APPLICATIONS

Business and Economics

Life Sciences

Social and Behavioral Sciences

continued on back endsheets

SEVENTH EDITION

BRIEF CALCULUS

an applied approach

Ron E. Larson
The Pennsylvania State University
The Behrend College

Bruce Edwards
University of Florida

with the assistance of
David C. Falvo
The Pennsylvania State University
The Behrend College

HOUGHTON MIFFLIN COMPANY
BOSTON NEW YORK

Publisher: Jack Shira
Associate Sponsoring Editor: Cathy Cantin
Development Manager: Maureen Ross
Development Editor: David George
Editorial Assistant: Elizabeth Kassab
Supervising Editor: Karen Carter
Senior Project Editor: Patty Bergin
Editorial Assistant: Julia Keller
Production Technology Supervisor: Gary Crespo
Senior Marketing Manager: Danielle Potvin Curran
Marketing Coordinator: Nicole Mollica
Senior Manufacturing Coordinator: Marie Barnes

We have included examples and exercises that use real-life data as well as technology output from a variety of software. This would not have been possible without the help of many people and organizations. Our wholehearted thanks goes to all for their time and effort.

Custom Publishing Editor: Todd Corbin
Custom Publishing Production Manager: Tina Kozik
Project Coordinator: Andrea Wagner

Cover Design: Galen B. Murphy
Cover Art: PhotoDisc

This book contains select works from existing Houghton Mifflin Company resources and was produced by Houghton Mifflin Custom Publishing for collegiate use. As such, those adopting and/or contributing to this work are responsible for editorial content, accuracy, continuity and completeness.

Printed in the United States of America.

ISBN-13: 978-0-618-66487-0
ISBN-10: 0-618-66487-4
N-05156

3 4 5 6 7 8 9 – WP – 07 06

Houghton Mifflin
Custom Publishing

222 Berkeley Street • Boston, MA 02116
Address all correspondence and order information to the above address.

Contents

*Available at the text-specific website at *college.hmco.com*.

A Word from the Authors

Welcome to *Brief Calculus: An Applied Approach*, Seventh Edition. In this revision, we have focused on making the text even more student-oriented. To encourage mastery and understanding, we have outlined a straightforward program of study with continual reinforcement and applicability to the real world.

Student-Oriented Approach

Each chapter begins with "What you should learn" and "Why you should learn it." The "What you should learn" is a list of *Objectives* that students will examine in the chapter. The "Why you should learn it" lists sample applications that appear throughout the chapter. Each section begins with a list of learning *Objectives*, enabling students to identify and focus on the key points of the section.

Following every example is a *Try It* exercise. The new problem allows for students to immediately practice the concept learned in the example.

It is crucial for a student to understand an algebraic concept before attempting to master a related calculus concept. To help students in this area, *Algebra Review* tips appear at point of use throughout the text. A two-page *Algebra Review* appears at the end of each chapter, which emphasizes key algebraic concepts discussed in the chapter.

Before students are exposed to selected topics, *Discovery* projects allow them to explore concepts on their own, making them more likely to remember the results. These optional boxed features can be omitted, if the instructor desires, with no loss of continuity in the coverage of the material.

Throughout the text, *Study Tips* address special cases, expand on concepts, and help students avoid common errors. *Side Comments* help explain the steps of a solution. State-of-the-art graphics help students with visualization, especially when working with functions of several variables.

Advances in *Technology* are helping to change the world around us. We have updated and increased technology coverage to be even more readily available at point of use. Students are encouraged to use a graphing utility, computer program, or spreadsheet software as a tool for exploration, discovery, and problem solving. Students are not required to have access to a graphing utility to use this text effectively. In addition to describing the benefits of using technology, the text also pays special attention to its possible misuse or misinterpretation.

Just before each section exercise set, the *Take Another Look* feature asks students to look back at one or more concepts presented in the section, using questions designed to enhance understanding of key ideas.

Each chapter presents many opportunities for students to assess their progress, both at the end of each section (*Prerequisite Review* and *Section Exercises*) and at the end of each chapter (*Chapter Summary*, *Study Strategies*, *Study Tools*, and *Review Exercises*). The test items in *Sample Post-Graduation Exam Questions* show the relevance of calculus. The test questions are representative of types of questions on several common post-graduation exams.

Business Capsules appear at the ends of numerous sections. These capsules and their accompanying exercises deal with business situations that are related to the mathematical concepts covered in the chapter.

Application to the Changing World Around Us

Students studying calculus need to understand how the subject matter relates to the real world. In this edition, we have focused on increasing the variety of applications, especially in the life sciences, economics, and finance. All real-data applications have been revised to use the most current information available. Exercises containing material from textbooks in other disciplines have been included to show the relevance of calculus in other areas. In addition, exercises involving the use of spreadsheets have been incorporated throughout.

We hope you enjoy the Seventh Edition. A readable text with a straightforward approach, it provides effective study tools and direct application to the lives and futures of calculus students.

Ron Larson

Bruce H. Edwards

Supplements

The integrated learning system for *Brief Calculus: An Applied Approach*, Seventh Edition, addresses the changing needs of today's instructors and students, offering dynamic teaching tools for instructors and interactive learning resources for students in print, CD-ROM, and online formats.

Resources

Eduspace®, Houghton Mifflin's Online Learning Tool

Eduspace® is an online learning environment that combines algorithmic tutorials, homework capabilities, and testing. Text-specific content, organized by section, is available to help students understand the mathematics covered in this text.

For the Instructor

Instructor ClassPrep CD-ROM with HM Testing (Windows, Macintosh)

ClassPrep offers complete instructor solutions and other instructor resources. *HM Testing* is a computerized test generator with algorithmically generated test items.

Instructor Website (math.college.hmco.com/instructors)

This website contains pdfs of the *Complete Solutions Guide* and *Test Item File and Instructor's Resource Guide*. Digital Figures and Lessons are available (ppts) for use as handouts or slides.

For the Student

HM mathSpace® Student CD-ROM

HM mathSpace contains a prerequisite algebra review, a link to our online graphing calculator, and graphing calculator programs.

Excel Made Easy: Video Instruction with Activities CD-ROM

Excel Made Easy uses easy-to-follow videos to help students master mathematical concepts introduced in class. The CD-ROM includes electronic spreadsheets and detailed tutorials.

SMARTHINKING™ Online Tutoring

Instructional Video and DVD Series by Dana Mosely

The video and DVD series complement the textbook topic coverage should a student struggle with the calculus concepts or miss a class.

Student Solutions Guide

This printed manual features step-by-step solutions to the odd-numbered exercises. A practice test with full solutions is available for each chapter.

Excel Guide for Finite Math and Applied Calculus

The *Excel Guide* provides useful information, including step-by-step examples and sample exercises.

Student Website (math.college.hmco.com/students)

The website contains self-quizzing content to help students strengthen their calculus skills, a link to our online graphing calculator, graphing calculator programs, and printable formula cards.

Acknowledgments

We would like to thank the many people who have helped us at various stages of this project during the past 24 years. Their encouragement, criticisms, and suggestions have been invaluable to us.

A special note of thanks goes to the instructors who responded to our survey and to all the students who have used the previous editions of the text.

Reviewers of the Seventh Edition

Scott Perkins
Lake Sumter Community College

Jose Gimenez
Temple University

Keng Deng
University of Louisiana at Lafayette

George Anastassiou
University of Memphis

Peggy Luczak
Camden County College

Bernadette Kocyba
J. Sergeant Reynolds Community College

Shane Goodwin
Brigham Young University of Idaho

Harvey Greenwald
California Polytechnic State University

Randall McNiece
San Jacinto College

Reviewers of Previous Editions

Carol Achs, *Mesa Community College*; David Bregenzer, *Utah State University*; Mary Chabot, *Mt. San Antonio College*; Joseph Chance, *University of Texas—Pan American*; John Chuchel, *University of California*; Miriam E. Connellan, *Marquette University*; William Conway, *University of Arizona*; Karabi Datta, *Northern Illinois University*; Roger A. Engle, *Clarion University of Pennsylvania*; Betty Givan, *Eastern Kentucky University*; Mark Greenhalgh, *Fullerton College*; Karen Hay, *Mesa Community College*; Raymond Heitmann, *University of Texas at Austin*; William C. Huffman, *Loyola University of Chicago*; Arlene Jesky, *Rose State College*; Ronnie Khuri, *University of Florida*; Duane Kouba, *University of California—Davis*; James A. Kurre, *The Pennsylvania State University*; Melvin Lax, *California State University—Long Beach*; Norbert Lerner, *State University of New York at Cortland*; Yuhlong Lio, *University of South Dakota*; Peter J. Livorsi, *Oakton Community College*; Samuel A. Lynch, *Southwest Missouri State University*; Kevin McDonald, *Mt. San Antonio College*; Earl H. McKinney, *Ball State University*; Philip R. Montgomery, *University of Kansas*; Mike Nasab, *Long Beach City College*; Karla Neal, *Louisiana State University*; James Osterburg, *University of Cincinnati*; Rita Richards, *Scottsdale Community College*; Stephen B. Rodi, *Austin Community College*; Yvonne Sandoval-Brown, *Pima Community College*; Richard Semmler, *Northern Virginia Community College—Annandale*; Bernard Shapiro, *University of Massachusetts, Lowell*; Jane Y. Smith, *University of Florida*; DeWitt L. Sumners, *Florida State University*; Jonathan Wilkin, *Northern Virginia Community College*; Carol G. Williams, *Pepperdine University*; Melvin R. Woodard, *Indiana University of Pennsylvania*; Carlton Woods, *Auburn University at Montgomery*; Jan E. Wynn, *Brigham Young University*; Robert A. Yawin, *Springfield Technical Community College*; Charles W. Zimmerman, *Robert Morris College*

Our thanks to David Falvo, The Behrend College, The Pennsylvania State University, for his contributions to this project. Our thanks also to Robert Hostetler, The Behrend College, The Pennsylvania State University, for his significant contributions to previous editions of this text.

We would also like to thank the staff at Larson Texts, Inc. who assisted with proofreading the manuscript, preparing and proofreading the art package, and checking and typesetting the supplements.

On a personal level, we are grateful to our spouses, Deanna Gilbert Larson and Consuelo Edwards, for their love, patience, and support. Also, a special thanks goes to R. Scott O'Neil.

If you have suggestions for improving this text, please feel free to write to us. Over the past two decades we have received many useful comments from both instructors and students, and we value these comments very highly.

Ron Larson

Ron Larson

Bruce H. Edwards

Bruce H. Edwards

Features

CHAPTER OPENERS

Each chapter opens with *Strategies for Success*, a checklist that outlines what students should learn and lists several applications of those objectives. Each chapter opener also contains a list of the section topics and a photo referring students to an interesting application in the section exercises.

chapter

2 Differentiation

2.1 The Derivative and the Slope of a Graph
2.2 Some Rules for Differentiation
2.3 Rates of Change: Velocity and Marginals
2.4 The Product and Quotient Rules
2.5 The Chain Rule
2.6 Higher-Order Derivatives
2.7 Implicit Differentiation
2.8 Related Rates

© Martyn Goddard/CORBIS

Higher-order derivatives are used to determine the acceleration function of a sports car. The acceleration function shows the changes in the car's velocity. As the car reaches its "cruising" speed, is the acceleration increasing or decreasing?

STRATEGIES FOR SUCCESS

WHAT YOU SHOULD LEARN:

- How to find the slope of a graph and calculate derivatives using the limit definition
- How to use the Constant Rule, Power Rule, Constant Multiple Rule, and Sum and Difference Rules
- How to find rates of change: velocity, marginal profit, marginal revenue, and marginal cost
- How to use the Product, Quotient, Chain, and General Power Rules
- How to calculate higher-order derivatives and derivatives using implicit differentiation
- How to solve related-rate problems and applications

WHY YOU SHOULD LEARN IT:

Derivatives have many applications in real life, as can be seen by the examples below, which represent a small sample of the applications in this chapter.

- Increasing Revenue, Example 10 on page 101
- Psychology: Migraine Prevalence, Exercise 62 on page 104
- Average Velocity, Exercises 15 and 16 on page 117
- Demand Function, Exercises 53 and 54 on page 129
- Quality Control, Exercise 58 on page 129
- Velocity and Acceleration, Exercises 41–44 and 50 on pages 145 and 146

81

SECTION 2.5 The Chain Rule 131

2.5 THE CHAIN RULE

- Find derivatives using the Chain Rule.
- Find derivatives using the General Power Rule.
- Write derivatives in simplified form.
- Use derivatives to answer questions about real-life situations.
- Use the differentiation rules to differentiate algebraic functions.

The Chain Rule

In this section, you will study one of the most powerful rules of differential calculus—the **Chain Rule.** This differentiation rule deals with composite functions and adds versatility to the rules presented in Sections 2.2 and 2.4. For example, compare the functions below. Those on the left can be differentiated without the Chain Rule, whereas those on the right are best done with the Chain Rule.

Without the Chain Rule	*With the Chain Rule*
$y = x^2 + 1$	$y = \sqrt{x^2 + 1}$
$y = x + 1$	$y = (x + 1)^{-1/2}$
$y = 3x + 2$	$y = (3x + 2)^5$
$y = \frac{x + 5}{x^2 + 2}$	$y = \left(\frac{x + 5}{x^2 + 2}\right)^2$

The Chain Rule

If $y = f(u)$ is a differentiable function of u, and $u = g(x)$ is a differentiable function of x, then $y = f(g(x))$ is a differentiable function of x, and

$$\frac{dy}{dx} = \frac{dy}{du} \cdot \frac{du}{dx}$$

or, equivalently,

$$\frac{d}{dx}[f(g(x))] = f'(g(x))g'(x).$$

Basically, the Chain Rule states that if y changes dy/du times as fast as u, and u changes du/dx times as fast as x, then y changes

$$\frac{dy}{du} \cdot \frac{du}{dx}$$

times as fast as x, as illustrated in Figure 2.28. One advantage of the dy/dx notation for derivatives is that it helps you remember differentiation rules, such as the Chain Rule. For instance, in the formula

$$dy/dx = (dy/du)(du/dx)$$

you can imagine that the du's divide out.

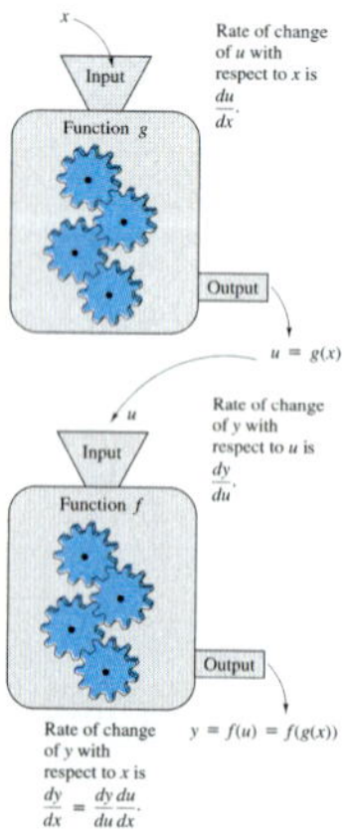

FIGURE 2.28

SECTION OBJECTIVES

Each section begins with a list of objectives covered in that section. This outline helps instructors with class planning and students in studying the material in the section.

DEFINITIONS AND THEOREMS

All definitions and theorems are highlighted for emphasis and easy reference.

EXAMPLES

To increase the usefulness of the text as a study tool, the Seventh Edition presents a wide variety of examples, each titled for easy reference. Many of these detailed examples display solutions that are presented graphically, analytically, and/or numerically to provide further insight into mathematical concepts. Side comments clarify the steps of the solution as necessary. Examples using real-life data are identified with a globe icon and are accompanied by the types of illustrations that students are used to seeing in newspapers and magazines.

TRY ITS

Appearing after every example, these new problems help students reinforce concepts right after they are presented.

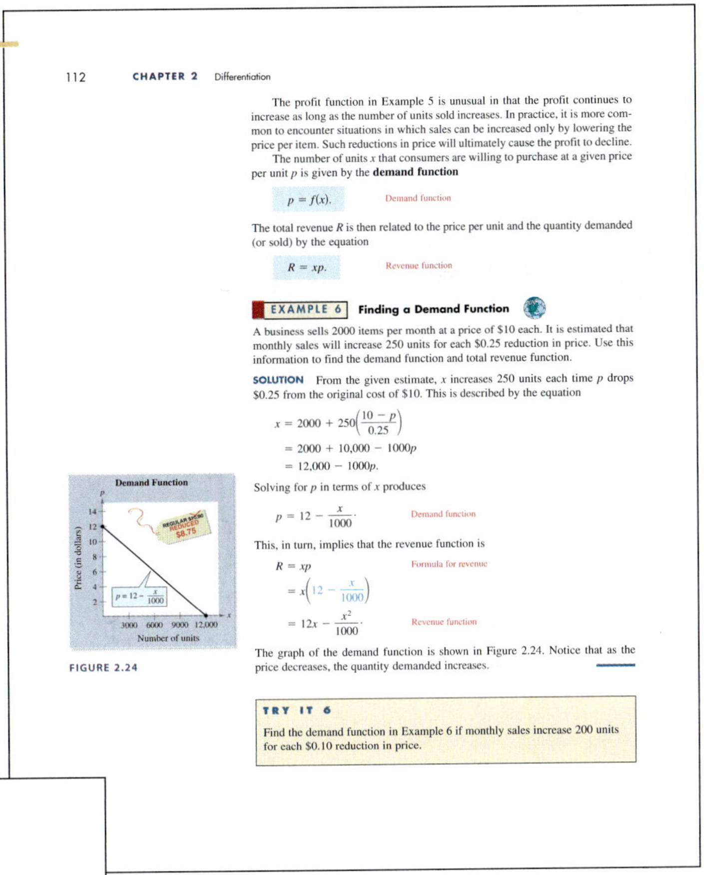

112 CHAPTER 2 Differentiation

The profit function in Example 5 is unusual in that the profit continues to increase as long as the number of units sold increases. In practice, it is more common to encounter situations in which sales can be increased only by lowering the price per item. Such reductions in price will ultimately cause the profit to decline.

The number of units x that consumers are willing to purchase at a given price per unit p is given by the **demand function**

$p = f(x).$ Demand function

The total revenue R is then related to the price per unit and the quantity demanded (or sold) by the equation

$R = xp.$ Revenue function

EXAMPLE 6 Finding a Demand Function

A business sells 2000 items per month at a price of \$10 each. It is estimated that monthly sales will increase 250 units for each \$0.25 reduction in price. Use this information to find the demand function and total revenue function.

SOLUTION From the given estimate, x increases 250 units each time p drops \$0.25 from the original cost of \$10. This is described by the equation

$$x = 2000 + 250\left(\frac{10-p}{0.25}\right)$$
$$= 2000 + 10{,}000 - 1000p$$
$$= 12{,}000 - 1000p.$$

Solving for p in terms of x produces

$$p = 12 - \frac{x}{1000}.$$ Demand function

This, in turn, implies that the revenue function is

$$R = xp$$ Formula for revenue
$$= x\left(12 - \frac{x}{1000}\right)$$
$$= 12x - \frac{x^2}{1000}.$$ Revenue function

The graph of the demand function is shown in Figure 2.24. Notice that as the price decreases, the quantity demanded increases.

FIGURE 2.24

TRY IT 6

Find the demand function in Example 6 if monthly sales increase 200 units for each \$0.10 reduction in price.

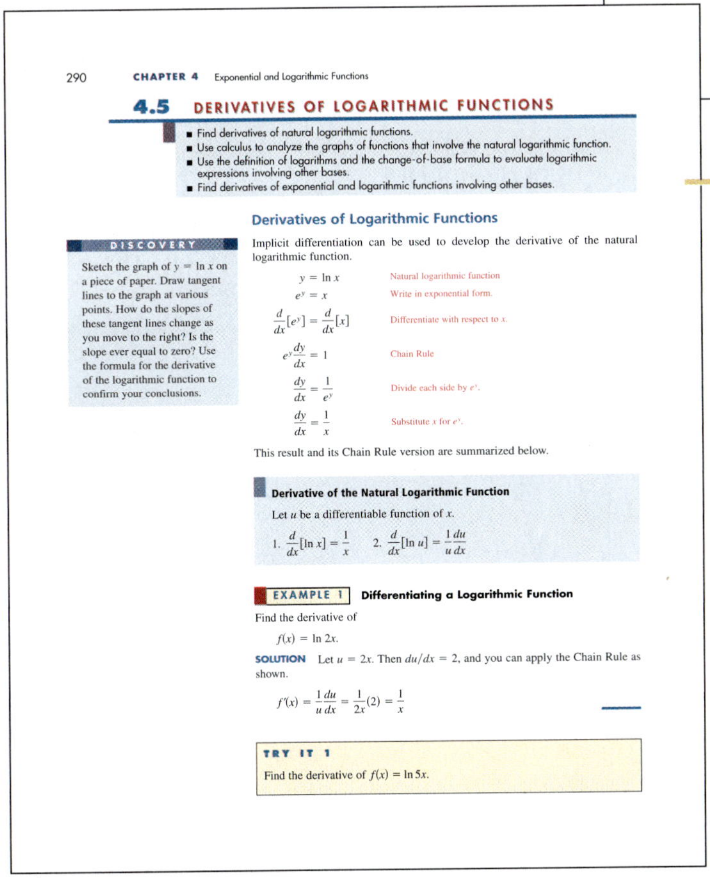

290 CHAPTER 4 Exponential and Logarithmic Functions

4.5 DERIVATIVES OF LOGARITHMIC FUNCTIONS

- Find derivatives of natural logarithmic functions.
- Use calculus to analyze the graphs of functions that involve the natural logarithmic function.
- Use the definition of logarithms and the change-of-base formula to evaluate logarithmic expressions involving other bases.
- Find derivatives of exponential and logarithmic functions involving other bases.

DISCOVERY

Sketch the graph of $y = \ln x$ on a piece of paper. Draw tangent lines to the graph at various points. How do the slopes of these tangent lines change as you move to the right? Is the slope ever equal to zero? Use the formula for the derivative of the logarithmic function to confirm your conclusions.

Derivatives of Logarithmic Functions

Implicit differentiation can be used to develop the derivative of the natural logarithmic function.

$y = \ln x$ Natural logarithmic function

$e^y = x$ Write in exponential form.

$\frac{d}{dx}[e^y] = \frac{d}{dx}[x]$ Differentiate with respect to x.

$e^y \frac{dy}{dx} = 1$ Chain Rule

$\frac{dy}{dx} = \frac{1}{e^y}$ Divide each side by e^y.

$\frac{dy}{dx} = \frac{1}{x}$ Substitute x for e^y.

This result and its Chain Rule version are summarized below.

Derivative of the Natural Logarithmic Function

Let u be a differentiable function of x.

1. $\frac{d}{dx}[\ln x] = \frac{1}{x}$ 2. $\frac{d}{dx}[\ln u] = \frac{1}{u}\frac{du}{dx}$

EXAMPLE 1 Differentiating a Logarithmic Function

Find the derivative of

$f(x) = \ln 2x.$

SOLUTION Let $u = 2x$. Then $du/dx = 2$, and you can apply the Chain Rule as shown.

$$f'(x) = \frac{1}{u}\frac{du}{dx} = \frac{1}{2x}(2) = \frac{1}{x}$$

TRY IT 1

Find the derivative of $f(x) = \ln 5x$.

DISCOVERY

Before students are exposed to selected topics, *Discovery* projects allow them to explore concepts on their own, making them more likely to remember the results. These optional boxed features can be omitted, if the instructor desires, with no loss of continuity in the coverage of material.

ALGEBRA REVIEWS

Algebra Reviews appear throughout each chapter and offer students algebraic support at point of use. These smaller reviews are then revisited in the Algebra Review at the end of each chapter, where additional details of examples with solutions and explanations are provided.

Not only is the function in Example 3 continuous on the entire real line, it is also differentiable there. For such functions, the only critical numbers are those for which $f'(x) = 0$. The next example considers a continuous function that has *both* types of critical numbers—those for which $f'(x) = 0$ and those for which f' is undefined.

ALGEBRA REVIEW

For help on the algebra in Example 4, see Example 2(d) in the *Chapter 3 Algebra Review*, on page 249.

EXAMPLE 4 Finding Increasing and Decreasing Intervals

Find the open intervals on which the function

$$f(x) = (x^2 - 4)^{2/3}$$

is increasing or decreasing.

SOLUTION Begin by finding the derivative of the function.

$$f'(x) = \frac{2}{3}(x^2 - 4)^{-1/3}(2x) \quad \text{Differentiate.}$$

$$= \frac{4x}{3(x^2 - 4)^{1/3}} \quad \text{Simplify.}$$

From this, you can see that the derivative is zero when $x = 0$ and the derivative is undefined when $x = \pm 2$. So, the critical numbers are

$$x = -2, \quad x = 0, \quad \text{and} \quad x = 2. \quad \text{Critical numbers}$$

This implies that the test intervals are

$$(-\infty, -2), \quad (-2, 0), \quad (0, 2), \quad \text{and} \quad (2, \infty). \quad \text{Test intervals}$$

The table summarizes the testing of these four intervals, and the graph of the function is shown in Figure 3.6.

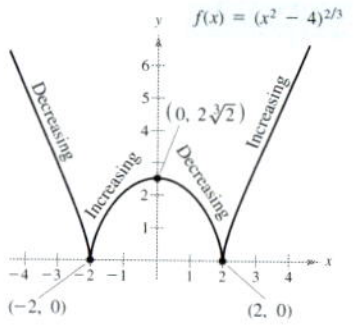

FIGURE 3.6

Interval	$-\infty < x < -2$	$-2 < x < 0$	$0 < x < 2$	$2 < x < \infty$
Test value	$x = -3$	$x = -1$	$x = 1$	$x = 3$
Sign of $f'(x)$	$f'(-3) < 0$	$f'(-1) > 0$	$f'(1) < 0$	$f'(3) > 0$
Conclusion	Decreasing	Increasing	Decreasing	Increasing

TRY IT 4

Find the open intervals on which the function $f(x) = x^{2/3}$ is increasing or decreasing.

ALGEBRA REVIEW

To test the intervals in the table, it is not necessary to *evaluate* $f'(x)$ at each test value—you only need to determine its sign. For example, you can determine the sign of $f'(-3)$ as shown.

$$f'(-3) = \frac{4(-3)}{3(9 - 4)^{1/3}} = \frac{\text{negative}}{\text{positive}} = \text{negative}$$

STUDY TIPS

Throughout the text, *Study Tips* help students avoid common errors, address special cases, and expand on theoretical concepts.

Finding Antiderivatives

The inverse relationship between the operations of integration and differentiation can be shown symbolically, as shown.

$$\frac{d}{dx}\left[\int f(x)\,dx\right] = f(x) \quad \text{Differentiation is the inverse of integration.}$$

$$\int f'(x)\,dx = f(x) + C \quad \text{Integration is the inverse of differentiation.}$$

This inverse relationship between integration and differentiation allows you to obtain integration formulas directly from differentiation formulas. The following summary lists the integration formulas that correspond to some of the differentiation formulas you have studied.

Basic Integration Rules

1. $\int k\,dx = kx + C$, k is a constant. — Constant Rule
2. $\int kf(x)\,dx = k\int f(x)\,dx$ — Constant Multiple Rule
3. $\int [f(x) + g(x)]\,dx = \int f(x)\,dx + \int g(x)\,dx$ — Sum Rule
4. $\int [f(x) - g(x)]\,dx = \int f(x)\,dx - \int g(x)\,dx$ — Difference Rule
5. $\int x^n\,dx = \frac{x^{n+1}}{n+1} + C, \quad n \neq -1$ — Simple Power Rule

STUDY TIP

You will study the General Power Rule for integration in Section 5.2 and the Exponential and Log Rules in Section 5.3.

STUDY TIP

In Example 2(b), the integral $\int 1\,dx$ is usually shortened to the form $\int dx$.

Be sure you see that the Simple Power Rule has the restriction that n cannot be -1. So, you *cannot* use the Simple Power Rule to evaluate the integral

$$\int \frac{1}{x}\,dx.$$

To evaluate this integral, you need the Log Rule, which is described in Section 5.3.

TRY IT 2

Find each indefinite integral.

(a) $\int 5\,dx$

(b) $\int -1\,dr$

(c) $\int 2\,dt$

EXAMPLE 2 Finding Indefinite Integrals

Find each indefinite integral.

(a) $\int \frac{1}{2}\,dx$ (b) $\int 1\,dx$ (c) $\int -5\,dt$

SOLUTION

(a) $\int \frac{1}{2}\,dx = \frac{1}{2}x + C$ (b) $\int 1\,dx = x + C$ (c) $\int -5\,dt = -5t + C$

TAKE ANOTHER LOOK

Starting with Chapter 1, each section in the text closes with a *Take Another Look* problem asking students to look back at one or more concepts presented in the section, using questions designed to enhance understanding of key ideas. These problems can be completed as group projects in class or as homework assignments. Because these problems encourage students to think, reason, and write about calculus, they emphasize the synthesis or the further exploration of the concepts presented in the section.

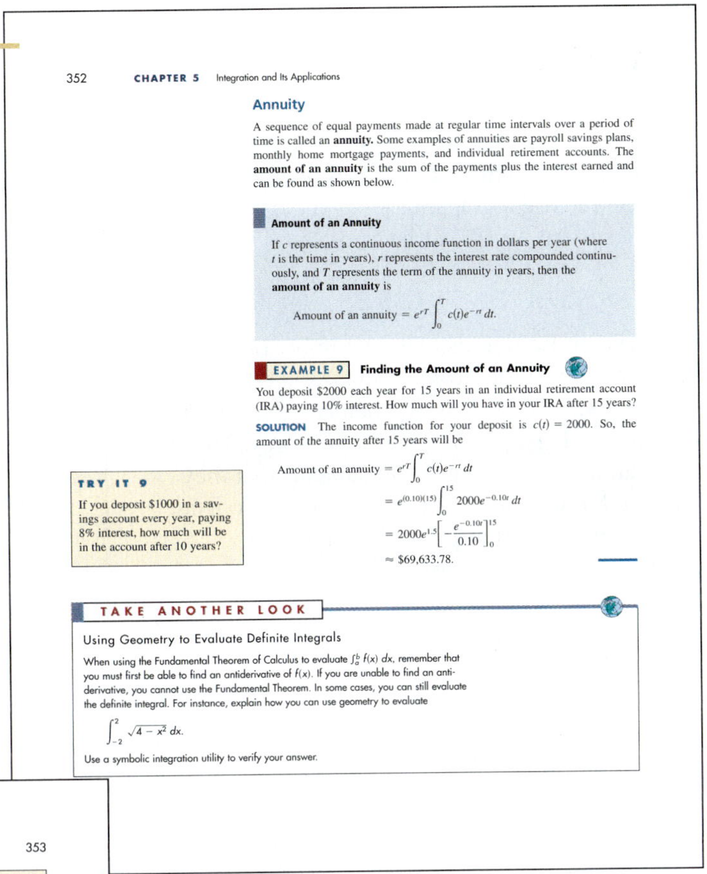
352 **CHAPTER 5** Integration and Its Applications

Annuity

A sequence of equal payments made at regular time intervals over a period of time is called an **annuity.** Some examples of annuities are payroll savings plans, monthly home mortgage payments, and individual retirement accounts. The **amount of an annuity** is the sum of the payments plus the interest earned and can be found as shown below.

Amount of an Annuity

If c represents a continuous income function in dollars per year (where t is the time in years), r represents the interest rate compounded continuously, and T represents the term of the annuity in years, then the **amount of an annuity** is

$$\text{Amount of an annuity} = e^{rT}\int_0^T c(t)e^{-rt}\,dt.$$

EXAMPLE 9 **Finding the Amount of an Annuity**

You deposit \$2000 each year for 15 years in an individual retirement account (IRA) paying 10% interest. How much will you have in your IRA after 15 years?

SOLUTION The income function for your deposit is $c(t) = 2000$. So, the amount of the annuity after 15 years will be

$$\begin{aligned}\text{Amount of an annuity} &= e^{rT}\int_0^T c(t)e^{-rt}\,dt\\ &= e^{(0.10)(15)}\int_0^{15} 2000e^{-0.10t}\,dt\\ &= 2000e^{1.5}\left[-\frac{e^{-0.10t}}{0.10}\right]_0^{15}\\ &\approx \$69{,}633.78.\end{aligned}$$

TRY IT 9

If you deposit \$1000 in a savings account every year, paying 8% interest, how much will be in the account after 10 years?

TAKE ANOTHER LOOK

Using Geometry to Evaluate Definite Integrals

When using the Fundamental Theorem of Calculus to evaluate $\int_a^b f(x)\,dx$, remember that you must first be able to find an antiderivative of $f(x)$. If you are unable to find an antiderivative, you cannot use the Fundamental Theorem. In some cases, you can still evaluate the definite integral. For instance, explain how you can use geometry to evaluate

$$\int_{-2}^{2}\sqrt{4-x^2}\,dx.$$

Use a symbolic integration utility to verify your answer.

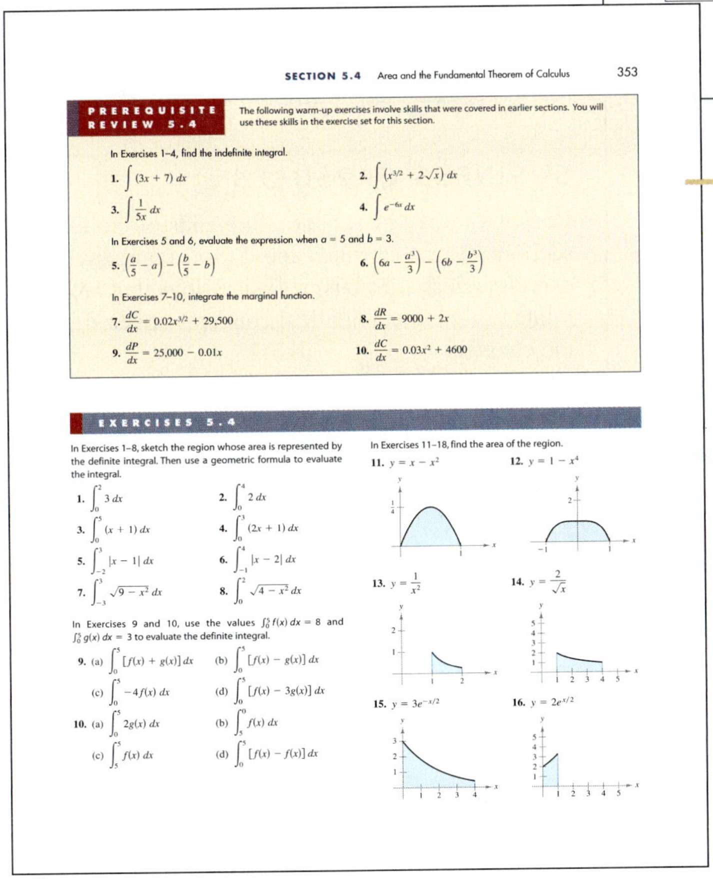
SECTION 5.4 Area and the Fundamental Theorem of Calculus 353

PREREQUISITE REVIEW 5.4 The following warm-up exercises involve skills that were covered in earlier sections. You will use these skills in the exercise set for this section.

In Exercises 1–4, find the indefinite integral.

1. $\int (3x + 7)\,dx$
2. $\int \left(x^{3/2} + 2\sqrt{x}\right)dx$
3. $\int \frac{1}{5x}\,dx$
4. $\int e^{-6x}\,dx$

In Exercises 5 and 6, evaluate the expression when $a = 5$ and $b = 3$.

5. $\left(\frac{a}{5} - a\right) - \left(\frac{b}{5} - b\right)$
6. $\left(6a - \frac{a^3}{3}\right) - \left(6b - \frac{b^3}{3}\right)$

In Exercises 7–10, integrate the marginal function.

7. $\frac{dC}{dx} = 0.02x^{3/2} + 29{,}500$
8. $\frac{dR}{dx} = 9000 + 2x$
9. $\frac{dP}{dx} = 25{,}000 - 0.01x$
10. $\frac{dC}{dx} = 0.03x^2 + 4600$

EXERCISES 5.4

In Exercises 1–8, sketch the region whose area is represented by the definite integral. Then use a geometric formula to evaluate the integral.

1. $\int_0^2 3\,dx$
2. $\int_0^4 2\,dx$
3. $\int_0^5 (x + 1)\,dx$
4. $\int_0^3 (2x + 1)\,dx$
5. $\int_{-2}^{3} |x - 1|\,dx$
6. $\int_{-1}^{4} |x - 2|\,dx$
7. $\int_{-3}^{3} \sqrt{9 - x^2}\,dx$
8. $\int_0^2 \sqrt{4 - x^2}\,dx$

In Exercises 9 and 10, use the values $\int_0^5 f(x)\,dx = 8$ and $\int_0^5 g(x)\,dx = 3$ to evaluate the definite integral.

9. (a) $\int_0^5 [f(x) + g(x)]\,dx$ (b) $\int_0^5 [f(x) - g(x)]\,dx$
(c) $\int_0^5 -4f(x)\,dx$ (d) $\int_0^5 [f(x) - 3g(x)]\,dx$
10. (a) $\int_0^5 2g(x)\,dx$ (b) $\int_5^0 f(x)\,dx$
(c) $\int_5^5 f(x)\,dx$ (d) $\int_0^5 [f(x) - f(x)]\,dx$

In Exercises 11–18, find the area of the region.

11. $y = x - x^2$
12. $y = 1 - x^4$
13. $y = \frac{1}{x^2}$
14. $y = \frac{2}{\sqrt{x}}$
15. $y = 3e^{-x/2}$
16. $y = 2e^{x/2}$

PREREQUISITE REVIEW

Starting with Chapter 1, each text section has a set of *Prerequisite Review* exercises. The exercises enable students to review and practice the previously learned skills necessary to master the new skills presented in the section. Answers to these sections appear in the back of the text.

EXERCISES

The text now contains almost 6000 exercises. Each exercise set is graded, progressing from skill-development problems to more challenging problems, to build confidence, skill, and understanding. The wide variety of types of exercises include many technology-oriented, real, and engaging problems. Answers to all odd-numbered exercises are included in the back of the text. To help instructors make homework assignments, many of the exercises in the text are labeled to indicate the area of application.

GRAPHING UTILITIES

Many exercises in the text can be solved using technology; however, the ⊕ symbol identifies all exercises for which students are specifically instructed to use a graphing utility, computer algebra system, or spreadsheet software.

TEXTBOOK EXERCISES

The Seventh Edition includes a number of exercises that contain material from textbooks in other disciplines, such as biology, chemistry, economics, finance, geology, physics, and psychology. These applications make the point to students that they will need to use calculus in future courses outside of the math curriculum. These exercises are identified by the icon and are labeled to indicate the subject area.

(a) Use a graphing utility to decide whether the board of trustees expects the gift income to increase or decrease over the five-year period.

(b) Find the expected total gift income over the five-year period.

(c) Determine the average annual gift income over the five-year period. Compare the result with the income given when $t = 3$.

61. *Learning Theory* A model for the ability M of a child to memorize, measured on a scale from 0 to 10, is

$$M = 1 + 1.6t \ln t, \quad 0 < t \le 4$$

where t is the child's age in years. Find the average value of this model between

(a) the child's first and second birthdays.

(b) the child's third and fourth birthdays.

62. *Revenue* A company sells a seasonal product. The revenue R (in dollars per year) generated by sales of the product can be modeled by

$$R = 410.5t^2e^{-t/30} + 25{,}000, \quad 0 \le t \le 365$$

where t is the time in days.

(a) Find the average daily receipts during the first quarter, which is given by $0 \le t \le 90$.

(b) Find the average daily receipts during the fourth quarter, which is given by $274 \le t \le 365$.

(c) Find the total daily receipts during the year.

Present Value In Exercises 63–68, find the present value of the income c (measured in dollars) over t_1 years at the given annual inflation rate r.

63. $c = 5000,\ r = 5\%,\ t_1 = 4$ years

64. $c = 450,\ r = 4\%,\ t_1 = 10$ years

65. $c = 150{,}000 + 2500t,\ r = 4\%,\ t_1 = 10$ years

66. $c = 30{,}000 + 500t,\ r = 7\%,\ t_1 = 6$ years

67. $c = 1000 + 50e^{t/2},\ r = 6\%,\ t_1 = 4$ years

68. $c = 5000 + 25te^{t/10},\ r = 6\%,\ t_1 = 10$ years

69. *Present Value* A company expects its income c during the next 4 years to be modeled by

$$c = 150{,}000 + 75{,}000t.$$

(a) Find the actual income for the business over the 4 years.

(b) Assuming an annual inflation rate of 4%, what is the present value of this income?

70. *Present Value* A professional athlete signs a three-year contract in which the earnings can be modeled by

$$c = 300{,}000 + 125{,}000t.$$

(a) Find the actual value of the athlete's contract.

(b) Assuming an annual inflation rate of 5%, what is the present value of the contract?

Future Value In Exercises 71 and 72, find the future value of the income (in dollars) given by $f(t)$ over t_1 years at the annual interest rate of r. If the function f represents a continuous investment over a period of t_1 years at an annual interest rate of r (compounded continuously), then the future value of the investment is given by

$$\text{Future value} = e^{rt_1}\int_0^{t_1} f(t)e^{-rt}\,dt.$$

71. $f(t) = 3000,\ r = 8\%,\ t_1 = 10$ years

72. $f(t) = 3000e^{0.05t},\ r = 10\%,\ t_1 = 5$ years

73. *Finance: Future Value* Use the equation from Exercises 71 and 72 to calculate the following. *(Source: Adapted from Garman/Forgue, Personal Finance, Fifth Edition)*

(a) The future value of \$1200 saved each year for 10 years earning 7% interest.

(b) A person who wishes to invest \$1200 each year finds one investment choice that is expected to pay 9% interest per year and another, riskier choice that may pay 10% interest per year. What is the difference in return (future value) if the investment is made for 15 years?

74. *Consumer Awareness* In 2004, the total cost to attend Pennsylvania State University for 1 year was estimated to be \$19,843. If your grandparents had continuously invested in a college fund according to the model

$$f(t) = 400t$$

for 18 years, at an annual interest rate of 10%, would the fund have grown enough to allow you to cover 4 years of expenses at Pennsylvania State University? *(Source: Pennsylvania State University)*

75. Use a program similar to the Midpoint Rule program on page 366 with $n = 10$ to approximate

$$\int_1^4 \frac{4}{\sqrt{x} + \sqrt[3]{x}}\,dx.$$

76. Use a program similar to the Midpoint Rule program on page 366 with $n = 12$ to approximate the volume of the solid generated by revolving the region bounded by the graphs of

$$y = \frac{10}{\sqrt{x}e^x},\ y = 0,\ x = 1, \text{ and } x = 4$$

about the x-axis.

36. *Minimum Cost* The ordering and transportation cost C of the components used in manufacturing a product is modeled by

$$C = 100\left(\frac{200}{x^2} + \frac{x}{x + 30}\right), \quad x \ge 1$$

where C is measured in thousands of dollars and x is the order size in hundreds. Find the order size that minimizes the cost. (*Hint:* Use the *root* feature of a graphing utility.)

37. *Revenue* The demand for a car wash is

$$x = 600 - 50p$$

where the current price is \$5.00. Can revenue be increased by lowering the price and thus attracting more customers? Use price elasticity of demand to determine your answer.

38. *Revenue* Repeat Exercise 37 for a demand function of

$$x = 800 - 40p.$$

39. *Demand* A demand function is modeled by $x = a/p^m$, where a is a constant and $m > 1$. Show that $\eta = -m$. In other words, show that a 1% increase in price results in an m% decrease in the quantity demanded.

40. *Sales* The sales S (in millions of dollars per year) for Lowe's for the years 1994 through 2003 can be modeled by

$$S = 201.556t^2 - 502.29t + 2622.8 + \frac{9286}{t},$$
$$4 \le t \le 13$$

where $t = 4$ corresponds to 1994. *(Source: Lowe's Companies)*

(a) During which year, from 1994 to 2003, were Lowe's sales increasing most rapidly?

(b) During which year were the sales increasing at the lowest rate?

(c) Find the rate of increase or decrease for each year in parts (a) and (b).

(d) Use a graphing utility to graph the sales function. Then use the *zoom* and *trace* features to confirm the results in parts (a), (b), and (c).

41. *Revenue* The revenue R (in millions of dollars per year) for Papa John's for the years 1994 through 2003 can be modeled by

$$R = \frac{-18.0 + 24.74t}{1 - 0.16t + 0.008t^2}, \quad 4 \le t \le 13$$

where $t = 4$ corresponds to 1994. *(Source: Papa John's Int'l.)*

(a) During which year, from 1994 to 2003, was Papa John's revenue the greatest? the least?

(b) During which year was the revenue increasing at the greatest rate? decreasing at the greatest rate?

(c) Use a graphing utility to graph the revenue function, and confirm your results in parts (a) and (b).

42. Match each graph with the function it best represents—a demand function, a revenue function, a cost function, or a profit function. Explain your reasoning. (The graphs are labeled a–d.)

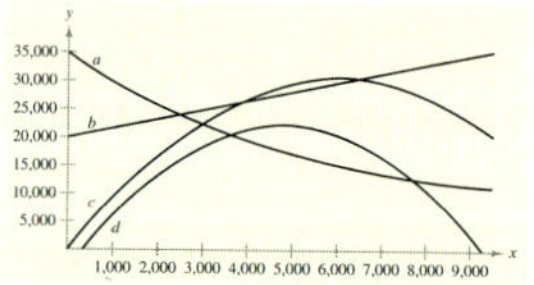

BUSINESS CAPSULE

Courtesy of Transperfect Translations

While graduate students, Elizabeth Elting and Phil Shawe co-founded TransPerfect Translations in 1992. They used a rented computer and a \$5000 credit card cash advance to market their service-oriented translation firm, now one of the largest in the country. Currently, they have a network of 4000 certified language specialists in North America, Europe, and Asia, which translates technical, legal, business, and marketing materials. In 2004, the company estimates its gross sales will be \$35 million.

43. *Research Project* Choose an innovative product like the one described above. Use your school's library, the Internet, or some other reference source to research the history of the product or service. Collect data about the revenue that the product or service has generated, and find a mathematical model of the data. Summarize your findings.

BUSINESS CAPSULES

Business Capsules appear at the ends of numerous sections. These capsules and their accompanying exercises deal with business situations that are related to the mathematical concepts covered in the chapter.

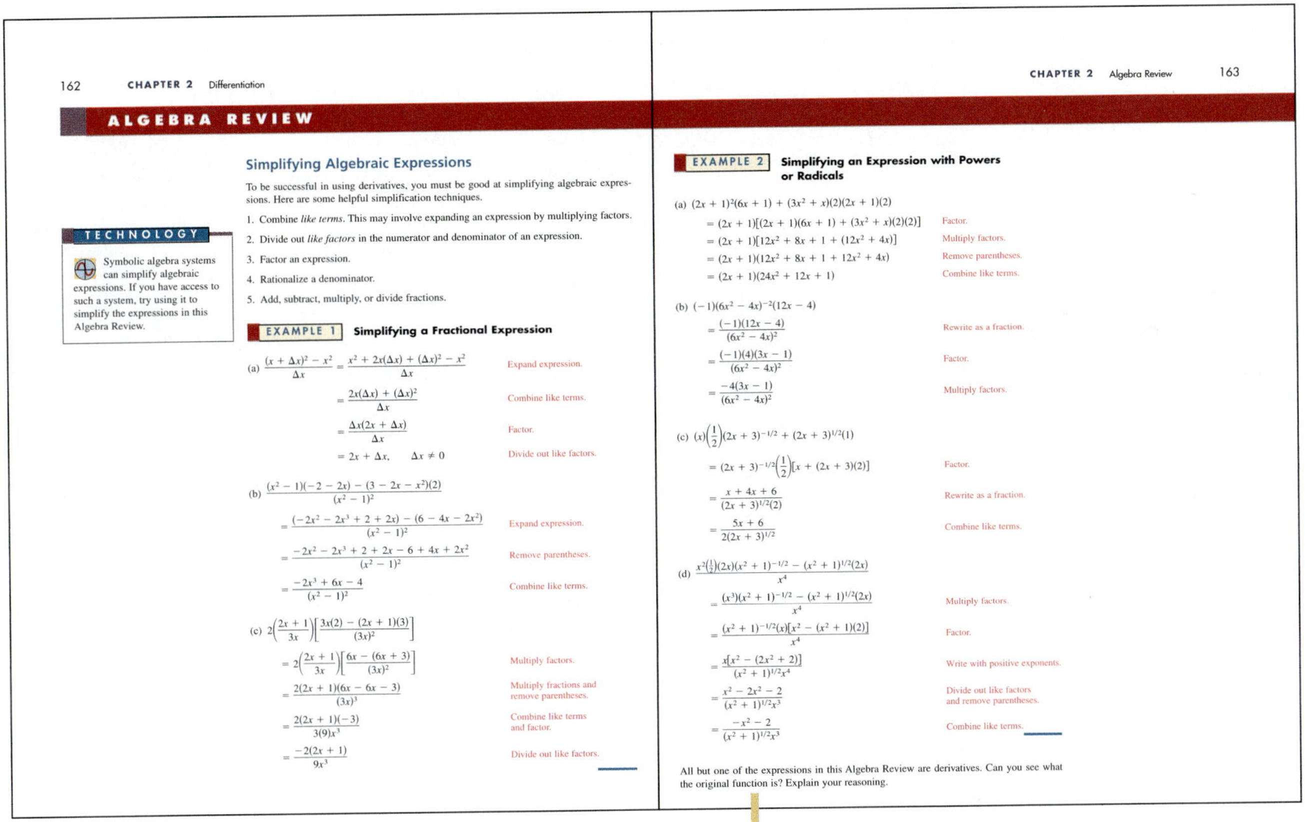

162 CHAPTER 2 Differentiation

CHAPTER 2 Algebra Review 163

ALGEBRA REVIEW

Simplifying Algebraic Expressions

To be successful in using derivatives, you must be good at simplifying algebraic expressions. Here are some helpful simplification techniques.

1. Combine *like terms*. This may involve expanding an expression by multiplying factors.
2. Divide out *like factors* in the numerator and denominator of an expression.
3. Factor an expression.
4. Rationalize a denominator.
5. Add, subtract, multiply, or divide fractions.

TECHNOLOGY

Symbolic algebra systems can simplify algebraic expressions. If you have access to such a system, try using it to simplify the expressions in this Algebra Review.

EXAMPLE 1 Simplifying a Fractional Expression

(a) $\dfrac{(x+\Delta x)^2 - x^2}{\Delta x} = \dfrac{x^2 + 2x(\Delta x) + (\Delta x)^2 - x^2}{\Delta x}$ Expand expression.

$= \dfrac{2x(\Delta x) + (\Delta x)^2}{\Delta x}$ Combine like terms.

$= \dfrac{\Delta x(2x + \Delta x)}{\Delta x}$ Factor.

$= 2x + \Delta x, \quad \Delta x \neq 0$ Divide out like factors.

(b) $\dfrac{(x^2-1)(-2-2x) - (3-2x-x^2)(2)}{(x^2-1)^2}$

$= \dfrac{(-2x^2 - 2x^3 + 2 + 2x) - (6 - 4x - 2x^2)}{(x^2-1)^2}$ Expand expression.

$= \dfrac{-2x^2 - 2x^3 + 2 + 2x - 6 + 4x + 2x^2}{(x^2-1)^2}$ Remove parentheses.

$= \dfrac{-2x^3 + 6x - 4}{(x^2-1)^2}$ Combine like terms.

(c) $2\left(\dfrac{2x+1}{3x}\right)\left[\dfrac{3x(2) - (2x+1)(3)}{(3x)^2}\right]$

$= 2\left(\dfrac{2x+1}{3x}\right)\left[\dfrac{6x - (6x+3)}{(3x)^2}\right]$ Multiply factors.

$= \dfrac{2(2x+1)(6x-6x-3)}{(3x)^3}$ Multiply fractions and remove parentheses.

$= \dfrac{2(2x+1)(-3)}{3(9)x^3}$ Combine like terms and factor.

$= \dfrac{-2(2x+1)}{9x^3}$ Divide out like factors.

EXAMPLE 2 Simplifying an Expression with Powers or Radicals

(a) $(2x+1)^2(6x+1) + (3x^2+x)(2)(2x+1)(2)$

$= (2x+1)[(2x+1)(6x+1) + (3x^2+x)(2)(2)]$ Factor.

$= (2x+1)[12x^2 + 8x + 1 + (12x^2 + 4x)]$ Multiply factors.

$= (2x+1)(12x^2 + 8x + 1 + 12x^2 + 4x)$ Remove parentheses.

$= (2x+1)(24x^2 + 12x + 1)$ Combine like terms.

(b) $(-1)(6x^2-4x)^{-2}(12x-4)$

$= \dfrac{(-1)(12x-4)}{(6x^2-4x)^2}$ Rewrite as a fraction.

$= \dfrac{(-1)(4)(3x-1)}{(6x^2-4x)^2}$ Factor.

$= \dfrac{-4(3x-1)}{(6x^2-4x)^2}$ Multiply factors.

(c) $(x)\left(\dfrac{1}{2}\right)(2x+3)^{-1/2} + (2x+3)^{1/2}(1)$

$= (2x+3)^{-1/2}\left(\dfrac{1}{2}\right)[x + (2x+3)(2)]$ Factor.

$= \dfrac{x + 4x + 6}{(2x+3)^{1/2}(2)}$ Rewrite as a fraction.

$= \dfrac{5x+6}{2(2x+3)^{1/2}}$ Combine like terms.

(d) $\dfrac{x^2\left(\frac{1}{2}\right)(2x)(x^2+1)^{-1/2} - (x^2+1)^{1/2}(2x)}{x^4}$

$= \dfrac{(x^3)(x^2+1)^{-1/2} - (x^2+1)^{1/2}(2x)}{x^4}$ Multiply factors.

$= \dfrac{(x^2+1)^{-1/2}(x)[x^2 - (x^2+1)(2)]}{x^4}$ Factor.

$= \dfrac{x[x^2 - (2x^2+2)]}{(x^2+1)^{1/2}x^4}$ Write with positive exponents.

$= \dfrac{x^2 - 2x^2 - 2}{(x^2+1)^{1/2}x^3}$ Divide out like factors and remove parentheses.

$= \dfrac{-x^2 - 2}{(x^2+1)^{1/2}x^3}$ Combine like terms.

All but one of the expressions in this Algebra Review are derivatives. Can you see what the original function is? Explain your reasoning.

ALGEBRA REVIEW

At the end of each chapter, the *Algebra Review* illustrates the key algebraic concepts used in the chapter. Often, rudimentary steps are provided in detail for selected examples from the chapter. This review offers additional support to those students who have trouble following examples as a result of poor algebra skills.

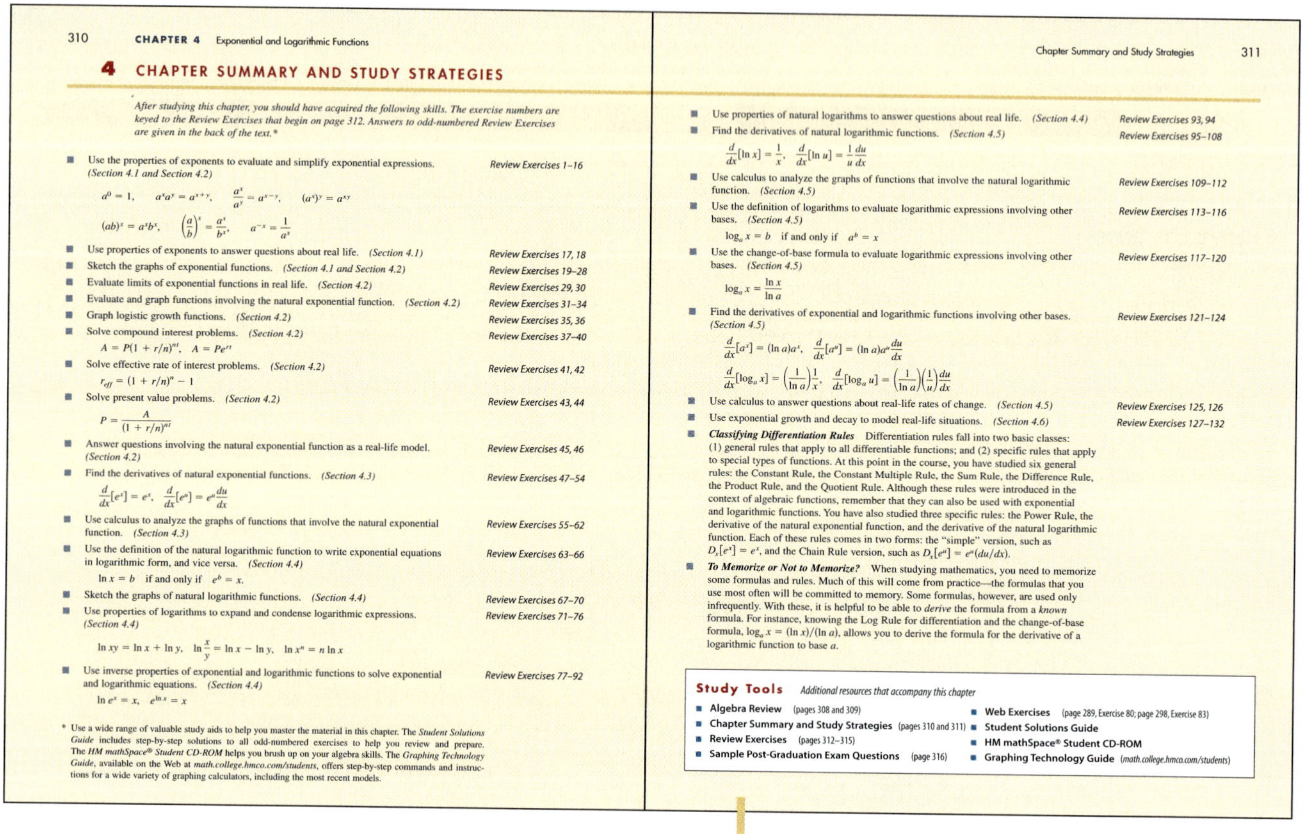

310 CHAPTER 4 Exponential and Logarithmic Functions

4 CHAPTER SUMMARY AND STUDY STRATEGIES

*After studying this chapter, you should have acquired the following skills. The exercise numbers are keyed to the Review Exercises that begin on page 312. Answers to odd-numbered Review Exercises are given in the back of the text.**

- Use the properties of exponents to evaluate and simplify exponential expressions. *(Section 4.1 and Section 4.2)* — *Review Exercises 1–16*

  $a^0 = 1, \quad a^x a^y = a^{x+y}, \quad \frac{a^x}{a^y} = a^{x-y}, \quad (a^x)^y = a^{xy}$

  $(ab)^x = a^x b^x, \quad \left(\frac{a}{b}\right)^x = \frac{a^x}{b^x}, \quad a^{-x} = \frac{1}{a^x}$
- Use properties of exponents to answer questions about real life. *(Section 4.1)* — *Review Exercises 17, 18*
- Sketch the graphs of exponential functions. *(Section 4.1 and Section 4.2)* — *Review Exercises 19–28*
- Evaluate limits of exponential functions in real life. *(Section 4.2)* — *Review Exercises 29, 30*
- Evaluate and graph functions involving the natural exponential function. *(Section 4.2)* — *Review Exercises 31–34*
- Graph logistic growth functions. *(Section 4.2)* — *Review Exercises 35, 36*
- Solve compound interest problems. *(Section 4.2)* — *Review Exercises 37–40*

  $A = P(1 + r/n)^{nt}, \quad A = Pe^{rt}$
- Solve effective rate of interest problems. *(Section 4.2)* — *Review Exercises 41, 42*

  $r_{eff} = (1 + r/n)^n - 1$
- Solve present value problems. *(Section 4.2)* — *Review Exercises 43, 44*

  $P = \frac{A}{(1 + r/n)^{nt}}$
- Answer questions involving the natural exponential function as a real-life model. *(Section 4.2)* — *Review Exercises 45, 46*
- Find the derivatives of natural exponential functions. *(Section 4.3)* — *Review Exercises 47–54*

  $\frac{d}{dx}[e^x] = e^x, \quad \frac{d}{dx}[e^u] = e^u \frac{du}{dx}$
- Use calculus to analyze the graphs of functions that involve the natural exponential function. *(Section 4.3)* — *Review Exercises 55–62*
- Use the definition of the natural logarithmic function to write exponential equations in logarithmic form, and vice versa. *(Section 4.4)* — *Review Exercises 63–66*

  $\ln x = b$ if and only if $e^b = x$.
- Sketch the graphs of natural logarithmic functions. *(Section 4.4)* — *Review Exercises 67–70*
- Use properties of logarithms to expand and condense logarithmic expressions. *(Section 4.4)* — *Review Exercises 71–76*

  $\ln xy = \ln x + \ln y, \quad \ln \frac{x}{y} = \ln x - \ln y, \quad \ln x^n = n \ln x$
- Use inverse properties of exponential and logarithmic functions to solve exponential and logarithmic equations. *(Section 4.4)* — *Review Exercises 77–92*

  $\ln e^x = x, \quad e^{\ln x} = x$

* Use a wide range of valuable study aids to help you master the material in this chapter. The *Student Solutions Guide* includes step-by-step solutions to all odd-numbered exercises to help you review and prepare. The *HM mathSpace® Student CD-ROM* helps you brush up on your algebra skills. The *Graphing Technology Guide*, available on the Web at *math.college.hmco.com/students*, offers step-by-step commands and instructions for a wide variety of graphing calculators, including the most recent models.

Chapter Summary and Study Strategies 311

- Use properties of natural logarithms to answer questions about real life. *(Section 4.4)* — *Review Exercises 93, 94*
- Find the derivatives of natural logarithmic functions. *(Section 4.5)* — *Review Exercises 95–108*

  $\frac{d}{dx}[\ln x] = \frac{1}{x}, \quad \frac{d}{dx}[\ln u] = \frac{1}{u}\frac{du}{dx}$
- Use calculus to analyze the graphs of functions that involve the natural logarithmic function. *(Section 4.5)* — *Review Exercises 109–112*
- Use the definition of logarithms to evaluate logarithmic expressions involving other bases. *(Section 4.5)* — *Review Exercises 113–116*

  $\log_a x = b$ if and only if $a^b = x$
- Use the change-of-base formula to evaluate logarithmic expressions involving other bases. *(Section 4.5)* — *Review Exercises 117–120*

  $\log_a x = \frac{\ln x}{\ln a}$
- Find the derivatives of exponential and logarithmic functions involving other bases. *(Section 4.5)* — *Review Exercises 121–124*

  $\frac{d}{dx}[a^x] = (\ln a)a^x, \quad \frac{d}{dx}[a^u] = (\ln a)a^u \frac{du}{dx}$

  $\frac{d}{dx}[\log_a x] = \left(\frac{1}{\ln a}\right)\frac{1}{x}, \quad \frac{d}{dx}[\log_a u] = \left(\frac{1}{\ln a}\right)\left(\frac{1}{u}\right)\frac{du}{dx}$
- Use calculus to answer questions about real-life rates of change. *(Section 4.5)* — *Review Exercises 125, 126*
- Use exponential growth and decay to model real-life situations. *(Section 4.6)* — *Review Exercises 127–132*
- ***Classifying Differentiation Rules*** Differentiation rules fall into two basic classes: (1) general rules that apply to all differentiable functions; and (2) specific rules that apply to special types of functions. At this point in the course, you have studied six general rules: the Constant Rule, the Constant Multiple Rule, the Sum Rule, the Difference Rule, the Product Rule, and the Quotient Rule. Although these rules were introduced in the context of algebraic functions, remember that they can also be used with exponential and logarithmic functions. You have also studied three specific rules: the Power Rule, the derivative of the natural exponential function, and the derivative of the natural logarithmic function. Each of these rules comes in two forms: the "simple" version, such as $D_x[e^x] = e^x$, and the Chain Rule version, such as $D_x[e^u] = e^u(du/dx)$.
- ***To Memorize or Not to Memorize?*** When studying mathematics, you need to memorize some formulas and rules. Much of this will come from practice—the formulas that you use most often will be committed to memory. Some formulas, however, are used only infrequently. With these, it is helpful to be able to *derive* the formula from a *known* formula. For instance, knowing the Log Rule for differentiation and the change-of-base formula, $\log_a x = (\ln x)/(\ln a)$, allows you to derive the formula for the derivative of a logarithmic function to base a.

Study Tools *Additional resources that accompany this chapter*

- **Algebra Review** (pages 308 and 309)
- **Chapter Summary and Study Strategies** (pages 310 and 311)
- **Review Exercises** (pages 312–315)
- **Sample Post-Graduation Exam Questions** (page 316)
- **Web Exercises** (page 289, Exercise 80; page 298, Exercise 83)
- **Student Solutions Guide**
- **HM mathSpace® Student CD-ROM**
- **Graphing Technology Guide** (*math.college.hmco.com/students*)

CHAPTER SUMMARY AND STUDY STRATEGIES

The *Chapter Summary* reviews the skills covered in the chapter and correlates each skill to the *Review Exercises* that test those skills. Following each *Chapter Summary* is a short list of *Study Strategies* for addressing topics or situations specific to the chapter, and a list of *Study Tools* that accompany each chapter.

REVIEW EXERCISES

The *Review Exercises* offer students opportunities for additional practice as they complete each chapter. Answers to all odd-numbered *Review Exercises* appear at the end of the text.

7 CHAPTER REVIEW EXERCISES

In Exercises 1 and 2, plot the points.

1. $(2, -1, 4), (-1, 3, -3)$
2. $(1, -2, -3), (-4, -3, 5)$

In Exercises 3 and 4, find the distance between the two points.

3. $(0, 0, 0), (2, 5, 9)$
4. $(-4, 1, 5), (1, 3, 7)$

In Exercises 5 and 6, find the midpoint of the line segment joining the two points.

5. $(2, 6, 4), (-4, 2, 8)$
6. $(5, 0, 7), (-1, -2, 9)$

In Exercises 7–10, find the standard form of the equation of the sphere.

7. Center: $(0, 1, 0)$; radius: 5
8. Center: $(4, -5, 3)$; radius: 10
9. Diameter endpoints: $(3, 4, 0), (5, 8, 2)$
10. Diameter endpoints: $(-2, 5, 1), (4, -3, 3)$

In Exercises 11 and 12, find the center and radius of the sphere.

11. $x^2 + y^2 + z^2 + 4x - 2y - 8z + 5 = 0$
12. $x^2 + y^2 + z^2 + 4y - 10z - 7 = 0$

In Exercises 13 and 14, sketch the *xy*-trace of the sphere.

13. $(x + 2)^2 + (y - 1)^2 + (z - 3)^2 = 25$
14. $(x - 1)^2 + (y + 3)^2 + (z - 6)^2 = 72$

In Exercises 15–18, find the intercepts and sketch the graph of the plane.

15. $x + 2y + 3z = 6$
16. $2y + z = 4$
17. $6x + 3y - 6z = 12$
18. $4x - y + 2z = 8$

In Exercises 19–26, identify the surface.

19. $x^2 + y^2 + z^2 - 2x + 4y - 6z + 5 = 0$
20. $16x^2 + 16y^2 - 9z^2 = 0$
21. $x^2 + \frac{y^2}{16} + \frac{z^2}{9} = 1$
22. $-x^2 + \frac{y^2}{16} + \frac{z^2}{9} = 1$

$\frac{x^2}{9} + y^2$

24. $-4x^2 + y^2 + z^2 = 4$
25. $z = \sqrt{x^2 + y^2}$
26. $z = 9x + 3y - 5$

In Exercises 27 and 28, find the function values.

27. $f(x, y) = xy^2$
    (a) $f(2, 3)$ (b) $f(0, 1)$
    (c) $f(-5, 7)$ (d) $f(-2, -4)$
28. $f(x, y) = \frac{x^2}{y}$
    (a) $f(6, 9)$ (b) $f(8, 4)$
    (c) $f(t, 2)$ (d) $f(r, r)$

In Exercises 29 and 30, describe the region *R* in the *xy*-plane that corresponds to the domain of the function. Then find the range of the function.

29. $f(x, y) = \sqrt{1 - x^2 - y^2}$
30. $f(x, y) = \frac{1}{x + y}$

In Exercises 31–34, describe the level curves of the function. Sketch the level curves for the given *c*-values.

31. $z = 10 - 2x - 5y$, $c = 0, 2, 4, 5, 10$
32. $z = \sqrt{9 - x^2 - y^2}$, $c = 0, 1, 2, 3$
33. $z = (xy)^2$, $c = 1, 4, 9, 12, 16$
34. $z = 2e^{xy}$, $c = 1, 2, 3, 4, 5$

35. ***Meteorology*** The contour map shown below represents the average yearly precipitation for Iowa. *(Source: U.S. National Oceanic and Atmospheric Administration)*
    (a) Discuss the use of color to represent the level curves.
    (b) Which part of Iowa receives the most precipitation?
    (c) Which part of Iowa receives the least precipitation?

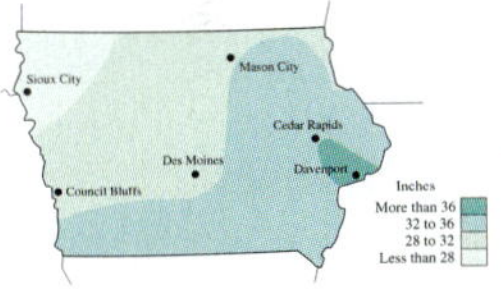

4 SAMPLE POST-GRADUATION EXAM QUESTIONS

CPA
GMAT
GRE
Actuarial
CLAST

The following questions represent the types of questions that appear on certified public accountant (CPA) exams, Graduate Management Admission Tests (GMAT), Graduate Records Exams (GRE), actuarial exams, and College-Level Academic Skills Tests (CLAST). The answers to the questions are given in the back of the book.

1. 10^x means that 10 is to be used as a factor x times, and 10^{-x} is equal to
   $$\frac{1}{10^x}.$$
   A very large or very small number, therefore, is frequently written as a decimal multiplied by 10^x, where x is an integer. Which, if any, are false?
   (a) $470{,}000 = 4.7 \times 10^5$
   (b) 450 billion $= 4.5 \times 10^{11}$
   (c) $0.00000000075 = 7.5 \times 10^{-10}$
   (d) 86 hundred-thousandths $= 8.6 \times 10^2$
2. The rate of decay of a radioactive substance is proportional to the amount of the substance present. Three years ago there was 6 grams of substance. Now there is 5 grams. How many grams will there be 3 years from now?
   (a) 4 (b) $\frac{25}{6}$ (c) $\frac{125}{36}$ (d) $\frac{75}{36}$
3. In a certain town, 45% of the people have brown hair, 30% have brown eyes, and 15% have both brown hair and brown eyes. What percent of the people in the town have neither brown hair nor brown eyes?
   (a) 25% (b) 35% (c) 40% (d) 50%
4. You deposit \$900 in a savings account that is compounded continuously at 4.76%. After 16 years, the amount in the account will be
   (a) \$1927.53 (b) \$1077.81 (c) \$943.88 (d) \$2827.53
5. A bookstore orders 75 books. Each book costs the bookstore \$29 and is sold for \$42. The bookstore must pay a \$4 service charge for each unsold book returned. If the bookstore returns seven books, how much profit will the bookstore make?
   (a) \$975 (b) \$947 (c) \$856 (d) \$681

Figure for 6–9

For Questions 6–9, use the data given in the graph.

6. In how many of the years were expenses greater than in the preceding year?
   (a) 2 (b) 4 (c) 1 (d) 3
7. In which year was the profit the greatest?
   (a) 1997 (b) 2000 (c) 1996 (d) 1998
8. In 1999, profits decreased by x percent from 1998 with x equal to
   (a) 60% (b) 140% (c) 340% (d) 40%
9. In 2000, profits increased by y percent from 1999 with y equal to
   (a) 64% (b) 136% (c) 178% (d) 378%

POST-GRADUATION EXAM QUESTIONS

To emphasize the relevance of calculus, every chapter concludes with sample questions representative of the types of questions on certified public accountant (CPA) exams, Graduate Management Admission Tests® (GMAT®), Graduate Record Examinations® (GRE®), actuarial exams, and College-Level Academic Skills Tests (CLAST). The answers to all *Post-Graduation Exam Questions* are given in the back of the text.

A Plan for You as a Student

Study Strategies

Your success in mathematics depends on your active participation both in class and outside of class. Because the material you learn each day builds on the material you have learned previously, it is important that you keep up with your course work every day and develop a clear plan of study. This set of guidelines highlights key study strategies to help you learn how to study mathematics.

Preparing for Class The syllabus your instructor provides is an invaluable resource that outlines the major topics to be covered in the course. Use it to help you prepare. As a general rule, you should set aside two to four hours of study time for each hour spent in class. Being prepared is the first step toward success. Before class:

- Review your notes from the previous class.
- Read the portion of the text that will be covered in class.
- Use the objectives listed at the beginning of each section to keep you focused on the main ideas of the section.
- Pay special attention to the definitions, rules, and concepts highlighted in boxes. Also, be sure you understand the meanings of mathematical symbols and terms written in boldface type. Keep a vocabulary journal for easy reference.
- Read through the solved examples. Use the side comments given in the solution steps to help you in the solution process. Also, read the *Study Tips* given in the margins.
- Make notes of anything you do not understand as you read through the text. If you still do not understand after your instructor covers the topic in question, ask questions before your instructor moves on to a new topic.
- Try the *Discovery* and *Technology* exercises to get a better grasp of the material before the instructor presents it.

Keeping Up Another important step toward success in mathematics involves your ability to keep up with the work. It is very easy to fall behind, especially if you miss a class. To keep up with the course work, be sure to:

- Attend every class. Bring your text, a notebook, a pen or pencil, and a calculator (scientific or graphing). If you miss a class, get the notes from a classmate as soon as possible and review them carefully.
- Participate in class. As mentioned above, if there is a topic you do not understand, ask about it before the instructor moves on to a new topic.
- Take notes in class. After class, read through your notes and add explanations so that your notes make sense to *you*. Fill in any gaps and note any questions you might have.
- Reread the portion of the text that was covered in class. This time, work each example *before* reading through the solution.
- Do your homework as soon as possible, while concepts are still fresh in your mind. Allow at least two hours of homework time for each hour spent in class so you do not fall behind. Learning mathematics is a step-by-step process, and you must understand each topic in order to learn the next one.
- When you are working problems for homework assignments, show every step in your solution. Then, if you make an error, it will be easier to find where the error occurred.
- Use your notes from class, the text discussion, the examples, and the *Study Tips* as you do your homework. Many exercises are keyed to specific examples in the text for easy reference.

Getting Extra Help It can be very frustrating when you do not understand concepts and are unable to complete homework assignments. However, there are many resources available to help you with your studies.

- Your instructor may have office hours. If you are feeling overwhelmed and need help, make an appointment to discuss your difficulties with your instructor.
- Find a study partner or a study group. Sometimes it helps to work through problems with another person.
- Arrange to get regular assistance from a tutor. Many colleges have math resource centers available on campus as well.
- Consult one of the many ancillaries available with this text: the *HM mathSpace® Student CD-ROM*, the *Student Solutions Guide*, videotapes, and additional study resources available at this text's website at *college.hmco.com.*
- Special assistance with algebra appears in the *Algebra Reviews*, which appear throughout each chapter. These short reviews are tied together in the larger *Algebra Review* section at the end of each chapter.

Preparing for an Exam The last step toward success in mathematics lies in how you prepare for and complete exams. If you have followed the suggestions given above, then you are almost ready for exams. Do not assume that you can cram for the exam the night before—this seldom works. As a final preparation for the exam:

- Read the *Chapter Summary* and *Study Strategies* keyed to each section, and review the concepts and terms.
- Work through the *Review Exercises* if you need extra practice on material from a particular section. You can practice for an exam by first trying to work through the exercises with your book and notebook closed.
- Take practice tests offered online at this text's website at *college.hmco.com.*
- When you study for an exam, first look at all definitions, properties, and formulas until you know them. Review your notes and the portion of the text that will be covered on the exam. Then work as many exercises as you can, especially any kinds of exercises that have given you trouble in the past, reworking homework problems as necessary.
- Start studying for your exam well in advance (at least a week). The first day or two, study only about two hours. Gradually increase your study time each day. Be completely prepared for the exam two days in advance. Spend the final day just building confidence so you can be relaxed during the exam.
- Avoid studying up until the last minute. This will only make you anxious. Allow yourself plenty of time to get to the testing location. When you take the exam, go in with a clear mind and a positive attitude.
- Once the exam begins, read through the directions and the entire exam before beginning. Work the problems that you know how to do first to avoid spending too much time on any one problem. Time management is extremely important when taking an exam.
- If you finish early, use the remaining time to go over your work.
- When you get an exam back, review it carefully and go over your errors. Rework the problems you answered incorrectly. Discovering the mistakes you made will help you improve your test-taking ability. Understanding how to correct your errors will help you build on the knowledge you have gained before you move on to the next topic.

chapter

0 A Precalculus Review

Richard Megna/Fundamental Photographs

The period of a pendulum is dependent only on the length of the pendulum. Changing the length of the pendulum can correct a slow running grandfather clock, regardless of the weight of the bob or the amplitude.

STRATEGIES FOR SUCCESS

WHAT YOU SHOULD LEARN:

This initial chapter, A Precalculus Review, is just that—a review chapter. Make sure that you have a solid understanding of all the material in this chapter before beginning Chapter 1. You can also use it as a reference as you progress through the text, coming back to brush up on an algebra skill that you may have forgotten over the course of the semester. As with all math courses, calculus is a building process—that is, you use what you know to go on to the next topic.

WHY YOU SHOULD LEARN IT:

Precalculus concepts have many applications in real life, as can be seen by the examples below, which represent a small sample of the applications in this chapter.

- Biology: pH Values, Exercise 29 on page 0-7
- Budget Variance, Exercises 47-50 on page 0-12
- Compound Interest, Exercises 57-60 on page 0-18
- Period of a Pendulum, Exercise 61 on page 0-18
- Chemistry: Finding Concentrations, Exercise 71 on page 0-24
- Installment Loan, Exercise 47 on page 0-32

0.1 THE REAL NUMBER LINE AND ORDER

- Represent, classify, and order real numbers.
- Use inequalities to represent sets of real numbers.
- Solve inequalities.
- Use inequalities to model and solve real-life problems.

The Real Number Line

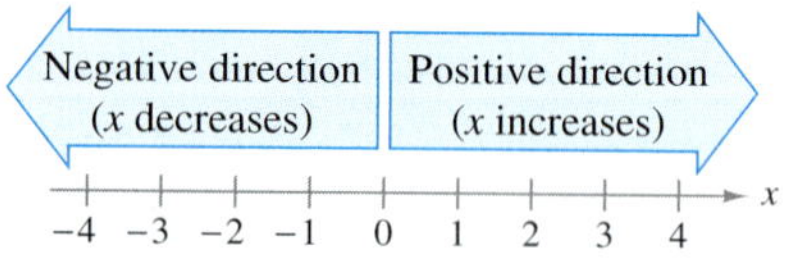

FIGURE 0.1 The Real Number Line

Real numbers can be represented with a coordinate system called the **real number line** (or x-axis), as shown in Figure 0.1. The **positive direction** (to the right) is denoted by an arrowhead and indicates the direction of increasing values of x. The real number corresponding to a particular point on the real number line is called the **coordinate** of the point. As shown in Figure 0.1, it is customary to label those points whose coordinates are integers.

The point on the real number line corresponding to zero is called the **origin.** Numbers to the right of the origin are **positive,** and numbers to the left of the origin are **negative.** The term **nonnegative** describes a number that is either positive or zero.

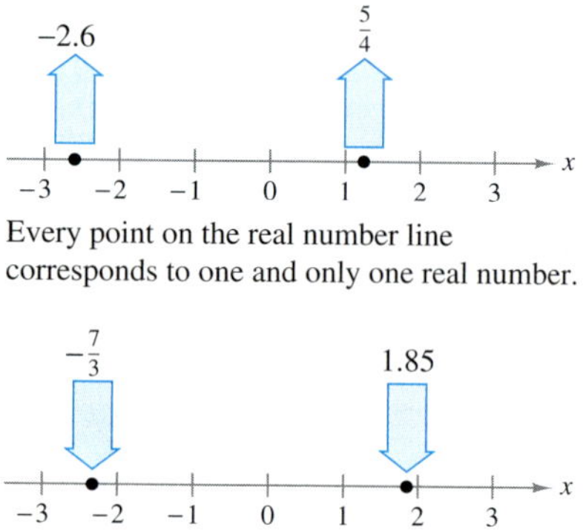

Every point on the real number line corresponds to one and only one real number.

Every real number corresponds to one and only one point on the real number line.

FIGURE 0.2

The importance of the real number line is that it provides you with a conceptually perfect picture of the real numbers. That is, each point on the real number line corresponds to one and only one real number, and each real number corresponds to one and only one point on the real number line. This type of relationship is called a **one-to-one correspondence** and is illustrated in Figure 0.2.

Each of the four points in Figure 0.2 corresponds to a real number that can be expressed as the ratio of two integers.

$-2.6 = -\frac{13}{5}$ $\qquad$ $\frac{5}{4}$

$-\frac{7}{3}$ $\qquad$ $1.85 = \frac{37}{20}$

Such numbers are called **rational.** Rational numbers have either terminating or infinitely repeating decimal representations.

Terminating Decimals	*Infinitely Repeating Decimals*
$\frac{2}{5} = 0.4$	$\frac{1}{3} = 0.333\ldots = 0.\overline{3}$*
$\frac{7}{8} = 0.875$	$\frac{12}{7} = 1.714285714285\ldots = 1.\overline{714285}$

Real numbers that are not rational are called **irrational,** and they cannot be represented as the ratio of two integers (or as terminating or infinitely repeating decimals). So, a decimal approximation is used to represent an irrational number. Some irrational numbers occur so frequently in applications that mathematicians have invented special symbols to represent them. For example, the symbols $\sqrt{2}$, π, and e represent irrational numbers whose decimal approximations are as shown. (See Figure 0.3.)

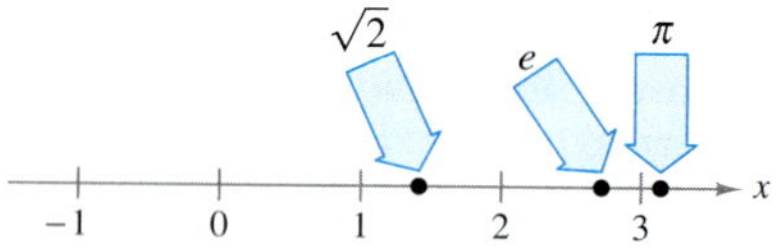

FIGURE 0.3

$$\sqrt{2} \approx 1.4142135623$$

$$\pi \approx 3.1415926535$$

$$e \approx 2.7182818284$$

*The bar indicates which digit or digits repeat infinitely.

Order and Intervals on the Real Number Line

One important property of the real numbers is that they are **ordered:** 0 is less than 1, -3 is less than -2.5, π is less than $\frac{22}{7}$, and so on. You can visualize this property on the real number line by observing that a is less than b if and only if a lies to the left of b on the real number line. Symbolically, "a is less than b" is denoted by the inequality

$$a < b.$$

For example, the inequality $\frac{3}{4} < 1$ follows from the fact that $\frac{3}{4}$ lies to the left of 1 on the real number line, as shown in Figure 0.4.

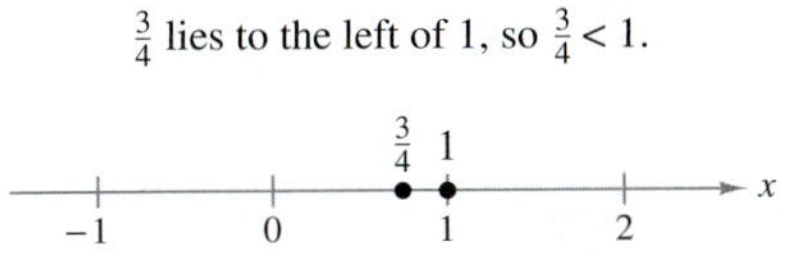

FIGURE 0.4

When three real numbers a, x, and b are ordered such that $a < x$ and $x < b$, we say that x is **between** a and b and write

$$a < x < b.$$ x is between a and b.

The set of *all* real numbers between a and b is called the **open interval** between a and b and is denoted by (a, b). An interval of the form (a, b) does not contain the "endpoints" a and b. Intervals that include their endpoints are called **closed** and are denoted by $[a, b]$. Intervals of the form $[a, b)$ and $(a, b]$ are neither open nor closed. Figure 0.5 shows the nine types of intervals on the real number line.

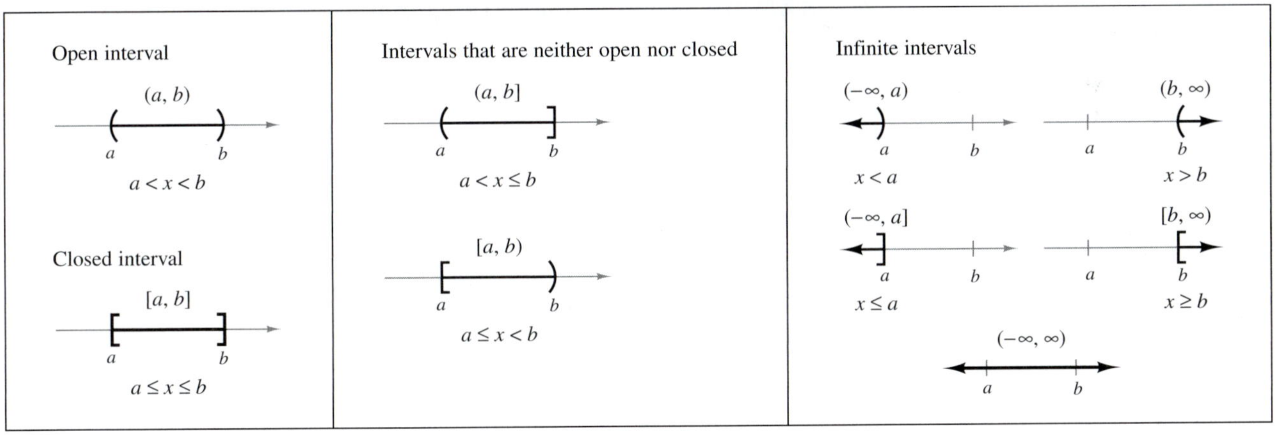

FIGURE 0.5 Intervals on the Real Number Line

STUDY TIP

Note that a square bracket is used to denote "less than or equal to" ($\leq$) or "greater than or equal to" ($\geq$). Furthermore, the symbols ∞ and $-\infty$ denote **positive** and **negative infinity.** These symbols do not denote real numbers; they merely let you describe unbounded conditions more concisely. For instance, the interval $[b, \infty)$ is unbounded to the right because it includes *all* real numbers that are greater than or equal to b.

Solving Inequalities

In calculus, you are frequently required to "solve inequalities" involving variable expressions such as $3x - 4 < 5$. The number a is a **solution** of an inequality if the inequality is true when a is substituted for x. The set of all values of x that satisfy an equality is called the **solution set** of the inequality. The following properties are useful for solving inequalities. (Similar properties are obtained if $<$ is replaced by $\leq$ and $>$ is replaced by $\geq$.)

STUDY TIP

Notice the differences between Properties 3 and 4. For example,

$$-3 < 4 \Rightarrow (-3)(2) < (4)(2)$$

and

$$-3 < 4 \Rightarrow (-3)(-2) > (4)(-2).$$

Properties of Inequalities

Let a, b, c, and d be real numbers.

1. Transitive property: $a < b$ and $b < c$ $\Rightarrow$ $a < c$
2. Adding inequalities: $a < b$ and $c < d$ $\Rightarrow$ $a + c < b + d$
3. Multiplying by a (positive) constant: $a < b$ $\Rightarrow$ $ac < bc$, $c > 0$
4. Multiplying by a (negative) constant: $a < b$ $\Rightarrow$ $ac > bc$, $c < 0$
5. Adding a constant: $a < b$ $\Rightarrow$ $a + c < b + c$
6. Subtracting a constant: $a < b$ $\Rightarrow$ $a - c < b - c$

ALGEBRA REVIEW

Once you have solved an inequality, it is a good idea to check some x-values in your solution set to see whether they satisfy the original inequality. You might also check some values outside your solution set to verify that they do *not* satisfy the inequality. For example, Figure 0.6 shows that when $x = 0$ or $x = 2$ the inequality is satisfied, but when $x = 4$ the inequality is not satisfied.

Note that you *reverse the inequality* when you multiply by a negative number. For example, if $x < 3$, then $-4x > -12$. This principle also applies to division by a negative number. So, if $-2x > 4$, then $x < -2$.

EXAMPLE 1 Solving an Inequality

Find the solution set of the inequality $3x - 4 < 5$.

SOLUTION

$3x - 4 < 5$	Write original inequality.
$3x - 4 + 4 < 5 + 4$	Add 4 to each side.
$3x < 9$	Simplify.
$\frac{1}{3}(3x) < \frac{1}{3}(9)$	Multiply each side by $\frac{1}{3}$.
$x < 3$	Simplify.

So, the solution set is the interval $(-\infty, 3)$, as shown in Figure 0.6.

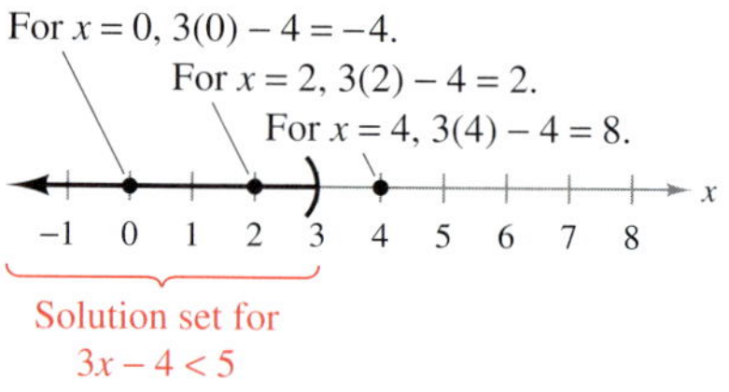

FIGURE 0.6

TRY IT 1

Find the solution set of the inequality $2x - 3 < 7$.

In Example 1, all five inequalities listed as steps in the solution have the same solution set, and they are called **equivalent inequalities.**

The inequality in Example 1 involves a first-degree polynomial. To solve inequalities involving polynomials of higher degree, you can use the fact that a polynomial can change signs *only* at its real zeros (the real numbers that make the polynomial zero). Between two consecutive real zeros, a polynomial must be entirely positive or entirely negative. This means that when the real zeros of a polynomial are put in order, they divide the real number line into **test intervals** in which the polynomial has no sign changes. That is, if a polynomial has the factored form

$$(x - r_1)(x - r_2), \ldots, (x - r_n), \qquad r_1 < r_2 < r_3 < \cdot \cdot \cdot < r_n$$

then the test intervals are

$$(-\infty, r_1), \quad (r_1, r_2), \quad \ldots, \quad (r_{n-1}, r_n), \quad \text{and} \quad (r_n, \infty).$$

For example, the polynomial

$$x^2 - x - 6 = (x - 3)(x + 2)$$

can change signs only at $x = -2$ and $x = 3$. To determine the sign of the polynomial in the intervals $(-\infty, -2)$, $(-2, 3)$, and $(3, \infty)$, you need to test only *one value* from each interval.

EXAMPLE 2 Solving a Polynomial Inequality

Find the solution set of the inequality $x^2 < x + 6$.

SOLUTION

$$x^2 < x + 6 \qquad \text{Write original inequality.}$$
$$x^2 - x - 6 < 0 \qquad \text{Polynomial form}$$
$$(x - 3)(x + 2) < 0 \qquad \text{Factor.}$$

So, the polynomial $x^2 - x - 6$ has $x = -2$ and $x = 3$ as its zeros. You can solve the inequality by testing the sign of the polynomial in each of the following intervals.

$$x < -2, \quad -2 < x < 3, \quad x > 3$$

To test an interval, choose a representative number in the interval and compute the sign of each factor. For example, for any $x < -2$, both of the factors $(x - 3)$ and $(x + 2)$ are negative. Consequently, the product (of two negative numbers) is positive, and the inequality is *not* satisfied in the interval

$$x < -2.$$

A convenient testing format is shown in Figure 0.7. Because the inequality is satisfied only by the center test interval, you can conclude that the solution set is given by the interval

$$-2 < x < 3. \qquad \text{Solution set}$$

Sign of $(x - 3)(x + 2)$

x	Sign	< 0?
-3	$(-)(-)$	No
-2	$(-)(0)$	No
-1	$(-)(+)$	Yes
0	$(-)(+)$	Yes
1	$(-)(+)$	Yes
2	$(-)(+)$	Yes
3	$(0)(+)$	No
4	$(+)(+)$	No

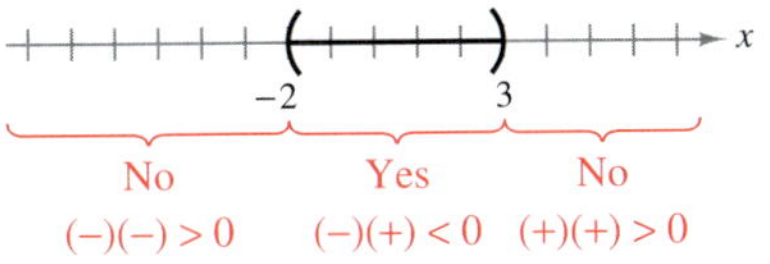

FIGURE 0.7 Is $(x - 3)(x + 2) < 0$?

TRY IT 2

Find the solution set of the inequality $x^2 > 3x + 10$.

Application

Inequalities are frequently used to describe conditions that occur in business and science. For instance, the inequality

$$144 \le W \le 180$$

describes the recommended weight W for a man whose height is 5 feet 10 inches. Example 3 shows how an inequality can be used to describe the production level of a manufacturing plant.

EXAMPLE 3 Production Levels

In addition to fixed overhead costs of \$500 per day, the cost of producing x units of an item is \$2.50 per unit. During the month of August, the total cost of production varied from a high of \$1325 to a low of \$1200 per day. Find the high and low *production levels* during the month.

SOLUTION Because it costs \$2.50 to produce one unit, it costs $2.5x$ to produce x units. Furthermore, because the fixed cost per day is \$500, the total daily cost of producing x units is

$$C = 2.5x + 500.$$

Now, because the cost ranged from \$1200 to \$1325, you can write the following.

$$1200 \le 2.5x + 500 \le 1325 \quad \text{Write original inequality.}$$

$$1200 - 500 \le 2.5x + 500 - 500 \le 1325 - 500 \quad \text{Subtract 500 from each side.}$$

$$700 \le 2.5x \le 825 \quad \text{Simplify.}$$

$$\frac{700}{2.5} \le \frac{2.5x}{2.5} \le \frac{825}{2.5} \quad \text{Divide each side by 2.5.}$$

$$280 \le x \le 330 \quad \text{Simplify.}$$

So, the daily production levels during the month of August varied from a low of 280 units to a high of 330 units, as shown in Figure 0.8.

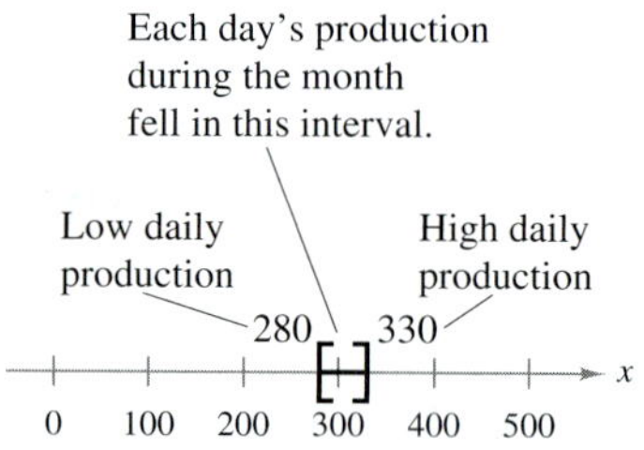

FIGURE 0.8

TRY IT 3

Use the information in Example 3 to find the high and low production levels if, during October, the total cost of production varied from a high of \$1500 to a low of \$1000 per day.

EXERCISES 0.1

In Exercises 1–10, determine whether the real number is rational or irrational.

*1. 0.7

2. -3678

3. $\frac{3\pi}{2}$

4. $3\sqrt{2} - 1$

5. $4.3\overline{451}$

6. $\frac{22}{7}$

7. $\sqrt[3]{64}$

8. $0.\overline{8177}$

9. $\sqrt[3]{60}$

10. $2e$

In Exercises 11–14, determine whether each given value of x satisfies the inequality.

11. $5x - 12 > 0$
    (a) $x = 3$ (b) $x = -3$
    (c) $x = \frac{5}{2}$ (d) $x = \frac{3}{2}$

12. $x + 1 < \frac{2x}{3}$
    (a) $x = 0$ (b) $x = 4$
    (c) $x = -4$ (d) $x = -3$

13. $0 < \frac{x - 2}{4} < 2$
    (a) $x = 4$ (b) $x = 10$
    (c) $x = 0$ (d) $x = \frac{7}{2}$

14. $-1 < \frac{3 - x}{2} \leq 1$
    (a) $x = 0$ (b) $x = \sqrt{5}$
    (c) $x = 1$ (d) $x = 5$

In Exercises 15–28, solve the inequality and sketch the graph of the solution on the real number line.

15. $x - 5 \geq 7$

16. $2x > 3$

17. $4x + 1 < 2x$

18. $2x + 7 < 3$

19. $4 - 2x < 3x - 1$

20. $x - 4 \leq 2x + 1$

21. $-4 < 2x - 3 < 4$

22. $0 \leq x + 3 < 5$

23. $\frac{3}{4} > x + 1 > \frac{1}{4}$

24. $-1 < -\frac{x}{3} < 1$

25. $\frac{x}{2} + \frac{x}{3} > 5$

26. $\frac{x}{2} - \frac{x}{3} > 5$

27. $2x^2 - x < 6$

28. $2x^2 + 1 < 9x - 3$

29. ***Biology: pH Values*** The pH scale measures the concentration of hydrogen ions in a solution. Strong acids produce low pH values, while strong bases produce high pH values. Represent the following approximate pH values on a real number line: hydrochloric acid, 0.0; lemon juice, 2.0; oven cleaner, 13.0; baking soda, 9.0; pure water, 7.0; black coffee, 5.0. *(Source: Adapted from Levine/Miller,* Biology: Discovering Life, *Second Edition)*

30. ***Physiology*** The maximum heart rate of a person in normal health is related to the person's age by the equation

$$r = 220 - A$$

where r is the maximum heart rate in beats per minute and A is the person's age in years. Some physiologists recommend that during physical activity a person should strive to increase his or her heart rate to at least 60% of the maximum heart rate for sedentary people and at most 90% of the maximum heart rate for highly fit people. Express as an interval the range of the target heart rate for a 20-year-old.

31. ***Profit*** The revenue for selling x units of a product is

$$R = 115.95x$$

and the cost of producing x units is

$$C = 95x + 750.$$

To obtain a profit, the revenue must be *greater than* the cost. For what values of x will this product return a profit?

32. ***Sales*** A doughnut shop at a shopping mall sells a dozen doughnuts for \$3.50. Beyond the fixed cost (for rent, utilities, and insurance) of \$170 per day, it costs \$1.75 for enough materials (flour, sugar, etc.) and labor to produce each dozen doughnuts. If the daily profit *varies between* \$40 and \$250, between what levels (in dozens) do the daily sales vary?

33. ***Reimbursement*** A pharmaceutical company reimburses their sales representatives \$0.35 per mile driven and \$100 for meals per week. The company allocates from \$200 to \$250 per sales representative each week. What are the minimum and maximum numbers of miles the company expects each representative to drive each week?

34. ***Area*** A square region is to have an area of *at least* 500 square meters. What must the length of the sides of the region be?

In Exercises 35 and 36, determine whether each statement is true or false, given $a < b$.

35. (a) $-2a < -2b$
    (b) $a + 2 < b + 2$
    (c) $6a < 6b$
    (d) $\frac{1}{a} < \frac{1}{b}$

36. (a) $a - 4 < b - 4$
    (b) $4 - a < 4 - b$
    (c) $-3b < -3a$
    (d) $\frac{a}{4} < \frac{b}{4}$

* The answers to the odd-numbered and selected even exercises are given in the back of the text. Worked-out solutions to the odd-numbered exercises are given in the *Student Solutions Guide*.

0.2 ABSOLUTE VALUE AND DISTANCE ON THE REAL NUMBER LINE

- Find the absolute values of real numbers and understand the properties of absolute value.
- Find the distance between two numbers on the real number line.
- Define intervals on the real number line.
- Find the midpoint of an interval and use intervals to model and solve real-life problems.

Absolute Value of a Real Number

TECHNOLOGY

Absolute value expressions can be evaluated on a graphing utility. When an expression such as $|3 - 8|$ is evaluated, parentheses should surround the expression, as in abs(3 − 8).

Definition of Absolute Value

The **absolute value** of a real number a is

$$|a| = \begin{cases} a, & \text{if } a \geq 0 \\ -a, & \text{if } a < 0. \end{cases}$$

At first glance, it may appear from this definition that the absolute value of a real number can be negative, but this is not possible. For example, let $a = -3$. Then, because $-3 < 0$, you have

$$\begin{aligned} |a| &= |-3| \\ &= -(-3) \\ &= 3. \end{aligned}$$

The following properties are useful for working with absolute values.

Properties of Absolute Value

1. Multiplication: $|ab| = |a||b|$
2. Division: $\left|\dfrac{a}{b}\right| = \dfrac{|a|}{|b|}, \quad b \neq 0$
3. Power: $|a^n| = |a|^n$
4. Square root: $\sqrt{a^2} = |a|$

Be sure you understand the fourth property in this list. A common error in algebra is to imagine that by squaring a number and then taking the square root, you come back to the original number. But this is true only if the original number is nonnegative. For instance, if $a = 2$, then

$$\sqrt{2^2} = \sqrt{4} = 2$$

but if $a = -2$, then

$$\sqrt{(-2)^2} = \sqrt{4} = 2.$$

The reason for this is that (by definition) the square root symbol $\sqrt{}$ denotes only the nonnegative root.

Distance on the Real Number Line

Consider two distinct points on the real number line, as shown in Figure 0.9.

1. The **directed distance from a to b** is $b - a$.
2. The **directed distance from b to a** is $a - b$.
3. The **distance between a and b** is $|a - b|$ or $|b - a|$.

In Figure 0.9, note that because b is to the right of a, the directed distance from a to b (moving to the right) is positive. Moreover, because a is to the left of b, the directed distance from b to a (moving to the left) is negative. The distance *between* two points on the real number line can never be negative.

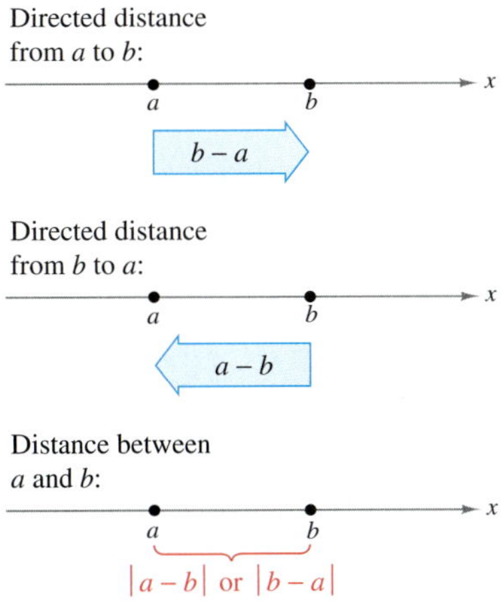

FIGURE 0.9

Distance Between Two Points on the Real Number Line

The distance d between points x_1 and x_2 on the real number line is given by

$$d = |x_2 - x_1| = \sqrt{(x_2 - x_1)^2}.$$

Note that the order of subtraction with x_1 and x_2 does not matter because

$$|x_2 - x_1| = |x_1 - x_2| \quad \text{and} \quad (x_2 - x_1)^2 = (x_1 - x_2)^2.$$

EXAMPLE 1 Finding Distance on the Real Number Line

Determine the distance between -3 and 4 on the real number line. What is the directed distance from -3 to 4? What is the directed distance from 4 to -3?

SOLUTION The distance between -3 and 4 is given by

$$|-3 - 4| = |-7| = 7 \quad \text{or} \quad |4 - (-3)| = |7| = 7 \qquad |a - b| \text{ or } |b - a|$$

as shown in Figure 0.10.

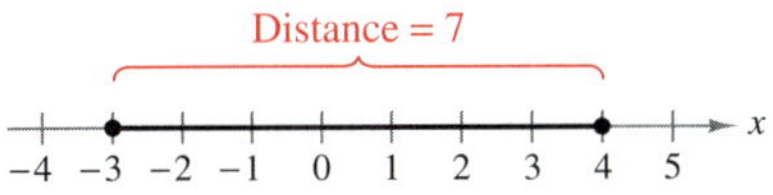

FIGURE 0.10

The directed distance from -3 to 4 is

$$4 - (-3) = 7. \qquad b - a$$

The directed distance from 4 to -3 is

$$-3 - 4 = -7. \qquad a - b$$

TRY IT 1

Determine the distance between -2 and 6 on the real number line. What is the directed distance from -2 to 6? What is the directed distance from 6 to -2?

Intervals Defined by Absolute Value

EXAMPLE 2 Defining an Interval on the Real Number Line

Find the interval on the real number line that contains all numbers that lie no more than two units from 3.

SOLUTION Let x be any point in this interval. You need to find all x such that the distance between x and 3 is less than or equal to 2. This implies that

$$|x - 3| \le 2.$$

Requiring the absolute value of $x - 3$ to be less than or equal to 2 means that $x - 3$ must lie between -2 and 2. So, you can write

$$-2 \le x - 3 \le 2.$$

Solving this pair of inequalities, you have

$$-2 + 3 \le x - 3 + 3 \le 2 + 3$$

$$1 \le x \le 5. \quad \text{Solution set}$$

So, the interval is $[1, 5]$, as shown in Figure 0.11.

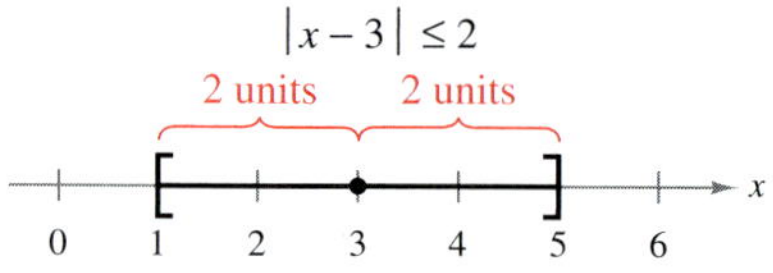

FIGURE 0.11

TRY IT 2

Find the interval on the real number line that contains all numbers that lie no more than four units from 6.

Two Basic Types of Inequalities Involving Absolute Value

Let a and d be real numbers, where $d > 0$.

$|x - a| \le d$ if and only if $a - d \le x \le a + d$.

$|x - a| \ge d$ if and only if $x \le a - d$ or $a + d \le x$.

Inequality	*Interpretation*	*Graph*
$\lvert x - a\rvert \le d$	All numbers x whose distance from a is less than or equal to d.	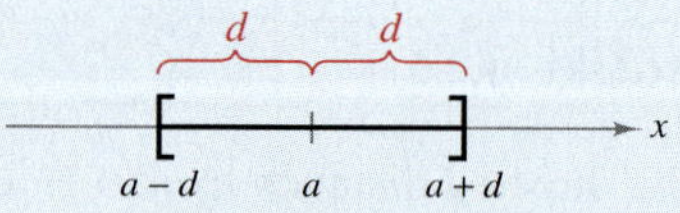
$\lvert x - a\rvert \ge d$	All numbers x whose distance from a is greater than or equal to d.	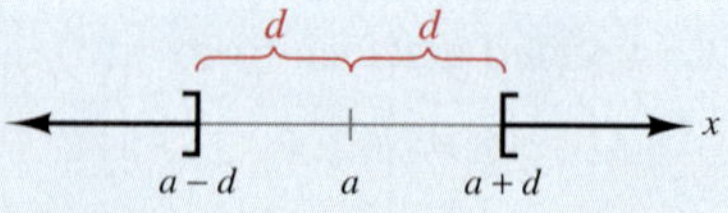

ALGEBRA REVIEW

Be sure you see that inequalities of the form $|x - a| \ge d$ have solution sets consisting of two intervals. To describe the two intervals without using absolute values, you must use *two* separate inequalities, connected by an "or" to indicate union.

Application

EXAMPLE 3 Quality Control

A large manufacturer hired a quality control firm to determine the reliability of a product. Using statistical methods, the firm determined that the manufacturer could expect $0.35\% \pm 0.17\%$ of the units to be defective. If the manufacturer offers a money-back guarantee on this product, how much should be budgeted to cover the refunds on 100,000 units? (Assume that the retail price is \$8.95.)

SOLUTION Let r represent the percent of defective units (written in decimal form). You know that r will differ from 0.0035 by at most 0.0017.

$$0.0035 - 0.0017 \le r \le 0.0035 + 0.0017$$

$$0.0018 \le r \le 0.0052 \qquad \text{Figure 0.12(a)}$$

Now, letting x be the number of defective units out of 100,000, it follows that $x = 100{,}000r$ and you have

$$0.0018(100{,}000) \le 100{,}000r \le 0.0052(100{,}000)$$

$$180 \le x \le 520. \qquad \text{Figure 0.12(b)}$$

Finally, letting C be the cost of refunds, you have $C = 8.95x$. So, the total cost of refunds for 100,000 units should fall within the interval given by

$$180(8.95) \le 8.95x \le 520(8.95)$$

$$\$1611 \le C \le \$4654. \qquad \text{Figure 0.12(c)}$$

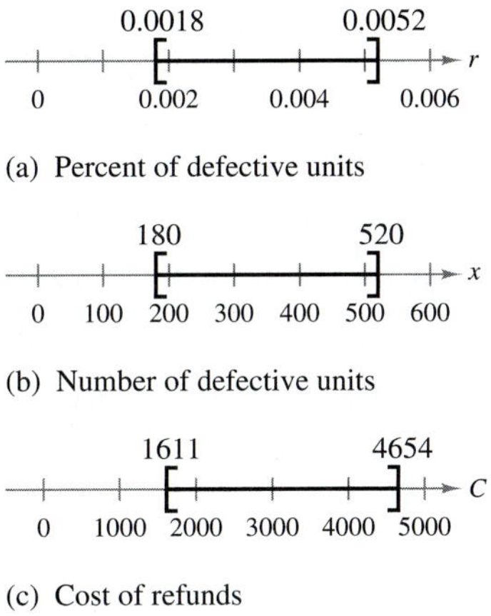

FIGURE 0.12

TRY IT 3

Use the information in Example 3 to determine how much should be budgeted to cover refunds on 250,000 units.

In Example 3, the manufacturer should expect to spend between \$1611 and \$4654 for refunds. Of course, the safer budget figure for refunds would be the higher of these estimates. However, from a statistical point of view, the most representative estimate would be the average of these two extremes. Graphically, the average of two numbers is the **midpoint** of the interval with the two numbers as endpoints, as shown in Figure 0.13.

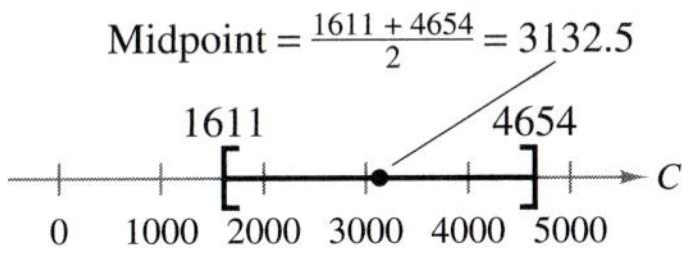

FIGURE 0.13

Midpoint of an Interval

The **midpoint** of the interval with endpoints a and b is found by taking the average of the endpoints.

$$\text{Midpoint} = \frac{a + b}{2}$$

EXERCISES 0.2

In Exercises 1–6, find (a) the directed distance from a to b, (b) the directed distance from b to a, and (c) the distance between a and b.

1. $a = 126, b = 75$ **2.** $a = -126, b = -75$

3. $a = 9.34, b = -5.65$ **4.** $a = -2.05, b = 4.25$

5. $a = \frac{16}{5}, b = \frac{112}{75}$ **6.** $a = -\frac{18}{5}, b = \frac{61}{15}$

In Exercises 7–18, use absolute values to describe the given interval (or pair of intervals) on the real number line.

7. $[-2, 2]$ **8.** $(-3, 3)$

9. $(-\infty, -2) \cup (2, \infty)$ **10.** $(-\infty, -3] \cup [3, \infty)$

11. $[2, 6]$ **12.** $(-7, -1)$

13. $(-\infty, 0) \cup (4, \infty)$ **14.** $(-\infty, 20) \cup (24, \infty)$

15. All numbers *less than* two units from 4

16. All numbers *more than* six units from 3

17. y is *at most* two units from a.

18. y is *less than* h units from c.

In Exercises 19–34, solve the inequality and sketch the graph of the solution on the real number line.

19. $|x| < 5$ **20.** $|2x| < 6$

21. $\left|\frac{x}{2}\right| > 3$ **22.** $|5x| > 10$

23. $|x + 2| < 5$ **24.** $|3x + 1| \geq 4$

25. $\left|\frac{x - 3}{2}\right| \geq 5$ **26.** $|2x + 1| < 5$

27. $|10 - x| > 4$ **28.** $|25 - x| \geq 20$

29. $|9 - 2x| < 1$

30. $\left|1 - \frac{2x}{3}\right| < 1$

31. $|x - a| \leq b, \ b > 0$

32. $|2x - a| \geq b, \ b > 0$

33. $\left|\frac{3x - a}{4}\right| < 2b, \ b > 0$

34. $\left|a - \frac{5x}{2}\right| > b, \ b > 0$

In Exercises 35–40, find the midpoint of the given interval.

35. $[7, 21]$ **36.** $[8.6, 11.4]$

37. $[-6.85, 9.35]$ **38.** $[-4.6, -1.3]$

39. $\left[-\frac{1}{2}, \frac{3}{4}\right]$ **40.** $\left[\frac{5}{6}, \frac{5}{2}\right]$

41. ***Chemistry*** Copper has a melting point M within 0.2°C of 1083.4°C. Use absolute values to write the range as an inequality.

42. ***Stock Price*** A stock market analyst predicts that over the next year the price p of a stock will not change from its current price of $\$33\frac{1}{8}$ by more than \$2. Use absolute values to write this prediction as an inequality.

43. ***Statistics*** The heights h of two-thirds of the members of a population satisfy the inequality

$$\left|\frac{h - 68.5}{2.7}\right| \leq 1$$

where h is measured in inches. Determine the interval on the real number line in which these heights lie.

44. ***Biology*** The American Kennel Club has developed guidelines for judging the features of various breeds of dogs. For collies, the guidelines specify that the weights for males satisfy the inequality

$$\left|\frac{w - 57.5}{7.5}\right| \leq 1$$

where w is measured in pounds. Determine the interval on the real number line in which these weights lie.

45. ***Production*** The estimated daily production x at a refinery is given by

$$|x - 200{,}000| \leq 25{,}000$$

where x is measured in barrels of oil. Determine the high and low production levels.

46. ***Manufacturing*** The acceptable weights for a 20-ounce cereal box are given by

$$|x - 20| \leq 0.75$$

where x is measured in ounces. Determine the high and low weights for the cereal box.

Budget Variance In Exercises 47–50, (a) use absolute value notation to represent the two intervals in which expenses must lie if they are to be within \$500 and within 5% of the specified budget amount and (b) using the more stringent constraint, determine whether the given expense is at variance with the budget restriction.

	Item	*Budget*	*Expense*
47.	Utilities	\$4750.00	\$5116.37
48.	Insurance	\$15,000.00	\$14,695.00
49.	Maintenance	\$20,000.00	\$22,718.35
50.	Taxes	\$7500.00	\$8691.00

0.3 EXPONENTS AND RADICALS

- Evaluate expressions involving exponents or radicals.
- Simplify expressions with exponents.
- Find the domains of algebraic expressions.

Expressions Involving Exponents or Radicals

Properties of Exponents

1. Whole-number exponents: $x^n = \underbrace{x \cdot x \cdot x \cdot \cdot \cdot x}_{n \text{ factors}}$
2. Zero exponent: $x^0 = 1, \quad x \neq 0$
3. Negative exponents: $x^{-n} = \dfrac{1}{x^n}, \quad x \neq 0$
4. Radicals (principal nth root): $\sqrt[n]{x} = a \Rightarrow x = a^n$
5. Rational exponents $(1/n)$: $x^{1/n} = \sqrt[n]{x}$
6. Rational exponents (m/n): $x^{m/n} = (x^{1/n})^m = \left(\sqrt[n]{x}\right)^m$
   $x^{m/n} = (x^m)^{1/n} = \sqrt[n]{x^m}$
7. Special convention (square root): $\sqrt[2]{x} = \sqrt{x}$

ALGEBRA REVIEW

If n is even, then the principal nth root is positive. For example, $\sqrt{4} = +2$ and $\sqrt[4]{81} = +3$.

EXAMPLE 1 Evaluating Expressions

	Expression	*x-Value*	*Substitution*
(a)	$y = -2x^2$	$x = 4$	$y = -2(4^2) = -2(16) = -32$
(b)	$y = 3x^{-3}$	$x = -1$	$y = 3(-1)^{-3} = \dfrac{3}{(-1)^3} = \dfrac{3}{-1} = -3$
(c)	$y = (-x)^2$	$x = \dfrac{1}{2}$	$y = \left(-\dfrac{1}{2}\right)^2 = \dfrac{1}{4}$
(d)	$y = \dfrac{2}{x^{-2}}$	$x = 3$	$y = \dfrac{2}{3^{-2}} = 2(3^2) = 18$

TRY IT 1

Evaluate $y = 4x^{-2}$ for $x = 3$.

EXAMPLE 2 Evaluating Expressions

	Expression	*x-Value*	*Substitution*
(a)	$y = 2x^{1/2}$	$x = 4$	$y = 2\sqrt{4} = 2(2) = 4$
(b)	$y = \sqrt[3]{x^2}$	$x = 8$	$y = 8^{2/3} = (8^{1/3})^2 = 2^2 = 4$

TRY IT 2

Evaluate $y = 4x^{1/3}$ for $x = 8$.

Operations with Exponents

TECHNOLOGY

Graphing utilities perform the established order of operations when evaluating an expression. To see this, try entering the expressions

$$1200\left(1 + \frac{0.09}{12}\right)^{12 \cdot 6}$$

and

$$1200 \times 1 + \left(\frac{0.09}{12}\right)^{12 \cdot 6}$$

into your graphing utility to see that the expressions result in different values.

Operations with Exponents

1. Multiplying like bases: $x^n x^m = x^{n+m}$ Add exponents.
2. Dividing like bases: $\dfrac{x^n}{x^m} = x^{n-m}$ Subtract exponents.
3. Removing parentheses: $(xy)^n = x^n y^n$

   $\left(\dfrac{x}{y}\right)^n = \dfrac{x^n}{y^n}$

   $(x^n)^m = x^{nm}$
4. Special conventions: $-x^n = -(x^n), \quad -x^n \neq (-x)^n$

   $cx^n = c(x^n), \quad cx^n \neq (cx)^n$

   $x^{n^m} = x^{(n^m)}, \quad x^{n^m} \neq (x^n)^m$

EXAMPLE 3 Simplifying Expressions with Exponents

Simplify each expression.

(a) $2x^2(x^3)$ (b) $(3x)^2\sqrt[3]{x}$ (c) $\dfrac{3x^2}{(x^{1/2})^3}$

(d) $\dfrac{5x^4}{(x^2)^3}$ (e) $x^{-1}(2x^2)$ (f) $\dfrac{-\sqrt{x}}{5x^{-1}}$

SOLUTION

(a) $2x^2(x^3) = 2x^{2+3} = 2x^5$ $\qquad x^n x^m = x^{n+m}$

(b) $(3x)^2\sqrt[3]{x} = 9x^2x^{1/3} = 9x^{2+(1/3)} = 9x^{7/3}$ $\qquad x^n x^m = x^{n+m}$

(c) $\dfrac{3x^2}{(x^{1/2})^3} = 3\left(\dfrac{x^2}{x^{3/2}}\right) = 3x^{2-(3/2)} = 3x^{1/2}$ $\qquad (x^n)^m = x^{nm}, \ \dfrac{x^n}{x^m} = x^{n-m}$

(d) $\dfrac{5x^4}{(x^2)^3} = \dfrac{5x^4}{x^6} = 5x^{4-6} = 5x^{-2} = \dfrac{5}{x^2}$ $\qquad (x^n)^m = x^{nm}, \ \dfrac{x^n}{x^m} = x^{n-m}$

(e) $x^{-1}(2x^2) = 2x^{-1}x^2 = 2x^{2-1} = 2x$ $\qquad x^n x^m = x^{n+m}$

(f) $\dfrac{-\sqrt{x}}{5x^{-1}} = -\dfrac{1}{5}\left(\dfrac{x^{1/2}}{x^{-1}}\right) = -\dfrac{1}{5}x^{(1/2)+1} = -\dfrac{1}{5}x^{3/2}$ $\qquad \dfrac{x^n}{x^m} = x^{n-m}$

TRY IT 3

Simplify each expression.

(a) $3x^2(x^4)$ (b) $(2x)^3\sqrt{x}$ (c) $\dfrac{4x^2}{(x^{1/3})^2}$

Note in Example 3 that one characteristic of simplified expressions is the absence of negative exponents. Another characteristic of simplified expressions is that sums and differences are written in *factored form*. To do this, you can use the **Distributive Property.**

$$abx^n + acx^{n+m} = ax^n(b + cx^m)$$

Study the next example carefully to be sure that you understand the concepts involved in the factoring process.

EXAMPLE 4 **Simplifying by Factoring**

Simplify each expression by factoring.

(a) $2x^2 - x^3$ (b) $2x^3 + x^2$ (c) $2x^{1/2} + 4x^{5/2}$ (d) $2x^{-1/2} + 3x^{5/2}$

SOLUTION

(a) $2x^2 - x^3 = x^2(2 - x)$

(b) $2x^3 + x^2 = x^2(2x + 1)$

(c) $2x^{1/2} + 4x^{5/2} = 2x^{1/2}(1 + 2x^2)$

(d) $2x^{-1/2} + 3x^{5/2} = x^{-1/2}(2 + 3x^3) = \dfrac{2 + 3x^3}{\sqrt{x}}$

TRY IT 4

Simplify each expression by factoring.

(a) $x^3 - 2x$

(b) $2x^{1/2} + 8x^{3/2}$

Many algebraic expressions obtained in calculus occur in unsimplified form. For instance, the two expressions shown in the following example are the result of an operation in calculus called **differentiation.** [The first is the derivative of $2(x + 1)^{3/2}(2x - 3)^{5/2}$, and the second is the derivative of $2(x + 1)^{1/2}(2x - 3)^{5/2}$.]

STUDY TIP

To check that the simplified expression is equivalent to the original expression, try substituting values for x into each expression.

EXAMPLE 5 **Simplifying by Factoring**

Simplify each expression by factoring.

(a) $3(x + 1)^{1/2}(2x - 3)^{5/2} + 10(x + 1)^{3/2}(2x - 3)^{3/2}$

(b) $(x + 1)^{-1/2}(2x - 3)^{5/2} + 10(x + 1)^{1/2}(2x - 3)^{3/2}$

SOLUTION

(a)
$$\begin{aligned} 3(x + 1)^{1/2}(2x - 3)^{5/2} &+ 10(x + 1)^{3/2}(2x - 3)^{3/2} \\ &= (x + 1)^{1/2}(2x - 3)^{3/2}[3(2x - 3) + 10(x + 1)] \\ &= (x + 1)^{1/2}(2x - 3)^{3/2}(6x - 9 + 10x + 10) \\ &= (x + 1)^{1/2}(2x - 3)^{3/2}(16x + 1) \end{aligned}$$

(b)
$$\begin{aligned} (x + 1)^{-1/2}(2x - 3)^{5/2} &+ 10(x + 1)^{1/2}(2x - 3)^{3/2} \\ &= (x + 1)^{-1/2}(2x - 3)^{3/2}[(2x - 3) + 10(x + 1)] \\ &= (x + 1)^{-1/2}(2x - 3)^{3/2}(2x - 3 + 10x + 10) \\ &= (x + 1)^{-1/2}(2x - 3)^{3/2}(12x + 7) \\ &= \frac{(2x - 3)^{3/2}(12x + 7)}{(x + 1)^{1/2}} \end{aligned}$$

TRY IT 5

Simplify the expression by factoring.

$$(x + 2)^{1/2}(3x - 1)^{3/2} + 4(x + 2)^{-1/2}(3x - 1)^{5/2}$$

Example 6 shows some additional types of expressions that can occur in calculus. [The expression in Example 6(d) is an antiderivative of $(x + 1)^{2/3}(2x + 3)$, and the expression in Example 6(e) is the derivative of $(x + 2)^3/(x - 1)^3$.]

TECHNOLOGY

A graphing utility offers several ways to calculate rational exponents and radicals. You should be familiar with the x-squared key [x^2]. This key squares the value of an expression.

For rational exponents or exponents other than 2, use the [^] key.

For rational expressions, you can use the square root key [$\sqrt{\ }$], the cube root key [$\sqrt[3]{\ }$], or the xth root key [$\sqrt[x]{\ }$].

Use a graphing utility to evaluate each expression.

(a) $(-8)^{2/3}$ (b) $(16 - 5)^4$

(c) $\sqrt{576}$ (d) $\sqrt[3]{729}$

(e) $\sqrt[4]{(16)^3}$

EXAMPLE 6 Factors Involving Quotients

Simplify each expression by factoring.

(a) $\dfrac{3x^2 + x^4}{2x}$

(b) $\dfrac{\sqrt{x} + x^{3/2}}{x}$

(c) $(9x + 2)^{-1/3} + 18(9x + 2)$

(d) $\dfrac{3}{5}(x + 1)^{5/3} + \dfrac{3}{4}(x + 1)^{8/3}$

(e) $\dfrac{3(x + 2)^2(x - 1)^3 - 3(x + 2)^3(x - 1)^2}{[(x - 1)^3]^2}$

SOLUTION

(a) $\dfrac{3x^2 + x^4}{2x} = \dfrac{x^2(3 + x^2)}{2x} = \dfrac{x^{2-1}(3 + x^2)}{2} = \dfrac{x(3 + x^2)}{2}$

(b) $\dfrac{\sqrt{x} + x^{3/2}}{x} = \dfrac{x^{1/2}(1 + x)}{x} = \dfrac{1 + x}{x^{1-(1/2)}} = \dfrac{1 + x}{\sqrt{x}}$

(c)
$$\begin{aligned}(9x + 2)^{-1/3} + 18(9x + 2) &= (9x + 2)^{-1/3}[1 + 18(9x + 2)^{4/3}]\\ &= \frac{1 + 18(9x + 2)^{4/3}}{\sqrt[3]{9x + 2}}\end{aligned}$$

(d)
$$\begin{aligned}\frac{3}{5}(x + 1)^{5/3} + \frac{3}{4}(x + 1)^{8/3} &= \frac{12}{20}(x + 1)^{5/3} + \frac{15}{20}(x + 1)^{8/3}\\ &= \frac{3}{20}(x + 1)^{5/3}[4 + 5(x + 1)]\\ &= \frac{3}{20}(x + 1)^{5/3}(4 + 5x + 5)\\ &= \frac{3}{20}(x + 1)^{5/3}(5x + 9)\end{aligned}$$

(e)
$$\begin{aligned}&\frac{3(x + 2)^2(x - 1)^3 - 3(x + 2)^3(x - 1)^2}{[(x - 1)^3]^2}\\ &= \frac{3(x + 2)^2(x - 1)^2[(x - 1) - (x + 2)]}{(x - 1)^6}\\ &= \frac{3(x + 2)^2(x - 1 - x - 2)}{(x - 1)^{6-2}}\\ &= \frac{-9(x + 2)^2}{(x - 1)^4}\end{aligned}$$

TRY IT 6

Simplify the expression by factoring.

$$\frac{2(x + 1)^2(x - 4) - 2(x + 1)(x - 4)^2}{[(x + 1)^2]^2}$$

Domain of an Algebraic Expression

When working with algebraic expressions involving x, you face the potential difficulty of substituting a value of x for which the expression is not defined (does not produce a real number). For example, the expression $\sqrt{2x+3}$ is *not defined* when $x = -2$ because $\sqrt{2(-2)+3}$ is not a real number.

The set of all values for which an expression is defined is called its **domain.** So, the domain of $\sqrt{2x+3}$ is the set of all values of x such that $\sqrt{2x+3}$ is a real number. In order for $\sqrt{2x+3}$ to represent a real number, it is necessary that $2x + 3 \geq 0$. In other words, $\sqrt{2x+3}$ is defined only for those values of x that lie in the interval $\left[-\frac{3}{2}, \infty\right)$, as shown in Figure 0.14.

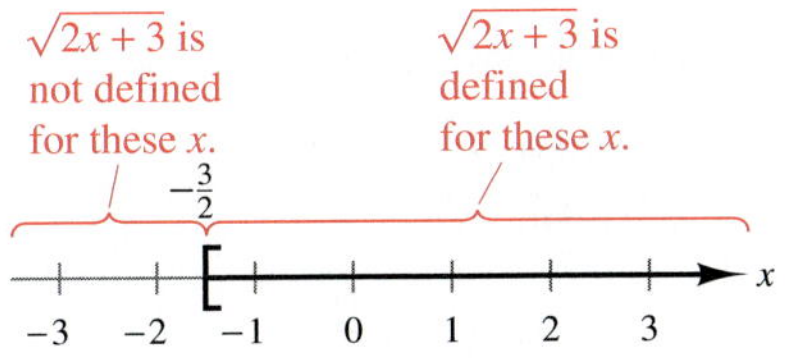

FIGURE 0.14

EXAMPLE 7 **Finding the Domain of an Expression**

Find the domain of each expression.

(a) $\sqrt{3x-2}$

(b) $\dfrac{1}{\sqrt{3x-2}}$

(c) $\sqrt[3]{9x+1}$

SOLUTION

(a) The domain of $\sqrt{3x-2}$ consists of all x such that

$$3x - 2 \geq 0 \qquad \text{Expression must be nonnegative.}$$

which implies that $x \geq \frac{2}{3}$. So, the domain is $\left[\frac{2}{3}, \infty\right)$.

(b) The domain of $1/\sqrt{3x-2}$ is the same as the domain of $\sqrt{3x-2}$, except that $1/\sqrt{3x-2}$ is not defined when $3x - 2 = 0$. Because this occurs when $x = \frac{2}{3}$, the domain is $\left(\frac{2}{3}, \infty\right)$.

(c) Because $\sqrt[3]{9x+1}$ is defined for all real numbers, its domain is $(-\infty, \infty)$.

TRY IT 7

Find the domain of each expression.

(a) $\sqrt{x-2}$

(b) $\dfrac{1}{\sqrt{x-2}}$

(c) $\sqrt[3]{x-2}$

EXERCISES 0.3

In Exercises 1–20, evaluate the expression for the given value of *x*.

	Expression	*x-Value*		*Expression*	*x-Value*
1.	$-3x^3$	$x = 2$	**2.**	$\frac{x^2}{2}$	$x = 6$
3.	$4x^{-3}$	$x = 2$	**4.**	$7x^{-2}$	$x = 4$
5.	$\frac{1 + x^{-1}}{x^{-1}}$	$x = 2$	**6.**	$x - 4x^{-2}$	$x = 3$
7.	$3x^2 - 4x^3$	$x = -2$	**8.**	$5(-x)^3$	$x = 3$
9.	$6x^0 - (6x)^0$	$x = 10$	**10.**	$\frac{1}{(-x)^{-3}}$	$x = 4$
11.	$\sqrt[3]{x^2}$	$x = 27$	**12.**	$\sqrt{x^3}$	$x = \frac{1}{9}$
13.	$x^{-1/2}$	$x = 4$	**14.**	$x^{-3/4}$	$x = 16$
15.	$x^{-2/5}$	$x = -32$	**16.**	$(x^{2/3})^3$	$x = 10$
17.	$500x^{60}$	$x = 1.01$	**18.**	$\frac{10{,}000}{x^{120}}$	$x = 1.075$
19.	$\sqrt[3]{x}$	$x = -154$	**20.**	$\sqrt[6]{x}$	$x = 325$

In Exercises 21–30, simplify the expression.

21. $6y^{-2}(2y^4)^{-3}$

22. $z^{-3}(3z^4)$

23. $10(x^2)^2$

24. $(4x^3)^2$

25. $\frac{7x^2}{x^{-3}}$

26. $\frac{x^{-3}}{\sqrt{x}}$

27. $\frac{12(x + y)^3}{9(x + y)^{-2}}$

28. $\left(\frac{12s^2}{9s}\right)^3$

29. $\frac{3x\sqrt{x}}{x^{1/2}}$

30. $\left(\sqrt[3]{x^2}\right)^3$

In Exercises 31–36, simplify by removing all possible factors from the radical.

31. (a) $\sqrt{8}$ (b) $\sqrt{18}$

32. (a) $\sqrt[3]{\frac{16}{27}}$ (b) $\sqrt[3]{\frac{24}{125}}$

33. (a) $\sqrt[3]{16x^5}$ (b) $\sqrt[4]{32x^4z^5}$

34. (a) $\sqrt[4]{(3x^2y^3)^4}$ (b) $\sqrt[3]{54x^7}$

35. (a) $\sqrt[3]{144x^9y^{-4}z^5}$ (b) $\sqrt{12(3x + 5)^7}$

36. (a) $\sqrt[4]{32xy^5z^{-8}}$ (b) $\sqrt{90(2x - 3y)^6}$

In Exercises 37–46, simplify each expression by factoring.

37. $4x^3 - 6x$

38. $8x^4 - 6x^2$

39. $2x^{5/2} + x^{-1/2}$

40. $5x^{3/2} - x^{-3/2}$

41. $3x(x + 1)^{3/2} - 6(x + 1)^{1/2}$

42. $2x(x - 1)^{5/2} - 4(x - 1)^{3/2}$

43. $\frac{(x + 1)(x - 1)^2 - (x - 1)^3}{(x + 1)^2}$

44. $\frac{(x - 4)(2x - 1)^3 - (2x - 1)^4}{(x - 4)^2}$

45. $(x^2 + 1)^2(x - 1)^{-1/2} + 2x(x - 1)^{1/2}(x^2 + 1)$

46. $(x^4 + 2)^3(x + 3)^{-1/2} + 4x^3(x^4 + 2)^2(x + 3)^{1/2}$

The symbol indicates an exercise in which you are instructed to use graphing technology or a symbolic computer algebra system. The solutions of other exercises may also be facilitated by use of appropriate technology.

In Exercises 47–56, find the domain of the given expression.

47. $\sqrt{x - 1}$

48. $\sqrt{5 - 2x}$

49. $\sqrt{x^2 + 3}$

50. $\sqrt{4x^2 + 1}$

51. $\frac{1}{\sqrt[3]{x - 1}}$

52. $\frac{1}{\sqrt[3]{x + 4}}$

53. $\frac{\sqrt{x + 2}}{x - 4}$

54. $\frac{\sqrt{x - 1}}{x + 1}$

55. $\sqrt{x - 1} + \sqrt{5 - x}$

56. $\frac{1}{\sqrt{2x + 3}} + \sqrt{6 - 4x}$

Compound Interest In Exercises 57–60, a certificate of deposit has a principal of P and an annual percentage rate of r (expressed as a decimal) compounded n times per year. Enter the compound interest formula

$$A = P\left(1 + \frac{r}{n}\right)^N$$

into a graphing utility and use it to find the balance after N compoundings.

57. $P = \$10{,}000, \quad r = 6.5\%, \quad n = 12, \quad N = 120$

58. $P = \$7000, \quad r = 5\%, \quad n = 365, \quad N = 1000$

59. $P = \$5000, \quad r = 5.5\%, \quad n = 4, \quad N = 60$

60. $P = \$8000, \quad r = 7\%, \quad n = 12, \quad N = 180$

61. ***The Period of a Pendulum*** The period of a pendulum is

$$T = 2\pi\sqrt{\frac{L}{32}}$$

where T is the period in seconds and L is the length of the pendulum in feet. Find the period of a pendulum whose length is 4 feet.

62. ***Annuity*** A balance A, after n annual payments of P dollars have been made into an annuity earning an annual percentage rate of r compounded annually, is given by

$$A = P(1 + r) + P(1 + r)^2 + \cdots + P(1 + r)^n.$$

Rewrite this formula by completing the following factorization: $A = P(1 + r)(\quad)$.

0.4 FACTORING POLYNOMIALS

- Use special products and factorization techniques to factor polynomials.
- Find the domains of radical expressions.
- Use synthetic division to factor polynomials of degree three or more.
- Use the Rational Zero Theorem to find the real zeros of polynomials.

Factorization Techniques

The **Fundamental Theorem of Algebra** states that every nth-degree polynomial

$$a_n x^n + a_{n-1}x^{n-1} + \cdot \cdot \cdot + a_1 x + a_0, \quad a_n \neq 0$$

has precisely n **zeros.** (The zeros may be repeated or imaginary.) The problem of finding the zeros of a polynomial is equivalent to the problem of factoring the polynomial into linear factors.

Special Products and Factorization Techniques

Quadratic Formula

$$ax^2 + bx + c = 0 \Rightarrow x = \frac{-b \pm \sqrt{b^2 - 4ac}}{2a}$$

Example

$$x^2 + 3x - 1 = 0 \Rightarrow x = \frac{-3 \pm \sqrt{13}}{2}$$

Special Products

$$x^2 - a^2 = (x - a)(x + a)$$
$$x^3 - a^3 = (x - a)(x^2 + ax + a^2)$$
$$x^3 + a^3 = (x + a)(x^2 - ax + a^2)$$
$$x^4 - a^4 = (x - a)(x + a)(x^2 + a^2)$$

Examples

$$x^2 - 9 = (x - 3)(x + 3)$$
$$x^3 - 8 = (x - 2)(x^2 + 2x + 4)$$
$$x^3 + 64 = (x + 4)(x^2 - 4x + 16)$$
$$x^4 - 16 = (x - 2)(x + 2)(x^2 + 4)$$

Binomial Theorem

$$(x + a)^2 = x^2 + 2ax + a^2$$
$$(x - a)^2 = x^2 - 2ax + a^2$$
$$(x + a)^3 = x^3 + 3ax^2 + 3a^2x + a^3$$
$$(x - a)^3 = x^3 - 3ax^2 + 3a^2x - a^3$$
$$(x + a)^4 = x^4 + 4ax^3 + 6a^2x^2 + 4a^3x + a^4$$
$$(x - a)^4 = x^4 - 4ax^3 + 6a^2x^2 - 4a^3x + a^4$$

Examples

$$(x + 3)^2 = x^2 + 6x + 9$$
$$(x^2 - 5)^2 = x^4 - 10x^2 + 25$$
$$(x + 2)^3 = x^3 + 6x^2 + 12x + 8$$
$$(x - 1)^3 = x^3 - 3x^2 + 3x - 1$$
$$(x + 2)^4 = x^4 + 8x^3 + 24x^2 + 32x + 16$$
$$(x - 4)^4 = x^4 - 16x^3 + 96x^2 - 256x + 256$$

$$(x + a)^n = x^n + nax^{n-1} + \frac{n(n-1)}{2!}a^2x^{n-2} + \frac{n(n-1)(n-2)}{3!}a^3x^{n-3} + \cdot \cdot \cdot + na^{n-1}x + a^n{}^*$$

$$(x - a)^n = x^n - nax^{n-1} + \frac{n(n-1)}{2!}a^2x^{n-2} - \frac{n(n-1)(n-2)}{3!}a^3x^{n-3} + \cdot \cdot \cdot \pm na^{n-1}x \mp a^n$$

Factoring by Grouping

$$acx^3 + adx^2 + bcx + bd = ax^2(cx + d) + b(cx + d)$$
$$= (ax^2 + b)(cx + d)$$

Example

$$3x^3 - 2x^2 - 6x + 4 = x^2(3x - 2) - 2(3x - 2)$$
$$= (x^2 - 2)(3x - 2)$$

* The factorial symbol ! is defined as follows:
$0! = 1,\ 1! = 1,\ 2! = 2 \cdot 1 = 2,\ 3! = 3 \cdot 2 \cdot 1 = 6,\ 4! = 4 \cdot 3 \cdot 2 \cdot 1 = 24$, and so on.

EXAMPLE 1 Applying the Quadratic Formula

Use the Quadratic Formula to find all real zeros of each polynomial.

(a) $4x^2 + 6x + 1$ (b) $x^2 + 6x + 9$ (c) $2x^2 - 6x + 5$

SOLUTION

(a) Using $a = 4$, $b = 6$, and $c = 1$, you can write

$$\begin{aligned} x = \frac{-b \pm \sqrt{b^2 - 4ac}}{2a} &= \frac{-6 \pm \sqrt{36 - 16}}{8} \\ &= \frac{-6 \pm \sqrt{20}}{8} \\ &= \frac{-6 \pm 2\sqrt{5}}{8} \\ &= \frac{2\left(-3 \pm \sqrt{5}\right)}{2(4)} \\ &= \frac{-3 \pm \sqrt{5}}{4}. \end{aligned}$$

So, there are two real zeros:

$$x = \frac{-3 - \sqrt{5}}{4} \approx -1.309 \quad \text{and} \quad x = \frac{-3 + \sqrt{5}}{4} \approx -0.191.$$

STUDY TIP

Try solving Example 1(b) by factoring. Do you obtain the same answer?

(b) In this case, $a = 1$, $b = 6$, and $c = 9$, and the Quadratic Formula yields

$$x = \frac{-b \pm \sqrt{b^2 - 4ac}}{2a} = \frac{-6 \pm \sqrt{36 - 36}}{2} = -\frac{6}{2} = -3.$$

So, there is one (repeated) real zero: $x = -3$.

(c) For this quadratic equation, $a = 2$, $b = -6$, and $c = 5$. So,

$$x = \frac{-b \pm \sqrt{b^2 - 4ac}}{2a} = \frac{6 \pm \sqrt{36 - 40}}{4} = \frac{6 \pm \sqrt{-4}}{4}.$$

Because $\sqrt{-4}$ is imaginary, there are no real zeros.

TRY IT 1

Use the Quadratic Formula to find all real zeros of each polynomial.

(a) $2x^2 + 4x + 1$ (b) $x^2 - 8x + 16$ (c) $2x^2 - x + 5$

The zeros in Example 1(a) are irrational, and the zeros in Example 1(c) are imaginary. In both of these cases the quadratic is said to be **irreducible** because it cannot be factored into linear factors with rational coefficients. The next example shows how to find the zeros associated with *reducible* quadratics. In this example, factoring is used to find the zeros of each quadratic. Try using the Quadratic Formula to obtain the same zeros.

EXAMPLE 2 Factoring Quadratics

Find the zeros of each quadratic polynomial.

(a) $x^2 - 5x + 6$ (b) $x^2 - 6x + 9$ (c) $2x^2 + 5x - 3$

SOLUTION

(a) Because

$$x^2 - 5x + 6 = (x - 2)(x - 3)$$

the zeros are $x = 2$ and $x = 3$.

(b) Because

$$x^2 - 6x + 9 = (x - 3)^2$$

the only zero is $x = 3$.

(c) Because

$$2x^2 + 5x - 3 = (2x - 1)(x + 3)$$

the zeros are $x = \frac{1}{2}$ and $x = -3$.

ALGEBRA REVIEW

The zeros of a polynomial in x are the values of x that make the polynomial zero. To find the zeros, factor the polynomial into linear factors and set each factor equal to zero. For instance, the zeros of $(x - 2)(x - 3)$ occur when $x - 2 = 0$ and $x - 3 = 0$.

TRY IT 2

Find the zeros of each quadratic polynomial.

(a) $x^2 - 2x - 15$ (b) $x^2 + 2x + 1$ (c) $2x^2 - 7x + 6$

EXAMPLE 3 Finding the Domain of a Radical Expression

Find the domain of $\sqrt{x^2 - 3x + 2}$.

SOLUTION Because

$$x^2 - 3x + 2 = (x - 1)(x - 2)$$

you know that the zeros of the quadratic are $x = 1$ and $x = 2$. So, you need to test the sign of the quadratic in the three intervals $(-\infty, 1)$, $(1, 2)$, and $(2, \infty)$, as shown in Figure 0.15. After testing each of these intervals, you can see that the quadratic is negative in the center interval and positive in the outer two intervals. Moreover, because the quadratic is zero when $x = 1$ and $x = 2$, you can conclude that the domain of $\sqrt{x^2 - 3x + 2}$ is

$$(-\infty, 1] \cup [2, \infty). \quad \text{Domain}$$

Values of $\sqrt{x^2 - 3x + 2}$

x	$\sqrt{x^2 - 3x + 2}$
0	$\sqrt{2}$
1	0
1.5	Undefined
2	0
3	$\sqrt{2}$

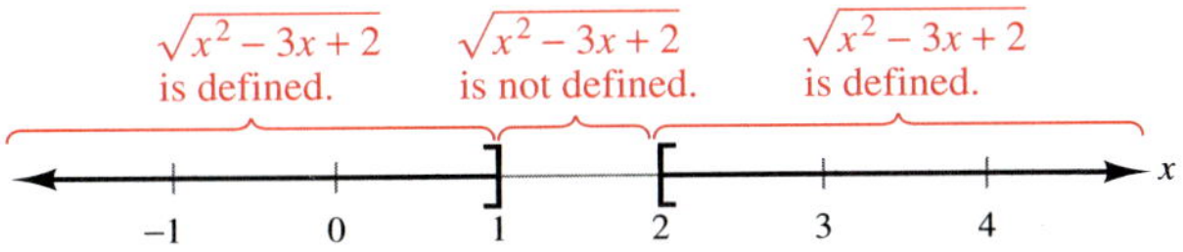

FIGURE 0.15

TRY IT 3

Find the domain of

$$\sqrt{x^2 + x - 2}.$$

Factoring Polynomials of Degree Three or More

It can be difficult to find the zeros of polynomials of degree three or more. However, if one of the zeros of a polynomial is known, then you can use that zero to reduce the degree of the polynomial. For example, if you know that $x = 2$ is a zero of $x^3 - 4x^2 + 5x - 2$, then you know that $(x - 2)$ is a factor, and you can use long division to factor the polynomial as shown.

$$\begin{aligned} x^3 - 4x^2 + 5x - 2 &= (x - 2)(x^2 - 2x + 1) \\ &= (x - 2)(x - 1)(x - 1) \end{aligned}$$

As an alternative to long division, many people prefer to use **synthetic division** to reduce the degree of a polynomial.

Synthetic Division for a Cubic Polynomial

Given: $x = x_1$ is a zero of $ax^3 + bx^2 + cx + d$.

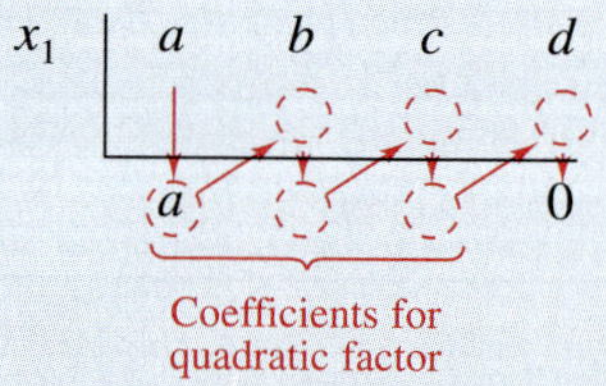

Vertical pattern: *Add terms.*

Diagonal pattern: *Multiply by x_1.*

Performing synthetic division on the polynomial

$$x^3 - 4x^2 + 5x - 2$$

using the given zero, $x = 2$, produces the following.

$$\begin{array}{r|rrrr} 2 & 1 & -4 & 5 & -2 \\ & & 2 & -4 & 2 \\ \hline & 1 & -2 & 1 & 0 \end{array}$$

$(x - 2)(x^2 - 2x + 1) = x^3 - 4x^2 + 5x - 2$

When you use synthetic division, remember to take *all* coefficients into account—*even if some of them are zero*. For instance, if you know that $x = -2$ is a zero of

$$x^3 + 3x + 14$$

you can apply synthetic division as shown.

$$\begin{array}{r|rrrr} -2 & 1 & 0 & 3 & 14 \\ & & -2 & 4 & -14 \\ \hline & 1 & -2 & 7 & 0 \end{array}$$

$(x + 2)(x^2 - 2x + 7) = x^3 + 3x + 14$

STUDY TIP

Note that synthetic division works *only* for divisors of the form $x - x_1$. [Remember that $x + x_1 = x - (-x_1)$.] You cannot use synthetic division to divide a polynomial by a quadratic such as $x^2 - 3$.

The Rational Zero Theorem

There is a systematic way to find the *rational* zeros of a polynomial. You can use the **Rational Zero Theorem** (also called the Rational Root Theorem).

Rational Zero Theorem

If a polynomial

$$a_nx^n + a_{n-1}x^{n-1} + \cdot \cdot \cdot + a_1x + a_0$$

has integer coefficients, then every *rational* zero is of the form $x = p/q$, where p is a factor of a_0, and q is a factor of a_n.

EXAMPLE 4 Using the Rational Zero Theorem

Find all real zeros of the polynomial.

$$2x^3 + 3x^2 - 8x + 3$$

SOLUTION

$$2x^3 + 3x^2 - 8x + 3$$

Factors of constant term: $\pm 1, \pm 3$

Factors of leading coefficient: $\pm 1, \pm 2$

The possible rational zeros are the factors of the constant term divided by the factors of the leading coefficient.

$$1, -1, 3, -3, \frac{1}{2}, -\frac{1}{2}, \frac{3}{2}, -\frac{3}{2}$$

By testing these possible zeros, you can see that $x = 1$ works.

$$2(1)^3 + 3(1)^2 - 8(1) + 3 = 2 + 3 - 8 + 3 = 0$$

Now, by synthetic division you have the following.

$$\begin{array}{r|rrrr} 1 & 2 & 3 & -8 & 3 \\ & & 2 & 5 & -3 \\ \hline & 2 & 5 & -3 & 0 \end{array}$$

$(x - 1)(2x^2 + 5x - 3) = 2x^3 + 3x^2 - 8x + 3$

Finally, by factoring the quadratic, $2x^2 + 5x - 3 = (2x - 1)(x + 3)$, you have

$$2x^3 + 3x^2 - 8x + 3 = (x - 1)(2x - 1)(x + 3)$$

and you can conclude that the zeros are $x = 1$, $x = \frac{1}{2}$, and $x = -3$.

STUDY TIP

In Example 4, you can check that the zeros are correct by substituting into the original polynomial.

Check that $x = 1$ is a zero.

$$\begin{aligned} &2(1)^3 + 3(1)^2 - 8(1) + 3 \\ &= 2 + 3 - 8 + 3 \\ &= 0 \end{aligned}$$

Check that $x = \frac{1}{2}$ is a zero.

$$\begin{aligned} &2\left(\frac{1}{2}\right)^3 + 3\left(\frac{1}{2}\right)^2 - 8\left(\frac{1}{2}\right) + 3 \\ &= \frac{1}{4} + \frac{3}{4} - 4 + 3 \\ &= 0 \end{aligned}$$

Check that $x = -3$ is a zero.

$$\begin{aligned} &2(-3)^3 + 3(-3)^2 - 8(-3) + 3 \\ &= -54 + 27 + 24 + 3 \\ &= 0 \end{aligned}$$

TRY IT 4

Find all real zeros of the polynomial.

$$2x^3 - 3x^2 - 3x + 2$$

EXERCISES 0.4

In Exercises 1–8, use the Quadratic Formula to find all real zeros of the second-degree polynomial.

1. $6x^2 - x - 1$

2. $8x^2 - 2x - 1$

3. $4x^2 - 12x + 9$

4. $9x^2 + 12x + 4$

5. $y^2 + 4y + 1$

6. $x^2 + 6x - 1$

7. $2x^2 + 3x - 4$

8. $3x^2 - 8x - 4$

In Exercises 9–18, write the second-degree polynomial as the product of two linear factors.

9. $x^2 - 4x + 4$

10. $x^2 + 10x + 25$

11. $4x^2 + 4x + 1$

12. $9x^2 - 12x + 4$

13. $x^2 + x - 2$

14. $2x^2 - x - 1$

15. $3x^2 - 5x + 2$

16. $x^2 - xy - 2y^2$

17. $x^2 - 4xy + 4y^2$

18. $a^2b^2 - 2abc + c^2$

In Exercises 19–34, completely factor the polynomial.

19. $81 - y^4$

20. $x^4 - 16$

21. $x^3 - 8$

22. $y^3 - 64$

23. $y^3 + 64$

24. $z^3 + 125$

25. $x^3 - 27$

26. $(x - a)^3 + b^3$

27. $x^3 - 4x^2 - x + 4$

28. $x^3 - x^2 - x + 1$

29. $2x^3 - 3x^2 + 4x - 6$

30. $x^3 - 5x^2 - 5x + 25$

31. $2x^3 - 4x^2 - x + 2$

32. $x^3 - 7x^2 - 4x + 28$

33. $x^4 - 15x^2 - 16$

34. $2x^4 - 49x^2 - 25$

In Exercises 35–52, find all real zeros of the polynomial.

35. $x^2 - 5x$

36. $2x^2 - 3x$

37. $x^2 - 9$

38. $x^2 - 25$

39. $x^2 - 3$

40. $x^2 - 8$

41. $(x - 3)^2 - 9$

42. $(x + 1)^2 - 8$

43. $x^2 + x - 2$

44. $x^2 + 5x + 6$

45. $x^2 - 5x + 6$

46. $x^2 + x - 20$

47. $x^3 + 64$

48. $x^3 - 216$

49. $x^4 - 16$

50. $x^4 - 625$

51. $x^3 - x^2 - 4x + 4$

52. $2x^3 + x^2 + 6x + 3$

In Exercises 53–56, find the interval (or intervals) on which the given expression is defined.

53. $\sqrt{x^2 - 4}$

54. $\sqrt{4 - x^2}$

55. $\sqrt{x^2 - 7x + 12}$

56. $\sqrt{x^2 - 8x + 15}$

In Exercises 57–60, use synthetic division to complete the indicated factorization.

57. $x^3 - 3x^2 - 6x - 2 = (x + 1)(\quad)$

58. $x^3 - 2x^2 - x + 2 = (x + 1)(\quad)$

59. $2x^3 - x^2 - 2x + 1 = (x - 1)(\quad)$

60. $x^4 - 16x^3 + 96x^2 - 256x + 256 = (x - 4)(\quad)$

In Exercises 61–68, use the Rational Zero Theorem as an aid in finding all real zeros of the polynomial.

61. $x^3 - x^2 - 10x - 8$

62. $x^3 - 7x - 6$

63. $x^3 - 6x^2 + 11x - 6$

64. $x^3 + 2x^2 - 5x - 6$

65. $6x^3 - 11x^2 - 19x - 6$

66. $18x^3 - 9x^2 - 8x + 4$

67. $x^3 - 3x^2 - 3x - 4$

68. $2x^3 - x^2 - 13x - 6$

69. ***Average Cost*** The minimum average cost of producing x units of a product occurs when the production level is set at the (positive) solution of

$$0.0003x^2 - 1200 = 0.$$

Determine this production level.

70. ***Profit*** The profit P from sales is given by

$$P = -200x^2 + 2000x - 3800$$

where x is the number of units sold per day (in hundreds). Determine the interval for x such that the profit will be greater than 1000.

71. ***Chemistry: Finding Concentrations*** Use the Quadratic Formula to solve the expression

$$1.8 \times 10^{-5} = \frac{x^2}{1.0 \times 10^{-4} - x}$$

which is needed to determine the quantity of hydrogen ions ($[H^+]$) in a solution of 1.0×10^{-4}M acetic acid. Because x represents a concentration of $[H^+]$, only positive values of x are possible solutions. *(Source: Adapted from Zumdahl,* Chemistry, *Sixth Edition)*

72. ***Finance*** After 2 years, an investment of \$1200 is made at an interest rate r, compounded annually, that will yield an amount of

$$A = 1200(1 + r)^2.$$

Determine the interest rate if $A = \$1300$.

0.5 FRACTIONS AND RATIONALIZATION

- Add and subtract rational expressions.
- Simplify rational expressions involving radicals.
- Rationalize numerators and denominators of rational expressions.

Operations with Fractions

In this section, you will review operations involving fractional expressions such as

$$\frac{2}{x}, \quad \frac{x^2 + 2x - 4}{x + 6}, \quad \text{and} \quad \frac{1}{\sqrt{x^2 + 1}}.$$

The first two expressions have polynomials as both numerator and denominator and are called **rational expressions.** A rational expression is **proper** if the degree of the numerator is less than the degree of the denominator. For example,

$$\frac{x}{x^2 + 1}$$

is proper. If the degree of the numerator is greater than or equal to the degree of the denominator, then the rational expression is **improper.** For example,

$$\frac{x^2}{x^2 + 1}, \quad \text{and} \quad \frac{x^3 + 2x + 1}{x + 1}$$

are both improper.

Operations with Fractions

1. Add fractions (find a common denominator):

$$\frac{a}{b} + \frac{c}{d} = \frac{a}{b}\left(\frac{d}{d}\right) + \frac{c}{d}\left(\frac{b}{b}\right) = \frac{ad}{bd} + \frac{bc}{bd} = \frac{ad + bc}{bd}, \quad b \neq 0, d \neq 0$$

2. Subtract fractions (find a common denominator):

$$\frac{a}{b} - \frac{c}{d} = \frac{a}{b}\left(\frac{d}{d}\right) - \frac{c}{d}\left(\frac{b}{b}\right) = \frac{ad}{bd} - \frac{bc}{bd} = \frac{ad - bc}{bd}, \quad b \neq 0, d \neq 0$$

3. Multiply fractions:

$$\left(\frac{a}{b}\right)\left(\frac{c}{d}\right) = \frac{ac}{bd}, \quad b \neq 0, d \neq 0$$

4. Divide fractions (invert and multiply):

$$\frac{a/b}{c/d} = \left(\frac{a}{b}\right)\left(\frac{d}{c}\right) = \frac{ad}{bc}, \quad \frac{a/b}{c} = \frac{a/b}{c/1} = \left(\frac{a}{b}\right)\left(\frac{1}{c}\right) = \frac{a}{bc}, \quad b \neq 0,$$

$c \neq 0, d \neq 0$

5. Divide out like factors:

$$\frac{\not{a}b}{\not{a}c} = \frac{b}{c}, \quad \frac{ab + ac}{ad} = \frac{\not{a}(b + c)}{\not{a}d} = \frac{b + c}{d}, \quad a \neq 0, c \neq 0, d \neq 0$$

EXAMPLE 1 Adding and Subtracting Rational Expressions

Perform the indicated operation and simplify.

(a) $x + \dfrac{1}{x}$ (b) $\dfrac{1}{x+1} - \dfrac{2}{2x-1}$

SOLUTION

(a) $x + \dfrac{1}{x} = \dfrac{x^2}{x} + \dfrac{1}{x}$ Write with common denominator.

$= \dfrac{x^2+1}{x}$ Add fractions.

(b) $$\frac{1}{x+1} - \frac{2}{2x-1} = \frac{(2x-1)}{(x+1)(2x-1)} - \frac{2(x+1)}{(x+1)(2x-1)}$$

$$= \frac{2x-1-2x-2}{2x^2+x-1} = \frac{-3}{2x^2+x-1}$$

TRY IT 1

Perform each indicated operation and simplify.

(a) $x + \dfrac{2}{x}$ (b) $\dfrac{2}{x+1} - \dfrac{1}{2x+1}$

In adding (or subtracting) fractions whose denominators have no common factors, it is convenient to use the following pattern.

$$\frac{a}{b} + \frac{c}{d} = \frac{a}{b} \times \frac{c}{d} = \frac{ad+bc}{bd}$$

For instance, in Example 1(b), you could have used this pattern as shown.

$$\frac{1}{x+1} - \frac{2}{2x-1} = \frac{(2x-1) - 2(x+1)}{(x+1)(2x-1)}$$

$$= \frac{2x-1-2x-2}{(x+1)(2x-1)} = \frac{-3}{2x^2+x-1}$$

In Example 1, the denominators of the rational expressions have no common factors. When the denominators do have common factors, it is best to find the least common denominator before adding or subtracting. For instance, when adding $1/x$ and $2/x^2$, you can recognize that the least common denominator is x^2 and write

$\dfrac{1}{x} + \dfrac{2}{x^2} = \dfrac{x}{x^2} + \dfrac{2}{x^2}$ Write with common denominator.

$= \dfrac{x+2}{x^2}.$ Add fractions.

This is further demonstrated in Example 2.

EXAMPLE 2 **Adding and Subtracting Rational Expressions**

Perform the indicated operation and simplify.

(a) $\dfrac{x}{x^2 - 1} + \dfrac{3}{x + 1}$ (b) $\dfrac{1}{2(x^2 + 2x)} - \dfrac{1}{4x}$

SOLUTION

(a) Because $x^2 - 1 = (x + 1)(x - 1)$, the least common denominator is $x^2 - 1$.

$$\frac{x}{x^2 - 1} + \frac{3}{x + 1} = \frac{x}{(x - 1)(x + 1)} + \frac{3}{x + 1} \qquad \text{Factor.}$$

$$= \frac{x}{(x - 1)(x + 1)} + \frac{3(x - 1)}{(x - 1)(x + 1)} \qquad \text{Write with common denominator.}$$

$$= \frac{x + 3x - 3}{(x - 1)(x + 1)} \qquad \text{Add fractions.}$$

$$= \frac{4x - 3}{x^2 - 1} \qquad \text{Simplify.}$$

(b) In this case, the least common denominator is $4x(x + 2)$.

$$\frac{1}{2(x^2 + 2x)} - \frac{1}{4x} = \frac{1}{2x(x + 2)} - \frac{1}{2(2x)} \qquad \text{Factor.}$$

$$= \frac{2}{2(2x)(x + 2)} - \frac{x + 2}{2(2x)(x + 2)} \qquad \text{Write with common denominator.}$$

$$= \frac{2 - x - 2}{4x(x + 2)} \qquad \text{Subtract fractions.}$$

$$= \frac{-\cancel{x}}{4\cancel{x}(x + 2)} \qquad \text{Divide out like factor.}$$

$$= \frac{-1}{4(x + 2)}, \quad x \neq 0 \qquad \text{Simplify.}$$

TRY IT 2

Perform each indicated operation and simplify.

(a) $\dfrac{x}{x^2 - 4} + \dfrac{2}{x - 2}$ (b) $\dfrac{1}{3(x^2 + 2x)} - \dfrac{1}{3x}$

ALGEBRA REVIEW

To add more than two fractions, you must find a denominator that is common to all the fractions. For instance, to add $\frac{1}{2}$, $\frac{1}{3}$, and $\frac{1}{5}$, use a (least) common denominator of 30 and write

$$\frac{1}{2} + \frac{1}{3} + \frac{1}{5} = \frac{15}{30} + \frac{10}{30} + \frac{6}{30} \qquad \text{Write with common denominator.}$$

$$= \frac{31}{30}. \qquad \text{Add fractions.}$$

To add more than two rational expressions, use a similar procedure, as shown in Example 3. (Expressions such as those shown in this example are used in calculus to perform an integration technique called integration by partial fractions.)

EXAMPLE 3 **Adding More than Two Rational Expressions**

Perform the indicated addition of rational expressions.

(a) $\frac{A}{x+2} + \frac{B}{x-3} + \frac{C}{x+4}$

(b) $\frac{A}{x+2} + \frac{B}{(x+2)^2} + \frac{C}{x-1}$

SOLUTION

(a) The least common denominator is $(x+2)(x-3)(x+4)$.

$$\frac{A}{x+2} + \frac{B}{x-3} + \frac{C}{x+4}$$
$$= \frac{A(x-3)(x+4) + B(x+2)(x+4) + C(x+2)(x-3)}{(x+2)(x-3)(x+4)}$$
$$= \frac{A(x^2+x-12) + B(x^2+6x+8) + C(x^2-x-6)}{(x+2)(x-3)(x+4)}$$
$$= \frac{Ax^2 + Bx^2 + Cx^2 + Ax + 6Bx - Cx - 12A + 8B - 6C}{(x+2)(x-3)(x+4)}$$
$$= \frac{(A+B+C)x^2 + (A+6B-C)x + (-12A+8B-6C)}{(x+2)(x-3)(x+4)}$$

(b) Here the least common denominator is $(x+2)^2(x-1)$.

$$\frac{A}{x+2} + \frac{B}{(x+2)^2} + \frac{C}{x-1}$$
$$= \frac{A(x+2)(x-1) + B(x-1) + C(x+2)^2}{(x+2)^2(x-1)}$$
$$= \frac{A(x^2+x-2) + B(x-1) + C(x^2+4x+4)}{(x+2)^2(x-1)}$$
$$= \frac{Ax^2 + Cx^2 + Ax + Bx + 4Cx - 2A - B + 4C}{(x+2)^2(x-1)}$$
$$= \frac{(A+C)x^2 + (A+B+4C)x + (-2A-B+4C)}{(x+2)^2(x-1)}$$

TRY IT 3

Perform each indicated addition of rational expressions.

(a) $\frac{A}{x+1} + \frac{B}{x-1} + \frac{C}{x+2}$

(b) $\frac{A}{x+1} + \frac{B}{(x+1)^2} + \frac{C}{x-2}$

Expressions Involving Radicals

In calculus, the operation of differentiation tends to produce "messy" expressions when applied to fractional expressions. This is especially true when the fractional expression involves radicals. When differentiation is used, it is important to be able to simplify these expressions so that you can obtain more manageable forms. All of the expressions in Examples 4 and 5 are the results of differentiation. In each case, note how much *simpler* the simplified form is than the original form.

EXAMPLE 4 Simplifying an Expression with Radicals

Simplify each expression.

(a) $\dfrac{\sqrt{x+1} - \dfrac{x}{2\sqrt{x+1}}}{x+1}$ (b) $\left(\dfrac{1}{x+\sqrt{x^2+1}}\right)\left(1 + \dfrac{2x}{2\sqrt{x^2+1}}\right)$

SOLUTION

(a) $$\frac{\sqrt{x+1} - \dfrac{x}{2\sqrt{x+1}}}{x+1} = \frac{\dfrac{2(x+1)}{2\sqrt{x+1}} - \dfrac{x}{2\sqrt{x+1}}}{x+1}$$ Write with common denominator.

$$= \frac{\dfrac{2x+2-x}{2\sqrt{x+1}}}{\dfrac{x+1}{1}}$$ Subtract fractions.

$$= \frac{x+2}{2\sqrt{x+1}}\left(\frac{1}{x+1}\right)$$ To divide, invert and multiply

$$= \frac{x+2}{2(x+1)^{3/2}}$$ Multiply.

(b) $$\left(\frac{1}{x+\sqrt{x^2+1}}\right)\left(1 + \frac{2x}{2\sqrt{x^2+1}}\right)$$

$$= \left(\frac{1}{x+\sqrt{x^2+1}}\right)\left(1 + \frac{x}{\sqrt{x^2+1}}\right)$$

$$= \left(\frac{1}{x+\sqrt{x^2+1}}\right)\left(\frac{\sqrt{x^2+1}}{\sqrt{x^2+1}} + \frac{x}{\sqrt{x^2+1}}\right)$$

$$= \left(\frac{1}{x+\sqrt{x^2+1}}\right)\left(\frac{x+\sqrt{x^2+1}}{\sqrt{x^2+1}}\right)$$

$$= \frac{1}{\sqrt{x^2+1}}$$

TRY IT 4

Simplify each expression.

(a) $\dfrac{\sqrt{x+2} - \dfrac{x}{4\sqrt{x+2}}}{x+2}$ (b) $\left(\dfrac{1}{x+\sqrt{x^2+4}}\right)\left(1 + \dfrac{x}{\sqrt{x^2+4}}\right)$

EXAMPLE 5 Simplifying an Expression with Radicals

Simplify the expression.

$$\frac{-x\left(\dfrac{2x}{2\sqrt{x^2+1}}\right)+\sqrt{x^2+1}}{x^2}+\left(\frac{1}{x+\sqrt{x^2+1}}\right)\left(1+\frac{2x}{2\sqrt{x^2+1}}\right)$$

SOLUTION From Example 4(b), you already know that the second part of this sum simplifies to $1/\sqrt{x^2+1}$. The first part simplifies as shown.

$$\begin{aligned}\frac{-x\left(\dfrac{2x}{2\sqrt{x^2+1}}\right)+\sqrt{x^2+1}}{x^2} &= \frac{-x^2}{x^2\sqrt{x^2+1}}+\frac{\sqrt{x^2+1}}{x^2}\\ &= \frac{-x^2}{x^2\sqrt{x^2+1}}+\frac{x^2+1}{x^2\sqrt{x^2+1}}\\ &= \frac{-x^2+x^2+1}{x^2\sqrt{x^2+1}}\\ &= \frac{1}{x^2\sqrt{x^2+1}}\end{aligned}$$

So, the sum is

$$\begin{aligned}\frac{-x\left(\dfrac{2x}{2\sqrt{x^2+1}}\right)+\sqrt{x^2+1}}{x^2}&+\left(\frac{1}{x+\sqrt{x^2+1}}\right)\left(1+\frac{2x}{2\sqrt{x^2+1}}\right)\\ &= \frac{1}{x^2\sqrt{x^2+1}}+\frac{1}{\sqrt{x^2+1}}\\ &= \frac{1}{x^2\sqrt{x^2+1}}+\frac{x^2}{x^2\sqrt{x^2+1}}\\ &= \frac{x^2+1}{x^2\sqrt{x^2+1}}\\ &= \frac{\sqrt{x^2+1}}{x^2}.\end{aligned}$$

TRY IT 5

Simplify the expression.

$$\frac{-x\left(\dfrac{3x}{3\sqrt{x^2+4}}\right)+\sqrt{x^2+4}}{x^2}+\left(\frac{1}{x+\sqrt{x^2+4}}\right)\left(1+\frac{3x}{3\sqrt{x^2+4}}\right)$$

ALGEBRA REVIEW

To check that the simplified expression in Example 5 is equivalent to the original expression, try substituting values of x into each expression. For instance, when you substitute $x = 1$ into each expression, you obtain $\sqrt{2}$.

Rationalization Techniques

In working with quotients involving radicals, it is often convenient to move the radical expression from the denominator to the numerator, or vice versa. For example, you can move $\sqrt{2}$ from the denominator to the numerator in the following quotient by multiplying by $\sqrt{2}/\sqrt{2}$.

Radical in Denominator		*Rationalize*		*Radical in Numerator*
$\dfrac{1}{\sqrt{2}}$	⇒	$\dfrac{1}{\sqrt{2}}\left(\dfrac{\sqrt{2}}{\sqrt{2}}\right)$	⇒	$\dfrac{\sqrt{2}}{2}$

This process is called **rationalizing the denominator.** A similar process is used to **rationalize the numerator.**

Rationalizing Techniques

1. If the denominator is $\sqrt{a}$, multiply by $\dfrac{\sqrt{a}}{\sqrt{a}}$.
2. If the denominator is $\sqrt{a} - \sqrt{b}$, multiply by $\dfrac{\sqrt{a} + \sqrt{b}}{\sqrt{a} + \sqrt{b}}$.
3. If the denominator is $\sqrt{a} + \sqrt{b}$, multiply by $\dfrac{\sqrt{a} - \sqrt{b}}{\sqrt{a} - \sqrt{b}}$.

The same guidelines apply to rationalizing numerators.

ALGEBRA REVIEW

The success of the second and third rationalizing techniques stems from the following.

$$(\sqrt{a} - \sqrt{b})(\sqrt{a} + \sqrt{b})$$
$$= a - b$$

EXAMPLE 6 Rationalizing Denominators and Numerators

Rationalize the denominator or numerator.

(a) $\dfrac{3}{\sqrt{12}}$ (b) $\dfrac{\sqrt{x+1}}{2}$ (c) $\dfrac{1}{\sqrt{5}+\sqrt{2}}$ (d) $\dfrac{1}{\sqrt{x}-\sqrt{x+1}}$

SOLUTION

(a) $$\frac{3}{\sqrt{12}} = \frac{3}{2\sqrt{3}} = \frac{3}{2\sqrt{3}}\left(\frac{\sqrt{3}}{\sqrt{3}}\right) = \frac{3\sqrt{3}}{2(3)} = \frac{\sqrt{3}}{2}$$

(b) $$\frac{\sqrt{x+1}}{2} = \frac{\sqrt{x+1}}{2}\left(\frac{\sqrt{x+1}}{\sqrt{x+1}}\right)$$
$$= \frac{x+1}{2\sqrt{x+1}}$$

(c) $$\frac{1}{\sqrt{5}+\sqrt{2}} = \frac{1}{\sqrt{5}+\sqrt{2}}\left(\frac{\sqrt{5}-\sqrt{2}}{\sqrt{5}-\sqrt{2}}\right) = \frac{\sqrt{5}-\sqrt{2}}{5-2} = \frac{\sqrt{5}-\sqrt{2}}{3}$$

(d) $$\frac{1}{\sqrt{x}-\sqrt{x+1}} = \frac{1}{\sqrt{x}-\sqrt{x+1}}\left(\frac{\sqrt{x}+\sqrt{x+1}}{\sqrt{x}+\sqrt{x+1}}\right)$$
$$= \frac{\sqrt{x}+\sqrt{x+1}}{x-(x+1)}$$
$$= -\sqrt{x}-\sqrt{x+1}$$

TRY IT 6

Rationalize the denominator or numerator.

(a) $\dfrac{5}{\sqrt{8}}$

(b) $\dfrac{\sqrt{x+2}}{4}$

(c) $\dfrac{1}{\sqrt{6}-\sqrt{3}}$

(d) $\dfrac{1}{\sqrt{x}+\sqrt{x+2}}$

EXERCISES 0.5

In Exercises 1–16, perform the indicated operations and simplify your answer.

1. $\dfrac{5}{x-1} + \dfrac{x}{x-1}$

2. $\dfrac{2x-1}{x+3} + \dfrac{1-x}{x+3}$

3. $\dfrac{2x}{x^2+2} - \dfrac{1-3x}{x^2+2}$

4. $\dfrac{5x+10}{2x-1} - \dfrac{2x+10}{2x-1}$

5. $\dfrac{2}{x^2-4} - \dfrac{1}{x-2}$

6. $\dfrac{x}{x^2+x-2} - \dfrac{1}{x+2}$

7. $\dfrac{5}{x-3} + \dfrac{3}{3-x}$

8. $\dfrac{x}{2-x} + \dfrac{2}{x-2}$

9. $\dfrac{A}{x+1} + \dfrac{B}{(x+1)^2} + \dfrac{C}{x-2}$

10. $\dfrac{A}{x-5} + \dfrac{B}{x+5} + \dfrac{C}{(x+5)^2}$

11. $\dfrac{A}{x-6} + \dfrac{Bx+C}{x^2+3}$

12. $\dfrac{Ax+B}{x^2+2} + \dfrac{C}{x-4}$

13. $-\dfrac{1}{x} + \dfrac{2}{x^2+1}$

14. $\dfrac{2}{x+1} + \dfrac{1-x}{x^2-2x+3}$

15. $\dfrac{1}{x^2-x-2} - \dfrac{x}{x^2-5x+6}$

16. $\dfrac{x-1}{x^2+5x+4} + \dfrac{2}{x^2-x-2} + \dfrac{10}{x^2+2x-8}$

In Exercises 17–30, simplify each expression.

17. $\dfrac{-x}{(x+1)^{3/2}} + \dfrac{2}{(x+1)^{1/2}}$

18. $2\sqrt{x}(x-2) + \dfrac{(x-2)^2}{2\sqrt{x}}$

19. $\dfrac{2-t}{2\sqrt{1+t}} - \sqrt{1+t}$

20. $-\dfrac{\sqrt{x^2+1}}{x^2} + \dfrac{1}{\sqrt{x^2+1}}$

21. $\left(2x\sqrt{x^2+1} - \dfrac{x^3}{\sqrt{x^2+1}}\right) \div (x^2+1)$

22. $\left(\sqrt{x^3+1} - \dfrac{3x^3}{2\sqrt{x^3+1}}\right) \div (x^3+1)$

23. $\dfrac{(x^2+2)^{1/2} - x^2(x^2+2)^{-1/2}}{x^2}$

24. $\dfrac{x(x+1)^{-1/2} - (x+1)^{1/2}}{x^2}$

25. $\dfrac{\dfrac{\sqrt{x+1}}{\sqrt{x}} - \dfrac{\sqrt{x}}{\sqrt{x+1}}}{2(x+1)}$

26. $\dfrac{\dfrac{2x^2}{3(x^2-1)^{2/3}} - (x^2-1)^{1/3}}{x^2}$

27. $\dfrac{x}{2(x+2)^{1/2}} + (x+2)^{1/2}$

28. $\dfrac{x}{(x-5)^{1/2}} + 2(x-5)^{1/2}$

29. $\dfrac{-x^2}{(2x+3)^{3/2}} + \dfrac{2x}{(2x+3)^{1/2}}$

30. $\dfrac{-x}{2(3+x^2)^{3/2}} + \dfrac{3}{(3+x^2)^{1/2}}$

In Exercises 31–44, rationalize the numerator or denominator and simplify.

31. $\dfrac{3}{\sqrt{6}}$

32. $\dfrac{5}{\sqrt{10}}$

33. $\dfrac{x}{\sqrt{x-4}}$

34. $\dfrac{4y}{\sqrt{y+8}}$

35. $\dfrac{49(x-3)}{\sqrt{x^2-9}}$

36. $\dfrac{10(x+2)}{\sqrt{x^2-x-6}}$

37. $\dfrac{5}{\sqrt{14}-2}$

38. $\dfrac{13}{6+\sqrt{10}}$

39. $\dfrac{2x}{5-\sqrt{3}}$

40. $\dfrac{x}{\sqrt{2}+\sqrt{3}}$

41. $\dfrac{1}{\sqrt{6}+\sqrt{5}}$

42. $\dfrac{\sqrt{15}+3}{12}$

43. $\dfrac{2}{\sqrt{x}+\sqrt{x-2}}$

44. $\dfrac{10}{\sqrt{x}+\sqrt{x+5}}$

In Exercises 45 and 46, perform the indicated operations and rationalize as needed.

45. $\dfrac{\dfrac{\sqrt{4-x^2}}{x^4} - \dfrac{2}{x^2\sqrt{4-x^2}}}{4-x^2}$

46. $\dfrac{\dfrac{\sqrt{x^2+1}}{x^2} - \dfrac{1}{x\sqrt{x^2+1}}}{x^2+1}$

47. *Installment Loan* The monthly payment M for an installment loan is given by the formula

$$M = P\left[\frac{r/12}{1-\left(\dfrac{1}{(r/12)+1}\right)^N}\right]$$

where P is the amount of the loan, r is the annual percentage rate, and N is the number of monthly payments. Enter the formula into a graphing utility, and use it to find the monthly payment for a loan of \$10,000 at an annual percentage rate of 14% ($r = 0.14$) for 5 years ($N = 60$ monthly payments).

48. *Inventory* A retailer has determined that the cost C of ordering and storing x units of a product is

$$C = 6x + \frac{900{,}000}{x}.$$

(a) Write the expression for cost as a single fraction.

(b) Determine the cost for ordering and storing $x = 240$ units of this product.

chapter

1 Functions, Graphs, and Limits

© Jeff Zaruba/CORBIS

A graph showing changes in a company's earnings and other financial indicators can depict the company's general financial trends over time.

STRATEGIES FOR SUCCESS

WHAT YOU SHOULD LEARN:

- How to plot points in the Cartesian plane and find the distance between two points
- How to sketch the graph of an equation, and find the *x*- and *y*-intercepts
- How to write equations of lines and sketch the lines
- How to evaluate, simplify, and find inverse functions
- How to find the limit of a function and discuss the continuity of a variety of functions

WHY YOU SHOULD LEARN IT:

Functions have many applications in real life, as can be seen by the examples below, which represent a small sample of the applications in this chapter.

- Dow Jones Industrial Average, Exercises 35 and 36 on page 9
- Break-Even Analysis, Exercises 63–68 on page 22
- Linear Depreciation, Exercises 86 and 87 on page 35
- Demand Function, Exercise 71 on page 47
- Market Equilibrium, Exercise 74 on page 48

1.1 THE CARTESIAN PLANE AND THE DISTANCE FORMULA

- Plot points in a coordinate plane and read data presented graphically.
- Find the distance between two points in a coordinate plane.
- Find the midpoints of line segments connecting two points.
- Translate points in a coordinate plane.

The Cartesian Plane

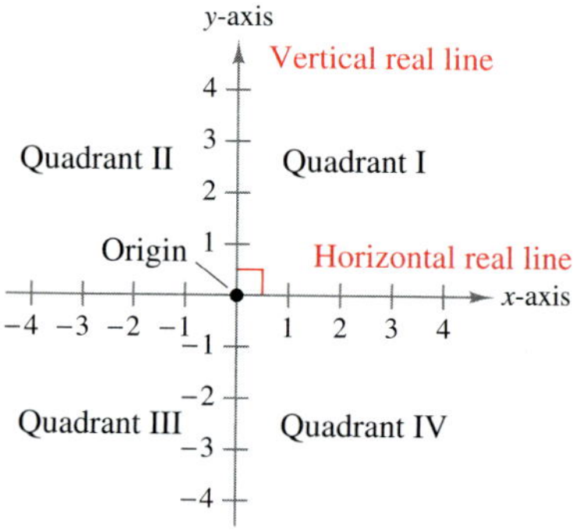

FIGURE 1.1 The Cartesian Plane

Just as you can represent real numbers by points on a real number line, you can represent ordered pairs of real numbers by points in a plane called the **rectangular coordinate system,** or the **Cartesian plane,** after the French mathematician René Descartes (1596–1650).

The Cartesian plane is formed by using two real number lines intersecting at right angles, as shown in Figure 1.1. The horizontal real number line is usually called the ***x*-axis,** and the vertical real number line is usually called the ***y*-axis.** The point of intersection of these two axes is the **origin,** and the two axes divide the plane into four parts called **quadrants.**

Each point in the plane corresponds to an **ordered pair** (x, y) of real numbers x and y, called **coordinates** of the point. The ***x*-coordinate** represents the directed distance from the y-axis to the point, and the ***y*-coordinate** represents the directed distance from the x-axis to the point, as shown in Figure 1.2.

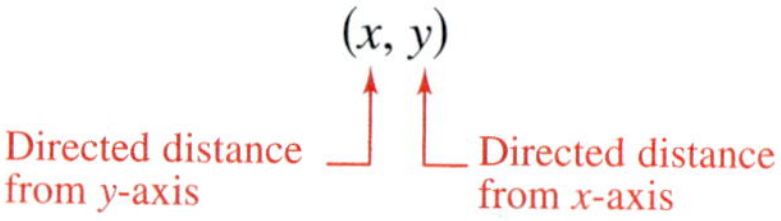

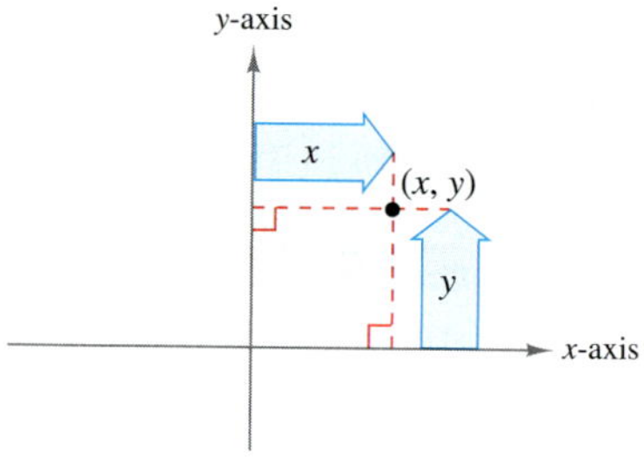

FIGURE 1.2

STUDY TIP

The notation (x, y) denotes both a point in the plane and an open interval on the real number line. The context will tell you which meaning is intended.

EXAMPLE 1 **Plotting Points in the Cartesian Plane**

Plot the points $(-1, 2)$, $(3, 4)$, $(0, 0)$, $(3, 0)$, and $(-2, -3)$.

SOLUTION To plot the point

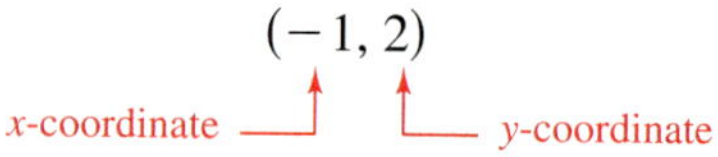

imagine a vertical line through -1 on the x-axis and a horizontal line through 2 on the y-axis. The intersection of these two lines is the point $(-1, 2)$. The other four points can be plotted in a similar way and are shown in Figure 1.3.

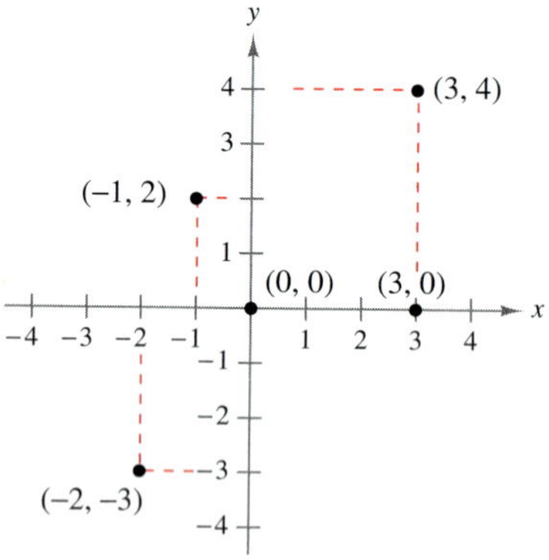

FIGURE 1.3

TRY IT 1

Plot the points $(-3, 2)$, $(4, -2)$, $(3, 1)$, $(0, -2)$, and $(-1, -2)$.

Using a rectangular coordinate system allows you to visualize relationships between two variables. It would be difficult to overestimate the importance of Descartes's introduction of coordinates to the plane. Today his ideas are in common use in virtually every scientific and business-related field. In Example 2, notice how much your intuition is enhanced by the use of a graphical presentation.

EXAMPLE 2 Sketching a Scatter Plot

The amounts A (in millions of dollars) spent on snowmobiles in the United States from 1993 through 2002 are shown in the table, where t represents the year. Sketch a scatter plot of the data. *(Source: National Sporting Goods Association)*

t	1993	1994	1995	1996	1997	1998	1999	2000	2001	2002
A	515	715	910	974	975	883	820	894	784	808

SOLUTION To sketch a *scatter plot* of the data given in the table, you simply represent each pair of values by an ordered pair (t, A), and plot the resulting points, as shown in Figure 1.4. For instance, the first pair of values is represented by the ordered pair (1993, 515). Note that the break in the t-axis indicates that the numbers between 0 and 1992 have been omitted.

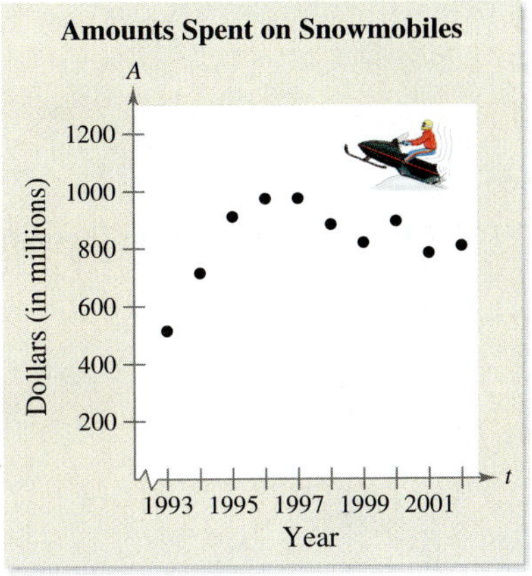

FIGURE 1.4

TRY IT 2

From 1991 to 2000, the enrollments E (in millions) of students in U.S. public colleges are shown, where t represents the year. Sketch a scatter plot of the data. *(Source: U.S. National Center for Education Statistics)*

t	1991	1992	1993	1994	1995	1996	1997	1998	1999	2000
E	11.3	11.4	11.2	11.1	11.1	11.1	11.2	11.1	11.3	11.8

STUDY TIP

In Example 2, you could let $t = 1$ represent the year 1993. In that case, the horizontal axis would not have been broken, and the tick marks would have been labeled 1 through 10 (instead of 1993 through 2002).

TECHNOLOGY

The scatter plot in Example 2 is only one way to represent the given data graphically. Two other techniques are shown at the right. The first is a *bar graph* and the second is a *line graph*. All three graphical representations were created with a computer. If you have access to computer graphing software, try using it to represent graphically the data given in Example 2.

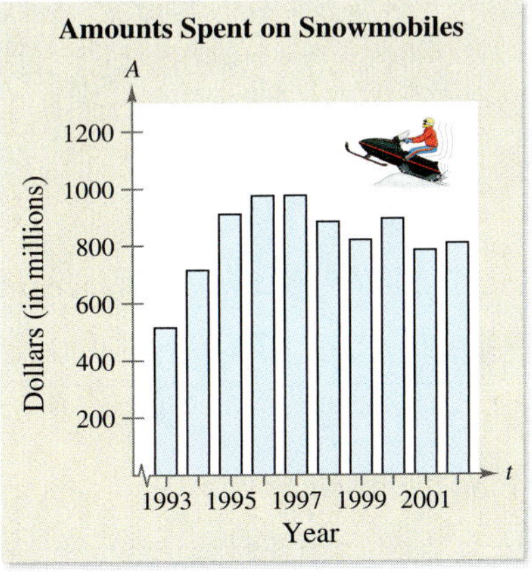

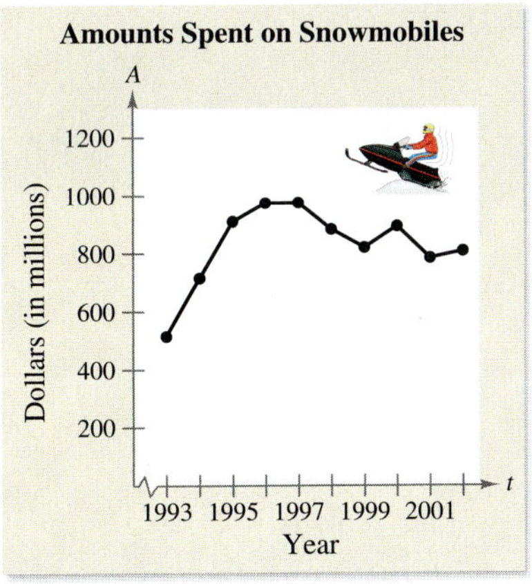

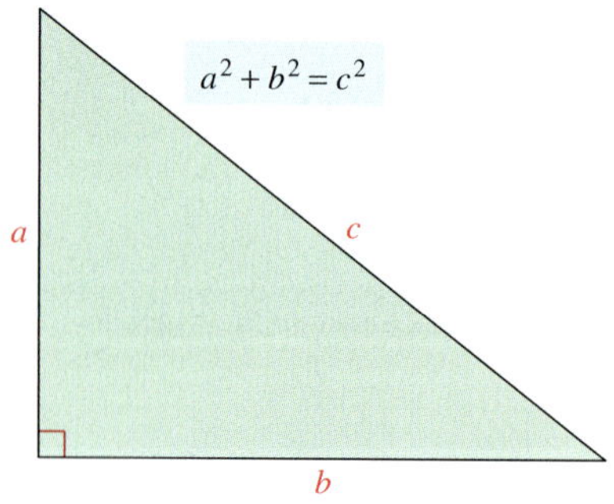

FIGURE 1.5 Pythagorean Theorem

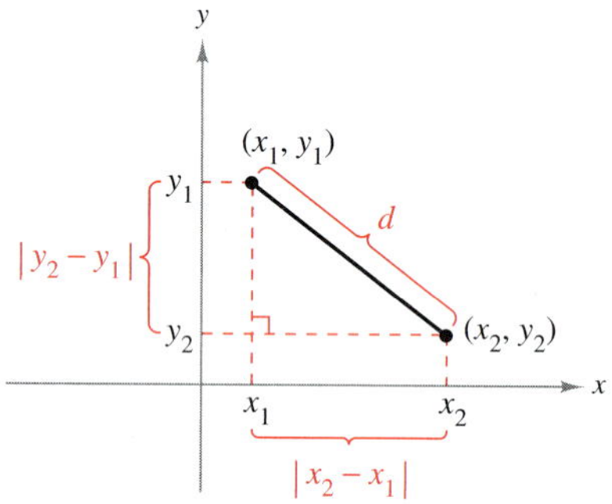

FIGURE 1.6 Distance Between Two Points

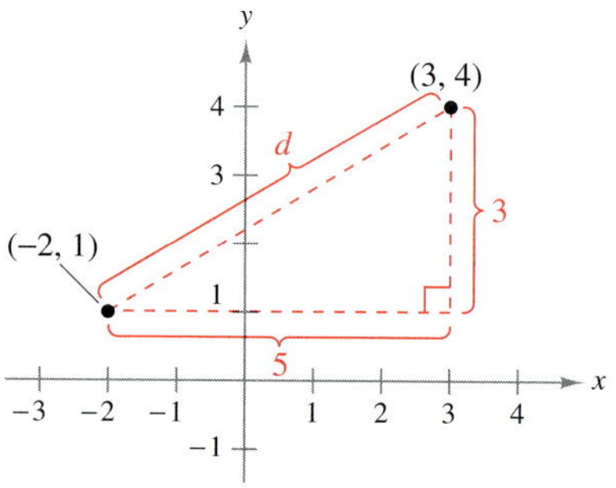

FIGURE 1.7

The Distance Formula

Recall from the Pythagorean Theorem that, for a right triangle with hypotenuse of length c and sides of lengths a and b, you have

$$a^2 + b^2 = c^2 \qquad \text{Pythagorean Theorem}$$

as shown in Figure 1.5. (The converse is also true. That is, if $a^2 + b^2 = c^2$, then the triangle is a right triangle.)

Suppose you want to determine the distance d between two points (x_1, y_1) and (x_2, y_2) in the plane. With these two points, a right triangle can be formed, as shown in Figure 1.6. The length of the vertical side of the triangle is

$$|y_2 - y_1|$$

and the length of the horizontal side is

$$|x_2 - x_1|.$$

By the Pythagorean Theorem, you can write

$$d^2 = |x_2 - x_1|^2 + |y_2 - y_1|^2$$
$$d = \sqrt{|x_2 - x_1|^2 + |y_2 - y_1|^2}$$
$$d = \sqrt{(x_2 - x_1)^2 + (y_2 - y_1)^2}.$$

This result is the **Distance Formula.**

The Distance Formula

The distance d between the points (x_1, y_1) and (x_2, y_2) in the plane is

$$d = \sqrt{(x_2 - x_1)^2 + (y_2 - y_1)^2}.$$

EXAMPLE 3 Finding a Distance

Find the distance between the points $(-2, 1)$ and $(3, 4)$.

SOLUTION Let $(x_1, y_1) = (-2, 1)$ and $(x_2, y_2) = (3, 4)$. Then apply the Distance Formula as shown.

$$d = \sqrt{(x_2 - x_1)^2 + (y_2 - y_1)^2} \qquad \text{Distance Formula}$$
$$= \sqrt{[3 - (-2)]^2 + (4 - 1)^2} \qquad \text{Substitute for } x_1, y_1, x_2, \text{ and } y_2.$$
$$= \sqrt{(5)^2 + (3)^2} \qquad \text{Simplify.}$$
$$= \sqrt{34}$$
$$\approx 5.83 \qquad \text{Use a calculator.}$$

Note in Figure 1.7 that a distance of 5.83 looks about right.

TRY IT 3

Find the distance between the points $(-2, 1)$ and $(2, 4)$.

EXAMPLE 4 Verifying a Right Triangle

Use the Distance Formula to show that the points (2, 1), (4, 0), and (5, 7) are vertices of a right triangle.

SOLUTION The three points are plotted in Figure 1.8. Using the Distance Formula, you can find the lengths of the three sides as shown below.

$$d_1 = \sqrt{(5-2)^2 + (7-1)^2} = \sqrt{9+36} = \sqrt{45}$$
$$d_2 = \sqrt{(4-2)^2 + (0-1)^2} = \sqrt{4+1} = \sqrt{5}$$
$$d_3 = \sqrt{(5-4)^2 + (7-0)^2} = \sqrt{1+49} = \sqrt{50}$$

Because

$$d_1^2 + d_2^2 = 45 + 5 = 50 = d_3^2$$

you can apply the converse of the Pythagorean Theorem to conclude that the triangle must be a right triangle.

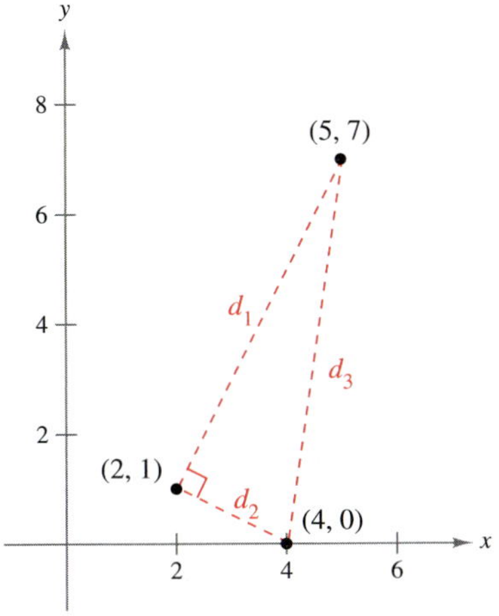

FIGURE 1.8

TRY IT 4

Use the Distance Formula to show that the points (2, −1), (5, 5), and (6, −3) are vertices of a right triangle.

The figures provided with Examples 3 and 4 were not really essential to the solution. *Nevertheless*, we strongly recommend that you develop the habit of including sketches with your solutions—even if they are not required.

EXAMPLE 5 Finding the Length of a Pass

In a football game, a quarterback throws a pass from the five-yard line, 20 yards from the sideline. The pass is caught by a wide receiver on the 45-yard line, 50 yards from the same sideline, as shown in Figure 1.9. How long was the pass?

SOLUTION You can find the length of the pass by finding the distance between the points (20, 5) and (50, 45).

$$d = \sqrt{(50-20)^2 + (45-5)^2} \quad \text{Distance Formula}$$
$$= \sqrt{900 + 1600}$$
$$= 50 \quad \text{Simplify.}$$

So, the pass was 50 yards long.

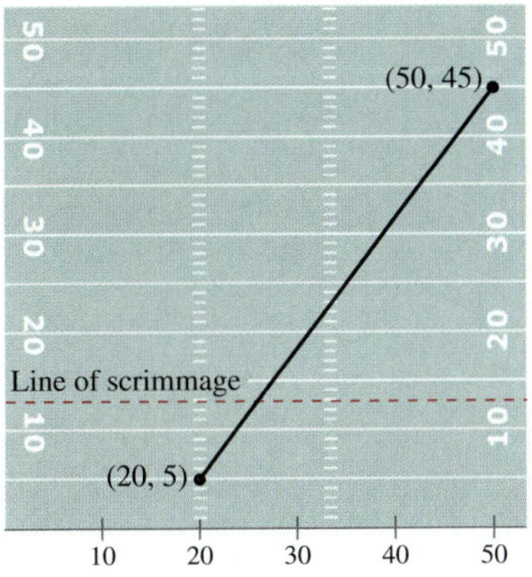

FIGURE 1.9

STUDY TIP

In Example 5, the scale along the goal line showing distance from the sideline does not normally appear on a football field. However, when you use coordinate geometry to solve real-life problems, you are free to place the coordinate system in any way that is convenient to the solution of the problem.

TRY IT 5

A quarterback throws a pass from the 10-yard line, 10 yards from the sideline. The pass is caught by a wide receiver on the 30-yard line, 25 yards from the same sideline. How long was the pass?

The Midpoint Formula

To find the **midpoint** of the line segment that joins two points in a coordinate plane, you can simply find the average values of the respective coordinates of the two endpoints.

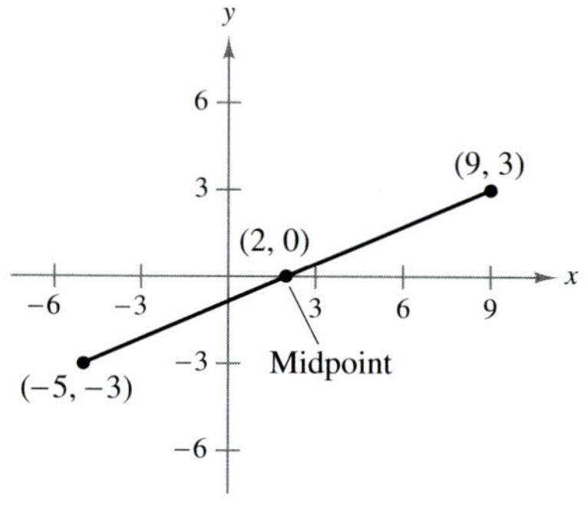

FIGURE 1.10

The Midpoint Formula

The midpoint of the segment joining the points (x_1, y_1) and (x_2, y_2) is

$$\text{Midpoint} = \left(\frac{x_1 + x_2}{2}, \frac{y_1 + y_2}{2}\right).$$

EXAMPLE 6 Finding a Segment's Midpoint

Find the midpoint of the line segment joining the points $(-5, -3)$ and $(9, 3)$, as shown in Figure 1.10.

SOLUTION Let $(x_1, y_1) = (-5, -3)$ and $(x_2, y_2) = (9, 3)$.

$$\text{Midpoint} = \left(\frac{x_1 + x_2}{2}, \frac{y_1 + y_2}{2}\right) = \left(\frac{-5 + 9}{2}, \frac{-3 + 3}{2}\right) = (2, 0)$$

TRY IT 6

Find the midpoint of the line segment joining $(-6, 2)$ and $(2, 8)$.

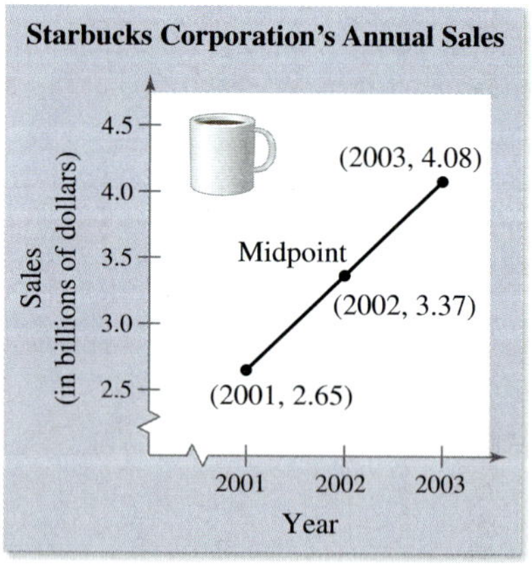

FIGURE 1.11

EXAMPLE 7 Estimating Annual Sales

Starbucks Corporation had annual sales of \$2.65 billion in 2001 and \$4.08 billion in 2003. Without knowing any additional information, what would you estimate the 2002 sales to have been? *(Source: Starbucks Corp.)*

SOLUTION One solution to the problem is to assume that sales followed a linear pattern. With this assumption, you can estimate the 2002 sales by finding the midpoint of the segment connecting the points (2001, 2.65) and (2003, 4.08).

$$\text{Midpoint} = \left(\frac{2001 + 2003}{2}, \frac{2.65 + 4.08}{2}\right) \approx (2002, 3.37)$$

So, you would estimate the 2002 sales to have been about \$3.37 billion, as shown in Figure 1.11. (The actual 2002 sales were \$3.29 billion.)

TRY IT 7

Maytag Corporation had annual sales of \$4.32 billion in 2001 and \$4.79 billion in 2003. What would you estimate the 2002 annual sales to have been? *(Source: Maytag Corp.)*

Translating Points in the Plane

EXAMPLE 8 **Translating Points in the Plane**

Figure 1.12(a) shows the vertices of a parallelogram. Find the vertices of the parallelogram after it has been translated two units down and four units to the right.

SOLUTION To translate each vertex two units down, subtract 2 from each y-coordinate. To translate each vertex four units to the right, add 4 to each x-coordinate.

Original Point	*Translated Point*
$(1, 0)$	$(1 + 4, 0 - 2) = (5, -2)$
$(3, 2)$	$(3 + 4, 2 - 2) = (7, 0)$
$(3, 6)$	$(3 + 4, 6 - 2) = (7, 4)$
$(1, 4)$	$(1 + 4, 4 - 2) = (5, 2)$

The translated parallelogram is shown in Figure 1.12(b).

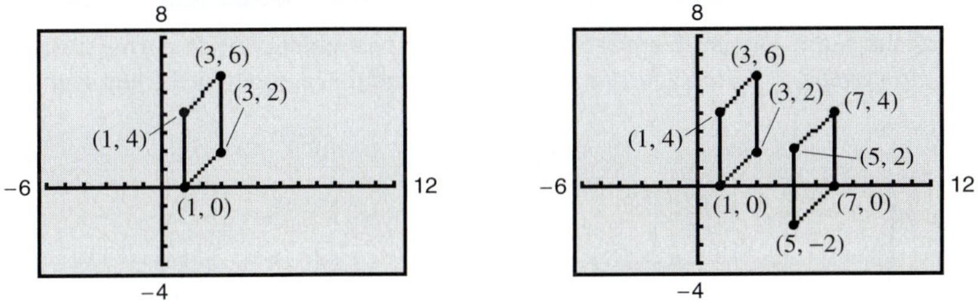

(a) (b)

FIGURE 1.12

DREAMWORKS/THE KOBAL COLLECTION

Many movies now use extensive computer graphics, much of which consists of transformations of points in two- and three-dimensional space. The photo above shows a scene from Shrek. *The movie's animators used computer graphics to design the scenery, characters, motion, and even the lighting in each scene.*

TRY IT 8

Find the vertices of the parallelogram in Example 8 after it has been translated two units to the left and four units down.

TAKE ANOTHER LOOK

Transforming Points in a Coordinate Plane

Example 8 illustrates points that have been translated (or *slid*) in a coordinate plane. The translated parallelogram is congruent to (has the same size and shape as) the original parallelogram. Try using a graphing utility to graph the transformed parallelogram for each of the following transformations. Describe the transformation. Is it a translation, a reflection, or a rotation? Is the transformed parallelogram congruent to the original parallelogram?

a. $(x, y) \Rightarrow (-x, y)$

b. $(x, y) \Rightarrow (x, -y)$

c. $(x, y) \Rightarrow (-x, -y)$

PREREQUISITE REVIEW 1.1

The following warm-up exercises involve skills that were covered in earlier sections. You will use these skills in the exercise set for this section.

In Exercises 1–6, simplify each expression.

1. $\sqrt{(3-6)^2 + [1-(-5)]^2}$

2. $\sqrt{(-2-0)^2 + [-7-(-3)]^2}$

3. $\dfrac{5+(-4)}{2}$

4. $\dfrac{-3+(-1)}{2}$

5. $\sqrt{27} + \sqrt{12}$

6. $\sqrt{8} - \sqrt{18}$

In Exercises 7–10, solve for x or y.

7. $\sqrt{(3-x)^2 + (7-4)^2} = \sqrt{45}$

8. $\sqrt{(6-2)^2 + (-2-y)^2} = \sqrt{52}$

9. $\dfrac{x+(-5)}{2} = 7$

10. $\dfrac{-7+y}{2} = -3$

EXERCISES 1.1

In Exercises 1–6, (a) find the length of each side of the right triangle and (b) show that these lengths satisfy the Pythagorean Theorem.

***1.**

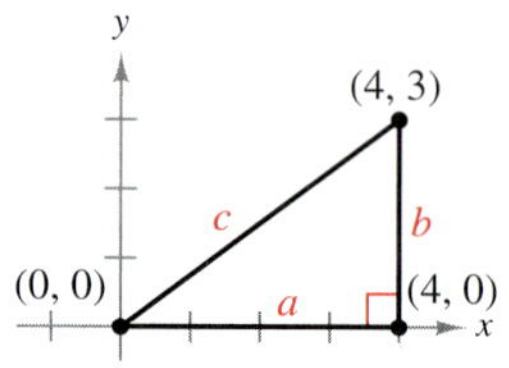

2.

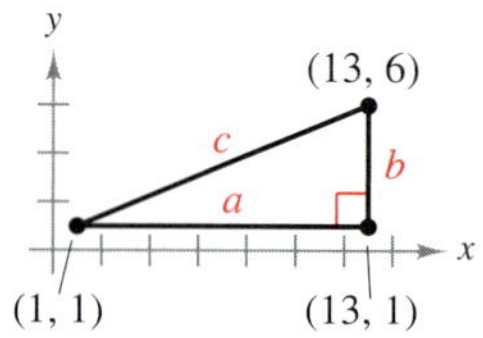

3.

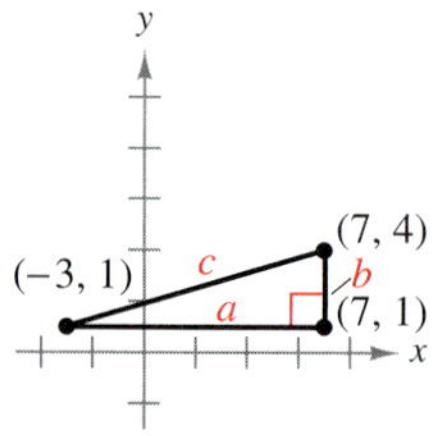

4.

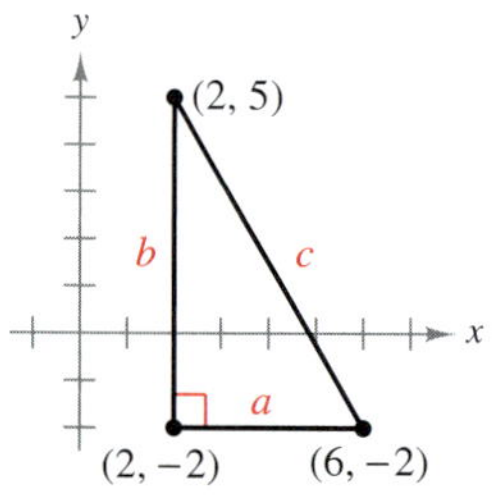

5.

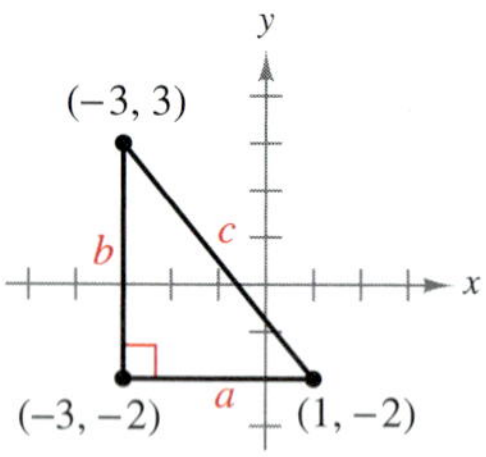

6.

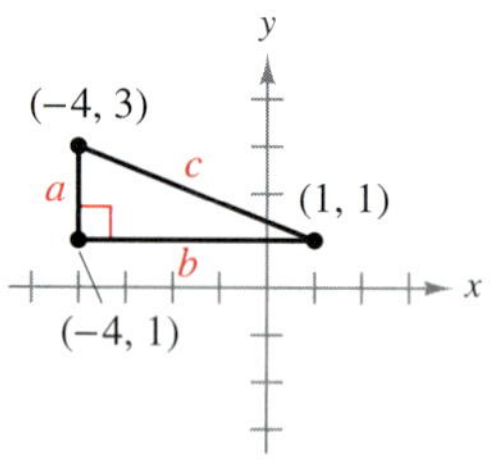

* The answers to the odd-numbered and selected even exercises are given in the back of the text. Worked-out solutions to the odd-numbered exercises are given in the *Student Solutions Guide*.

In Exercises 7–14, (a) plot the points, (b) find the distance between the points, and (c) find the midpoint of the line segment joining the points.

7. $(3, 1), (5, 5)$

8. $(-3, 2), (3, -2)$

9. $\left(\frac{1}{2}, 1\right), \left(-\frac{3}{2}, -5\right)$

10. $\left(\frac{2}{3}, -\frac{1}{3}\right), \left(\frac{5}{6}, 1\right)$

11. $(2, 2), (4, 14)$

12. $(-3, 7), (1, -1)$

13. $\left(1, \sqrt{3}\right), (-1, 1)$

14. $(-2, 0), \left(0, \sqrt{2}\right)$

In Exercises 15–18, show that the points form the vertices of the given figure. (A rhombus is a quadrilateral whose sides have the same length.)

	Vertices	*Figure*
15.	$(0, 1), (3, 7), (4, -1)$	Right triangle
16.	$(1, -3), (3, 2), (-2, 4)$	Isosceles triangle
17.	$(0, 0), (1, 2), (2, 1), (3, 3)$	Rhombus
18.	$(0, 1), (3, 7), (4, 4), (1, -2)$	Parallelogram

In Exercises 19–22, use the Distance Formula to determine whether the points are collinear (lie on the same line).

19. $(0, -4), (2, 0), (3, 2)$

20. $(0, 4), (7, -6), (-5, 11)$

21. $(-2, -6), (1, -3), (5, 2)$

22. $(-1, 1), (3, 3), (5, 5)$

In Exercises 23 and 24, find x such that the distance between the points is 5.

23. $(1, 0), (x, -4)$

24. $(2, -1), (x, 2)$

In Exercises 25 and 26, find y such that the distance between the points is 8.

25. $(0, 0), (3, y)$

26. $(5, 1), (5, y)$

27. Use the Midpoint Formula repeatedly to find the three points that divide the segment joining (x_1, y_1) and (x_2, y_2) into four equal parts.

28. Show that $\left(\frac{1}{3}[2x_1 + x_2], \frac{1}{3}[2y_1 + y_2]\right)$ is one of the points of trisection of the line segment joining (x_1, y_1) and (x_2, y_2). Then, find the second point of trisection by finding the midpoint of the segment joining

$$\left(\frac{1}{3}[2x_1 + x_2], \frac{1}{3}[2y_1 + y_2]\right) \text{ and } (x_2, y_2).$$

29. Use Exercise 27 to find the points that divide the line segment joining the given points into four equal parts.

(a) $(1, -2), (4, -1)$

(b) $(-2, -3), (0, 0)$

30. Use Exercise 28 to find the points of trisection of the line segment joining the given points.

(a) $(1, -2), (4, 1)$

(b) $(-2, -3), (0, 0)$

31. ***Building Dimensions*** The base and height of the trusses for the roof of a house are 32 feet and 5 feet, respectively (see figure).

(a) Find the distance d from the eaves to the peak of the roof.

(b) The length of the house is 40 feet. Use the result of part (a) to find the number of square feet of roofing.

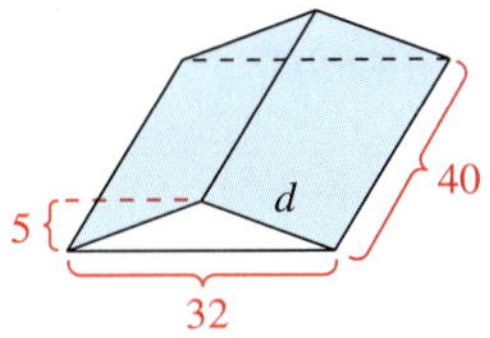

32. ***Wire Length*** A guy wire is stretched from a broadcasting tower at a point 200 feet above the ground to an anchor 125 feet from the base (see figure). How long is the wire?

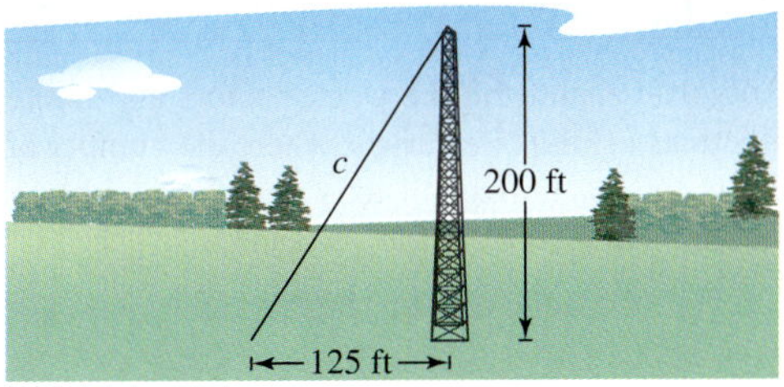

The symbol indicates an exercise in which you are instructed to use graphing technology or a symbolic computer algebra system. The solutions of other exercises may also be facilitated by use of appropriate technology.

In Exercises 33 and 34, use a graphing utility to graph a scatter plot, a bar graph, or a line graph to represent the data. Describe any trends that appear.

33. ***Consumer Trends*** The numbers (in millions) of cable television subscribers in the United States for 1992–2001 are shown in the table. *(Source: Nielsen Media Research)*

Year	1992	1993	1994	1995	1996
Subscribers	57.2	58.8	60.5	63.0	64.6

Year	1997	1998	1999	2000	2001
Subscribers	65.9	67.0	68.5	69.3	73.0

34. ***Consumer Trends*** The numbers (in millions) of cellular telephone subscribers in the United States for 1993–2002 are shown in the table. *(Source: Cellular Telecommunications & Internet Association)*

Year	1993	1994	1995	1996	1997
Subscribers	16.0	24.1	33.8	44.0	55.3

Year	1998	1999	2000	2001	2002
Subscribers	69.2	86.0	109.5	128.4	140.8

Dow Jones Industrial Average In Exercises 35 and 36, use the figure below showing the Dow Jones Industrial Average for common stocks. *(Source: Dow Jones, Inc.)*

35. Estimate the Dow Jones Industrial Average for each date.

(a) March 2002 (b) December 2002

(c) May 2003 (d) January 2004

36. Estimate the percent increase or decrease in the Dow Jones Industrial Average (a) from April 2002 to November 2002 and (b) from June 2003 to February 2004.

Figure for 35 and 36

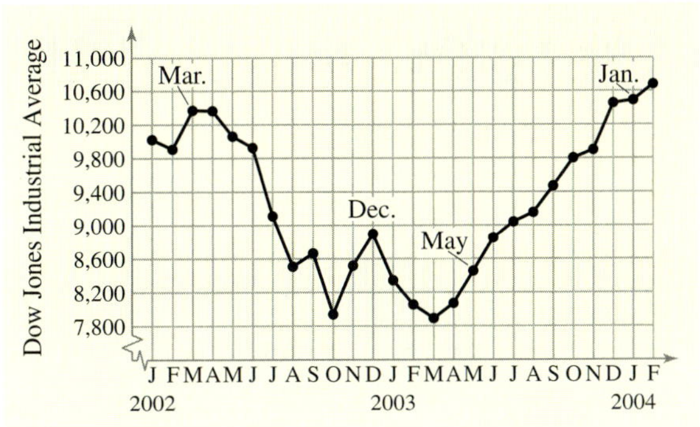

Construction In Exercises 37 and 38, use the figure, which shows the median sales prices of existing one-family homes sold (in thousands of dollars) in the United States from 1987 to 2002. *(Source: National Association of Realtors)*

37. Estimate the median sales price of existing one-family homes for each year.

(a) 1987 (b) 1992

(c) 1997 (d) 2002

38. Estimate the percent increases in the value of existing one-family homes (a) from 1993 to 1994 and (b) from 2001 to 2002.

Figure for 37 and 38

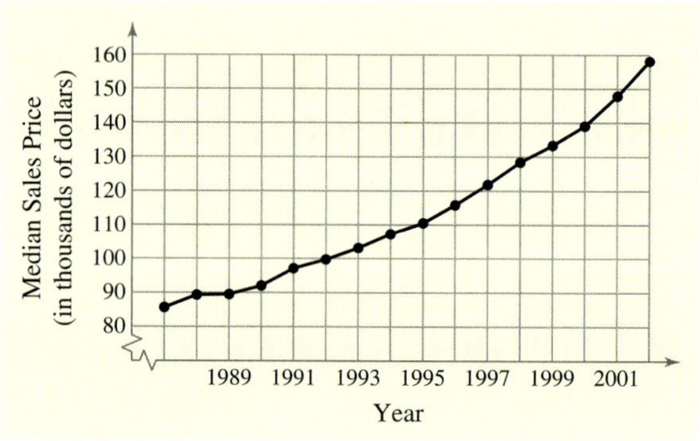

Research Project In Exercises 39 and 40, (a) use the Midpoint Formula to estimate the revenue and profit of the company in 2001. (b) Then use your school's library, the Internet, or some other reference source to find the actual revenue and profit for 2001. (c) Did the revenue and profit increase in a linear pattern from 1999 to 2003? Explain your reasoning. (d) What were the company's expenses during each of the given years? (e) How would you rate the company's growth from 1999 to 2003? *(Source: Walgreen Company and The Yankee Candle Company)*

39. Walgreen Company

Year	1999	2001	2003
Revenue (millions of $)	17,839		32,505
Profit (millions of $)	624.1		1157.3

40. The Yankee Candle Company

Year	1999	2001	2003
Revenue (millions of $)	256.6		508.6
Profit (millions of $)	34.3		74.8

Computer Graphics In Exercises 41 and 42, the red figure is translated to a new position in the plane to form the blue figure. (a) Find the vertices of the transformed figure. (b) Then use a graphing utility to draw both figures.

41.

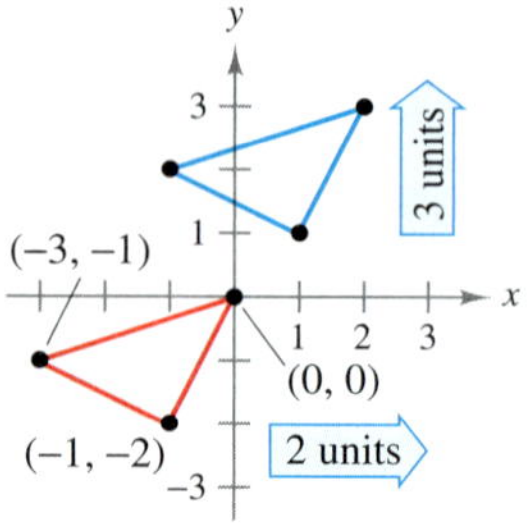

42.

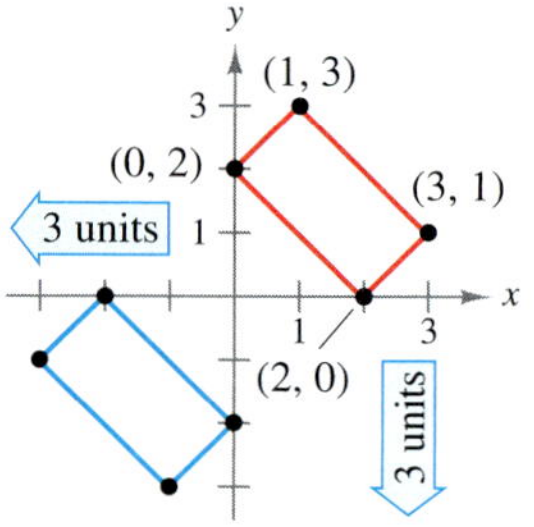

43. ***Economics*** The table shows the numbers of ear infections treated by doctors at HMO clinics of three different sizes: small, medium, and large.

Cases per small clinic	Cases per medium clinic	Cases per large clinic	Number of doctors
0	0	0	0
20	30	35	1
28	42	49	2
35	53	62	3
40	60	70	4

(a) Show the relationship between doctors and treated ear infections using *three* curves, where the number of doctors is on the horizontal axis and the number of ear infections treated is on the vertical axis.

(b) Compare the three relationships.

(Source: Adapted from Taylor, Economics, *Fourth Edition)*

The symbol indicates an exercise that contains material from textbooks in other disciplines.

1.2 GRAPHS OF EQUATIONS

- Sketch graphs of equations by hand.
- Find the x- and y-intercepts of graphs of equations.
- Write the standard forms of equations of circles.
- Find the points of intersection of two graphs.
- Use mathematical models to model and solve real-life problems.

The Graph of an Equation

In Section 1.1, you used a coordinate system to represent graphically the relationship between two quantities. There, the graphical picture consisted of a collection of points in a coordinate plane (see Example 2 in Section 1.1).

Frequently, a relationship between two quantities is expressed as an equation. For instance, degrees on the Fahrenheit scale are related to degrees on the Celsius scale by the equation

$$F = \tfrac{9}{5}C + 32.$$

In this section, you will study some basic procedures for sketching the graphs of such equations. The **graph** of an equation is the set of all points that are solutions of the equation.

EXAMPLE 1 Sketching the Graph of an Equation

Sketch the graph of $y = 7 - 3x$.

SOLUTION The simplest way to sketch the graph of an equation is the *point-plotting method.* With this method, you construct a table of values that consists of several solution points of the equation, as shown in the table below. For instance, when $x = 0$

$$y = 7 - 3(0) = 7$$

which implies that $(0, 7)$ is a solution point of the graph.

x	0	1	2	3	4
$y = 7 - 3x$	7	4	1	-2	-5

From the table, it follows that

$$(0, 7), (1, 4), (2, 1), (3, -2), \text{ and } (4, -5)$$

are solution points of the equation. After plotting these points, you can see that they appear to lie on a line, as shown in Figure 1.13. The graph of the equation is the line that passes through the five plotted points.

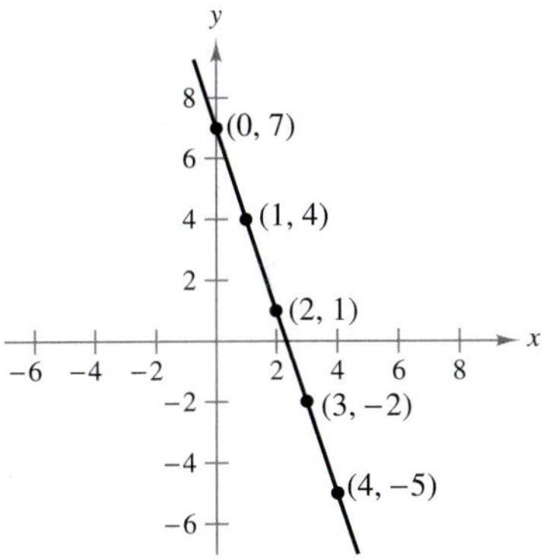

FIGURE 1.13 Solution Points for $y = 7 - 3x$

TRY IT 1

Sketch the graph of $y = 2x + 1$.

STUDY TIP

Even though we refer to the sketch shown in Figure 1.13 as the graph of $y = 7 - 3x$, it actually represents only a *portion* of the graph. The entire graph is a line that would extend off the page.

EXAMPLE 2 Sketching the Graph of an Equation

Sketch the graph of $y = x^2 - 2$.

SOLUTION Begin by constructing a table of values, as shown below.

x	-2	-1	0	1	2	3
$y = x^2 - 2$	2	-1	-2	-1	2	7

Next, plot the points given in the table, as shown in Figure 1.14(a). Finally, connect the points with a smooth curve, as shown in Figure 1.14(b).

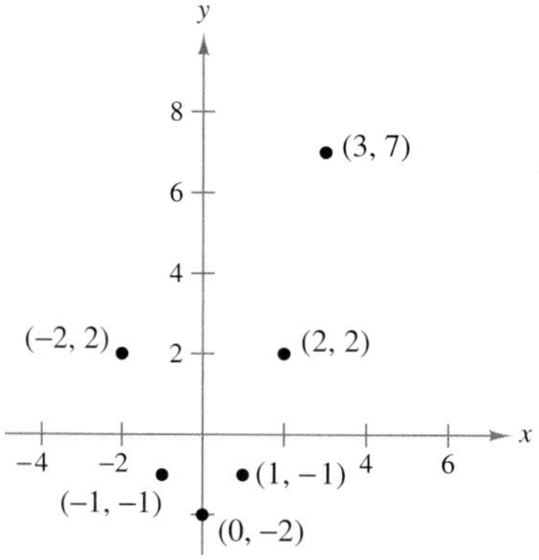

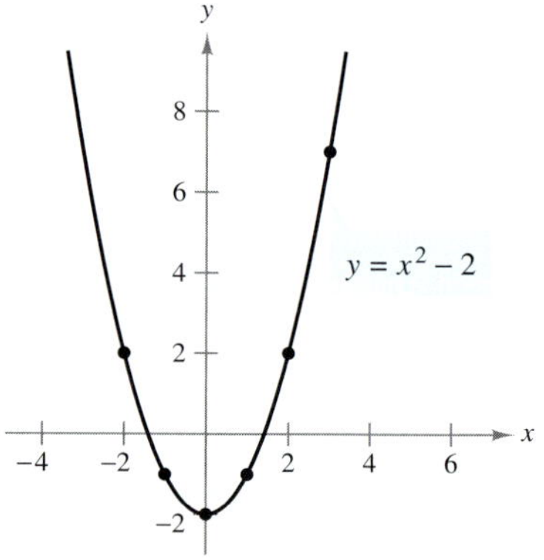

FIGURE 1.14

STUDY TIP

The graph shown in Example 2 is a **parabola.** The graph of any second-degree equation of the form

$$y = ax^2 + bx + c, \quad a \neq 0$$

has a similar shape. If $a > 0$, the parabola opens upward, as in Figure 1.14(b), and if $a < 0$, the parabola opens downward.

TRY IT 2

Sketch the graph of $y = x^2 - 4$.

The point-plotting technique demonstrated in Examples 1 and 2 is easy to use, but it does have some shortcomings. With too few solution points, you can badly misrepresent the graph of a given equation. For instance, how would you connect the four points in Figure 1.15? Without further information, any one of the three graphs in Figure 1.16 would be reasonable.

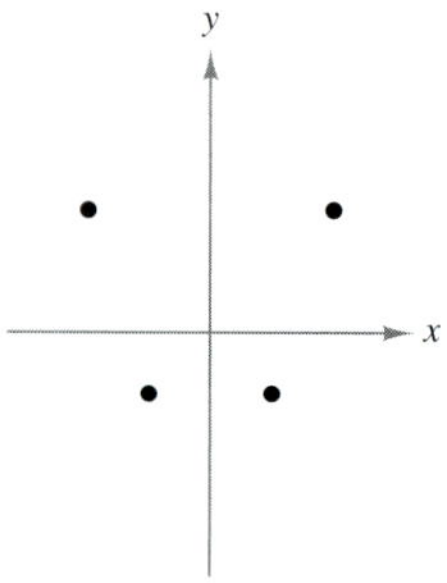

FIGURE 1.15

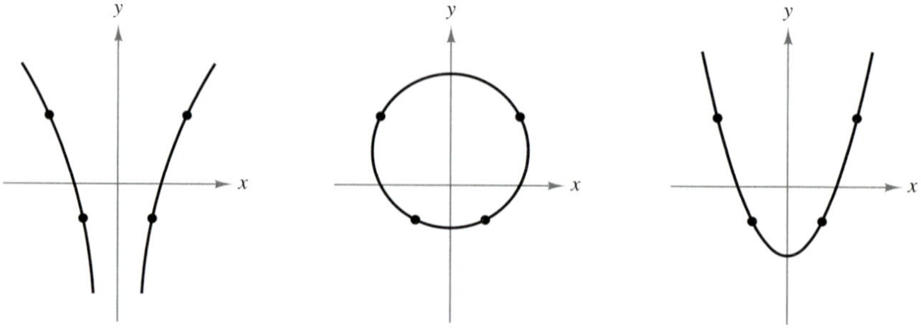

FIGURE 1.16

Intercepts of a Graph

It is often easy to determine the solution points that have zero as either the x-coordinate or the y-coordinate. These points are called **intercepts** because they are the points at which the graph intersects the x- or y-axis.

Some texts denote the x-intercept as the x-coordinate of the point $(a, 0)$ rather than the point itself. Unless it is necessary to make a distinction, we will use the term **intercept** to mean either the point or the coordinate.

A graph may have no intercepts or several intercepts, as shown in Figure 1.17.

ALGEBRA REVIEW

Finding intercepts involves solving equations. For a review of some techniques for solving equations, see page 73.

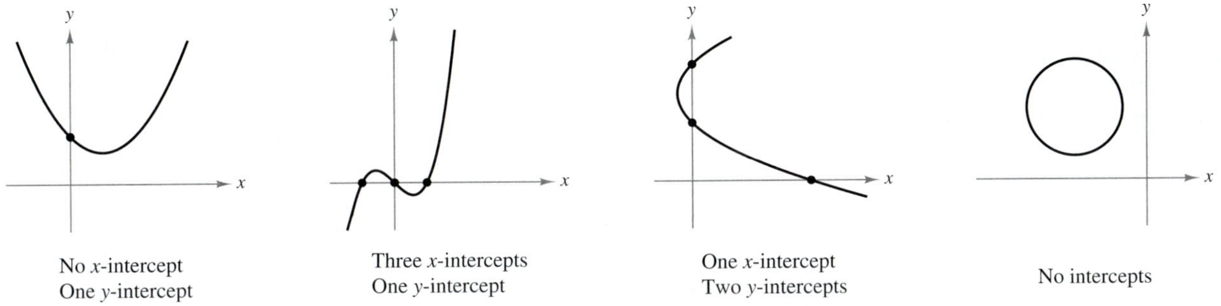

FIGURE 1.17

Finding Intercepts

1. To find **x-intercepts,** let y be zero and solve the equation for x.
2. To find **y-intercepts,** let x be zero and solve the equation for y.

EXAMPLE 3 **Finding x- and y-Intercepts**

Find the x- and y-intercepts of the graph of each equation.

(a) $y = x^3 - 4x$ (b) $x = y^2 - 3$

SOLUTION

(a) Let $y = 0$. Then $0 = x(x^2 - 4) = x(x + 2)(x - 2)$ has solutions $x = 0$ and $x = \pm 2$. Let $x = 0$. Then $y = (0)^3 - 4(0) = 0$.

x-intercepts: $(0, 0), (2, 0), (-2, 0)$ y-intercept: $(0, 0)$ See Figure 1.18.

(b) Let $y = 0$. Then $x = (0)^2 - 3 = -3$. Let $x = 0$. Then $y^2 - 3 = 0$ has solutions $y = \pm\sqrt{3}$.

x-intercept: $(-3, 0)$ y-intercepts: $(0, \sqrt{3}), (0, -\sqrt{3})$ See Figure 1.19.

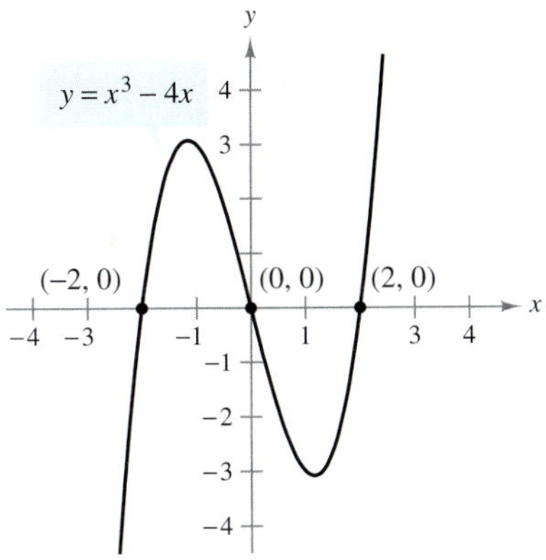

FIGURE 1.18

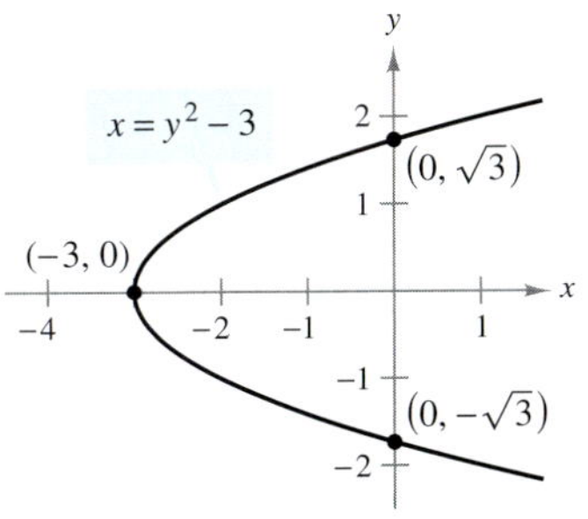

FIGURE 1.19

TRY IT 3

Find the x- and y-intercepts of the graph of each equation.

(a) $y = x^2 - 2x - 3$ (b) $y^2 - 4 = x$

TECHNOLOGY

Zooming in to Find Intercepts

You can use the *zoom* feature of a graphing utility to approximate the x-intercepts of a graph. Suppose you want to approximate the x-intercept(s) of the graph of

$$y = 2x^3 - 3x + 2.$$

Begin by graphing the equation, as shown below in part (a). From the viewing window shown, the graph appears to have only one x-intercept. This intercept lies between -2 and -1. By zooming in on the intercept, you can improve the approximation, as shown in part (b). To three decimal places, the solution is $x \approx -1.476$.

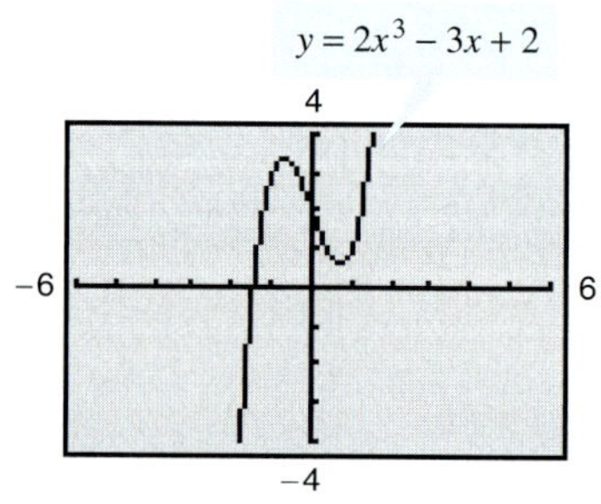

(a)

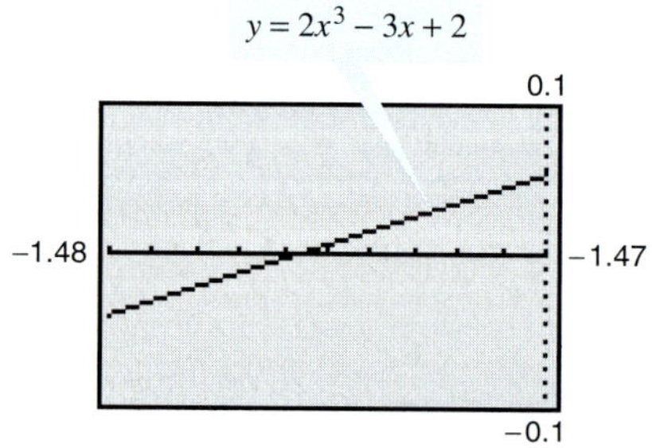

(b)

Here are some suggestions for using the *zoom* feature.

1. With each successive zoom-in, adjust the x-scale so that the viewing window shows at least one tick mark on each side of the x-intercept.
2. The error in your approximation will be less than the distance between two scale marks.
3. The *trace* feature can usually be used to add one more decimal place of accuracy without changing the viewing window.

Part (a) below shows the graph of $y = x^2 - 5x + 3$. Parts (b) and (c) show "zoom-in views" of the two intercepts. From these views, you can approximate the x-intercepts to be $x \approx 0.697$ and $x \approx 4.303$.

STUDY TIP

Some graphing utilities have a built-in program that can find the x-intercepts of a graph. If your graphing utility has this feature, try using it to find the x-intercept of the graph shown on the left. (Your calculator may call this the *root* or *zero* feature.)

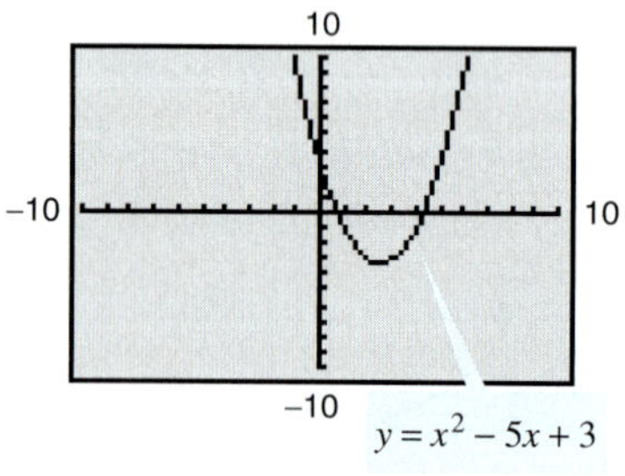

(a)

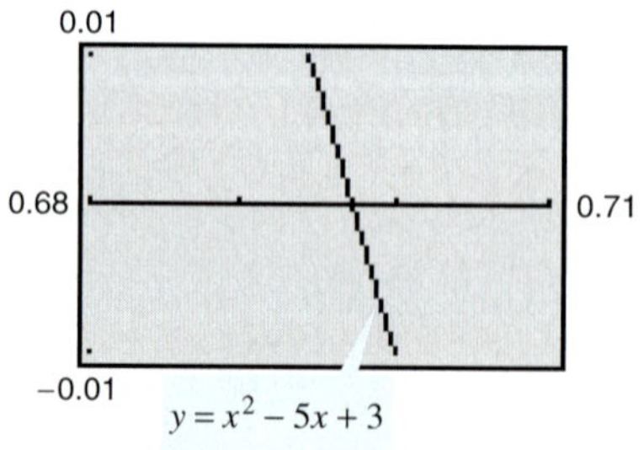

(b)

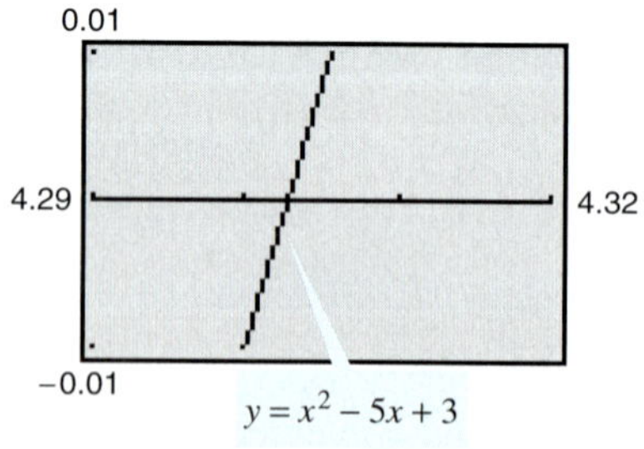

(c)

Circles

Throughout this course, you will learn to recognize several types of graphs from their equations. For instance, you should recognize that the graph of a second-degree equation of the form

$$y = ax^2 + bx + c, \quad a \neq 0$$

is a parabola (see Example 2). Another easily recognized graph is that of a **circle.**

Consider the circle shown in Figure 1.20. A point (x, y) is on the circle if and only if its distance from the center (h, k) is r. By the Distance Formula,

$$\sqrt{(x - h)^2 + (y - k)^2} = r.$$

By squaring both sides of this equation, you obtain the **standard form of the equation of a circle.**

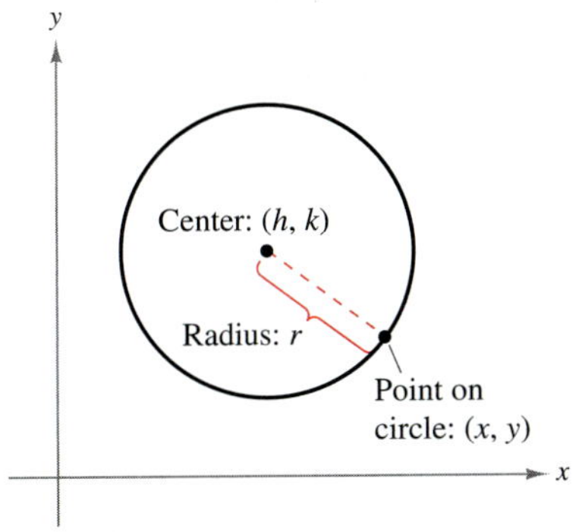

FIGURE 1.20

Standard Form of the Equation of a Circle

The point (x, y) lies on the circle of **radius** r and **center** (h, k) if and only if

$$(x - h)^2 + (y - k)^2 = r^2.$$

From this result, you can see that the standard form of the equation of a circle *with its center at the origin*, $(h, k) = (0, 0)$, is simply

$$x^2 + y^2 = r^2. \qquad \text{Circle with center at origin}$$

EXAMPLE 4 Finding the Equation of a Circle

The point $(3, 4)$ lies on a circle whose center is at $(-1, 2)$, as shown in Figure 1.21. Find the standard form of the equation of this circle.

SOLUTION The radius of the circle is the distance between $(-1, 2)$ and $(3, 4)$.

$$r = \sqrt{[3 - (-1)]^2 + (4 - 2)^2} \qquad \text{Distance Formula}$$
$$= \sqrt{16 + 4} \qquad \text{Simplify.}$$
$$= \sqrt{20} \qquad \text{Radius}$$

Using $(h, k) = (-1, 2)$, the standard form of the equation of the circle is

$$(x - h)^2 + (y - k)^2 = r^2$$
$$[x - (-1)]^2 + (y - 2)^2 = (\sqrt{20})^2 \qquad \text{Substitute for } h, k, \text{ and } r.$$
$$(x + 1)^2 + (y - 2)^2 = 20. \qquad \text{Write in standard form.}$$

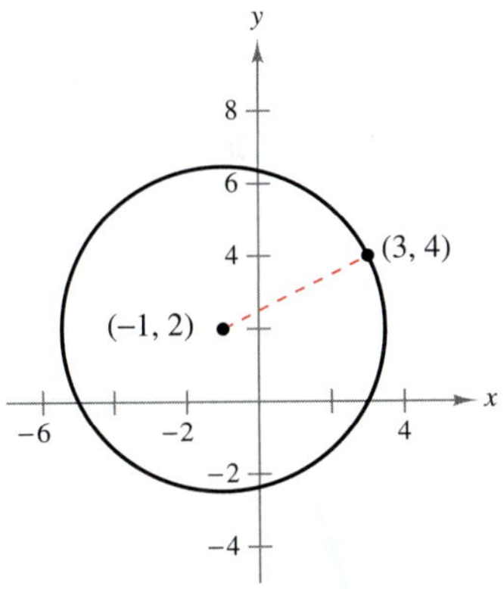

$(x + 1)^2 + (y - 2)^2 = 20$

FIGURE 1.21

TRY IT 4

The point $(1, 5)$ lies on a circle whose center is at $(-2, 1)$. Find the standard form of the equation of this circle.

TECHNOLOGY

To graph a circle on a graphing utility, you can solve its equation for y and graph the top and bottom halves of the circle separately. For instance, you can graph the circle $(x + 1)^2 + (y - 2)^2 = 20$ by graphing the following equations.

$$y = 2 + \sqrt{20 - (x + 1)^2}$$

$$y = 2 - \sqrt{20 - (x + 1)^2}$$

If you want the result to appear circular, you need to use a square setting, as shown below.

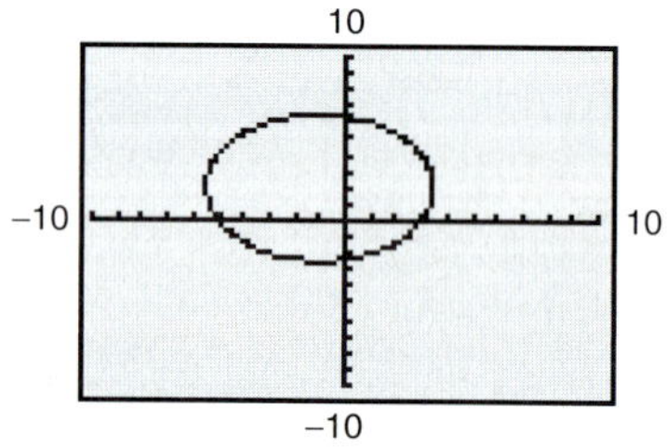

Standard setting

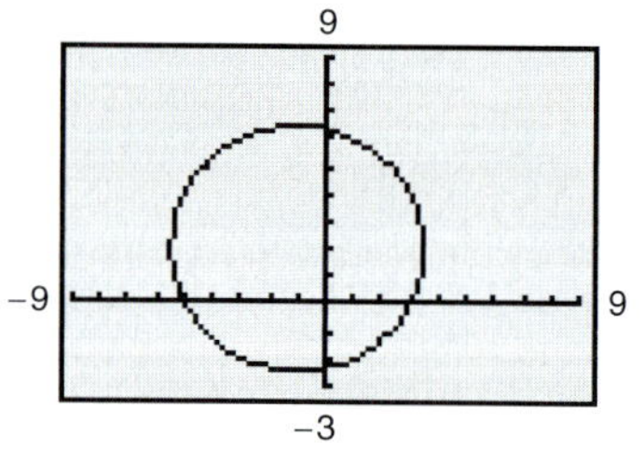

Square setting

TRY IT 5

Write the equation of the circle $x^2 + y^2 - 4x + 2y + 1 = 0$ in standard form and sketch its graph.

General Form of the Equation of a Circle

$$Ax^2 + Ay^2 + Dx + Ey + F = 0, \qquad A \neq 0$$

To change from general form to standard form, you can use a process called **completing the square,** as demonstrated in Example 5.

EXAMPLE 5 Completing the Square

Sketch the graph of the circle whose general equation is

$$4x^2 + 4y^2 + 20x - 16y + 37 = 0.$$

SOLUTION First divide by 4 so that the coefficients of x^2 and y^2 are both 1.

$$4x^2 + 4y^2 + 20x - 16y + 37 = 0 \qquad \text{Write original equation.}$$

$$x^2 + y^2 + 5x - 4y + \tfrac{37}{4} = 0 \qquad \text{Divide each side by 4.}$$

$$(x^2 + 5x + \quad) + (y^2 - 4y + \quad) = -\tfrac{37}{4} \qquad \text{Group terms.}$$

$$\left(x^2 + 5x + \tfrac{25}{4}\right) + (y^2 - 4y + 4) = -\tfrac{37}{4} + \tfrac{25}{4} + 4 \qquad \text{Complete the square.}$$

(Half)^2 (Half)^2

$$\left(x + \tfrac{5}{2}\right)^2 + (y - 2)^2 = 1 \qquad \text{Write in standard form.}$$

From the standard form, you can see that the circle is centered at $\left(-\frac{5}{2}, 2\right)$ and has a radius of 1, as shown in Figure 1.22.

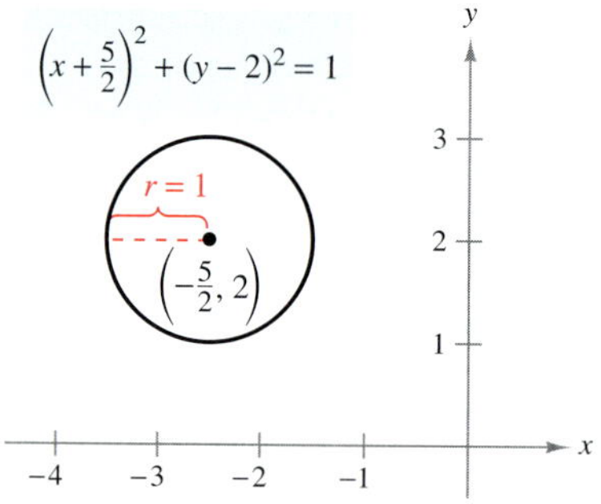

FIGURE 1.22

The general equation $Ax^2 + Ay^2 + Dx + Ey + F = 0$ may not always represent a circle. In fact, such an equation will have no solution points if the procedure of completing the square yields the impossible result

$$(x - h)^2 + (y - k)^2 = \text{negative number.} \qquad \text{No solution points}$$

Further assistance with the calculus and algebra used in this example is available on the CD that accompanies this text.

Points of Intersection

A **point of intersection** of two graphs is an ordered pair that is a solution point of both graphs. For instance, Figure 1.23 shows that the graphs of

$$y = x^2 - 3 \quad \text{and} \quad y = x - 1$$

have two points of intersection: $(2, 1)$ and $(-1, -2)$. To find the points analytically, set the two y-values equal to each other and solve the equation

$$x^2 - 3 = x - 1$$

for x.

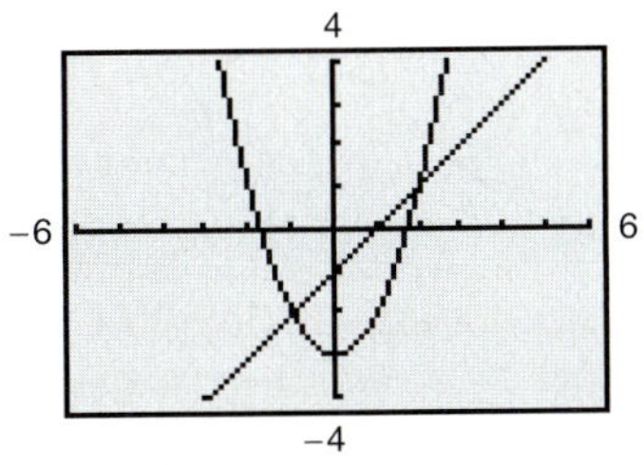

FIGURE 1.23

A common business application that involves points of intersection is **break-even analysis.** The marketing of a new product typically requires an initial investment. When sufficient units have been sold so that the total revenue has offset the total cost, the sale of the product has reached the **break-even point.** The **total cost** of producing x units of a product is denoted by C, and the **total revenue** from the sale of x units of the product is denoted by R. So, you can find the break-even point by setting the cost C equal to the revenue R, and solving for x.

STUDY TIP

The *Technology* note on page 14 describes how to use a graphing utility to find the x-intercepts of a graph. A similar procedure can be used to find the points of intersection of two graphs. (Your calculator may call this the *intersect* feature.)

EXAMPLE 6 Finding a Break-Even Point

A business manufactures a product at a cost of \$0.65 per unit and sells the product for \$1.20 per unit. The company's initial investment to produce the product was \$10,000. How many units must the company sell to break even?

SOLUTION The total cost of producing x units of the product is given by

$$C = 0.65x + 10{,}000. \qquad \text{Cost equation}$$

The total revenue from the sale of x units is given by

$$R = 1.2x. \qquad \text{Revenue equation}$$

To find the break-even point, set the cost equal to the revenue and solve for x.

$$R = C \qquad \text{Set revenue equal to cost.}$$

$$1.2x = 0.65x + 10{,}000 \qquad \text{Substitute for } R \text{ and } C.$$

$$0.55x = 10{,}000 \qquad \text{Subtract } 0.65x \text{ from each side.}$$

$$x = \frac{10{,}000}{0.55} \qquad \text{Divide each side by 0.55.}$$

$$x \approx 18{,}182 \qquad \text{Use a calculator.}$$

So, the company must sell 18,182 units before it breaks even. This result is shown graphically in Figure 1.24.

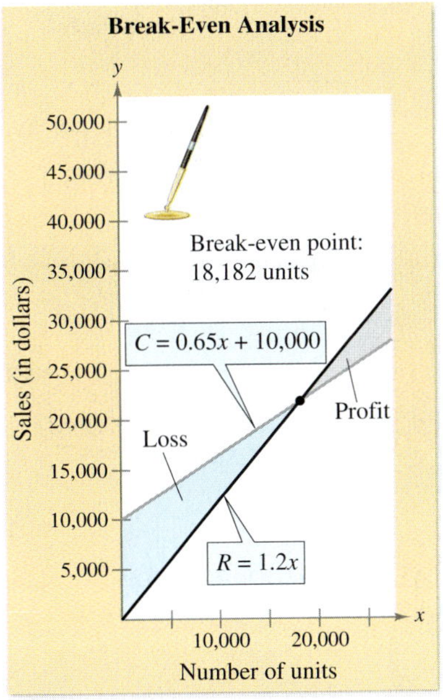

FIGURE 1.24

TRY IT 6

How many units must the company in Example 6 sell to break even if the selling price is \$1.45 per unit?

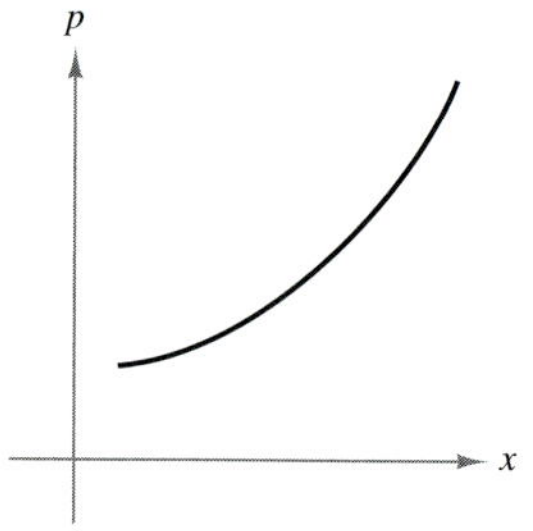

FIGURE 1.25 Supply Curve

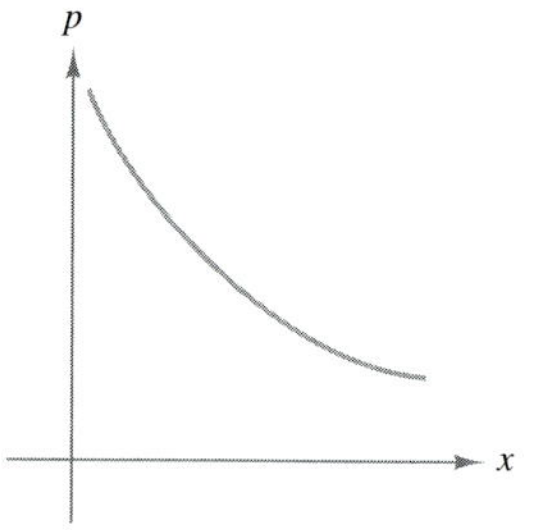

FIGURE 1.26 Demand Curve

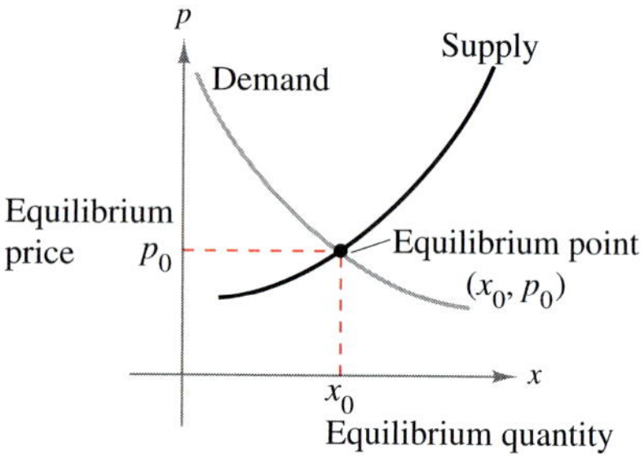

FIGURE 1.27 Equilibrium Point

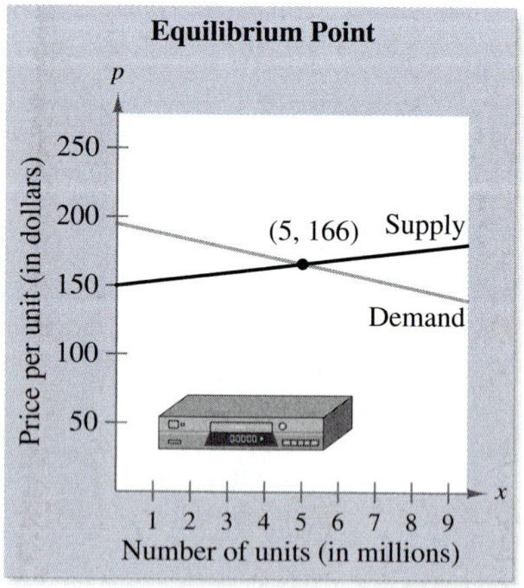

FIGURE 1.28

Two types of applications that economists use to analyze a market are supply and demand equations. A **supply equation** shows the relationship between the unit price p of a product and the quantity supplied x. The graph of a supply equation is called a **supply curve.** (See Figure 1.25.) A typical supply curve rises because producers of a product want to sell more units if the unit price is higher.

A **demand equation** shows the relationship between the unit price p of a product and the quantity demanded x. The graph of a demand equation is called a **demand curve.** (See Figure 1.26.) A typical demand curve tends to show a decrease in the quantity demanded with each increase in price.

In an ideal situation, with no other factors present to influence the market, the production level should stabilize at the point of intersection of the graphs of the supply and demand equations. This point is called the **equilibrium point.** The x-coordinate of the equilibrium point is called the **equilibrium quantity** and the p-coordinate is called the **equilibrium price.** (See Figure 1.27.) You can find the equilibrium point by setting the demand equation equal to the supply equation and solving for x.

EXAMPLE 7 Finding the Equilibrium Point

The demand and supply equations for a DVD player are given by

$p = 195 - 5.8x$ Demand equation

$p = 150 + 3.2x$ Supply equation

where p is the price in dollars and x represents the number of units in millions. Find the equilibrium point for this market.

SOLUTION Begin by setting the demand equation equal to the supply equation.

$195 - 5.8x = 150 + 3.2x$ Set equations equal to each other.

$45 - 5.8x = 3.2x$ Subtract 150 from each side.

$45 = 9x$ Add $5.8x$ to each side.

$5 = x$ Divide each side by 9.

So, the equilibrium point occurs when the demand and supply are each five million units. (See Figure 1.28.) The price that corresponds to this x-value is obtained by substituting $x = 5$ into either of the original equations. For instance, substituting into the demand equation produces

$$p = 195 - 5.8(5) = 195 - 29 = \$166.$$

Substitute $x = 5$ into the supply equation to see that you obtain the same price.

TRY IT 7

The demand and supply equations for a calculator are $p = 136 - 3.5x$ and $p = 112 + 2.5x$, respectively, where p is the price in dollars and x represents the number of units in millions. Find the equilibrium point for this market.

Mathematical Models

In this text, you will see many examples of the use of equations as **mathematical models** of real-life phenomena. In developing a mathematical model to represent actual data, you should strive for two (often conflicting) goals—accuracy and simplicity.

EXAMPLE 8 Using Mathematical Models

The table shows the annual sales (in millions of dollars) for Dillard's and Kohl's for 1999 through 2003. In the spring of 2004, the publication *Value Line* listed the projected 2004 sales for the companies as \$7740 million and \$11,975 million, respectively. How do you think these projections were obtained? *(Source: Dillard's Inc. and Kohl's Corp.)*

Year	1999	2000	2001	2002	2003
t	9	10	11	12	13
Dillard's	8677	8567	8155	7911	7599
Kohl's	4557	6152	7489	9120	10,282

SOLUTION The projections were obtained by using past sales to predict future sales. The past sales were modeled by equations that were found by a statistical procedure called least squares regression analysis.

$$S = -16.86t^2 + 89.7t + 9269, \quad 9 \le t \le 13 \qquad \text{Dillard's}$$

$$S = -40.86t^2 + 2340.7t - 13{,}202, \quad 9 \le t \le 13 \qquad \text{Kohl's}$$

Using $t = 14$ to represent 2004, you can predict the 2004 sales to be

$$S = -16.86(14)^2 + 89.7(14) + 9269 \approx 7220 \qquad \text{Dillard's}$$

$$S = -40.86(14)^2 + 2340.7(14) - 13{,}202 \approx 11{,}559. \qquad \text{Kohl's}$$

These two projections are close to those projected by *Value Line*. The graphs of the two models are shown in Figure 1.29.

ALGEBRA REVIEW

For help in evaluating the expressions in Example 8, see the review of order of operations on page 72.

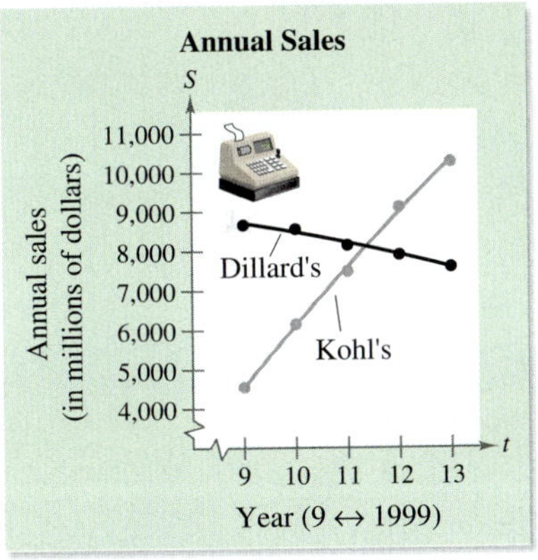

FIGURE 1.29

TRY IT 8

The table shows the annual sales (in millions of dollars) for Dollar General for 1995 through 2002. In the winter of 2004, the publication *Value Line* listed projected 2004 sales for the company as \$7800 million. How does this projection compare with the projection obtained using the model below? *(Source: Dollar General Corp.)*

$$S = 32.326t^2 + 78.23t + 530.9, \quad 5 \le t \le 12$$

Year	1995	1996	1997	1998	1999	2000	2001	2002
t	5	6	7	8	9	10	11	12
Sales	1764.2	2134.4	2627.3	3221.0	3888.0	4550.6	5322.9	6100.4

STUDY TIP

To test the accuracy of a model, you can compare the actual data with the values given by the model. For instance, the table below compares the actual Kohl's sales with those given by the model.

Year	1999	2000	2001
Actual	4557	6152	7489
Model	4554.6	6119	7601.6

Year	2002	2003
Actual	9120	10,282
Model	9002.6	10,322

Much of your study of calculus will center around the behavior of the graphs of mathematical models. Figure 1.30 shows the graphs of six basic algebraic equations. Familiarity with these graphs will help you in the creation and use of mathematical models.

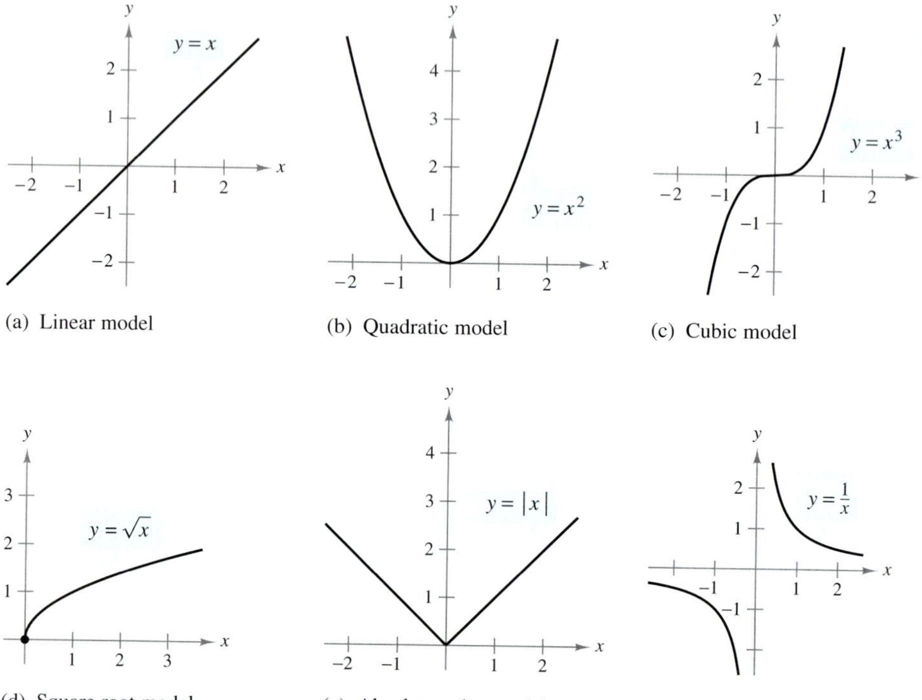

(a) Linear model (b) Quadratic model (c) Cubic model

(d) Square root model (e) Absolute value model (f) Rational model

FIGURE 1.30

TAKE ANOTHER LOOK

Graphical, Numerical, and Analytic Solutions

Most problems in calculus can be solved in a variety of ways. Often, you can solve a problem graphically, numerically (using a table), and analytically. For instance, Example 6 compares graphical and analytic approaches to finding points of intersection.

In Example 8, suppose you were asked to find the point in time at which Kohl's sales exceeded Dillard's sales. Explain how to use *each* of the three approaches to answer the question. For this question, which approach do you think is best? Explain. Suppose you answered the question and obtained $t = 11.36$. What date does this represent—April 2001 or April 2002? Explain.

PREREQUISITE REVIEW 1.2

The following warm-up exercises involve skills that were covered in earlier sections. You will use these skills in the exercise set for this section.

In Exercises 1–6, solve for y.

1. $5y - 12 = x$

2. $-y = 15 - x$

3. $x^3y + 2y = 1$

4. $x^2 + x - y^2 - 6 = 0$

5. $(x - 2)^2 + (y + 1)^2 = 9$

6. $(x + 6)^2 + (y - 5)^2 = 81$

In Exercises 7–10, complete the square to write the expression as a perfect square trinomial.

7. $x^2 - 4x + \square$

8. $x^2 + 6x + \square$

9. $x^2 - 5x + \square$

10. $x^2 + 3x + \square$

In Exercises 11–14, factor the expression.

11. $x^2 - 3x + 2$

12. $x^2 + 5x + 6$

13. $y^2 - 3y + \frac{9}{4}$

14. $y^2 - 7y + \frac{49}{4}$

EXERCISES 1.2

In Exercises 1–6, determine whether the points are solution points of the given equation.

1. $2x - y - 3 = 0$

(a) $(1, 2)$ (b) $(1, -1)$ (c) $(4, 5)$

2. $7x + 4y - 6 = 0$

(a) $(6, -9)$ (b) $(-5, 10)$ (c) $\left(\frac{1}{2}, \frac{5}{8}\right)$

3. $x^2 + y^2 = 4$

(a) $\left(1, -\sqrt{3}\right)$ (b) $\left(\frac{1}{2}, -1\right)$ (c) $\left(\frac{3}{2}, \frac{7}{2}\right)$

4. $x^2y + x^2 - 5y = 0$

(a) $\left(0, \frac{1}{5}\right)$ (b) $(2, 4)$ (c) $(-2, -4)$

5. $x^2 - xy + 4y = 3$

(a) $(0, 2)$ (b) $\left(-2, -\frac{1}{6}\right)$ (c) $(3, -6)$

6. $3y + 2xy - x^2 = 5$

(a) $(-7, -5)$ (b) $(-1, 6)$ (c) $\left(1, \frac{6}{5}\right)$

In Exercises 7–12, match the equation with its graph. Use a graphing utility, set for a square setting, to confirm your result. [The graphs are labeled (a)–(f).]

7. $y = x - 2$

8. $y = -\frac{1}{2}x + 2$

9. $y = x^2 + 2x$

10. $y = \sqrt{9 - x^2}$

11. $y = |x| - 2$

12. $y = x^3 - x$

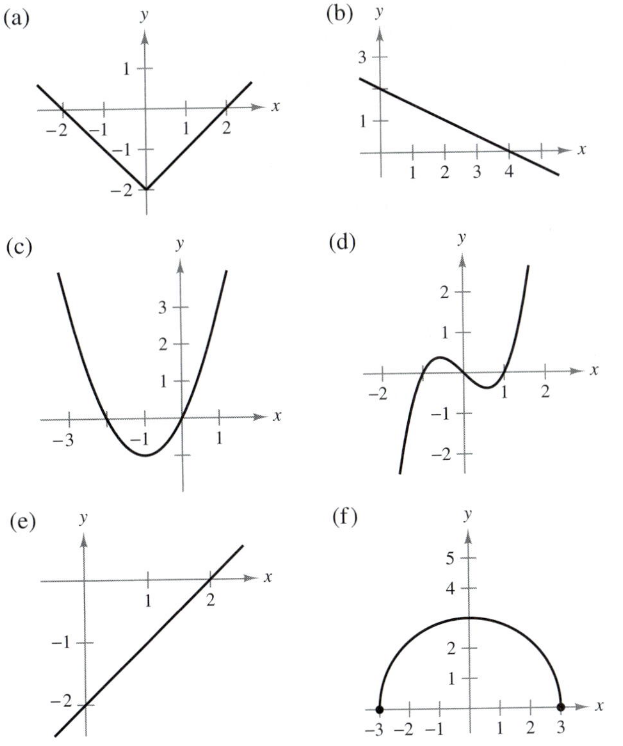

In Exercises 13–22, find the x- and y-intercepts of the graph of the equation.

13. $2x - y - 3 = 0$

14. $4x - 2y - 5 = 0$

15. $y = x^2 + x - 2$

16. $y = x^2 - 4x + 3$

17. $y = x^2\sqrt{9 - x^2}$

18. $y^2 = x^3 - 4x$

19. $y = \dfrac{x^2 - 4}{x - 2}$

20. $y = \dfrac{x^2 + 3x}{(3x + 1)^2}$

21. $x^2y - x^2 + 4y = 0$

22. $2x^2y + 8y - x^2 = 1$

In Exercises 23–38, sketch the graph of the equation and label the intercepts. Use a graphing utility to verify your results.

23. $y = 2x + 3$

24. $y = -3x + 2$

25. $y = x^2 - 3$

26. $y = x^2 + 6$

27. $y = (x - 1)^2$

28. $y = (5 - x)^2$

29. $y = x^3 + 2$

30. $y = 1 - x^3$

31. $y = -\sqrt[3]{x + 1}$

32. $y = \sqrt{x + 1}$

33. $y = |x + 1|$

34. $y = -|x - 2|$

35. $y = 1/(x - 3)$

36. $y = 1/(x^2 + 1)$

37. $x = y^2 - 4$

38. $x = 4 - y^2$

In Exercises 39–46, write the general form of the equation of the circle.

39. Center: $(0, 0)$; radius: 3

40. Center: $(0, 0)$; radius: 5

41. Center: $(2, -1)$; radius: 4

42. Center: $(-4, 3)$; radius: 3

43. Center: $(-1, 2)$; solution point: $(0, 0)$

44. Center: $(3, -2)$; solution point: $(-1, 1)$

45. Endpoints of a diameter: $(3, 3), (-3, 3)$

46. Endpoints of a diameter: $(-4, -1), (4, 1)$

In Exercises 47–54, complete the square to write the equation of the circle in standard form. Then use a graphing utility to graph the circle.

47. $x^2 + y^2 - 2x + 6y + 6 = 0$

48. $x^2 + y^2 - 2x + 6y - 15 = 0$

49. $x^2 + y^2 + 4x + 6y - 3 = 0$

50. $x^2 + y^2 - 4x + 2y + 3 = 0$

51. $2x^2 + 2y^2 - 2x - 2y - 3 = 0$

52. $4x^2 + 4y^2 - 4x + 2y - 1 = 0$

53. $16x^2 + 16y^2 + 16x + 40y - 7 = 0$

54. $3x^2 + 3y^2 - 6y - 1 = 0$

In Exercises 55–62, find the points of intersection (if any) of the graphs of the equations. Use a graphing utility to check your results.

55. $x + y = 2, 2x - y = 1$

56. $x + y = 7, 3x - 2y = 11$

57. $x^2 + y^2 = 25, 2x + y = 10$

58. $x^2 + y = 4, 2x - y = 1$

59. $y = x^3, y = 2x$

60. $y = \sqrt{x}, y = x$

61. $y = x^4 - 2x^2 + 1, y = 1 - x^2$

62. $y = x^3 - 2x^2 + x - 1, y = -x^2 + 3x - 1$

63. ***Break-Even Analysis*** You are setting up a part-time business with an initial investment of \$15,000. The unit cost of the product is \$11.80, and the selling price is \$19.30.

(a) Find equations for the total cost C and total revenue R for x units.

(b) Find the break-even point by finding the point of intersection of the cost and revenue equations.

(c) How many units would yield a profit of \$1000?

64. ***Break-Even Analysis*** A 2004 Chevrolet Malibu costs \$20,930 with a gasoline engine. A 2004 Toyota Prius costs \$22,052 with a hybrid engine. The Malibu gets 16 miles per gallon of gasoline and the Prius gets 35 miles per gallon of gasoline. Assume that the price of gasoline is \$1.759. *(Source: Adapted from* Consumer Reports, *May 2004)*

(a) Show that the cost C_g of driving the Chevrolet Malibu x miles is

$$C_g = 20{,}930 + 1.759x/16$$

and the cost C_h of driving the Toyota Prius x miles is

$$C_h = 22{,}052 + 1.759x/35.$$

(b) Find the break-even point. That is, find the mileage at which the hybrid-powered Toyota Prius becomes more economical than the gasoline-powered Chevrolet Malibu.

Break-Even Analysis In Exercises 65–68, find the sales necessary to break even for the given cost and revenue equations. (Round your answer up to the nearest whole unit.) Use a graphing utility to graph the equations and then find the break-even point.

65. $C = 0.85x + 35{,}000, R = 1.55x$

66. $C = 6x + 500{,}000, R = 35x$

67. $C = 8650x + 250{,}000, R = 9950x$

68. $C = 5.5\sqrt{x} + 10{,}000, R = 3.29x$

69. ***Supply and Demand*** The demand and supply equations for an electronic organizer are given by

$p = 180 - 4x$ Demand equation

$p = 75 + 3x$ Supply equation

where p is the price in dollars and x represents the number of units, in thousands. Find the equilibrium point for this market.

70. ***Supply and Demand*** The demand and supply equations for a portable CD player are given by

$p = 190 - 15x$ Demand equation

$p = 75 + 8x$ Supply equation

where p is the price in dollars and x represents the number of units, in hundreds of thousands. Find the equilibrium point for this market.

71. *Consumer Trends* The amounts of money y (in millions of dollars) spent on college textbooks in the United States in the years 1995 to 2002 are shown in the table. *(Source: Book Industry Study Group, Inc.)*

Year	1995	1996	1997	1998
Expense	2708	2920	3110	3365

Year	1999	2000	2001	2002
Expense	3773	3905	4187	4706

A mathematical model for the data is given by

$$y = 2.177t^3 - 41.99t^2 + 497.1t + 985$$

where t represents the year, with $t = 5$ corresponding to 1995.

(a) Compare the actual expenses with those given by the model. How good is the model? Explain your reasoning.

(b) Use the model to predict the expenses in 2010.

72. *Farm Work Force* The numbers of workers in farm work force in the United States for selected years from 1955 to 2000, as percents of the total work force, are shown in the table. *(Source: Department of Commerce)*

Year	1955	1960	1965	1970	1975
Percent	9.9	7.8	5.9	4.2	3.6

Year	1980	1985	1990	1995	2000
Percent	3.1	2.8	2.6	2.6	1.7

A mathematical model for the data is given by

$$y = \frac{-4.97 + 0.021t}{1 - 0.025t}$$

where y represents the percent and t represents the year, with $t = 55$ corresponding to 1955.

(a) Compare the actual percents with those given by the model. How good is the model?

(b) Use the model to predict the farm work force as a percent of the total work force in 2010.

(c) Discuss the validity of your prediction in part (b).

73. *Weekly Salary* A mathematical model for the average weekly salary y of a person in finance, insurance, or real estate is given by

$$y = \frac{292.48 + 37.72t}{1 + 0.02t}$$

where t represents the year, with $t = 7$ corresponding to 1997. *(Source: U.S. Bureau of Labor Statistics)*

(a) Use the model to complete the table.

Year	1997	1998	1999	2000	2001	2004
Salary						

(b) This model was created using actual data from 1997 through 2002. How accurate do you think the model is in predicting the 2004 average weekly salary? Explain your reasoning.

(c) What does this model predict the average weekly salary to be in 2006? Do you think this prediction is valid?

74. *Medicine* A mathematical model for the numbers of kidney transplants performed in the United States in the years 1998 to 2002 is given by

$$y = 60.64t^2 - 544.0t + 12{,}624$$

where y is the number of transplants and t is the time in years, with $t = 8$ corresponding to 1998. *(Source: United Network for Organ Sharing)*

(a) Enter the model into a graphing utility and use it to complete the table.

Year	1998	1999	2000	2001	2002
Transplants					

(b) Use your school's library, the Internet, or some other reference source to find the actual numbers of kidney transplants for the years 1998 to 2002. Compare the actual numbers with those given by the model. How good is the model? Explain your reasoning.

(c) Using this model, what is the prediction for the number of transplants in the year 2008? How valid do you think the prediction is? What factors could affect this model's accuracy?

75. Use a graphing utility to graph the equation $y = cx + 1$ for $c = 1, 2, 3, 4,$ and 5. Then make a conjecture about the x-coefficient and the graph of the equation.

76. Define the break-even point for a business marketing a new product. Give examples of a linear cost equation and a linear revenue equation for which the break-even point is 10,000 units.

In Exercises 77–82, use a graphing utility to graph the equation. Use the graphing utility to approximate the x- and y-intercepts of the graph.

77. $y = 0.24x^2 + 1.32x + 5.36$

78. $y = -0.56x^2 - 5.34x + 6.25$

79. $y = \sqrt{0.3x^2 - 4.3x + 5.7}$

80. $y = \sqrt{-1.21x^2 + 2.34x + 5.6}$

81. $y = \dfrac{0.2x^2 + 1}{0.1x + 2.4}$

82. $y = \dfrac{0.4x - 5.3}{0.4x^2 + 5.3}$

1.3 LINES IN THE PLANE AND SLOPE

- Use the slope-intercept form of a linear equation to sketch graphs.
- Find slopes of lines passing through two points.
- Use the point-slope form to write equations of lines.
- Find equations of parallel and perpendicular lines.
- Use linear equations to model and solve real-life problems.

TECHNOLOGY

On most graphing utilities, the display screen is two-thirds as high as it is wide. On such screens, you can obtain a graph with a true geometric perspective by using a **square setting**—one in which

$$\frac{Y_{max} - Y_{min}}{X_{max} - X_{min}} = \frac{2}{3}.$$

One such setting is shown below. Notice that the x and y tick marks are equally spaced on a square setting, but not on a standard setting.

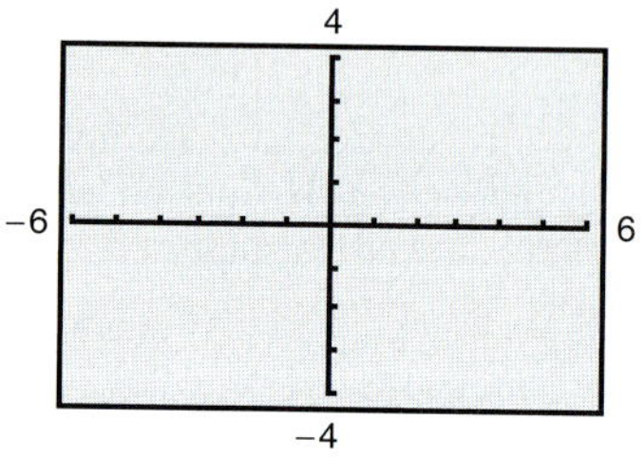

Using Slope

The simplest mathematical model for relating two variables is the **linear equation** $y = mx + b$. The equation is called *linear* because its graph is a line. (In this text, the term *line* is used to mean *straight line*.) By letting $x = 0$, you can see that the line crosses the y-axis at $y = b$, as shown in Figure 1.31. In other words, the y-intercept is $(0, b)$. The steepness or slope of the line is m.

$$y = mx + b$$

Slope ↑ (m) y-intercept ↑ (b)

The **slope** of a line is the number of units the line rises (or falls) vertically for each unit of horizontal change from left to right, as shown in Figure 1.31.

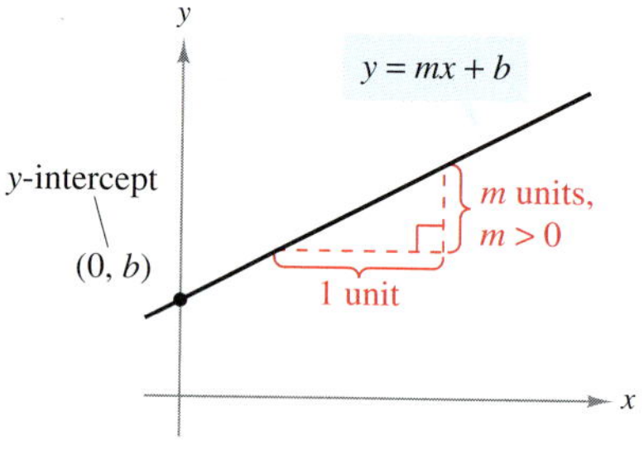

Positive slope, line rises.

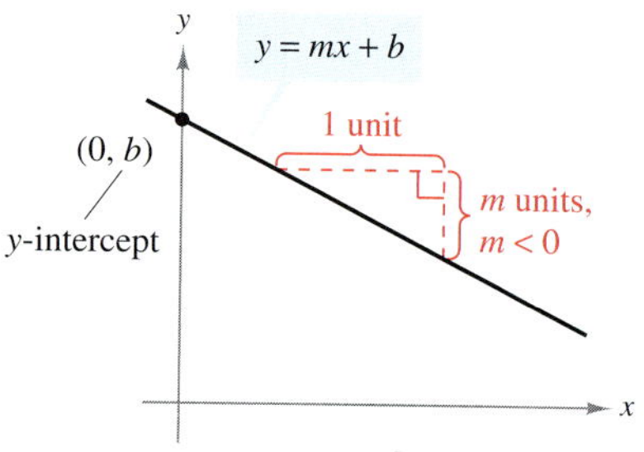

Negative slope, line falls.

FIGURE 1.31

DISCOVERY

Use a graphing utility to compare the slopes of the lines $y = mx$, where $m = 0.5, 1, 2,$ and 4. Which line rises most quickly? Now, let $m = -0.5, -1, -2,$ and -4. Which line falls most quickly? Let $m = 0.01, 0.001,$ and 0.0001. What is the slope of a horizontal line? Use a square setting to obtain a true geometric perspective.

A linear equation that is written in the form $y = mx + b$ is said to be written in **slope-intercept form.**

The Slope-Intercept Form of the Equation of a Line

The graph of the equation

$$y = mx + b$$

is a line whose slope is m and whose y-intercept is $(0, b)$.

Once you have determined the slope and the y-intercept of a line, it is a relatively simple matter to sketch its graph.

In the following example, note that none of the lines is vertical. A vertical line has an equation of the form

$x = a.$ Vertical line

Because such an equation cannot be written in the form $y = mx + b$, it follows that the slope of a vertical line is undefined, as indicated in Figure 1.32.

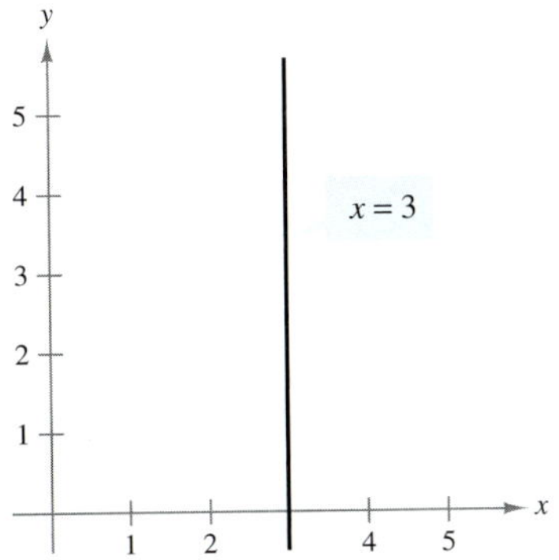

FIGURE 1.32 When the line is vertical, the slope is undefined.

EXAMPLE 1 Graphing a Linear Equation

Sketch the graph of each linear equation.

(a) $y = 2x + 1$

(b) $y = 2$

(c) $x + y = 2$

SOLUTION

(a) Because $b = 1$, the y-intercept is $(0, 1)$. Moreover, because the slope is $m = 2$, the line *rises* two units for each unit the line moves to the right, as shown in Figure 1.33(a).

(b) By writing this equation in the form $y = (0)x + 2$, you can see that the y-intercept is $(0, 2)$ and the slope is zero. A zero slope implies that the line is horizontal—that is, it doesn't rise *or* fall, as shown in Figure 1.33(b).

(c) By writing this equation in slope-intercept form

$x + y = 2$ Write original equation.

$y = -x + 2$ Subtract x from each side.

$y = (-1)x + 2$ Write in slope-intercept form.

you can see that the y-intercept is $(0, 2)$. Moreover, because the slope is $m = -1$, this line *falls* one unit for each unit the line moves to the right, as shown in Figure 1.33(c).

TRY IT 1

Sketch the graph of each linear equation.

(a) $y = 4x - 2$

(b) $x = 1$

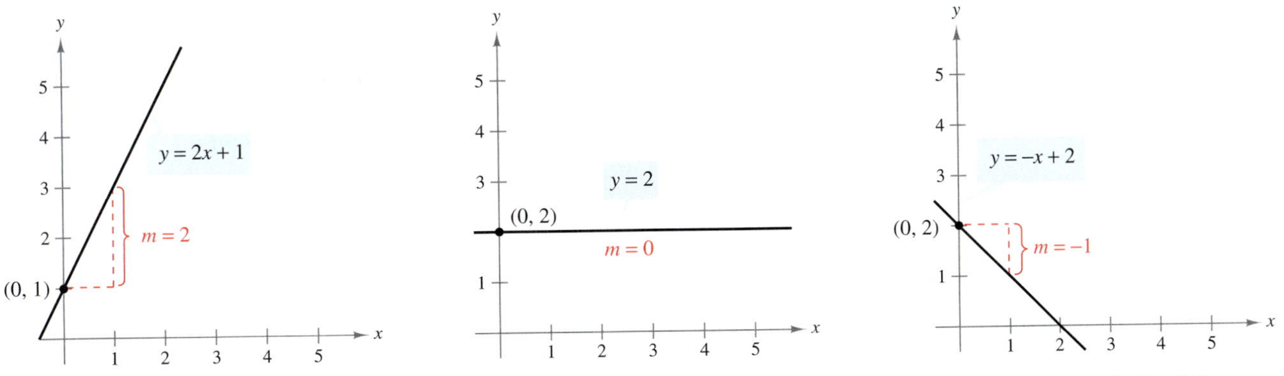

(a) When m is positive, the line rises.

(b) When m is zero, the line is horizontal.

(c) When m is negative, the line falls.

FIGURE 1.33

In real-life problems, the slope of a line can be interpreted as either a *ratio* or a *rate*. If the x-axis and y-axis have the same unit of measure, then the slope has no units and is a **ratio.** If the x-axis and y-axis have different units of measure, then the slope is a **rate** or **rate of change.**

EXAMPLE 2 Using Slope as a Ratio

The maximum recommended slope of a wheelchair ramp is $\frac{1}{12} \approx 0.083$. A business is installing a wheelchair ramp that rises 22 inches over a horizontal length of 24 feet, as shown in Figure 1.34. Is the ramp steeper than recommended? *(Source: American Disabilities Act Handbook)*

SOLUTION The horizontal length of the ramp is 24 feet or $12(24) = 288$ inches. So, the slope of the ramp is

$$\text{Slope} = \frac{\text{vertical change}}{\text{horizontal change}} = \frac{22 \text{ in.}}{288 \text{ in.}} \approx 0.076.$$

So, the slope is not steeper than recommended.

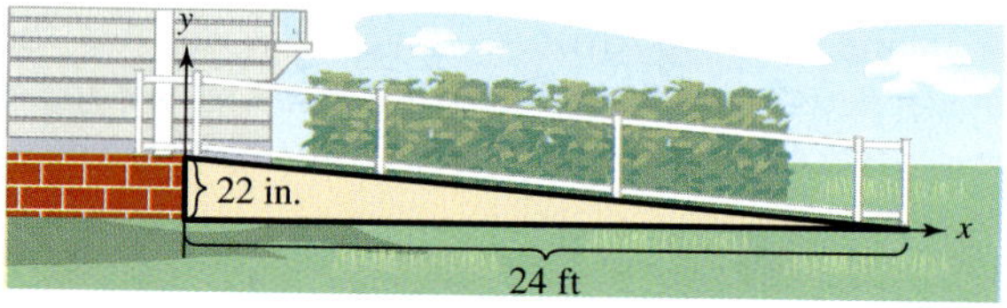

FIGURE 1.34

TRY IT 2

If the ramp in Example 2 rises 27 inches over a horizontal length of 26 feet, is it steeper than recommended?

EXAMPLE 3 Using Slope as a Rate of Change

A manufacturing company determines that the total cost in dollars of producing x units of a product is $C = 25x + 3500$. Describe the practical significance of the y-intercept and slope of the line given by this equation.

SOLUTION The y-intercept $(0, 3500)$ tells you that the cost of producing zero units is \$3500. This is the **fixed cost** of production—it includes costs that must be paid regardless of the number of units produced. The slope of $m = 25$ tells you that the cost of producing each unit is \$25, as shown in Figure 1.35. Economists call the cost per unit the **marginal cost.** If the production increases by one unit, then the "margin" or extra amount of cost is \$25.

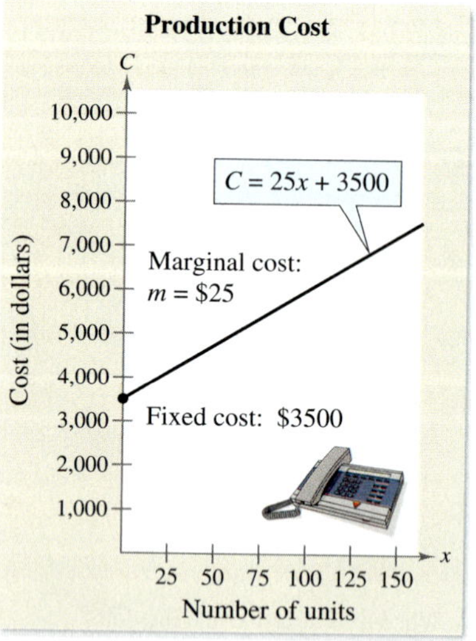

FIGURE 1.35

TRY IT 3

A small business purchases a copier and determines that the value of the copier t years after its purchase is $V = -175t + 875$. Describe the practical significance of the y-intercept and slope of the line given by this equation.

Finding the Slope of a Line

Given an equation of a nonvertical line, you can find its slope by writing the equation in slope-intercept form. If you are not given an equation, you can still find the slope of a line. For instance, suppose you want to find the slope of the line passing through the points (x_1, y_1) and (x_2, y_2), as shown in Figure 1.36. As you move from left to right along this line, a change of $(y_2 - y_1)$ units in the vertical direction corresponds to a change of $(x_2 - x_1)$ units in the horizontal direction. These two changes are denoted by the symbols

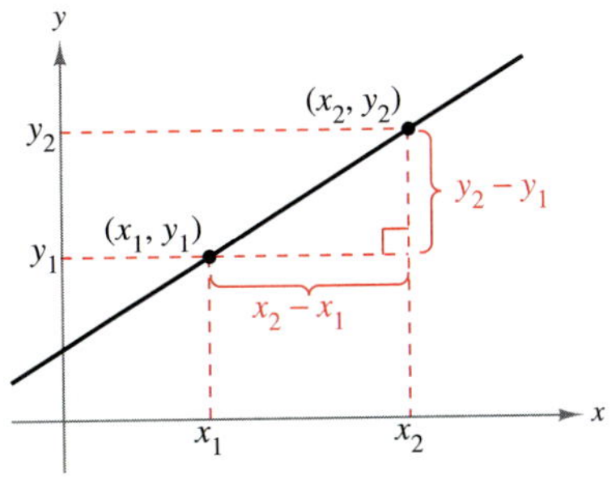

FIGURE 1.36

$$\Delta y = y_2 - y_1 = \text{the change in } y$$

and

$$\Delta x = x_2 - x_1 = \text{the change in } x.$$

(The symbol Δ is the Greek capital letter delta, and the symbols Δy and Δx are read as "delta y" and "delta x.") The ratio of Δy to Δx represents the slope of the line that passes through the points (x_1, y_1) and (x_2, y_2).

$$\text{Slope} = \frac{\Delta y}{\Delta x} = \frac{y_2 - y_1}{x_2 - x_1}$$

Be sure you see that Δx represents a single number, not the product of two numbers (Δ and x). The same is true for Δy.

The Slope of a Line Passing Through Two Points

The **slope** m of the line passing through (x_1, y_1) and (x_2, y_2) is

$$m = \frac{\Delta y}{\Delta x} = \frac{y_2 - y_1}{x_2 - x_1}$$

where $x_1 \neq x_2$.

When this formula is used for slope, the *order of subtraction* is important. Given two points on a line, you are free to label either one of them as (x_1, y_1) and the other as (x_2, y_2). However, once you have done this, you must form the numerator and denominator using the same order of subtraction.

$$\underbrace{m = \frac{y_2 - y_1}{x_2 - x_1}}_{\text{Correct}} \qquad \underbrace{m = \frac{y_1 - y_2}{x_1 - x_2}}_{\text{Correct}} \qquad \underbrace{\cancel{m = \frac{y_2 - y_1}{x_1 - x_2}}}_{\text{Incorrect}}$$

For instance, the slope of the line passing through the points (3, 4) and (5, 7) can be calculated as

$$m = \frac{7 - 4}{5 - 3} = \frac{3}{2}$$

or

$$m = \frac{4 - 7}{3 - 5} = \frac{-3}{-2} = \frac{3}{2}.$$

EXAMPLE 4 Finding the Slope of a Line

Find the slope of the line passing through each pair of points.

(a) $(-2, 0)$ and $(3, 1)$ (b) $(-1, 2)$ and $(2, 2)$

(c) $(0, 4)$ and $(1, -1)$ (d) $(3, 4)$ and $(3, 1)$

SOLUTION

(a) Letting $(x_1, y_1) = (-2, 0)$ and $(x_2, y_2) = (3, 1)$, you obtain a slope of

$$m = \frac{y_2 - y_1}{x_2 - x_1} = \frac{1 - 0}{3 - (-2)} = \frac{1}{5}$$

Difference in y-values

Difference in x-values

as shown in Figure 1.37(a).

(b) The slope of the line passing through $(-1, 2)$ and $(2, 2)$ is

$$m = \frac{2 - 2}{2 - (-1)} = \frac{0}{3} = 0.$$

See Figure 1.37(b).

(c) The slope of the line passing through $(0, 4)$ and $(1, -1)$ is

$$m = \frac{-1 - 4}{1 - 0} = \frac{-5}{1} = -5.$$

See Figure 1.37(c).

(d) The slope of the vertical line passing through $(3, 4)$ and $(3, 1)$ is not defined because division by zero is undefined. [See Figure 1.37(d).]

DISCOVERY

The line in Example 4(b) is a horizontal line. Find an equation for this line. The line in Example 4(d) is a vertical line. Find an equation for this line.

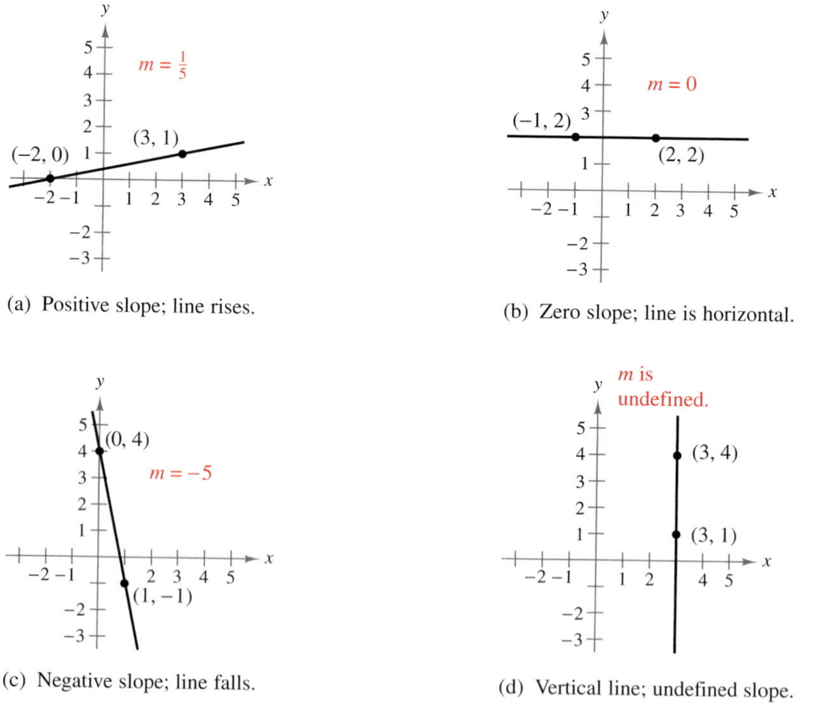

(a) Positive slope; line rises. (b) Zero slope; line is horizontal.

(c) Negative slope; line falls. (d) Vertical line; undefined slope.

FIGURE 1.37

TRY IT 4

Find the slope of the line passing through each pair of points.

(a) $(-3, 2)$ and $(5, 18)$

(b) $(-2, 1)$ and $(-4, 2)$

Writing Linear Equations

If (x_1, y_1) is a point lying on a nonvertical line of slope m and (x, y) is *any other* point on the line, then

$$\frac{y - y_1}{x - x_1} = m.$$

This equation, involving the variables x and y, can be rewritten in the form $y - y_1 = m(x - x_1)$, which is the **point-slope form** of the equation of a line.

Point-Slope Form of the Equation of a Line

The equation of the line with slope m passing through the point (x_1, y_1) is

$$y - y_1 = m(x - x_1).$$

The point-slope form is most useful for *finding* the equation of a nonvertical line. You should remember this formula—it is used throughout the text.

EXAMPLE 5 Using the Point-Slope Form

Find the equation of the line that has a slope of 3 and passes through the point $(1, -2)$.

SOLUTION Use the point-slope form with $m = 3$ and $(x_1, y_1) = (1, -2)$.

$$y - y_1 = m(x - x_1) \qquad \text{Point-slope form}$$
$$y - (-2) = 3(x - 1) \qquad \text{Substitute for } m, x_1, \text{ and } y_1.$$
$$y + 2 = 3x - 3 \qquad \text{Simplify.}$$
$$y = 3x - 5 \qquad \text{Write in slope-intercept form.}$$

The slope-intercept form of the equation of the line is $y = 3x - 5$. The graph of this line is shown in Figure 1.38.

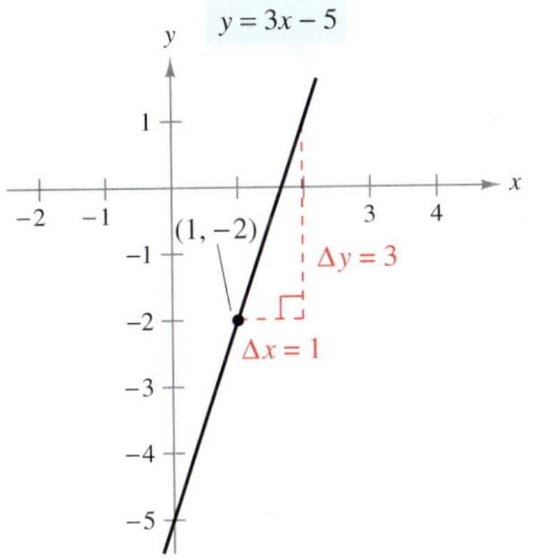

FIGURE 1.38

TRY IT 5

Find the equation of the line that has a slope of 2 and passes through the point $(-1, 2)$.

The point-slope form can be used to find an equation of the line passing through points (x_1, y_1) and (x_2, y_2). To do this, first find the slope of the line

$$m = \frac{y_2 - y_1}{x_2 - x_1}, \qquad x_1 \neq x_2$$

and then use the point-slope form to obtain the equation

$$y - y_1 = \frac{y_2 - y_1}{x_2 - x_1}(x - x_1). \qquad \text{Two-point form}$$

This is sometimes called the **two-point form** of the equation of a line.

STUDY TIP

The two-point form of a line is similar to the slope-intercept form. What is the slope of a line given in two-point form

$$y - y_1 = \frac{y_2 - y_1}{x_2 - x_1}(x - x_1)?$$

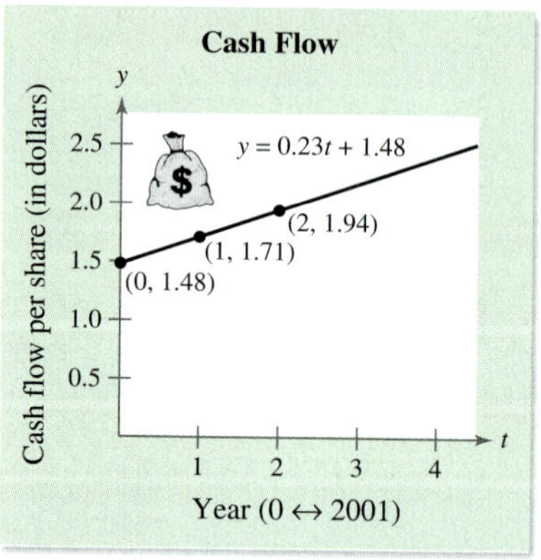

FIGURE 1.39

EXAMPLE 6 Predicting Cash Flow Per Share

The cash flow per share for Ruby Tuesday, Inc. was \$1.48 in 2001 and \$1.71 in 2002. Using only this information, write a linear equation that gives the cash flow per share in terms of the year. Then predict the cash flow for 2003. *(Source: Ruby Tuesday, Inc.)*

SOLUTION Let $t = 0$ represent 2001. Then the two given values are represented by the data points (0, 1.48) and (1, 1.71). The slope of the line through these points is

$$m = \frac{1.71 - 1.48}{1 - 0} = 0.23.$$

Using the point-slope form, you can find the equation that relates the cash flow y and the year t to be $y = 0.23t + 1.48$. According to this equation, the cash flow in 2003 was \$1.94, as shown in Figure 1.39. (In this case, the prediction is fairly good—the actual cash flow in 2003 was \$2.09.)

TRY IT 6

The sales per share for Clorox Company were \$16.49 in 2001 and \$18.21 in 2002. Write a linear equation that gives the sales per share in terms of the year. Let $t = 0$ represent 2001. Then predict the sales per share for 2003. *(Source: Clorox Company)*

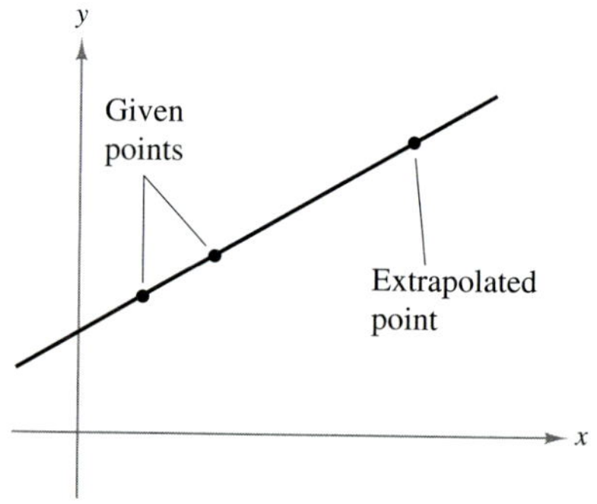

(a) Linear Extrapolation

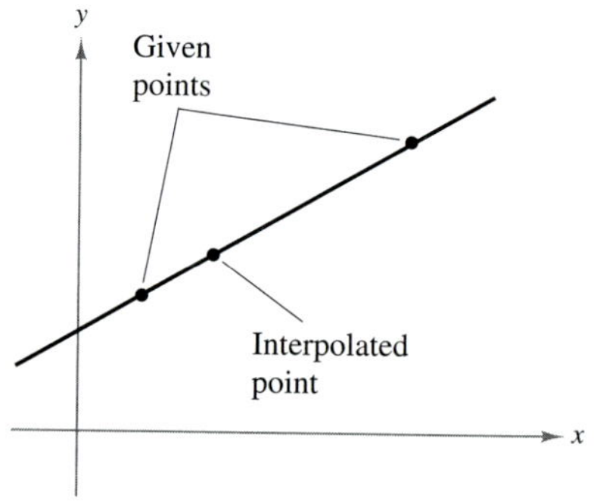

(b) Linear Interpolation
FIGURE 1.40

The prediction method illustrated in Example 6 is called **linear extrapolation.** Note in Figure 1.40(a) that an extrapolated point does not lie between the given points. When the estimated point lies between two given points, as shown in Figure 1.40(b), the procedure is called **linear interpolation.**

Because the slope of a vertical line is not defined, its equation cannot be written in slope-intercept form. However, every line has an equation that can be written in the **general form**

$$Ax + By + C = 0$$ General form

where A and B are not both zero. For instance, the vertical line given by $x = a$ can be represented by the general form $x - a = 0$. The five most common forms of equations of lines are summarized below.

Equations of Lines

1. General form: $Ax + By + C = 0$
2. Vertical line: $x = a$
3. Horizontal line: $y = b$
4. Slope-intercept form: $y = mx + b$
5. Point-slope form: $y - y_1 = m(x - x_1)$

Parallel and Perpendicular Lines

Slope can be used to decide whether two nonvertical lines in a plane are parallel, perpendicular, or neither.

> **Parallel and Perpendicular Lines**
>
> 1. Two distinct nonvertical lines are **parallel** if and only if their slopes are equal. That is,
>
> $$m_1 = m_2.$$
>
> 2. Two nonvertical lines are **perpendicular** if and only if their slopes are negative reciprocals of each other. That is,
>
> $$m_1 = -1/m_2.$$

TECHNOLOGY

On a graphing utility, lines will not appear to have the correct slope unless you use a viewing window that has a "square setting." For instance, try graphing the lines in Example 7 using the standard setting $-10 \le x \le 10$ and $-10 \le y \le 10$. Then reset the viewing window with the square setting $-9 \le x \le 9$ and $-6 \le y \le 6$. On which setting do the lines $y = \frac{2}{3}x - \frac{5}{3}$ and $y = -\frac{3}{2}x + 2$ appear to be perpendicular?

EXAMPLE 7 Finding Parallel and Perpendicular Lines

Find equations of the lines that pass through the point $(2, -1)$ and are

(a) parallel to the line $2x - 3y = 5$.

(b) perpendicular to the line $2x - 3y = 5$.

SOLUTION By writing the given equation in slope-intercept form

$2x - 3y = 5$	Write original equation.
$-3y = -2x + 5$	Subtract $2x$ from each side.
$y = \frac{2}{3}x - \frac{5}{3}$	Write in slope-intercept form.

you can see that it has a slope of $m = \frac{2}{3}$, as shown in Figure 1.41.

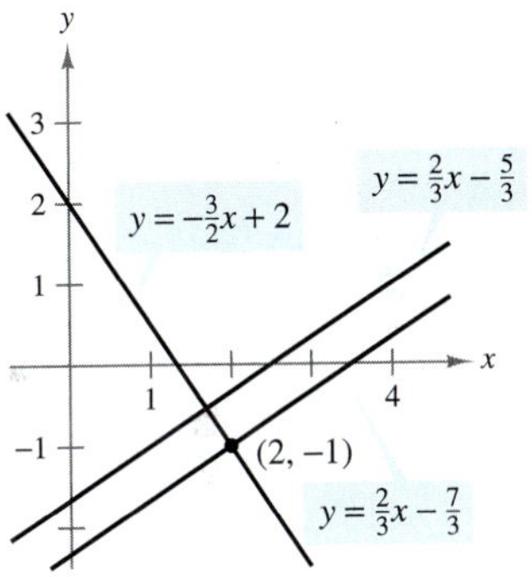

FIGURE 1.41

(a) Any line parallel to the given line must also have a slope of $\frac{2}{3}$. So, the line through $(2, -1)$ that is parallel to the given line has the following equation.

$y - (-1) = \frac{2}{3}(x - 2)$	Write in point-slope form.
$3(y + 1) = 2(x - 2)$	Multiply each side by 3.
$3y + 3 = 2x - 4$	Distributive Property
$2x - 3y - 7 = 0$	Write in general form.
$y = \frac{2}{3}x - \frac{7}{3}$	Write in slope-intercept form.

(b) Any line perpendicular to the given line must have a slope of

$$\frac{-1}{\frac{2}{3}} \text{ or } -\frac{3}{2}.$$

So, the line through $(2, -1)$ that is perpendicular to the given line has the following equation.

$y - (-1) = -\frac{3}{2}(x - 2)$	Write in point-slope form.
$2(y + 1) = -3(x - 2)$	Multiply each side by 2.
$2y + 2 = -3x + 6$	Distributive Property
$3x + 2y - 4 = 0$	Write in general form.
$y = -\frac{3}{2}x + 2$	Write in slope-intercept form.

TRY IT 7

Find equations of the lines that pass through the point $(2, 1)$ and are

(a) parallel to the line $2x - 4y = 5$.

(b) perpendicular to the line $2x - 4y = 5$.

Extended Application: Linear Depreciation

Most business expenses can be deducted the same year they occur. One exception to this is the cost of property that has a useful life of more than 1 year, such as buildings, cars, or equipment. Such costs must be **depreciated** over the useful life of the property. If the *same amount* is depreciated each year, the procedure is called **linear depreciation** or **straight-line depreciation.** The *book value* is the difference between the original value and the total amount of depreciation accumulated to date.

TRY IT 8

Write a linear equation for the machine in Example 8 if the salvage value at the end of 8 years is $1000.

EXAMPLE 8 Depreciating Equipment

Your company has purchased a $12,000 machine that has a useful life of 8 years. The salvage value at the end of 8 years is $2000. Write a linear equation that describes the book value of the machine each year.

SOLUTION Let V represent the value of the machine at the end of year t. You can represent the initial value of the machine by the ordered pair (0, 12,000) and the salvage value of the machine by the ordered pair (8, 2000). The slope of the line is

$$m = \frac{2000 - 12{,}000}{8 - 0} = -\$1250 \qquad m = \frac{y_2 - y_1}{t_2 - t_1}$$

which represents the annual depreciation in *dollars per year*. Using the point-slope form, you can write the equation of the line as shown.

$$V - 12{,}000 = -1250(t - 0) \qquad \text{Write in point-slope form.}$$

$$V = -1250t + 12{,}000 \qquad \text{Write in slope-intercept form.}$$

The table shows the book value of the machine at the end of each year.

t	0	1	2	3	4	5	6	7	8
V	12,000	10,750	9500	8250	7000	5750	4500	3250	2000

The graph of this equation is shown in Figure 1.42.

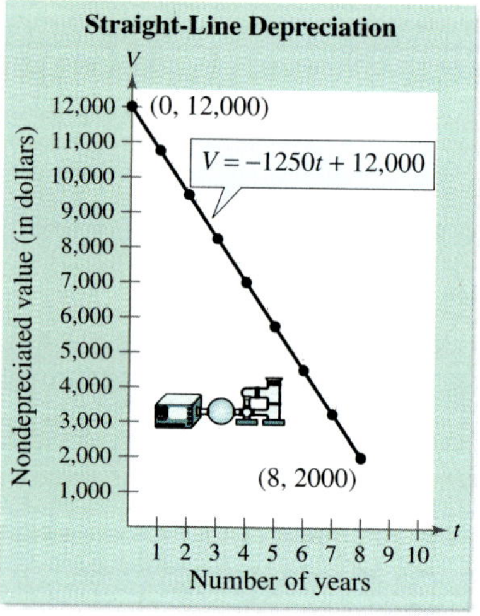

FIGURE 1.42

TAKE ANOTHER LOOK

Comparing Different Types of Depreciation

The Internal Revenue Service allows businesses to choose different types of depreciation. Another type is

$$\textit{Uniform Declining Balances: } V = 12{,}000\left(\frac{n - 1.605}{n}\right)^t, \quad n = 8.$$

Construct a table that compares this type of depreciation with linear depreciation. What are the advantages of each type?

PREREQUISITE REVIEW 1.3

The following warm-up exercises involve skills that were covered in earlier sections. You will use these skills in the exercise set for this section.

In Exercises 1 and 2, simplify the expression.

1. $\dfrac{5-(-2)}{-3-4}$

2. $\dfrac{-7-(-0)}{4-1}$

3. Evaluate $-\dfrac{1}{m}$ when $m = -3$.

4. Evaluate $-\dfrac{1}{m}$ when $m = \dfrac{6}{7}$.

In Exercises 5–10, solve for y in terms of x.

5. $-4x + y = 7$

6. $3x - y = 7$

7. $y - 2 = 3(x - 4)$

8. $y - (-5) = -1[x - (-2)]$

9. $y - (-3) = \dfrac{4-(-3)}{2-1}(x - 2)$

10. $y - 1 = \dfrac{-3-1}{-7-(-1)}[x - (-1)]$

EXERCISES 1.3

In Exercises 1–4, estimate the slope of the line.

1.

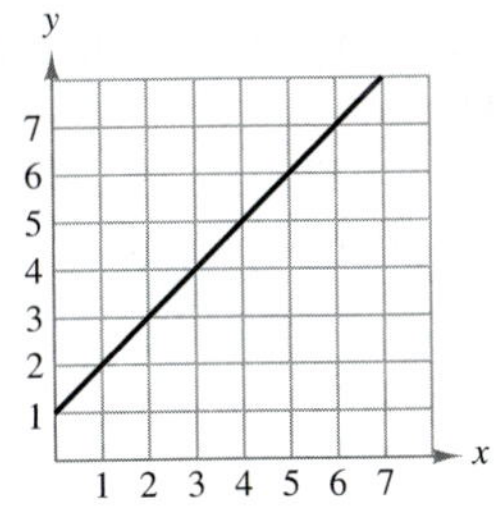

2.

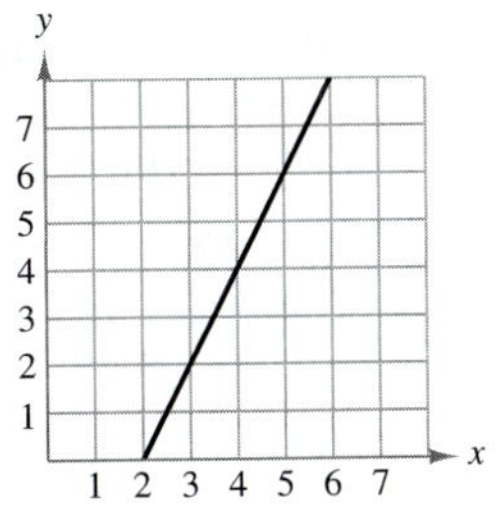

3.

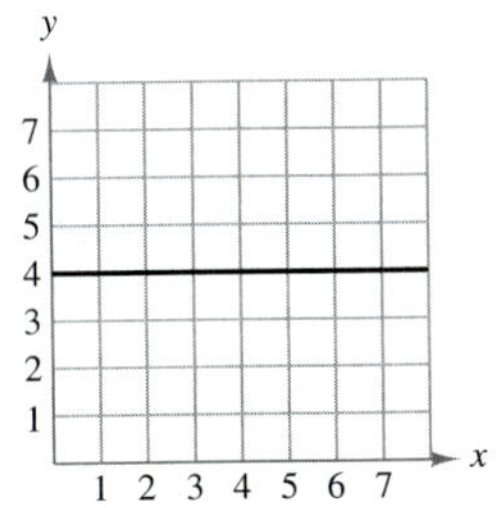

4.

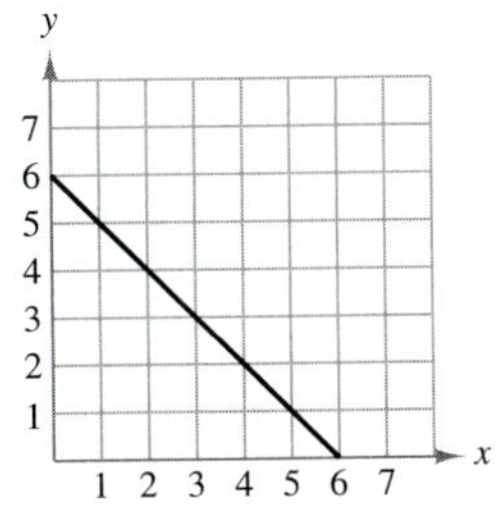

In Exercises 5–16, plot the points and find the slope of the line passing through the pair of points.

5. $(3, -4), (5, 2)$

6. $(1, 2), (-2, 2)$

7. $(\frac{1}{2}, 2), (6, 2)$

8. $(\frac{11}{3}, -2), (\frac{11}{3}, -10)$

9. $(-8, -3), (-8, -5)$

10. $(2, -1), (-2, -5)$

11. $(-2, 1), (4, -3)$

12. $(3, -5), (-2, -5)$

13. $(\frac{1}{4}, -2), (-\frac{3}{8}, 1)$

14. $(-\frac{3}{2}, -5), (\frac{5}{6}, 4)$

15. $(\frac{2}{3}, \frac{5}{2}), (\frac{1}{4}, -\frac{5}{6})$

16. $(\frac{7}{8}, \frac{3}{4}), (\frac{5}{4}, -\frac{1}{4})$

In Exercises 17–24, use the point on the line and the slope of the line to find three additional points through which the line passes. (There are many correct answers.)

	Point	*Slope*		*Point*	*Slope*
17.	$(2, 1)$	$m = 0$	**18.**	$(-3, -1)$	$m = 0$
19.	$(6, -4)$	$m = \frac{2}{3}$	**20.**	$(-2, -2)$	$m = \frac{5}{2}$
21.	$(1, 7)$	$m = -3$	**22.**	$(10, -6)$	$m = -1$
23.	$(-8, 1)$	m is undefined.			
24.	$(-3, 4)$	m is undefined.			

In Exercises 25–34, find the slope and y-intercept (if possible) of the equation of the line.

25. $x + 5y = 20$

26. $2x + y = 40$

27. $7x - 5y = 15$

28. $6x - 5y = 15$

29. $3x - y = 15$

30. $2x - 3y = 24$

31. $x = 4$

32. $x + 5 = 0$

33. $y - 4 = 0$

34. $y + 1 = 0$

In Exercises 35–46, write an equation of the line that passes through the points. Then use the equation to sketch the line.

35. $(4, 3), (0, -5)$

36. $(-3, -4), (1, 4)$

37. $(0, 0), (-1, 3)$

38. $(-3, 6), (1, 2)$

39. $(2, 3), (2, -2)$

40. $(6, 1), (10, 1)$

41. $(3, -1), (-2, -1)$

42. $(2, 5), (2, -10)$

43. $(-\frac{1}{3}, 1), (-\frac{2}{3}, \frac{5}{6})$

44. $(\frac{7}{8}, \frac{3}{4}), (\frac{5}{4}, -\frac{1}{4})$

45. $(-\frac{1}{2}, 4), (\frac{1}{2}, 8)$

46. $(4, -1), (\frac{1}{4}, -5)$

In Exercises 47–56, write an equation of the line that passes through the given point and has the given slope. Then use a graphing utility to graph the line.

	Point	*Slope*		*Point*	*Slope*
47.	$(0, 3)$	$m = \frac{3}{4}$	**48.**	$(0, 0)$	$m = \frac{2}{3}$
49.	$(-1, 2)$	m is undefined.			
50.	$(0, 4)$	m is undefined.			
51.	$(-2, 7)$	$m = 0$	**52.**	$(-2, 4)$	$m = 0$
53.	$(0, -2)$	$m = -4$	**54.**	$(-1, -4)$	$m = -2$
55.	$\left(0, \frac{2}{3}\right)$	$m = \frac{3}{4}$	**56.**	$\left(0, -\frac{2}{3}\right)$	$m = \frac{1}{6}$

In Exercises 57 and 58, explain how to use the concept of slope to determine whether the three points are collinear. Then explain how to use the Distance Formula to determine whether the points are collinear.

57. $(-2, 1), (-1, 0), (2, -2)$

58. $(0, 4), (7, -6), (-5, 11)$

59. Write an equation of the vertical line with x-intercept at 3.

60. Write an equation of the horizontal line through $(0, -5)$.

61. Write an equation of the line with y-intercept at -10 and parallel to all horizontal lines.

62. Write an equation of the line with x-intercept at -5 and parallel to all vertical lines.

In Exercises 63–70, write the equations of the lines through the given point (a) parallel to the given line and (b) perpendicular to the given line. Then use a graphing utility to graph all three equations in the same viewing window.

	Point	*Line*
63.	$(-3, 2)$	$x + y = 7$
64.	$(2, 1)$	$4x - 2y = 3$
65.	$\left(-\frac{2}{3}, \frac{7}{8}\right)$	$3x + 4y = 7$
66.	$\left(\frac{7}{8}, \frac{3}{4}\right)$	$5x + 3y = 0$
67.	$(-1, 0)$	$y + 3 = 0$
68.	$(2, 5)$	$y + 4 = 0$
69.	$(1, 1)$	$x - 2 = 0$
70.	$(12, -3)$	$x + 4 = 0$

In Exercises 71–78, sketch the graph of the equation. Use a graphing utility to verify your result.

71. $y = -2$

72. $y = -4$

73. $2x - y - 3 = 0$

74. $x + 2y + 6 = 0$

75. $y = -2x + 1$

76. $4x + 5y = 20$

77. $y + 2 = -4(x + 1)$

78. $y - 1 = 3(x + 4)$

79. ***Population*** The resident population of South Carolina (in thousands) was 3860 in 1997 and 4107 in 2002. Assume that the relationship between the population y and the year t is linear. Let $t = 7$ represent 1997. *(Source: U.S. Census Bureau)*

(a) Write a linear model for the data. What is the slope and what does it tell you about the population?

(b) Estimate the population in 1999.

(c) Use your model to estimate the population in 2001.

(d) Use your school's library, the Internet, or some other reference source to find the actual populations in 1999 and 2001. How close were your estimates?

(e) Do you think your model could be used to predict the population in 2006? Explain.

80. ***Annual Salary*** Your annual salary was \$26,300 in 2002 and \$29,700 in 2004. Assume your salary can be modeled by a linear equation.

(a) Write a linear equation giving your salary S in terms of the year. Let $t = 2$ represent 2002.

(b) Use the linear model to predict your salary in 2008.

81. ***Temperature Conversion*** Write a linear equation that expresses the relationship between the temperature in degrees Celsius C and degrees Fahrenheit F. Use the fact that water freezes at 0°C (32°F) and boils at 100°C (212°F).

82. ***Chemistry*** Use the result of Exercise 81 to answer the following:

(a) A person has a temperature of 102.5°F. What is this temperature on the Celsius scale?

(b) If the temperature in a room is 74°F, what is this temperature on the Celsius scale?

(Source: Adapted from Zumdahl, Chemistry, *Sixth Edition)*

83. ***Reimbursed Expenses*** A company reimburses its sales representatives \$150 per day for lodging and meals, plus \$0.34 per mile driven. Write a linear equation giving the daily cost C in terms of x, the number of miles driven.

84. ***Union Negotiation*** You are on a negotiating panel in a union hearing for a large corporation. The union is asking for a base pay of \$9.25 per hour plus an additional piecework rate of \$0.80 per unit produced. The corporation is offering a base pay of \$6.85 per hour plus a piecework rate of \$1.15.

(a) Write a linear equation for the hourly wages W in terms of x, the number of units produced per hour, for each pay schedule.

(b) Use a graphing utility to graph each linear equation and find the point of intersection.

(c) Interpret the meaning of the point of intersection of the graphs. How would you use this information to advise the corporation and the union?

85. *Chemistry* Ethylene glycol is the main component in automobile antifreeze. To monitor the temperature of an auto cooling system, you intend to use a meter that reads from 0 to 100. You devise a new temperature scale (°A) based on the approximate melting and boiling points of a typical antifreeze solution (−45°C and 115°C). You wish these points to correspond to 0°A and 100°A, respectively.

(a) Derive an expression for converting between °A and °C.

(b) Derive an expression for converting between °F and °A.

(c) At what temperature would your thermometer and a Celsius thermometer give the same numerical reading?

(d) Your thermometer reads 86°A. What is the temperature in °C and in °F?

(e) What is a temperature of 45°C in °A?

(Source: Zumdahl, Chemistry, *Sixth Edition)*

86. *Linear Depreciation* A company constructs a warehouse for $825,000. The warehouse has an estimated useful life of 25 years, after which its value is expected to be $75,000. Write a linear equation giving the value y of the warehouse during its 25 years of useful life. (Let t represent the time in years.)

87. *Linear Depreciation* A small business purchases a piece of equipment for $1025. After 5 years the equipment will be outdated, having no value.

(a) Write a linear equation giving the value y of the equipment in terms of the time t in years, $0 \le t \le 5$.

(b) Use a graphing utility to graph the equation.

(c) Move the cursor along the graph and estimate (to two-decimal-place accuracy) the value of the equipment when $t = 3$.

(d) Move the cursor along the graph and estimate (to two-decimal-place accuracy) the time when the value of the equipment will be $600.

88. *College Enrollment* A small college had 2546 students in 2002 and 2702 students in 2004. If the enrollment follows a linear growth pattern, how many students will the college have in 2008?

89. *Consumer Awareness* A real estate office handles an apartment complex with 50 units. When the rent is $380 per month, all 50 units are occupied. When the rent is $425, however, the average number of occupied units drops to 47. Assume that the relationship between the monthly rent p and the demand x is linear. (The term *demand* refers to the number of occupied units.)

(a) Write a linear equation expressing x in terms of p.

(b) *Linear Extrapolation* Predict the number of occupied units when the rent is set at $455.

(c) *Linear Interpolation* Predict the number of occupied units when the rent is set at $395.

90. *Profit* You are a contractor and have purchased a piece of equipment for $26,500. The equipment costs an average of $5.25 per hour for fuel and maintenance, and the operator is paid $9.50 per hour.

(a) Write a linear equation giving the total cost C of operating the equipment for t hours.

(b) You charge your customers $25 per hour of machine use. Write an equation for the revenue R derived from t hours of use.

(c) Use the formula for profit, $P = R - C$, to write an equation for the profit derived from t hours of use.

(d) Find the number of hours you must operate the equipment before you break even.

91. *Personal Income* Personal income (in billions of dollars) in the United States was 6937 in 1997 and 8685 in 2001. Assume that the relationship between the personal income Y and the time t (in years) is linear. Let $t = 0$ correspond to 1990. *(Source: U.S. Bureau of Economic Analysis)*

(a) Write a linear model for the data.

(b) *Linear Interpolation* Estimate the personal income in 1999.

(c) *Linear Extrapolation* Estimate the personal income in 2002.

(d) Use your school's library, the Internet, or some other reference source to find the actual personal income in 1999 and 2002. How close were your estimates?

92. *Sales Commission* As a salesperson, you receive a monthly salary of $2000, plus a commission of 7% of sales. You are offered a new job at $2300 per month, plus a commission of 5% of sales.

(a) Write a linear equation for your current monthly wage W in terms of your monthly sales S.

(b) Write a linear equation for the monthly wage W of your job offer in terms of the monthly sales S.

(c) Use a graphing utility to graph both equations in the same viewing window. Find the point of intersection. What does it signify?

(d) You think you can sell $20,000 worth of a product per month. Should you change jobs? Explain.

In Exercises 93–102, use a graphing utility to graph the cost function. Determine the maximum production level x, given that the cost C cannot exceed $100,000.

93. $C = 23{,}500 + 3100x$

94. $C = 30{,}000 + 575x$

95. $C = 18{,}375 + 1150x$

96. $C = 24{,}900 + 1785x$

97. $C = 75{,}500 + 89x$

98. $C = 83{,}620 + 67x$

99. $C = 32{,}000 + 650x$

100. $C = 53{,}500 + 495x$

101. $C = 50{,}000 + 0.25x$

102. $C = 75{,}500 + 1.50x$

1.4 FUNCTIONS

- Decide whether relations between two variables are functions.
- Find the domains and ranges of functions.
- Use function notation and evaluate functions.
- Combine functions to create other functions.
- Find inverse functions algebraically.

Functions

In many common relationships between two variables, the value of one of the variables depends on the value of the other variable. For example, the sales tax on an item depends on its selling price, the distance an object moves in a given amount of time depends on its speed, the price of mailing a package with an overnight delivery service depends on the package's weight, and the area of a circle depends on its radius.

Consider the relationship between the area of a circle and its radius. This relationship can be expressed by the equation

$$A = \pi r^2.$$

In this equation, the value of A depends on the choice of r. Because of this, A is the **dependent variable** and r is the **independent variable.**

Most of the relationships that you will study in this course have the property that for a given value of the independent variable, there corresponds exactly one value of the dependent variable. Such a relationship is a **function.**

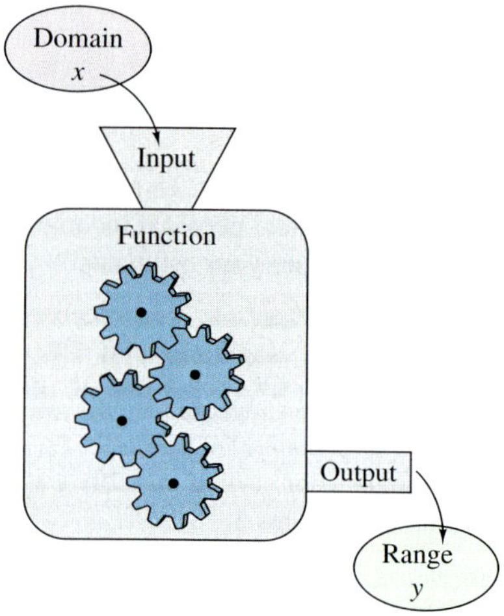

FIGURE 1.43

Definition of Function

A **function** is a relationship between two variables such that to each value of the independent variable there corresponds exactly one value of the dependent variable.

The **domain** of the function is the set of all values of the independent variable for which the function is defined. The **range** of the function is the set of all values taken on by the dependent variable.

In Figure 1.43, notice that you can think of a function as a machine that inputs values of the independent variable and outputs values of the dependent variable.

Although functions can be described by various means such as tables, graphs, and diagrams, they are most often specified by formulas or equations. For instance, the equation

$$y = 4x^2 + 3$$

describes y as a function of x. For this function, x is the independent variable and y is the dependent variable.

EXAMPLE 1 Deciding Whether Relations Are Functions

Which of the equations below define y as a function of x?

(a) $x + y = 1$ (b) $x^2 + y^2 = 1$

(c) $x^2 + y = 1$ (d) $x + y^2 = 1$

SOLUTION To decide whether an equation defines a function, it is helpful to isolate the dependent variable on the left side. For instance, to decide whether the equation $x + y = 1$ defines y as a function of x, write the equation in the form

$$y = 1 - x.$$

From this form, you can see that for any value of x, there is exactly one value of y. So, y is a function of x.

Original Equation	*Rewritten Equation*	*Test: Is y a function of x?*
(a) $x + y = 1$	$y = 1 - x$	Yes, each value of x determines exactly one value of y.
(b) $x^2 + y^2 = 1$	$y = \pm\sqrt{1 - x^2}$	No, some values of x determine two values of y.
(c) $x^2 + y = 1$	$y = 1 - x^2$	Yes, each value of x determines exactly one value of y.
(d) $x + y^2 = 1$	$y = \pm\sqrt{1 - x}$	No, some values of x determine two values of y.

Note that the equations that assign two values ($\pm$) to the dependent variable for a given value of the independent variable do not define functions of x. For instance, in part (b), when $x = 0$, the equation $y = \pm\sqrt{1 - x^2}$ indicates that $y = +1$ or $y = -1$. Figure 1.44 shows the graphs of the four equations.

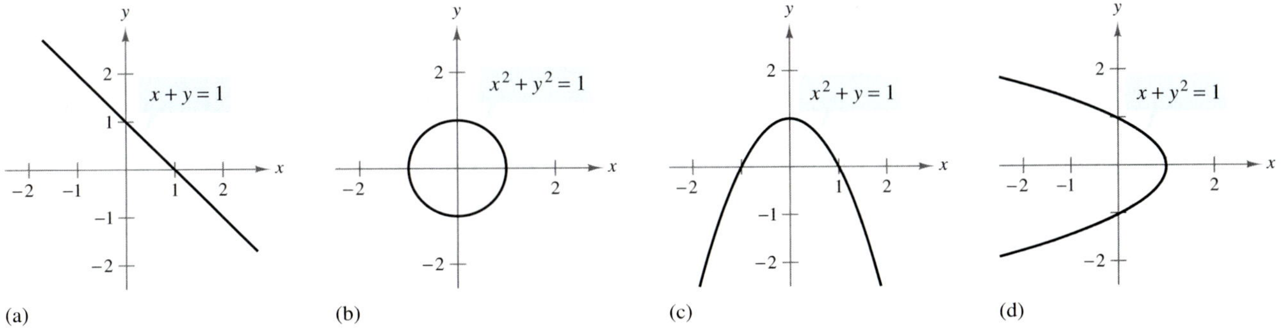

FIGURE 1.44

TECHNOLOGY

The procedure used in Example 1, isolating the dependent variable on the left side, is also useful for graphing equations with a graphing utility. In fact, the standard graphing program on most graphing utilities is called a "function grapher." To graph an equation in which y is not a function of x, such as a circle, you usually have to enter two or more equations into the graphing utility.

TRY IT 1

Which of the equations below define y as a function of x? Explain your answer.

(a) $x - y = 1$ (b) $x^2 + y^2 = 4$ (c) $y^2 + x = 2$ (d) $x^2 - y = 0$

The Graph of a Function

When the graph of a function is sketched, the standard convention is to let the horizontal axis represent the independent variable. When this convention is used, the test described in Example 1 has a nice graphical interpretation called the **vertical line test.** This test states that if every vertical line intersects the graph of an equation at most once, then the equation defines y as a function of x. For instance, in Figure 1.44, the graphs in parts (a) and (c) pass the vertical line test, but those in parts (b) and (d) do not.

The domain of a function may be described explicitly, or it may be *implied* by an equation used to define the function. For example, the function given by

$$y = \frac{1}{x^2 - 4}$$

has an implied domain that consists of all real x except $x = \pm 2$. These two values are excluded from the domain because division by zero is undefined.

Another type of implied domain is that used to avoid even roots of negative numbers, as indicated in Example 2.

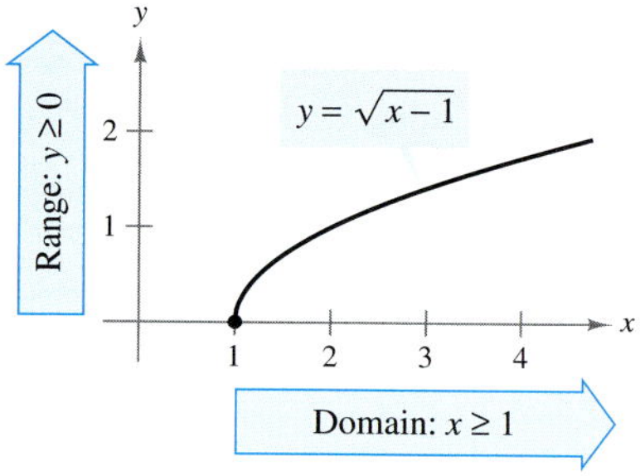

(a)

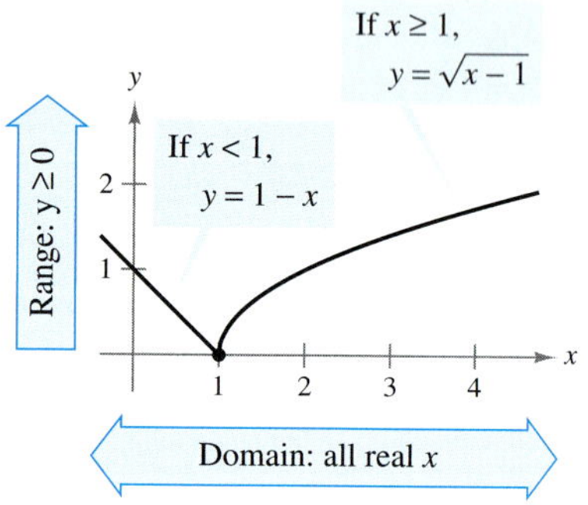

(b)

FIGURE 1.45

EXAMPLE 2 **Finding the Domain and Range of a Function**

Find the domain and range of each function.

(a) $y = \sqrt{x - 1}$ (b) $y = \begin{cases} 1 - x, & x < 1 \\ \sqrt{x - 1}, & x \geq 1 \end{cases}$

SOLUTION

(a) Because $\sqrt{x - 1}$ is not defined for $x - 1 < 0$ (that is, for $x < 1$), it follows that the domain of the function is the interval $x \geq 1$ or $[1, \infty)$. To find the range, observe that $\sqrt{x - 1}$ is never negative. Moreover, as x takes on the various values in the domain, y takes on all nonnegative values. So, the range is the interval $y \geq 0$ or $[0, \infty)$. The graph of the function, shown in Figure 1.45(a), confirms these results.

(b) Because this function is defined for $x < 1$ *and* for $x \geq 1$, the domain is the entire set of real numbers. This function is called a **piecewise-defined function** because it is defined by two or more equations over a specified domain. When $x \geq 1$, the function behaves as in part (a). For $x < 1$, the value of $1 - x$ is positive, and therefore the range of the function is $y \geq 0$ or $[0, \infty)$, as shown in Figure 1.45(b).

TRY IT 2

Find the domain and range of each function.

(a) $y = \sqrt{x + 1}$

(b) $y = \begin{cases} x^2, & x \leq 0 \\ \sqrt{x}, & x > 0 \end{cases}$

A function is **one-to-one** if to each value of the dependent variable in the range there corresponds exactly one value of the independent variable. For instance, the function in Example 2(a) is one-to-one, whereas the function in Example 2(b) is not one-to-one.

Geometrically, a function is one-to-one if every horizontal line intersects the graph of the function at most once. This geometrical interpretation is the **horizontal line test** for one-to-one functions. So, a graph that represents a one-to-one function must satisfy *both* the vertical line test and the horizontal line test.

Function Notation

When using an equation to define a function, you generally isolate the dependent variable on the left. For instance, writing the equation $x + 2y = 1$ as

$$y = \frac{1 - x}{2}$$

indicates that y is the dependent variable. In **function notation,** this equation has the form

$$f(x) = \frac{1 - x}{2}. \qquad \text{Function notation}$$

The independent variable is x, and the name of the function is "f." The symbol $f(x)$ is read as "f of x," and it denotes the value of the dependent variable. For instance, the value of f when $x = 3$ is

$$f(3) = \frac{1 - (3)}{2} = \frac{-2}{2} = -1.$$

The value $f(3)$ is called a **function value,** and it lies in the range of f. This means that the point $(3, f(3))$ lies on the graph of f. One of the advantages of function notation is that it allows you to be less wordy. For instance, instead of asking "What is the value of y when $x = 3$?" you can ask "What is $f(3)$?"

EXAMPLE 3 Evaluating a Function

Find the value of the function

$$f(x) = 2x^2 - 4x + 1$$

when x is -1, 0, and 2. Is f one-to-one?

SOLUTION When $x = -1$, the value of f is

$$f(-1) = 2(-1)^2 - 4(-1) + 1 = 2 + 4 + 1 = 7.$$

When $x = 0$, the value of f is

$$f(0) = 2(0)^2 - 4(0) + 1 = 0 - 0 + 1 = 1.$$

When $x = 2$, the value of f is

$$f(2) = 2(2)^2 - 4(2) + 1 = 8 - 8 + 1 = 1.$$

Because two different values of x yield the same value of $f(x)$, the function is *not* one-to-one, as shown in Figure 1.46.

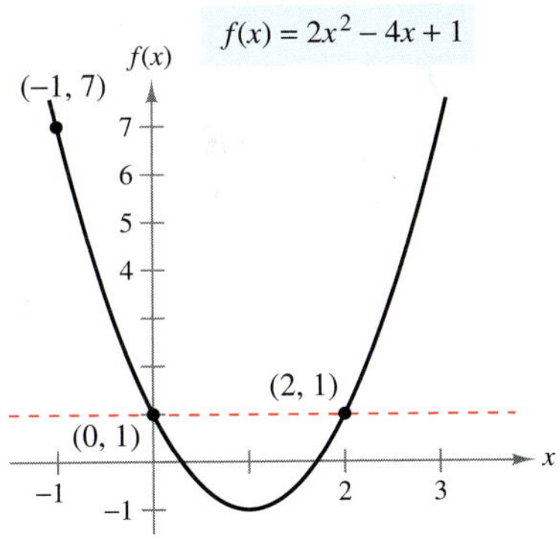

FIGURE 1.46

> **TRY IT 3**
>
> Find the value of $f(x) = x^2 - 5x + 1$ when x is 0, 1, and 4. Is f one-to-one?

STUDY TIP

You can use the horizontal line test to determine whether the function in Example 3 is one-to-one. Because the line $y = 1$ intersects the graph of the function twice, the function is *not* one-to-one.

Example 3 suggests that the role of the variable x in the equation

$$f(x) = 2x^2 - 4x + 1$$

is simply that of a placeholder. Informally, f could be defined by the equation

$$f(\quad) = 2(\quad)^2 - 4(\quad) + 1.$$

To evaluate $f(-2)$, simply place -2 in each set of parentheses.

$$f(-2) = 2(-2)^2 - 4(-2) + 1 = 8 + 8 + 1 = 17$$

The ratio in Example 4(b) is called a *difference quotient*. In Section 2.1, you will see that it has special significance in calculus.

TECHNOLOGY

Most graphing utilities can be programmed to evaluate functions. The program depends on the calculator used. The pseudocode below can be translated into a program for a graphing utility. (Appendix E lists the program for several models of graphing utilities.)

Program
- *Label a.*
- *Input x.*
- *Display function value.*
- *Goto a.*

To use this program, enter a function. Then run the program—it will allow you to evaluate the function at several values of x.

EXAMPLE 4 Evaluating a Function

Let $f(x) = x^2 - 4x + 7$, and find

(a) $f(x + \Delta x)$ (b) $\dfrac{f(x + \Delta x) - f(x)}{\Delta x}$.

SOLUTION

(a) To evaluate f at $x + \Delta x$, substitute $x + \Delta x$ for x in the original function, as shown.

$$\begin{aligned} f(x + \Delta x) &= (x + \Delta x)^2 - 4(x + \Delta x) + 7 \\ &= x^2 + 2x\,\Delta x + (\Delta x)^2 - 4x - 4\,\Delta x + 7 \end{aligned}$$

(b) Using the result of part (a), you can write

$$\begin{aligned} &\frac{f(x + \Delta x) - f(x)}{\Delta x} \\ &\quad = \frac{[(x + \Delta x)^2 - 4(x + \Delta x) + 7] - [x^2 - 4x + 7]}{\Delta x} \\ &\quad = \frac{x^2 + 2x\,\Delta x + (\Delta x)^2 - 4x - 4\,\Delta x + 7 - x^2 + 4x - 7}{\Delta x} \\ &\quad = \frac{2x\,\Delta x + (\Delta x)^2 - 4\,\Delta x}{\Delta x} \\ &\quad = 2x + \Delta x - 4, \quad \Delta x \neq 0. \end{aligned}$$

TRY IT 4

Let $f(x) = x^2 - 2x + 3$, and find (a) $f(x + \Delta x)$ and (b) $\dfrac{f(x + \Delta x) - f(x)}{\Delta x}$.

Although f is often used as a convenient function name and x as the independent variable, you can use other symbols. For instance, the following equations all define the same function.

$$f(x) = x^2 - 4x + 7$$
$$f(t) = t^2 - 4t + 7$$
$$g(s) = s^2 - 4s + 7$$

Combinations of Functions

Two functions can be combined in various ways to create new functions. For instance, if $f(x) = 2x - 3$ and $g(x) = x^2 + 1$, you can form the following functions.

$$f(x) + g(x) = (2x - 3) + (x^2 + 1) = x^2 + 2x - 2 \quad \text{Sum}$$
$$f(x) - g(x) = (2x - 3) - (x^2 + 1) = -x^2 + 2x - 4 \quad \text{Difference}$$
$$f(x)g(x) = (2x - 3)(x^2 + 1) = 2x^3 - 3x^2 + 2x - 3 \quad \text{Product}$$
$$\frac{f(x)}{g(x)} = \frac{2x - 3}{x^2 + 1} \quad \text{Quotient}$$

You can combine two functions in yet another way called a **composition.** The resulting function is a **composite function.**

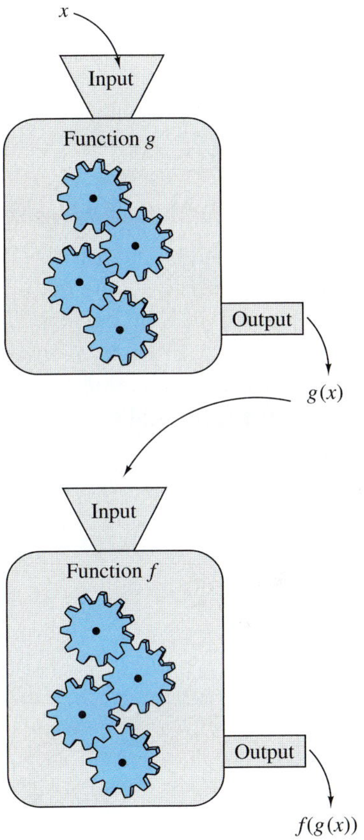

FIGURE 1.47

Definition of Composite Function

The function given by $(f \circ g)(x) = f(g(x))$ is the **composite** of f with g. The **domain** of $(f \circ g)$ is the set of all x in the domain of g such that $g(x)$ is in the domain of f, as indicated in Figure 1.47.

The composite of f with g may not be equal to the composite of g with f, as shown in the next example.

EXAMPLE 5 Forming Composite Functions

Let $f(x) = 2x - 3$ and $g(x) = x^2 + 1$, and find

(a) $f(g(x))$ (b) $g(f(x))$.

SOLUTION

(a) The composite of f with g is given by

$$f(g(x)) = 2(g(x)) - 3 \quad \text{Evaluate } f \text{ at } g(x).$$
$$= 2(x^2 + 1) - 3 \quad \text{Substitute } x^2 + 1 \text{ for } g(x).$$
$$= 2x^2 - 1. \quad \text{Simplify.}$$

(b) The composite of g with f is given by

$$g(f(x)) = (f(x))^2 + 1 \quad \text{Evaluate } g \text{ at } f(x).$$
$$= (2x - 3)^2 + 1 \quad \text{Substitute } 2x - 3 \text{ for } f(x).$$
$$= 4x^2 - 12x + 10. \quad \text{Simplify.}$$

STUDY TIP

The results of $f(g(x))$ and $g(f(x))$ are different in Example 5. You can verify this by substituting specific values of x into each function and comparing the results.

TRY IT 5

Let $f(x) = 2x + 1$ and $g(x) = x^2 + 2$, and find

(a) $f(g(x))$ (b) $g(f(x))$.

Inverse Functions

Informally, the inverse function of f is another function g that "undoes" what f has done.

$$x \xrightarrow{f} f(x) \xrightarrow{g} g(f(x)) = x$$

ALGEBRA REVIEW

Don't be confused by the use of the superscript -1 to denote the inverse function f^{-1}. In this text, whenever f^{-1} is written, it *always* refers to the inverse function of f and *not* to the reciprocal of $f(x)$.

Definition of Inverse Function

Let f and g be two functions such that

$$f(g(x)) = x \text{ for each } x \text{ in the domain of } g$$

and

$$g(f(x)) = x \text{ for each } x \text{ in the domain of } f.$$

Under these conditions, the function g is the **inverse function** of f. The function g is denoted by f^{-1}, which is read as "f-inverse." So,

$$f(f^{-1}(x)) = x \text{ and } f^{-1}(f(x)) = x.$$

The domain of f must be equal to the range of f^{-1}, and the range of f must be equal to the domain of f^{-1}.

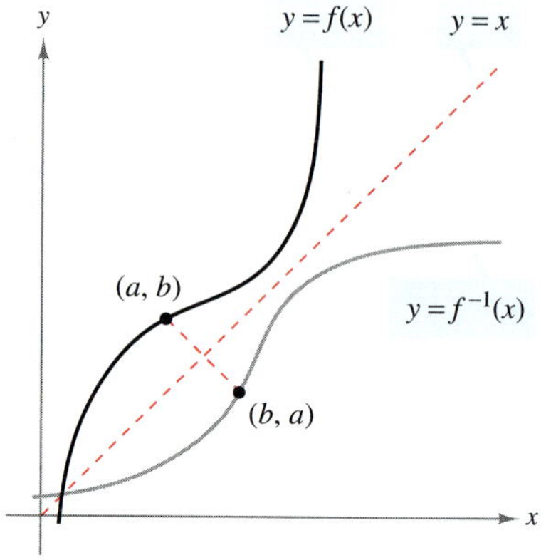

FIGURE 1.48 The graph of f^{-1} is a reflection of the graph of f in the line $y = x$.

EXAMPLE 6 Finding Inverse Functions

Several functions and their inverse functions are shown below. In each case, note that the inverse function "undoes" the original function. For instance, to undo multiplication by 2, you should divide by 2.

(a) $f(x) = 2x$ $\quad f^{-1}(x) = \frac{1}{2}x$

(b) $f(x) = \frac{1}{3}x$ $\quad f^{-1}(x) = 3x$

(c) $f(x) = x + 4$ $\quad f^{-1}(x) = x - 4$

(d) $f(x) = 2x - 5$ $\quad f^{-1}(x) = \frac{1}{2}(x + 5)$

(e) $f(x) = x^3$ $\quad f^{-1}(x) = \sqrt[3]{x}$

(f) $f(x) = \dfrac{1}{x}$ $\quad f^{-1}(x) = \dfrac{1}{x}$

STUDY TIP

You can verify that the functions in Example 6 are inverse functions by substituting specific values of x.

TRY IT 6

Informally find the inverse function of each function.

(a) $f(x) = \frac{1}{5}x$ (b) $f(x) = 3x + 2$

The graphs of f and f^{-1} are mirror images of each other (with respect to the line $y = x$), as shown in Figure 1.48. Try using a graphing utility to confirm this for each of the functions given in Example 6.

The functions in Example 6 are simple enough so that their inverse functions can be found by inspection. The next example demonstrates a strategy for finding the inverse functions of more complicated functions.

EXAMPLE 7 Finding an Inverse Function

Find the inverse function of $f(x) = \sqrt{2x - 3}$.

SOLUTION Begin by replacing $f(x)$ with y. Then, interchange x and y and solve for y.

$$f(x) = \sqrt{2x-3} \quad \text{Write original function.}$$
$$y = \sqrt{2x-3} \quad \text{Replace } f(x) \text{ with } y.$$
$$x = \sqrt{2y-3} \quad \text{Interchange } x \text{ and } y.$$
$$x^2 = 2y - 3 \quad \text{Square each side.}$$
$$x^2 + 3 = 2y \quad \text{Add 3 to each side.}$$
$$\frac{x^2+3}{2} = y \quad \text{Divide each side by 2.}$$

So, the inverse function has the form

$$f^{-1}(\quad) = \frac{(\quad)^2 + 3}{2}.$$

Using x as the independent variable, you can write

$$f^{-1}(x) = \frac{x^2+3}{2}, \quad x \geq 0.$$

In Figure 1.49, note that the domain of f^{-1} coincides with the range of f.

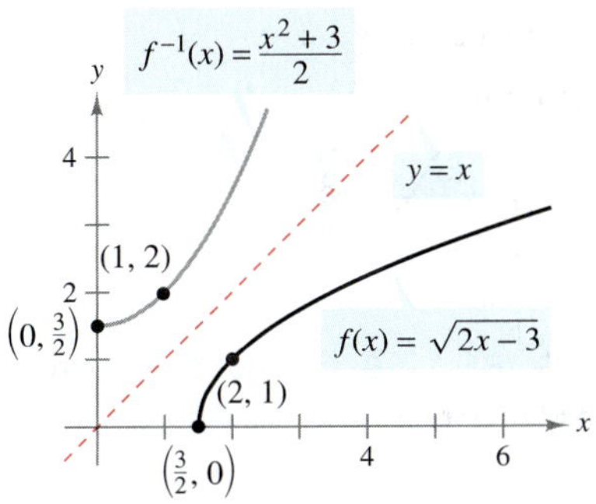

FIGURE 1.49

TRY IT 7

Find the inverse function of $f(x) = x^2 + 2$ for $x \geq 0$.

After you have found an inverse function, you should check your results. You can check your results *graphically* by observing that the graphs of f and f^{-1} are reflections of each other in the line $y = x$. You can check your results *algebraically* by evaluating $f(f^{-1}(x))$ and $f^{-1}(f(x))$—both should be equal to x.

Check that $f(f^{-1}(x)) = x$

$$f(f^{-1}(x)) = f\left(\frac{x^2+3}{2}\right) = \sqrt{2\left(\frac{x^2+3}{2}\right) - 3} = \sqrt{x^2} = x, \quad x \geq 0$$

Check that $f^{-1}(f(x)) = x$

$$f^{-1}(f(x)) = f^{-1}\left(\sqrt{2x-3}\right) = \frac{\left(\sqrt{2x-3}\right)^2 + 3}{2} = \frac{2x}{2} = x, \quad x \geq \frac{3}{2}$$

TECHNOLOGY

A graphing utility can help you check that the graphs of f and f^{-1} are reflections of each other in the line $y = x$. To do this, graph $y = f(x)$, $y = f^{-1}(x)$, and $y = x$ in the same viewing window, using a *square setting*.

Not every function has an inverse function. In fact, for a function to have an inverse function, it must be one-to-one.

EXAMPLE 8 A Function That Has No Inverse Function

Show that the function

$$f(x) = x^2 - 1$$

has no inverse function. (Assume that the domain of f is the set of all real numbers.)

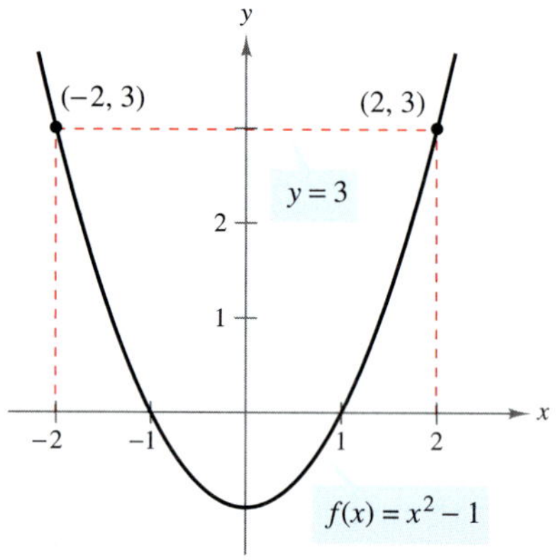

FIGURE 1.50 f is not one-to-one and has no inverse function.

SOLUTION Begin by sketching the graph of f, as shown in Figure 1.50. Note that

$$f(2) = (2)^2 - 1 = 3$$

and

$$f(-2) = (-2)^2 - 1 = 3.$$

So, f does not pass the horizontal line test, which implies that it is not one-to-one, and therefore has no inverse function. The same conclusion can be obtained by trying to find the inverse function of f algebraically.

$f(x) = x^2 - 1$	Write original function.
$y = x^2 - 1$	Replace $f(x)$ with y.
$x = y^2 - 1$	Interchange x and y.
$x + 1 = y^2$	Add 1 to each side.
$\pm\sqrt{x + 1} = y$	Take square root of each side.

The last equation does not define y as a function of x, and so f has no inverse function.

TRY IT 8

Show that the function

$$f(x) = x^2 + 4$$

has no inverse function.

TAKE ANOTHER LOOK

Comparing Composition Orders

You are buying an automobile whose price is $22,500. Which of the following options would you choose? Explain.

a. You are given a factory rebate of $2000, followed by a dealer discount of 10%.

b. You are given a dealer discount of 10%, followed by a factory rebate of $2000.

Let $f(x) = x - 2000$ and let $g(x) = 0.9x$. Which option is represented by the composite $f(g(x))$? Which is represented by the composite $g(f(x))$?

PREREQUISITE REVIEW 1.4

The following warm-up exercises involve skills that were covered in earlier sections. You will use these skills in the exercise set for this section.

In Exercises 1–6, simplify the expression.

1. $5(-1)^2 - 6(-1) + 9$

2. $(-2)^3 + 7(-2)^2 - 10$

3. $(x - 2)^2 + 5x - 10$

4. $(3 - x) + (x + 3)^3$

5. $\dfrac{1}{1 - (1 - x)}$

6. $1 + \dfrac{x - 1}{x}$

In Exercises 7–12, solve for y in terms of x.

7. $2x + y - 6 = 11$

8. $5y - 6x^2 - 1 = 0$

9. $(y - 3)^2 = 5 + (x + 1)^2$

10. $y^2 - 4x^2 = 2$

11. $x = \dfrac{2y - 1}{4}$

12. $x = \sqrt[3]{2y - 1}$

EXERCISES 1.4

In Exercises 1–8, decide whether the equation defines y as a function of x.

1. $x^2 + y^2 = 4$

2. $x + y^2 = 4$

3. $\frac{1}{2}x - 6y = -3$

4. $3x - 2y + 5 = 0$

5. $x^2 + y = 4$

6. $x^2 + y^2 - 2x - 4y + 1 = 0$

7. $y^2 = x^2 - 1$

8. $x^2y - x^2 + 4y = 0$

In Exercises 9–16, use a graphing utility to graph the function. Then determine the domain and range of the function.

9. $f(x) = 2x^2 - 5x + 1$

10. $f(x) = 5x^3 + 6x^2 - 1$

11. $f(x) = \dfrac{|x|}{x}$

12. $f(x) = \sqrt{9 - x^2}$

13. $f(x) = \dfrac{x}{\sqrt{x - 4}}$

14. $f(x) = \dfrac{2x}{\sqrt{x + 1}}$

15. $f(x) = \dfrac{x - 2}{x + 4}$

16. $f(x) = \dfrac{x^2}{1 - x}$

In Exercises 17–20, find the domain and range of the function. Use interval notation to write your result.

17. $f(x) = x^3$

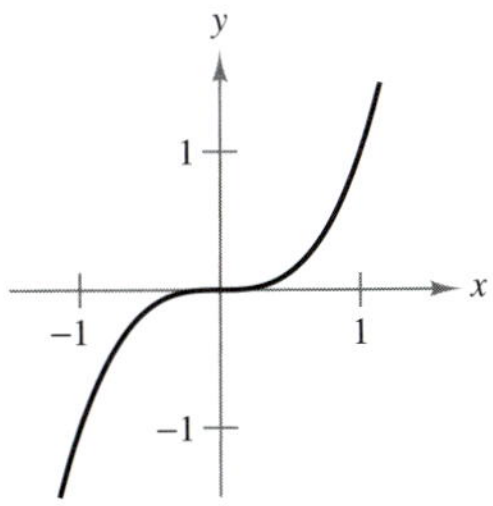

18. $f(x) = \sqrt{2x - 3}$

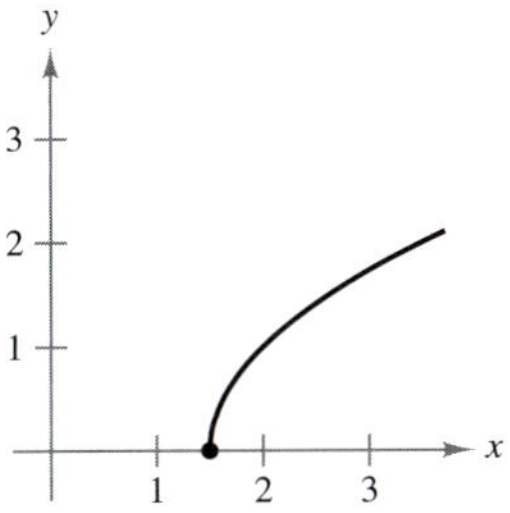

19. $f(x) = 4 - x^2$

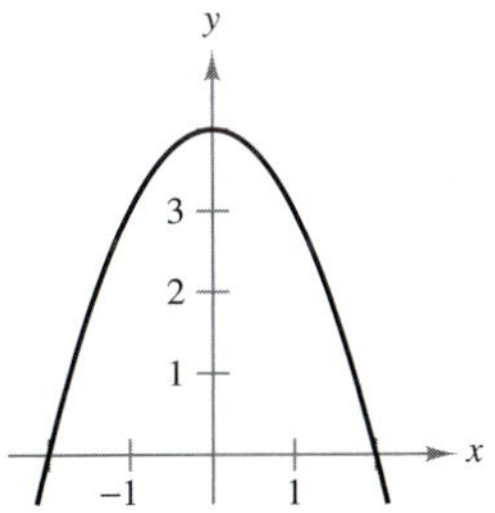

20. $f(x) = |x - 2|$

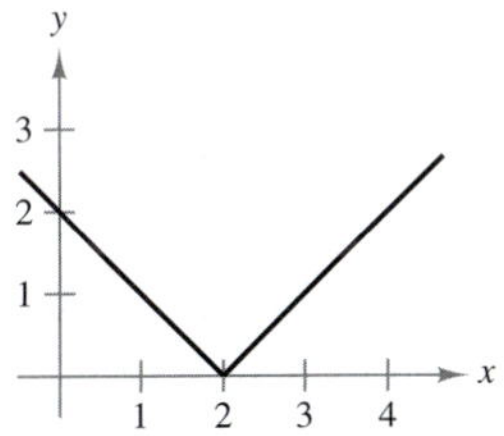

In Exercises 21–24, evaluate the function at the specified values of the independent variable. Simplify the result.

21. $f(x) = 2x - 3$

(a) $f(0)$ (b) $f(-3)$

(c) $f(x - 1)$ (d) $f(x + \Delta x)$

22. $f(x) = x^2 - 2x + 2$

(a) $f\left(\frac{1}{2}\right)$ (b) $f(-1)$

(c) $f(c + 2)$ (d) $f(x + \Delta x)$

23. $g(x) = 1/x$

(a) $g(2)$ (b) $g\left(\frac{1}{4}\right)$

(c) $g(x + 4)$ (d) $g(x + \Delta x) - g(x)$

24. $f(x) = |x| + 4$

(a) $f(2)$ (b) $f(-2)$

(c) $f(x + 2)$ (d) $f(x + \Delta x) - f(x)$

In Exercises 25–30, evaluate the difference quotient and simplify the result.

25. $f(x) = x^2 - 4x + 1$

$$\frac{f(x + \Delta x) - f(x)}{\Delta x}$$

26. $h(x) = x^2 - x + 1$

$$\frac{h(2 + \Delta x) - h(2)}{\Delta x}$$

27. $g(x) = \sqrt{x+3}$

$\dfrac{g(x+\Delta x) - g(x)}{\Delta x}$

28. $f(x) = \dfrac{1}{\sqrt{x-1}}$

$\dfrac{f(x) - f(2)}{x-2}$

29. $f(x) = \dfrac{1}{x-2}$

$\dfrac{f(x+\Delta x) - f(x)}{\Delta x}$

30. $f(x) = \dfrac{1}{x+4}$

$\dfrac{f(x+\Delta x) - f(x)}{\Delta x}$

In Exercises 31–34, use the vertical line test to determine whether y is a function of x.

31. $x^2 + y^2 = 9$

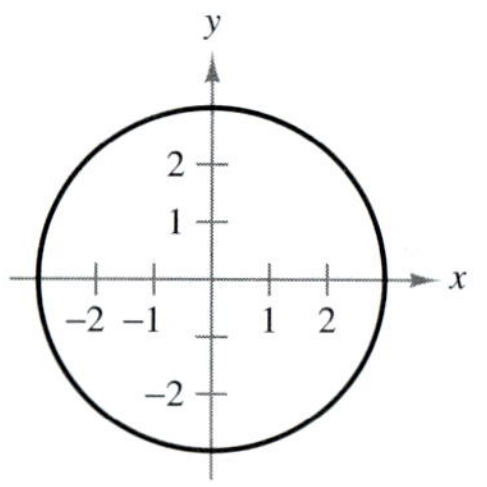

32. $x - xy + y + 1 = 0$

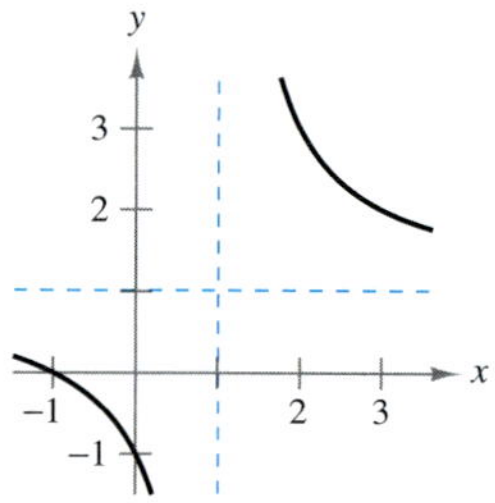

33. $x^2 = xy - 1$

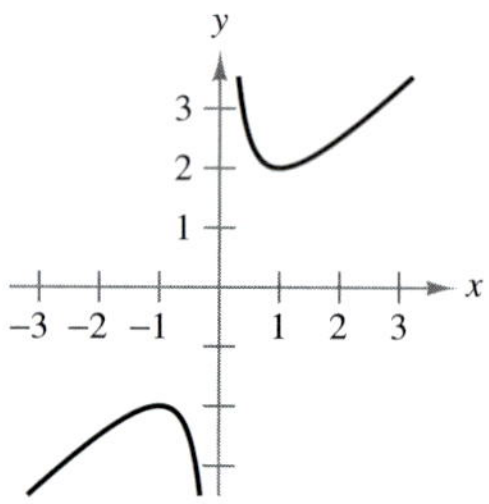

34. $x = |y|$

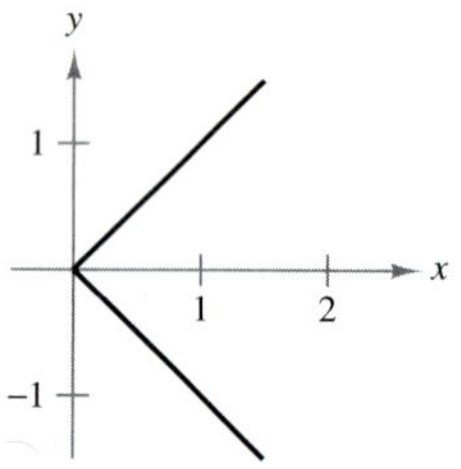

In Exercises 35–40, find (a) $f(x) + g(x)$, (b) $f(x) \cdot g(x)$, (c) $f(x)/g(x)$, (d) $f(g(x))$, and (e) $g(f(x))$ if defined.

35. $f(x) = 2x - 5$

$g(x) = 5$

36. $f(x) = 2x - 5$

$g(x) = 2 - x$

37. $f(x) = x^2 + 1$

$g(x) = x - 1$

38. $f(x) = x^2 + 5$

$g(x) = \sqrt{1-x}$

39. $f(x) = \dfrac{1}{x}$

$g(x) = \dfrac{1}{x^2}$

40. $f(x) = \dfrac{x}{x+1}$

$g(x) = x^3$

41. Given $f(x) = \sqrt{x}$ and $g(x) = x^2 - 1$, find the composite functions.

(a) $f(g(1))$ (b) $g(f(1))$

(c) $g(f(0))$ (d) $f(g(-4))$

(e) $f(g(x))$ (f) $g(f(x))$

42. Given $f(x) = 1/x$ and $g(x) = x^2 - 1$, find the composite functions.

(a) $f(g(2))$ (b) $g(f(2))$

(c) $f\left(g\left(1/\sqrt{2}\right)\right)$ (d) $g\left(f\left(1/\sqrt{2}\right)\right)$

(e) $f(g(x))$ (f) $g(f(x))$

In Exercises 43–46, select a function from (a) $f(x) = cx$, (b) $g(x) = cx^2$, (c) $h(x) = c\sqrt{|x|}$, and (d) $r(x) = c/x$ and determine the value of the constant c such that the function fits the data in the table.

43.

x	-4	-1	0	1	4
y	-32	-2	0	-2	-32

44.

x	-4	-1	0	1	4
y	-1	$-\frac{1}{4}$	0	$\frac{1}{4}$	1

45.

x	-4	-1	0	1	4
y	-8	-32	Undefined	32	8

46.

x	-4	-1	0	1	4
y	6	3	0	3	6

In Exercises 47–50, show that f and g are inverse functions by showing that $f(g(x)) = x$ and $g(f(x)) = x$. Then sketch the graphs of f and g on the same coordinate axes.

47. $f(x) = 5x + 1,\quad g(x) = \dfrac{x-1}{5}$

48. $f(x) = \dfrac{1}{x},\quad g(x) = \dfrac{1}{x}$

49. $f(x) = 9 - x^2,\ x \ge 0,\quad g(x) = \sqrt{9-x},\ x \le 9$

50. $f(x) = 1 - x^3,\quad g(x) = \sqrt[3]{1-x}$

In Exercises 51–58, find the inverse function of f. Then sketch the graphs of f and f^{-1} on the same coordinate axes.

51. $f(x) = 2x - 3$

52. $f(x) = 6 - 3x$

53. $f(x) = x^5$

54. $f(x) = x^3 + 1$

55. $f(x) = \sqrt{9 - x^2},\quad 0 \le x \le 3$

56. $f(x) = \sqrt{x^2 - 4},\quad x \ge 2$

57. $f(x) = x^{2/3},\quad x \ge 0$

58. $f(x) = x^{3/5}$

In Exercises 59–64, use a graphing utility to graph the function. Then use the horizontal line test to determine whether the function is one-to-one. If it is, find its inverse function.

59. $f(x) = 3 - 7x$

60. $f(x) = \sqrt{x-2}$

61. $f(x) = x^2$

62. $f(x) = x^4$

63. $f(x) = |x - 2|$

64. $f(x) = 3$

65. Use the graph of $f(x) = \sqrt{x}$ below to sketch the graph of each function.

(a) $y = \sqrt{x} + 2$

(b) $y = -\sqrt{x}$

(c) $y = \sqrt{x - 2}$

(d) $y = \sqrt{x + 3}$

(e) $y = \sqrt{x} - 4$

(f) $y = 2\sqrt{x}$

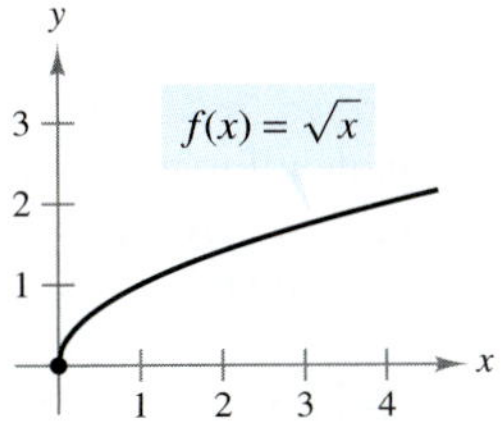

66. Use the graph of $f(x) = |x|$ below to sketch the graph of each function.

(a) $y = |x| + 3$

(b) $y = -\frac{1}{2}|x|$

(c) $y = |x - 2|$

(d) $y = |x + 1| - 1$

(e) $y = 2|x|$

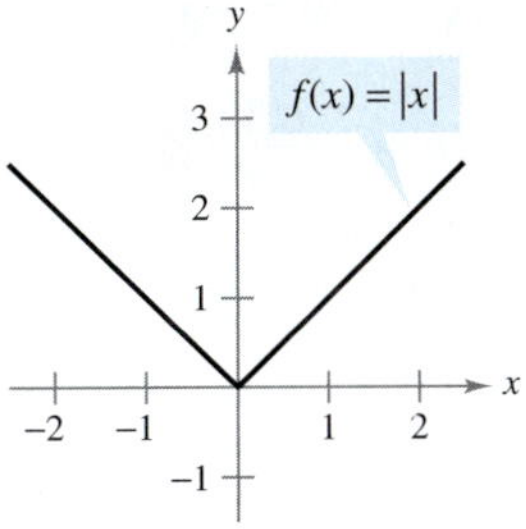

67. Use the graph of $f(x) = x^2$ to find a formula for each of the functions whose graphs are shown.

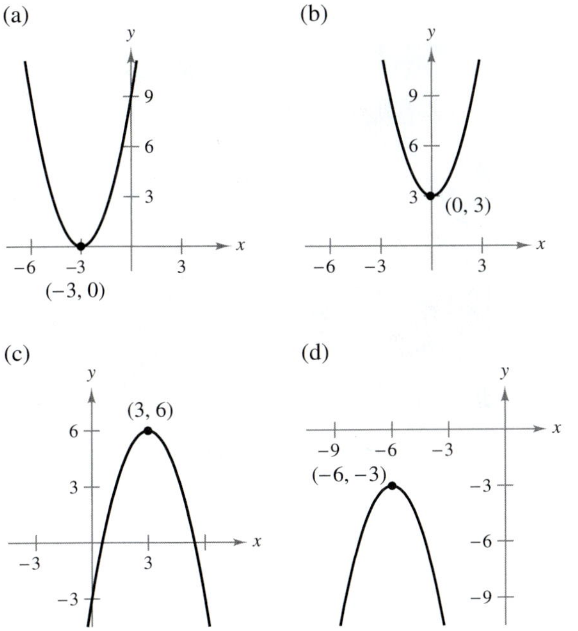

68. *Real Estate* Express the value V of a real estate firm in terms of x, the number of acres of property owned. Each acre is valued at \$2500 and other company assets total \$750,000.

69. *Owning a Business* You own two restaurants. From 1998 to 2004, the sales R_1 (in thousands of dollars) for one restaurant can be modeled by

$$R_1 = 480 - 8t - 0.8t^2, \quad t = 0, 1, 2, 3, 4, 5, 6$$

where $t = 0$ represents 1998. During the same seven-year period, the sales R_2 (in thousands of dollars) for the second restaurant can be modeled by

$$R_2 = 254 + 0.78t, \quad t = 0, 1, 2, 3, 4, 5, 6.$$

Write a function that represents the total sales for the two restaurants. Use a graphing utility to graph the total sales function.

70. *Cost* The inventor of a new game believes that the variable cost for producing the game is \$0.95 per unit. The fixed cost is \$6000.

(a) Express the total cost C as a function of x, the number of games sold.

(b) Find a formula for the average cost per unit $\overline{C} = C/x$.

(c) The selling price for each game is \$1.69. How many units must be sold before the average cost per unit falls below the selling price?

71. *Demand* The demand function for a commodity is

$$p = \frac{14.75}{1 + 0.01x}, \quad x \geq 0$$

where p is the price per unit and x is the number of units sold.

(a) Find x as a function of p.

(b) Find the number of units sold when the price is \$10.

72. *Cost* A power station is on one side of a river that is $\frac{1}{2}$ mile wide. A factory is 3 miles downstream on the other side of the river (see figure). It costs \$10/ft to run the power lines on land and \$15/ft to run them under water. Express the cost C of running the lines from the power station to the factory as a function of x.

73. *Cost* The weekly cost of producing x units in a manufacturing process is given by the function

$$C(x) = 70x + 375.$$

The number of units produced in t hours is given by $x(t) = 40t$. Find and interpret $C(x(t))$.

74. *Market Equilibrium* The supply function for a product relates the number of units x that producers are willing to supply for a given price per unit p. The supply and demand functions for a market are

$$p = \frac{2}{5}x + 4 \qquad \text{Supply}$$

$$p = -\frac{16}{25}x + 30. \qquad \text{Demand}$$

(a) Use a graphing utility to graph the supply and demand functions in the same viewing window.

(b) Use the *trace* feature of the graphing utility to find the *equilibrium point* for the market.

(c) For what values of x does the demand exceed the supply?

(d) For what values of x does the supply exceed the demand?

75. *Profit* A radio manufacturer charges \$90 per unit for units that cost \$60 to produce. To encourage large orders from distributors, the manufacturer will reduce the price by \$0.01 per unit for each unit in excess of 100 units. (For example, an order of 101 units would have a price of \$89.99 per unit, and an order of 102 units would have a price of \$89.98 per unit.) This price reduction is discontinued when the price per unit drops to \$75.

(a) Express the price per unit p as a function of the order size x.

(b) Express the profit P as a function of the order size x.

76. *Cost, Revenue, and Profit* A company invests \$98,000 for equipment to produce a new product. Each unit of the product costs \$12.30 and is sold for \$17.98. Let x be the number of units produced and sold.

(a) Write the total cost C as a function of x.

(b) Write the revenue R as a function of x.

(c) Write the profit P as a function of x.

77. *Revenue* For groups of 80 or more people, a charter bus company determines the rate r (in dollars per person) according to the formula

$$r = 8 - 0.05(n - 80), \quad n \geq 80$$

where n is the number of people.

(a) Express the revenue R for the bus company as a function of n.

(b) Complete the table.

n	90	100	110	120	130	140	150
R							

(c) Criticize the formula for the rate. Would you use this formula? Explain your reasoning.

78. *Medicine* The temperature of a patient after being given a fever-reducing drug is given by

$$F(t) = 98 + \frac{3}{t + 1}$$

where F is the temperature in degrees Fahrenheit and t is the time in hours since the drug was administered. Use a graphing utility to graph the function. Be sure to choose an appropriate viewing window. For what values of t do you think this function would be valid? Explain.

In Exercises 79–86, use a graphing utility to graph the function. Then use the *zoom* and *trace* features to find the zeros of the function. Is the function one-to-one?

79. $f(x) = 9x - 4x^2$

80. $f(x) = 2\left(3x^2 - \dfrac{6}{x}\right)$

81. $g(t) = \dfrac{t + 3}{1 - t}$

82. $h(x) = 6x^3 - 12x^2 + 4$

83. $f(x) = \dfrac{4 - x^2}{x}$

84. $g(x) = \left|\dfrac{1}{2}x^2 - 4\right|$

85. $g(x) = x^2\sqrt{x^2 - 4}$

86. $f(x) = \dfrac{\sqrt{x^2 - 16}}{x^2}$

BUSINESS CAPSULE

Robert L. Boverie/CardSenders

CardSenders is a home-based greeting card service for businesses. Phyllis and Robert Boverie bought the company in 1990, which has expanded into the United Kingdom, Canada, Asia, and Mexico. Currently there are over 200 licensees and consultants. Start-up costs run from \$995.00 for consultants and \$6,900.00 for licensees.

87. *Research Project* Use your school's library, the Internet, or some other reference source to find information about the start-up costs of beginning a business, such as the example above. Write a short paper about the company.

1.5 LIMITS

- Find limits of functions graphically and numerically.
- Use the properties of limits to evaluate limits of functions.
- Use different analytic techniques to evaluate limits of functions.
- Evaluate one-sided limits.
- Recognize unbounded behavior of functions.

The Limit of a Function

In everyday language, people refer to a speed limit, a wrestler's weight limit, the limit of one's endurance, or stretching a spring to its limit. These phrases all suggest that a limit is a bound, which on some occasions may not be reached but on other occasions may be reached or exceeded.

Consider a spring that will break only if a weight of 10 pounds or more is attached. To determine how far the spring will stretch without breaking, you could attach increasingly heavier weights and measure the spring length s for each weight w, as shown in Figure 1.51. If the spring length approaches a value of L, then it is said that "the limit of s as w approaches 10 is L." A mathematical limit is much like the limit of a spring. The notation for a limit is

$$\lim_{x \to c} f(x) = L$$

which is read as "the limit of $f(x)$ as x approaches c is L."

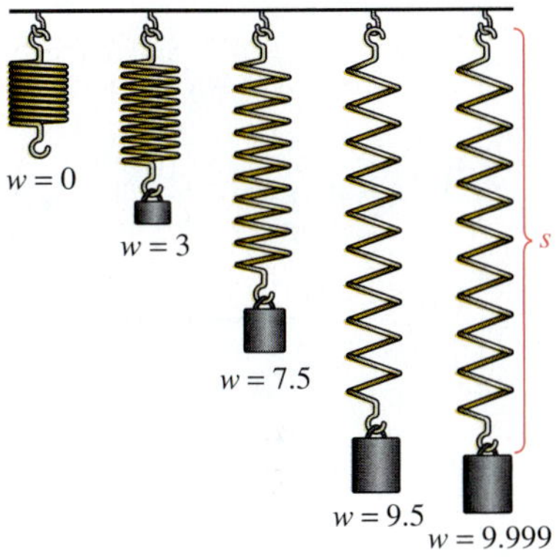

FIGURE 1.51 What is the limit of s as w approaches 10 pounds?

EXAMPLE 1 Finding a Limit

Find the limit: $\lim_{x \to 1} (x^2 + 1)$.

SOLUTION Let $f(x) = x^2 + 1$. From the graph of f in Figure 1.52, it appears that $f(x)$ approaches 2 as x approaches 1 from either side, and you can write

$$\lim_{x \to 1} (x^2 + 1) = 2.$$

The table yields the same conclusion. Notice that as x gets closer and closer to 1, $f(x)$ gets closer and closer to 2.

x approaches 1. → ← x approaches 1.

x	0.900	0.990	0.999	1.000	1.001	1.010	1.100
$f(x)$	1.810	1.980	1.998	2.000	2.002	2.020	2.210

$f(x)$ approaches 2. → ← $f(x)$ approaches 2.

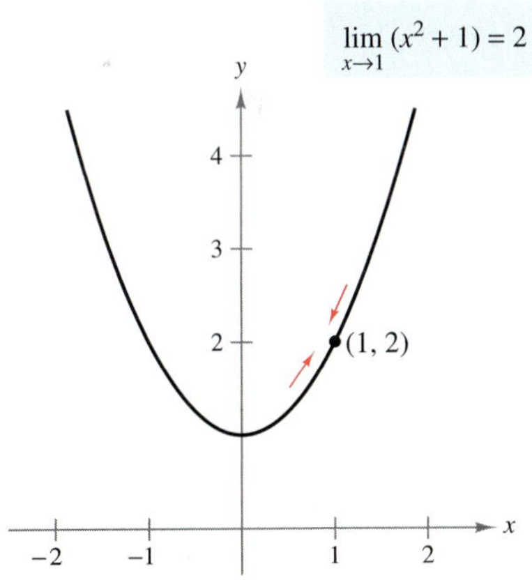

FIGURE 1.52

TRY IT 1

Find the limit: $\lim_{x \to 1} (2x + 4)$.

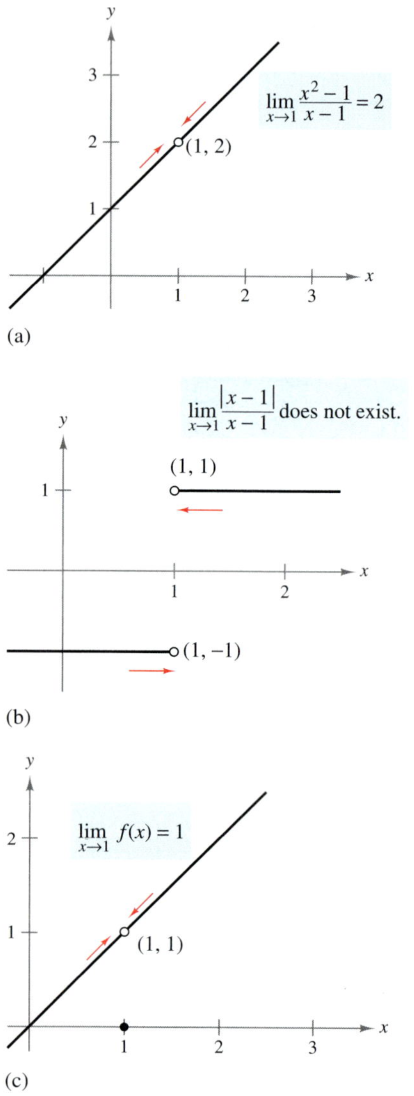

FIGURE 1.53

TRY IT 2

Find the limit: $\lim_{x\to 2} f(x)$.

(a) $f(x) = \dfrac{x^2 - 4}{x - 2}$

(b) $f(x) = \dfrac{|x - 2|}{x - 2}$

(c) $f(x) = \begin{cases} x^2, & x \neq 2 \\ 0, & x = 2 \end{cases}$

EXAMPLE 2 Finding Limits Graphically and Numerically

Find the limit: $\lim_{x\to 1} f(x)$.

(a) $f(x) = \dfrac{x^2 - 1}{x - 1}$ (b) $f(x) = \dfrac{|x - 1|}{x - 1}$ (c) $f(x) = \begin{cases} x, & x \neq 1 \\ 0, & x = 1 \end{cases}$

SOLUTION

(a) From the graph of f, in Figure 1.53(a), it appears that $f(x)$ approaches 2 as x approaches 1 from either side. A missing point is denoted by the open dot on the graph. This conclusion is reinforced by the table. Be sure you see that *it does not matter that $f(x)$ is undefined when $x = 1$*. The limit depends only on values of $f(x)$ near 1, not at 1.

x approaches 1. → ← x approaches 1.

x	0.900	0.990	0.999	1.000	1.001	1.010	1.100
$f(x)$	1.900	1.990	1.999	?	2.001	2.010	2.100

$f(x)$ approaches 2. → ← $f(x)$ approaches 2.

(b) From the graph of f, in Figure 1.53(b), you can see that $f(x) = -1$ for all values to the left of $x = 1$ and $f(x) = 1$ for all values to the right of $x = 1$. So, $f(x)$ is approaching a different value from the left of $x = 1$ than it is from the right of $x = 1$. In such situations, we say that the *limit does not exist*. This conclusion is reinforced by the table.

x approaches 1. → ← x approaches 1.

x	0.900	0.990	0.999	1.000	1.001	1.010	1.100
$f(x)$	−1.000	−1.000	−1.000	?	1.000	1.000	1.000

$f(x)$ approaches −1. → ← $f(x)$ approaches 1.

(c) From the graph of f, in Figure 1.53(c), it appears that $f(x)$ approaches 1 as x approaches 1 from either side. This conclusion is reinforced by the table. It does not matter that $f(1) = 0$. The limit depends only on values of $f(x)$ near 1, not at 1.

x approaches 1. → ← x approaches 1.

x	0.900	0.990	0.999	1.000	1.001	1.010	1.100
$f(x)$	0.900	0.990	0.999	0.000	1.001	1.010	1.100

$f(x)$ approaches 1. → ← $f(x)$ approaches 1.

There are three important ideas to learn from Examples 1 and 2.

1. Saying that the limit of $f(x)$ approaches L as x approaches c means that the value of $f(x)$ may be made *arbitrarily close* to the number L by choosing x closer and closer to c.
2. For a limit to exist, you must allow x to approach c from *either side* of c. If $f(x)$ approaches a different number as x approaches c from the left than it does as x approaches c from the right, then the limit *does not exist*. [See Example 2(b).]
3. The value of $f(x)$ when $x = c$ has no bearing on the existence or nonexistence of the limit of $f(x)$ as x approaches c. For instance, in Example 2(a), the limit of $f(x)$ exists as x approaches 1 even though the function f is not defined at $x = 1$.

Definition of the Limit of a Function

If $f(x)$ becomes arbitrarily close to a single number L as x approaches c from either side, then

$$\lim_{x \to c} f(x) = L$$

which is read as "the **limit** of $f(x)$ as x approaches c is L."

TECHNOLOGY

Try using a graphing utility to determine the following limit.

$$\lim_{x \to 1} \frac{x^3 + 4x - 5}{x - 1}$$

You can do this by graphing

$$f(x) = \frac{x^3 + 4x - 5}{x - 1}$$

and zooming in near $x = 1$. From the graph, what does the limit appear to be?

Properties of Limits

Many times the limit of $f(x)$ as x approaches c is simply $f(c)$, as shown in Example 1. Whenever the limit of $f(x)$ as x approaches c is

$$\lim_{x \to c} f(x) = f(c) \qquad \text{Substitute } c \text{ for } x.$$

the limit can be evaluated by **direct substitution.** (In the next section, you will learn that a function that has this property is *continuous at c*.) It is important that you learn to recognize the types of functions that have this property. Some basic ones are given in the following list.

Properties of Limits

Let b and c be real numbers, and let n be a positive integer.

1. $\lim_{x \to c} b = b$
2. $\lim_{x \to c} x = c$
3. $\lim_{x \to c} x^n = c^n$
4. $\lim_{x \to c} \sqrt[n]{x} = \sqrt[n]{c}$

In Property 4, if n is even, then c must be positive.

By combining the properties of limits with the rules for operating with limits shown below, you can find limits for a wide variety of algebraic functions.

TECHNOLOGY

Symbolic computer algebra systems are capable of evaluating limits. Try using a computer algebra system to evaluate the limit given in Example 3.

Operations with Limits

Let b and c be real numbers, let n be a positive integer, and let f and g be functions with the following limits.

$$\lim_{x\to c} f(x) = L \quad \text{and} \quad \lim_{x\to c} g(x) = K$$

1. Scalar multiple: $\lim_{x\to c} [bf(x)] = bL$
2. Sum or difference: $\lim_{x\to c} [f(x) \pm g(x)] = L \pm K$
3. Product: $\lim_{x\to c} [f(x) \cdot g(x)] = LK$
4. Quotient: $\lim_{x\to c} \dfrac{f(x)}{g(x)} = \dfrac{L}{K}$, provided $K \neq 0$
5. Power: $\lim_{x\to c} [f(x)]^n = L^n$
6. Radical: $\lim_{x\to c} \sqrt[n]{f(x)} = \sqrt[n]{L}$

In Property 6, if n is even, then L must be positive.

DISCOVERY

Use a graphing utility to graph $y_1 = 1/x^2$. Does y_1 approach a limit as x approaches 0? Evaluate $y_1 = 1/x^2$ at several positive and negative values of x near 0 to confirm your answer. Does $\lim_{x\to 1} 1/x^2$ exist?

EXAMPLE 3 Finding the Limit of a Polynomial Function

Find the limit: $\lim_{x\to 2} (x^2 + 2x - 3)$.

$$\begin{aligned}
\lim_{x\to 2} (x^2 + 2x - 3) &= \lim_{x\to 2} x^2 + \lim_{x\to 2} 2x - \lim_{x\to 2} 3 && \text{Apply Property 2.}\\
&= 2^2 + 2(2) - 3 && \text{Use direct substitution.}\\
&= 4 + 4 - 3 && \text{Simplify.}\\
&= 5
\end{aligned}$$

TRY IT 3

Find the limit: $\lim_{x\to 1} (2x^2 - x + 4)$.

Example 3 is an illustration of the following important result, which states that the limit of a polynomial can be evaluated by direct substitution.

The Limit of a Polynomial Function

If p is a polynomial function and c is any real number, then

$$\lim_{x\to c} p(x) = p(c).$$

Techniques for Evaluating Limits

Many techniques for evaluating limits are based on the following important theorem. Basically, the theorem states that if two functions agree at all but a single point c, then they have identical limit behavior at $x = c$.

> **The Replacement Theorem**
>
> Let c be a real number and let $f(x) = g(x)$ for all $x \neq c$. If the limit of $g(x)$ exists as $x \to c$, then the limit of $f(x)$ also exists and
>
> $$\lim_{x \to c} f(x) = \lim_{x \to c} g(x).$$

To apply the Replacement Theorem, you can use a result from algebra which states that for a polynomial function p, $p(c) = 0$ if and only if $(x - c)$ is a factor of $p(x)$. This concept is demonstrated in Example 4.

EXAMPLE 4 Finding the Limit of a Function

Find the limit: $\lim_{x \to 1} \dfrac{x^3 - 1}{x - 1}$.

SOLUTION Note that the numerator and denominator are zero when $x = 1$. This implies that $x - 1$ is a factor of both, and you can divide out this like factor.

$$\frac{x^3 - 1}{x - 1} = \frac{(x - 1)(x^2 + x + 1)}{x - 1} \qquad \text{Factor numerator.}$$

$$= \frac{\cancel{(x - 1)}(x^2 + x + 1)}{\cancel{x - 1}} \qquad \text{Divide out like factor.}$$

$$= x^2 + x + 1, \quad x \neq 1 \qquad \text{Simplify.}$$

So, the rational function $(x^3 - 1)/(x - 1)$ and the polynomial function $x^2 + x + 1$ agree for all values of x other than $x = 1$, and you can apply the Replacement Theorem.

$$\lim_{x \to 1} \frac{x^3 - 1}{x - 1} = \lim_{x \to 1} (x^2 + x + 1) = 1^2 + 1 + 1 = 3$$

Figure 1.54 illustrates this result graphically. Note that the two graphs are identical except that the graph of g contains the point $(1, 3)$, whereas this point is missing on the graph of f. (In the graph of f in Figure 1.54, the missing point is denoted by an open dot.)

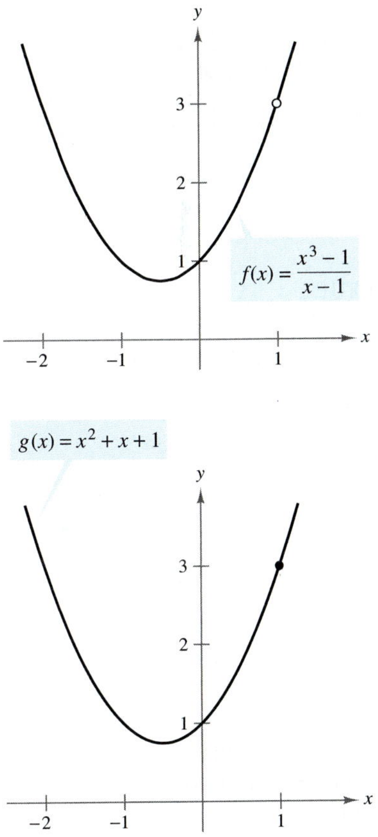

FIGURE 1.54

> **TRY IT 4**
>
> Find the limit: $\lim_{x \to 2} \dfrac{x^3 - 8}{x - 2}$.

The technique used to evaluate the limit in Example 4 is called the **dividing out** technique. This technique is further demonstrated in the next example.

DISCOVERY

Use a graphing utility to graph

$$y = \frac{x^2 + x - 6}{x + 3}.$$

Is the graph a line? Why or why not?

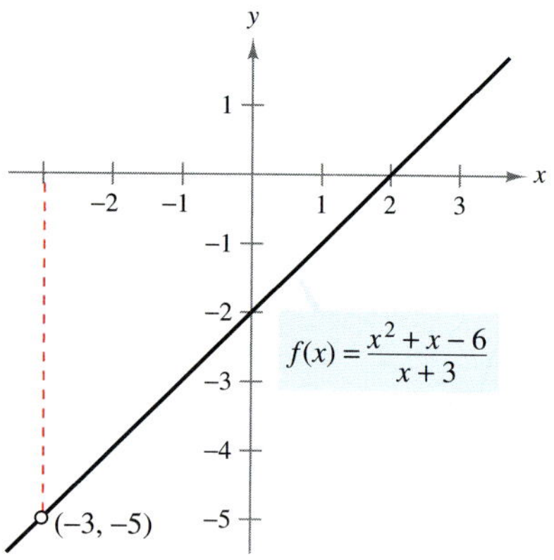

FIGURE 1.55 f is undefined when $x = -3$.

EXAMPLE 5 Using the Dividing Out Technique

Find the limit: $\lim_{x \to -3} \frac{x^2 + x - 6}{x + 3}$.

SOLUTION Direct substitution fails because both the numerator and the denominator are zero when $x = -3$.

$$\lim_{x \to -3} \frac{x^2 + x - 6}{x + 3} \qquad \begin{array}{l} \leftarrow \lim_{x \to -3} (x^2 + x - 6) = 0 \\ \leftarrow \lim_{x \to -3} (x + 3) = 0 \end{array}$$

However, because the limits of both the numerator and denominator are zero, you know that they have a *common factor* of $x + 3$. So, for all $x \neq -3$, you can divide out this factor to obtain the following.

$$\begin{aligned} \lim_{x \to -3} \frac{x^2 + x - 6}{x + 3} &= \lim_{x \to -3} \frac{(x - 2)(x + 3)}{x + 3} && \text{Factor numerator.} \\ &= \lim_{x \to -3} \frac{(x - 2)\cancel{(x + 3)}}{\cancel{x + 3}} && \text{Divide out like factor.} \\ &= \lim_{x \to -3} (x - 2) && \text{Simplify.} \\ &= -5 && \text{Direct substitution} \end{aligned}$$

This result is shown graphically in Figure 1.55. Note that the graph of f coincides with the graph of $g(x) = x - 2$, except that the graph of f has a hole at $(-3, -5)$.

TRY IT 5

Find the limit: $\lim_{x \to 3} \frac{x^2 + x - 12}{x - 3}$.

STUDY TIP

When you try to evaluate a limit and both the numerator and denominator are zero, remember that you must rewrite the fraction so that the new denominator does not have 0 as its limit. One way to do this is to divide out like factors, as shown in Example 5. Another technique is to rationalize the numerator, as shown in Example 6.

EXAMPLE 6 Finding a Limit of a Function

Find the limit: $\lim_{x \to 0} \frac{\sqrt{x + 1} - 1}{x}$.

SOLUTION Direct substitution fails because both the numerator and the denominator are zero when $x = 0$. In this case, you can rewrite the fraction by rationalizing the numerator.

$$\begin{aligned} \frac{\sqrt{x + 1} - 1}{x} &= \left(\frac{\sqrt{x + 1} - 1}{x}\right)\left(\frac{\sqrt{x + 1} + 1}{\sqrt{x + 1} + 1}\right) \\ &= \frac{(x + 1) - 1}{x(\sqrt{x + 1} + 1)} \\ &= \frac{\cancel{x}}{\cancel{x}(\sqrt{x + 1} + 1)} = \frac{1}{\sqrt{x + 1} + 1}, \quad x \neq 0 \end{aligned}$$

Now, using the Replacement Theorem, you can evaluate the limit as shown.

$$\lim_{x \to 0} \frac{\sqrt{x + 1} - 1}{x} = \lim_{x \to 0} \frac{1}{\sqrt{x + 1} + 1} = \frac{1}{1 + 1} = \frac{1}{2}$$

TRY IT 6

Find the limit: $\lim_{x \to 0} \frac{\sqrt{x + 4} - 2}{x}$.

One-Sided Limits

In Example 2(b), you saw that one way in which a limit can fail to exist is when a function approaches a different value from the left of c than it approaches from the right of c. This type of behavior can be described more concisely with the concept of a **one-sided limit.**

$$\lim_{x\to c^-} f(x) = L \qquad \text{Limit from the left}$$

$$\lim_{x\to c^+} f(x) = L \qquad \text{Limit from the right}$$

The first of these two limits is read as "the limit of $f(x)$ as x approaches c from the left is L." The second is read as "the limit of $f(x)$ as x approaches c from the right is L."

EXAMPLE 7 **Finding One-Sided Limits**

Find the limit as $x \to 0$ from the left and the limit as $x \to 0$ from the right for the function

$$f(x) = \frac{|2x|}{x}.$$

SOLUTION From the graph of f, shown in Figure 1.56, you can see that $f(x) = -2$ for all $x < 0$. So, the limit from the left is

$$\lim_{x\to 0^-} \frac{|2x|}{x} = -2. \qquad \text{Limit from the left}$$

Because $f(x) = 2$ for all $x > 0$, the limit from the right is

$$\lim_{x\to 0^+} \frac{|2x|}{x} = 2. \qquad \text{Limit from the right}$$

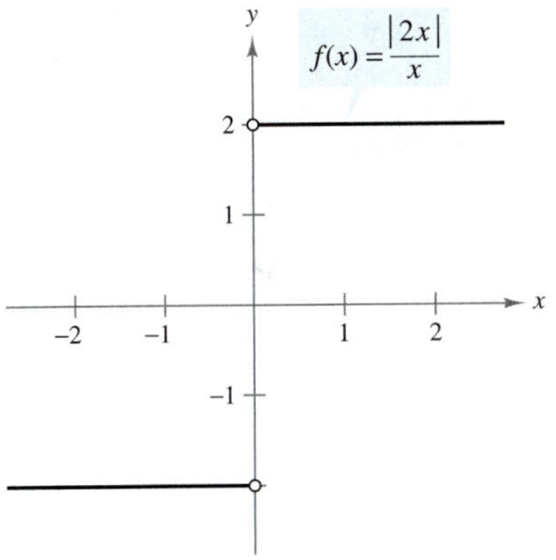

FIGURE 1.56

TRY IT 7

Find each limit. (a) $\lim_{x\to 2^-} \frac{|x-2|}{x-2}$ (b) $\lim_{x\to 2^+} \frac{|x-2|}{x-2}$

TECHNOLOGY

On most graphing utilities, the absolute value function is denoted by *abs*. You can verify the result in Example 7 by graphing

$$y = \frac{\text{abs}(2x)}{x}$$

in the viewing window $-3 \le x \le 3$ and $-3 \le y \le 3$.

In Example 7, note that the function approaches different limits from the left and from the right. In such cases, the limit of $f(x)$ as $x \to c$ does not exist. For the limit of a function to exist as $x \to c$, *both* one-sided limits must exist and must be equal.

Existence of a Limit

If f is a function and c and L are real numbers, then

$$\lim_{x\to c} f(x) = L$$

if and only if both the left and right limits are equal to L.

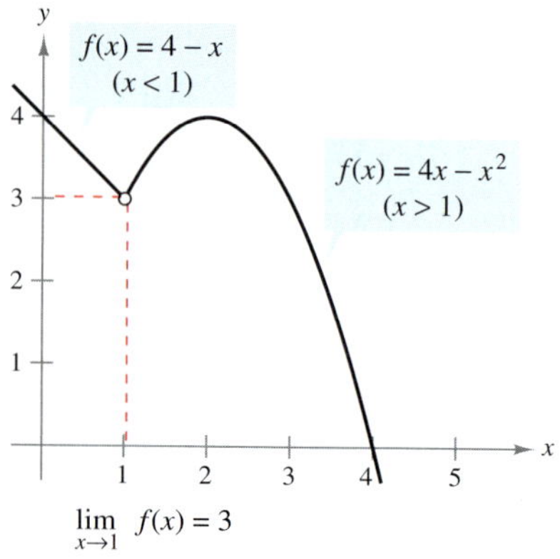

FIGURE 1.57

TRY IT 8

Find the limit of $f(x)$ as x approaches 0.

$$f(x) = \begin{cases} x^2 + 1, & x < 0 \\ 2x + 1, & x > 0 \end{cases}$$

EXAMPLE 8 Finding One-Sided Limits

Find the limit of $f(x)$ as x approaches 1.

$$f(x) = \begin{cases} 4 - x, & x < 1 \\ 4x - x^2, & x > 1 \end{cases}$$

SOLUTION Remember that you are concerned about the value of f near $x = 1$ rather than at $x = 1$. So, for $x < 1$, $f(x)$ is given by $4 - x$, and you can use direct substitution to obtain

$$\lim_{x\to 1^-} f(x) = \lim_{x\to 1^-} (4 - x)$$
$$= 4 - 1 = 3.$$

For $x > 1$, $f(x)$ is given by $4x - x^2$, and you can use direct substitution to obtain

$$\lim_{x\to 1^+} f(x) = \lim_{x\to 1^+} (4x - x^2)$$
$$= 4(1) - 1^2 = 4 - 1 = 3.$$

Because both one-sided limits exist and are equal to 3, it follows that

$$\lim_{x\to 1} f(x) = 3.$$

The graph in Figure 1.57 confirms this conclusion.

EXAMPLE 9 Comparing One-Sided Limits

An overnight delivery service charges \$8 for the first pound and \$2 for each additional pound. Let x represent the weight of a parcel and let $f(x)$ represent the shipping cost.

$$f(x) = \begin{cases} 8, & 0 < x \le 1 \\ 10, & 1 < x \le 2 \\ 12, & 2 < x \le 3 \end{cases}$$

Show that the limit of $f(x)$ as $x \to 2$ does not exist.

SOLUTION The graph of f is shown in Figure 1.58. The limit of $f(x)$ as x approaches 2 from the left is

$$\lim_{x\to 2^-} f(x) = 10$$

whereas the limit of $f(x)$ as x approaches 2 from the right is

$$\lim_{x\to 2^+} f(x) = 12.$$

Because these one-sided limits are not equal, the limit of $f(x)$ as $x \to 2$ does not exist.

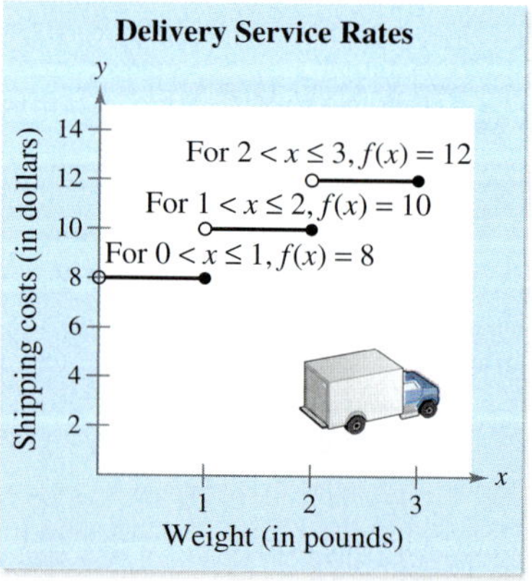

FIGURE 1.58

TRY IT 9

Show that the limit of $f(x)$ as $x \to 1$ does not exist in Example 9.

Unbounded Behavior

Example 9 shows a limit that fails to exist because the limits from the left and right differ. Another important way in which a limit can fail to exist is when $f(x)$ increases or decreases without bound as x approaches c.

EXAMPLE 10 An Unbounded Function

Find the limit (if possible).

$$\lim_{x \to 2} \frac{3}{x-2}$$

SOLUTION From Figure 1.59, you can see that $f(x)$ decreases without bound as x approaches 2 from the left and $f(x)$ increases without bound as x approaches 2 from the right. Symbolically, you can write this as

$$\lim_{x \to 2^-} \frac{3}{x-2} = -\infty$$

and

$$\lim_{x \to 2^+} \frac{3}{x-2} = \infty.$$

Because f is unbounded as x approaches 2, the limit does not exist.

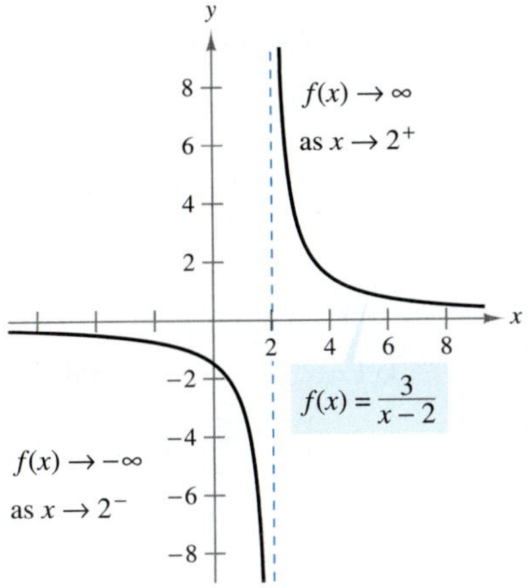

FIGURE 1.59

TRY IT 10

Find the limit (if possible): $\lim_{x \to -2} \dfrac{5}{x+2}$.

STUDY TIP

The equal sign in the statement $\lim_{x \to c^+} f(x) = \infty$ does not mean that the limit exists. On the contrary, it tells you how the limit *fails to exist* by denoting the unbounded behavior of $f(x)$ as x approaches c.

TAKE ANOTHER LOOK

Evaluating a Limit

Consider the limit from Example 6.

$$\lim_{x \to 0} \frac{\sqrt{x+1} - 1}{x}$$

a. Approximate the limit graphically, using a graphing utility.

b. Approximate the limit numerically, by constructing a table.

c. Find the limit analytically, using the Replacement Theorem, as shown in Example 6.

Of the three methods, which do you prefer? Explain your reasoning.

PREREQUISITE REVIEW 1.5

The following warm-up exercises involve skills that were covered in earlier sections. You will use these skills in the exercise set for this section.

In Exercises 1–4, evaluate the expression and simplify.

1. $f(x) = x^2 - 3x + 3$

(a) $f(-1)$ (b) $f(c)$ (c) $f(x + h)$

2. $f(x) = \begin{cases} 2x - 2, & x < 1 \\ 3x + 1, & x \geq 1 \end{cases}$

(a) $f(-1)$ (b) $f(3)$ (c) $f(t^2 + 1)$

3. $f(x) = x^2 - 2x + 2$ $\quad \dfrac{f(1 + h) - f(1)}{h}$

4. $f(x) = 4x$ $\quad \dfrac{f(2 + h) - f(2)}{h}$

In Exercises 5–8, find the domain and range of the function and sketch its graph.

5. $h(x) = -\dfrac{5}{x}$

6. $g(x) = \sqrt{25 - x^2}$

7. $f(x) = |x - 3|$

8. $f(x) = \dfrac{|x|}{x}$

In Exercises 9 and 10, determine whether y is a function of x.

9. $9x^2 + 4y^2 = 49$

10. $2x^2y + 8x = 7y$

EXERCISES 1.5

In Exercises 1–8, complete the table and use the result to estimate the limit. Use a graphing utility to graph the function to confirm your result.

1. $\lim_{x \to 2} (5x + 4)$

x	1.9	1.99	1.999	2	2.001	2.01	2.1
$f(x)$				?			

2. $\lim_{x \to 2} (x^2 - 3x + 1)$

x	1.9	1.99	1.999	2	2.001	2.01	2.1
$f(x)$				?			

3. $\lim_{x \to 2} \dfrac{x - 2}{x^2 - 4}$

x	1.9	1.99	1.999	2	2.001	2.01	2.1
$f(x)$				?			

4. $\lim_{x \to 2} \dfrac{x^5 - 32}{x - 2}$

x	1.9	1.99	1.999	2	2.001	2.01	2.1
$f(x)$				?			

5. $\lim_{x \to 0} \dfrac{\sqrt{x + 3} - \sqrt{3}}{x}$

x	-0.1	-0.01	-0.001	0	0.001	0.01	0.1
$f(x)$				?			

6. $\lim_{x \to 0} \dfrac{\sqrt{x + 2} - \sqrt{2}}{x}$

x	-0.1	-0.01	-0.001	0	0.001	0.01	0.1
$f(x)$				?			

7. $\lim_{x\to 0^-} \dfrac{\dfrac{1}{x+4} - \dfrac{1}{4}}{x}$

x	-0.5	-0.1	-0.01	-0.001	0
$f(x)$					?

8. $\lim_{x\to 0^+} \dfrac{\dfrac{1}{2+x} - \dfrac{1}{2}}{2x}$

x	0.5	0.1	0.01	0.001	0
$f(x)$					?

In Exercises 9–12, use the graph to find the limit (if it exists).

9.

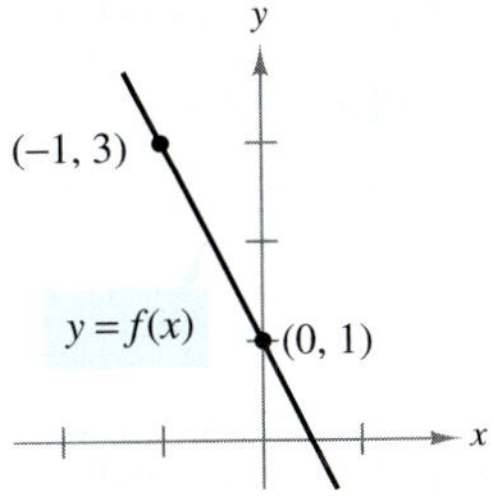

(a) $\lim_{x\to 0} f(x)$

(b) $\lim_{x\to -1} f(x)$

10.

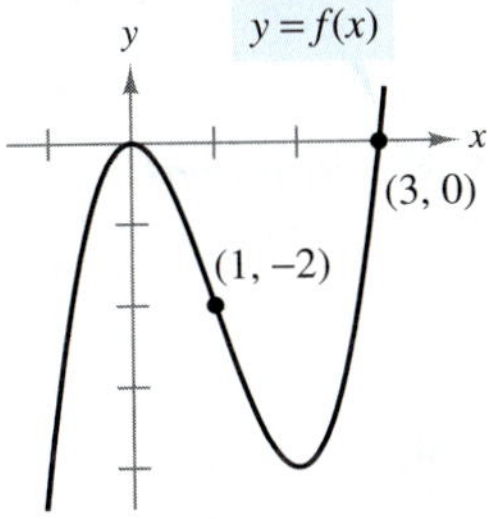

(a) $\lim_{x\to 1} f(x)$

(b) $\lim_{x\to 3} f(x)$

11.

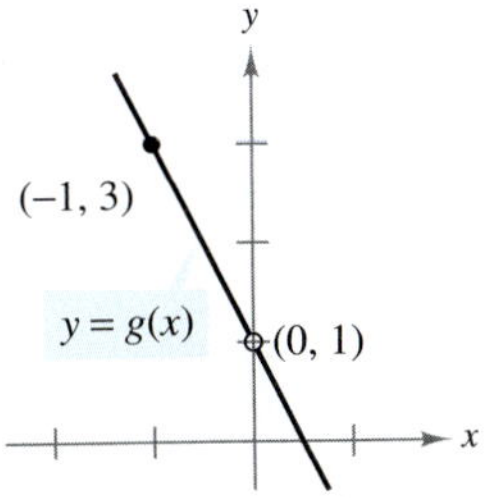

(a) $\lim_{x\to 0} g(x)$

(b) $\lim_{x\to -1} g(x)$

12.

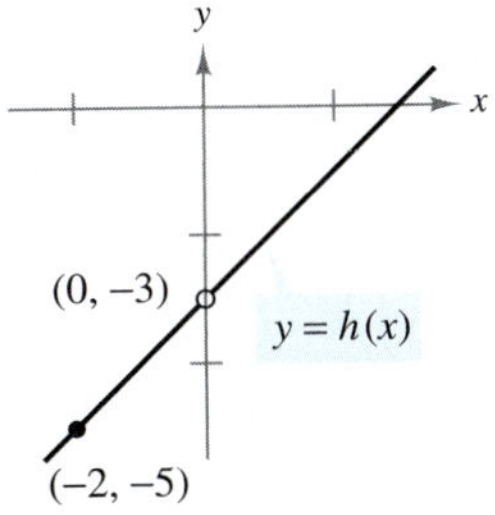

(a) $\lim_{x\to -2} h(x)$

(b) $\lim_{x\to 0} h(x)$

In Exercises 13 and 14, find the limit of (a) $f(x) + g(x)$, (b) $f(x)g(x)$, and (c) $f(x)/g(x)$ as x approaches c.

13. $\lim_{x\to c} f(x) = 3$

$\lim_{x\to c} g(x) = 9$

14. $\lim_{x\to c} f(x) = \frac{3}{2}$

$\lim_{x\to c} g(x) = \frac{1}{2}$

In Exercises 15 and 16, find the limit of (a) $\sqrt{f(x)}$, (b) $[3f(x)]$, and (c) $[f(x)]^2$ as x approaches c.

15. $\lim_{x\to c} f(x) = 16$

16. $\lim_{x\to c} f(x) = 9$

In Exercises 17–22, use the graph to find the limit (if it exists).

(a) $\lim_{x\to c^+} f(x)$

(b) $\lim_{x\to c^-} f(x)$

(c) $\lim_{x\to c} f(x)$

17.

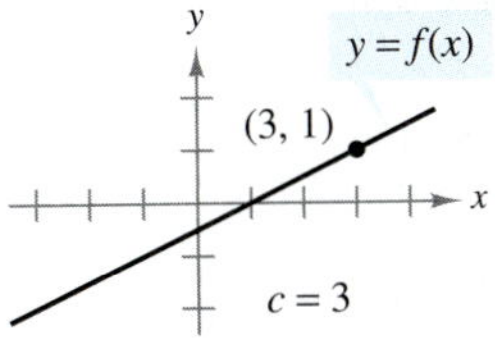

18.

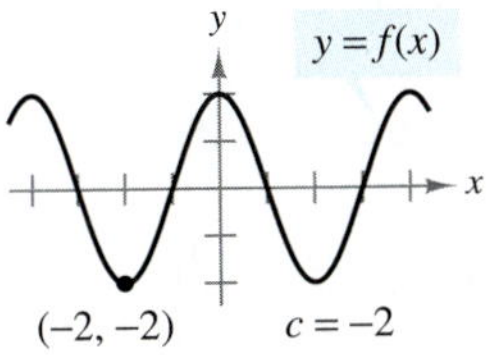

19.

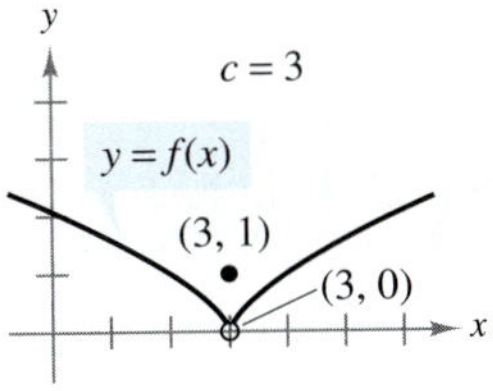

20.

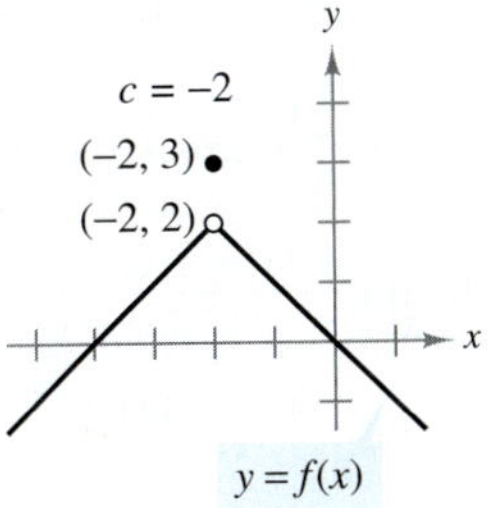

21.

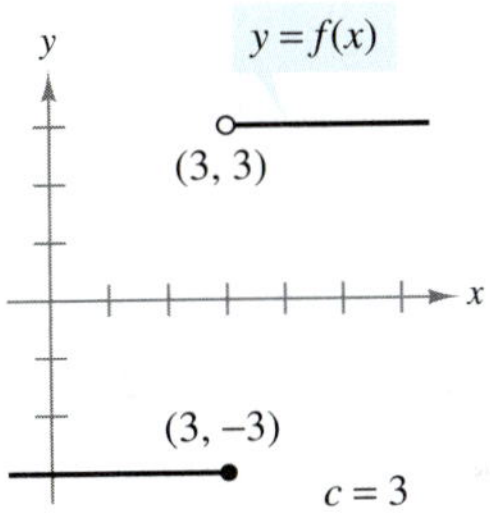

22.

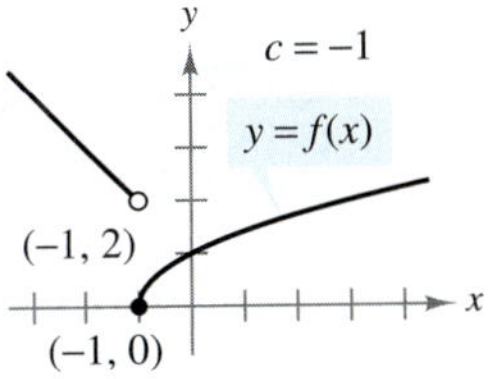

In Exercises 23–40, find the limit.

23. $\lim_{x\to 2} x^4$

24. $\lim_{x\to -2} x^3$

25. $\lim_{x\to -3} (3x + 2)$

26. $\lim_{x\to 0} (2x - 3)$

27. $\lim_{x\to 1} (1 - x^2)$

28. $\lim_{x\to 2} (-x^2 + x - 2)$

29. $\lim_{x\to 3} \sqrt{x+1}$

30. $\lim_{x\to 4} \sqrt[3]{x+4}$

31. $\lim_{x\to -3} \dfrac{2}{x+2}$

32. $\lim_{x\to -2} \dfrac{3x+1}{2-x}$

33. $\lim_{x\to -2} \dfrac{x^2-1}{2x}$

34. $\lim_{x\to -1} \dfrac{4x-5}{3-x}$

35. $\lim_{x\to 7} \dfrac{5x}{x+2}$

36. $\lim_{x\to 3} \dfrac{\sqrt{x+1}}{x-4}$

37. $\lim_{x\to 3} \dfrac{\sqrt{x+1}-1}{x}$

38. $\lim_{x\to 5} \dfrac{\sqrt{x+4}-2}{x}$

39. $\lim_{x\to 1} \dfrac{\dfrac{1}{x+4} - \dfrac{1}{4}}{x}$

40. $\lim_{x\to 2} \dfrac{\dfrac{1}{x+2} - \dfrac{1}{2}}{x}$

In Exercises 41–58, find the limit (if it exists).

41. $\lim_{x\to -1} \dfrac{x^2-1}{x+1}$

42. $\lim_{x\to -1} \dfrac{2x^2-x-3}{x+1}$

43. $\lim_{x\to 2} \dfrac{x-2}{x^2-4x+4}$

44. $\lim_{x\to 2} \dfrac{2-x}{x^2-4}$

45. $\lim_{t\to 5} \dfrac{t-5}{t^2-25}$

46. $\lim_{t\to 1} \dfrac{t^2+t-2}{t^2-1}$

47. $\lim_{x\to -2} \dfrac{x^3+8}{x+2}$

48. $\lim_{x\to 1} \dfrac{x^3-1}{x-1}$

49. $\lim_{x\to -2} \dfrac{|x+2|}{x+2}$

50. $\lim_{x\to 2} \dfrac{|x-2|}{x-2}$

51. $\lim_{x\to 3} f(x)$, where $f(x) = \begin{cases} \frac{1}{3}x - 2, & x \le 3 \\ -2x + 5, & x > 3 \end{cases}$

52. $\lim_{s\to 1} f(s)$, where $f(s) = \begin{cases} s, & s \le 1 \\ 1 - s, & s > 1 \end{cases}$

53. $\lim_{\Delta x\to 0} \dfrac{2(x+\Delta x) - 2x}{\Delta x}$

54. $\lim_{\Delta x\to 0} \dfrac{4(x+\Delta x) - 5 - (4x-5)}{\Delta x}$

55. $\lim_{\Delta x\to 0} \dfrac{\sqrt{x+2+\Delta x} - \sqrt{x+2}}{\Delta x}$

56. $\lim_{\Delta x\to 0} \dfrac{\sqrt{x+\Delta x} - \sqrt{x}}{\Delta x}$

57. $\lim_{\Delta t\to 0} \dfrac{(t+\Delta t)^2 - 5(t+\Delta t) - (t^2-5t)}{\Delta t}$

58. $\lim_{\Delta t\to 0} \dfrac{(t+\Delta t)^2 - 4(t+\Delta t) + 2 - (t^2-4t+2)}{\Delta t}$

Graphical, Numerical, and Analytic Analysis In Exercises 59–62, use a graphing utility to graph the function and estimate the limit. Use a table to reinforce your conclusion. Then find the limit by analytic methods.

59. $\lim_{x\to 1^-} \dfrac{2}{x^2-1}$

60. $\lim_{x\to 1^+} \dfrac{5}{1-x}$

61. $\lim_{x\to -2^-} \dfrac{1}{x+2}$

62. $\lim_{x\to 0^-} \dfrac{x+1}{x}$

In Exercises 63–66, use a graphing utility to estimate the limit (if it exists).

63. $\lim_{x\to 2} \dfrac{x^2-5x+6}{x^2-4x+4}$

64. $\lim_{x\to 1} \dfrac{x^2+6x-7}{x^3-x^2+2x-2}$

65. $\lim_{x\to -4} \dfrac{x^3+4x^2+x+4}{2x^2+7x-4}$

66. $\lim_{x\to -2} \dfrac{4x^3+7x^2+x+6}{3x^2-x-14}$

67. The limit of

$$f(x) = (1+x)^{1/x}$$

is a natural base for many business applications, as you will see in Section 4.2.

$$\lim_{x\to 0} (1+x)^{1/x} = e \approx 2.718$$

(a) Show the reasonableness of this limit by completing the table.

x	-0.01	-0.001	-0.0001	0	0.0001	0.001	0.01
$f(x)$							

(b) Use a graphing utility to graph f and to confirm the answer in part (a).

(c) Find the domain and range of the function.

68. Find $\lim_{x\to 0} f(x)$, given

$$4 - x^2 \le f(x) \le 4 + x^2, \text{ for all } x.$$

69. ***Environment*** The cost (in dollars) of removing $p\%$ of the pollutants from the water in a small lake is given by

$$C = \frac{25{,}000p}{100-p}, \quad 0 \le p < 100$$

where C is the cost and p is the percent of pollutants.

(a) Find the cost of removing 50% of the pollutants.

(b) What percent of the pollutants can be removed for \$100,000?

(c) Evaluate $\lim_{p\to 100^-} C$. Explain your results.

70. ***Compound Interest*** You deposit \$1000 in an account that is compounded quarterly at an annual rate of r (in decimal form). The balance A after 10 years is

$$A = 1000\left(1 + \frac{r}{4}\right)^{40}.$$

Does the limit of A exist as the interest rate approaches 6%? If so, what is the limit?

71. ***Compound Interest*** Consider a certificate of deposit that pays 10% (annual percentage rate) on an initial deposit of \$500. The balance A after 10 years is

$$A = 500(1 + 0.1x)^{10/x}$$

where x is the length of the compounding period (in years).

(a) Use a graphing utility to graph A, where $0 \le x \le 1$.

(b) Use the *zoom* and *trace* features to estimate the balance for quarterly compounding and daily compounding.

(c) Use the *zoom* and *trace* features to estimate

$$\lim_{x\to 0^+} A.$$

What do you think this limit represents? Explain your reasoning.

1.6 CONTINUITY

- Determine the continuity of functions.
- Determine the continuity of functions on a closed interval.
- Use the greatest integer function to model and solve real-life problems.
- Use compound interest models to solve real-life problems.

Continuity

In mathematics, the term "continuous" has much the same meaning as it does in everyday use. To say that a function is continuous at $x = c$ means that there is no interruption in the graph of f at c. The graph of f is unbroken at c, and there are no holes, jumps, or gaps. As simple as this concept may seem, its precise definition eluded mathematicians for many years. In fact, it was not until the early 1800s that a precise definition was finally developed.

Before looking at this definition, consider the function whose graph is shown in Figure 1.60. This figure identifies three values of x at which the function f is not continuous.

1. At $x = c_1$, $f(c_1)$ is not defined.
2. At $x = c_2$, $\lim\limits_{x \to c_2} f(x)$ does not exist.
3. At $x = c_3$, $f(c_3) \neq \lim\limits_{x \to c_3} f(x)$.

At all other points in the interval (a, b), the graph of f is uninterrupted, which implies that the function f is continuous at all other points in the interval (a, b).

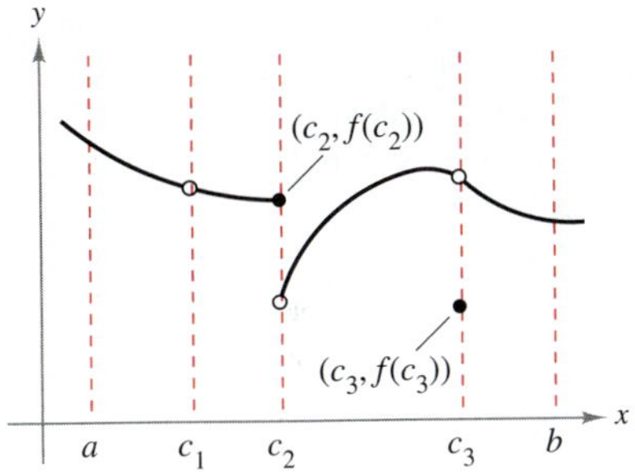

FIGURE 1.60 f is not continuous when $x = c_1, c_2, c_3$.

Definition of Continuity

Let c be a number in the interval (a, b), and let f be a function whose domain contains the interval (a, b). The function f is **continuous at the point c** if the following conditions are true.

1. $f(c)$ is defined.
2. $\lim\limits_{x \to c} f(x)$ exists.
3. $\lim\limits_{x \to c} f(x) = f(c)$.

If f is continuous at every point in the interval (a, b), then it is **continuous on an open interval (a, b).**

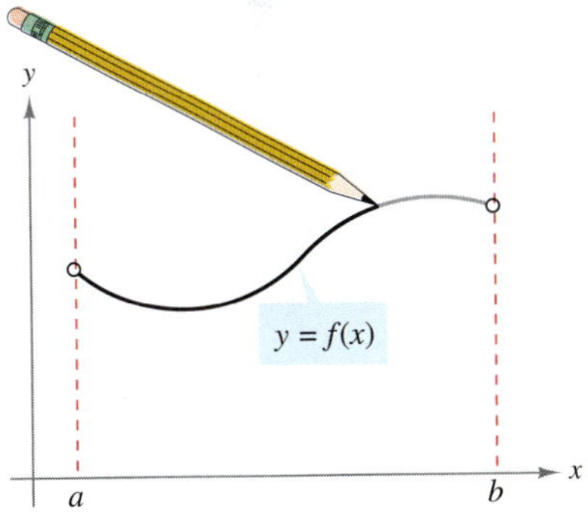

FIGURE 1.61 On the interval (a, b), the graph of f can be traced with a pencil.

Roughly, you can say that a function is continuous on an interval if its graph on the interval can be traced using a pencil and paper without lifting the pencil from the paper, as shown in Figure 1.61.

TECHNOLOGY

Most graphing utilities can draw graphs in two different modes: *connected mode* and *dot mode*. The *connected mode* works well as long as the function is continuous on the entire interval represented by the viewing window. If, however, the function is not continuous at one or more x-values in the viewing window, then the *connected mode* may try to "connect" parts of the graphs that should not be connected. For instance, try graphing the function $y_1 = (x + 3)/(x - 2)$ on the viewing window $-8 \le x \le 8$ and $-6 \le y \le 6$. Do you notice any problems?

In Section 1.5, you studied several types of functions that meet the three conditions for continuity. Specifically, if *direct substitution* can be used to evaluate the limit of a function at c, then the function is continuous at c. Two types of functions that have this property are polynomial functions and rational functions.

Continuity of Polynomial and Rational Functions

1. A polynomial function is continuous at every real number.
2. A rational function is continuous at every number in its domain.

EXAMPLE 1 **Determining Continuity of a Polynomial Function**

Discuss the continuity of each function.

(a) $f(x) = x^2 - 2x + 3$

(b) $f(x) = x^3 - x$

SOLUTION Each of these functions is a *polynomial function*. So, each is continuous on the entire real line, as indicated in Figure 1.62.

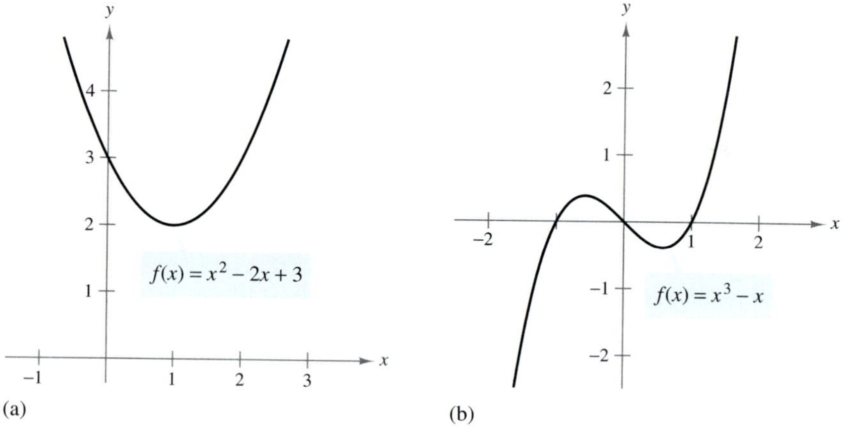

FIGURE 1.62 Both functions are continuous on $(-\infty, \infty)$.

STUDY TIP

A graphing utility can give misleading information about the continuity of a function. Graph the function

$$f(x) = \frac{x^3 + 8}{x + 2}$$

in the standard viewing window. Does the graph appear to be continuous? For what values of x is the function continuous?

TRY IT 1

Discuss the continuity of each function.

(a) $f(x) = x^2 + x + 1$ (b) $f(x) = x^3 + x$

Polynomial functions are one of the most important types of functions used in calculus. Be sure you see from Example 1 that the graph of a polynomial function is continuous on the entire real line, and therefore has no holes, jumps, or gaps. Rational functions, on the other hand, need not be continuous on the entire real line, as shown in Example 2.

EXAMPLE 2 Determining Continuity of a Rational Function

Discuss the continuity of each function.

(a) $f(x) = 1/x$ (b) $f(x) = (x^2 - 1)/(x - 1)$ (c) $f(x) = 1/(x^2 + 1)$

SOLUTION Each of these functions is a rational function and is therefore continuous at every number in its domain.

(a) The domain of $f(x) = 1/x$ consists of all real numbers except $x = 0$. So, this function is continuous on the intervals $(-\infty, 0)$ and $(0, \infty)$. [See Figure 1.63(a).]

(b) The domain of $f(x) = (x^2 - 1)/(x - 1)$ consists of all real numbers except $x = 1$. So, this function is continuous on the intervals $(-\infty, 1)$ and $(1, \infty)$. [See Figure 1.63(b).]

(c) The domain of $f(x) = 1/(x^2 + 1)$ consists of all real numbers. So, this function is continuous on the entire real line. [See Figure 1.63(c).]

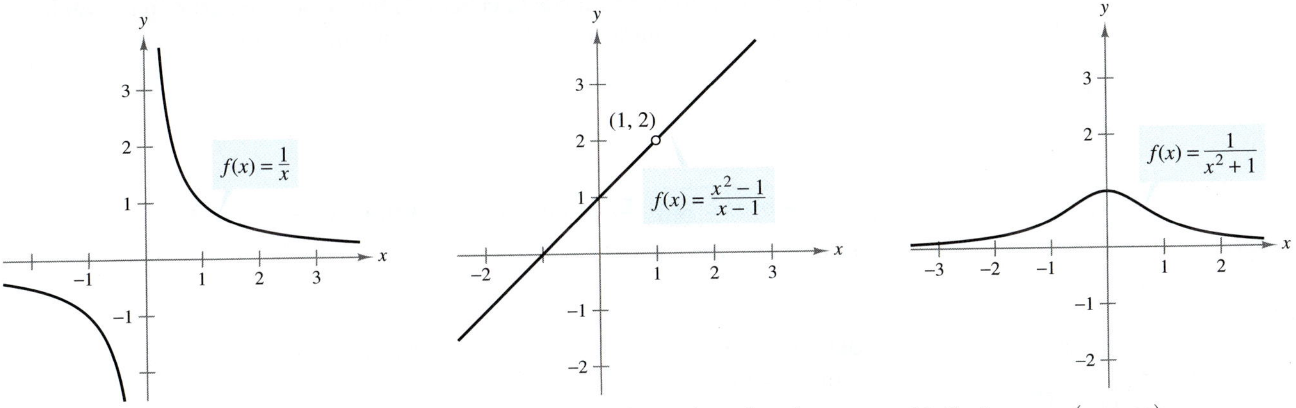

(a) Continuous on $(-\infty, 0)$ and $(0, \infty)$ (b) Continuous on $(-\infty, 1)$ and $(1, \infty)$ (c) Continuous on $(-\infty, \infty)$

FIGURE 1.63

TRY IT 2

Discuss the continuity of each function.

(a) $f(x) = \dfrac{1}{x - 1}$ (b) $f(x) = \dfrac{x^2 - 4}{x - 2}$ (c) $f(x) = \dfrac{1}{x^2 + 2}$

Consider an open interval I that contains a real number c. If a function f is defined on I (except possibly at c), and f is not continuous at c, then f is said to have a **discontinuity** at c. Discontinuities fall into two categories: **removable** and **nonremovable.** A discontinuity at c is called removable if f can be made continuous by appropriately defining (or redefining) $f(c)$. For instance, the function in Example 2(b) has a removable discontinuity at $(1, 2)$. To remove the discontinuity, all you need to do is redefine the function so that $f(1) = 2$.

A discontinuity at $x = c$ is nonremovable if the function cannot be made continuous at $x = c$ by defining or redefining the function at $x = c$. For instance, the function in Example 2(a) has a nonremovable discontinuity at $x = 0$.

Continuity on a Closed Interval

The intervals discussed in Examples 1 and 2 are open. To discuss continuity on a closed interval, you can use the concept of one-sided limits, as defined in Section 1.5.

Definition of Continuity on a Closed Interval

Let f be defined on a closed interval $[a, b]$. If f is continuous on the open interval (a, b) and

$$\lim_{x\to a^+} f(x) = f(a) \qquad \text{and} \qquad \lim_{x\to b^-} f(x) = f(b)$$

then f is **continuous on the closed interval** $[a, b]$. Moreover, f is **continuous from the right** at a and **continuous from the left** at b.

Similar definitions can be made to cover continuity on intervals of the form $(a, b]$ and $[a, b)$, or on infinite intervals. For example, the function

$$f(x) = \sqrt{x}$$

is continuous on the infinite interval $[0, \infty)$.

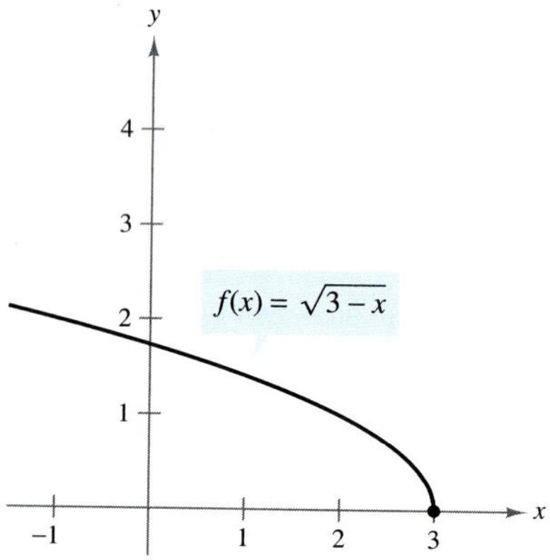

FIGURE 1.64

EXAMPLE 3 Examining Continuity at an Endpoint

Discuss the continuity of

$$f(x) = \sqrt{3 - x}.$$

SOLUTION Notice that the domain of f is the set $(-\infty, 3]$. Moreover, f is continuous from the left at $x = 3$ because

$$\begin{aligned}\lim_{x\to 3^-} f(x) &= \lim_{x\to 3^-} \sqrt{3 - x}\\ &= 0\\ &= f(3).\end{aligned}$$

For all $x < 3$, the function f satisfies the three conditions for continuity. So, you can conclude that f is continuous on the interval $(-\infty, 3]$, as shown in Figure 1.64.

TRY IT 3

Discuss the continuity of $f(x) = \sqrt{x - 2}$.

STUDY TIP

When working with radical functions of the form

$$f(x) = \sqrt{g(x)}$$

remember that the domain of f coincides with the solution of $g(x) \geq 0$.

EXAMPLE 4 Examining Continuity on a Closed Interval

Discuss the continuity of $g(x) = \begin{cases} 5 - x, & -1 \le x \le 2 \\ x^2 - 1, & 2 < x \le 3 \end{cases}$.

SOLUTION The polynomial functions $5 - x$ and $x^2 - 1$ are continuous on the intervals $[-1, 2]$ and $(2, 3]$, respectively. So, to conclude that g is continuous on the entire interval $[-1, 3]$, you only need to check the behavior of g when $x = 2$. You can do this by taking the one-sided limits when $x = 2$.

$$\lim_{x \to 2^-} g(x) = \lim_{x \to 2^-} (5 - x) = 3$$ Limit from the left

and

$$\lim_{x \to 2^+} g(x) = \lim_{x \to 2^+} (x^2 - 1) = 3$$ Limit from the right

Because these two limits are equal,

$$\lim_{x \to 2} g(x) = g(2) = 3.$$

So, g is continuous at $x = 2$ and, consequently, it is continuous on the entire interval $[-1, 3]$. The graph of g is shown in Figure 1.65.

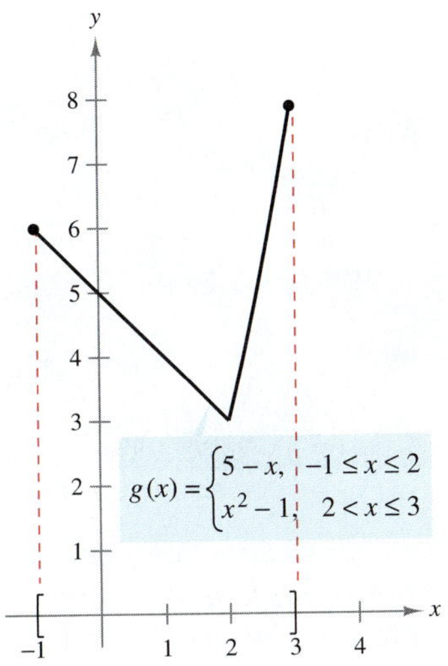

FIGURE 1.65

TRY IT 4

Discuss the continuity of $f(x) = \begin{cases} x + 2, & -1 \le x < 3 \\ 14 - x^2, & 3 \le x \le 5 \end{cases}$.

The Greatest Integer Function

Many functions that are used in business applications are **step functions.** For instance, the function in Example 9 in Section 1.5 is a step function. The **greatest integer function** is another example of a step function. This function is denoted by

$[\![x]\!]$ = greatest integer less than or equal to x.

For example,

$[\![-2.1]\!]$ = greatest integer less than or equal to $-2.1 = -3$

$[\![-2]\!]$ = greatest integer less than or equal to $-2 = -2$

$[\![1.5]\!]$ = greatest integer less than or equal to $1.5 = 1$.

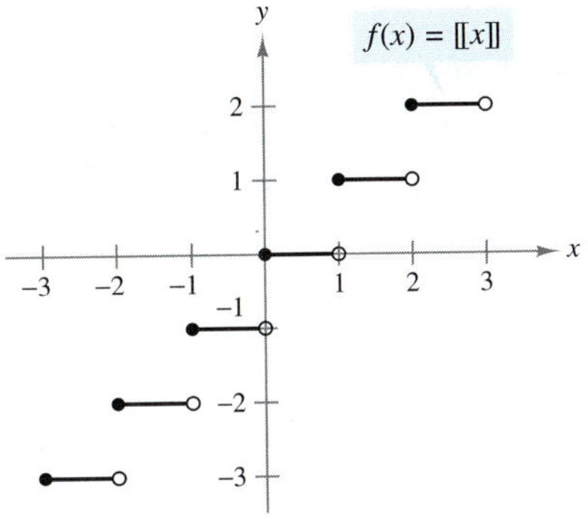

FIGURE 1.66 Greatest Integer Function

Note that the graph of the greatest integer function (Figure 1.66) jumps up one unit at each integer. This implies that the function is not continuous at each integer.

In real-life applications, the domain of the greatest integer function is often restricted to nonnegative values of x. In such cases this function serves the purpose of **truncating** the decimal portion of x. For example, 1.345 is truncated to 1 and 3.57 is truncated to 3. That is,

$$[\![1.345]\!] = 1 \quad \text{and} \quad [\![3.57]\!] = 3.$$

TECHNOLOGY

Use a graphing utility to calculate the following.

(a) $[\![3.5]\!]$ (b) $[\![-3.5]\!]$ (c) $[\![0]\!]$

EXAMPLE 5 **Modeling a Cost Function**

A bookbinding company produces 10,000 books in an eight-hour shift. The fixed cost *per shift* amounts to \$5000, and the unit cost per book is \$3. Using the greatest integer function, you can write the cost of producing x books as

$$C = 5000\left(1 + \left[\!\left[\frac{x-1}{10{,}000}\right]\!\right]\right) + 3x.$$

Sketch the graph of this cost function.

SOLUTION Note that during the first eight-hour shift

$$\left[\!\left[\frac{x-1}{10{,}000}\right]\!\right] = 0, \quad 1 \le x \le 10{,}000$$

which implies

$$C = 5000\left(1 + \left[\!\left[\frac{x-1}{10{,}000}\right]\!\right]\right) + 3x = 5000 + 3x.$$

During the second eight-hour shift

$$\left[\!\left[\frac{x-1}{10{,}000}\right]\!\right] = 1, \quad 10{,}001 \le x \le 20{,}000$$

which implies

$$C = 5000\left(1 + \left[\!\left[\frac{x-1}{10{,}000}\right]\!\right]\right) + 3x$$
$$= 10{,}000 + 3x.$$

The graph of C is shown in Figure 1.67. Note the graph's discontinuities.

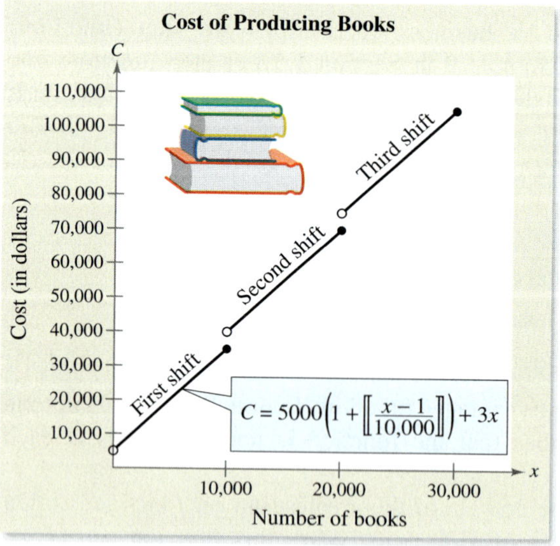

FIGURE 1.67

R.R. Donnelley & Sons

R. R. Donnelley & Sons Company is one of the world's largest commercial printers. It prints and binds a major share of the national publications in the United States, including Time, Newsweek, and TV Guide. In 2004, the printing and binding of books accounted for 15% of Donnelley's business. The other part of its business came from the printing and binding of catalogs, magazines, directories, and other types of publications. In the photo, an employee is aligning a roll of paper on a printing press in one of Donnelley's plants.

TRY IT 5

Use a graphing utility to graph the cost function in Example 5.

TECHNOLOGY

Step Functions and Compound Functions

To graph a step function or compound function with a graphing utility, you must be familiar with the utility's programming language. For instance, different graphing utilities have different "integer truncation" functions. One is IPart(x), and it yields the truncated integer part of x. For example, IPart$(-1.2) = -1$ and IPart$(3.4) = 3$. The other function is Int(x), which is the greatest integer function. The graphs of these two functions are shown below. When graphing a step function, you should set your graphing utility to *dot mode*.

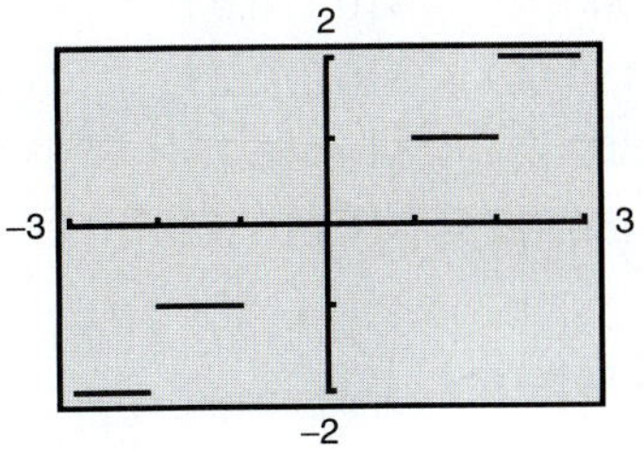

Graph of $f(x) = \text{IPart}(x)$

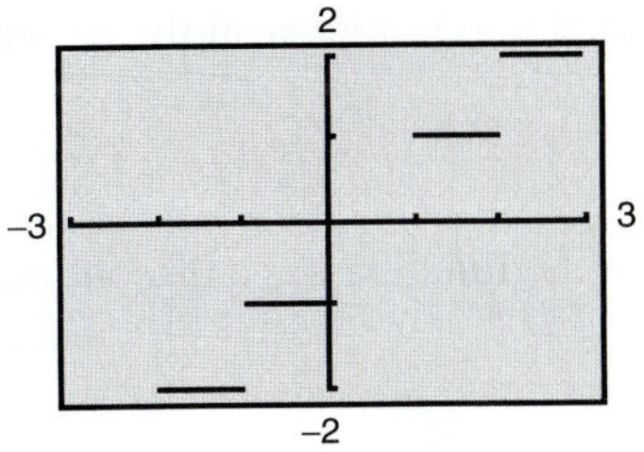

Graph of $f(x) = \text{Int}(x)$

On some graphing utilities, you can graph a piecewise-defined function such as

$$f(x) = \begin{cases} x^2 - 4, & x \le 2 \\ -x + 2, & 2 < x \end{cases}.$$

The graph of this function is shown below.

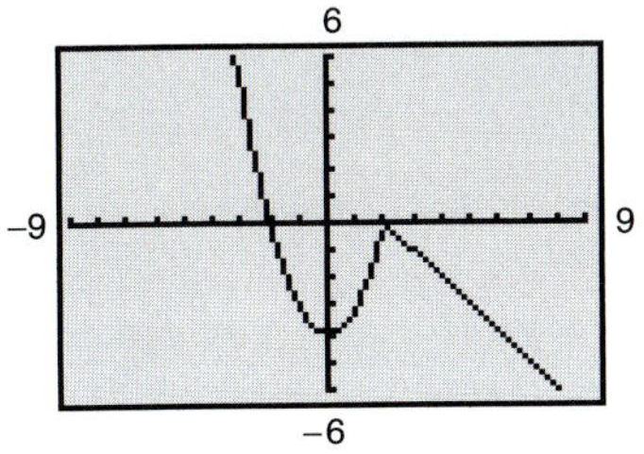

Consult the user's guide for your graphing utility for specific keystrokes you can use to graph these functions.

TECHNOLOGY

You can use a spreadsheet or the *table* feature of a graphing utility to create a table. Try doing this for the data shown at the right. (Consult the user's manual of a spreadsheet software program for specific instructions on how to create a table.)

Extended Application: Compound Interest

Banks and other financial institutions differ on how interest is paid to an account. If the interest is added to the account so that future interest is paid on previously earned interest, then the interest is said to be compounded. Suppose, for example, that you deposited \$10,000 in an account that pays 6% interest, compounded quarterly. Because the 6% is the annual interest rate, the quarterly rate is $\frac{1}{4}(0.06) = 0.015$ or 1.5%. The balances during the first five quarters are shown below.

Quarter	*Balance*
1st	\$10,000.00
2nd	$10{,}000.00 + (0.015)(10{,}000.00) = \$10{,}150.00$
3rd	$10{,}150.00 + (0.015)(10{,}150.00) = \$10{,}302.25$
4th	$10{,}302.25 + (0.015)(10{,}302.25) = \$10{,}456.78$
5th	$10{,}456.78 + (0.015)(10{,}456.78) = \$10{,}613.63$

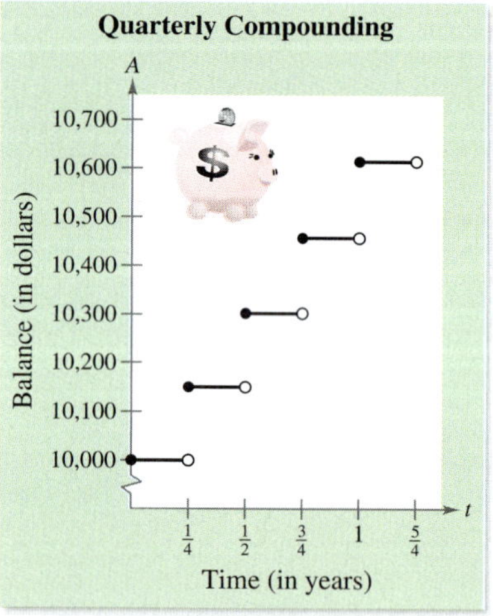

FIGURE 1.68

EXAMPLE 6 Graphing Compound Interest

Sketch the graph of the balance in the account described above.

SOLUTION Let A represent the balance in the account and let t represent the time, in years. You can use the greatest integer function to represent the balance, as shown.

$$A = 10{,}000(1 + 0.015)^{[\![4t]\!]}$$

From the graph shown in Figure 1.68, notice that the function has a discontinuity at each quarter.

TRY IT 6

Write an equation that gives the balance of the account in Example 6 if the annual interest rate is 8%.

TAKE ANOTHER LOOK

Compound Interest

If P dollars is deposited in an account, compounded n times per year, with an annual rate of r (in decimal form), then the balance A after t years is given by

$$A = P\left(1 + \frac{r}{n}\right)^{nt}.$$

Sketch the graph of each function. Which function is continuous? Describe the differences in policy between a bank that uses the first formula and a bank that uses the second formula.

a. $A = P\left(1 + \frac{r}{n}\right)^{nt}$ b. $A = P\left(1 + \frac{r}{n}\right)^{[\![nt]\!]}$

PREREQUISITE REVIEW 1.6

The following warm-up exercises involve skills that were covered in earlier sections. You will use these skills in the exercise set for this section.

In Exercises 1–4, simplify the expression.

1. $\dfrac{x^2 + 6x + 8}{x^2 - 6x - 16}$

2. $\dfrac{x^2 - 5x - 6}{x^2 - 9x + 18}$

3. $\dfrac{2x^2 - 2x - 12}{4x^2 - 24x + 36}$

4. $\dfrac{x^3 - 16x}{x^3 + 2x^2 - 8x}$

In Exercises 5–8, solve for *x*.

5. $x^2 + 7x = 0$

6. $x^2 + 4x - 5 = 0$

7. $3x^2 + 8x + 4 = 0$

8. $x^3 + 5x^2 - 24x = 0$

In Exercises 9 and 10, find the limit.

9. $\lim_{x \to 3} (2x^2 - 3x + 4)$

10. $\lim_{x \to -2} (3x^3 - 8x + 7)$

EXERCISES 1.6

In Exercises 1–10, determine whether the function is continuous on the entire real line. Explain your reasoning.

1. $f(x) = 5x^3 - x^2 + 2$

2. $f(x) = (x^2 - 1)^3$

3. $f(x) = \dfrac{1}{x^2 - 4}$

4. $f(x) = \dfrac{1}{9 - x^2}$

5. $f(x) = \dfrac{1}{4 + x^2}$

6. $f(x) = \dfrac{3x}{x^2 + 1}$

7. $f(x) = \dfrac{2x - 1}{x^2 - 8x + 15}$

8. $f(x) = \dfrac{x + 4}{x^2 - 6x + 5}$

9. $g(x) = \dfrac{x^2 - 4x + 4}{x^2 - 4}$

10. $g(x) = \dfrac{x^2 - 9x + 20}{x^2 - 16}$

In Exercises 11–34, describe the interval(s) on which the function is continuous.

11. $f(x) = \dfrac{x^2 - 1}{x}$

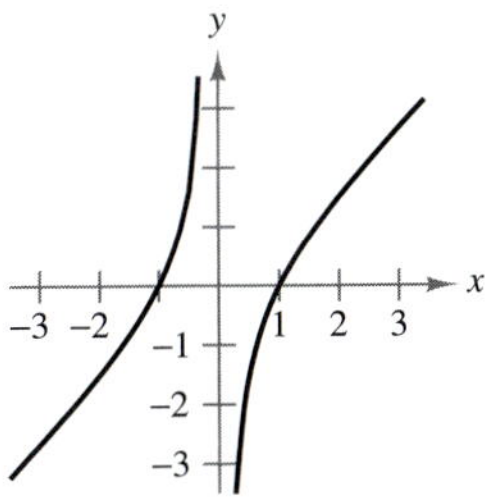

12. $f(x) = \dfrac{1}{x^2 - 4}$

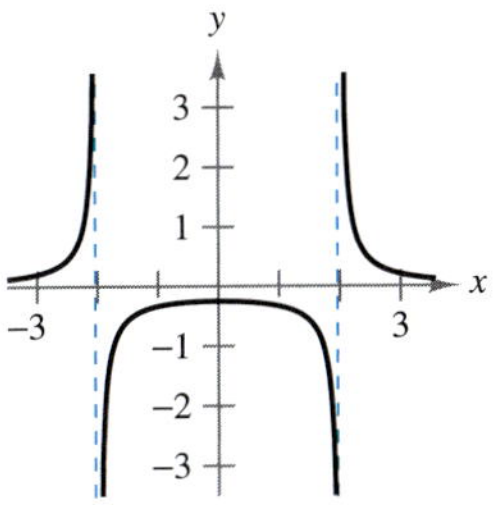

13. $f(x) = \dfrac{x^2 - 1}{x + 1}$

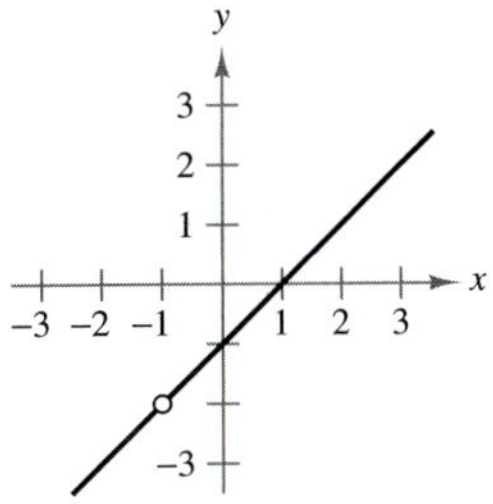

14. $f(x) = \dfrac{x^3 - 8}{x - 2}$

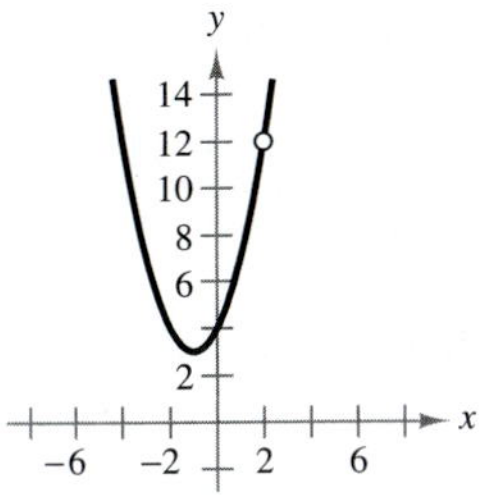

15. $f(x) = x^2 - 2x + 1$

16. $f(x) = 3 - 2x - x^2$

17. $f(x) = \dfrac{x}{x^2 - 1}$

18. $f(x) = \dfrac{x - 3}{x^2 - 9}$

19. $f(x) = \dfrac{x}{x^2 + 1}$

20. $f(x) = \dfrac{1}{x^2 + 1}$

21. $f(x) = \dfrac{x - 5}{x^2 - 9x + 20}$

22. $f(x) = \dfrac{x - 1}{x^2 + x - 2}$

23. $f(x) = [\![2x]\!] + 1$

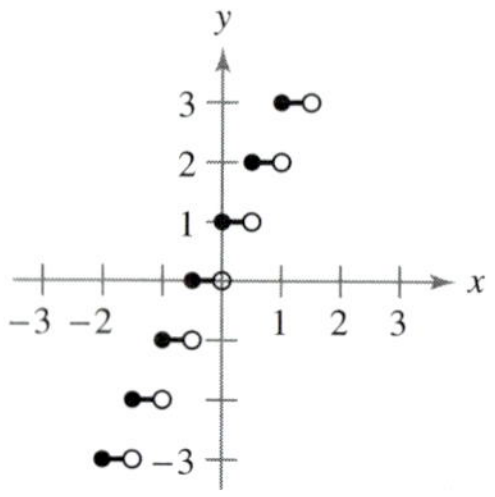

24. $f(x) = \dfrac{[\![x]\!]}{2} + x$

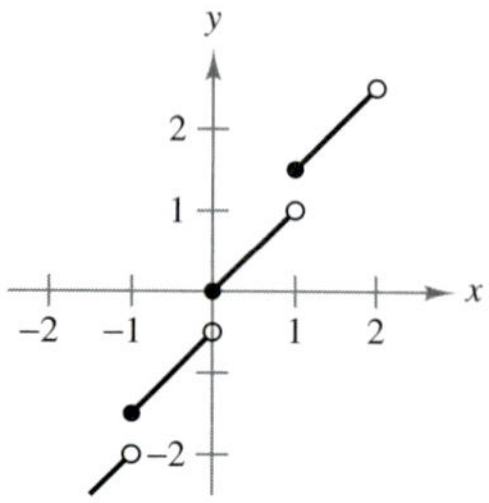

25. $f(x) = \begin{cases} -2x + 3, & x < 1 \\ x^2, & x \geq 1 \end{cases}$

26. $f(x) = \begin{cases} 3 + x, & x \leq 2 \\ x^2 + 1, & x > 2 \end{cases}$

27. $f(x) = \begin{cases} \frac{1}{2}x + 1, & x \leq 2 \\ 3 - x, & x > 2 \end{cases}$

28. $f(x) = \begin{cases} x^2 - 4, & x \leq 0 \\ 3x + 1, & x > 0 \end{cases}$

29. $f(x) = \dfrac{|x + 1|}{x + 1}$

30. $f(x) = \dfrac{|4 - x|}{4 - x}$

31. $f(x) = [\![x - 1]\!]$

32. $f(x) = x - [\![x]\!]$

33. $h(x) = f(g(x)),\quad f(x) = \dfrac{1}{\sqrt{x}},\quad g(x) = x - 1, x > 1$

34. $h(x) = f(g(x)),\quad f(x) = \dfrac{1}{x - 1},\quad g(x) = x^2 + 5$

In Exercises 35–38, discuss the continuity of the function on the closed interval. If there are any discontinuities, determine whether they are removable.

Function	*Interval*
35. $f(x) = x^2 - 4x - 5$	$[-1, 5]$
36. $f(x) = \dfrac{5}{x^2 + 1}$	$[-2, 2]$
37. $f(x) = \dfrac{1}{x - 2}$	$[1, 4]$
38. $f(x) = \dfrac{x}{x^2 - 4x + 3}$	$[0, 4]$

In Exercises 39–44, sketch the graph of the function and describe the interval(s) on which the function is continuous.

39. $f(x) = \dfrac{x^2 - 16}{x - 4}$

40. $f(x) = \dfrac{2x^2 + x}{x}$

41. $f(x) = \dfrac{x^3 + x}{x}$

42. $f(x) = \dfrac{x - 3}{4x^2 - 12x}$

43. $f(x) = \begin{cases} x^2 + 1, & x < 0 \\ x - 1, & x \geq 0 \end{cases}$

44. $f(x) = \begin{cases} x^2 - 4, & x \leq 0 \\ 2x + 4, & x > 0 \end{cases}$

In Exercises 45 and 46, find the constant *a* (Exercise 45) and the constants *a* and *b* (Exercise 46) such that the function is continuous on the entire real line.

45. $f(x) = \begin{cases} x^3, & x \leq 2 \\ ax^2, & x > 2 \end{cases}$

46. $f(x) = \begin{cases} 2, & x \leq -1 \\ ax + b, & -1 < x < 3 \\ -2, & x \geq 3 \end{cases}$

In Exercises 47–52, use a graphing utility to graph the function. Use the graph to determine any *x*-values at which the function is not continuous.

47. $h(x) = \dfrac{1}{x^2 - x - 2}$

48. $k(x) = \dfrac{x - 4}{x^2 - 5x + 4}$

49. $f(x) = \begin{cases} 2x - 4, & x \leq 3 \\ x^2 - 2x, & x > 3 \end{cases}$

50. $f(x) = \begin{cases} 3x - 1, & x \leq 1 \\ x + 1, & x > 1 \end{cases}$

51. $f(x) = x - 2[\![x]\!]$

52. $f(x) = [\![2x - 1]\!]$

In Exercises 53–56, describe the interval(s) on which the function is continuous.

53. $f(x) = \dfrac{x}{x^2 + 1}$

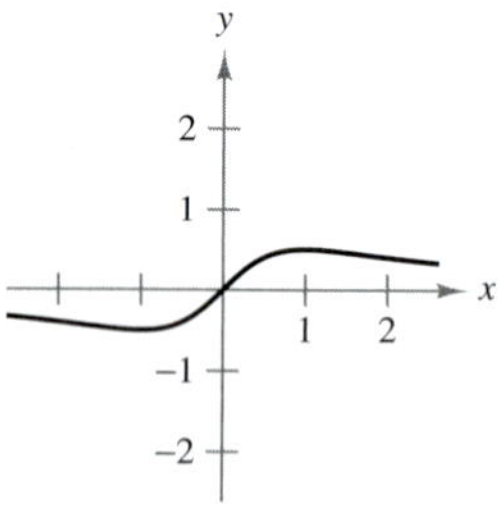

54. $f(x) = x\sqrt{x + 3}$

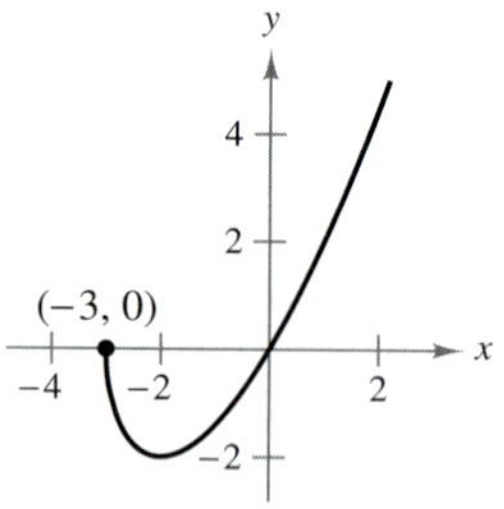

55. $f(x) = \frac{1}{2}[\![2x]\!]$

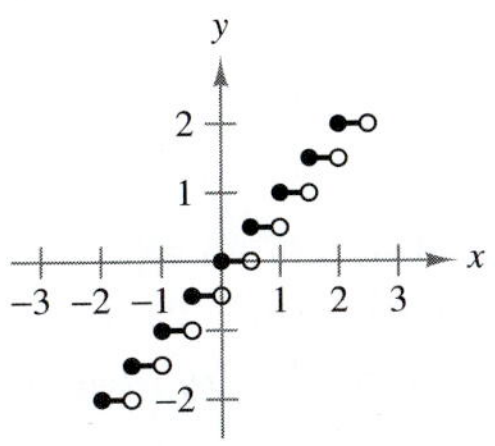

56. $f(x) = \dfrac{x+1}{\sqrt{x}}$

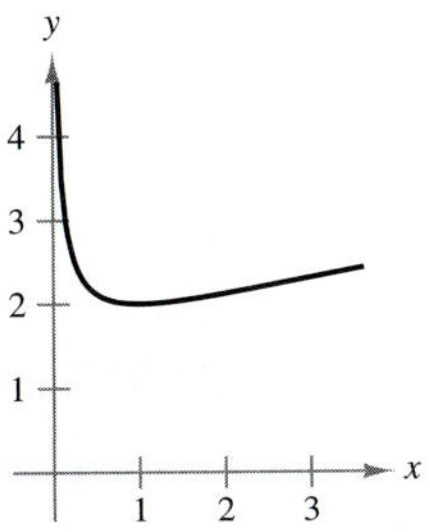

Writing In Exercises 57 and 58, use a graphing utility to graph the function on the interval $[-4, 4]$. Does the graph of the function appear to be continuous on this interval? Is the function in fact continuous on $[-4, 4]$? Write a short paragraph about the importance of examining a function analytically as well as graphically.

57. $f(x) = \dfrac{x^2 + x}{x}$

58. $f(x) = \dfrac{x^3 - 8}{x - 2}$

59. ***Compound Interest*** A deposit of \$7500 is made in an account that pays 6% compounded quarterly. The amount A in the account after t years is

$$A = 7500(1.015)^{[\![4t]\!]}, \quad t \geq 0.$$

(a) Sketch the graph of A. Is the graph continuous? Explain your reasoning.

(b) What is the balance after 7 years?

60. ***Environmental Cost*** The cost C (in millions of dollars) of removing x percent of the pollutants emitted from the smokestack of a factory can be modeled by

$$C = \frac{2x}{100 - x}.$$

(a) What is the implied domain of C? Explain your reasoning.

(b) Use a graphing utility to graph the cost function. Is the function continuous on its domain? Explain your reasoning.

(c) Find the cost of removing 75% of the pollutants from the smokestack.

61. ***Consumer Awareness*** A shipping company's charge for sending an overnight package from New York to Atlanta is \$9.80 for the first pound and \$2.50 for each additional pound or fraction thereof. Use the greatest integer function to create a model for the charge C for overnight delivery of a package weighing x pounds. Use a graphing utility to graph the function, and discuss its continuity.

62. ***Consumer Awareness*** A cab company charges \$3 for the first mile and \$0.25 for each additional mile or fraction thereof. Use the greatest integer function to create a model for the cost C of a cab ride n miles long. Use a graphing utility to graph the function, and discuss its continuity.

63. ***Consumer Awareness*** A dial-direct long distance call between two cities costs \$1.04 for the first 2 minutes and \$0.36 for each additional minute or fraction thereof.

(a) Use the greatest integer function to write the cost C of a call in terms of the time t (in minutes). Sketch the graph of the cost function and discuss its continuity.

(b) Find the cost of a nine-minute call.

64. ***Salary Contract*** A union contract guarantees a 9% yearly increase for 5 years. For a current salary of \$28,500, the salary for the next 5 years is given by

$$S = 28{,}500(1.09)^{[\![t]\!]}$$

where $t = 0$ represents the present year.

(a) Use the greatest integer function of a graphing utility to graph the salary function, and discuss its continuity.

(b) Find the salary during the fifth year (when $t = 5$).

65. ***Inventory Management*** The number of units in inventory in a small company is

$$N = 25\left(2\left[\!\!\left[\frac{t+2}{2}\right]\!\!\right] - t\right), \quad 0 \leq t \leq 12$$

where the real number t is the time in months.

(a) Use the greatest integer function of a graphing utility to graph this function, and discuss its continuity.

(b) How often must the company replenish its inventory?

66. ***Owning a Franchise*** You have purchased a franchise. You have determined a linear model for your revenue as a function of time. Is the model a continuous function? Would your actual revenue be a continuous function of time? Explain your reasoning.

67. ***Biology*** The gestation period of rabbits is only 26 to 30 days. Therefore, the population of a form (rabbits' home) can increase dramatically in a short period of time. The table gives the population of a form, where t is the time in months and N is the rabbit population.

t	0	1	2	3	4	5	6
N	2	8	10	14	10	15	12

Graph the population as a function of time. Find any points of discontinuity in the function. Explain your reasoning.

ALGEBRA REVIEW

Order of Operations

Much of the algebra in this chapter involves evaluation of algebraic expressions. When you evaluate an algebraic expression, you need to know the priorities assigned to different operations. These priorities are called the *order of operations.*

1. Perform operations inside *symbols of grouping or absolute value symbols,* starting with the innermost symbol.
2. Evaluate all *exponential* expressions.
3. Perform all *multiplications* and *divisions* from left to right.
4. Perform all *additions* and *subtractions* from left to right.

EXAMPLE 1 Using Order of Operations

Evaluate each expression.

(a) $7 - [(5 \cdot 3) + 2^3]$

(b) $[36 \div (3^2 \cdot 2)] + 6$

(c) $36 - [3^2 \cdot (2 \div 6)]$

(d) $10 - 2(8 + |5 - 7|)$

SOLUTION

(a)
$$\begin{aligned} 7 - [(5 \cdot 3) + 2^3] &= 7 - [15 + 2^3] && \text{Multiply inside parentheses.} \\ &= 7 - [15 + 8] && \text{Evaluate exponential expression.} \\ &= 7 - 23 && \text{Add inside brackets.} \\ &= -16 && \text{Subtract.} \end{aligned}$$

(b)
$$\begin{aligned} [36 \div (3^2 \cdot 2)] + 6 &= [36 \div (9 \cdot 2)] + 6 && \text{Evaluate exponential expression inside parentheses.} \\ &= [36 \div 18] + 6 && \text{Multiply inside parentheses.} \\ &= 2 + 6 && \text{Divide inside brackets.} \\ &= 8 && \text{Add.} \end{aligned}$$

(c)
$$\begin{aligned} 36 - [3^2 \cdot (2 \div 6)] &= 36 - \left[3^2 \cdot \tfrac{1}{3}\right] && \text{Divide inside parentheses.} \\ &= 36 - \left[9 \cdot \tfrac{1}{3}\right] && \text{Evaluate exponential expression.} \\ &= 36 - 3 && \text{Multiply inside brackets.} \\ &= 33 && \text{Subtract.} \end{aligned}$$

(d)
$$\begin{aligned} 10 - 2(8 + |5 - 7|) &= 10 - 2(8 + |-2|) && \text{Subtract inside absolute value symbols.} \\ &= 10 - 2(8 + 2) && \text{Evaluate absolute value.} \\ &= 10 - 2(10) && \text{Add inside parentheses.} \\ &= 10 - 20 && \text{Multiply.} \\ &= -10 && \text{Subtract.} \end{aligned}$$

TECHNOLOGY

Most scientific and graphing calculators use the same order of operations listed above. Try entering the expressions in Example 1 into your calculator. Do you get the same results?

Solving Equations

A second algebraic skill in this chapter is solving an equation in one variable.

1. To solve a *linear equation*, you can add or subtract the same quantity from each side of the equation. You can also multiply or divide each side of the equation by the same *nonzero* quantity.
2. To solve a *quadratic equation*, you can take the square root of each side, use factoring, or use the Quadratic Formula.
3. To solve a *radical equation*, isolate the radical on one side of the equation and square each side of the equation.
4. To solve an *absolute value equation*, use the definition of absolute value to rewrite the equation as two equations.

TECHNOLOGY

The equations in Example 2 are solved algebraically. Most graphing utilities have a "solve" key that allows you to solve the equation graphically. If you have a graphing utility, try using it to solve graphically the equations in Example 2.

EXAMPLE 2 Solving Equations

Solve each equation.

(a) $3x - 3 = 5x - 7$

(b) $2x^2 = 10$

(c) $2x^2 + 5x - 6 = 6$

(d) $\sqrt{2x - 7} = 5$

SOLUTION

(a) $3x - 3 = 5x - 7$ Write original (linear) equation.

$-3 = 2x - 7$ Subtract $3x$ from each side.

$4 = 2x$ Add 7 to each side.

$2 = x$ Divide each side by 2.

(b) $2x^2 = 10$ Write original (quadratic) equation.

$x^2 = 5$ Divide each side by 2.

$x = \pm\sqrt{5}$ Take the square root of each side.

(c) $2x^2 + 5x - 6 = 6$ Write original (quadratic) equation.

$2x^2 + 5x - 12 = 0$ Write in general form.

$(2x - 3)(x + 4) = 0$ Factor.

$2x - 3 = 0 \Rightarrow x = \frac{3}{2}$ Set first factor equal to zero.

$x + 4 = 0 \Rightarrow x = -4$ Set second factor equal to zero.

(d) $\sqrt{2x - 7} = 5$ Write original (radical) equation.

$2x - 7 = 25$ Square each side.

$2x = 32$ Add 7 to each side.

$x = 16$ Divide each side by 2.

ALGEBRA REVIEW

You should be aware that solving radical equations can sometimes lead to *extraneous solutions* (those that do not satisfy the original equation). For example, squaring both sides of the following equation yields two possible solutions, one of which is extraneous.

$$\sqrt{x} = x - 2$$
$$x = x^2 - 4x + 4$$
$$0 = x^2 - 5x + 4$$
$$= (x - 4)(x - 1)$$

$x - 4 = 0 \Rightarrow x = 4$ (solution)

$x - 1 = 0 \Rightarrow x = 1$ (extraneous)

1 CHAPTER SUMMARY AND STUDY STRATEGIES

*After studying this chapter, you should have acquired the following skills. The exercise numbers are keyed to the view Exercises that begin on page 76. Answers to odd-numbered Review Exercises are given in the back of the text.**

- Plot points in a coordinate plane and read data presented graphically. *(Section 1.1)* — *Review Exercises 1–4*
- Find the distance between two points in a coordinate plane. *(Section 1.1)* — *Review Exercises 5–8*

  $d = \sqrt{(x_2 - x_1)^2 + (y_2 - y_1)^2}$
- Find the midpoints of line segments connecting two points. *(Section 1.1)* — *Review Exercises 9–12*

  $\text{Midpoint} = \left(\frac{x_1 + x_2}{2}, \frac{y_1 + y_2}{2}\right)$
- Interpret real-life data that is presented graphically. *(Section 1.1)* — *Review Exercises 13, 14*
- Translate points in a coordinate plane. *(Section 1.1)* — *Review Exercises 15, 16*
- Construct a bar graph from real-life data. *(Section 1.1)* — *Review Exercise 17*
- Sketch graphs of equations by hand. *(Section 1.2)* — *Review Exercises 18–27*
- Find the x- and y-intercepts of graphs of equations algebraically *and* graphically using a graphing utility. *(Section 1.2)* — *Review Exercises 28, 29*
- Write the standard forms of equations of circles, given the center and a point on the circle. *(Section 1.2)* — *Review Exercises 30, 31*

  $(x - h)^2 + (y - k)^2 = r^2$
- Convert equations of circles from general form to standard form by completing the square, and sketch the circles. *(Section 1.2)* — *Review Exercises 32, 33*
- Find the points of intersection of two graphs algebraically *and* graphically using a graphing utility. *(Section 1.2)* — *Review Exercises 34–37*
- Find the break-even point for a business. *(Section 1.2)* — *Review Exercises 38, 39*

  The break-even point occurs when the revenue R is equal to the cost C.
- Find the equilibrium points of supply equations and demand equations. *(Section 1.2)* — *Review Exercise 40*

  The equilibrium point is the point of intersection of the graphs of the supply and demand equations.
- Use the slope-intercept form of a linear equation to sketch graphs of lines. *(Section 1.3)* — *Review Exercises 41–46*

  $y = mx + b$
- Find slopes of lines passing through two points. *(Section 1.3)* — *Review Exercises 47–50*

  $m = \frac{y_2 - y_1}{x_2 - x_1}$
- Use the point-slope form to write equations of lines and graph equations using a graphing utility. *(Section 1.3)* — *Review Exercises 51, 52*

  $y - y_1 = m(x - x_1)$

* Use a wide range of valuable study aids to help you master the material in this chapter. The *Student Solutions Guide* includes step-by-step solutions to all odd-numbered exercises to help you review and prepare. The *HM mathSpace® Student CD-ROM* helps you brush up on your algebra skills. The *Graphing Technology Guide*, available on the Web at *math.college.hmco.com/students*, offers step-by-step commands and instructions for a wide variety of graphing calculators, including the most recent models.

- Find equations of parallel and perpendicular lines. *(Section 1.3)* — *Review Exercises 53, 54*

  Parallel lines: $m_1 = m_2$ Perpendicular lines: $m_1 = -\dfrac{1}{m_2}$
- Use linear equations to solve real-life problems such as predicting future sales or creating a linear depreciation schedule. *(Section 1.3)* — *Review Exercises 55, 56*
- Use the vertical line test to decide whether equations define functions. *(Section 1.4)* — *Review Exercises 57–60*
- Use function notation to evaluate functions. *(Section 1.4)* — *Review Exercises 61, 62*
- Use a graphing utility to graph functions and find the domains and ranges of functions. *(Section 1.4)* — *Review Exercises 63–68*
- Combine functions to create other functions. *(Section 1.4)* — *Review Exercises 69, 70*
- Use the horizontal line test to determine whether functions have inverse functions. If they do, find the inverse functions. *(Section 1.4)* — *Review Exercises 71–74*
- Determine whether limits exist. If they do, find the limits. *(Section 1.5)* — *Review Exercises 75–92*
- Use a table to estimate one-sided limits. *(Section 1.5)* — *Review Exercises 93, 94*
- Determine whether statements about limits are true or false. *(Section 1.5)* — *Review Exercises 95–100*
- Determine whether functions are continuous at a point, on an open interval, and on a closed interval. *(Section 1.6)* — *Review Exercises 101–108*
- Determine the constant such that f is continuous. *(Section 1.6)* — *Review Exercises 109, 110*
- Use analytic and graphical models of real-life data to solve real-life problems. *(Section 1.6)* — *Review Exercises 111–114*

On pages xxv–xxviii of the preface, we included a feature called A Plan for You as a Student. *If you have not already read this feature, we encourage you to do so now. Here are some other strategies that can help you succeed in this course.*

- ***Use a Graphing Utility*** A graphing calculator or graphing software for a computer can help you in this course in two important ways. As an *exploratory device*, a graphing utility allows you to learn concepts by allowing you to compare graphs of equations. For instance, sketching the graphs of $y = x^2$, $y = x^2 + 1$, and $y = x^2 - 1$ helps confirm that adding (or subtracting) a constant to (or from) a function shifts the graph of the function vertically. As a *problem-solving tool*, a graphing utility frees you of some of the drudgery of sketching complicated graphs by hand. The time that you save can be spent using mathematics to solve real-life problems.
- ***Use the Warm-Up Exercises*** Each exercise set in this text begins with a set of warm-up exercises. We urge you to begin each homework session by quickly working all of the warm-up exercises (all are answered in the back of the text). The "old" skills covered in the warm-up exercises are needed to master the "new" skills in the section exercise set. The warm-up exercises remind you that mathematics is cumulative—to be successful in this course, you must retain "old" skills.
- ***Use the Additional Study Aids*** The additional study aids were prepared specifically to help you master the concepts discussed in the text. They are the *Student Solutions Guide*, the *Calculus: An Applied Approach Learning Tools Student CD-ROM*, *The Algebra of Calculus*, and the *Graphing Technology Guide*.

Study Tools *Additional resources that accompany this chapter*

- **Algebra Review** (pages 72 and 73)
- **Chapter Summary and Study Strategies** (pages 74 and 75)
- **Review Exercises** (pages 76–79)
- **Sample Post-Graduation Exam Questions** (page 80)
- **Web Exercises** (page 10, Exercises 39 and 40; page 23, Exercise 74; page 34, Exercise 79; page 35, Exercise 91; page 48, Exercise 87)
- **Student Solutions Guide**
- **HM mathSpace® Student CD-ROM**
- **Graphing Technology Guide** (math.college.hmco.com/students)

1 CHAPTER REVIEW EXERCISES

In Exercises 1–4, match the data with the real-life situation that it represents. [The graphs are labeled (a)–(d).]

1. Population of Texas

2. Population of California

3. Number of unemployed workers in the United States

4. Best Buy sales

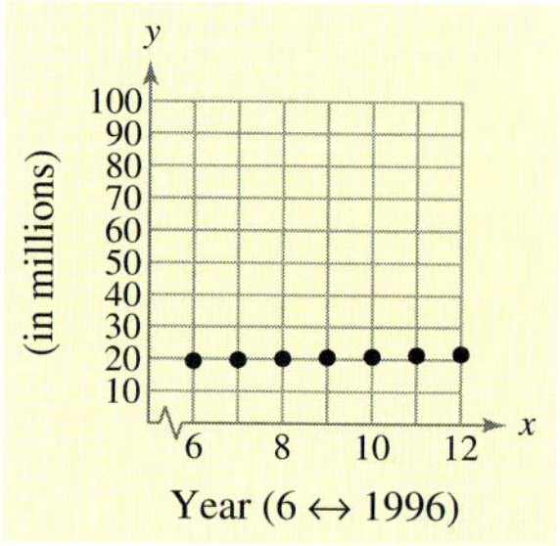

(a)

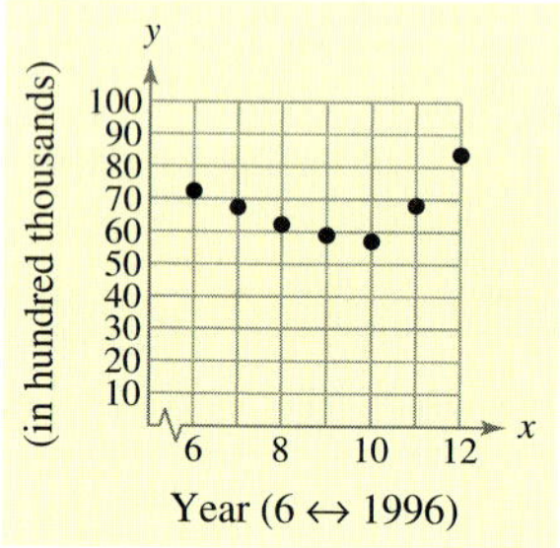

(b)

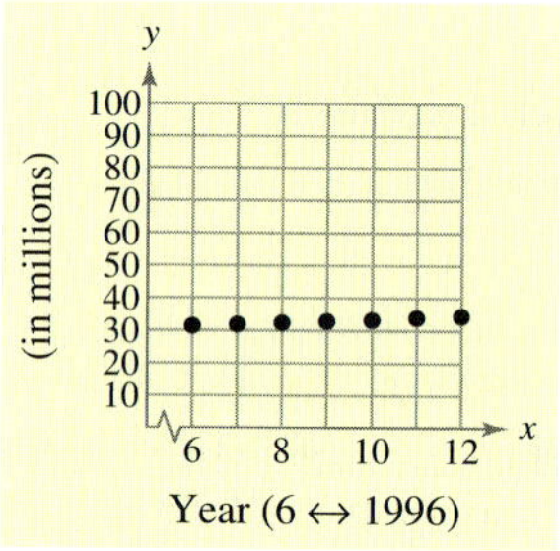

(c)

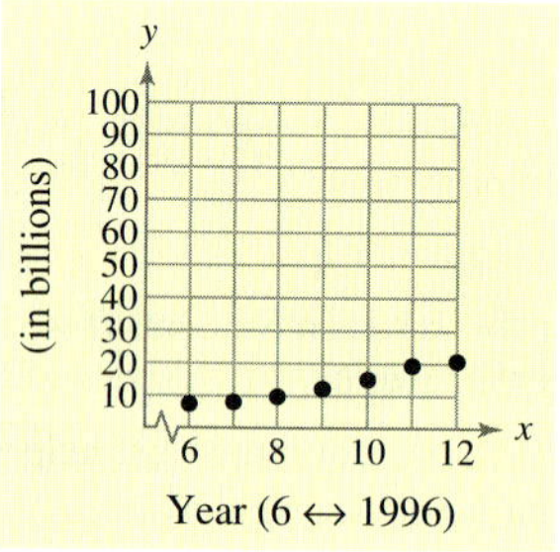

(d)

In Exercises 5–8, find the distance between the two points.

5. $(0, 0), (5, 2)$

6. $(1, 2), (4, 3)$

7. $(-1, 3), (-4, 6)$

8. $(6, 8), (-3, 7)$

In Exercises 9–12, find the midpoint of the line segment connecting the two points.

9. $(5, 6), (9, 2)$

10. $(0, 0), (-4, 8)$

11. $(-10, 4), (-6, 8)$

12. $(7, -9), (-3, 5)$

In Exercises 13 and 14, use the graph below, which gives the revenues, costs, and profits for Pixar from 1999 through 2003. (Pixar develops and produces animated feature films.) *(Source: Pixar)*

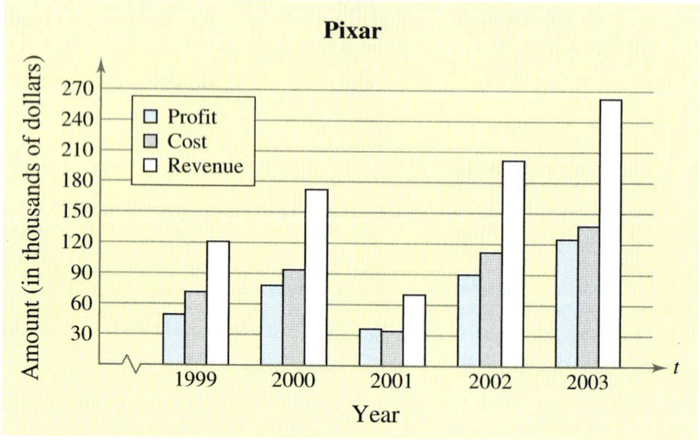

13. Write an equation that relates the revenue R, cost C, and profit P. Explain the relationship between the heights of the bars and the equation.

14. Estimate the revenue, cost, and profit for Pixar for each year.

15. Translate the triangle whose vertices are $(1, 3)$, $(2, 4)$, and $(5, 6)$ three units to the right and four units up. Find the coordinates of the translated vertices.

16. Translate the rectangle whose vertices are $(-2, 1)$, $(-1, 2)$, $(1, 0)$, and $(0, -1)$ four units to the right and one unit down.

17. ***Biology*** The following data represent six intertidal invertebrate species collected from four stations along the Maine coast.

Mytilus	107	*Gammarus*	78
Littorina	65	*Arbacia*	6
Nassarius	112	*Mya*	18

Use a graphing utility to construct a bar graph that represents the data. *(Source: Adapted from Haefner,* Exploring Marine Biology: Laboratory and Field Exercises*)*

In Exercises 18–27, sketch the graph of the equation.

18. $y = 4x - 12$

19. $y = 4 - 3x$

20. $y = x^2 + 5$

21. $y = 1 - x^2$

22. $y = |4 - x|$

23. $y = |2x - 3|$

24. $y = x^3 + 4$

25. $y = x^3 + 2x^2 - x + 2$

26. $y = \sqrt{4x + 1}$

27. $y = \sqrt{2x}$

In Exercises 28 and 29, find the x- and y-intercepts of the graph of the equation algebraically. Use a graphing utility to verify your results.

28. $4x + y + 3 = 0$

29. $y = (x - 1)^3 + 2(x - 1)^2$

In Exercises 30 and 31, write the standard form of the equation of the circle.

30. Center: $(0, 0)$

Solution point: $(2, \sqrt{5})$

31. Center: $(2, -1)$

Solution point: $(-1, 7)$

In Exercises 32 and 33, complete the square to write the equation of the circle in standard form. Determine the radius and center of the circle. Then sketch the circle.

32. $x^2 + y^2 - 6x + 8y = 0$

33. $x^2 + y^2 + 10x + 4y - 7 = 0$

In Exercises 34–37, find the point(s) of intersection of the graphs algebraically. Then use a graphing utility to verify your results.

34. $x + y = 2,\ 2x - y = 1$

35. $x^2 + y^2 = 5,\ x - y = 1$

36. $y = x^3,\ y = x$

37. $y = \sqrt{x},\ y = x$

38. ***Break-Even Analysis*** The student government association wants to raise money by having a T-shirt sale. Each shirt costs \$8. The silk screening costs \$200 for the design, plus \$2 per shirt. Each shirt will sell for \$14.

(a) Find equations for the total cost C and the total revenue R for selling x shirts.

(b) Find the break-even point.

39. ***Break-Even Analysis*** You are starting a part-time business. You make an initial investment of \$6000. The unit cost of the product is \$6.50, and the selling price is \$13.90.

(a) Find equations for the total cost C and the total revenue R for selling x units of the product.

(b) Find the break-even point.

40. ***Supply and Demand*** The demand and supply equations for a cordless screwdriver are given by

$p = 91.4 - 0.009x$ Demand equation

$p = 6.4 + 0.008x$ Supply equation

where p is the price in dollars and x represents the number of units. Find the equilibrium point for this market.

In Exercises 41–46, find the slope and y-intercept (if possible) of the linear equation. Then sketch the graph of the equation.

41. $3x + y = -2$

42. $-\frac{1}{3}x + \frac{5}{6}y = 1$

43. $y = -\frac{5}{3}$

44. $x = -3$

45. $-2x - 5y - 5 = 0$

46. $3.2x - 0.8y + 5.6 = 0$

In Exercises 47–50, find the slope of the line passing through the two points.

47. $(0, 0), (7, 6)$

48. $(-1, 5), (-5, 7)$

49. $(10, 17), (-11, -3)$

50. $(-11, -3), (-1, -3)$

In Exercises 51 and 52, find an equation of the line that passes through the point and has the given slope. Then use a graphing utility to graph the line.

51. Point: $(3, -1)$; slope: $m = -2$

52. Point: $(-3, -3)$; slope: $m = \frac{1}{2}$

In Exercises 53 and 54, find the general form of the equation of the line passing through the point and satisfying the given condition.

53. Point: $(-3, 6)$

(a) Slope is $\frac{7}{8}$.

(b) Parallel to the line $4x + 2y = 7$.

(c) Passes through the origin.

(d) Perpendicular to the line $3x - 2y = 2$.

54. Point: $(1, -3)$

(a) Parallel to the x-axis.

(b) Perpendicular to the x-axis.

(c) Parallel to the line $-4x + 5y = -3$.

(d) Perpendicular to the line $5x - 2y = 3$.

55. ***Demand*** When a wholesaler sold a product at \$32 per unit, sales were 750 units per week. After a price increase of \$5 per unit, however, the sales dropped to 700 units per week.

(a) Write the quantity demanded x as a linear function of the price p.

(b) *Linear Interpolation* Predict the number of units sold at a price of \$34.50 per unit.

(c) *Linear Extrapolation* Predict the number of units sold at a price of \$42.00 per unit.

56. ***Linear Depreciation*** A small business purchases a typesetting system for \$117,000. After 9 years, the system will be obsolete and have no value.

(a) Write a linear equation giving the value v of the system in terms of the time t.

(b) Use a graphing utility to graph the function.

(c) Use a graphing utility to estimate the value of the system after 4 years.

(d) Use a graphing utility to estimate the time when the system's value will be \$84,000.

In Exercises 57–60, use the vertical line test to determine whether y is a function of x.

57. $y = -x^2 + 2$

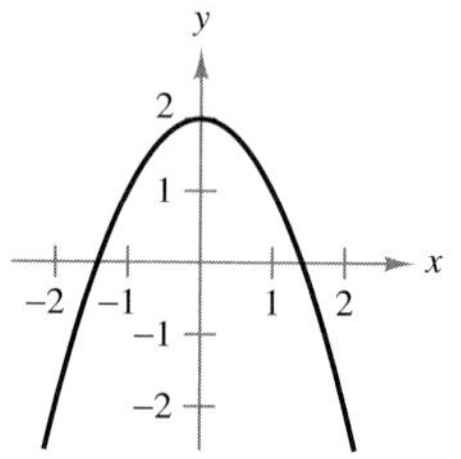

58. $x^2 + y^2 = 4$

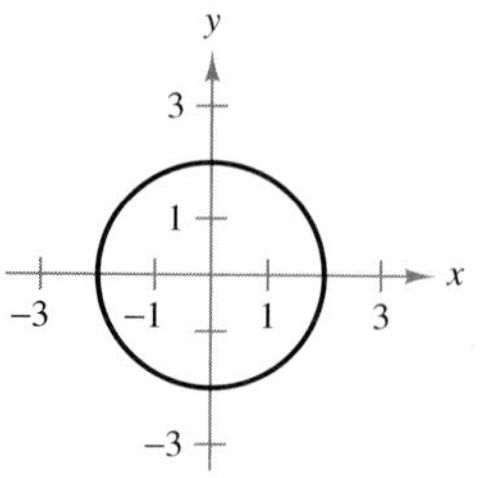

59. $y^2 - \frac{1}{4}x^2 = 4$

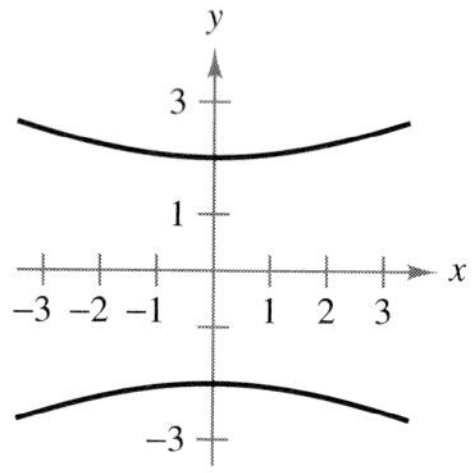

60. $y = |x + 4|$

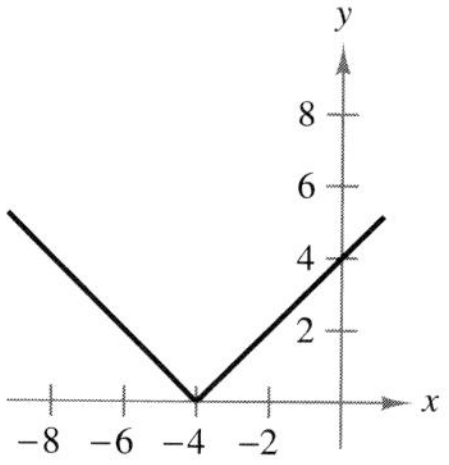

In Exercises 61 and 62, evaluate the function at the specified values of the independent variable. Simplify the result.

61. $f(x) = 3x + 4$

(a) $f(1)$ (b) $f(x + 1)$ (c) $f(2 + \Delta x)$

62. $f(x) = x^2 + 4x + 3$

(a) $f(0)$ (b) $f(x - 1)$ (c) $f(x + \Delta x) - f(x)$

In Exercises 63–68, use a graphing utility to graph the function. Then find the domain and range of the function.

63. $f(x) = x^2 + 3x + 2$

64. $f(x) = 2$

65. $f(x) = \sqrt{x + 1}$

66. $f(x) = \dfrac{x - 3}{x^2 + x - 12}$

67. $f(x) = -|x| + 3$

68. $f(x) = -\frac{12}{13}x - \frac{7}{8}$

In Exercises 69 and 70, use f and g to find the combinations of the functions.

(a) $f(x) + g(x)$ (b) $f(x) - g(x)$ (c) $f(x)g(x)$

(d) $\dfrac{f(x)}{g(x)}$ (e) $f(g(x))$ (f) $g(f(x))$

69. $f(x) = 1 + x^2, \quad g(x) = 2x - 1$

70. $f(x) = 2x - 3, \quad g(x) = \sqrt{x + 1}$

In Exercises 71–74, find the inverse function of f (if it exists).

71. $f(x) = \frac{3}{2}x$

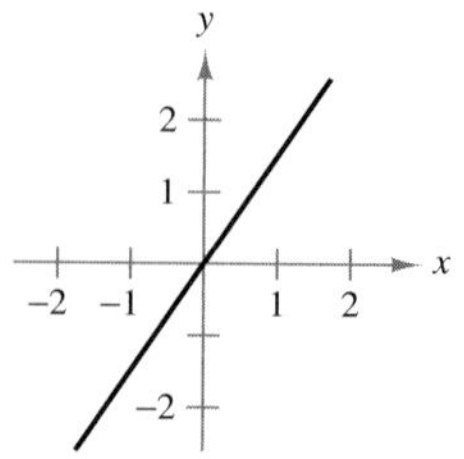

72. $f(x) = |x + 1|$

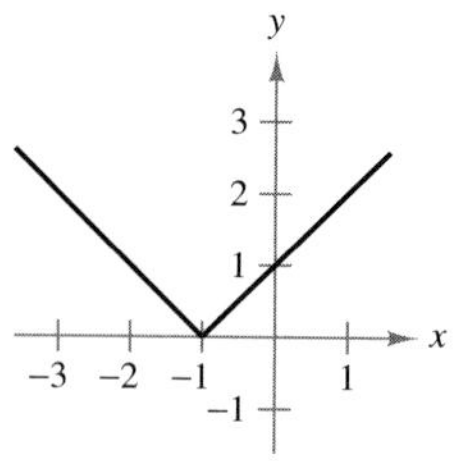

73. $f(x) = -x^2 + \frac{1}{2}$

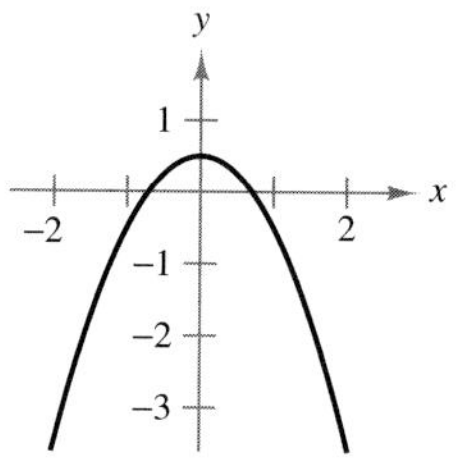

74. $f(x) = x^3 - 1$

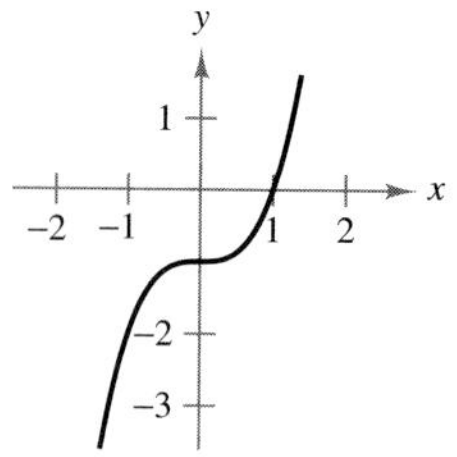

In Exercises 75–92, find the limit (if it exists).

75. $\lim_{x\to 2} (5x - 3)$

76. $\lim_{x\to 2} (2x + 9)$

77. $\lim_{x\to 2} (5x - 3)(2x + 3)$

78. $\lim_{x\to 2} \dfrac{5x - 3}{2x + 9}$

79. $\lim_{t\to 3} \dfrac{t^2 + 1}{t}$

80. $\lim_{t\to 0} \dfrac{t^2 + 1}{t}$

81. $\lim_{t\to 1} \dfrac{t + 1}{t - 2}$

82. $\lim_{t\to 2} \dfrac{t + 1}{t - 2}$

83. $\lim_{x\to -2} \dfrac{x + 2}{x^2 - 4}$

84. $\lim_{x\to 3^-} \dfrac{x^2 - 9}{x - 3}$

85. $\lim_{x\to 0^+} \left(x - \dfrac{1}{x}\right)$

86. $\lim_{x\to 1/2} \dfrac{2x - 1}{6x - 3}$

87. $\lim_{x\to 0} \dfrac{[1/(x - 2)] - 1}{x}$

88. $\lim_{x\to 0} \dfrac{[1/(x - 4)] - (1/4)}{x}$

89. $\lim_{t\to 0} \dfrac{\left(1/\sqrt{t + 4}\right) - (1/2)}{t}$

90. $\lim_{s\to 0} \dfrac{\left(1/\sqrt{1 + s}\right) - 1}{s}$

91. $\lim_{\Delta x\to 0} \dfrac{(x + \Delta x)^3 - (x + \Delta x) - (x^3 - x)}{\Delta x}$

92. $\lim_{\Delta x\to 0} \dfrac{1 - (x + \Delta x)^2 - (1 - x^2)}{\Delta x}$

In Exercises 93 and 94, use a table to estimate the limit.

93. $\lim_{x\to 1^+} \dfrac{\sqrt{2x + 1} - \sqrt{3}}{x - 1}$

94. $\lim_{x\to 1^+} \dfrac{1 - \sqrt[3]{x}}{x - 1}$

True or False? In Exercises 95–100, determine whether the statement is true or false. If it is false, explain why or give an example that shows it is false.

95. $\lim_{x\to 0} \frac{|x|}{x} = 1$

96. $\lim_{x\to 0} x^3 = 0$

97. $\lim_{x\to 0} \sqrt{x} = 0$

98. $\lim_{x\to 0} \sqrt[3]{x} = 0$

99. $\lim_{x\to 2} f(x) = 3, \quad f(x) = \begin{cases} 3, & x \le 2 \\ 0, & x > 2 \end{cases}$

100. $\lim_{x\to 3} f(x) = 1, \quad f(x) = \begin{cases} x - 2, & x \le 3 \\ -x^2 + 8x - 14, & x > 3 \end{cases}$

In Exercises 101–108, describe the interval(s) on which the function is continuous.

101. $f(x) = \dfrac{1}{(x+4)^2}$

102. $f(x) = \dfrac{x+2}{x}$

103. $f(x) = \dfrac{3}{x+1}$

104. $f(x) = \dfrac{x+1}{2x+2}$

105. $f(x) = [\![x+3]\!]$

106. $f(x) = [\![x]\!] - 2$

107. $f(x) = \begin{cases} x, & x \le 0 \\ x + 1, & x > 0 \end{cases}$

108. $f(x) = \begin{cases} x, & x \le 0 \\ x^2, & x > 0 \end{cases}$

In Exercises 109 and 110, find the constant a such that f is continuous on the entire real line.

109. $f(x) = \begin{cases} -x + 1, & x \le 3 \\ ax - 8, & x > 3 \end{cases}$

110. $f(x) = \begin{cases} x + 1, & x < 1 \\ 2x + a, & x \ge 1 \end{cases}$

111. ***National Debt*** The table lists the national debt D (in billions of dollars) for selected years. A mathematical model for the national debt is

$$D = 2.7502t^3 - 61.061t^2 + 598.79t + 3103.6,$$
$$0 \le t \le 13$$

where $t = 0$ represents 1990. *(Source: U.S. Department of the Treasury)*

t	0	1	2	3	4
D	3206.3	3598.2	4001.8	4351.0	4643.3

t	5	6	7	8	9
D	4920.6	5181.5	5369.2	5478.2	5605.5

t	10	11	12	13
D	5628.7	5769.9	6198.4	6752.0

(a) Use a graphing utility to graph the model.

(b) Create a table that compares the values given by the model with the actual data.

(c) Use the model to estimate the national debt in 2008.

112. ***Consumer Awareness*** A cellular phone company charges \$2 for the first minute and \$0.10 for each additional minute or fraction thereof. Use the greatest integer function to create a model for the cost C of a phone call lasting t minutes. Use a graphing utility to graph the function, and discuss its continuity.

113. ***Recycling*** A recycling center pays \$0.25 for each pound of aluminum cans. Twenty-four aluminum cans weigh one pound. A mathematical model for the amount A paid by the recycling center is

$$A = \frac{1}{4}\left[\!\!\left[\frac{x}{24}\right]\!\!\right]$$

where x is the number of cans.

(a) Use a graphing utility to graph the function and then discuss its continuity.

(b) How much does a recycling center pay out for 1500 cans?

114. ***Biology*** A researcher experimenting with strains of corn produced the results in the figure below. From the figure, visually estimate the x- and y-intercepts, and use these points to write an equation for each line.

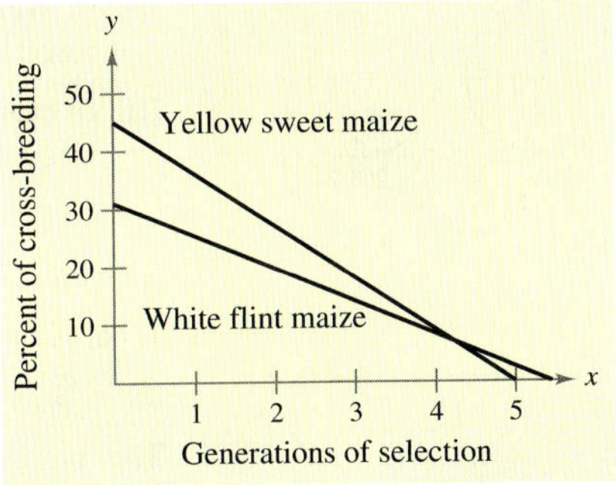

(Source: Adapted from Levine/Miller, Biology: Discovering Life, *Second Edition)*

1 SAMPLE POST-GRADUATION EXAM QUESTIONS

CPA
GMAT
GRE
Actuarial
CLAST

The following questions represent the types of questions that appear on certified public accountant (CPA) exams, Graduate Management Admission Tests (GMAT), Graduate Records Exams (GRE), actuarial exams, and College-Level Academic Skills Tests (CLAST). The answers to the questions are given in the back of the book.

In Questions 1–5, use the data given in the graphs. *(Source: U.S. Bureau of Labor Statistics)*

Figure for 1–5

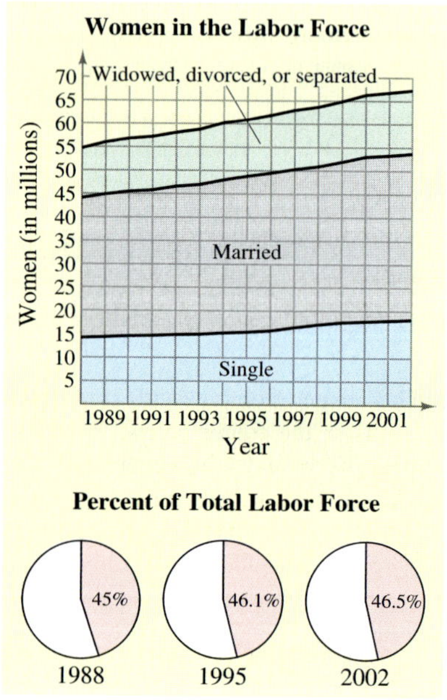

1. The total labor force in 2002 was about y million with y equal to

(a) 100 (b) 118 (c) 129 (d) 145 (e) 154

2. In 1995, the percent of married women in the labor force was about

(a) 19 (b) 32 (c) 45 (d) 55 (e) 82

3. What was the first year when more than 60 million women were in the labor force?

(a) 1992 (b) 1994 (c) 1996 (d) 1998 (e) 2000

4. Between 1990 and 2000, the number of women in the labor force

(a) increased by about 17% (b) increased by about 25%
(c) increased by about 50% (d) increased by about 100%
(e) increased by about 125%

5. Which of the statements about the labor force can be inferred from the graphs?

I. Between 1988 and 2002, there were no years when more than 20 million widowed, divorced, or separated women were in the labor force.

II. In every year between 1988 and 2002, the number of married women in the labor force increased.

III. In every year between 1988 and 2002, women made up at least $\frac{2}{5}$ of the total labor force.

(a) I only (b) II only (c) I and II only
(d) II and III only (e) I, II, and III

6. What is the length of the line segment connecting $(1, 3)$ and $(-1, 5)$?

(a) $\sqrt{3}$ (b) 2 (c) $2\sqrt{2}$ (d) 4 (e) 8

7. The interest charged on a loan is p dollars per \$1000 for the first month and q dollars per \$1000 for each month after the first month. How much interest will be charged during the first 3 months on a loan of \$10,000?

(a) $30p$ (b) $30q$ (c) $p + 2q$
(d) $20p + 10q$ (e) $10p + 20q$

8. If $x + y > 5$ and $x - y > 3$, which of the following describes the x solutions?

(a) $x > 3$ (b) $x > 4$ (c) $x > 5$ (d) $x < 5$ (e) $x < 3$

Figure for 9

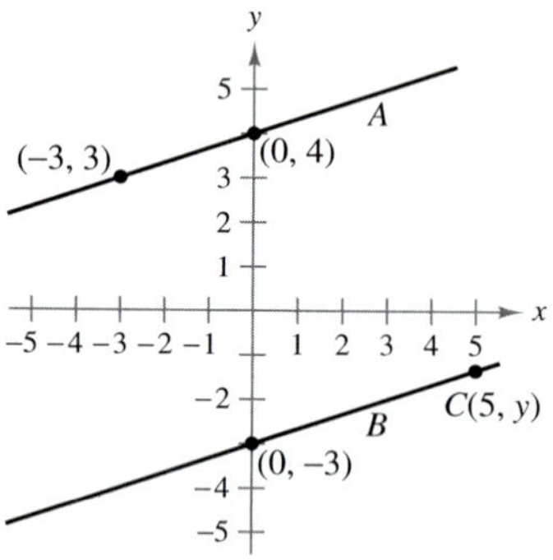

9. In the figure at the left, in order for line A to be parallel to line B, the coordinates of C must be $(5, y)$ with y equal to which of the following?

(a) -4 (b) $-\frac{4}{3}$ (c) 0 (d) $\frac{1}{3}$ (e) 5

chapter

2 Differentiation

Higher-order derivatives are used to determine the acceleration function of a sports car. The acceleration function shows the changes in the car's velocity. As the car reaches its "cruising" speed, is the acceleration increasing or decreasing?

STRATEGIES FOR SUCCESS

WHAT YOU SHOULD LEARN:

- How to find the slope of a graph and calculate derivatives using the limit definition
- How to use the Constant Rule, Power Rule, Constant Multiple Rule, and Sum and Difference Rules
- How to find rates of change: velocity, marginal profit, marginal revenue, and marginal cost
- How to use the Product, Quotient, Chain, and General Power Rules
- How to calculate higher-order derivatives and derivatives using implicit differentiation
- How to solve related-rate problems and applications

WHY YOU SHOULD LEARN IT:

Derivatives have many applications in real life, as can be seen by the examples below, which represent a small sample of the applications in this chapter.

- Increasing Revenue, Example 10 on page 101
- Psychology: Migraine Prevalence, Exercise 62 on page 104
- Average Velocity, Exercises 15 and 16 on page 117
- Demand Function, Exercises 53 and 54 on page 129
- Quality Control, Exercise 58 on page 129
- Velocity and Acceleration, Exercises 41–44 and 50 on pages 145 and 146

2.1 THE DERIVATIVE AND THE SLOPE OF A GRAPH

- Identify tangent lines to a graph at a point.
- Approximate the slopes of tangent lines to graphs at points.
- Use the limit definition to find the slopes of graphs at points.
- Use the limit definition to find the derivatives of functions.
- Describe the relationship between differentiability and continuity.

Tangent Line to a Graph

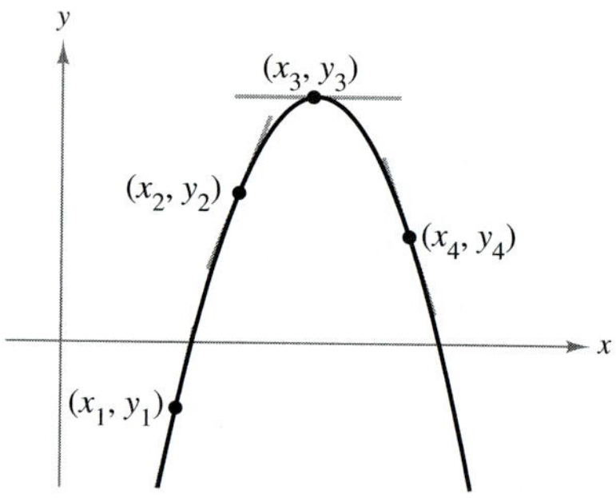

FIGURE 2.1 The slope of a nonlinear graph changes from one point to another.

Calculus is a branch of mathematics that studies rates of change of functions. In this course, you will learn that rates of change have many applications in real life. In Section 1.3, you learned how the slope of a line indicates the rate at which the line rises or falls. For a line, this rate (or slope) is the same at every point on the line. For graphs other than lines, the rate at which the graph rises or falls changes from point to point. For instance, in Figure 2.1, the parabola is rising more quickly at the point (x_1, y_1) than it is at the point (x_2, y_2). At the vertex (x_3, y_3), the graph levels off, and at the point (x_4, y_4), the graph is falling.

To determine the rate at which a graph rises or falls at a *single point*, you can find the slope of the **tangent line** at the point. In simple terms, the tangent line to the graph of a function f at a point $P(x_1, y_1)$ is the line that best approximates the graph at that point, as shown in Figure 2.1. Figure 2.2 shows other examples of tangent lines.

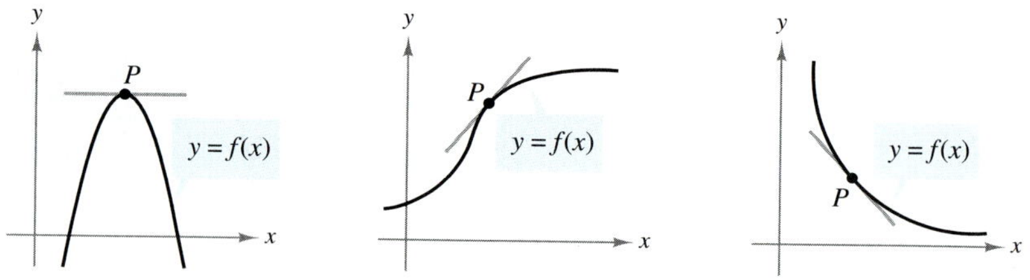

FIGURE 2.2 Tangent Line to a Graph at a Point

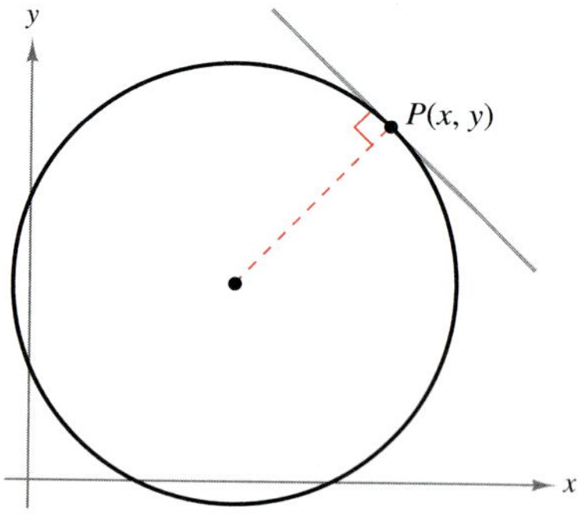

FIGURE 2.3 Tangent Line to a Circle

When Isaac Newton (1642–1727) was working on the "tangent line problem," he realized that it is difficult to define precisely what is meant by a tangent to a general curve. From geometry, you know that a line is tangent to a circle if the line intersects the circle at only one point, as shown in Figure 2.3. Tangent lines to a noncircular graph, however, can intersect the graph at more than one point. For instance, in the second graph in Figure 2.2, if the tangent line were extended, it would intersect the graph at a point other than the point of tangency. In this section, you will see how the notion of a limit can be used to define a general tangent line.

DISCOVERY

Use a graphing utility to sketch the graph of $f(x) = 2x^3 - 4x^2 + 3x - 5$. On the same screen, sketch the graphs of $y = x - 5$, $y = 2x - 5$, and $y = 3x - 5$. Which of these lines, if any, appears to be tangent to the graph of f at the point $(0, -5)$? Explain your reasoning.

Slope of a Graph

Because a tangent line approximates the graph at a point, the problem of finding the slope of a graph at a point becomes one of finding the slope of the tangent line at the point.

EXAMPLE 1 Approximating the Slope of a Graph

Use the graph in Figure 2.4 to approximate the slope of the graph of $f(x) = x^2$ at the point $(1, 1)$.

SOLUTION From the graph of $f(x) = x^2$, you can see that the tangent line at $(1, 1)$ rises approximately two units for each unit change in x. So, the slope of the tangent line at $(1, 1)$ is given by

$$\text{Slope} = \frac{\text{change in } y}{\text{change in } x} \approx \frac{2}{1} = 2.$$

Because the tangent line at the point $(1, 1)$ has a slope of about 2, you can conclude that the graph has a slope of about 2 at the point $(1, 1)$.

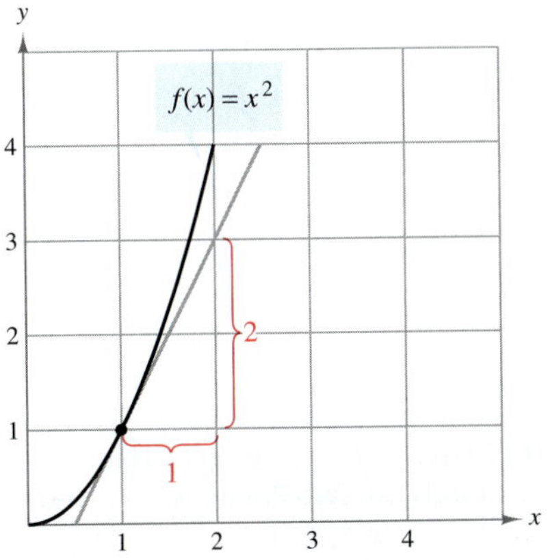

FIGURE 2.4

STUDY TIP

When visually approximating the slope of a graph, note that the scales on the horizontal and vertical axes may differ. When this happens (as it frequently does in applications), the slope of the tangent line is distorted, and you must be careful to account for the difference in scales.

TRY IT 1

Use the graph to approximate the slope of the graph of $f(x) = x^3$ at the point $(1, 1)$.

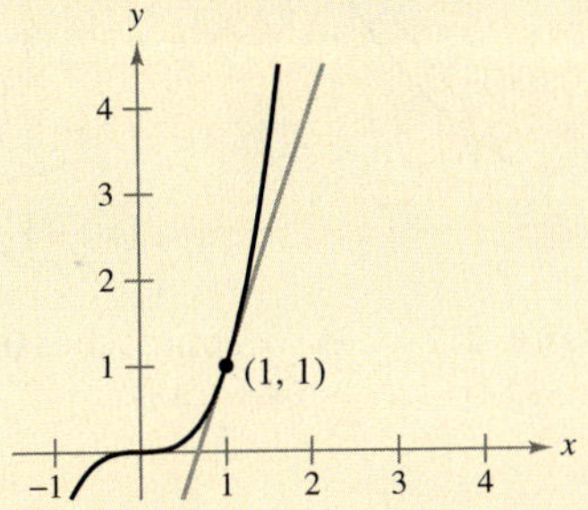

EXAMPLE 2 Interpreting Slope

Figure 2.5 graphically depicts the average daily temperature (in degrees Fahrenheit) in Duluth, Minnesota. Estimate the slope of this graph at the indicated point and give a physical interpretation of the result. *(Source: National Oceanic and Atmospheric Administration)*

SOLUTION From the graph, you can see that the tangent line at the given point falls approximately 27 units for each two-unit change in x. So, you can estimate the slope at the given point to be

$$\begin{aligned}\text{Slope} &= \frac{\text{change in } y}{\text{change in } x} \approx \frac{-27}{2} \\ &= -13.5 \text{ degrees per month.}\end{aligned}$$

This means that you can expect the average daily temperatures in November to be about 13.5 degrees *lower* than the corresponding temperatures in October.

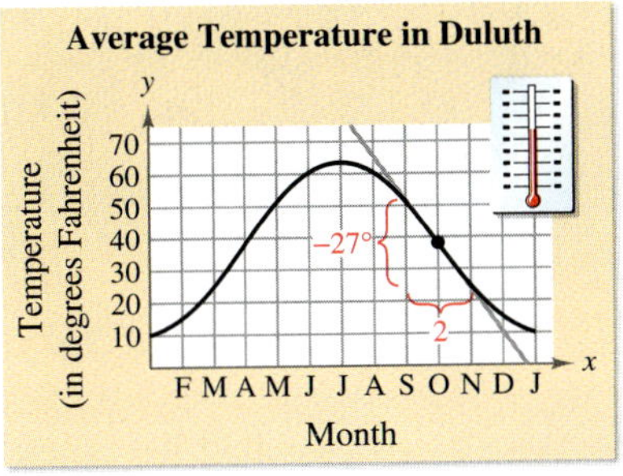

FIGURE 2.5

TRY IT 2

For which months do the slopes of the tangent lines appear to be positive? Negative? Interpret these slopes in the context of the problem.

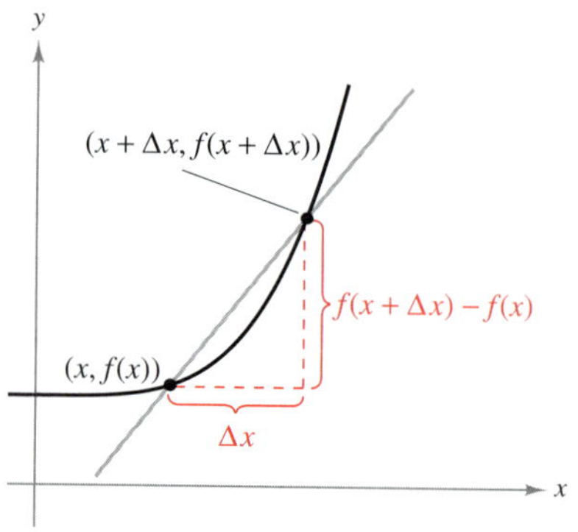

FIGURE 2.6 The Secant Line Through the Two Points $(x, f(x))$ and $(x + \Delta x, f(x + \Delta x))$

Slope and the Limit Process

In Examples 1 and 2, you approximated the slope of a graph at a point by making a careful graph and then "eyeballing" the tangent line at the point of tangency. A more precise method of approximating tangent lines makes use of a **secant line** through the point of tangency and a second point on the graph, as shown in Figure 2.6. If $(x, f(x))$ is the point of tangency and $(x + \Delta x, f(x + \Delta x))$ is a second point on the graph of f, then the slope of the secant line through the two points is

$$m_{\text{sec}} = \frac{f(x + \Delta x) - f(x)}{\Delta x}. \qquad \text{Slope of secant line}$$

The right side of this equation is called the **difference quotient.** The denominator Δx is the **change in x,** and the numerator is the **change in y.** The beauty of this procedure is that you obtain better and better approximations of the slope of the tangent line by choosing the second point closer and closer to the point of tangency, as shown in Figure 2.7.

Using the limit process, you can find the *exact* slope of the tangent line at $(x, f(x))$.

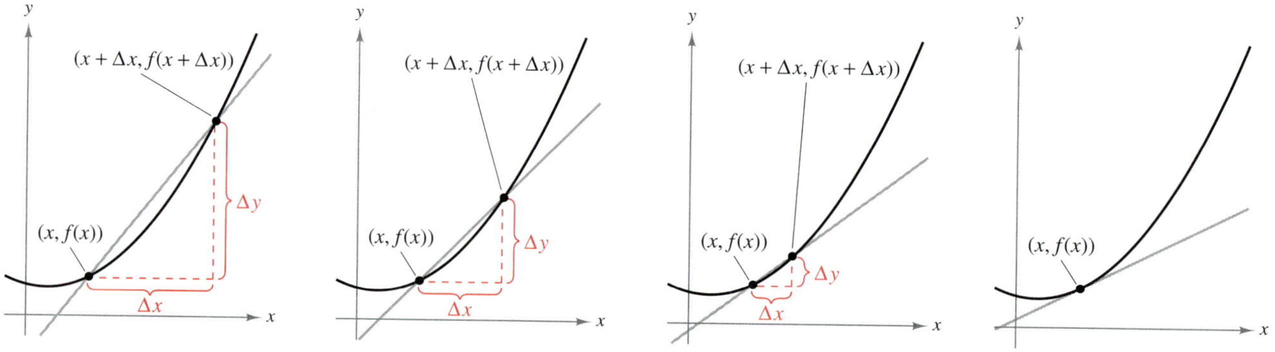

FIGURE 2.7 As Δx approaches 0, the secant lines approach the tangent line.

Definition of the Slope of a Graph

The **slope** m of the graph of f at the point $(x, f(x))$ is equal to the slope of its tangent line at $(x, f(x))$, and is given by

$$m = \lim_{\Delta x \to 0} m_{\text{sec}} = \lim_{\Delta x \to 0} \frac{f(x + \Delta x) - f(x)}{\Delta x}$$

provided this limit exists.

STUDY TIP

Δx is used as a variable to represent the change in x in the definition of the slope of a graph. Other variables may also be used. For instance, this definition is sometimes written as

$$m = \lim_{h \to 0} \frac{f(x + h) - f(x)}{h}.$$

EXAMPLE 3 Finding Slope by the Limit Process

Find the slope of the graph of $f(x) = x^2$ at the point $(-2, 4)$.

SOLUTION Begin by finding an expression that represents the slope of a secant line at the point $(-2, 4)$.

$$\begin{aligned}
m_{sec} &= \frac{f(-2 + \Delta x) - f(-2)}{\Delta x} && \text{Set up difference quotient.}\\
&= \frac{(-2 + \Delta x)^2 - (-2)^2}{\Delta x} && \text{Use } f(x) = x^2.\\
&= \frac{4 - 4\,\Delta x + (\Delta x)^2 - 4}{\Delta x} && \text{Expand terms.}\\
&= \frac{-4\,\Delta x + (\Delta x)^2}{\Delta x} && \text{Simplify.}\\
&= \frac{\Delta x(-4 + \Delta x)}{\Delta x} && \text{Factor and divide out.}\\
&= -4 + \Delta x, \quad \Delta x \neq 0 && \text{Simplify.}
\end{aligned}$$

Next, take the limit of m_{sec} as $\Delta x \to 0$.

$$m = \lim_{\Delta x \to 0} m_{sec} = \lim_{\Delta x \to 0} (-4 + \Delta x) = -4$$

So, the graph of f has a slope of -4 at the point $(-2, 4)$, as shown in Figure 2.8.

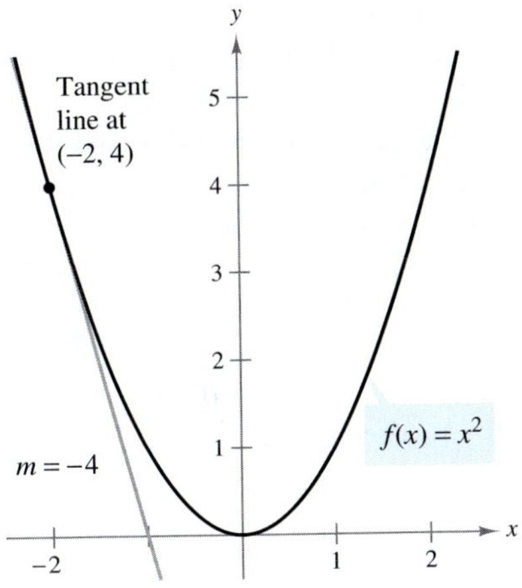

FIGURE 2.8

TRY IT 3

Find the slope of the graph of $f(x) = x^2$ at the point $(2, 4)$.

EXAMPLE 4 Finding the Slope of a Graph

Find the slope of $f(x) = -2x + 4$.

SOLUTION You know from your study of linear functions that the line given by $f(x) = -2x + 4$ has a slope of -2, as shown in Figure 2.9. This conclusion is consistent with the limit definition of slope.

$$\begin{aligned}
m &= \lim_{\Delta x \to 0} \frac{f(x + \Delta x) - f(x)}{\Delta x}\\
&= \lim_{\Delta x \to 0} \frac{[-2(x + \Delta x) + 4] - [-2x + 4]}{\Delta x}\\
&= \lim_{\Delta x \to 0} \frac{-2x - 2\,\Delta x + 4 + 2x - 4}{\Delta x}\\
&= \lim_{\Delta x \to 0} \frac{-2\Delta x}{\Delta x} = -2
\end{aligned}$$

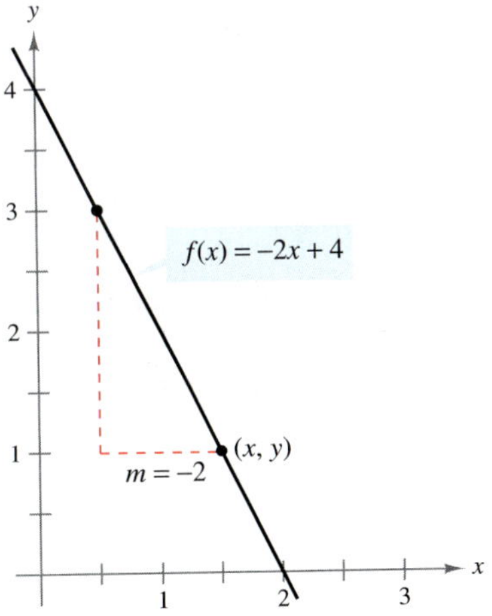

FIGURE 2.9

TRY IT 4

Find the slope of the graph of $f(x) = 2x + 5$.

DISCOVERY

Use a graphing utility to graph the function $y_1 = x^2 + 1$ and the three lines $y_2 = 3x - 1$, $y_3 = 4x - 3$, and $y_4 = 5x - 5$. Which of these lines appears to be tangent to y_1 at the point $(2, 5)$? Confirm your answer by showing that the graphs of y_1 and its tangent line have only one point of intersection, whereas the graphs of y_1 and the other lines each have two points of intersection.

It is important that you see the distinction between the ways the difference quotients were set up in Examples 3 and 4. In Example 3, you were finding the slope of a graph at a specific point $(c, f(c))$. To find the slope, you can use the following form of a difference quotient.

$$m = \lim_{\Delta x \to 0} \frac{f(c + \Delta x) - f(c)}{\Delta x} \qquad \text{Slope at specific point}$$

In Example 4, however, you were finding a formula for the slope at *any* point on the graph. In such cases, you should use x, rather than c, in the difference quotient.

$$m = \lim_{\Delta x \to 0} \frac{f(x + \Delta x) - f(x)}{\Delta x} \qquad \text{Formula for slope}$$

Except for linear functions, this form will always produce a function of x, which can then be evaluated to find the slope at any desired point.

EXAMPLE 5 Finding a Formula for the Slope of a Graph

Find a formula for the slope of the graph of $f(x) = x^2 + 1$. What are the slopes at the points $(-1, 2)$ and $(2, 5)$?

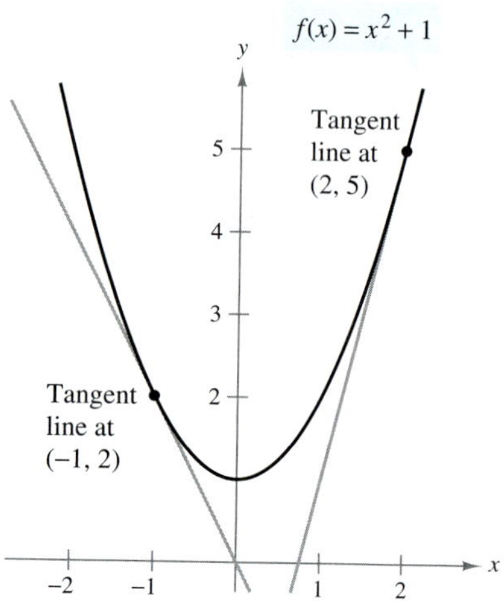

FIGURE 2.10

SOLUTION

$$\begin{aligned}
m_{sec} &= \frac{f(x + \Delta x) - f(x)}{\Delta x} && \text{Set up difference quotient.}\\
&= \frac{[(x + \Delta x)^2 + 1] - (x^2 + 1)}{\Delta x} && \text{Use } f(x) = x^2 + 1.\\
&= \frac{x^2 + 2x\,\Delta x + (\Delta x)^2 + 1 - x^2 - 1}{\Delta x} && \text{Expand terms.}\\
&= \frac{2x\,\Delta x + (\Delta x)^2}{\Delta x} && \text{Simplify.}\\
&= \frac{\Delta x(2x + \Delta x)}{\Delta x} && \text{Factor and divide out.}\\
&= 2x + \Delta x, \quad \Delta x \neq 0 && \text{Simplify.}
\end{aligned}$$

Next, take the limit of m_{sec} as $\Delta x \to 0$.

$$\begin{aligned}
m &= \lim_{\Delta x \to 0} m_{sec}\\
&= \lim_{\Delta x \to 0} (2x + \Delta x)\\
&= 2x
\end{aligned}$$

Using the formula $m = 2x$, you can find the slopes at the specified points. At $(-1, 2)$ the slope is $m = 2(-1) = -2$, and at $(2, 5)$ the slope is $m = 2(2) = 4$. The graph of f is shown in Figure 2.10.

TRY IT 5

Find a formula for the slope of the graph of $f(x) = 4x^2 + 1$. What are the slopes at the points $(0, 1)$ and $(1, 5)$?

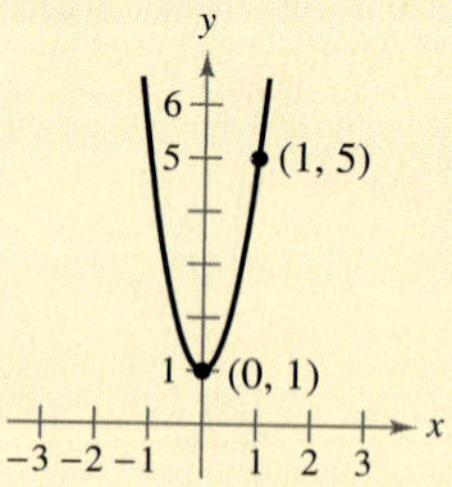

STUDY TIP

The slope of the graph of $f(x) = x^2 + 1$ varies for different values of x. For what value of x is the slope equal to 0?

The Derivative of a Function

In Example 5, you started with the function $f(x) = x^2 + 1$ and used the limit process to derive another function, $m = 2x$, that represents the slope of the graph of f at the point $(x, f(x))$. This derived function is called the **derivative** of f at x. It is denoted by $f'(x)$, which is read as "f prime of x."

Definition of the Derivative

The **derivative of f at x** is given by

$$f'(x) = \lim_{\Delta x \to 0} \frac{f(x + \Delta x) - f(x)}{\Delta x}$$

provided this limit exists. A function is **differentiable** at x if its derivative exists at x. The process of finding derivatives is called **differentiation.**

STUDY TIP

The notation dy/dx is read as "the derivative of y with respect to x," and using limit notation, you can write

$$\frac{dy}{dx} = \lim_{\Delta x \to 0} \frac{\Delta y}{\Delta x}$$
$$= \lim_{\Delta x \to 0} \frac{f(x + \Delta x) - f(x)}{\Delta x}$$
$$= f'(x).$$

In addition to $f'(x)$, other notations can be used to denote the derivative of $y = f(x)$. The most common are

$$\frac{dy}{dx}, \quad y', \quad \frac{d}{dx}[f(x)], \quad \text{and} \quad D_x[y].$$

EXAMPLE 6 **Finding a Derivative**

Find the derivative of $f(x) = 3x^2 - 2x$.

SOLUTION

$$f'(x) = \lim_{\Delta x \to 0} \frac{f(x + \Delta x) - f(x)}{\Delta x}$$
$$= \lim_{\Delta x \to 0} \frac{[3(x + \Delta x)^2 - 2(x + \Delta x)] - (3x^2 - 2x)}{\Delta x}$$
$$= \lim_{\Delta x \to 0} \frac{3x^2 + 6x\,\Delta x + 3(\Delta x)^2 - 2x - 2\,\Delta x - 3x^2 + 2x}{\Delta x}$$
$$= \lim_{\Delta x \to 0} \frac{6x\,\Delta x + 3(\Delta x)^2 - 2\,\Delta x}{\Delta x}$$
$$= \lim_{\Delta x \to 0} \frac{\cancel{\Delta x}(6x + 3\,\Delta x - 2)}{\cancel{\Delta x}}$$
$$= \lim_{\Delta x \to 0} (6x + 3\,\Delta x - 2)$$
$$= 6x - 2$$

So, the derivative of $f(x) = 3x^2 - 2x$ is $f'(x) = 6x - 2$.

TRY IT 6

Find the derivative of $f(x) = x^2 - 5x$.

In many applications, it is convenient to use a variable other than x as the independent variable. Example 7 shows a function that uses t as the independent variable.

TECHNOLOGY

You can use a graphing utility to confirm the result given in Example 7. One way to do this is to choose a point on the graph of $y = 2/t$, such as $(1, 2)$, and find the equation of the tangent line at that point. Using the derivative found in the example, you know that the slope of the tangent line when $t = 1$ is $m = -2$. This means that the tangent line at the point $(1, 2)$ is

$$y - y_1 = m(t - t_1)$$
$$y - 2 = -2(t - 1) \text{ or}$$
$$y = -2t + 4.$$

By graphing $y = 2/t$ and $y = -2t + 4$ in the same viewing window, as shown below, you can confirm that the line is tangent to the graph at the point $(1, 2)$.

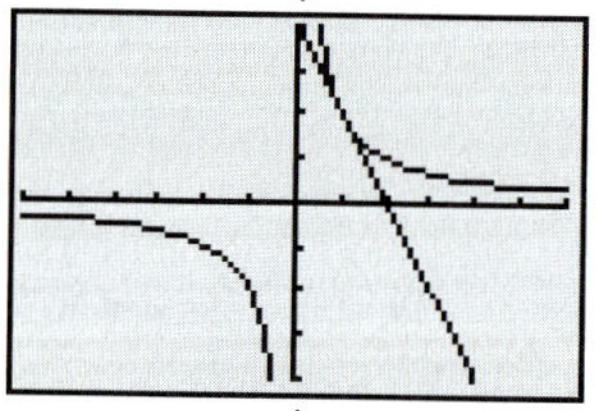

EXAMPLE 7 Finding a Derivative

Find the derivative of y with respect to t for the function

$$y = \frac{2}{t}.$$

SOLUTION Consider $y = f(t)$, and use the limit process as shown.

$$\frac{dy}{dt} = \lim_{\Delta t \to 0} \frac{f(t + \Delta t) - f(t)}{\Delta t} \qquad \text{Set up difference quotient.}$$

$$= \lim_{\Delta t \to 0} \frac{\dfrac{2}{t + \Delta t} - \dfrac{2}{t}}{\Delta t} \qquad \text{Use } f(t) = 2/t.$$

$$= \lim_{\Delta t \to 0} \frac{\dfrac{2t - 2t - 2\,\Delta t}{t(t + \Delta t)}}{\Delta t} \qquad \text{Expand terms.}$$

$$= \lim_{\Delta t \to 0} \frac{-2\,\Delta t}{t(\Delta t)(t + \Delta t)} \qquad \text{Factor and divide out.}$$

$$= \lim_{\Delta t \to 0} \frac{-2}{t(t + \Delta t)} \qquad \text{Simplify.}$$

$$= -\frac{2}{t^2} \qquad \text{Evaluate the limit.}$$

So, the derivative of y with respect to t is

$$\frac{dy}{dt} = -\frac{2}{t^2}.$$

Remember that the derivative of a function gives you a formula for finding the slope of the tangent line at any point on the graph of the function. For example, the slope of the tangent line to the graph of f at the point $(1, 2)$ is given by

$$f'(1) = -\frac{2}{1^2} = -2.$$

To find the slopes of the graph at other points, substitute the t-coordinate of the point into the derivative, as shown below.

Point	*t-Coordinate*	*Slope*
$(2, 1)$	$t = 2$	$m = f'(2) = -\frac{2}{2^2} = -\frac{1}{2}$
$(-2, -1)$	$t = -2$	$m = f'(-2) = -\frac{2}{(-2)^2} = -\frac{1}{2}$

TRY IT 7

Find the derivative of y with respect to t for the function $y = 4/t$.

Differentiability and Continuity

Not every function is differentiable. Figure 2.11 shows some common situations in which a function will not be differentiable at a point—vertical tangent lines, discontinuities, and sharp turns in the graph. Each of the functions shown in Figure 2.11 is differentiable at every value of x *except* $x = 0$.

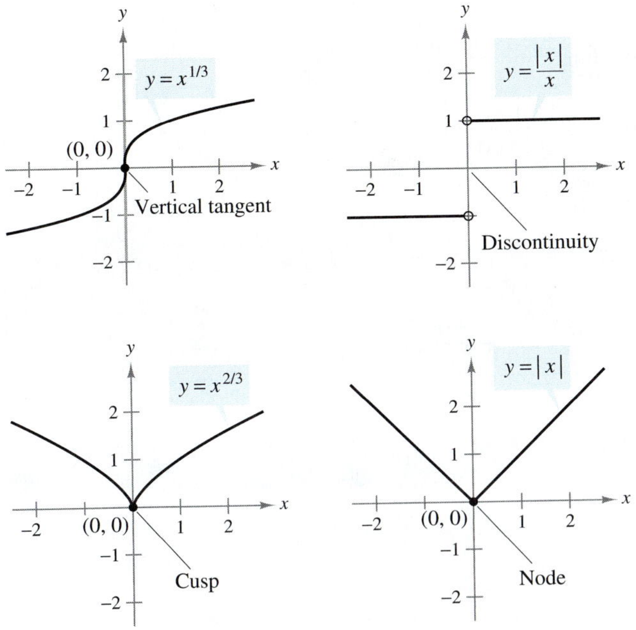

FIGURE 2.11 Functions That Are Not Differentiable at $x = 0$

In Figure 2.11, you can see that all but one of the functions are continuous at $x = 0$ but none are differentiable there. This shows that continuity is not a strong enough condition to guarantee differentiability. On the other hand, if a function is differentiable at a point, then it must be continuous at that point. This important result is stated in the following theorem.

Differentiability Implies Continuity

If a function is differentiable at $x = c$, then it is continuous at $x = c$.

TAKE ANOTHER LOOK

A Graphing Utility Experiment

Use a graphing utility to graph $f(x) = x^3 - 3x$ and the lines below.

a. $y = 2$ b. $y = x - 3$ c. $y = -3x$ d. $y = -x + 1$

Do any of these lines appear to be tangent? If so, what is the point of tangency? Explain how you can verify your conclusion analytically.

PREREQUISITE REVIEW 2.1

The following warm-up exercises involve skills that were covered in earlier sections. You will use these skills in the exercise set for this section.

In Exercises 1 and 2, find an equation of the line containing P and Q.

1. $P(2, 1),\ Q(2, 4)$

2. $P(2, 2),\ Q(-5, 2)$

In Exercises 3–6, find the limit.

3. $\displaystyle\lim_{\Delta x \to 0} \frac{2x\Delta x + (\Delta x)^2}{\Delta x}$

4. $\displaystyle\lim_{\Delta x \to 0} \frac{3x^2\Delta x + 3x(\Delta x)^2 + (\Delta x)^3}{\Delta x}$

5. $\displaystyle\lim_{\Delta x \to 0} \frac{1}{x(x + \Delta x)}$

6. $\displaystyle\lim_{\Delta x \to 0} \frac{(x + \Delta x)^2 - x^2}{\Delta x}$

In Exercises 7–10, find the domain of the function.

7. $f(x) = \dfrac{1}{x - 1}$

8. $f(x) = \dfrac{1}{5}x^3 - 2x^2 + \dfrac{1}{3}x - 1$

9. $f(x) = \dfrac{6x}{x^3 + x}$

10. $f(x) = \dfrac{x^2 - 2x - 24}{x^2 + x - 12}$

EXERCISES 2.1

In Exercises 1–4, trace the graph and sketch the tangent lines at (x_1, y_1) and (x_2, y_2).

1.

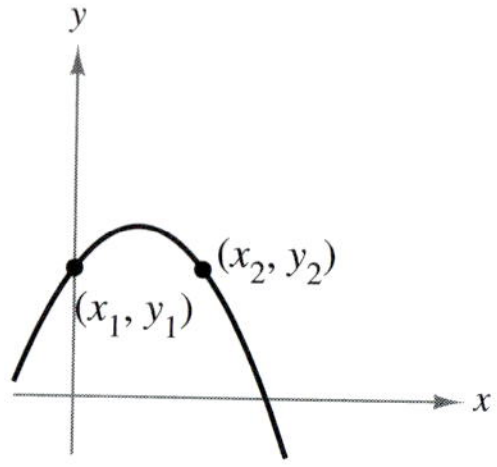

2.

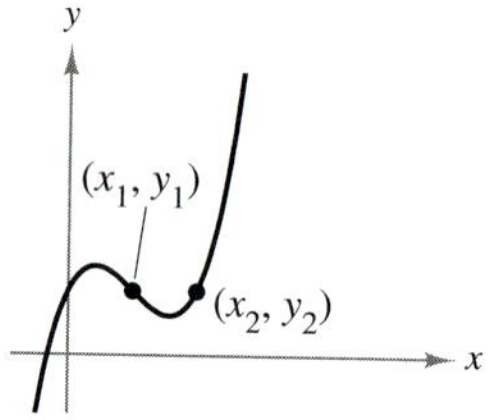

3.

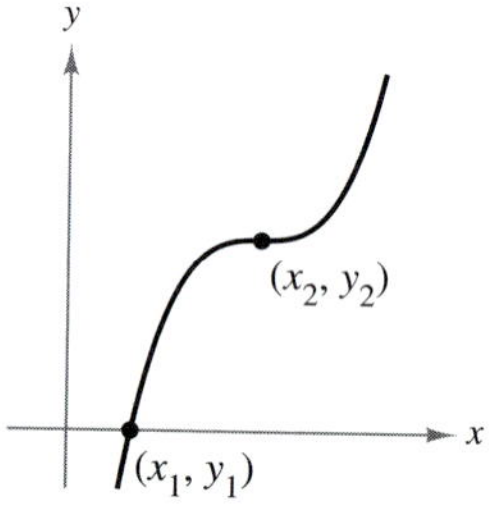

4.

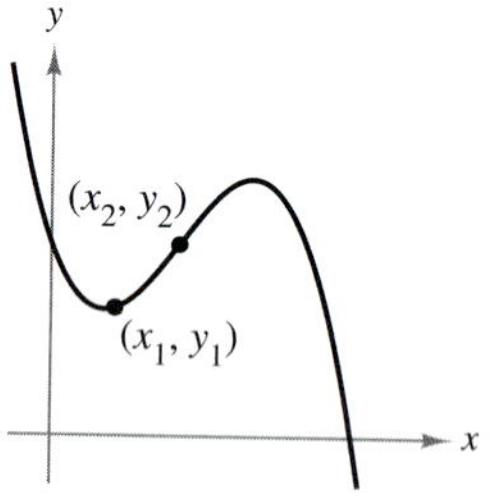

In Exercises 5–10, estimate the slope of the graph at the point (x, y). (Each square on the grid is 1 unit by 1 unit.)

5.

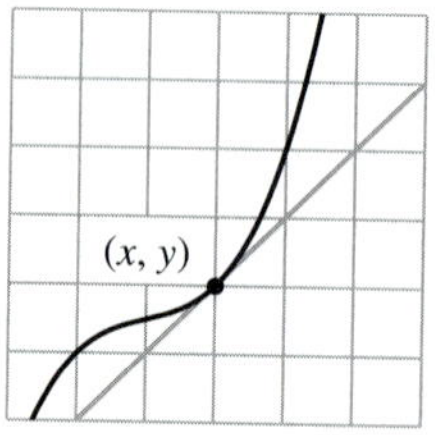

6.

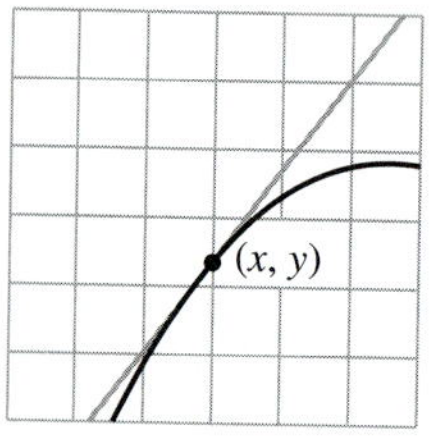

7.

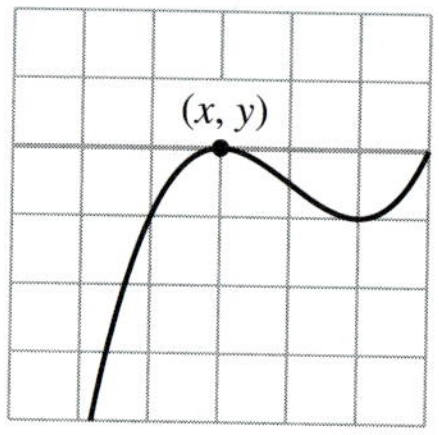

8.

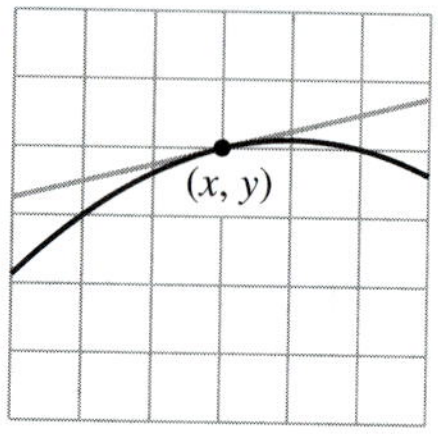

9.

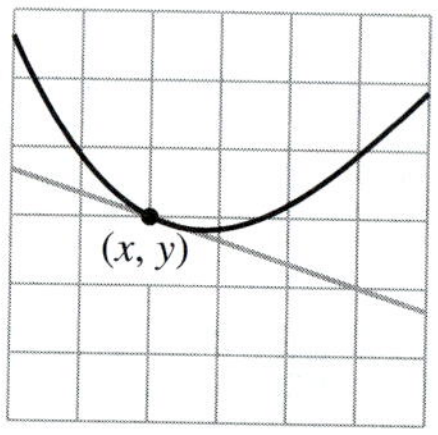

10.

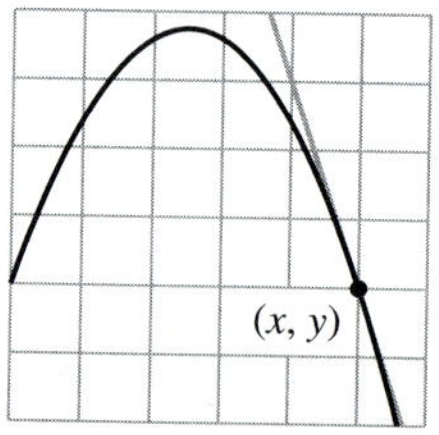

11. ***Revenue*** The graph (on page 91) represents the revenue R (in millions of dollars per year) for Polo Ralph Lauren from 1996 through 2002, where $t = 6$ corresponds to 1996. Estimate the slopes of the graph for the years 1997, 2000, and 2002. *(Source: Polo Ralph Lauren Corp.)*

Figure for 11

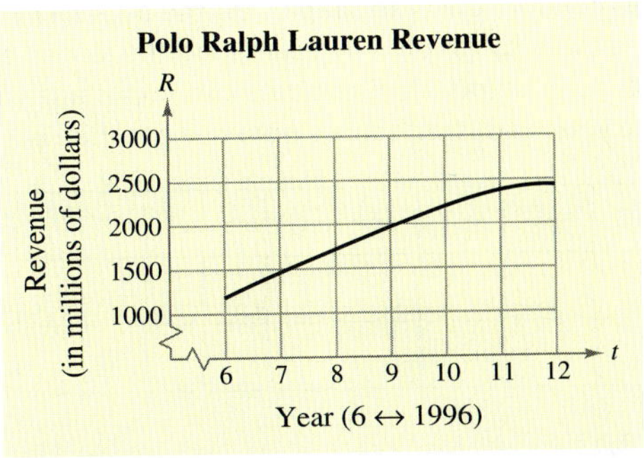

12. *Sales* The graph represents the sales S (in millions of dollars per year) for Scotts Company from 1997 through 2003, where $t = 7$ corresponds to 1997. Estimate the slopes of the graph for the years 1998, 2001, and 2003. *(Source: Scotts Company)*

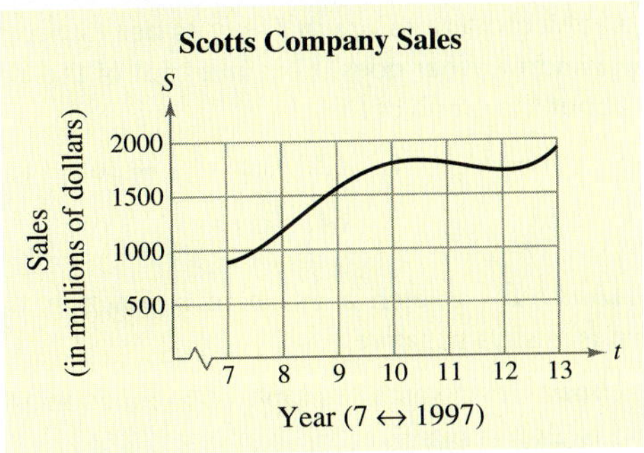

13. *Consumer Trends* The graph shows the number of visitors V to a national park in hundreds of thousands during a one-year period, where $t = 1$ corresponds to January. Estimate the slopes of the graph at $t = 1$, 8, and 12.

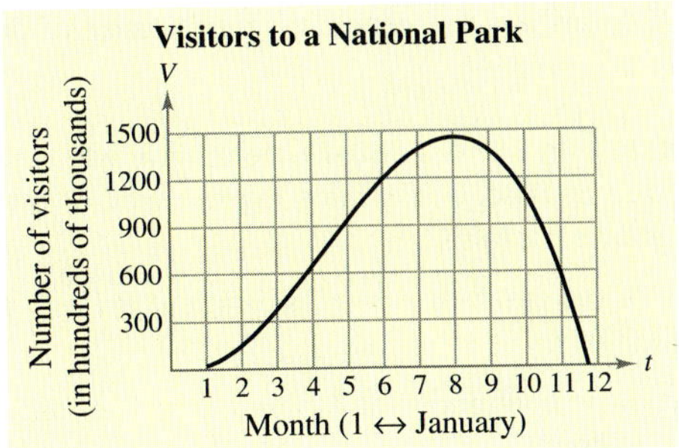

14. *Athletics* Two long distance runners starting out side by side begin a 10,000-meter run. Their distances are given by $s = f(t)$ and $s = g(t)$, respectively, where s is measured in thousands of meters and t is measured in minutes.

Figure for 14

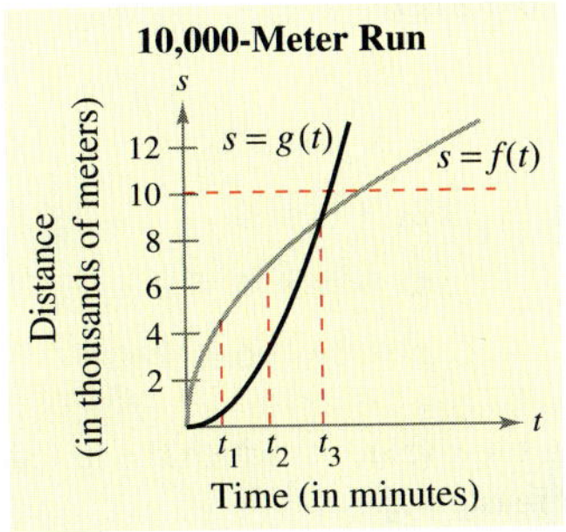

(a) Which runner is running faster at t_1?

(b) What conclusion can you make regarding their rates at t_2?

(c) What conclusion can you make regarding their rates at t_3?

(d) Which runner finishes the race first? Explain.

In Exercises 15–26, use the limit definition to find the derivative of the function.

15. $f(x) = 3$

16. $f(x) = -4$

17. $f(x) = -5x + 3$

18. $f(x) = \frac{1}{2}x + 5$

19. $f(x) = x^2 - 4$

20. $f(x) = 1 - x^2$

21. $h(t) = \sqrt{t - 1}$

22. $f(x) = \sqrt{x + 2}$

23. $f(t) = t^3 - 12t$

24. $f(t) = t^3 + t^2$

25. $f(x) = \dfrac{1}{x + 2}$

26. $g(s) = \dfrac{1}{s - 1}$

In Exercises 27–36, find the slope of the tangent line to the graph of f at the given point.

27. $f(x) = 6 - 2x$; $(2, 2)$

28. $f(x) = 2x + 4$; $(1, 6)$

29. $f(x) = -1$; $(0, -1)$

30. $f(x) = 6$; $(-2, 6)$

31. $f(x) = x^2 - 2$; $(2, 2)$

32. $f(x) = x^2 + 2x + 1$; $(-3, 4)$

33. $f(x) = x^3 - x$; $(2, 6)$

34. $f(x) = x^3 + 2x$; $(1, 3)$

35. $f(x) = \sqrt{1 - 2x}$; $(-4, 3)$

36. $f(x) = \sqrt{2x - 2}$; $(9, 4)$

In Exercises 37–44, find an equation of the tangent line to the graph of f at the given point. Then verify your result by sketching the graph of f and the tangent line.

37. $f(x) = \frac{1}{2}x^2$; $(2, 2)$

38. $f(x) = -x^2$; $(-1, -1)$

39. $f(x) = (x - 1)^2$; $(-2, 9)$

40. $f(x) = 2x^2 - 1$; $(0, -1)$

41. $f(x) = \sqrt{x} + 1$; $(4, 3)$

42. $f(x) = \sqrt{x + 2}$; $(7, 3)$

43. $f(x) = \dfrac{1}{x}$; $(1, 1)$

44. $f(x) = \dfrac{1}{x - 1}$; $(2, 1)$

In Exercises 45–48, find an equation of the line that is tangent to the graph of f and parallel to the given line.

Function	*Line*
45. $f(x) = -\frac{1}{4}x^2$	$x + y = 0$
46. $f(x) = x^2 + 1$	$2x + y = 0$
47. $f(x) = -\frac{1}{2}x^3$	$6x + y + 4 = 0$
48. $f(x) = x^2 - x$	$x + 2y - 6 = 0$

In Exercises 49–56, describe the x-values at which the function is differentiable. Explain your reasoning.

49. $y = |x + 3|$

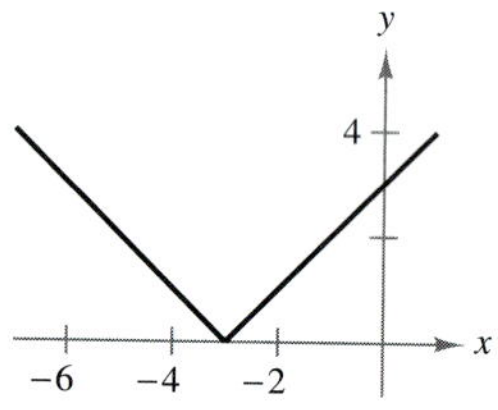

50. $y = |x^2 - 9|$

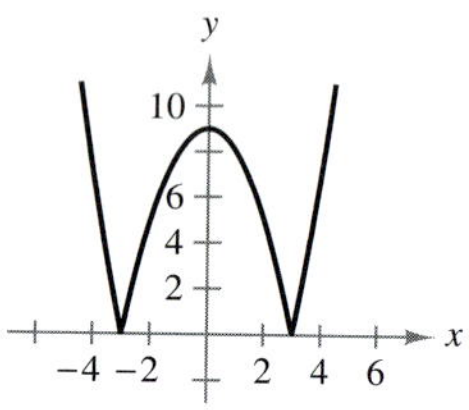

51. $y = (x - 3)^{2/3}$

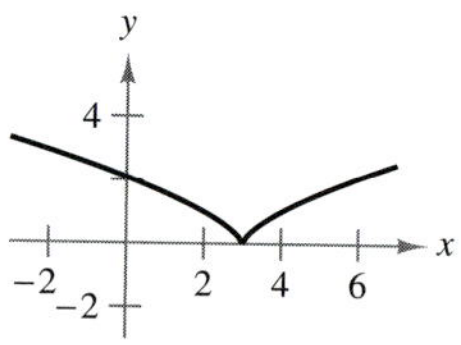

52. $y = x^{2/5}$

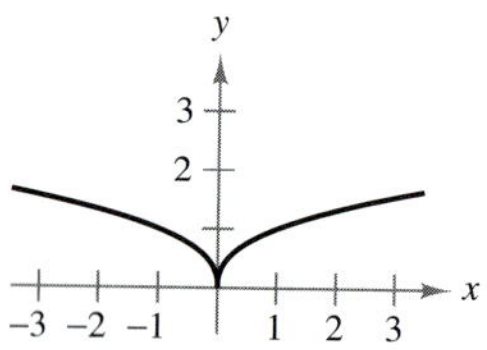

53. $y = \sqrt{x - 1}$

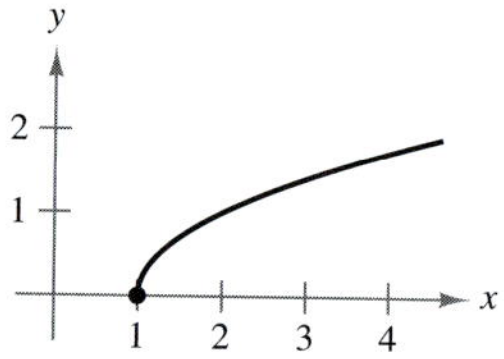

54. $y = \dfrac{x^2}{x^2 - 4}$

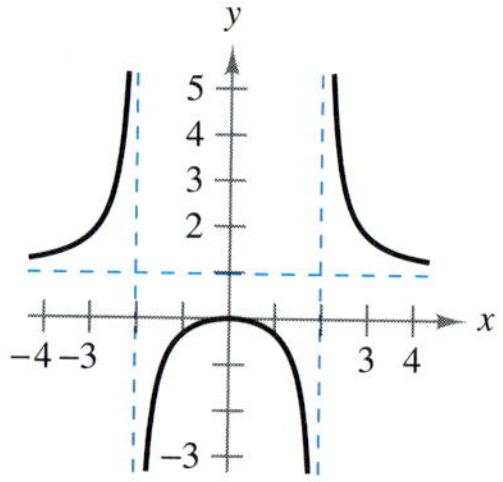

55. $y = \begin{cases} x^3 + 3, & x < 0 \\ x^3 - 3, & x \geq 0 \end{cases}$

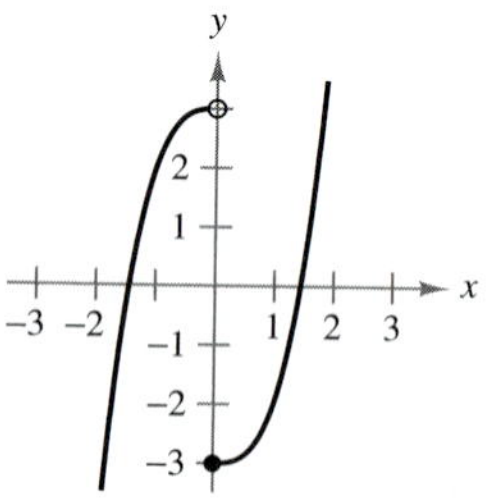

56. $y = \begin{cases} x^2, & x \leq 1 \\ -x^2, & x > 1 \end{cases}$

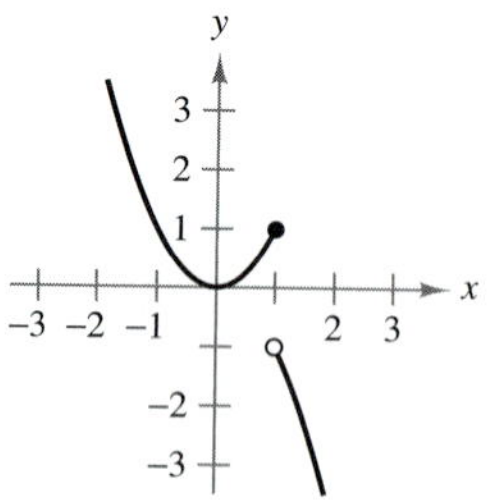

Graphical, Numerical, and Analytic Analysis In Exercises 57–60, use a graphing utility to graph f on the interval $[-2, 2]$. Complete the table by graphically estimating the slopes of the graph at the given points. Then evaluate the slopes analytically and compare your results with those obtained graphically.

x	-2	$-\frac{3}{2}$	-1	$-\frac{1}{2}$	0	$\frac{1}{2}$	1	$\frac{3}{2}$	2
$f(x)$									
$f'(x)$									

57. $f(x) = \frac{1}{4}x^3$

58. $f(x) = \frac{1}{2}x^2$

59. $f(x) = -\frac{1}{2}x^3$

60. $f(x) = -\frac{3}{2}x^2$

In Exercises 61–64, find the derivative of the given function f. Then use a graphing utility to graph f and its derivative in the same viewing window. What does the x-intercept of the derivative indicate about the graph of f?

61. $f(x) = x^2 - 4x$

62. $f(x) = 2 + 6x - x^2$

63. $f(x) = x^3 - 3x$

64. $f(x) = x^3 - 6x^2$

65. ***Think About It*** Sketch a graph of a function whose derivative is always negative.

66. ***Think About It*** Sketch a graph of a function whose derivative is always positive.

67. ***Writing*** Use a graphing utility to graph the two functions $f(x) = x^2 + 1$ and $g(x) = |x| + 1$ in the same viewing window. Use the *zoom* and *trace* features to analyze the graphs near the point $(0, 1)$. What do you observe? Which function is differentiable at this point? Write a short paragraph describing the geometric significance of differentiability at a point.

True or False? In Exercises 68–71, determine whether the statement is true or false. If it is false, explain why or give an example that shows it is false.

68. The slope of the graph of $y = x^2$ is different at every point on the graph of f.

69. If a function is continuous at a point, then it is differentiable at that point.

70. If a function is differentiable at a point, then it is continuous at that point.

71. A tangent line to a graph can intersect the graph at more than one point.

2.2 SOME RULES FOR DIFFERENTIATION

- Find the derivatives of functions using the Constant Rule.
- Find the derivatives of functions using the Power Rule.
- Find the derivatives of functions using the Constant Multiple Rule.
- Find the derivatives of functions using the Sum and Difference Rules.
- Use derivatives to answer questions about real-life situations.

The Constant Rule

In Section 2.1, you found derivatives by the limit process. This process is tedious, even for simple functions, but fortunately there are rules that greatly simplify differentiation. These rules allow you to calculate derivatives without the *direct* use of limits.

The Constant Rule

The derivative of a constant function is zero. That is,

$$\frac{d}{dx}[c] = 0, \qquad c \text{ is a constant.}$$

PROOF Let $f(x) = c$. Then, by the limit definition of the derivative, you can write

$$f'(x) = \lim_{\Delta x \to 0} \frac{f(x + \Delta x) - f(x)}{\Delta x} = \lim_{\Delta x \to 0} \frac{c - c}{\Delta x} = \lim_{\Delta x \to 0} 0 = 0.$$

So,

$$\frac{d}{dx}[c] = 0.$$

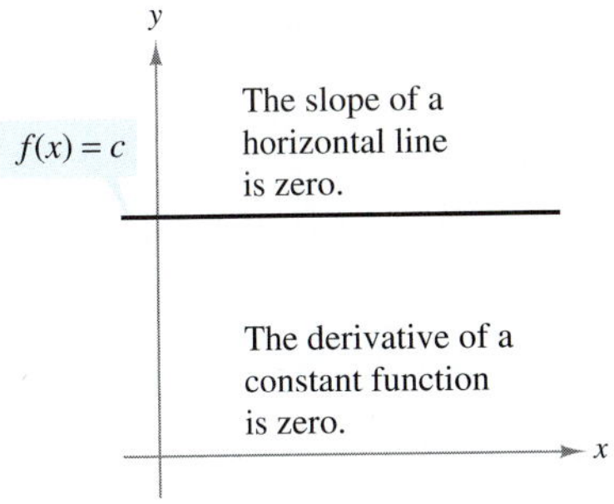

FIGURE 2.12

STUDY TIP

Note in Figure 2.12 that the Constant Rule is equivalent to saying that the slope of a horizontal line is zero.

EXAMPLE 1 **Finding Derivatives of Constant Functions**

(a) $\frac{d}{dx}[7] = 0$

(b) If $f(x) = 0$, then $f'(x) = 0$.

(c) If $y = 2$, then $\frac{dy}{dx} = 0$.

(d) If $g(t) = -\frac{3}{2}$, then $g'(t) = 0$.

STUDY TIP

An interpretation of the Constant Rule says that the tangent line to a constant function is the function itself. Find an equation of the tangent line to $f(x) = -4$ at $x = 3$.

TRY IT 1

Find the derivative of each function.

(a) $f(x) = -2$ (b) $y = \pi$ (c) $g(w) = \sqrt{5}$ (d) $s(t) = 320.5$

The Power Rule

The binomial expansion process is used to prove the Power Rule.

$$(x + \Delta x)^2 = x^2 + 2x\,\Delta x + (\Delta x)^2$$
$$(x + \Delta x)^3 = x^3 + 3x^2\,\Delta x + 3x(\Delta x)^2 + (\Delta x)^3$$
$$(x + \Delta x)^n = x^n + nx^{n-1}\,\Delta x + \underbrace{\frac{n(n-1)x^{n-2}}{2}(\Delta x)^2 + \cdots + (\Delta x)^n}_{(\Delta x)^2 \text{ is a factor of these terms.}}$$

The (Simple) Power Rule

$$\frac{d}{dx}[x^n] = nx^{n-1}, \qquad n \text{ is any real number.}$$

PROOF We prove only the case in which n is a positive integer. Let $f(x) = x^n$. Using the binomial expansion, you can write

$$f'(x) = \lim_{\Delta x \to 0} \frac{f(x + \Delta x) - f(x)}{\Delta x} \qquad \text{Definition of derivative}$$
$$= \lim_{\Delta x \to 0} \frac{(x + \Delta x)^n - x^n}{\Delta x}$$
$$= \lim_{\Delta x \to 0} \frac{x^n + nx^{n-1}\,\Delta x + \frac{n(n-1)x^{n-2}}{2}(\Delta x)^2 + \cdots + (\Delta x)^n - x^n}{\Delta x}$$
$$= \lim_{\Delta x \to 0} \left[nx^{n-1} + \frac{n(n-1)x^{n-2}}{2}(\Delta x) + \cdots + (\Delta x)^{n-1}\right]$$
$$= nx^{n-1} + 0 + \cdots + 0 = nx^{n-1}.$$

For the Power Rule, the case in which $n = 1$ is worth remembering as a separate differentiation rule. That is,

$$\frac{d}{dx}[x] = 1. \qquad \text{The derivative of } x \text{ is 1.}$$

This rule is consistent with the fact that the slope of the line given by $y = x$ is 1. (See Figure 2.13.)

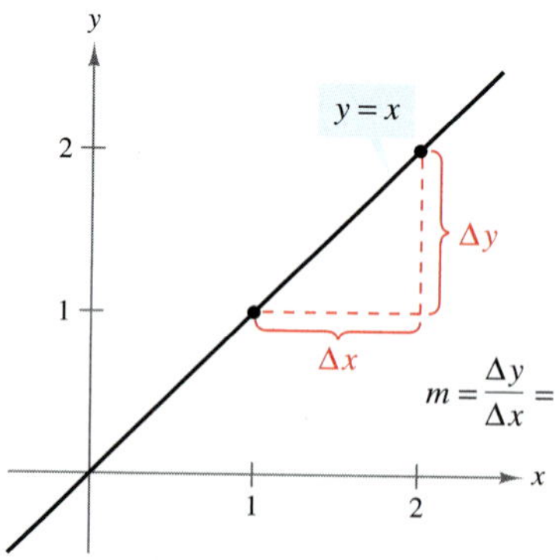

FIGURE 2.13 The slope of the line $y = x$ is 1.

EXAMPLE 2 Applying the Power Rule

Find the derivative of each function.

Function	*Derivative*
(a) $f(x) = x^3$	$f'(x) = 3x^2$
(b) $y = \dfrac{1}{x^2} = x^{-2}$	$\dfrac{dy}{dx} = (-2)x^{-3} = -\dfrac{2}{x^3}$
(c) $g(t) = t$	$g'(t) = 1$
(d) $R = x^4$	$\dfrac{dR}{dx} = 4x^3$

TRY IT 2

Find the derivative of each function.

(a) $f(x) = x^4$ (b) $y = \dfrac{1}{x^3}$

(c) $g(w) = w^2$ (d) $s(t) = \dfrac{1}{t}$

In Example 2(b), note that *before* differentiating, you should rewrite $1/x^2$ as x^{-2}. Rewriting is the first step in *many* differentiation problems.

Original Function: $y = \dfrac{1}{x^2}$ → Rewrite: $y = x^{-2}$ → Differentiate: $\dfrac{dy}{dx} = (-2)x^{-3}$ → Simplify: $\dfrac{dy}{dx} = -\dfrac{2}{x^3}$

Remember that the derivative of a function f is another function that gives the slope of the graph of f at any point at which f is differentiable. So, you can use the derivative to find slopes, as shown in Example 3.

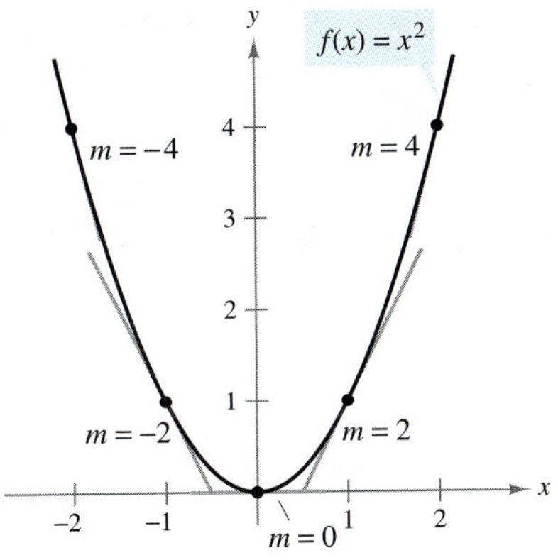

FIGURE 2.14

EXAMPLE 3 Finding the Slope of a Graph

Find the slopes of the graph of

$f(x) = x^2$ Original function

when $x = -2, -1, 0, 1,$ and 2.

SOLUTION Begin by using the Power Rule to find the derivative of f.

$f'(x) = 2x$ Derivative

You can use the derivative to find the slopes of the graph of f, as shown.

x-Value	*Slope of Graph of f*
$x = -2$	$m = f'(-2) = 2(-2) = -4$
$x = -1$	$m = f'(-1) = 2(-1) = -2$
$x = 0$	$m = f'(0) = 2(0) = 0$
$x = 1$	$m = f'(1) = 2(1) = 2$
$x = 2$	$m = f'(2) = 2(2) = 4$

The graph of f is shown in Figure 2.14.

TRY IT 3

Find the slopes of the graph of $f(x) = x^3$ when $x = -1, 0,$ and 1.

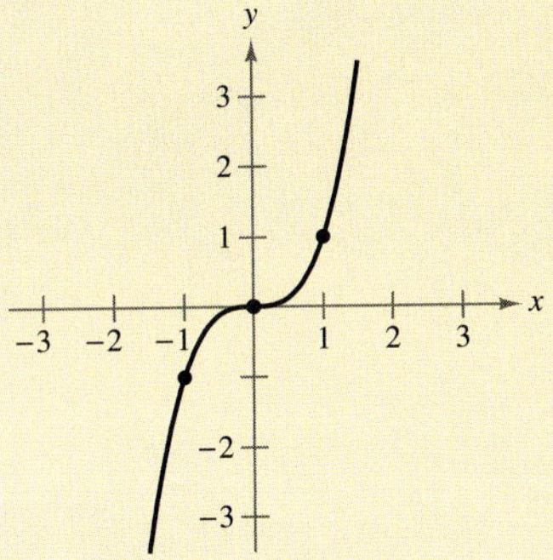

The Constant Multiple Rule

To prove the Constant Multiple Rule, the following property of limits is used.

$$\lim_{x \to a} cg(x) = c\left[\lim_{x \to a} g(x)\right]$$

The Constant Multiple Rule

If f is a differentiable function of x, and c is a real number, then

$$\frac{d}{dx}[cf(x)] = cf'(x), \qquad c \text{ is a constant.}$$

PROOF Apply the definition of the derivative to produce

$$\frac{d}{dx}[cf(x)] = \lim_{\Delta x \to 0} \frac{cf(x + \Delta x) - cf(x)}{\Delta x} \qquad \text{Definition of derivative}$$

$$= \lim_{\Delta x \to 0} c\left[\frac{f(x + \Delta x) - f(x)}{\Delta x}\right]$$

$$= c\left[\lim_{\Delta x \to 0} \frac{f(x + \Delta x) - f(x)}{\Delta x}\right] = cf'(x).$$

Informally, the Constant Multiple Rule states that constants can be factored out of the differentiation process.

$$\frac{d}{dx}[cf(x)] = c\frac{d}{dx}[\;f(x)] = cf'(x)$$

The usefulness of this rule is often overlooked, especially when the constant appears in the denominator, as shown below.

$$\frac{d}{dx}\left[\frac{f(x)}{c}\right] = \frac{d}{dx}\left[\frac{1}{c}f(x)\right] = \frac{1}{c}\left(\frac{d}{dx}[\;f(x)]\right) = \frac{1}{c}f'(x)$$

To use the Constant Multiple Rule efficiently, look for constants that can be factored out *before* differentiating. For example,

$$\frac{d}{dx}[5x^2] = 5\frac{d}{dx}[x^2] \qquad \text{Factor out 5.}$$

$$= 5(2x) \qquad \text{Differentiate.}$$

$$= 10x \qquad \text{Simplify.}$$

and

$$\frac{d}{dx}\left[\frac{x^2}{5}\right] = \frac{1}{5}\left(\frac{d}{dx}[x^2]\right) \qquad \text{Factor out } \tfrac{1}{5}.$$

$$= \frac{1}{5}(2x) \qquad \text{Differentiate.}$$

$$= \frac{2}{5}x. \qquad \text{Simplify.}$$

EXAMPLE 4 Using the Power and Constant Multiple Rules

Differentiate each function.

(a) $y = 2x^{1/2}$ (b) $f(t) = \dfrac{4t^2}{5}$

SOLUTION

(a) Using the Constant Multiple Rule and the Power Rule, you can write

$$\frac{dy}{dx} = \frac{d}{dx}[2x^{1/2}] = \underbrace{2\frac{d}{dx}[x^{1/2}]}_{\text{Constant Multiple Rule}} = \underbrace{2\left(\frac{1}{2}x^{-1/2}\right)}_{\text{Power Rule}} = x^{-1/2} = \frac{1}{\sqrt{x}}.$$

(b) Begin by rewriting $f(t)$ as

$$f(t) = \frac{4t^2}{5} = \frac{4}{5}t^2.$$

Then, use the Constant Multiple Rule and the Power Rule to obtain

$$f'(t) = \frac{d}{dt}\left[\frac{4}{5}t^2\right] = \frac{4}{5}\left[\frac{d}{dt}(t^2)\right] = \frac{4}{5}(2t) = \frac{8}{5}t.$$

TECHNOLOGY

If you have access to a symbolic differentiation utility, try using it to confirm the derivatives shown in this section.

TRY IT 4

Differentiate each function.

(a) $y = 4x^2$

(b) $f(x) = 16x^{1/2}$

You may find it helpful to combine the Constant Multiple Rule and the Power Rule into one combined rule.

$$\frac{d}{dx}[cx^n] = cnx^{n-1}, \qquad n \text{ is a real number, } c \text{ is a constant.}$$

For instance, in Example 4(b), you can apply this combined rule to obtain

$$\frac{d}{dt}\left[\frac{4}{5}t^2\right] = \left(\frac{4}{5}\right)(2)(t) = \frac{8}{5}t.$$

The three functions in the next example are simple, yet errors are frequently made in differentiating functions involving constant multiples of the first power of x. Keep in mind that

$$\frac{d}{dx}[cx] = c, \qquad c \text{ is a constant.}$$

EXAMPLE 5 Applying the Constant Multiple Rule

Find the derivative of each function.

Original Function	*Derivative*
(a) $y = -\dfrac{3x}{2}$	$y' = -\dfrac{3}{2}$
(b) $y = 3\pi x$	$y' = 3\pi$
(c) $y = -\dfrac{x}{2}$	$y' = -\dfrac{1}{2}$

TRY IT 5

Find the derivative of each function.

(a) $y = \dfrac{t}{4}$

(b) $y = -\dfrac{2x}{5}$

TECHNOLOGY

Most graphing utilities have a built-in program to approximate the derivative of a function at a specific point. The pseudocode below can be translated into a program for a graphing utility. (Appendix E lists the program for several models of graphing utilities.)

Program
- Begin loop.
- Input x.
- Find the derivative of a function with respect to the variable x at a given value of x and store in d.
- Display d.
- End loop.

This program lets you conveniently evaluate the derivative of a function at several values of x.

Parentheses can play an important role in the use of the Constant Multiple Rule and the Power Rule. In Example 6, be sure you understand the mathematical conventions involving the use of parentheses.

EXAMPLE 6 Using Parentheses When Differentiating

Find the derivative of each function.

(a) $y = \dfrac{5}{2x^3}$ (b) $y = \dfrac{5}{(2x)^3}$ (c) $y = \dfrac{7}{3x^{-2}}$ (d) $y = \dfrac{7}{(3x)^{-2}}$

SOLUTION

Function	*Rewrite*	*Differentiate*	*Simplify*
(a) $y = \dfrac{5}{2x^3}$	$y = \dfrac{5}{2}(x^{-3})$	$y' = \dfrac{5}{2}(-3x^{-4})$	$y' = -\dfrac{15}{2x^4}$
(b) $y = \dfrac{5}{(2x)^3}$	$y = \dfrac{5}{8}(x^{-3})$	$y' = \dfrac{5}{8}(-3x^{-4})$	$y' = -\dfrac{15}{8x^4}$
(c) $y = \dfrac{7}{3x^{-2}}$	$y = \dfrac{7}{3}(x^2)$	$y' = \dfrac{7}{3}(2x)$	$y' = \dfrac{14x}{3}$
(d) $y = \dfrac{7}{(3x)^{-2}}$	$y = 63(x^2)$	$y' = 63(2x)$	$y' = 126x$

TRY IT 6

Find the derivative of each function.

(a) $y = \dfrac{9}{4x^2}$ (b) $y = \dfrac{9}{(4x)^2}$

ALGEBRA REVIEW

When differentiating functions involving radicals, you should rewrite the function with rational exponents. For instance, you should rewrite $y = \sqrt[3]{x}$ as $y = x^{1/3}$, and you should rewrite

$$y = \frac{1}{\sqrt[3]{x^4}} \quad \text{as} \quad y = x^{-4/3}.$$

EXAMPLE 7 Differentiating Radical Functions

Find the derivative of each function.

(a) $y = \sqrt{x}$ (b) $y = \dfrac{1}{2\sqrt[3]{x^2}}$ (c) $y = \sqrt{2x}$

SOLUTION

Function	*Rewrite*	*Differentiate*	*Simplify*
(a) $y = \sqrt{x}$	$y = x^{1/2}$	$y' = \left(\dfrac{1}{2}\right)x^{-1/2}$	$y' = \dfrac{1}{2\sqrt{x}}$
(b) $y = \dfrac{1}{2\sqrt[3]{x^2}}$	$y = \dfrac{1}{2}x^{-2/3}$	$y' = \dfrac{1}{2}\left(-\dfrac{2}{3}\right)x^{-5/3}$	$y' = -\dfrac{1}{3x^{5/3}}$
(c) $y = \sqrt{2x}$	$y = \sqrt{2}(x^{1/2})$	$y' = \sqrt{2}\left(\dfrac{1}{2}\right)x^{-1/2}$	$y' = \dfrac{1}{\sqrt{2x}}$

TRY IT 7

Find the derivative of each function.

(a) $y = \sqrt{5x}$

(b) $y = \sqrt[3]{x}$

The Sum and Difference Rules

The next two rules are ones that you might expect to be true, and you may have used them without thinking about it. For instance, if you were asked to differentiate $y = 3x + 2x^3$, you would probably write

$$y' = 3 + 6x^2$$

without questioning your answer. The validity of differentiating a sum of functions term by term is given by the Sum and Difference Rules.

The Sum and Difference Rules

The derivative of the sum (or difference) of two differentiable functions is the sum (or difference) of their derivatives.

$$\frac{d}{dx}[f(x) + g(x)] = f'(x) + g'(x) \qquad \text{Sum Rule}$$

$$\frac{d}{dx}[f(x) - g(x)] = f'(x) - g'(x) \qquad \text{Difference Rule}$$

PROOF Let $h(x) = f(x) + g(x)$. Then, you can prove the Sum Rule as shown.

$$\begin{aligned}
h'(x) &= \lim_{\Delta x \to 0} \frac{h(x + \Delta x) - h(x)}{\Delta x} \qquad \text{Definition of derivative} \\
&= \lim_{\Delta x \to 0} \frac{f(x + \Delta x) + g(x + \Delta x) - f(x) - g(x)}{\Delta x} \\
&= \lim_{\Delta x \to 0} \frac{f(x + \Delta x) - f(x) + g(x + \Delta x) - g(x)}{\Delta x} \\
&= \lim_{\Delta x \to 0} \left[\frac{f(x + \Delta x) - f(x)}{\Delta x} + \frac{g(x + \Delta x) - g(x)}{\Delta x}\right] \\
&= \lim_{\Delta x \to 0} \frac{f(x + \Delta x) - f(x)}{\Delta x} + \lim_{\Delta x \to 0} \frac{g(x + \Delta x) - g(x)}{\Delta x} \\
&= f'(x) + g'(x)
\end{aligned}$$

So,

$$\frac{d}{dx}[f(x) + g(x)] = f'(x) + g'(x).$$

The Difference Rule can be proved in a similar manner.

The Sum and Difference Rules can be extended to the sum or difference of any finite number of functions. For instance, if $y = f(x) + g(x) + h(x)$, then $y' = f'(x) + g'(x) + h'(x)$.

STUDY TIP

Look back at Example 6 on page 87. Notice that the example asks for the derivative of the difference of two functions. Verify this result by using the Difference Rule.

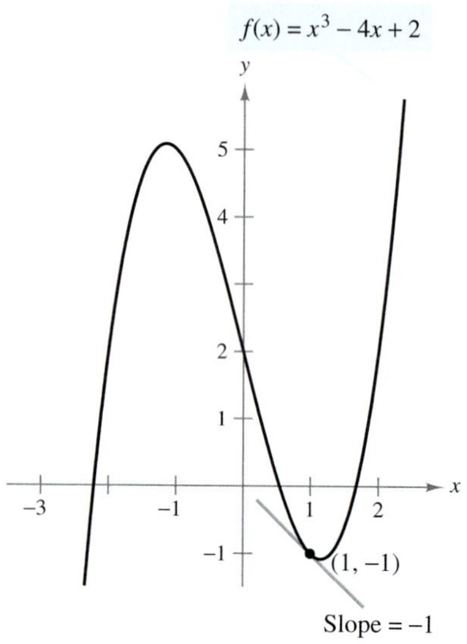

FIGURE 2.15

With the four differentiation rules listed in this section, you can differentiate *any* polynomial function.

EXAMPLE 8 Using the Sum and Difference Rules

Find the slope of the graph of $f(x) = x^3 - 4x + 2$ at the point $(1, -1)$.

SOLUTION The derivative of $f(x)$ is

$$f'(x) = 3x^2 - 4.$$

So, the slope of the graph of f at $(1, -1)$ is

$$\text{Slope} = f'(1) = 3(1)^2 - 4 = -1$$

as shown in Figure 2.15.

TRY IT 8

Find the slope of the graph of $f(x) = x^2 - 5x + 1$ at the point $(2, -5)$.

Example 8 illustrates the use of the derivative for determining the shape of a graph. A rough sketch of the graph of $f(x) = x^3 - 4x + 2$ might lead you to think that the point $(1, -1)$ is a minimum point of the graph. After finding the slope at this point to be -1, however, you can conclude that the minimum point (where the slope is 0) is farther to the right. (You will study techniques for finding minimum and maximum points in Section 3.2.)

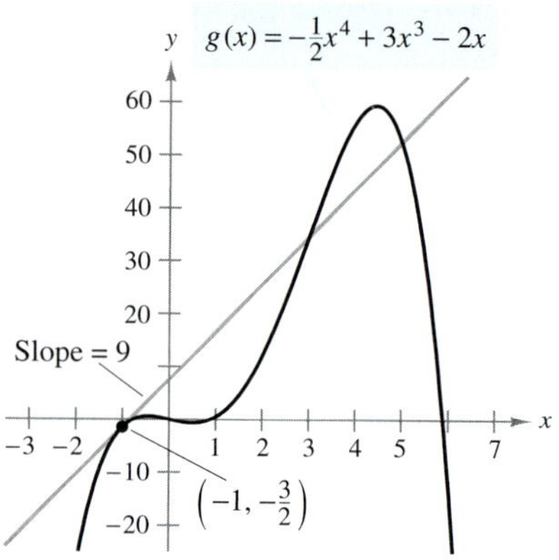

FIGURE 2.16

EXAMPLE 9 Using the Sum and Difference Rules

Find an equation of the tangent line to the graph of

$$g(x) = -\frac{1}{2}x^4 + 3x^3 - 2x$$

at the point $\left(-1, -\frac{3}{2}\right)$.

SOLUTION The derivative of $g(x)$ is $g'(x) = -2x^3 + 9x^2 - 2$, which implies that the slope of the graph at the point $\left(-1, -\frac{3}{2}\right)$ is

$$\begin{aligned}\text{Slope} = g'(-1) &= -2(-1)^3 + 9(-1)^2 - 2\\ &= 2 + 9 - 2\\ &= 9\end{aligned}$$

as shown in Figure 2.16. Using the point-slope form, you can write the equation of the tangent line at $\left(-1, -\frac{3}{2}\right)$ as shown.

$$y - \left(-\frac{3}{2}\right) = 9[x - (-1)] \qquad \text{Point-slope form}$$

$$y = 9x + \frac{15}{2} \qquad \text{Equation of tangent line}$$

TRY IT 9

Find an equation of the tangent line to the graph of $f(x) = -x^2 + 3x - 2$ at the point $(2, 0)$.

Application

EXAMPLE 10 Modeling Revenue

From 1998 through 2003, the revenue R (in millions of dollars per year) for Microsoft Corporation can be modeled by

$$R = 174.343t^3 - 5630.45t^2 + 63{,}029.8t - 218{,}635, \qquad 8 \le t \le 13$$

where $t = 8$ represents 1998. At what rate was Microsoft's revenue changing in 1999? *(Source: Microsoft Corporation)*

SOLUTION One way to answer this question is to find the derivative of the revenue model with respect to time.

$$\frac{dR}{dt} = 523.029t^2 - 11{,}260.90t + 63{,}029.8, \qquad 8 \le t \le 13$$

In 1999 (when $t = 9$), the rate of change of the revenue with respect to time is given by

$$523.029(9)^2 - 11{,}260.90(9) + 63{,}029.8 \approx 4047.$$

Because R is measured in millions of dollars and t is measured in years, it follows that the derivative dR/dt is measured in millions of dollars per year. So, at the end of 1999, Microsoft's revenues were increasing at a rate of about \$4047 million per year, as shown in Figure 2.17.

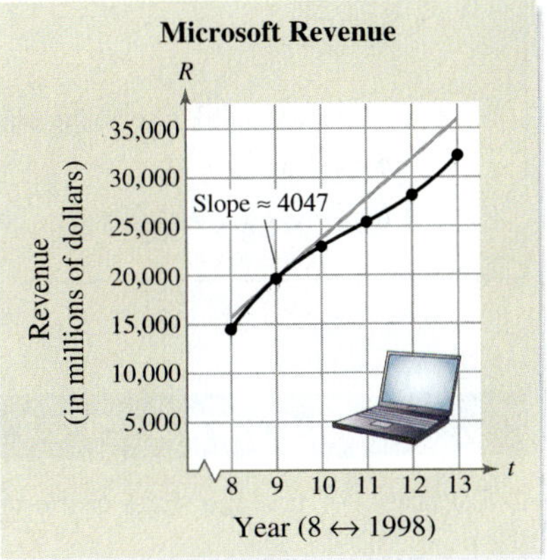

FIGURE 2.17

TRY IT 10

From 1995 through 2002, the revenue per share R (in dollars) for McDonald's Corporation can be modeled by

$$R = 0.0043t^2 + 0.689t + 3.39, \qquad 5 \le t \le 12$$

where $t = 5$ represents 1995. At what rate was McDonald's revenue per share changing in 1998? *(Source: McDonald's Corporation)*

TAKE ANOTHER LOOK

Units for Rates of Change

In Example 10, the units for R are millions of dollars and the units for dR/dt are millions of dollars per year. State the units for the derivatives of each model.

a. A population model that gives the population P (in millions of people) of the United States in terms of the year t. What are the units for dP/dt?

b. A position model that gives the height s (in feet) of an object in terms of the time t (in seconds). What are the units for ds/dt?

c. A demand model that gives the price per unit p (in dollars) of a product in terms of the number of units x sold. What are the units for dp/dx?

PREREQUISITE REVIEW 2.2

The following warm-up exercises involve skills that were covered in earlier sections. You will use these skills in the exercise set for this section.

In Exercises 1 and 2, evaluate each expression when $x = 2$.

1. (a) $2x^2$ (b) $(2x)^2$ (c) $2x^{-2}$

2. (a) $\dfrac{1}{(3x)^2}$ (b) $\dfrac{1}{4x^3}$ (c) $\dfrac{(2x)^{-3}}{4x^{-2}}$

In Exercises 3–6, simplify the expression.

3. $4(3)x^3 + 2(2)x$

4. $\frac{1}{2}(3)x^2 - \frac{3}{2}x^{1/2}$

5. $\left(\frac{1}{4}\right)x^{-3/4}$

6. $\frac{1}{3}(3)x^2 - 2\left(\frac{1}{2}\right)x^{-1/2} + \frac{1}{3}x^{-2/3}$

In Exercises 7–10, solve the equation.

7. $3x^2 + 2x = 0$

8. $x^3 - x = 0$

9. $x^2 + 8x - 20 = 0$

10. $x^2 - 10x - 24 = 0$

EXERCISES 2.2

In Exercises 1–4, find the slope of the tangent line to $y = x^n$ at the point (1, 1).

1. (a) $y = x^2$

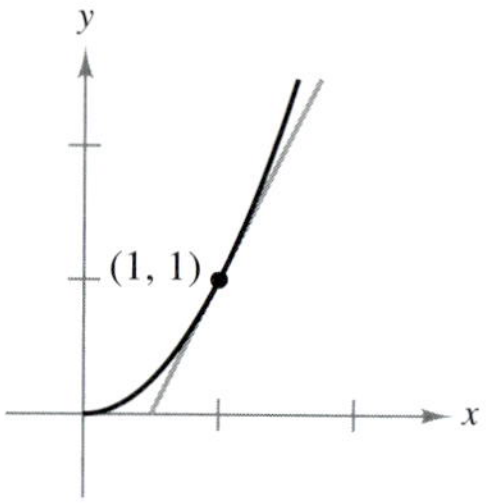

(b) $y = x^{1/2}$

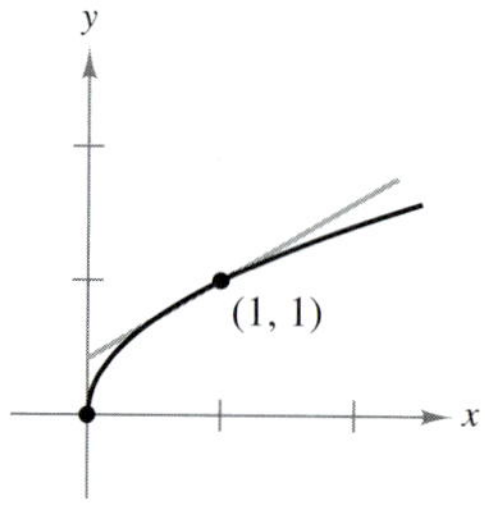

2. (a) $y = x^{3/2}$

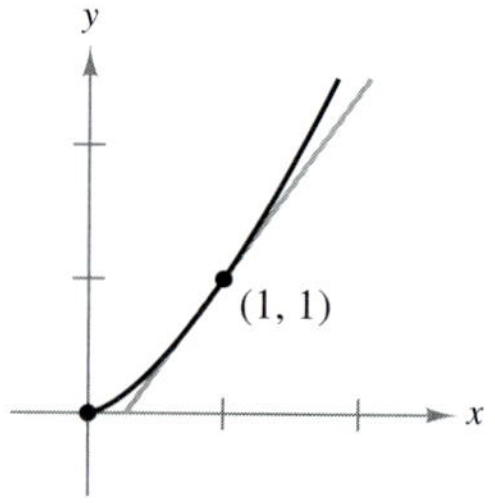

(b) $y = x^3$

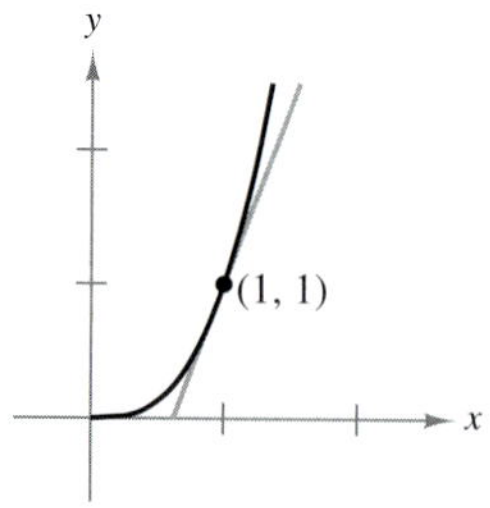

3. (a) $y = x^{-1}$

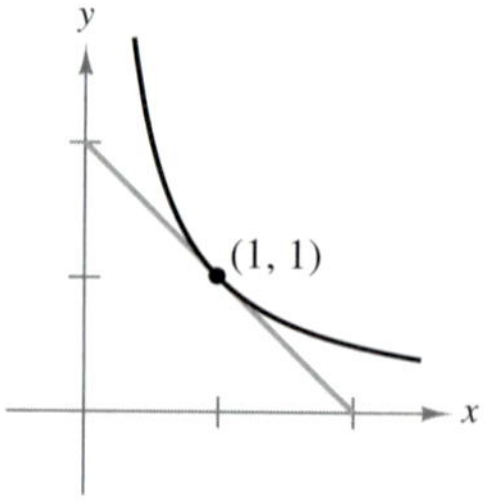

(b) $y = x^{-1/3}$

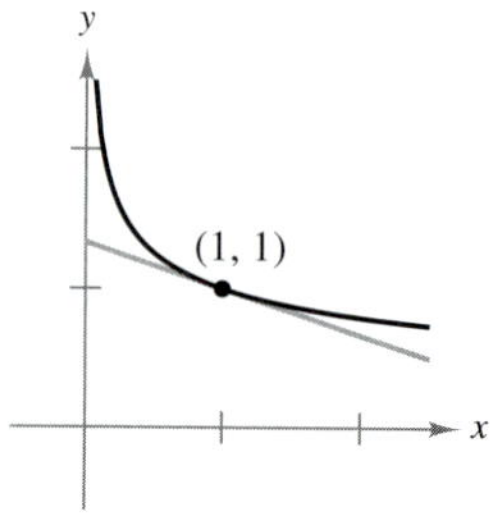

4. (a) $y = x^{-1/2}$

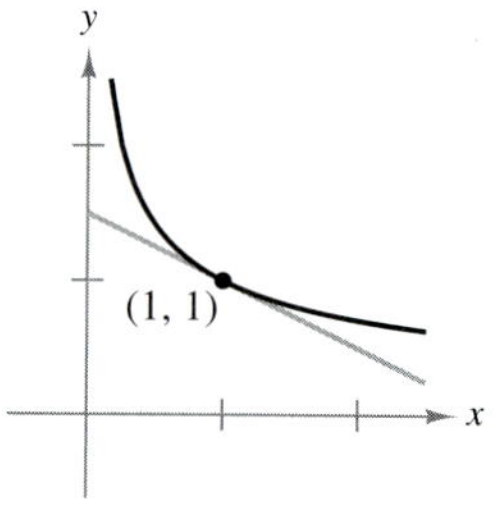

(b) $y = x^{-2}$

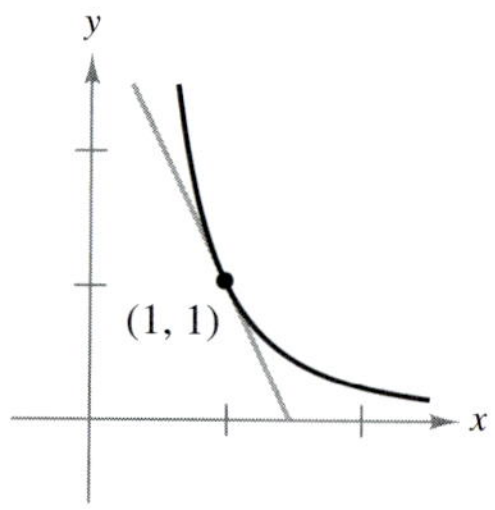

In Exercises 5–20, find the derivative of the function.

5. $y = 3$

6. $f(x) = -2$

7. $f(x) = 4x + 1$

8. $g(x) = 3x - 1$

9. $g(x) = x^2 + 4x - 1$

10. $y = t^2 + 2t - 3$

11. $f(t) = -3t^2 + 2t - 4$

12. $y = x^3 - 9x^2 + 2$

13. $s(t) = t^3 - 2t + 4$

14. $y = 2x^3 - x^2 + 3x - 1$

15. $y = 4t^{4/3}$

16. $h(x) = x^{5/2}$

17. $f(x) = 4\sqrt{x}$

18. $g(x) = 4\sqrt[3]{x} + 2$

19. $y = 4x^{-2} + 2x^2$

20. $s(t) = 4t^{-1} + 1$

In Exercises 21–26, use Example 6 as a model to find the derivative.

Function	*Rewrite*	*Differentiate*	*Simplify*
21. $y = \dfrac{1}{4x^3}$			
22. $y = \dfrac{2}{3x^2}$			
23. $y = \dfrac{1}{(4x)^3}$			
24. $y = \dfrac{\pi}{(3x)^2}$			
25. $y = \dfrac{\sqrt{x}}{x}$			
26. $y = \dfrac{4x}{x^{-3}}$			

In Exercises 27–32, find the value of the derivative of the function at the given point.

Function	*Point*
27. $f(x) = \dfrac{1}{x}$	$(1, 1)$
28. $f(t) = 4 - \dfrac{4}{3t}$	$\left(\dfrac{1}{2}, \dfrac{4}{3}\right)$
29. $f(x) = -\frac{1}{2}x(1 + x^2)$	$(1, -1)$
30. $y = 3x\left(x^2 - \dfrac{2}{x}\right)$	$(2, 18)$
31. $y = (2x + 1)^2$	$(0, 1)$
32. $f(x) = 3(5 - x)^2$	$(5, 0)$

In Exercises 33–46, find $f'(x)$.

33. $f(x) = x^2 - \dfrac{4}{x} - 3x^{-2}$

34. $f(x) = x^2 - 3x - 3x^{-2} + 5x^{-3}$

35. $f(x) = x^2 - 2x - \dfrac{2}{x^4}$

36. $f(x) = x^2 + 4x + \dfrac{1}{x}$

37. $f(x) = x(x^2 + 1)$

38. $f(x) = (x^2 + 2x)(x + 1)$

39. $f(x) = (x + 4)(2x^2 - 1)$

40. $f(x) = (3x^2 - 5x)(x^2 + 2)$

41. $f(x) = \dfrac{2x^3 - 4x^2 + 3}{x^2}$

42. $f(x) = \dfrac{2x^2 - 3x + 1}{x}$

43. $f(x) = \dfrac{4x^3 - 3x^2 + 2x + 5}{x^2}$

44. $f(x) = \dfrac{-6x^3 + 3x^2 - 2x + 1}{x}$

45. $f(x) = x^{4/5} + x$

46. $f(x) = x^{1/3} - 1$

In Exercises 47–50, find an equation of the tangent line to the graph of the function at the given point.

Function	*Point*
47. $y = -2x^4 + 5x^2 - 3$	$(1, 0)$
48. $y = x^3 + x$	$(-1, -2)$
49. $f(x) = \sqrt[3]{x} + \sqrt[5]{x}$	$(1, 2)$
50. $f(x) = \dfrac{1}{\sqrt[3]{x^2}} - x$	$(-1, 2)$

In Exercises 51–54, determine the point(s), if any, at which the graph of the function has a horizontal tangent line.

51. $y = -x^4 + 3x^2 - 1$

52. $y = x^3 + 3x^2$

53. $y = \frac{1}{2}x^2 + 5x$

54. $y = x^2 + 2x$

In Exercises 55 and 56,

(a) Sketch the graphs of f, g, and h on the same set of coordinate axes.

(b) Find $f'(1)$, $g'(1)$, and $h'(1)$.

(c) Sketch the graph of the tangent line to each graph when $x = 1$.

55. $f(x) = x^3$
$g(x) = x^3 + 3$
$h(x) = x^3 - 2$

56. $f(x) = \sqrt{x}$
$g(x) = \sqrt{x} + 4$
$h(x) = \sqrt{x} - 2$

57. Use the Constant Rule, the Constant Multiple Rule, and the Sum Rule to find $h'(1)$ given that $f'(1) = 3$.

(a) $h(x) = f(x) - 2$

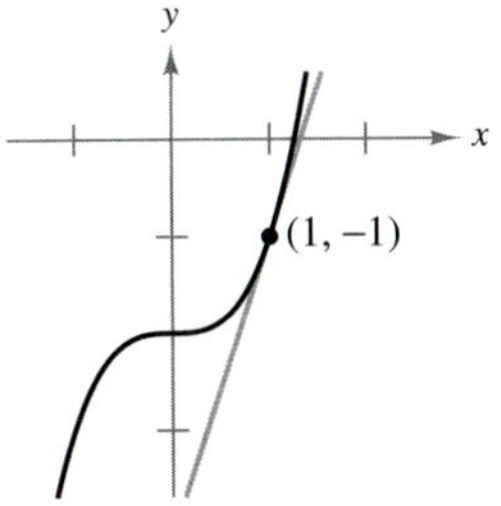

(b) $h(x) = 2f(x)$

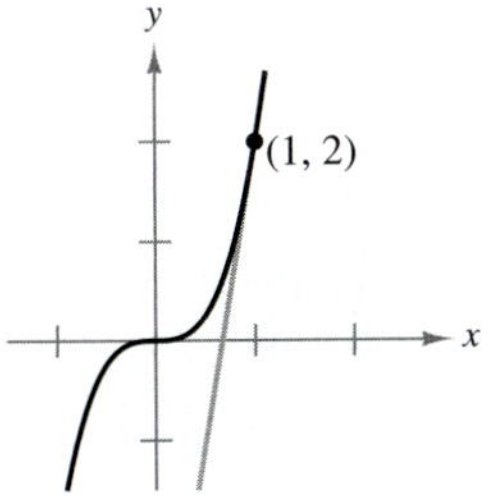

(c) $h(x) = -f(x)$

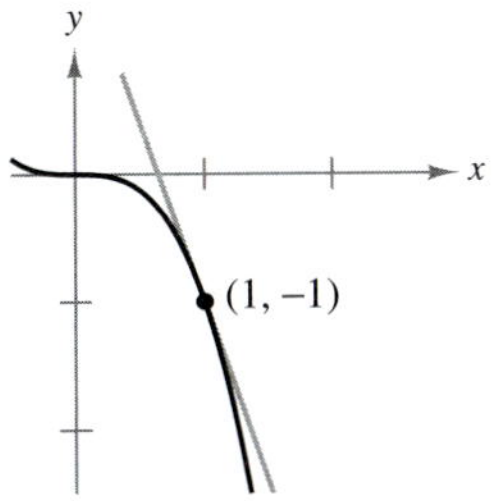

(d) $h(x) = -1 + 2f(x)$

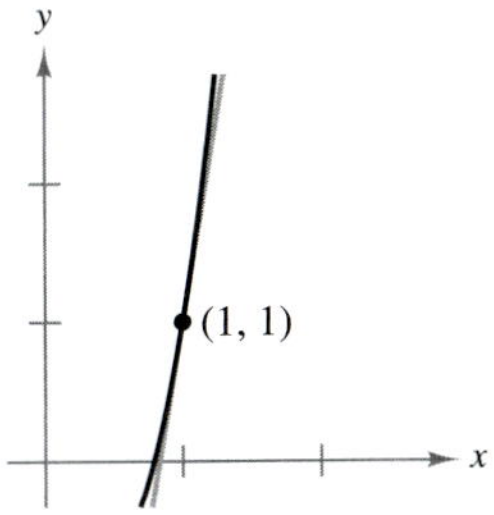

58. *Revenue* The revenue R (in millions of dollars per year) for Polo Ralph Lauren from 1996 through 2002 can be modeled by

$$R = -1.17879t^4 + 38.3641t^3 - 469.994t^2 + 2820.22t - 5577.7$$

where $t = 6$ corresponds to 1996. *(Source: Polo Ralph Lauren Corp.)*

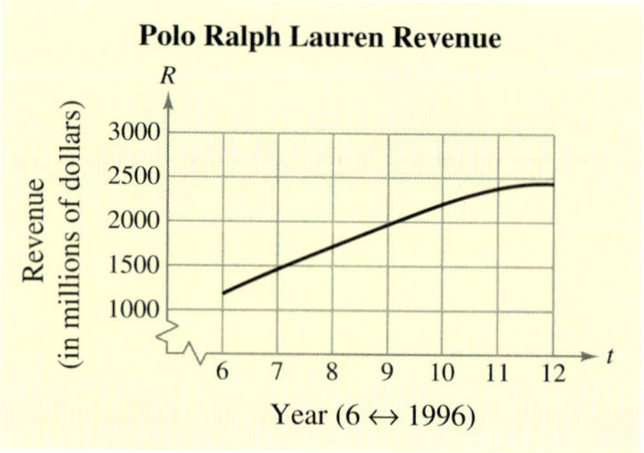

(a) Find the slopes of the graph for the years 1997, 2000, and 2002.

(b) Compare your results with those obtained in Exercise 11 in Section 2.1.

(c) What are the units for the slope of the graph? Interpret the slope of the graph in the context of the problem.

59. *Sales* The sales S (in millions of dollars per year) for Scotts Company from 1997 through 2003 can be modeled by

$$S = 8.70947t^4 - 341.0927t^3 + 4885.752t^2 - 30{,}118.17t + 68{,}395.3$$

where $t = 7$ corresponds to 1997. *(Source: Scotts Company)*

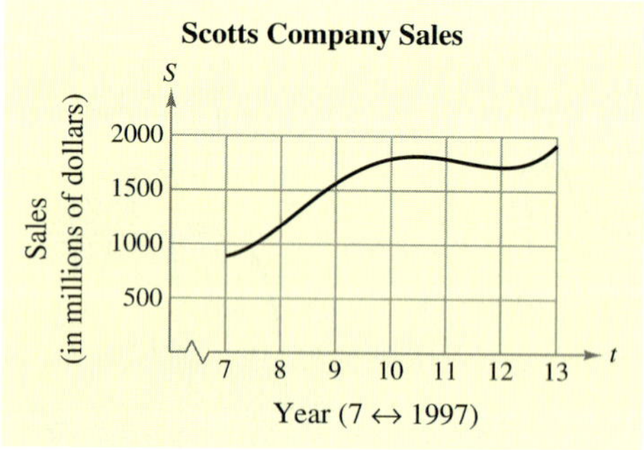

(a) Find the slopes of the graph for the years 1998, 2001, and 2003.

(b) Compare your results with those obtained in Exercise 12 in Section 2.1.

(c) What are the units for the slope of the graph? Interpret the slope of the graph in the context of the problem.

60. *Cost* The variable cost for manufacturing an electrical component is \$7.75 per unit, and the fixed cost is \$500. Write the cost C as a function of x, the number of units produced. Show that the derivative of this cost function is a constant and is equal to the variable cost.

61. *Profit* A college club raises funds by selling candy bars for \$1.00 each. The club pays \$0.60 for each candy bar and has annual fixed costs of \$250. Write the profit P as a function of x, the number of candy bars sold. Show that the derivative of the profit function is a constant and is equal to the profit on each candy bar sold.

62. *Psychology: Migraine Prevalence* The graph illustrates the prevalence of migraine headaches in males and females in selected income groups. *(Source: Adapted from Sue/Sue/Sue,* Understanding Abnormal Behavior, *Seventh Edition)*

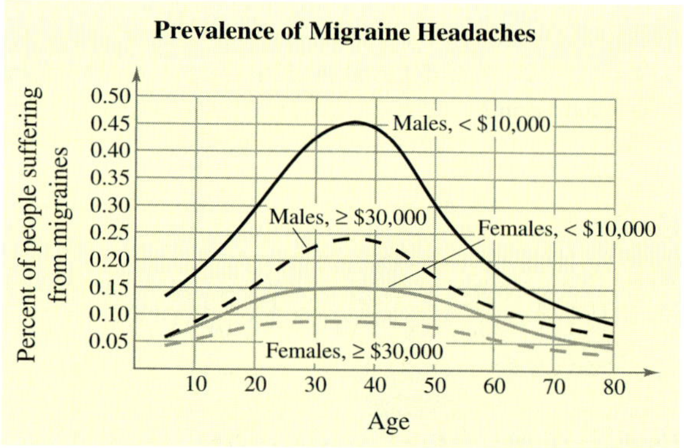

(a) Write a short paragraph describing your general observations about the prevalence of migraines in females and males with respect to age group and income bracket.

(b) Describe the graph of the derivative of each curve, and explain the significance of each derivative. Include an explanation of the units of the derivatives, and indicate the time intervals in which the derivatives would be positive and negative.

In Exercises 63 and 64, use a graphing utility to graph f and f' over the given interval. Determine any points at which the graph of f has horizontal tangents.

	Function	*Interval*
63.	$f(x) = 4.1x^3 - 12x^2 + 2.5x$	$[0, 3]$
64.	$f(x) = x^3 - 1.4x^2 - 0.96x + 1.44$	$[-2, 2]$

True or False? In Exercises 65 and 66, determine whether the statement is true or false. If it is false, explain why or give an example that shows it is false.

65. If $f'(x) = g'(x)$, then $f(x) = g(x)$.

66. If $f(x) = g(x) + c$, then $f'(x) = g'(x)$.

2.3 RATES OF CHANGE: VELOCITY AND MARGINALS

- Find the average rates of change of functions over intervals.
- Find the instantaneous rates of change of functions at points.
- Find the marginal revenues, marginal costs, and marginal profits for products.

Average Rate of Change

In Sections 2.1 and 2.2, you studied the two primary applications of derivatives.

1. **Slope** The derivative of f is a function that gives the slope of the graph of f at a point $(x, f(x))$.
2. **Rate of Change** The derivative of f is a function that gives the rate of change of $f(x)$ with respect to x at the point $(x, f(x))$.

In this section, you will see that there are many real-life applications of rates of change. A few are velocity, acceleration, population growth rates, unemployment rates, production rates, and water flow rates. Although rates of change often involve change with respect to time, you can investigate the rate of change of one variable with respect to any other related variable.

When determining the rate of change of one variable with respect to another, you must be careful to distinguish between *average* and *instantaneous* rates of change. The distinction between these two rates of change is comparable to the distinction between the slope of the secant line through two points on a graph and the slope of the tangent line at one point on the graph.

Definition of Average Rate of Change

If $y = f(x)$, then the **average rate of change** of y with respect to x on the interval $[a, b]$ is

$$\text{Average rate of change} = \frac{f(b) - f(a)}{b - a} = \frac{\Delta y}{\Delta x}.$$

Note that $f(a)$ is the value of the function at the *left* endpoint of the interval, $f(b)$ is the value of the function at the *right* endpoint of the interval, and $b - a$ is the width of the interval, as shown in Figure 2.18.

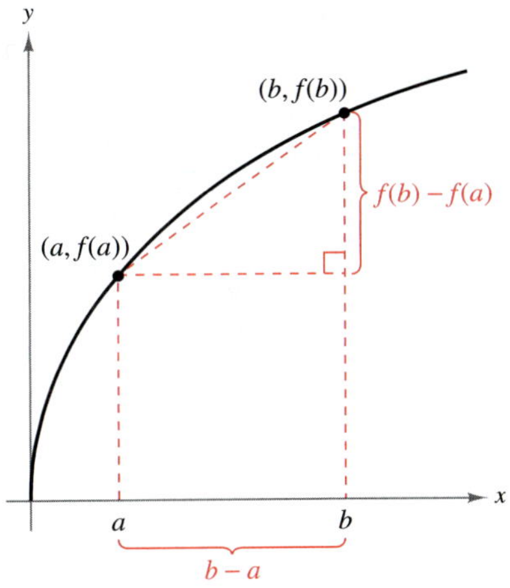

FIGURE 2.18

STUDY TIP

In real-life problems, it is important to list the units of measure for a rate of change. The units for $\Delta y/\Delta x$ are "y-units" per "x-units." For example, if y is measured in miles and x is measured in hours, then $\Delta y/\Delta x$ is measured in *miles per hour.*

STUDY TIP

In Example 1, the average rate of change is positive when the concentration increases and negative when the concentration decreases, as shown in Figure 2.19.

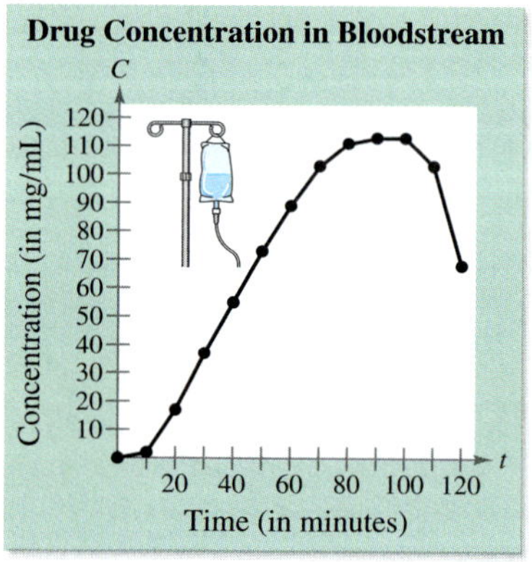

FIGURE 2.19

EXAMPLE 1 Medicine

The concentration C (in milligrams per milliliter) of a drug in a patient's bloodstream is monitored over 10-minute intervals for 2 hours, where t is measured in minutes, as shown in the table. Find the average rate of change over each interval.

(a) $[0, 10]$ (b) $[0, 20]$ (c) $[100, 110]$

t	0	10	20	30	40	50	60	70	80	90	100	110	120
C	0	2	17	37	55	73	89	103	111	113	113	103	68

SOLUTION

(a) For the interval $[0, 10]$, the average rate of change is

Value of C at right endpoint

Value of C at left endpoint

$$\frac{\Delta C}{\Delta t} = \frac{2 - 0}{10 - 0} = \frac{2}{10} = 0.2 \text{ milligram per milliliter per minute.}$$

Width of interval

(b) For the interval $[0, 20]$, the average rate of change is

$$\frac{\Delta C}{\Delta t} = \frac{17 - 0}{20 - 0} = \frac{17}{20} = 0.85 \text{ milligram per milliliter per minute.}$$

(c) For the interval $[100, 110]$, the average rate of change is

$$\frac{\Delta C}{\Delta t} = \frac{103 - 113}{110 - 100} = \frac{-10}{10} = -1 \text{ milligram per milliliter per minute.}$$

TRY IT 1

Use the table in Example 1 to find the average rates of change over each interval.

(a) $[0, 120]$ (b) $[90, 100]$ (c) $[90, 120]$

The rates of change in Example 1 are in milligrams per milliliter per minute because the concentration is measured in milligrams per milliliter and the time is measured in minutes.

Concentration is measured in milligrams per milliliter.

Rate of change is measured in milligrams per milliliter per minute.

$$\frac{\Delta C}{\Delta t} = \frac{2 - 0}{10 - 0} = \frac{2}{10} = 0.2 \text{ milligram per milliliter per minute}$$

Time is measured in minutes.

A common application of an average rate of change is to find the **average velocity** of an object that is moving in a straight line. That is,

$$\text{Average velocity} = \frac{\text{change in distance}}{\text{change in time}}.$$

This formula is demonstrated in Example 2.

EXAMPLE 2 Finding an Average Velocity

If a free-falling object is dropped from a height of 100 feet, and *air resistance is neglected*, the height h (in feet) of the object at time t (in seconds) is given by

$$h = -16t^2 + 100.$$ (See Figure 2.20.)

Find the average velocity of the object over each interval.

(a) $[1, 2]$ (b) $[1, 1.5]$ (c) $[1, 1.1]$

SOLUTION You can use the position equation $h = -16t^2 + 100$ to determine the heights at $t = 1$, $t = 1.1$, $t = 1.5$, and $t = 2$, as shown in the table.

t (in seconds)	0	1	1.1	1.5	2
h (in feet)	100	84	80.64	64	36

(a) For the interval $[1, 2]$, the object falls from a height of 84 feet to a height of 36 feet. So, the average velocity is

$$\frac{\Delta h}{\Delta t} = \frac{36 - 84}{2 - 1} = \frac{-48}{1} = -48 \text{ feet per second.}$$

(b) For the interval $[1, 1.5]$, the average velocity is

$$\frac{\Delta h}{\Delta t} = \frac{64 - 84}{1.5 - 1} = \frac{-20}{0.5} = -40 \text{ feet per second.}$$

(c) For the interval $[1, 1.1]$, the average velocity is

$$\frac{\Delta h}{\Delta t} = \frac{80.64 - 84}{1.1 - 1} = \frac{-3.36}{0.1} = -33.6 \text{ feet per second.}$$

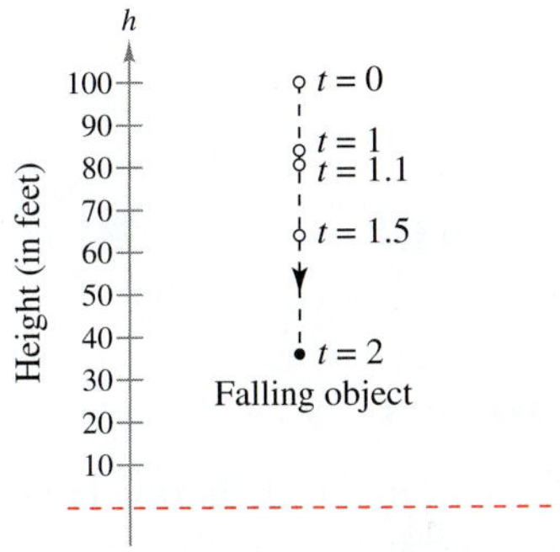

FIGURE 2.20 Some falling objects have considerable air resistance. Other falling objects have negligible air resistance. When modeling a falling-body problem, you must decide whether to account for air resistance or neglect it.

TRY IT 2

The height h (in feet) of a free-falling object at time t (in seconds) is given by $h = -16t^2 + 180$. Find the average velocity of the object over each interval.

(a) $[0, 1]$ (b) $[1, 2]$ (c) $[2, 3]$

STUDY TIP

In Example 2, the average velocities are negative because the object is moving downward.

Instantaneous Rate of Change and Velocity

Suppose in Example 2 you wanted to find the rate of change of h at the instant $t = 1$ second. Such a rate is called an **instantaneous rate of change.** You can approximate the instantaneous rate of change at $t = 1$ by calculating the average rate of change over smaller and smaller intervals of the form $[1, 1 + \Delta t]$, as shown in the table. From the table, it seems reasonable to conclude that the instantaneous rate of change of the height when $t = 1$ is -32 feet per second.

Δt approaches 0.

Δt	1	0.5	0.1	0.01	0.001	0.0001	0
$\dfrac{\Delta h}{\Delta t}$	-48	-40	-33.6	-32.16	-32.016	-32.0016	-32

$\dfrac{\Delta h}{\Delta t}$ approaches -32.

STUDY TIP

The limit in this definition is the same as the limit in the definition of the derivative of f at x. This is the second major interpretation of the derivative—as an *instantaneous rate of change in one variable with respect to another*. Recall that the first interpretation of the derivative is as the slope of the graph of f at x.

Definition of Instantaneous Rate of Change

The **instantaneous rate of change** (or simply **rate of change**) of $y = f(x)$ at x is the limit of the average rate of change on the interval $[x, x + \Delta x]$, as Δx approaches 0.

$$\lim_{\Delta x \to 0} \frac{\Delta y}{\Delta x} = \lim_{\Delta x \to 0} \frac{f(x + \Delta x) - f(x)}{\Delta x}$$

If y is a distance and x is time, then the rate of change is a **velocity.**

EXAMPLE 3 Finding an Instantaneous Rate of Change

Find the velocity of the object in Example 2 when $t = 1$.

SOLUTION From Example 2, you know that the height of the falling object is given by

$$h = -16t^2 + 100. \qquad \text{Position function}$$

By taking the derivative of this position function, you obtain the velocity function.

$$h'(t) = -32t \qquad \text{Velocity function}$$

The velocity function gives the velocity at *any* time. So, when $t = 1$, the velocity is

$$\begin{aligned} h'(1) &= -32(1) \\ &= -32 \text{ feet per second.} \end{aligned}$$

TRY IT 3

Find the velocity of the object in Try It 2 when $t = 1.75$ and $t = 2$.

The general **position function** for a free-falling object, neglecting air resistance, is

$$h = -16t^2 + v_0 t + h_0$$ Position function

where h is the height (in feet), t is the time (in seconds), v_0 is the initial velocity (in feet per second), and h_0 is the initial height (in feet). Remember that the model assumes that positive velocities indicate upward motion and negative velocities indicate downward motion. The derivative $h' = -32t + v_0$ is the **velocity function.** The absolute value of the velocity is the **speed** of the object.

DISCOVERY

Graph the polynomial function $h = -16t^2 + 16t + 32$ from Example 4 on the domain $0 \le t \le 2$. What is the maximum value of h? What is the derivative of h at this maximum point? In general, discuss how the derivative can be used to find the maximum or minimum values of a function.

EXAMPLE 4 Finding the Velocity of a Diver

At time $t = 0$, a diver jumps from a diving board that is 32 feet high, as shown in Figure 2.21. Because the diver's initial velocity is 16 feet per second, his position function is

$$h = -16t^2 + 16t + 32.$$ Position function

(a) When does the diver hit the water?

(b) What is the diver's velocity at impact?

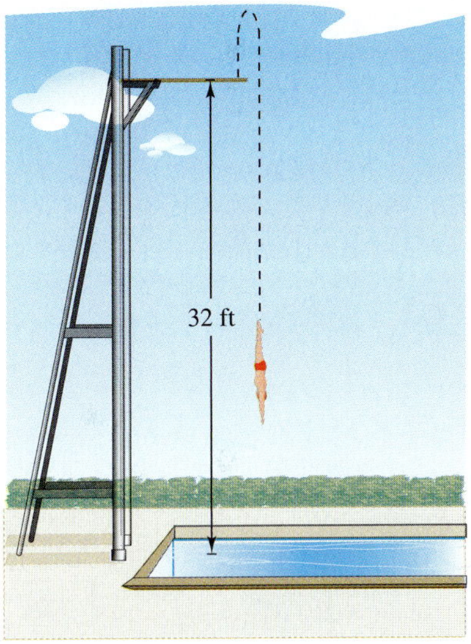

FIGURE 2.21

SOLUTION

(a) To find the time at which the diver hits the water, let $h = 0$ and solve for t.

$-16t^2 + 16t + 32 = 0$ Set h equal to 0.

$-16(t^2 - t - 2) = 0$ Factor out common factor.

$-16(t + 1)(t - 2) = 0$ Factor.

$t = -1$ or $t = 2$ Solve for t.

The solution $t = -1$ does not make sense in the problem because it would mean the diver hits the water 1 second before he jumps. So, you can conclude that the diver hits the water when $t = 2$ seconds.

(b) The velocity at time t is given by the derivative

$h' = -32t + 16.$ Velocity function

The velocity at time $t = 2$ is $-32(2) + 16 = -48$ feet per second.

TRY IT 4

Give the position function of a diver who jumps from a board 12 feet high with initial velocity 16 feet per second. Then find the diver's velocity function.

In Example 4, note that the diver's initial velocity is $v_0 = 16$ feet per second (upward) and his initial height is $h_0 = 32$ feet.

Initial velocity is 16 feet per second.

Initial height is 32 feet.

$$h = -16t^2 + 16t + 32$$

Rates of Change in Economics: Marginals

Another important use of rates of change is in the field of economics. Economists refer to *marginal profit*, *marginal revenue*, and *marginal cost* as the rates of change of the profit, revenue, and cost with respect to the number x of units produced or sold. An equation that relates these three quantities is

$$P = R - C$$

where P, R, and C represent the following quantities.

$P =$ total profit

$R =$ total revenue

and

$C =$ total cost

The derivatives of these quantities are called the **marginal profit, marginal revenue,** and **marginal cost,** respectively.

$$\frac{dP}{dx} = \text{marginal profit}$$

$$\frac{dR}{dx} = \text{marginal revenue}$$

$$\frac{dC}{dx} = \text{marginal cost}$$

In many business and economics problems, the number of units produced or sold is restricted to positive integer values, as indicated in Figure 2.22(a). (Of course, it could happen that a sale involves half or quarter units, but it is hard to conceive of a sale involving $\sqrt{2}$ units.) The variable that denotes such units is called a **discrete variable.** To analyze a function of a discrete variable x, you can temporarily assume that x is a **continuous variable** and is able to take on any real value in a given interval, as indicated in Figure 2.22(b). Then, you can use the methods of calculus to find the x-value that corresponds to the marginal revenue, maximum profit, minimum cost, or whatever is called for. Finally, you should round the solution to the nearest sensible x-value—cents, dollars, units, or days, depending on the context of the problem.

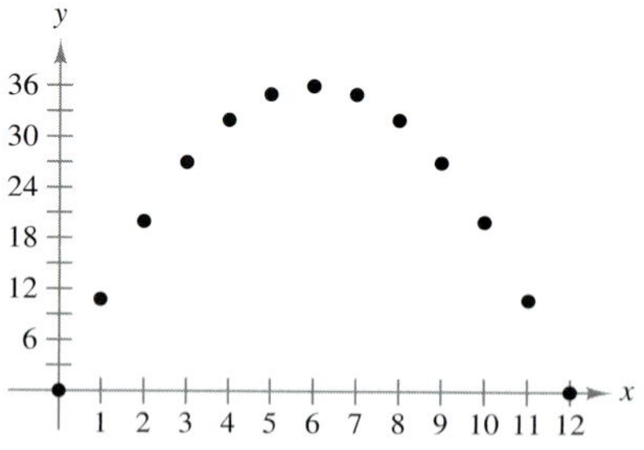

(a) Function of a Discrete Variable

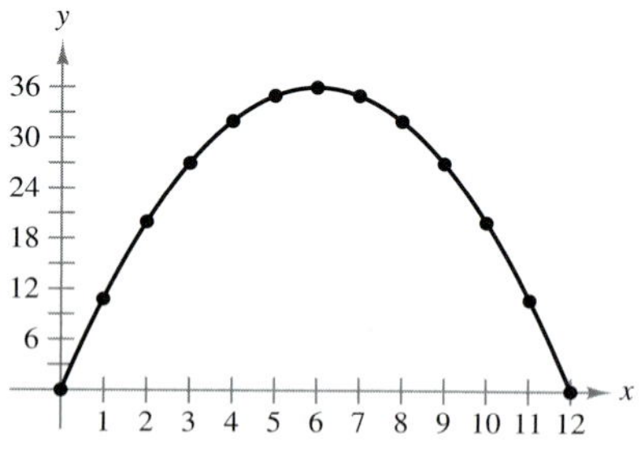

(b) Function of a Continuous Variable

FIGURE 2.22

EXAMPLE 5 Finding the Marginal Profit

The profit derived from selling x units of an alarm clock is given by

$$P = 0.0002x^3 + 10x.$$

(a) Find the marginal profit for a production level of 50 units.

(b) Compare this with the actual gain in profit obtained by increasing the production level from 50 to 51 units.

SOLUTION

(a) Because the profit is $P = 0.0002x^3 + 10x$, the marginal profit is given by the derivative

$$dP/dx = 0.0006x^2 + 10.$$

When $x = 50$, the marginal profit is

$$0.0006(50)^2 + 10 = 1.5 + 10$$
$$= \$11.50 \text{ per unit.}$$ Marginal profit for $x = 50$

(b) For $x = 50$, the actual profit is

$$P = (0.0002)(50)^3 + 10(50)$$ Substitute 50 for x.
$$= 25 + 500$$
$$= \$525.00$$ Actual profit for $x = 50$

and for $x = 51$, the actual profit is

$$P = (0.0002)(51)^3 + 10(51)$$ Substitute 51 for x.
$$\approx 26.53 + 510$$
$$= \$536.53.$$ Actual profit for $x = 51$

So, the additional profit obtained by increasing the production level from 50 to 51 units is

$$536.53 - 525.00 = \$11.53.$$ Extra profit for one unit

Note that the actual profit increase of \$11.53 (when x increases from 50 to 51 units) can be approximated by the marginal profit of \$11.50 per unit (when $x = 50$), as shown in Figure 2.23.

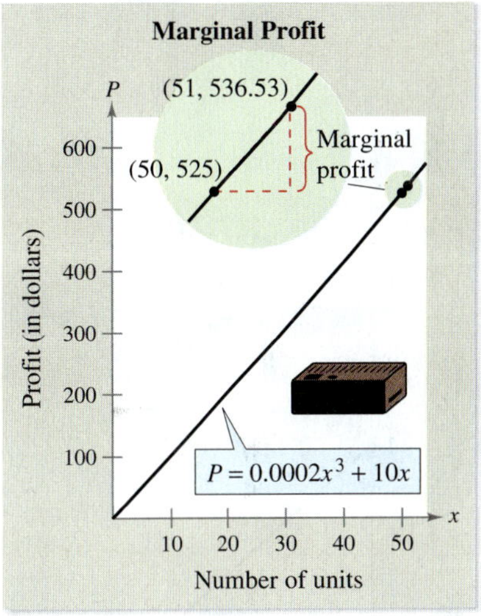

FIGURE 2.23

TRY IT 5

Use the profit function in Example 5 to find the marginal profit for a production level of 100 units. Compare this with the actual gain in profit by increasing production from 100 to 101 units.

STUDY TIP

The reason the marginal profit gives a good approximation of the actual change in profit is that the graph of P is nearly straight over the interval $50 \le x \le 51$. You will study more about the use of marginals to approximate actual changes in Section 3.8.

The profit function in Example 5 is unusual in that the profit continues to increase as long as the number of units sold increases. In practice, it is more common to encounter situations in which sales can be increased only by lowering the price per item. Such reductions in price will ultimately cause the profit to decline.

The number of units x that consumers are willing to purchase at a given price per unit p is given by the **demand function**

$$p = f(x).$$ Demand function

The total revenue R is then related to the price per unit and the quantity demanded (or sold) by the equation

$$R = xp.$$ Revenue function

EXAMPLE 6 Finding a Demand Function

A business sells 2000 items per month at a price of \$10 each. It is estimated that monthly sales will increase 250 units for each \$0.25 reduction in price. Use this information to find the demand function and total revenue function.

SOLUTION From the given estimate, x increases 250 units each time p drops \$0.25 from the original cost of \$10. This is described by the equation

$$\begin{aligned} x &= 2000 + 250\left(\frac{10 - p}{0.25}\right) \\ &= 2000 + 10{,}000 - 1000p \\ &= 12{,}000 - 1000p. \end{aligned}$$

Solving for p in terms of x produces

$$p = 12 - \frac{x}{1000}.$$ Demand function

This, in turn, implies that the revenue function is

$$\begin{aligned} R &= xp && \text{Formula for revenue} \\ &= x\left(12 - \frac{x}{1000}\right) \\ &= 12x - \frac{x^2}{1000}. && \text{Revenue function} \end{aligned}$$

The graph of the demand function is shown in Figure 2.24. Notice that as the price decreases, the quantity demanded increases.

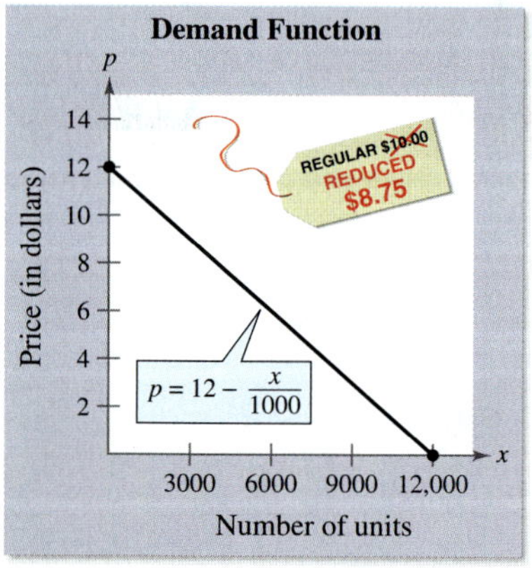

FIGURE 2.24

TRY IT 6

Find the demand function in Example 6 if monthly sales increase 200 units for each \$0.10 reduction in price.

TECHNOLOGY

Modeling a Demand Function

To model a demand function, you need data that indicate how many units of a product will sell at a given price. As you might imagine, such data are not easy to obtain for a new product. After a product has been on the market awhile, however, its sales history can provide the necessary data.

As an example, consider the two bar graphs shown below. From these graphs, you can see that from 1996 through 1999, the price of prerecorded videocassettes dropped from an average price of about \$13 to an average price of about \$10. During that time, the number of units sold increased from about 600 million to about 800 million. *(Source: Paul Kagan Associates, Inc.)*

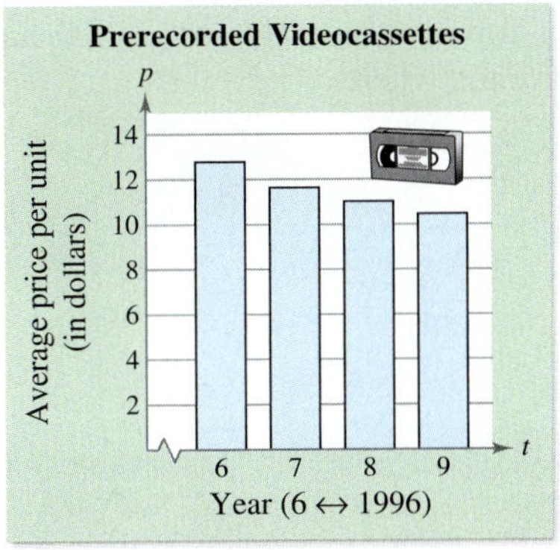

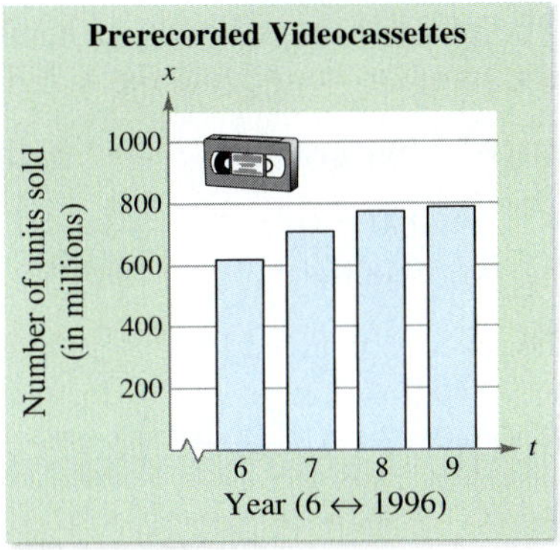

The information in the two bar graphs is combined in the table, where p represents the price (in dollars) and x represents the units sold (in millions).

t	6	7	8	9
p	12.75	11.61	10.99	10.44
x	617.0	707.8	772.0	786.9

By entering the ordered pairs (x, p) into a calculator or computer that has a power regression program, you can find that the power model for the demand for prerecorded videocassettes is: $p = 1673.35x^{-0.758}$, $617.0 \le x \le 786.9$. A graph of this demand function and its data points is shown below.

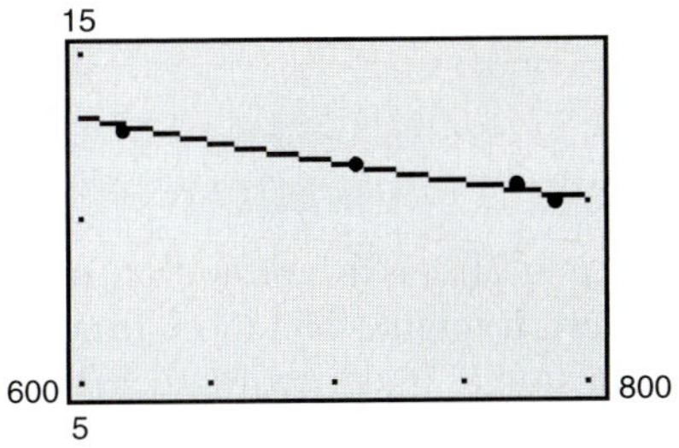

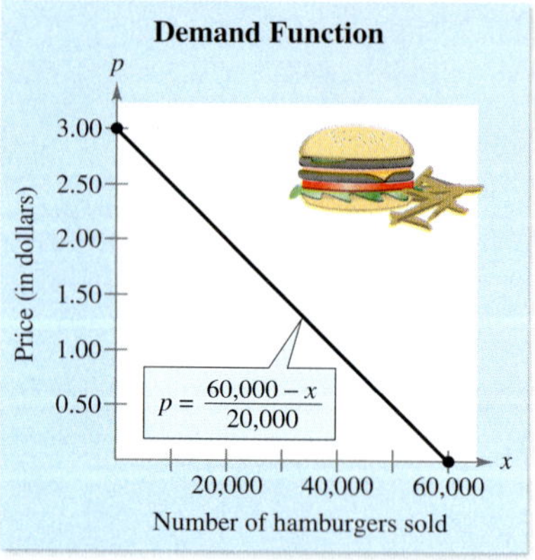

FIGURE 2.25 As the price decreases, more hamburgers are sold.

EXAMPLE 7 Finding the Marginal Revenue

A fast-food restaurant has determined that the monthly demand for its hamburgers is given by

$$p = \frac{60{,}000 - x}{20{,}000}.$$

Figure 2.25 shows that as the price decreases, the quantity demanded increases. The table shows the demands for hamburgers at various prices.

x	60,000	50,000	40,000	30,000	20,000	10,000	0
p	\$0.00	\$0.50	\$1.00	\$1.50	\$2.00	\$2.50	\$3.00

Find the increase in revenue per hamburger for monthly sales of 20,000 hamburgers. In other words, find the marginal revenue when $x = 20{,}000$.

SOLUTION Because the demand is given by

$$p = \frac{60{,}000 - x}{20{,}000}$$

and the revenue is given by $R = xp$, you have

$$R = xp = x\left(\frac{60{,}000 - x}{20{,}000}\right)$$

$$= \frac{1}{20{,}000}(60{,}000x - x^2).$$

By differentiating, you can find the marginal revenue to be

$$\frac{dR}{dx} = \frac{1}{20{,}000}(60{,}000 - 2x).$$

So, when $x = 20{,}000$, the marginal revenue is

$$\frac{1}{20{,}000}[60{,}000 - 2(20{,}000)] = \frac{20{,}000}{20{,}000}$$

$$= \$1 \text{ per unit.}$$

TRY IT 7

Find the revenue function and marginal revenue for a demand function of $p = 2000 - 4x$.

STUDY TIP

Writing a demand function in the form $p = f(x)$ is a convention used in economics. From a consumer's point of view, it might seem more reasonable to think that the quantity demanded is a function of the price. Mathematically, however, the two points of view are equivalent because a typical demand function is one-to-one and so has an inverse function. For instance, in Example 7, you could write the demand function as $x = 60{,}000 - 20{,}000p$.

EXAMPLE 8 Finding the Marginal Profit

Suppose in Example 7 that the cost of producing x hamburgers is

$$C = 5000 + 0.56x, \quad 0 \le x \le 50{,}000.$$

Find the profit and the marginal profit for each production level.

(a) $x = 20{,}000$ (b) $x = 24{,}400$ (c) $x = 30{,}000$

SOLUTION From Example 7, you know that the total revenue from selling x hamburgers is

$$R = \frac{1}{20{,}000}(60{,}000x - x^2).$$

Because the total profit is given by $P = R - C$, you have

$$\begin{aligned} P &= \frac{1}{20{,}000}(60{,}000x - x^2) - (5000 + 0.56x) \\ &= 3x - \frac{x^2}{20{,}000} - 5000 - 0.56x \\ &= 2.44x - \frac{x^2}{20{,}000} - 5000. \end{aligned}$$

See Figure 2.26.

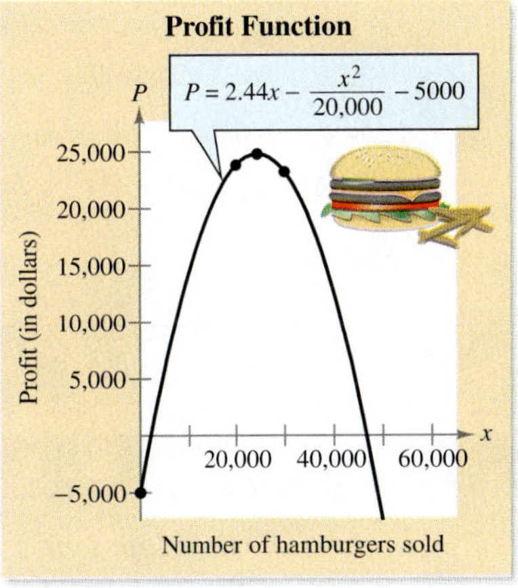

FIGURE 2.26

So, the marginal profit is

$$\frac{dP}{dx} = 2.44 - \frac{x}{10{,}000}.$$

Using these formulas, you can compute the profit and marginal profit.

Production	*Profit*	*Marginal Profit*
(a) $x = 20{,}000$	$P = \$23{,}800.00$	$2.44 - \frac{20{,}000}{10{,}000} = \$0.44$ per unit
(b) $x = 24{,}400$	$P = \$24{,}768.00$	$2.44 - \frac{24{,}400}{10{,}000} = \$0.00$ per unit
(c) $x = 30{,}000$	$P = \$23{,}200.00$	$2.44 - \frac{30{,}000}{10{,}000} = -\$0.56$ per unit

TRY IT 8

From Example 8, compare the marginal profit when 10,000 units are produced with the actual increase in profit from 10,000 units to 10,001 units.

TAKE ANOTHER LOOK

Interpreting Marginal Profit

In Example 8 and Figure 2.26, notice that when 20,000 hamburgers are sold, the marginal profit is positive, when 24,400 are sold, the marginal profit is zero, and when 30,000 are sold, the marginal profit is negative. What does this mean? If you owned this business, how much would you charge for hamburgers? Explain your reasoning.

PREREQUISITE REVIEW 2.3

The following warm-up exercises involve skills that were covered in earlier sections. You will use these skills in the exercise set for this section.

In Exercises 1 and 2, evaluate the expression.

1. $\dfrac{-63 - (-105)}{21 - 7}$

2. $\dfrac{-37 - 54}{16 - 3}$

In Exercises 3–10, find the derivative of the function.

3. $y = 4x^2 - 2x + 7$

4. $y = -3t^3 + 2t^2 - 8$

5. $s = -16t^2 + 24t + 30$

6. $y = -16x^2 + 54x + 70$

7. $A = \frac{1}{10}(-2r^3 + 3r^2 + 5r)$

8. $y = \frac{1}{9}(6x^3 - 18x^2 + 63x - 15)$

9. $y = 12x - \dfrac{x^2}{5000}$

10. $y = 138 + 74x - \dfrac{x^3}{10{,}000}$

EXERCISES 2.3

1. ***Research and Development*** The graph shows the amounts A (in billions of dollars per year) spent on R&D in the United States from 1980 through 2002. Approximate the average rate of change of A during each period. *(Source: U.S. National Science Foundation)*

(a) 1980–1985 (b) 1985–1990 (c) 1990–1995

(d) 1995–2000 (e) 1980–2002 (f) 1990–2002

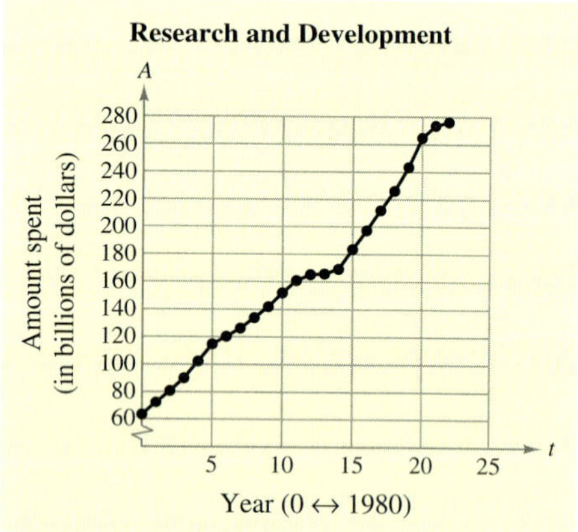

2. ***Trade Deficit*** The graph shows the values I (in billions of dollars per year) of goods imported to the United States and the value E (in billions of dollars per year) of goods exported from the United States from 1980 through 2002. Approximate each indicated average rate of change. *(Source: U.S. International Trade Administration)*

(a) Imports: 1980–1990 (b) Exports: 1980–1990

(c) Imports: 1990–2000 (d) Exports: 1990–2000

(e) Imports: 1980–2002 (f) Exports: 1980–2002

Figure for 2

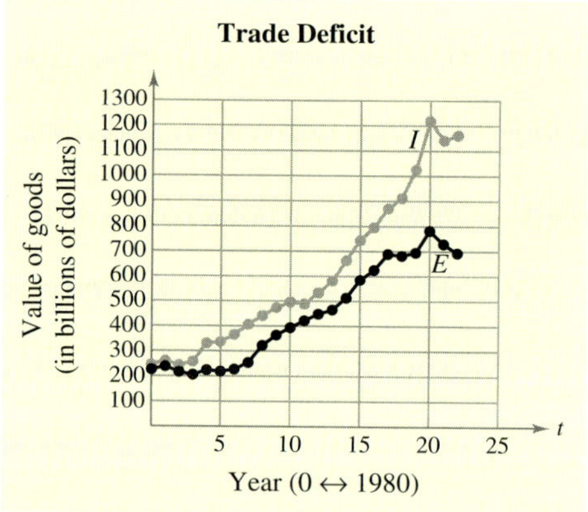

In Exercises 3–8, sketch the graph of the function and find its average rate of change on the interval. Compare this rate with the instantaneous rates of change at the endpoints of the interval.

3. $f(t) = 2t + 7;\ [1, 2]$

4. $h(x) = 1 - x;\ [0, 1]$

5. $h(x) = x^2 - 4x + 2;\ [-2, 2]$

6. $f(x) = x^2 - 6x - 1;\ [-1, 3]$

7. $f(x) = \dfrac{1}{x};\ [1, 4]$

8. $f(x) = \dfrac{1}{\sqrt{x}};\ [1, 4]$

In Exercises 9 and 10, use a graphing utility to graph the function and find its average rate of change on the interval. Compare this rate with the instantaneous rates of change at the endpoints of the interval.

9. $g(x) = x^4 - x^2 + 2;\ [1, 3]$

10. $g(x) = x^3 - 1;\ [-1, 1]$

11. *Consumer Trends* The graph shows the number of visitors V to a national park in hundreds of thousands during a one-year period, where $t = 1$ represents January.

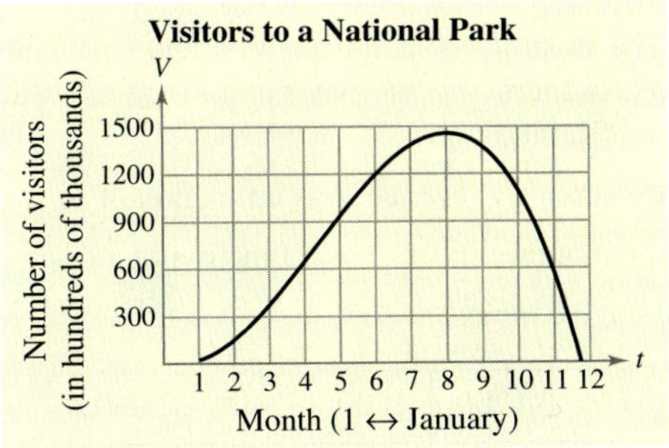

(a) Estimate the rate of change of V over the interval $[9, 12]$ and explain your results.

(b) Over what interval is the average rate of change approximately equal to the rate of change at $t = 8$? Explain your reasoning.

12. *Medicine* The graph shows the estimated number of milligrams of a pain medication M in the bloodstream t hours after a 1000-milligram dose of the drug has been given.

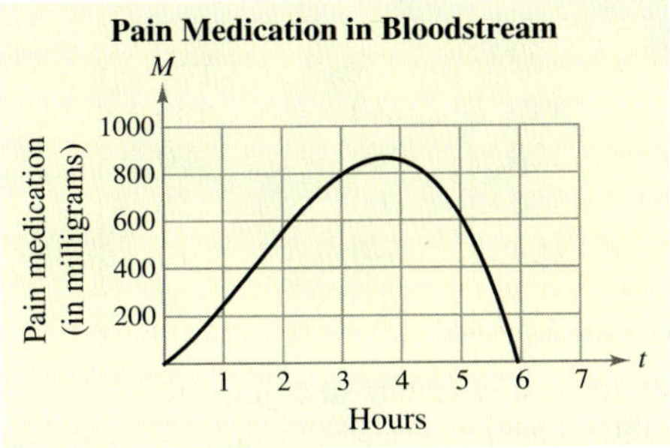

(a) Estimate the one-hour interval over which the average rate of change is the greatest.

(b) Over what interval is the average rate of change approximately equal to the rate of change at $t = 4$? Explain your reasoning.

13. *Medicine* The effectiveness E (on a scale from 0 to 1) of a pain-killing drug t hours after entering the bloodstream is given by

$$E = \frac{1}{27}(9t + 3t^2 - t^3), \quad 0 \le t \le 4.5.$$

Find the average rate of change of E on each indicated interval and compare this rate with the instantaneous rates of change at the endpoints of the interval.

(a) $[0, 1]$ (b) $[1, 2]$ (c) $[2, 3]$ (d) $[3, 4]$

14. *Chemistry: Wind Chill* At 0° Celsius, the heat loss H (in kilocalories per square meter per hour) from a person's body can be modeled by

$$H = 33(10\sqrt{v} - v + 10.45)$$

where v is the wind speed (in meters per second).

(a) Find $\frac{dH}{dv}$ and interpret its meaning in this situation.

(b) Find the rates of change of H when $v = 2$ and when $v = 5$.

15. *Velocity* The height s (in feet) at time t (in seconds) of a silver dollar dropped from the top of the Washington Monument is given by

$$s = -16t^2 + 555.$$

(a) Find the average velocity on the interval $[2, 3]$.

(b) Find the instantaneous velocities when $t = 2$ and when $t = 3$.

(c) How long will it take the dollar to hit the ground?

(d) Find the velocity of the dollar when it hits the ground.

16. *Physics: Velocity* A racecar travels northward on a straight, level track at a constant speed, traveling 0.750 kilometer in 20.0 seconds. The return trip over the same track is made in 25.0 seconds.

(a) What is the average velocity of the car in meters per second for the first leg of the run?

(b) What is the average velocity for the total trip?

(Source: Shipman/Wilson/Todd, An Introduction to Physical Science, *Tenth Edition)*

Marginal Cost In Exercises 17–20, find the marginal cost for producing x units. (The cost is measured in dollars.)

17. $C = 4500 + 1.47x$

18. $C = 104{,}000 + 7200x$

19. $C = 55{,}000 + 470x - 0.25x^2, \quad 0 \le x \le 940$

20. $C = 100(9 + 3\sqrt{x})$

Marginal Revenue In Exercises 21–24, find the marginal revenue for producing x units. (The revenue is measured in dollars.)

21. $R = 50x - 0.5x^2$

22. $R = 30x - x^2$

23. $R = -6x^3 + 8x^2 + 200x$

24. $R = 50(20x - x^{3/2})$

Marginal Profit In Exercises 25–28, find the marginal profit for producing x units. (The profit is measured in dollars.)

25. $P = -2x^2 + 72x - 145$

26. $P = -0.25x^2 + 2000x - 1{,}250{,}000$

27. $P = -0.00025x^2 + 12.2x - 25{,}000$

28. $P = -0.5x^3 + 30x^2 - 164.25x - 1000$

29. *Marginal Cost* The cost (in dollars) of producing x units of a product is given by

$$C = 3.6\sqrt{x} + 500.$$

(a) Find the additional cost when the production increases from 9 to 10 units.

(b) Find the marginal cost when $x = 9$.

(c) Compare the results of parts (a) and (b).

30. *Marginal Revenue* The revenue (in dollars) from renting x apartments can be modeled by

$$R = 2x(900 + 32x - x^2).$$

(a) Find the additional revenue when the number of rentals is increased from 14 to 15.

(b) Find the marginal revenue when $x = 14$.

(c) Compare the results of parts (a) and (b).

31. *Marginal Profit* The profit (in dollars) from selling x units of calculus textbooks is given by

$$p = -0.05x^2 + 20x - 1000.$$

(a) Find the additional profit when the sales increase from 150 to 151 units.

(b) Find the marginal profit when $x = 150$.

(c) Compare the results of parts (a) and (b).

32. *Population Growth* The population of a developing rural area has been growing according to the model

$$P = 22t^2 + 52t + 10{,}000$$

where t is time in years, with $t = 0$ corresponding to 1990.

(a) Evaluate P for $t = 0$, 10, 15, 20, and 25. Explain these values.

(b) Determine the population growth rate, dP/dt.

(c) Evaluate dP/dt for the same values as in part (a). Explain your results.

33. *Health* The temperature of a person during an illness is given by $T = -0.0375t^2 + 0.3t + 100.4$, where t is time in hours since the person started to show signs of a fever.

(a) Use a graphing utility to graph the function. Be sure to choose an appropriate window.

(b) Do the slopes of the tangent lines appear to be positive or negative? What does this tell you?

(c) Evaluate the function for $t = 0$, 4, 8, and 12.

(d) Find dT/dt and explain its meaning in this situation.

(e) Evaluate dT/dt for $t = 0$, 4, 8, and 12.

34. *Marginal Profit* The profit (in dollars) from selling x units of a product is given by

$$P = 36{,}000 + 2048\sqrt{x} - \frac{1}{8x^2}, \quad 150 \le x \le 275.$$

Find the marginal profit for each of the following sales.

(a) $x = 150$ (b) $x = 175$ (c) $x = 200$

(d) $x = 225$ (e) $x = 250$ (f) $x = 275$

35. *Profit* The monthly demand function and cost function for x newspapers at a newsstand are given by $p = 5 - 0.001x$ and $C = 35 + 1.5x$.

(a) Find the monthly revenue R as a function of x.

(b) Find the monthly profit P as a function of x.

(c) Complete the table.

x	600	1200	1800	2400	3000
dR/dx					
dP/dx					
P					

36. *Economics* Use the table to answer the questions below.

Quantity produced and sold (Q)	Price (p)	Total revenue (TR)	Marginal revenue (MR)
0	160	0	—
2	140	280	130
4	120	480	90
6	100	600	50
8	80	640	10
10	60	600	−30

(a) Use the *regression* feature of a graphing utility to find a quadratic model that relates the total revenue (TR) to the quantity produced and sold (Q).

(b) Using derivatives, find a model for marginal revenue from the model you found in part (a).

(c) Calculate the marginal revenue for all values of Q using your model in part (b), and compare these values with the actual values given. How good is your model?

(Source: Adapted from Taylor, Economics, *Fourth Edition)*

37. *Marginal Profit* When a glass of lemonade at a lemonade stand was \$0.75, 400 glasses were sold. When the price was lowered to \$0.50, 500 glasses were sold. Assume that the demand function is linear and that the variable and fixed costs are \$0.05 and \$20, respectively.

(a) Find the profit P as a function of x, the number of glasses of lemonade sold.

(b) Use a graphing utility to graph P, and comment about the slopes of P when $x = 200$ and when $x = 400$.

(c) Find the marginal profits when 200 glasses of lemonade are sold and when 400 glasses of lemonade are sold.

38. ***Marginal Profit*** When the admission price for a baseball game was \$6 per ticket, 36,000 tickets were sold. When the price was raised to \$7, only 33,000 tickets were sold. Assume that the demand function is linear and that the variable and fixed costs for the ballpark owners are \$0.20 and \$85,000, respectively.

(a) Find the profit P as a function of x, the number of tickets sold.

(b) Use a graphing utility to graph P, and comment about the slopes of P when $x = 18{,}000$ and when $x = 36{,}000$.

(c) Find the marginal profits when 18,000 tickets are sold and when 36,000 tickets are sold.

39. ***Marginal Profit*** In Exercise 38, suppose ticket sales decreased to 30,000 when the price increased to \$7. How would this change the answers?

40. ***Marginal Cost*** The cost C of producing x units is modeled by $C = v(x) + k$, where v represents the variable cost and k represents the fixed cost. Show that the marginal cost is independent of the fixed cost.

41. ***Profit*** The demand function for a product is given by $p = 50/\sqrt{x}$ for $1 \le x \le 8000$, and the cost function is given by $C = 0.5x + 500$ for $0 \le x \le 8000$.

Find the marginal profits for (a) $x = 900$, (b) $x = 1600$, (c) $x = 2500$, and (d) $x = 3600$.

If you were in charge of setting the price for this product, what price would you set? Explain your reasoning.

42. ***Inventory Management*** The annual inventory cost for a manufacturer is given by $C = 1{,}008{,}000/Q + 6.3Q$, where Q is the order size when the inventory is replenished. Find the change in annual cost when Q is increased from 350 to 351, and compare this with the instantaneous rate of change when $Q = 350$.

43. ***Center of Population*** Since 1790, the center of population of the United States has been gradually moving westward. Use the figure to estimate the rate (in miles per year) at which the center of population was moving *westward* during the given period. *(Source: U.S. Census Bureau)*

(a) From 1790 to 1900 (b) From 1900 to 2000

(c) From 1790 to 2000

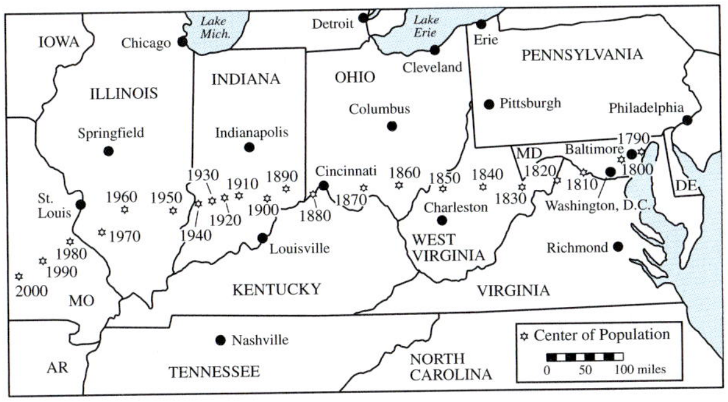

44. ***Consumer Awareness*** A car is driven 15,000 miles a year and gets x miles per gallon. The average fuel cost is \$1.30 per gallon. Find the annual cost C of fuel as a function of x and use this function to complete the table.

x	10	15	20	25	30	35	40
C							
dC/dx							

Who would benefit more from a 1 mile per gallon increase in fuel efficiency—the driver who gets 15 miles per gallon or the driver who gets 35 miles per gallon? Explain.

45. ***Writing*** The number N of gallons of regular unleaded gasoline sold by a gasoline station at a price of p dollars per gallon is given by $N = f(p)$.

(a) Describe the meaning of $f'(1.479)$.

(b) Is $f'(1.479)$ usually positive or negative? Explain.

46. Consider the function given by $f(x) = \dfrac{4}{x}$, $\quad 0 < x \le 5$.

(a) Use a graphing utility to graph f and f' in the same viewing window.

(b) Complete the table.

x	$\frac{1}{8}$	$\frac{1}{4}$	$\frac{1}{2}$	1	2	3	4	5
$f(x)$								
$f'(x)$								

(c) Find the average rate of change of f over the intervals determined by consecutive x-values in the table.

In Exercises 47 and 48, use a graphing utility to graph f and f'. Then determine the points (if any) at which f has a horizontal tangent.

47. $f(x) = \frac{1}{4}x^3, \quad -2 \le x \le 2$

48. $f(x) = x^4 - 12x^3 + 52x^2 - 96x + 64, \quad 1 \le x \le 5$

49. ***Biology*** Many populations in nature exhibit logistic growth, which consists of four phases, as shown in the figure. Describe the rate of growth of the population in each phase, and give possible reasons as to why the rates might be changing from phase to phase. *(Source: Adapted from Levine/Miller,* Biology: Discovering Life, *Second Edition)*

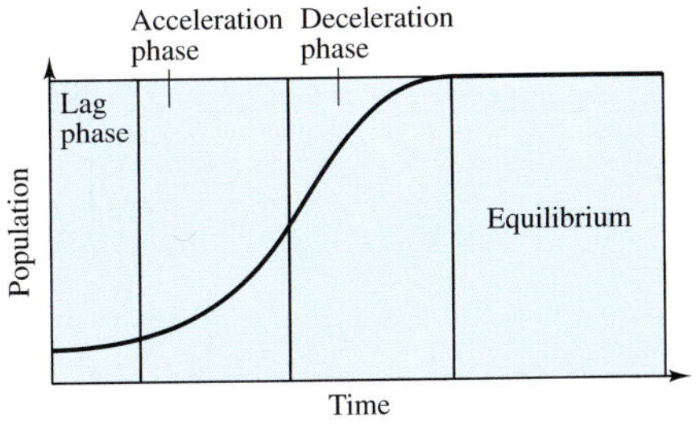

2.4 THE PRODUCT AND QUOTIENT RULES

- Find the derivatives of functions using the Product Rule.
- Find the derivatives of functions using the Quotient Rule.
- Simplify derivatives.
- Use derivatives to answer questions about real-life situations.

The Product Rule

In Section 2.2, you saw that the derivative of a sum or difference of two functions is simply the sum or difference of their derivatives. The rules for the derivative of a product or quotient of two functions are not as simple.

The Product Rule

The derivative of the product of two differentiable functions is equal to the first function times the derivative of the second plus the second function times the derivative of the first.

$$\frac{d}{dx}[f(x)g(x)] = f(x)g'(x) + g(x)f'(x)$$

STUDY TIP

Rather than trying to remember the formula for the Product Rule, it can be more helpful to remember its verbal statement: *the first function times the derivative of the second plus the second function times the derivative of the first.*

PROOF Some mathematical proofs, such as the proof of the Sum Rule, are straightforward. Others involve clever steps that may not appear to follow clearly from a prior step. The proof below involves such a step—adding and subtracting the same quantity. (This step is shown in color.) Let $F(x) = f(x)g(x)$.

$$\begin{aligned}
F'(x) &= \lim_{\Delta x \to 0} \frac{F(x + \Delta x) - F(x)}{\Delta x} \\
&= \lim_{\Delta x \to 0} \frac{f(x + \Delta x)g(x + \Delta x) - f(x)g(x)}{\Delta x} \\
&= \lim_{\Delta x \to 0} \frac{f(x + \Delta x)g(x + \Delta x) - f(x + \Delta x)g(x) + f(x + \Delta x)g(x) - f(x)g(x)}{\Delta x} \\
&= \lim_{\Delta x \to 0} \left[f(x + \Delta x)\frac{g(x + \Delta x) - g(x)}{\Delta x} + g(x)\frac{f(x + \Delta x) - f(x)}{\Delta x} \right] \\
&= \lim_{\Delta x \to 0} f(x + \Delta x)\frac{g(x + \Delta x) - g(x)}{\Delta x} + \lim_{\Delta x \to 0} g(x)\frac{f(x + \Delta x) - f(x)}{\Delta x} \\
&= \left[\lim_{\Delta x \to 0} f(x + \Delta x)\right]\left[\lim_{\Delta x \to 0} \frac{g(x + \Delta x) - g(x)}{\Delta x}\right] \\
&\quad + \left[\lim_{\Delta x \to 0} g(x)\right]\left[\lim_{\Delta x \to 0} \frac{f(x + \Delta x) - f(x)}{\Delta x}\right] \\
&= f(x)g'(x) + g(x)f'(x)
\end{aligned}$$

EXAMPLE 1 Finding the Derivative of a Product

Find the derivative of $y = (3x - 2x^2)(5 + 4x)$.

SOLUTION Using the Product Rule, you can write

$$\frac{dy}{dx} = \overbrace{(3x - 2x^2)}^{\text{First}} \overbrace{\frac{d}{dx}[5 + 4x]}^{\text{Derivative of second}} + \overbrace{(5 + 4x)}^{\text{Second}} \overbrace{\frac{d}{dx}[3x - 2x^2]}^{\text{Derivative of first}}$$

$$= (3x - 2x^2)(4) + (5 + 4x)(3 - 4x)$$

$$= (12x - 8x^2) + (15 - 8x - 16x^2)$$

$$= 15 + 4x - 24x^2.$$

TRY IT 1

Find the derivative of $y = (4x + 3x^2)(6 - 3x)$.

STUDY TIP

In general, the derivative of the product of two functions is not equal to the product of the derivatives of the two functions. To see this, compare the product of the derivatives of $f(x) = 3x - 2x^2$ and $g(x) = 5 + 4x$ with the derivative found in Example 1.

In the next example, notice that the first step in differentiating is *rewriting the original function.*

EXAMPLE 2 Finding the Derivative of a Product

Find the derivative of

$$f(x) = \left(\frac{1}{x} + 1\right)(x - 1). \qquad \text{Original function}$$

SOLUTION Rewrite the function. Then use the Product Rule to find the derivative.

$$f(x) = (x^{-1} + 1)(x - 1) \qquad \text{Rewrite function.}$$

$$f'(x) = (x^{-1} + 1)\frac{d}{dx}[x - 1] + (x - 1)\frac{d}{dx}[x^{-1} + 1] \qquad \text{Product Rule}$$

$$= (x^{-1} + 1)(1) + (x - 1)(-x^{-2})$$

$$= \frac{1}{x} + 1 - \frac{x - 1}{x^2}$$

$$= \frac{x + x^2 - x + 1}{x^2} \qquad \text{Write with common denominator.}$$

$$= \frac{x^2 + 1}{x^2} \qquad \text{Simplify.}$$

TECHNOLOGY

If you have access to a symbolic differentiation utility, try using it to confirm several of the derivatives in this section. The form of the derivative can depend on how you use software.

TRY IT 2

Find the derivative of

$$f(x) = \left(\frac{1}{x} + 1\right)(2x + 1).$$

You now have two differentiation rules that deal with products—the Constant Multiple Rule and the Product Rule. The difference between these two rules is that the Constant Multiple Rule deals with the product of a constant and a variable quantity:

Constant (c) — Variable quantity ($f(x)$)

$$F(x) = c\,f(x)$$ Use Constant Multiple Rule.

whereas the Product Rule deals with the product of two variable quantities:

Variable quantity ($f(x)$) — Variable quantity ($g(x)$)

$$F(x) = f(x)\,g(x).$$ Use Product Rule.

The next example compares these two rules.

STUDY TIP

You could calculate the derivatives in Example 3 without the Product Rule. For Example 3(a),

$$y = 2x(x^2 + 3x) = 2x^3 + 6x^2$$

and

$$\frac{dy}{dx} = 6x^2 + 12x.$$

EXAMPLE 3 Comparing Differentiation Rules

Find the derivative of each function.

(a) $y = 2x(x^2 + 3x)$

(b) $y = 2(x^2 + 3x)$

SOLUTION

(a) By the Product Rule,

$$\begin{aligned}\frac{dy}{dx} &= (2x)\frac{d}{dx}[x^2 + 3x] + (x^2 + 3x)\frac{d}{dx}[2x] && \text{Product Rule}\\ &= (2x)(2x + 3) + (x^2 + 3x)(2)\\ &= 4x^2 + 6x + 2x^2 + 6x\\ &= 6x^2 + 12x.\end{aligned}$$

(b) By the Constant Multiple Rule,

$$\begin{aligned}\frac{dy}{dx} &= 2\frac{d}{dx}[x^2 + 3x] && \text{Constant Multiple Rule}\\ &= 2(2x + 3)\\ &= 4x + 6.\end{aligned}$$

TRY IT 3

Find the derivative of each function.

(a) $y = 3x(2x^2 + 5x)$

(b) $y = 3(2x^2 + 5x)$

The Product Rule can be extended to products that have more than two factors. For example, if f, g, and h are differentiable functions of x, then

$$\frac{d}{dx}[f(x)g(x)h(x)] = f'(x)g(x)h(x) + f(x)g'(x)h(x) + f(x)g(x)h'(x).$$

The Quotient Rule

In Section 2.2, you saw that by using the Constant Rule, Power Rule, Constant Multiple Rule, and Sum and Difference Rules, you were able to differentiate any polynomial function. By combining these rules with the Quotient Rule, you can now differentiate any *rational* function.

The Quotient Rule

The derivative of the quotient of two differentiable functions is equal to the denominator times the derivative of the numerator minus the numerator times the derivative of the denominator, all divided by the square of the denominator.

$$\frac{d}{dx}\left[\frac{f(x)}{g(x)}\right] = \frac{g(x)f'(x) - f(x)g'(x)}{[g(x)]^2}, \qquad g(x) \neq 0$$

STUDY TIP

From this differentiation rule, you can see that the derivative of a quotient is not, in general, the quotient of the derivatives. That is,

$$\frac{d}{dx}\left[\frac{f(x)}{g(x)}\right] \neq \frac{f'(x)}{g'(x)}.$$

PROOF Let $F(x) = f(x)/g(x)$. As in the proof of the Product Rule, a key step in this proof is adding and subtracting the same quantity.

$$F'(x) = \lim_{\Delta x \to 0} \frac{F(x + \Delta x) - F(x)}{\Delta x}$$

$$= \lim_{\Delta x \to 0} \frac{\dfrac{f(x + \Delta x)}{g(x + \Delta x)} - \dfrac{f(x)}{g(x)}}{\Delta x}$$

$$= \lim_{\Delta x \to 0} \frac{g(x)f(x + \Delta x) - f(x)g(x + \Delta x)}{\Delta x g(x)g(x + \Delta x)}$$

$$= \lim_{\Delta x \to 0} \frac{g(x)f(x + \Delta x) - f(x)g(x) + f(x)g(x) - f(x)g(x + \Delta x)}{\Delta x g(x)g(x + \Delta x)}$$

$$= \frac{\displaystyle\lim_{\Delta x \to 0} \frac{g(x)[f(x + \Delta x) - f(x)]}{\Delta x} - \lim_{\Delta x \to 0} \frac{f(x)[g(x + \Delta x) - g(x)]}{\Delta x}}{\displaystyle\lim_{\Delta x \to 0} [g(x)g(x + \Delta x)]}$$

$$= \frac{g(x)\left[\displaystyle\lim_{\Delta x \to 0} \frac{f(x + \Delta x) - f(x)}{\Delta x}\right] - f(x)\left[\displaystyle\lim_{\Delta x \to 0} \frac{g(x + \Delta x) - g(x)}{\Delta x}\right]}{\displaystyle\lim_{\Delta x \to 0} [g(x)g(x + \Delta x)]}$$

$$= \frac{g(x)f'(x) - f(x)g'(x)}{[g(x)]^2}$$

STUDY TIP

As suggested for the Product Rule, it can be more helpful to remember the verbal statement of the Quotient Rule rather than trying to remember the formula for the rule.

EXAMPLE 4 **Finding the Derivative of a Quotient**

Find the derivative of $y = \dfrac{x - 1}{2x + 3}$.

SOLUTION Apply the Quotient Rule, as shown.

$$\begin{aligned}\frac{dy}{dx} &= \frac{(2x + 3)\frac{d}{dx}[x - 1] - (x - 1)\frac{d}{dx}[2x + 3]}{(2x + 3)^2}\\ &= \frac{(2x + 3)(1) - (x - 1)(2)}{(2x + 3)^2}\\ &= \frac{2x + 3 - 2x + 2}{(2x + 3)^2}\\ &= \frac{5}{(2x + 3)^2}\end{aligned}$$

ALGEBRA REVIEW

When applying the Quotient Rule, it is suggested that you enclose all factors and derivatives in symbols of grouping, such as parentheses. Also, pay special attention to the subtraction required in the numerator.

TRY IT 4

Find the derivative of $y = \dfrac{x + 4}{5x - 2}$.

EXAMPLE 5 **Finding an Equation of a Tangent Line**

Find the equation of the tangent line to the graph of

$$y = \frac{2x^2 - 4x + 3}{2 - 3x}$$

when $x = 1$.

SOLUTION Apply the Quotient Rule, as shown.

$$\begin{aligned}\frac{dy}{dx} &= \frac{(2 - 3x)\frac{d}{dx}[2x^2 - 4x + 3] - (2x^2 - 4x + 3)\frac{d}{dx}[2 - 3x]}{(2 - 3x)^2}\\ &= \frac{(2 - 3x)(4x - 4) - (2x^2 - 4x + 3)(-3)}{(2 - 3x)^2}\\ &= \frac{-12x^2 + 20x - 8 - (-6x^2 + 12x - 9)}{(2 - 3x)^2}\\ &= \frac{-12x^2 + 20x - 8 + 6x^2 - 12x + 9}{(2 - 3x)^2}\\ &= \frac{-6x^2 + 8x + 1}{(2 - 3x)^2}\end{aligned}$$

When $x = 1$, the value of the function is $y = -1$ and the slope is $m = 3$. Using the point-slope form of a line, you can find the equation of the tangent line to be $y = 3x - 4$. The graph of the function and the tangent line is shown in Figure 2.27.

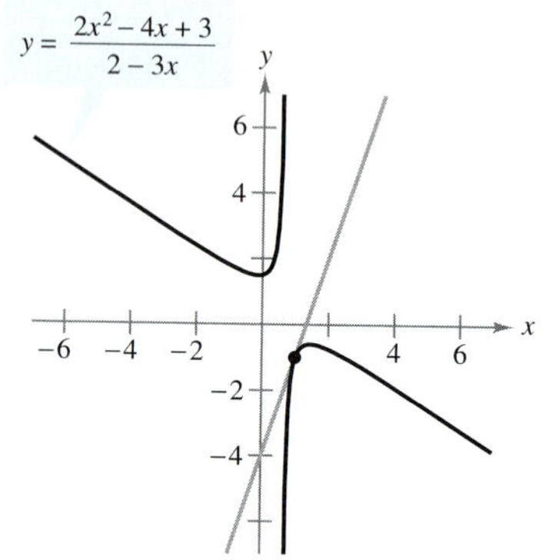

FIGURE 2.27

TRY IT 5

Find the equation of the tangent line to the graph of

$$y = \frac{x^2 - 4}{2x + 5} \text{ when } x = 0.$$

Sketch the line tangent to the graph at $x = 0$.

EXAMPLE 6 **Finding the Derivative of a Quotient**

Find the derivative of

$$y = \frac{3 - (1/x)}{x + 5}.$$

SOLUTION Begin by rewriting the original function. Then apply the Quotient Rule and simplify the result.

$$y = \frac{3 - (1/x)}{x + 5}$$ Write original function.

$$= \frac{3x - 1}{x(x + 5)}$$ Multiply numerator and denominator by x.

$$= \frac{3x - 1}{x^2 + 5x}$$ Rewrite.

$$\frac{dy}{dx} = \frac{(x^2 + 5x)(3) - (3x - 1)(2x + 5)}{(x^2 + 5x)^2}$$ Apply Quotient Rule.

$$= \frac{(3x^2 + 15x) - (6x^2 + 13x - 5)}{(x^2 + 5x)^2}$$

$$= \frac{-3x^2 + 2x + 5}{(x^2 + 5x)^2}$$ Simplify.

TRY IT 6

Find the derivative of $y = \dfrac{3 - (2/x)}{x + 4}$.

Not every quotient needs to be differentiated by the Quotient Rule. For instance, each of the quotients in the next example can be considered as the product of a constant and a function of x. In such cases, the Constant Multiple Rule is more efficient than the Quotient Rule.

EXAMPLE 7 **Rewriting Before Differentiating**

Find the derivative of each function.

	Original Function	*Rewrite*	*Differentiate*	*Simplify*
(a)	$y = \dfrac{x^2 + 3x}{6}$	$y = \dfrac{1}{6}(x^2 + 3x)$	$y' = \dfrac{1}{6}(2x + 3)$	$y' = \dfrac{1}{3}x + \dfrac{1}{2}$
(b)	$y = \dfrac{5x^4}{8}$	$y = \dfrac{5}{8}x^4$	$y' = \dfrac{5}{8}(4x^3)$	$y' = \dfrac{5}{2}x^3$
(c)	$y = \dfrac{-3(3x - 2x^2)}{7x}$	$y = -\dfrac{3}{7}(3 - 2x)$	$y' = -\dfrac{3}{7}(-2)$	$y' = \dfrac{6}{7}$
(d)	$y = \dfrac{9}{5x^2}$	$y = \dfrac{9}{5}(x^{-2})$	$y' = \dfrac{9}{5}(-2x^{-3})$	$y' = -\dfrac{18}{5x^3}$

STUDY TIP

To see the efficiency of using the Constant Multiple Rule in Example 7, try using the Quotient Rule to find the derivatives of the four functions.

TRY IT 7

Find the derivative of each function.

(a) $y = \dfrac{x^2 + 4x}{5}$ (b) $y = \dfrac{3x^4}{4}$

Simplifying Derivatives

EXAMPLE 8 Combining the Product and Quotient Rules

Find the derivative of

$$y = \frac{(1 - 2x)(3x + 2)}{5x - 4}.$$

SOLUTION This function contains a product within a quotient. You could first multiply the factors in the numerator and then apply the Quotient Rule. However, to gain practice in using the Product Rule within the Quotient Rule, try differentiating as shown.

$$y' = \frac{(5x - 4)\frac{d}{dx}[(1 - 2x)(3x + 2)] - (1 - 2x)(3x + 2)\frac{d}{dx}[5x - 4]}{(5x - 4)^2}$$

$$= \frac{(5x - 4)[(1 - 2x)(3) + (3x + 2)(-2)] - (1 - 2x)(3x + 2)(5)}{(5x - 4)^2}$$

$$= \frac{(5x - 4)(-12x - 1) - (1 - 2x)(15x + 10)}{(5x - 4)^2}$$

$$= \frac{(-60x^2 + 43x + 4) - (-30x^2 - 5x + 10)}{(5x - 4)^2}$$

$$= \frac{-30x^2 + 48x - 6}{(5x - 4)^2}$$

TRY IT 8

Find the derivative of $y = \dfrac{(1 + x)(2x - 1)}{x - 1}$.

In the examples in this section, much of the work in obtaining the final form of the derivative occurs *after* the differentiation. As summarized in the list below, direct application of differentiation rules often yields results that are not in simplified form. Note that two characteristics of simplified form are the absence of negative exponents and the combining of like terms.

	f′(x) After Differentiating	*f′(x) After Simplifying*
Example 1	$(3x - 2x^2)(4) + (5 + 4x)(3 - 4x)$	$15 + 4x - 24x^2$
Example 2	$(x^{-1} + 1)(1) + (x - 1)(-x^{-2})$	$\dfrac{x^2 + 1}{x^2}$
Example 5	$\dfrac{(2 - 3x)(4x - 4) - (2x^2 - 4x + 3)(-3)}{(2 - 3x)^2}$	$\dfrac{-6x^2 + 8x + 1}{(2 - 3x)^2}$
Example 8	$\dfrac{(5x - 4)[(1 - 2x)(3) + (3x + 2)(-2)] - (1 - 2x)(3x + 2)(5)}{(5x - 4)^2}$	$\dfrac{-30x^2 + 48x - 6}{(5x - 4)^2}$

Application

EXAMPLE 9 Rate of Change of Blood Pressure

As blood moves from the heart through the major arteries out to the capillaries and back through the veins, the systolic pressure continuously drops. Consider a person whose blood pressure P (in millimeters of mercury) is given by

$$P = \frac{25t^2 + 125}{t^2 + 1}, \qquad 0 \le t \le 10$$

where t is measured in seconds. At what rate is the blood pressure changing 5 seconds after blood leaves the heart?

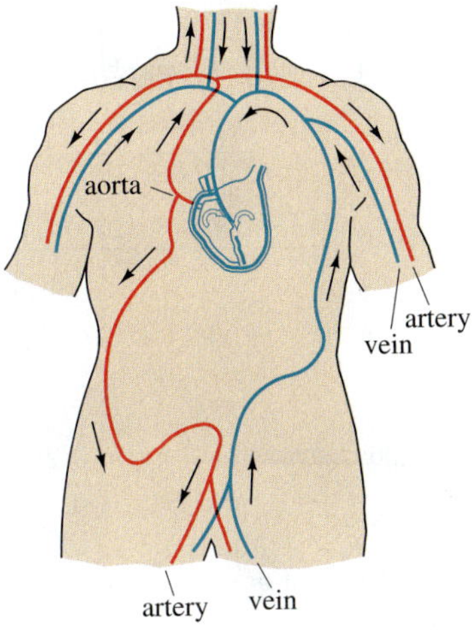

SOLUTION Begin by applying the Quotient Rule.

$$\frac{dP}{dt} = \frac{(t^2 + 1)(50t) - (25t^2 + 125)(2t)}{(t^2 + 1)^2} \qquad \text{Quotient Rule}$$

$$= \frac{50t^3 + 50t - 50t^3 - 250t}{(t^2 + 1)^2}$$

$$= -\frac{200t}{(t^2 + 1)^2} \qquad \text{Simplify.}$$

When $t = 5$, the rate of change is

$$-\frac{200(5)}{26^2} \approx -1.48 \text{ millimeters per second.}$$

So, the pressure is *dropping* at a rate of 1.48 millimeters per second when $t = 5$ seconds.

TRY IT 9

In Example 9, find the rate at which blood pressure is changing at each time shown in the table below. Describe the changes in blood pressure as the blood moves away from the heart.

t	0	1	2	3	4	5	6	7
$\frac{dP}{dt}$								

TAKE ANOTHER LOOK

Testing Differentiation Rules

What is wrong with the differentiations at the right? If you were teaching a calculus class, how could you convince your class that these derivatives are incorrect?

~~$f(x) = (x^3)(x^4)$~~
~~$f'(x) = (3x^2)(4x^3)$~~

~~$g(x) = \frac{x^3}{x^4}$~~
~~$g'(x) = \frac{3x^2}{4x^3}$~~

PREREQUISITE REVIEW 2.4

The following warm-up exercises involve skills that were covered in earlier sections. You will use these skills in the exercise set for this section.

In Exercises 1–10, simplify the expression.

1. $(x^2 + 1)(2) + (2x + 7)(2x)$

2. $(2x - x^3)(8x) + (4x^2)(2 - 3x^2)$

3. $x(4)(x^2 + 2)^3(2x) + (x^2 + 4)(1)$

4. $x^2(2)(2x + 1)(2) + (2x + 1)^4(2x)$

5. $\dfrac{(2x + 7)(5) - (5x + 6)(2)}{(2x + 7)^2}$

6. $\dfrac{(x^2 - 4)(2x + 1) - (x^2 + x)(2x)}{(x^2 - 4)^2}$

7. $\dfrac{(x^2 + 1)(2) - (2x + 1)(2x)}{(x^2 + 1)^2}$

8. $\dfrac{(1 - x^4)(4) - (4x - 1)(-4x^3)}{(1 - x^4)^2}$

9. $(x^{-1} + x)(2) + (2x - 3)(-x^{-2} + 1)$

10. $\dfrac{(1 - x^{-1})(1) - (x - 4)(x^{-2})}{(1 - x^{-1})^2}$

In Exercises 11–14, find $f'(2)$.

11. $f(x) = 3x^2 - x + 4$

12. $f(x) = -x^3 + x^2 + 8x$

13. $f(x) = \dfrac{1}{x}$

14. $f(x) = x^2 - \dfrac{1}{x^2}$

EXERCISES 2.4

In Exercises 1–14, find the value of the derivative of the function at the given point.

	Function	*Point*
1.	$f(x) = x^2(3x^3 - 1)$	$(1, 2)$
2.	$f(x) = (x^2 + 1)(2x + 5)$	$(-1, 6)$
3.	$f(x) = \frac{1}{3}(2x^3 - 4)$	$\left(0, -\frac{4}{3}\right)$
4.	$f(x) = \frac{1}{7}(5 - 6x^2)$	$\left(1, -\frac{1}{7}\right)$
5.	$g(x) = (x^2 - 4x + 3)(x - 2)$	$(4, 6)$
6.	$g(x) = (x^2 - 2x + 1)(x^3 - 1)$	$(1, 0)$
7.	$h(x) = \dfrac{x}{x - 5}$	$(6, 6)$
8.	$h(x) = \dfrac{x^2}{x + 3}$	$\left(-1, \dfrac{1}{2}\right)$
9.	$f(t) = \dfrac{2t^2 - 3}{3t + 1}$	$\left(3, \dfrac{3}{2}\right)$
10.	$f(x) = \dfrac{3x}{x^2 + 4}$	$\left(-1, -\dfrac{3}{5}\right)$
11.	$g(x) = \dfrac{2x + 1}{x - 5}$	$(6, 13)$
12.	$f(x) = \dfrac{x + 1}{x - 1}$	$(2, 3)$
13.	$f(t) = \dfrac{t^2 - 1}{t + 4}$	$(1, 0)$
14.	$g(x) = \dfrac{4x - 5}{x^2 - 1}$	$(0, 5)$

In Exercises 15–22, find the derivative of the function. Use Example 7 as a model.

	Function	*Rewrite*	*Differentiate*	*Simplify*
15.	$y = \dfrac{x^2 + 2x}{x}$			
16.	$y = \dfrac{4x^{3/2}}{x}$			
17.	$y = \dfrac{7}{3x^3}$			
18.	$y = \dfrac{4}{5x^2}$			
19.	$y = \dfrac{4x^2 - 3x}{8\sqrt{x}}$			
20.	$y = \dfrac{3x^2 - 4x}{6x}$			
21.	$y = \dfrac{x^2 - 4x + 3}{x - 1}$			
22.	$y = \dfrac{x^2 - 4}{x + 2}$			

In Exercises 23–38, find the derivative of the function.

23. $f(x) = (x^3 - 3x)(2x^2 + 3x + 5)$

24. $h(t) = (t^5 - 1)(4t^2 - 7t - 3)$

25. $g(t) = (2t^3 - 1)^2$

26. $h(p) = (p^3 - 2)^2$

27. $f(x) = \sqrt[3]{x}\left(\sqrt{x} + 3\right)$

28. $f(x) = \sqrt[3]{x}(x + 1)$

29. $f(x) = \dfrac{3x - 2}{2x - 3}$

30. $f(x) = \dfrac{x^3 + 3x + 2}{x^2 - 1}$

31. $f(x) = \dfrac{3 - 2x - x^2}{x^2 - 1}$

32. $f(x) = (x^5 - 3x)\left(\dfrac{1}{x^2}\right)$

33. $f(x) = x\left(1 - \dfrac{2}{x + 1}\right)$

34. $h(t) = \dfrac{t + 2}{t^2 + 5t + 6}$

35. $g(s) = \dfrac{s^2 - 2s + 5}{\sqrt{s}}$

36. $f(x) = \dfrac{x + 1}{\sqrt{x}}$

37. $g(x) = \left(\dfrac{x - 3}{x + 4}\right)(x^2 + 2x + 1)$

38. $f(x) = (3x^3 + 4x)(x - 5)(x + 1)$

In Exercises 39–44, find an equation of the tangent line to the graph of the function at the given point. Then use a graphing utility to graph the function and the tangent line in the same viewing window.

Function	*Point*
39. $f(x) = (x - 1)^2(x - 2)$	$(0, -2)$
40. $h(x) = (x^2 - 1)^2$	$(-2, 9)$
41. $f(x) = \dfrac{x - 2}{x + 1}$	$\left(1, -\frac{1}{2}\right)$
42. $f(x) = \dfrac{2x + 1}{x - 1}$	$(2, 5)$
43. $f(x) = \left(\dfrac{x + 5}{x - 1}\right)(2x + 1)$	$(0, -5)$
44. $g(x) = (x + 2)\left(\dfrac{x - 5}{x + 1}\right)$	$(0, -10)$

In Exercises 45–48, find the point(s), if any, at which the graph of f has a horizontal tangent.

45. $f(x) = \dfrac{x^2}{x - 1}$

46. $f(x) = \dfrac{x^2}{x^2 + 1}$

47. $f(x) = \dfrac{x^4}{x^3 + 1}$

48. $f(x) = \dfrac{x^4 + 3}{x^2 + 1}$

In Exercises 49–52, use a graphing utility to graph f and f' on the interval $[-2, 2]$.

49. $f(x) = x(x + 1)$

50. $f(x) = x^2(x + 1)$

51. $f(x) = x(x + 1)(x - 1)$

52. $f(x) = x^2(x + 1)(x - 1)$

Demand In Exercises 53 and 54, use the demand function to find the rate of change in the demand x for the given price p.

53. $x = 275\left(1 - \dfrac{3p}{5p + 1}\right)$, $p = \$4$

54. $x = 300 - p - \dfrac{2p}{p + 1}$, $p = \$3$

55. ***Environment*** The model

$$f(t) = \frac{t^2 - t + 1}{t^2 + 1}$$

measures the percent of the normal level of oxygen in a pond, where t is the time (in weeks) after organic waste is dumped into the pond. Find the rates of change of f with respect to t when (a) $t = 0.5$, (b) $t = 2$, and (c) $t = 8$.

56. ***Physical Science*** The temperature T of food placed in a refrigerator is modeled by

$$T = 10\left(\frac{4t^2 + 16t + 75}{t^2 + 4t + 10}\right)$$

where t is the time (in hours). What is the initial temperature of the food? Find the rates of change of T with respect to t when (a) $t = 1$, (b) $t = 3$, (c) $t = 5$, and (d) $t = 10$.

57. ***Population Growth*** A population of bacteria is introduced into a culture. The number of bacteria P can be modeled by

$$P = 500\left(1 + \frac{4t}{50 + t^2}\right)$$

where t is the time (in hours). Find the rate of change of the population when $t = 2$.

58. ***Quality Control*** The percent P of defective parts produced by a new employee t days after the employee starts work can be modeled by

$$P = \frac{t + 1750}{50(t + 2)}.$$

Find the rates of change of P when (a) $t = 1$ and (b) $t = 10$.

59. ***Profit*** You decide to form a partnership with another business. Your business determines that the demand x for your product is inversely proportional to the square of the price for $x \geq 5$.

(a) The price is \$1000 and the demand is 16 units. Find the demand function.

(b) Your partner determines that the product costs \$250 per unit and the fixed cost is \$10,000. Find the cost function.

(c) Find the profit function and use a graphing utility to graph it. From the graph, what price would you negotiate with your partner for this product? Explain your reasoning.

60. *Profit* You are managing a store and have been adjusting the price of an item. You have found that you make a profit of \$50 when 10 units are sold, \$60 when 12 units are sold, and \$65 when 14 units are sold.

(a) Fit these data to the model $P = ax^2 + bx + c$.

(b) Use a graphing utility to graph P.

(c) Find the point on the graph at which the marginal profit is zero. Interpret this point in the context of the problem.

61. *Demand Function* Given $f(x) = x^2 + 1$, which function would most likely represent a demand function?

(a) $p = f(x)$

(b) $p = xf(x)$

(c) $p = 1/f(x)$

Explain your reasoning. Use a graphing utility to graph each function, and use each graph as part of your explanation.

62. *Cost* The cost of producing x units of a product is given by

$$C = x^3 - 15x^2 + 87x - 73, \qquad 4 \le x \le 9.$$

(a) Use a graphing utility to graph the marginal cost function and the average cost function, C/x, in the same viewing window.

(b) Find the point of intersection of the graphs of dC/dx and C/x. Does this point have any significance?

63. *Inventory Replenishment* The ordering and transportation cost C (in thousands of dollars) of the components used in manufacturing a product is given by

$$C = 100\left(\frac{200}{x^2} + \frac{x}{x + 30}\right), \qquad 1 \le x$$

where x is the order size (in hundreds). Find the rate of change of C with respect to x for each order size.

(a) $x = 10$

(b) $x = 15$

(c) $x = 20$

What do these rates of change imply about increasing the size of an order?

64. *Sales Analysis* The monthly sales of memberships M at a newly built fitness center are modeled by

$$M(t) = \frac{300t}{t^2 + 1} + 8$$

where t is the number of months since the center opened.

(a) Find $M'(t)$.

(b) Find $M(3)$ and $M'(3)$ and interpret the results.

(c) Find $M(24)$ and $M'(24)$ and interpret the results.

65. *Consumer Awareness* The prices of 1 pound of 100% ground beef in the United States from 1995 to 2002 can be modeled by

$$P = \frac{1.47 - 0.311t + 0.0173t^2}{1 - 0.206t + 0.0112t^2}$$

where t is the year, with $t = 5$ corresponding to 1995. Find dP/dt and evaluate it for $t = 5, 7, 9$, and 11. Interpret the meaning of these values. *(Source: U.S. Bureau of Labor Statistics)*

BUSINESS CAPSULE

The Blackwood Centre for Adolescent Development, a state-sponsored secondary school for young people at risk in Victoria, Australia, joined forces in 2000 with the Centre for Executive Development (CED). With the CED providing fundraising and logistical support, the Blackwood Centre has been able to transform its program into a model of success, offering their students training and educational opportunities for entry into the business world. The CED has gained team-building skills, improved workplace assessment methods, and a stronger connection to the community.

66. *Research Project* Use your school's library, the Internet, or some other reference source to find information about partnerships between companies and federal, state, or local government that have benefited their communities. (One such partnership is described above.) Write a short paper about the partnership.

2.5 THE CHAIN RULE

- Find derivatives using the Chain Rule.
- Find derivatives using the General Power Rule.
- Write derivatives in simplified form.
- Use derivatives to answer questions about real-life situations.
- Use the differentiation rules to differentiate algebraic functions.

The Chain Rule

In this section, you will study one of the most powerful rules of differential calculus—the **Chain Rule.** This differentiation rule deals with composite functions and adds versatility to the rules presented in Sections 2.2 and 2.4. For example, compare the functions below. Those on the left can be differentiated without the Chain Rule, whereas those on the right are best done with the Chain Rule.

Without the Chain Rule	*With the Chain Rule*
$y = x^2 + 1$	$y = \sqrt{x^2 + 1}$
$y = x + 1$	$y = (x + 1)^{-1/2}$
$y = 3x + 2$	$y = (3x + 2)^5$
$y = \dfrac{x + 5}{x^2 + 2}$	$y = \left(\dfrac{x + 5}{x^2 + 2}\right)^2$

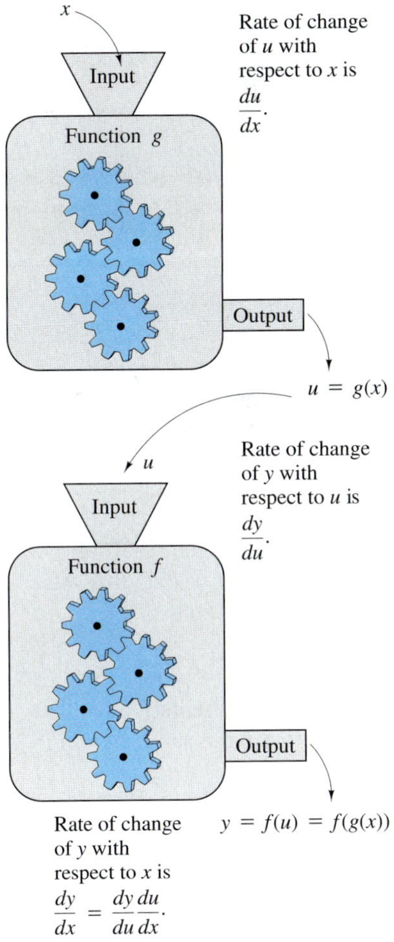

FIGURE 2.28

The Chain Rule

If $y = f(u)$ is a differentiable function of u, and $u = g(x)$ is a differentiable function of x, then $y = f(g(x))$ is a differentiable function of x, and

$$\frac{dy}{dx} = \frac{dy}{du} \cdot \frac{du}{dx}$$

or, equivalently,

$$\frac{d}{dx}[f(g(x))] = f'(g(x))g'(x).$$

Basically, the Chain Rule states that if y changes dy/du times as fast as u, and u changes du/dx times as fast as x, then y changes

$$\frac{dy}{du} \cdot \frac{du}{dx}$$

times as fast as x, as illustrated in Figure 2.28. One advantage of the dy/dx notation for derivatives is that it helps you remember differentiation rules, such as the Chain Rule. For instance, in the formula

$$dy/dx = (dy/du)(du/dx)$$

you can imagine that the du's divide out.

When applying the Chain Rule, it helps to think of the composite function $y = f(g(x))$ or $y = f(u)$ as having two parts—an *inside* and an *outside*—as illustrated below.

Inside

$$y = f(g(x)) = f(u)$$

Outside

The Chain Rule tells you that the derivative of $y = f(u)$ is the derivative of the outer function (at the inner function u) *times* the derivative of the inner function. That is,

$$y' = f'(u) \cdot u'.$$

TRY IT 1

Write each function as the composition of two functions, where $y = f(g(x))$.

(a) $y = \dfrac{1}{\sqrt{x+1}}$

(b) $y = (x^2 + 2x + 5)^3$

EXAMPLE 1 Decomposing Composite Functions

Write each function as the composition of two functions.

(a) $y = \dfrac{1}{x+1}$ (b) $y = \sqrt{3x^2 - x + 1}$

SOLUTION There is more than one correct way to decompose each function. One way for each is shown below.

$y = f(g(x))$	$u = g(x)$ *(inside)*	$y = f(u)$ *(outside)*
(a) $y = \dfrac{1}{x+1}$	$u = x + 1$	$y = \dfrac{1}{u}$
(b) $y = \sqrt{3x^2 - x + 1}$	$u = 3x^2 - x + 1$	$y = \sqrt{u}$

EXAMPLE 2 Using the Chain Rule

Find the derivative of $y = (x^2 + 1)^3$.

SOLUTION To apply the Chain Rule, you need to identify the inside function u.

$$y = (\underbrace{x^2 + 1}_{u})^3 = u^3$$

By the Chain Rule, you can write the derivative as shown.

$$\frac{dy}{dx} = \underbrace{3(x^2 + 1)^2}_{\frac{dy}{du}}\underbrace{(2x)}_{\frac{du}{dx}} = 6x(x^2 + 1)^2$$

STUDY TIP

Try checking the result of Example 2 by expanding the function to obtain

$$y = x^6 + 3x^4 + 3x^2 + 1$$

and finding the derivative. Do you obtain the same answer?

TRY IT 2

Find the derivative of $y = (x^3 + 1)^2$.

The General Power Rule

The function in Example 2 illustrates one of the most common types of composite functions—a power function of the form

$$y = [u(x)]^n.$$

The rule for differentiating such functions is called the **General Power Rule,** and it is a special case of the Chain Rule.

The General Power Rule

If $y = [u(x)]^n$, where u is a differentiable function of x and n is a real number, then

$$\frac{dy}{dx} = n[u(x)]^{n-1}\frac{du}{dx}$$

or, equivalently,

$$\frac{d}{dx}[u^n] = nu^{n-1}u'.$$

PROOF Apply the Chain Rule and the Simple Power Rule as shown.

$$\begin{aligned}\frac{dy}{dx} &= \frac{dy}{du} \cdot \frac{du}{dx} \\ &= \frac{d}{du}[u^n]\frac{du}{dx} \\ &= nu^{n-1}\frac{du}{dx}\end{aligned}$$

EXAMPLE 3 **Using the General Power Rule**

Find the derivative of

$$f(x) = (3x - 2x^2)^3.$$

SOLUTION The inside function is $u = 3x - 2x^2$. So, by the General Power Rule,

$$\begin{aligned}f'(x) &= \overset{n}{3}\overbrace{(3x - 2x^2)^2}^{u^{n-1}}\overbrace{\frac{d}{dx}[3x - 2x^2]}^{u'} \\ &= 3(3x - 2x^2)^2(3 - 4x) \\ &= (9 - 12x)(3x - 2x^2)^2.\end{aligned}$$

TECHNOLOGY

If you have access to a symbolic differentiation utility, try using it to confirm the result of Example 3.

TRY IT 3

Find the derivative of $y = (x^2 + 3x)^4$.

EXAMPLE 4 Rewriting Before Differentiating

Find the tangent line to the graph of

$$y = \sqrt[3]{(x^2 + 4)^2} \qquad \text{Original function}$$

when $x = 2$.

SOLUTION Begin by rewriting the function in rational exponent form.

$$y = (x^2 + 4)^{2/3} \qquad \text{Rewrite original function.}$$

Then, using the inside function, $u = x^2 + 4$, apply the General Power Rule.

$$\frac{dy}{dx} = \underbrace{\frac{2}{3}}_{n}\underbrace{(x^2 + 4)^{-1/3}}_{u^{n-1}}\underbrace{(2x)}_{u'} \qquad \text{Apply General Power Rule.}$$

$$= \frac{4x(x^2 + 4)^{-1/3}}{3}$$

$$= \frac{4x}{3\sqrt[3]{x^2 + 4}} \qquad \text{Simplify.}$$

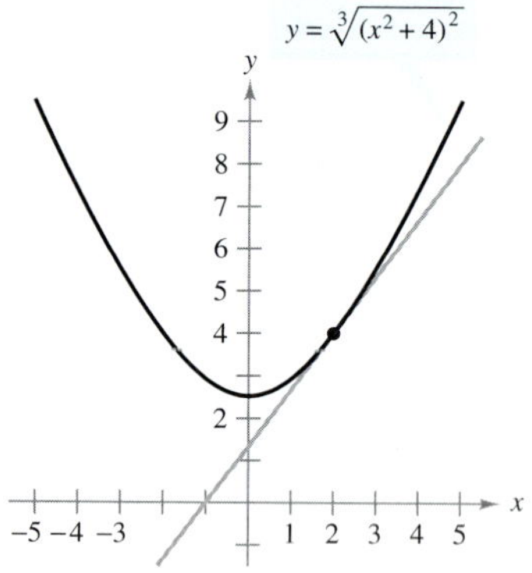

FIGURE 2.29

When $x = 2$, $y = 4$ and the slope of the line tangent to the graph at $(2, 4)$ is $\frac{4}{3}$. Using the point-slope form, you can find the equation of the tangent line to be $y = \frac{4}{3}x + \frac{4}{3}$. The graph of the function and the tangent line is shown in Figure 2.29.

TRY IT 4

Find the tangent line to the graph of $y = \sqrt[3]{(x + 4)^2}$ when $x = 4$. Sketch the line tangent to the graph at $x = 4$.

STUDY TIP

The derivative of a quotient can sometimes be found more easily with the General Power Rule than with the Quotient Rule. This is especially true when the numerator is a constant, as shown in Example 5.

EXAMPLE 5 Finding the Derivative of a Quotient

Find the derivative of each function.

(a) $y = \dfrac{3}{x^2 + 1}$ (b) $y = \dfrac{3}{(x + 1)^2}$

SOLUTION

(a) Begin by rewriting the function as

$$y = 3(x^2 + 1)^{-1}. \qquad \text{Rewrite original function.}$$

Then apply the General Power Rule to obtain

$$\frac{dy}{dx} = -3(x^2 + 1)^{-2}(2x) = -\frac{6x}{(x^2 + 1)^2}. \qquad \text{Apply General Power Rule.}$$

(b) Begin by rewriting the function as

$$y = 3(x + 1)^{-2}. \qquad \text{Rewrite original function.}$$

Then apply the General Power Rule to obtain

$$\frac{dy}{dx} = -6(x + 1)^{-3}(1) = -\frac{6}{(x + 1)^3}. \qquad \text{Apply General Power Rule.}$$

TRY IT 5

Find the derivative of each function.

(a) $y = \dfrac{4}{2x + 1}$

(b) $y = \dfrac{2}{(x - 1)^3}$

Simplification Techniques

Throughout this chapter, writing derivatives in simplified form has been emphasized. The reason for this is that most applications of derivatives require a simplified form. The next two examples illustrate some useful simplification techniques.

EXAMPLE 6 Simplifying by Factoring Out Least Powers

Find the derivative of $y = x^2\sqrt{1 - x^2}$.

$$\begin{aligned}
y &= x^2\sqrt{1-x^2} && \text{Write original function.}\\
&= x^2(1-x^2)^{1/2} && \text{Rewrite function.}\\
y' &= x^2\frac{d}{dx}\left[(1-x^2)^{1/2}\right] + (1-x^2)^{1/2}\frac{d}{dx}\left[x^2\right] && \text{Product Rule}\\
&= x^2\left[\frac{1}{2}(1-x^2)^{-1/2}(-2x)\right] + (1-x^2)^{1/2}(2x) && \text{Power Rule}\\
&= -x^3(1-x^2)^{-1/2} + 2x(1-x^2)^{1/2}\\
&= x(1-x^2)^{-1/2}\left[-x^2(1) + 2(1-x^2)\right] && \text{Factor.}\\
&= x(1-x^2)^{-1/2}(2-3x^2)\\
&= \frac{x(2-3x^2)}{\sqrt{1-x^2}} && \text{Simplify.}
\end{aligned}$$

ALGEBRA REVIEW

In Example 6, note that you subtract exponents when factoring. That is, when $(1 - x^2)^{-1/2}$ is factored out of $(1 - x^2)^{1/2}$, the *remaining* factor has an exponent of $\frac{1}{2} - \left(-\frac{1}{2}\right) = 1$. So,

$$(1-x^2)^{1/2} = (1-x^2)^{-1/2}(1-x^2)^1.$$

TRY IT 6

Find and simplify the derivative of $y = x^2\sqrt{x^2 + 1}$.

EXAMPLE 7 Differentiating a Quotient Raised to a Power

Find the derivative of

$$f(x) = \left(\frac{3x-1}{x^2+3}\right)^2.$$

SOLUTION

$$\begin{aligned}
f'(x) &= \underbrace{2}_{n}\underbrace{\left(\frac{3x-1}{x^2+3}\right)}_{u^{n-1}}\underbrace{\frac{d}{dx}\left[\frac{3x-1}{x^2+3}\right]}_{u'}\\
&= \left[\frac{2(3x-1)}{x^2+3}\right]\left[\frac{(x^2+3)(3)-(3x-1)(2x)}{(x^2+3)^2}\right]\\
&= \frac{2(3x-1)(3x^2+9-6x^2+2x)}{(x^2+3)^3}\\
&= \frac{2(3x-1)(-3x^2+2x+9)}{(x^2+3)^3}
\end{aligned}$$

STUDY TIP

In Example 7, try to find $f'(x)$ by applying the Quotient Rule to

$$f(x) = \frac{(3x-1)^2}{(x^2+3)^2}.$$

Which method do you prefer?

TRY IT 7

Find the derivative of

$$f(x) = \left(\frac{x+1}{x-5}\right)^2.$$

U.S. Cellular is in the process of converting its wireless network to an advanced wireless digital technology. The benefits of wireless digital technology are fewer blocked calls, reduced background noise and interference, improved security and privacy, greater capacity, and expanded coverage.

Application

EXAMPLE 8 Finding Rates of Change

From 1993 through 2002, the revenue per share R (in dollars) for U.S. Cellular can be modeled by $R = (-0.003t^2 + 0.42t + 0.5)^2$ for $3 \le t \le 12$, where $t = 3$ corresponds to 1993. Use the model to approximate the rates of change in the revenue per share in 1994, 1996, and 2000. If you had been a U.S. Cellular stockholder from 1993 through 2002, would you have been satisfied with the performance of this stock? *(Source: U.S. Cellular)*

SOLUTION The rate of change in R is given by the derivative dR/dt. You can use the General Power Rule to find the derivative.

$$\frac{dR}{dt} = 2(-0.003t^2 + 0.42t + 0.5)^1(-0.006t + 0.42)$$

$$= (-0.012t + 0.84)(-0.003t^2 + 0.42t + 0.5)$$

In 1994, the revenue per share was changing at a rate of

$$[-0.012(4) + 0.84][-0.003(4)^2 + 0.42(4) + 0.5] \approx \$1.69 \text{ per year.}$$

In 1996, the revenue per share was changing at a rate of

$$[-0.012(6) + 0.84][-0.003(6)^2 + 0.42(6) + 0.5] \approx \$2.24 \text{ per year.}$$

In 2000, the revenue per share was changing at a rate of

$$[-0.012(10) + 0.84][-0.003(10)^2 + 0.42(10) + 0.5] \approx \$3.17 \text{ per year.}$$

The graph of the revenue per share function is shown in Figure 2.30. For most investors, the performance of U.S. Cellular stock would be considered to be very good.

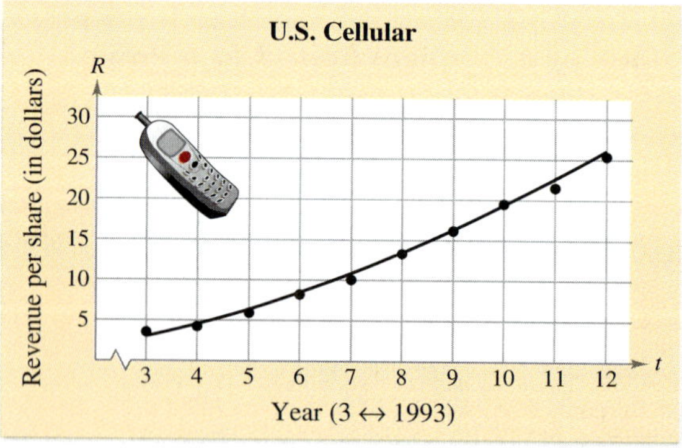

FIGURE 2.30

TRY IT 8

From 1994 through 2002, the sales per share (in dollars) for Dollar Tree can be modeled by $S = (-0.009t^2 + 0.52t - 0.4)^2$ for $4 \le t \le 12$, where $t = 4$ corresponds to 1994. Use the model to approximate the rate of change in sales per share in 2000. *(Source: Dollar Tree Stores, Inc.)*

Summary of Differentiation Rules

You now have all the rules you need to differentiate *any* algebraic function. For your convenience, they are summarized below.

Summary of Differentiation Rules

1. Constant Rule $\frac{d}{dx}[c] = 0,$ c is a constant.
2. Constant Multiple Rule $\frac{d}{dx}[cu] = c\frac{du}{dx},$ c is a constant.
3. Sum and Difference Rules $\frac{d}{dx}[u \pm v] = \frac{du}{dx} \pm \frac{dv}{dx}$
4. Product Rule $\frac{d}{dx}[uv] = u\frac{dv}{dx} + v\frac{du}{dx}$
5. Quotient Rule $\frac{d}{dx}\left[\frac{u}{v}\right] = \frac{v\frac{du}{dx} - u\frac{dv}{dx}}{v^2}$
6. Power Rules $\frac{d}{dx}[x^n] = nx^{n-1}$

   $\frac{d}{dx}[u^n] = nu^{n-1}\frac{du}{dx}$
7. Chain Rule $\frac{dy}{dx} = \frac{dy}{du} \cdot \frac{du}{dx}$

TAKE ANOTHER LOOK

Comparing Power Rules

You now have two power rules for differentiation.

$\frac{d}{dx}[x^n] = nx^{n-1}$ Simple Power Rule

$\frac{d}{dx}[u^n] = nu^{n-1}\frac{du}{dx}$ General Power Rule

Explain how you can tell which rule to use. Then state whether you would use the simple or general rule for each function.

a. $y = x^4$ b. $y = (x^2 + 1)$

c. $y = (x^2 + 1)^2$ d. $y = \frac{1}{x^4}$

e. $y = \frac{1}{x - 1}$ f. $y = \frac{1}{\sqrt{2x - 1}}$

PREREQUISITE REVIEW 2.5

The following warm-up exercises involve skills that were covered in earlier sections. You will use these skills in the exercise set for this section.

In Exercises 1–6, rewrite the expression with rational exponents.

1. $\sqrt[5]{(1-5x)^2}$ **2.** $\sqrt[4]{(2x-1)^3}$ **3.** $\dfrac{1}{\sqrt{4x^2+1}}$

4. $\dfrac{1}{\sqrt[3]{x-6}}$ **5.** $\dfrac{\sqrt{x}}{\sqrt[3]{1-2x}}$ **6.** $\dfrac{\sqrt{(3-7x)^3}}{2x}$

In Exercises 7–10, factor the expression.

7. $3x^3 - 6x^2 + 5x - 10$ **8.** $5x\sqrt{x} - x - 5\sqrt{x} + 1$

9. $4(x^2+1)^2 - x(x^2+1)^3$ **10.** $-x^5 + 3x^3 + x^2 - 3$

EXERCISES 2.5

In Exercises 1–8, identify the inside function, $u = g(x)$, and the outside function, $y = f(u)$.

$y = f(g(x))$	$u = g(x)$	$y = f(u)$
1. $y = (6x-5)^4$		
2. $y = (x^2 - 2x + 3)^3$		
3. $y = (4 - x^2)^{-1}$		
4. $y = (x^2+1)^{4/3}$		
5. $y = \sqrt{5x-2}$		
6. $y = \sqrt{9-x^2}$		
7. $y = (3x+1)^{-1}$		
8. $y = (x+1)^{-1/2}$		

In Exercises 9–16, match the function with the rule that you would use to find the derivative *most efficiently*.

(a) Simple Power Rule (b) Constant Rule

(c) General Power Rule (d) Quotient Rule

9. $f(x) = \dfrac{2}{1-x^3}$ **10.** $f(x) = \dfrac{2x}{1-x^3}$

11. $f(x) = \sqrt[3]{8^2}$ **12.** $f(x) = \sqrt[3]{x^2}$

13. $f(x) = \dfrac{x^2+2}{x}$ **14.** $f(x) = \dfrac{x^4 - 2x + 1}{\sqrt{x}}$

15. $f(x) = \dfrac{2}{x-2}$ **16.** $f(x) = \dfrac{5}{x^2+1}$

In Exercises 17–34, use the General Power Rule to find the derivative of the function.

17. $y = (2x-7)^3$ **18.** $y = (3x^2+1)^4$

19. $g(x) = (4-2x)^3$ **20.** $h(t) = (1-t^2)^4$

21. $h(x) = (6x - x^3)^2$ **22.** $f(x) = (4x - x^2)^3$

23. $f(x) = (x^2-9)^{2/3}$ **24.** $f(t) = (9t+2)^{2/3}$

25. $f(t) = \sqrt{t+1}$ **26.** $g(x) = \sqrt{2x+3}$

27. $s(t) = \sqrt{2t^2 + 5t + 2}$ **28.** $y = \sqrt[3]{3x^3 + 4x}$

29. $y = \sqrt[3]{9x^2+4}$ **30.** $y = 2\sqrt{4-x^2}$

31. $f(x) = -3\sqrt[4]{2-9x}$ **32.** $f(x) = (25 + x^2)^{-1/2}$

33. $h(x) = (4 - x^3)^{-4/3}$ **34.** $f(x) = (4-3x)^{-5/2}$

In Exercises 35–40, find an equation of the tangent line to the graph of f at the point $(2, f(2))$. Use a graphing utility to check your result by graphing the original function and the tangent line in the same viewing window.

35. $f(x) = 2(x^2-1)^3$ **36.** $f(x) = 3(9x-4)^4$

37. $f(x) = \sqrt{4x^2-7}$ **38.** $f(x) = x\sqrt{x^2+5}$

39. $f(x) = \sqrt{x^2 - 2x + 1}$ **40.** $f(x) = (4 - 3x^2)^{-2/3}$

In Exercises 41–44, use a symbolic differentiation utility to find the derivative of the function. Graph the function and its derivative in the same viewing window. Describe the behavior of the function when the derivative is zero.

41. $f(x) = \dfrac{\sqrt{x}+1}{x^2+1}$ **42.** $f(x) = \sqrt{\dfrac{2x}{x+1}}$

43. $f(x) = \sqrt{\dfrac{x+1}{x}}$ **44.** $f(x) = \sqrt{x}(2-x^2)$

In Exercises 45–64, find the derivative of the function.

45. $y = \dfrac{1}{x-2}$ **46.** $s(t) = \dfrac{1}{t^2+3t-1}$

47. $y = -\dfrac{4}{(t+2)^2}$ **48.** $f(x) = \dfrac{3}{(x^3-4)^2}$

49. $f(x) = \dfrac{1}{(x^2-3x)^2}$ **50.** $y = \dfrac{1}{\sqrt{x+2}}$

51. $g(t) = \frac{1}{t^2 - 2}$

52. $g(x) = \frac{3}{\sqrt[3]{x^3 - 1}}$

53. $f(x) = x(3x - 9)^3$

54. $f(x) = x^3(x - 4)^2$

55. $y = x\sqrt{2x + 3}$

56. $y = t\sqrt{t + 1}$

57. $y = t^2\sqrt{t - 2}$

58. $y = \sqrt{x}(x - 2)^2$

59. $f(x) = \sqrt{\frac{3 - 2x}{4x}}$

60. $g(t) = \frac{3t^2}{\sqrt{t^2 + 2t - 1}}$

61. $f(x) = \sqrt{x^2 + 1} - \sqrt{x^2 - 1}$

62. $y = \sqrt{x - 1} + \sqrt{x + 1}$

63. $y = \left(\frac{6 - 5x}{x^2 - 1}\right)^2$

64. $y = \left(\frac{4x^2}{3 - x}\right)^3$

In Exercises 65–70, find an equation of the tangent line to the graph of the function at the given point. Then use a graphing utility to graph the function and the tangent line in the same viewing window.

Function	*Point*
65. $f(t) = \frac{36}{(3 - t)^2}$	$(0, 4)$
66. $s(x) = \frac{1}{\sqrt{x^2 - 3x + 4}}$	$\left(3, \frac{1}{2}\right)$
67. $f(t) = (t^2 - 9)\sqrt{t + 2}$	$(-1, -8)$
68. $y = \frac{2x}{\sqrt{x + 1}}$	$(3, 3)$
69. $f(x) = \frac{x + 1}{\sqrt{2x - 3}}$	$(2, 3)$
70. $y = \frac{x}{\sqrt{25 + x^2}}$	$(0, 0)$

71. ***Compound Interest*** You deposit \$1000 in an account with an annual interest rate of r (in decimal form) compounded monthly. At the end of 5 years, the balance is

$$A = 1000\left(1 + \frac{r}{12}\right)^{60}.$$

Find the rates of change of A with respect to r when (a) $r = 0.08$, (b) $r = 0.10$, and (c) $r = 0.12$.

72. ***Environment*** An environmental study indicates that the average daily level P of a certain pollutant in the air in parts per million can be modeled by the equation

$$P = 0.25\sqrt{0.5n^2 + 5n + 25}$$

where n is the number of residents of the community in thousands. Find the rate at which the level of pollutant is increasing when the population of the community is 12,000.

73. ***Biology*** The number N of bacteria in a culture after t days is modeled by

$$N = 400\left[1 - \frac{3}{(t^2 + 2)^2}\right].$$

Complete the table. What can you conclude?

t	0	1	2	3	4
dN/dt					

74. ***Depreciation*** The value V of a machine t years after it is purchased is inversely proportional to the square root of $t + 1$. The initial value of the machine is \$10,000.

(a) Write V as a function of t.

(b) Find the rate of depreciation when $t = 1$.

(c) Find the rate of depreciation when $t = 3$.

75. ***Depreciation*** Repeat Exercise 74 given that the value of the machine t years after it is purchased is inversely proportional to the cube root of $t + 1$.

76. ***Credit Card Rate*** The average annual rate r (in percent form) for commercial bank credit cards from 1994 through 2002 can be modeled by

$$r = \sqrt{-0.14239t^4 + 3.939t^3 - 39.0835t^2 + 161.037t + 22.13}$$

where $t = 4$ corresponds to 1994. *(Source: Federal Reserve Bulletin)*

(a) Find the derivative of this model. Which differentiation rule(s) did you use?

(b) Use a graphing utility to graph the derivative. Use the interval $4 \le t \le 12$.

(c) Use the *trace* feature to find the years during which the finance rate was changing the most.

(d) Use the *trace* feature to find the years during which the finance rate was changing the least.

True or False? In Exercises 77 and 78, determine whether the statement is true or false. If it is false, explain why or give an example that shows it is false.

77. If $y = (1 - x)^{1/2}$, then $y' = \frac{1}{2}(1 - x)^{-1/2}$.

78. If y is a differentiable function of u, u is a differentiable function of v, and v is a differentiable function of x, then

$$\frac{dy}{dx} = \frac{dy}{du} \cdot \frac{du}{dv} \cdot \frac{dv}{dx}.$$

2.6 HIGHER-ORDER DERIVATIVES

- Find higher-order derivatives.
- Find and use the position functions to determine the velocity and acceleration of moving objects.

STUDY TIP

In the context of higher-order derivatives, the "standard" derivative f' is often called the **first derivative** of f.

Second, Third, and Higher-Order Derivatives

The derivative of f' is the **second derivative** of f and is denoted by f''.

$$\frac{d}{dx}[f'(x)] = f''(x) \qquad \text{Second derivative}$$

The derivative of f'' is the **third derivative** of f and is denoted by f'''.

$$\frac{d}{dx}[f''(x)] = f'''(x) \qquad \text{Third derivative}$$

By continuing this process, you obtain **higher-order derivatives** of f. Higher-order derivatives are denoted as follows.

DISCOVERY

For each function, find the indicated higher-order derivative.

(a) $y = x^2$ y''

(b) $y = x^3$ y'''

(c) $y = x^4$ $y^{(4)}$

(d) $y = x^n$ $y^{(n)}$

Notation for Higher-Order Derivatives

1. 1st derivative:	y',	$f'(x)$,	$\frac{dy}{dx}$,	$\frac{d}{dx}[f(x)]$,	$D_x[y]$
2. 2nd derivative:	y'',	$f''(x)$,	$\frac{d^2y}{dx^2}$,	$\frac{d^2}{dx^2}[f(x)]$,	$D_x^2[y]$
3. 3rd derivative:	y''',	$f'''(x)$,	$\frac{d^3y}{dx^3}$,	$\frac{d^3}{dx^3}[f(x)]$,	$D_x^3[y]$
4. 4th derivative:	$y^{(4)}$,	$f^{(4)}(x)$,	$\frac{d^4y}{dx^4}$,	$\frac{d^4}{dx^4}[f(x)]$,	$D_x^4[y]$
5. *n*th derivative:	$y^{(n)}$,	$f^{(n)}(x)$,	$\frac{d^ny}{dx^n}$,	$\frac{d^n}{dx^n}[f(x)]$,	$D_x^n[y]$

EXAMPLE 1 Finding Higher-Order Derivatives

Find the first five derivatives of $f(x) = 2x^4 - 3x^2$.

$$f(x) = 2x^4 - 3x^2 \qquad \text{Write original function.}$$
$$f'(x) = 8x^3 - 6x \qquad \text{First derivative}$$
$$f''(x) = 24x^2 - 6 \qquad \text{Second derivative}$$
$$f'''(x) = 48x \qquad \text{Third derivative}$$
$$f^{(4)}(x) = 48 \qquad \text{Fourth derivative}$$
$$f^{(5)}(x) = 0 \qquad \text{Fifth derivative}$$

TRY IT 1

Find the first four derivatives of

$$f(x) = 6x^3 - 2x^2 + 1.$$

EXAMPLE 2 Finding Higher-Order Derivatives

Find the value of $g'''(2)$ for the function

$$g(t) = -t^4 + 2t^3 + t + 4. \qquad \text{Original function}$$

SOLUTION Begin by differentiating three times.

$$g'(t) = -4t^3 + 6t^2 + 1 \qquad \text{First derivative}$$
$$g''(t) = -12t^2 + 12t \qquad \text{Second derivative}$$
$$g'''(t) = -24t + 12 \qquad \text{Third derivative}$$

Then, evaluate the third derivative of g at $t = 2$.

$$g'''(2) = -24(2) + 12$$
$$= -36 \qquad \text{Value of third derivative}$$

TRY IT 2

Find the value of $g'''(1)$ for $g(x) = x^4 - x^3 + 2x$.

TECHNOLOGY

Higher-order derivatives of nonpolynomial functions can be difficult to find by hand. If you have access to a symbolic differentiation utility, try using it to find higher-order derivatives.

Examples 1 and 2 show how to find higher-order derivatives of *polynomial* functions. Note that with each successive differentiation, the degree of the polynomial drops by one. Eventually, higher-order derivatives of polynomial functions degenerate to a constant function. Specifically, the nth-order derivative of an nth-degree polynomial function

$$f(x) = a_n x^n + a_{n-1}x^{n-1} + \cdot \cdot \cdot + a_1 x + a_0$$

is the constant function

$$f^{(n)}(x) = n!a_n$$

where $n! = 1 \cdot 2 \cdot 3 \cdot \cdot \cdot n$. Each derivative of order higher than n is the zero function. Polynomial functions are the *only* functions with this characteristic. For other functions, successive differentiation never produces a constant function.

EXAMPLE 3 Finding Higher-Order Derivatives

Find the first four derivatives of $y = x^{-1}$.

$$y = x^{-1} = \frac{1}{x} \qquad \text{Write original function.}$$

$$y' = (-1)x^{-2} = -\frac{1}{x^2} \qquad \text{First derivative}$$

$$y'' = (-1)(-2)x^{-3} = \frac{2}{x^3} \qquad \text{Second derivative}$$

$$y''' = (-1)(-2)(-3)x^{-4} = -\frac{6}{x^4} \qquad \text{Third derivative}$$

$$y^{(4)} = (-1)(-2)(-3)(-4)x^{-5} = \frac{24}{x^5} \qquad \text{Fourth derivative}$$

TRY IT 3

Find the fourth derivative of

$$y = \frac{1}{x^2}.$$

Acceleration

STUDY TIP

Acceleration is measured in units of length per unit of time squared. For instance, if the velocity is measured in feet per second, then the acceleration is measured in "feet per second squared," or, more formally, in "feet per second per second."

In Section 2.3, you saw that the velocity of an object moving in a straight path (neglecting air resistance) is given by the derivative of its position function. In other words, the rate of change of the position with respect to time is defined to be the velocity. In a similar way, the rate of change of the velocity with respect to time is defined to be the **acceleration** of the object.

$$s = f(t) \qquad \text{Position function}$$

$$\frac{ds}{dt} = f'(t) \qquad \text{Velocity function}$$

$$\frac{d^2s}{dt^2} = f''(t) \qquad \text{Acceleration function}$$

To find the position, velocity, or acceleration at a particular time t, substitute the given value of t into the appropriate function, as illustrated in Example 4.

EXAMPLE 4 Finding Acceleration

A ball is thrown into the air from the top of a 160-foot cliff, as shown in Figure 2.31. The initial velocity of the ball is 48 feet per second, which implies that the position function is

$$s = -16t^2 + 48t + 160$$

where the time t is measured in seconds. Find the height, the velocity, and the acceleration of the ball when $t = 3$.

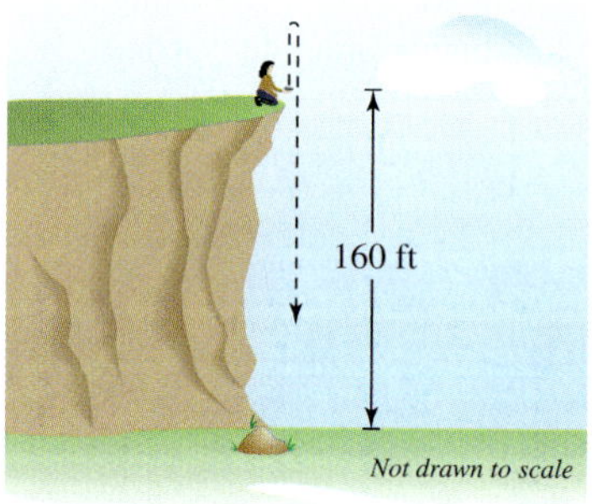

FIGURE 2.31

SOLUTION Begin by differentiating to find the velocity and acceleration functions.

$$s = -16t^2 + 48t + 160 \qquad \text{Position function}$$

$$\frac{ds}{dt} = -32t + 48 \qquad \text{Velocity function}$$

$$\frac{d^2s}{dt^2} = -32 \qquad \text{Acceleration function}$$

To find the height, velocity, and acceleration when $t = 3$, substitute $t = 3$ into each of the functions above.

$$\text{Height} = -16(3)^2 + 48(3) + 160 = 160 \text{ feet}$$

$$\text{Velocity} = -32(3) + 48 = -48 \text{ feet per second}$$

$$\text{Acceleration} = -32 \text{ feet per second squared}$$

TRY IT 4

A ball is thrown upward from the top of an 80-foot cliff with an initial velocity of 64 feet per second. Give the position function. Then find the velocity and acceleration functions.

In Example 4, notice that the acceleration of the ball is -32 feet per second squared at any time t. This constant acceleration is due to the gravitational force of Earth and is called the **acceleration due to gravity.** Note that the negative value indicates that the ball is being pulled *down*—toward Earth.

Although the acceleration exerted on a falling object is relatively constant near Earth's surface, it varies greatly throughout our solar system. Large planets exert a much greater gravitational pull than do small planets or moons. The next example describes the motion of a free-falling object on the moon.

NASA

The acceleration due to gravity on the surface of the moon is only about one-sixth that exerted by Earth. So, if you were on the moon and threw an object into the air, it would rise to a greater height than it would on Earth's surface.

EXAMPLE 5 Finding Acceleration on the Moon

An astronaut standing on the surface of the moon throws a rock into the air. The height s (in feet) of the rock is given by

$$s = -\frac{27}{10}t^2 + 27t + 6$$

where t is measured in seconds. How does the acceleration due to gravity on the moon compare with that on Earth?

SOLUTION

$$s = -\frac{27}{10}t^2 + 27t + 6 \qquad \text{Position function}$$

$$\frac{ds}{dt} = -\frac{27}{5}t + 27 \qquad \text{Velocity function}$$

$$\frac{d^2s}{dt^2} = -\frac{27}{5} \qquad \text{Acceleration function}$$

So, the acceleration at any time is

$$-\frac{27}{5} = -5.4 \text{ feet per second squared}$$

—about one-sixth of the acceleration due to gravity on Earth.

TRY IT 5

The position function on Earth, where s is measured in meters, t is measured in seconds, v_0 is the initial velocity in meters per second, and h_0 is the initial height in meters, is

$$s = -4.9t^2 + v_0t + h_0.$$

If the initial velocity is 2.2 and the initial height is 3.6, what is the acceleration due to gravity on Earth in meters per second per second?

The position function described in Example 5 neglects air resistance, which is appropriate because the moon has no atmosphere—and *no air resistance*. This means that the position function for any free-falling object on the moon is given by

$$s = -\frac{27}{10}t^2 + v_0t + h_0$$

where s is the height (in feet), t is the time (in seconds), v_0 is the initial velocity, and h_0 is the initial height. For instance, the rock in Example 5 was thrown upward with an initial velocity of 27 feet per second and had an initial height of 6 feet. This position function is valid for all objects, whether heavy ones such as hammers or light ones such as feathers.

In 1971, astronaut David R. Scott demonstrated the lack of atmosphere on the moon by dropping a hammer and a feather from the same height. Both took exactly the same time to fall to the ground. If they were dropped from a height of 6 feet, how long did each take to hit the ground?

EXAMPLE 6 **Finding Velocity and Acceleration**

The velocity v (in feet per second) of a certain automobile starting from rest is

$$v = \frac{80t}{t+5} \qquad \text{Velocity function}$$

where t is the time (in seconds). The positions of the automobile at 10-second intervals are shown in Figure 2.32. Find the velocity and acceleration of the automobile at 10-second intervals from $t = 0$ to $t = 60$.

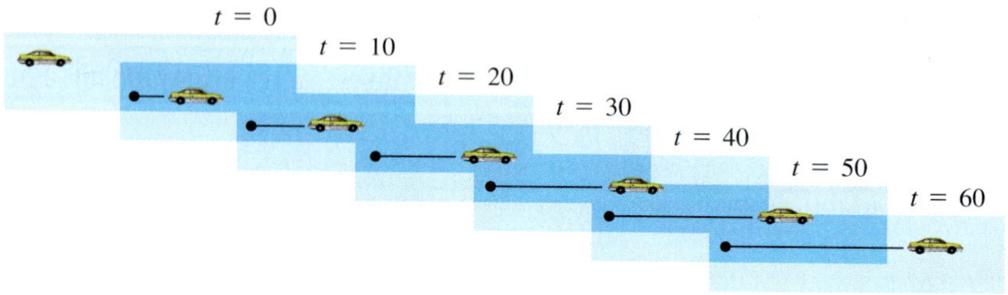

FIGURE 2.32

SOLUTION To find the acceleration function, differentiate the velocity function.

$$\frac{dv}{dt} = \frac{(t+5)(80) - (80t)(1)}{(t+5)^2}$$

$$= \frac{400}{(t+5)^2} \qquad \text{Acceleration function}$$

t (seconds)	0	10	20	30	40	50	60
v (ft/sec)	0	53.3	64.0	68.6	71.1	72.7	73.8
$\frac{dv}{dt}$ (ft/sec^2)	16	1.78	0.64	0.33	0.20	0.13	0.09

In the table, note that the acceleration approaches zero as the velocity levels off. This observation should agree with your experience—when riding in an accelerating automobile, you do not feel the velocity, but you do feel the acceleration. In other words, you feel changes in velocity.

TRY IT 6

Use a graphing utility to graph the velocity function and acceleration function in Example 6 in the same viewing window. Compare the graphs with the table at the right. As the velocity levels off, what does the acceleration approach?

TAKE ANOTHER LOOK

Acceleration Due to Gravity

Newton's Law of Universal Gravitation states that the gravitational attraction of two objects is directly proportional to their masses and inversely proportional to the square of the distance between their centers. The Earth has a mass of 5.979×10^{24} kilograms and a radius of 6371 kilometers. The moon has a mass of 7.354×10^{22} kilograms and a radius of 1738 kilometers. Discuss how you could find the ratio of the Earth's gravity to the moon's gravity. What is this ratio?

PREREQUISITE REVIEW 2.6

The following warm-up exercises involve skills that were covered in earlier sections. You will use these skills in the exercise set for this section.

In Exercises 1–4, solve the equation.

1. $-16t^2 + 24t = 0$

2. $-16t^2 + 80t + 224 = 0$

3. $-16t^2 + 128t + 320 = 0$

4. $-16t^2 + 9t + 1440 = 0$

In Exercises 5–8, find dy/dx.

5. $y = x^2(2x + 7)$

6. $y = (x^2 + 3x)(2x^2 - 5)$

7. $y = \dfrac{x^2}{2x + 7}$

8. $y = \dfrac{x^2 + 3x}{2x^2 - 5}$

In Exercises 9 and 10, find the domain and range of f.

9. $f(x) = x^2 - 4$

10. $f(x) = \sqrt{x - 7}$

EXERCISES 2.6

In Exercises 1–14, find the second derivative of the function.

1. $f(x) = 5 - 4x$

2. $f(x) = 3x - 1$

3. $f(x) = x^2 + 7x - 4$

4. $f(x) = 3x^2 + 4x$

5. $g(t) = \frac{1}{3}t^3 - 4t^2 + 2t$

6. $f(x) = 4(x^2 - 1)^2$

7. $f(t) = \dfrac{3}{4t^2}$

8. $g(t) = t^{-1/3}$

9. $f(x) = 3(2 - x^2)^3$

10. $f(x) = x\sqrt[3]{x}$

11. $f(x) = \dfrac{x + 1}{x - 1}$

12. $g(t) = -\dfrac{4}{(t + 2)^2}$

13. $y = x^2(x^2 + 4x + 8)$

14. $h(s) = s^3(s^2 - 2s + 1)$

In Exercises 15–20, find the third derivative of the function.

15. $f(x) = x^5 - 3x^4$

16. $f(x) = x^4 - 2x^3$

17. $f(x) = 5x(x + 4)^3$

18. $f(x) = (x - 1)^2$

19. $f(x) = \dfrac{3}{16x^2}$

20. $f(x) = \dfrac{1}{x}$

In Exercises 21–26, find the given value.

	Function	*Value*
21.	$g(t) = 5t^4 + 10t^2 + 3$	$g''(2)$
22.	$f(x) = 9 - x^2$	$f''(-\sqrt{5})$
23.	$f(x) = \sqrt{4 - x}$	$f'''(-5)$
24.	$f(t) = \sqrt{2t + 3}$	$f'''(\frac{1}{2})$
25.	$f(x) = x^2(3x^2 + 3x - 4)$	$f'''(-2)$
26.	$g(x) = 2x^3(x^2 - 5x + 4)$	$g'''(0)$

In Exercises 27–32, find the higher-order derivative.

	Given	*Derivative*
27.	$f'(x) = 2x^2$	$f''(x)$
28.	$f''(x) = 20x^3 - 36x^2$	$f'''(x)$
29.	$f''(x) = (2x - 2)/x$	$f'''(x)$
30.	$f'''(x) = 2\sqrt{x - 1}$	$f^{(4)}(x)$
31.	$f^{(4)}(x) = (x + 1)^2$	$f^{(6)}(x)$
32.	$f(x) = x^3 - 2x$	$f''(x)$

In Exercises 33–40, find the second derivative and solve the equation $f''(x) = 0$.

33. $f(x) = x^3 - 9x^2 + 27x - 27$

34. $f(x) = 3x^3 - 9x + 1$

35. $f(x) = (x + 3)(x - 4)(x + 5)$

36. $f(x) = (x + 2)(x - 2)(x + 3)(x - 3)$

37. $f(x) = x\sqrt{x^2 - 1}$

38. $f(x) = x\sqrt{4 - x^2}$

39. $f(x) = \dfrac{x}{x^2 + 3}$

40. $f(x) = \dfrac{x}{x^2 + 1}$

41. ***Velocity and Acceleration*** A ball is propelled straight upward from ground level with an initial velocity of 144 feet per second.

(a) Write the position function of the ball.

(b) Write the velocity and acceleration functions.

(c) When is the ball at its highest point? How high is this point?

(d) How fast is the ball traveling when it hits the ground? How is this speed related to the initial velocity?

42. ***Velocity and Acceleration*** A brick becomes dislodged from the top of the Empire State Building (at a height of 1250 feet) and falls to the sidewalk below.

(a) Write the position function of the brick.

(b) Write the velocity and acceleration functions.

(c) How long does it take the brick to hit the sidewalk?

(d) How fast is the brick traveling when it hits the sidewalk?

43. ***Velocity and Acceleration*** The velocity (in feet per second) of an automobile starting from rest is modeled by

$$\frac{ds}{dt} = \frac{90t}{t + 10}.$$

Create a table showing the velocity and acceleration at 10-second intervals during the first minute of travel. What can you conclude?

44. ***Stopping Distance*** A car is traveling at a rate of 66 feet per second (45 miles per hour) when the brakes are applied. The position function for the car is given by $s = -8.25t^2 + 66t$, where s is measured in feet and t is measured in seconds. Create a table showing the position, velocity, and acceleration for each given value of t. What can you conclude?

In Exercises 45 and 46, use a graphing utility to graph f, f', and f'' in the same viewing window. What is the relationship among the degree of f and the degrees of its successive derivatives? In general, what is the relationship among the degree of a polynomial function and the degrees of its successive derivatives?

45. $f(x) = x^2 - 6x + 6$

46. $f(x) = 3x^3 - 9x$

In Exercises 47 and 48, the graphs of f, f', and f'' are shown on the same set of coordinate axes. Which is which? Explain your reasoning.

47.

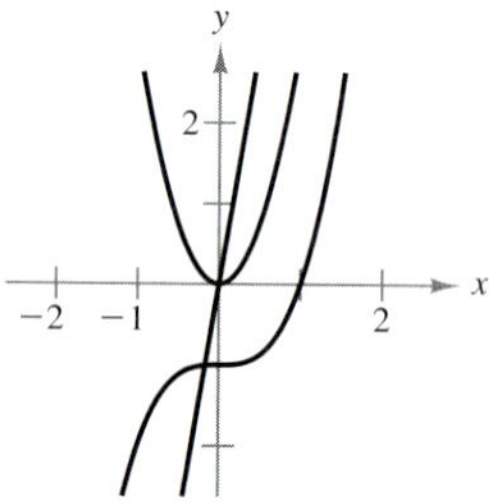

48.

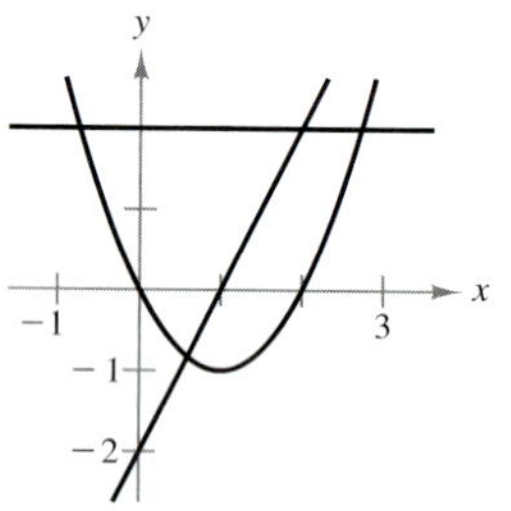

49. ***Data Analysis*** The table shows the median prices y (in thousands of dollars) of new privately owned U.S. homes in the South for 1995 to 2002. *(Source: U.S. Census Bureau)*

t	5	6	7	8	9
y	124.5	126.2	129.6	135.8	145.9

t	10	11	12
y	148.0	155.4	163.4

A model for the data is

$$y = -0.0828t^3 + 2.443t^2 - 17.06t + 158.7$$

where t is the year, with $t = 5$ corresponding to 1995.

(a) Use a graphing utility to graph the model and plot the data in the same viewing window.

(b) Find the first and second derivatives of the function.

(c) Show that the price of homes was increasing from 1995 to 2002.

(d) Find the year when the price was increasing at the greatest rate.

(e) Explain the relationship among your answers for parts (b), (c), and (d).

50. ***Projectile Motion*** An object is thrown upward from the top of a 64-foot building with an initial velocity of 48 feet per second.

(a) Write the position function of the object.

(b) Find the velocity and acceleration functions.

(c) When will the object hit the ground?

(d) When is the velocity of the object zero?

(e) How high does the object go?

(f) Use a graphing utility to graph the position, velocity, and acceleration functions in the same viewing window. Write a short paragraph that describes the relationship among these functions.

True or False? In Exercises 51–56, determine whether the statement is true or false. If it is false, explain why or give an example that shows it is false.

51. If $y = f(x)g(x)$, then $y' = f'(x)g'(x)$.

52. If $y = (x + 1)(x + 2)(x + 3)(x + 4)$, then $\dfrac{d^5y}{dx^5} = 0$.

53. If $f'(c)$ and $g'(c)$ are zero and $h(x) = f(x)g(x)$, then $h'(c) = 0$.

54. If $f(x)$ is an nth-degree polynomial, then $f^{(n+1)}(x) = 0$.

55. The second derivative represents the rate of change of the first derivative.

56. If the velocity of an object is constant, then its acceleration is zero.

57. ***Finding a Pattern*** Develop a general rule for $[xf(x)]^{(n)}$ where f is a differentiable function of x.

2.7 IMPLICIT DIFFERENTIATION

- Find derivatives explicitly.
- Find derivatives implicitly.
- Use derivatives to answer questions about real-life situations.

Implicit and Explicit Functions

So far in the text, functions involving two variables have generally been expressed in the **explicit form** $y = f(x)$. That is, one of the two variables has been explicitly given in terms of the other. For example,

$$y = 3x - 5, \quad s = -16t^2 + 20t, \quad \text{and} \quad u = 3w - w^2$$

are each written in explicit form, and we say that y, s, and u are functions of x, t, and w, explicitly. Many functions, however, are not given explicitly and are only implied by a given equation, as shown in Example 1.

EXAMPLE 1 Finding a Derivative Explicitly

Find dy/dx for the equation

$$xy = 1.$$

SOLUTION In this equation, y is **implicitly** defined as a function of x. One way to find dy/dx is first to solve the equation for y, then differentiate as usual.

$$xy = 1 \qquad \text{Write original equation.}$$

$$y = \frac{1}{x} \qquad \text{Solve for } y.$$

$$= x^{-1} \qquad \text{Rewrite.}$$

$$\frac{dy}{dx} = -x^{-2} \qquad \text{Differentiate with respect to } x.$$

$$= -\frac{1}{x^2} \qquad \text{Simplify.}$$

TRY IT 1

Find dy/dx for the equation $x^2y = 1$.

The procedure shown in Example 1 works well whenever you can easily write the given function explicitly. You cannot, however, use this procedure when you are unable to solve for y as a function of x. For instance, how would you find dy/dx in the equation

$$x^2 - 2y^3 + 4y = 2$$

where it is very difficult to express y as a function of x explicitly? To do this, you can use a procedure called **implicit differentiation.**

Implicit Differentiation

To understand how to find dy/dx implicitly, you must realize that the differentiation is taking place *with respect to x*. This means that when you differentiate terms involving x alone, you can differentiate as usual. *But* when you differentiate terms involving y, you must apply the Chain Rule because you are assuming that y is defined implicitly as a function of x. Study the next example carefully. Note in particular how the Chain Rule is used to introduce the dy/dx factors in Examples 2(b) and 2(d).

EXAMPLE 2 Applying the Chain Rule

Differentiate each expression with respect to x.

(a) $3x^2$ (b) $2y^3$ (c) $x + 3y$ (d) xy^2

SOLUTION

(a) The only variable in this expression is x. So, to differentiate with respect to x, you can use the Simple Power Rule and the Constant Multiple Rule to obtain

$$\frac{d}{dx}[3x^2] = 6x.$$

(b) This case is different. The variable in the expression is y, and yet you are asked to differentiate with respect to x. To do this, assume that y is a differentiable function of x and use the Chain Rule.

$$\frac{d}{dx}\overbrace{[2y^3]}^{cu^n} = \overbrace{2}^{c}\,\overbrace{(3)}^{n}\,\overbrace{y^2}^{u^{n-1}}\,\overbrace{\frac{dy}{dx}}^{u'} \qquad \text{Chain Rule}$$

$$= 6y^2\frac{dy}{dx}$$

(c) This expression involves both x and y. By the Sum Rule and the Constant Multiple Rule, you can write

$$\frac{d}{dx}[x + 3y] = 1 + 3\frac{dy}{dx}.$$

(d) By the Product Rule and the Chain Rule, you can write

$$\frac{d}{dx}[xy^2] = x\frac{d}{dx}[y^2] + y^2\frac{d}{dx}[x] \qquad \text{Product Rule}$$

$$= x\left(2y\frac{dy}{dx}\right) + y^2(1) \qquad \text{Chain Rule}$$

$$= 2xy\frac{dy}{dx} + y^2.$$

TRY IT 2

Differentiate each expression with respect to x.

(a) $4x^3$ (b) $3y^2$ (c) $x + 5y$ (d) xy^3

Implicit Differentiation

Consider an equation involving x and y in which y is a differentiable function of x. You can use the steps below to find dy/dx.

1. Differentiate both sides of the equation *with respect to x*.
2. Write the result so that all terms involving dy/dx are on the left side of the equation and all other terms are on the right side of the equation.
3. Factor dy/dx out of the terms on the left side of the equation.
4. Solve for dy/dx by dividing both sides of the equation by the left-hand factor that does not contain dy/dx.

In Example 3, note that implicit differentiation can produce an expression for dy/dx that contains both x and y.

EXAMPLE 3 Finding the Slope of a Graph Implicitly

Find the slope of the tangent line to the ellipse given by $x^2 + 4y^2 = 4$ at the point $(\sqrt{2}, -1/\sqrt{2})$, as shown in Figure 2.33.

SOLUTION

$$x^2 + 4y^2 = 4 \qquad \text{Write original equation.}$$

$$\frac{d}{dx}[x^2 + 4y^2] = \frac{d}{dx}[4] \qquad \text{Differentiate with respect to } x.$$

$$2x + 8y\left(\frac{dy}{dx}\right) = 0 \qquad \text{Implicit differentiation}$$

$$8y\left(\frac{dy}{dx}\right) = -2x \qquad \text{Subtract } 2x \text{ from each side.}$$

$$\frac{dy}{dx} = \frac{-2x}{8y} \qquad \text{Divide each side by } 8y.$$

$$\frac{dy}{dx} = -\frac{x}{4y} \qquad \text{Simplify.}$$

To find the slope at the given point, substitute $x = \sqrt{2}$ and $y = -1/\sqrt{2}$ into the derivative, as shown below.

$$-\frac{\sqrt{2}}{4(-1/\sqrt{2})} = \frac{1}{2}$$

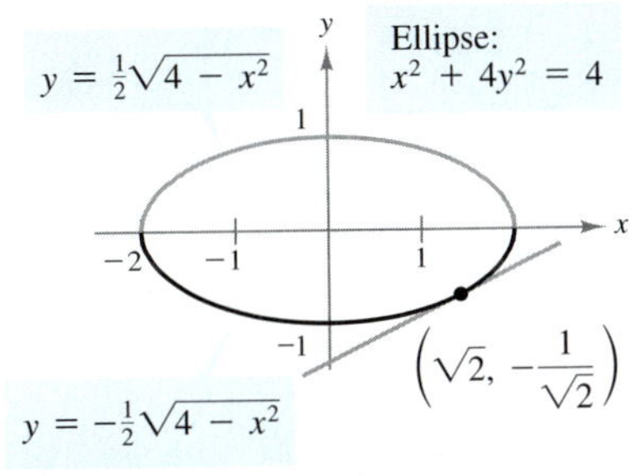

FIGURE 2.33 Slope of tangent line is $\frac{1}{2}$.

STUDY TIP

To see the benefit of implicit differentiation, try reworking Example 3 using the explicit function

$$y = -\frac{1}{2}\sqrt{4 - x^2}.$$

The graph of this function is the lower half of the ellipse.

TRY IT 3

Find the slope of the tangent line to the circle $x^2 + y^2 = 25$ at the point $(3, -4)$.

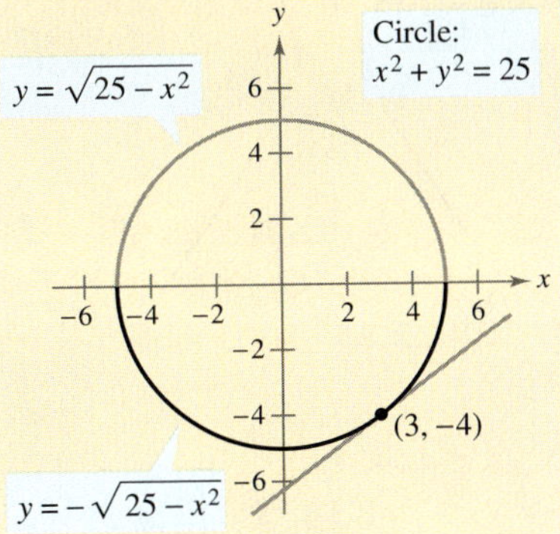

EXAMPLE 4 Using Implicit Differentiation

Find dy/dx for the equation $y^3 + y^2 - 5y - x^2 = -4$.

SOLUTION

$$y^3 + y^2 - 5y - x^2 = -4 \quad \text{Write original equation.}$$

$$\frac{d}{dx}[y^3 + y^2 - 5y - x^2] = \frac{d}{dx}[-4] \quad \text{Differentiate with respect to } x.$$

$$3y^2\frac{dy}{dx} + 2y\frac{dy}{dx} - 5\frac{dy}{dx} - 2x = 0 \quad \text{Implicit differentiation}$$

$$3y^2\frac{dy}{dx} + 2y\frac{dy}{dx} - 5\frac{dy}{dx} = 2x \quad \text{Collect } dy/dx \text{ terms.}$$

$$\frac{dy}{dx}(3y^2 + 2y - 5) = 2x \quad \text{Factor.}$$

$$\frac{dy}{dx} = \frac{2x}{3y^2 + 2y - 5}$$

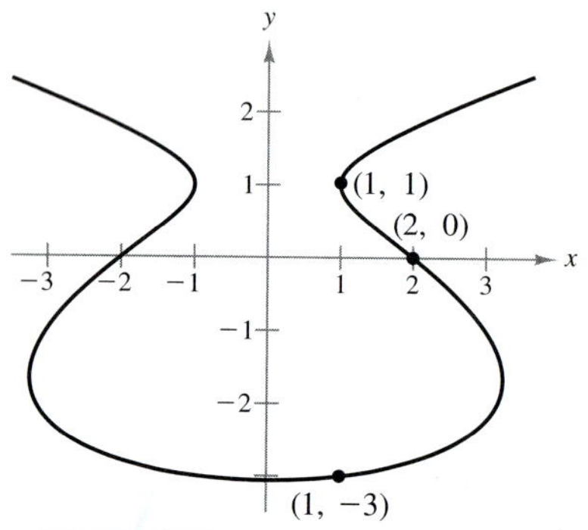

$y^3 + y^2 - 5y - x^2 = -4$

FIGURE 2.34

The graph of the original equation is shown in Figure 2.34. What are the slopes of the graph at the points $(1, -3)$, $(2, 0)$, and $(1, 1)$?

TRY IT 4

Find dy/dx for the equation $y^2 + x^2 - 2y - 4x = 4$.

EXAMPLE 5 Finding the Slope of a Graph Implicitly

Find the slope of the graph of $2x^2 - y^2 = 1$ at the point $(1, 1)$.

SOLUTION Begin by finding dy/dx implicitly.

$$2x^2 - y^2 = 1 \quad \text{Write original equation.}$$

$$4x - 2y\left(\frac{dy}{dx}\right) = 0 \quad \text{Differentiate with respect to } x.$$

$$-2y\left(\frac{dy}{dx}\right) = -4x \quad \text{Subtract } 4x \text{ from each side.}$$

$$\frac{dy}{dx} = \frac{2x}{y} \quad \text{Divide each side by } -2y.$$

At the point $(1, 1)$, the slope of the graph is

$$\frac{2(1)}{1} = 2$$

as shown in Figure 2.35. The graph is called a **hyperbola.**

$2x^2 - y^2 = 1$

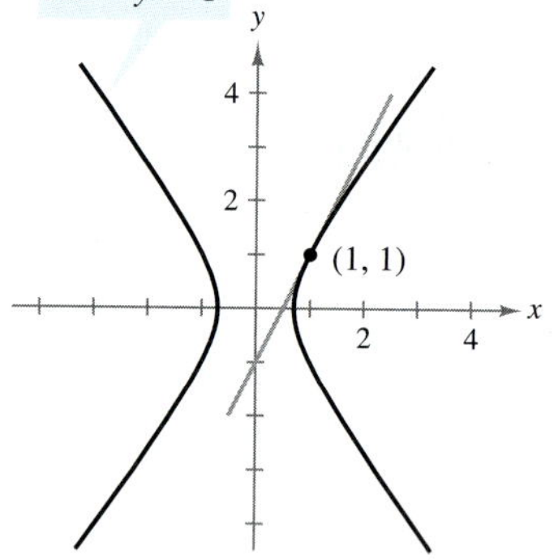

FIGURE 2.35 Hyperbola

TRY IT 5

Find the slope of the graph of $x^2 - 9y^2 = 16$ at the point $(5, 1)$.

Application

EXAMPLE 6 **Using a Demand Function**

The demand function for a product is modeled by

$$p = \frac{3}{0.000001x^3 + 0.01x + 1}$$

where p is measured in dollars and x is measured in thousands of units, as shown in Figure 2.36. Find the rate of change of the demand x with respect to the price p when $x = 100$.

SOLUTION To simplify the differentiation, begin by rewriting the function. Then, differentiate *with respect to p*.

$$p = \frac{3}{0.000001x^3 + 0.01x + 1}$$

$$0.000001x^3 + 0.01x + 1 = \frac{3}{p}$$

$$0.000003x^2\frac{dx}{dp} + 0.01\frac{dx}{dp} = -\frac{3}{p^2}$$

$$(0.000003x^2 + 0.01)\frac{dx}{dp} = -\frac{3}{p^2}$$

$$\frac{dx}{dp} = -\frac{3}{p^2(0.000003x^2 + 0.01)}$$

When $x = 100$, the price is

$$p = \frac{3}{0.000001(100)^3 + 0.01(100) + 1} = \$1.$$

So, when $x = 100$ and $p = 1$, the rate of change of the demand with respect to the price is

$$-\frac{3}{(1)^2[0.000003(100)^2 + 0.01]} = -75.$$

This means that when $x = 100$, the demand is dropping at the rate of 75 thousand units for each dollar increase in price.

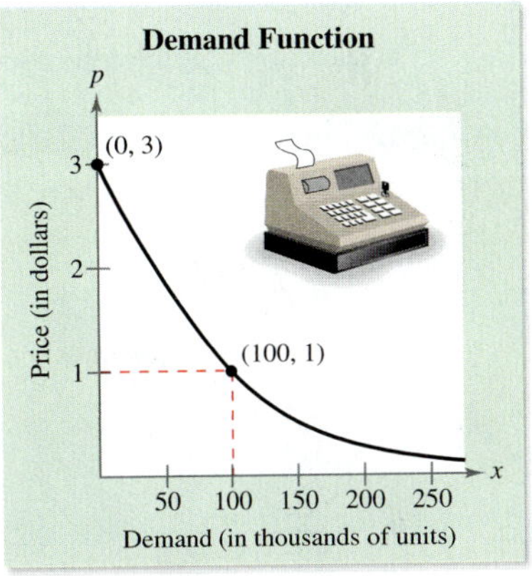

FIGURE 2.36

TRY IT 6

The demand function for a product is given by

$$p = \frac{2}{0.001x^2 + 1}.$$

Find dx/dp implicitly.

TAKE ANOTHER LOOK

Comparing Derivatives

In Example 6, the derivative dx/dp does not represent the slope of the graph of the demand function. Because the demand function is given by $p = f(x)$, the slope of the graph is given by dp/dx. Find dp/dx. Show that the two derivatives are related by $dx/dp = 1/(dp/dx)$. What does dp/dx represent?

PREREQUISITE REVIEW 2.7

The following warm-up exercises involve skills that were covered in earlier sections. You will use these skills in the exercise set for this section.

In Exercises 1–6, solve the equation for y.

1. $x - \dfrac{y}{x} = 2$

2. $\dfrac{4}{x-3} = \dfrac{1}{y}$

3. $xy - x + 6y = 6$

4. $12 + 3y = 4x^2 + x^2y$

5. $x^2 + y^2 = 5$

6. $x = \pm\sqrt{6 - y^2}$

In Exercises 7–10, evaluate the expression at the given point.

7. $\dfrac{3x^2 - 4}{3y^2}$, $(2, 1)$

8. $\dfrac{x^2 - 2}{1 - y}$, $(0, -3)$

9. $\dfrac{5x}{3y^2 - 12y + 5}$, $(-1, 2)$

10. $\dfrac{1}{y^2 - 2xy + x^2}$, $(4, 3)$

EXERCISES 2.7

In Exercises 1–12, find dy/dx.

1. $5xy = 1$

2. $\frac{1}{2}x^2 - y = 6x$

3. $y^2 = 1 - x^2,\ 0 \le x \le 1$

4. $4x^2y - \dfrac{3}{y} = 0$

5. $x^2y^2 - 4y = 1$

6. $xy^2 + 4xy = 10$

7. $4y^2 - xy = 2$

8. $2xy^3 - x^2y = 2$

9. $\dfrac{2y - x}{y^2 - 3} = 5$

10. $\dfrac{xy - y^2}{y - x} = 1$

11. $\dfrac{x + y}{2x - y} = 1$

12. $\dfrac{2x + y}{x - 5y} = 1$

In Exercises 13–24, find dy/dx by implicit differentiation and evaluate the derivative at the given point.

Equation	*Point*
13. $x^2 + y^2 = 49$	$(0, 7)$
14. $x^2 - y^2 = 16$	$(4, 0)$
15. $y + xy = 4$	$(-5, -1)$
16. $x^2 - y^3 = 3$	$(2, 1)$
17. $x^3 - xy + y^2 = 4$	$(0, -2)$
18. $x^2y + y^2x = -2$	$(2, -1)$
19. $x^3y^3 - y = x$	$(0, 0)$
20. $x^3 + y^3 = 2xy$	$(1, 1)$
21. $x^{1/2} + y^{1/2} = 9$	$(16, 25)$
22. $\sqrt{xy} = x - 2y$	$(4, 1)$
23. $x^{2/3} + y^{2/3} = 5$	$(8, 1)$
24. $(x + y)^3 = x^3 + y^3$	$(-1, 1)$

In Exercises 25–30, find the slope of the graph at the given point.

25. $3x^2 - 2y + 5 = 0$

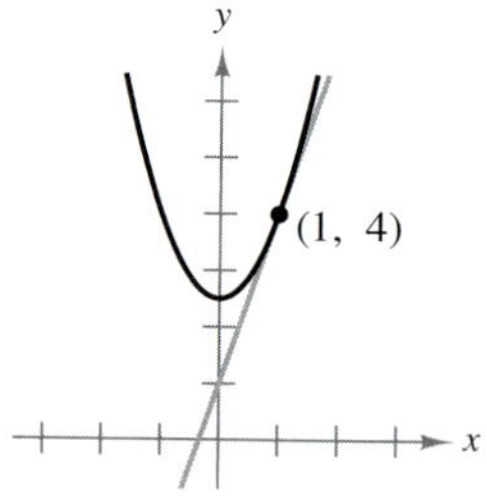

26. $4x^2 + 2y - 1 = 0$

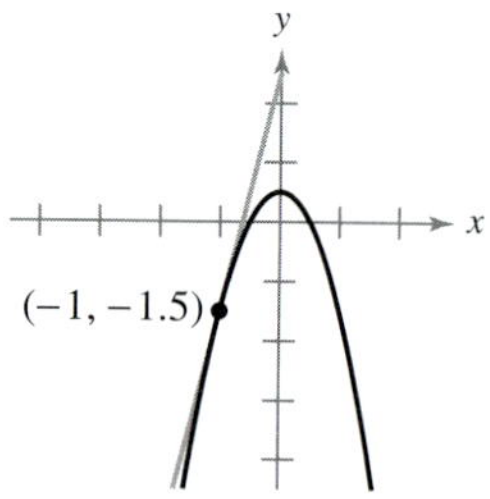

27. $x^2 + y^2 = 4$

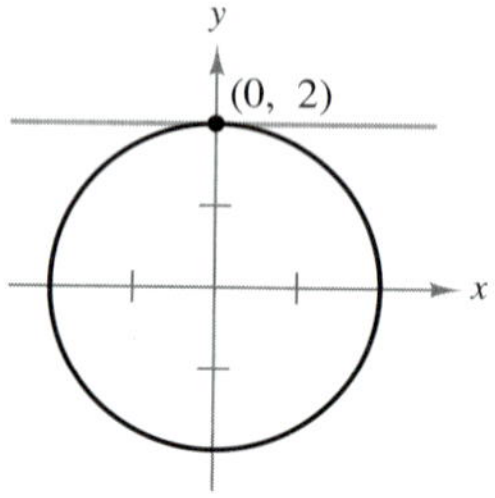

28. $4x^2 + y^2 = 4$

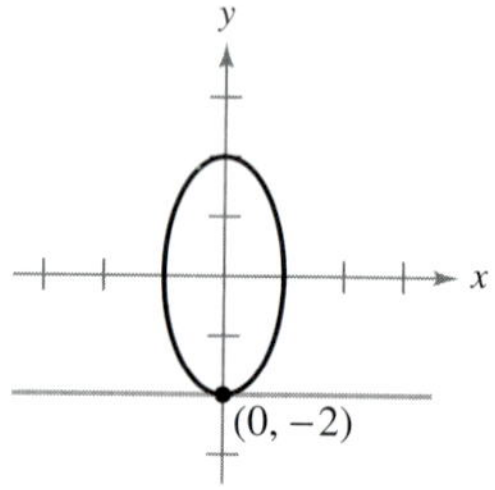

29. $4x^2 + 9y^2 = 36$

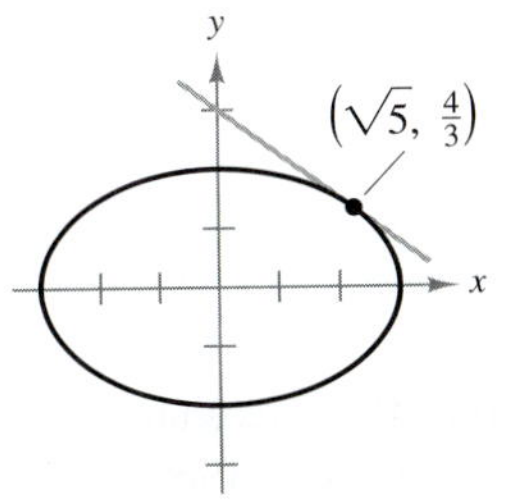

30. $x^2 - y^3 = 0$

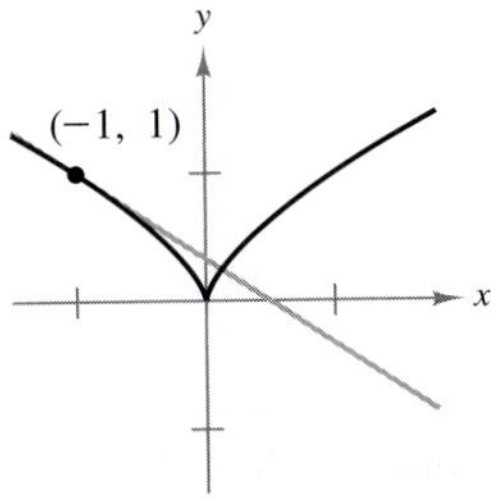

In Exercises 31–34, find dy/dx implicitly and explicitly (the explicit functions are shown on the graph) and show that the results are equivalent. Use the graph to estimate the slope of the tangent line at the labeled point. Then verify your result analytically by evaluating dy/dx at the point.

31. $x^2 + y^2 = 25$

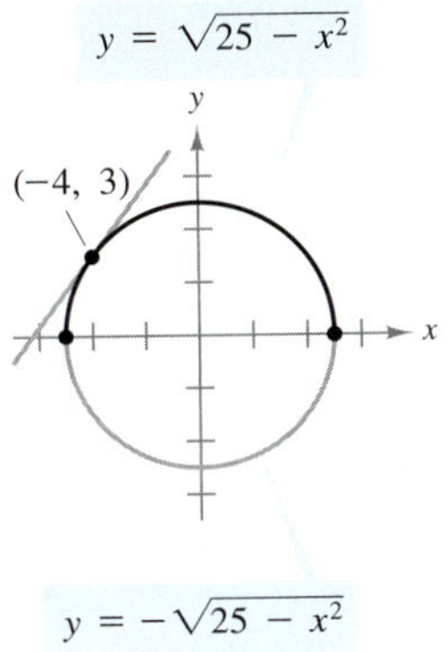

32. $9x^2 + 16y^2 = 144$

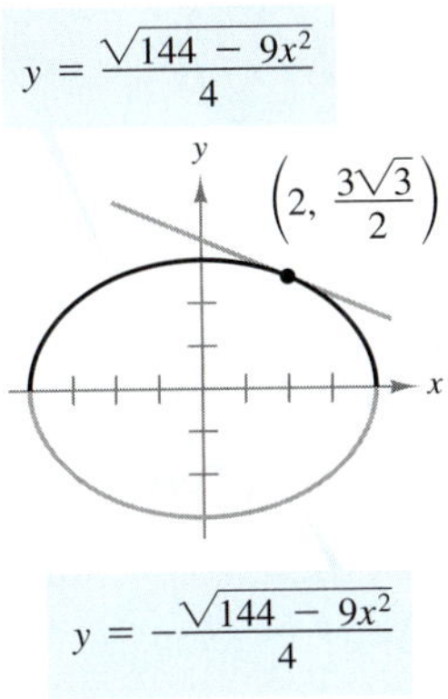

33. $x - y^2 - 1 = 0$

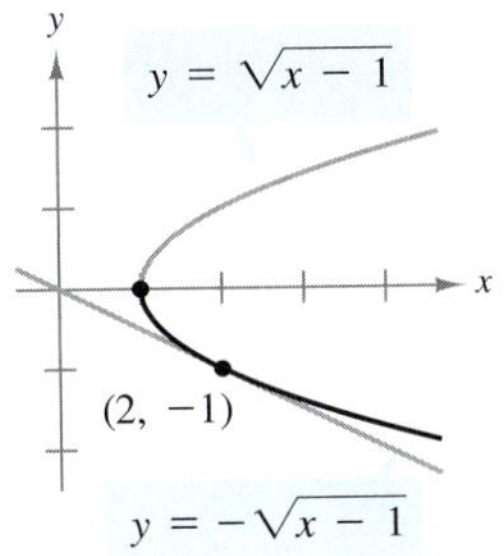

34. $4y^2 - x^2 = 7$

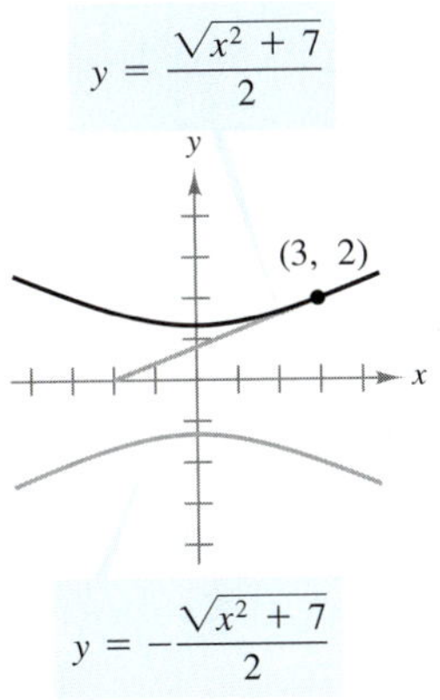

In Exercises 35–40, find equations of the tangent lines to the graph at the given points. Use a graphing utility to graph the equation and the tangent lines in the same viewing window.

	Equation	*Points*
35.	$x^2 + y^2 = 169$	$(5, 12)$ and $(-12, 5)$
36.	$x^2 + y^2 = 9$	$(0, 3)$ and $(2, \sqrt{5})$
37.	$y^2 = 5x^3$	$(1, \sqrt{5})$ and $(1, -\sqrt{5})$
38.	$4xy + x^2 = 5$	$(1, 1)$ and $(5, -1)$
39.	$x^3 + y^3 = 8$	$(0, 2)$ and $(2, 0)$
40.	$y^2 = \frac{x^3}{4 - x}$	$(2, 2)$ and $(2, -2)$

Demand In Exercises 41–44, find the rate of change of x with respect to p.

41. $p = 0.006x^4 + 0.02x^2 + 10, \quad x \ge 0$

42. $p = 0.002x^4 + 0.01x^2 + 5, \quad x \ge 0$

43. $p = \sqrt{\frac{200 - x}{2x}}, \quad 0 < x \le 200$

44. $p = \sqrt{\frac{500 - x}{2x}}, \quad 0 < x \le 500$

45. ***Production*** Let x represent the units of labor and y the capital invested in a manufacturing process. When 135,540 units are produced, the relationship between labor and capital can be modeled by $100x^{0.75}y^{0.25} = 135{,}540$.

(a) Find the rate of change of y with respect to x when $x = 1500$ and $y = 1000$.

(b) The model used in the problem is called the *Cobb-Douglas production function*. Graph the model on a graphing utility and describe the relationship between labor and capital.

46. ***Health: U.S. AIDS Epidemic*** The numbers (in millions) of cases y of AIDS reported in the years 1994 to 2001 can be modeled by

$$y^2 + 4436 = -4.2460t^4 + 146.821t^3 - 1728.00t^2 + 7456.6t$$

where $t = 4$ corresponds to 1994. *(Source: U.S. Centers for Disease Control and Prevention)*

(a) Use a graphing utility to graph the model and describe the results.

(b) Use the graph to determine the year during which the number of reported cases decreasing most rapidly.

(c) Complete the table to confirm your estimate.

t	4	5	6	7	8	9	10	11
y								
y'								

2.8 RELATED RATES

- Examine related variables.
- Solve related-rate problems.

Related Variables

In this section, you will study problems involving variables that are changing with respect to time. If two or more such variables are related to each other, then their rates of change with respect to time are also related.

For instance, suppose that x and y are related by the equation $y = 2x$. If both variables are changing with respect to time, then their rates of change will also be related.

x and y are related. The rates of change of x and y are related.

$$y = 2x \quad \Rightarrow \quad \frac{dy}{dt} = 2\frac{dx}{dt}$$

In this simple example, you can see that because y always has twice the value of x, it follows that the rate of change of y with respect to time is always twice the rate of change of x with respect to time.

EXAMPLE 1 Examining Two Rates That Are Related

The variables x and y are differentiable functions of t and are related by the equation

$$y = x^2 + 3.$$

When $x = 1$, $dx/dt = 2$. Find dy/dt when $x = 1$.

SOLUTION Use the Chain Rule to differentiate both sides of the equation with respect to t.

$$y = x^2 + 3 \qquad \text{Write original equation.}$$

$$\frac{d}{dt}[y] = \frac{d}{dt}[x^2 + 3] \qquad \text{Differentiate with respect to } t.$$

$$\frac{dy}{dt} = 2x\frac{dx}{dt} \qquad \text{Apply Chain Rule.}$$

When $x = 1$ and $dx/dt = 2$, you have

$$2x\frac{dx}{dt} = 2(1)(2)$$

$$= 4.$$

TRY IT 1

When $x = 1$, $dx/dt = 3$. Find dy/dt when $x = 1$ if $y = x^3 + 2$.

Solving Related-Rate Problems

In Example 1, you are *given* the mathematical model.

Given equation: $y = x^2 + 3$

Given rate: $\dfrac{dx}{dt} = 2$ when $x = 1$

Find: $\dfrac{dy}{dt}$ when $x = 1$

In the next example, you are asked to *create* a similar mathematical model.

EXAMPLE 2 Changing Area

A pebble is dropped into a calm pool of water, causing ripples in the form of concentric circles, as shown in Figure 2.37. The radius r of the outer ripple is increasing at a constant rate of 1 foot per second. When the radius is 4 feet, at what rate is the total area A of the disturbed water changing?

FIGURE 2.37

SOLUTION The variables r and A are related by the equation for the area of a circle, $A = \pi r^2$. To solve this problem, use the fact that the rate of change of the radius is given by dr/dt.

Equation: $A = \pi r^2$

Given rate: $\dfrac{dr}{dt} = 1$ when $r = 4$

Find: $\dfrac{dA}{dt}$ when $r = 4$

Using this model, you can proceed as in Example 1.

$$A = \pi r^2 \quad \text{Write original equation.}$$

$$\frac{d}{dt}[A] = \frac{d}{dt}[\pi r^2] \quad \text{Differentiate with respect to } t.$$

$$\frac{dA}{dt} = 2\pi r\frac{dr}{dt} \quad \text{Apply Chain Rule.}$$

When $r = 4$ and $dr/dt = 1$, you have

$$2\pi r\frac{dr}{dt} = 2\pi(4)(1)$$

$$= 8\pi \text{ feet squared per second}$$

TRY IT 2

If the radius r of the outer ripple in Example 2 is increasing at a rate of 2 feet per second, at what rate is the total area changing when the radius is 3 feet?

STUDY TIP

In Example 2, note that the radius changes at a *constant* rate ($dr/dt = 1$ for all t), but the area changes at a *nonconstant* rate.

When $r = 1$ *ft*	*When* $r = 2$ *ft*	*When* $r = 3$ *ft*	*When* $r = 4$ *ft*
$\dfrac{dA}{dt} = 2\pi \text{ ft}^2/\text{sec}$	$\dfrac{dA}{dt} = 4\pi \text{ ft}^2/\text{sec}$	$\dfrac{dA}{dt} = 6\pi \text{ ft}^2/\text{sec}$	$\dfrac{dA}{dt} = 8\pi \text{ ft}^2/\text{sec}$

The solution shown in Example 2 illustrates the steps for solving a related-rate problem.

Guidelines for Solving a Related-Rate Problem

1. Identify all *given* quantities and all quantities *to be determined.* If possible, make a sketch and label the quantities.
2. Write an equation that relates all variables whose rates of change are either given or to be determined.
3. Use the Chain Rule to differentiate both sides of the equation *with respect to time.*
4. Substitute into the resulting equation all known values of the variables and their rates of change. Then solve for the required rate of change.

STUDY TIP

Be sure you notice the order of Steps 3 and 4 in the guidelines. Do not substitute the known values for the variables until after you have differentiated.

In Step 2 of the guidelines, note that you must write an equation that relates the given variables. To help you with this step, reference tables that summarize many common formulas are included in the appendices. For instance, the volume of a sphere of radius r is given by the formula

$$V = \frac{4}{3}\pi r^3$$

as listed in Appendix D.

The table below shows the mathematical models for some common rates of change that can be used in the first step of the solution of a related-rate problem.

Verbal statement	Mathematical model
The velocity of a car after traveling 1 hour is 50 miles per hour.	$x =$ distance traveled $\frac{dx}{dt} = 50$ when $t = 1$
Water is being pumped into a swimming pool at the rate of 10 cubic feet per minute.	$V =$ volume of water in pool $\frac{dV}{dt} = 10\ \text{ft}^3/\text{min}$
A population of bacteria is increasing at the rate of 2000 per hour.	$x =$ number in population $\frac{dx}{dt} = 2000$ bacteria per hour
Revenue is increasing at the rate of \$4000 per month.	$R =$ revenue $\frac{dR}{dt} = 4000$ dollars per month

EXAMPLE 3 Changing Volume

Air is being pumped into a spherical balloon at the rate of 4.5 cubic inches per minute. See Figure 2.38. Find the rate of change of the radius when the radius is 2 inches.

SOLUTION Let V represent the volume of the balloon and let r represent the radius. Because the volume is increasing at the rate of 4.5 cubic inches per minute, you know that $dV/dt = 4.5$. An equation that relates V and r is $V = \frac{4}{3}\pi r^3$. So, the problem can be represented by the model shown below.

Equation: $V = \dfrac{4}{3}\pi r^3$

Given rate: $\dfrac{dV}{dt} = 4.5$

Find: $\dfrac{dr}{dt}$ when $r = 2$

By differentiating the equation, you obtain

$$V = \frac{4}{3}\pi r^3 \qquad \text{Write original equation.}$$

$$\frac{d}{dt}[V] = \frac{d}{dt}\left[\frac{4}{3}\pi r^3\right] \qquad \text{Differentiate with respect to } t.$$

$$\frac{dV}{dt} = \frac{4}{3}\pi(3r^2)\frac{dr}{dt} \qquad \text{Apply Chain Rule.}$$

$$\frac{1}{4\pi r^2}\frac{dV}{dt} = \frac{dr}{dt}. \qquad \text{Solve for } dr/dt.$$

When $r = 2$ and $dV/dt = 4.5$, the rate of change of the radius is

$$\frac{1}{4\pi r^2}\frac{dV}{dt} = \frac{1}{4\pi(2^2)}(4.5)$$

$$\approx 0.09 \text{ inch per minute.}$$

In Example 3, note that the volume is increasing at a *constant rate* but the radius is increasing at a *variable* rate. In this particular example, the radius is increasing more and more slowly as t increases. This is illustrated in Figure 2.38 and the table below.

t	1	3	5	7	9	11
$V = 4.5t$	4.5	13.5	22.5	31.5	40.5	49.5
$r = \sqrt[3]{\dfrac{3V}{4\pi}}$	1.02	1.48	1.75	1.96	2.13	2.28
$\dfrac{dr}{dt}$	0.34	0.16	0.12	0.09	0.08	0.07

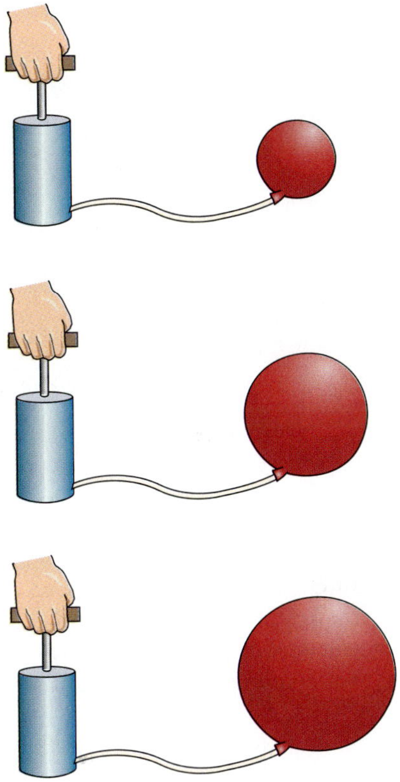

FIGURE 2.38 Expanding Balloon

TRY IT 3

If the radius of a spherical balloon increases at a rate of 1.5 inches per minute, find the rate at which the surface area changes when the radius is 6 inches. (Formula for surface area of a sphere: $S = 4\pi r^2$)

EXAMPLE 4 Analyzing a Profit Function

A company's profit P (in dollars) from selling x units of a product can be modeled by

$$P = 500x - \left(\frac{1}{4}\right)x^2. \qquad \text{Model for profit}$$

The sales are increasing at a rate of 10 units per day. Find the rate of change in the profit (in dollars per day) when 500 units have been sold.

SOLUTION Because you are asked to find the rate of change in dollars per day, you should differentiate the given equation with respect to the time t.

$$P = 500x - \left(\frac{1}{4}\right)x^2 \qquad \text{Write model for profit.}$$

$$\frac{dP}{dt} = 500\left(\frac{dx}{dt}\right) - 2\left(\frac{1}{4}\right)(x)\left(\frac{dx}{dt}\right) \qquad \text{Differentiate with respect to } t.$$

The sales are increasing at a constant rate of 10 units per day, so

$$\frac{dx}{dt} = 10.$$

When $x = 500$ units and $dx/dt = 10$, the rate of change in the profit is

$$500\left(\frac{dx}{dt}\right) - 2\left(\frac{1}{4}\right)(x)\left(\frac{dx}{dt}\right) = 500(10) - 2\left(\frac{1}{4}\right)(500)(10) \qquad \text{Substitute for } dx/dt \text{ and } x.$$

$$= 5000 - 2500$$

$$= \$2500 \text{ per day.} \qquad \text{Simplify.}$$

The graph of the profit function (in terms of x) is shown in Figure 2.39.

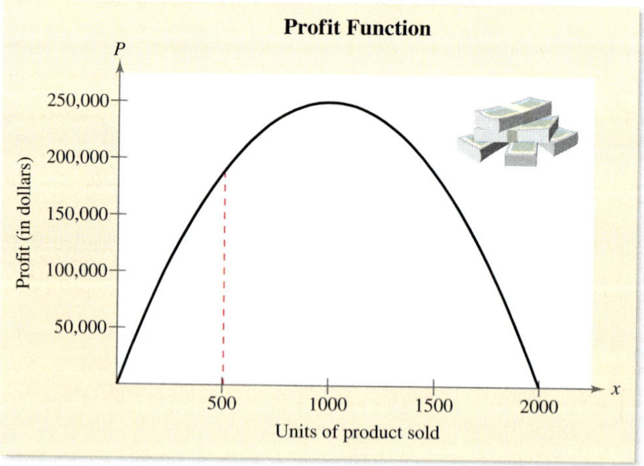

FIGURE 2.39

STUDY TIP

In Example 4, note that one of the keys to successful use of calculus in applied problems is the interpretation of a rate of change as a derivative.

TRY IT 4

Find the rate of change in profit (in dollars per day) when 50 units have been sold, sales have increased at a rate of 10 units per day, and $P = 200x - \frac{1}{2}x^2$.

EXAMPLE 5 Increasing Production

A company is increasing the production of a product at the rate of 200 units per week. The weekly demand function is modeled by

$$p = 100 - 0.001x$$

where p is the price per unit and x is the number of units produced in a week. Find the rate of change of the revenue with respect to time when the weekly production is 2000 units.

SOLUTION

Equation: $R = xp = x(100 - 0.001x) = 100x - 0.001x^2$

Given rate: $\dfrac{dx}{dt} = 200$

Find: $\dfrac{dR}{dt}$ when $x = 2000$

By differentiating the equation, you obtain

$$R = 100x - 0.001x^2 \qquad \text{Write original equation.}$$

$$\frac{d}{dt}[R] = \frac{d}{dt}[100x - 0.001x^2] \qquad \text{Differentiate with respect to } t.$$

$$\frac{dR}{dt} = (100 - 0.002x)\frac{dx}{dt}. \qquad \text{Apply Chain Rule.}$$

Using $x = 2000$ and $dx/dt = 200$, you have

$$(100 - 0.002x)\frac{dx}{dt} = [100 - 0.002(2000)](200)$$

$$= \$19{,}200 \text{ per week.}$$

TRY IT 5

Find the rate of change of revenue with respect to time for the company in Example 5 if the weekly demand function is

$$p = 150 - 0.002x.$$

TAKE ANOTHER LOOK

Increasing Revenue

In Example 5, the company is increasing its revenue for the product by decreasing the price. The number of units produced and the price per unit during the first 10 weeks are shown in the table.

t	1	2	3	4	5	6	7	8	9	10
x	200	400	600	800	1000	1200	1400	1600	1800	2000
p	\$99.8	\$99.6	\$99.4	\$99.2	\$99.0	\$98.8	\$98.6	\$98.4	\$98.2	\$98.0

Is the revenue increasing during this 10-week period? For how many weeks can the company continue to drop the price before the revenue begins to decline? Do you think companies use methods such as this to set prices? Why or why not?

PREREQUISITE REVIEW 2.8

The following warm-up exercises involve skills that were covered in earlier sections. You will use these skills in the exercise set for this section.

In Exercises 1–6, write a formula for the given quantity.

1. Area of a circle

2. Volume of a sphere

3. Surface area of a cube

4. Volume of a cube

5. Volume of a cone

6. Area of a triangle

In Exercises 7–10, find dy/dx by implicit differentiation.

7. $x^2 + y^2 = 9$

8. $3xy - x^2 = 6$

9. $x^2 + 2y + xy = 12$

10. $x + xy^2 - y^2 = xy$

EXERCISES 2.8

In Exercises 1–4, find the given values of dy/dt and dx/dt.

	Equation	*Find*	*Given*
1.	$y = x^2 - \sqrt{x}$	(a) $\dfrac{dy}{dt}$	$x = 4, \dfrac{dx}{dt} = 8$
		(b) $\dfrac{dx}{dt}$	$x = 16, \dfrac{dy}{dt} = 12$
2.	$y = x^2 - 3x$	(a) $\dfrac{dy}{dt}$	$x = 3, \dfrac{dx}{dt} = 2$
		(b) $\dfrac{dx}{dt}$	$x = 1, \dfrac{dy}{dt} = 5$
3.	$xy = 4$	(a) $\dfrac{dy}{dt}$	$x = 8, \dfrac{dx}{dt} = 10$
		(b) $\dfrac{dx}{dt}$	$x = 1, \dfrac{dy}{dt} = -6$
4.	$x^2 + y^2 = 25$	(a) $\dfrac{dy}{dt}$	$x = 3, y = 4, \dfrac{dx}{dt} = 8$
		(b) $\dfrac{dx}{dt}$	$x = 4, y = 3, \dfrac{dy}{dt} = -2$

5. ***Area*** The radius r of a circle is increasing at a rate of 2 inches per minute. Find the rates of change of the area when (a) $r = 6$ inches and (b) $r = 24$ inches.

6. ***Volume*** The radius r of a sphere is increasing at a rate of 2 inches per minute. Find the rates of change of the volume when (a) $r = 6$ inches and (b) $r = 24$ inches.

7. ***Area*** Let A be the area of a circle of radius r that is changing with respect to time. If dr/dt is constant, is dA/dt constant? Explain your reasoning.

8. ***Volume*** Let V be the volume of a sphere of radius r that is changing with respect to time. If dr/dt is constant, is dV/dt constant? Explain your reasoning.

9. ***Volume*** A spherical balloon is inflated with gas at a rate of 20 cubic feet per minute. How fast is the radius of the balloon changing at the instant the radius is (a) 1 foot and (b) 2 feet?

10. ***Volume*** The radius r of a right circular cone is increasing at a rate of 2 inches per minute. The height h of the cone is related to the radius by $h = 3r$. Find the rates of change of the volume when (a) $r = 6$ inches and (b) $r = 24$ inches.

11. ***Cost, Revenue, and Profit*** A company that manufactures sport supplements calculates that its costs and revenue can be modeled by the equations

$$C = 125{,}000 + 0.75x \quad \text{and} \quad R = 250x - \frac{1}{10}x^2$$

where x is the number of units of sport supplements produced in 1 week. If production in one particular week is 1000 units and is increasing at a rate of 150 units per week, find:

(a) the rate at which the cost is changing.

(b) the rate at which the revenue is changing.

(c) the rate at which the profit is changing.

12. ***Cost, Revenue, and Profit*** A company that manufactures pet toys calculates that its costs and revenue can be modeled by the equations

$$C = 75{,}000 + 1.05x \quad \text{and} \quad R = 500x - \frac{x^2}{25}$$

where x is the number of toys produced in 1 week. If production in one particular week is 5000 toys and is increasing at a rate of 250 toys per week, find:

(a) the rate at which the cost is changing.

(b) the rate at which the revenue is changing.

(c) the rate at which the profit is changing.

13. ***Expanding Cube*** All edges of a cube are expanding at a rate of 3 centimeters per second. How fast is the volume changing when each edge is (a) 1 centimeter and (b) 10 centimeters?

14. ***Expanding Cube*** All edges of a cube are expanding at a rate of 3 centimeters per second. How fast is the surface area changing when each edge is (a) 1 centimeter and (b) 10 centimeters?

15. ***Moving Point*** A point is moving along the graph of $y = x^2$ such that dx/dt is 2 centimeters per minute. Find dy/dt for each value of x.

(a) $x = -3$ (b) $x = 0$ (c) $x = 1$ (d) $x = 3$

16. ***Moving Point*** A point is moving along the graph of $y = 1/(1 + x^2)$ such that dx/dt is 2 centimeters per minute. Find dy/dt for each value of x.

(a) $x = -2$ (b) $x = 2$ (c) $x = 0$ (d) $x = 10$

17. ***Speed*** A 25-foot ladder is leaning against a house (see figure). The base of the ladder is pulled away from the house at a rate of 2 feet per second. How fast is the top of the ladder moving down the wall when the base is (a) 7 feet, (b) 15 feet, and (c) 24 feet from the house?

Figure for 17

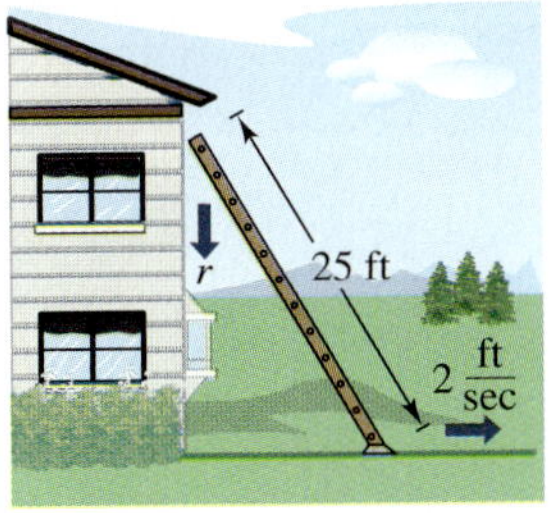

Figure for 18

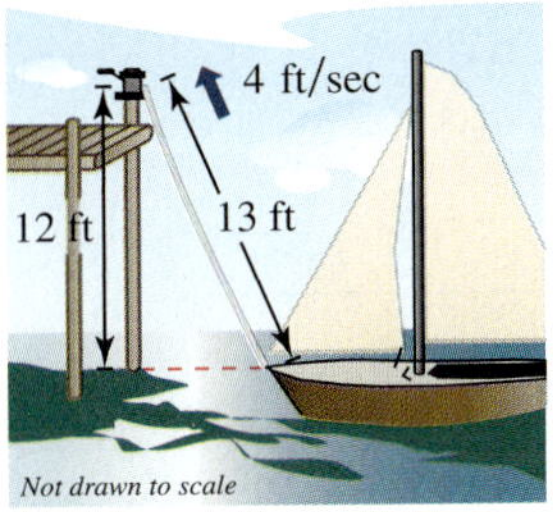

18. ***Speed*** A boat is pulled by a winch on a dock, and the winch is 12 feet above the deck of the boat (see figure). The winch pulls the rope at a rate of 4 feet per second. Find the speed of the boat when 13 feet of rope is out. What happens to the speed of the boat as it gets closer and closer to the dock?

19. ***Air Traffic Control*** An air traffic controller spots two airplanes at the same altitude converging to a point as they fly at right angles to each other. One airplane is 150 miles from the point and has a speed of 450 miles per hour. The other is 200 miles from the point and has a speed of 600 miles per hour.

(a) At what rate is the distance between the planes changing?

(b) How much time does the controller have to get one of the airplanes on a different flight path?

20. ***Speed*** An airplane flying at an altitude of 6 miles passes directly over a radar antenna (see figure). When the airplane is 10 miles away ($s = 10$), the radar detects that the distance s is changing at a rate of 240 miles per hour. What is the speed of the airplane?

Figure for 20

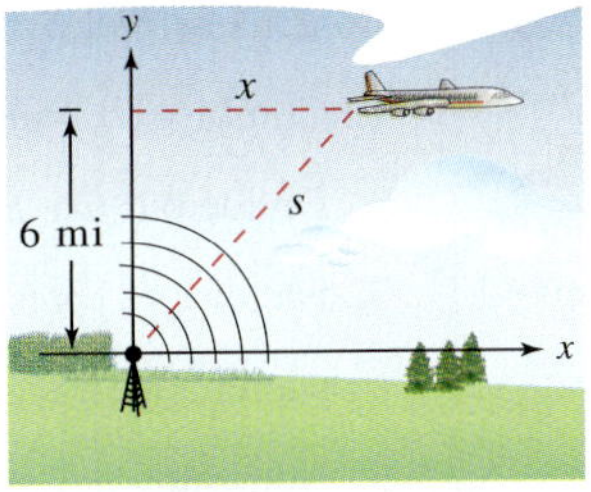

Figure for 21

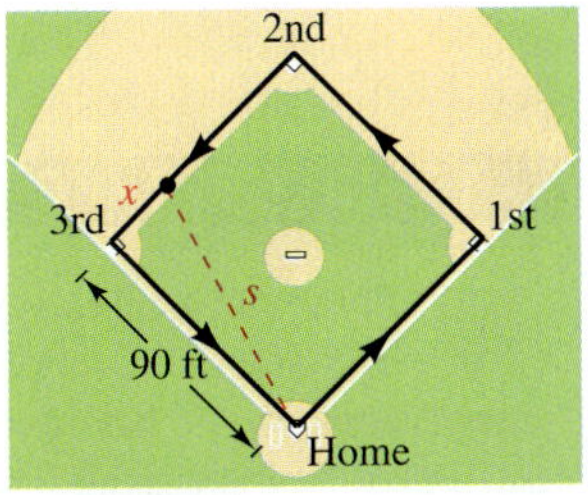

21. ***Athletics*** A (square) baseball diamond has sides that are 90 feet long (see figure). A player 26 feet from third base is running at a speed of 30 feet per second. At what rate is the player's distance from home plate changing?

22. ***Advertising Costs*** A retail sporting goods store estimates that weekly sales S and weekly advertising costs x are related by the equation $S = 2250 + 50x + 0.35x^2$. The current weekly advertising costs are \$1500, and these costs are increasing at a rate of \$125 per week. Find the current rate of change of weekly sales.

23. ***Environment*** An accident at an oil drilling platform is causing a circular oil slick. The slick is 0.08 foot thick, and when the radius is 750 feet, the radius of the slick is increasing at the rate of 0.5 foot per minute. At what rate (in cubic feet per minute) is oil flowing from the site of the accident?

24. ***Profit*** A company is increasing the production of a product at the rate of 25 units per week. The demand and cost functions for the product are given by $p = 50 - 0.01x$ and $C = 4000 + 40x - 0.02x^2$. Find the rate of change of the profit with respect to time when the weekly sales are $x = 800$ units. Use a graphing utility to graph the profit function, and use the *zoom* and *trace* features of the graphing utility to verify your result.

25. ***Sales*** The profit for a product is increasing at a rate of \$6384 per week. The demand and cost functions for the product are given by $p = 6000 - 0.4x^2$ and $C = 2400x + 5200$. Find the rate of change of sales with respect to time when the weekly sales are $x = 44$ units.

26. ***Cost*** The annual cost (in millions of dollars) for a government agency to seize $p\%$ of an illegal drug is given by

$$C = \frac{528p}{100 - p}, \qquad 0 \le p < 100.$$

The agency's goal is to increase p by 5% per year. Find the rates of change of the cost when (a) $p = 30\%$ and (b) $p = 60\%$. Use a graphing utility to graph C. What happens to the graph of C as p approaches 100?

ALGEBRA REVIEW

Simplifying Algebraic Expressions

To be successful in using derivatives, you must be good at simplifying algebraic expressions. Here are some helpful simplification techniques.

1. Combine *like terms*. This may involve expanding an expression by multiplying factors.
2. Divide out *like factors* in the numerator and denominator of an expression.
3. Factor an expression.
4. Rationalize a denominator.
5. Add, subtract, multiply, or divide fractions.

TECHNOLOGY

Symbolic algebra systems can simplify algebraic expressions. If you have access to such a system, try using it to simplify the expressions in this Algebra Review.

EXAMPLE 1 Simplifying a Fractional Expression

(a) $\dfrac{(x + \Delta x)^2 - x^2}{\Delta x} = \dfrac{x^2 + 2x(\Delta x) + (\Delta x)^2 - x^2}{\Delta x}$ Expand expression.

$= \dfrac{2x(\Delta x) + (\Delta x)^2}{\Delta x}$ Combine like terms.

$= \dfrac{\Delta x(2x + \Delta x)}{\Delta x}$ Factor.

$= 2x + \Delta x, \quad \Delta x \neq 0$ Divide out like factors.

(b) $\dfrac{(x^2 - 1)(-2 - 2x) - (3 - 2x - x^2)(2)}{(x^2 - 1)^2}$

$= \dfrac{(-2x^2 - 2x^3 + 2 + 2x) - (6 - 4x - 2x^2)}{(x^2 - 1)^2}$ Expand expression.

$= \dfrac{-2x^2 - 2x^3 + 2 + 2x - 6 + 4x + 2x^2}{(x^2 - 1)^2}$ Remove parentheses.

$= \dfrac{-2x^3 + 6x - 4}{(x^2 - 1)^2}$ Combine like terms.

(c) $2\left(\dfrac{2x + 1}{3x}\right)\left[\dfrac{3x(2) - (2x + 1)(3)}{(3x)^2}\right]$

$= 2\left(\dfrac{2x + 1}{3x}\right)\left[\dfrac{6x - (6x + 3)}{(3x)^2}\right]$ Multiply factors.

$= \dfrac{2(2x + 1)(6x - 6x - 3)}{(3x)^3}$ Multiply fractions and remove parentheses.

$= \dfrac{2(2x + 1)(-3)}{3(9)x^3}$ Combine like terms and factor.

$= \dfrac{-2(2x + 1)}{9x^3}$ Divide out like factors.

EXAMPLE 2 **Simplifying an Expression with Powers or Radicals**

(a) $(2x + 1)^2(6x + 1) + (3x^2 + x)(2)(2x + 1)(2)$

$= (2x + 1)[(2x + 1)(6x + 1) + (3x^2 + x)(2)(2)]$ — Factor.

$= (2x + 1)[12x^2 + 8x + 1 + (12x^2 + 4x)]$ — Multiply factors.

$= (2x + 1)(12x^2 + 8x + 1 + 12x^2 + 4x)$ — Remove parentheses.

$= (2x + 1)(24x^2 + 12x + 1)$ — Combine like terms.

(b) $(-1)(6x^2 - 4x)^{-2}(12x - 4)$

$= \dfrac{(-1)(12x - 4)}{(6x^2 - 4x)^2}$ — Rewrite as a fraction.

$= \dfrac{(-1)(4)(3x - 1)}{(6x^2 - 4x)^2}$ — Factor.

$= \dfrac{-4(3x - 1)}{(6x^2 - 4x)^2}$ — Multiply factors.

(c) $(x)\left(\frac{1}{2}\right)(2x + 3)^{-1/2} + (2x + 3)^{1/2}(1)$

$= (2x + 3)^{-1/2}\left(\frac{1}{2}\right)[x + (2x + 3)(2)]$ — Factor.

$= \dfrac{x + 4x + 6}{(2x + 3)^{1/2}(2)}$ — Rewrite as a fraction.

$= \dfrac{5x + 6}{2(2x + 3)^{1/2}}$ — Combine like terms.

(d) $\dfrac{x^2\left(\frac{1}{2}\right)(2x)(x^2 + 1)^{-1/2} - (x^2 + 1)^{1/2}(2x)}{x^4}$

$= \dfrac{(x^3)(x^2 + 1)^{-1/2} - (x^2 + 1)^{1/2}(2x)}{x^4}$ — Multiply factors.

$= \dfrac{(x^2 + 1)^{-1/2}(x)[x^2 - (x^2 + 1)(2)]}{x^4}$ — Factor.

$= \dfrac{x[x^2 - (2x^2 + 2)]}{(x^2 + 1)^{1/2}x^4}$ — Write with positive exponents.

$= \dfrac{x^2 - 2x^2 - 2}{(x^2 + 1)^{1/2}x^3}$ — Divide out like factors and remove parentheses.

$= \dfrac{-x^2 - 2}{(x^2 + 1)^{1/2}x^3}$ — Combine like terms.

All but one of the expressions in this Algebra Review are derivatives. Can you see what the original function is? Explain your reasoning.

2 CHAPTER SUMMARY AND STUDY STRATEGIES

*After studying this chapter, you should have acquired the following skills. The exercise numbers are keyed to the Review Exercises that begin on page 166. Answers to odd-numbered Review Exercises are given in the back of the text.**

Skills

- Approximate the slope of the tangent line to a graph at a point. *(Section 2.1)* — *Review Exercises 1–4*
- Interpret the slope of a graph in a real-life setting. *(Section 2.1)* — *Review Exercises 5–8*
- Use the limit definition to find the derivative of a function and the slope of a graph at a point. *(Section 2.1)* — *Review Exercises 9–16*

  $$f'(x) = \lim_{\Delta x \to 0} \frac{f(x + \Delta x) - f(x)}{\Delta x}$$

- Use the derivative to find the slope of a graph at a point. *(Section 2.1)* — *Review Exercises 17–24*
- Use the graph of a function to recognize points at which the function is not differentiable. *(Section 2.1)* — *Review Exercises 25–28*
- Use the Constant Multiple Rule for differentiation. *(Section 2.2)* — *Review Exercises 29, 30*

  $$\frac{d}{dx}[cf(x)] = cf'(x)$$

- Use the Sum and Difference Rules for differentiation. *(Section 2.2)* — *Review Exercises 31–38*

  $$\frac{d}{dx}[f(x) \pm g(x)] = f'(x) \pm g'(x)$$

- Find the average rate of change of a function over an interval and the instantaneous rate of change at a point. *(Section 2.3)* — *Review Exercises 39, 40*

  $$\text{Average rate of change} = \frac{f(b) - f(a)}{b - a}$$

  $$\text{Instantaneous rate of change} = \lim_{\Delta x \to 0} \frac{f(x + \Delta x) - f(x)}{\Delta x}$$

- Find the average and instantaneous rates of change of a quantity in a real-life problem. *(Section 2.3)* — *Review Exercises 41–44*

* Use a wide range of valuable study aids to help you master the material in this chapter. The *Student Solutions Guide* includes step-by-step solutions to all odd-numbered exercises to help you review and prepare. The *HM mathSpace® Student CD-ROM* helps you brush up on your algebra skills. The *Graphing Technology Guide*, available on the Web at *math.college.hmco.com/students*, offers step-by-step commands and instructions for a wide variety of graphing calculators, including the most recent models.

- Find the velocity of an object that is moving in a straight line. *(Section 2.3)* — *Review Exercises 45, 46*
- Create mathematical models for the revenue, cost, and profit for a product. *(Section 2.3)* — *Review Exercises 47, 48*

  $P = R - C, \qquad R = xp$
- Find the marginal revenue, marginal cost, and marginal profit for a product. *(Section 2.3)* — *Review Exercises 49–58*
- Use the Product Rule for differentiation. *(Section 2.4)* — *Review Exercises 59–62*

  $$\frac{d}{dx}[f(x)g(x)] = f(x)g'(x) + g(x)f'(x)$$
- Use the Quotient Rule for differentiation. *(Section 2.4)* — *Review Exercises 63, 64*

  $$\frac{d}{dx}\left[\frac{f(x)}{g(x)}\right] = \frac{g(x)f'(x) - f(x)g'(x)}{[g(x)]^2}$$
- Use the General Power Rule for differentiation. *(Section 2.5)* — *Review Exercises 65–68*

  $$\frac{d}{dx}[u^n] = nu^{n-1}u'$$
- Use differentiation rules efficiently to find the derivative of any algebraic function, then simplify the result. *(Section 2.5)* — *Review Exercises 69–78*
- Use derivatives to answer questions about real-life situations. *(Sections 2.1–2.5)* — *Review Exercises 79, 80*
- Find higher-order derivatives. *(Section 2.6)* — *Review Exercises 81–88*
- Find and use the position function to determine the velocity and acceleration of a moving object. *(Section 2.6)* — *Review Exercises 89, 90*
- Find derivatives implicitly. *(Section 2.7)* — *Review Exercises 91–98*
- Solve related-rate problems. *(Section 2.8)* — *Review Exercises 99, 100*
- ***Simplify Your Derivatives*** Often our students ask if they have to simplify their derivatives. Our answer is "Yes, if you expect to use them." In the next chapter, you will see that almost all applications of derivatives require that the derivatives be written in simplified form. It is not difficult to see the advantage of a derivative in simplified form. Consider, for instance, the derivative of $f(x) = x/\sqrt{x^2+1}$. The "raw form" produced by the Quotient and Chain Rules

  $$f'(x) = \frac{(x^2+1)^{1/2}(1) - (x)\left(\frac{1}{2}\right)(x^2+1)^{-1/2}(2x)}{\left(\sqrt{x^2+1}\right)^2}$$

  is obviously much more difficult to use than the simplified form

  $$f'(x) = \frac{1}{(x^2+1)^{3/2}}.$$
- ***List Units of Measure in Applied Problems*** When using derivatives in real-life applications, be sure to list the units of measure for each variable. For instance, if R is measured in dollars and t is measured in years, then the derivative dR/dt is measured in dollars per year.

Study Tools *Additional resources that accompany this chapter*

- **Algebra Review** (pages 162 and 163)
- **Chapter Summary and Study Strategies** (pages 164 and 165)
- **Review Exercises** (pages 166–169)
- **Sample Post-Graduation Exam Questions** (page 170)
- **Web Exercises** (page 130, Exercise 66)
- **Student Solutions Guide**
- **HM mathSpace® Student CD-ROM**
- **Graphing Technology Guide** (*math.college.hmco.com/students*)

2 CHAPTER REVIEW EXERCISES

In Exercises 1–4, approximate the slope of the tangent line to the graph at (x, y).

1.

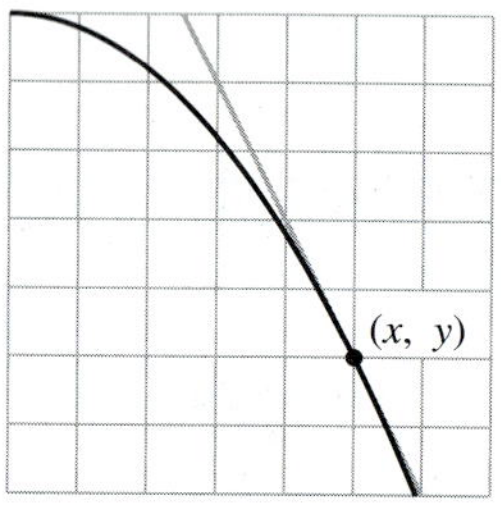

2.

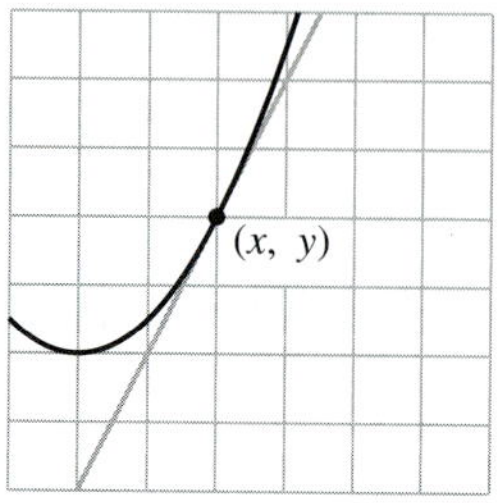

3.

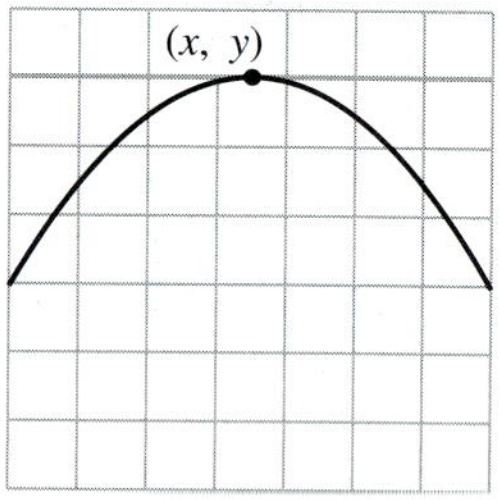

4.

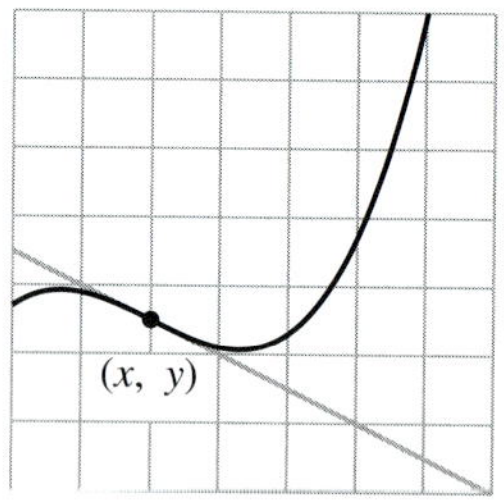

5. *Sales* The graph approximates the annual sales S (in millions of dollars per year) of Home Depot for the years 1996 to 2002, with $t = 6$ corresponding to 1996. Estimate the slopes of the graph when $t = 7$, $t = 10$, and $t = 12$. Interpret each slope in the context of the problem. *(Source: The Home Depot, Inc.)*

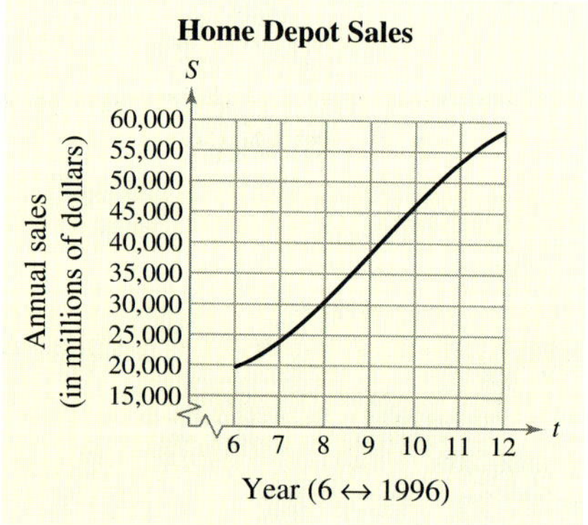

6. *Consumer Trends* The graph approximates the number of subscribers S (in thousands per year) of cellular telephones for 1993 to 2002, with $t = 3$ corresponding to 1993. Estimate the slopes of the graph when $t = 4$, $t = 8$, and $t = 12$. Interpret each slope in the context of the problem. *(Source: Cellular Telecommunications & Internet Association)*

Figure for 6

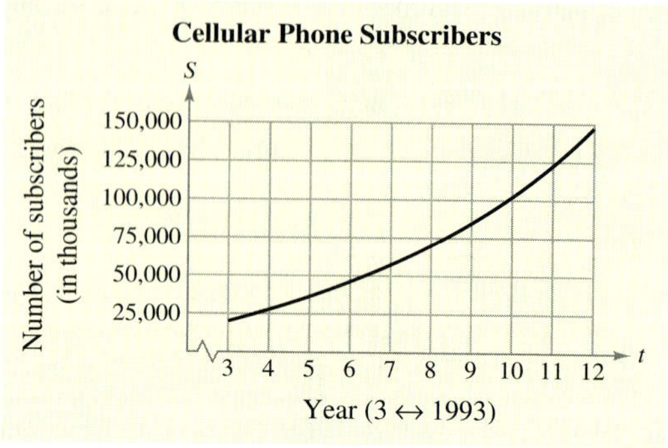

7. *Medicine* The graph shows the estimated number of milligrams of a pain medication M in the bloodstream t hours after a 1000-milligram dose of the drug has been given. Estimate the slopes of the graph at $t = 0$, 4, and 6.

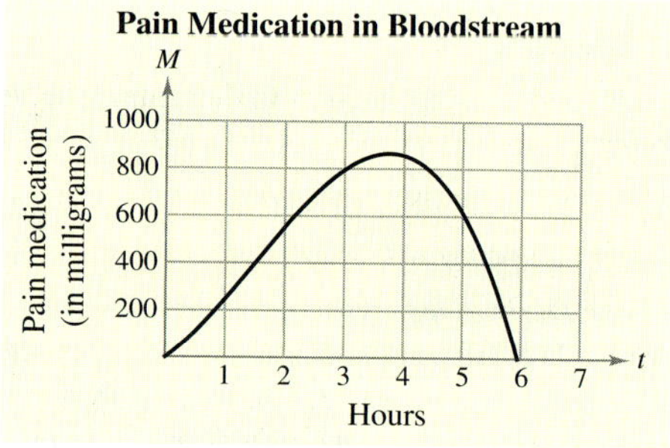

8. *Athletics* Two white-water rafters leave a campsite simultaneously and start downstream. Their distances from the campsite are given by $s = f(t)$ and $s = g(t)$, where s is measured in miles and t is measured in hours.

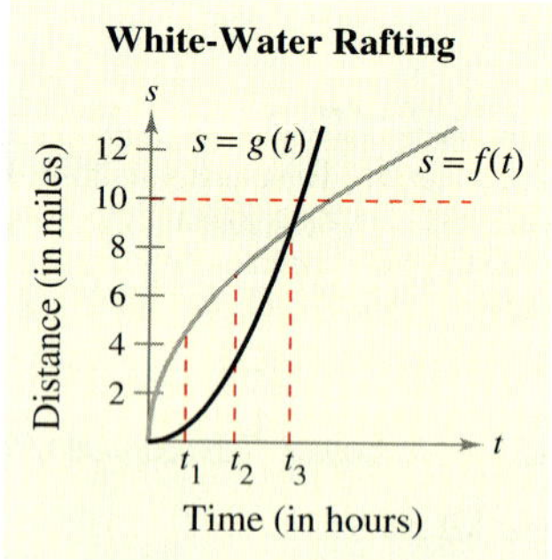

(a) Which rafter is traveling at a greater rate at t_1?

(b) What can you conclude about their rates at t_2?

(c) What can you conclude about their rates at t_3?

(d) Which rafter finishes the trip first? Explain your reasoning.

In Exercises 9–16, use the limit definition to find the derivative of the function. Then use the limit definition to find the slope of the tangent line to the graph of f at the given point.

9. $f(x) = -3x - 5;\ (-2, 1)$

10. $f(x) = 7x + 3;\ (-1, 4)$

11. $f(x) = x^2 - 4x;\ (1, -3)$

12. $f(x) = x^2 + 10;\ (2, 14)$

13. $f(x) = \sqrt{x + 9};\ (-5, 2)$

14. $f(x) = \sqrt{x - 1};\ (10, 3)$

15. $f(x) = \dfrac{1}{x - 5};\ (6, 1)$

16. $f(x) = \dfrac{1}{x + 4};\ (-3, 1)$

In Exercises 17–24, find the slope of the graph of f at the given point.

17. $f(x) = 8 - 5x;\ (3, -7)$

18. $f(x) = 2 - 3x;\ (1, -1)$

19. $f(x) = -\frac{1}{2}x^2 + 2x;\ (2, 2)$

20. $f(x) = 4 - x^2;\ (-1, 3)$

21. $f(x) = \sqrt{x} + 2;\ (9, 5)$

22. $f(x) = 2\sqrt{x} + 1;\ (4, 5)$

23. $f(x) = \dfrac{5}{x};\ (1, 5)$

24. $f(x) = \dfrac{2}{x} - 1;\ \left(\dfrac{1}{2}, 3\right)$

In Exercises 25–28, determine the x-value at which the function is not differentiable.

25. $y = \dfrac{x + 1}{x - 1}$

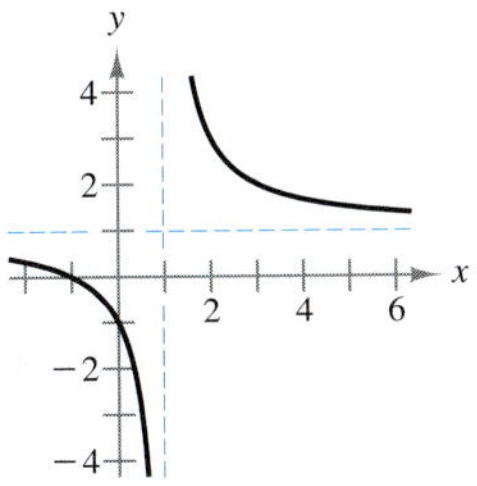

26. $y = -|x| + 3$

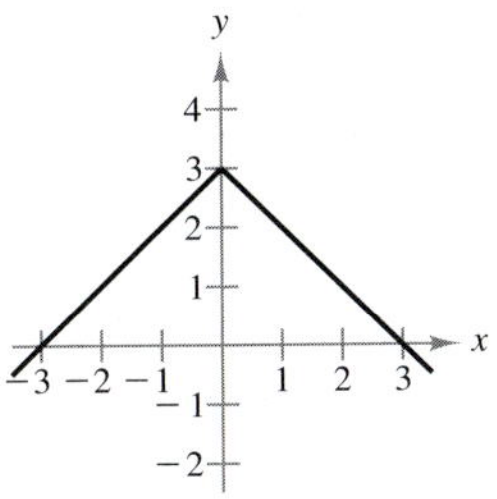

27. $y = \begin{cases} -x - 2, & x \le 0 \\ x^3 + 2, & x > 0 \end{cases}$

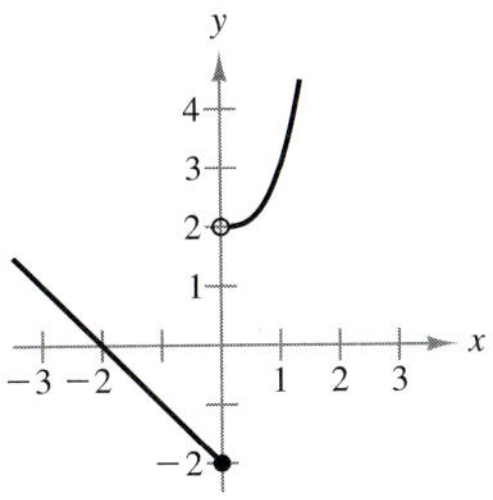

28. $y = (x + 1)^{2/3}$

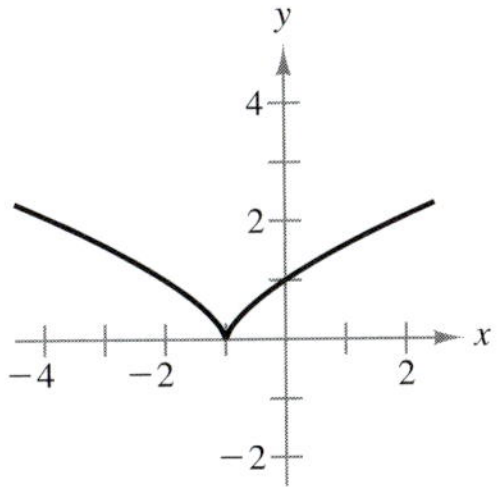

In Exercises 29–38, find the equation of the tangent line at the given point. Then use a graphing utility to graph the function and the equation of the tangent line in the same viewing window.

	Function	*Point*
29.	$g(t) = \dfrac{2}{3t^2}$	$\left(1, \dfrac{2}{3}\right)$
30.	$h(x) = \dfrac{2}{(3x)^2}$	$\left(2, \dfrac{1}{18}\right)$
31.	$f(x) = x^2 + 3$	$(1, 4)$
32.	$f(x) = 2x^2 - 3x + 1$	$(2, 3)$
33.	$y = 11x^4 - 5x^2 + 1$	$(-1, 7)$
34.	$y = x^3 - 5 + \dfrac{3}{x^3}$	$(-1, -9)$
35.	$f(x) = \sqrt{x} - \dfrac{1}{\sqrt{x}}$	$(1, 0)$
36.	$f(x) = 2x^{-3} + 4 - \sqrt{x}$	$(1, 5)$
37.	$f(x) = \dfrac{x^2 + 3}{x}$	$(1, 4)$
38.	$f(x) = -x^2 - 4x - 4$	$(-4, -4)$

In Exercises 39 and 40, find the average rate of change of the function over the indicated interval. Then compare the average rate of change with the instantaneous rates of change at the endpoints of the interval.

39. $f(x) = x^2 + 3x - 4;\quad [0, 1]$

40. $f(x) = x^3 + x;\quad [-2, 2]$

41. ***Sales*** The annual sales S (in millions of dollars per year) of Home Depot for the years 1996 to 2002 can be modeled by

$$S = -172.361t^3 + 4740.75t^2 - 35{,}441.6t + 98{,}831$$

where $t = 6$ corresponds to 1996. A graph of this model appears in Exercise 5. *(Source: The Home Depot, Inc.)*

(a) Find the average rate of change for the interval from 1998 to 2002.

(b) Find the instantaneous rates of change of the model for 1998 and 2002.

(c) Interpret the results of parts (a) and (b) in the context of the problem.

42. ***Consumer Trends*** The numbers of subscribers S (in thousands per year) of cellular telephones for the years 1993 to 2002 can be modeled by

$$S = \frac{1168.2751 + 5366.2569t}{1 - 0.046307t}$$

where $t = 3$ corresponds to 1993. A graph of this model appears in Exercise 6. *(Source: Cellular Telecommunications & Internet Association)*

(a) Find the average rate of change for the interval from 1997 to 2002.

(b) Find the instantaneous rates of change of the model for 1997 and 2002.

(c) Interpret the results of parts (a) and (b) in the context of the problem.

43. *Retail Price* The average retail price P (in dollars) of 1 pound of 100% ground beef from 1996 to 2002 can be modeled by the equation

$$P = -0.001059t^4 + 0.03015t^3 - 0.2850t^2 + 1.007t + 0.50$$

where t is the year, with $t = 6$ corresponding to 1996. *(Source: U.S. Bureau of Labor Statistics)*

(a) Find the rate of change of the price with respect to the year.

(b) At what rate was the price of 100% ground beef changing in 1997? in 2000? in 2002?

(c) Use a graphing utility to graph the function for $6 \le t \le 12$. During which years was the price increasing? decreasing?

(d) For what years do the slopes of the tangent lines appear to be positive? negative?

(e) Compare your answers for parts (c) and (d).

44. *Recycling* The amount T of recycled paper products in millions of tons from 1993 to 2001 can be modeled by the equation

$$T = \sqrt{2.4890t^3 - 62.062t^2 + 553.16t - 509.4}$$

where t is the year, with $t = 3$ corresponding to 1993. *(Source: Franklin Associates, Ltd.)*

(a) Use a graphing utility to graph the equation. Be sure to choose an appropriate window.

(b) Determine dT/dt. Evaluate dT/dt for 1993, 1998, and 2001.

(c) Is dT/dt positive for $t \ge 3$? Does this agree with the graph of the function? What does this tell you about this situation? Explain your reasoning.

45. *Velocity* A rock is dropped from a tower on the Brooklyn Bridge, 276 feet above the East River. Let t represent the time in seconds.

(a) Write a model for the position function (assume that air resistance is negligible).

(b) Find the average velocity during the first 2 seconds.

(c) Find the instantaneous velocities when $t = 2$ and $t = 3$.

(d) How long will it take for the rock to hit the water?

(e) When it hits the water, what is the rock's speed?

46. *Velocity* The straight-line distance s (in feet) traveled by an accelerating bicyclist can be modeled by

$$s = 2t^{3/2}, \quad 0 \le t \le 8$$

where t is the time (in seconds). Complete the table, showing the velocity of the bicyclist at two-second intervals.

Table for 46

Time, t	0	2	4	6	8
Velocity					

47. *Cost, Revenue, and Profit* The fixed cost of operating a small flower shop is \$2500 per month. The average cost of a floral arrangement is \$15 and the average price is \$27.50. Write the monthly revenue, cost, and profit functions for the floral shop in terms of x, the number of arrangements sold.

48. *Profit* The weekly demand and cost functions for a product are given by

$$p = 1.89 - 0.0083x \quad \text{and} \quad C = 21 + 0.65x.$$

Write the profit function for this product.

Marginal Cost In Exercises 49–52, find the marginal cost function.

49. $C = 2500 + 320x$

50. $C = 225x + 4500$

51. $C = 370 + 2.55\sqrt{x}$

52. $C = 475 + 5.25x^{2/3}$

Marginal Revenue In Exercises 53–56, find the marginal revenue function.

53. $R = 200x - \frac{1}{5}x^2$

54. $R = 150x - \frac{3}{4}x^2$

55. $R = \dfrac{35x}{\sqrt{x - 2}}, \quad x \ge 6$

56. $R = x\left(5 + \dfrac{10}{\sqrt{x}}\right)$

Marginal Profit In Exercises 57 and 58, find the marginal profit function.

57. $P = -0.0002x^3 + 6x^2 - x - 2000$

58. $P = -\frac{1}{15}x^3 + 4000x^2 - 120x - 144{,}000$

In Exercises 59–78, find the derivative of the function. Simplify your result.

59. $f(x) = x^3(5 - 3x^2)$

60. $y = (3x^2 + 7)(x^2 - 2x)$

61. $y = (4x - 3)(x^3 - 2x^2)$

62. $s = \left(4 - \dfrac{1}{t^2}\right)(t^2 - 3t)$

63. $f(x) = \dfrac{6x - 5}{x^2 + 1}$

64. $f(x) = \dfrac{x^2 + x - 1}{x^2 - 1}$

65. $f(x) = (5x^2 + 2)^3$

66. $f(x) = \sqrt[3]{x^2 - 1}$

67. $h(x) = \dfrac{2}{\sqrt{x + 1}}$

68. $g(x) = \sqrt{x^6 - 12x^3 + 9}$

69. $g(x) = x\sqrt{x^2 + 1}$

70. $g(t) = \dfrac{t}{(1 - t)^3}$

71. $f(x) = x(1 - 4x^2)^2$

72. $f(x) = \left(x^2 + \dfrac{1}{x}\right)^5$

73. $h(x) = [x^2(2x + 3)]^3$

74. $f(x) = [(x - 2)(x + 4)]^2$

75. $f(x) = x^2(x - 1)^5$

76. $f(s) = s^3(s^2 - 1)^{5/2}$

77. $h(t) = \dfrac{\sqrt{3t + 1}}{(1 - 3t)^2}$

78. $g(x) = \dfrac{(3x + 1)^2}{(x^2 + 1)^2}$

79. ***Physical Science*** The temperature T (in degrees Fahrenheit) of food placed in a freezer can be modeled by

$$T = \frac{1300}{t^2 + 2t + 25}$$

where t is the time (in hours).

(a) Find the rates of change of T when $t = 1, t = 3, t = 5$, and $t = 10$.

(b) Graph the model on a graphing utility and describe the rate at which the temperature is changing.

80. ***Forestry*** According to the *Doyle Log Rule*, the volume V (in board-feet) of a log of length L (feet) and diameter D (inches) at the small end is

$$V = \left(\frac{D - 4}{4}\right)^2 L.$$

Find the rates at which the volume is changing with respect to D for a 12-foot-long log whose smallest diameter is (a) 8 inches, (b) 16 inches, (c) 24 inches, and (d) 36 inches.

In Exercises 81–88, find the given derivative.

81. Given $f(x) = 3x^2 + 7x + 1$, find $f''(x)$.

82. Given $f'(x) = 5x^4 - 6x^2 + 2x$, find $f'''(x)$.

83. Given $f'''(x) = -\dfrac{6}{x^4}$, find $f^{(5)}(x)$.

84. Given $f(x) = \sqrt{x}$, find $f^{(4)}(x)$.

85. Given $f'(x) = 7x^{5/2}$, find $f''(x)$.

86. Given $f(x) = x^2 + \dfrac{3}{x}$, find $f''(x)$.

87. Given $f''(x) = 6\sqrt[3]{x}$, find $f'''(x)$.

88. Given $f'''(x) = 20x^4 - \dfrac{2}{x^3}$, find $f^{(5)}(x)$.

89. ***Athletics*** A person dives from a 30-foot platform with an initial velocity of 5 feet per second (upward).

(a) Find the position function of the diver.

(b) How long will it take for the diver to hit the water?

(c) What is the diver's velocity at impact?

(d) What is the diver's acceleration at impact?

90. ***Velocity and Acceleration*** The position function of a particle is given by

$$s = \frac{1}{t^2 + 2t + 1}$$

where s is the height (in feet) and t is the time (in seconds). Find the velocity and acceleration functions.

In Exercises 91–94, use implicit differentiation to find dy/dx.

91. $x^2 + 3xy + y^3 = 10$

92. $x^2 + 9xy + y^2 = 0$

93. $y^2 - x^2 + 8x - 9y - 1 = 0$

94. $y^2 + x^2 - 6y - 2x - 5 = 0$

In Exercises 95–98, use implicit differentiation to find an equation of the tangent line at the given point.

	Equation	*Point*
95.	$y^2 = x - y$	$(2, 1)$
96.	$2\sqrt[3]{x} + 3\sqrt{y} = 10$	$(8, 4)$
97.	$y^2 - 2x = xy$	$(1, 2)$
98.	$y^3 - 2x^2y + 3xy^2 = -1$	$(0, -1)$

99. ***Water Level*** A swimming pool is 40 feet long, 20 feet wide, 4 feet deep at the shallow end, and 9 feet deep at the deep end (see figure). Water is being pumped into the pool at the rate of 10 cubic feet per minute. How fast is the water level rising when there is 4 feet of water in the deep end?

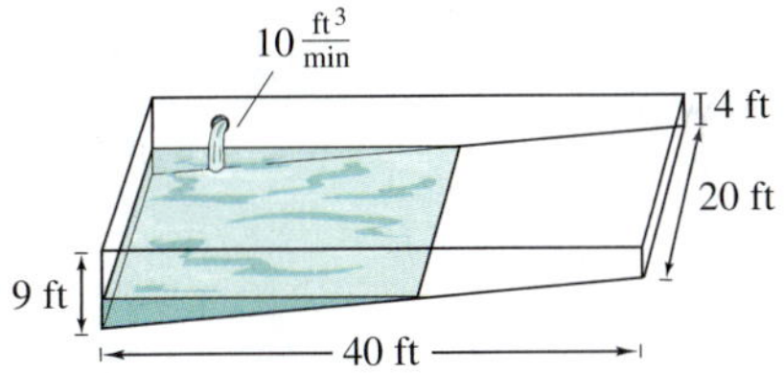

100. ***Profit*** The demand and cost functions for a product can be modeled by

$$p = 211 - 0.002x$$

and

$$C = 30x + 1{,}500{,}000$$

where x is the number of units produced.

(a) Write the profit function for this product.

(b) Find the marginal profit when 80,000 units are produced.

(c) Graph the profit function on a graphing utility and use the graph to determine the price you would charge for the product. Explain your reasoning.

2 SAMPLE POST-GRADUATION EXAM QUESTIONS

CPA
GMAT
GRE
Actuarial
CLAST

The following questions represent the types of questions that appear on certified public accountant (CPA) exams, Graduate Management Admission Tests (GMAT), Graduate Records Exams (GRE), actuarial exams, and College-Level Academic Skills Tests (CLAST). The answers to the questions are given in the back of the book.

Figure for 1

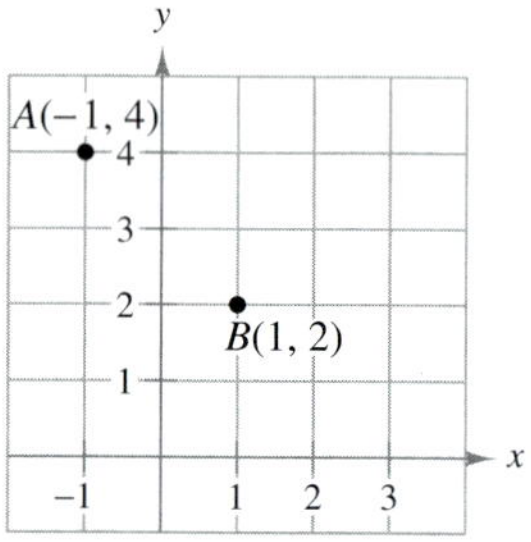

1. What is the length of the line segment that connects A to B (see graph)?

(a) 2 (b) 4 (c) $2\sqrt{2}$ (d) 6 (e) $\sqrt{3}$

For Questions 2–4, refer to the following table. *(Source: U.S. Census Bureau)*

Participation in National Elections (millions of persons)

Characteristic	1992 Persons of voting age	1992 Percent voted	1996 Persons of voting age	1996 Percent voted	2000 Persons of voting age	2000 Percent voted
Total	185.7	61.3	193.7	54.2	202.6	54.7
Male	88.6	60.2	92.6	52.8	97.1	53.1
Female	97.1	62.3	101.0	55.5	105.5	56.2
Age 18 to 20	9.7	38.5	10.8	31.2	11.9	28.4
21 to 24	14.6	33.4	13.9	33.4	14.9	35.4
25 to 34	41.6	53.2	40.1	43.1	37.3	43.7
35 to 44	39.7	63.6	43.3	54.9	44.5	55.0
45 to 64	49.1	70.0	53.7	64.4	61.4	64.1
65 years and over	30.8	70.1	31.9	67.0	32.8	67.6

2. Which of the following groups had the highest percent of voters in 1996?

(a) Male (b) Age 35 to 44 (c) Age 25 to 34

(d) Age 18 to 20 (e) Female

3. In 2000, what percent (to the nearest percent) of persons of voting age were female?

(a) 56 (b) 48 (c) 53 (d) 57 (e) 52

4. In 1992, how many males of voting age voted?

(a) 60,493,300 (b) 111,791,400 (c) 53,337,200

(d) 35,262,800 (e) 48,892,800

For Questions 5 and 6, refer to the following example.
The position s at time t of an object in motion is given by $s(t) = t(t^2 - 2)^2$.

5. The acceleration function is

(a) $5t^4 - 12t^2 + 4$ (b) $5t^4 - 8t$ (c) $20t^3 - 24t$

(d) $20t^3 - 8$ (e) $t^5 - 4t^2 + 4$

6. The velocity of the object at time $t = 3$ is

(a) 301 (b) 381 (c) 147 (d) 532 (e) 468

chapter

3 Applications of the Derivative

Economists use the derivative to measure the increase or decrease in demand for products, such as food, when the price is lowered or raised.

STRATEGIES FOR SUCCESS

WHAT YOU SHOULD LEARN:

- How to find the open intervals on which a function is increasing or decreasing
- How to determine relative and absolute extrema of a function
- How to determine the concavity and points of inflection of a graph
- How to solve real-life optimization problems
- How to determine vertical and horizontal asymptotes of a graph
- How to use calculus to analyze the shape of the graph of a function

WHY YOU SHOULD LEARN IT:

Derivatives have many applications in real life, as can be seen by the examples below, which represent a small sample of the applications in this chapter.

- Chemistry: Molecular Velocity, Exercise 36 on page 180
- Diminishing Returns, Exercises 51 and 52 on page 199
- Price Elasticity, Exercises 33-35 on page 218
- Learning Curve, Exercises 61 and 62 on page 230
- Marginal Analysis, Exercises 27-36 on page 247

3.1 INCREASING AND DECREASING FUNCTIONS

- Test for increasing and decreasing functions.
- Find the critical numbers of functions and find the open intervals on which functions are increasing or decreasing.
- Use increasing and decreasing functions to model and solve real-life problems.

Increasing and Decreasing Functions

A function is **increasing** if its graph moves up as x moves to the right and **decreasing** if its graph moves down as x moves to the right. The following definition states this more formally.

Definition of Increasing and Decreasing Functions

A function f is **increasing** on an interval if for any x_1 and x_2 in the interval

$$x_2 > x_1 \quad \text{implies} \quad f(x_2) > f(x_1).$$

A function f is **decreasing** on an interval if for any x_1 and x_2 in the interval

$$x_2 > x_1 \quad \text{implies} \quad f(x_2) < f(x_1).$$

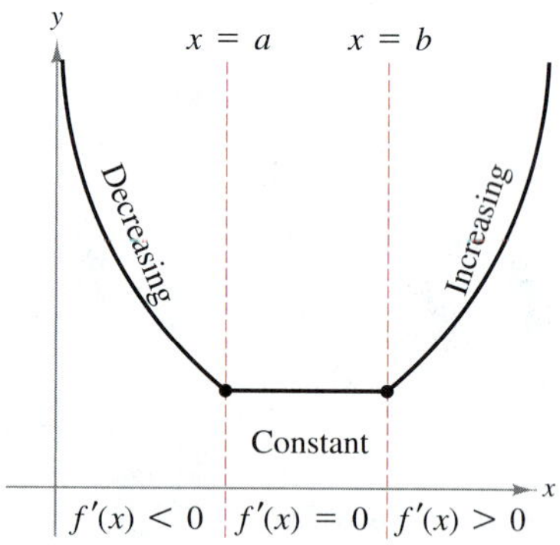

FIGURE 3.1

The function in Figure 3.1 is decreasing on the interval $(-\infty, a)$, constant on the interval (a, b), and increasing on the interval (b, ∞). Actually, from the definition of increasing and decreasing functions, the function shown in Figure 3.1 is decreasing on the interval $(-\infty, a]$ and increasing on the interval $[b, \infty)$. This text restricts the discussion to finding *open* intervals on which a function is increasing or decreasing.

The derivative of a function can be used to determine whether the function is increasing or decreasing on an interval.

Test for Increasing and Decreasing Functions

Let f be differentiable on the interval (a, b).

1. If $f'(x) > 0$ for all x in (a, b), then f is increasing on (a, b).
2. If $f'(x) < 0$ for all x in (a, b), then f is decreasing on (a, b).
3. If $f'(x) = 0$ for all x in (a, b), then f is constant on (a, b).

STUDY TIP

The conclusions in the first two cases of testing for increasing and decreasing functions are valid even if $f'(x) = 0$ at a finite number of x-values in (a, b).

EXAMPLE 1 Testing for Increasing and Decreasing Functions

Show that the function

$$f(x) = x^2$$

is decreasing on the open interval $(-\infty, 0)$ and increasing on the open interval $(0, \infty)$.

SOLUTION The derivative of f is

$$f'(x) = 2x.$$

On the open interval $(-\infty, 0)$, the fact that x is negative implies that $f'(x) = 2x$ is also negative. So, by the test for a decreasing function, you can conclude that f is *decreasing* on this interval. Similarly, on the open interval $(0, \infty)$, the fact that x is positive implies that $f'(x) = 2x$ is also positive. So, it follows that f is *increasing* on this interval, as shown in Figure 3.2.

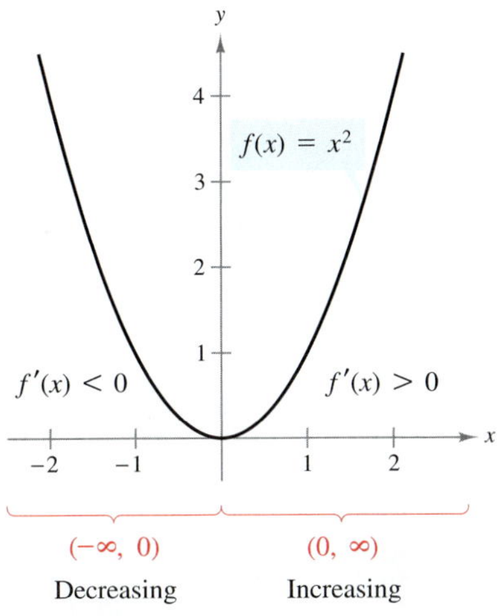

FIGURE 3.2

TRY IT 1

Show that the function $f(x) = x^4$ is decreasing on the open interval $(-\infty, 0)$ and increasing on the open interval $(0, \infty)$.

DISCOVERY

Use a graphing utility to graph $f(x) = 2 - x^2$ and $f'(x) = -2x$ in the same viewing window. On what interval is f increasing? On what interval is f' positive? Describe how the first derivative can be used to determine where a function is increasing and decreasing. Repeat this analysis for $g(x) = x^3 - x$ and $g'(x) = 3x^2 - 1$.

EXAMPLE 2 Modeling Consumption

From 1993 through 2001, the consumption M of mozzarella cheese (in pounds per person per year) can be modeled by

$$M = 0.007t^2 + 0.15t + 7.1, \qquad 3 \le t \le 11$$

where $t = 3$ corresponds to 1993 (see Figure 3.3). Show that the consumption of mozzarella cheese was increasing from 1993 through 2001. *(Source: U.S. Department of Agriculture)*

SOLUTION The derivative of this model is $dM/dt = 0.014t + 0.15$. As long as t is positive, the derivative is also positive. So, the function is increasing, which implies that the consumption of mozzarella cheese was increasing from 1993 through 2001.

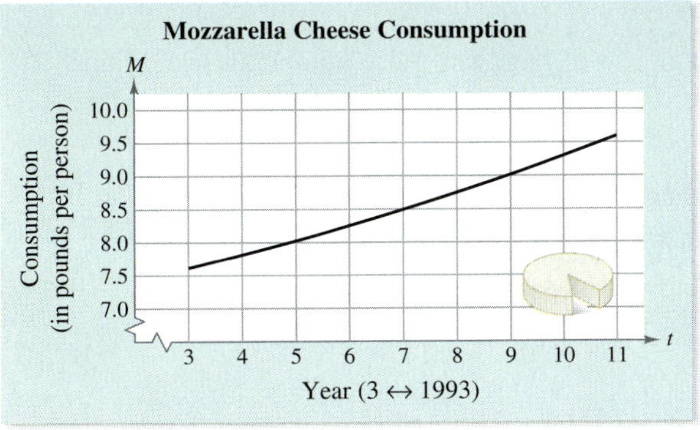

FIGURE 3.3

TRY IT 2

From 1990 through 1999, the consumption of bottled water (in gallons per person per year) can be modeled by

$$W = 0.099t^2 + 0.17t + 7.9, \quad 0 \le t \le 9$$

where $t = 0$ corresponds to 1990. Show that the consumption of bottled water was increasing from 1990 to 1999. *(Source: U.S. Department of Agriculture)*

Critical Numbers and Their Use

In Example 1, you were given two intervals: one on which the function was decreasing and one on which it was increasing. Suppose you had been asked to determine these intervals. To do this, you could have used the fact that for a continuous function, $f'(x)$ can change signs only at x-values where $f'(x) = 0$ or at x-values where $f'(x)$ is undefined, as shown in Figure 3.4. These two types of numbers are called the **critical numbers** of f.

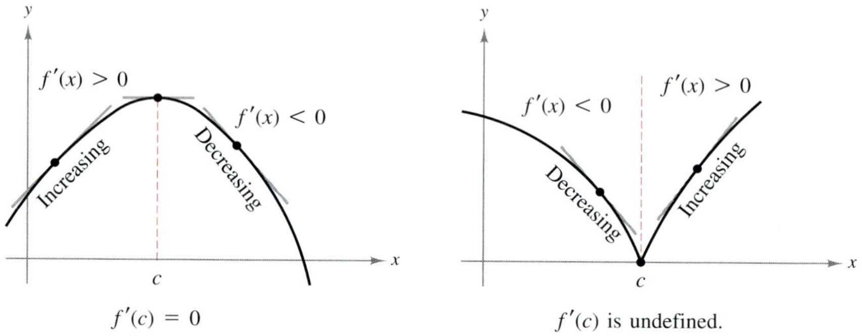

FIGURE 3.4

Definition of Critical Number

If f is defined at c, then c is a critical number of f if $f'(c) = 0$ or if f' is undefined at c.

STUDY TIP

This definition requires that a critical number be in the domain of the function. For example, $x = 0$ is not a critical number of the function $f(x) = 1/x$.

To determine the intervals on which a continuous function is increasing or decreasing, you can use the guidelines below.

Guidelines for Applying Increasing/Decreasing Test

1. Find the derivative of f.
2. Locate the critical numbers of f and use these numbers to determine test intervals. That is, find all x for which $f'(x) = 0$ or $f'(x)$ is undefined.
3. Test the sign of $f'(x)$ at an arbitrary number in each of the test intervals.
4. Use the test for increasing and decreasing functions to decide whether f is increasing or decreasing on each interval.

EXAMPLE 3 Finding Increasing and Decreasing Intervals

Find the open intervals on which the function is increasing or decreasing.

$$f(x) = x^3 - \frac{3}{2}x^2$$

SOLUTION Begin by finding the derivative of f. Then set the derivative equal to zero and solve for the critical numbers.

$$f'(x) = 3x^2 - 3x \quad \text{Differentiate original function.}$$
$$3x^2 - 3x = 0 \quad \text{Set derivative equal to 0.}$$
$$3(x)(x - 1) = 0 \quad \text{Factor.}$$
$$x = 0, x = 1 \quad \text{Critical numbers}$$

Because there are no x-values for which f' is undefined, it follows that $x = 0$ and $x = 1$ are the *only* critical numbers. So, the intervals that need to be tested are $(-\infty, 0)$, $(0, 1)$, and $(1, \infty)$. The table summarizes the testing of these three intervals.

Interval	$-\infty < x < 0$	$0 < x < 1$	$1 < x < \infty$
Test value	$x = -1$	$x = \frac{1}{2}$	$x = 2$
Sign of $f'(x)$	$f'(-1) = 6 > 0$	$f'(\frac{1}{2}) = -\frac{3}{4} < 0$	$f'(2) = 6 > 0$
Conclusion	Increasing	Decreasing	Increasing

The graph of f is shown in Figure 3.5. Note that the test values in the intervals were chosen for convenience—other x-values could have been used.

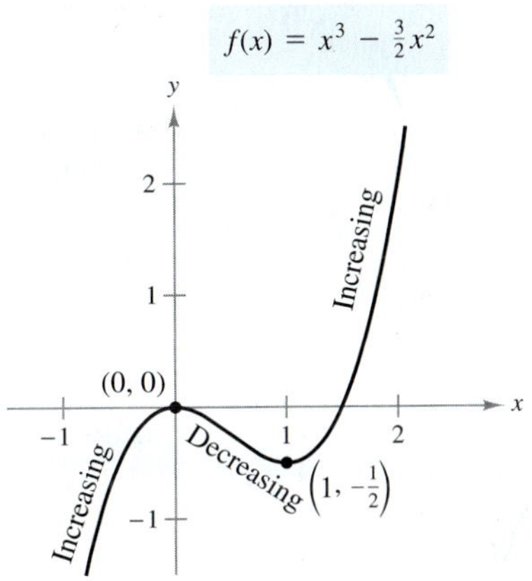

FIGURE 3.5

TRY IT 3

Find the open intervals on which the function $f(x) = x^3 - 12x$ is increasing or decreasing.

TECHNOLOGY

You can use the *trace* feature of a graphing utility to confirm the result of Example 3. Begin by graphing the function, as shown at the right. Then activate the *trace* feature and move the cursor from left to right. In intervals on which the function is increasing, note that the y-values increase as the x-values increase, whereas in intervals on which the function is decreasing, the y-values decrease as the x-values increase.

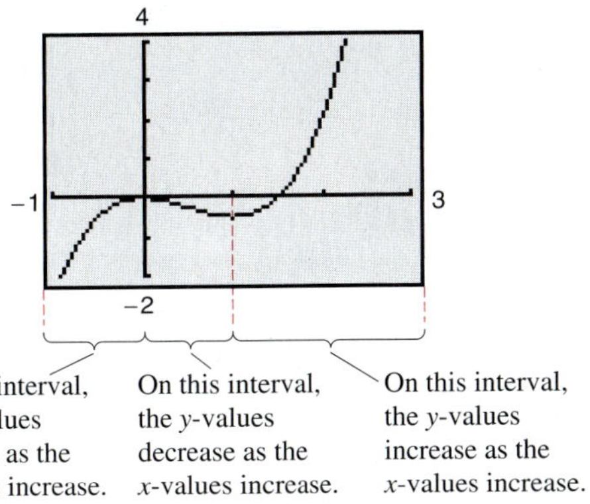

Not only is the function in Example 3 continuous on the entire real line, it is also differentiable there. For such functions, the only critical numbers are those for which $f'(x) = 0$. The next example considers a continuous function that has *both* types of critical numbers—those for which $f'(x) = 0$ and those for which f' is undefined.

ALGEBRA REVIEW

For help on the algebra in Example 4, see Example 2(d) in the *Chapter 3 Algebra Review*, on page 249.

EXAMPLE 4 Finding Increasing and Decreasing Intervals

Find the open intervals on which the function

$$f(x) = (x^2 - 4)^{2/3}$$

is increasing or decreasing.

SOLUTION Begin by finding the derivative of the function.

$$f'(x) = \frac{2}{3}(x^2 - 4)^{-1/3}(2x) \qquad \text{Differentiate.}$$

$$= \frac{4x}{3(x^2 - 4)^{1/3}} \qquad \text{Simplify.}$$

From this, you can see that the derivative is zero when $x = 0$ and the derivative is undefined when $x = \pm 2$. So, the critical numbers are

$$x = -2, \quad x = 0, \quad \text{and} \quad x = 2. \qquad \text{Critical numbers}$$

This implies that the test intervals are

$$(-\infty, -2), \quad (-2, 0), \quad (0, 2), \quad \text{and} \quad (2, \infty). \qquad \text{Test intervals}$$

The table summarizes the testing of these four intervals, and the graph of the function is shown in Figure 3.6.

Interval	$-\infty < x < -2$	$-2 < x < 0$	$0 < x < 2$	$2 < x < \infty$
Test value	$x = -3$	$x = -1$	$x = 1$	$x = 3$
Sign of $f'(x)$	$f'(-3) < 0$	$f'(-1) > 0$	$f'(1) < 0$	$f'(3) > 0$
Conclusion	Decreasing	Increasing	Decreasing	Increasing

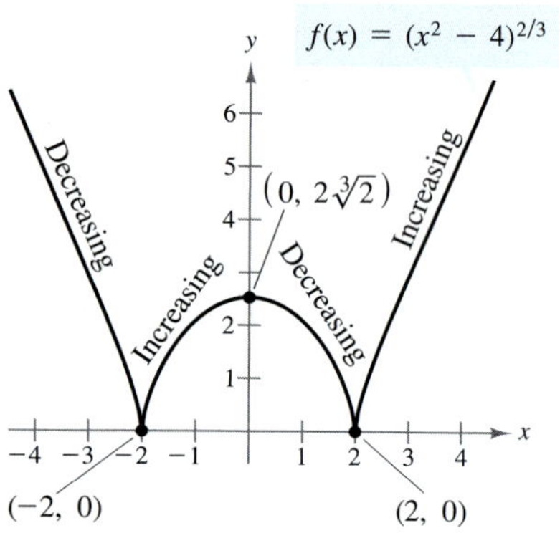

FIGURE 3.6

TRY IT 4

Find the open intervals on which the function $f(x) = x^{2/3}$ is increasing or decreasing.

ALGEBRA REVIEW

To test the intervals in the table, it is not necessary to *evaluate* $f'(x)$ at each test value—you only need to determine its sign. For example, you can determine the sign of $f'(-3)$ as shown.

$$f'(-3) = \frac{4(-3)}{3(9 - 4)^{1/3}} = \frac{\text{negative}}{\text{positive}} = \text{negative}$$

The functions in Examples 1 through 4 are continuous on the entire real line. If there are isolated x-values at which a function is not continuous, then these x-values should be used along with the critical numbers to determine the test intervals. For example, the function

$$f(x) = \frac{x^4 + 1}{x^2}$$

is not continuous when $x = 0$. Because the derivative of f

$$f'(x) = \frac{2(x^4 - 1)}{x^3}$$

is zero when $x = \pm 1$, you should use the following numbers to determine the test intervals.

$x = -1, x = 1$ Critical numbers

$x = 0$ Discontinuity

After testing $f'(x)$, you can determine that the function is decreasing on the intervals $(-\infty, -1)$ and $(0, 1)$, and increasing on the intervals $(-1, 0)$ and $(1, \infty)$, as shown in Figure 3.7.

The converse of the test for increasing and decreasing functions is *not* true. For instance, it is possible for a function to be increasing on an interval even though its derivative is not positive at every point in the interval.

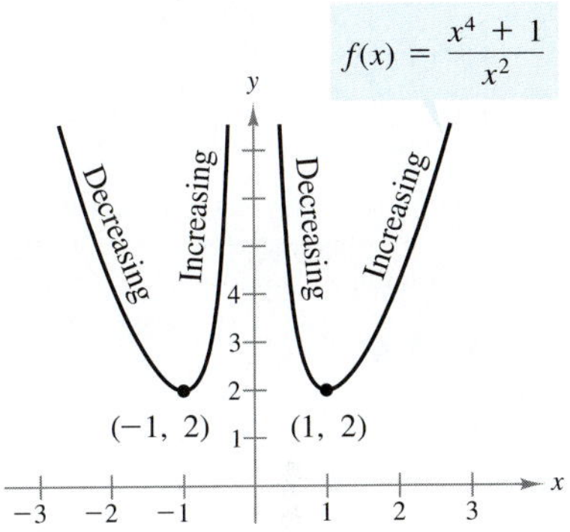

FIGURE 3.7

EXAMPLE 5 Testing an Increasing Function

Show that

$$f(x) = x^3 - 3x^2 + 3x$$

is increasing on the entire real line.

SOLUTION From the derivative of f

$$f'(x) = 3x^2 - 6x + 3 = 3(x - 1)^2$$

you can see that the only critical number is $x = 1$. So, the test intervals are $(-\infty, 1)$ and $(1, \infty)$. The table summarizes the testing of these two intervals. From Figure 3.8, you can see that f is increasing on the entire real line, even though $f'(1) = 0$. To convince yourself of this, look back at the definition of an increasing function.

Interval	$-\infty < x < 1$	$1 < x < \infty$
Test value	$x = 0$	$x = 2$
Sign of $f'(x)$	$f'(0) = 3(-1)^2 > 0$	$f'(2) = 3(1)^2 > 0$
Conclusion	Increasing	Increasing

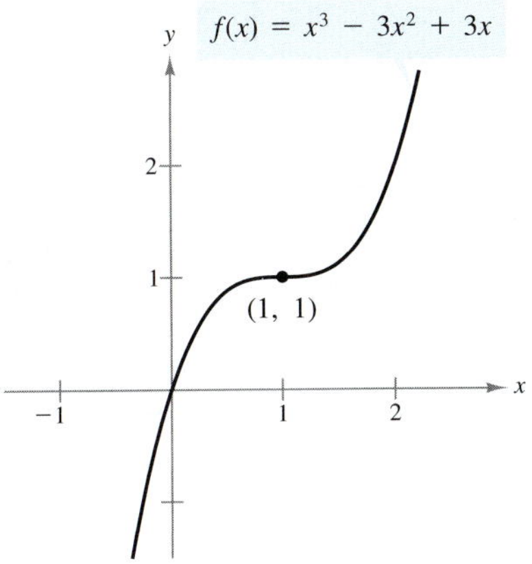

FIGURE 3.8

TRY IT 5

Show that $f(x) = -x^3 + 2$ is decreasing on the entire real line.

Application

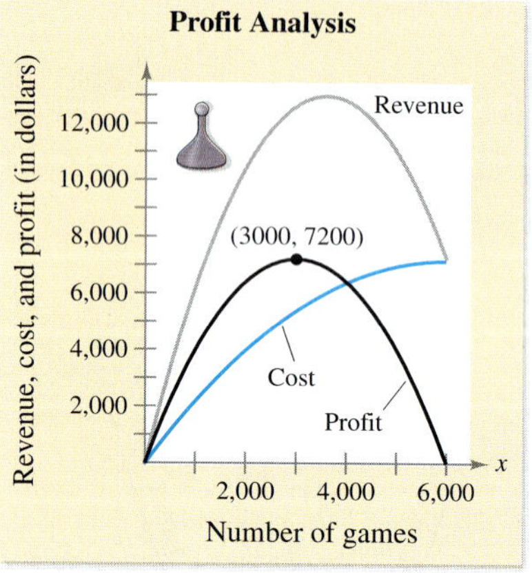

FIGURE 3.9

TRY IT 6

A national distributor of pet toys determines the cost and revenue functions for one of its toys.

$$C = 1.2x - 0.0001x^2, \quad 0 \le x \le 6000$$

$$R = 3.6x - 0.0005x^2, \quad 0 \le x \le 6000$$

Determine the interval on which the profit function is increasing.

EXAMPLE 6 Profit Analysis

A national toy distributor determines the cost and revenue models for one of its games.

$$C = 2.4x - 0.0002x^2, \qquad 0 \le x \le 6000$$

$$R = 7.2x - 0.001x^2, \qquad 0 \le x \le 6000$$

Determine the interval on which the profit function is increasing.

SOLUTION The profit for producing x games is

$$\begin{aligned} P &= R - C \\ &= (7.2x - 0.001x^2) - (2.4x - 0.0002x^2) \\ &= 4.8x - 0.0008x^2. \end{aligned}$$

To find the interval on which the profit is increasing, set the marginal profit P' equal to zero and solve for x.

$$P' = 4.8 - 0.0016x \qquad \text{Differentiate profit function.}$$

$$4.8 - 0.0016x = 0 \qquad \text{Set } P' \text{ equal to 0.}$$

$$-0.0016x = -4.8 \qquad \text{Subtract 4.8 from each side.}$$

$$x = \frac{-4.8}{-0.0016} \qquad \text{Divide each side by } -0.0016.$$

$$x = 3000 \text{ games} \qquad \text{Simplify.}$$

On the interval (0, 3000), P' is positive and the profit is *increasing*. On the interval (3000, 6000), P' is negative and the profit is *decreasing*. The graphs of the cost, revenue, and profit functions are shown in Figure 3.9.

TAKE ANOTHER LOOK

Comparing Cost, Revenue, and Profit

Use the models from Example 6 to answer the questions.

a. What is the demand function for the product described in the example?

b. What price would you set to obtain a maximum profit?

c. What price would you set to obtain a maximum revenue?

d. Why doesn't the maximum revenue occur at the same x-value as the maximum profit?

PREREQUISITE REVIEW 3.1

The following warm-up exercises involve skills that were covered in earlier sections. You will use these skills in the exercise set for this section.

In Exercises 1–4, solve the equation.

1. $x^2 = 8x$

2. $15x = \frac{5}{8}x^2$

3. $\frac{x^2 - 25}{x^3} = 0$

4. $\frac{2x}{\sqrt{1 - x^2}} = 0$

In Exercises 5–8, find the domain of the expression.

5. $\frac{x + 3}{x - 3}$

6. $\frac{2}{\sqrt{1 - x}}$

7. $\frac{2x + 1}{x^2 - 3x - 10}$

8. $\frac{3x}{\sqrt{9 - 3x^2}}$

In Exercises 9–12, evaluate the expression when $x = -2$, 0, and 2.

9. $-2(x + 1)(x - 1)$

10. $4(2x + 1)(2x - 1)$

11. $\frac{2x + 1}{(x - 1)^2}$

12. $\frac{-2(x + 1)}{(x - 4)^2}$

EXERCISES 3.1

In Exercises 1–4, evaluate the derivative of the function at the indicated points on the graph.

1. $f(x) = \frac{x^2}{x^2 + 4}$

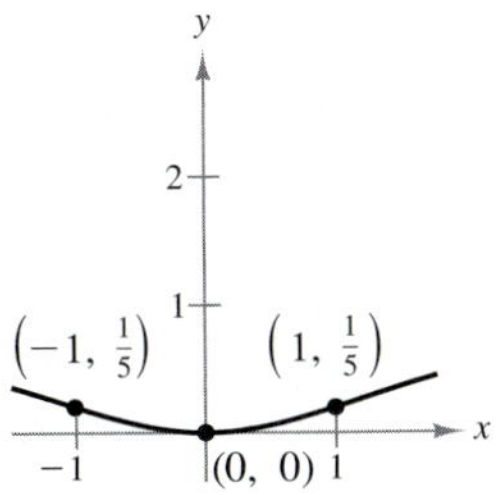

2. $f(x) = x + \frac{32}{x^2}$

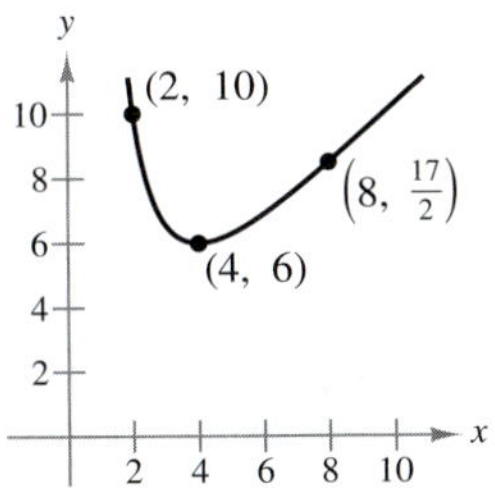

3. $f(x) = (x + 2)^{2/3}$

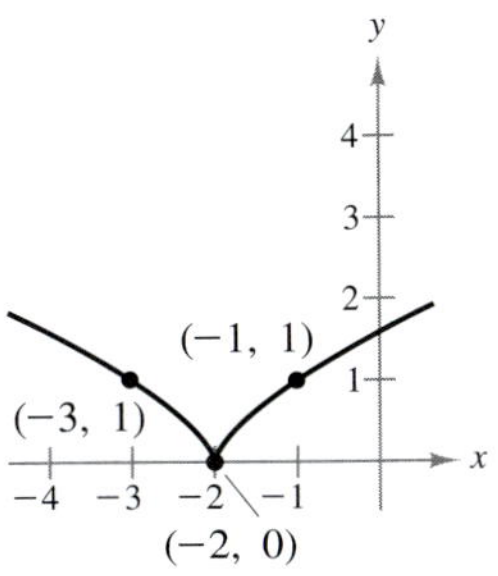

4. $f(x) = -3x\sqrt{x + 1}$

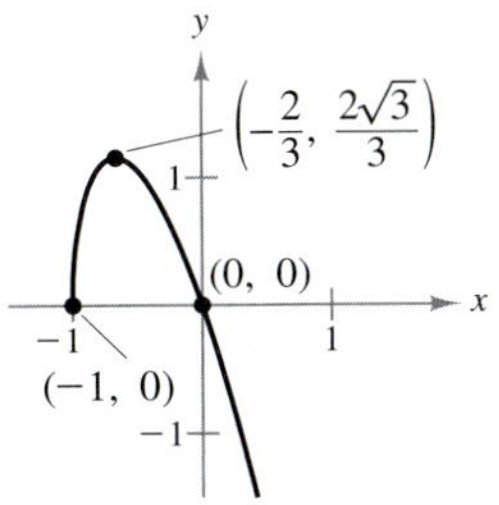

In Exercises 5–8, use the derivative to identify the open intervals on which the function is increasing or decreasing. Verify your result with the graph of the function.

5. $f(x) = -(x + 1)^2$

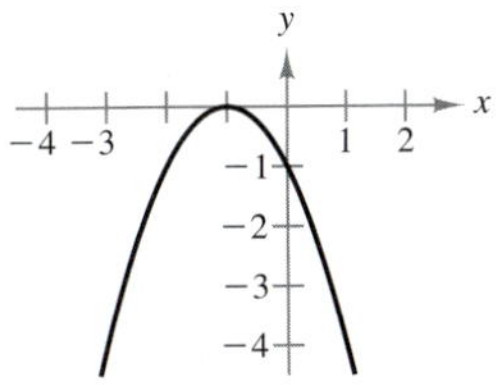

6. $f(x) = \frac{x^3}{4} - 3x$

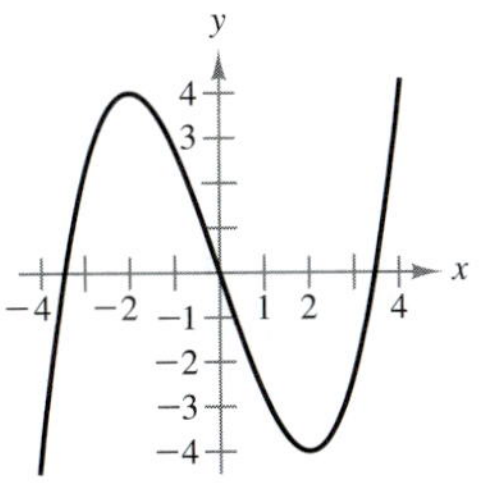

7. $f(x) = x^4 - 2x^2$

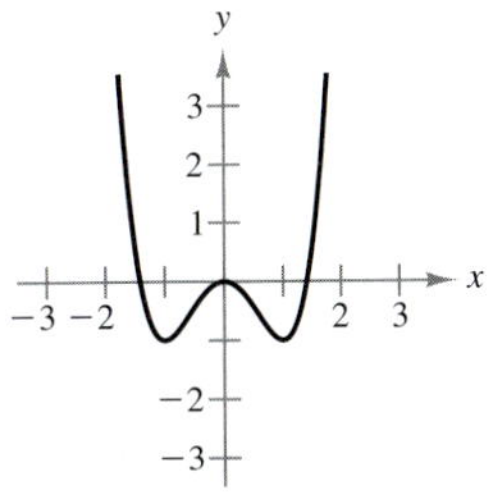

8. $f(x) = \frac{x^2}{x + 1}$

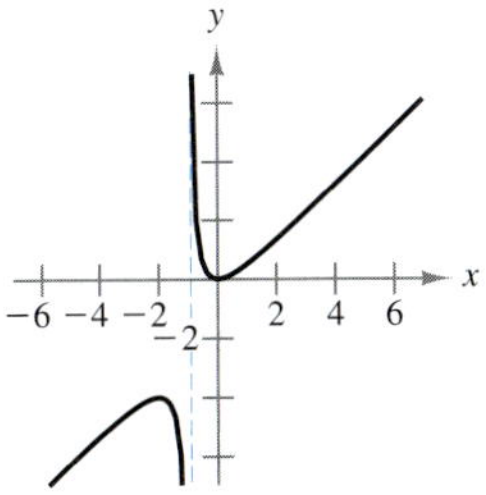

In Exercises 9–18, find the critical numbers and the open intervals on which the function is increasing or decreasing. Sketch the graph of the function.

9. $f(x) = 2x - 3$

10. $f(x) = 5 - 3x$

11. $g(x) = -(x - 1)^2$

12. $g(x) = (x + 2)^2$

13. $y = x^2 - 5x$

14. $y = -x^2 + 2x$

15. $y = x^3 - 6x^2$

16. $y = (x - 2)^3$

17. $f(x) = \sqrt{x^2 - 1}$

18. $f(x) = \sqrt{4 - x^2}$

In Exercises 19–28, find the critical numbers and the open intervals on which the function is increasing or decreasing. Then use a graphing utility to graph the function.

19. $f(x) = -2x^2 + 4x + 3$

20. $f(x) = x^2 + 8x + 10$

21. $y = 3x^3 + 12x^2 + 15x$

22. $y = x^3 - 3x + 2$

23. $f(x) = x\sqrt{x + 1}$

24. $h(x) = x\sqrt[3]{x - 1}$

25. $f(x) = x^4 - 2x^3$

26. $f(x) = \frac{1}{4}x^4 - 2x^2$

27. $f(x) = \dfrac{x}{x^2 + 4}$

28. $f(x) = \dfrac{x^2}{x^2 + 4}$

In Exercises 29–34, find the critical numbers and the open intervals on which the function is increasing or decreasing. (*Hint:* Check for discontinuities.) Sketch the graph of the function.

29. $f(x) = \dfrac{2x}{16 - x^2}$

30. $f(x) = \dfrac{x}{x + 1}$

31. $y = \begin{cases} 4 - x^2, & x \le 0 \\ -2x, & x > 0 \end{cases}$

32. $y = \begin{cases} 2x + 1, & x \le -1 \\ x^2 - 2, & x > -1 \end{cases}$

33. $y = \begin{cases} 3x + 1, & x \le 1 \\ 5 - x^2, & x > 1 \end{cases}$

34. $y = \begin{cases} -x^3 + 1, & x \le 0 \\ -x^2 + 2x, & x > 0 \end{cases}$

35. ***Cost*** The ordering and transportation cost C (in hundreds of dollars) for an automobile dealership is modeled by

$$C = 10\left(\frac{1}{x} + \frac{x}{x + 3}\right), \quad 1 \le x$$

where x is the number of automobiles ordered.

(a) Find the intervals on which C is increasing or decreasing.

(b) Use a graphing utility to graph the cost function.

(c) Use the *trace* feature to determine the order sizes for which the cost is \$900. Assuming that the revenue function is increasing for $x \ge 0$, which order size would you use? Explain your reasoning.

36. ***Chemistry: Molecular Velocity*** Plots of the relative numbers of N_2 (nitrogen) molecules that have a given velocity at each of three temperatures (in degrees Kelvin) are shown in the figure. Identify the differences in the average velocities (indicated by the peaks of the curves) for the three temperatures, and describe the intervals on which the velocity is increasing and decreasing for each of the three temperatures. *(Source: Adapted from Zumdahl, Chemistry, Sixth Edition)*

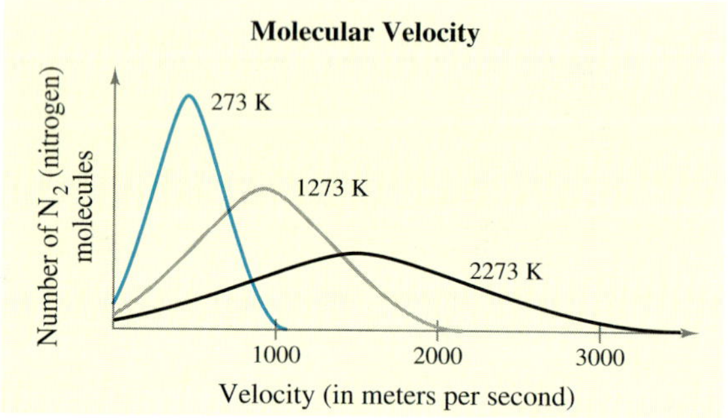

Position Function In Exercises 37 and 38, the position function gives the height s (in feet) of a ball, where the time t is measured in seconds. Find the time interval on which the ball is rising and the interval on which it is falling.

37. $s = 96t - 16t^2, \quad 0 \le t \le 6$

38. $s = -16t^2 + 64t, \quad 0 \le t \le 4$

39. ***Law Degrees*** The number y of law degrees conferred in the United States from 1970 to 2000 can be modeled by

$$y = 2.743t^3 - 171.55t^2 + 3462.3t + 15{,}265, \quad 0 \le t \le 30$$

where t is the time in years, with $t = 0$ corresponding to 1970. *(Source: U.S. National Center for Education Statistics)*

(a) Use a graphing utility to graph the model. Then graphically estimate the years during which the model is increasing and the years during which it is decreasing.

(b) Use the test for increasing and decreasing functions to verify the result of part (a).

40. ***Profit*** The profit P made by a cinema from selling x bags of popcorn can be modeled by

$$P = 2.36x - \frac{x^2}{25{,}000} - 3500, \quad 0 \le x \le 50{,}000.$$

(a) Find the intervals on which P is increasing and decreasing.

(b) If you owned the cinema, what price would you charge to obtain a maximum profit for popcorn? Explain your reasoning.

3.2 EXTREMA AND THE FIRST-DERIVATIVE TEST

- Recognize the occurrence of relative extrema of functions.
- Use the First-Derivative Test to find the relative extrema of functions.
- Find absolute extrema of continuous functions on a closed interval.
- Find minimum and maximum values of real-life models and interpret the results in context.

Relative Extrema

You have used the derivative to determine the intervals on which a function is increasing or decreasing. In this section, you will examine the points at which a function changes from increasing to decreasing, or vice versa. At such a point, the function has a **relative extremum.** (The plural of extremum is *extrema.*) The **relative extrema** of a function include the **relative minima** and **relative maxima** of the function. For instance, the function shown in Figure 3.10 has two relative extrema—the left point is a relative maximum and the right point is a relative minimum.

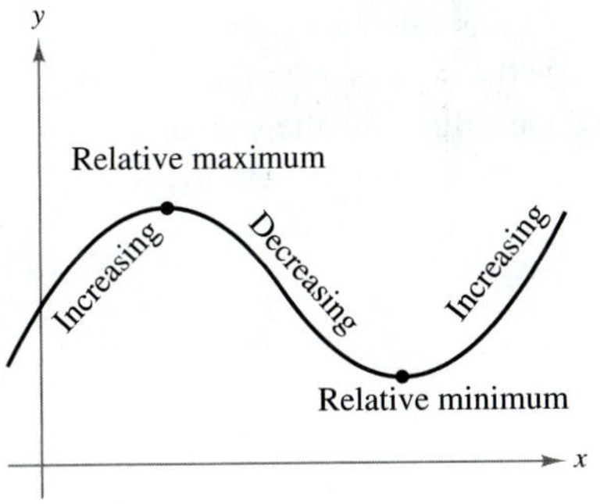

FIGURE 3.10

Definition of Relative Extrema

Let f be a function defined at c.

1. $f(c)$ is a **relative maximum** of f if there exists an interval (a, b) containing c such that $f(x) \leq f(c)$ for all x in (a, b).
2. $f(c)$ is a **relative minimum** of f if there exists an interval (a, b) containing c such that $f(x) \geq f(c)$ for all x in (a, b).

If $f(c)$ is a relative extremum of f, then the relative extremum is said to *occur* at $x = c$.

For a continuous function, the relative extrema must occur at critical numbers of the function, as shown in Figure 3.11.

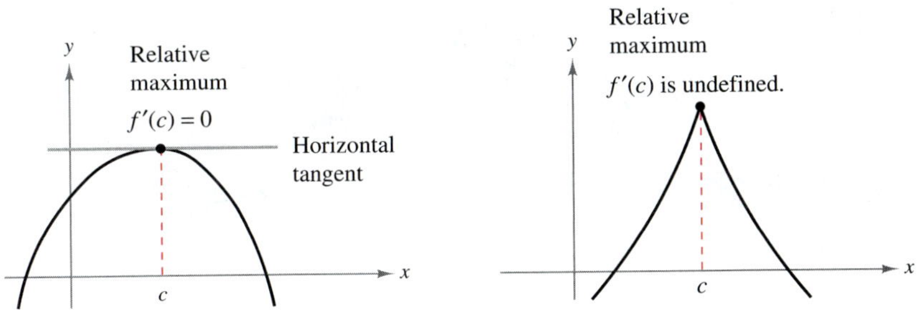

FIGURE 3.11

Occurrence of Relative Extrema

If f has a relative minimum or relative maximum when $x = c$, then c is a critical number of f. That is, either $f'(c) = 0$ or $f'(c)$ is undefined.

The First-Derivative Test

DISCOVERY

Use a graphing utility to graph the function $f(x) = x^2$ and its first derivative $f'(x) = 2x$ in the same viewing window. Where does f have a relative minimum? What is the sign of f' to the left of this relative minimum? What is the sign of f' to the right? Describe how the sign of f' can be used to determine the relative extrema of a function.

The discussion on the preceding page implies that in your search for relative extrema of a continuous function, you only need to test the critical numbers of the function. Once you have determined that c is a critical number of a function f, the **First-Derivative Test** for relative extrema enables you to classify $f(c)$ as a relative minimum, a relative maximum, or neither.

First-Derivative Test for Relative Extrema

Let f be continuous on the interval (a, b) in which c is the only critical number. If f is differentiable on the interval (except possibly at c), then $f(c)$ can be classified as a relative minimum, a relative maximum, or neither, as shown.

1. On the interval (a, b), if $f'(x)$ is negative to the left of $x = c$ and positive to the right of $x = c$, then $f(c)$ is a relative minimum.
2. On the interval (a, b), if $f'(x)$ is positive to the left of $x = c$ and negative to the right of $x = c$, then $f(c)$ is a relative maximum.
3. On the interval (a, b), if $f'(x)$ has the same sign to the left and right of $x = c$, then $f(c)$ is not a relative extremum of f.

A graphical interpretation of the First-Derivative Test is shown in Figure 3.12.

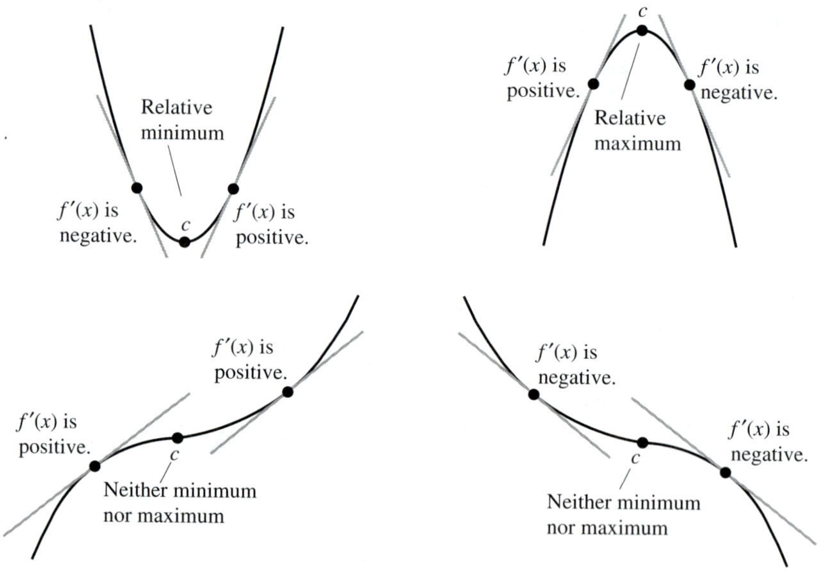

FIGURE 3.12

EXAMPLE 1 Finding Relative Extrema

Find all relative extrema of the function

$$f(x) = 2x^3 - 3x^2 - 36x + 14.$$

SOLUTION Begin by finding the critical numbers of f.

$f'(x) = 6x^2 - 6x - 36$ — Find derivative of f.

$6x^2 - 6x - 36 = 0$ — Set derivative equal to 0.

$6(x^2 - x - 6) = 0$ — Factor out common factor.

$6(x - 3)(x + 2) = 0$ — Factor.

$x = -2, x = 3$ — Critical numbers

Because $f'(x)$ is defined for all x, the only critical numbers of f are $x = -2$ and $x = 3$. Using these numbers, you can form the three test intervals $(-\infty, -2)$, $(-2, 3)$, and $(3, \infty)$. The testing of the three intervals is shown in the table.

Interval	$-\infty < x < -2$	$-2 < x < 3$	$3 < x < \infty$
Test value	$x = -3$	$x = 0$	$x = 4$
Sign of $f'(x)$	$f'(-3) = 36 > 0$	$f'(0) = -36 < 0$	$f'(4) = 36 > 0$
Conclusion	Increasing	Decreasing	Increasing

Using the First-Derivative Test, you can conclude that the critical number -2 yields a relative maximum [$f'(x)$ changes sign from positive to negative], and the critical number 3 yields a relative minimum [$f'(x)$ changes sign from negative to positive].

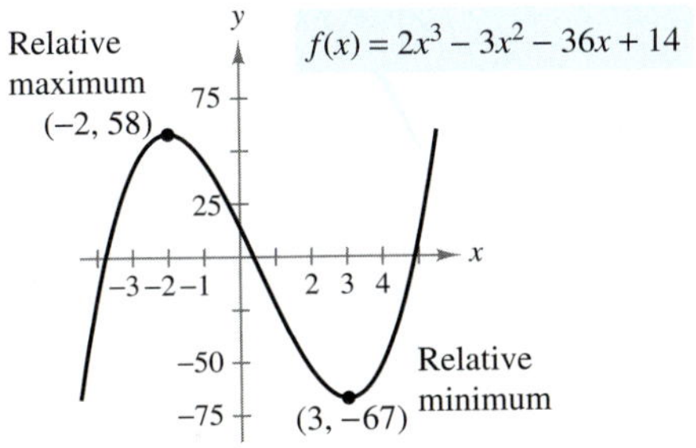

FIGURE 3.13

The graph of f is shown in Figure 3.13. To find the y-coordinates of the relative extrema, substitute the x-coordinates into the function. For instance, the relative maximum is $f(-2) = 58$ and the relative minimum is $f(3) = -67$.

STUDY TIP

In Section 2.2, Example 8, you examined the graph of the function $f(x) = x^3 - 4x + 2$ and discovered that it does *not* have a relative minimum at the point $(1, -1)$. Try using the First-Derivative Test to find the point at which the graph *does* have a relative minimum.

TRY IT 1

Find all relative extrema of $f(x) = 2x^3 - 6x + 1$. Sketch a graph of the function and label the relative extrema.

In Example 1, both critical numbers yielded relative extrema. In the next example, only one of the two critical numbers yields a relative extremum.

ALGEBRA REVIEW

For help on the algebra in Example 2, see Example 2(c) in the *Chapter 3 Algebra Review*, on page 249.

EXAMPLE 2 Finding Relative Extrema

Find all relative extrema of the function $f(x) = x^4 - x^3$.

SOLUTION From the derivative of the function

$$f'(x) = 4x^3 - 3x^2 = x^2(4x - 3)$$

you can see that the function has only two critical numbers: $x = 0$ and $x = \frac{3}{4}$. These numbers produce the test intervals $(-\infty, 0)$, $\left(0, \frac{3}{4}\right)$, and $\left(\frac{3}{4}, \infty\right)$, which are tested in the table.

Interval	$-\infty < x < 0$	$0 < x < \frac{3}{4}$	$\frac{3}{4} < x < \infty$
Test value	$x = -1$	$x = \frac{1}{2}$	$x = 1$
Sign of $f'(x)$	$f'(-1) = -7 < 0$	$f'\left(\frac{1}{2}\right) = -\frac{1}{4} < 0$	$f'(1) = 1 > 0$
Conclusion	Decreasing	Decreasing	Increasing

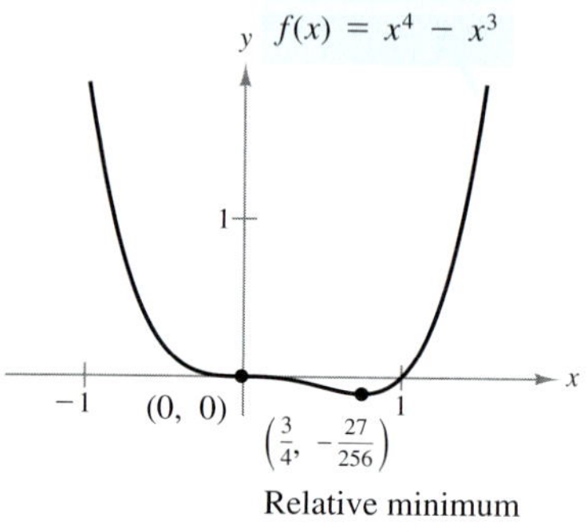

FIGURE 3.14

By the First-Derivative Test, it follows that f has a relative minimum when $x = \frac{3}{4}$, as shown in Figure 3.14. Note that the critical number $x = 0$ does not yield a relative extremum.

TRY IT 2

Find all relative extrema of $f(x) = x^4 - 4x^3$.

EXAMPLE 3 Finding Relative Extrema

Find all relative extrema of the function

$$f(x) = 2x - 3x^{2/3}.$$

SOLUTION From the derivative of the function

$$f'(x) = 2 - \frac{2}{x^{1/3}} = \frac{2(x^{1/3} - 1)}{x^{1/3}}$$

you can see that $f'(1) = 0$ and f' is undefined at $x = 0$. So, the function has two critical numbers: $x = 1$ and $x = 0$. These numbers produce the test intervals $(-\infty, 0)$, $(0, 1)$, and $(1, \infty)$. By testing these intervals, you can conclude that f has a relative maximum at $(0, 0)$ and a relative minimum at $(1, -1)$, as shown in Figure 3.15.

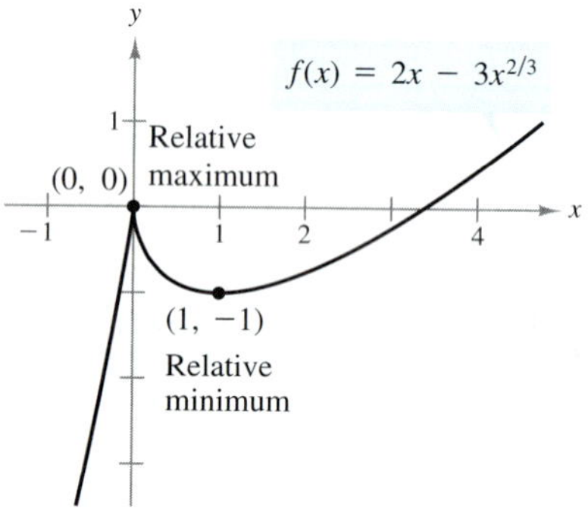

FIGURE 3.15

TRY IT 3

Find all relative extrema of $f(x) = 3x^{2/3} - 2x$.

TECHNOLOGY

Finding Relative Extrema

There are several ways to use technology to find relative extrema of a function. One way is to use a graphing utility to graph the function, and then use the *zoom* and *trace* features to find the relative minimum and relative maximum points. For instance, consider the graph of

$$f(x) = 3.1x^3 - 7.3x^2 + 1.2x + 2.5,$$

as shown below.

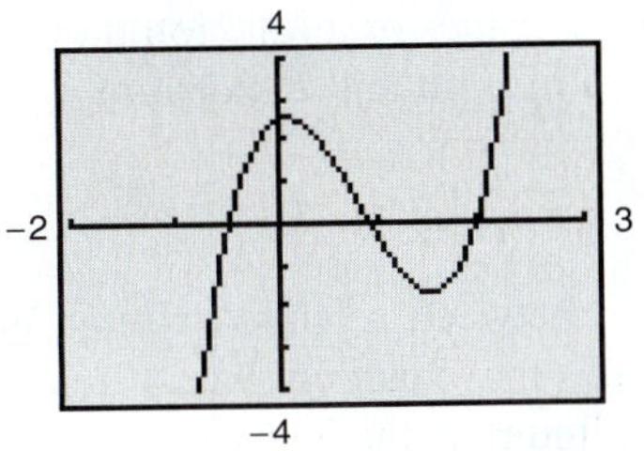

From the graph, you can see that the function has one relative maximum and one relative minimum. You can approximate the coordinates of these points by zooming in and using the *trace* feature, as shown below.

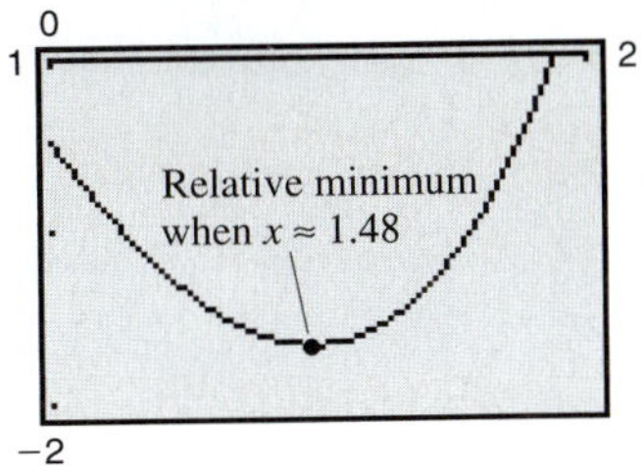

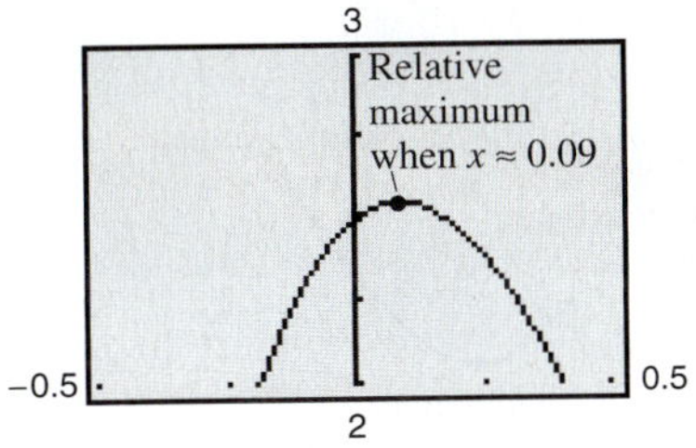

A second way to use technology to find relative extrema is to perform the First-Derivative Test with a symbolic differentiation utility. You can use the utility to differentiate the function, set the derivative equal to zero, and then solve the resulting equation. After obtaining the critical numbers, 1.48287 and 0.0870148, you can graph the function and observe that the first yields a relative minimum and the second yields a relative maximum.

$$f(x) = 3.1x^3 - 7.3x^2 + 1.2x + 2.5 \qquad \text{Write original function.}$$

$$f'(x) = \frac{d}{dx}[3.1x^3 - 7.3x^2 + 1.2x + 2.5] \qquad \text{Differentiate with respect to } x.$$

$$f'(x) = 9.3x^2 - 14.6x + 1.2 \qquad \text{First derivative}$$

$$9.3x^2 - 14.6x + 1.2 = 0 \qquad \text{Set derivative equal to 0.}$$

$$x = \frac{73 \pm \sqrt{4213}}{93} \qquad \text{Solve for } x.$$

$$x \approx 1.48288,\ x \approx 0.0870148 \qquad \text{Approximate.}$$

STUDY TIP

Some graphing calculators have a special feature that allows you to find the minimum or maximum of a function on an interval. Consult the user's manual for information on the *minimum value* and *maximum value* features of your graphing utility.

Absolute Extrema

The terms *relative minimum* and *relative maximum* describe the *local* behavior of a function. To describe the *global* behavior of the function on an entire interval, you can use the terms **absolute maximum** and **absolute minimum.**

Definition of Absolute Extrema

Let f be defined on an interval I containing c.

1. $f(c)$ is an **absolute minimum of f** on I if $f(c) \leq f(x)$ for every x in I.
2. $f(c)$ is an **absolute maximum of f** on I if $f(c) \geq f(x)$ for every x in I.

The absolute minimum and absolute maximum values of a function on an interval are sometimes simply called the **minimum** and **maximum** of f on I.

Be sure that you understand the distinction between relative extrema and absolute extrema. For instance, in Figure 3.16, the function has a relative minimum that also happens to be an absolute minimum on the interval $[a, b]$. The relative maximum of f, however, is not the absolute maximum on the interval $[a, b]$. The next theorem points out that if a continuous function has a closed interval as its domain, then it *must* have both an absolute minimum and an absolute maximum on the interval. From Figure 3.16, note that these extrema can occur at endpoints of the interval.

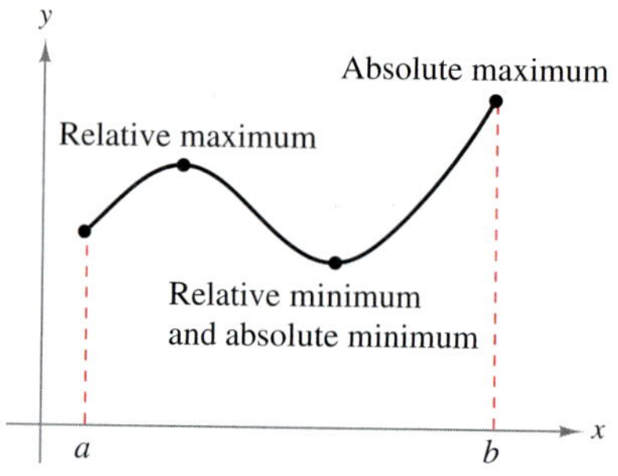

FIGURE 3.16

Extreme Value Theorem

If f is continuous on $[a, b]$, then f has both a minimum value and a maximum value on $[a, b]$.

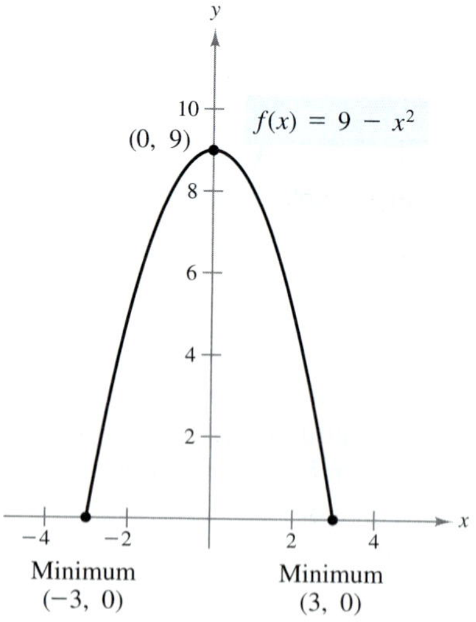

FIGURE 3.17

Although a continuous function has just one minimum and one maximum value on a closed interval, either of these values can occur for more than one x-value. For instance, on the interval $[-3, 3]$, the function $f(x) = 9 - x^2$ has a minimum value of zero when $x = -3$ *and* when $x = 3$, as shown in Figure 3.17.

When looking for the extreme value of a function on a *closed* interval, remember that you must consider the values of the function at the endpoints as well as at the critical numbers of the function. You can use the guidelines below to find extrema on a closed interval.

TECHNOLOGY

A graphing utility can help you locate the extrema of a function on a closed interval. For instance, try using a graphing utility to confirm the results of Example 4. (Set the viewing window to $-1 \le x \le 6$ and $-8 \le y \le 4$.) Use the *trace* feature to check that the minimum y-value occurs when $x = 3$ and the maximum y-value occurs when $x = 0$.

Guidelines for Finding Extrema on a Closed Interval

To find the extrema of a continuous function f on a closed interval $[a, b]$, use the steps below.

1. Evaluate f at each of its critical numbers in (a, b).
2. Evaluate f at each endpoint, a and b.
3. The least of these values is the minimum, and the greatest is the maximum.

EXAMPLE 4 Finding Extrema on a Closed Interval

Find the minimum and maximum values of

$$f(x) = x^2 - 6x + 2$$

on the interval $[0, 5]$.

SOLUTION Begin by finding the critical numbers of the function.

$$f'(x) = 2x - 6 \qquad \text{Find derivative of } f.$$
$$2x - 6 = 0 \qquad \text{Set derivative equal to 0.}$$
$$2x = 6 \qquad \text{Add 6 to each side.}$$
$$x = 3 \qquad \text{Solve for } x.$$

From this, you can see that the only critical number of f is $x = 3$. Because this number lies in the interval under question, you should test the values of $f(x)$ at this number *and* at the endpoints of the interval, as shown in the table.

x-value	Endpoint: $x = 0$	Critical number: $x = 3$	Endpoint: $x = 5$
$f(x)$	$f(0) = 2$	$f(3) = -7$	$f(5) = -3$
Conclusion	Maximum	Minimum	Neither maximum nor minimum

From the table, you can see that the minimum of f on the interval $[0, 5]$ is $f(3) = -7$. Moreover, the maximum of f on the interval $[0, 5]$ is $f(0) = 2$. This is confirmed by the graph of f, as shown in Figure 3.18.

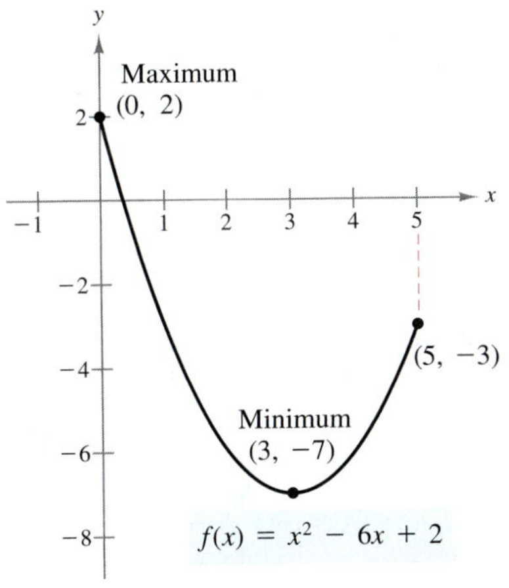

FIGURE 3.18

TRY IT 4

Find the minimum and maximum values of $f(x) = x^2 - 8x + 10$ on the interval $[0, 7]$. Sketch the graph of $f(x)$ and label the minimum and maximum values.

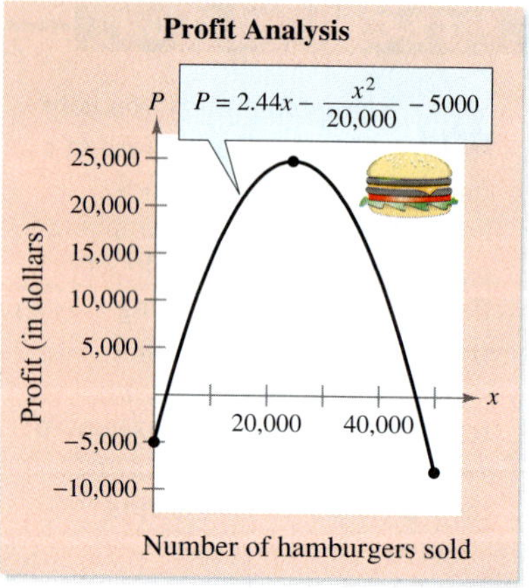

FIGURE 3.19

Applications of Extrema

Finding the minimum and maximum values of a function is one of the most common applications of calculus.

EXAMPLE 5 Finding the Maximum Profit

Recall the fast-food restaurant in Examples 7 and 8 in Section 2.3. The restaurant's profit function for hamburgers is given by

$$P = 2.44x - \frac{x^2}{20,000} - 5000, \quad 0 \le x \le 50,000.$$

Find the production level that produces a maximum profit.

SOLUTION To begin, find an equation for marginal profit. Then set the marginal profit equal to zero and solve for x.

$$P' = 2.44 - \frac{x}{10,000} \qquad \text{Find marginal profit.}$$

$$2.44 - \frac{x}{10,000} = 0 \qquad \text{Set marginal profit equal to 0.}$$

$$-\frac{x}{10,000} = -2.44 \qquad \text{Subtract 2.44 from each side.}$$

$$x = 24,400 \text{ hamburgers} \qquad \text{Critical number}$$

From Figure 3.19, you can see that the critical number $x = 24,400$ corresponds to the production level that produces a maximum profit. To find the maximum profit, substitute $x = 24,400$ into the profit function.

$$P = 2.44x - \frac{x^2}{20,000} - 5000$$
$$= 2.44(24,400) - \frac{(24,400)^2}{20,000} - 5000$$
$$= \$24,768$$

TRY IT 5

Verify the results of Example 5 by completing the table.

x (units)	24,000	24,200	24,300
P (profit)			

x (units)	24,400	24,500	24,600
P (profit)			

x (units)	24,800	25,000
P (profit)		

TAKE ANOTHER LOOK

Setting the Price of a Product

In Example 5, you discovered that a production level of 24,400 hamburgers corresponds to a maximum profit. Remember that this model assumes that the quantity demanded and the price per unit are related by a demand function. So, the only way to sell more hamburgers is to lower the price, and consequently lower the profit. What is the demand function for Example 5? What is the price per unit that produces a maximum profit?

PREREQUISITE REVIEW 3.2

The following warm-up exercises involve skills that were covered in earlier sections. You will use these skills in the exercise set for this section.

In Exercises 1–6, solve the equation $f'(x) = 0$.

1. $f(x) = 4x^4 - 2x^2 + 1$

2. $f(x) = \frac{1}{3}x^3 - \frac{3}{2}x^2 - 10x$

3. $f(x) = 5x^{4/5} - 4x$

4. $f(x) = \frac{1}{2}x^2 - 3x^{5/3}$

5. $f(x) = \dfrac{x + 4}{x^2 + 1}$

6. $f(x) = \dfrac{x - 1}{x^2 + 4}$

In Exercises 7–10, use $g(x) = -x^5 - 2x^4 + 4x^3 + 2x - 1$ to determine the sign of the derivative.

7. $g'(-4)$

8. $g'(0)$

9. $g'(1)$

10. $g'(3)$

In Exercises 11 and 12, decide whether the function is increasing or decreasing on the given interval.

11. $f(x) = 2x^2 - 11x - 6, \quad (3, 6)$

12. $f(x) = x^3 + 2x^2 - 4x - 8, \quad (-2, 0)$

EXERCISES 3.2

In Exercises 1–4, use a table similar to that in Example 1 to find all relative extrema of the function.

1. $f(x) = -2x^2 + 4x + 3$

2. $f(x) = x^2 + 8x + 10$

3. $f(x) = x^2 - 6x$

4. $f(x) = -4x^2 + 4x + 1$

In Exercises 5–12, find all relative extrema of the function.

5. $g(x) = 6x^3 - 15x^2 + 12x$

6. $g(x) = \frac{1}{5}x^5 - x$

7. $h(x) = -(x + 4)^3$

8. $h(x) = 2(x - 3)^3$

9. $f(x) = x^3 - 6x^2 + 15$

10. $f(x) = x^4 - 32x + 4$

11. $f(x) = x^4 - 2x^3$

12. $f(x) = x^4 - 12x^3$

In Exercises 13–18, use a graphing utility to graph the function. Then find all relative extrema of the function.

13. $f(x) = (x - 1)^{2/3}$

14. $f(t) = (t - 1)^{1/3}$

15. $g(t) = t - \dfrac{1}{2t^2}$

16. $f(x) = x + \dfrac{1}{x}$

17. $f(x) = \dfrac{x}{x + 1}$

18. $h(x) = \dfrac{4}{x^2 + 1}$

In Exercises 19–26, find the absolute extrema of the function on the closed interval.

	Function	*Interval*
19.	$f(x) = 2(3 - x)$	$[-1, 2]$
20.	$f(x) = \frac{1}{3}(2x + 5)$	$[0, 5]$
21.	$f(x) = 5 - 2x^2$	$[0, 3]$
22.	$f(x) = x^2 + 2x - 4$	$[-1, 1]$
23.	$f(x) = x^3 - 3x^2$	$[-1, 3]$
24.	$f(x) = x^3 - 12x$	$[0, 4]$
25.	$h(s) = \dfrac{1}{3 - s}$	$[0, 2]$
26.	$h(t) = \dfrac{t}{t - 2}$	$[3, 5]$

In Exercises 27–30, find the absolute extrema of the function on the closed interval. Use a graphing utility to verify your results.

	Function	*Interval*
27.	$f(x) = 3x^{2/3} - 2x$	$[-1, 2]$
28.	$g(t) = \dfrac{t^2}{t^2 + 3}$	$[-1, 1]$
29.	$h(t) = (t - 1)^{2/3}$	$[-7, 2]$
30.	$g(x) = 4\left(1 + \dfrac{1}{x} + \dfrac{1}{x^2}\right)$	$[-4, 5]$

In Exercises 31–34, use a graphing utility to find graphically the absolute extrema of the function on the closed interval.

	Function	*Interval*
31.	$f(x) = 0.4x^3 - 1.8x^2 + x - 3$	$[0, 5]$
32.	$f(x) = 3.2x^5 + 5x^3 - 3.5x$	$[0, 1]$
33.	$f(x) = \frac{4}{3}x\sqrt{3 - x}$	$[0, 3]$
34.	$f(x) = 4\sqrt{x} - 2x + 1$	$[0, 6]$

In Exercises 35–38, find the absolute extrema of the function on the interval $[0, \infty)$.

35. $f(x) = \dfrac{4x}{x^2 + 1}$

36. $f(x) = \dfrac{8}{x + 1}$

37. $f(x) = \dfrac{2x}{x^2 + 4}$

38. $f(x) = 8 - \dfrac{4x}{x^2 + 1}$

In Exercises 39 and 40, find the maximum value of $|f''(x)|$ on the closed interval. (You will use this skill in Section 6.5 to estimate the error in the Trapezoidal Rule.)

Function	*Interval*
39. $f(x) = x^3(3x^2 - 10)$	$[0, 1]$
40. $f(x) = \dfrac{1}{x^2 + 1}$	$[0, 3]$

In Exercises 41 and 42, find the maximum value of $|f^{(4)}(x)|$ on the closed interval. (You will use this skill in Section 6.5 to estimate the error in Simpson's Rule.) Use a graphing utility to verify your answer.

Function	*Interval*
41. $f(x) = 15x^4 - \left(\dfrac{2x - 1}{2}\right)^6$	$[0, 1]$
42. $f(x) = \dfrac{1}{x^2}$	$[1, 2]$

43. ***Cost*** A retailer has determined the cost C for ordering and storing x units of a product to be modeled by

$$C = 3x + \frac{20{,}000}{x}, \quad 0 < x \le 200.$$

The delivery truck can bring at most 200 units per order. Find the order size that will minimize the cost. Use a graphing utility to verify your result.

44. ***Profit*** The quantity demanded x for a product is inversely proportional to the cube of the price p for $p > 1$. When the price is \$10 per unit, the quantity demanded is eight units. The initial cost is \$100 and the cost per unit is \$4. What price will yield a maximum profit?

45. ***Profit*** When soft drinks were sold for \$0.80 per can at football games, approximately 6000 cans were sold. When the price was raised to \$1.00 per can, the quantity demanded dropped to 5600. The initial cost is \$5000 and the cost per unit is \$0.40. Assuming that the demand function is linear, use the *table* feature of a graphing utility to determine the price that will yield a maximum profit.

46. ***Medical Science*** Coughing forces the trachea (windpipe) to contract, which in turn affects the velocity of the air through the trachea. The velocity of the air during coughing can be modeled by

$$v = k(R - r)r^2, \quad 0 \le r < R$$

where k is a constant, R is the normal radius of the trachea, and r is the radius during coughing. What radius r will produce the maximum air velocity?

47. ***Population*** The resident population P (in millions) of the United States from 1790 to 2000 can be modeled by

$$P = 0.00000583t^3 + 0.005003t^2 + 0.13775t + 4.658,$$
$$-10 \le t \le 200$$

where $t = 0$ corresponds to 1800. *(Source: U.S. Census Bureau)*

(a) Make a conjecture about the maximum and minimum populations in the U.S. from 1790 to 2000.

(b) Analytically find the maximum and minimum populations over the interval.

(c) Write a brief paragraph comparing your conjecture with your results in part (b).

48. ***Biology: Fertility Rates*** The graph of the United States fertility rate shows the number of births per 1000 women in their lifetime according to the birth rate in that particular year. *(Source: U.S. National Center for Health Statistics)*

(a) Around what year was the fertility rate the highest, and to how many births per 1000 women did this rate correspond?

(b) During which time periods was the fertility rate increasing most rapidly? Most slowly?

(c) During which time periods was the fertility rate decreasing most rapidly? Most slowly?

(d) Give some possible real-life reasons for fluctuations in the fertility rate.

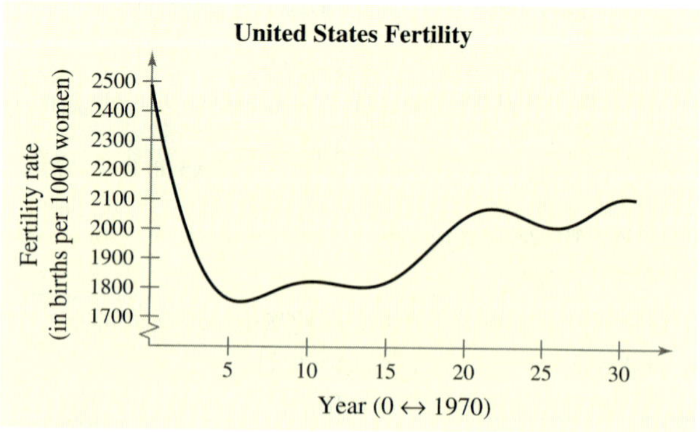

3.3 CONCAVITY AND THE SECOND-DERIVATIVE TEST

- Determine the intervals on which the graphs of functions are concave upward or concave downward.
- Find the points of inflection of the graphs of functions.
- Use the Second-Derivative Test to find the relative extrema of functions.
- Find the points of diminishing returns of input-output models.

Concavity

You already know that locating the intervals over which a function f increases or decreases is helpful in determining its graph. In this section, you will see that locating the intervals on which f' increases or decreases can determine where the graph of f is curving upward or curving downward. This property of curving upward or downward is defined formally as the **concavity** of the graph of the function.

Definition of Concavity

Let f be differentiable on an open interval I. The graph of f is

1. **concave upward** on I if f' is increasing on the interval.
2. **concave downward** on I if f' is decreasing on the interval.

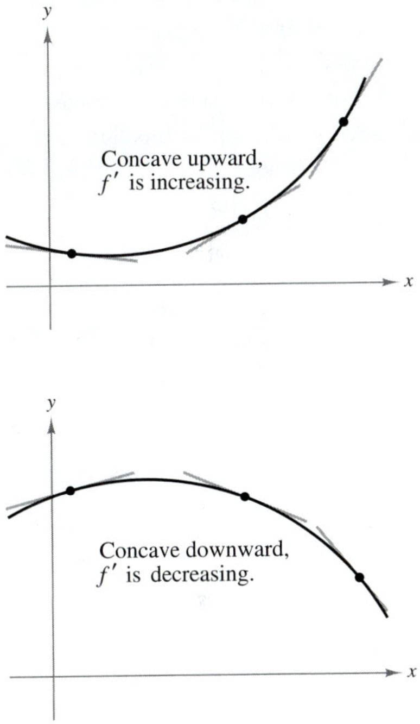

FIGURE 3.20

From Figure 3.20, you can observe the following graphical interpretation of concavity.

1. A curve that is concave upward lies *above* its tangent line.
2. A curve that is concave downward lies *below* its tangent line.

This visual test for concavity is useful when the graph of a function is given. To determine concavity without seeing a graph, you need an analytic test. It turns out that you can use the second derivative to determine these intervals in much the same way that you use the first derivative to determine the intervals on which f is increasing or decreasing.

Test for Concavity

Let f be a function whose second derivative exists on an open interval I.

1. If $f''(x) > 0$ for all x in I, then f is concave upward on I.
2. If $f''(x) < 0$ for all x in I, then f is concave downward on I.

For a *continuous* function f, you can find the intervals on which the graph of f is concave upward and concave downward as follows. [For a function that is not continuous, the test intervals should be formed using points of discontinuity, along with the points at which $f''(x)$ is zero or undefined.]

DISCOVERY

Use a graphing utility to graph the function $f(x) = x^3 - x$ and its second derivative $f''(x) = 6x$ in the same viewing window. On what interval is f concave upward? On what interval is f'' positive? Describe how the second derivative can be used to determine where a function is concave upward and concave downward. Repeat this analysis for the functions $g(x) = x^4 - 6x^2$ and $g''(x) = 12x^2 - 12$.

Guidelines for Applying Concavity Test

1. Locate the x-values at which $f''(x) = 0$ or $f''(x)$ is undefined.
2. Use these x-values to determine the test intervals.
3. Test the sign of $f''(x)$ in each test interval.

EXAMPLE 1 **Applying the Test for Concavity**

(a) The graph of the function

$$f(x) = x^2 \quad \text{Original function}$$

is concave upward on the entire real line because its second derivative

$$f''(x) = 2 \quad \text{Second derivative}$$

is positive for all x. (See Figure 3.21.)

(b) The graph of the function

$$f(x) = \sqrt{x} \quad \text{Original function}$$

is concave downward for $x > 0$ because its second derivative

$$f''(x) = -\frac{1}{4}x^{-3/2} \quad \text{Second derivative}$$

is negative for all $x > 0$. (See Figure 3.22.)

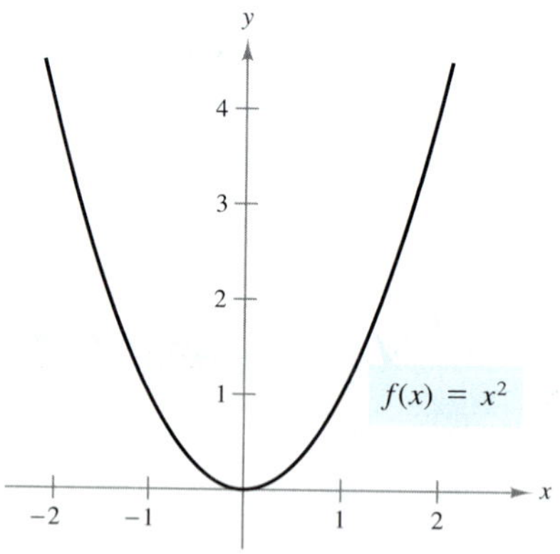

FIGURE 3.21 Concave Upward

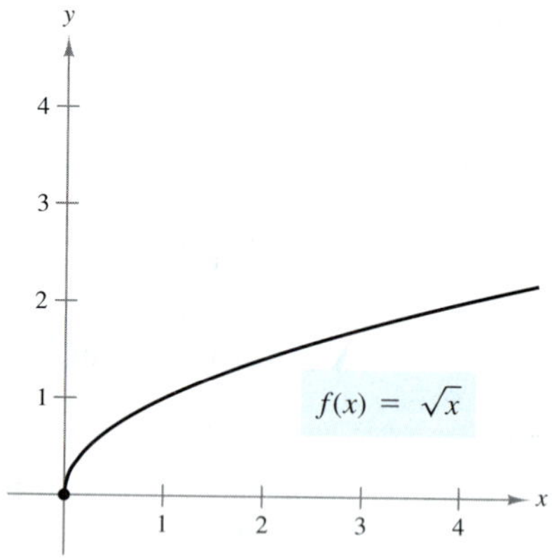

FIGURE 3.22 Concave Downward

TRY IT 1

(a) Find the second derivative of $f(x) = -2x^2$ and discuss the concavity of the graph.

(b) Find the second derivative of $f(x) = -2\sqrt{x}$ and discuss the concavity of the graph.

EXAMPLE 2 Determining Concavity

Determine the intervals on which the graph of the function is concave upward or concave downward.

$$f(x) = \frac{6}{x^2 + 3}$$

SOLUTION Begin by finding the second derivative of f.

$$f(x) = 6(x^2 + 3)^{-1} \quad \text{Rewrite original function.}$$

$$f'(x) = (-6)(2x)(x^2 + 3)^{-2} \quad \text{Chain Rule}$$

$$= \frac{-12x}{(x^2 + 3)^2} \quad \text{Simplify.}$$

$$f''(x) = \frac{(x^2 + 3)^2(-12) - (-12x)(2)(2x)(x^2 + 3)}{(x^2 + 3)^4} \quad \text{Quotient Rule}$$

$$= \frac{-12(x^2 + 3) + (48x^2)}{(x^2 + 3)^3} \quad \text{Simplify.}$$

$$= \frac{36(x^2 - 1)}{(x^2 + 3)^3} \quad \text{Simplify.}$$

From this, you can see that $f''(x)$ is defined for all real numbers and $f''(x) = 0$ when $x = \pm 1$. So, you can test the concavity of f by testing the intervals $(-\infty, -1)$, $(-1, 1)$, and $(1, \infty)$, as shown in the table. The graph of f is shown in Figure 3.23.

Interval	$-\infty < x < -1$	$-1 < x < 1$	$1 < x < \infty$
Test value	$x = -2$	$x = 0$	$x = 2$
Sign of $f''(x)$	$f''(-2) > 0$	$f''(0) < 0$	$f''(2) > 0$
Conclusion	Concave upward	Concave downward	Concave upward

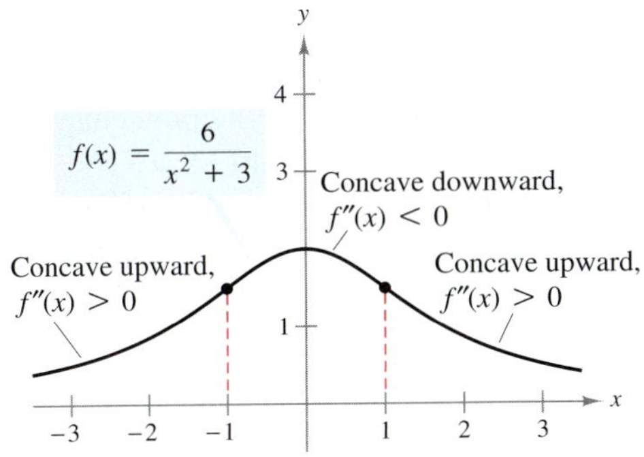

FIGURE 3.23

ALGEBRA REVIEW

For help on the algebra in Example 2, see Example 1(a) in the *Chapter 3 Algebra Review*, on page 248.

STUDY TIP

In Example 2, f' is increasing on the interval $(1, \infty)$ even though f is decreasing there. Be sure you see that the increasing or decreasing of f' does not necessarily correspond to the increasing or decreasing of f.

TRY IT 2

Determine the intervals on which the graph of the function is concave upward and concave downward.

$$f(x) = \frac{24}{x^2 + 12}$$

Points of Inflection

If the tangent line to a graph exists at a point at which the concavity changes, then the point is a **point of inflection.** Three examples of inflection points are shown in Figure 3.24. (Note that the third graph has a vertical tangent line at its point of inflection.)

STUDY TIP

As shown in Figure 3.24, a graph crosses its tangent line at a point of inflection.

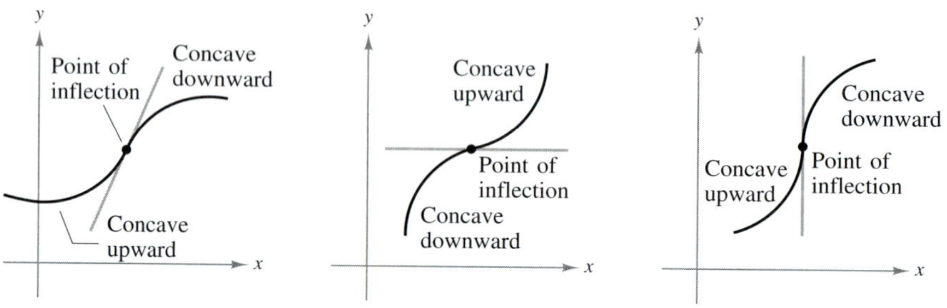

FIGURE 3.24 The graph *crosses* its tangent line at a point of inflection.

Definition of Point of Inflection

If the graph of a continuous function has a tangent line at a point where its concavity changes from upward to downward (or vice versa), then the point is a **point of inflection.**

DISCOVERY

Use a graphing utility to graph $f(x) = x^3 - 6x^2 + 12x - 6$ and $f''(x) = 6x - 12$ in the same viewing window. At what x-value does $f''(x) = 0$? At what x-value does the point of inflection occur? Repeat this analysis for $g(x) = x^4 - 5x^2 + 7$ and $g''(x) = 12x^2 - 10$. Make a general statement about the relationship of the point of inflection of a function and the second derivative of a function.

Because a point of inflection occurs where the concavity of a graph changes, it must be true that at such points the sign of f'' changes. So, to locate possible points of inflection, you only need to determine the values of x for which $f''(x) = 0$ or for which $f''(x)$ does not exist. This parallels the procedure for locating the relative extrema of f by determining the critical numbers of f.

Property of Points of Inflection

If $(c, f(c))$ is a point of inflection of the graph of f, then either $f''(c) = 0$ or $f''(c)$ is undefined at c.

EXAMPLE 3 Finding Points of Inflection

Discuss the concavity of the graph of f and find its points of inflection.

$$f(x) = x^4 + x^3 - 3x^2 + 1$$

SOLUTION Begin by finding the second derivative of f.

$$f'(x) = 4x^3 + 3x^2 - 6x \quad \text{Find first derivative.}$$
$$f''(x) = 12x^2 + 6x - 6 \quad \text{Find second derivative.}$$
$$= 6(2x - 1)(x + 1) \quad \text{Factor.}$$

From this, you can see that the possible points of inflection occur at $x = \frac{1}{2}$ and $x = -1$. After testing the intervals $(-\infty, -1)$, $\left(-1, \frac{1}{2}\right)$, and $\left(\frac{1}{2}, \infty\right)$, you can determine that the graph is concave upward on $(-\infty, -1)$, concave downward on $\left(-1, \frac{1}{2}\right)$, and concave upward on $\left(\frac{1}{2}, \infty\right)$. Because the concavity changes at $x = -1$ and $x = \frac{1}{2}$, you can conclude that the graph has points of inflection at these x-values, as shown in Figure 3.25.

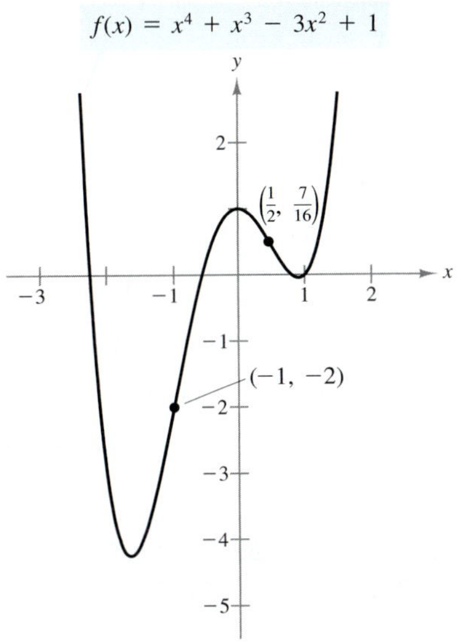

FIGURE 3.25 Two Points of Inflection

TRY IT 3

Discuss the concavity of the graph of f and find its points of inflection.

$$f(x) = x^4 - 2x^3 + 1$$

It is possible for the second derivative to be zero at a point that is *not* a point of inflection. For example, compare the graphs of $f(x) = x^3$ and $g(x) = x^4$, as shown in Figure 3.26. Both second derivatives are zero when $x = 0$, but only the graph of f has a point of inflection at $x = 0$. This shows that before concluding that a point of inflection exists at a value of x for which $f''(x) = 0$, you must test to be certain that the concavity actually changes at that point.

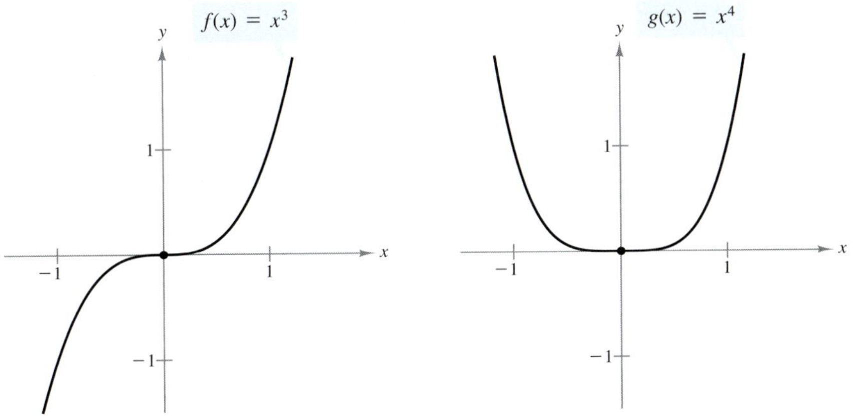

$f''(0) = 0$, and $(0, 0)$ is a point of inflection.

$g''(0) = 0$, but $(0, 0)$ is not a point of inflection.

FIGURE 3.26

The Second-Derivative Test

The second derivative can be used to perform a simple test for relative minima and relative maxima. If f is a function such that $f'(c) = 0$ and the graph of f is concave upward at $x = c$, then $f(c)$ is a relative minimum of f. Similarly, if f is a function such that $f'(c) = 0$ and the graph of f is concave downward at $x = c$, then $f(c)$ is a relative maximum of f, as shown in Figure 3.27.

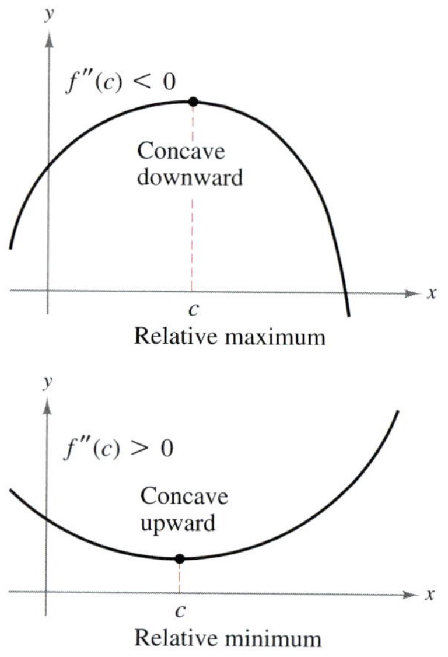

FIGURE 3.27

Second-Derivative Test

Let $f'(c) = 0$, and let f'' exist on an open interval containing c.

1. If $f''(c) > 0$, then $f(c)$ is a relative minimum.
2. If $f''(c) < 0$, then $f(c)$ is a relative maximum.
3. If $f''(c) = 0$, then the test fails. In such cases, you can use the First-Derivative Test to determine whether $f(c)$ is a relative minimum, a relative maximum, or neither.

EXAMPLE 4 **Using the Second-Derivative Test**

Find the relative extrema of

$$f(x) = -3x^5 + 5x^3.$$

SOLUTION Begin by finding the first derivative of f.

$$f'(x) = -15x^4 + 15x^2$$
$$= 15x^2(1 - x^2)$$

From this derivative, you can see that $x = 0$, $x = -1$, and $x = 1$ are the only critical numbers of f. Using the second derivative

$$f''(x) = -60x^3 + 30x$$

you can apply the Second-Derivative Test, as shown.

Point	*Sign of $f''(x)$*	*Conclusion*
$(-1, -2)$	$f''(-1) = 30 > 0$	Relative minimum
$(0, 0)$	$f''(0) = 0$	Test fails.
$(1, 2)$	$f''(1) = -30 < 0$	Relative maximum

Because the test fails at $(0, 0)$, you can apply the First-Derivative Test to conclude that the point $(0, 0)$ is neither a relative minimum nor a relative maximum—a test for concavity would show that this point is a point of inflection. The graph of f is shown in Figure 3.28.

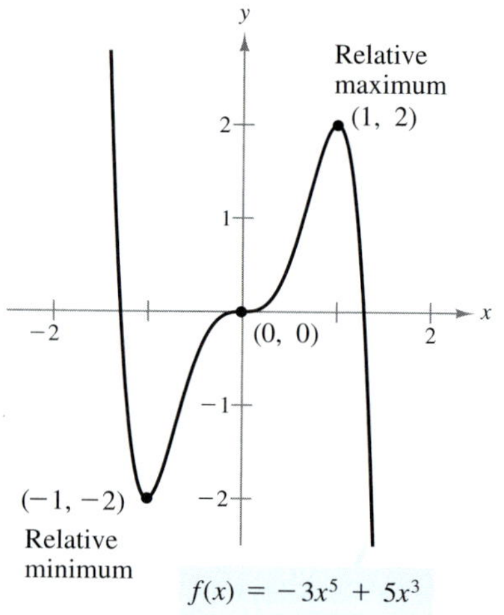

FIGURE 3.28

TRY IT 4

Find all relative extrema of $f(x) = x^4 - 4x^3 + 1$.

Extended Application: Diminishing Returns

In economics, the notion of concavity is related to the concept of **diminishing returns.** Consider a function

$$\underset{\text{Output}}{y} = f(\underset{\text{Input}}{x})$$

where x measures input (in dollars) and y measures output (in dollars). In Figure 3.29, notice that the graph of this function is concave upward on the interval (a, c) and is concave downward on the interval (c, b). On the interval (a, c), each additional dollar of input returns more than the previous input dollar. By contrast, on the interval (c, b), each additional dollar of input returns less than the previous input dollar. The point $(c, f(c))$ is called the **point of diminishing returns.** An increased investment beyond this point is usually considered a poor use of capital.

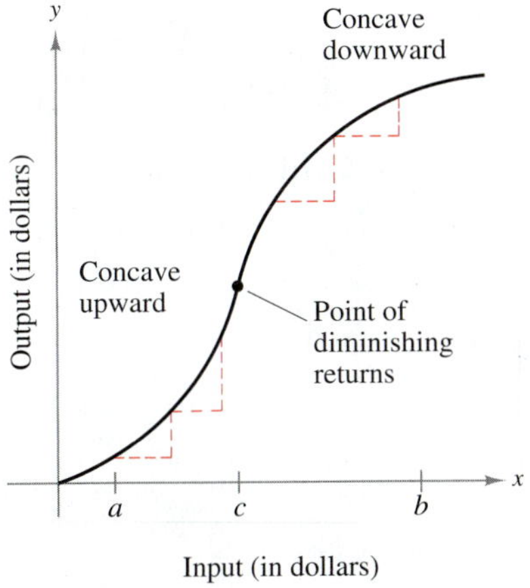

FIGURE 3.29

EXAMPLE 5 Exploring Diminishing Returns

By increasing its advertising cost x (in thousands of dollars) for a product, a company discovers that it can increase the sales y (in thousands of dollars) according to the model

$$y = -\frac{1}{10}x^3 + 6x^2 + 400, \qquad 0 \le x \le 40.$$

Find the point of diminishing returns for this product.

SOLUTION Begin by finding the first and second derivatives.

$$y' = 12x - \frac{3x^2}{10} \quad \text{First derivative} \qquad y'' = 12 - \frac{3x}{5} \quad \text{Second derivative}$$

The second derivative is zero only when $x = 20$. By testing the intervals $(0, 20)$ and $(20, 40)$, you can conclude that the graph has a point of diminishing returns when $x = 20$, as shown in Figure 3.30.

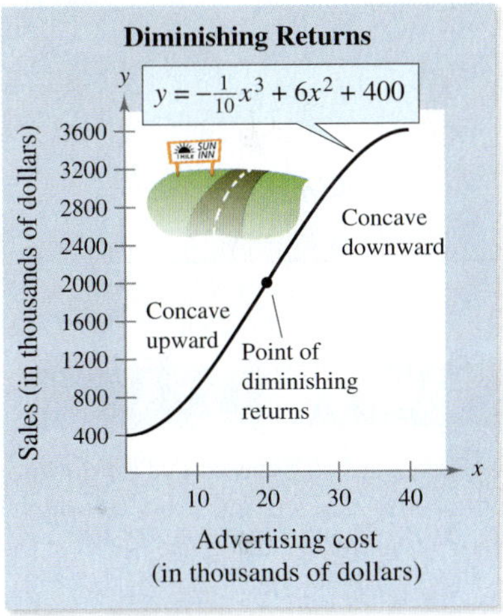

FIGURE 3.30

TRY IT 5

Find the point of diminishing returns for the model below, where R is the revenue (in thousands of dollars) and x is the advertising cost (in thousands of dollars).

$$R = \frac{1}{20{,}000}(450x^2 - x^3), \qquad 0 \le x \le 300$$

TAKE ANOTHER LOOK

Diminishing Returns

Use the function in Example 5 to verify that additional dollars of input return more than the previous input dollars on the interval $(0, 20)$ and less than the previous input dollars on the interval $(20, 40)$.

PREREQUISITE REVIEW 3.3

The following warm-up exercises involve skills that were covered in earlier sections. You will use these skills in the exercise set for this section.

In Exercises 1–6, find the second derivative of the function.

1. $f(x) = 4x^4 - 9x^3 + 5x - 1$
2. $g(s) = (s^2 - 1)(s^2 - 3s + 2)$
3. $g(x) = (x^2 + 1)^4$
4. $f(x) = (x - 3)^{4/3}$
5. $h(x) = \dfrac{4x + 3}{5x - 1}$
6. $f(x) = \dfrac{2x - 1}{3x + 2}$

In Exercises 7–10, find the critical numbers of the function.

7. $f(x) = 5x^3 - 5x + 11$
8. $f(x) = x^4 - 4x^3 - 10$
9. $g(t) = \dfrac{16 + t^2}{t}$
10. $h(x) = \dfrac{x^4 - 50x^2}{8}$

EXERCISES 3.3

In Exercises 1–8, analytically find the intervals on which the graph is concave upward and those on which it is concave downward. Verify your results using the graph of the function.

1. $y = x^2 - x - 2$

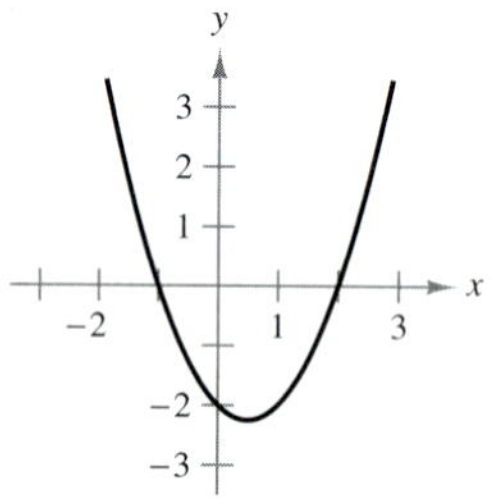

2. $y = -x^3 + 3x^2 - 2$

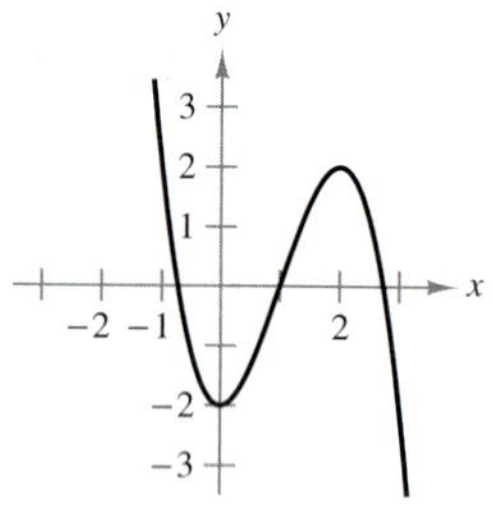

3. $f(x) = \dfrac{x^2 - 1}{2x + 1}$

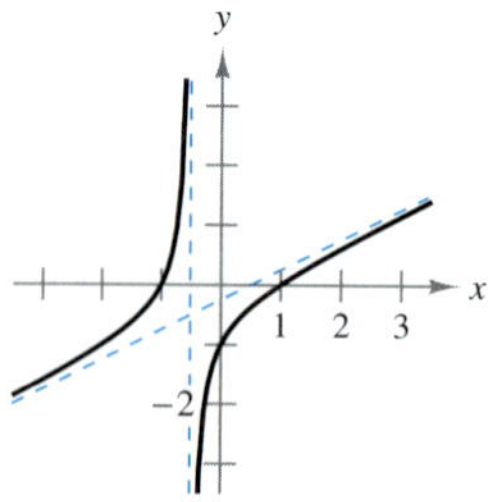

4. $f(x) = \dfrac{x^2 + 4}{4 - x^2}$

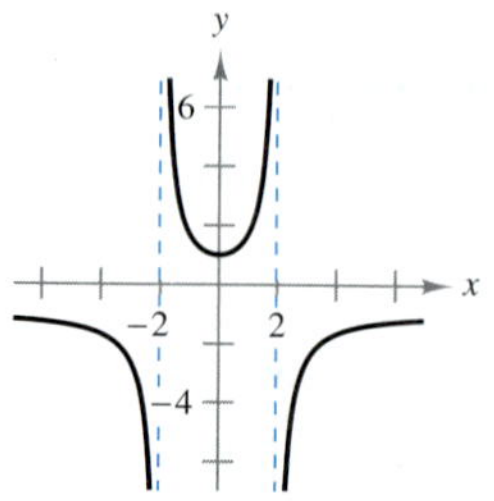

5. $f(x) = \dfrac{24}{x^2 + 12}$

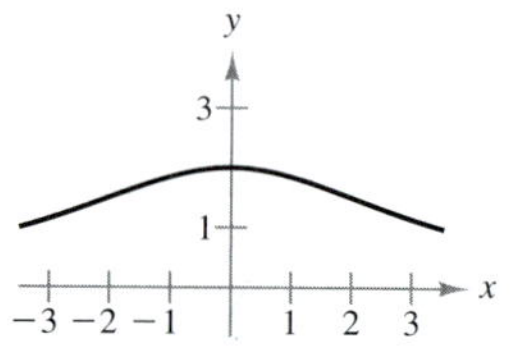

6. $f(x) = \dfrac{x^2}{x^2 + 1}$

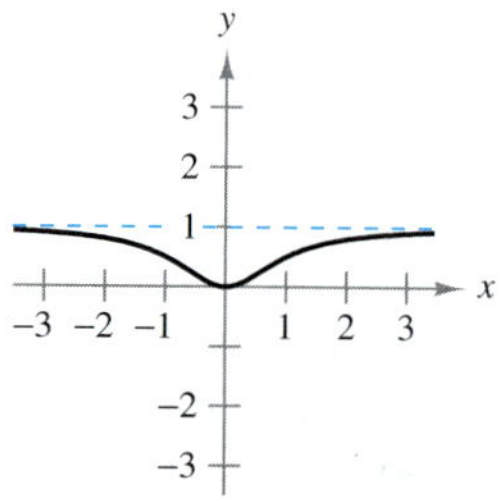

7. $y = -x^3 + 6x^2 - 9x - 1$

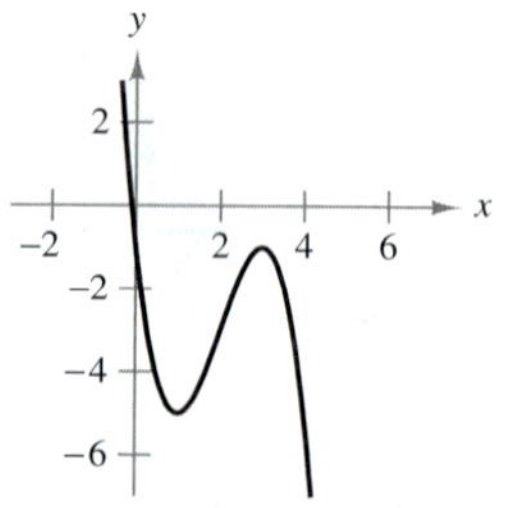

8. $y = x^5 + 5x^4 - 40x^2$

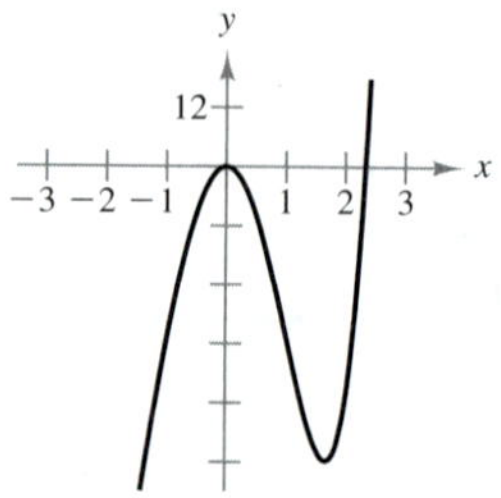

In Exercises 9–18, find all relative extrema of the function. Use the Second-Derivative Test when applicable.

9. $f(x) = 6x - x^2$

10. $f(x) = (x - 5)^2$

11. $f(x) = x^3 - 5x^2 + 7x$

12. $f(x) = x^4 - 4x^3 + 2$

13. $f(x) = x^{2/3} - 3$

14. $f(x) = x + \dfrac{4}{x}$

15. $f(x) = \sqrt{x^2 + 1}$

16. $f(x) = \sqrt{4 - x^2}$

17. $f(x) = \dfrac{x}{x - 1}$

18. $f(x) = \dfrac{x}{x^2 - 1}$

In Exercises 19–22, use a graphing utility to estimate graphically all relative extrema of the function.

19. $f(x) = \frac{1}{2}x^4 - \frac{1}{3}x^3 - \frac{1}{2}x^2$

20. $f(x) = -\frac{1}{3}x^5 - \frac{1}{2}x^4 + x$

21. $f(x) = 5 + 3x^2 - x^3$

22. $f(x) = 3x^3 + 5x^2 - 2$

In Exercises 23–26, state the signs of $f'(x)$ and $f''(x)$ on the interval $(0, 2)$.

23.

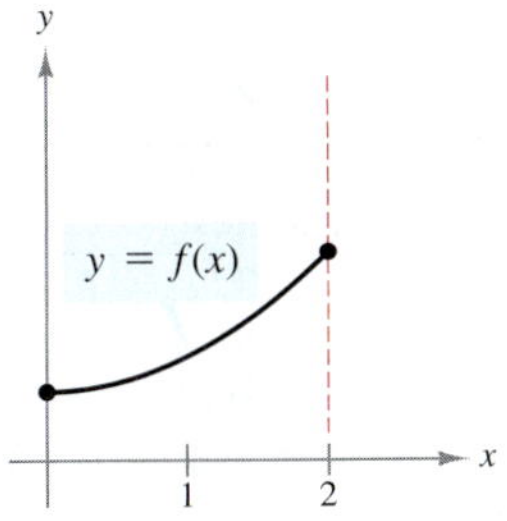

24.

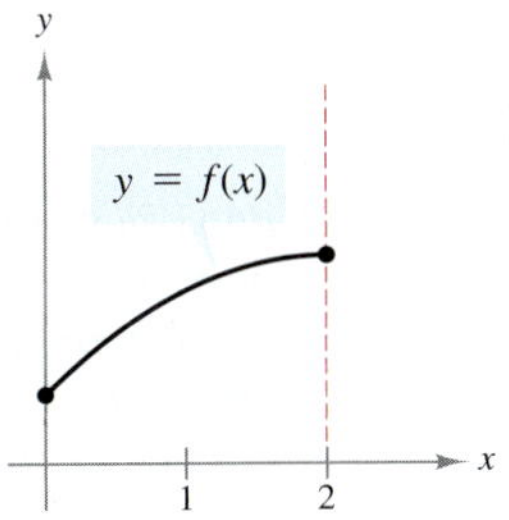

25.

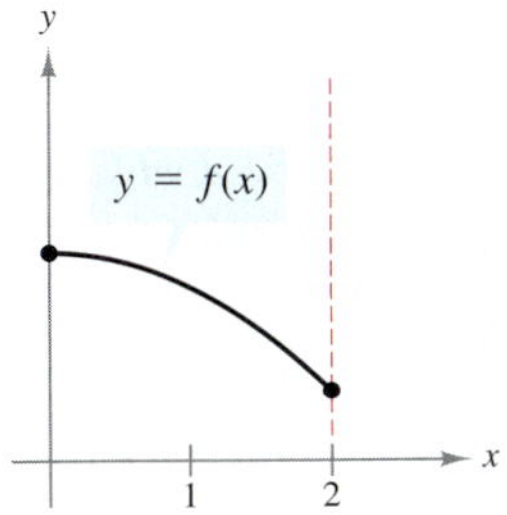

26.

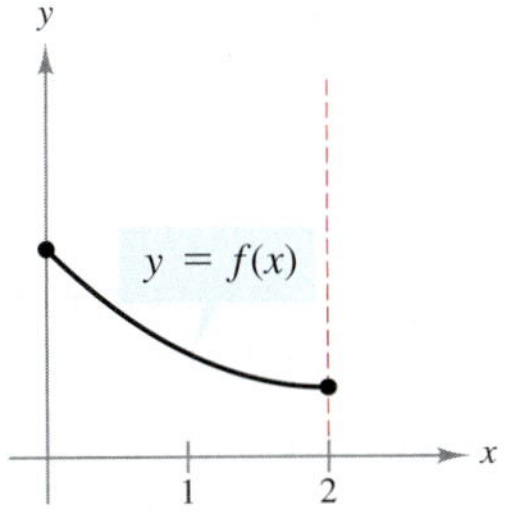

In Exercises 27–34, find the point(s) of inflection of the graph of the function.

27. $f(x) = x^3 - 9x^2 + 24x - 18$

28. $f(x) = x(6 - x)^2$

29. $f(x) = (x - 1)^3(x - 5)$

30. $f(x) = x^4 - 18x^2 + 5$

31. $g(x) = 2x^4 - 8x^3 + 12x^2 + 12x$

32. $f(x) = -4x^3 - 8x^2 + 32$

33. $h(x) = (x - 2)^3(x - 1)$

34. $f(t) = (1 - t)(t - 4)(t^2 - 4)$

In Exercises 35–46, use a graphing utility to graph the function and identify all relative extrema and points of inflection.

35. $f(x) = x^3 - 12x$

36. $f(x) = x^3 - 3x$

37. $f(x) = x^3 - 6x^2 + 12x$

38. $f(x) = x^3 - \frac{3}{2}x^2 - 6x$

39. $f(x) = \frac{1}{4}x^4 - 2x^2$

40. $f(x) = 2x^4 - 8x + 3$

41. $g(x) = (x - 2)(x + 1)^2$

42. $g(x) = (x - 6)(x + 2)^3$

43. $g(x) = x\sqrt{x + 3}$

44. $g(x) = x\sqrt{9 - x}$

45. $f(x) = \dfrac{4}{1 + x^2}$

46. $f(x) = \dfrac{2}{x^2 - 1}$

In Exercises 47 and 48, sketch a graph of a function f having the given characteristics.

	Function	First Derivative	Second Derivative
47.	$f(2) = 0$	$f'(x) < 0,\ x < 3$	$f''(x) > 0$
	$f(4) = 0$	$f'(3) = 0$	
		$f'(x) > 0,\ x > 3$	
48.	$f(2) = 0$	$f'(x) > 0,\ x < 3$	$f''(x) > 0,\ x \neq 3$
	$f(4) = 0$	$f'(3)$ is undefined.	
		$f'(x) < 0,\ x > 3$	

In Exercises 49 and 50, use the graph to sketch the graph of f'. Find the intervals on which (a) $f'(x)$ is positive, (b) $f'(x)$ is negative, (c) f' is increasing, and (d) f' is decreasing. For each of these intervals, describe the corresponding behavior of f.

49.

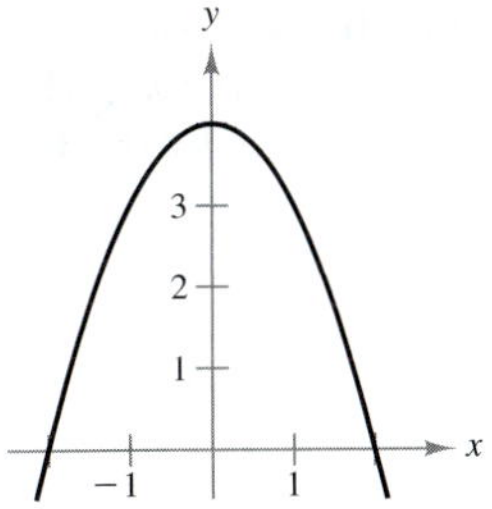

50.

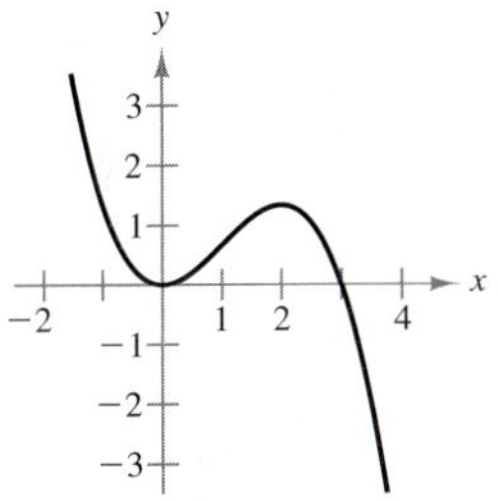

Point of Diminishing Returns In Exercises 51 and 52, identify the point of diminishing returns for the input-output function. For each function, R is the revenue and x is the amount spent on advertising. Use a graphing utility to verify your results.

51. $R = \dfrac{1}{50{,}000}(600x^2 - x^3), \quad 0 \le x \le 400$

52. $R = -\frac{4}{9}(x^3 - 9x^2 - 27), \quad 0 \le x \le 5$

Average Cost In Exercises 53 and 54, you are given the total cost of producing x units. Find the production level that minimizes the average cost per unit. Use a graphing utility to verify your results.

53. $C = 0.5x^2 + 15x + 5000$

54. $C = 0.002x^3 + 20x + 500$

Productivity In Exercises 55 and 56, consider a college student who works from 7 P.M. to 11 P.M. assembling mechanical components. The number N of components assembled after t hours is given by the function. At what time is the student assembling components at the greatest rate?

55. $N = -0.12t^3 + 0.54t^2 + 8.22t, \quad 0 \le t \le 4$

56. $N = \dfrac{20t^2}{4 + t^2}, \quad 0 \le t \le 4$

Sales Growth In Exercises 57 and 58, find the time t in years when the annual sales x of a new product are increasing at the greatest rate. Use a graphing utility to verify your results.

57. $x = \dfrac{10{,}000t^2}{9 + t^2}$ **58.** $x = \dfrac{500{,}000t^2}{36 + t^2}$

In Exercises 59–62, use a graphing utility to graph f, f', and f'' in the same viewing window. Graphically locate the relative extrema and points of inflection of the graph.

	Function	*Interval*
59.	$f(x) = \frac{1}{2}x^3 - x^2 + 3x - 5$	$[0, 3]$
60.	$f(x) = -\frac{1}{20}x^5 - \frac{1}{12}x^2 - \frac{1}{3}x + 1$	$[-2, 2]$
61.	$f(x) = \dfrac{2}{x^2 + 1}$	$[-3, 3]$
62.	$f(x) = \dfrac{x^2}{x^2 + 1}$	$[-3, 3]$

63. ***Dow Jones Industrial Average*** The graph shows the Dow Jones Industrial Average y on Black Monday, October 19, 1987, where $t = 0$ corresponds to 8:30 A.M., when the market opens, and $t = 7.5$ corresponds to 4 P.M., the closing time. *(Source: Wall Street Journal)*

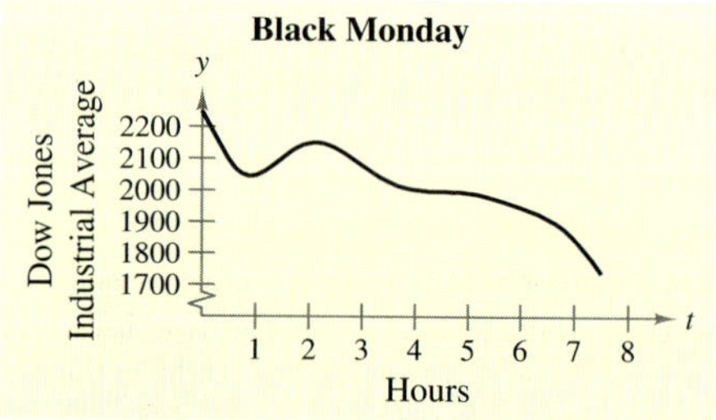

(a) Estimate the relative extrema and absolute extrema of the graph. Interpret your results in the context of the problem.

(b) Estimate the point of inflection of the graph on the interval $[3, 6]$. Interpret your result in the context of the problem.

64. ***Think About It*** Let S represent monthly sales of a new model of MP3 player. Write a statement describing S' and S'' for each of the following.

(a) The rate of change of sales is increasing.

(b) Sales are increasing, but at a greater rate.

(c) The rate of change of sales is steady.

(d) Sales are steady.

(e) Sales are declining, but at a lower rate.

(f) Sales have bottomed out and have begun to rise.

65. ***Medicine*** The spread of a virus can be modeled by

$$N = -t^3 + 12t^2, \quad 0 \le t \le 12$$

where N is the number of people infected in hundreds, and t is the time in weeks.

(a) What is the maximum number of people projected to be infected?

(b) When will the virus be spreading most rapidly?

(c) Use a graphing utility to verify your results.

BUSINESS CAPSULE

Gordon Weinberger won a baking contest two years in a row and decided to start his own pie making company. He raised money and opened Gordon's Top of the Tree Baking Company in Londonderry, NH in 1994. His pies sold well in small stores, but not in the larger markets. He was in debt, so he painted a schoolbus in psychedelic patterns and drove 1500 miles a week to visit as many supermarkets as he could. He eventually found a buyer who ordered 40 tractor-trailers full, which meant $1 million in sales. Gordon's yearly sales hit $5 million. In 2002, he sold the company to Mrs. Smith's Bakeries.

66. ***Research Project*** Use your school's library, the Internet, or some other reference source to research the financial history of a small company like the one above. Gather the data on the company's costs and revenues over a period of time, and use a graphing utility to graph a scatter plot of the data. Fit models to the data. Do the models appear to be concave upward or downward? Do they appear to be increasing or decreasing? Discuss the implications of your answers.

3.4 OPTIMIZATION PROBLEMS

■ Solve real-life optimization problems.

Solving Optimization Problems

One of the most common applications of calculus is the determination of optimum (minimum or maximum) values. Before learning a general method for solving optimization problems, consider the next example.

EXAMPLE 1 Finding the Maximum Volume

A manufacturer wants to design an open box that has a square base and a surface area of 108 square inches, as shown in Figure 3.31. What dimensions will produce a box with a maximum volume?

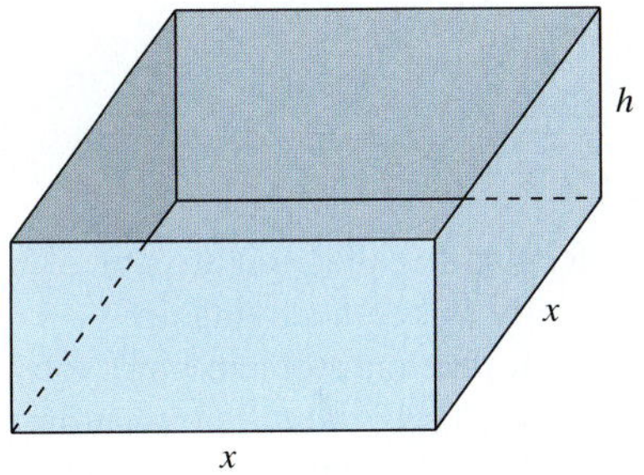

FIGURE 3.31 Open Box with Square Base: $S = x^2 + 4xh = 108$

SOLUTION Because the base of the box is square, the volume is

$$V = x^2h. \qquad \text{Primary equation}$$

This equation is called the **primary equation** because it gives a formula for the quantity to be optimized. The surface area of the box is

$$S = (\text{area of base}) + (\text{area of four sides})$$
$$108 = x^2 + 4xh. \qquad \text{Secondary equation}$$

Because V is to be optimized, it helps to express V as a function of just one variable. To do this, solve the secondary equation for h in terms of x to obtain

$$h = \frac{108 - x^2}{4x}$$

and substitute into the primary equation.

$$V = x^2h = x^2\left(\frac{108 - x^2}{4x}\right) = 27x - \frac{1}{4}x^3 \qquad \text{Function of one variable}$$

Before finding which x-value yields a maximum value of V, you need to determine the **feasible domain** of the function. That is, what values of x make sense in the problem? Because x must be nonnegative and the area of the base $(A = x^2)$ is at most 108, you can conclude that the feasible domain is

$$0 \le x \le \sqrt{108}. \qquad \text{Feasible domain}$$

Using the techniques described in the first three sections of this chapter, you can determine that $\left(\text{on the interval } 0 \le x \le \sqrt{108}\right)$ this function has an absolute maximum when $x = 6$ inches and $h = 3$ inches.

TRY IT 1

Use a graphing utility to graph the volume function $V = 27x - \frac{1}{4}x^3$ on $0 \le x \le \sqrt{108}$ from Example 1. Verify that the function has an absolute maximum when $x = 6$. What is the maximum volume?

In studying Example 1, be sure that you understand the basic question that it asks. Some students have trouble with optimization problems because they are too eager to start solving the problem by using a standard formula. For instance, in Example 1, you should realize that there are infinitely many open boxes having 108 square inches of surface area. You might begin to solve this problem by asking yourself which basic shape would seem to yield a maximum volume. Should the box be tall, cubical, or squat? You might even try calculating a few volumes, as shown in Figure 3.32, to see if you can get a good feeling for what the optimum dimensions should be.

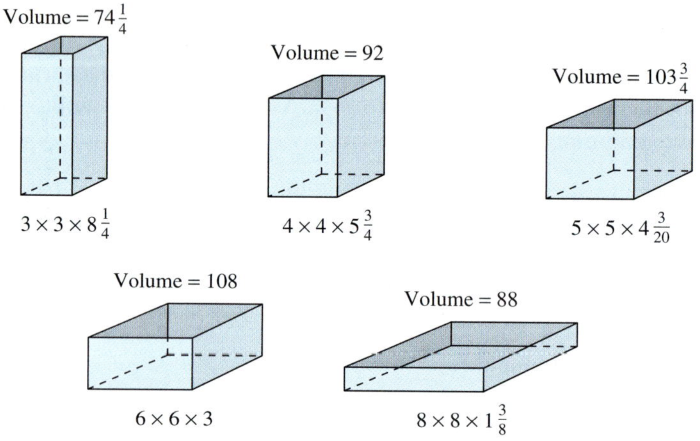

FIGURE 3.32 Which box has the maximum volume?

STUDY TIP

Remember that you are not ready to begin solving an optimization problem until you have clearly identified what the problem is. Once you are sure you understand what is being asked, you are ready to begin considering a method for solving the problem.

There are several steps in the solution of Example 1. The first step is to sketch a diagram and identify all *known* quantities and all quantities *to be determined.* The second step is to write a primary equation for the quantity to be optimized. Then, a secondary equation is used to rewrite the primary equation as a function of one variable. Finally, calculus is used to determine the optimum value. These steps are summarized below.

STUDY TIP

When performing Step 5, remember that to determine the maximum or minimum value of a continuous function f on a closed interval, you need to compare the values of f at its critical numbers with the values of f at the endpoints of the interval. The largest of these values is the desired maximum and the smallest is the desired minimum.

Guidelines for Solving Optimization Problems

1. Identify all given quantities and all quantities to be determined. When feasible, sketch a diagram.
2. Write a **primary equation** for the quantity that is to be maximized or minimized. (A summary of several common formulas is given in Appendix D.)
3. Reduce the primary equation to one having a single independent variable. This may involve the use of a **secondary equation** that relates the independent variables of the primary equation.
4. Determine the **feasible domain** of the primary equation. That is, determine the values for which the stated problem makes sense.
5. Determine the desired maximum or minimum value by the calculus techniques discussed in Sections 3.1 through 3.3.

EXAMPLE 2 Finding a Minimum Sum

The product of two positive numbers is 288. Minimize the sum of the second number and twice the first number.

SOLUTION

1. Let x be the first number, y the second, and S the sum to be minimized.
2. Because you want to minimize S, the primary equation is

$$S = 2x + y. \qquad \text{Primary equation}$$

3. Because the product of the two numbers is 288, you can write the secondary equation as

$$xy = 288 \qquad \text{Secondary equation}$$
$$y = \frac{288}{x}.$$

Using this result, you can rewrite the primary equation as a function of one variable.

$$S = 2x + \frac{288}{x} \qquad \text{Function of one variable}$$

4. Because the numbers are positive, the feasible domain is

$$0 < x. \qquad \text{Feasible domain}$$

5. To find the minimum value of S, begin by finding its critical numbers.

$$\frac{dS}{dx} = 2 - \frac{288}{x^2} \qquad \text{Find derivative of } S.$$
$$0 = 2 - \frac{288}{x^2} \qquad \text{Set derivative equal to 0.}$$
$$x^2 = 144 \qquad \text{Simplify.}$$
$$x = \pm 12 \qquad \text{Critical numbers}$$

Choosing the positive x-value, you can use the First-Derivative Test to conclude that S is decreasing on the interval $(0, 12)$ and increasing on the interval $(12, \infty)$, as shown in the table. So, $x = 12$ yields a minimum, and the two numbers are

$$x = 12 \quad \text{and} \quad y = \frac{288}{12} = 24.$$

Interval	$0 < x < 12$	$12 < x < \infty$
Test value	$x = 1$	$x = 13$
Sign of $\frac{dS}{dx}$	$\frac{dS}{dx} < 0$	$\frac{dS}{dx} > 0$
Conclusion	S is decreasing.	S is increasing.

ALGEBRA REVIEW

For help on the algebra in Example 2, see Example 1(b) in the *Chapter 3 Algebra Review,* on page 248.

TECHNOLOGY

After you have written the primary equation as a function of a single variable, you can estimate the optimum value by graphing the function. For instance, the graph of

$$S = 2x + \frac{288}{x}$$

shown below indicates that the minimum value of S occurs when x is about 12.

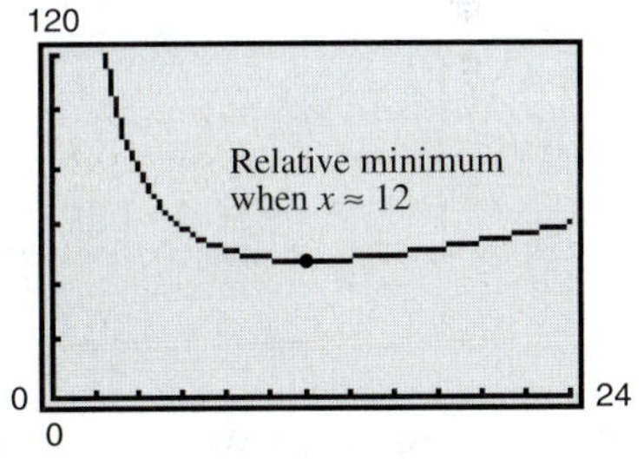

TRY IT 2

The product of two numbers is 72. Minimize the sum of the second number and twice the first number.

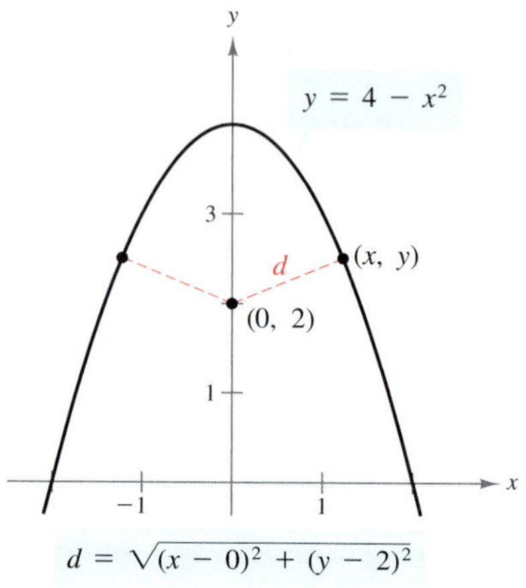

FIGURE 3.33

EXAMPLE 3 Finding a Minimum Distance

Find the points on the graph of

$$y = 4 - x^2$$

that are closest to $(0, 2)$.

SOLUTION

1. Figure 3.33 indicates that there are two points at a minimum distance from the point $(0, 2)$.
2. You are asked to minimize the distance d. So, you can use the Distance Formula to obtain a primary equation.

$$d = \sqrt{(x - 0)^2 + (y - 2)^2} \qquad \text{Primary equation}$$

3. Using the secondary equation $y = 4 - x^2$, you can rewrite the primary equation as a function of a single variable.

$$d = \sqrt{x^2 + (4 - x^2 - 2)^2} \qquad \text{Substitute } 4 - x^2 \text{ for } y.$$
$$= \sqrt{x^4 - 3x^2 + 4} \qquad \text{Simplify.}$$

Because d is smallest when the expression under the radical is smallest, you simplify the problem by finding the minimum value of

$$f(x) = x^4 - 3x^2 + 4.$$

4. The domain of f is the entire real line.
5. To find the minimum value of $f(x)$, first find the critical numbers of f.

$$f'(x) = 4x^3 - 6x \qquad \text{Find derivative of } f.$$
$$0 = 4x^3 - 6x \qquad \text{Set derivative equal to 0.}$$
$$0 = 2x(2x^2 - 3) \qquad \text{Factor.}$$
$$x = 0,\ x = \sqrt{\tfrac{3}{2}},\ x = -\sqrt{\tfrac{3}{2}} \qquad \text{Critical numbers}$$

By the First-Derivative Test, you can conclude that $x = 0$ yields a relative maximum, whereas both $\sqrt{3/2}$ and $-\sqrt{3/2}$ yield a minimum. So, on the graph of $y = 4 - x^2$, the points that are closest to the point $(0, 2)$ are

$$\left(\sqrt{\tfrac{3}{2}}, \tfrac{5}{2}\right) \quad \text{and} \quad \left(-\sqrt{\tfrac{3}{2}}, \tfrac{5}{2}\right).$$

TRY IT 3

Find the points of the graph of $y = 4 - x^2$ that are closest to $(0, 3)$.

ALGEBRA REVIEW

For help on the algebra in Example 3, see Example 1(c) in the *Chapter 3 Algebra Review*, on page 248.

STUDY TIP

To confirm the result in Example 3, try computing the distances between several points on the graph of $y = 4 - x^2$ and the point $(0, 2)$. For instance, the distance between $(1, 3)$ and $(0, 2)$ is

$$d = \sqrt{(0 - 1)^2 + (2 - 3)^2}$$
$$= \sqrt{2} \approx 1.414.$$

Note that this is greater than the distance between $\left(\sqrt{3/2}, 5/2\right)$ and $(0, 2)$, which is

$$d = \sqrt{\left(0 - \sqrt{\tfrac{3}{2}}\right)^2 + \left(2 - \tfrac{5}{2}\right)^2}$$
$$= \sqrt{\tfrac{7}{4}} \approx 1.323.$$

EXAMPLE 4 Finding a Minimum Area

A rectangular page will contain 24 square inches of print. The margins at the top and bottom of the page are $1\frac{1}{2}$ inches wide. The margins on each side are 1 inch wide. What should the dimensions of the page be to minimize the amount of paper used?

SOLUTION

1. A diagram of the page is shown in Figure 3.34.

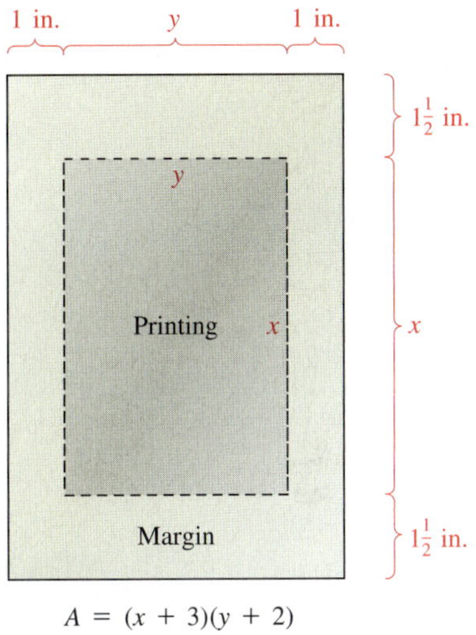

FIGURE 3.34

2. Letting A be the area to be minimized, the primary equation is

$$A = (x + 3)(y + 2). \qquad \text{Primary equation}$$

3. The printed area inside the margins is given by

$$24 = xy. \qquad \text{Secondary equation}$$

Solving this equation for y produces

$$y = \frac{24}{x}.$$

By substituting this into the primary equation, you obtain

$$A = (x + 3)\left(\frac{24}{x} + 2\right) \qquad \text{Write as a function of one variable.}$$

$$= (x + 3)\left(\frac{24 + 2x}{x}\right) \qquad \text{Rewrite as fraction.}$$

$$= \frac{2x^2}{x} + \frac{30x}{x} + \frac{72}{x} \qquad \text{Multiply and separate into terms.}$$

$$= 2x + 30 + \frac{72}{x}. \qquad \text{Simplify.}$$

4. Because x must be positive, the feasible domain is $x > 0$.

5. To find the minimum area, begin by finding the critical numbers of A.

$$\frac{dA}{dx} = 2 - \frac{72}{x^2} \qquad \text{Find derivative of } A.$$

$$0 = 2 - \frac{72}{x^2} \qquad \text{Set derivative equal to 0.}$$

$$-2 = -\frac{72}{x^2} \qquad \text{Subtract 2 from each side.}$$

$$x^2 = 36 \qquad \text{Simplify.}$$

$$x = \pm 6 \qquad \text{Critical numbers}$$

Because $x = -6$ is not in the feasible domain, you only need to consider the critical number $x = 6$. Using the First-Derivative Test, it follows that A is a minimum when $x = 6$. So, the dimensions of the page should be

$$x + 3 = 6 + 3 = 9 \text{ inches by } y + 2 = \frac{24}{6} + 2 = 6 \text{ inches.}$$

TRY IT 4

A rectangular page will contain 54 square inches of print. The margins at the top and bottom of the page are $1\frac{1}{2}$ inches wide. The margins on each side are 1 inch wide. What should the dimensions of the page be to minimize the amount of paper used?

As applications go, the four examples described in this section are fairly simple, and yet the resulting primary equations are quite complicated. Real-life applications often involve equations that are at least as complex as these four. Remember that one of the main goals of this course is to enable you to use the power of calculus to analyze equations that at first glance seem formidable.

Also remember that once you have found the primary equation, you can use the graph of the equation to help solve the problem. For instance, the graphs of the primary equations in Examples 1 through 4 are shown in Figure 3.35.

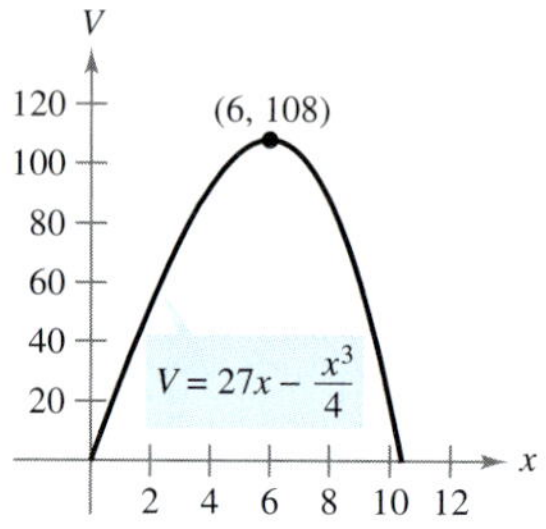

Example 1

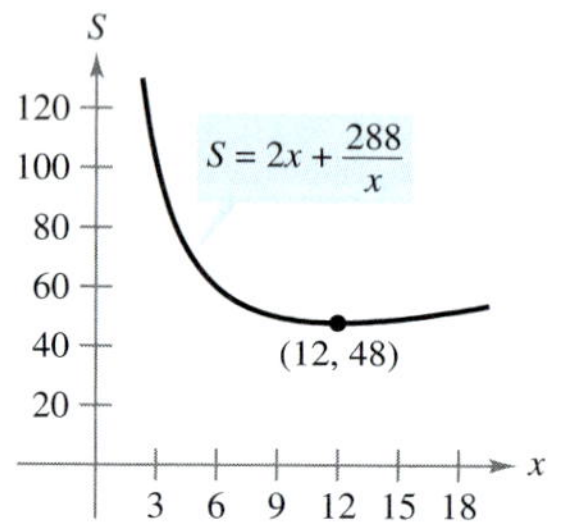

Example 2

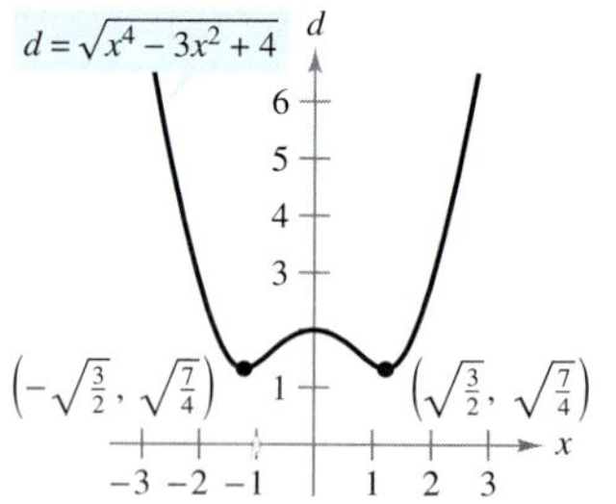

Example 3

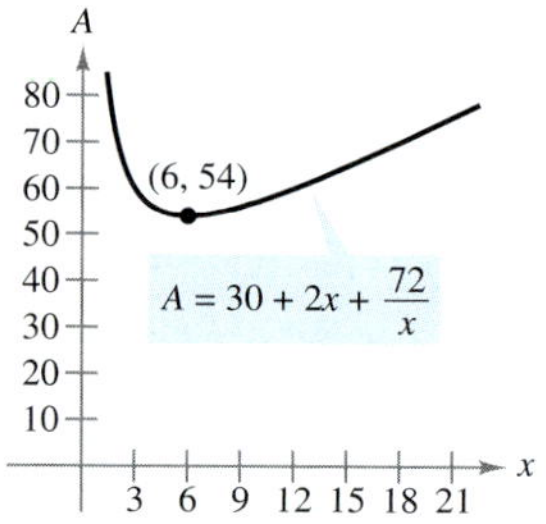

Example 4

FIGURE 3.35

TAKE ANOTHER LOOK

Comparing Graphical, Numerical, and Analytic Approaches

In Examples 1, 2, 3, and 4, an analytic approach was used to find the minimum or maximum value. To compare this type of analysis with other approaches, divide into at most eight groups and solve the examples employing a numerical approach (using a table) or a graphical approach (using a graphing utility).

Example 1 – *numerical* approach
Example 1 – *graphical* approach
Example 2 – *numerical* approach
Example 2 – *graphical* approach
Example 3 – *numerical* approach
Example 3 – *graphical* approach
Example 4 – *numerical* approach
Example 4 – *graphical* approach

When finished, compare all three approaches for each example. What are the advantages of each approach? What are the disadvantages?

PREREQUISITE REVIEW 3.4

The following warm-up exercises involve skills that were covered in earlier sections. You will use these skills in the exercise set for this section.

In Exercises 1–4, write a formula for the written statement.

1. The sum of one number and half a second number is 12.
2. The product of one number and twice another is 24.
3. The area of a rectangle is 24 square units.
4. The distance between two points is 10 units.

In Exercises 5–10, find the critical numbers of the function.

5. $y = x^2 + 6x - 9$
6. $y = 2x^3 - x^2 - 4x$
7. $y = 5x + \dfrac{125}{x}$
8. $y = 3x + \dfrac{96}{x^2}$
9. $y = \dfrac{x^2 + 1}{x}$
10. $y = \dfrac{x}{x^2 + 9}$

EXERCISES 3.4

In Exercises 1–6, find two positive numbers satisfying the given requirements.

1. The sum is 110 and the product is a maximum.
2. The sum is S and the product is a maximum.
3. The sum of the first and twice the second is 36 and the product is a maximum.
4. The sum of the first and twice the second is 100 and the product is a maximum.
5. The product is 192 and the sum is a minimum.
6. The product is 192 and the sum of the first plus three times the second is a minimum.
7. What positive number x minimizes the sum of x and its reciprocal?
8. The difference of two numbers is 50. Find the two numbers such that their product is a minimum.

In Exercises 9 and 10, find the length and width of a rectangle that has the given perimeter and a maximum area.

9. Perimeter: 100 meters
10. Perimeter: P units

In Exercises 11 and 12, find the length and width of the rectangle that has the given area and a minimum perimeter.

11. Area: 64 square feet
12. Area: A square centimeters

13. ***Maximum Area*** A rancher has 200 feet of fencing to enclose two adjacent rectangular corrals (see figure). What dimensions should be used so that the enclosed area will be a maximum?

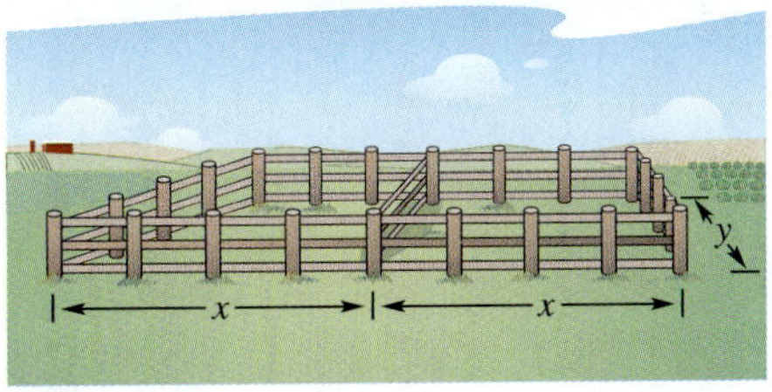

Figure for 13

14. ***Area*** A dairy farmer plans to enclose a rectangular pasture adjacent to a river. To provide enough grass for the herd, the pasture must contain 180,000 square meters. No fencing is required along the river. What dimensions will use the smallest amount of fencing?

15. ***Maximum Volume***
    (a) Verify that each of the rectangular solids shown in the figure has a surface area of 150 square inches.
    (b) Find the volume of each solid.
    (c) Determine the dimensions of a rectangular solid (with a square base) of maximum volume if its surface area is 150 square inches.

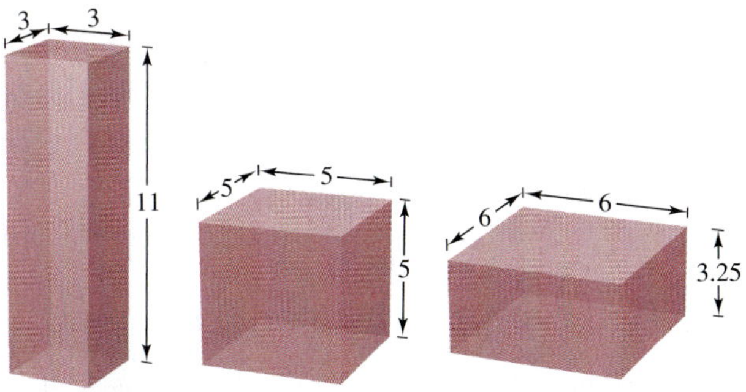

16. *Maximum Volume* Determine the dimensions of a rectangular solid (with a square base) with maximum volume if its surface area is 337.5 square centimeters.

17. *Maximum Area* A Norman window is constructed by adjoining a semicircle to the top of an ordinary rectangular window (see figure). Find the dimensions of a Norman window of maximum area if the total perimeter is 16 feet.

18. *Volume* An open box is to be made from a six-inch by six-inch square piece of material by cutting equal squares from the corners and turning up the sides (see figure). Find the volume of the largest box that can be made.

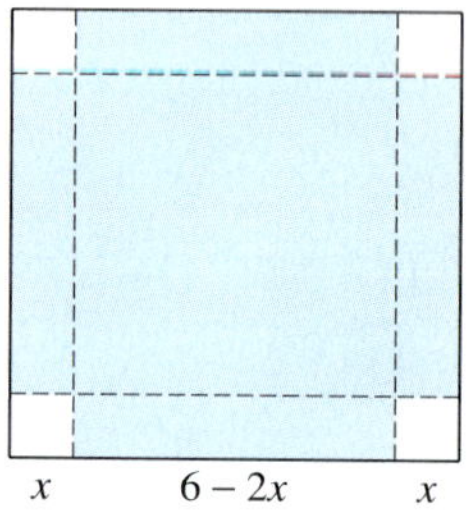

19. *Volume* An open box is to be made from a two-foot by three-foot rectangular piece of material by cutting equal squares from the corners and turning up the sides. Find the volume of the largest box that can be made in this manner.

20. *Minimum Surface Area* A net enclosure for golf practice is open at one end (see figure). The volume of the enclosure is $83\frac{1}{3}$ cubic meters. Find the dimensions that require the smallest amount of netting.

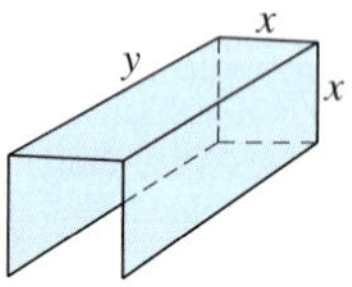

21. *Gardening* A home gardener estimates that if she plants 16 apple trees, the average yield will be 80 apples per tree. But because of the size of the garden, for each additional tree planted the yield will decrease by four apples per tree. How many trees should be planted to maximize the total yield of apples? What is the maximum yield?

22. *Area* A rectangular page is to contain 36 square inches of print. The margins at the top and bottom and on each side are to be $1\frac{1}{2}$ inches. Find the dimensions of the page that will minimize the amount of paper used.

23. *Area* A rectangular page is to contain 30 square inches of print. The margins at the top and bottom of the page are to be 2 inches wide. The margins on each side are to be 1 inch wide. Find the dimensions of the page such that the least amount of paper is used.

24. *Maximum Area* A rectangle is bounded by the x- and y-axes and the graph of

$$y = \frac{6 - x}{2}$$

(see figure). What length and width should the rectangle have so that its area is a maximum?

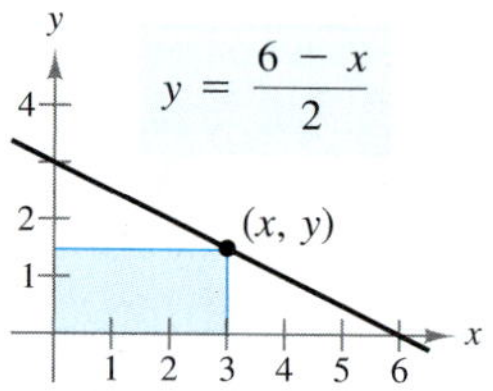

Figure for 24

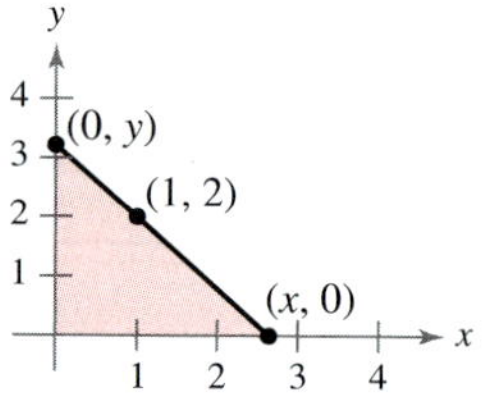

Figure for 25

25. *Minimum Length* A right triangle is formed in the first quadrant by the x- and y-axes and a line through the point (1, 2) (see figure).

(a) Write the length L of the hypotenuse as a function of x.

(b) Use a graphing utility to approximate x graphically such that the length of the hypotenuse is a minimum.

(c) Find the vertices of the triangle such that its area is a minimum.

26. *Maximum Area* A rectangle is bounded by the x-axis and the semicircle

$$y = \sqrt{25 - x^2}$$

(see figure). What length and width should the rectangle have so that its area is a maximum?

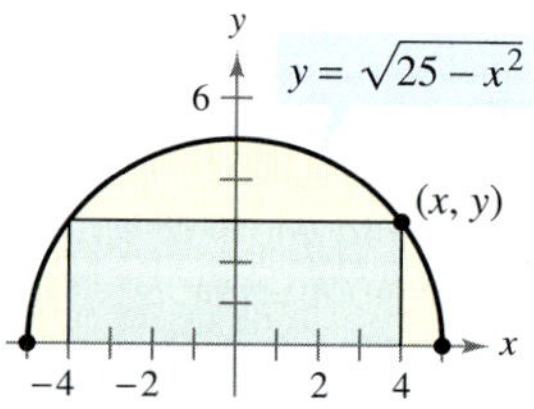

27. *Area* Find the dimensions of the largest rectangle that can be inscribed in a semicircle of radius r. (See Exercise 26.)

28. *Volume* You are designing a soft drink container that has the shape of a right circular cylinder. The container is supposed to hold 12 fluid ounces (1 fluid ounce is approximately 1.80469 cubic inches). Find the dimensions that will use a minimum amount of construction material.

29. *Volume* Find the volume of the largest right circular cylinder that can be inscribed in a sphere of radius r (see figure on next page).

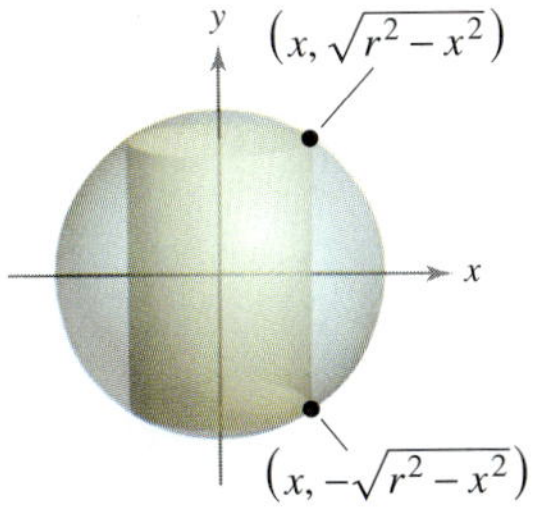

Figure for 29

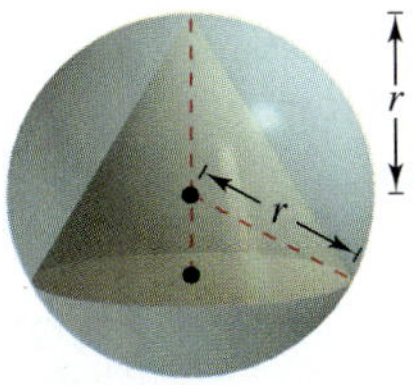

Figure for 30

30. *Maximum Volume* Find the volume of the largest right circular cone that can be inscribed in a sphere of radius r.

In Exercises 31 and 32, find the points on the graph of the function that are closest to the given point.

Function	*Point*
31. $f(x) = x^2 + 1$	$(0, 4)$
32. $f(x) = x^2$	$\left(2, \frac{1}{2}\right)$

33. *Maximum Volume* A rectangular package to be sent by a postal service can have a maximum combined length and girth of 108 inches. Find the dimensions of the package with maximum volume. Assume that the package's dimensions are x by x by y (see figure).

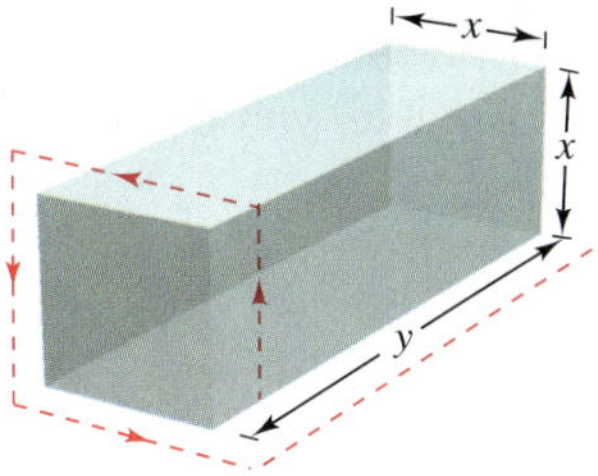

34. *Minimum Surface Area* A solid is formed by adjoining two hemispheres to the ends of a right circular cylinder. The total volume of the solid is 12 cubic inches. Find the radius of the cylinder that produces the minimum surface area.

35. *Minimum Area* The combined perimeter of a circle and a square is 16. Find the dimensions of the circle and square that produce a minimum total area.

36. *Minimum Area* The combined perimeter of an equilateral triangle and a square is 10. Find the dimensions of the triangle and square that produce a minimum total area.

37. *Minimum Time* You are in a boat 2 miles from the nearest point on the coast. You are to go to point Q, located 3 miles down the coast and 1 mile inland (see figure). You can row at a rate of 2 miles per hour and you can walk at a rate of 4 miles per hour. Toward what point on the coast should you row in order to reach point Q in the least time?

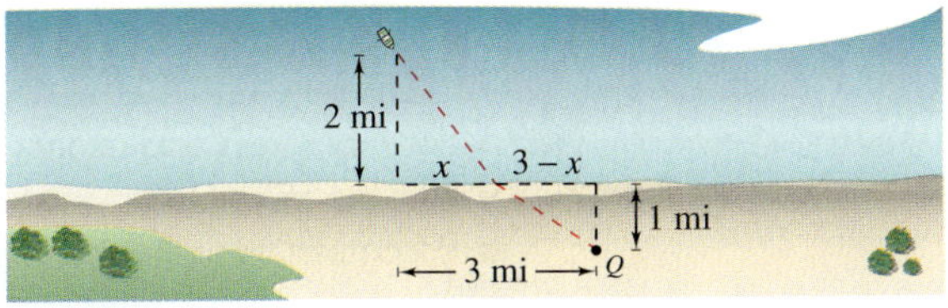

Figure for 37

38. *Maximum Area* An indoor physical fitness room consists of a rectangular region with a semicircle on each end. The perimeter of the room is to be a 200-meter running track. Find the dimensions that will make the area of the rectangular region as large as possible.

39. *Farming* A strawberry farmer will receive \$4 per bushel of strawberries during the first week of harvesting. Each week after that, the value will drop \$0.10 per bushel. The farmer estimates that there are approximately 120 bushels of strawberries in the fields, and that the crop is increasing at a rate of four bushels per week. When should the farmer harvest the strawberries to maximize their value? How many bushels of strawberries will yield the maximum value? What is the maximum value of the strawberries?

40. *Beam Strength* A wooden beam has a rectangular cross section of height h and width w (see figure). The strength S of the beam is directly proportional to its width and the square of its height. What are the dimensions of the strongest beam that can be cut from a round log of diameter 24 inches? (*Hint:* $S = kh^2w$, where k is the proportionality constant.)

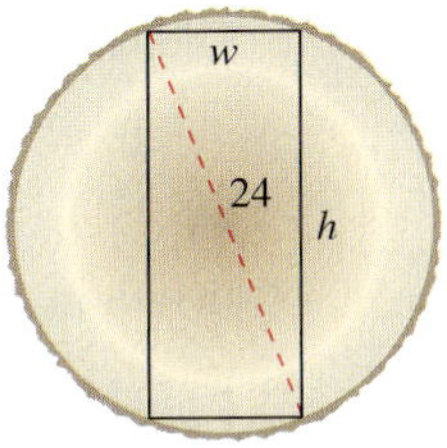

41. *Maximum Area* Use a graphing utility to graph the primary equation and its first derivative to find the dimensions of the rectangle of maximum area that can be inscribed in a semicircle of radius 10.

42. *Area* Four feet of wire is to be used to form a square and a circle.

(a) Express the sum of the areas of the square and the circle as a function A of the side of the square x.

(b) What is the domain of A?

(c) Use a graphing utility to graph A on its domain.

(d) How much wire should be used for the square and how much for the circle in order to enclose the smallest total area? the greatest total area?

3.5 BUSINESS AND ECONOMICS APPLICATIONS

- Solve business and economics optimization problems.
- Find the price elasticity of demand for demand functions.
- Recognize basic business terms and formulas.

Optimization in Business and Economics

The problems in this section are primarily optimization problems. So, the five-step procedure used in Section 3.4 is an appropriate strategy to follow.

EXAMPLE 1 Finding the Maximum Revenue

A company has determined that its total revenue (in dollars) for a product can be modeled by

$$R = -x^3 + 450x^2 + 52{,}500x$$

where x is the number of units produced (and sold). What production level will yield a maximum revenue?

SOLUTION

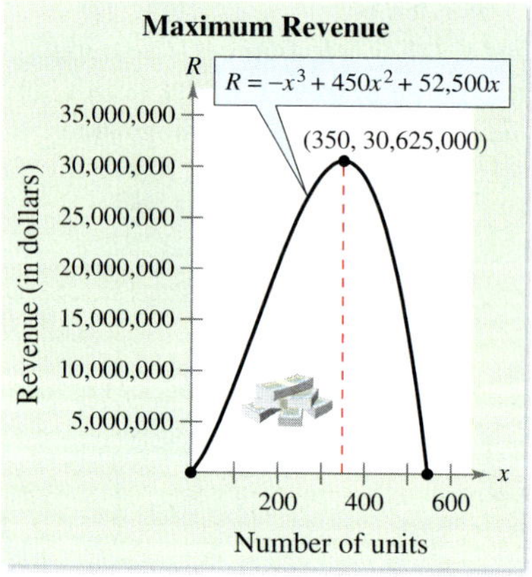

FIGURE 3.36 Maximum revenue occurs when $dR/dx = 0$.

1. A sketch of the revenue function is shown in Figure 3.36.
2. The primary equation is the given revenue function.

   $$R = -x^3 + 450x^2 + 52{,}500x \qquad \text{Primary equation}$$

3. Because R is already given as a function of one variable, you do not need a secondary equation.
4. The feasible domain of the primary equation is

   $$0 \le x \le 546. \qquad \text{Feasible domain}$$

   This is determined by finding the x-intercepts of the revenue function, as shown in Figure 3.36.
5. To maximize the revenue, find the critical numbers.

   $$\frac{dR}{dx} = -3x^2 + 900x + 52{,}500 = 0 \qquad \text{Set derivative equal to 0.}$$

   $$-3(x - 350)(x + 50) = 0 \qquad \text{Factor.}$$

   $$x = 350, x = -50 \qquad \text{Critical numbers}$$

   The only critical number in the feasible domain is $x = 350$. From the graph of the function, you can see that the production level of 350 units corresponds to a maximum revenue.

TRY IT 1

Find the number of units that must be produced to maximize the revenue function $R = -x^3 + 150x^2 + 9375x$. What is the maximum revenue?

To study the effects of production levels on cost, economists use the **average cost function** $\overline{C}$, which is defined as

$$\overline{C} = \frac{C}{x}$$ Average cost function

where $C = f(x)$ is the total cost function and x is the number of units produced.

EXAMPLE 2 Finding the Minimum Average Cost

A company estimates that the cost (in dollars) of producing x units of a product can be modeled by $C = 800 + 0.04x + 0.0002x^2$. Find the production level that minimizes the average cost per unit.

SOLUTION

1. C represents the total cost, x represents the number of units produced, and $\overline{C}$ represents the average cost per unit.
2. The primary equation is

$$\overline{C} = \frac{C}{x}.$$ Primary equation

3. Substituting the given equation for C produces

$$\overline{C} = \frac{800 + 0.04x + 0.0002x^2}{x}$$ Substitute for C.

$$= \frac{800}{x} + 0.04 + 0.0002x.$$ Function of one variable

4. The feasible domain for this function is

$$0 < x.$$ Feasible domain

5. You can find the critical numbers as shown.

$$\frac{d\overline{C}}{dx} = -\frac{800}{x^2} + 0.0002 = 0$$ Set derivative equal to 0.

$$0.0002 = \frac{800}{x^2}$$

$$x^2 = \frac{800}{0.0002}$$ Multiply each side by x^2 and divide each side by 0.0002.

$$x^2 = 4{,}000{,}000$$

$$x = \pm 2000$$ Critical numbers

By choosing the positive value of x and sketching the graph of $\overline{C}$, as shown in Figure 3.37, you can see that a production level of $x = 2000$ minimizes the average cost per unit.

STUDY TIP

To see that $x = 2000$ corresponds to a minimum average cost in Example 2, try evaluating $\overline{C}$ for several values of x. For instance, when $x = 400$, the average cost per unit is $\overline{C} = \$2.12$, but when $x = 2000$, the average cost per unit is $\overline{C} = \$0.84$.

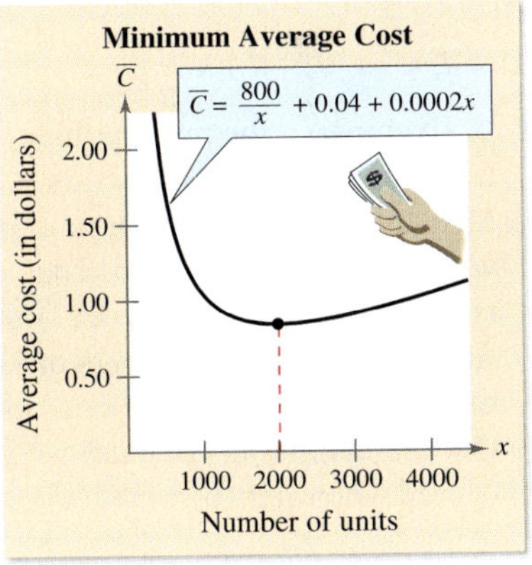

FIGURE 3.37 Minimum average cost occurs when $d\overline{C}/dx = 0$.

TRY IT 2

Find the production level that minimizes the average cost per unit for the cost function $C = 400 + 0.05x + 0.0025x^2$.

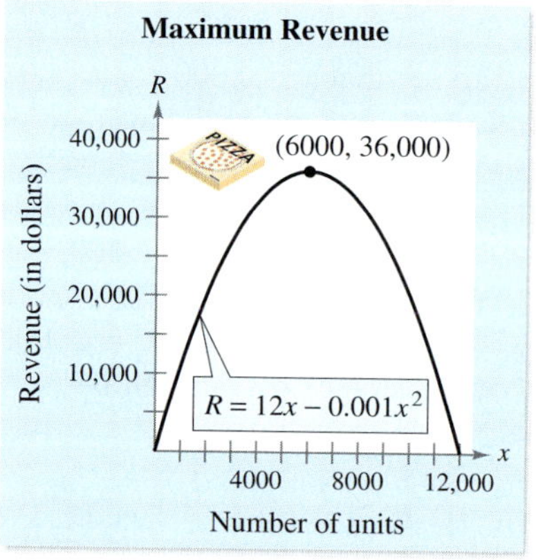

FIGURE 3.38

STUDY TIP

In Example 3, the revenue function was written as a function of x. It could also have been written as a function of p. That is, $R = 1000(12p - p^2)$. By finding the critical numbers of this function, you can determine that the maximum revenue occurs when $p = 6$.

EXAMPLE 3 Finding the Maximum Revenue

A business sells 2000 units of a product per month at a price of \$10 each. It can sell 250 more items per month for each \$0.25 reduction in price. What price per unit will maximize the monthly revenue?

SOLUTION

1. Let x represent the number of units sold in a month, let p represent the price per unit, and let R represent the monthly revenue.

2. Because the revenue is to be maximized, the primary equation is

$$R = xp. \qquad \text{Primary equation}$$

3. A price of $p = \$10$ corresponds to $x = 2000$, and a price of $p = \$9.75$ corresponds to $x = 2250$. Using this information, you can use the point-slope form to create the demand equation.

$$p - 10 = \frac{10 - 9.75}{2000 - 2250}(x - 2000) \qquad \text{Point-slope form}$$

$$p - 10 = -0.001(x - 2000) \qquad \text{Simplify.}$$

$$p = -0.001x + 12 \qquad \text{Secondary equation}$$

Substituting this valuc into thc revenue equation produces

$$R = x(-0.001x + 12) \qquad \text{Substitute for } p.$$

$$= -0.001x^2 + 12x. \qquad \text{Function of one variable}$$

4. The feasible domain of the revenue function is

$$0 \le x \le 12{,}000. \qquad \text{Feasible domain}$$

5. To maximize the revenue, find the critical numbers.

$$\frac{dR}{dx} = 12 - 0.002x = 0 \qquad \text{Set derivative equal to 0.}$$

$$-0.002x = -12$$

$$x = 6000 \qquad \text{Critical number}$$

From the graph of R in Figure 3.38, you can see that this production level yields a maximum revenue. The price that corresponds to this production level is

$$p = 12 - 0.001x \qquad \text{Demand function}$$

$$= 12 - 0.001(6000) \qquad \text{Substitute 6000 for } x.$$

$$= \$6. \qquad \text{Price per unit}$$

TRY IT 3

Find the price per unit that will maximize the monthly revenue for the business in Example 3 if it can sell only 200 more items per month for each \$0.25 reduction in price.

EXAMPLE 4 Finding the Maximum Profit

The marketing department of a business has determined that the demand for a product can be modeled by

$$p = \frac{50}{\sqrt{x}}.$$

The cost of producing x units is given by

$$C = 0.5x + 500.$$

What price will yield a maximum profit?

SOLUTION

1. Let R represent the revenue, P the profit, p the price per unit, x the number of units, and C the total cost of producing x units.
2. Because you are maximizing the profit, the primary equation is

$$P = R - C. \qquad \text{Primary equation}$$

3. Because the revenue is $R = xp$, you can write the profit function as

$$\begin{aligned} P &= R - C \\ &= xp - (0.5x + 500) && \text{Substitute for } C. \\ &= x\left(\frac{50}{\sqrt{x}}\right) - 0.5x - 500 && \text{Substitute for } p. \\ &= 50\sqrt{x} - 0.5x - 500. && \text{Function of one variable} \end{aligned}$$

4. The feasible domain of the function is $127 < x \le 7872$. (When x is less than 127 or greater than 7872, the profit is negative.)
5. To maximize the profit, find the critical numbers.

$$\begin{aligned} \frac{dP}{dx} = \frac{25}{\sqrt{x}} - 0.5 &= 0 && \text{Set derivative equal to 0.} \\ \sqrt{x} &= 50 && \text{Isolate } x\text{-term on one side.} \\ x &= 2500 && \text{Critical number} \end{aligned}$$

From the graph of the profit function shown in Figure 3.39, you can see that $x = 2500$ corresponds to a maximum profit. The price that corresponds to $x = 2500$ is

$$p = \frac{50}{\sqrt{x}} = \frac{50}{\sqrt{2500}} = \frac{50}{50} = \$1.00. \qquad \text{Price per unit}$$

ALGEBRA REVIEW

For help on the algebra in Example 4, see Example 2(b) in the *Chapter 3 Algebra Review*, on page 249.

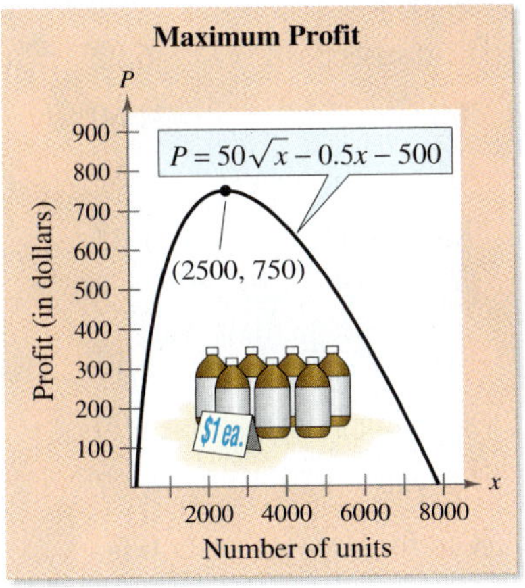

FIGURE 3.39

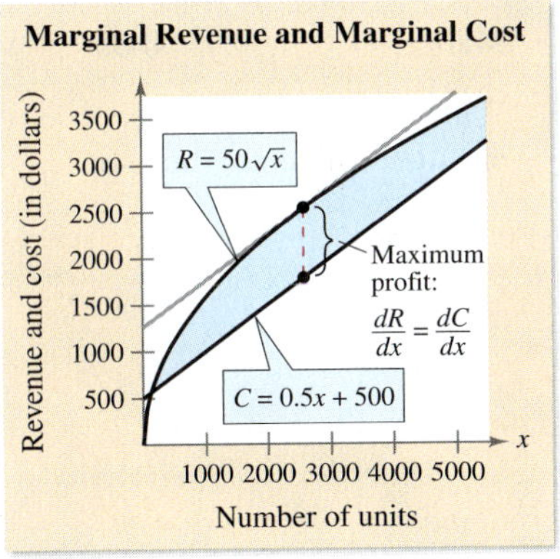

FIGURE 3.40

STUDY TIP

To find the maximum profit in Example 4, the equation $P = R - C$ was differentiated and set equal to zero. From the equation

$$\frac{dP}{dx} = \frac{dR}{dx} - \frac{dC}{dx} = 0$$

it follows that the maximum profit occurs when the marginal revenue is equal to the marginal cost, as shown in Figure 3.40.

TRY IT 4

Find the price that will maximize profit for the demand and cost functions.

$$p = \frac{40}{\sqrt{x}} \text{ and } C = 2x + 50$$

STUDY TIP

The list below shows some estimates of elasticities of demand for common products. *(Source: James Kearl, Principles of Economics)*

Item	*Absolute Value of Elasticity*
Cottonseed oil	6.92
Tomatoes	4.60
Restaurant meals	1.63
Automobiles	1.35
Cable TV	1.20
Beer	1.13
Housing	1.00
Movies	0.87
Clothing	0.60
Cigarettes	0.51
Coffee	0.25
Gasoline	0.15
Newspapers	0.10
Mail	0.05

Which of these items are elastic? Which are inelastic?

Price Elasticity of Demand

One way economists measure the responsiveness of consumers to a change in the price of a product is with **price elasticity of demand.** For example, a drop in the price of vegetables might result in a much greater demand for vegetables; such a demand is called **elastic.** On the other hand, the demand for items such as milk and water is relatively unresponsive to changes in price; the demand for such items is called **inelastic.**

More formally, the elasticity of demand is the percent change of a quantity demanded x, divided by the percent change in its price p. You can develop a formula for price elasticity of demand using the approximation

$$\frac{\Delta p}{\Delta x} \approx \frac{dp}{dx}$$

which is based on the definition of the derivative. Using this approximation, you can write

$$\begin{aligned}\text{Price elasticity of demand} &= \frac{\text{rate of change in demand}}{\text{rate of change in price}}\\ &= \frac{\Delta x/x}{\Delta p/p}\\ &= \frac{p/x}{\Delta p/\Delta x}\\ &\approx \frac{p/x}{dp/dx}.\end{aligned}$$

Definition of Price Elasticity of Demand

If $p = f(x)$ is a differentiable function, then the **price elasticity of demand** is given by

$$\eta = \frac{p/x}{dp/dx}$$

where η is the lowercase Greek letter eta. For a given price, the demand is **elastic** if $|\eta| > 1$, the demand is **inelastic** if $|\eta| < 1$, and the demand has **unit elasticity** if $|\eta| = 1$.

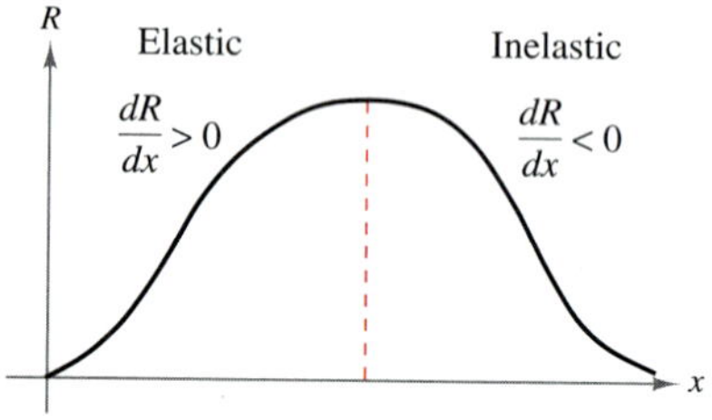

FIGURE 3.41 Revenue Curve

Price elasticity of demand is related to the total revenue function, as indicated in Figure 3.41 and the list below.

1. If the demand is *elastic*, then a decrease in price is accompanied by an increase in unit sales sufficient to increase the total revenue.
2. If the demand is *inelastic*, then a decrease in price is not accompanied by an increase in unit sales sufficient to increase the total revenue.

EXAMPLE 5 Comparing Elasticity and Revenue

The demand function for a product is modeled by

$$p = 21 - 1.5\sqrt{x}, \quad 0 \le x \le 196$$

as shown in Figure 3.42(a).

(a) Find the intervals on which the demand is elastic, inelastic, and of unit elasticity.

(b) Use the result of part (a) to describe the behavior of the revenue function.

SOLUTION

(a) The price elasticity of demand is given by

$$\eta = \frac{p/x}{dp/dx} \qquad \text{Formula for price elasticity of demand}$$

$$= \frac{\dfrac{21 - 1.5\sqrt{x}}{x}}{\dfrac{-3}{4\sqrt{x}}} \qquad \text{Substitute for } p/x \text{ and } dp/dx.$$

$$= \frac{-28\sqrt{x} + 2x}{x} \qquad \text{Multiply numerator and denominator by } -\frac{4\sqrt{x}}{3}.$$

$$= -\frac{28\sqrt{x}}{x} + 2. \qquad \text{Rewrite as two fractions and simplify.}$$

The demand is of unit elasticity when $|\eta| = 1$. In the interval $[0, 196]$, the only solution of the equation

$$|\eta| = \left| -\frac{28\sqrt{x}}{x} + 2 \right| = 1 \qquad \text{Unit elasticity}$$

is $x = 784/9$. So, the demand is of unit elasticity when $x = 784/9$. For x-values in the interval $(0, 784/9)$,

$$|\eta| = \left| -\frac{28\sqrt{x}}{x} + 2 \right| > 1, \quad 0 < x < 784/9 \qquad \text{Elastic}$$

which implies that the demand is elastic when $0 < x < 784/9$. For x-values in the interval $(784/9, 196)$,

$$|\eta| = \left| -\frac{28\sqrt{x}}{x} + 2 \right| < 1, \quad 784/9 < x < 196 \qquad \text{Inelastic}$$

which implies that the demand is inelastic when $784/9 < x < 196$.

(b) From part (a), you can conclude that the revenue function R is increasing on the open interval $(0, 784/9)$, is decreasing on the open interval $(784/9, 196)$, and is a maximum when $x = 784/9$, as indicated in Figure 3.42(b).

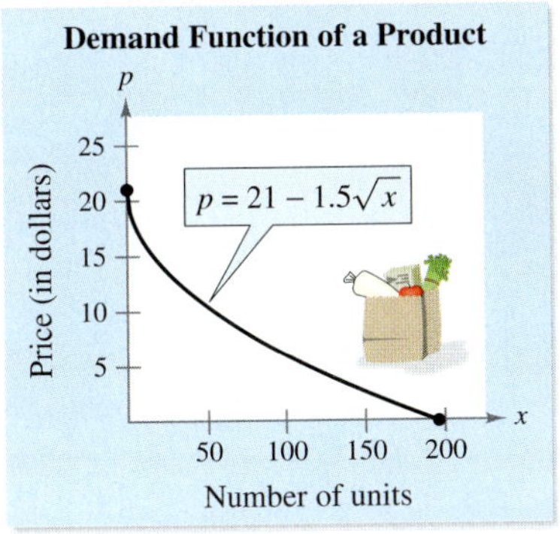

(a)

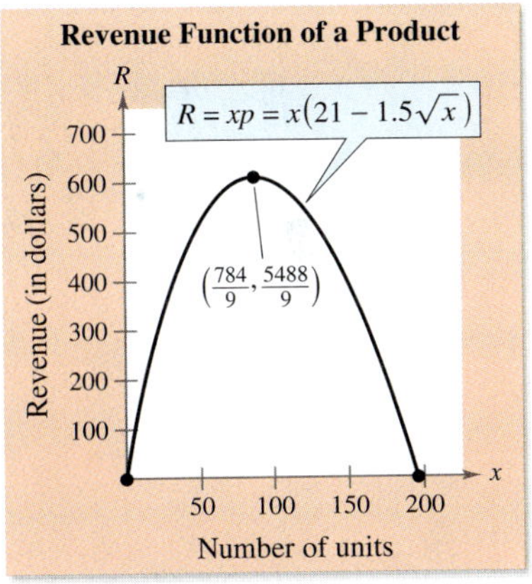

(b)

FIGURE 3.42

TRY IT 5

Find the intervals on which the demand function $p = 34 - 2\sqrt{x}$, $0 \le x \le 289$, is elastic, inelastic, and of unit elasticity.

STUDY TIP

In the discussion of price elasticity of demand, the price is assumed to decrease as the quantity demanded increases. So, the demand function $p = f(x)$ is decreasing and dp/dx is negative.

Business Terms and Formulas

This section concludes with a summary of the basic business terms and formulas used in this section. A summary of the graphs of the demand, revenue, cost, and profit functions is shown in Figure 3.43.

Summary of Business Terms and Formulas

x = number of units produced (or sold)	η = price elasticity of demand $= (p/x)/(dp/dx)$
p = price per unit	dR/dx = marginal revenue
R = total revenue from selling x units $= xp$	dC/dx = marginal cost
C = total cost of producing x units	dP/dx = marginal profit
P = total profit from selling x units $= R - C$	
$\overline{C}$ = average cost per unit $= \dfrac{C}{x}$	

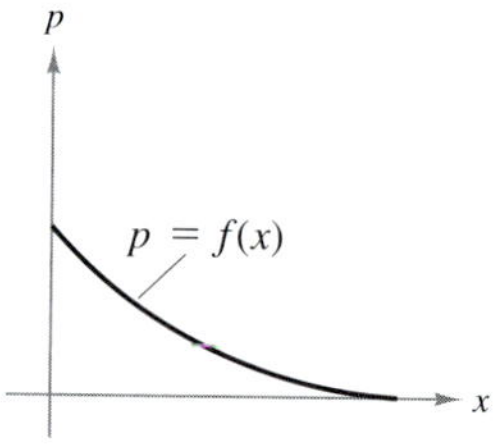

Demand function

Quantity demanded increases as price decreases.

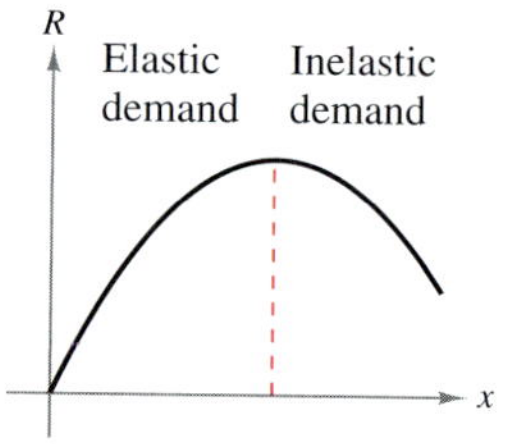

Revenue function

The low prices required to sell more units eventually result in a decreasing revenue.

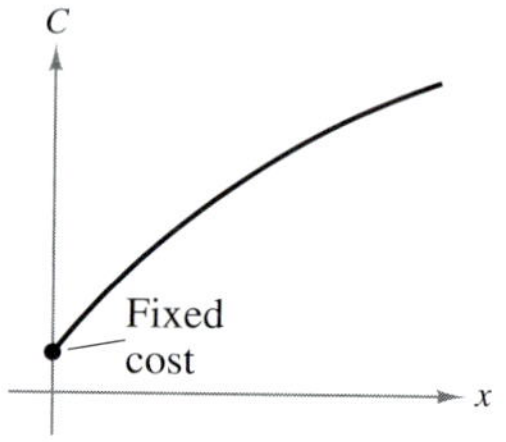

Cost function

The total cost to produce x units includes the fixed cost.

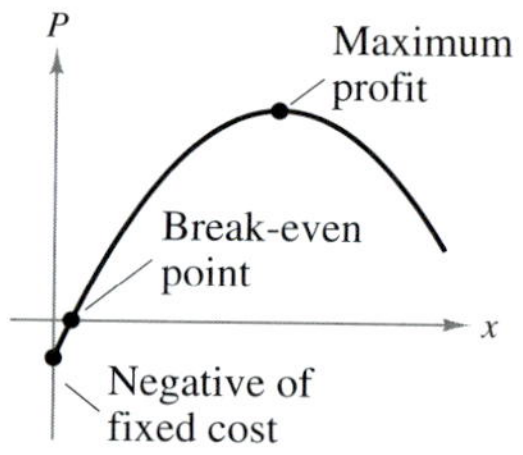

Profit function

The break-even point occurs when $R = C$.

FIGURE 3.43

TAKE ANOTHER LOOK

Demand Function

Throughout this text, it is assumed that demand functions are decreasing. Can you think of a product that has an increasing demand function? That is, can you think of a product that becomes more in demand as its price increases? Explain your reasoning, and sketch a graph of the function.

PREREQUISITE REVIEW 3.5

The following warm-up exercises involve skills that were covered in earlier sections. You will use these skills in the exercise set for this section.

In Exercises 1–4, evaluate the expression for $x = 150$.

1. $\left|-\dfrac{300}{x} + 3\right|$
2. $\left|-\dfrac{600}{5x} + 2\right|$
3. $\left|\dfrac{(20x^{-1/2})/x}{-10x^{-3/2}}\right|$
4. $\left|\dfrac{(4000/x^2)/x}{-8000x^{-3}}\right|$

In Exercises 5–10, find the marginal revenue, marginal cost, or marginal profit.

5. $C = 650 + 1.2x + 0.003x^2$
6. $P = 0.01x^2 + 11x$
7. $R = 14x - \dfrac{x^2}{2000}$
8. $R = 3.4x - \dfrac{x^2}{1500}$
9. $P = -0.7x^2 + 7x - 50$
10. $C = 1700 + 4.2x + 0.001x^3$

EXERCISES 3.5

In Exercises 1–4, find the number of units x that produces a maximum revenue R.

1. $R = 800x - 0.2x^2$
2. $R = 48x^2 - 0.02x^3$
3. $R = 400x - x^2$
4. $R = 30x^{2/3} - 2x$

In Exercises 5–8, find the number of units x that produces the minimum average cost per unit $\overline{C}$.

5. $C = 1.25x^2 + 25x + 8000$
6. $C = 0.001x^3 + 5x + 250$
7. $C = 2x^2 + 255x + 5000$
8. $C = 0.02x^3 + 55x^2 + 1250$

In Exercises 9–12, find the price per unit p that produces the maximum profit P.

	Cost Function	*Demand Function*
9.	$C = 100 + 30x$	$p = 90 - x$
10.	$C = 0.5x + 600$	$p = \dfrac{60}{\sqrt{x}}$
11.	$C = 8000 + 50x + 0.03x^2$	$p = 70 - 0.001x$
12.	$C = 35x + 500$	$p = 50 - 0.1\sqrt{x}$

Average Cost In Exercises 13 and 14, use the cost function to find the production level for which the average cost is a minimum. For this production level, show that the marginal cost and average cost are equal. Use a graphing utility to graph the average cost function and verify your results.

13. $C = 2x^2 + 5x + 18$
14. $C = x^3 - 6x^2 + 13x$

15. ***Maximum Profit*** A commodity has a demand function modeled by

$$p = 100 - 0.5x^2$$

and a total cost function modeled by $C = 40x + 37.5$.

(a) What price yields a maximum profit?

(b) When the profit is maximized, what is the average cost per unit?

16. ***Maximum Profit*** How would the answer to Exercise 15 change if the marginal cost rose from \$40 per unit to \$50 per unit? In other words, rework Exercise 15 using the cost function $C = 50x + 37.5$.

Maximum Profit In Exercises 17 and 18, find the amount s of advertising that maximizes the profit P. (s and P are measured in thousands of dollars.) Find the point of diminishing returns.

17. $P = -2s^3 + 35s^2 - 100s + 200$
18. $P = -0.1s^3 + 6s^2 + 400$

19. ***Maximum Profit*** The cost per unit of producing a type of radio is \$60. The manufacturer charges \$90 per unit for orders of 100 or less. To encourage large orders, however, the manufacturer reduces the charge by \$0.10 per radio for each order in excess of 100 units. For instance, an order of 101 radios would be \$89.90 per radio, an order of 102 radios would be \$89.80 per radio, and so on. Find the largest order the manufacturer should allow to obtain a maximum profit.

20. *Maximum Profit* A real estate office handles a 50-unit apartment complex. When the rent is \$580 per month, all units are occupied. For each \$40 increase in rent, however, an average of one unit becomes vacant. Each occupied unit requires an average of \$45 per month for service and repairs. What rent should be charged to obtain a maximum profit?

21. *Maximum Revenue* When a wholesaler sold a product at \$40 per unit, sales were 300 units per week. After a price increase of \$5, however, the average number of units sold dropped to 275 per week. Assuming that the demand function is linear, what price per unit will yield a maximum total revenue?

22. *Maximum Profit* Assume that the amount of money deposited in a bank is proportional to the square of the interest rate the bank pays on the money. Furthermore, the bank can reinvest the money at 12% simple interest. Find the interest rate the bank should pay to maximize its profit.

23. *Minimum Cost* A power station is on one side of a river that is 0.5 mile wide, and a factory is 6 miles downstream on the other side of the river (see figure). It costs \$6 per foot to run overland power lines and \$8 per foot to run underwater power lines. Write a cost function for running the power lines from the power station to the factory. Use a graphing utility to graph your function. Estimate the value of x that minimizes the cost. Explain your results.

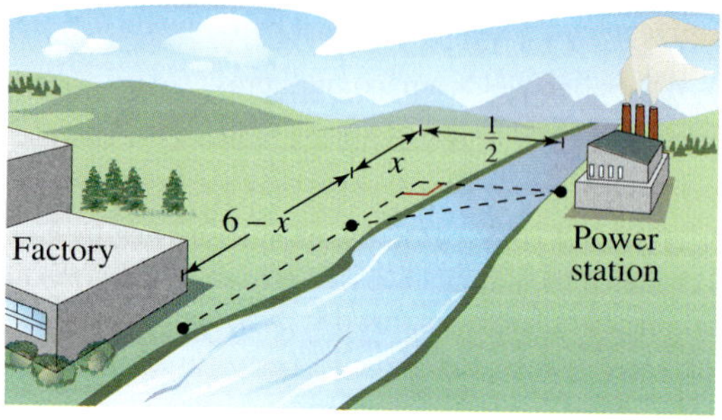

24. *Minimum Cost* An offshore oil well is 1 mile off the coast. The oil refinery is 2 miles down the coast. Laying pipe in the ocean is twice as expensive as laying it on land. Find the most economical path for the pipe from the well to the oil refinery.

25. *Minimum Cost* A small business uses a minivan to make deliveries. The cost per hour for fuel is $C = v^2/600$, where v is the speed of the minivan (in miles per hour). The driver is paid \$10 per hour. Find the speed that minimizes the cost of a 110-mile trip. (Assume there are no costs other than fuel and wages.)

26. *Minimum Cost* Repeat Exercise 25 for a fuel cost per hour of

$$C = \frac{v^2 + 360}{720}$$

and a wage of \$8 per hour.

Elasticity In Exercises 27–32, find the price elasticity of demand for the demand function at the indicated x-value. Is the demand elastic, inelastic, or of unit elasticity at the indicated x-value? Use a graphing utility to graph the revenue function, and identify the intervals of elasticity and inelasticity.

	Demand Function	*Quantity Demanded*
27.	$p = 400 - 3x$	$x = 20$
28.	$p = 5 - 0.03x$	$x = 100$
29.	$p = 20 - 0.0002x$	$x = 30$
30.	$p = \dfrac{500}{x + 2}$	$x = 23$
31.	$p = \dfrac{100}{x^2} + 2$	$x = 10$
32.	$p = 100 - \sqrt{0.2x}$	$x = 125$

33. *Elasticity* The demand function for a product is given by

$$x = p^2 - 20p + 100.$$

(a) Consider a price of \$2. If the price increases by 5%, what is the corresponding percent change in the quantity demanded?

(b) Average elasticity of demand is defined to be the percent change in quantity divided by the percent change in price. Use the percent in part (a) to find the average elasticity over the interval [2, 2.1].

(c) Find the elasticity for a price of \$2 and compare the result with that in part (b).

(d) Find an expression for the total revenue and find the values of x and p that maximize the total revenue.

34. *Elasticity* The demand function for a product is given by

$$p^3 + x^3 = 9.$$

(a) Find the price elasticity of demand when $x = 2$.

(b) Find the values of x and p that maximize the total revenue.

(c) For the value of x found in part (b), show that the price elasticity of demand has unit elasticity.

35. *Elasticity* The demand function for a product is given by

$$p = 20 - 0.02x, \quad 0 < x < 1000.$$

(a) Find the price elasticity of demand when $x = 560$.

(b) Find the values of x and p that maximize the total revenue.

(c) For the value of x found in part (b), show that the price elasticity of demand has unit elasticity.

36. *Minimum Cost* The ordering and transportation cost C of the components used in manufacturing a product is modeled by

$$C = 100\left(\frac{200}{x^2} + \frac{x}{x+30}\right), \quad x \geq 1$$

where C is measured in thousands of dollars and x is the order size in hundreds. Find the order size that minimizes the cost. (*Hint:* Use the *root* feature of a graphing utility.)

37. *Revenue* The demand for a car wash is

$$x = 600 - 50p$$

where the current price is \$5.00. Can revenue be increased by lowering the price and thus attracting more customers? Use price elasticity of demand to determine your answer.

38. *Revenue* Repeat Exercise 37 for a demand function of

$$x = 800 - 40p.$$

39. *Demand* A demand function is modeled by $x = a/p^m$, where a is a constant and $m > 1$. Show that $\eta = -m$. In other words, show that a 1% increase in price results in an m% decrease in the quantity demanded.

40. *Sales* The sales S (in millions of dollars per year) for Lowe's for the years 1994 through 2003 can be modeled by

$$S = 201.556t^2 - 502.29t + 2622.8 + \frac{9286}{t},$$

$$4 \leq t \leq 13$$

where $t = 4$ corresponds to 1994. *(Source: Lowe's Companies)*

(a) During which year, from 1994 to 2003, were Lowe's sales increasing most rapidly?

(b) During which year were the sales increasing at the lowest rate?

(c) Find the rate of increase or decrease for each year in parts (a) and (b).

(d) Use a graphing utility to graph the sales function. Then use the *zoom* and *trace* features to confirm the results in parts (a), (b), and (c).

41. *Revenue* The revenue R (in millions of dollars per year) for Papa John's for the years 1994 through 2003 can be modeled by

$$R = \frac{-18.0 + 24.74t}{1 - 0.16t + 0.008t^2}, \quad 4 \leq t \leq 13$$

where $t = 4$ corresponds to 1994. *(Source: Papa John's Int'l.)*

(a) During which year, from 1994 to 2003, was Papa John's revenue the greatest? the least?

(b) During which year was the revenue increasing at the greatest rate? decreasing at the greatest rate?

(c) Use a graphing utility to graph the revenue function, and confirm your results in parts (a) and (b).

42. Match each graph with the function it best represents—a demand function, a revenue function, a cost function, or a profit function. Explain your reasoning. (The graphs are labeled a–d.)

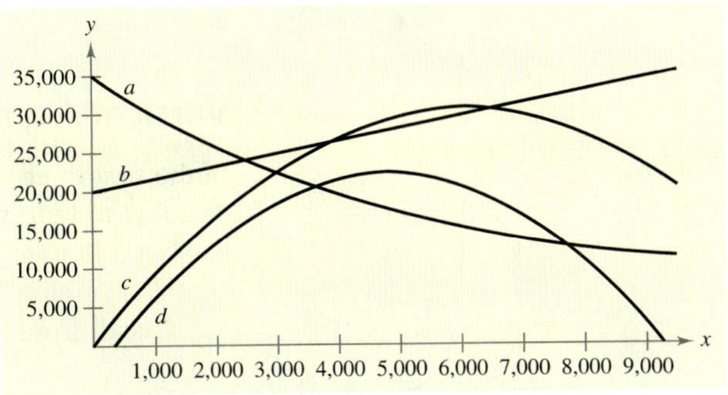

BUSINESS CAPSULE

Courtesy of Transperfect Translations

While graduate students, Elizabeth Elting and Phil Shawe co-founded TransPerfect Translations in 1992. They used a rented computer and a \$5000 credit card cash advance to market their service-oriented translation firm, now one of the largest in the country. Currently, they have a network of 4000 certified language specialists in North America, Europe, and Asia, which translates technical, legal, business, and marketing materials. In 2004, the company estimates its gross sales will be \$35 million.

43. *Research Project* Choose an innovative product like the one described above. Use your school's library, the Internet, or some other reference source to research the history of the product or service. Collect data about the revenue that the product or service has generated, and find a mathematical model of the data. Summarize your findings.

3.6 ASYMPTOTES

- Find the vertical asymptotes of functions and find infinite limits.
- Find the horizontal asymptotes of functions and find limits at infinity.
- Use asymptotes to answer questions about real-life situations.

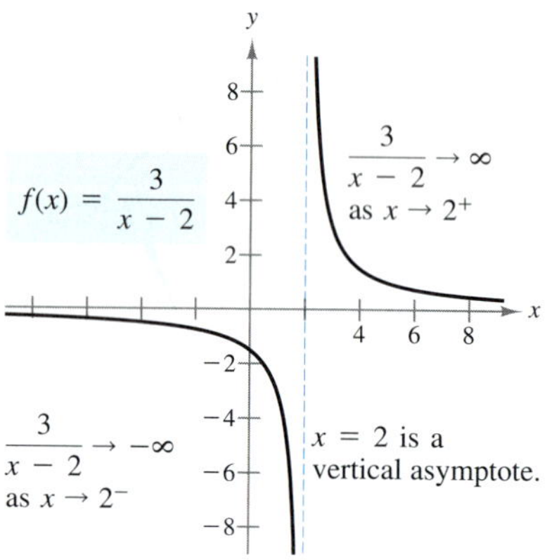

FIGURE 3.44

Vertical Asymptotes and Infinite Limits

In the first three sections of this chapter, you studied ways in which you can use calculus to help analyze the graph of a function. In this section, you will study another valuable aid to curve sketching: the determination of vertical and horizontal asymptotes.

Recall from Section 1.5, Example 10, that the function

$$f(x) = \frac{3}{x-2}$$

is unbounded as x approaches 2 (see Figure 3.44). This type of behavior is described by saying that the line $x = 2$ is a **vertical asymptote** of the graph of f. The type of limit in which $f(x)$ approaches infinity (or negative infinity) as x approaches c from the left or from the right is an **infinite limit.** The infinite limits for the function $f(x) = 3/(x-2)$ can be written as

$$\lim_{x \to 2^-} \frac{3}{x-2} = -\infty$$

and

$$\lim_{x \to 2^+} \frac{3}{x-2} = \infty.$$

Definition of Vertical Asymptote

If $f(x)$ approaches infinity (or negative infinity) as x approaches c from the right or from the left, then the line $x = c$ is a **vertical asymptote** of the graph of f.

TECHNOLOGY

When you use a graphing utility to graph a function that has a vertical asymptote, the utility may try to connect separate branches of the graph. For instance, the figure at the right shows the graph of

$$f(x) = \frac{3}{x-2}$$

on a graphing calculator.

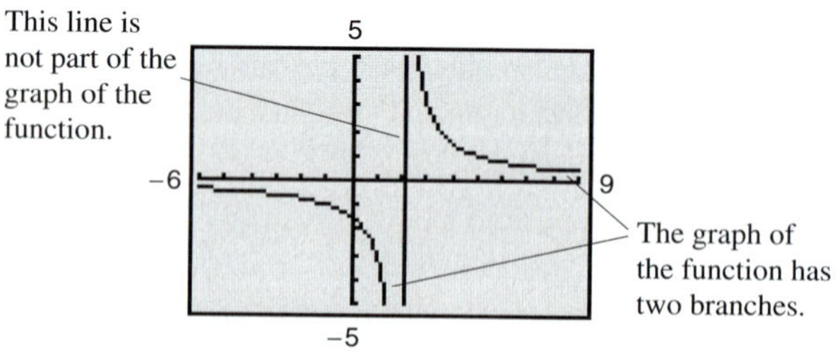

One of the most common instances of a vertical asymptote is the graph of a *rational function*—that is, a function of the form $f(x) = p(x)/q(x)$, where $p(x)$ and $q(x)$ are polynomials. If c is a real number such that $q(c) = 0$ and $p(c) \neq 0$, the graph of f has a vertical asymptote at $x = c$. Example 1 shows four cases.

EXAMPLE 1 **Finding Infinite Limits**

Find each limit.

	Limit from the left	*Limit from the right*	
(a)	$\lim_{x\to 1^-} \dfrac{1}{x-1} = -\infty$	$\lim_{x\to 1^+} \dfrac{1}{x-1} = \infty$	See Figure 3.45(a).
(b)	$\lim_{x\to 1^-} \dfrac{-1}{x-1} = \infty$	$\lim_{x\to 1^+} \dfrac{-1}{x-1} = -\infty$	See Figure 3.45(b).
(c)	$\lim_{x\to 1^-} \dfrac{-1}{(x-1)^2} = -\infty$	$\lim_{x\to 1^+} \dfrac{-1}{(x-1)^2} = -\infty$	See Figure 3.45(c).
(d)	$\lim_{x\to 1^-} \dfrac{1}{(x-1)^2} = \infty$	$\lim_{x\to 1^+} \dfrac{1}{(x-1)^2} = \infty$	See Figure 3.45(d).

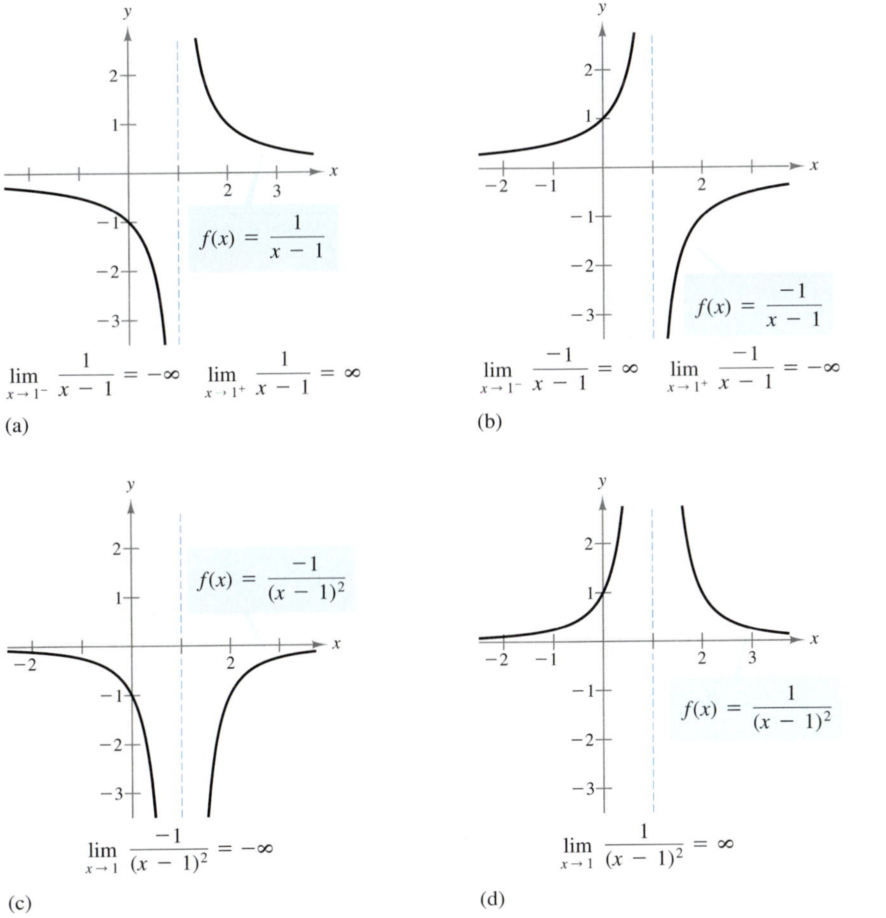

FIGURE 3.45

TECHNOLOGY

Use a spreadsheet or table to verify the results shown in Example 1. (Consult the user's manual of a spreadsheet software program for specific instructions on how to create a table.) For instance, in Example 1(a), notice that the values of $f(x) = 1/(x-1)$ decrease and increase without bound as x gets closer and closer to 1 from the left and the right.

x Approaches 1 from the Left

x	$f(x) = 1/(x-1)$
0	-1
0.9	-10
0.99	-100
0.999	-1000
0.9999	$-10{,}000$

x Approaches 1 from the Right

x	$f(x) = 1/(x-1)$
2	1
1.1	10
1.01	100
1.001	1000
1.0001	10,000

TRY IT 1

Find each limit.

(a) *Limit from the left*

$$\lim_{x\to 2^-} \frac{1}{x-2}$$

Limit from the right

$$\lim_{x\to 2^+} \frac{1}{x-2}$$

(b) *Limit from the left*

$$\lim_{x\to -3^-} \frac{-1}{x+3}$$

Limit from the right

$$\lim_{x\to -3^+} \frac{-1}{x+3}$$

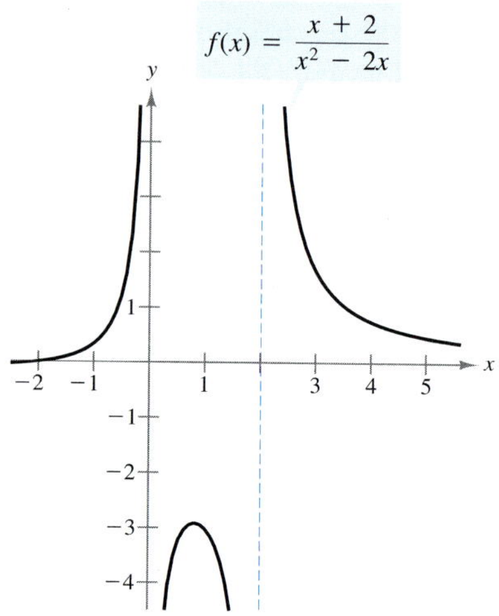

FIGURE 3.46 Vertical Asymptotes at $x = 0$ and $x = 2$

Each of the graphs in Example 1 has only one vertical asymptote. As shown in the next example, the graph of a rational function can have more than one vertical asymptote.

EXAMPLE 2 Finding Vertical Asymptotes

Find the vertical asymptotes of the graph of

$$f(x) = \frac{x+2}{x^2-2x}.$$

SOLUTION The possible vertical asymptotes correspond to the x-values for which the denominator is zero.

$$x^2 - 2x = 0 \qquad \text{Set denominator equal to 0.}$$

$$x(x-2) = 0 \qquad \text{Factor.}$$

$$x = 0, x = 2 \qquad \text{Zeros of denominator}$$

Because the numerator of f is not zero at either of these x-values, you can conclude that the graph of f has two vertical asymptotes—one at $x = 0$ and one at $x = 2$, as shown in Figure 3.46.

TRY IT 2

Find the vertical asymptote(s) of the graph of

$$f(x) = \frac{x+4}{x^2-4x}.$$

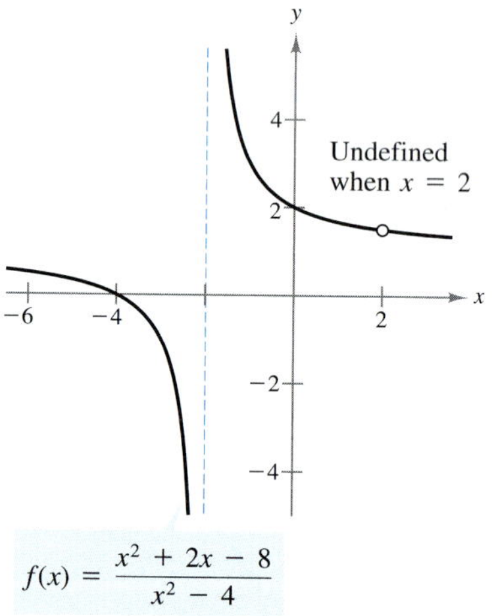

FIGURE 3.47 Vertical Asymptote at $x = -2$

EXAMPLE 3 Finding Vertical Asymptotes

Find the vertical asymptotes of the graph of

$$f(x) = \frac{x^2+2x-8}{x^2-4}.$$

SOLUTION First factor the numerator and denominator. Then divide out like factors.

$$f(x) = \frac{x^2+2x-8}{x^2-4} \qquad \text{Write original function.}$$

$$= \frac{(x+4)(x-2)}{(x+2)(x-2)} \qquad \text{Factor numerator and denominator.}$$

$$= \frac{(x+4)\cancel{(x-2)}}{(x+2)\cancel{(x-2)}} \qquad \text{Divide out like factors.}$$

$$= \frac{x+4}{x+2}, \quad x \neq 2 \qquad \text{Simplify.}$$

For all values of x other than $x = 2$, the graph of this simplified function is the same as the graph of f. So, you can conclude that the graph of f has only one vertical asymptote. This occurs at $x = -2$, as shown in Figure 3.47.

TRY IT 3

Find the vertical asymptotes of the graph of

$$f(x) = \frac{x^2+4x+3}{x^2-9}.$$

From Example 3, you know that the graph of

$$f(x) = \frac{x^2 + 2x - 8}{x^2 - 4}$$

has a vertical asymptote at $x = -2$. This implies that the limit of $f(x)$ as $x \to -2$ from the right (or from the left) is either ∞ or $-\infty$. But without looking at the graph, how can you determine that the limit from the left is *negative* infinity and the limit from the right is *positive* infinity? That is, why is the limit from the left

$$\lim_{x \to -2^-} \frac{x^2 + 2x - 8}{x^2 - 4} = -\infty \qquad \text{Limit from the left}$$

and why is the limit from the right

$$\lim_{x \to -2^+} \frac{x^2 + 2x - 8}{x^2 - 4} = \infty? \qquad \text{Limit from the right}$$

It is cumbersome to determine these limits analytically, and you may find the graphical method shown in Example 4 to be more efficient.

From the left, $f(x)$ approaches positive infinity.

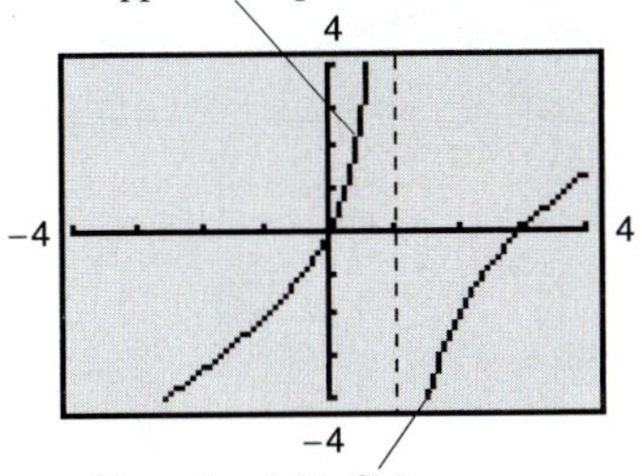

From the right, $f(x)$ approaches negative infinity.

FIGURE 3.48

EXAMPLE 4 Determining Infinite Limits

Find the limits.

$$\lim_{x \to 1^-} \frac{x^2 - 3x}{x - 1} \quad \text{and} \quad \lim_{x \to 1^+} \frac{x^2 - 3x}{x - 1}$$

SOLUTION Begin by considering the function

$$f(x) = \frac{x^2 - 3x}{x - 1}.$$

Because the denominator is zero when $x = 1$ and the numerator is not zero when $x = 1$, it follows that the graph of the function has a vertical asymptote at $x = 1$. This implies that each of the given limits is either ∞ or $-\infty$. To determine which, use a graphing utility to graph the function, as shown in Figure 3.48. From the graph, you can see that the limit from the left is positive infinity and the limit from the right is negative infinity. That is,

$$\lim_{x \to 1^-} \frac{x^2 - 3x}{x - 1} = \infty \qquad \text{Limit from the left}$$

and

$$\lim_{x \to 1^+} \frac{x^2 - 3x}{x - 1} = -\infty. \qquad \text{Limit from the right}$$

STUDY TIP

In Example 4, try evaluating $f(x)$ at x-values that are just barely to the left of 1. You will find that you can make the values of $f(x)$ arbitrarily large by choosing x sufficiently close to 1. For instance, $f(0.99999) = 199{,}999$.

TRY IT 4

Find the limits.

$$\lim_{x \to 2^-} \frac{x^2 - 4x}{x - 2} \quad \text{and} \quad \lim_{x \to 2^+} \frac{x^2 - 4x}{x - 2}$$

Then verify your solution by graphing the function.

Horizontal Asymptotes and Limits at Infinity

Another type of limit, called a **limit at infinity,** specifies a finite value approached by a function as x increases (or decreases) without bound.

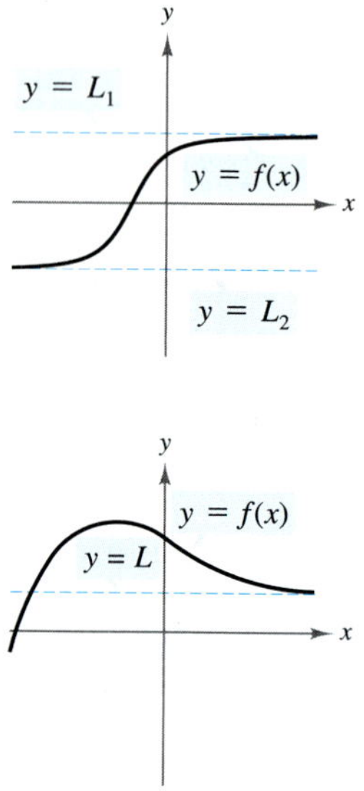

FIGURE 3.49

Definition of Horizontal Asymptote

If f is a function and L_1 and L_2 are real numbers, the statements

$$\lim_{x\to\infty} f(x) = L_1 \quad \text{and} \quad \lim_{x\to-\infty} f(x) = L_2$$

denote **limits at infinity.** The lines $y = L_1$ and $y = L_2$ are **horizontal asymptotes** of the graph of f.

Figure 3.49 shows two ways in which the graph of a function can approach one or more horizontal asymptotes. Note that it is possible for the graph of a function to cross its horizontal asymptote.

Limits at infinity share many of the properties of limits discussed in Section 1.5. When finding horizontal asymptotes, you can use the property that

$$\lim_{x\to\infty} \frac{1}{x^r} = 0, \quad r > 0 \quad \text{and} \quad \lim_{x\to-\infty} \frac{1}{x^r} = 0, \quad r > 0.$$

(The second limit assumes that x^r is defined when $x < 0$.)

EXAMPLE 5 Finding Limits at Infinity

Find the limit: $\lim_{x\to\infty} \left(5 - \frac{2}{x^2}\right)$.

SOLUTION

$$\lim_{x\to\infty} \left(5 - \frac{2}{x^2}\right) = \lim_{x\to\infty} 5 - \lim_{x\to\infty} \frac{2}{x^2} \qquad \lim_{x\to\infty} [f(x) - g(x)] = \lim_{x\to\infty} f(x) - \lim_{x\to\infty} g(x)$$

$$= \lim_{x\to\infty} 5 - 2\left(\lim_{x\to\infty} \frac{1}{x^2}\right) \qquad \lim_{x\to\infty} cf(x) = c \lim_{x\to\infty} f(x)$$

$$= 5 - 2(0)$$

$$= 5$$

You can verify this limit by sketching the graph of

$$f(x) = 5 - \frac{2}{x^2}$$

as shown in Figure 3.50. Note that the graph has $y = 5$ as a horizontal asymptote to the right. By evaluating the limit of $f(x)$ as $x \to -\infty$, you can show that this line is also a horizontal asymptote to the left.

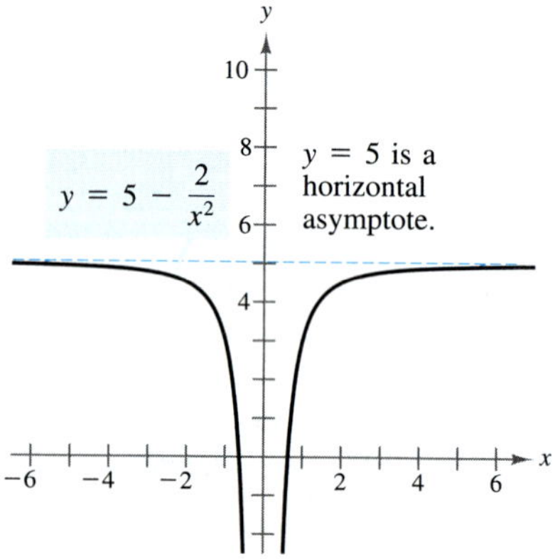

FIGURE 3.50

TRY IT 5

Find the limit: $\lim_{x\to\infty} \left(2 + \frac{5}{x^2}\right)$.

There is an easy way to determine whether the graph of a *rational* function has a horizontal asymptote. This shortcut is based on a comparison of the degrees of the numerator and denominator of the rational function.

Horizontal Asymptotes of Rational Functions

Let $f(x) = p(x)/q(x)$ be a rational function.

1. If the degree of the numerator is less than the degree of the denominator, then $y = 0$ is a horizontal asymptote of the graph of f (to the left and to the right).
2. If the degree of the numerator is equal to the degree of the denominator, then $y = a/b$ is a horizontal asymptote of the graph of f (to the left and to the right), where a and b are the leading coefficients of $p(x)$ and $q(x)$, respectively.
3. If the degree of the numerator is greater than the degree of the denominator, then the graph of f has no horizontal asymptote.

TECHNOLOGY

Some functions have two horizontal asymptotes: one to the right and one to the left. For instance, try sketching the graph of

$$f(x) = \frac{x}{\sqrt{x^2 + 1}}.$$

What horizontal asymptotes does the function appear to have?

EXAMPLE 6 Finding Horizontal Asymptotes

Find the horizontal asymptote of the graph of each function.

(a) $y = \dfrac{-2x + 3}{3x^2 + 1}$ (b) $y = \dfrac{-2x^2 + 3}{3x^2 + 1}$ (c) $y = \dfrac{-2x^3 + 3}{3x^2 + 1}$

SOLUTION

(a) Because the degree of the numerator is less than the degree of the denominator, $y = 0$ is a horizontal asymptote. [See Figure 3.51(a).]

(b) Because the degree of the numerator is equal to the degree of the denominator, the line $y = -\frac{2}{3}$ is a horizontal asymptote. [See Figure 3.51(b).]

(c) Because the degree of the numerator is greater than the degree of the denominator, the graph has no horizontal asymptote. [See Figure 3.51(c).]

TRY IT 6

Find the horizontal asymptote of the graph of each function.

(a) $y = \dfrac{2x + 1}{4x^2 + 5}$

(b) $y = \dfrac{2x^2 + 1}{4x^2 + 5}$

(c) $y = \dfrac{2x^3 + 1}{4x^2 + 5}$

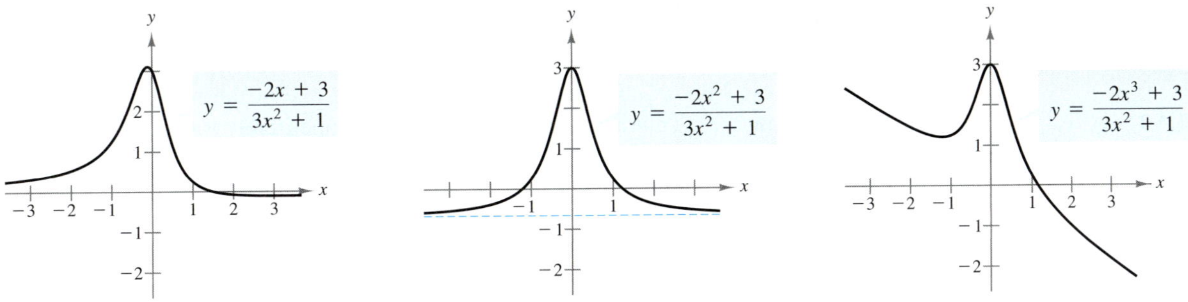

(a) $y = 0$ is a horizontal asymptote. (b) $y = -\frac{2}{3}$ is a horizontal asymptote. (c) No horizontal asymptote

FIGURE 3.51

Applications of Asymptotes

There are many examples of asymptotic behavior in real life. For instance, Example 7 describes the asymptotic behavior of an average cost function.

STUDY TIP

In Example 7, suppose that the small business had made an initial investment of \$50,000. How would this change the answers to the questions? Would it change the average cost of producing x units? Would it change the limiting average cost per unit?

EXAMPLE 7 Modeling Average Cost

A small business invests \$5000 in a new product. In addition to this initial investment, the product will cost \$0.50 per unit to produce. Find the average cost per unit if 1000 units are produced, if 10,000 units are produced, and if 100,000 units are produced. What is the limit of the average cost as the number of units produced increases?

SOLUTION From the given information, you can model the total cost C (in dollars) by

$$C = 0.5x + 5000 \qquad \text{Total cost function}$$

where x is the number of units produced. This implies that the average cost function is

$$\overline{C} = \frac{C}{x} = 0.5 + \frac{5000}{x}. \qquad \text{Average cost function}$$

If only 1000 units are produced, then the average cost per unit is

$$\overline{C} = 0.5 + \frac{5000}{1000} = \$5.50. \qquad \text{Average cost for 1000 units}$$

If 10,000 units are produced, then the average cost per unit is

$$\overline{C} = 0.5 + \frac{5000}{10{,}000} = \$1.00. \qquad \text{Average cost for 10,000 units}$$

If 100,000 units are produced, then the average cost per unit is

$$\overline{C} = 0.5 + \frac{5000}{100{,}000} = \$0.55. \qquad \text{Average cost for 100,000 units}$$

As x approaches infinity, the limiting average cost per unit is

$$\lim_{x\to\infty}\left(0.5 + \frac{5000}{x}\right) = \$0.50.$$

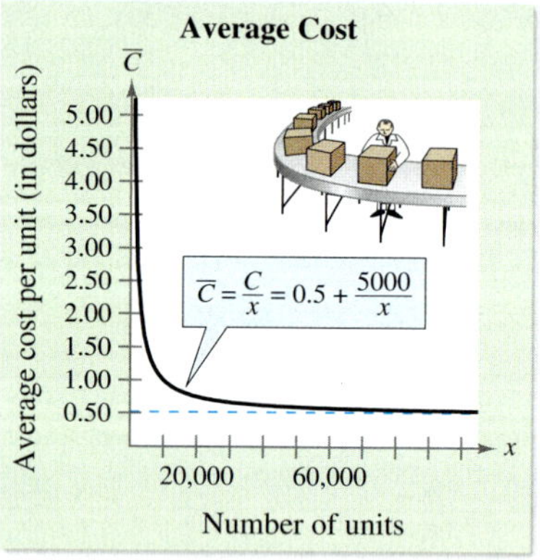

FIGURE 3.52 As $x \to \infty$, the average cost per unit approaches \$0.50.

As shown in Figure 3.52, this example points out one of the major problems of small businesses. That is, it is difficult to have competitively low prices when the production level is low.

TRY IT 7

A small business invests \$25,000 in a new product. In addition, the product will cost \$0.75 per unit to produce. Find the cost function and the average cost function. What is the limit of the average cost function as production increases?

EXAMPLE 8 Modeling Smokestack Emission

A manufacturing plant has determined that the cost C (in dollars) of removing $p\%$ of the smokestack pollutants of its main smokestack is modeled by

$$C = \frac{80{,}000p}{100 - p}, \qquad 0 \le p < 100.$$

What is the vertical asymptote of this function? What does the vertical asymptote mean to the plant owners?

SOLUTION The graph of the cost function is shown in Figure 3.53. From the graph, you can see that $p = 100$ is the vertical asymptote. This means that as the plant attempts to remove higher and higher percents of the pollutants, the cost increases dramatically. For instance, the cost of removing 85% of the pollutants is

$$C = \frac{80{,}000(85)}{100 - 85} \approx \$453{,}333 \qquad \text{Cost for 85\% removal}$$

but the cost of removing 90% is

$$C = \frac{80{,}000(90)}{100 - 90} = \$720{,}000. \qquad \text{Cost for 90\% removal}$$

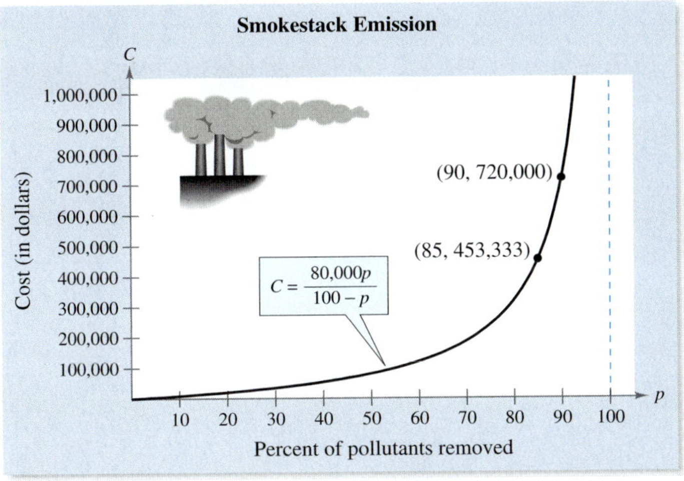

FIGURE 3.53

Rob Crandall/Rainbow

Since the 1980s, industries in the United States have spent billions of dollars to reduce air pollution.

TRY IT 8

According to the cost function in Example 8, is it possible to remove 100% of the smokestack pollutants? Why or why not?

TAKE ANOTHER LOOK

Indirect Costs

In Example 8, the given cost model considers only the direct cost for the manufacturing plant. Describe the possible indirect costs incurred by society as the plant attempts to remove more and more of the pollutants from its smokestack emission.

PREREQUISITE REVIEW 3.6

The following warm-up exercises involve skills that were covered in earlier sections. You will use these skills in the exercise set for this section.

In Exercises 1–8, find the limit.

1. $\lim_{x \to 2} (x + 1)$

2. $\lim_{x \to -1} (3x + 4)$

3. $\lim_{x \to -3} \frac{2x^2 + x - 15}{x + 3}$

4. $\lim_{x \to 2} \frac{3x^2 - 8x + 4}{x - 2}$

5. $\lim_{x \to 2^+} \frac{x^2 - 5x + 6}{x^2 - 4}$

6. $\lim_{x \to 1^-} \frac{x^2 - 6x + 5}{x^2 - 1}$

7. $\lim_{x \to 0^+} \sqrt{x}$

8. $\lim_{x \to 1^+} \left(x + \sqrt{x - 1}\right)$

In Exercises 9–12, find the average cost and the marginal cost.

9. $C = 150 + 3x$

10. $C = 1900 + 1.7x + 0.002x^2$

11. $C = 0.005x^2 + 0.5x + 1375$

12. $C = 760 + 0.05x$

EXERCISES 3.6

In Exercises 1–8, find the vertical and horizontal asymptotes. Write the asymptotes as equations of lines.

1. $f(x) = \frac{x^2 + 1}{x^2}$

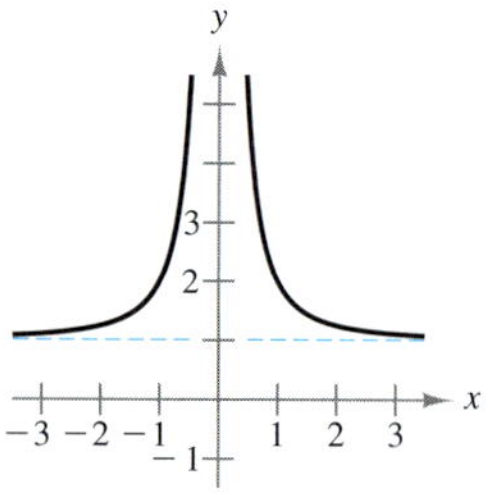

2. $f(x) = \frac{4}{(x - 2)^3}$

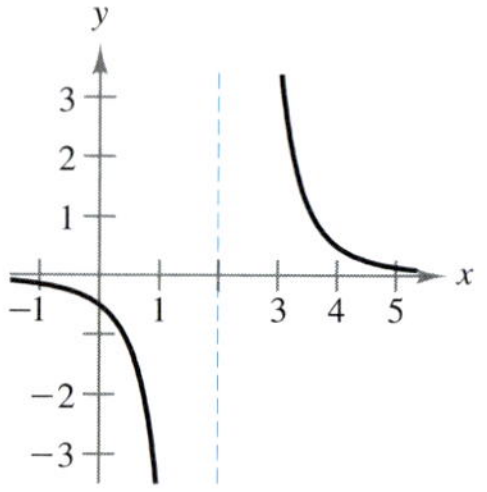

3. $f(x) = \frac{x^2 - 2}{x^2 - x - 2}$

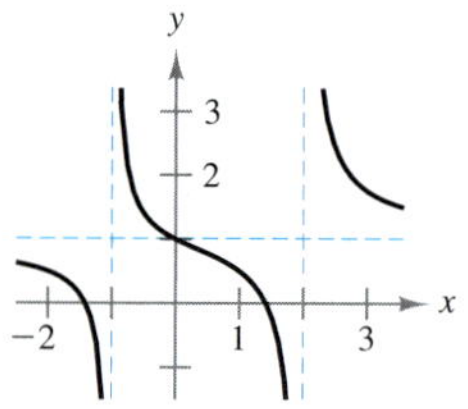

4. $f(x) = \frac{2 + x}{1 - x}$

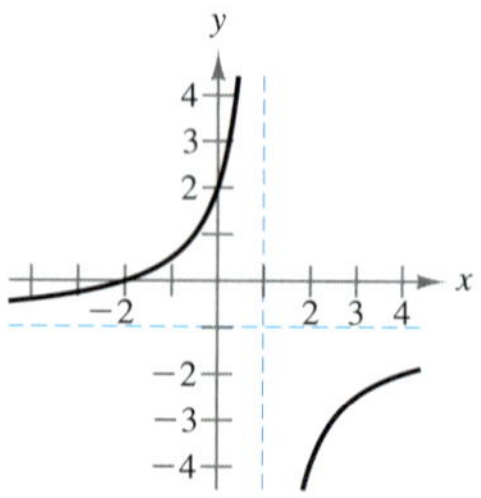

5. $f(x) = \frac{3x^2}{2(x^2 + 1)}$

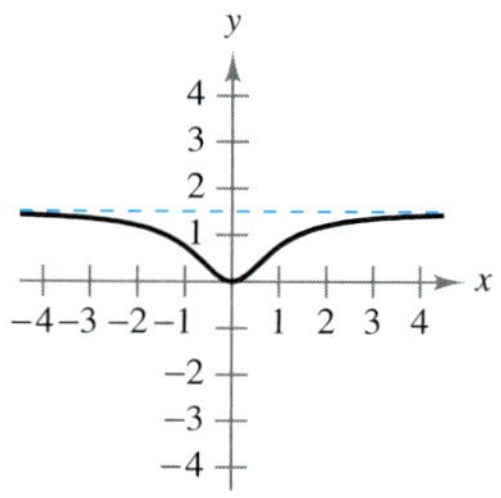

6. $f(x) = \frac{-4x}{x^2 + 4}$

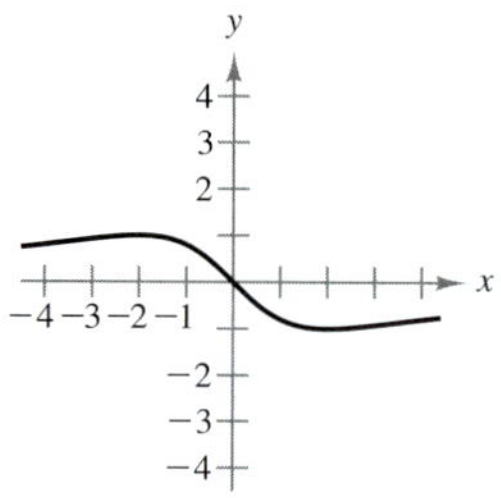

7. $f(x) = \frac{x^2 - 1}{2x^2 - 8}$

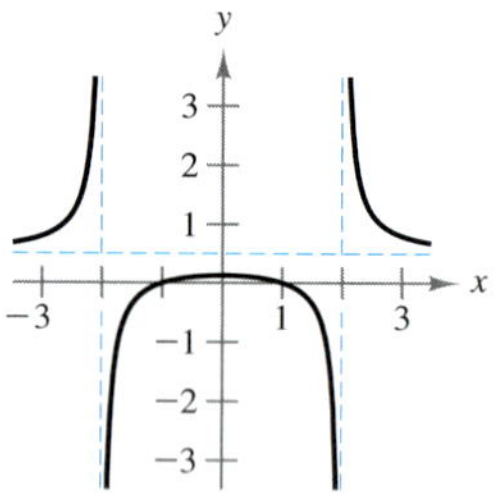

8. $f(x) = \frac{x^2 + 1}{x^3 - 8}$

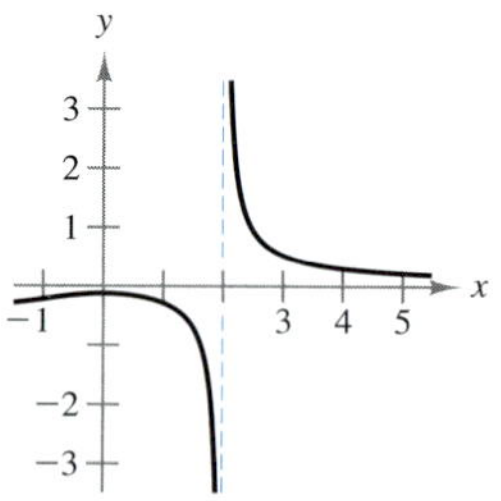

In Exercises 9–14, match the function with its graph. Use horizontal asymptotes as an aid. [The graphs are labeled (a)–(f).]

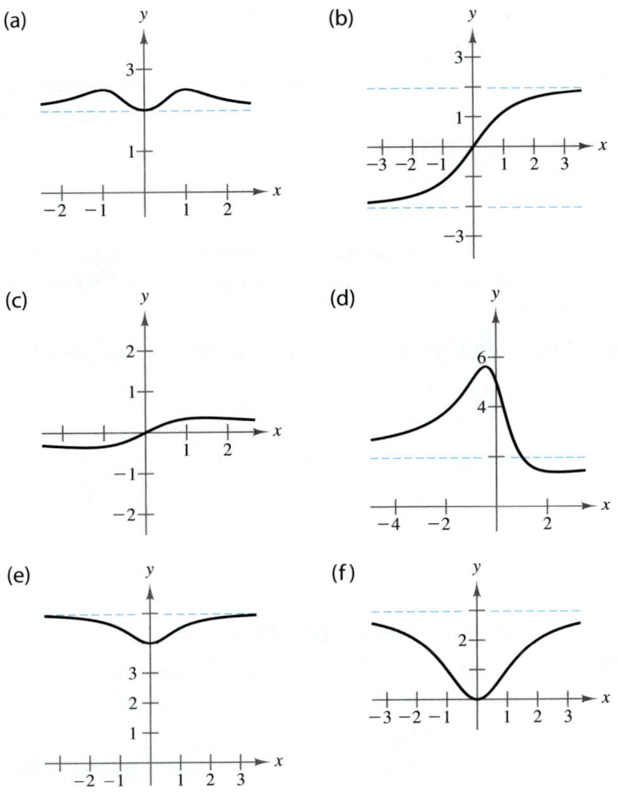

9. $f(x) = \dfrac{3x^2}{x^2 + 2}$

10. $f(x) = \dfrac{2x}{\sqrt{x^2 + 2}}$

11. $f(x) = \dfrac{x}{x^2 + 2}$

12. $f(x) = 2 + \dfrac{x^2}{x^4 + 1}$

13. $f(x) = 5 - \dfrac{1}{x^2 + 1}$

14. $f(x) = \dfrac{2x^2 - 3x + 5}{x^2 + 1}$

In Exercises 15–22, find the limit.

15. $\displaystyle\lim_{x \to -2^-} \frac{1}{(x + 2)^2}$

16. $\displaystyle\lim_{x \to -2^-} \frac{1}{x + 2}$

17. $\displaystyle\lim_{x \to 3^+} \frac{x - 4}{x - 3}$

18. $\displaystyle\lim_{x \to 1^+} \frac{2 + x}{1 - x}$

19. $\displaystyle\lim_{x \to 4^-} \frac{x^2}{x^2 - 16}$

20. $\displaystyle\lim_{x \to 4} \frac{x^2}{x^2 + 16}$

21. $\displaystyle\lim_{x \to 0^-} \left(1 + \frac{1}{x}\right)$

22. $\displaystyle\lim_{x \to 0^-} \left(x^2 - \frac{1}{x}\right)$

In Exercises 23–32, find the limit.

23. $\displaystyle\lim_{x \to \infty} \frac{2x - 1}{3x + 2}$

24. $\displaystyle\lim_{x \to \infty} \frac{5x^3 + 1}{10x^3 - 3x^2 + 7}$

25. $\displaystyle\lim_{x \to \infty} \frac{3x}{4x^2 - 1}$

26. $\displaystyle\lim_{x \to \infty} \frac{2x^{10} - 1}{10x^{11} - 3}$

27. $\displaystyle\lim_{x \to -\infty} \frac{5x^2}{x + 3}$

28. $\displaystyle\lim_{x \to \infty} \frac{x^3 - 2x^2 + 3x + 1}{x^2 - 3x + 2}$

29. $\displaystyle\lim_{x \to \infty} (2x - x^{-2})$

30. $\displaystyle\lim_{x \to \infty} (2 - x^{-3})$

31. $\displaystyle\lim_{x \to -\infty} \left(\frac{2x}{x - 1} + \frac{3x}{x + 1}\right)$

32. $\displaystyle\lim_{x \to \infty} \left(\frac{2x^2}{x - 1} + \frac{3x}{x + 1}\right)$

In Exercises 33 and 34, complete the table. Then use the result to estimate the limit of $f(x)$ as x approaches infinity.

33. $f(x) = \dfrac{x + 1}{x\sqrt{x}}$

x	10^0	10^1	10^2	10^3	10^4	10^5	10^6
$f(x)$							

34. $f(x) = x^2 - x\sqrt{x(x - 1)}$

x	10^0	10^1	10^2	10^3	10^4	10^5	10^6
$f(x)$							

In Exercises 35 and 36, use a spreadsheet software program to complete the table and use the result to estimate the limit of $f(x)$ as x approaches infinity.

35. $f(x) = \dfrac{x^2 - 1}{0.02x^2}$

x	10^0	10^1	10^2	10^3	10^4	10^5	10^6
$f(x)$							

36. $f(x) = \dfrac{3x^2}{0.1x^2 + 1}$

x	10^0	10^1	10^2	10^3	10^4	10^5	10^6
$f(x)$							

In Exercises 37 and 38, use a graphing utility to complete the table and use the result to estimate the limit of $f(x)$ as x approaches infinity and as x approaches negative infinity.

37. $f(x) = \dfrac{2x}{\sqrt{x^2 + 4}}$

x	-10^6	-10^4	-10^2	10^0	10^2	10^4	10^6
$f(x)$							

38. $f(x) = x - \sqrt{x(x-1)}$

x	-10^6	-10^4	-10^2	10^0	10^2	10^4	10^6
$f(x)$							

In Exercises 39–56, sketch the graph of the equation. Use intercepts, extrema, and asymptotes as sketching aids.

39. $y = \dfrac{2+x}{1-x}$

40. $y = \dfrac{x-3}{x-2}$

41. $f(x) = \dfrac{x^2}{x^2+9}$

42. $f(x) = \dfrac{x}{x^2+4}$

43. $g(x) = \dfrac{x^2}{x^2-16}$

44. $g(x) = \dfrac{x}{x^2-4}$

45. $xy^2 = 4$

46. $x^2y = 4$

47. $y = \dfrac{2x}{1-x}$

48. $y = \dfrac{2x}{1-x^2}$

49. $y = 3(1 - x^{-2})$

50. $y = 1 + x^{-1}$

51. $f(x) = \dfrac{1}{x^2-x-2}$

52. $f(x) = \dfrac{x-2}{x^2-4x+3}$

53. $g(x) = \dfrac{x^2-x-2}{x-2}$

54. $g(x) = \dfrac{x^2-9}{x+3}$

55. $y = \dfrac{2x^2-6}{(x-1)^2}$

56. $y = \dfrac{x}{(x+1)^2}$

57. ***Cost*** The cost C (in dollars) of producing x units of a product is $C = 1.35x + 4570$.

(a) Find the average cost function $\overline{C}$.

(b) Find $\overline{C}$ when $x = 100$ and when $x = 1000$.

(c) What is the limit of $\overline{C}$ as x approaches infinity?

58. ***Average Cost*** A business has a cost (in dollars) of $C = 0.5x + 500$ for producing x units.

(a) Find the average cost function $\overline{C}$.

(b) Find $\overline{C}$ when $x = 250$ and when $x = 1250$.

(c) What is the limit of $\overline{C}$ as x approaches infinity?

59. ***Cost*** The cost C (in millions of dollars) for the federal government to seize $p\%$ of a type of illegal drug as it enters the country is modeled by

$$C = 528p/(100 - p), \qquad 0 \le p < 100.$$

(a) Find the cost of seizing 25%, 50%, and 75%.

(b) Find the limit of C as $p \to 100^-$.

(c) Use a graphing utility to verify the result of part (b).

60. ***Cost*** The cost C (in dollars) of removing $p\%$ of the air pollutants in the stack emission of a utility company that burns coal is modeled by

$$C = 80{,}000p/(100 - p), \qquad 0 \le p < 100.$$

(a) Find the cost of removing 15%, 50%, and 90%.

(b) Find the limit of C as $p \to 100^-$.

(c) Use a graphing utility to verify the result of part (b).

61. ***Learning Curve*** Psychologists have developed mathematical models to predict performance P (the percent of correct responses) as a function of n, the number of times a task is performed. One such model is

$$P = \frac{b + \theta a(n-1)}{1 + \theta(n-1)}$$

where a, b, and θ are constants that depend on the actual learning situation. Find the limit of P as n approaches infinity.

62. ***Learning Curve*** Consider the learning curve given by

$$P = \frac{0.5 + 0.9(n-1)}{1 + 0.9(n-1)}, \qquad 0 < n.$$

(a) Complete the table for the model.

n	1	2	3	4	5	6	7	8	9	10
P										

(b) Find the limit as n approaches infinity.

(c) Use a graphing utility to graph this learning curve, and interpret the graph in the context of the problem.

63. ***Biology: Wildlife Management*** The state game commission introduces 30 elk into a new state park. The population N of the herd is modeled by

$$N = [10(3 + 4t)]/(1 + 0.1t)$$

where t is the time in years.

(a) Find the size of the herd after 5, 10, and 25 years.

(b) According to this model, what is the limiting size of the herd as time progresses?

64. ***Average Profit*** The cost and revenue functions for a product are $C = 34.5x + 15{,}000$ and $R = 69.9x$.

(a) Find the average profit function

$$\overline{P} = (R - C)/x.$$

(b) Find the average profit when x is 1000, 10,000, and 100,000.

(c) What is the limit of the average profit function as x approaches infinity? Explain your reasoning.

65. ***Average Profit*** The cost and revenue functions for a product are $C = 25.5x + 1000$ and $R = 75.5x$.

(a) Find the average profit function $\overline{P} = \dfrac{R - C}{x}$.

(b) Find the average profit when x is 100, 500, and 1000.

(c) What is the limit of the average profit function as x approaches infinity? Explain your reasoning.

3.7 CURVE SKETCHING: A SUMMARY

- Analyze the graphs of functions.
- Recognize the graphs of simple polynomial functions.

Summary of Curve-Sketching Techniques

It would be difficult to overstate the importance of using graphs in mathematics. Descartes's introduction of analytic geometry contributed significantly to the rapid advances in calculus that began during the mid-seventeenth century.

So far, you have studied several concepts that are useful in analyzing the graph of a function.

- x-intercepts and y-intercepts (Section 1.2)
- Domain and range (Section 1.4)
- Continuity (Section 1.6)
- Differentiability (Section 2.1)
- Relative extrema (Section 3.2)
- Concavity (Section 3.3)
- Points of inflection (Section 3.3)
- Vertical asymptotes (Section 3.6)
- Horizontal asymptotes (Section 3.6)

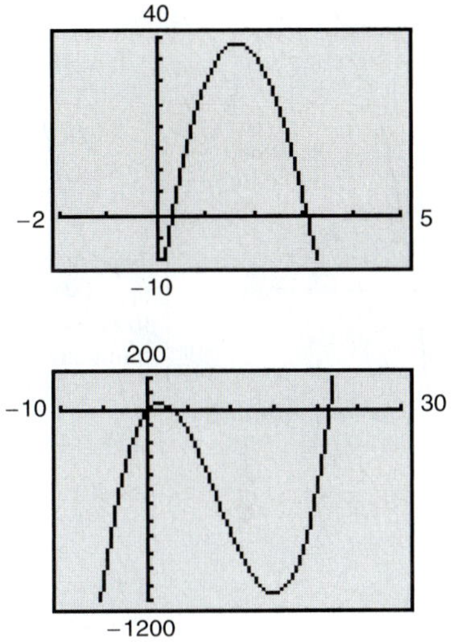

FIGURE 3.54

When you are sketching the graph of a function, either by hand or with a graphing utility, remember that you cannot normally show the *entire* graph. The decision as to which part of the graph to show is crucial. For instance, which of the viewing windows in Figure 3.54 better represents the graph of

$$f(x) = x^3 - 25x^2 + 74x - 20?$$

The lower viewing window gives a more complete view of the graph, but the context of the problem might indicate that the upper view is better. Here are some guidelines for analyzing the graph of a function.

Guidelines for Analyzing the Graph of a Function

1. Determine the domain and range of the function. If the function models a real-life situation, consider the context.
2. Determine the intercepts and asymptotes of the graph.
3. Locate the x-values where $f'(x)$ and $f''(x)$ are zero or undefined. Use the results to determine the relative extrema and points of inflection.

In these guidelines, note the importance of *algebra* (as well as calculus) for solving the equations $f(x) = 0$, $f'(x) = 0$, and $f''(x) = 0$.

TECHNOLOGY

Which of the viewing windows best represents the graph of the function

$$f(x) = \frac{x^3 + 8x^2 - 33x}{5}?$$

(a) Xmin $= -15$, Xmax $= 1$, Ymin $= -10$, Ymax $= 60$

(b) Xmin $= -10$, Xmax $= 10$, Ymin $= -10$, Ymax $= 10$

(c) Xmin $= -13$, Xmax $= 5$, Ymin $= -10$, Ymax $= 60$

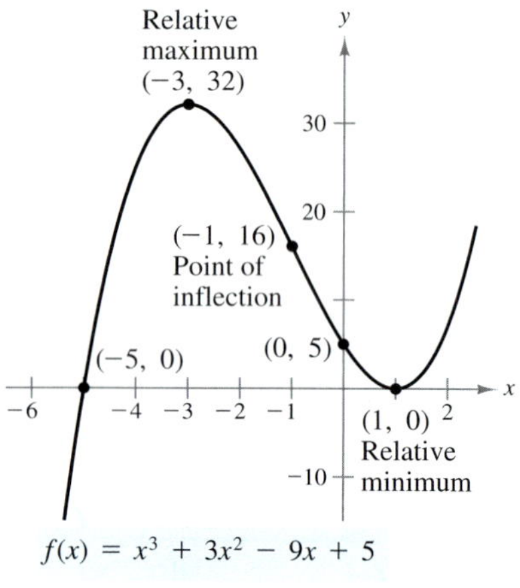

FIGURE 3.55

EXAMPLE 1 **Analyzing a Graph**

Analyze the graph of

$$f(x) = x^3 + 3x^2 - 9x + 5. \quad \text{Original function}$$

SOLUTION Begin by finding the intercepts of the graph. This function factors as

$$f(x) = (x-1)^2(x+5). \quad \text{Factored form}$$

So, the x-intercepts occur when $x = 1$ and $x = -5$. The derivative is

$$f'(x) = 3x^2 + 6x - 9 \quad \text{First derivative}$$
$$= 3(x-1)(x+3). \quad \text{Factored form}$$

So, the critical numbers of f are $x = 1$ and $x = -3$. The second derivative of f is

$$f''(x) = 6x + 6 \quad \text{Second derivative}$$
$$= 6(x+1) \quad \text{Factored form}$$

which implies that the second derivative is zero when $x = -1$. By testing the values of $f'(x)$ and $f''(x)$, as shown in the table, you can see that f has one relative minimum, one relative maximum, and one point of inflection. The graph of f is shown in Figure 3.55.

	$f(x)$	$f'(x)$	$f''(x)$	Characteristics of graph
x in $(-\infty, -3)$		+	−	Increasing, concave downward
$x = -3$	32	0	−	Relative maximum
x in $(-3, -1)$		−	−	Decreasing, concave downward
$x = -1$	16	−	0	Point of inflection
x in $(-1, 1)$		−	+	Decreasing, concave upward
$x = 1$	0	0	+	Relative minimum
x in $(1, \infty)$		+	+	Increasing, concave upward

TRY IT 1

Analyze the graph of $f(x) = -x^3 + 3x^2 + 9x - 27$.

TECHNOLOGY

In Example 1, you are able to find the zeros of f, f', and f'' algebraically (by factoring). When this is not feasible, you can use a graphing utility to find the zeros. For instance, the function

$$g(x) = x^3 + 3x^2 - 9x + 6$$

is similar to the function in the example, but it does not factor with integer coefficients. Using a graphing utility, you can determine that the function has only one x-intercept, $x \approx -5.0275$.

EXAMPLE 2 Analyzing a Graph

Analyze the graph of

$$f(x) = x^4 - 12x^3 + 48x^2 - 64x. \qquad \text{Original function}$$

SOLUTION Begin by finding the intercepts of the graph. This function factors as

$$f(x) = x(x^3 - 12x^2 + 48x - 64)$$
$$= x(x - 4)^3. \qquad \text{Factored form}$$

So, the x-intercepts occur when $x = 0$ and $x = 4$. The derivative is

$$f'(x) = 4x^3 - 36x^2 + 96x - 64 \qquad \text{First derivative}$$
$$= 4(x - 1)(x - 4)^2. \qquad \text{Factored form}$$

So, the critical numbers of f are $x = 1$ and $x = 4$. The second derivative of f is

$$f''(x) = 12x^2 - 72x + 96 \qquad \text{Second derivative}$$
$$= 12(x - 4)(x - 2) \qquad \text{Factored form}$$

which implies that the second derivative is zero when $x = 2$ and $x = 4$. By testing the values of $f'(x)$ and $f''(x)$, as shown in the table, you can see that f has one relative minimum and two points of inflection. The graph is shown in Figure 3.56.

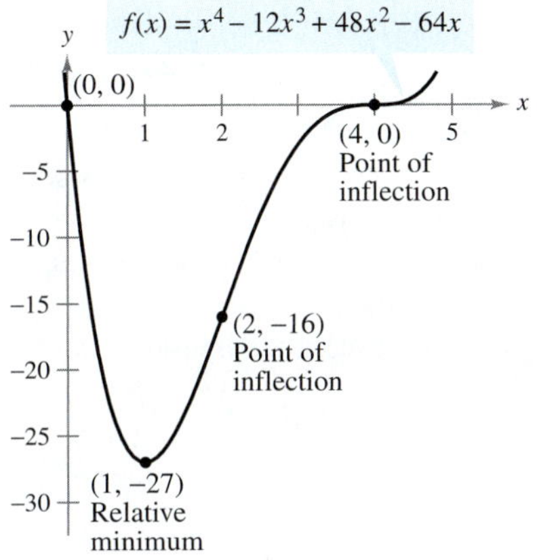

FIGURE 3.56

	$f(x)$	$f'(x)$	$f''(x)$	Characteristics of graph
x in $(-\infty, 1)$		−	+	Decreasing, concave upward
$x = 1$	−27	0	+	Relative minimum
x in $(1, 2)$		+	+	Increasing, concave upward
$x = 2$	−16	+	0	Point of inflection
x in $(2, 4)$		+	−	Increasing, concave downward
$x = 4$	0	0	0	Point of inflection
x in $(4, \infty)$		+	+	Increasing, concave upward

TRY IT 2

Analyze the graph of $f(x) = x^4 - 4x^3 + 5$.

DISCOVERY

A polynomial function of degree n can have at most $n - 1$ relative extrema and at most $n - 2$ points of inflection. For instance, the third-degree polynomial in Example 1 has two relative extrema and one point of inflection. Similarly, the fourth-degree polynomial function in Example 2 has one relative extremum and two points of inflection. Is it possible for a third-degree function to have no relative extrema? Is it possible for a fourth-degree function to have no relative extrema?

DISCOVERY

Show that the function in Example 3 can be rewritten as

$$f(x) = \frac{x^2 - 2x + 4}{x - 2} = x + \frac{4}{x - 2}.$$

Use a graphing utility to graph f together with the line $y = x$. How do the two graphs compare as you zoom out? Describe what is meant by a "slant asymptote." Find the slant asymptote of the function $g(x) = \dfrac{x^2 - x - 1}{x - 1}$.

EXAMPLE 3 Analyzing a Graph

Analyze the graph of

$$f(x) = \frac{x^2 - 2x + 4}{x - 2}. \qquad \text{Original function}$$

SOLUTION The y-intercept occurs at $(0, -2)$. Using the Quadratic Formula on the numerator, you can see that there are no x-intercepts. Because the denominator is zero when $x = 2$ (and the numerator is not zero when $x = 2$), it follows that $x = 2$ is a vertical asymptote of the graph. There are no horizontal asymptotes because the degree of the numerator is greater than the degree of the denominator. The derivative is

$$f'(x) = \frac{(x - 2)(2x - 2) - (x^2 - 2x + 4)}{(x - 2)^2} \qquad \text{First derivative}$$

$$= \frac{x(x - 4)}{(x - 2)^2}. \qquad \text{Factored form}$$

So, the critical numbers of f are $x = 0$ and $x = 4$. The second derivative is

$$f''(x) = \frac{(x - 2)^2(2x - 4) - (x^2 - 4x)(2)(x - 2)}{(x - 2)^4} \qquad \text{Second derivative}$$

$$= \frac{(x - 2)(2x^2 - 8x + 8 - 2x^2 + 8x)}{(x - 2)^4}$$

$$= \frac{8}{(x - 2)^3}. \qquad \text{Factored form}$$

Because the second derivative has no zeros and because $x = 2$ is not in the domain of the function, you can conclude that the graph has no points of inflection. By testing the values of $f'(x)$ and $f''(x)$, as shown in the table, you can see that f has one relative minimum and one relative maximum. The graph of f is shown in Figure 3.57.

	$f(x)$	$f'(x)$	$f''(x)$	Characteristics of graph
x in $(-\infty, 0)$		+	−	Increasing, concave downward
$x = 0$	-2	0	−	Relative maximum
x in $(0, 2)$		−	−	Decreasing, concave downward
$x = 2$	Undef.	Undef.	Undef.	Vertical asymptote
x in $(2, 4)$		−	+	Decreasing, concave upward
$x = 4$	6	0	+	Relative minimum
x in $(4, \infty)$		+	+	Increasing, concave upward

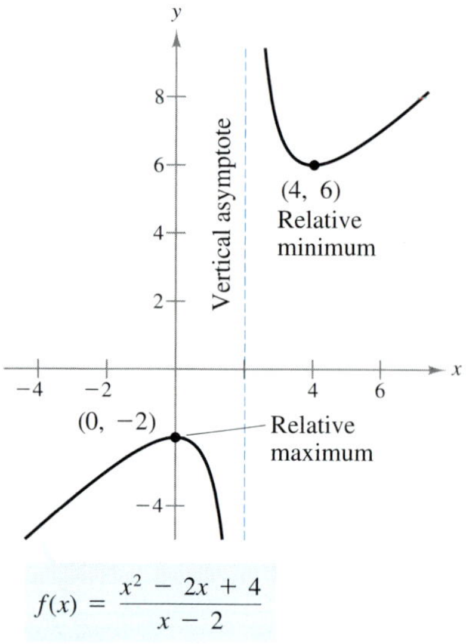

FIGURE 3.57

TRY IT 3

Analyze the graph of $f(x) = \dfrac{x^2}{x - 1}$.

EXAMPLE 4 Analyzing a Graph

Analyze the graph of

$$f(x) = \frac{2(x^2 - 9)}{x^2 - 4}.$$ Original function

SOLUTION Begin by writing the function in factored form.

$$f(x) = \frac{2(x - 3)(x + 3)}{(x - 2)(x + 2)}$$ Factored form

The y-intercept is $\left(0, \frac{9}{2}\right)$, and the x-intercepts are $(-3, 0)$ and $(3, 0)$. There are vertical asymptotes at $x = \pm 2$ and a horizontal asymptote at $y = 2$. The first derivative is

$$f'(x) = \frac{2[(x^2 - 4)(2x) - (x^2 - 9)(2x)]}{(x^2 - 4)^2}$$ First derivative

$$= \frac{20x}{(x^2 - 4)^2}.$$ Factored form

So, the critical number of f is $x = 0$. The second derivative of f is

$$f''(x) = \frac{(x^2 - 4)^2(20) - (20x)(2)(2x)(x^2 - 4)}{(x^2 - 4)^4}$$ Second derivative

$$= \frac{20(x^2 - 4)(x^2 - 4 - 4x^2)}{(x^2 - 4)^4}$$

$$= -\frac{20(3x^2 + 4)}{(x^2 - 4)^3}.$$ Factored form

Because the second derivative has no zeros and $x = \pm 2$ are not in the domain of the function, you can conclude that the graph has no points of inflection. By testing the values of $f'(x)$ and $f''(x)$, as shown in the table, you can see that f has one relative minimum. The graph of f is shown in Figure 3.58.

	$f(x)$	$f'(x)$	$f''(x)$	Characteristics of graph
x in $(-\infty, -2)$		$-$	$-$	Decreasing, concave downward
$x = -2$	Undef.	Undef.	Undef.	Vertical asymptote
x in $(-2, 0)$		$-$	$+$	Decreasing, concave upward
$x = 0$	$\frac{9}{2}$	0	$+$	Relative minimum
x in $(0, 2)$		$+$	$+$	Increasing, concave upward
$x = 2$	Undef.	Undef.	Undef.	Vertical asymptote
x in $(2, \infty)$		$+$	$-$	Increasing, concave downward

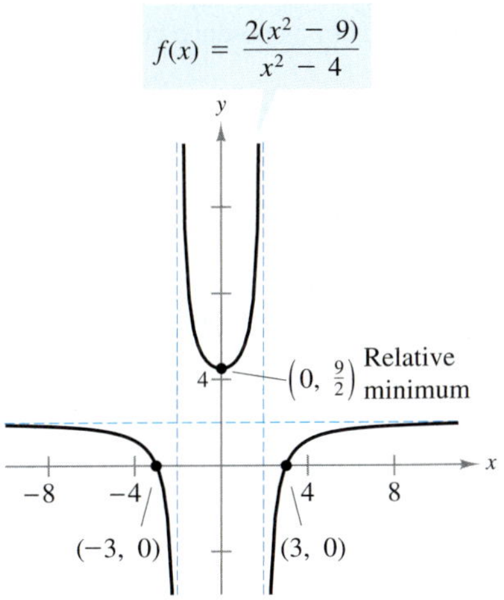

FIGURE 3.58

TRY IT 4

Analyze the graph of $f(x) = \dfrac{x^2 + 1}{x^2 - 1}$.

TECHNOLOGY

Some graphing utilities will not graph the function in Example 5 properly if the function is entered as

f(x) = 2x^(5/3) − 5x^(4/3).

To correct for this, you can enter the function as

$f(x) = 2(\sqrt[3]{x})$^5 − $5(\sqrt[3]{x})$^4.

Try entering both functions into a graphing utility to see whether both functions produce correct graphs.

ALGEBRA REVIEW

For help on the algebra in Example 5, see Example 2(a) in the *Chapter 3 Algebra Review*, on page 249.

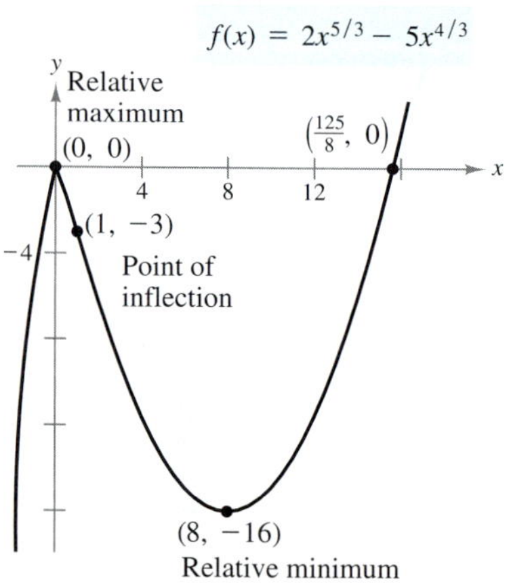

FIGURE 3.59

EXAMPLE 5 Analyzing a Graph

Analyze the graph of

$$f(x) = 2x^{5/3} - 5x^{4/3}.\qquad \text{Original function}$$

SOLUTION Begin by writing the function in factored form.

$$f(x) = x^{4/3}(2x^{1/3} - 5)\qquad \text{Factored form}$$

One of the intercepts is (0, 0). A second x-intercept occurs when $2x^{1/3} = 5$.

$$2x^{1/3} = 5$$
$$x^{1/3} = \tfrac{5}{2}$$
$$x = \left(\tfrac{5}{2}\right)^3$$
$$x = \tfrac{125}{8}$$

The first derivative is

$$f'(x) = \tfrac{10}{3}x^{2/3} - \tfrac{20}{3}x^{1/3}\qquad \text{First derivative}$$
$$= \tfrac{10}{3}x^{1/3}(x^{1/3} - 2).\qquad \text{Factored form}$$

So, the critical numbers of f are $x = 0$ and $x = 8$. The second derivative is

$$f''(x) = \tfrac{20}{9}x^{-1/3} - \tfrac{20}{9}x^{-2/3}\qquad \text{Second derivative}$$
$$= \tfrac{20}{9}x^{-2/3}(x^{1/3} - 1)$$
$$= \frac{20(x^{1/3} - 1)}{9x^{2/3}}.\qquad \text{Factored form}$$

So, possible points of inflection occur when $x = 1$ and when $x = 0$. By testing the values of $f'(x)$ and $f''(x)$, as shown in the table, you can see that f has one relative maximum, one relative minimum, and one point of inflection. The graph of f is shown in Figure 3.59.

	$f(x)$	$f'(x)$	$f''(x)$	Characteristics of graph
x in $(-\infty, 0)$		+	−	Increasing, concave downward
$x = 0$	0	0	Undef.	Relative maximum
x in $(0, 1)$		−	−	Decreasing, concave downward
$x = 1$	−3	−	0	Point of inflection
x in $(1, 8)$		−	+	Decreasing, concave upward
$x = 8$	−16	0	+	Relative minimum
x in $(8, \infty)$		+	+	Increasing, concave upward

TRY IT 5

Analyze the graph of

$$f(x) = 2x^{3/2} - 6x^{1/2}.$$

Summary of Simple Polynomial Graphs

A summary of the graphs of polynomial functions of degrees 0, 1, 2, and 3 is shown in Figure 3.60. Because of their simplicity, lower-degree polynomial functions are commonly used as mathematical models.

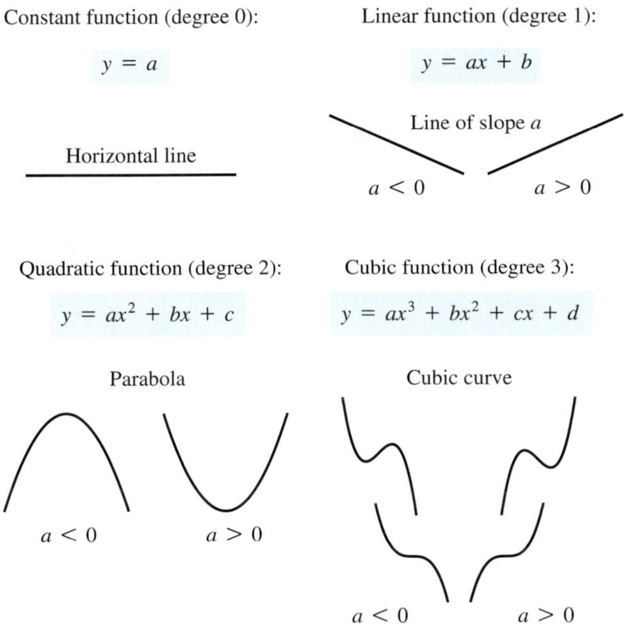

FIGURE 3.60

STUDY TIP

The graph of any cubic polynomial has one point of inflection. The slope of the graph at the point of inflection may be zero or nonzero.

TAKE ANOTHER LOOK

Graphs of Fourth-Degree Polynomial Functions

In the summary presented above, the graphs of cubic functions are classified into four basic types. How many basic types of graphs are possible with fourth-degree polynomial functions? Make a rough sketch of each type. Then use a graphing utility to classify each of the following. (In each case, use a viewing window that shows all the basic characteristics of the graph.)

a. $y = x^4$

b. $y = -x^4 + 5x^2$

c. $y = x^4 - x^3 + x$

d. $y = -x^4 - 4x^3 - 3x^2 + x$

e. $y = -x^4 + 2x^2$

f. $y = -x^4 + x^3$

g. $y = x^4 - 5x$

h. $y = x^4 - 8x^3 + 8x^2$

PREREQUISITE REVIEW 3.7

The following warm-up exercises involve skills that were covered in earlier sections. You will use these skills in the exercise set for this section.

In Exercises 1–4, find the vertical and horizontal asymptotes of the graph.

1. $f(x) = \dfrac{1}{x^2}$

2. $f(x) = \dfrac{8}{(x - 2)^2}$

3. $f(x) = \dfrac{40x}{x + 3}$

4. $f(x) = \dfrac{x^2 - 3}{x^2 - 4x + 3}$

In Exercises 5–10, determine the open intervals on which the function is increasing or decreasing.

5. $f(x) = x^2 + 4x + 2$

6. $f(x) = -x^2 - 8x + 1$

7. $f(x) = x^3 - 3x + 1$

8. $f(x) = \dfrac{-x^3 + x^2 - 1}{x^2}$

9. $f(x) = \dfrac{x - 2}{x - 1}$

10. $f(x) = -x^3 - 4x^2 + 3x + 2$

EXERCISES 3.7

In Exercises 1–20, sketch the graph of the function. Choose a scale that allows all relative extrema and points of inflection to be identified on the graph.

1. $y = -x^2 - 2x + 3$

2. $y = 2x^2 - 4x + 1$

3. $y = x^3 - 4x^2 + 6$

4. $y = -\frac{1}{3}(x^3 - 3x + 2)$

5. $y = 2 - x - x^3$

6. $y = x^3 + 3x^2 + 3x + 2$

7. $y = 3x^3 - 9x + 1$

8. $y = -4x^3 + 6x^2$

9. $y = 3x^4 + 4x^3$

10. $y = 3x^4 - 6x^2$

11. $y = x^3 - 6x^2 + 3x + 10$

12. $y = -x^3 + 3x^2 + 9x - 2$

13. $y = x^4 - 8x^3 + 18x^2 - 16x + 5$

14. $y = x^4 - 4x^3 + 16x - 16$

15. $y = x^4 - 4x^3 + 16x$

16. $y = x^5 + 1$

17. $y = x^5 - 5x$

18. $y = (x - 1)^5$

19. $y = \begin{cases} x^2 + 1, & x \le 0 \\ 1 - 2x, & x > 0 \end{cases}$

20. $y = \begin{cases} x^2 + 4, & x < 0 \\ 4 - x, & x \ge 0 \end{cases}$

In Exercises 21–32, use a graphing utility to graph the function. Choose a window that allows all relative extrema and points of inflection to be identified on the graph.

21. $y = \dfrac{x^2 + 2}{x^2 + 1}$

22. $y = \dfrac{x}{x^2 + 1}$

23. $y = 3x^{2/3} - 2x$

24. $y = 3x^{2/3} - x^2$

25. $y = 1 - x^{2/3}$

26. $y = (1 - x)^{2/3}$

27. $y = x^{1/3} + 1$

28. $y = x^{-1/3}$

29. $y = x^{5/3} - 5x^{2/3}$

30. $y = x^{4/3}$

31. $y = x\sqrt{x^2 - 9}$

32. $y = \dfrac{x}{\sqrt{x^2 - 4}}$

In Exercises 33–42, sketch the graph of the function. Label the intercepts, relative extrema, points of inflection, and asymptotes. Then state the domain of the function.

33. $y = \dfrac{5 - 3x}{x - 2}$

34. $y = \dfrac{x^2 + 1}{x^2 - 2}$

35. $y = \dfrac{2x}{x^2 - 1}$

36. $y = \dfrac{x^2 - 6x + 12}{x - 4}$

37. $y = x\sqrt{4 - x}$

38. $y = x\sqrt{4 - x^2}$

39. $y = \dfrac{x - 3}{x}$

40. $y = x + \dfrac{32}{x^2}$

41. $y = \dfrac{x^3}{x^3 - 1}$

42. $y = \dfrac{x^4}{x^4 - 1}$

In Exercises 43–46, find values of a, b, c, and d such that the graph of $f(x) = ax^3 + bx^2 + cx + d$ will resemble the given graph. Then use a graphing utility to verify your result. (There are many correct answers.)

43.

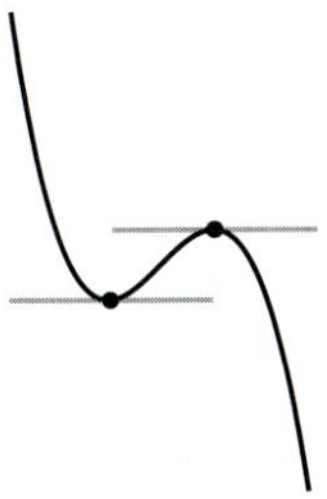

44.

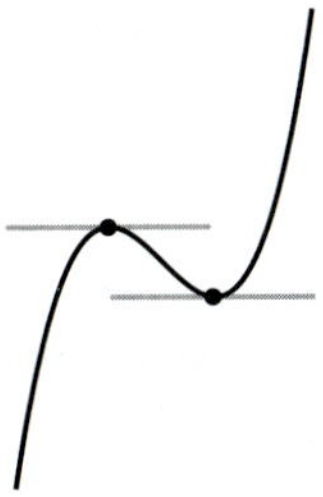

45.

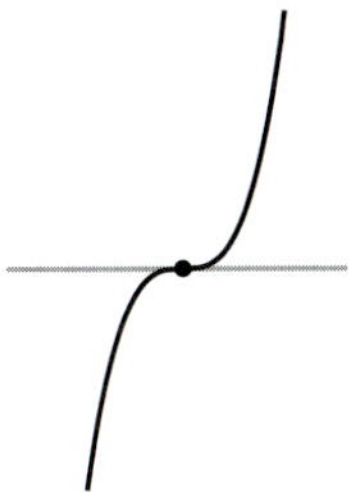

46.

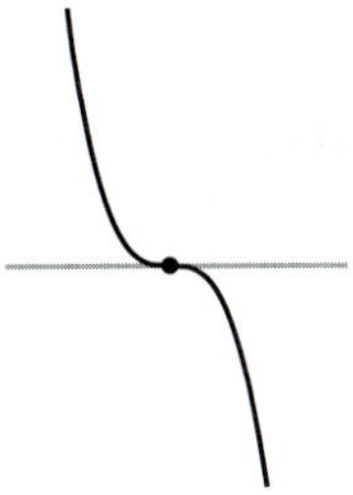

In Exercises 47–50, use the graph of f' or f'' to sketch the graph of f. (There are many correct answers.)

47.

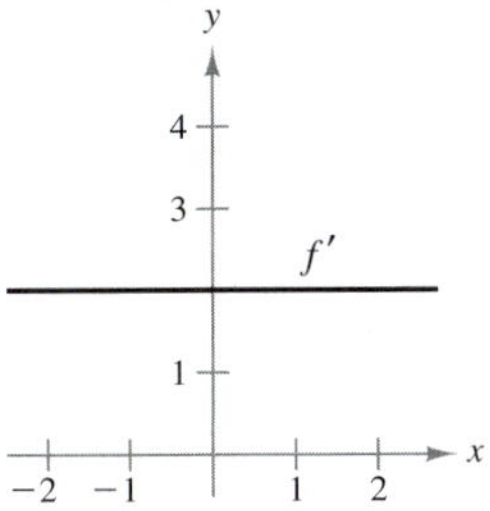

48.

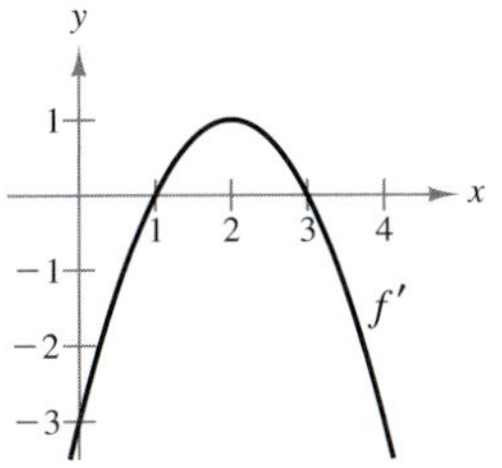

49.

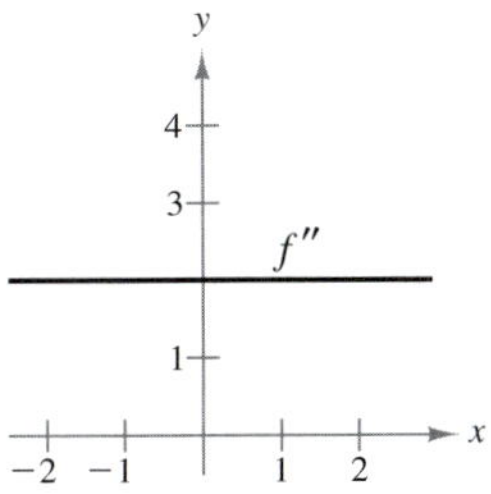

50.

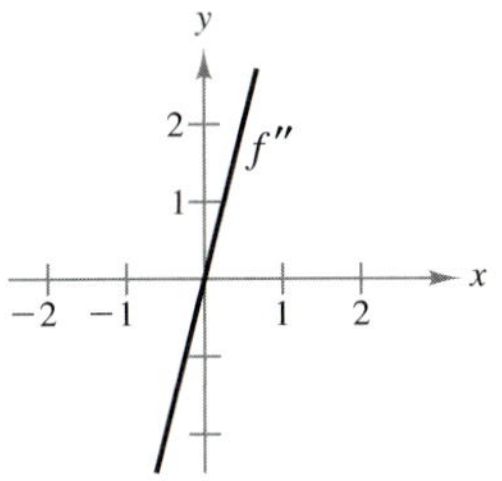

In Exercises 51 and 52, sketch a graph of a function f having the given characteristics. (There are many correct answers.)

51. $f(-2) = 0$

$f(0) = 0$

$f'(x) > 0, \quad -\infty < x < -1$

$f'(-1) = 0$

$f'(x) < 0, \quad -1 < x < 0$

$f'(0) = 0$

$f'(x) > 0, \quad 0 < x < \infty$

52. $f(-1) = 0$

$f(3) = 0$

$f'(1)$ is undefined.

$f'(x) < 0, \quad -\infty < x < 1$

$f'(x) > 0, \quad 1 < x < \infty$

$f''(x) < 0, \quad x \neq 1$

$\lim_{x\to\infty} f(x) = 4$

53. *Cost* An employee of a delivery company earns \$9 per hour driving a delivery van in an area where gasoline costs \$1.80 per gallon. When the van is driven at a constant speed s (in miles per hour, with $40 \leq s \leq 65$), the van gets $500/s$ miles per gallon.

(a) Find the cost C as a function of s for a 100-mile trip on an interstate highway.

(b) Use a graphing utility to graph the function found in part (a) and determine the most economical speed.

54. *Profit* The management of a company is considering three possible models for predicting the company's profits from 2001 through 2006. Model I gives the expected annual profits if the current trends continue. Models II and III give the expected annual profits for various combinations of increased labor and energy costs. In each model, p is the profit (in billions of dollars) and $t = 0$ corresponds to 2001.

Model I: $p = 0.03t^2 - 0.01t + 3.39$

Model II: $p = 0.08t + 3.36$

Model III: $p = -0.07t^2 + 0.05t + 3.38$

(a) Use a graphing utility to graph all three models in the same viewing window.

(b) For which models are profits increasing during the interval from 2001 through 2006?

(c) Which model is the most optimistic? Which is the most pessimistic?

55. *Meteorology* The monthly normal temperature T (in degrees Fahrenheit) for Pittsburgh, Pennsylvania can be modeled by

$$T = \frac{23.011 - 1.0t + 0.048t^2}{1 - 0.204t + 0.014t^2}, \quad 1 \leq t \leq 12$$

where t is the month, with $t = 1$ corresponding to January. Use a graphing utility to graph the model and find all absolute extrema. Explain the meaning of those values. *(Source: National Climatic Data Center)*

Writing In Exercises 56 and 57, use a graphing utility to graph the function. Explain why there is no vertical asymptote when a superficial examination of the function may indicate that there should be one.

56. $h(x) = \dfrac{6 - 2x}{3 - x}$

57. $g(x) = \dfrac{x^2 + x - 2}{x - 1}$

3.8 DIFFERENTIALS AND MARGINAL ANALYSIS

- Find the differentials of functions.
- Use differentials to approximate changes in functions.
- Use differentials to approximate changes in real-life models.

Differentials

When the derivative was defined in Section 2.1 as the limit of the ratio $\Delta y/\Delta x$, it seemed natural to retain the quotient symbolism for the limit itself. So, the derivative of y with respect to x was denoted by

$$\frac{dy}{dx} = \lim_{\Delta x \to 0} \frac{\Delta y}{\Delta x}$$

even though we did not interpret dy/dx as the quotient of two separate quantities. In this section, you will see that the quantities dy and dx can be assigned meanings in such a way that their quotient, when $dx \neq 0$, is equal to the derivative of y with respect to x.

STUDY TIP

In this definition, dx can have any nonzero value. In most applications, however, dx is chosen to be small and this choice is denoted by $dx = \Delta x$.

Definition of Differentials

Let $y = f(x)$ represent a differentiable function. The **differential of x** (denoted by dx) is any nonzero real number. The **differential of y** (denoted by dy) is

$$dy = f'(x)\,dx.$$

One use of differentials is in approximating the change in $f(x)$ that corresponds to a change in x, as shown in Figure 3.61. This change is denoted by

$$\Delta y = f(x + \Delta x) - f(x). \qquad \text{Change in } y$$

STUDY TIP

Note in Figure 3.61 that near the point of tangency, the graph of f is very close to the tangent line. This is the essence of the approximations used in this section. In other words, near the point of tangency, $dy \approx \Delta y$.

In Figure 3.61, notice that as Δx gets smaller and smaller, the values of dy and Δy get closer and closer. That is, when Δx is small, $dy \approx \Delta y$.

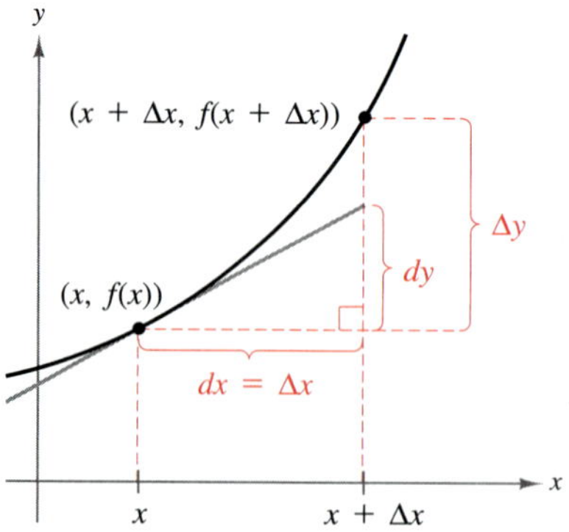

FIGURE 3.61

This **tangent line approximation** is the basis for most applications of differentials.

EXAMPLE 1 Interpreting Differentials Graphically

Consider the function given by

$f(x) = x^2$. Original function

Find the value of dy when $x = 1$ and $dx = 0.01$. Compare this with the value of Δy when $x = 1$ and $\Delta x = 0.01$. Interpret the results graphically.

SOLUTION Begin by finding the derivative of f.

$f'(x) = 2x$ Derivative of f

When $x = 1$ and $dx = 0.01$, the value of the differential dy is

$dy = f'(x)\,dx$ Differential of y

$= f'(1)(0.01)$ Substitute 1 for x and 0.01 for dx.

$= 2(1)(0.01)$ Use $f'(x) = 2x$.

$= 0.02.$ Simplify.

When $x = 1$ and $\Delta x = 0.01$, the value of Δy is

$\Delta y = f(x + \Delta x) - f(x)$ Change in y

$= f(1.01) - f(1)$ Substitute 1 for x and 0.01 for Δx.

$= (1.01)^2 - (1)^2$

$= 0.0201.$ Simplify.

Note that $dy \approx \Delta y$, as shown in Figure 3.62.

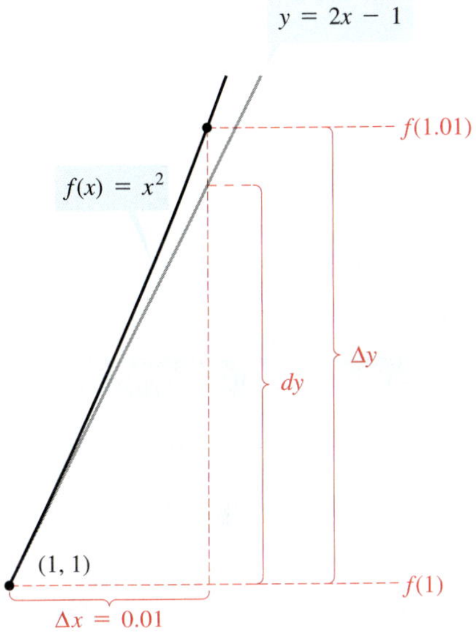

FIGURE 3.62

TRY IT 1

Find the value of dy when $x = 2$ and $dx = 0.01$ for $f(x) = x^4$. Compare this with the value of Δy when $x = 2$ and $\Delta x = 0.01$.

The validity of the approximation

$$dy \approx \Delta y, \quad dx \neq 0$$

stems from the definition of the derivative. That is, the existence of the limit

$$f'(x) = \lim_{\Delta x \to 0} \frac{f(x + \Delta x) - f(x)}{\Delta x}$$

implies that when Δx is close to zero, then $f'(x)$ is close to the difference quotient. So, you can write

$$\frac{f(x + \Delta x) - f(x)}{\Delta x} \approx f'(x)$$

$$f(x + \Delta x) - f(x) \approx f'(x)\,\Delta x$$

$$\Delta y \approx f'(x)\,\Delta x.$$

Substituting dx for Δx and dy for $f'(x)\,dx$ produces

$$\Delta y \approx dy.$$

STUDY TIP

Find an equation of the tangent line $y = g(x)$ to the graph of $f(x) = x^2$ at the point $x = 1$. Evaluate $g(1.01)$ and $f(1.01)$.

Marginal Analysis

Differentials are used in economics to approximate changes in revenue, cost, and profit. Suppose that $R = f(x)$ is the total revenue for selling x units of a product. When the number of units increases by 1, the change in x is $\Delta x = 1$, and the change in R is

$$\Delta R = f(x + \Delta x) - f(x) \approx dR = \frac{dR}{dx}\,dx.$$

In other words, you can use the differential dR to approximate the change in the revenue that accompanies the sale of one additional unit. Similarly, the differentials dC and dP can be used to approximate the changes in cost and profit that accompany the sale (or production) of one additional unit.

EXAMPLE 2 Using Marginal Analysis

The demand function for a product is modeled by

$$p = \sqrt{400 - x}, \quad 0 \le x \le 400.$$

Use differentials to approximate the change in revenue as sales increase from 256 units to 257 units. Compare this with the actual change in revenue.

SOLUTION Begin by finding the marginal revenue, dR/dx.

$$R = xp \qquad \text{Formula for revenue}$$

$$= x\sqrt{400 - x} \qquad \text{Use } p = \sqrt{400 - x}.$$

$$\frac{dR}{dx} = x\left(\frac{1}{2}\right)(400 - x)^{-1/2}(-1) + (400 - x)^{1/2}(1) \qquad \text{Product Rule}$$

$$= \frac{800 - 3x}{2\sqrt{400 - x}} \qquad \text{Simplify.}$$

When $x = 256$ and $dx = \Delta x = 1$, you can approximate the change in the revenue to be

$$\frac{800 - 3(256)}{2\sqrt{400 - 256}}(1) \approx \$1.33.$$

When x increases from 256 to 257, the actual change in revenue is

$$\Delta R = 257\sqrt{400 - 257} - 256\sqrt{400 - 256}$$

$$\approx 3073.27 - 3072.00$$

$$= \$1.27.$$

TECHNOLOGY

Use a graphing utility to graph the revenue function $R(x) = x\sqrt{400 - x}$ and the tangent line approximation $y = \frac{4}{3}(x - 256) + 3072$ in the viewing window $250 \le x \le 260$, $3065 \le y \le 3080$. Explain why the two curves appear to be almost identical. Where do the curves intersect? Use the graphs to verify the solution to Example 2.

TRY IT 2

The demand function for a product is modeled by

$$p = \sqrt{200 - x}.$$

Use differentials to approximate the change in revenue as sales increase from 100 to 101 units. Compare this with the actual change in revenue.

EXAMPLE 3 Using Marginal Analysis

The profit derived from selling x units of an item is modeled by

$$P = 0.0002x^3 + 10x.$$

Use the differential dP to approximate the change in profit when the production level changes from 50 to 51 units. Compare this with the actual gain in profit obtained by increasing the production level from 50 to 51 units.

SOLUTION The marginal profit is

$$\frac{dP}{dx} = 0.0006x^2 + 10.$$

When $x = 50$ and $dx = 1$, the differential is

$$[0.0006(50)^2 + 10](1) = \$11.50.$$

When x changes from 50 to 51 units, the actual change in profit is

$$\begin{aligned} \Delta P &= [(0.0002)(51)^3 + 10(51)] - [(0.0002)(50)^3 + 10(50)] \\ &\approx 536.53 - 525.00 \\ &= \$11.53. \end{aligned}$$

These values are shown graphically in Figure 3.63.

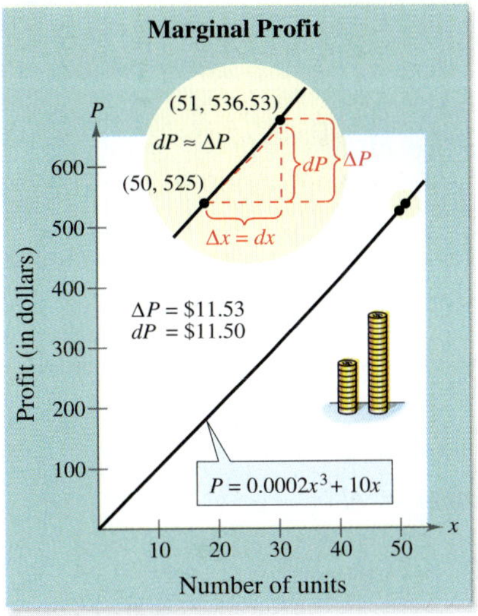

FIGURE 3.63

STUDY TIP

Example 3 uses differentials to solve the same problem that was solved in Example 5 in Section 2.3. Look back at that solution. Which approach do you prefer?

STUDY TIP

Find an equation of the tangent line $y = f(x)$ to the graph of $P = 0.0002x^3 + 10x$ at the point $x = 50$. Evaluate $f(51)$ and $p(51)$.

TRY IT 3

Use the differential dP to approximate the change in profit for the profit function in Example 3 when the production level changes from 40 to 41 units. Compare this with the actual gain in profit obtained by increasing the production level from 40 to 41 units.

Formulas for Differentials

You can use the definition of differentials to rewrite each differentiation rule in **differential form.** For example, if u and v are differentiable functions of x, then $du = (du/dx)\,dx$ and $dv = (dv/dx)\,dx$, which implies that you can write the Product Rule in the following differential form.

$$d[uv] = \frac{d}{dx}[uv]\,dx \qquad \text{Differential of } uv$$

$$= \left[u\frac{dv}{dx} + v\frac{du}{dx}\right]dx \qquad \text{Product Rule}$$

$$= u\frac{dv}{dx}dx + v\frac{du}{dx}dx$$

$$= u\,dv + v\,du \qquad \text{Differential form of Product Rule}$$

The following summary gives the differential forms of the differentiation rules presented so far in the text.

Differential Forms of Differentiation Rules

Constant Multiple Rule:	$d[cu] = c\,du$
Sum or Difference Rule:	$d[u \pm v] = du \pm dv$
Product Rule:	$d[uv] = u\,dv + v\,du$
Quotient Rule:	$d\left[\dfrac{u}{v}\right] = \dfrac{v\,du - u\,dv}{v^2}$
Constant Rule:	$d[c] = 0$
Power Rule:	$d[x^n] = nx^{n-1}\,dx$

The next example compares the derivatives and differentials of several simple functions.

TRY IT 4

Find the differential dy of each function.

(a) $y = 4x^3$

(b) $y = \dfrac{2x + 1}{3}$

(c) $y = 3x^2 - 2x$

(d) $y = \dfrac{1}{x^2}$

EXAMPLE 4 Finding Differentials

Find the differential dy of each function.

Function	*Derivative*	*Differential*
(a) $y = x^2$	$\dfrac{dy}{dx} = 2x$	$dy = 2x\,dx$
(b) $y = \dfrac{3x + 2}{5}$	$\dfrac{dy}{dx} = \dfrac{3}{5}$	$dy = \dfrac{3}{5}\,dx$
(c) $y = 2x^2 - 3x$	$\dfrac{dy}{dx} = 4x - 3$	$dy = (4x - 3)\,dx$
(d) $y = \dfrac{1}{x}$	$\dfrac{dy}{dx} = -\dfrac{1}{x^2}$	$dy = -\dfrac{1}{x^2}\,dx$

Error Propagation

A common use of differentials is the estimation of errors that result from inaccuracies of physical measuring devices. This is shown in Example 5.

EXAMPLE 5 **Estimating Measurement Errors**

The radius of a ball bearing is measured to be 0.7 inch, as shown in Figure 3.64. This implies that the volume of the ball bearing is $\frac{4}{3}\pi(0.7)^3 \approx 1.4368$ cubic inches. You are told that the measurement of the radius is correct to within 0.01 inch. How far off could the calculation of the volume be?

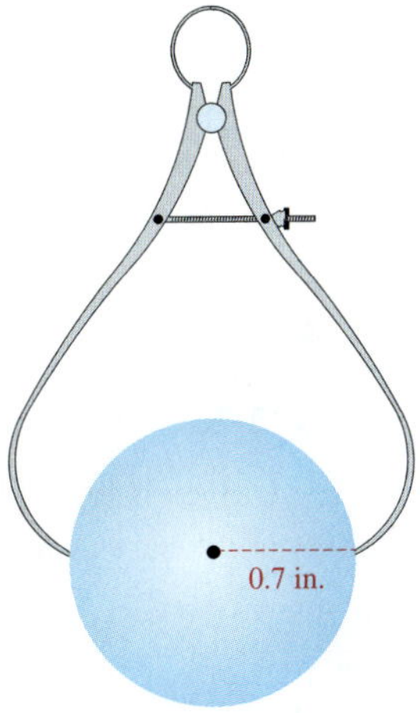

FIGURE 3.64

SOLUTION Because the value of r can be off by 0.01 inch, it follows that

$$-0.01 \leq \Delta r \leq 0.01. \qquad \text{Possible error in measuring}$$

Using $\Delta r = dr$, you can estimate the possible error in the volume.

$$V = \frac{4}{3}\pi r^3 \qquad \text{Formula for volume}$$

$$dV = \frac{dV}{dr}\,dr \qquad \text{Formula for differential of } V$$

The possible error in the volume is

$$4\pi r^2\,dr = 4\pi(0.7)^2(\pm 0.01) \qquad \text{Substitute for } r \text{ and } dr.$$

$$\approx \pm 0.0616 \text{ cubic inch.} \qquad \text{Possible error}$$

So, the volume of the ball bearing could range between

$$(1.4368 - 0.0616) = 1.3752 \text{ cubic inches}$$

and

$$(1.4368 + 0.0616) = 1.4984 \text{ cubic inches.}$$

In Example 5, the **relative error** in the volume is defined to be the ratio of dV to V. This ratio is

$$\frac{dV}{V} \approx \frac{\pm 0.0616}{1.4368} \approx \pm 0.0429.$$

This corresponds to a **percentage error** of 4.29%.

TRY IT 5

Find the surface area of the ball bearing in Example 5. How far off could your calculation of the surface area be? The surface area of a sphere is given by $S = 4\pi r^2$.

TAKE ANOTHER LOOK

Finding Propagated Errors

a. In Example 5, if the radius of the ball bearing were measured to be 1.5 inches, correct to within 0.01 inch, would the percentage error of the volume be greater or smaller than in the example?

b. In Example 5, if the radius of the ball bearing were measured to be 0.7 inch, correct to within 0.02 inch, would the percentage error of the volume be greater or smaller than in the example?

PREREQUISITE REVIEW 3.8

The following warm-up exercises involve skills that were covered in earlier sections. You will use these skills in the exercise set for this section.

In Exercises 1–12, find the derivative.

1. $C = 44 + 0.09x^2$

2. $C = 250 + 0.15x$

3. $R = x(1.25 + 0.02\sqrt{x})$

4. $R = x(15.5 - 1.55x)$

5. $P = -0.03x^{1/3} + 1.4x - 2250$

6. $P = -0.02x^2 + 25x - 1000$

7. $A = \frac{1}{4}\sqrt{3}x^2$

8. $A = 6x^2$

9. $C = 2\pi r$

10. $P = 4w$

11. $S = 4\pi r^2$

12. $P = 2x + \sqrt{2}x$

In Exercises 13–16, write a formula for the quantity.

13. Area A of a circle of radius r

14. Area A of a square of side x

15. Volume V of a cube of edge x

16. Volume V of a sphere of radius r

EXERCISES 3.8

In Exercises 1–6, find the differential dy.

1. $y = 3x^2 - 4$

2. $y = 2x^{3/2}$

3. $y = (4x - 1)^3$

4. $y = (1 - 2x^2)^4$

5. $y = \sqrt{x^2 + 1}$

6. $y = \sqrt[3]{6x^2}$

In Exercises 7–10, let $x = 1$ and $\Delta x = 0.01$. Find Δy.

7. $f(x) = 5x^2 - 1$

8. $f(x) = \sqrt{3x}$

9. $f(x) = \dfrac{4}{\sqrt[3]{x}}$

10. $f(x) = \dfrac{x}{x^2 + 1}$

In Exercises 11–14, compare the values of dy and Δy.

11.	$y = x^3$	$x = 1$	$\Delta x = dx = 0.1$
12.	$y = 1 - 2x^2$	$x = 0$	$\Delta x = dx = -0.1$
13.	$y = x^4 + 1$	$x = -1$	$\Delta x = dx = 0.01$
14.	$y = 2x^3 + 1$	$x = 2$	$\Delta x = dx = 0.01$

In Exercises 15–20, let $x = 2$ and complete the table for the function.

$dx = \Delta x$	dy	Δy	$\Delta y - dy$	$dy/\Delta y$
1.000				
0.500				
0.100				
0.010				
0.001				

15. $y = x^2$

16. $y = x^5$

17. $y = \dfrac{1}{x^2}$

18. $y = \dfrac{1}{x}$

19. $y = \sqrt[4]{x}$

20. $y = \sqrt{x}$

In Exercises 21–24, find an equation of the tangent line to the function at the given point. Then find the function value and the tangent line value at $f(x + \Delta x)$ and $y(x + \Delta x)$ for $\Delta x = -0.01$ and 0.01.

	Function	*Point*
21.	$f(x) = 2x^3 - x^2 + 1$	$(-2, -19)$
22.	$f(x) = 3x^2 - 1$	$(2, 11)$
23.	$f(x) = \dfrac{x}{x^2 + 1}$	$(0, 0)$
24.	$f(x) = \sqrt{25 - x^2}$	$(3, 4)$

25. *Demand* The demand function for a product is modeled by

$$p = 75 - 0.25x.$$

(a) If x changes from 7 to 8, what is the corresponding change in p? Compare the values of Δp and dp.

(b) Repeat part (a) when x changes from 70 to 71 units.

26. *Biology: Wildlife Management* A state game commission introduces 50 deer into newly acquired state game lands. The population N of the herd can be modeled by

$$N = \frac{10(5 + 3t)}{1 + 0.04t}$$

where t is the time in years. Use differentials to approximate the change in the herd size from $t = 5$ to $t = 6$.

Marginal Analysis In Exercises 27–32, use differentials to approximate the change in cost, revenue, or profit corresponding to an increase in sales of one unit. For instance, in Exercise 27, approximate the change in cost as x increases from 12 units to 13 units. Then use a graphing utility to graph the function, and use the *trace* feature to verify your result.

	Function	*x-Value*
27.	$C = 0.05x^2 + 4x + 10$	$x = 12$
28.	$C = 0.025x^2 + 8x + 5$	$x = 10$
29.	$R = 30x - 0.15x^2$	$x = 75$
30.	$R = 50x - 1.5x^2$	$x = 15$
31.	$P = -0.5x^3 + 2500x - 6000$	$x = 50$
32.	$P = -x^2 + 60x - 100$	$x = 25$

33. *Marginal Analysis* A retailer has determined that the monthly sales x of a watch is 150 units when the price is \$50, but decreases to 120 units when the price is \$60. Assume that the demand is a linear function of the price. Find the revenue R as a function of x and approximate the change in revenue for a one-unit increase in sales when $x = 141$. Make a sketch showing dR and ΔR.

34. *Marginal Analysis* A manufacturer determines that the demand x for a product is inversely proportional to the square of the price p. When the price is \$10, the demand is 2500. Find the revenue R as a function of x and approximate the change in revenue for a one-unit increase in sales when $x = 3000$. Make a sketch showing dR and ΔR.

35. *Marginal Analysis* The demand x for a radio is 30,000 units per week when the price is \$25 and 40,000 units when the price is \$20. The initial investment is \$275,000 and the cost per unit is \$17. Assume that the demand is a linear function of the price. Find the profit P as a function of x and approximate the change in profit for a one-unit increase in sales when $x = 28{,}000$. Make a sketch showing dP and ΔP.

36. *Marginal Analysis* The variable cost for the production of a calculator is \$14.25 and the initial investment is \$110,000. Find the total cost C as a function of x, the number of units produced. Then use differentials to approximate the change in the cost for a one-unit increase in production when $x = 50{,}000$. Make a sketch showing dC and ΔC. Explain why $dC = \Delta C$ in this problem.

37. *Area* The area A of a square of side x is $A = x^2$.

(a) Compute dA and ΔA in terms of x and Δx.

(b) In the figure, identify the region whose area is dA.

(c) Identify the region whose area is $\Delta A - dA$.

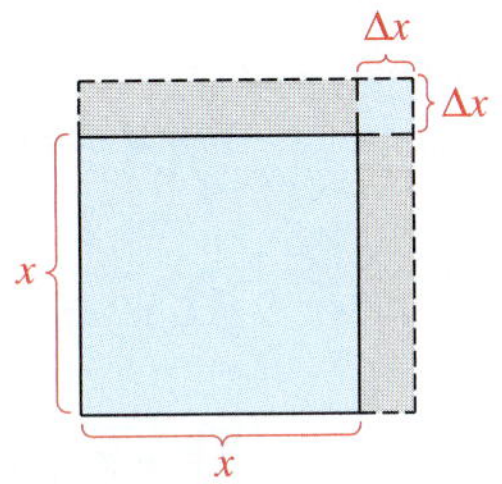

38. *Area* The side of a square is measured to be 12 inches, with a possible error of $\frac{1}{64}$ inch. Use differentials to approximate the possible error and the relative error in computing the area of the square.

39. *Area* The radius of a circle is measured to be 10 inches, with a possible error of $\frac{1}{8}$ inch. Use differentials to approximate the possible error and the relative error in computing the area of the circle.

40. *Volume and Surface Area* The edge of a cube is measured to be 12 inches, with a possible error of 0.03 inch. Use differentials to approximate the possible error and the relative error in computing (a) the volume of the cube and (b) the surface area of the cube.

41. *Volume* The radius of a sphere is measured to be 6 inches, with a possible error of 0.02 inch. Use differentials to approximate the possible error and the relative error in computing the volume of the sphere.

42. *Medical Science* The concentration C (in milligrams per milliliter) of a drug in a patient's bloodstream t hours after injection into muscle tissue is modeled by

$$C = \frac{3t}{27 + t^3}.$$

Use differentials to approximate the change in the concentration when t changes from $t = 1$ to $t = 1.5$.

True or False? In Exercises 43 and 44, determine whether the statement is true or false. If it is false, explain why or give an example that shows it is false.

43. If $y = x + c$, then $dy = dx$.

44. If $y = ax + b$, then $\Delta y/\Delta x = dy/dx$.

ALGEBRA REVIEW

Solving Equations

Much of the algebra in Chapter 3 involves simplifying algebraic expressions (see pages 162 and 163) and solving algebraic equations (see page 73). The Algebra Review on page 73 illustrates some of the basic techniques for solving equations. On these two pages, you can review some of the more complicated techniques for solving equations.

When solving an equation, remember that your basic goal is to isolate the variable on one side of the equation. To do this, you use inverse operations. For instance, to get rid of the *subtract* 2 in

$$x - 2 = 0$$

you *add* 2 to each side of the equation. Similarly, to get rid of the *square root* in

$$\sqrt{x + 3} = 2$$

you *square* both sides of the equation.

EXAMPLE 1 Solving an Equation

Solve each equation.

(a) $\dfrac{36(x^2 - 1)}{(x^2 + 3)^3} = 0$ (b) $0 = 2 - \dfrac{288}{x^2}$ (c) $0 = 2x(2x^2 - 3)$

SOLUTION

(a) $\dfrac{36(x^2 - 1)}{(x^2 + 3)^3} = 0$ — Example 2, page 193

$36(x^2 - 1) = 0$ — A fraction is zero only if its numerator is zero.

$x^2 - 1 = 0$ — Divide each side by 36.

$x^2 = 1$ — Add 1 to each side.

$x = \pm 1$ — Take the square root of each side.

(b) $0 = 2 - \dfrac{288}{x^2}$ — Example 2, page 203

$-2 = -\dfrac{288}{x^2}$ — Subtract 2 from each side.

$1 = \dfrac{144}{x^2}$ — Divide each side by -2.

$x^2 = 144$ — Multiply each side by x^2.

$x = \pm 12$ — Take the square root of each side.

(c) $0 = 2x(2x^2 - 3)$ — Example 3, page 204

$2x = 0 \Rightarrow x = 0$ — Set first factor equal to zero.

$2x^2 - 3 = 0 \Rightarrow x = \pm\sqrt{\frac{3}{2}}$ — Set second factor equal to zero.

EXAMPLE 2 **Solving an Equation**

Solve each equation.

(a) $\dfrac{20(x^{1/3}-1)}{9x^{2/3}}=0$ (b) $\dfrac{25}{\sqrt{x}}-0.5=0$

(c) $x^2(4x-3)=0$ (d) $\dfrac{4x}{3(x^2-4)^{1/3}}=0$

(e) $g'(x)=0$, where $g(x)=(x-2)(x+1)^2$

SOLUTION

(a) $\dfrac{20(x^{1/3}-1)}{9x^{2/3}}=0$ — Example 5, page 236

$20(x^{1/3}-1)=0$ — A fraction is zero only if its numerator is zero.

$x^{1/3}-1=0$ — Divide each side by 20.

$x^{1/3}=1$ — Add 1 to each side.

$x=1$ — Cube each side.

(b) $\dfrac{25}{\sqrt{x}}-0.5=0$ — Example 4, page 213

$\dfrac{25}{\sqrt{x}}=0.5$ — Add 0.5 to each side.

$25=0.5\sqrt{x}$ — Multiply each side by $\sqrt{x}$.

$50=\sqrt{x}$ — Divide each side by 0.5.

$2500=x$ — Square both sides.

(c) $x^2(4x-3)=0$ — Example 2, page 184

$x^2=0 \Rightarrow x=0$ — Set first factor equal to zero.

$4x-3=0 \Rightarrow x=\frac{3}{4}$ — Set second factor equal to zero.

(d) $\dfrac{4x}{3(x^2-4)^{1/3}}=0$ — Example 4, page 176

$4x=0$ — A fraction is zero only if its numerator is zero.

$x=0$ — Divide each side by 4.

(e) $g(x)=(x-2)(x+1)^2$ — Exercise 41, page 199

$(x-2)(2)(x+1)+(x+1)^2(1)=0$ — Find derivative and set equal to zero.

$(x+1)[2(x-2)+(x+1)]=0$ — Factor.

$(x+1)(2x-4+x+1)=0$ — Multiply factors.

$(x+1)(3x-3)=0$ — Combine like terms.

$x+1=0 \Rightarrow x=-1$ — Set first factor equal to zero.

$3x-3=0 \Rightarrow x=1$ — Set second factor equal to zero.

3 CHAPTER SUMMARY AND STUDY STRATEGIES

*After studying this chapter, you should have acquired the following skills. The exercise numbers are keyed to the Review Exercises that begin on page 252. Answers to odd-numbered Review Exercises are given in the back of the text.**

Skills

- Find the critical numbers of a function. *(Section 3.1)* — *Review Exercises 1–4*
  c is a critical number of f if $f'(c) = 0$ or $f'(c)$ is undefined.
- Find the open intervals on which a function is increasing or decreasing. *(Section 3.1)* — *Review Exercises 5–8*
  Increasing if $f'(x) > 0$
  Decreasing if $f'(x) < 0$
- Find intervals on which a real-life model is increasing or decreasing, and interpret the results in context. *(Section 3.1)* — *Review Exercises 9, 10, 95*
- Use the First-Derivative Test to find the relative extrema of a function. *(Section 3.2)* — *Review Exercises 11–20*
- Find the absolute extrema of a continuous function on a closed interval. *(Section 3.2)* — *Review Exercises 21–30*
- Find minimum and maximum values of a real-life model and interpret the results in context. *(Section 3.2)* — *Review Exercises 31, 32*
- Find the open intervals on which the graph of a function is concave upward or concave downward. *(Section 3.3)* — *Review Exercises 33–36*
  Concave upward if $f''(x) > 0$
  Concave downward if $f''(x) < 0$
- Find the points of inflection of the graph of a function. *(Section 3.3)* — *Review Exercises 37–40*
- Use the Second-Derivative Test to find the relative extrema of a function. *(Section 3.3)* — *Review Exercises 41–44*
- Find the point of diminishing returns of an input-output model. *(Section 3.3)* — *Review Exercises 45, 46*
- Solve real-life optimization problems. *(Section 3.4)* — *Review Exercises 47–53, 96*
- Solve business and economics optimization problems. *(Section 3.5)* — *Review Exercises 54–58, 99*
- Find the price elasticity of demand for a demand function. *(Section 3.5)* — *Review Exercises 59–62*
- Find the vertical and horizontal asymptotes of a function and sketch its graph. *(Section 3.6)* — *Review Exercises 63–68*
- Find infinite limits and limits at infinity. *(Section 3.6)* — *Review Exercises 69–76*
- Use asymptotes to answer questions about real life. *(Section 3.6)* — *Review Exercises 77, 78*
- Analyze the graph of a function. *(Section 3.7)* — *Review Exercises 79–86*
- Find the differential of a function. *(Section 3.8)* — *Review Exercises 87–90*
- Use differentials to approximate changes in a function. *(Section 3.8)* — *Review Exercises 91–94*
- Use differentials to approximate changes in real-life models. *(Section 3.8)* — *Review Exercises 97, 98*

* Use a wide range of valuable study aids to help you master the material in this chapter. The *Student Solutions Guide* includes step-by-step solutions to all odd-numbered exercises to help you review and prepare. The *HM mathSpace® Student CD-ROM* helps you brush up on your algebra skills. The *Graphing Technology Guide*, available on the Web at *math.college.hmco.com/students*, offers step-by-step commands and instructions for a wide variety of graphing calculators, including the most recent models.

- ***Solve Problems Graphically, Analytically, and Numerically*** When analyzing the graph of a function, use a variety of problem-solving strategies. For instance, if you were asked to analyze the graph of

  $$f(x) = x^3 - 4x^2 + 5x - 4$$

  you could begin *graphically*. That is, you could use a graphing utility to find a viewing window that appears to show the important characteristics of the graph. From the graph shown below, the function appears to have one relative minimum, one relative maximum, and one point of inflection.

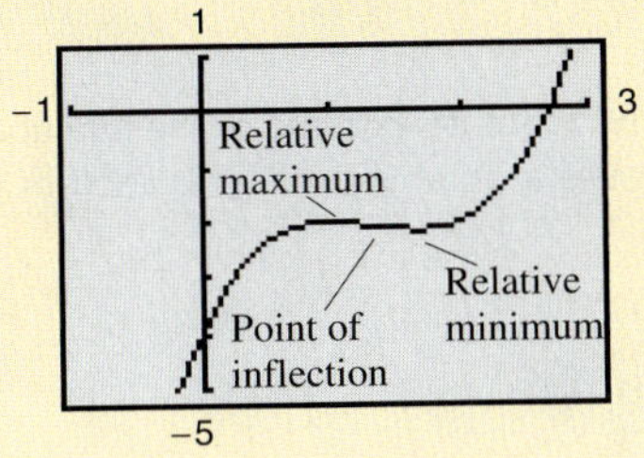

  Next, you could use calculus to *analyze* the graph. Because the derivative of f is

  $$f'(x) = 3x^2 - 8x + 5 = (3x - 5)(x - 1)$$

  the critical numbers of f are $x = \frac{5}{3}$ and $x = 1$. By the First-Derivative Test, you can conclude that $x = \frac{5}{3}$ yields a relative minimum and $x = 1$ yields a relative maximum. Because

  $$f''(x) = 6x - 8$$

  you can conclude that $x = \frac{4}{3}$ yields a point of inflection. Finally, you could analyze the graph *numerically*. For instance, you could construct a table of values and observe that f is increasing on the interval $(-\infty, 1)$, decreasing on the interval $\left(1, \frac{5}{3}\right)$, and increasing on the interval $\left(\frac{5}{3}, \infty\right)$.

- ***Problem-Solving Strategies*** If you get stuck when trying to solve an optimization problem, consider the strategies below.

  1. *Draw a Diagram.* If feasible, draw a diagram that represents the problem. Label all known values and unknown values on the diagram.
  2. *Solve a Simpler Problem.* Simplify the problem, or write several simple examples of the problem. For instance, if you are asked to find the dimensions that will produce a maximum area, try calculating the areas of several examples.
  3. *Rewrite the Problem in Your Own Words.* Rewriting a problem can help you understand it better.
  4. *Guess and Check.* Try guessing the answer, then check your guess in the statement of the original problem. By refining your guesses, you may be able to think of a general strategy for solving the problem.

Study Tools *Additional resources that accompany this chapter*

- **Algebra Review** (pages 248 and 249)
- **Chapter Summary and Study Strategies** (pages 250 and 251)
- **Review Exercises** (pages 252–255)
- **Sample Post-Graduation Exam Questions** (page 256)
- **Web Exercises** (page 200, Exercise 66; page 219, Exercise 43)
- **Student Solutions Guide**
- **HM mathSpace® Student CD-ROM**
- **Graphing Technology Guide** (*math.college.hmco.com/students*)

3 CHAPTER REVIEW EXERCISES

In Exercises 1–4, find the critical numbers of the function.

1. $f(x) = -x^2 + 2x + 4$
2. $g(x) = (x - 1)^2(x - 3)$
3. $h(x) = \sqrt{x}(x - 3)$
4. $f(x) = (x + 1)^3$

In Exercises 5–8, determine the open intervals on which the function is increasing or decreasing. Solve the problem analytically and graphically.

5. $f(x) = x^2 + x - 2$
6. $g(x) = -x^2 + 7x - 12$
7. $h(x) = \dfrac{x^2 - 3x - 4}{x - 3}$
8. $f(x) = -x^3 + 6x^2 - 2$

9. ***Meteorology*** The monthly normal temperature T (in degrees Fahrenheit) for New York City can be modeled by

$$T = 0.0385t^4 - 1.122t^3 + 9.67t^2 - 21.8t + 47$$

where $1 \le t \le 12$ and $t = 1$ corresponds to January. *(Source: National Climatic Data Center)*

(a) Find the interval(s) on which the model is increasing.

(b) Find the interval(s) on which the model is decreasing.

(c) Interpret the results of parts (a) and (b).

(d) Use a graphing utility to graph the model.

10. ***CD Shipments*** The number S of manufacturer unit shipments (in millions) of CDs in the United States from 1998 through 2002 can be modeled by

$$S = 5.8583t^3 - 28.943t^2 - 34.36t + 940.6$$

where $-2 \le t \le 2$ and $t = 0$ corresponds to 2000. *(Source: Recording Industry Association of America)*

(a) Find the interval(s) on which the model is increasing.

(b) Find the interval(s) on which the model is decreasing.

(c) Interpret the results of parts (a) and (b).

(d) Use a graphing utility to graph the model.

In Exercises 11–20, use the First-Derivative Test to find the relative extrema of the function. Then use a graphing utility to verify your result.

11. $f(x) = 4x^3 - 6x^2 - 2$
12. $f(x) = \frac{1}{4}x^4 - 8x$
13. $g(x) = x^2 - 16x + 12$
14. $h(x) = 4 + 10x - x^2$
15. $h(x) = 2x^2 - x^4$
16. $s(x) = x^4 - 8x^2 + 3$
17. $f(x) = \dfrac{6}{x^2 + 1}$
18. $f(x) = \dfrac{2}{x^2 - 1}$
19. $h(x) = \dfrac{x^2}{x - 2}$
20. $g(x) = x - 6\sqrt{x}, \quad x > 0$

In Exercises 21–30, find the absolute extrema of the function on the closed interval. Then use a graphing utility to confirm your result.

21. $f(x) = x^2 + 5x + 6; \quad [-3, 0]$
22. $f(x) = x^4 - 2x^3; \quad [0, 2]$
23. $f(x) = x^3 - 12x + 1; \quad [-4, 4]$
24. $f(x) = x^3 + 2x^2 - 3x + 4; \quad [-3, 2]$
25. $f(x) = 4\sqrt{x} - x^2; \quad [0, 3]$
26. $f(x) = 2\sqrt{x} - x; \quad [0, 9]$
27. $f(x) = 3x^4 - 6x^2 + 2; \quad [0, 2]$
28. $f(x) = -x^4 + x^2 + 2; \quad [0, 2]$
29. $f(x) = \dfrac{2x}{x^2 + 1}; \quad [-1, 2]$
30. $f(x) = \dfrac{8}{x} + x; \quad [1, 4]$

31. ***Surface Area*** A right circular cylinder of radius r and height h has a volume of 25 cubic inches. The total surface area of the cylinder in terms of r is given by

$$S = 2\pi r\left(r + \frac{25}{\pi r^2}\right).$$

Use a graphing utility to graph S and S' and find the value of r that yields the minimum surface area.

32. ***Environment*** When organic waste is dumped into a pond, the decomposition of the waste consumes oxygen. A model for the oxygen level O (where 1 is the normal level) of a pond as waste material oxidizes is

$$O = \frac{t^2 - t + 1}{t^2 + 1}, \quad 0 \le t$$

where t is the time in weeks.

(a) When is the oxygen level lowest? What is this level?

(b) When is the oxygen level highest? What is this level?

(c) Describe the oxygen level as t increases.

In Exercises 33–36, determine the open intervals on which the graph of the function is concave upward or concave downward. Then use a graphing utility to confirm your result.

33. $f(x) = (x - 2)^3$

34. $h(x) = x^5 - 10x^2$

35. $g(x) = \frac{1}{4}(-x^4 + 8x^2 - 12)$

36. $h(x) = x^3 - 6x$

In Exercises 37–40, find the points of inflection of the graph of the function.

37. $f(x) = \frac{1}{2}x^4 - 4x^3$

38. $f(x) = \frac{1}{4}x^4 - 2x^2 - x$

39. $f(x) = x^3(x - 3)^2$

40. $f(x) = (x + 2)^2(x - 4)$

In Exercises 41–44, use the Second-Derivative Test to find the relative extrema of the function.

41. $f(x) = x^5 - 5x^3$

42. $f(x) = x(x^2 - 3x - 9)$

43. $f(x) = (x - 1)^3(x + 4)^2$

44. $f(x) = (x - 2)^2(x + 2)^2$

Point of Diminishing Returns In Exercises 45 and 46, identify the point of diminishing returns for the input-output function. For each function, R is the revenue (in thousands of dollars) and x is the amount spent on advertising (in thousands of dollars).

45. $R = \frac{1}{1500}(150x^2 - x^3), \quad 0 \le x \le 100$

46. $R = -\frac{2}{3}(x^3 - 12x^2 - 6), \quad 0 \le x \le 8$

47. ***Minimum Sum*** Find two positive numbers whose product is 169 and whose sum is a minimum. Solve the problem analytically, and use a graphing utility to solve the problem graphically.

48. ***Length*** The wall of a building is to be braced by a beam that must pass over a five-foot fence that is parallel to the building and 4 feet from the building. Find the length of the shortest beam that can be used.

49. ***Newspaper Circulation*** The total number N of daily newspapers in circulation (in millions) in the United States from 1970 through 2000 can be modeled by

$$N = 0.020t^3 - 1.19t^2 + 9.0t + 1746$$

where $0 \le t \le 30$ and $t = 0$ corresponds to 1970. *(Source: Editor and Publisher Company)*

(a) Find the absolute maximum and minimum over the time period.

(b) Find the year when the circulation was changing at the greatest rate.

(c) Briefly explain your results for parts (a) and (b).

50. ***Minimum Cost*** A fence is to be built to enclose a rectangular region of 4800 square feet. The fencing material along three sides costs \$3 per foot. The fencing material along the fourth side costs \$4 per foot.

(a) Find the most economical dimensions of the region.

(b) How would the result of part (a) change if the fencing material costs for all sides increased by \$1 per foot?

51. ***Biology*** The growth of a red oak tree is approximated by the model

$$y = -0.003x^3 + 0.137x^2 + 0.458x - 0.839, \quad 2 \le x \le 34$$

where y is the height of the tree in feet and x is its age in years. Find the age of the tree when it is growing most rapidly. Then use a graphing utility to graph the function to verify your result. (*Hint:* Use the viewing window $2 \le x \le 34$ and $-10 \le y \le 60$.)

52. ***Consumer Trends*** The average number of hours N (per person per year) of TV usage in the United States from 1996 through 2001 can be modeled by

$$N = -2.870t^3 + 79.62t^2 - 639.1t + 3473$$

where $t = 6$ corresponds to 1996. *(Source: Veronis Suhler Stevenson)*

(a) Find the intervals on which dN/dt is increasing and decreasing.

(b) Find the limit of N as $t \to 0$.

(c) Briefly explain your results for parts (a) and (b).

53. ***Medicine: Poiseuille's Law*** The speed of blood that is r centimeters from the center of an artery is modeled by

$$s(r) = c(R^2 - r^2)$$

where c is a constant, R is the radius of the artery, and s is measured in centimeters per second. Show that the speed is a maximum at the center of an artery.

54. ***Profit*** The demand and cost functions for a product are $p = 36 - 4x$ and $C = 2x^2 + 6$.

(a) What level of production will produce a maximum profit?

(b) What level of production will produce a minimum average cost per unit?

55. ***Revenue*** For groups of 20 or more, a theater determines the ticket price p according to the formula

$$p = 15 - 0.1(n - 20), \quad 20 \le n \le N$$

where n is the number in the group. What should the value of N be? Explain your reasoning.

56. ***Minimum Cost*** The cost of fuel to run a locomotive is proportional to the $\frac{3}{2}$ power of the speed. At a speed of 25 miles per hour, the cost of fuel is \$50 per hour. Other costs amount to \$100 per hour. Find the speed that will minimize the cost per mile.

57. ***Inventory Cost*** The cost C of inventory modeled by

$$C = \left(\frac{Q}{x}\right)s + \left(\frac{x}{2}\right)r$$

depends on ordering and storage costs, where Q is the number of units sold per year, r is the cost of storing one unit for 1 year, s is the cost of placing an order, and x is the number of units in the order. Determine the order size that will minimize the cost when $Q = 10{,}000$, $s = 4.5$, and $r = 5.76$.

58. ***Profit*** The demand and cost functions for a product are given by

$$p = 600 - 3x$$

and

$$C = 0.3x^2 + 6x + 600$$

where p is the price per unit, x is the number of units, and C is the total cost. The profit for producing x units is given by

$$P = xp - C - xt$$

where t is the excise tax per unit. Find the maximum profits for excise taxes of $t = \$5$, $t = \$10$, and $t = \$20$.

In Exercises 59–62, find the intervals on which the demand is elastic, inelastic, and of unit elasticity.

59. $p = 30 - 0.2x, \quad 0 \le x \le 150$

60. $p = 60 - 0.04x, \quad 0 \le x \le 1500$

61. $p = \sqrt{300 - x}, \quad 0 \le x \le 300$

62. $p = \sqrt{960 - x}, \quad 0 \le x \le 960$

In Exercises 63–68, find the vertical and horizontal asymptotes of the graph. Then use a graphing utility to graph the function.

63. $h(x) = \dfrac{2x + 3}{x - 4}$

64. $g(x) = \dfrac{5x^2}{x^2 + 2}$

65. $f(x) = \dfrac{\sqrt{9x^2 + 1}}{x}$

66. $h(x) = \dfrac{3x}{\sqrt{x^2 + 2}}$

67. $f(x) = \dfrac{3}{x^2 - 5x + 4}$

68. $h(x) = \dfrac{2x^2 + 3x - 5}{x - 1}$

In Exercises 69–76, find the limit, if it exists.

69. $\displaystyle\lim_{x \to 0^+} \left(x - \frac{1}{x^3}\right)$

70. $\displaystyle\lim_{x \to 0^-} \left(3 + \frac{1}{x}\right)$

71. $\displaystyle\lim_{x \to -1^+} \frac{x^2 - 2x + 1}{x + 1}$

72. $\displaystyle\lim_{x \to 3^-} \frac{3x^2 + 1}{x^2 - 9}$

73. $\displaystyle\lim_{x \to \infty} \frac{5x^2 + 3}{2x^2 - x + 1}$

74. $\displaystyle\lim_{x \to \infty} \frac{3x^2 - 2x + 3}{x + 1}$

75. $\displaystyle\lim_{x \to -\infty} \frac{3x^2}{x + 2}$

76. $\displaystyle\lim_{x \to -\infty} \left(\frac{x}{x - 2} + \frac{2x}{x + 2}\right)$

77. ***Health*** For a person with sensitive skin, the maximum amount T (in hours) of exposure to the sun that can be tolerated before skin damage occurs can be modeled by

$$T = \frac{0.37s + 23.8}{s}, \quad 0 < s \le 120$$

where s is the Sunsor Scale reading. *(Source: Sunsor, Inc.)*

(a) Use a graphing utility to graph the model. Compare your result with the graph below.

(b) Describe the value of T as s increases.

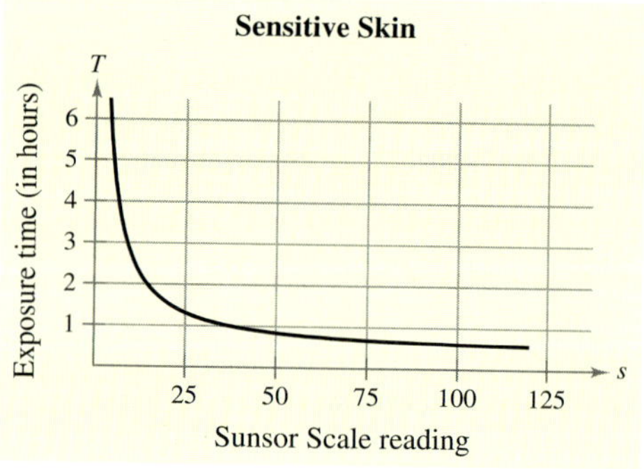

78. ***Average Cost and Profit*** The cost and revenue functions for a product are given by

$$C = 10{,}000 + 48.9x$$

and

$$R = 68.5x.$$

(a) Find the average cost function.

(b) What is the limit of the average cost as x approaches infinity?

(c) Find the average profits when x is 1 million, 2 million, and 10 million.

(d) What is the limit of the average profit as x increases without bound?

In Exercises 79–86, sketch the graph of the function. Label the intercepts, relative extrema, points of inflection, and asymptotes. State the domain of the function.

79. $f(x) = 4x - x^2$

80. $f(x) = 4x^3 - x^4$

81. $f(x) = x\sqrt{16 - x^2}$

82. $f(x) = x^2\sqrt{9 - x^2}$

83. $f(x) = \dfrac{x + 1}{x - 1}$

84. $f(x) = \dfrac{2x}{1 + x^2}$

85. $f(x) = x^2 + \dfrac{2}{x}$

86. $f(x) = x^{4/5}$

In Exercises 87–90, find the differential *dy*.

87. $y = 6x^2 - 5$

88. $y = (3x^2 - 2)^3$

89. $y = -\dfrac{5}{\sqrt[3]{x}}$

90. $y = \dfrac{2 - x}{x + 5}$

In Exercises 91–94, use differentials to approximate the change in cost, revenue, or profit corresponding to an increase in sales of one unit.

91. $C = 40x^2 + 1225, \quad x = 10$

92. $C = 1.5\sqrt[3]{x} + 500, \quad x = 125$

93. $R = 6.25x + 0.4x^{3/2}, \quad x = 225$

94. $P = 0.003x^2 + 0.019x - 1200, \quad x = 750$

95. ***Revenue Per Share*** The revenues per share R (in dollars) for the Walt Disney Company for the years 1992 through 2003 are shown in the table. *(Source: The Walt Disney Company)*

Year, t	2	3	4	5	6	7
Revenue per share, R	4.77	5.31	6.40	7.70	10.50	11.10

Year, t	8	9	10	11	12	13
Revenue per share, R	11.21	11.34	12.09	12.52	12.40	13.23

(a) Use a graphing utility to create a scatter plot of the data, where t is the time in years, with $t = 2$ corresponding to 1992.

(b) Describe any trends and/or patterns of the data.

(c) A model for the data is

$$R = \frac{4.72 - 1.605t + 0.1741t^2}{1 - 0.356t + 0.0420t^2 - 0.00112t^3},$$

$2 \le t \le 13.$

Graph the model and the data in the same viewing window.

(d) Find the years when the revenue per share was increasing and decreasing.

(e) Find the years when the rate of change of the revenue per share was increasing and decreasing.

(f) Briefly explain your results for parts (d) and (e).

96. ***Medicine*** The effectiveness E of a pain-killing drug t hours after entering the bloodstream is modeled by

$$E = 22.5t + 7.5t^2 - 2.5t^3, \quad 0 \le t \le 4.5.$$

(a) Use a graphing utility to graph the equation. Choose an appropriate window.

(b) Find the maximum effectiveness the pain-killing drug attains over the interval $[0, 4.5]$.

97. ***Surface Area and Volume*** The diameter of a sphere is measured to be 18 inches with a possible error of 0.05 inch. Use differentials to approximate the possible error in the surface area and the volume of the sphere.

98. ***Demand*** A company finds that the demand for its product is modeled by

$$p = 85 - 0.125x.$$

If x changes from 7 to 8, what is the corresponding change in p? Compare the values of Δp and dp.

99. ***Economics: Revenue*** Consider the following cost and demand information for a monopoly (in dollars). Complete the table, and then use the information to answer the questions. *(Source: Adapted from Taylor,* Economics, *Fourth Edition)*

Quantity of output	Price	Total revenue	Marginal revenue
1	14.00		
2	12.00		
3	10.00		
4	8.50		
5	7.00		
6	5.50		

(a) Use the *regression* feature of a graphing utility to find a quadratic model for the total revenue data.

(b) From the total revenue model you found in part (a), use derivatives to find an equation for the marginal revenue. Now use the values for output in the table and compare the results with the values in the marginal revenue column of the table. How close was your model?

(c) What quantity maximizes total revenue for the monopoly?

3 SAMPLE POST-GRADUATION EXAM QUESTIONS

CPA
GMAT
GRE
Actuarial
CLAST

The following questions represent the types of questions that appear on certified public accountant (CPA) exams, Graduate Management Admission Tests (GMAT), Graduate Records Exams (GRE), actuarial exams, and College-Level Academic Skills Tests (CLAST). The answers to the questions are given in the back of the book.

Figure for 1–3

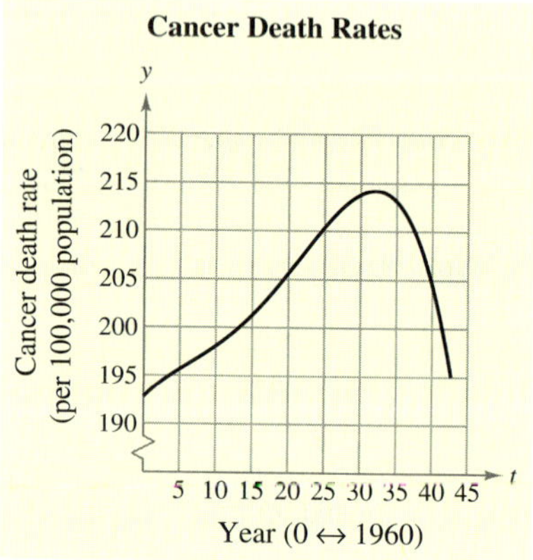

For Questions 1–3, use the data shown in the graph.
(Source: U.S. National Center for Health Statistics)

1. The percent decrease in the cancer death rate between 1990 and 2000 was about

(a) 3% (b) 6.5% (c) 4% (d) 16%

2. Which of the following statements about the cancer death rate can be inferred from the graph?

I. The cancer death rate reached a maximum around 1990.

II. The cancer death rate was increasing at a maximum rate around 1980.

III. Between 1960 and 1990, the average rate of change of the cancer death rate was about 0.75 per year.

(a) I and II (b) I and III (c) II and III (d) I, II, and III

3. Let f represent the cancer death rate function. In 1990, $f'(x) =$

(a) 208 (b) 0 (c) 1 (d) undefined

4. In 2000, a company issued 75,000 shares of stock. Each share of stock was worth \$85.75. Five years later, each share of stock was worth \$72.21. How much less were the shares worth in 2005 than in 2000?

(a) \$1,115,500 (b) \$1,051,500 (c) \$1,155,000 (d) \$1,015,500

5. Which of the figures below most resembles the graph of $f(x) = \dfrac{x}{x^2 - 4}$?

(a)

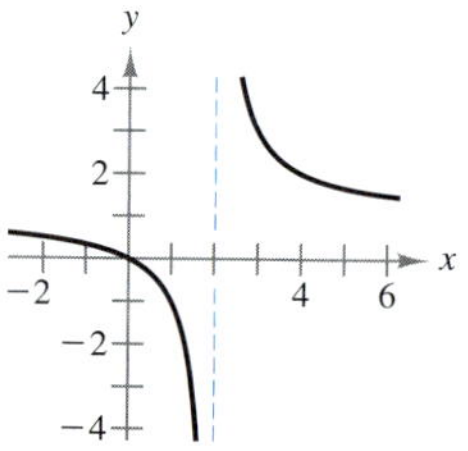

(b)

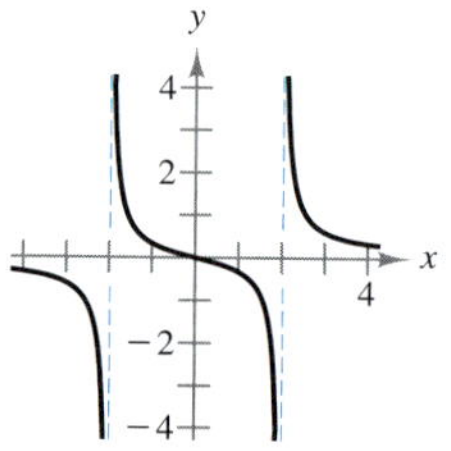

(c)

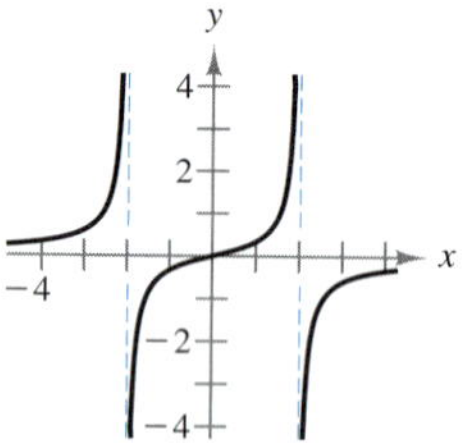

(d)

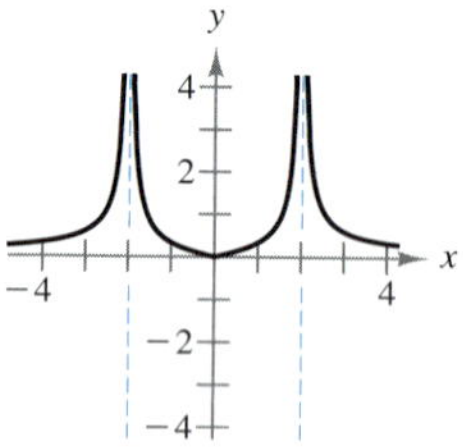

chapter

4 Exponential and Logarithmic Functions

On January 2, 2004, the Indonesian islands of Bali and Lombok experienced an earthquake measuring 6.1 on the Richter scale, a logarithmic function that serves as one way to calculate an earthquake's magnitude.

STRATEGIES FOR SUCCESS

WHAT YOU SHOULD LEARN:

- How to graph the exponential function $f(x) = a^x$ and how to graph the natural exponential function $f(x) = e^x$
- How to calculate derivatives of exponential functions
- How to graph the logarithmic function $f(x) = \ln x$ and use it to solve exponential and logarithmic equations
- How to calculate the derivatives of logarithmic functions
- How to solve exponential growth and decay applications

WHY YOU SHOULD LEARN IT:

Exponential and logarithmic functions have many applications in real life, as can be seen by the examples below, which represent a small sample of the applications in this chapter.

- Property Value, Exercise 37 on page 263
- Present Value, Exercises 41 and 42 on page 271
- Normal Probability Density Function, Exercises 45–48 on page 280
- Human Memory Model, Exercise 79 on page 289
- Effective Yield, Exercises 31–34 on page 306
- Earthquake Intensity, Exercise 48 on page 307

4.1 EXPONENTIAL FUNCTIONS

- Use the properties of exponents to evaluate and simplify exponential expressions.
- Sketch the graphs of exponential functions.

Exponential Functions

You are already familiar with the behavior of algebraic functions such as

$$f(x) = x^2, \quad g(x) = \sqrt{x} = x^{1/2}, \quad \text{and} \quad h(x) = \frac{1}{x} = x^{-1}$$

each of which involves a variable raised to a constant power. By interchanging roles and raising a constant to a variable power, you obtain another important class of functions called **exponential functions.** Some simple examples are

$$f(x) = 2^x, \quad g(x) = \left(\frac{1}{10}\right)^x = \frac{1}{10^x}, \quad \text{and} \quad h(x) = 3^{2x} = 9^x.$$

In general, you can use any positive base $a \neq 1$ as the base of an exponential function.

Definition of Exponential Function

If $a > 0$ and $a \neq 1$, then the **exponential function** with base a is given by

$$f(x) = a^x.$$

STUDY TIP

In the definition above, the base $a = 1$ is excluded because it yields

$$f(x) = 1^x = 1.$$

This is a constant function, not an exponential function.

When working with exponential functions, the properties of exponents, shown below, are useful.

Properties of Exponents

Let a and b be positive numbers.

1. $a^0 = 1$
2. $a^x a^y = a^{x+y}$
3. $\dfrac{a^x}{a^y} = a^{x-y}$
4. $(a^x)^y = a^{xy}$
5. $(ab)^x = a^x b^x$
6. $\left(\dfrac{a}{b}\right)^x = \dfrac{a^x}{b^x}$
7. $a^{-x} = \dfrac{1}{a^x}$

EXAMPLE 1 Applying Properties of Exponents

Simplify each expression using the properties of exponents.

(a) $(2^2)(2^3)$ (b) $(2^2)(2^{-3})$ (c) $(3^2)^3$

(d) $\left(\frac{1}{3}\right)^{-2}$ (e) $\frac{3^2}{3^3}$ (f) $(2^{1/2})(3^{1/2})$

SOLUTION

(a) $(2^2)(2^3) = 2^{2+3} = 2^5 = 32$ Apply Property 2.

(b) $(2^2)(2^{-3}) = 2^{2-3} = 2^{-1} = \frac{1}{2}$ Apply Properties 2 and 7.

(c) $(3^2)^3 = 3^{2(3)} = 3^6 = 729$ Apply Property 4.

(d) $\left(\frac{1}{3}\right)^{-2} = \frac{1}{(1/3)^2} = \left(\frac{1}{1/3}\right)^2 = 3^2 = 9$ Apply Properties 6 and 7.

(e) $\frac{3^2}{3^3} = 3^{2-3} = 3^{-1} = \frac{1}{3}$ Apply Properties 3 and 7.

(f) $(2^{1/2})(3^{1/2}) = [(2)(3)]^{1/2} = 6^{1/2} = \sqrt{6}$ Apply Property 5.

TRY IT 1

Simplify each expression using the properties of exponents.

(a) $(3^2)(3^3)$ (b) $(3^2)(3^{-1})$

(c) $(2^3)^2$ (d) $\left(\frac{1}{2}\right)^{-3}$

(e) $\frac{2^2}{2^3}$ (f) $(2^{1/2})(5^{1/2})$

Although Example 1 demonstrates the properties of exponents with integer and rational exponents, it is important to realize that the properties hold for *all* real exponents. With a calculator, you can obtain approximations of a^x for any base a and any real exponent x. Here are some examples.

$$2^{-0.6} \approx 0.660, \qquad \pi^{0.75} \approx 2.360, \qquad (1.56)^{\sqrt{2}} \approx 1.876$$

EXAMPLE 2 Dating Organic Material

In living organic material, the ratio of radioactive carbon isotopes to the total number of carbon atoms is about 1 to 10^{12}. When organic material dies, its radioactive carbon isotopes begin to decay, with a half-life of about 5715 years. This means that after 5715 years, the ratio of isotopes to atoms will have decreased to one-half the original ratio, after a second 5715 years the ratio will have decreased to one-fourth of the original, and so on. Figure 4.1 shows this decreasing ratio. The formula for the ratio R of carbon isotopes to carbon atoms is

$$R = \left(\frac{1}{10^{12}}\right)\left(\frac{1}{2}\right)^{t/5715}$$

where t is the time in years. Find the value of R for each period of time.

(a) 10,000 years (b) 20,000 years (c) 25,000 years

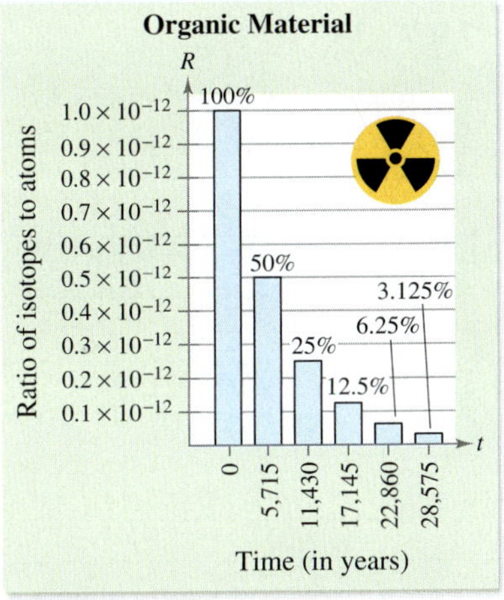

FIGURE 4.1

SOLUTION

(a) $R = \left(\frac{1}{10^{12}}\right)\left(\frac{1}{2}\right)^{10{,}000/5715} \approx 2.973 \times 10^{-13}$ Ratio for 10,000 years

(b) $R = \left(\frac{1}{10^{12}}\right)\left(\frac{1}{2}\right)^{20{,}000/5715} \approx 8.842 \times 10^{-14}$ Ratio for 20,000 years

(c) $R = \left(\frac{1}{10^{12}}\right)\left(\frac{1}{2}\right)^{25{,}000/5715} \approx 4.821 \times 10^{-14}$ Ratio for 25,000 years

TRY IT 2

Use the formula for the ratio of carbon isotopes to carbon atoms in Example 2 to find the value of R for each period of time.

(a) 5,000 years

(b) 15,000 years

(c) 30,000 years

Graphs of Exponential Functions

The basic nature of the graph of an exponential function can be determined by the point-plotting method or by using a graphing utility.

EXAMPLE 3 **Graphing Exponential Functions**

Sketch the graph of each exponential function.

(a) $f(x) = 2^x$ (b) $g(x) = \left(\frac{1}{2}\right)^x = 2^{-x}$ (c) $h(x) = 3^x$

SOLUTION To sketch these functions by hand, you can begin by constructing a table of values, as shown below.

x	-3	-2	-1	0	1	2	3	4
$f(x) = 2^x$	$\frac{1}{8}$	$\frac{1}{4}$	$\frac{1}{2}$	1	2	4	8	16
$g(x) = 2^{-x}$	8	4	2	1	$\frac{1}{2}$	$\frac{1}{4}$	$\frac{1}{8}$	$\frac{1}{16}$
$h(x) = 3^x$	$\frac{1}{27}$	$\frac{1}{9}$	$\frac{1}{3}$	1	3	9	27	81

The graphs of the three functions are shown in Figure 4.2. Note that the graphs of $f(x) = 2^x$ and $h(x) = 3^x$ are increasing, whereas the graph of $g(x) = 2^{-x}$ is decreasing.

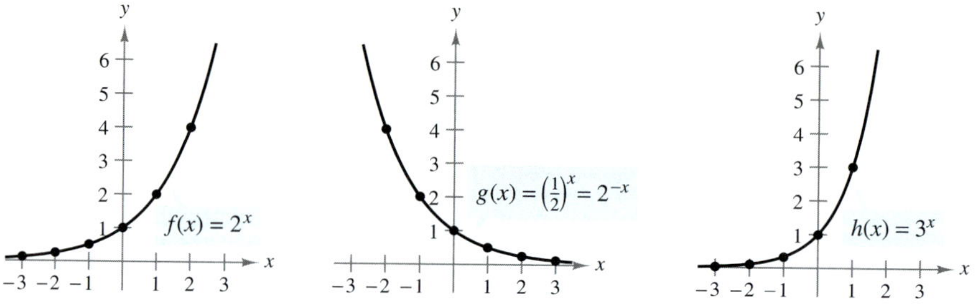

(a) (b) (c)

FIGURE 4.2

STUDY TIP

Note that a graph of the form $f(x) = a^x$, as shown in Example 3(a), is a reflection in the y-axis of the graph of the form $f(x) = a^{-x}$, as shown in Example 3(b).

TRY IT 3

Complete the table of values for $f(x) = 5^x$. Sketch the graph of the exponential function.

x	-3	-2	-1	0
$f(x)$				

x	1	2	3
$f(x)$			

TECHNOLOGY

Try graphing the functions $f(x) = 2^x$ and $h(x) = 3^x$ in the same viewing window, as shown at the right. From the display, you can see that the graph of h is increasing more rapidly than the graph of f.

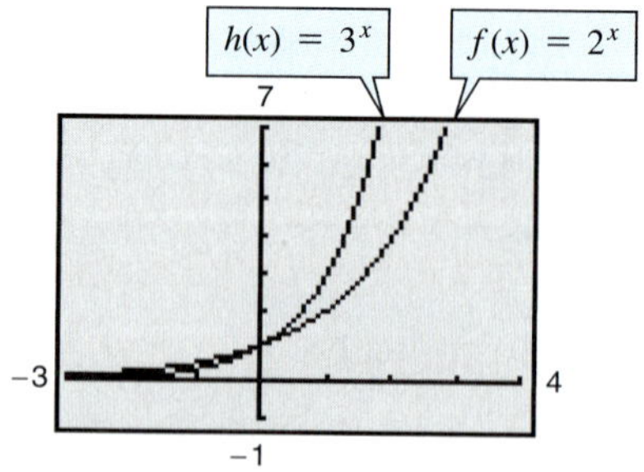

The forms of the graphs in Figure 4.2 are typical of the graphs of the exponential functions $y = a^{-x}$ and $y = a^x$, where $a > 1$. The basic characteristics of such graphs are summarized in Figure 4.3.

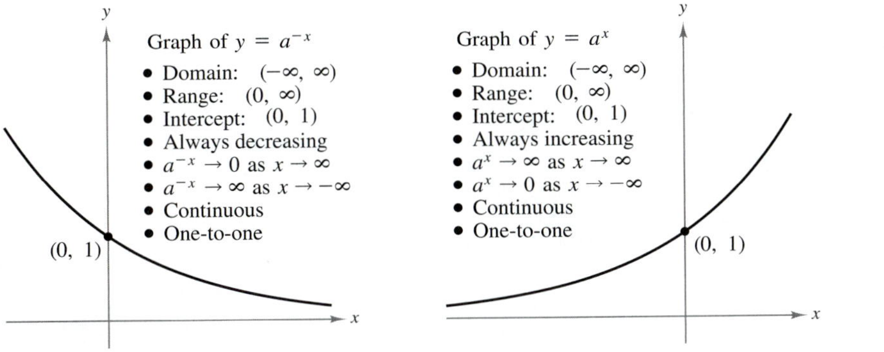

FIGURE 4.3 Characteristics of the Exponential Functions $y = a^{-x}$ and $y = a^x$ $(a > 1)$

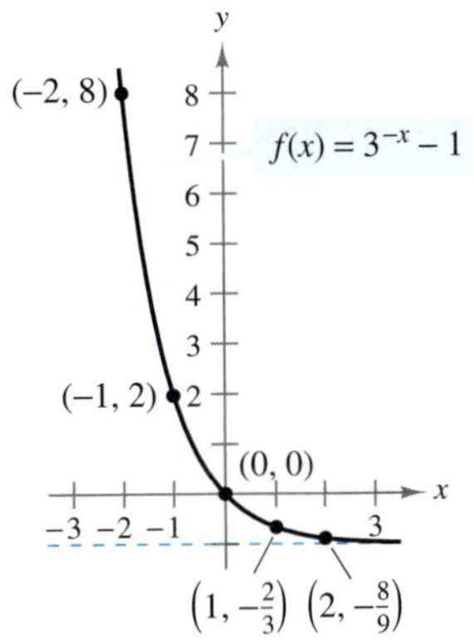

FIGURE 4.4

EXAMPLE 4 Graphing an Exponential Function

Sketch the graph of

$$f(x) = 3^{-x} - 1.$$

SOLUTION Begin by creating a table of values, as shown below.

x	-2	-1	0	1	2
$f(x)$	$3^2 - 1 = 8$	$3^1 - 1 = 2$	$3^0 - 1 = 0$	$3^{-1} - 1 = -\frac{2}{3}$	$3^{-2} - 1 = -\frac{8}{9}$

From the limit

$$\lim_{x\to\infty} (3^{-x} - 1) = \lim_{x\to\infty} 3^{-x} - \lim_{x\to\infty} 1$$
$$= \lim_{x\to\infty} \frac{1}{3^x} - \lim_{x\to\infty} 1$$
$$= 0 - 1$$
$$= -1$$

you can see that $y = -1$ is a horizontal asymptote of the graph. The graph is shown in Figure 4.4.

TRY IT 4

Complete the table of values for $f(x) = 2^{-x} + 1$. Sketch the graph of the function. Determine the horizontal asymptote of the graph.

x	-3	-2	-1	0
$f(x)$				

x	1	2	3
$f(x)$			

TAKE ANOTHER LOOK

Finding a Pattern

Use a graphing utility to investigate the function $f(x) = a^x$ for $0 < a < 1$, $a = 1$, and $a > 1$. Discuss the effect that a has on the shape of the graph.

PREREQUISITE REVIEW 4.1

The following warm-up exercises involve skills that were covered in earlier sections. You will use these skills in the exercise set for this section.

In Exercises 1–6, describe how the graph of g is related to the graph of f.

1. $g(x) = f(x + 2)$

2. $g(x) = -f(x)$

3. $g(x) = -1 + f(x)$

4. $g(x) = f(-x)$

5. $g(x) = f(x - 1)$

6. $g(x) = f(x) + 2$

In Exercises 7–10, discuss the continuity of the function.

7. $f(x) = \dfrac{x^2 + 2x - 1}{x + 4}$

8. $f(x) = \dfrac{x^2 - 3x + 1}{x^2 + 2}$

9. $f(x) = \dfrac{x^2 - 3x - 4}{x^2 - 1}$

10. $f(x) = \dfrac{x^2 - 5x + 4}{x^2 + 1}$

In Exercises 11–16, solve for x.

11. $2x - 6 = 4$

12. $3x + 1 = 5$

13. $(x + 4)^2 = 25$

14. $(x - 2)^2 = 8$

15. $x^2 + 4x - 5 = 0$

16. $2x^2 - 3x + 1 = 0$

EXERCISES 4.1

In Exercises 1 and 2, evaluate each expression.

1. (a) $5(5^3)$ (b) $27^{2/3}$
(c) $64^{3/4}$ (d) $81^{1/2}$
(e) $25^{3/2}$ (f) $32^{2/5}$

2. (a) $\left(\frac{1}{5}\right)^3$ (b) $\left(\frac{1}{8}\right)^{1/3}$
(c) $64^{2/3}$ (d) $\left(\frac{5}{8}\right)^2$
(e) $100^{3/2}$ (f) $4^{5/2}$

In Exercises 3–6, use the properties of exponents to simplify the expression.

3. (a) $(5^2)(5^3)$ (b) $(5^2)(5^{-3})$
(c) $(5^2)^2$ (d) 5^{-3}

4. (a) $\dfrac{5^3}{5^6}$ (b) $\left(\dfrac{1}{5}\right)^{-2}$
(c) $(8^{1/2})(2^{1/2})$ (d) $(32^{3/2})\left(\frac{1}{2}\right)^{3/2}$

5. (a) $\dfrac{5^3}{25^2}$ (b) $(9^{2/3})(3)(3^{2/3})$
(c) $[(25^{1/2})(5^2)]^{1/3}$ (d) $(8^2)(4^3)$

6. (a) $(4^3)(4^2)$ (b) $\left(\frac{1}{4}\right)^2(4^2)$
(c) $(4^6)^{1/2}$ (d) $[(8^{-1})(8^{2/3})]^3$

In Exercises 7–10, evaluate the function. If necessary, use a graphing utility, rounding your answers to three decimal places.

7. $f(x) = 2^{x-1}$
(a) $f(3)$ (b) $f\left(\frac{1}{2}\right)$
(c) $f(-2)$ (d) $f\left(-\frac{3}{2}\right)$

8. $f(x) = 3^{x+2}$
(a) $f(-4)$ (b) $f\left(-\frac{1}{2}\right)$
(c) $f(2)$ (d) $f\left(-\frac{5}{2}\right)$

9. $g(x) = 1.05^x$
(a) $g(-2)$ (b) $g(120)$
(c) $g(12)$ (d) $g(5.5)$

10. $g(x) = 1.075^x$
(a) $g(1.2)$ (b) $g(180)$
(c) $g(60)$ (d) $g(12.5)$

In Exercises 11–18, solve the equation for x.

11. $3^x = 81$

12. $5^{x+1} = 125$

13. $\left(\frac{1}{3}\right)^{x-1} = 27$

14. $\left(\frac{1}{5}\right)^{2x} = 625$

15. $4^3 = (x + 2)^3$

16. $4^2 = (x + 2)^2$

17. $x^{3/4} = 8$

18. $(x + 3)^{4/3} = 16$

In Exercises 19–24, match the function with its graph. [The graphs are labeled (a)–(f).]

(a)

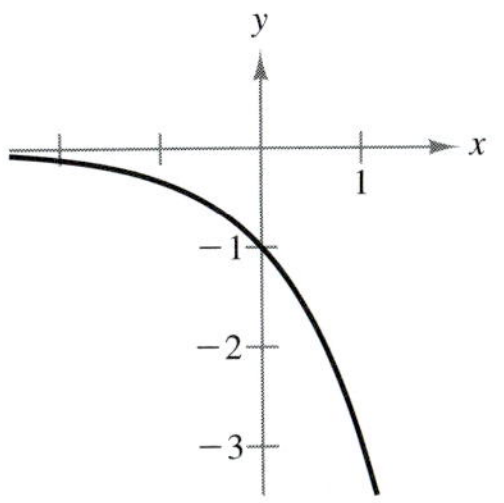

(b)

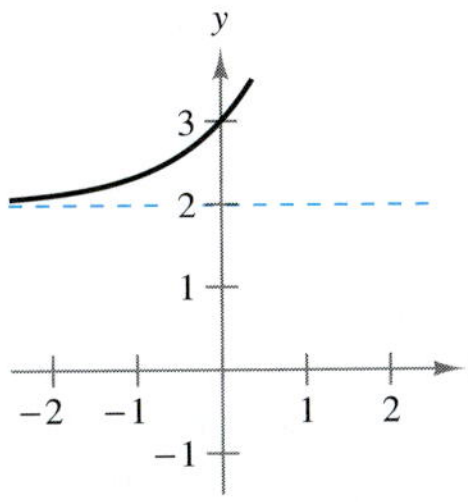

(c)

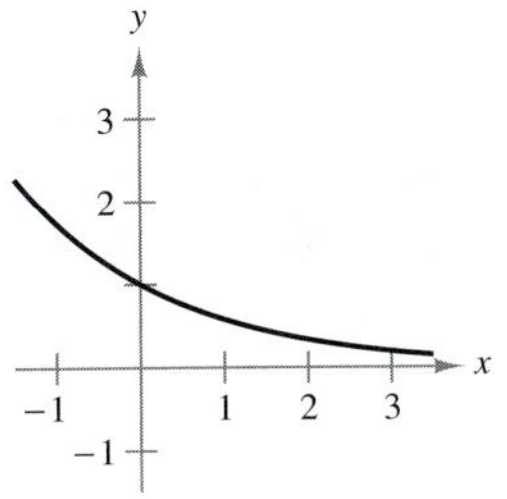

(d)

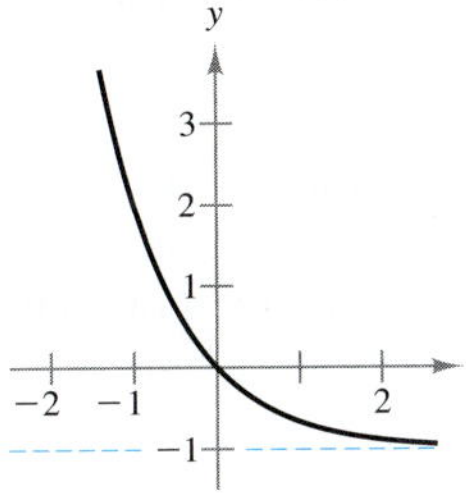

(e)

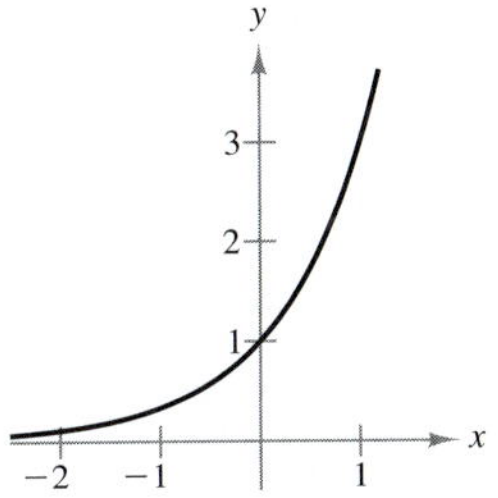

(f)

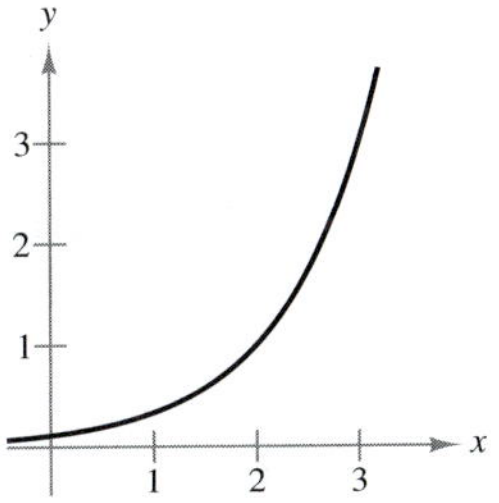

19. $f(x) = 3^x$

20. $f(x) = 3^{-x/2}$

21. $f(x) = -3^x$

22. $f(x) = 3^{x-2}$

23. $f(x) = 3^{-x} - 1$

24. $f(x) = 3^x + 2$

In Exercises 25–34, sketch the graph of the function.

25. $f(x) = 6^x$

26. $f(x) = 4^x$

27. $f(x) = \left(\frac{1}{5}\right)^x = 5^{-x}$

28. $f(x) = \left(\frac{1}{4}\right)^x = 4^{-x}$

29. $y = 3^{-x^2}$

30. $y = 2^{-x^2}$

31. $y = 3^{-|x|}$

32. $y = 3^{|x|}$

33. $s(t) = \frac{1}{4}(3^{-t})$

34. $s(t) = 2^{-t} + 3$

35. ***Population Growth*** The population P (in millions) of the United States from 1992 through 2002 can be modeled by the exponential function

$$P(t) = 251.27(1.0118)^t$$

where t is the time in years, with $t = 2$ corresponding to 1992. Use the model to estimate the population in the years (a) 2006 and (b) 2012. *(Source: U.S. Census Bureau)*

36. ***Sales*** The sales S (in millions of dollars) for Starbucks from 1994 through 2003 can be modeled by the exponential function

$$S(t) = 116.59(1.3295)^t$$

where t is the time in years, with $t = 4$ corresponding to 1994. Use the model to estimate the sales in the years (a) 2006 and (b) 2012. *(Source: Starbucks Corp.)*

37. ***Property Value*** Suppose that the value of a piece of property doubles every 15 years. If you buy the property for \$64,000, its value t years after the date of purchase should be

$$V(t) = 64{,}000(2)^{t/15}.$$

Use the model to approximate the value of the property (a) 5 years and (b) 20 years after it is purchased.

38. ***Inflation Rate*** Suppose that the annual rate of inflation averages 5% over the next 10 years. With this rate of inflation, the approximate cost C of goods or services during any year in that decade will be given by

$$C(t) = P(1.05)^t, \quad 0 \le t \le 10$$

where t is time in years and P is the present cost. If the price of a movie theater ticket is presently \$6.95, estimate the price 10 years from now.

39. ***Depreciation*** After t years, the value of a car that originally cost \$16,000 depreciates so that each year it is worth $\frac{3}{4}$ of its value for the previous year. Find a model for $V(t)$, the value of the car after t years. Sketch a graph of the model and determine the value of the car 4 years after it was purchased.

40. ***Radioactive Decay*** After t years, the initial mass of 16 grams of a radioactive element whose half-life is 30 years is given by

$$y = 16\left(\tfrac{1}{2}\right)^{t/30}, \quad t \ge 0.$$

(a) Use a graphing utility to graph the function.

(b) How much of the initial mass remains after 50 years?

(c) Use the *zoom* and *trace* features of a graphing utility to find the time required for the mass to decay to an amount of 1 gram.

41. ***Radioactive Decay*** After t years, the initial mass of 23 grams of a radioactive element whose half-life is 45 years is given by

$$y = 23\left(\tfrac{1}{2}\right)^{t/45}, \quad t \ge 0.$$

(a) Use a graphing utility to graph the function.

(b) How much of the initial mass remains after 75 years?

(c) Use the *zoom* and *trace* features of a graphing utility to find the time required for the mass to decay to an amount of 1 gram.

4.2 NATURAL EXPONENTIAL FUNCTIONS

- Evaluate and graph functions involving the natural exponential function.
- Solve compound interest problems.
- Solve present value problems.

Natural Exponential Functions

In Section 4.1, exponential functions were introduced using an unspecified base a. In calculus, the most convenient (or natural) choice for a base is the irrational number e, whose decimal approximation is

$$e \approx 2.71828182846.$$

Although this choice of base may seem unusual, its convenience will become apparent as the rules for differentiating exponential functions are developed in Section 4.3. In that development, you will encounter the limit used in the definition of e.

TECHNOLOGY

Try graphing $y = (1 + x)^{1/x}$ with a graphing utility. Then use the *zoom* and *trace* features to find values of y near $x = 0$. You will find that the y-values get closer and closer to the number $e \approx 2.71828$.

Limit Definition of e

The irrational number e is defined to be the limit of $(1 + x)^{1/x}$ as $x \to 0$. That is,

$$\lim_{x\to 0} (1 + x)^{1/x} = e.$$

EXAMPLE 1 Graphing the Natural Exponential Function

Sketch the graph of $f(x) = e^x$.

SOLUTION Begin by evaluating the function for several values of x, as shown in the table.

x	-2	-1	0	1	2
$f(x)$	$e^{-2} \approx 0.135$	$e^{-1} \approx 0.368$	$e^0 = 1$	$e^1 \approx 2.718$	$e^2 \approx 7.389$

The graph of $f(x) = e^x$ is shown in Figure 4.5. Note that e^x is positive for all values of x. Moreover, the graph has the x-axis as a horizontal asymptote to the left. That is,

$$\lim_{x\to -\infty} e^x = 0.$$

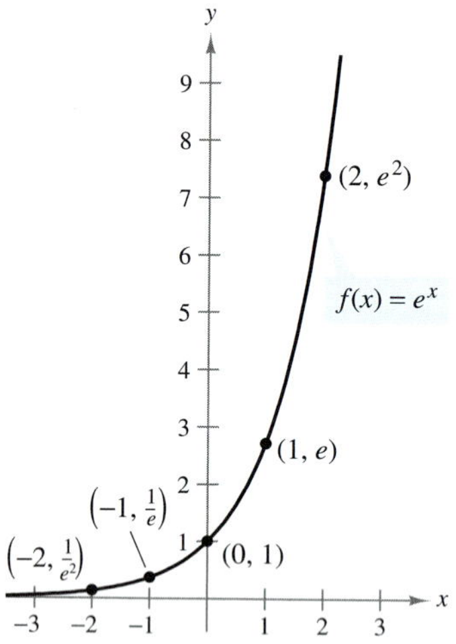

FIGURE 4.5

TRY IT 1

Complete the table of values for $f(x) = e^{-x}$. Sketch the graph of the function.

x	-2	-1	0	1	2
$f(x)$					

Exponential functions are often used to model the growth of a quantity or a population. When the quantity's growth *is not* restricted, an exponential model is often used. When the quantity's growth *is* restricted, the best model is often a **logistic growth function** of the form

$$f(t) = \frac{a}{1 + be^{-kt}}.$$

Graphs of both types of population growth models are shown in Figure 4.6.

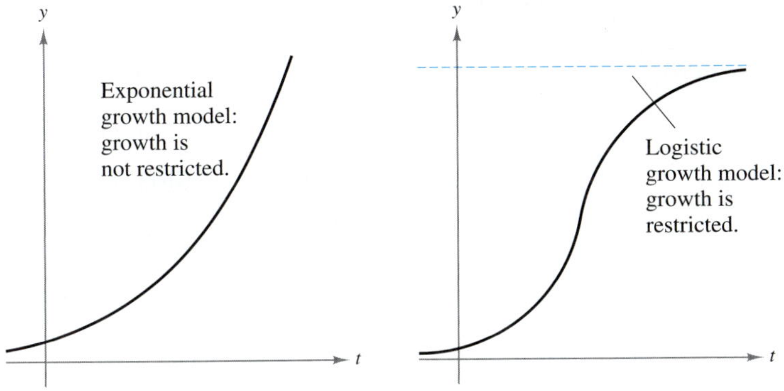

FIGURE 4.6

When a culture is grown in a dish, the size of the dish and the available food limit the culture's growth.

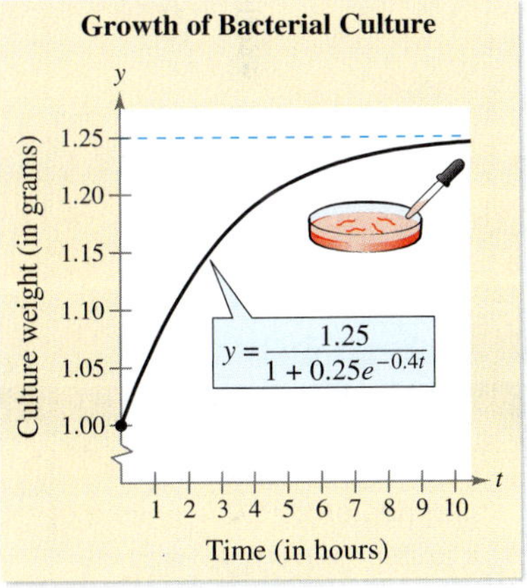

FIGURE 4.7

EXAMPLE 2 Modeling a Population

A bacterial culture is growing according to the *logistic growth model*

$$y = \frac{1.25}{1 + 0.25e^{-0.4t}}, \quad t \geq 0$$

where y is the culture weight (in grams) and t is the time (in hours). Find the weight of the culture after 0 hours, 1 hour, and 10 hours. What is the limit of the model as t increases without bound?

SOLUTION

$$y = \frac{1.25}{1 + 0.25e^{-0.4(0)}} = 1 \text{ gram} \qquad \text{Weight when } t = 0$$

$$y = \frac{1.25}{1 + 0.25e^{-0.4(1)}} \approx 1.071 \text{ grams} \qquad \text{Weight when } t = 1$$

$$y = \frac{1.25}{1 + 0.25e^{-0.4(10)}} \approx 1.244 \text{ grams} \qquad \text{Weight when } t = 10$$

As t approaches infinity, the limit of y is

$$\lim_{t\to\infty} \frac{1.25}{1 + 0.25e^{-0.4t}} = \lim_{t\to\infty} \frac{1.25}{1 + (0.25/e^{0.4t})}$$

$$= \frac{1.25}{1 + 0}$$

$$= 1.25.$$

So, as t increases without bound, the weight of the culture approaches 1.25 grams. The graph of the model is shown in Figure 4.7.

TRY IT 2

A bacterial culture is growing according to the model

$$y = \frac{1.50}{1 + 0.2e^{-0.5t}}, \quad t \geq 0$$

where y is the culture weight (in grams) and t is the time (in hours). Find the weight of the culture after 0 hours, 1 hour, and 10 hours. What is the limit of the model as t increases without bound?

Extended Application: Compound Interest

TECHNOLOGY

Use a spreadsheet software program or the *table* feature of a graphing utility to reproduce the table at the right. (Consult the user's manual for a spreadsheet software program for specific instructions on how to create a table.) Do you get the same results as those shown in the table?

If P dollars is deposited in an account at an annual interest rate of r (in decimal form), what is the balance after 1 year? The answer depends on the number of times the interest is compounded, according to the formula

$$A = P\left(1 + \frac{r}{n}\right)^n$$

where n is the number of compoundings per year. The balances for a deposit of \$1000 at 8%, at various compounding periods, are shown in the table.

Number of times compounded per year, n	Balance (in dollars), A
Annually, $n = 1$	$A = 1000\left(1 + \frac{0.08}{1}\right)^1 = \$1080.00$
Semiannually, $n = 2$	$A = 1000\left(1 + \frac{0.08}{2}\right)^2 = \$1081.60$
Quarterly, $n = 4$	$A = 1000\left(1 + \frac{0.08}{4}\right)^4 \approx \$1082.43$
Monthly, $n = 12$	$A = 1000\left(1 + \frac{0.08}{12}\right)^{12} \approx \$1083.00$
Daily, $n = 365$	$A = 1000\left(1 + \frac{0.08}{365}\right)^{365} \approx \$1083.28$

You may be surprised to discover that as n increases, the balance A approaches a limit, as indicated in the following development. In this development, let $x = r/n$. Then $x \to 0$ as $n \to \infty$, and you have

$$\begin{aligned} A &= \lim_{n\to\infty} P\left(1 + \frac{r}{n}\right)^n \\ &= P \lim_{n\to\infty}\left[\left(1 + \frac{r}{n}\right)^{n/r}\right]^r \\ &= P\left[\lim_{x\to 0}(1 + x)^{1/x}\right]^r \qquad \text{Substitute } x \text{ for } r/n. \\ &= Pe^r. \end{aligned}$$

This limit is the balance after 1 year of **continuous compounding.** So, for a deposit of \$1000 at 8%, compounded continuously, the balance at the end of the year would be

$$\begin{aligned} A &= 1000e^{0.08} \\ &\approx \$1083.29. \end{aligned}$$

DISCOVERY

Use a spreadsheet software program or the *table* feature of a graphing utility to evaluate the expression

$$\left(1 + \frac{1}{n}\right)^n$$

for each value of n.

n	$(1 + 1/n)^n$
10	
100	
1000	
10,000	
100,000	

What can you conclude? Try the same thing for negative values of n.

Summary of Compound Interest Formulas

Let P be the amount deposited, t the number of years, A the balance, and r the annual interest rate (in decimal form).

1. Compounded n times per year: $A = P\left(1 + \frac{r}{n}\right)^{nt}$
2. Compounded continuously: $A = Pe^{rt}$

The average interest rates paid by banks on savings accounts have varied greatly during the past 30 years. At times, savings accounts have earned as much as 12% annual interest and at times they have earned as little as 3%. The next example shows how the annual interest rate can affect the balance of an account.

EXAMPLE 3 Finding Account Balances

You are creating a trust fund for your newborn nephew. You deposit \$12,000 in an account, with instructions that the account be turned over to your nephew on his 25th birthday. Compare the balances in the account for each situation.

(a) 7%, compounded continuously

(b) 7%, compounded quarterly

(c) 11%, compounded continuously

(d) 11%, compounded quarterly

SOLUTION

(a) $12{,}000e^{0.07(25)} \approx 69{,}055.23$ — 7%, compounded continuously

(b) $12{,}000\left(1 + \frac{0.07}{4}\right)^{4(25)} \approx 68{,}017.87$ — 7%, compounded quarterly

(c) $12{,}000e^{0.11(25)} \approx 187{,}711.58$ — 11%, compounded continuously

(d) $12{,}000\left(1 + \frac{0.11}{4}\right)^{4(25)} \approx 180{,}869.07$ — 11%, compounded quarterly

The growth of the account for parts (a) and (c) is shown in Figure 4.8. Notice the dramatic difference between the balances at 7% and 11%.

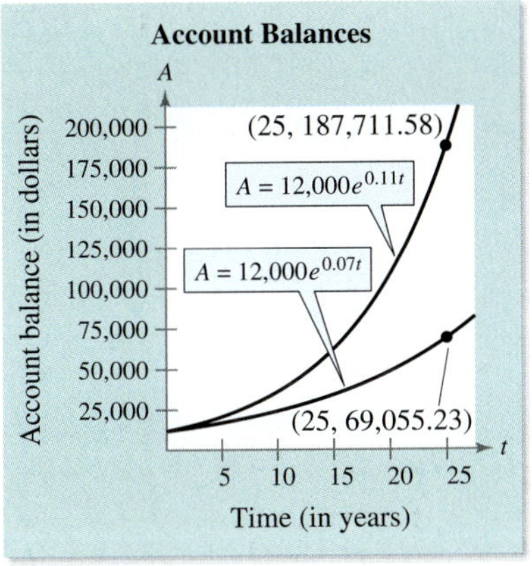

FIGURE 4.8

TRY IT 3

Find the balance in an account if \$2000 is deposited for 10 years at an interest of 9%, compounded as follows. Compare the results and make a general statement about compounding.

(a) quarterly (b) monthly

(c) daily (d) continuously

In Example 3, note that the interest earned depends on the frequency with which the interest is compounded. The annual percentage rate is called the **stated rate** or **nominal rate.** However, the nominal rate does not reflect the actual rate at which interest is earned, which means that the compounding produced an **effective rate** that is larger than the normal rate. In general, the effective rate corresponding to a nominal rate of r that is compounded n times per year is

$$\text{Effective rate} = r_{eff} = \left(1 + \frac{r}{n}\right)^n - 1.$$

EXAMPLE 4 Finding the Effective Rate of Interest

Find the effective rate of interest corresponding to a nominal rate of 6% per year compounded (a) annually, (b) quarterly, and (c) monthly.

SOLUTION

(a) $r_{eff} = \left(1 + \frac{r}{n}\right)^n - 1$ Formula for effective rate of interest

$= \left(1 + \frac{0.06}{1}\right)^1 - 1$ Substitute for r and n.

$= 1.06 - 1$ Simplify.

$= 0.06$

So, the effective rate is 6% per year.

(b) $r_{eff} = \left(1 + \frac{r}{n}\right)^n - 1$ Formula for effective rate of interest

$= \left(1 + \frac{0.06}{4}\right)^4 - 1$ Substitute for r and n.

$= (1.015)^4 - 1$ Simplify.

≈ 0.0614

So, the effective rate is about 6.14% per year.

(c) $r_{eff} = \left(1 + \frac{r}{n}\right)^n - 1$ Formula for effective rate of interest

$= \left(1 + \frac{0.06}{12}\right)^{12} - 1$ Substitute for r and n.

$\approx (1.005)^{12} - 1$ Simplify.

≈ 0.0618

So, the effective rate is about 6.18% per year.

TRY IT 4

Find the effective rate of interest corresponding to a nominal rate of 7% per year compounded (a) semiannually and (b) daily.

Present Value

In planning for the future, this problem often arises: "How much money P should be deposited now, at a fixed rate of interest r, in order to have a balance of A, t years from now?" The answer to this question is given by the **present value** of A.

To find the present value of a future investment, use the formula for compound interest as shown.

$A = P\left(1 + \frac{r}{n}\right)^{nt}$ Formula for compound interest

Solving for P gives a present value of

$$P = \frac{A}{\left(1 + \frac{r}{n}\right)^{nt}} \quad \text{or} \quad P = \frac{A}{(1 + i)^N}$$

where $i = r/n$ is the interest rate per compounding period and $N = nt$ is the total number of compounding periods. You will learn another way to find the present value of a future investment in Section 6.2.

EXAMPLE 5 Finding Present Value

An investor is purchasing a 12-year certificate of deposit that pays an annual percentage rate of 8%, compounded monthly. How much should the person invest in order to obtain a balance of \$15,000 at maturity?

SOLUTION Here, $A = 15{,}000$, $r = 0.08$, $n = 12$, and $t = 12$. Using the formula for present value, you obtain

$$P = \frac{15{,}000}{\left(1 + \frac{0.08}{12}\right)^{12(12)}} \qquad \text{Substitute for } A, r, n, \text{ and } t.$$

$$\approx 5761.72. \qquad \text{Simplify.}$$

So, the person should invest \$5761.72 in the certificate of deposit.

TRY IT 5

How much money should be deposited in an account paying 6% interest compounded monthly in order to have a balance of \$20,000 after 3 years?

TAKE ANOTHER LOOK

Compound Interest

You want to invest \$5000 in a certificate of deposit for 10 years. You are given the options below. Which would you choose?

a. You can buy a 10-year certificate of deposit that earns 7%, compounded continuously. You are guaranteed a 7% rate for the entire 10 years, but you cannot withdraw the money early without paying a substantial penalty.

b. You can buy a five-year certificate of deposit that earns 6%, compounded continuously. After 5 years, you can reinvest your money at whatever the current interest rate is at that time.

c. You can buy a two-year certificate of deposit that earns 5%, compounded continuously. After 2 years, you can reinvest your money at whatever the current interest rate is at that time.

PREREQUISITE REVIEW 4.2

The following warm-up exercises involve skills that were covered in earlier sections. You will use these skills in the exercise set for this section.

In Exercises 1–4, discuss the continuity of the function.

1. $f(x) = \dfrac{3x^2 + 2x + 1}{x^2 + 1}$

2. $f(x) = \dfrac{x + 1}{x^2 - 4}$

3. $f(x) = \dfrac{x^2 - 6x + 5}{x^2 - 3}$

4. $g(x) = \dfrac{x^2 - 9x + 20}{x - 4}$

In Exercises 5–12, find the limit.

5. $\displaystyle\lim_{x\to\infty} \frac{25}{1 + 4x}$

6. $\displaystyle\lim_{x\to\infty} \frac{16x}{3 + x^2}$

7. $\displaystyle\lim_{x\to\infty} \frac{8x^3 + 2}{2x^3 + x}$

8. $\displaystyle\lim_{x\to\infty} \frac{x}{2x}$

9. $\displaystyle\lim_{x\to\infty} \frac{3}{2 + (1/x)}$

10. $\displaystyle\lim_{x\to\infty} \frac{6}{1 + x^{-2}}$

11. $\displaystyle\lim_{x\to\infty} 2^{-x}$

12. $\displaystyle\lim_{x\to\infty} \frac{7}{1 + 5x}$

EXERCISES 4.2

In Exercises 1–4, use the properties of exponents to simplify the expression.

1. (a) $(e^3)(e^4)$ (b) $(e^3)^4$
(c) $(e^3)^{-2}$ (d) e^0

2. (a) $\left(\dfrac{1}{e}\right)^{-2}$ (b) $\left(\dfrac{e^5}{e^2}\right)^{-1}$
(c) $\dfrac{e^5}{e^3}$ (d) $\dfrac{1}{e^{-3}}$

3. (a) $(e^2)^{5/2}$ (b) $(e^2)(e^{1/2})$
(c) $(e^{-2})^{-3}$ (d) $\dfrac{e^5}{e^{-2}}$

4. (a) $(e^{-3})^{2/3}$ (b) $\dfrac{e^4}{e^{-1/2}}$
(c) $(e^{-2})^{-4}$ (d) $(e^{-4})(e^{-3/2})$

In Exercises 5–12, solve the equation for *x*.

5. $e^{-3x} = e$

6. $e^x = 1$

7. $e^{\sqrt{x}} = e^3$

8. $e^{-1/x} = \sqrt{e}$

9. $x^{2/3} = \sqrt[3]{e^2}$

10. $\dfrac{x^2}{2} = e^2$

11. $3x^3 = 9e^3$

12. $x^{-2} = \dfrac{2}{e^2}$

In Exercises 13–18, match the function with its graph. [The graphs are labeled (a)–(f).]

13. $f(x) = e^{2x+1}$

14. $f(x) = e^{-x/2}$

15. $f(x) = e^{x^2}$

16. $f(x) = e^{-1/x}$

17. $f(x) = e^{\sqrt{x}}$

18. $f(x) = -e^x + 1$

(a)

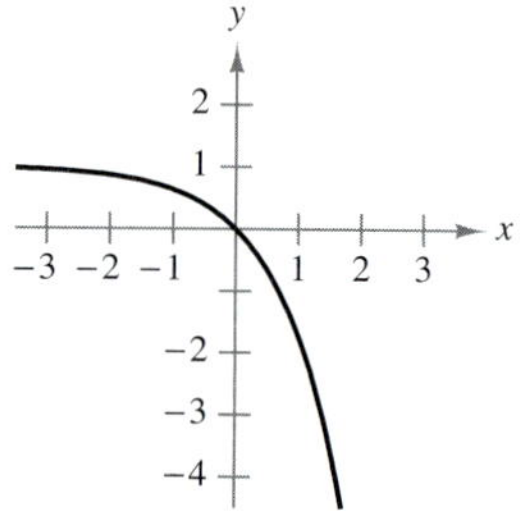

(b)

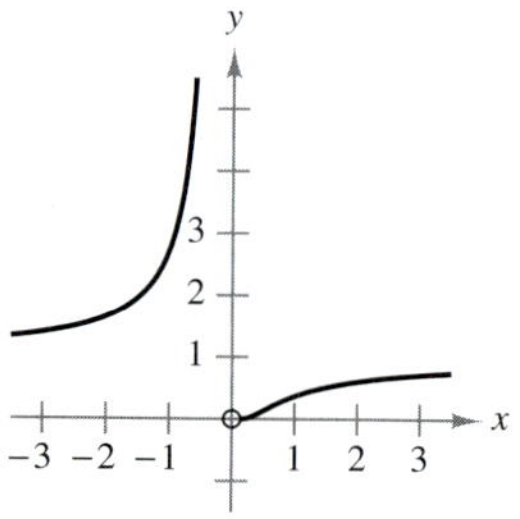

(c)

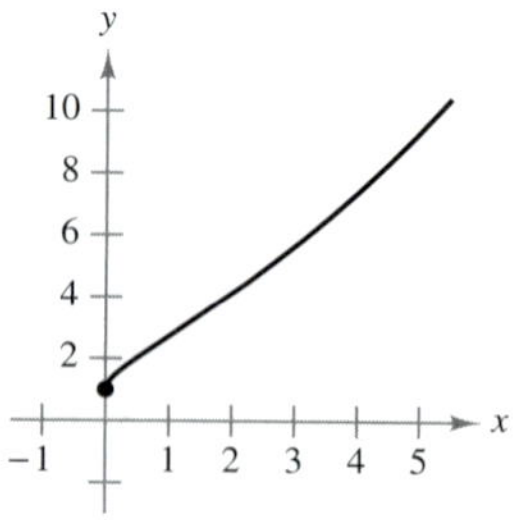

(d)

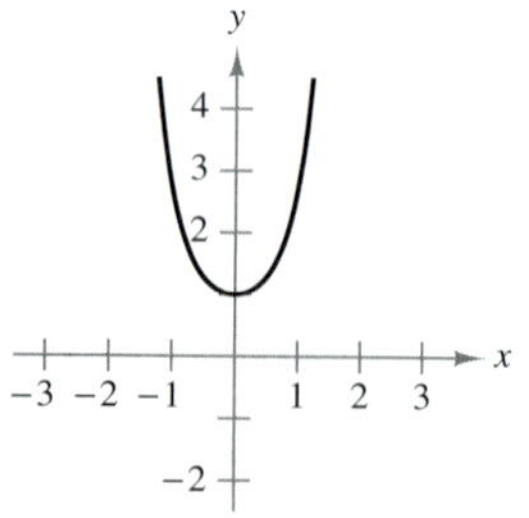

(e)

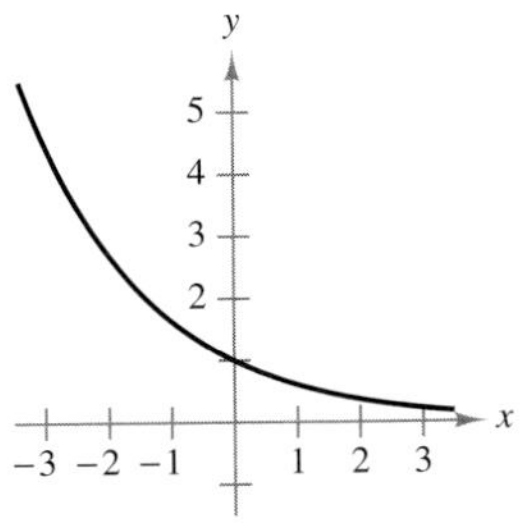

(f)

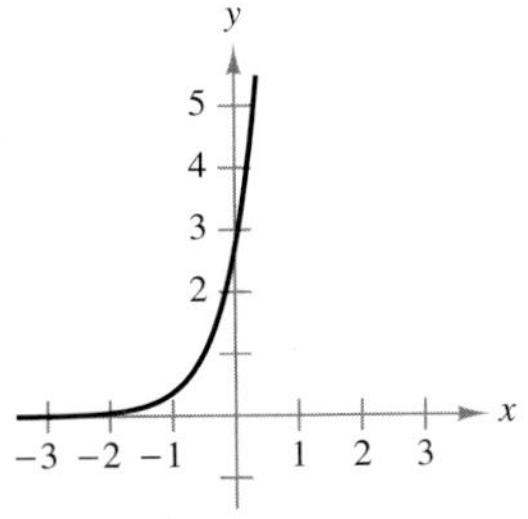

In Exercises 19–22, sketch the graph of the function.

19. $h(x) = e^{x-2}$

20. $f(x) = e^{2x}$

21. $g(x) = e^{1-x}$

22. $j(x) = e^{-x+2}$

In Exercises 23–26, use a graphing utility to graph the function. Be sure to choose an appropriate viewing window.

23. $N(t) = 500e^{-0.2t}$

24. $A(t) = 500e^{0.15t}$

25. $g(x) = \dfrac{2}{1 + e^{x^2}}$

26. $g(x) = \dfrac{10}{1 + e^{-x}}$

In Exercises 27–30, use a graphing utility to graph the function. Determine whether the function has any horizontal asymptotes and discuss the continuity of the function.

27. $f(x) = \dfrac{e^x + e^{-x}}{2}$

28. $f(x) = \dfrac{e^x - e^{-x}}{2}$

29. $f(x) = \dfrac{2}{1 + e^{1/x}}$

30. $f(x) = \dfrac{2}{1 + 2e^{-0.2x}}$

Compound Interest In Exercises 31–34, complete the table to determine the balance A for P dollars invested at rate r for t years, compounded n times per year.

n	1	2	4	12	365	Continuous compounding
A						

31. $P = \$1000,\ r = 3\%,\ t = 10$ years

32. $P = \$2500,\ r = 5\%,\ t = 20$ years

33. $P = \$1000,\ r = 3\%,\ t = 40$ years

34. $P = \$2500,\ r = 5\%,\ t = 40$ years

Compound Interest In Exercises 35–38, complete the table to determine the amount of money P that should be invested at rate r to produce a final balance of \$100,000 in t years.

t	1	10	20	30	40	50
P						

35. $r = 4\%$, compounded continuously

36. $r = 3\%$, compounded continuously

37. $r = 5\%$, compounded monthly

38. $r = 6\%$, compounded daily

39. ***Effective Rate*** Find the effective rate of interest corresponding to a nominal rate of 9% per year compounded (a) annually, (b) semiannually, (c) quarterly, and (d) monthly.

40. ***Effective Rate*** Find the effective rate of interest corresponding to a nominal rate of 7.5% per year compounded (a) annually, (b) semiannually, (c) quarterly, and (d) monthly.

41. ***Present Value*** How much should be deposited in an account paying 7.2% interest compounded monthly in order to have a balance of \$15,503.77 three years from now?

42. ***Present Value*** How much should be deposited in an account paying 7.8% interest compounded monthly in order to have a balance of \$21,154.03 four years from now?

43. ***Future Value*** Find the future value of an \$8000 investment if the interest rate is 4.5% compounded monthly for 2 years.

44. ***Future Value*** Find the future value of a \$6000 investment if the interest rate is 6.25% compounded monthly for 3 years.

45. ***Demand*** The demand function for a product is modeled by

$$p = 5000\left(1 - \frac{4}{4 + e^{-0.002x}}\right).$$

Find the price of the product if the quantity demanded is (a) $x = 100$ units and (b) $x = 500$ units. What is the limit of the price as x increases without bound?

46. ***Demand*** The demand function for a product is modeled by

$$p = 10{,}000\left(1 - \frac{3}{3 + te^{-0.001t}}\right).$$

Find the price of the product if the quantity demanded is (a) $x = 1000$ units and (b) $x = 1500$ units. What is the limit of the price as x increases without bound?

47. ***Probability*** The average time between incoming calls at a switchboard is 3 minutes. If a call has just come in, the probability that the next call will come within the next t minutes is

$$P(t) = 1 - e^{-t/3}.$$

Find the probability of each situation.

(a) A call comes in within $\frac{1}{2}$ minute.

(b) A call comes in within 2 minutes.

(c) A call comes in within 5 minutes.

48. ***Consumer Awareness*** An automobile gets 28 miles per gallon at speeds of up to and including 50 miles per hour. At speeds greater than 50 miles per hour, the number of miles per gallon drops at the rate of 12% for each 10 miles per hour. If s is the speed (in miles per hour) and y is the number of miles per gallon, then

$$y = 28e^{0.6-0.012s}, \quad s > 50.$$

Use this information to complete the table. What can you conclude?

Speed (s)	50	55	60	65	70
Miles per gallon (y)					

49. ***Sales*** The sales S (in millions of dollars) for Avon Products from 1994 through 2003 can be modeled by

$$S = 3557.12e^{0.0475t}$$

where t is time in years, with $t = 4$ corresponding to 1994. *(Source: Avon Products Inc.)*

(a) Find the sales in 1995, 2000, and 2003.

(b) Using the data points from part (a), would a linear model fit the data? Explain your reasoning.

(c) Use the exponential growth model to estimate when the sales will exceed 10 billion dollars.

50. ***Population*** The population P (in thousands) of Las Vegas, Nevada from 1970 to 2000 can be modeled by

$$P = 115.49e^{0.0445t}$$

where t is the time in years, with $t = 0$ corresponding to 1970. *(Source: U.S. Census Bureau)*

(a) Find the populations in 1970, 1980, 1990, and 2000.

(b) Explain why the data do not fit a linear model.

(c) Use the model to estimate when the population will exceed 750,000.

51. ***Biology*** The population y of a bacterial culture is modeled by the logistic growth function

$$y = \frac{925}{1 + e^{-0.3t}}$$

where t is the time in days.

(a) Use a graphing utility to graph the model.

(b) Does the population have a limit as t increases without bound? Explain your answer.

(c) How would the limit change if the model were

$$y = \frac{1000}{1 + e^{-0.3t}}?$$

Explain your answer. Draw some conclusions about this type of model.

52. ***Biology: Cell Division*** Suppose that you have a single imaginary bacterium able to divide to form two new cells every 30 seconds. Make a table of values for the number of individuals in the population over 30-second intervals up to 5 minutes. Graph the points and use a graphing utility to fit an exponential model to the data. *(Source: Adapted from Levine/Miller,* Biology: Discovering Life, *Second Edition)*

53. ***Learning Theory*** In a learning theory project, the proportion P of correct responses after n trials can be modeled by

$$P = \frac{0.83}{1 + e^{-0.2n}}.$$

(a) Use a graphing utility to estimate the proportion of correct responses after 10 trials. Verify your result analytically.

(b) Use a graphing utility to estimate the number of trials required to have a proportion of correct responses of 0.75.

(c) Does the proportion of correct responses have a limit as n increases without bound? Explain your answer.

54. ***Learning Theory*** In a typing class, the average number N of words per minute typed after t weeks of lessons can be modeled by

$$N = \frac{95}{1 + 8.5e^{-0.12t}}.$$

(a) Use a graphing utility to estimate the average number of words per minute typed after 10 weeks. Verify your result analytically.

(b) Use a graphing utility to estimate the number of weeks required to achieve an average of 70 words per minute.

(c) Does the number of words per minute have a limit as t increases without bound? Explain your answer.

4.3 DERIVATIVES OF EXPONENTIAL FUNCTIONS

- Find the derivatives of natural exponential functions.
- Use calculus to analyze the graphs of functions that involve the natural exponential function.
- Explore the normal probability density function.

Derivatives of Exponential Functions

In Section 4.2, it was stated that the most convenient base for exponential functions is the irrational number e. The convenience of this base stems primarily from the fact that the function $f(x) = e^x$ *is its own derivative*. You will see that this is not true of other exponential functions of the form $y = a^x$ where $a \neq e$. To verify that $f(x) = e^x$ is its own derivative, notice that the limit

$$\lim_{\Delta x \to 0} (1 + \Delta x)^{1/\Delta x} = e$$

implies that for small values of Δx, $e \approx (1 + \Delta x)^{1/\Delta x}$, or $e^{\Delta x} \approx 1 + \Delta x$. This approximation is used in the following derivation.

$$\begin{aligned}
f'(x) &= \lim_{\Delta x \to 0} \frac{f(x + \Delta x) - f(x)}{\Delta x} && \text{Definition of derivative} \\
&= \lim_{\Delta x \to 0} \frac{e^{x+\Delta x} - e^x}{\Delta x} && \text{Use } f(x) = e^x. \\
&= \lim_{\Delta x \to 0} \frac{e^x(e^{\Delta x} - 1)}{\Delta x} && \text{Factor numerator.} \\
&= \lim_{\Delta x \to 0} \frac{e^x[(1 + \Delta x) - 1]}{\Delta x} && \text{Substitute } 1 + \Delta x \text{ for } e^{\Delta x}. \\
&= \lim_{\Delta x \to 0} \frac{e^x(\cancel{\Delta x})}{\cancel{\Delta x}} && \text{Divide out like factor.} \\
&= \lim_{\Delta x \to 0} e^x && \text{Simplify.} \\
&= e^x && \text{Evaluate limit.}
\end{aligned}$$

If u is a function of x, you can apply the Chain Rule to obtain the derivative of e^u with respect to x. Both formulas are summarized below.

Derivative of the Natural Exponential Function

Let u be a differentiable function of x.

1. $\dfrac{d}{dx}[e^x] = e^x$ 2. $\dfrac{d}{dx}[e^u] = e^u \dfrac{du}{dx}$

DISCOVERY

Use a spreadsheet software program to compare the expressions $e^{\Delta x}$ and $1 + \Delta x$ for values of Δx near 0.

Δx	$e^{\Delta x}$	$1 + \Delta x$
0.1		
0.01		
0.001		

What can you conclude? Explain how this result is used in the development of the derivative of $f(x) = e^x$.

TECHNOLOGY

Let $f(x) = e^x$. Use a graphing utility to evaluate $f(x)$ and the numerical derivative of $f(x)$ at each x-value. Explain the results.

(a) $x = -2$ (b) $x = 0$ (c) $x = 2$

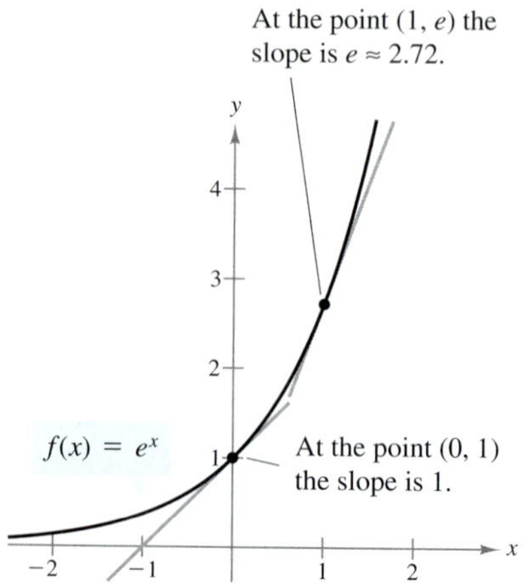

FIGURE 4.9

EXAMPLE 1 Interpreting a Derivative Graphically

Find the slopes of the tangent lines to

$$f(x) = e^x \qquad \text{Original function}$$

at the points (0, 1) and (1, e). What conclusion can you make?

SOLUTION Because the derivative of f is

$$f'(x) = e^x \qquad \text{Derivative}$$

it follows that the slope of the tangent line to the graph of f is

$$f'(0) = e^0 = 1 \qquad \text{Slope at point (0, 1)}$$

at the point (0, 1) and

$$f'(1) = e^1 = e \qquad \text{Slope at point } (1, e)$$

at the point (1, e), as shown in Figure 4.9. From this pattern, you can see that the slope of the tangent line to the graph of $f(x) = e^x$ at any point (x, e^x) is equal to the y-coordinate of the point.

TRY IT 1

Find the equations of the tangent lines to $f(x) = e^x$ at the points (0, 1) and (1, e).

STUDY TIP

In Example 2, notice that when you differentiate an exponential function, the exponent does not change. For instance, the derivative of $y = e^{3x}$ is $y' = 3e^{3x}$. In both the function and its derivative, the exponent is $3x$.

EXAMPLE 2 Differentiating Exponential Functions

Differentiate each function.

(a) $f(x) = e^{2x}$ (b) $f(x) = e^{-3x^2}$

(c) $f(x) = 6e^{x^3}$ (d) $f(x) = e^{-x}$

SOLUTION

(a) Let $u = 2x$. Then $du/dx = 2$, and you can apply the Chain Rule.

$$f'(x) = e^u \frac{du}{dx} = e^{2x}(2) = 2e^{2x}$$

(b) Let $u = -3x^2$. Then $du/dx = -6x$, and you can apply the Chain Rule.

$$f'(x) = e^u \frac{du}{dx} = e^{-3x^2}(-6x) = -6xe^{-3x^2}$$

(c) Let $u = x^3$. Then $du/dx = 3x^2$, and you can apply the Chain Rule.

$$f'(x) = 6e^u \frac{du}{dx} = 6e^{x^3}(3x^2) = 18x^2e^{x^3}$$

(d) Let $u = -x$. Then $du/dx = -1$, and you can apply the Chain Rule.

$$f'(x) = e^u \frac{du}{dx} = e^{-x}(-1) = -e^{-x}$$

TRY IT 2

Differentiate each function.

(a) $f(x) = e^{3x}$

(b) $f(x) = e^{-2x^3}$

(c) $f(x) = 4e^{x^2}$

(d) $f(x) = e^{-2x}$

The differentiation rules that you studied in Chapter 2 can be used with exponential functions, as shown in Example 3.

EXAMPLE 3 Differentiating Exponential Functions

Differentiate each function.

(a) $f(x) = xe^x$ (b) $f(x) = \dfrac{e^x - e^{-x}}{2}$ (c) $f(x) = \dfrac{e^x}{x}$ (d) $f(x) = xe^x - e^x$

SOLUTION

(a) $f(x) = xe^x$ — Write original function.

$f'(x) = xe^x + e^x(1)$ — Product Rule

$= xe^x + e^x$ — Simplify.

(b) $f(x) = \dfrac{e^x - e^{-x}}{2}$ — Write original function.

$= \frac{1}{2}(e^x - e^{-x})$ — Rewrite.

$f'(x) = \frac{1}{2}(e^x + e^{-x})$ — Constant Multiple Rule

(c) $f(x) = \dfrac{e^x}{x}$ — Write original function.

$f'(x) = \dfrac{xe^x - e^x(1)}{x^2}$ — Quotient Rule

$= \dfrac{e^x(x - 1)}{x^2}$ — Simplify.

(d) $f(x) = xe^x - e^x$ — Write original function.

$f'(x) = [xe^x + e^x(1)] - e^x$ — Product and Difference Rules

$= xe^x + e^x - e^x$

$= xe^x$ — Simplify.

TRY IT 3

Differentiate each function.

(a) $f(x) = x^2e^x$ (b) $f(x) = \dfrac{e^x + e^{-x}}{2}$

(c) $f(x) = \dfrac{e^x}{x^2}$ (d) $f(x) = x^2e^x - e^x$

TECHNOLOGY

If you have access to a symbolic differentiation utility, try using it to find the derivatives of the functions in Example 3.

Applications

In Chapter 3, you learned how to use derivatives to analyze the graphs of functions. The next example applies those techniques to a function composed of exponential functions. In the example, notice that $e^a = e^b$ implies that $a = b$.

David Buffington/PhotoDisc

Utility wires strung between poles have the shape of a catenary.

EXAMPLE 4 Analyzing a Catenary

When a telephone wire is hung between two poles, the wire forms a U-shaped curve called a **catenary.** For instance, the function

$$y = 30(e^{x/60} + e^{-x/60}), \qquad -30 \le x \le 30$$

models the shape of a telephone wire strung between two poles that are 60 feet apart (x and y are measured in feet). Show that the lowest point on the wire is midway between the two poles. How much does the wire sag between the two poles?

SOLUTION The derivative of the function is

$$\begin{aligned} y' &= 30\left[e^{x/60}\left(\tfrac{1}{60}\right) + e^{-x/60}\left(-\tfrac{1}{60}\right)\right] \\ &= \tfrac{1}{2}(e^{x/60} - e^{-x/60}). \end{aligned}$$

To find the critical numbers, set the derivative equal to zero.

$$\tfrac{1}{2}(e^{x/60} - e^{-x/60}) = 0 \qquad \text{Set derivative equal to 0.}$$
$$e^{x/60} - e^{-x/60} = 0 \qquad \text{Multiply each side by 2.}$$
$$e^{x/60} = e^{-x/60} \qquad \text{Add } e^{-x/60} \text{ to each side.}$$
$$\frac{x}{60} = -\frac{x}{60} \qquad \text{If } e^a = e^b, \text{ then } a = b.$$
$$x = -x \qquad \text{Multiply each side by 60.}$$
$$2x = 0 \qquad \text{Add } x \text{ to each side.}$$
$$x = 0 \qquad \text{Divide each side by 2.}$$

Using the First-Derivative Test, you can determine that the critical number $x = 0$ yields a relative minimum of the function. From the graph in Figure 4.10, you can see that this relative minimum is actually a minimum on the interval $[-30, 30]$. To find how much the wire sags between the two poles, you can compare its height at each pole with its height at the midpoint.

$$y = 30(e^{-30/60} + e^{-(-30)/60}) \approx 67.7 \text{ feet} \qquad \text{Height at left pole}$$
$$y = 30(e^{0/60} + e^{-(0)/60}) = 60 \text{ feet} \qquad \text{Height at midpoint}$$
$$y = 30(e^{30/60} + e^{-(30)/60}) \approx 67.7 \text{ feet} \qquad \text{Height at right pole}$$

From this, you can see that the wire sags about 7.7 feet.

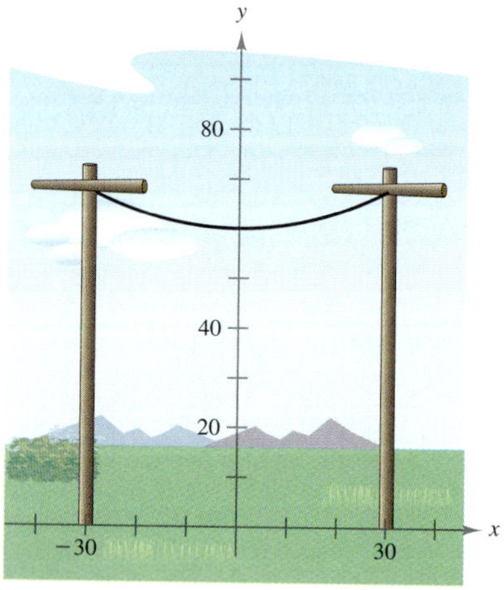

FIGURE 4.10

TRY IT 4

Use a graphing utility to graph the function in Example 4. Verify the minimum value. Use the information in the example to choose an appropriate viewing window.

EXAMPLE 5 Finding a Maximum Revenue

The demand function for a product is modeled by

$$p = 56e^{-0.000012x} \qquad \text{Demand function}$$

where p is the price per unit (in dollars) and x is the number of units. What price will yield a maximum revenue?

SOLUTION The revenue function is

$$R = xp = 56xe^{-0.000012x}. \qquad \text{Revenue function}$$

To find the maximum revenue *analytically*, you would set the marginal revenue, dR/dx, equal to zero and solve for x. In this problem, it is easier to use a *graphical* approach. After experimenting to find a reasonable viewing window, you can obtain a graph of R that is similar to that shown in Figure 4.11. Using the *zoom* and *trace* features, you can conclude that the maximum revenue occurs when x is about 83,300 units. To find the price that corresponds to this production level, substitute $x \approx 83{,}300$ into the demand function.

$$p \approx 56e^{-0.000012(83{,}300)} \approx \$20.61.$$

So, a price of about \$20.61 will yield a maximum revenue.

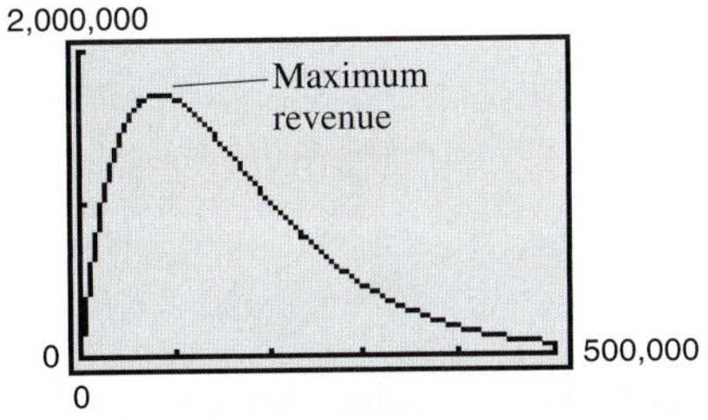

FIGURE 4.11 Use the *zoom* and *trace* features to approximate the x-value that corresponds to the maximum revenue.

TRY IT 5

The demand function for a product is modeled by

$$p = 50e^{-0.0000125x}$$

where p is the price per unit in dollars and x is the number of units. What price will yield a maximum revenue?

STUDY TIP

Try solving the problem in Example 5 analytically. When you do this, you obtain

$$\frac{dR}{dx} = 56xe^{-0.000012x}(-0.000012) + e^{-0.000012x}(56) = 0.$$

Explain how you would solve this equation. What is the solution?

The Normal Probability Density Function

If you take a course in statistics or quantitative business analysis, you will spend quite a bit of time studying the characteristics and use of the **normal probability density function** given by

$$f(x) = \frac{1}{\sigma\sqrt{2\pi}}e^{-(x-\mu)^2/2\sigma^2}$$

where σ is the lowercase Greek letter sigma, and μ is the lowercase Greek letter mu. In this formula, σ represents the *standard deviation* of the probability distribution, and μ represents the *mean* of the probability distribution.

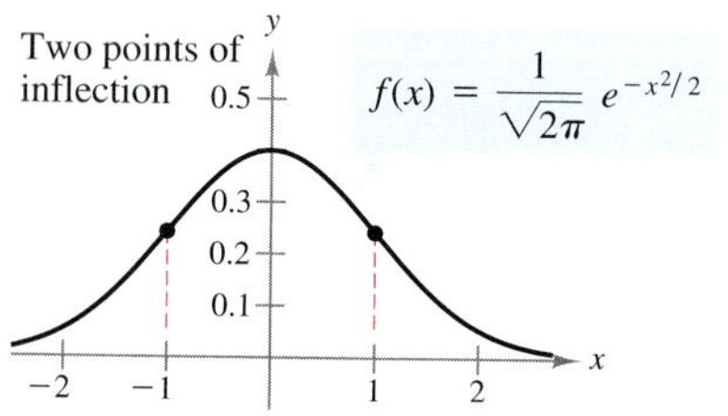

FIGURE 4.12 The graph of the normal probability density function is bell-shaped.

EXAMPLE 6 Exploring a Probability Density Function

Show that the graph of the normal probability density function

$$f(x) = \frac{1}{\sqrt{2\pi}}e^{-x^2/2} \qquad \text{Original function}$$

has points of inflection at $x = \pm 1$.

SOLUTION Begin by finding the second derivative of the function.

$$f'(x) = \frac{1}{\sqrt{2\pi}}(-x)e^{-x^2/2} \qquad \text{First derivative}$$

$$f''(x) = \frac{1}{\sqrt{2\pi}}[(-x)(-x)e^{-x^2/2} + (-1)e^{-x^2/2}] \qquad \text{Second derivative}$$

$$= \frac{1}{\sqrt{2\pi}}(e^{-x^2/2})(x^2 - 1) \qquad \text{Simplify.}$$

By setting the second derivative equal to 0, you can determine that $x = \pm 1$. By testing the concavity of the graph, you can then conclude that these x-values yield points of inflection, as shown in Figure 4.12.

TRY IT 6

Graph the normal probability density function

$$f(x) = \frac{1}{4\sqrt{2\pi}}e^{-x^2/32}$$

and approximate the points of inflection.

TAKE ANOTHER LOOK

The Normal Probability Density Function

Use Example 6 as a model to show that the graph of the normal probability density function with $\mu = 0$

$$f(x) = \frac{1}{\sigma\sqrt{2\pi}}e^{-x^2/2\sigma^2}$$

has points of inflection at $x = \pm\sigma$. What is the maximum of the function? Use a graphing utility to verify your answer by graphing the function for several values of σ.

PREREQUISITE REVIEW 4.3

The following warm-up exercises involve skills that were covered in earlier sections. You will use these skills in the exercise set for this section.

In Exercises 1–4, factor the expression.

1. $x^2e^x - \frac{1}{2}e^x$

2. $(xe^{-x})^{-1} + e^x$

3. $xe^x - e^{2x}$

4. $e^x - xe^{-x}$

In Exercises 5–8, find the derivative of the function.

5. $f(x) = \dfrac{3}{7x^2}$

6. $g(x) = 3x^2 - \dfrac{x}{6}$

7. $f(x) = (4x - 3)(x^2 + 9)$

8. $f(t) = \dfrac{t - 2}{\sqrt{t}}$

In Exercises 9 and 10, find the relative extrema of the function.

9. $f(x) = \frac{1}{8}x^3 - 2x$

10. $f(x) = x^4 - 2x^2 + 5$

EXERCISES 4.3

In Exercises 1–4, find the slope of the tangent line to the exponential function at the point (0, 1).

1. $y = e^{3x}$

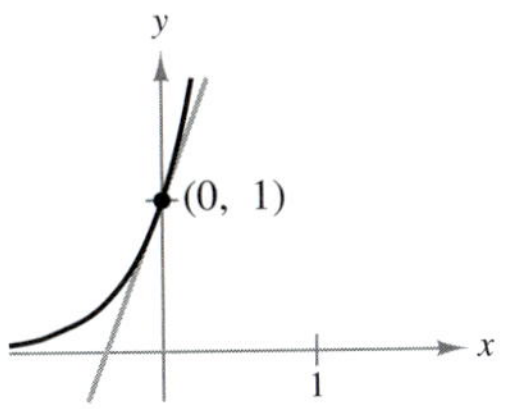

2. $y = e^{2x}$

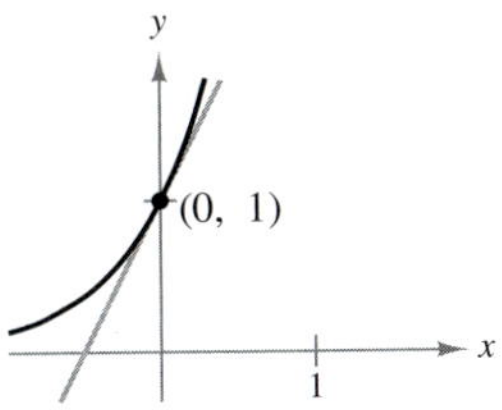

3. $y = e^{-x}$

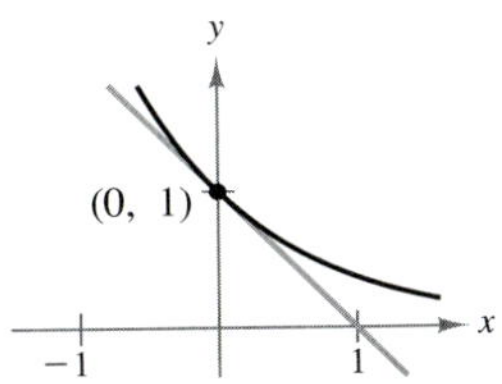

4. $y = e^{-2x}$

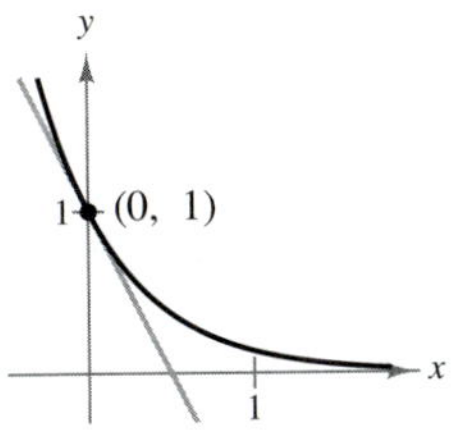

In Exercises 5–16, find the derivative of the function.

5. $y = e^{4x}$

6. $y = e^{1-x}$

7. $y = e^{-x^2}$

8. $f(x) = e^{1/x}$

9. $f(x) = e^{-1/x^2}$

10. $g(x) = e^{\sqrt{x}}$

11. $f(x) = (x^2 + 1)e^{4x}$

12. $y = 4x^3e^{-x}$

13. $f(x) = \dfrac{2}{(e^x + e^{-x})^3}$

14. $f(x) = \dfrac{(e^x + e^{-x})^4}{2}$

15. $y = xe^x - 4e^{-x}$

16. $y = x^2e^x - 2xe^x + 2e^x$

In Exercises 17–22, determine an equation of the tangent line to the function at the given point.

Function	*Point*
17. $y = e^{-2x+x^2}$	$(0, 1)$
18. $g(x) = e^{x^3}$	$\left(-1, \dfrac{1}{e}\right)$
19. $y = x^2e^{-x}$	$\left(2, \dfrac{4}{e^2}\right)$
20. $y = \dfrac{x}{e^{2x}}$	$\left(1, \dfrac{1}{e^2}\right)$
21. $y = (e^{2x} + 1)^3$	$(0, 8)$
22. $y = (e^{4x} - 2)^2$	$(0, 1)$

In Exercises 23–26, find *dy/dx* implicitly.

23. $xe^x + 2ye^x = 0$

24. $x^2y - xe^x + 2 = 0$

25. $x^2e^{-x} + 2y^2 - xy = 0$

26. $e^{xy} + x^2 - y^2 = 10$

In Exercises 27–30, find the second derivative.

27. $f(x) = 2e^{3x} + 3e^{-2x}$

28. $f(x) = (1 + 2x)e^{4x}$

29. $f(x) = 5e^{-x} - 2e^{-5x}$

30. $f(x) = (3 + 2x)e^{-3x}$

In Exercises 31–34, graph and analyze the function. Include extrema, points of inflection, and asymptotes in your analysis.

31. $f(x) = \dfrac{1}{2 - e^{-x}}$

32. $f(x) = \dfrac{e^x - e^{-x}}{2}$

33. $f(x) = x^2e^{-x}$

34. $f(x) = xe^{-x}$

In Exercises 35 and 36, use a graphing utility to graph the function. Determine any asymptotes of the graph.

35. $f(x) = \dfrac{8}{1 + e^{-0.5x}}$ **36.** $g(x) = \dfrac{8}{1 + e^{-0.5/x}}$

Depreciation In Exercises 37 and 38, the value V (in dollars) of an item is a function of the time t (in years).

(a) Sketch the function over the interval $[0, 10]$. Use a graphing utility to verify your graph.

(b) Find the rate of change of V when $t = 1$.

(c) Find the rate of change of V when $t = 5$.

(d) Use the values $(0, V(0))$ and $(10, V(10))$ to find the linear depreciation model for the item.

(e) Compare the exponential function and the model from part (d). What are the advantages of each?

37. $V = 15{,}000e^{-0.6286t}$ **38.** $V = 500{,}000e^{-0.2231t}$

39. ***Forestry*** To estimate the defoliation p (in percent of foliage) caused by gypsy moths during a year, a forester counts the number x (in thousands) of egg masses on $\frac{1}{40}$ of an acre the preceding fall. The defoliation is modeled by

$$p = \frac{300}{3 + 17e^{-1.57x}}.$$

(Source: National Forest Service)

(a) Use a graphing utility to graph the model.

(b) Estimate the percent of defoliation if 2000 egg masses are counted.

(c) Estimate the number of egg masses for which the amount of defoliation is increasing most rapidly.

40. ***Learning Theory*** The average typing speed N (in words per minute) after t weeks of lessons is modeled by

$$N = \frac{95}{1 + 8.5e^{-0.12t}}.$$

Find the rates at which the typing speed is changing when (a) $t = 5$ weeks, (b) $t = 10$ weeks, and (c) $t = 30$ weeks.

41. ***Compound Interest*** The balance A (in dollars) in a savings account is given by $A = 5000e^{0.08t}$, where t is measured in years. Find the rates at which the balance is changing when (a) $t = 1$ year, (b) $t = 10$ years, and (c) $t = 50$ years.

42. ***Ebbinghaus Model*** The *Ebbinghaus Model* for human memory is $p = (100 - a)e^{-bt} + a$, where p is the percent retained after t weeks. (The constants a and b vary from one person to another.) If $a = 20$ and $b = 0.5$, at what rate is information being retained after 1 week? After 3 weeks?

43. ***Agriculture*** The yield V (in pounds per acre) for an orchard at age t (in years) is modeled by

$$V = 7955.6e^{-0.0458/t}.$$

At what rate is the yield changing when $t = 5$ years? When $t = 10$ years? When $t = 25$ years?

44. ***Employment*** From 1995 to 2002, the numbers y (in millions) of employed people in the United States can be modeled by

$$y = 115.46 + 1.592t + 0.0552t^2 - 0.00004e^t$$

where $t = 5$ corresponds to 1995. *(Source: U.S. Bureau of Labor Statistics)*

(a) Use a graphing utility to graph the model.

(b) Use the graph to estimate the rates of change in the number of employed people in 1995, 1998, and 2002.

(c) Confirm the results of part (b) analytically.

45. ***Probability*** A survey of high school seniors from a certain school district who took the SAT has determined that the mean score on the mathematics portion was 650 with a standard deviation of 12.5.

(a) Assuming the data can be modeled by a normal probability density function, find a model for these data.

(b) Use a graphing utility to graph the model. Be sure to choose an appropriate viewing window.

(c) Find the derivative of the model.

(d) Show that $f' > 0$ for $x < \mu$ and $f' < 0$ for $x > \mu$.

46. ***Probability*** A survey of a college freshman class has determined that the mean height of females in the class is 64 inches with a standard deviation of 3.2 inches.

(a) Assuming the data can be modeled by a normal probability density function, find a model for these data.

(b) Use a graphing utility to graph the model. Be sure to choose an appropriate viewing window.

(c) Find the derivative of the model.

(d) Show that $f' > 0$ for $x < \mu$ and $f' < 0$ for $x > \mu$.

47. Use a graphing utility to graph the normal probability density function with $\mu = 0$ and $\sigma = 2, 3$, and 4 in the same viewing window. What effect does the standard deviation σ have on the function? Explain your reasoning.

48. Use a graphing utility to graph the normal probability density function with $\sigma = 1$ and $\mu = -2, 1$, and 3 in the same viewing window. What effect does the mean μ have on the function? Explain your reasoning.

49. ***Athletics*** A parachutist jumps from a plane and opens the parachute at a height of 2000 feet. The height of the parachutist is $h = 1950 + 50e^{-1.6t} - 20t$, where h is the height (in feet) and t is the time (in seconds) since the parachute was opened.

(a) Find dh/dt and use a graphing utility to graph dh/dt.

(b) Evaluate dh/dt for $t = 0, 1, 5, 10$, and 20.

(c) Interpret your results for parts (a) and (b).

4.4 LOGARITHMIC FUNCTIONS

- Sketch the graphs of natural logarithmic functions.
- Use properties of logarithms to simplify, expand, and condense logarithmic expressions.
- Use inverse properties of exponential and logarithmic functions to solve exponential and logarithmic equations.
- Use properties of natural logarithms to answer questions about real-life situations.

The Natural Logarithmic Function

From your previous algebra courses, you should be somewhat familiar with logarithms. For instance, the **common logarithm** $\log_{10} x$ is defined as

$$\log_{10} x = b \quad \text{if and only if} \quad 10^b = x.$$

The base of common logarithms is 10. In calculus, the most useful base for logarithms is the number e.

Definition of the Natural Logarithmic Function

The **natural logarithmic function,** denoted by $\ln x$, is defined as

$$\ln x = b \quad \text{if and only if} \quad e^b = x.$$

$\ln x$ is read as "el en of x" or as "the natural log of x."

This definition implies that the natural logarithmic function and the natural exponential function are inverse functions. So, every logarithmic equation can be written in an equivalent exponential form and every exponential equation can be written in logarithmic form. Here are some examples.

Logarithmic form:	*Exponential form:*
$\ln 1 = 0$	$e^0 = 1$
$\ln e = 1$	$e^1 = e$
$\ln \frac{1}{e} = -1$	$e^{-1} = \frac{1}{e}$
$\ln 2 \approx 0.693$	$e^{0.693} \approx 2$

Because the functions $f(x) = e^x$ and $g(x) = \ln x$ are inverse functions, their graphs are reflections of each other in the line $y = x$. This reflective property is illustrated in Figure 4.13. The figure also contains a summary of several properties of the graph of the natural logarithmic function.

Notice that the domain of the natural logarithmic function is the set of *positive real numbers*—be sure you see that $\ln x$ is not defined for zero or for negative numbers. You can test this on your calculator. If you try evaluating $\ln(-1)$ or $\ln 0$, your calculator should indicate that the value is not a real number.

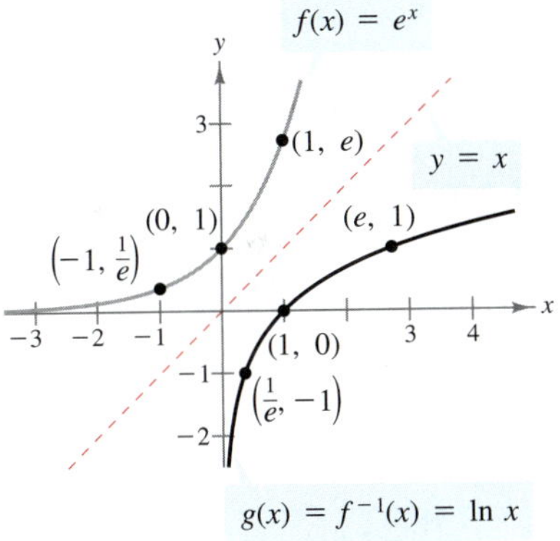

$g(x) = \ln x$

- Domain: $(0, \infty)$
- Range: $(-\infty, \infty)$
- Intercept: $(1, 0)$
- Always increasing
- $\ln x \to \infty$ as $x \to \infty$
- $\ln x \to -\infty$ as $x \to 0^+$
- Continuous
- One-to-one

FIGURE 4.13

EXAMPLE 1 Graphing Logarithmic Functions

Sketch the graph of each function.

(a) $f(x) = \ln(x + 1)$ (b) $f(x) = 2\ln(x - 2)$

SOLUTION

(a) Because the natural logarithmic function is defined only for positive values, the domain of the function is $0 < x + 1$, or

$-1 < x.$ Domain

To sketch the graph, begin by constructing a table of values, as shown below. Then plot the points in the table and connect them with a smooth curve, as shown in Figure 4.14(a).

x	-0.5	0	0.5	1	1.5	2
$\ln(x + 1)$	-0.693	0	0.405	0.693	0.916	1.099

(b) The domain of this function is $0 < x - 2$, or

$2 < x.$ Domain

A table of values for the function is shown below, and its graph is shown in Figure 4.14(b).

x	2.5	3	3.5	4	4.5	5
$2\ln(x - 2)$	-1.386	0	0.811	1.386	1.833	2.197

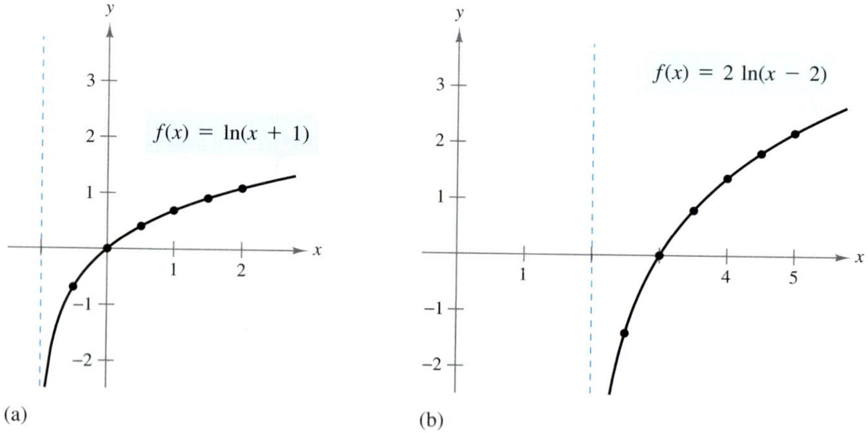

FIGURE 4.14

TECHNOLOGY

What happens when you take the logarithm of a negative number? Some graphing utilities do not give an error message for $\ln(-1)$. Instead, the graphing utility displays a complex number. For the purpose of this text, however, it is assumed that the domain of the logarithmic function is the set of positive real numbers.

TRY IT 1

Use a graphing utility to complete the table and graph the function.

$f(x) = \ln(x + 2)$

x	-1.5	-1	-0.5
$f(x)$			

x	0	0.5	1
$f(x)$			

STUDY TIP

How does the graph of $f(x) = \ln(x + 1)$ relate to the graph of $y = \ln x$? The graph of f is a translation of the graph of $y = \ln x$ one unit to the left.

Properties of Logarithmic Functions

Recall from Section 1.4 that inverse functions have the property that

$$f(f^{-1}(x)) = x \quad \text{and} \quad f^{-1}(f(x)) = x.$$

The properties listed below follow from the fact that the natural logarithmic function and the natural exponential function are inverse functions.

Inverse Properties of Logarithms and Exponents

1. $\ln e^x = x$ 2. $e^{\ln x} = x$

EXAMPLE 2 Applying Inverse Properties

Simplify each expression.

(a) $\ln e^{\sqrt{2}}$ (b) $e^{\ln 3x}$

SOLUTION

(a) Because $\ln e^x = x$, it follows that

$$\ln e^{\sqrt{2}} = \sqrt{2}.$$

(b) Because $e^{\ln x} = x$, it follows that

$$e^{\ln 3x} = 3x.$$

TRY IT 2

Simplify each expression.

(a) $\ln e^3$ (b) $e^{\ln(x+1)}$

Most of the properties of exponential functions can be rewritten in terms of logarithmic functions. For instance, the property

$$e^x e^y = e^{x+y}$$

states that you can multiply two exponential expressions by adding their exponents. In terms of logarithms, this property becomes

$$\ln xy = \ln x + \ln y.$$

This property and two other properties of logarithms are summarized below.

Properties of Logarithms

1. $\ln xy = \ln x + \ln y$ 2. $\ln \dfrac{x}{y} = \ln x - \ln y$

3. $\ln x^n = n \ln x$

STUDY TIP

There is no general property that can be used to rewrite $\ln(x + y)$. Specifically, $\ln(x + y)$ is not equal to $\ln x + \ln y$.

Rewriting a logarithm of a single quantity as the sum, difference, or multiple of logarithms is called *expanding* the logarithmic expression. The reverse procedure is called *condensing* a logarithmic expression.

EXAMPLE 3 Expanding Logarithmic Expressions

Use the properties of logarithms to rewrite each expression as a sum, difference, or multiple of logarithms. (Assume $x > 0$ and $y > 0$.)

(a) $\ln \dfrac{10}{9}$ (b) $\ln \sqrt{x^2 + 1}$ (c) $\ln \dfrac{xy}{5}$ (d) $\ln[x^2(x + 1)]$

SOLUTION

(a) $\ln \frac{10}{9} = \ln 10 - \ln 9$ — Property 2

(b) $\ln \sqrt{x^2 + 1} = \ln(x^2 + 1)^{1/2}$ — Rewrite with rational exponent.

$= \frac{1}{2}\ln(x^2 + 1)$ — Property 3

(c) $\ln \dfrac{xy}{5} = \ln(xy) - \ln 5$ — Property 2

$= \ln x + \ln y - \ln 5$ — Property 1

(d) $\ln[x^2(x + 1)] = \ln x^2 + \ln(x + 1)$ — Property 1

$= 2 \ln x + \ln(x + 1)$ — Property 3

TECHNOLOGY

Try using a graphing utility to verify the results of Example 3(b). That is, try graphing the functions

$y = \ln \sqrt{x^2 + 1}$

and

$y = \dfrac{1}{2}\ln(x^2 + 1).$

Because these two functions are equivalent, their graphs should coincide.

TRY IT 3

Use the properties of logarithms to rewrite each expression as a sum, difference, or multiple of logarithms. (Assume $x > 0$ and $y > 0$.)

(a) $\ln \dfrac{2}{5}$ (b) $\ln \sqrt[3]{x + 2}$ (c) $\ln \dfrac{x}{5y}$ (d) $\ln x(x + 1)^2$

EXAMPLE 4 Condensing Logarithmic Expressions

Use the properties of logarithms to rewrite each expression as the logarithm of a single quantity. (Assume $x > 0$ and $y > 0$.)

(a) $\ln x + 2 \ln y$

(b) $2 \ln(x + 2) - 3 \ln x$

SOLUTION

(a) $\ln x + 2 \ln y = \ln x + \ln y^2$ — Property 3

$= \ln xy^2$ — Property 1

(b) $2 \ln(x + 2) - 3 \ln x = \ln(x + 2)^2 - \ln x^3$ — Property 3

$= \ln \dfrac{(x + 2)^2}{x^3}$ — Property 2

TRY IT 4

Use the properties of logarithms to rewrite each expression as the logarithm of a single quantity. (Assume $x > 0$ and $y > 0$.)

(a) $4 \ln x + 3 \ln y$

(b) $\ln(x + 1) - 2 \ln(x + 3)$

Solving Exponential and Logarithmic Equations

The inverse properties of logarithms and exponents can be used to solve exponential and logarithmic equations, as shown in the next two examples.

EXAMPLE 5 **Solving Exponential Equations**

Solve each equation.

(a) $e^x = 5$ (b) $10 + e^{0.1t} = 14$

SOLUTION

(a) $e^x = 5$ Write original equation.

$\ln e^x = \ln 5$ Take natural log of each side.

$x = \ln 5$ Inverse property: $\ln e^x = x$

(b) $10 + e^{0.1t} = 14$ Write original equation.

$e^{0.1t} = 4$ Subtract 10 from each side.

$\ln e^{0.1t} = \ln 4$ Take natural log of each side.

$0.1t = \ln 4$ Inverse property: $\ln e^{0.1t} = 0.1t$

$t = 10 \ln 4$ Multiply each side by 10.

ALGEBRA REVIEW

In the examples on this page, note that the key step in solving an exponential equation is to take the log of each side, and the key step in solving a logarithmic equation is to exponentiate each side.

TRY IT 5

Solve each equation.

(a) $e^x = 6$

(b) $5 + e^{0.2t} = 10$

EXAMPLE 6 **Solving Logarithmic Equations**

Solve each equation.

(a) $\ln x = 5$ (b) $3 + 2 \ln x = 7$

SOLUTION

(a) $\ln x = 5$ Write original equation.

$e^{\ln x} = e^5$ Exponentiate each side.

$x = e^5$ Inverse property: $e^{\ln x} = x$

(b) $3 + 2 \ln x = 7$ Write original equation.

$2 \ln x = 4$ Subtract 3 from each side.

$\ln x = 2$ Divide each side by 2.

$e^{\ln x} = e^2$ Exponentiate each side.

$x = e^2$ Inverse property: $e^{\ln x} = x$

TRY IT 6

Solve each equation.

(a) $\ln x = 4$

(b) $4 + 5 \ln x = 19$

Application

EXAMPLE 7 Finding Doubling Time

You deposit P dollars in an account whose annual interest rate is r, compounded continuously. How long will it take for your balance to double?

SOLUTION The balance in the account after t years is

$$A = Pe^{rt}.$$

So, the balance will have doubled when $Pe^{rt} = 2P$. To find the "doubling time," solve this equation for t.

$Pe^{rt} = 2P$ Balance in account has doubled.

$e^{rt} = 2$ Divide each side by P.

$\ln e^{rt} = \ln 2$ Take natural log of each side.

$rt = \ln 2$ Inverse property: $\ln e^{rt} = rt$

$t = \frac{1}{r}\ln 2$ Divide each side by r.

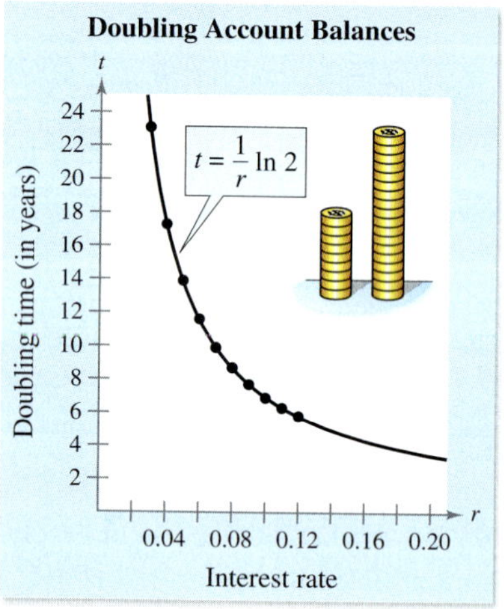

FIGURE 4.15

From this result, you can see that the time it takes for the balance to double is inversely proportional to the interest rate r. The table shows the doubling times for several interest rates. Notice that the doubling time decreases as the rate increases. The relationship between doubling time and the interest rate is shown graphically in Figure 4.15.

r	3%	4%	5%	6%	7%	8%	9%	10%	11%	12%
t	23.1	17.3	13.9	11.6	9.9	8.7	7.7	6.9	6.3	5.8

TRY IT 7

Use the equation found in Example 7 to determine the amount of time it would take for your balance to double at an interest rate of 8.75%.

TAKE ANOTHER LOOK

Tripling Time

You deposit P dollars in an account whose annual interest rate is r, compounded continuously. Use Example 7 as a model to determine how long it will take for your balance to triple. Complete the table below and use a graph to illustrate your answer.

Rate, r	3%	4%	5%	6%	7%	8%	9%	10%	11%	12%
Tripling time, t										

PREREQUISITE REVIEW 4.4

The following warm-up exercises involve skills that were covered in earlier sections. You will use these skills in the exercise set for this section.

In Exercises 1–8, use the properties of exponents to simplify the expression.

1. $(4^2)(4^{-3})$ **2.** $(2^3)^2$ **3.** $\dfrac{3^4}{3^{-2}}$ **4.** $\left(\dfrac{3}{2}\right)^{-3}$

5. e^0 **6.** $(3e)^4$ **7.** $\left(\dfrac{2}{e^3}\right)^{-1}$ **8.** $\left(\dfrac{4e^2}{25}\right)^{-3/2}$

In Exercises 9–12, solve for *x*.

9. $0 < x + 4$ **10.** $0 < x^2 + 1$

11. $0 < \sqrt{x^2 - 1}$ **12.** $0 < x - 5$

In Exercises 13 and 14, find the balance in the account after 10 years.

13. $P = \$1900$, $r = 6\%$, compounded continuously

14. $P = \$2500$, $r = 3\%$, compounded continuously

EXERCISES 4.4

In Exercises 1–8, write the logarithmic equation as an exponential equation, or vice versa.

1. $\ln 2 = 0.6931 . . .$

2. $\ln 8.4 = 2.1282 . . .$

3. $\ln 0.2 = -1.6094 . . .$

4. $\ln 0.056 = -2.8824 . . .$

5. $e^0 = 1$

6. $e^2 = 7.3891 . . .$

7. $e^{-3} = 0.0498 . . .$

8. $e^{0.25} = 1.2840 . . .$

In Exercises 9–12, match the function with its graph. [The graphs are labeled (a)–(d).]

9. $f(x) = 2 + \ln x$ **10.** $f(x) = -\ln x$

11. $f(x) = \ln(x + 2)$ **12.** $f(x) = -\ln(x - 1)$

(a)

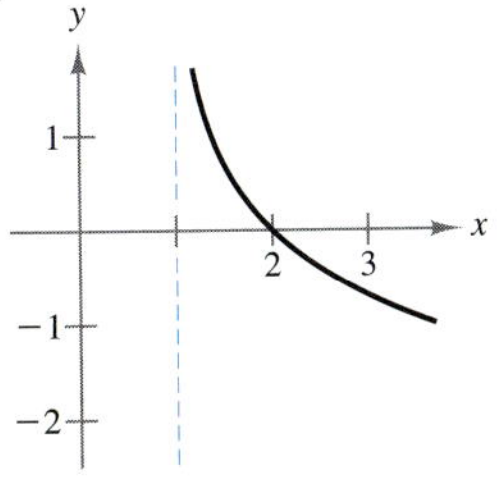

(b)

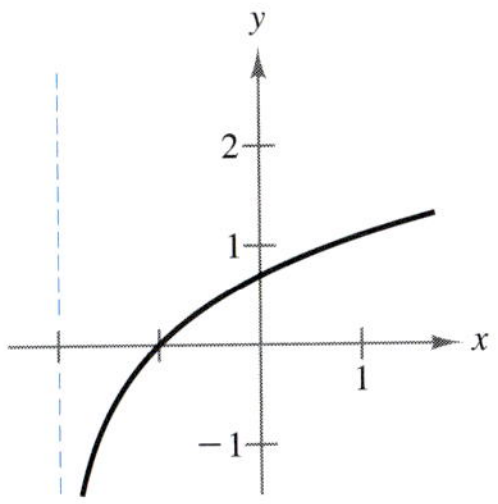

(c)

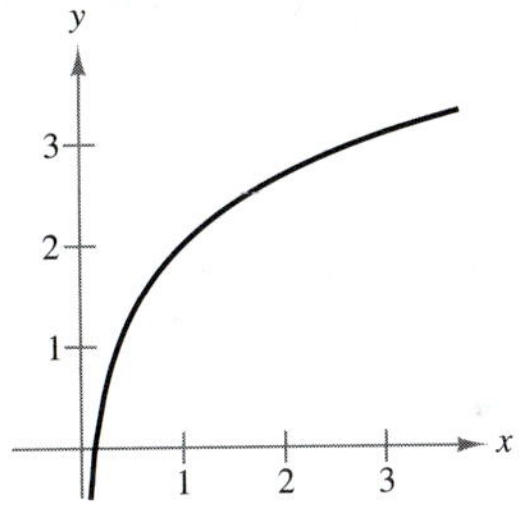

(d)

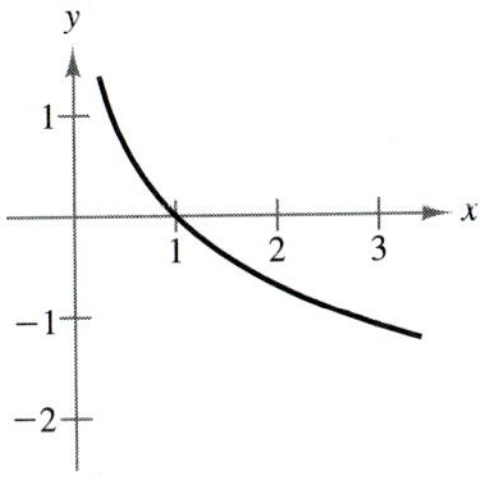

In Exercises 13–18, sketch the graph of the function.

13. $y = \ln(x - 1)$ **14.** $y = \ln|x|$

15. $y = \ln 2x$ **16.** $y = 5 + \ln x$

17. $y = 3 \ln x$ **18.** $y = \frac{1}{4} \ln x$

In Exercises 19–22, analytically show that the functions are inverse functions. Then use a graphing utility to show this graphically.

19. $f(x) = e^{2x}$

$g(x) = \ln \sqrt{x}$

20. $f(x) = e^x - 1$

$g(x) = \ln(x + 1)$

21. $f(x) = e^{2x-1}$

$g(x) = \frac{1}{2} + \ln \sqrt{x}$

22. $f(x) = e^{x/3}$

$g(x) = \ln x^3$

In Exercises 23–28, apply the inverse properties of logarithmic and exponential functions to simplify the expression.

23. $\ln e^{x^2}$

24. $\ln e^{2x-1}$

25. $e^{\ln(5x+2)}$

26. $e^{\ln\sqrt{x}}$

27. $-1 + \ln e^{2x}$

28. $-8 + e^{\ln x^3}$

In Exercises 29 and 30, use the properties of logarithms and the fact that $\ln 2 \approx 0.6931$ and $\ln 3 \approx 1.0986$ to approximate the logarithm. Then use a calculator to confirm your approximation.

29. (a) $\ln 6$ (b) $\ln \frac{3}{2}$

(c) $\ln 81$ (d) $\ln \sqrt{3}$

30. (a) $\ln 0.25$ (b) $\ln 24$

(c) $\ln \sqrt[3]{12}$ (d) $\ln \frac{1}{72}$

In Exercises 31–40, use the properties of logarithms to write the expression as a sum, difference, or multiple of logarithms.

31. $\ln \frac{2}{3}$

32. $\ln \frac{1}{5}$

33. $\ln xyz$

34. $\ln \dfrac{xy}{z}$

35. $\ln \sqrt{x^2 + 1}$

36. $\ln \sqrt{\dfrac{x^3}{x+1}}$

37. $\ln [z(z-1)^2]$

38. $\ln\left(x\sqrt[3]{x^2+1}\right)$

39. $\ln \dfrac{3x(x+1)}{(2x+1)^2}$

40. $\ln \dfrac{2x}{\sqrt{x^2-1}}$

In Exercises 41–50, write the expression as the logarithm of a single quantity.

41. $\ln(x-2) - \ln(x+2)$

42. $\ln(2x+1) + \ln(2x-1)$

43. $3\ln x + 2\ln y - 4\ln z$

44. $2\ln 3 - \frac{1}{2}\ln(x^2+1)$

45. $3[\ln x + \ln(x+3) - \ln(x+4)]$

46. $\frac{1}{3}[2\ln(x+3) + \ln x - \ln(x^2-1)]$

47. $\frac{3}{2}[\ln x(x^2+1) - \ln(x+1)]$

48. $2\left[\ln x + \frac{1}{4}\ln(x+1)\right]$

49. $\frac{1}{3}\ln(x+1) - \frac{2}{3}\ln(x-1)$

50. $\frac{1}{2}\ln(x-2) + \frac{3}{2}\ln(x+2)$

In Exercises 51–68, solve for x or t.

51. $e^{\ln x} = 4$

52. $e^{\ln x^2} - 9 = 0$

53. $\ln x = 0$

54. $2\ln x = 4$

55. $e^{x+1} = 4$

56. $e^{-0.5x} = 0.075$

57. $300e^{-0.2t} = 700$

58. $400e^{-0.0174t} = 1000$

59. $4e^{2x-1} - 1 = 5$

60. $2e^{-x+1} - 5 = 9$

61. $\dfrac{10}{1 + 4e^{-0.01x}} = 2.5$

62. $\dfrac{50}{1 + 12e^{-0.02x}} = 10.5$

63. $5^{2x} = 15$

64. $2^{1-x} = 6$

65. $500(1.07)^t = 1000$

66. $400(1.06)^t = 1300$

67. $1000\left(1 + \dfrac{0.07}{12}\right)^{12t} = 3000$

68. $2000\left(1 + \dfrac{0.06}{12}\right)^{12t} = 10{,}000$

69. ***Compound Interest*** A deposit of \$1000 is made in an account that earns interest at an annual rate of 5%. How long will it take for the balance to double if the interest is compounded (a) annually, (b) monthly, (c) daily, and (d) continuously?

70. ***Compound Interest*** Complete the table, which shows the time t necessary for P dollars to triple if the interest is compounded continuously at the rate of r.

r	2%	4%	6%	8%	10%	12%	14%
t							

71. ***Chemistry: Carbon Dating*** The remnants of an ancient fire in a cave in Africa showed a $^{14}_{6}C$ (carbon-14) decay rate of 3.1 counts per minute per gram of carbon. Assuming that the decay rate of $^{14}_{6}C$ in freshly cut wood (corrected for changes in the $^{14}_{6}C$ content of the atmosphere) is 13.6 counts per minute per gram of carbon, calculate the age of the remnants. The half-life of $^{14}_{6}C$ is 5715 years. Use the integrated first-order rate law,

$$\ln(N/N_0) = -kt$$

where N is the number of nuclides present at time t, N_0 is the number of nuclides present at time 0, $k = 0.693/5715$, and t is the time of the fire. *(Source: Adapted from Zumdahl,* Chemistry, *Sixth Edition)*

72. ***Demand*** The demand function for a product is given by

$$p = 250 - 0.8e^{0.005x}$$

where p is the price per unit and x is the number of units sold. Find the numbers of units sold for prices of (a) $p = \$200$ and (b) $p = \$125$.

73. ***Population Growth*** The population P (in thousands) of Orlando, Florida from 1980 through 2000 can be modeled by

$$P = 821.95e^{0.0358t}$$

where $t = 0$ corresponds to 1980. *(Source: U.S. Census Bureau)*

(a) According to this model, what was the population of Orlando in 2000?

(b) According to this model, in what year will Orlando have a population of 2,500,000?

74. *Population Growth* The population P (in thousands) of Houston, Texas from 1980 through 2000 can be modeled by

$$P = 2734.07e^{0.0210t}$$

where $t = 0$ corresponds to 1980. *(Source: U.S. Census Bureau)*

(a) According to this model, what was the population of Houston in 2000?

(b) According to this model, in what year will Houston have a population of 6,000,000?

Carbon Dating In Exercises 75–78, you are given the ratio of carbon atoms in a fossil. Use the information to estimate the age of the fossil. In living organic material, the ratio of radioactive carbon isotopes to the total number of carbon atoms is about 1 to 10^{12}. (See Example 2 in Section 4.1.) When organic material dies, its radioactive carbon isotopes begin to decay, with a half-life of about 5715 years. So, the ratio R of carbon isotopes to carbon-14 atoms is modeled by

$$R = 10^{-12}\left(\tfrac{1}{2}\right)^{t/5715}$$

where t is the time (in years) and $t = 0$ represents the time when the organic material died.

75. $R = 0.32 \times 10^{-12}$ **76.** $R = 0.27 \times 10^{-12}$

77. $R = 0.22 \times 10^{-12}$ **78.** $R = 0.13 \times 10^{-12}$

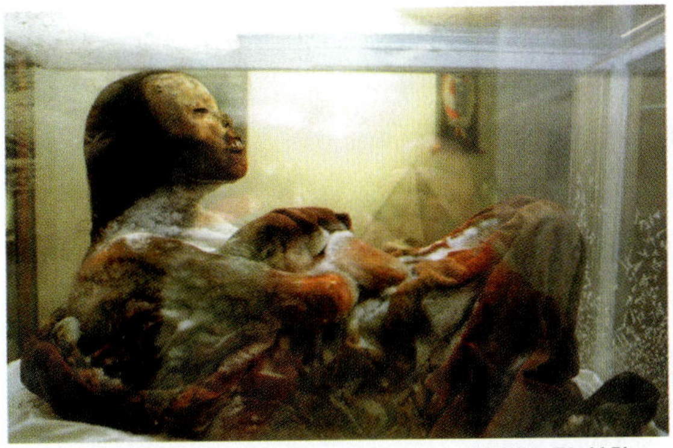

AP/Wide World Photos

In 1995, archeologist Johan Reinhard discovered the frozen remains of a young Incan woman atop Mt. Ampato in Peru. Carbon dating was used to estimate the age of the "Ice Maiden" at 500 years.

79. *Learning Theory* Students in a mathematics class were given an exam and then retested monthly with equivalent exams. The average score S (on a 100-point scale) for the class can be modeled by

$$S = 80 - 14\ln(t + 1), \qquad 0 \le t \le 12$$

where t is the time in months.

(a) What was the average score on the original exam?

(b) What was the average score after 4 months?

(c) After how many months was the average score 46?

80. *Research Project* Use a graphing utility to graph

$$y = 10\ln\left(\frac{10 + \sqrt{100 - x^2}}{10}\right) - \sqrt{100 - x^2}$$

over the interval $(0, 10]$. This graph is called a *tractrix* or *pursuit curve*. Use your school's library, the Internet, or some other reference source to find information about a tractrix. Explain how such a curve can arise in a real-life setting.

81. Demonstrate that

$$\frac{\ln x}{\ln y} \ne \ln\frac{x}{y} = \ln x - \ln y$$

by completing the table.

x	y	$\frac{\ln x}{\ln y}$	$\ln\frac{x}{y}$	$\ln x - \ln y$
1	2			
3	4			
10	5			
4	0.5			

82. Complete the table using $f(x) = \dfrac{\ln x}{x}$.

x	1	5	10	10^2	10^4	10^6
$f(x)$						

(a) Use the table to estimate the limit: $\lim_{x\to\infty} f(x)$.

(b) Use a graphing utility to estimate the relative extrema of f.

In Exercises 83 and 84, use a graphing utility to verify that the functions are equivalent for $x > 0$.

83. $f(x) = \ln\dfrac{x^2}{4}$

$g(x) = 2\ln x - \ln 4$

84. $f(x) = \ln\sqrt{x(x^2 + 1)}$

$g(x) = \frac{1}{2}[\ln x + \ln(x^2 + 1)]$

True or False? In Exercises 85–90, determine whether the statement is true or false given that $f(x) = \ln x$. If it is false, explain why or give an example that shows it is false.

85. $f(0) = 0$

86. $f(ax) = f(a) + f(x), \quad a > 0, x > 0$

87. $f(x - 2) = f(x) - f(2), \quad x > 2$

88. $\sqrt{f(x)} = \frac{1}{2}f(x)$

89. If $f(u) = 2f(v)$, then $v = u^2$.

90. If $f(x) < 0$, then $0 < x < 1$.

4.5 DERIVATIVES OF LOGARITHMIC FUNCTIONS

- Find derivatives of natural logarithmic functions.
- Use calculus to analyze the graphs of functions that involve the natural logarithmic function.
- Use the definition of logarithms and the change-of-base formula to evaluate logarithmic expressions involving other bases.
- Find derivatives of exponential and logarithmic functions involving other bases.

DISCOVERY

Sketch the graph of $y = \ln x$ on a piece of paper. Draw tangent lines to the graph at various points. How do the slopes of these tangent lines change as you move to the right? Is the slope ever equal to zero? Use the formula for the derivative of the logarithmic function to confirm your conclusions.

Derivatives of Logarithmic Functions

Implicit differentiation can be used to develop the derivative of the natural logarithmic function.

$$y = \ln x \qquad \text{Natural logarithmic function}$$

$$e^y = x \qquad \text{Write in exponential form.}$$

$$\frac{d}{dx}[e^y] = \frac{d}{dx}[x] \qquad \text{Differentiate with respect to } x.$$

$$e^y\frac{dy}{dx} = 1 \qquad \text{Chain Rule}$$

$$\frac{dy}{dx} = \frac{1}{e^y} \qquad \text{Divide each side by } e^y.$$

$$\frac{dy}{dx} = \frac{1}{x} \qquad \text{Substitute } x \text{ for } e^y.$$

This result and its Chain Rule version are summarized below.

Derivative of the Natural Logarithmic Function

Let u be a differentiable function of x.

1. $\dfrac{d}{dx}[\ln x] = \dfrac{1}{x}$ 2. $\dfrac{d}{dx}[\ln u] = \dfrac{1}{u}\dfrac{du}{dx}$

EXAMPLE 1 **Differentiating a Logarithmic Function**

Find the derivative of

$$f(x) = \ln 2x.$$

SOLUTION Let $u = 2x$. Then $du/dx = 2$, and you can apply the Chain Rule as shown.

$$f'(x) = \frac{1}{u}\frac{du}{dx} = \frac{1}{2x}(2) = \frac{1}{x}$$

TRY IT 1

Find the derivative of $f(x) = \ln 5x$.

EXAMPLE 2 Differentiating Logarithmic Functions

Find the derivative of each function.

(a) $f(x) = \ln(2x^2 + 4)$ (b) $f(x) = x \ln x$ (c) $f(x) = \dfrac{\ln x}{x}$

SOLUTION

(a) Let $u = 2x^2 + 4$. Then $du/dx = 4x$, and you can apply the Chain Rule.

$$f'(x) = \frac{1}{u}\frac{du}{dx} \qquad \text{Chain Rule}$$

$$= \frac{1}{2x^2 + 4}(4x)$$

$$= \frac{2x}{x^2 + 2} \qquad \text{Simplify.}$$

(b) Using the Product Rule, you can find the derivative.

$$f'(x) = x\frac{d}{dx}[\ln x] + (\ln x)\frac{d}{dx}[x] \qquad \text{Product Rule}$$

$$= x\left(\frac{1}{x}\right) + (\ln x)(1)$$

$$= 1 + \ln x \qquad \text{Simplify.}$$

(c) Using the Quotient Rule, you can find the derivative.

$$f'(x) = \frac{x\dfrac{d}{dx}[\ln x] - (\ln x)\dfrac{d}{dx}[x]}{x^2} \qquad \text{Quotient Rule}$$

$$= \frac{x\left(\dfrac{1}{x}\right) - \ln x}{x^2}$$

$$= \frac{1 - \ln x}{x^2} \qquad \text{Simplify.}$$

STUDY TIP

When you are differentiating logarithmic functions, it is often helpful to use the properties of logarithms to rewrite the function *before* differentiating. To see the advantage of rewriting before differentiating, try using the Chain Rule to differentiate $f(x) = \ln\sqrt{x + 1}$ and compare your work with that shown in Example 3.

TRY IT 2

Find the derivative of each function.

(a) $f(x) = \ln(x^2 - 4)$

(b) $f(x) = x^2 \ln x$

(c) $f(x) = -\dfrac{\ln x}{x^2}$

EXAMPLE 3 Rewriting Before Differentiating

Find the derivative of $f(x) = \ln\sqrt{x + 1}$.

SOLUTION

$$f(x) = \ln\sqrt{x + 1} \qquad \text{Write original function.}$$

$$= \ln(x + 1)^{1/2} \qquad \text{Rewrite with rational exponent.}$$

$$= \frac{1}{2}\ln(x + 1) \qquad \text{Property of logarithms}$$

$$f'(x) = \frac{1}{2}\left(\frac{1}{x + 1}\right) \qquad \text{Differentiate.}$$

$$= \frac{1}{2(x + 1)} \qquad \text{Simplify.}$$

TRY IT 3

Find the derivative of $f(x) = \ln \sqrt[3]{x + 1}$.

DISCOVERY

What is the domain of the function $f(x) = \ln\sqrt{x+1}$ in Example 3? What is the domain of the function $f'(x) = 1/[2(x+1)]$? In general, you must be careful to understand the domains of functions involving logarithms. For example, are the domains of the functions $y_1 = \ln x^2$ and $y_2 = 2\ln x$ the same? Try graphing them on your graphing utility.

The next example is an even more dramatic illustration of the benefit of rewriting a function before differentiating.

EXAMPLE 4 Rewriting Before Differentiating

Find the derivative of

$$f(x) = \ln[x(x^2+1)^2].$$

SOLUTION

$$\begin{aligned} f(x) &= \ln[x(x^2+1)^2] && \text{Write original function.}\\ &= \ln x + \ln(x^2+1)^2 && \text{Logarithmic properties}\\ &= \ln x + 2\ln(x^2+1) && \text{Logarithmic properties}\\ f'(x) &= \frac{1}{x} + 2\left(\frac{2x}{x^2+1}\right) && \text{Differentiate.}\\ &= \frac{1}{x} + \frac{4x}{x^2+1} && \text{Simplify.} \end{aligned}$$

TRY IT 4

Find the derivative of $f(x) = \ln[x^2\sqrt{x^2+1}]$.

ALGEBRA REVIEW

Finding the derivative of the function in Example 4 without first rewriting would be a formidable task.

$$f'(x) = \frac{1}{x(x^2+1)^2}\frac{d}{dx}[x(x^2+1)^2]$$

You might try showing that this yields the same result obtained in Example 4, but be careful—the algebra is messy.

TECHNOLOGY

A symbolic differentiation utility will not generally list the derivative of the logarithmic function in the form obtained in Example 4. Use a symbolic differentiation utility to find the derivative of the function in Example 4. Show that the two forms are equivalent by rewriting the answer obtained in Example 4.

Applications

EXAMPLE 5 Analyzing a Graph

Analyze the graph of the function $f(x) = \frac{x^2}{2} - \ln x$.

SOLUTION From Figure 4.16, it appears that the function has a minimum at $x = 1$. To find the minimum analytically, find the critical numbers by setting the derivative of f equal to zero and solving for x.

$$f(x) = \frac{x^2}{2} - \ln x \qquad \text{Write original function.}$$

$$f'(x) = x - \frac{1}{x} \qquad \text{Differentiate.}$$

$$x - \frac{1}{x} = 0 \qquad \text{Set derivative equal to 0.}$$

$$x = \frac{1}{x} \qquad \text{Add } 1/x \text{ to each side.}$$

$$x^2 = 1 \qquad \text{Multiply each side by } x.$$

$$x = \pm 1 \qquad \text{Take square root of each side.}$$

Of these two possible critical numbers, only the positive one lies in the domain of f. By applying the First-Derivative Test, you can confirm that the function has a relative minimum when $x = 1$.

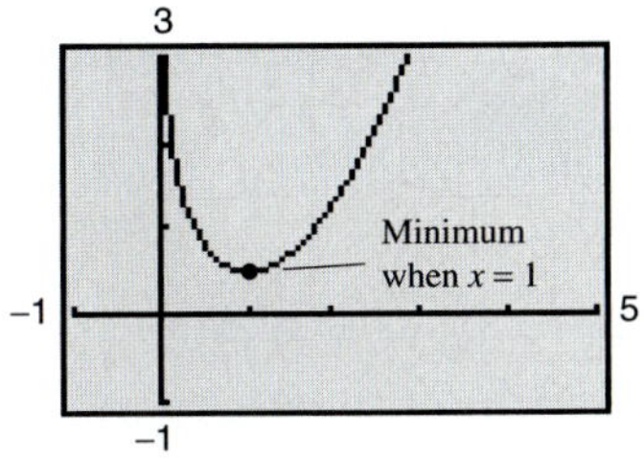

FIGURE 4.16

TRY IT 5

Determine the relative extrema of the function

$f(x) = x - 2\ln x$.

EXAMPLE 6 Finding a Rate of Change

A group of 200 college students was tested every 6 months over a four-year period. The group was composed of students who took Spanish during the fall semester of their freshman year and did not take subsequent Spanish courses. The average test score p (in percent) is modeled by

$$p = 91.6 - 15.6\ln(t + 1), \quad 0 \le t \le 48$$

where t is the time in months, as shown in Figure 4.17. At what rate was the average score changing after 1 year?

SOLUTION The rate of change is

$$\frac{dp}{dt} = -\frac{15.6}{t + 1}.$$

When $t = 12$, $dp/dt = -1.2$, which means that the average score was decreasing at the rate of 1.2% per month.

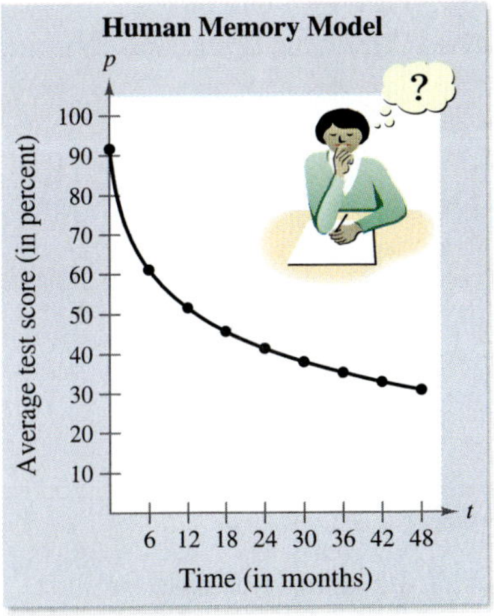

FIGURE 4.17

TRY IT 6

Suppose the average test score p in Example 6 was modeled by $p = 92.3 - 16.9\ln(t + 1)$, where t is the time in months. How would the rate at which the average test score changed after 1 year compare with that of the model in Example 6?

Other Bases

This chapter began with a definition of a general exponential function

$$f(x) = a^x$$

where a is a positive number such that $a \neq 1$. The corresponding **logarithm to the base a** is defined by

$$\log_a x = b \quad \text{if and only if} \quad a^b = x.$$

As with the natural logarithmic function, the domain of the logarithmic function to the base a is the set of positive numbers.

TECHNOLOGY

Use a graphing utility to graph the three functions $y_1 = \log_2 x = \ln x/\ln 2$, $y_2 = 2^x$, and $y_3 = x$ in the same viewing window. Explain why the graphs of y_1 and y_2 are reflections of each other in the line $y_3 = x$.

EXAMPLE 7 Evaluating Logarithms

Evaluate each logarithm without using a calculator.

(a) $\log_2 8$ (b) $\log_{10} 100$ (c) $\log_{10} \frac{1}{10}$ (d) $\log_3 81$

SOLUTION

(a) $\log_2 8 = 3$ — $2^3 = 8$

(b) $\log_{10} 100 = 2$ — $10^2 = 100$

(c) $\log_{10} \frac{1}{10} = -1$ — $10^{-1} = \frac{1}{10}$

(d) $\log_3 81 = 4$ — $3^4 = 81$

TRY IT 7

Evaluate each logarithm without using a calculator.

(a) $\log_2 16$

(b) $\log_{10} \frac{1}{100}$

(c) $\log_2 \frac{1}{32}$

(d) $\log_5 125$

Logarithms to the base 10 are called **common logarithms.** Most calculators have only two logarithm keys—a natural logarithm key denoted by [LN] and a common logarithm key denoted by [LOG]. Logarithms to other bases can be evaluated with the following change-of-base formula.

$$\log_a x = \frac{\ln x}{\ln a}$$ Change-of-base formula

EXAMPLE 8 Evaluating Logarithms

Use the change-of-base formula and a calculator to evaluate each logarithm.

(a) $\log_2 3$ (b) $\log_3 6$ (c) $\log_2(-1)$

SOLUTION In each case, use the change-of-base formula and a calculator.

(a) $\log_2 3 = \frac{\ln 3}{\ln 2} \approx 1.585$ — $\log_a x = \frac{\ln x}{\ln a}$

(b) $\log_3 6 = \frac{\ln 6}{\ln 3} \approx 1.631$ — $\log_a x = \frac{\ln x}{\ln a}$

(c) $\log_2(-1)$ is not defined.

TRY IT 8

Use the change-of-base formula and a calculator to evaluate each logarithm.

(a) $\log_2 5$

(b) $\log_3 18$

(c) $\log_4 80$

(d) $\log_{16} 0.25$

To find derivatives of exponential or logarithmic functions to bases other than e, you can either convert to base e or use the differentiation rules shown on the next page.

Other Bases and Differentiation

Let u be a differentiable function of x.

1. $\dfrac{d}{dx}[a^x] = (\ln a)a^x$
2. $\dfrac{d}{dx}[a^u] = (\ln a)a^u \dfrac{du}{dx}$
3. $\dfrac{d}{dx}[\log_a x] = \left(\dfrac{1}{\ln a}\right)\dfrac{1}{x}$
4. $\dfrac{d}{dx}[\log_a u] = \left(\dfrac{1}{\ln a}\right)\left(\dfrac{1}{u}\right)\dfrac{du}{dx}$

STUDY TIP

Remember that you can convert to base e using the formulas

$$a^x = e^{(\ln a)x}$$

and

$$\log_a x = \left(\frac{1}{\ln a}\right)\ln x.$$

PROOF By definition, $a^x = e^{(\ln a)x}$. So, you can prove the first rule by letting $u = (\ln a)x$ and differentiating with base e to obtain

$$\frac{d}{dx}[a^x] = \frac{d}{dx}[e^{(\ln a)x}] = e^u\frac{du}{dx} = e^{(\ln a)x}(\ln a) = (\ln a)a^x.$$

EXAMPLE 9 **Finding a Rate of Change**

Radioactive carbon isotopes have a half-life of 5715 years. If 1 gram of the isotopes is present in an object now, the amount A (in grams) that will be present after t years is

$$A = \left(\frac{1}{2}\right)^{t/5715}.$$

At what rate is the amount changing when $t = 10{,}000$ years?

SOLUTION The derivative of A with respect to t is

$$\frac{dA}{dt} = \left(\ln\frac{1}{2}\right)\left(\frac{1}{2}\right)^{t/5715}\left(\frac{1}{5715}\right).$$

When $t = 10{,}000$, the rate at which the amount is changing is

$$\left(\ln\frac{1}{2}\right)\left(\frac{1}{2}\right)^{10{,}000/5715}\left(\frac{1}{5715}\right) \approx -0.000036$$

which implies that the amount of isotopes in the object is decreasing at the rate of 0.000036 gram per year.

TRY IT 9

Use a graphing utility to graph the model in Example 9. Describe the rate at which the amount is changing as time t increases.

TAKE ANOTHER LOOK

Finding an Average Rate of Change

Suppose you forgot the differentiation rules for exponential functions. Could you still answer the question in Example 9 by calculating

$$\left(\frac{1}{2}\right)^{10{,}001/5715} - \left(\frac{1}{2}\right)^{10{,}000/5715}?$$

Explain your reasoning. Illustrate each technique graphically. Which technique involves the slope of a tangent line? A secant line?

PREREQUISITE REVIEW 4.5

The following warm-up exercises involve skills that were covered in earlier sections. You will use these skills in the exercise set for this section.

In Exercises 1–6, expand the logarithmic expression.

1. $\ln(x + 1)^2$

2. $\ln x(x + 1)$

3. $\ln \dfrac{x}{x + 1}$

4. $\ln\left(\dfrac{x}{x - 3}\right)^3$

5. $\ln \dfrac{4x(x - 7)}{x^2}$

6. $\ln x^3(x + 1)$

In Exercises 7 and 8, find dy/dx implicitly.

7. $y^2 + xy = 7$

8. $x^2y - xy^2 = 3x$

In Exercises 9 and 10, find the second derivative of f.

9. $f(x) = x^2(x + 1) - 3x^3$

10. $f(x) = -\dfrac{1}{x^2}$

EXERCISES 4.5

In Exercises 1–4, find the slope of the tangent line to the graph of the function at the point $(1, 0)$.

1. $y = \ln x^3$

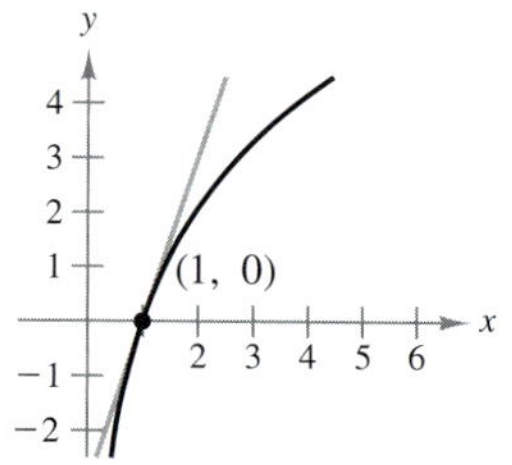

2. $y = \ln x^{5/2}$

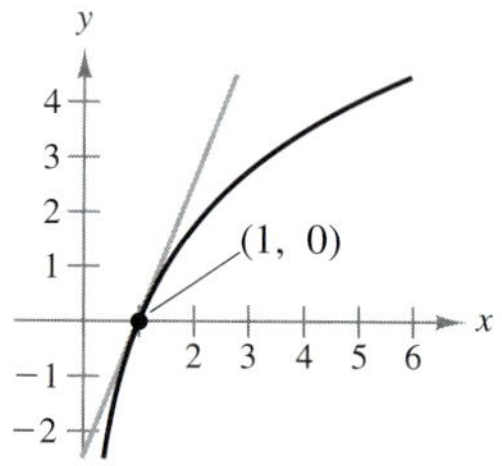

3. $y = \ln x^2$

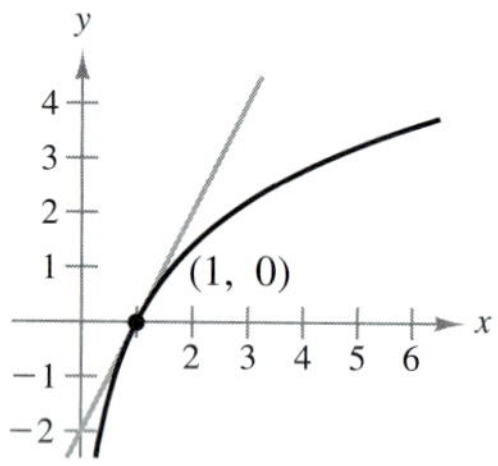

4. $y = \ln x^{1/2}$

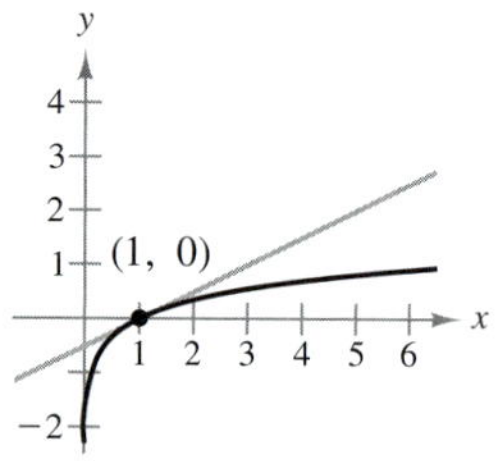

In Exercises 5–26, find the derivative of the function.

5. $y = \ln x^2$

6. $f(x) = \ln 2x$

7. $y = \ln(x^2 + 3)$

8. $f(x) = \ln(1 - x^2)$

9. $y = \ln\sqrt{x^4 - 4x}$

10. $y = \ln(1 - x)^{3/2}$

11. $y = \frac{1}{2}(\ln x)^6$

12. $y = (\ln x^2)^2$

13. $f(x) = x^2 \ln x$

14. $y = \dfrac{\ln x}{x^2}$

15. $y = \ln\left(x\sqrt{x^2 - 1}\right)$

16. $y = \ln \dfrac{x}{x^2 + 1}$

17. $y = \ln \dfrac{x}{x + 1}$

18. $y = \ln \dfrac{x^2}{x^2 + 1}$

19. $y = \ln \sqrt[3]{\dfrac{x - 1}{x + 1}}$

20. $y = \ln \sqrt{\dfrac{x + 1}{x - 1}}$

21. $y = \ln \dfrac{\sqrt{4 + x^2}}{x}$

22. $y = \ln\left(x\sqrt{4 + x^2}\right)$

23. $g(x) = e^{-x} \ln x$

24. $f(x) = x \ln e^{x^2}$

25. $g(x) = \ln \dfrac{e^x + e^{-x}}{2}$

26. $f(x) = \ln \dfrac{1 + e^x}{1 - e^x}$

In Exercises 27–30, write the expression with base e.

27. 2^x

28. 3^x

29. $\log_4 x$

30. $\log_3 x$

In Exercises 31–36, use a calculator to evaluate the logarithm. Round to three decimal places.

31. $\log_2 48$

32. $\log_5 12$

33. $\log_3 \frac{1}{2}$

34. $\log_7 \frac{2}{9}$

35. $\log_{1/5} 31$

36. $\log_{2/3} 32$

In Exercises 37–46, find the derivative of the function.

37. $y = 3^x$

38. $y = \left(\frac{1}{4}\right)^x$

39. $f(x) = \log_2 x$

40. $g(x) = \log_5 x$

41. $h(x) = 4^{2x-3}$

42. $y = 6^{5x}$

43. $y = \log_{10}(x^2 + 6x)$

44. $f(x) = 10^{x^2}$

45. $y = x2^x$

46. $y = x3^{x+1}$

In Exercises 47–50, determine an equation of the tangent line to the function at the given point.

Function	*Point*
47. $y = x \ln x$	$(1, 0)$
48. $y = \dfrac{\ln x}{x}$	$\left(e, \dfrac{1}{e}\right)$
49. $y = \log_3(3x + 7)$	$\left(\dfrac{2}{3}, 2\right)$
50. $g(x) = \log_2(3x - 1)$	$(11, 5)$

In Exercises 51–54, find dy/dx implicitly.

51. $x^2 - 3 \ln y + y^2 = 10$

52. $\ln xy + 5x = 30$

53. $4x^3 + \ln y^2 + 2y = 2x$

54. $4xy + \ln(x^2y) = 7$

In Exercises 55–58, find the second derivative of the function.

55. $f(x) = x \ln \sqrt{x} + 2x$

56. $f(x) = 3 + 2 \ln x$

57. $f(x) = 5^x$

58. $f(x) = \log_{10} x$

59. ***Sound Intensity*** The relationship between the number of decibels β and the intensity of a sound I in watts per square centimeter is given by

$$\beta = 10 \log_{10}\left(\frac{I}{10^{-16}}\right).$$

Find the rate of change in the number of decibels when the intensity is 10^{-4} watts per square centimeter.

60. ***Chemistry*** The temperatures T (°F) at which water boils at selected pressures p (pounds per square inch) can be modeled by

$$T = 87.97 + 34.96 \ln p + 7.91\sqrt{p}.$$

Find the rate of change of the temperature when the pressure is 60 pounds per square inch.

In Exercises 61–66, find the slope of the graph at the indicated point. Then write an equation of the tangent line at the point.

61. $f(x) = 1 + 2x \ln x$

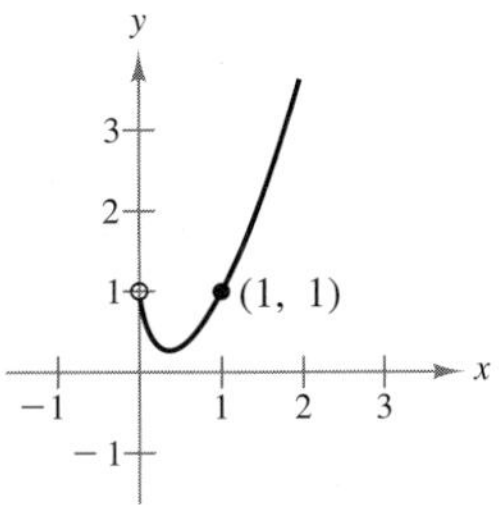

62. $f(x) = 2 \ln x^3$

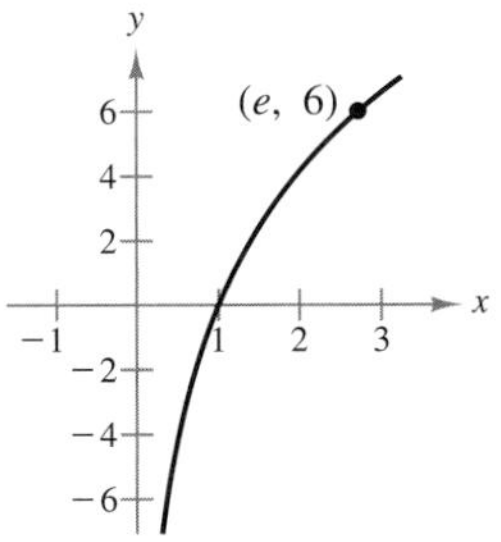

63. $f(x) = \ln \dfrac{5(x + 2)}{x}$

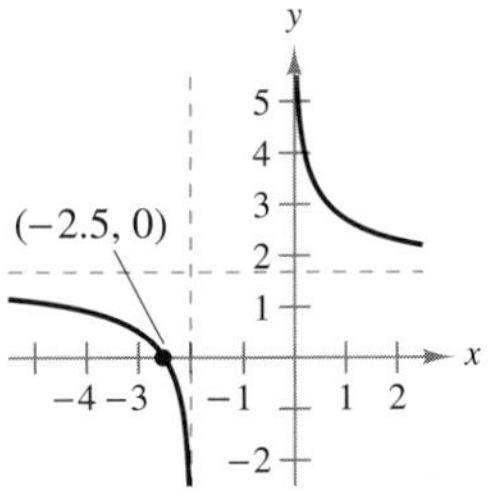

64. $f(x) = \ln\left(x\sqrt{x + 3}\right)$

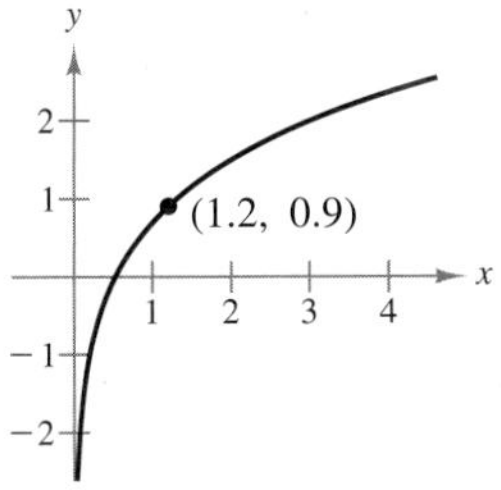

65. $f(x) = x \log_2 x$

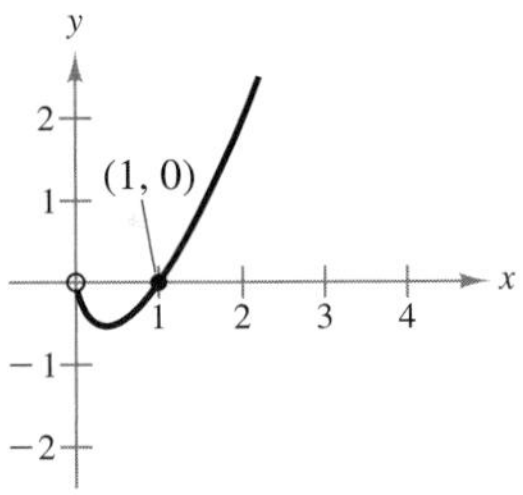

66. $f(x) = x^2 \log_3 x$

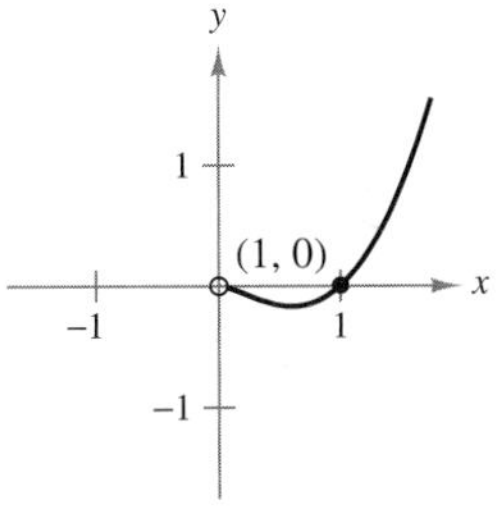

In Exercises 67–72, graph and analyze the function. Include any relative extrema and points of inflection in your analysis. Use a graphing utility to verify your results.

67. $y = x - \ln x$

68. $y = \frac{1}{2}x^2 - \ln x$

69. $y = \dfrac{\ln x}{x}$

70. $y = x \ln x$

71. $y = x^2 \ln x$

72. $y = (\ln x)^2$

In Exercises 73–76, find dx/dp for the demand function. Interpret this rate of change when the price is \$10.

73. $x = \ln\dfrac{1000}{p}$

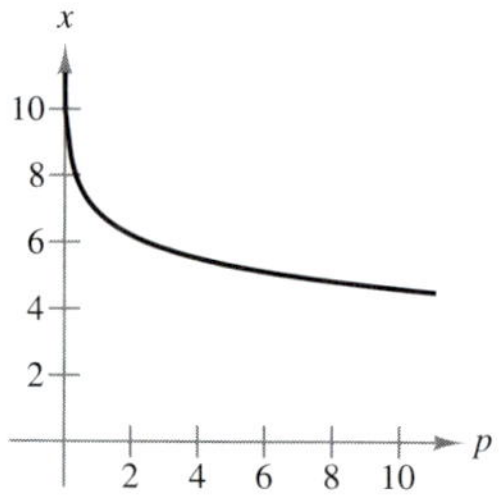

74. $x = 1000 - p \ln p$

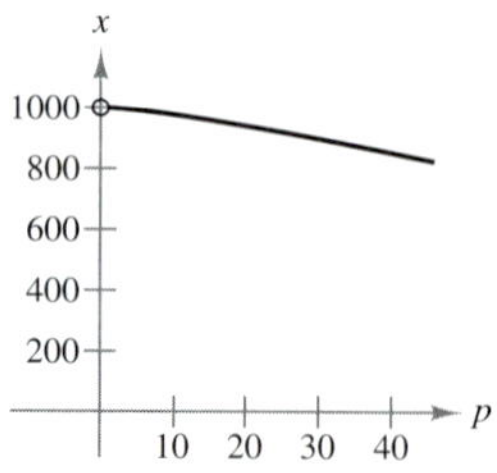

75. $x = \dfrac{500}{\ln(p^2 + 1)}$

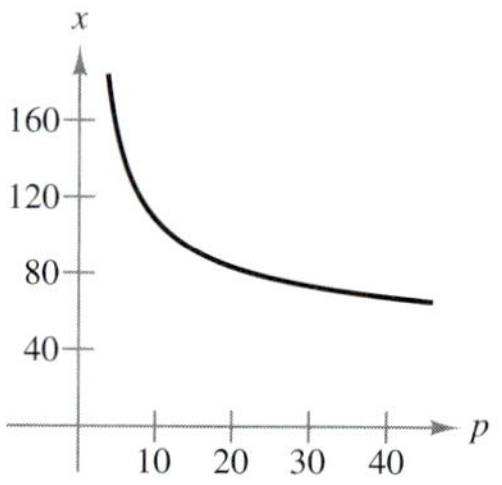

76. $x = 300 - 50 \ln(\ln p)$

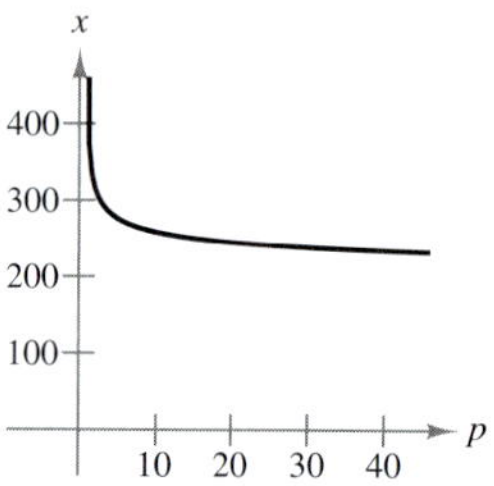

77. Solve the demand function in Exercise 73 for p. Use the result to find dp/dx. Then find the rate of change when $p = \$10$. What is the relationship between this derivative and dx/dp?

78. Solve the demand function in Exercise 75 for p. Use the result to find dp/dx. Then find the rate of change when $p = \$10$. What is the relationship between this derivative and dx/dp?

79. ***Minimum Average Cost*** The cost of producing x units of a product is modeled by

$$C = 500 + 300x - 300 \ln x, \qquad x \geq 1.$$

(a) Find the average cost function $\overline{C}$.

(b) Analytically find the minimum average cost.

(c) Use a graphing utility to confirm your results.

80. ***Minimum Average Cost*** The cost of producing x units of a product is modeled by

$$C = 100 + 25x - 120 \ln x, \quad x \geq 1.$$

(a) Find the average cost function $\overline{C}$.

(b) Analytically find the minimum average cost.

(c) Use a graphing utility to confirm your results.

81. ***Consumer Trends*** The retail sales S (in billions of dollars per year) of e-commerce companies in the United States from 1998 to 2002 can be modeled by $S = -210.3 + 103.30 \ln t$, where $t = 8$ corresponds to 1998. *(Source: Consumer Online Report)*

(a) Use a graphing utility to graph S over the interval $[8, 12]$.

(b) Estimate the amount of sales in 2000.

(c) At what rate were the sales changing in 2000?

82. ***Home Mortgage*** The term t (in years) of a \$120,000 home mortgage at 10% interest can be approximated by

$$t = \frac{5.315}{-6.7968 + \ln x}, \quad x > 1000$$

where x is the monthly payment in dollars.

(a) Use a graphing utility to graph the model.

(b) Use the model to approximate the term of a home mortgage for which the monthly payment is \$1167.41. What is the total amount paid?

(c) Use the model to approximate the term of a home mortgage for which the monthly payment is \$1068.45. What is the total amount paid?

(d) Find the instantaneous rate of change of t with respect to x when $x = 1167.41$ and $x = 1068.45$.

(e) Write a short paragraph describing the benefit of the higher monthly payment.

BUSINESS CAPSULE

Parrot Mountain Company

The Parrot Mountain Company is a mail-order business started by Angel Santiago in 1990. The company offers the famous Parrot Mountain roller skates used by bird trainers and carries more than 800 bird-related items.

83. ***Research Project*** Use your school's library, the Internet, or some other reference source to research information about a mail-order company, such as that mentioned above. Collect data about the company (sales or membership over a 20-year period, for example) and find a mathematical model to represent the data.

4.6 EXPONENTIAL GROWTH AND DECAY

- Use exponential growth and decay to model real-life situations.

Exponential Growth and Decay

In this section, you will learn to create models of *exponential growth and decay*. Real-life situations that involve exponential growth and decay deal with a substance or population whose *rate of change at any time t is proportional to the amount of the substance present at that time*. For example, the rate of decomposition of a radioactive substance is proportional to the amount of radioactive substance at a given instant. In its simplest form, this relationship is described by the equation below.

Rate of change of y — is — proportional to y.

$$\frac{dy}{dt} = ky$$

In this equation, k is a constant and y is a function of t. The solution of this equation is shown below.

Law of Exponential Growth and Decay

If y is a positive quantity whose rate of change with respect to time is proportional to the quantity present at any time t, then y is of the form

$$y = Ce^{kt}$$

where C is the **initial value** and k is the **constant of proportionality. Exponential growth** is indicated by $k > 0$ and **exponential decay** by $k < 0$.

PROOF Because the rate of change of y is proportional to y, you can write

$$\frac{dy}{dt} = ky.$$

You can see that $y = Ce^{kt}$ is a solution of this equation by differentiating to obtain $dy/dt = kCe^{kt}$ and substituting

$$\begin{aligned}\frac{dy}{dt} &= kCe^{kt}\\ &= k(Ce^{kt})\\ &= ky.\end{aligned}$$

DISCOVERY

Use a graphing utility to graph $y = Ce^{2t}$ for $C = 1, 2,$ and 5. How does the value of C affect the shape of the graph? Now graph $y = 2e^{kt}$ for $k = -2, -1, 0, 1,$ and 2. How does the value of k affect the shape of the graph? Which function grows faster, $y = e^x$ or $y = x^{10}$?

STUDY TIP

In the model $y = Ce^{kt}$, C is called the "initial value" because when $t = 0$

$$y = Ce^{k(0)} = C(1) = C.$$

Applications

Much of the cost of nuclear energy is the cost of disposing of radioactive waste. Because of the long half-life of the waste, it must be stored in containers that will remain undisturbed for thousands of years.

Radioactive decay is measured in terms of **half-life,** the number of years required for half of the atoms in a sample of radioactive material to decay. The half-lives of some common radioactive isotopes are as shown.

Uranium (^{238}U)	4,470,000,000 years
Plutonium (^{239}Pu)	24,100 years
Carbon (^{14}C)	5,715 years
Radium (^{226}Ra)	1,599 years
Einsteinium (^{254}Es)	276 days
Nobelium (^{257}No)	25 seconds

EXAMPLE 1 Modeling Radioactive Decay

A sample contains 1 gram of radium. How much radium will remain after 1000 years?

SOLUTION Let y represent the mass (in grams) of the radium in the sample. Because the rate of decay is proportional to y, you can apply the Law of Exponential Decay to conclude that y is of the form

$$y = Ce^{kt}$$

where t is the time in years. From the given information, you know that $y = 1$ when $t = 0$. Substituting these values into the model produces

$$1 = Ce^{k(0)} \qquad \text{Substitute 1 for } y \text{ and 0 for } t.$$

which implies that $C = 1$. Because radium has a half-life of 1599 years, you know that $y = \frac{1}{2}$ when $t = 1599$. Substituting these values into the model allows you to solve for k.

$$y = e^{kt} \qquad \text{Exponential decay model}$$

$$\frac{1}{2} = e^{k(1599)} \qquad \text{Substitute } \tfrac{1}{2} \text{ for } y \text{ and 1599 for } t.$$

$$\ln \frac{1}{2} = 1599k \qquad \text{Take natural log of each side.}$$

$$\frac{1}{1599} \ln \frac{1}{2} = k \qquad \text{Divide each side by 1599.}$$

So, $k \approx -0.0004335$, and the exponential decay model is

$$y = e^{-0.0004335t}.$$

To find the amount of radium remaining in the sample after 1000 years, substitute $t = 1000$ into the model. This produces

$$y = e^{-0.0004335(1000)} \approx 0.648 \text{ gram.}$$

The graph of the model is shown in Figure 4.18.

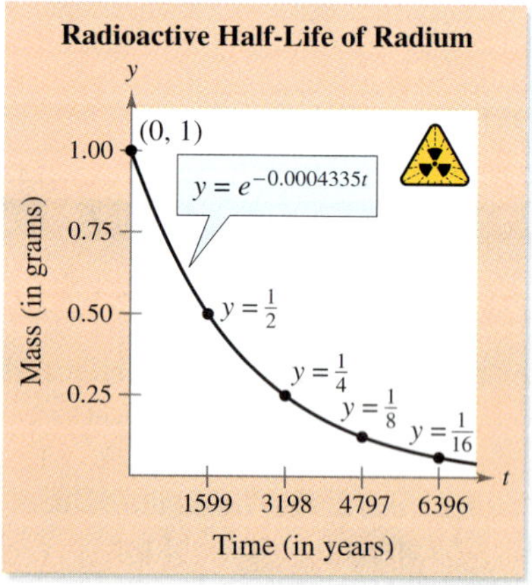

FIGURE 4.18

TRY IT 1

Use the model in Example 1 to determine the number of years required for a one-gram sample of radium to decay to 0.4 gram.

Note: Instead of approximating the value of k in Example 1, you could leave the value exact and obtain

$$y = e^{\ln[(1/2)^{(t/1599)}]} = \frac{1}{2}^{(t/1599)}.$$

This version of the model clearly shows the "half-life." When $t = 1599$, the value of y is $\frac{1}{2}$. When $t = 2(1599)$, the value of y is $\frac{1}{4}$, and so on.

The steps used in Example 1 are summarized below.

Guidelines for Modeling Exponential Growth and Decay

1. Use the given information to write *two* sets of conditions involving y and t.
2. Substitute the given conditions into the model $y = Ce^{kt}$ and use the results to solve for the constants C and k. (If one of the conditions involves $t = 0$, substitute that value first to solve for C.)
3. Use the model $y = Ce^{kt}$ to answer the question.

EXAMPLE 2 **Modeling Population Growth**

In a research experiment, a population of fruit flies is increasing in accordance with the exponential growth model. After 2 days, there are 100 flies, and after 4 days, there are 300 flies. How many flies will there be after 5 days?

SOLUTION Let y be the number of flies at time t. From the given information, you know that $y = 100$ when $t = 2$ and $y = 300$ when $t = 4$. Substituting this information into the model $y = Ce^{kt}$ produces

$$100 = Ce^{2k} \quad \text{and} \quad 300 = Ce^{4k}.$$

To solve for k, solve for C in the first equation and substitute the result into the second equation.

$$300 = Ce^{4k} \qquad \text{Second equation}$$

$$300 = \left(\frac{100}{e^{2k}}\right)e^{4k} \qquad \text{Substitute } 100/e^{2k} \text{ for } C.$$

$$\frac{300}{100} = e^{2k} \qquad \text{Divide each side by 100.}$$

$$\ln 3 = 2k \qquad \text{Take natural log of each side.}$$

$$\frac{1}{2}\ln 3 = k \qquad \text{Solve for } k.$$

Using $k = \frac{1}{2}\ln 3 \approx 0.5493$, you can determine that $C \approx 100/e^{2(0.5493)} \approx 33$. So, the exponential growth model is

$$y = 33e^{0.5493t}$$

as shown in Figure 4.19. This implies that, after 5 days, the population is

$$y = 33e^{0.5493(5)} \approx 514 \text{ flies.}$$

ALGEBRA REVIEW

For help with the algebra in Example 2, see Example 1(c) in the *Chapter 4 Algebra Review* on page 308.

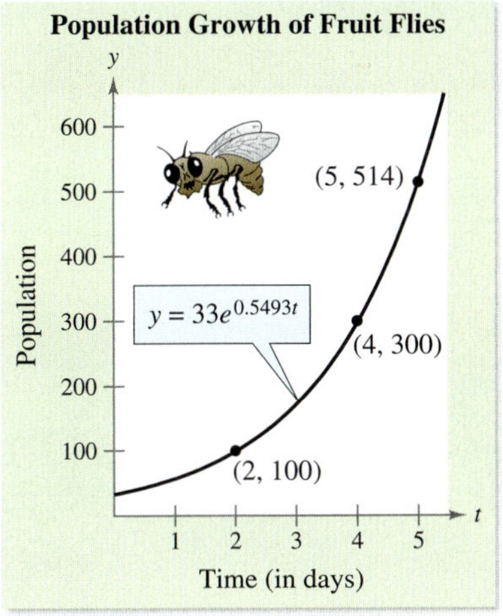

FIGURE 4.19

TRY IT 2

Find the exponential growth model if a population of fruit flies is 100 after 2 days and 400 after 4 days.

EXAMPLE 3 Modeling Compound Interest

Money is deposited in an account for which the interest is compounded continuously. The balance in the account doubles in 6 years. What is the annual interest rate?

SOLUTION The balance A in an account with continuously compounded interest is given by the exponential growth model

$$A = Pe^{rt} \qquad \text{Exponential growth model}$$

where P is the original deposit, r is the annual interest rate (in decimal form), and t is the time (in years). From the given information, you know that $A = 2P$ when $t = 6$, as shown in Figure 4.20. Use this information to solve for r.

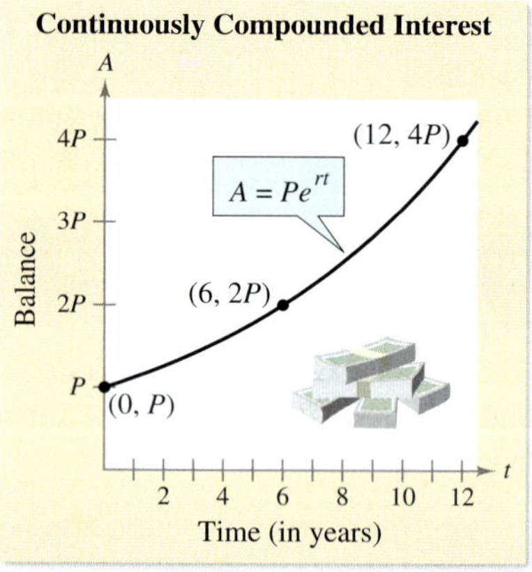

FIGURE 4.20

$$A = Pe^{rt} \qquad \text{Exponential growth model}$$

$$2P = Pe^{r(6)} \qquad \text{Substitute } 2P \text{ for } A \text{ and 6 for } t.$$

$$2 = e^{6r} \qquad \text{Divide each side by } P.$$

$$\ln 2 = 6r \qquad \text{Take natural log of each side.}$$

$$\frac{1}{6}\ln 2 = r \qquad \text{Divide each side by 6.}$$

So, the annual interest rate is

$$r = \frac{1}{6}\ln 2$$

$$\approx 0.1155$$

or about 11.55%.

TRY IT 3

Find the annual interest rate if the balance in an account doubles in 8 years where the interest is compounded continuously.

Each of the examples in this section uses the exponential growth model in which the base is e. Exponential growth, however, can be modeled with *any* base. That is, the model

$$y = Ca^{bt}$$

also represents exponential growth. (To see this, note that the model can be written in the form $y = Ce^{(\ln a)bt}$.) In some real-life settings, bases other than e are more convenient. For instance, in Example 1, knowing that the half-life of radium is 1599 years, you can immediately write the exponential decay model as

$$y = \left(\frac{1}{2}\right)^{t/1599}.$$

Using this model, the amount of radium left in the sample after 1000 years is

$$y = \left(\frac{1}{2}\right)^{1000/1599} \approx 0.648 \text{ gram}$$

which is the same answer obtained in Example 1.

STUDY TIP

Can you see why you can immediately write the model $y = \left(\frac{1}{2}\right)^{t/1599}$ for the radioactive decay described in Example 1? Notice that when $t = 1599$, the value of y is $\frac{1}{2}$, when $t = 3198$, the value of y is $\frac{1}{4}$, and so on.

TECHNOLOGY

Fitting an Exponential Model to Data

Most graphing utilities have programs that allow you to find the *least squares regression exponential model* for data. Depending on the type of graphing utility, the program may fit the data to a model of the form

$y = ab^x$ Exponential model with base b

or

$y = ae^{bx}$. Exponential model with base e

To see how to use such a program, consider the example below.

The cash flow per share y for Harley-Davidson, Inc. from 1995 through 2002 is shown in the table. *(Source: Harley-Davidson, Inc.)*

x	5	6	7	8	9	10	11	12
y	\$0.51	\$0.66	\$0.80	\$0.98	\$1.26	\$1.59	\$1.95	\$2.50

In the table, $x = 5$ corresponds to 1995. To fit an exponential model to these data, enter the coordinates listed below into the statistical data bank of a graphing utility.

(5, 0.51), (6, 0.66), (7, 0.80), (8, 0.98)

(9, 1.26), (10, 1.59), (11, 1.95), (12, 2.50)

After running the exponential regression program with a graphing utility that uses the model $y = ab^x$, the display should read $a \approx 0.167$ and $b \approx 1.252$. (The correlation of $r \approx 0.9996$ tells you that the fit is very good.) So, a model for the data is

$y = 0.167(1.252)^x$. Exponential model with base b

If you use a graphing utility that uses the model $y = ae^{bx}$, the display should read $a \approx 0.167$ and $b \approx 0.225$. The corresponding model is

$y = 0.167e^{0.225x}$. Exponential model with base e

The graph of the second model is shown at the right. Notice that one way to interpret the model is that the cash flow per share increased by about 22.5% each year from 1995 through 2002.

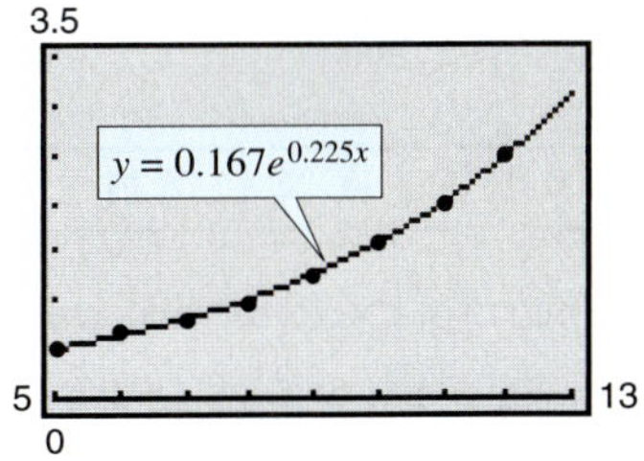

You can use either model to predict the cash flow per share in future years. For instance, in 2003 ($x = 13$), the cash flow per share is predicted to be

$$y = 0.167e^{(0.225)(13)}$$
$$\approx \$3.11$$

In 2003, *Value Line* predicted the cash flow for 2003 to be \$3.15.

ALGEBRA REVIEW

For help with the algebra in Example 4, see Example 1(b) in the *Chapter 4 Algebra Review* on page 308.

EXAMPLE 4 Modeling Sales

Four months after discontinuing advertising on national television, a manufacturer notices that sales have dropped from 100,000 MP3 players per month to 80,000 MP3 players. If the sales follow an exponential pattern of decline, what will they be after another 4 months?

SOLUTION Let y represent the number of MP3 players, let t represent the time (in months), and consider the exponential decay model

$$y = Ce^{kt}.\qquad \text{Exponential decay model}$$

From the given information, you know that $y = 100{,}000$ when $t = 0$. Using this information, you have

$$100{,}000 = Ce^0$$

which implies that $C = 100{,}000$. To solve for k, use the fact that $y = 80{,}000$ when $t = 4$.

$$y = 100{,}000e^{kt}\qquad \text{Exponential decay model}$$
$$80{,}000 = 100{,}000e^{k(4)}\qquad \text{Substitute 80,000 for } y \text{ and 4 for } t.$$
$$0.8 = e^{4k}\qquad \text{Divide each side by 100,000.}$$
$$\ln 0.8 = 4k\qquad \text{Take natural log of each side.}$$
$$\frac{1}{4}\ln 0.8 = k\qquad \text{Divide each side by 4.}$$

So, $k = \frac{1}{4}\ln 0.8 \approx -0.0558$, which means that the model is

$$y = 100{,}000e^{-0.0558t}.$$

After four more months ($t = 8$), you can expect sales to drop to

$$y = 100{,}000e^{-0.0558(8)}$$
$$\approx 64{,}000 \text{ MP3 players}$$

as shown in Figure 4.21.

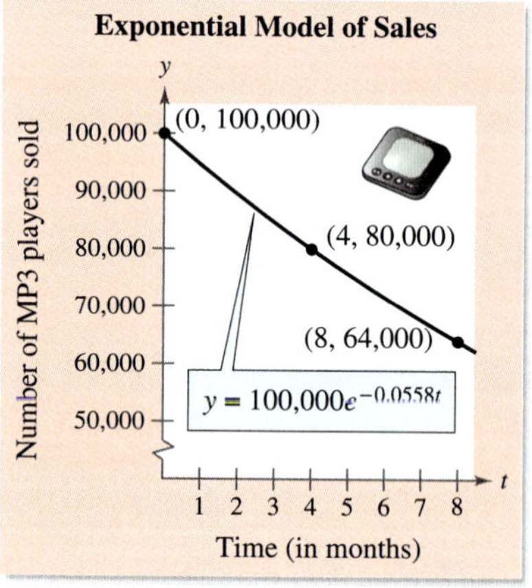

FIGURE 4.21

TRY IT 4

Use the model in Example 4 to determine when sales drop to 50,000 MP3 players.

TAKE ANOTHER LOOK

Comparing Exponential and Linear Models

In Example 4, it was assumed that the sales were following an exponential pattern of decline. This implies that the sales were dropping by the same *percent* each month. Suppose instead that the sales were following a linear pattern of decline (dropping by the same *amount* each month). How would this affect the answer to the question given in Example 4? Which of the two models do you think might be more realistic? Explain your reasoning.

PREREQUISITE REVIEW 4.6

The following warm-up exercises involve skills that were covered in earlier sections. You will use these skills in the exercise set for this section.

In Exercises 1–4, solve the equation for k.

1. $12 = 24e^{4k}$

2. $10 = 3e^{5k}$

3. $25 = 16e^{-0.01k}$

4. $22 = 32e^{-0.02k}$

In Exercises 5–8, find the derivative of the function.

5. $y = 32e^{0.23t}$

6. $y = 18e^{0.072t}$

7. $y = 24e^{-1.4t}$

8. $y = 25e^{-0.001t}$

In Exercises 9–12, simplify the expression.

9. $e^{\ln 4}$

10. $4e^{\ln 3}$

11. $e^{\ln(2x+1)}$

12. $e^{\ln(x^2+1)}$

EXERCISES 4.6

In Exercises 1–6, find the exponential function $y = Ce^{kt}$ that passes through the two given points.

1. $y = Ce^{kt}$

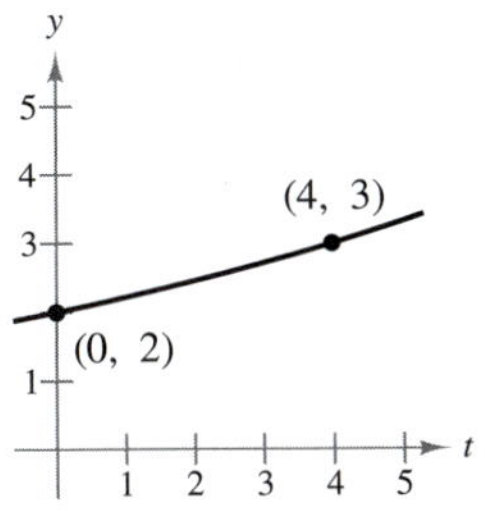

2. $y = Ce^{kt}$

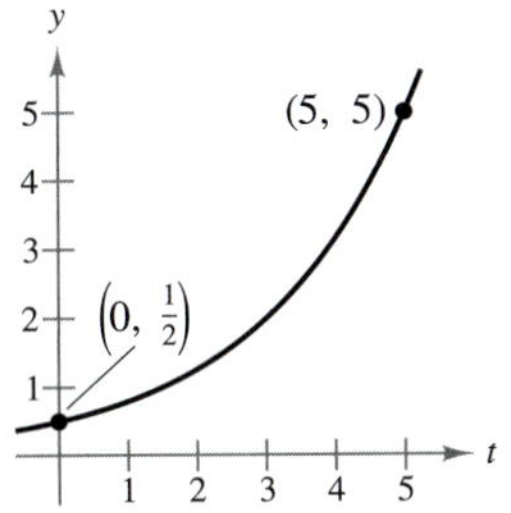

3. $y = Ce^{kt}$

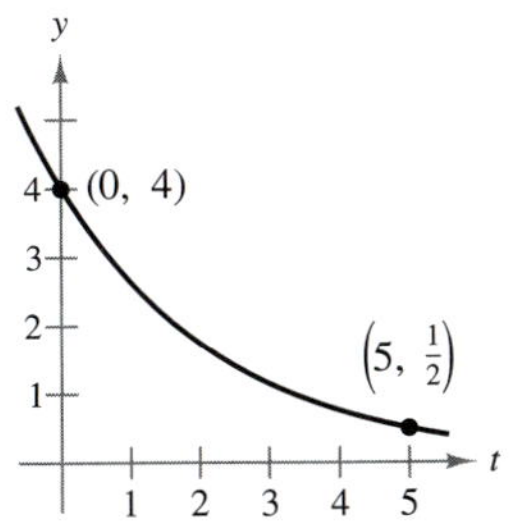

4. $y = Ce^{kt}$

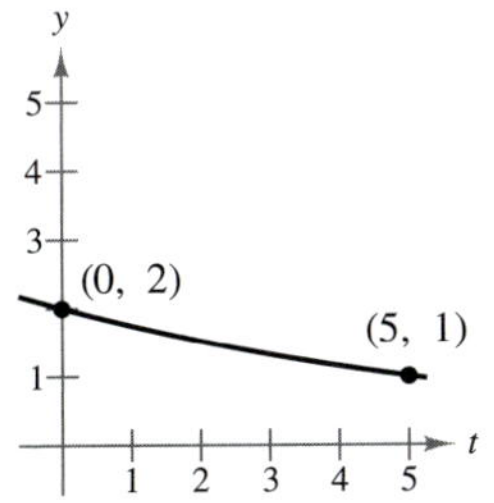

5. $y = Ce^{kt}$

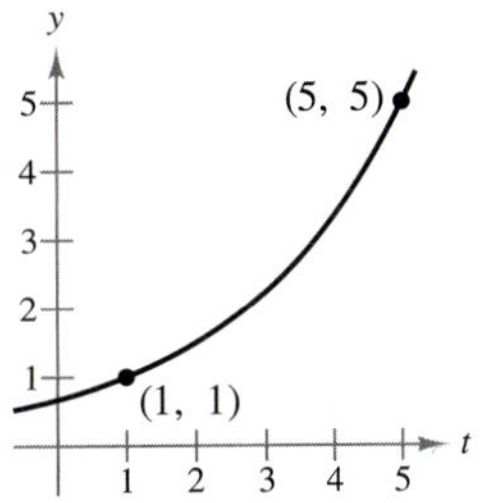

6. $y = Ce^{kt}$

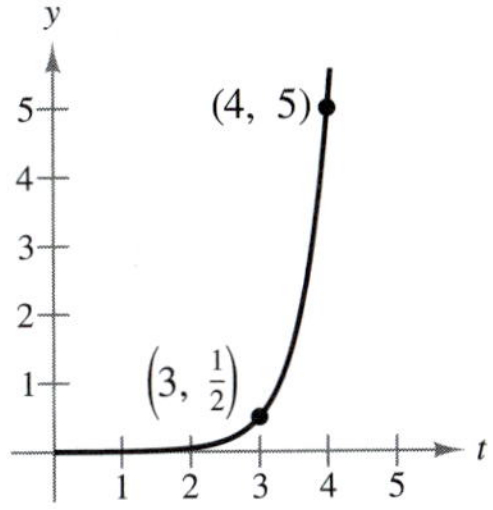

In Exercises 7–10, use the given information to write an equation for y. Confirm your result analytically by showing that the function satisfies the equation $dy/dt = Cy$. Does the function represent exponential growth or exponential decay?

7. $\frac{dy}{dt} = 2y, \quad y = 10$ when $t = 0$

8. $\frac{dy}{dt} = -\frac{2}{3}y, \quad y = 20$ when $t = 0$

9. $\frac{dy}{dt} = -4y, \quad y = 30$ when $t = 0$

10. $\frac{dy}{dt} = 5.2y, \quad y = 18$ when $t = 0$

Radioactive Decay In Exercises 11–16, complete the table for each radioactive isotope.

	Isotope	Half-life (in years)	Initial quantity	Amount after 1000 years	Amount after 10,000 years
11.	^{226}Ra	1599	10 grams		
12.	^{226}Ra	1599		1.5 grams	
13.	^{14}C	5715			2 grams
14.	^{14}C	5715	3 grams		
15.	^{239}Pu	24,100		2.1 grams	
16.	^{239}Pu	24,100			0.4 gram

17. ***Radioactive Decay*** What percent of a present amount of radioactive radium (^{226}Ra) will remain after 900 years?

18. ***Radioactive Decay*** Find the half-life of a radioactive material if after 1 year 99.57% of the initial amount remains.

19. ***Carbon Dating*** ^{14}C dating assumes that the carbon dioxide on the Earth today has the same radioactive content as it did centuries ago. If this is true, then the amount of ^{14}C absorbed by a tree that grew several centuries ago should be the same as the amount of ^{14}C absorbed by a similar tree today. A piece of ancient charcoal contains only 15% as much of the radioactive carbon as a piece of modern charcoal. How long ago was the tree burned to make the ancient charcoal? (The half-life of ^{14}C is 5715 years.)

20. ***Carbon Dating*** Repeat Exercise 19 for a piece of charcoal that contains 30% as much radioactive carbon as a modern piece.

In Exercises 21 and 22, find exponential models

$$y_1 = Ce^{k_1 t} \quad \text{and} \quad y_2 = C(2)^{k_2 t}$$

that pass through the points. Compare the values of k_1 and k_2. Briefly explain your results.

21. $(0, 5), (12, 20)$

22. $(0, 8), \left(20, \frac{1}{2}\right)$

23. ***Population Growth*** The number of a certain type of bacteria increases continuously at a rate proportional to the number present. There are 150 present at a given time and 450 present 5 hours later.

(a) How many will there be 10 hours after the initial time?

(b) How long will it take for the population to double?

(c) Does the answer to part (b) depend on the starting time? Explain your reasoning.

24. ***School Enrollment*** In 1960, the total enrollment in public universities and colleges in the United States was 2.3 million students. By 2000, enrollment had risen to 12.0 million students. Assume enrollment can be modeled by exponential growth. *(Source: U.S. Census Bureau)*

(a) Estimate the total enrollments in 1970, 1980, and 1990.

(b) How many years until the enrollment doubles from the 2000 figure?

(c) By what percent is the enrollment increasing each year?

Compound Interest In Exercises 25–30, complete the table for an account in which interest is compounded continuously.

	Initial investment	Annual rate	Time to double	Amount after 10 years	Amount after 25 years
25.	\$1,000	12%			
26.	\$20,000	$10\frac{1}{2}$%			
27.	\$750		$7\frac{3}{4}$ years		
28.	\$10,000		5 years		
29.	\$500			\$1292.85	
30.	\$2,000				\$6008.33

31. ***Effective Yield*** The effective yield is the annual rate i that will produce the same interest per year as the nominal rate r compounded n times per year.

(a) For a rate r that is compounded n times per year, show that the effective yield is

$$i = \left(1 + \frac{r}{n}\right)^n - 1.$$

(b) Find the effective yield for a nominal rate of 6%, compounded monthly.

32. ***Effective Yield*** The effective yield is the annual rate i that will produce the same interest per year as the nominal rate r.

(a) For a rate r that is compounded continuously, show that the effective yield is $i = e^r - 1$.

(b) Find the effective yield for a nominal rate of 6%, compounded continuously.

Effective Yield In Exercises 33 and 34, use the results of Exercises 31 and 32 to complete the table showing the effective yield for a nominal rate of r.

Number of compoundings per year	4	12	365	Continuous
Effective yield				

33. $r = 5\%$

34. $r = 7\frac{1}{2}\%$

35. ***Investment: Rule of 70*** Verify that the time necessary for an investment to double its value is approximately $70/r$, where r is the annual interest rate entered as a percent.

36. ***Investment: Rule of 70*** Use the Rule of 70 from Exercise 35 to approximate the times necessary for an investment to double in value if (a) $r = 10\%$ and (b) $r = 7\%$.

37. ***Revenue*** The revenues for Sonic Corporation were \$83.3 million in 1993 and \$446.6 million in 2003. *(Source: Sonic Corporation)*

(a) Use an exponential growth model to estimate the revenue in 2008.

(b) Use a linear model to estimate the 2008 revenue.

(c) Use a graphing utility to graph the models from parts (a) and (b). Which model is more accurate?

38. ***Sales*** The sales (in millions of dollars) for in-line skating and wheel sports in the United States were \$150 million in 1990 and \$1074 million in 2000. *(Source: National Sporting Goods Association)*

(a) Use the *regression* feature of a graphing utility to find an exponential growth model and a linear model for the data.

(b) Use the exponential growth model to estimate the sales in 2006.

(c) Use the linear model to estimate the sales in 2006.

(d) Use a graphing utility to graph the models from part (a). Which model is more accurate?

39. ***Sales*** The cumulative sales S (in thousands of units) of a new product after it has been on the market for t years are modeled by

$$S = Ce^{k/t}.$$

During the first year, 5000 units were sold. The saturation point for the market is 30,000 units. That is, the limit of S as $t \to \infty$ is 30,000.

(a) Solve for C and k in the model.

(b) How many units will be sold after 5 years?

(c) Use a graphing utility to graph the sales function.

40. ***Sales*** The cumulative sales S (in thousands of units) of a new product after it has been on the market for t years are modeled by

$$S = 30(1 - 3^{kt}).$$

During the first year, 5000 units were sold.

(a) Solve for k in the model.

(b) What is the saturation point for this product?

(c) How many units will be sold after 5 years?

(d) Use a graphing utility to graph the sales function.

41. ***Learning Curve*** The management of a factory finds that the maximum number of units a worker can produce in a day is 30. The learning curve for the number of units N produced per day after a new employee has worked t days is modeled by $N = 30(1 - e^{kt})$. After 20 days on the job, a worker is producing 19 units in a day. How many days should pass before this worker is producing 25 units per day?

42. ***Learning Curve*** The management in Exercise 41 requires that a new employee be producing at least 20 units per day after 30 days on the job.

(a) Find a learning curve model that describes this minimum requirement.

(b) Find the number of days before a minimal achiever is producing 25 units per day.

43. ***Profit*** Because of a slump in the economy, a company finds that its annual profits have dropped from \$742,000 in 1998 to \$632,000 in 2000. If the profit follows an exponential pattern of decline, what is the expected profit for 2003? (Let $t = 0$ correspond to 1998.)

44. ***Revenue*** A small business assumes that the demand function for one of its new products can be modeled by $p = Ce^{kx}$. When $p = \$45$, $x = 1000$ units, and when $p = \$40$, $x = 1200$ units.

(a) Solve for C and k.

(b) Find the values of x and p that will maximize the revenue for this product.

45. ***Revenue*** Repeat Exercise 44 given that when $p = \$5$, $x = 300$ units, and when $p = \$4$, $x = 400$ units.

46. ***Forestry*** The value V (in dollars) of a tract of timber can be modeled by $V = 100{,}000e^{0.75\sqrt{t}}$, where $t = 0$ corresponds to 1990. If money earns interest at a rate of 4%, compounded continuously, then the present value A of the timber at any time t is $A = Ve^{-0.04t}$. Find the year in which the timber should be harvested to maximize the present value.

47. ***Forestry*** Repeat Exercise 46 using the model

$$V = 100{,}000e^{0.6\sqrt{t}}.$$

48. ***Earthquake Intensity*** On the Richter scale, the magnitude R of an earthquake of intensity I is given by

$$R = \frac{\ln I - \ln I_0}{\ln 10}$$

where I_0 is the minimum intensity used for comparison. Assume $I_0 = 1$.

(a) Find the intensity of the 1906 San Francisco earthquake in which $R = 8.3$.

(b) Find the factor by which the intensity is increased when the value of R is doubled.

(c) Find dR/dI.

ALGEBRA REVIEW

Solving Exponential and Logarithmic Equations

To find the extrema or points of inflection of an exponential or logarithmic function, you must know how to solve exponential and logarithmic equations. A few examples are given on page 285. Some additional examples are presented in this Algebra Review.

As with all equations, remember that your basic goal is to isolate the variable on one side of the equation. To do this, you use inverse operations. For instance, to get rid of an exponential expression such as e^{2x}, take the natural log of each side and use the property $\ln e^{2x} = 2x$. Similarly, to get rid of a logarithmic expression such as $\log_2 3x$, exponentiate each side and use the property $2^{\log_2 3x} = 3x$.

EXAMPLE 1 Solving Exponential Equations

Solve each exponential equation.

(a) $25 = 5e^{7t}$ (b) $80{,}000 = 100{,}000e^{k(4)}$ (c) $300 = \left(\dfrac{100}{e^{2k}}\right)e^{4k}$

SOLUTION

(a)

$25 = 5e^{7t}$	Write original equation.
$5 = e^{7t}$	Divide each side by 5.
$\ln 5 = \ln e^{7t}$	Take natural log of each side.
$\ln 5 = 7t$	Apply the property $\ln e^a = a$.
$\frac{1}{7}\ln 5 = t$	Divide each side by 7.

(b)

$80{,}000 = 100{,}000e^{k(4)}$	Example 4, page 304
$0.8 = e^{4k}$	Divide each side by 100,000.
$\ln 0.8 = \ln e^{4k}$	Take natural log of each side.
$\ln 0.8 = 4k$	Apply the property $\ln e^a = a$.
$\frac{1}{4}\ln 0.8 = k$	Divide each side by 4.

(c)

$300 = \left(\dfrac{100}{e^{2k}}\right)e^{4k}$	Example 2, page 301
$300 = (100)\dfrac{e^{4k}}{e^{2k}}$	Rewrite product.
$300 = 100e^{4k-2k}$	To divide powers, subtract exponents.
$300 = 100e^{2k}$	Simplify.
$3 = e^{2k}$	Divide each side by 100.
$\ln 3 = \ln e^{2k}$	Take natural log of each side.
$\ln 3 = 2k$	Apply the property $\ln e^a = a$.
$\frac{1}{2}\ln 3 = k$	Divide each side by 2.

EXAMPLE 2 Solving Logarithmic Equations

Solve each logarithmic equation.

(a) $\ln x = 2$ (b) $5 + 2 \ln x = 4$

(c) $2 \ln 3x = 4$ (d) $\ln x - \ln(x - 1) = 1$

SOLUTION

(a)

$\ln x = 2$	Write original equation.
$e^{\ln x} = e^2$	Exponentiate each side.
$x = e^2$	Apply the property $e^{\ln a} = a$.

(b)

$5 + 2 \ln x = 4$	Write original equation.
$2 \ln x = -1$	Subtract 5 from each side.
$\ln x = -\frac{1}{2}$	Divide each side by 2.
$e^{\ln x} = e^{-1/2}$	Exponentiate each side.
$x = e^{-1/2}$	Apply the property $e^{\ln a} = a$.

(c)

$2 \ln 3x = 4$	Write original equation.
$\ln 3x = 2$	Divide each side by 2.
$e^{\ln 3x} = e^2$	Exponentiate each side.
$3x = e^2$	Apply the property $e^{\ln a} = a$.
$x = \frac{1}{3}e^2$	Divide each side by 3.

(d)

$\ln x - \ln(x - 1) = 1$	Write original equation.
$\ln \frac{x}{x - 1} = 1$	$\ln m - \ln n = \ln(m/n)$
$e^{\ln(x/x-1)} = e^1$	Exponentiate each side.
$\frac{x}{x - 1} = e^1$	Apply the property $e^{\ln a} = a$.
$x = ex - e$	Multiply each side by $x - 1$.
$x - ex = -e$	Subtract ex from each side.
$x(1 - e) = -e$	Factor.
$x = \frac{-e}{1 - e}$	Divide each side by $1 - e$.
$x = \frac{e}{e - 1}$	Simplify.

STUDY TIP

Because the domain of a logarithmic function generally does not include all real numbers, be sure to check for extraneous solutions.

4 CHAPTER SUMMARY AND STUDY STRATEGIES

*After studying this chapter, you should have acquired the following skills. The exercise numbers are keyed to the Review Exercises that begin on page 312. Answers to odd-numbered Review Exercises are given in the back of the text.**

- Use the properties of exponents to evaluate and simplify exponential expressions. *(Section 4.1 and Section 4.2)* — *Review Exercises 1–16*

  $$a^0 = 1, \quad a^x a^y = a^{x+y}, \quad \frac{a^x}{a^y} = a^{x-y}, \quad (a^x)^y = a^{xy}$$

  $$(ab)^x = a^x b^x, \quad \left(\frac{a}{b}\right)^x = \frac{a^x}{b^x}, \quad a^{-x} = \frac{1}{a^x}$$

- Use properties of exponents to answer questions about real life. *(Section 4.1)* — *Review Exercises 17, 18*
- Sketch the graphs of exponential functions. *(Section 4.1 and Section 4.2)* — *Review Exercises 19–28*
- Evaluate limits of exponential functions in real life. *(Section 4.2)* — *Review Exercises 29, 30*
- Evaluate and graph functions involving the natural exponential function. *(Section 4.2)* — *Review Exercises 31–34*
- Graph logistic growth functions. *(Section 4.2)* — *Review Exercises 35, 36*
- Solve compound interest problems. *(Section 4.2)* — *Review Exercises 37–40*

  $$A = P(1 + r/n)^{nt}, \quad A = Pe^{rt}$$

- Solve effective rate of interest problems. *(Section 4.2)* — *Review Exercises 41, 42*

  $$r_{eff} = (1 + r/n)^n - 1$$

- Solve present value problems. *(Section 4.2)* — *Review Exercises 43, 44*

  $$P = \frac{A}{(1 + r/n)^{nt}}$$

- Answer questions involving the natural exponential function as a real-life model. *(Section 4.2)* — *Review Exercises 45, 46*
- Find the derivatives of natural exponential functions. *(Section 4.3)* — *Review Exercises 47–54*

  $$\frac{d}{dx}[e^x] = e^x, \quad \frac{d}{dx}[e^u] = e^u \frac{du}{dx}$$

- Use calculus to analyze the graphs of functions that involve the natural exponential function. *(Section 4.3)* — *Review Exercises 55–62*
- Use the definition of the natural logarithmic function to write exponential equations in logarithmic form, and vice versa. *(Section 4.4)* — *Review Exercises 63–66*

  $\ln x = b$ if and only if $e^b = x$.

- Sketch the graphs of natural logarithmic functions. *(Section 4.4)* — *Review Exercises 67–70*
- Use properties of logarithms to expand and condense logarithmic expressions. *(Section 4.4)* — *Review Exercises 71–76*

  $$\ln xy = \ln x + \ln y, \quad \ln \frac{x}{y} = \ln x - \ln y, \quad \ln x^n = n \ln x$$

- Use inverse properties of exponential and logarithmic functions to solve exponential and logarithmic equations. *(Section 4.4)* — *Review Exercises 77–92*

  $$\ln e^x = x, \quad e^{\ln x} = x$$

* Use a wide range of valuable study aids to help you master the material in this chapter. The *Student Solutions Guide* includes step-by-step solutions to all odd-numbered exercises to help you review and prepare. The *HM mathSpace® Student CD-ROM* helps you brush up on your algebra skills. The *Graphing Technology Guide*, available on the Web at *math.college.hmco.com/students*, offers step-by-step commands and instructions for a wide variety of graphing calculators, including the most recent models.

- Use properties of natural logarithms to answer questions about real life. *(Section 4.4)* — *Review Exercises 93, 94*
- Find the derivatives of natural logarithmic functions. *(Section 4.5)* — *Review Exercises 95–108*

  $$\frac{d}{dx}[\ln x] = \frac{1}{x}, \quad \frac{d}{dx}[\ln u] = \frac{1}{u}\frac{du}{dx}$$

- Use calculus to analyze the graphs of functions that involve the natural logarithmic function. *(Section 4.5)* — *Review Exercises 109–112*
- Use the definition of logarithms to evaluate logarithmic expressions involving other bases. *(Section 4.5)* — *Review Exercises 113–116*

  $\log_a x = b$ if and only if $a^b = x$

- Use the change-of-base formula to evaluate logarithmic expressions involving other bases. *(Section 4.5)* — *Review Exercises 117–120*

  $$\log_a x = \frac{\ln x}{\ln a}$$

- Find the derivatives of exponential and logarithmic functions involving other bases. *(Section 4.5)* — *Review Exercises 121–124*

  $$\frac{d}{dx}[a^x] = (\ln a)a^x, \quad \frac{d}{dx}[a^u] = (\ln a)a^u\frac{du}{dx}$$

  $$\frac{d}{dx}[\log_a x] = \left(\frac{1}{\ln a}\right)\frac{1}{x}, \quad \frac{d}{dx}[\log_a u] = \left(\frac{1}{\ln a}\right)\left(\frac{1}{u}\right)\frac{du}{dx}$$

- Use calculus to answer questions about real-life rates of change. *(Section 4.5)* — *Review Exercises 125, 126*
- Use exponential growth and decay to model real-life situations. *(Section 4.6)* — *Review Exercises 127–132*
- ***Classifying Differentiation Rules*** Differentiation rules fall into two basic classes: (1) general rules that apply to all differentiable functions; and (2) specific rules that apply to special types of functions. At this point in the course, you have studied six general rules: the Constant Rule, the Constant Multiple Rule, the Sum Rule, the Difference Rule, the Product Rule, and the Quotient Rule. Although these rules were introduced in the context of algebraic functions, remember that they can also be used with exponential and logarithmic functions. You have also studied three specific rules: the Power Rule, the derivative of the natural exponential function, and the derivative of the natural logarithmic function. Each of these rules comes in two forms: the "simple" version, such as $D_x[e^x] = e^x$, and the Chain Rule version, such as $D_x[e^u] = e^u(du/dx)$.
- ***To Memorize or Not to Memorize?*** When studying mathematics, you need to memorize some formulas and rules. Much of this will come from practice—the formulas that you use most often will be committed to memory. Some formulas, however, are used only infrequently. With these, it is helpful to be able to *derive* the formula from a *known* formula. For instance, knowing the Log Rule for differentiation and the change-of-base formula, $\log_a x = (\ln x)/(\ln a)$, allows you to derive the formula for the derivative of a logarithmic function to base a.

Study Tools *Additional resources that accompany this chapter*

- **Algebra Review** (pages 308 and 309)
- **Chapter Summary and Study Strategies** (pages 310 and 311)
- **Review Exercises** (pages 312–315)
- **Sample Post-Graduation Exam Questions** (page 316)
- **Web Exercises** (page 289, Exercise 80; page 298, Exercise 83)
- **Student Solutions Guide**
- **HM mathSpace® Student CD-ROM**
- **Graphing Technology Guide** *(math.college.hmco.com/students)*

4 CHAPTER REVIEW EXERCISES

In Exercises 1–4, evaluate the expression.

1. $32^{3/5}$

2. $16^{3/2}$

3. $\left(\frac{1}{25}\right)^{-3/2}$

4. $\left(\frac{27}{8}\right)^{-1/3}$

In Exercises 5–8, use the properties of exponents to simplify the expression.

5. $\left(\frac{25}{4}\right)^{0}$

6. $(9^{1/3})(3^{1/3})$

7. $\dfrac{6^3}{36^2}$

8. $\dfrac{1}{4}\left(\dfrac{1}{2}\right)^{-3}$

In Exercises 9–16, solve the equation for *x*.

9. $5^x = 625$

10. $32^{-x} = 2^{3x+1}$

11. $x^{3/4} = 8$

12. $x^{5/2} = 243$

13. $e^{-1/2} = e^{x-1}$

14. $e^{-5} = e^{2x+1}$

15. $\dfrac{x^3}{3} = e^3$

16. $4x^2 = e^5$

17. ***Revenue*** The gross revenues R (in millions of dollars) of symphony orchestras in the United States from 1995 through 2001 can be modeled by

$$R = 378.914(1.067)^t$$

where $t = 5$ corresponds to 1995. *(Source: American Symphony Orchestra League, Inc.)*

(a) Use this model to find the gross revenues in 1995, 1998, and 2001.

(b) Do you think the model will be valid for years beyond 2001? Explain your reasoning.

18. ***Profit*** The net profits P (in millions of dollars) for California Pizza Kitchen from 1999 through 2003 can be modeled by

$$P = 36.182(1.193)^t$$

where $t = 9$ corresponds to 1999. *(Source: California Pizza Kitchen, Inc.)*

(a) Use this model to estimate the net profits in 1999, 2001, and 2003.

(b) Do you think the model will be valid for years beyond 2003? Explain your reasoning.

In Exercises 19–28, sketch the graph of the function.

19. $f(x) = 9^{x/2}$

20. $g(x) = 16^{3x/2}$

21. $f(t) = \left(\frac{1}{6}\right)^t$

22. $g(t) = \left(\frac{1}{3}\right)^{-t}$

23. $f(x) = \left(\frac{1}{2}\right)^{2x} + 4$

24. $g(x) = \left(\frac{2}{3}\right)^{2x} + 1$

25. $f(x) = e^{-x} + 1$

26. $g(x) = e^{2x} - 1$

27. $f(x) = 1 - e^x$

28. $g(x) = 2 + e^{x-1}$

29. ***Demand*** The demand function for a product is given by

$$p = 12{,}500 - \frac{10{,}000}{2 + e^{-0.001x}}$$

where p is the price per unit and x is the number of units produced (see figure). What is the limit of the price as x increases without bound? Explain what this means in the context of the problem.

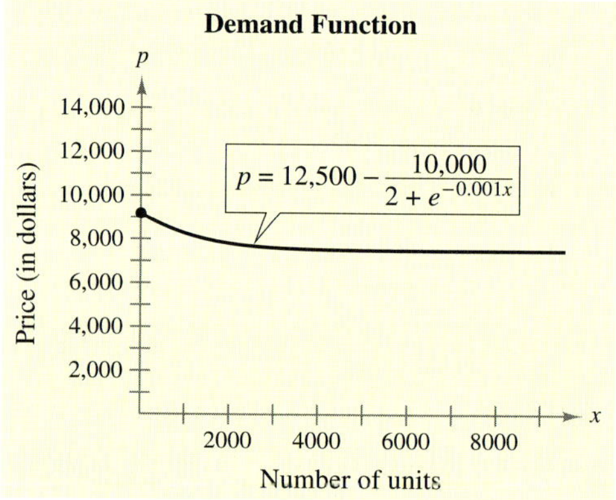

30. ***Biology: Endangered Species*** Biologists consider a species of a plant or animal to be endangered if it is expected to become extinct in less than 20 years. The population y of a certain species is modeled by

$$y = 1096e^{-0.39t}$$

(see figure). Is this species endangered? Explain your reasoning.

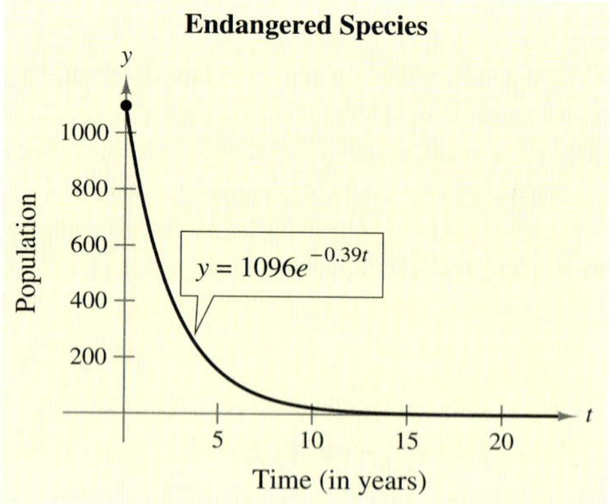

In Exercises 31–34, evaluate the function at each indicated value.

31. $f(x) = 2e^{x-1}$

(a) $x = 2$ (b) $x = \frac{1}{2}$ (c) $x = 10$

32. $f(t) = e^{4t} - 1$

(a) $t = 0$ (b) $t = 2$ (c) $t = -\frac{3}{4}$

33. $g(t) = 12e^{-0.2t}$

(a) $t = 17$ (b) $t = 50$ (c) $t = 100$

34. $g(x) = \dfrac{24}{1 + e^{-0.3x}}$

(a) $x = 0$ (b) $x = 300$ (c) $x = 1000$

35. ***Biology*** A lake is stocked with 500 fish and the fish population P begins to increase according to the logistic growth model

$$P = \frac{10{,}000}{1 + 19e^{-t/5}}, \quad 0 \le t$$

where t is measured in months.

(a) Use a graphing utility to graph the function.

(b) Estimate the number of fish in the lake after 4 months.

(c) Does the population have a limit as t increases without bound? Explain your reasoning.

(d) After how many months is the population increasing most rapidly? Explain your reasoning.

36. ***Medicine*** On a college campus of 5000 students, the spread of a flu virus through the student body is modeled by

$$P = \frac{5000}{1 + 4999e^{-0.8t}}, \quad 0 \le t$$

where P is the total number of infected people and t is the time, measured in days.

(a) Use a graphing utility to graph the function.

(b) How many students will be infected after 5 days?

(c) According to this model, will all the students on campus become infected with the flu? Explain your reasoning.

In Exercises 37 and 38, complete the table to determine the balance A when P dollars is invested at an annual rate of r for t years, compounded n times per year.

n	1	2	4	12	365	Continuous compounding
A						

37. $P = \$1000,\ r = 4\%,\ t = 5$ years

38. $P = \$5000,\ r = 6\%,\ t = 20$ years

In Exercises 39 and 40, \$2000 is deposited in an account. Decide which account, (a) or (b), will have the greater balance after 10 years.

39. (a) 5%, compounded continuously

(b) 6%, compounded quarterly

40. (a) $6\frac{1}{2}\%$, compounded monthly

(b) $6\frac{1}{4}\%$, compounded continuously

Effective Rate In Exercises 41 and 42, find the effective rate of interest corresponding to a nominal rate r, compounded (a) quarterly and (b) monthly.

41. $r = 6\%$ **42.** $r = 8.25\%$

43. ***Present Value*** How much should be deposited in an account paying 6.5% interest compounded quarterly in order to have a balance of \$12,000 three years from now?

44. ***Present Value*** How much should be deposited in an account paying 8% interest compounded monthly in order to have a balance of \$20,000 five years from now?

45. ***Vital Statistics*** The populations P (in millions) of people 65 years old and over in the United States from 1980 to 2000 can be modeled by

$$P = 22.9 + 0.87t - 0.019t^2 + 1.2960e^{-t}, \quad 0 \le t \le 20$$

where $t = 0$ corresponds to 1980. Use this model to estimate the populations of people 65 years old and over in 1980, 1995, and 2000. *(Source: U.S. Census Bureau)*

46. ***Revenue*** The revenues R (in millions of dollars per year) for Papa John's International from 1996 to 2003 can be modeled by

$$R = 2085.8 - 913.50t + 141.524t^2 - 6.1919t^3 + 0.00089e^t, \quad 6 \le t \le 13$$

where $t = 6$ corresponds to 1996. Use this model to estimate the revenues for Papa John's in 1996, 2000, and 2003. *(Source: Papa John's International)*

In Exercises 47–54, find the derivative of the function.

47. $y = 4e^{x^2}$ **48.** $y = 4e^{\sqrt{x}}$

49. $y = \dfrac{x}{e^{2x}}$ **50.** $y = x^2e^x$

51. $y = \sqrt{4e^{4x}}$ **52.** $y = \sqrt[3]{2e^{3x}}$

53. $y = \dfrac{5}{1 + e^{2x}}$ **54.** $y = \dfrac{10}{1 - 2e^x}$

In Exercises 55–62, graph and analyze the function. Include any relative extrema, points of inflection, and asymptotes in your analysis.

55. $f(x) = 4e^{-x}$ **56.** $f(x) = 2e^{x^2}$

57. $f(x) = x^3e^x$ **58.** $f(x) = \dfrac{e^x}{x^2}$

59. $f(x) = \frac{1}{xe^x}$

60. $f(x) = \frac{x^2}{e^x}$

61. $f(x) = xe^{2x}$

62. $f(x) = xe^{-2x}$

In Exercises 63 and 64, write the logarithmic equation as an exponential equation.

63. $\ln 12 = 2.4849 \ldots$

64. $\ln 0.6 = -0.5108 \ldots$

In Exercises 65 and 66, write the exponential equation as a logarithmic equation.

65. $e^{1.5} = 4.4816 \ldots$

66. $e^{-4} = 0.0183 \ldots$

In Exercises 67–70, sketch the graph of the function.

67. $y = \ln(4 - x)$

68. $y = 5 + \ln x$

69. $y = \ln \frac{x}{3}$

70. $y = -2 \ln x$

In Exercises 71–76, use the properties of logarithms to write the expression as a sum, difference, or multiple of logarithms.

71. $\ln \sqrt{x^2(x-1)}$

72. $\ln \sqrt[3]{x^2 - 1}$

73. $\ln \frac{x^2}{(x+1)^3}$

74. $\ln \frac{x^2}{x^2 + 1}$

75. $\ln\left(\frac{1-x}{3x}\right)^3$

76. $\ln\left(\frac{x-1}{x+1}\right)^2$

In Exercises 77–92, solve the equation for x.

77. $e^{\ln x} = 3$

78. $e^{\ln(x+2)} = 5$

79. $\ln x = 3e^{-1}$

80. $\ln x = 2e^5$

81. $\ln 2x - \ln(3x - 1) = 0$

82. $\ln x - \ln(x + 1) = 2$

83. $e^{2x-1} - 6 = 0$

84. $4e^{2x-3} - 5 = 0$

85. $\ln x + \ln(x - 3) = 0$

86. $2 \ln x + \ln(x - 2) = 0$

87. $e^{-1.386x} = 0.25$

88. $e^{-0.01x} - 5.25 = 0$

89. $100(1.21)^x = 110$

90. $500(1.075)^{120x} = 100{,}000$

91. $\frac{40}{1 - 5e^{-0.01x}} = 200$

92. $\frac{50}{1 - 2e^{-0.001x}} = 1000$

93. ***Home Mortgage*** The monthly payment M for a home mortgage of P dollars for t years at an annual interest rate of $r\%$ is given by

$$M = P\left\{\frac{\frac{r}{12}}{1 - \left[\frac{1}{(r/12) + 1}\right]^{12t}}\right\}.$$

(a) Use a graphing utility to graph the model when $P = \$100{,}000$ and $r = 8\%$.

(b) You are given a choice of a 20-year term or a 30-year term. Which would you choose? Explain your reasoning.

94. ***Hourly Wages*** The average hourly wages w in the United States from 1980 to 2000 can be modeled by

$$w = 7.60 + 0.168t + 0.0075t^2 - 0.76472e^{-t}$$

where $t = 0$ corresponds to 1980. *(Source: U.S. Bureau of Labor Statistics)*

(a) Use a graphing utility to graph the model.

(b) Use the model to determine the year in which the average hourly wage was \$12.

(c) For how many years past 2000 do you think this equation might be a good model for the average hourly wage? Explain your reasoning.

In Exercises 95–108, find the derivative of the function.

95. $f(x) = \ln 3x^2$

96. $y = \ln \sqrt{x}$

97. $y = \ln \frac{x(x-1)}{x-2}$

98. $y = \ln \frac{x^2}{x+1}$

99. $f(x) = \ln e^{2x+1}$

100. $f(x) = \ln e^{x^2}$

101. $y = \frac{\ln x}{x^3}$

102. $y = \frac{x^2}{\ln x}$

103. $y = \ln(x^2 - 2)^{2/3}$

104. $y = \ln \sqrt[3]{x^3 + 1}$

105. $f(x) = \ln\left(x^2\sqrt{x+1}\right)$

106. $f(x) = \ln \frac{x}{\sqrt{x+1}}$

107. $y = \ln \frac{e^x}{1 + e^x}$

108. $y = \ln\left(e^{2x}\sqrt{e^{2x} - 1}\right)$

In Exercises 109–112, graph and analyze the function. Include any relative extrema and points of inflection in your analysis.

109. $y = \ln(x + 3)$

110. $y = \frac{8 \ln x}{x^2}$

111. $y = \ln \frac{10}{x+2}$

112. $y = \ln \frac{x^2}{9 - x^2}$

In Exercises 113–116, evaluate the logarithm.

113. $\log_7 49$

114. $\log_2 32$

115. $\log_{10} 1$

116. $\log_4 \frac{1}{64}$

In Exercises 117–120, use the change-of-base formula to evaluate the logarithm. Round the result to three decimal places.

117. $\log_5 10$

118. $\log_4 12$

119. $\log_{16} 64$

120. $\log_4 125$

In Exercises 121–124, find the derivative of the function.

121. $y = \log_3(2x - 1)$

122. $y = \log_{10} \dfrac{3}{x}$

123. $y = \log_2 \dfrac{1}{x^2}$

124. $y = \log_{16}(x^2 - 3x)$

125. ***Depreciation*** After t years, the value V of a car purchased for \$20,000 is given by

$$V = 20{,}000(0.75)^t.$$

(a) Sketch a graph of the function and determine the value of the car 2 years after it was purchased.

(b) Find the rates of change of V with respect to t when $t = 1$ and when $t = 4$.

(c) After how many years will the car be worth \$5000?

126. ***Inflation Rate*** If the annual rate of inflation averages 5% over the next 10 years, then the approximate cost of goods or services C during any year in that decade is given by

$$C = P(1.05)^t$$

where t is the time in years and P is the present cost.

(a) The price of an oil change is presently \$24.95. Estimate the price of an oil change 10 years from now.

(b) Find the rate of change of C with respect to t when $t = 1$.

127. ***Medical Science*** A medical solution contains 500 milligrams of a drug per milliliter when the solution is prepared. After 40 days, it contains only 300 milligrams per milliliter. Assuming that the rate of decomposition is proportional to the concentration present, find an equation giving the concentration A after t days.

128. ***Population Growth*** A population is growing continuously at the rate of $2\frac{1}{2}\%$ per year. Find the time necessary for the population to (a) double in size and (b) triple in size.

129. ***Radioactive Decay*** A sample of radioactive waste is taken from a nuclear plant. The sample contains 50 grams of strontium-90 at time $t = 0$ years and 42.031 grams after 7 years. What is the half-life of strontium-90?

130. ***Radioactive Decay*** The half-life of cobalt-60 is 5.2 years. Find the time it would take for a sample of 0.5 gram of cobalt-60 to decay to 0.1 gram.

131. ***Profit*** The profit P (in millions of dollars) for Affiliated Computer Services, Inc. was \$17.6 million in 1995 and \$306.8 million in 2003 (see figure). Use an exponential growth model to predict the profit in 2006. *(Source: Affiliated Computer Services, Inc.)*

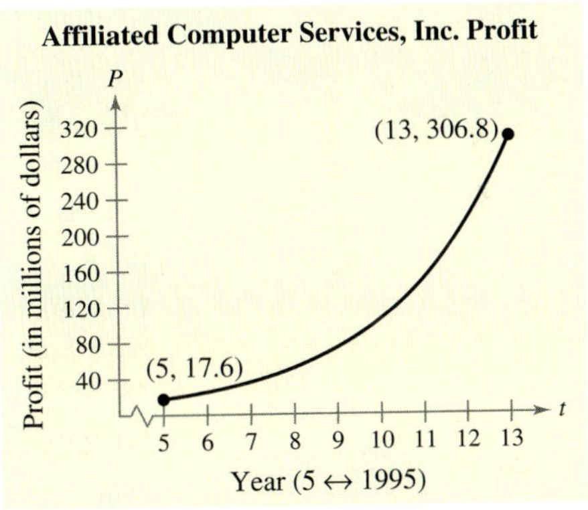

132. ***Profit*** The profit P (in millions of dollars) for MBNA Corporation was \$266.6 million in 1994 and \$2338.1 million in 2003 (see figure). Use an exponential growth model to predict the profit in 2006. *(Source: MBNA Corporation)*

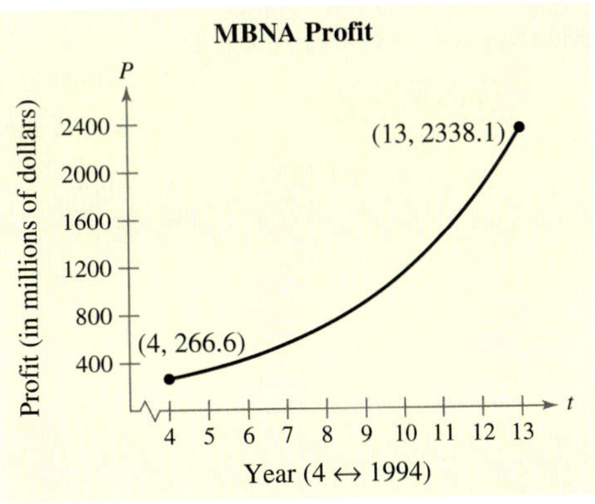

4 SAMPLE POST-GRADUATION EXAM QUESTIONS

CPA
GMAT
GRE
Actuarial
CLAST

The following questions represent the types of questions that appear on certified public accountant (CPA) exams, Graduate Management Admission Tests (GMAT), Graduate Records Exams (GRE), actuarial exams, and College-Level Academic Skills Tests (CLAST). The answers to the questions are given in the back of the book.

1. 10^x means that 10 is to be used as a factor x times, and 10^{-x} is equal to $\frac{1}{10^x}$.

A very large or very small number, therefore, is frequently written as a decimal multiplied by 10^x, where x is an integer. Which, if any, are false?

(a) $470{,}000 = 4.7 \times 10^5$

(b) $450 \text{ billion} = 4.5 \times 10^{11}$

(c) $0.00000000075 = 7.5 \times 10^{-10}$

(d) $86 \text{ hundred-thousandths} = 8.6 \times 10^2$

2. The rate of decay of a radioactive substance is proportional to the amount of the substance present. Three years ago there was 6 grams of substance. Now there is 5 grams. How many grams will there be 3 years from now?

(a) 4 (b) $\frac{25}{6}$ (c) $\frac{125}{36}$ (d) $\frac{75}{36}$

3. In a certain town, 45% of the people have brown hair, 30% have brown eyes, and 15% have both brown hair and brown eyes. What percent of the people in the town have neither brown hair nor brown eyes?

(a) 25% (b) 35% (c) 40% (d) 50%

4. You deposit $900 in a savings account that is compounded continuously at 4.76%. After 16 years, the amount in the account will be

(a) $1927.53 (b) $1077.81 (c) $943.88 (d) $2827.53

5. A bookstore orders 75 books. Each book costs the bookstore $29 and is sold for $42. The bookstore must pay a $4 service charge for each unsold book returned. If the bookstore returns seven books, how much profit will the bookstore make?

(a) $975 (b) $947 (c) $856 (d) $681

Figure for 6–9

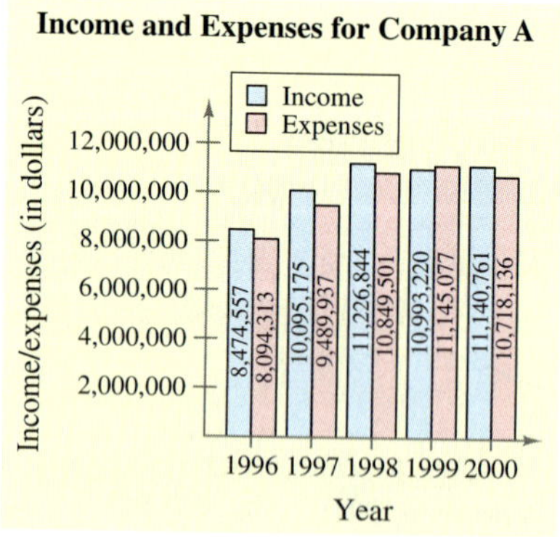

For Questions 6–9, use the data given in the graph.

6. In how many of the years were expenses greater than in the preceding year?

(a) 2 (b) 4 (c) 1 (d) 3

7. In which year was the profit the greatest?

(a) 1997 (b) 2000 (c) 1996 (d) 1998

8. In 1999, profits decreased by x percent from 1998 with x equal to

(a) 60% (b) 140% (c) 340% (d) 40%

9. In 2000, profits increased by y percent from 1999 with y equal to

(a) 64% (b) 136% (c) 178% (d) 378%

chapter

5 Integration and Its Applications

Jerry Alexander/Alamy

Integration can be used to solve physics problems, such as finding the time it takes for a sandbag to fall to the ground when dropped from a hot air balloon.

STRATEGIES FOR SUCCESS

WHAT YOU SHOULD LEARN:

- How to find the antiderivative F of a function—that is, $F'(x) = f(x)$
- How to use the General Power Rule, Exponential Rule, and Log Rule to calculate antiderivatives
- How to evaluate definite integrals and apply the Fundamental Theorem of Calculus to find the area bounded by two graphs
- How to use the Midpoint Rule to approximate definite integrals
- How to use integration to find the volume of a solid of revolution

WHY YOU SHOULD LEARN IT:

Integration has many applications in real life, as demonstrated by the examples below, which represent a small sample of the applications in this chapter.

- Demand Function, Exercises 67–70 on page 327
- Vertical Motion, Exercises 75–78 on page 328
- Marginal Propensity to Consume, Exercises 57 and 58 on page 336
- Annuity, Example 9 on page 352, Exercises 81–84 on page 355
- Capital Accumulation, Exercises 85–88 on page 355
- Consumer and Producer Surpluses, Exercises 41–46 on page 363
- Lorenz Curve, Exercises 58 and 59 on page 364

5.1 ANTIDERIVATIVES AND INDEFINITE INTEGRALS

- Understand the definition of antiderivative.
- Use indefinite integral notation for antiderivatives.
- Use basic integration rules to find antiderivatives.
- Use initial conditions to find particular solutions of indefinite integrals.
- Use antiderivatives to solve real-life problems.

Antiderivatives

Up to this point in the text, you have been concerned primarily with this problem: given a function, find its derivative. Many important applications of calculus involve the inverse problem: given the derivative of a function, find the function. For example, suppose you are given

$$f'(x) = 2, \quad g'(x) = 3x^2, \quad \text{and} \quad s'(t) = 4t.$$

Your goal is to determine the functions f, g, and s. By making educated guesses, you might come up with the following functions.

$$f(x) = 2x \quad \text{because} \quad \frac{d}{dx}[2x] = 2.$$

$$g(x) = x^3 \quad \text{because} \quad \frac{d}{dx}[x^3] = 3x^2.$$

$$s(t) = 2t^2 \quad \text{because} \quad \frac{d}{dt}[2t^2] = 4t.$$

This operation of determining the original function from its derivative is the inverse operation of differentiation. It is called **antidifferentiation.**

Definition of Antiderivative

A function F is an **antiderivative** of a function f if for every x in the domain of f, it follows that $F'(x) = f(x)$.

If $F(x)$ is an antiderivative of $f(x)$, then $F(x) + C$, where C is any constant, is also an antiderivative of $f(x)$. For example,

$$F(x) = x^3, \quad G(x) = x^3 - 5, \quad \text{and} \quad H(x) = x^3 + 0.3$$

are all antiderivatives of $3x^2$ because the derivative of each is $3x^2$. As it turns out, *all* antiderivatives of $3x^2$ are of the form $x^3 + C$. So, the process of antidifferentiation does not determine a single function, but rather a *family* of functions, each differing from the others by a constant.

STUDY TIP

In this text, the phrase "$F(x)$ is an antiderivative of $f(x)$" is used synonymously with "F is an antiderivative of f."

Notation for Antiderivatives and Indefinite Integrals

The antidifferentiation process is also called **integration** and is denoted by the symbol

$$\int \qquad \text{Integral sign}$$

which is called an **integral sign.** The symbol

$$\int f(x)\,dx \qquad \text{Indefinite integral}$$

is the **indefinite integral** of $f(x)$, and it denotes the family of antiderivatives of $f(x)$. That is, if $F'(x) = f(x)$ for all x, then you can write

Integral sign → $\int$; Differential → dx

$$\int f(x)\,dx = F(x) + C$$

Integrand → $f(x)$; Antiderivative → $F(x) + C$

where $f(x)$ is the **integrand** and C is the **constant of integration.** The differential dx in the indefinite integral identifies the variable of integration. That is, the symbol $\int f(x)\,dx$ denotes the "antiderivative of f *with respect to* x" just as the symbol dy/dx denotes the "derivative of y *with respect to* x."

Integral Notation of Antiderivatives

The notation

$$\int f(x)\,dx = F(x) + C$$

where C is an arbitrary constant, means that F is an antiderivative of f. That is, $F'(x) = f(x)$ for all x in the domain of f.

DISCOVERY

Verify that $F_1(x) = x^2 - 2x$, $F_2(x) = x^2 - 2x - 1$, and $F_3(x) = (x - 1)^2$ are all antiderivatives of $f(x) = 2x - 2$. Use a graphing utility to graph F_1, F_2, and F_3 in the same coordinate plane. How are their graphs related? What can you say about the graph of any other antiderivative of f?

EXAMPLE 1 **Notation for Antiderivatives**

Using integral notation, you can write the three antiderivatives from the beginning of this section as shown.

(a) $\int 2\,dx = 2x + C$ (b) $\int 3x^2\,dx = x^3 + C$ (c) $\int 4t\,dt = 2t^2 + C$

TRY IT 1

Rewrite each antiderivative using integral notation.

(a) $\frac{d}{dx}[3x] = 3$ (b) $\frac{d}{dx}[x^2] = 2x$ (c) $\frac{d}{dt}[3t^3] = 9t^2$

Finding Antiderivatives

The inverse relationship between the operations of integration and differentiation can be shown symbolically, as shown.

$$\frac{d}{dx}\left[\int f(x)\,dx\right] = f(x)$$ Differentiation is the inverse of integration.

$$\int f'(x)\,dx = f(x) + C$$ Integration is the inverse of differentiation.

This inverse relationship between integration and differentiation allows you to obtain integration formulas directly from differentiation formulas. The following summary lists the integration formulas that correspond to some of the differentiation formulas you have studied.

Basic Integration Rules

1. $\int k\,dx = kx + C$, k is a constant. — Constant Rule
2. $\int kf(x)\,dx = k\int f(x)\,dx$ — Constant Multiple Rule
3. $\int [f(x) + g(x)]\,dx = \int f(x)\,dx + \int g(x)\,dx$ — Sum Rule
4. $\int [f(x) - g(x)]\,dx = \int f(x)\,dx - \int g(x)\,dx$ — Difference Rule
5. $\int x^n\,dx = \dfrac{x^{n+1}}{n+1} + C,\quad n \neq -1$ — Simple Power Rule

STUDY TIP

You will study the General Power Rule for integration in Section 5.2 and the Exponential and Log Rules in Section 5.3.

STUDY TIP

In Example 2(b), the integral $\int 1\,dx$ is usually shortened to the form $\int dx$.

TRY IT 2

Find each indefinite integral.

(a) $\int 5\,dx$

(b) $\int -1\,dr$

(c) $\int 2\,dt$

Be sure you see that the Simple Power Rule has the restriction that n cannot be -1. So, you *cannot* use the Simple Power Rule to evaluate the integral

$$\int \frac{1}{x}\,dx.$$

To evaluate this integral, you need the Log Rule, which is described in Section 5.3.

EXAMPLE 2 **Finding Indefinite Integrals**

Find each indefinite integral.

(a) $\int \frac{1}{2}\,dx$ (b) $\int 1\,dx$ (c) $\int -5\,dt$

SOLUTION

(a) $\int \frac{1}{2}\,dx = \frac{1}{2}x + C$ (b) $\int 1\,dx = x + C$ (c) $\int -5\,dt = -5t + C$

EXAMPLE 3 Finding an Indefinite Integral

Find $\int 3x\,dx$.

SOLUTION

$$\begin{aligned}\int 3x\,dx &= 3\int x\,dx && \text{Constant Multiple Rule}\\ &= 3\int x^1\,dx && \text{Rewrite } x \text{ as } x^1.\\ &= 3\left(\frac{x^2}{2}\right) + C && \text{Simple Power Rule with } n = 1\\ &= \frac{3}{2}x^2 + C && \text{Simplify.}\end{aligned}$$

TECHNOLOGY

If you have access to a symbolic integration program, try using it to find antiderivatives.

TRY IT 3

Find $\int 5x\,dx$.

In finding indefinite integrals, a strict application of the basic integration rules tends to produce cumbersome constants of integration. For instance, in Example 3, you could have written

$$\int 3x\,dx = 3\int x\,dx = 3\left(\frac{x^2}{2} + C\right) = \frac{3}{2}x^2 + 3C.$$

However, because C represents *any* constant, it is unnecessary to write $3C$ as the constant of integration. You can simply write $\frac{3}{2}x^2 + C$.

In Example 3, note that the general pattern of integration is similar to that of differentiation.

Original Integral: $\int 3x\,dx$ → Rewrite: $3\int x^1\,dx$ → Integrate: $3\left(\frac{x^2}{2}\right) + C$ → Simplify: $\frac{3}{2}x^2 + C$

STUDY TIP

Remember that you can check your answer to an antidifferentiation problem by differentiating. For instance, in Example 4(b), you can check that $\frac{2}{3}x^{3/2}$ is the correct antiderivative by differentiating to obtain

$$\frac{d}{dx}\left[\frac{2}{3}x^{3/2}\right] = \left(\frac{2}{3}\right)\left(\frac{3}{2}\right)x^{1/2} = \sqrt{x}.$$

EXAMPLE 4 Rewriting Before Integrating

Find each indefinite integral.

(a) $\int \frac{1}{x^3}\,dx$

(b) $\int \sqrt{x}\,dx$

SOLUTION

	Original Integral	*Rewrite*	*Integrate*	*Simplify*
(a)	$\int \frac{1}{x^3}\,dx$	$\int x^{-3}\,dx$	$\frac{x^{-2}}{-2} + C$	$-\frac{1}{2x^2} + C$
(b)	$\int \sqrt{x}\,dx$	$\int x^{1/2}\,dx$	$\frac{x^{3/2}}{3/2} + C$	$\frac{2}{3}x^{3/2} + C$

TRY IT 4

Find each indefinite integral.

(a) $\int \frac{1}{x^2}\,dx$ (b) $\int \sqrt[3]{x}\,dx$

With the five basic integration rules, you can integrate *any* polynomial function, as demonstrated in the next example.

TRY IT 5

Find each indefinite integral.

(a) $\int (x + 4)\,dx$

(b) $\int (4x^3 - 5x + 2)\,dx$

EXAMPLE 5 Integrating Polynomial Functions

Find each indefinite integral.

(a) $\int (x + 2)\,dx$

(b) $\int (3x^4 - 5x^2 + x)\,dx$

SOLUTION

(a) Use the Sum Rule to integrate each part separately.

$$\int (x + 2)\,dx = \int x\,dx + \int 2\,dx$$
$$= \frac{x^2}{2} + 2x + C$$

(b) Try to identify each basic integration rule used to evaluate this integral.

$$\int (3x^4 - 5x^2 + x)\,dx = 3\left(\frac{x^5}{5}\right) - 5\left(\frac{x^3}{3}\right) + \frac{x^2}{2} + C$$
$$= \frac{3}{5}x^5 - \frac{5}{3}x^3 + \frac{1}{2}x^2 + C$$

STUDY TIP

When integrating quotients, remember *not* to integrate the numerator and denominator separately. For instance, in Example 6, be sure you see that

$$\int \frac{x+1}{\sqrt{x}}\,dx \neq \frac{\int (x+1)\,dx}{\int \sqrt{x}\,dx}.$$

EXAMPLE 6 Rewriting Before Integrating

Find $\int \frac{x+1}{\sqrt{x}}\,dx.$

SOLUTION Begin by rewriting the quotient in the integrand as a sum. Then rewrite each term using rational exponents.

$$\int \frac{x+1}{\sqrt{x}}\,dx = \int \left(\frac{x}{\sqrt{x}} + \frac{1}{\sqrt{x}}\right) dx \qquad \text{Rewrite as a sum.}$$
$$= \int (x^{1/2} + x^{-1/2})\,dx \qquad \text{Rewrite using rational exponents.}$$
$$= \frac{x^{3/2}}{3/2} + \frac{x^{1/2}}{1/2} + C \qquad \text{Apply Power Rule.}$$
$$= \frac{2}{3}x^{3/2} + 2x^{1/2} + C \qquad \text{Simplify.}$$

TRY IT 6

Find $\int \frac{x+2}{\sqrt{x}}\,dx.$

ALGEBRA REVIEW

For help on the algebra in Example 6, see Example 1(a) in the *Chapter 5 Algebra Review*, on page 378.

TECHNOLOGY

If you have access to a symbolic integration utility, try using it to solve Example 6. The utility may list the antiderivative as $2\sqrt{x}(x + 3)/3$, which is equivalent to the result listed above.

Particular Solutions

You have already seen that the **differential equation** (an equation that involves x, y, and derivatives of y) $y = \int f(x)\,dx$ has many solutions, each differing from the others by a constant. This means that the graphs of any two antiderivatives of f are vertical translations of each other. For instance, Figure 5.1 shows the graphs of several antiderivatives of the form

$$y = F(x) = \int (3x^2 - 1)\,dx = x^3 - x + C$$

where $F(x) = x^3 - x + C$ is called the **general solution** of the differential equation. Each of these antiderivatives is a solution of $dy/dx = 3x^2 - 1$.

In many applications of integration, you are given enough information to determine a **particular solution.** To do this, you only need to know the value of $F(x)$ for one value of x. (This information is called an **initial condition.**) For example, in Figure 5.1, there is only one curve that passes through the point (2, 4). To find this curve, use the information below.

$F(x) = x^3 - x + C$ General solution

$F(2) = 4$ Initial condition

By using the initial condition in the general solution, you can determine that $F(2) = 2^3 - 2 + C = 4$, which implies that $C = -2$. So, the particular solution is

$F(x) = x^3 - x - 2.$ Particular solution

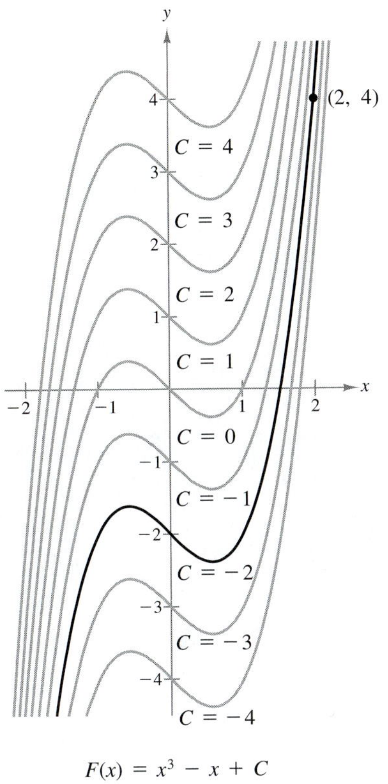

FIGURE 5.1

EXAMPLE 7 Finding a Particular Solution

Find the general solution of

$$F'(x) = 2x - 2$$

and find the particular solution that satisfies the initial condition $F(1) = 2$.

SOLUTION Begin by integrating to find the general solution.

$$F(x) = \int (2x - 2)\,dx \quad \text{Integrate } F'(x) \text{ to obtain } F(x).$$

$$= x^2 - 2x + C \quad \text{General solution}$$

Using the initial condition $F(1) = 2$, you can write

$$F(1) = 1^2 - 2(1) + C = 2$$

which implies that $C = 3$. So, the particular solution is

$F(x) = x^2 - 2x + 3.$ Particular solution

This solution is shown graphically in Figure 5.2. Note that each of the gray curves represents a solution of the equation $F'(x) = 2x - 2$. The black curve, however, is the only solution that passes through the point (1, 2), which means that $F(x) = x^2 - 2x + 3$ is the only solution that satisfies the initial condition.

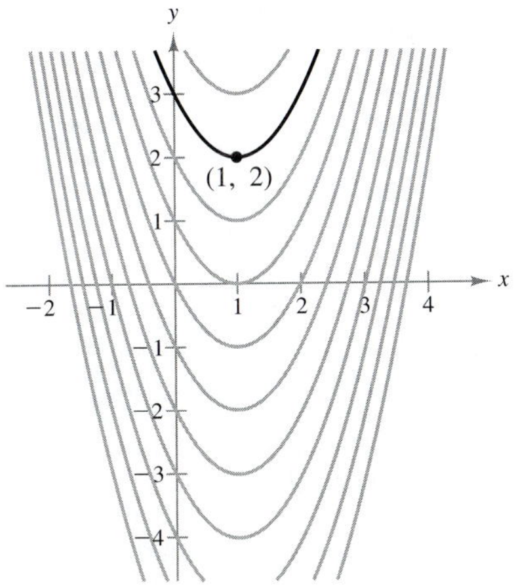

FIGURE 5.2

TRY IT 7

Find the general solution of $F'(x) = 4x + 2$, and find the particular solution that satisfies the initial condition $F(1) = 8$.

Applications

In Chapter 2, you used the general position function (neglecting air resistance) for a falling object

$$s(t) = -16t^2 + v_0t + s_0$$

where $s(t)$ is the height (in feet) and t is the time (in seconds). In the next example, integration is used to *derive* this function.

EXAMPLE 8 Deriving a Position Function

A ball is thrown upward with an initial velocity of 64 feet per second from an initial height of 80 feet, as shown in Figure 5.3. Derive the position function giving the height s (in feet) as a function of the time t (in seconds). When does the ball hit the ground?

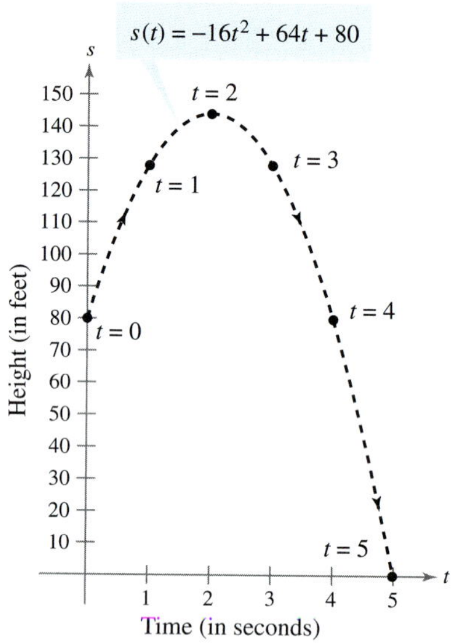

FIGURE 5.3

SOLUTION Let $t = 0$ represent the initial time. Then the two given conditions can be written as

$$s(0) = 80 \qquad \text{Initial height is 80 feet.}$$
$$s'(0) = 64. \qquad \text{Initial velocity is 64 feet per second.}$$

Because the acceleration due to gravity is -32 feet per second per second, you can integrate the acceleration function to find the velocity function as shown.

$$s''(t) = -32 \qquad \text{Acceleration due to gravity}$$
$$s'(t) = \int -32\, dt \qquad \text{Integrate } s''(t) \text{ to obtain } s'(t).$$
$$= -32t + C_1 \qquad \text{Velocity function}$$

Using the initial velocity, you can conclude that $C_1 = 64$.

$$s'(t) = -32t + 64 \qquad \text{Velocity function}$$
$$s(t) = \int (-32t + 64)\, dt \qquad \text{Integrate } s'(t) \text{ to obtain } s(t).$$
$$= -16t^2 + 64t + C_2 \qquad \text{Position function}$$

Using the initial height, it follows that $C_2 = 80$. So, the position function is given by

$$s(t) = -16t^2 + 64t + 80. \qquad \text{Position function}$$

To find the time when the ball hits the ground, set the position function equal to 0 and solve for t.

$$-16t^2 + 64t + 80 = 0 \qquad \text{Set } s(t) \text{ equal to zero.}$$
$$-16(t + 1)(t - 5) = 0 \qquad \text{Factor.}$$
$$t = -1, \quad t = 5 \qquad \text{Solve for } t.$$

Because the time must be positive, you can conclude that the ball hits the ground 5 seconds after it is thrown.

TRY IT 8

Derive the position function if a ball is thrown upward with an initial velocity of 32 feet per second from an initial height of 48 feet. When does the ball hit the ground? With what velocity does the ball hit the ground?

EXAMPLE 9 Finding a Cost Function

The marginal cost for producing x units of a product is modeled by

$$\frac{dC}{dx} = 32 - 0.04x. \qquad \text{Marginal cost}$$

It costs \$50 to produce one unit. Find the total cost of producing 200 units.

SOLUTION To find the cost function, integrate the marginal cost function.

$$C = \int (32 - 0.04x)\,dx \qquad \text{Integrate } \frac{dC}{dx} \text{ to obtain } C.$$

$$= 32x - 0.04\left(\frac{x^2}{2}\right) + K$$

$$= 32x - 0.02x^2 + K \qquad \text{Cost function}$$

To solve for K, use the initial condition that $C = 50$ when $x = 1$.

$$50 = 32(1) - 0.02(1)^2 + K \qquad \text{Substitute 50 for } C \text{ and 1 for } x.$$

$$18.02 = K \qquad \text{Solve for } K.$$

So, the total cost function is given by

$$C = 32x - 0.02x^2 + 18.02 \qquad \text{Cost function}$$

which implies that the cost of producing 200 units is

$$C = 32(200) - 0.02(200)^2 + 18.02$$
$$= \$5618.02.$$

TRY IT 9

The marginal cost function for producing x units of a product is modeled by

$$\frac{dC}{dx} = 28 - 0.02x.$$

It costs \$40 to produce one unit. Find the cost of producing 200 units.

TAKE ANOTHER LOOK

Investigating Marginal Cost

In Example 9, you were given a marginal cost function of

$$\frac{dC}{dx} = 32 - 0.04x.$$

This means that the cost of making each additional unit decreases by about \$0.04. You can confirm this by finding the costs of making different amounts of the product.

x	1	2	3	4	5	6
C	\$50.00	\$81.94	\$113.84	\$145.70	\$177.52	\$209.30

From this, you can see that the first unit cost \$50, the second cost \$31.94, the third cost \$31.90, the fourth cost \$31.86, the fifth cost \$31.82, and so on. At what production level would this costing scheme cease to make sense? Explain your reasoning.

PREREQUISITE REVIEW 5.1

The following warm-up exercises involve skills that were covered in earlier sections. You will use these skills in the exercise set for this section.

In Exercises 1–6, rewrite the expression using rational exponents.

1. $\dfrac{\sqrt{x}}{x}$

2. $\sqrt[3]{2x}(2x)$

3. $\sqrt{5x^3} + \sqrt{x^5}$

4. $\dfrac{1}{\sqrt{x}} + \dfrac{1}{\sqrt[3]{x^2}}$

5. $\dfrac{(x+1)^3}{\sqrt{x+1}}$

6. $\dfrac{\sqrt{x}}{\sqrt[3]{x}}$

In Exercises 7–10, let $(x, y) = (2, 2)$, and solve the equation for C.

7. $y = x^2 + 5x + C$

8. $y = 3x^3 - 6x + C$

9. $y = -16x^2 + 26x + C$

10. $y = -\frac{1}{4}x^4 - 2x^2 + C$

EXERCISES 5.1

In Exercises 1–8, verify the statement by showing that the derivative of the right side is equal to the integrand of the left side.

1. $\displaystyle\int \left(-\frac{9}{x^4}\right) dx = \frac{3}{x^3} + C$

2. $\displaystyle\int \frac{4}{\sqrt{x}}\, dx = 8\sqrt{x} + C$

3. $\displaystyle\int \left(4x^3 - \frac{1}{x^2}\right) dx = x^4 + \frac{1}{x} + C$

4. $\displaystyle\int \left(1 - \frac{1}{\sqrt[3]{x^2}}\right) dx = x - 3\sqrt[3]{x} + C$

5. $\displaystyle\int 2\sqrt{x}(x-3)\, dx = \frac{4x^{3/2}(x-5)}{5} + C$

6. $\displaystyle\int 4\sqrt{x}(x^2-2)\, dx = \frac{8x^{3/2}(3x^2-14)}{21} + C$

7. $\displaystyle\int \frac{x^2-1}{x^{3/2}}\, dx = \frac{2(x^2+3)}{3\sqrt{x}} + C$

8. $\displaystyle\int \frac{2x-1}{x^{4/3}}\, dx = \frac{3(x+1)}{\sqrt[3]{x}} + C$

In Exercises 9–20, find the indefinite integral and check your result by differentiation.

9. $\displaystyle\int 6\, dx$

10. $\displaystyle\int -4\, dx$

11. $\displaystyle\int 5t^2\, dt$

12. $\displaystyle\int 3t^4\, dt$

13. $\displaystyle\int 5x^{-3}\, dx$

14. $\displaystyle\int 4y^{-3}\, dy$

15. $\displaystyle\int du$

16. $\displaystyle\int dr$

17. $\displaystyle\int e\, dt$

18. $\displaystyle\int e^3\, dy$

19. $\displaystyle\int y^{3/2}\, dy$

20. $\displaystyle\int v^{-1/2}\, dv$

In Exercises 21–28, find the indefinite integral using the columns in Example 4 as a model. Use a symbolic integration utility to verify your results.

21. $\displaystyle\int \sqrt[3]{x}\, dx$

22. $\displaystyle\int \frac{1}{x^2}\, dx$

23. $\displaystyle\int \frac{1}{x\sqrt{x}}\, dx$

24. $\displaystyle\int \frac{1}{x^2\sqrt{x}}\, dx$

25. $\displaystyle\int x(x^2+3)\, dx$

26. $\displaystyle\int t(t^2+2)\, dt$

27. $\displaystyle\int \frac{1}{2x^3}\, dx$

28. $\displaystyle\int \frac{1}{8x^3}\, dx$

In Exercises 29–32, find two functions that have the given derivative, and sketch the graph of each. (There is more than one correct answer.)

29.

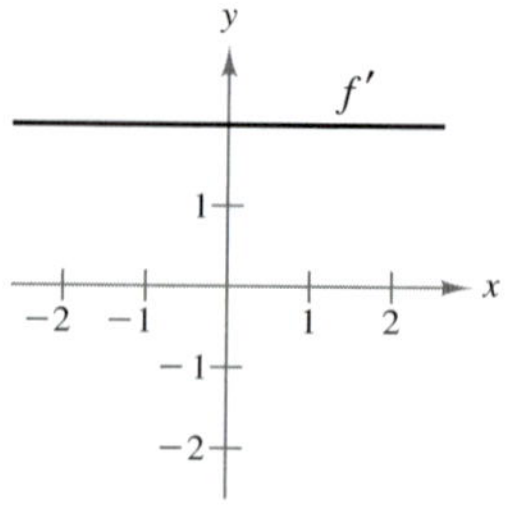

30.

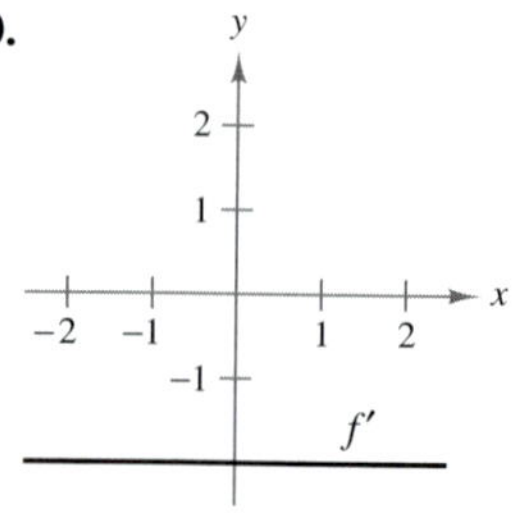

31.

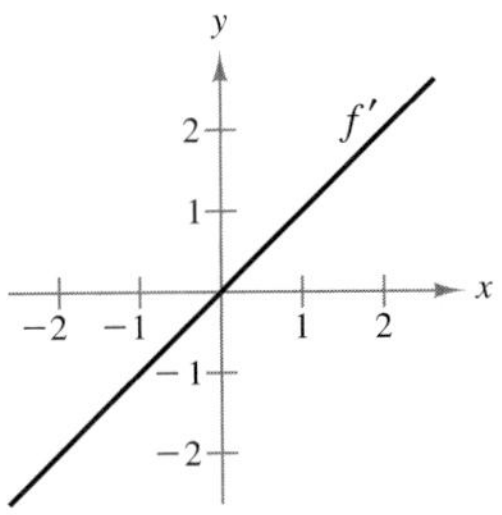

32.

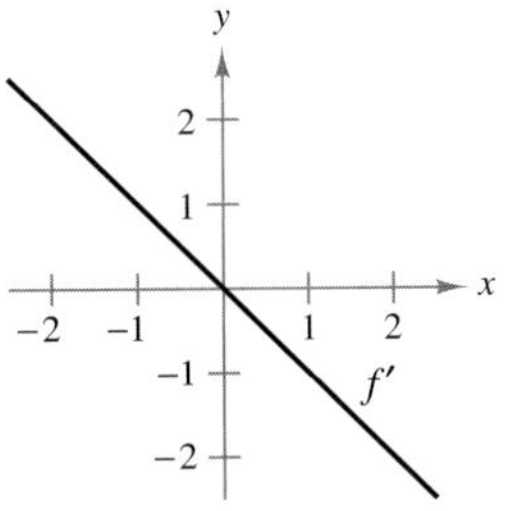

In Exercises 33–42, find the indefinite integral and check your result by differentiation.

33. $\int (x^3 + 2)\,dx$

34. $\int (x^2 - 2x + 3)\,dx$

35. $\int \left(\sqrt[3]{x} - \frac{1}{2\sqrt[3]{x}}\right) dx$

36. $\int \left(\sqrt{x} + \frac{1}{2\sqrt{x}}\right) dx$

37. $\int \left(\sqrt[3]{x^2} + 1\right) dx$

38. $\int (\sqrt[4]{x^3} + 1)\,dx$

39. $\int \frac{1}{3x^4}\,dx$

40. $\int \frac{1}{4x^2}\,dx$

41. $\int \frac{2x^3 + 1}{x^3}\,dx$

42. $\int \frac{t^2 + 2}{t^2}\,dt$

In Exercises 43–48, use a symbolic integration utility to find the indefinite integral.

43. $\int u(3u^2 + 1)\,du$

44. $\int \sqrt{x}(x + 1)\,dx$

45. $\int (x - 1)(6x - 5)\,dx$

46. $\int (2t^2 - 1)^2\,dt$

47. $\int y^2\sqrt{y}\,dy$

48. $\int (1 + 3t)t^2\,dt$

In Exercises 49–54, find the particular solution $y = f(x)$ that satisfies the differential equation and initial condition.

49. $f'(x) = 3\sqrt{x} + 3;\quad f(1) = 4$

50. $f'(x) = \frac{1}{5}x - 2;\quad f(10) = -10$

51. $f'(x) = 6x(x - 1);\quad f(1) = -1$

52. $f'(x) = (2x - 3)(2x + 3);\quad f(3) = 0$

53. $f'(x) = \frac{2 - x}{x^3},\ x > 0;\quad f(2) = \frac{3}{4}$

54. $f'(x) = \frac{x^2 - 5}{x^2},\ x > 0;\quad f(1) = 2$

In Exercises 55 and 56, you are shown a family of graphs, each of which is a general solution of the given differential equation. Find the equation of the particular solution that passes through the indicated point.

55. $\frac{dy}{dx} = -5x - 2$

56. $\frac{dy}{dx} = 2(x - 1)$

Figure for 55

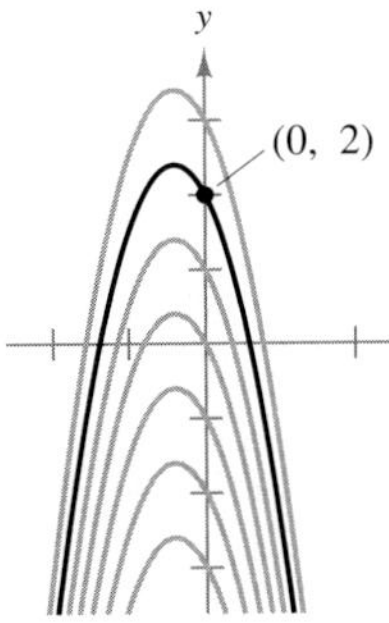

Figure for 56

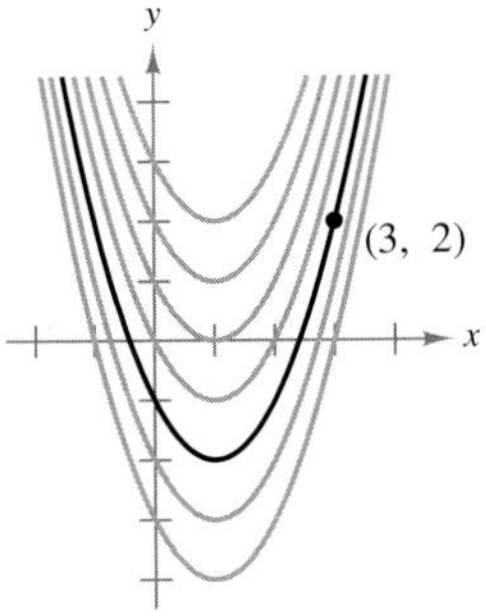

In Exercises 57 and 58, find the equation of the function f whose graph passes through the point.

	Derivative	*Point*
57.	$f'(x) = 6\sqrt{x} - 10$	(4, 2)
58.	$f'(x) = \frac{6}{x^2}$	(2, 5)

In Exercises 59–62, find a function f that satisfies the conditions.

59. $f''(x) = 2,\quad f'(2) = 5,\quad f(2) = 10$

60. $f''(x) = x^2,\quad f'(0) = 6,\quad f(0) = 3$

61. $f''(x) = x^{-2/3},\quad f'(8) = 6,\quad f(0) = 0$

62. $f''(x) = x^{-3/2},\quad f'(1) = 2,\quad f(9) = -4$

Cost In Exercises 63–66, find the cost function for the marginal cost and fixed cost.

	Marginal Cost	*Fixed Cost (x = 0)*
63.	$\frac{dC}{dx} = 85$	\$5500
64.	$\frac{dC}{dx} = \frac{1}{50}x + 10$	\$1000
65.	$\frac{dC}{dx} = \frac{1}{20\sqrt{x}} + 4$	\$750
66.	$\frac{dC}{dx} = \frac{\sqrt[4]{x}}{10} + 10$	\$2300

Demand Function In Exercises 67–70, find the revenue and demand functions for the given marginal revenue. (Use the fact that $R = 0$ when $x = 0$.)

67. $\frac{dR}{dx} = 225 - 3x$

68. $\frac{dR}{dx} = 310 - 4x$

69. $\frac{dR}{dx} = 225 + 2x - x^2$

70. $\frac{dR}{dx} = 100 - 6x - 2x^2$

Profit In Exercises 71–74, find the profit function for the given marginal profit and initial condition.

	Marginal Profit	*Initial Condition*
71.	$\frac{dP}{dx} = -18x + 1650$	$P(15) = \$22,725$
72.	$\frac{dP}{dx} = -40x + 250$	$P(5) = \$650$
73.	$\frac{dP}{dx} = -24x + 805$	$P(12) = \$8000$
74.	$\frac{dP}{dx} = -30x + 920$	$P(8) = \$6500$

Vertical Motion In Exercises 75–78, use $a(t) = -32$ feet per second per second as the acceleration due to gravity.

75. A ball is thrown vertically upward with an initial velocity of 60 feet per second. How high will the ball go?

76. The Grand Canyon is 6000 feet deep at the deepest part. A rock is dropped from this height. Express the height of the rock as a function of the time t (in seconds). How long will it take the rock to hit the canyon floor?

77. With what initial velocity must an object be thrown upward from the ground to reach the height of the Washington Monument (550 feet)?

78. A balloon, rising vertically with a velocity of 16 feet per second, releases a sandbag at an instant when the balloon is 64 feet above the ground.

(a) How many seconds after its release will the bag strike the ground?

(b) With what velocity will the bag strike the ground?

79. *Cost* A company produces a product for which the marginal cost of producing x units is modeled by

$$\frac{dC}{dx} = 2x - 12$$

and the fixed costs are \$125.

(a) Find the total cost function and the average cost function.

(b) Find the total cost of producing 50 units.

(c) In part (b), how much of the total cost is fixed? How much is variable? Give examples of fixed costs associated with the manufacturing of a product. Give examples of variable costs.

80. *Population Growth* The growth rate of Horry County in South Carolina can be modeled by

$$\frac{dP}{dt} = 105.74t + 2639.3$$

where t is the time in years, with $t = 0$ corresponding to 1970. The county's population was 196,629 in 2000. *(Source: U.S. Census Bureau)*

(a) Find the model for Horry County's population.

(b) Use the model to predict the population in 2010. Does your answer seem reasonable? Explain your reasoning.

81. *Vital Statistics* The rate of increase of the number of married couples M (in thousands) in the United States from 1970 to 2000 can be modeled by

$$\frac{dM}{dt} = 0.636t^2 - 28.48t + 632.7$$

where t is the time in years, with $t = 0$ corresponding to 1970. The number of married couples in 2000 was 56,497 thousand. *(Source: U.S. Census Bureau)*

(a) Find the model for the number of married couples in the United States.

(b) Use the model to predict the number of married couples in the United States in 2010. Does your answer seem reasonable? Explain your reasoning.

82. *Economics: Marginal Benefits and Costs* The table gives the marginal benefit and marginal cost of producing x products for a given company. Plot the points in each column and use the *regression* feature of a graphing utility to find a linear model for marginal benefit and a quadratic model for marginal cost. Then use integration to find the benefit B and cost C equations. Assume $B(0) = 0$ and $C(0) = 425$. Finally, find the intervals in which the benefit exceeds the cost of producing x products, and make a recommendation for how many products the company should produce based on your findings. *(Source: Adapted from Taylor,* Economics, *Fourth Edition)*

Number of products	1	2	3	4	5
Marginal benefit	330	320	290	270	250
Marginal cost	150	120	100	110	120

Number of products	6	7	8	9	10
Marginal benefit	230	210	190	170	160
Marginal cost	140	160	190	250	320

83. *Research Project* Use your school's library, the Internet, or some other reference source to research a company that markets a natural resource. Find data on the revenue of the company and on the consumption of the resource. Then find a model for each. Is the company's revenue related to the consumption of the resource? Explain your reasoning.

5.2 THE GENERAL POWER RULE

- Use the General Power Rule to find indefinite integrals.
- Use substitution to find indefinite integrals.
- Use the General Power Rule to solve real-life problems.

The General Power Rule

In Section 5.1, you used the Simple Power Rule

$$\int x^n\,dx = \frac{x^{n+1}}{n+1} + C, \qquad n \neq -1$$

to find antiderivatives of functions expressed as powers of x alone. In this section, you will study a technique for finding antiderivatives of more complicated functions.

To begin, consider how you might find the antiderivative of $2x(x^2 + 1)^3$. Because you are hunting for a function whose derivative is $2x(x^2 + 1)^3$, you might discover the antiderivative as shown.

$$\frac{d}{dx}[(x^2+1)^4] = 4(x^2+1)^3(2x) \qquad \text{Use Chain Rule.}$$

$$\frac{d}{dx}\left[\frac{(x^2+1)^4}{4}\right] = (x^2+1)^3(2x) \qquad \text{Divide both sides by 4.}$$

$$\frac{(x^2+1)^4}{4} + C = \int 2x(x^2+1)^3\,dx \qquad \text{Write in integral form.}$$

The key to this solution is the presence of the factor $2x$ in the integrand. In other words, this solution works because $2x$ is precisely the derivative of $(x^2 + 1)$. Letting $u = x^2 + 1$, you can write

$$\int \overbrace{(x^2+1)^3}^{u^3}\,\underbrace{2x\,dx}_{du} = \int u^3\,du$$

$$= \frac{u^4}{4} + C.$$

This is an example of the **General Power Rule** for integration.

General Power Rule for Integration

If u is a differentiable function of x, then

$$\int u^n \frac{du}{dx}\,dx = \int u^n\,du = \frac{u^{n+1}}{n+1} + C, \qquad n \neq -1.$$

When using the General Power Rule, you must first identify a factor u of the integrand that is raised to a power. Then, you must show that its derivative du/dx is also a factor of the integrand. This is demonstrated in Example 1.

EXAMPLE 1 Applying the General Power Rule

Find each indefinite integral.

(a) $\int 3(3x - 1)^4 \, dx$ (b) $\int (2x + 1)(x^2 + x) \, dx$

(c) $\int 3x^2\sqrt{x^3 - 2} \, dx$ (d) $\int \frac{-4x}{(1 - 2x^2)^2} \, dx$

SOLUTION

(a) $\int 3(3x - 1)^4 \, dx = \int \overbrace{(3x - 1)^4}^{u^n} \overbrace{(3)}^{\frac{du}{dx}} \, dx$ Let $u = 3x - 1$.

$= \frac{(3x - 1)^5}{5} + C$ General Power Rule

(b) $\int (2x + 1)(x^2 + x) \, dx = \int \overbrace{(x^2 + x)}^{u^n} \overbrace{(2x + 1)}^{\frac{du}{dx}} \, dx$ Let $u = x^2 + x$.

$= \frac{(x^2 + x)^2}{2} + C$ General Power Rule

(c) $\int 3x^2\sqrt{x^3 - 2} \, dx = \int \overbrace{(x^3 - 2)^{1/2}}^{u^n} \overbrace{(3x^2)}^{\frac{du}{dx}} \, dx$ Let $u = x^3 - 2$.

$= \frac{(x^3 - 2)^{3/2}}{3/2} + C$ General Power Rule

$= \frac{2}{3}(x^3 - 2)^{3/2} + C$ Simplify.

(d) $\int \frac{-4x}{(1 - 2x^2)^2} \, dx = \int \overbrace{(1 - 2x^2)^{-2}}^{u^n} \overbrace{(-4x)}^{\frac{du}{dx}} \, dx$ Let $u = 1 - 2x^2$.

$= \frac{(1 - 2x^2)^{-1}}{-1} + C$ General Power Rule

$= -\frac{1}{1 - 2x^2} + C$ Simplify.

STUDY TIP

Example 1(b) illustrates a case of the General Power Rule that is sometimes overlooked—when the power is $n = 1$. In this case, the rule takes the form

$$\int u \frac{du}{dx} \, dx = \frac{u^2}{2} + C.$$

STUDY TIP

Remember that you can verify the result of an indefinite integral by differentiating the function. Check the answer to Example 1(d) by differentiating the function

$$F(x) = -\frac{1}{1 - 2x^2} + C.$$

$$\frac{d}{dx}\left(-\frac{1}{1 - 2x^2} + C\right) = \frac{-4x}{(1 - 2x^2)^2}$$

TRY IT 1

Find each indefinite integral.

(a) $\int (3x^2 + 6)(x^3 + 6x)^2 \, dx$ (b) $\int 2x\sqrt{x^2 - 2} \, dx$

Many times, part of the derivative du/dx is missing from the integrand, and in *some* cases you can make the necessary adjustments to apply the General Power Rule.

EXAMPLE 2 Multiplying and Dividing by a Constant

Find $\int x(3 - 4x^2)^2\, dx$.

SOLUTION Let $u = 3 - 4x^2$. To apply the General Power Rule, you need to create $du/dx = -8x$ as a factor of the integrand. You can accomplish this by multiplying and dividing by the constant -8.

$$\int x(3 - 4x^2)^2\, dx = \int \left(-\frac{1}{8}\right)\overbrace{(3 - 4x^2)^2}^{u^n}\overbrace{(-8x)}^{\frac{du}{dx}}\, dx \qquad \text{Multiply and divide by } -8.$$

$$= -\frac{1}{8}\int (3 - 4x^2)^2(-8x)\, dx \qquad \text{Factor } -\tfrac{1}{8} \text{ out of integrand.}$$

$$= \left(-\frac{1}{8}\right)\frac{(3 - 4x^2)^3}{3} + C \qquad \text{General Power Rule}$$

$$= -\frac{(3 - 4x^2)^3}{24} + C \qquad \text{Simplify.}$$

ALGEBRA REVIEW

For help on the algebra in Example 2, see Example 1(b) in the *Chapter 5 Algebra Review*, on page 378.

STUDY TIP

Try using the Chain Rule to check the result of Example 2. After differentiating $-\frac{1}{24}(3 - 4x^2)^3$ and simplifying, you should obtain the original integrand.

TRY IT 2

Find $\int x^3(3x^4 + 1)^2\, dx$.

EXAMPLE 3 A Failure of the General Power Rule

Find $\int -8(3 - 4x^2)^2\, dx$.

SOLUTION Let $u = 3 - 4x^2$. As in Example 2, to apply the General Power Rule you must create $du/dx = -8x$ as a factor of the integrand. In Example 2, you could do this by multiplying and dividing by a constant, and then factoring that constant out of the integrand. This strategy doesn't work with variables. That is,

$$\int -8(3 - 4x^2)^2\, dx \neq \frac{1}{x}\int (3 - 4x^2)^2(-8x)\, dx.$$

To find this indefinite integral, you can expand the integrand and use the Simple Power Rule.

$$\int -8(3 - 4x^2)^2\, dx = \int (-72 + 192x^2 - 128x^4)\, dx$$

$$= -72x + 64x^3 - \frac{128}{5}x^5 + C$$

STUDY TIP

In Example 3, be sure you see that you cannot factor variable quantities outside the integral sign. After all, if this were permissible, then you could move the entire integrand outside the integral sign and eliminate the need for all integration rules except the rule $\int dx = x + C$.

TRY IT 3

Find $\int 2(3x^4 + 1)^2\, dx$.

When an integrand contains an extra constant factor that is not needed as part of du/dx, you can simply move the factor outside the integral sign, as shown in the next example.

EXAMPLE 4 Applying the General Power Rule

Find $\int 7x^2\sqrt{x^3+1}\,dx$.

SOLUTION Let $u = x^3 + 1$. Then you need to create $du/dx = 3x^2$ by multiplying and dividing by 3. The constant factor $\frac{7}{3}$ is not needed as part of du/dx, and can be moved outside the integral sign.

$$\begin{aligned}
\int 7x^2\sqrt{x^3+1}\,dx &= \int 7x^2(x^3+1)^{1/2}\,dx && \text{Rewrite with rational exponent.}\\
&= \int \frac{7}{3}(x^3+1)^{1/2}(3x^2)\,dx && \text{Multiply and divide by 3.}\\
&= \frac{7}{3}\int (x^3+1)^{1/2}(3x^2)\,dx && \text{Factor } \tfrac{7}{3} \text{ outside integral.}\\
&= \frac{7}{3}\,\frac{(x^3+1)^{3/2}}{3/2} + C && \text{General Power Rule}\\
&= \frac{14}{9}(x^3+1)^{3/2} + C && \text{Simplify.}
\end{aligned}$$

TRY IT 4

Find $\int 5x\sqrt{x^2+1}\,dx$.

ALGEBRA REVIEW

For help on the algebra in Example 4, see Example 1(c) in the *Chapter 5 Algebra Review*, on page 378.

TECHNOLOGY

If you use a symbolic integration utility to find indefinite integrals, you should be in for some surprises. This is true because integration is not nearly as straightforward as differentiation. By trying different integrands, you should be able to find several that the program cannot solve: in such situations, it may list a new indefinite integral. You should also be able to find several that have horrendous antiderivatives, some with functions that you may not recognize.

Substitution

The integration technique used in Examples 1, 2, and 4 depends on your ability to recognize or create an integrand of the form $u^n\,du/dx$. With more complicated integrands, it is difficult to recognize the steps needed to fit the integrand to a basic integration formula. When this occurs, an alternative procedure called **substitution** or **change of variables** can be helpful. With this procedure, you completely rewrite the integral in terms of u and du. That is, if $u = f(x)$, then $du = f'(x)\,dx$, and the General Power Rule takes the form

$$\int u^n \frac{du}{dx}\,dx = \int u^n\,du. \qquad \text{General Power Rule}$$

DISCOVERY

Calculate the derivative of each function. Which one is the antiderivative of $f(x) = \sqrt{1-3x}$?

$F(x) = (1-3x)^{3/2} + C$

$F(x) = \frac{2}{3}(1-3x)^{3/2} + C$

$F(x) = -\frac{2}{9}(1-3x)^{3/2} + C$

EXAMPLE 5 Integrating by Substitution

Find $\int \sqrt{1-3x}\,dx$.

SOLUTION Begin by letting $u = 1 - 3x$. Then, $du/dx = -3$ and $du = -3\,dx$. This implies that $dx = -\frac{1}{3}\,du$, and you can find the indefinite integral as shown.

$$\int \sqrt{1-3x}\,dx = \int (1-3x)^{1/2}\,dx \qquad \text{Rewrite with rational exponent.}$$

$$= \int u^{1/2}\left(-\frac{1}{3}\,du\right) \qquad \text{Substitute } u \text{ and } du.$$

$$= -\frac{1}{3}\int u^{1/2}\,du \qquad \text{Factor } -\tfrac{1}{3} \text{ out of integrand.}$$

$$= -\frac{1}{3}\frac{u^{3/2}}{3/2} + C \qquad \text{Apply Power Rule.}$$

$$= -\frac{2}{9}u^{3/2} + C \qquad \text{Simplify.}$$

$$= -\frac{2}{9}(1-3x)^{3/2} + C \qquad \text{Substitute } 1-3x \text{ for } u.$$

TRY IT 5

Find $\int \sqrt{1-2x}\,dx$ by the method of substitution.

To become efficient at integration, you should learn to use *both* techniques discussed in this section. For simpler integrals, you should use pattern recognition and create du/dx by multiplying and dividing by an appropriate constant. For more complicated integrals, you should use a formal change of variables, as shown in Example 5. (You will learn more about this technique in Chapter 6.) For the integrals in this section's exercise set, try working several of the problems twice—once with pattern recognition and once using formal substitution.

Extended Application: Propensity to Consume

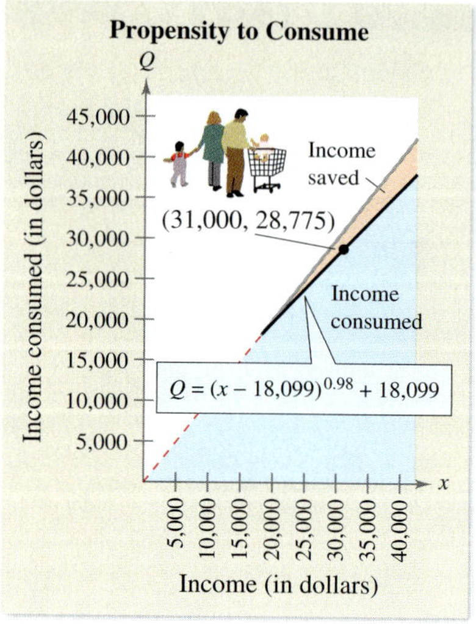

FIGURE 5.4

In 2001, the U.S. poverty level for a family of four was about \$18,100. Families at or below the poverty level tend to consume 100% of their income—that is, they use all their income to purchase necessities such as food, clothing, and shelter. As income level increases, the average consumption tends to drop below 100%. For instance, a family earning \$20,000 may be able to save \$400 and so consume only \$19,600 (98%) of their income. As the income increases, the ratio of consumption to savings tends to decrease. The rate of change of consumption with respect to income is called the **marginal propensity to consume.** *(Source: U.S. Census Bureau)*

EXAMPLE 6 Analyzing Consumption

For a family of four in 2001, the marginal propensity to consume income x can be modeled by

$$\frac{dQ}{dx} = \frac{0.98}{(x - 18{,}099)^{0.02}}, \qquad x \geq 18{,}100$$

where Q represents the income consumed. Use the model to estimate the amount consumed by a family of four whose 2001 income was \$31,000.

SOLUTION Begin by integrating dQ/dx to find a model for the consumption Q. Use the initial condition that $Q = 18{,}100$ when $x = 18{,}100$.

$$\frac{dQ}{dx} = \frac{0.98}{(x - 18{,}099)^{0.02}} \qquad \text{Marginal propensity to consume}$$

$$Q = \int \frac{0.98}{(x - 18{,}099)^{0.02}}\,dx \qquad \text{Integrate to obtain } Q.$$

$$= \int 0.98(x - 18{,}099)^{-0.02}\,dx \qquad \text{Rewrite.}$$

$$= (x - 18{,}099)^{0.98} + C \qquad \text{General Power Rule}$$

$$= (x - 18{,}099)^{0.98} + 18{,}099 \qquad \text{Use initial condition to find } C.$$

Using this model, you can estimate that a family of four with an income of $x = 31{,}000$ consumed about \$28,775. The graph of Q is shown in Figure 5.4.

ALGEBRA REVIEW

When you use the initial condition to find the value of C in Example 6, you substitute 18,100 for Q and 18,100 for x.

$$Q = (x - 18{,}099)^{0.98} + C$$
$$18{,}100 = (18{,}100 - 18{,}099)^{0.98} + C$$
$$18{,}100 = 1 + C$$
$$18{,}099 = C$$

TRY IT 6

According to the model in Example 6, at what income level would a family of four consume \$28,000?

TAKE ANOTHER LOOK

Propensity to Consume

According to the model in Example 6, at what income level would a family of four consume \$36,000? Use the graph in Figure 5.4 to verify your results.

PREREQUISITE REVIEW 5.2

The following warm-up exercises involve skills that were covered in earlier sections. You will use these skills in the exercise set for this section.

In Exercises 1–10, find the indefinite integral.

1. $\int (2x^3 + 1)\, dx$

2. $\int (x^{1/2} + 3x - 4)\, dx$

3. $\int \frac{1}{x^2}\, dx$

4. $\int \frac{1}{3t^3}\, dt$

5. $\int (1 + 2t)t^{3/2}\, dt$

6. $\int \sqrt{x}(2x - 1)\, dx$

7. $\int \frac{5x^3 + 2}{x^2}\, dx$

8. $\int \frac{2x^2 - 5}{x^4}\, dx$

9. $\int (x^2 + 1)^2\, dx$

10. $\int (x^3 - 2x + 1)^2\, dx$

In Exercises 11–14, simplify the expression.

11. $\left(-\frac{5}{4}\right)\frac{(x - 2)^4}{4}$

12. $\left(\frac{1}{6}\right)\frac{(x - 1)^{-2}}{-2}$

13. $(6)\frac{(x^2 + 3)^{2/3}}{2/3}$

14. $\left(\frac{5}{2}\right)\frac{(1 - x^3)^{-1/2}}{-1/2}$

EXERCISES 5.2

In Exercises 1–8, identify u and du/dx for the integral $\int u^n(du/dx)\, dx$.

1. $\int (5x^2 + 1)^2(10x)\, dx$

2. $\int (3 - 4x^2)^3(-8x)\, dx$

3. $\int \sqrt{1 - x^2}(-2x)\, dx$

4. $\int 3x^2\sqrt{x^3 + 1}\, dx$

5. $\int \left(4 + \frac{1}{x^2}\right)^5\left(\frac{-2}{x^3}\right) dx$

6. $\int \frac{1}{(1 + 2x)^2}(2)\, dx$

7. $\int (1 + \sqrt{x})^3\left(\frac{1}{2\sqrt{x}}\right) dx$

8. $\int (4 - \sqrt{x})^2\left(\frac{-1}{2\sqrt{x}}\right) dx$

In Exercises 9–28, find the indefinite integral and check the result by differentiation.

9. $\int (1 + 2x)^4(2)\, dx$

10. $\int (x^2 - 1)^3(2x)\, dx$

11. $\int \sqrt{5x^2 - 4}(10x)\, dx$

12. $\int \sqrt{3 - x^3}(3x^2)\, dx$

13. $\int (x - 1)^4\, dx$

14. $\int (x - 3)^{5/2}\, dx$

15. $\int x(x^2 - 1)^7\, dx$

16. $\int x(1 - 2x^2)^3\, dx$

17. $\int \frac{x^2}{(1 + x^3)^2}\, dx$

18. $\int \frac{x^2}{(x^3 - 1)^2}\, dx$

19. $\int \frac{x + 1}{(x^2 + 2x - 3)^2}\, dx$

20. $\int \frac{6x}{(1 + x^2)^3}\, dx$

21. $\int \frac{x - 2}{\sqrt{x^2 - 4x + 3}}\, dx$

22. $\int \frac{4x + 6}{(x^2 + 3x + 7)^3}\, dx$

23. $\int 5u\sqrt[3]{1 - u^2}\, du$

24. $\int u^3\sqrt{u^4 + 2}\, du$

25. $\int \frac{4y}{\sqrt{1 + y^2}}\, dy$

26. $\int \frac{x^2}{\sqrt{1 - x^3}}\, dx$

27. $\int \frac{-3}{\sqrt{2t + 3}}\, dt$

28. $\int \frac{t + 2t^2}{\sqrt{t}}\, dt$

In Exercises 29–34, use a symbolic integration utility to find the indefinite integral.

29. $\int \frac{x^3}{\sqrt{1 - x^4}}\, dx$

30. $\int \frac{3x}{\sqrt{1 - 4x^2}}\, dx$

31. $\int \left(1 + \frac{4}{t^2}\right)^2\left(\frac{1}{t^3}\right) dt$

32. $\int \left(1 + \frac{1}{t}\right)^3\left(\frac{1}{t^2}\right) dt$

33. $\int (x^3 + 3x)(x^2 + 1)\, dx$

34. $\int (3 - 2x - 4x^2)(1 + 4x)\, dx$

In Exercises 35–42, use formal substitution (as illustrated in Example 5) to find the indefinite integral.

35. $\int x(6x^2 - 1)^3\,dx$

36. $\int x^2(1 - x^3)^2\,dx$

37. $\int x^2(2 - 3x^3)^{3/2}\,dx$

38. $\int t\sqrt{t^2 + 1}\,dt$

39. $\int \frac{x}{\sqrt{x^2 + 25}}\,dx$

40. $\int \frac{3}{\sqrt{2x + 1}}\,dx$

41. $\int \frac{x^2 + 1}{\sqrt{x^3 + 3x + 4}}\,dx$

42. $\int \sqrt{x}(4 - x^{3/2})^2\,dx$

In Exercises 43–46, (a) perform the integration in two ways: once using the Simple Power Rule and once using the General Power Rule. (b) Explain the difference in the results. (c) Which method do you prefer? Explain your reasoning.

43. $\int (2x - 1)^2\,dx$

44. $\int (3 - 2x)^2\,dx$

45. $\int x(x^2 - 1)^2\,dx$

46. $\int x(2x^2 + 1)^2\,dx$

47. Find the equation of the function f whose graph passes through the point $\left(0, \frac{4}{3}\right)$ and whose derivative is

$$f'(x) = x\sqrt{1 - x^2}.$$

48. Find the equation of the function f whose graph passes through the point $\left(0, \frac{7}{3}\right)$ and whose derivative is

$$f'(x) = x\sqrt{1 - x^2}.$$

49. ***Cost*** The marginal cost of a product is modeled by

$$\frac{dC}{dx} = \frac{4}{\sqrt{x + 1}}.$$

When $x = 15$, $C = 50$.

(a) Find the cost function.

(b) Use a graphing utility to graph dC/dx and C in the same viewing window.

50. ***Cost*** The marginal cost of a product is modeled by

$$\frac{dC}{dx} = \frac{12}{\sqrt[3]{12x + 1}}.$$

When $x = 13$, $C = 100$.

(a) Find the cost function.

(b) Use a graphing utility to graph dC/dx and C in the same viewing window.

Supply Function In Exercises 51 and 52, find the supply function $x = f(p)$ that satisfies the initial conditions.

51. $\frac{dx}{dp} = p\sqrt{p^2 - 25}$, $x = 600$ when $p = \$13$

52. $\frac{dx}{dp} = \frac{10}{\sqrt{p - 3}}$, $x = 100$ when $p = \$3$

Demand Function In Exercises 53 and 54, find the demand function $x = f(p)$ that satisfies the initial conditions.

53. $\frac{dx}{dp} = -\frac{6000p}{(p^2 - 16)^{3/2}}$, $x = 5000$ when $p = \$5$

54. $\frac{dx}{dp} = -\frac{400}{(0.02p - 1)^3}$, $x = 10{,}000$ when $p = \$100$

55. ***Gardening*** An evergreen nursery usually sells a type of shrub after 5 years of growth and shaping. The growth rate during those 5 years is approximated by

$$\frac{dh}{dt} = \frac{17.6t}{\sqrt{17.6t^2 + 1}}$$

where t is time in years and h is height in inches. The seedlings are 6 inches tall when planted ($t = 0$).

(a) Find the height function.

(b) How tall are the shrubs when they are sold?

56. ***Cash Flow*** The rate of disbursement dQ/dt of a \$2 million federal grant is proportional to the square of $100 - t$, where t is the time (in days, $0 \le t \le 100$) and Q is the amount that remains to be disbursed. Find the amount that remains to be disbursed after 50 days. Assume that the entire grant will be disbursed after 100 days.

Marginal Propensity to Consume In Exercises 57 and 58, (a) use the marginal propensity to consume, dQ/dx, to write Q as a function of x, where x is the income (in dollars) and Q is the income consumed (in dollars). Assume that 100% of the income is consumed for families that have annual incomes of \$20,000 or less. (b) Use the result of part (a) to complete the table showing the income consumed and the income saved, $x - Q$, for various incomes. (c) Use a graphing utility to represent graphically the income consumed and saved.

x	20,000	50,000	100,000	150,000
Q				
$x - Q$				

57. $\frac{dQ}{dx} = \frac{0.95}{(x - 19{,}999)^{0.05}}$, $x \ge 20{,}000$

58. $\frac{dQ}{dx} = \frac{0.93}{(x - 19{,}999)^{0.07}}$, $x \ge 20{,}000$

In Exercises 59 and 60, use a symbolic integration utility to find the indefinite integral. Verify the result by differentiating.

59. $\int \frac{1}{\sqrt{x} + \sqrt{x + 1}}\,dx$

60. $\int \frac{x}{\sqrt{3x + 2}}\,dx$

5.3 EXPONENTIAL AND LOGARITHMIC INTEGRALS

- Use the Exponential Rule to find indefinite integrals.
- Use the Log Rule to find indefinite integrals.

Using the Exponential Rule

Each of the differentiation rules for exponential functions has its corresponding integration rule.

Integrals of Exponential Functions

Let u be a differentiable function of x.

$$\int e^x\,dx = e^x + C$$ Simple Exponential Rule

$$\int e^u \frac{du}{dx}\,dx = \int e^u\,du = e^u + C$$ General Exponential Rule

EXAMPLE 1 **Integrating Exponential Functions**

Find each indefinite integral.

(a) $\int 2e^x\,dx$

(b) $\int 2e^{2x}\,dx$

(c) $\int (e^x + x)\,dx$

SOLUTION

(a) $\int 2e^x\,dx = 2\int e^x\,dx$ Constant Multiple Rule

$= 2e^x + C$ Simple Exponential Rule

(b) $\int 2e^{2x}\,dx = \int e^{2x}(2)\,dx$ Let $u = 2x$, then $\frac{du}{dx} = 2.$

$= \int e^u \frac{du}{dx}\,dx$

$= e^{2x} + C$ General Exponential Rule

(c) $\int (e^x + x)\,dx = \int e^x\,dx + \int x\,dx$ Sum Rule

$= e^x + \frac{x^2}{2} + C$ Simple Exponential and Power Rules

You can check each of these results by differentiating.

TRY IT 1

Find each indefinite integral.

(a) $\int 3e^x\,dx$

(b) $\int 5e^{5x}\,dx$

(c) $\int (e^x - x)\,dx$

TECHNOLOGY

If you use a symbolic integration utility to find antiderivatives of exponential or logarithmic functions, you can easily obtain results that are beyond the scope of this course. For instance, the antiderivative of e^{x^2} involves the imaginary unit i and the probability function called "ERF." In this course, you are not expected to interpret or use such results. You can simply state that the function cannot be integrated using elementary functions.

EXAMPLE 2 Integrating an Exponential Function

Find $\int e^{3x+1}\,dx$.

SOLUTION Let $u = 3x + 1$, then $du/dx = 3$. You can introduce the missing factor of 3 in the integrand by multiplying and dividing by 3.

$$\int e^{3x+1}\,dx = \frac{1}{3}\int e^{3x+1}(3)\,dx \qquad \text{Multiply and divide by 3.}$$

$$= \frac{1}{3}\int e^{u}\frac{du}{dx}\,dx \qquad \text{Substitute } u \text{ and } du/dx.$$

$$= \frac{1}{3}e^{u} + C \qquad \text{General Exponential Rule}$$

$$= \frac{1}{3}e^{3x+1} + C \qquad \text{Substitute for } u.$$

TRY IT 2

Find $\int e^{2x+3}\,dx$.

ALGEBRA REVIEW

For help on the algebra in Example 3, see Example 1(d) in the *Chapter 5 Algebra Review*, on page 378.

EXAMPLE 3 Integrating an Exponential Function

Find $\int 5xe^{-x^2}\,dx$.

SOLUTION Let $u = -x^2$, then $du/dx = -2x$. You can create the factor $-2x$ in the integrand by multiplying and dividing by -2.

$$\int 5xe^{-x^2}\,dx = \int\left(-\frac{5}{2}\right)e^{-x^2}(-2x)\,dx \qquad \text{Multiply and divide by } -2.$$

$$= -\frac{5}{2}\int e^{-x^2}(-2x)\,dx \qquad \text{Factor } -\tfrac{5}{2} \text{ out of the integrand.}$$

$$= -\frac{5}{2}\int e^{u}\frac{du}{dx}\,dx \qquad \text{Substitute } u \text{ and } \frac{du}{dx}.$$

$$= -\frac{5}{2}e^{u} + C \qquad \text{General Exponential Rule}$$

$$= -\frac{5}{2}e^{-x^2} + C \qquad \text{Substitute for } u.$$

STUDY TIP

Remember that you cannot introduce a missing *variable* in the integrand. For instance, you cannot find $\int e^{x^2}\,dx$ by multiplying and dividing by $2x$ and then factoring $1/(2x)$ out of the integral. That is,

$$\int e^{x^2}\,dx \neq \frac{1}{2x}\int e^{x^2}(2x)\,dx.$$

TRY IT 3

Find $\int 4xe^{x^2}\,dx$.

Using the Log Rule

When the Power Rules for integration were introduced in Sections 5.1 and 5.2, you saw that they work for powers other than $n = -1$.

$$\int x^n\,dx = \frac{x^{n+1}}{n+1} + C, \quad n \neq -1 \qquad \text{Simple Power Rule}$$

$$\int u^n \frac{du}{dx}\,dx = \int u^n\,du = \frac{u^{n+1}}{n+1} + C, \quad n \neq -1 \qquad \text{General Power Rule}$$

The Log Rules for integration allow you to integrate functions of the form $\int x^{-1}\,dx$ and $\int u^{-1}\,du$.

DISCOVERY

The General Power Rule is not valid for $n = -1$. Can you find an antiderivative for u^{-1}?

Integrals of Logarithmic Functions

Let u be a differentiable function of x.

$$\int \frac{1}{x}\,dx = \ln|x| + C \qquad \text{Simple Logarithmic Rule}$$

$$\int \frac{du/dx}{u}\,dx = \int \frac{1}{u}\,du = \ln|u| + C \qquad \text{General Logarithmic Rule}$$

You can verify each of these rules by differentiating. For instance, to verify that $d/dx[\ln|x|] = 1/x$, notice that

$$\frac{d}{dx}[\ln x] = \frac{1}{x} \quad \text{and} \quad \frac{d}{dx}[\ln(-x)] = \frac{-1}{-x} = \frac{1}{x}.$$

STUDY TIP

Notice the absolute values in the Log Rules. For those special cases in which u or x cannot be negative, you can omit the absolute value. For instance, in Example 4(b), it is not necessary to write the antiderivative as $\ln|x^2| + C$ because x^2 cannot be negative.

EXAMPLE 4 Integrating Logarithmic Functions

Find each indefinite integral.

(a) $\int \frac{4}{x}\,dx$ (b) $\int \frac{2x}{x^2}\,dx$ (c) $\int \frac{3}{3x+1}\,dx$

SOLUTION

(a) $\int \frac{4}{x}\,dx = 4\int \frac{1}{x}\,dx$ — Constant Multiple Rule

$= 4\ln|x| + C$ — Simple Logarithmic Rule

(b) $\int \frac{2x}{x^2}\,dx = \int \frac{du/dx}{u}\,dx$ — Let $u = x^2$, then $\frac{du}{dx} = 2x$.

$= \ln|u| + C$ — General Logarithmic Rule

$= \ln x^2 + C$ — Substitute for u.

(c) $\int \frac{3}{3x+1}\,dx = \int \frac{du/dx}{u}\,dx$ — Let $u = 3x + 1$, then $\frac{du}{dx} = 3$.

$= \ln|u| + C$ — General Logarithmic Rule

$= \ln|3x+1| + C$ — Substitute for u.

TRY IT 4

Find each indefinite integral.

(a) $\int \frac{2}{x}\,dx$

(b) $\int \frac{3x^2}{x^3}\,dx$

(c) $\int \frac{2}{2x+1}\,dx$

EXAMPLE 5 **Using the Log Rule**

Find $\int \frac{1}{2x-1}\,dx$.

SOLUTION Let $u = 2x - 1$, then $du/dx = 2$. You can create the necessary factor of 2 in the integrand by multiplying and dividing by 2.

$$\int \frac{1}{2x-1}\,dx = \frac{1}{2}\int \frac{2}{2x-1}\,dx \qquad \text{Multiply and divide by 2.}$$

$$= \frac{1}{2}\int \frac{du/dx}{u}\,dx \qquad \text{Substitute } u \text{ and } \frac{du}{dx}.$$

$$= \frac{1}{2}\ln|u| + C \qquad \text{General Log Rule}$$

$$= \frac{1}{2}\ln|2x-1| + C \qquad \text{Substitute for } u.$$

TRY IT 5

Find $\int \frac{1}{4x+1}\,dx$.

EXAMPLE 6 **Using the Log Rule**

Find $\int \frac{6x}{x^2+1}\,dx$.

SOLUTION Let $u = x^2 + 1$, then $du/dx = 2x$. You can create the necessary factor of $2x$ in the integrand by factoring a 3 out of the integrand.

$$\int \frac{6x}{x^2+1}\,dx = 3\int \frac{2x}{x^2+1}\,dx \qquad \text{Factor 3 out of integrand.}$$

$$= 3\int \frac{du/dx}{u}\,dx \qquad \text{Substitute } u \text{ and } \frac{du}{dx}.$$

$$= 3\ln|u| + C \qquad \text{General Log Rule}$$

$$= 3\ln(x^2+1) + C \qquad \text{Substitute for } u.$$

TRY IT 6

Find $\int \frac{3x}{x^2+4}\,dx$.

ALGEBRA REVIEW

For help on the algebra in the integral at the right, see Example 2(d) in the *Chapter 5 Algebra Review*, on page 379.

Integrals to which the Log Rule can be applied are often given in disguised form. For instance, if a rational function has a numerator of degree greater than or equal to that of the denominator, you should use long division to rewrite the integrand. Here is an example.

$$\int \frac{x^2+6x+1}{x^2+1}\,dx = \int \left(1 + \frac{6x}{x^2+1}\right)dx$$

$$= x + 3\ln(x^2+1) + C$$

The next example summarizes some additional situations in which it is helpful to rewrite the integrand in order to recognize the antiderivative.

EXAMPLE 7 **Rewriting Before Integrating**

Find each indefinite integral.

(a) $\displaystyle\int \frac{3x^2 + 2x - 1}{x^2}\,dx$ (b) $\displaystyle\int \frac{1}{1 + e^{-x}}\,dx$ (c) $\displaystyle\int \frac{x^2 + x + 1}{x - 1}\,dx$

SOLUTION

(a) Begin by rewriting the integrand as the sum of three fractions.

$$\int \frac{3x^2 + 2x - 1}{x^2}\,dx = \int \left(\frac{3x^2}{x^2} + \frac{2x}{x^2} - \frac{1}{x^2}\right) dx$$
$$= \int \left(3 + \frac{2}{x} - \frac{1}{x^2}\right) dx$$
$$= 3x + 2\ln|x| + \frac{1}{x} + C$$

(b) Begin by rewriting the integrand by multiplying and dividing by e^x.

$$\int \frac{1}{1 + e^{-x}}\,dx = \int \left(\frac{e^x}{e^x}\right)\frac{1}{1 + e^{-x}}\,dx$$
$$= \int \frac{e^x}{e^x + 1}\,dx$$
$$= \ln(e^x + 1) + C$$

(c) Begin by dividing the numerator by the denominator.

$$\int \frac{x^2 + x + 1}{x - 1}\,dx = \int \left(x + 2 + \frac{3}{x - 1}\right) dx$$
$$= \frac{x^2}{2} + 2x + 3\ln|x - 1| + C$$

ALGEBRA REVIEW

For help on the algebra in Example 7, see Example 2(a)–(c) in the *Chapter 5 Algebra Review*, on page 379.

TRY IT 7

Find each indefinite integral.

(a) $\displaystyle\int \frac{4x^2 - 3x + 2}{x^2}\,dx$

(b) $\displaystyle\int \frac{2}{e^{-x} + 1}\,dx$

(c) $\displaystyle\int \frac{x^2 + 2x + 4}{x + 1}\,dx$

STUDY TIP

The Exponential and Log Rules are necessary to solve certain real-life problems, such as population growth. You will see such problems in the exercise set for this section.

TAKE ANOTHER LOOK

Using the General Log Rule

One of the most common applications of the Log Rule is to find indefinite integrals of the form

$$\int \frac{a}{bx + c}\,dx.$$

Describe a quick way to find this indefinite integral. Then apply your technique to each integral below.

a. $\displaystyle\int \frac{1}{2x - 5}\,dx$ b. $\displaystyle\int \frac{4}{3x + 2}\,dx$ c. $\displaystyle\int \frac{7}{8x - 3}\,dx$

Use a symbolic integration utility to verify your results.

PREREQUISITE REVIEW 5.3

The following warm-up exercises involve skills that were covered in earlier sections. You will use these skills in the exercise set for this section.

In Exercises 1 and 2, find the domain of the function.

1. $y = \ln(2x - 5)$

2. $y = \ln(x^2 - 5x + 6)$

In Exercises 3–6, use long division to rewrite the quotient.

3. $\dfrac{x^2 + 4x + 2}{x + 2}$

4. $\dfrac{x^2 - 6x + 9}{x - 4}$

5. $\dfrac{x^3 + 4x^2 - 30x - 4}{x^2 - 4x}$

6. $\dfrac{x^4 - x^3 + x^2 + 15x + 2}{x^2 + 5}$

In Exercises 7–10, evaluate the integral.

7. $\displaystyle\int \left(x^3 + \frac{1}{x^2}\right) dx$

8. $\displaystyle\int \frac{x^2 + 2x}{x}\, dx$

9. $\displaystyle\int \frac{x^3 + 4}{x^2}\, dx$

10. $\displaystyle\int \frac{x + 3}{x^3}\, dx$

EXERCISES 5.3

In Exercises 1–12, use the Exponential Rule to find the indefinite integral.

1. $\displaystyle\int 2e^{2x}\, dx$

2. $\displaystyle\int 3e^{-3x}\, dx$

3. $\displaystyle\int e^{4x}\, dx$

4. $\displaystyle\int e^{-0.25x}\, dx$

5. $\displaystyle\int 9xe^{-x^2}\, dx$

6. $\displaystyle\int 3xe^{0.5x^2}\, dx$

7. $\displaystyle\int 5x^2 e^{x^3}\, dx$

8. $\displaystyle\int (2x + 1)e^{x^2+x}\, dx$

9. $\displaystyle\int (x^2 + 2x)e^{x^3+3x^2-1}\, dx$

10. $\displaystyle\int 3(x - 4)e^{x^2-8x}\, dx$

11. $\displaystyle\int 5e^{2-x}\, dx$

12. $\displaystyle\int 3e^{-(x+1)}\, dx$

In Exercises 13–28, use the Log Rule to find the indefinite integral.

13. $\displaystyle\int \frac{1}{x + 1}\, dx$

14. $\displaystyle\int \frac{1}{x - 5}\, dx$

15. $\displaystyle\int \frac{1}{3 - 2x}\, dx$

16. $\displaystyle\int \frac{1}{6x - 5}\, dx$

17. $\displaystyle\int \frac{2}{3x + 5}\, dx$

18. $\displaystyle\int \frac{5}{2x - 1}\, dx$

19. $\displaystyle\int \frac{x}{x^2 + 1}\, dx$

20. $\displaystyle\int \frac{x^2}{3 - x^3}\, dx$

21. $\displaystyle\int \frac{x^2}{x^3 + 1}\, dx$

22. $\displaystyle\int \frac{x}{x^2 + 4}\, dx$

23. $\displaystyle\int \frac{x + 3}{x^2 + 6x + 7}\, dx$

24. $\displaystyle\int \frac{x^2 + 2x + 3}{x^3 + 3x^2 + 9x + 1}\, dx$

25. $\displaystyle\int \frac{1}{x \ln x}\, dx$

26. $\displaystyle\int \frac{1}{x(\ln x)^2}\, dx$

27. $\displaystyle\int \frac{e^{-x}}{1 - e^{-x}}\, dx$

28. $\displaystyle\int \frac{e^x}{1 + e^x}\, dx$

In Exercises 29–38, use a symbolic integration utility to find the indefinite integral.

29. $\displaystyle\int \frac{1}{x^2} e^{2/x}\, dx$

30. $\displaystyle\int \frac{1}{x^3} e^{1/4x^2}\, dx$

31. $\displaystyle\int \frac{1}{\sqrt{x}} e^{\sqrt{x}}\, dx$

32. $\displaystyle\int \frac{e^{1/\sqrt{x}}}{x^{3/2}}\, dx$

33. $\displaystyle\int (e^x - 2)^2\, dx$

34. $\displaystyle\int (e^x - e^{-x})^2\, dx$

35. $\displaystyle\int \frac{e^{-x}}{1 + e^{-x}}\, dx$

36. $\displaystyle\int \frac{3e^x}{2 + e^x}\, dx$

37. $\displaystyle\int \frac{4e^{2x}}{5 - e^{2x}}\, dx$

38. $\displaystyle\int \frac{-e^{3x}}{2 - e^{3x}}\, dx$

In Exercises 39–54, use any basic integration formula or formulas to find the indefinite integral.

39. $\displaystyle\int \frac{e^{2x} + 2e^x + 1}{e^x}\,dx$

40. $\displaystyle\int (6x + e^x)\sqrt{3x^2 + e^x}\,dx$

41. $\displaystyle\int e^x\sqrt{1 - e^x}\,dx$

42. $\displaystyle\int \frac{2(e^x - e^{-x})}{(e^x + e^{-x})^2}\,dx$

43. $\displaystyle\int \frac{1}{(x-1)^2}\,dx$

44. $\displaystyle\int \frac{1}{\sqrt{x+1}}\,dx$

45. $\displaystyle\int 4e^{2x-1}\,dx$

46. $\displaystyle\int (5e^{-2x} + 1)\,dx$

47. $\displaystyle\int \frac{x^3 - 8x}{2x^2}\,dx$

48. $\displaystyle\int \frac{x-1}{4x}\,dx$

49. $\displaystyle\int \frac{2}{1 + e^{-x}}\,dx$

50. $\displaystyle\int \frac{3}{1 + e^{-3x}}\,dx$

51. $\displaystyle\int \frac{x^2 + 2x + 5}{x - 1}\,dx$

52. $\displaystyle\int \frac{x-3}{x+3}\,dx$

53. $\displaystyle\int \frac{1 + e^{-x}}{1 + xe^{-x}}\,dx$

54. $\displaystyle\int \frac{5}{e^{-5x} + 7}\,dx$

In Exercises 55 and 56, find the equation of the function f whose graph passes through the point.

55. $f'(x) = \dfrac{x^2 + 4x + 3}{x - 1}$; $(2, 4)$

56. $f'(x) = \dfrac{x^3 - 4x^2 + 3}{x - 3}$; $(4, -1)$

57. ***Biology*** A population of bacteria is growing at the rate of

$$\frac{dP}{dt} = \frac{3000}{1 + 0.25t}$$

where t is the time in days. When $t = 0$, the population is 1000.

(a) Write an equation that models the population P in terms of the time t.

(b) What is the population after 3 days?

(c) After how many days will the population be 12,000?

58. ***Biology*** Because of an insufficient oxygen supply, the trout population in a lake is dying. The population's rate of change can be modeled by

$$\frac{dP}{dt} = -125e^{-t/20}$$

where t is the time in days. When $t = 0$, the population is 2500.

(a) Write an equation that models the population P in terms of the time t.

(b) What is the population after 15 days?

(c) According to this model, how long will it take for the entire trout population to die?

59. ***Demand*** The marginal price for the demand of a product can be modeled by $dp/dx = 0.1e^{-x/500}$, where x is the quantity demanded. When the demand is 600 units, the price is \$30.

(a) Find the demand function, $p = f(x)$.

(b) Use a graphing utility to graph the demand function. Does price increase or decrease as demand increases?

(c) Use the *zoom* and *trace* features of the graphing utility to find the quantity demanded when the price is \$22.

60. ***Revenue*** The marginal revenue for the sale of a product can be modeled by

$$\frac{dR}{dx} = 50 - 0.02x + \frac{100}{x+1}$$

where x is the quantity demanded.

(a) Find the revenue function.

(b) Use a graphing utility to graph the revenue function.

(c) Find the revenue when 1500 units are sold.

(d) Use the *zoom* and *trace* features of the graphing utility to find the number of units sold when the revenue is \$60,230.

61. ***Average Salary*** From 1995 through 2002, the average salary for superintendents S (in dollars) in the United States changed at the rate of

$$\frac{dS}{dt} = 2621.7e^{0.07t}$$

where $t = 5$ corresponds to 1995. In 2001, the average salary for superintendents was \$118,496. *(Source: Educational Research Service)*

(a) Write a model that gives the average salary for superintendents per year.

(b) Use the model to find the average salary for superintendents in 1999.

62. ***Sales*** The rate of change in sales for The Yankee Candle Company from 1998 through 2003 can be modeled by

$$\frac{dS}{dt} = 1.04t + \frac{544.694}{t}$$

where S is the sales (in millions) and $t = 8$ corresponds to 1998. In 1999, the sales for The Yankee Candle Company were \$256.6 million. *(Source: The Yankee Candle Company)*

(a) Find a model for sales from 1998 through 2003.

(b) Find The Yankee Candle Company's sales in 2002.

True or False? In Exercises 63 and 64, determine whether the statement is true or false. If it is false, explain why or give an example that shows it is false.

63. $(\ln x)^{1/2} = \frac{1}{2}(\ln x)$

64. $\displaystyle\int \ln x = \left(\frac{1}{x}\right) + C$

5.4 AREA AND THE FUNDAMENTAL THEOREM OF CALCULUS

- Evaluate definite integrals.
- Evaluate definite integrals using the Fundamental Theorem of Calculus.
- Use definite integrals to solve marginal analysis problems.
- Find the average values of functions over closed intervals.
- Use properties of even and odd functions to help evaluate definite integrals.
- Find the amounts of annuities.

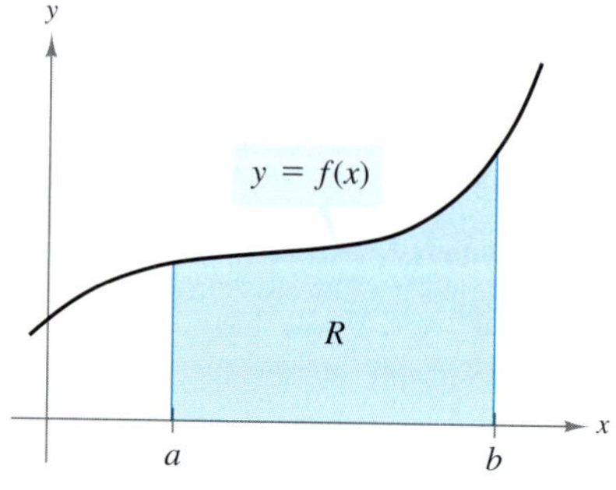

FIGURE 5.5 $\int_a^b f(x)\,dx = \text{Area}$

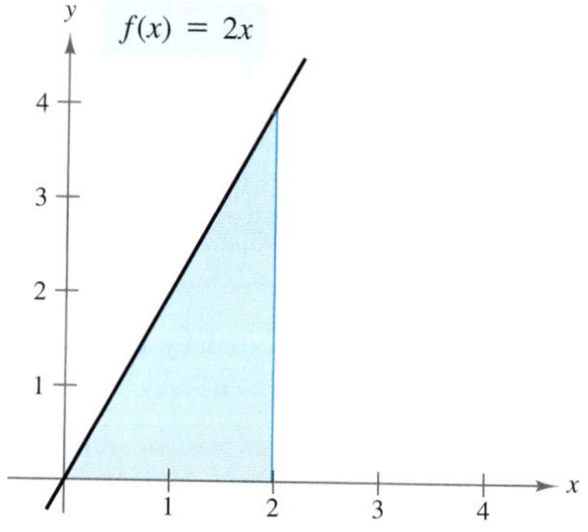

FIGURE 5.6

TRY IT 1

Evaluate the definite integral using a geometric formula. Illustrate your answer with an appropriate sketch.

$$\int_0^3 4x\,dx$$

Area and Definite Integrals

From your study of geometry, you know that area is a number that defines the size of a bounded region. For simple regions, such as rectangles, triangles, and circles, area can be found using geometric formulas.

In this section, you will learn how to use calculus to find the areas of nonstandard regions, such as the region R shown in Figure 5.5.

Definition of a Definite Integral

Let f be nonnegative and continuous on the closed interval $[a, b]$. The area of the region bounded by the graph of f, the x-axis, and the lines $x = a$ and $x = b$ is denoted by

$$\text{Area} = \int_a^b f(x)\,dx.$$

The expression $\int_a^b f(x)\,dx$ is called the **definite integral** from a to b, where a is the **lower limit of integration** and b is the **upper limit of integration.**

EXAMPLE 1 Evaluating a Definite Integral

Evaluate

$$\int_0^2 2x\,dx.$$

SOLUTION This definite integral represents the area of the region bounded by the graph of $f(x) = 2x$, the x-axis, and the line $x = 2$, as shown in Figure 5.6. The region is triangular, with a height of four units and a base of two units.

$$\int_0^2 2x\,dx = \frac{1}{2}(\text{base})(\text{height}) \qquad \text{Formula for area of triangle}$$

$$= \frac{1}{2}(2)(4) = 4 \qquad \text{Simplify.}$$

The Fundamental Theorem of Calculus

Consider the function A, which denotes the area of the region shown in Figure 5.7. To discover the relationship between A and f, let x increase by an amount Δx. This increases the area by ΔA. Let $f(m)$ and $f(M)$ denote the minimum and maximum values of f on the interval $[x, x + \Delta x]$.

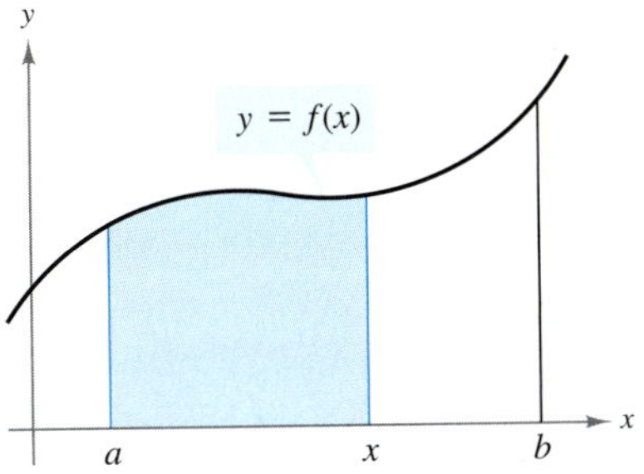

FIGURE 5.7 $A(x) =$ Area from a to x

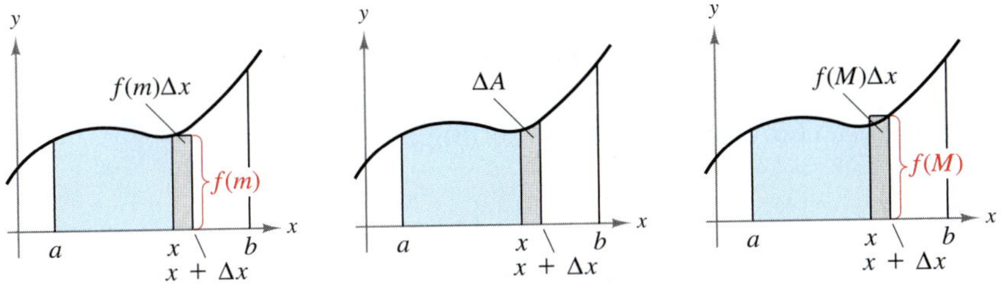

FIGURE 5.8

As indicated in Figure 5.8, you can write the inequality below.

$$f(m)\,\Delta x \le \Delta A \le f(M)\,\Delta x \qquad \text{See Figure 5.8.}$$

$$f(m) \le \frac{\Delta A}{\Delta x} \le f(M) \qquad \text{Divide each term by } \Delta x.$$

$$\lim_{\Delta x \to 0} f(m) \le \lim_{\Delta x \to 0} \frac{\Delta A}{\Delta x} \le \lim_{\Delta x \to 0} f(M) \qquad \text{Take limit of each term.}$$

$$f(x) \le A'(x) \le f(x) \qquad \text{Definition of derivative of } A(x)$$

So, $f(x) = A'(x)$, and $A(x) = F(x) + C$, where $F'(x) = f(x)$. Because $A(a) = 0$, it follows that $C = -F(a)$. So, $A(x) = F(x) - F(a)$, which implies that

$$A(b) = \int_a^b f(x)\,dx = F(b) - F(a).$$

This equation tells you that *if you can find an antiderivative for f*, then you can use the antiderivative to evaluate the definite integral $\int_a^b f(x)\,dx$. This result is called the **Fundamental Theorem of Calculus.**

The Fundamental Theorem of Calculus

If f is nonnegative and continuous on the closed interval $[a, b]$, then

$$\int_a^b f(x)\,dx = F(b) - F(a)$$

where F is any function such that $F'(x) = f(x)$ for all x in $[a, b]$.

STUDY TIP

There are two basic ways to introduce the Fundamental Theorem of Calculus. One way uses an area function, as shown here. The other uses a summation process, as shown in Appendix A.

Guidelines for Using the Fundamental Theorem of Calculus

1. The Fundamental Theorem of Calculus describes a way of *evaluating* a definite integral, not a procedure for finding antiderivatives.
2. In applying the Fundamental Theorem, it is helpful to use the notation

$$\int_a^b f(x)\,dx = F(x)\Big]_a^b = F(b) - F(a).$$

3. The constant of integration C can be dropped because

$$\begin{aligned}\int_a^b f(x)\,dx &= \Big[F(x) + C\Big]_a^b \\ &= [F(b) + C] - [F(a) + C] \\ &= F(b) - F(a) + C - C \\ &= F(b) - F(a).\end{aligned}$$

In the development of the Fundamental Theorem of Calculus, f was assumed to be nonnegative on the closed interval $[a, b]$. As such, the definite integral was defined as an area. Now, with the Fundamental Theorem, the definition can be extended to include functions that are negative on all or part of the closed interval $[a, b]$. Specifically, if f is *any* function that is continuous on a closed interval $[a, b]$, then the **definite integral** of $f(x)$ from a to b is defined to be

$$\int_a^b f(x)\,dx = F(b) - F(a)$$

where F is an antiderivative of f. Remember that definite integrals do not necessarily represent areas and can be negative, zero, or positive.

STUDY TIP

Be sure you see the distinction between indefinite and definite integrals. The *indefinite integral*

$$\int f(x)\,dx$$

denotes a family of *functions*, each of which is an antiderivative of f, whereas the *definite integral*

$$\int_a^b f(x)\,dx$$

is a *number*.

Properties of Definite Integrals

Let f and g be continuous on the closed interval $[a, b]$.

1. $\displaystyle\int_a^b kf(x)\,dx = k\int_a^b f(x)\,dx,\quad k$ is a constant.
2. $\displaystyle\int_a^b [f(x) \pm g(x)]\,dx = \int_a^b f(x)\,dx \pm \int_a^b g(x)\,dx$
3. $\displaystyle\int_a^b f(x)\,dx = \int_a^c f(x)\,dx + \int_c^b f(x)\,dx,\qquad a < c < b$
4. $\displaystyle\int_a^a f(x)\,dx = 0$
5. $\displaystyle\int_a^b f(x)\,dx = -\int_b^a f(x)\,dx$

EXAMPLE 2 Finding Area by the Fundamental Theorem

Find the area of the region bounded by the x-axis and the graph of

$$f(x) = x^2 - 1, \qquad 1 \le x \le 2.$$

SOLUTION Note that $f(x) \ge 0$ on the interval $1 \le x \le 2$, as shown in Figure 5.9. So, you can represent the area of the region by a definite integral. To find the area, use the Fundamental Theorem of Calculus.

$$\text{Area} = \int_1^2 (x^2 - 1)\,dx \qquad \text{Definition of definite integral}$$

$$= \left[\frac{x^3}{3} - x\right]_1^2 \qquad \text{Find antiderivative.}$$

$$= \left[\frac{(2)^3}{3} - 2\right] - \left[\frac{(1)^3}{3} - 1\right] \qquad \text{Apply Fundamental Theorem.}$$

$$= \frac{4}{3} \qquad \text{Simplify.}$$

So, the area of the region is $\frac{4}{3}$ square units.

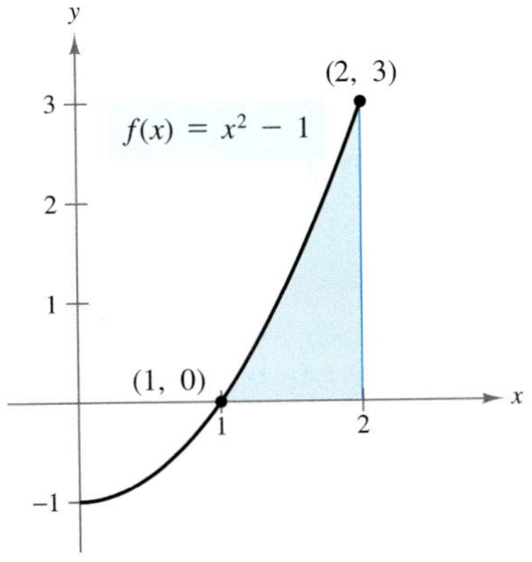

FIGURE 5.9 Area $= \int_1^2 (x^2 - 1)\,dx$

TRY IT 2

Find the area of the region bounded by the x-axis and the graph of $f(x) = x^2 + 1$, $2 \le x \le 3$.

STUDY TIP

It is easy to make errors in signs when evaluating definite integrals. To avoid such errors, enclose the values of the antiderivative at the upper and lower limits of integration in separate sets of parentheses, as shown above.

EXAMPLE 3 Evaluating a Definite Integral

Evaluate the definite integral

$$\int_0^1 (4t + 1)^2\,dt$$

and sketch the region whose area is represented by the integral.

SOLUTION

$$\int_0^1 (4t + 1)^2\,dt = \frac{1}{4}\int_0^1 (4t + 1)^2(4)\,dt \qquad \text{Multiply and divide by 4.}$$

$$= \frac{1}{4}\left[\frac{(4t + 1)^3}{3}\right]_0^1 \qquad \text{Find antiderivative.}$$

$$= \frac{1}{4}\left[\left(\frac{5^3}{3}\right) - \left(\frac{1}{3}\right)\right] \qquad \text{Apply Fundamental Theorem.}$$

$$= \frac{31}{3} \qquad \text{Simplify.}$$

The region is shown in Figure 5.10.

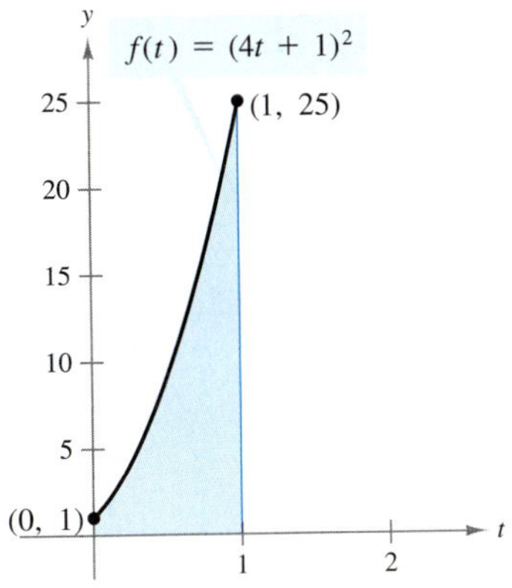

FIGURE 5.10

TRY IT 3

Evaluate $\int_0^1 (2t + 3)^3\,dt$.

EXAMPLE 4 **Evaluating Definite Integrals**

Evaluate each definite integral.

(a) $\int_0^3 e^{2x}\,dx$ (b) $\int_1^2 \frac{1}{x}\,dx$ (c) $\int_1^4 -3\sqrt{x}\,dx$

SOLUTION

(a) $\int_0^3 e^{2x}\,dx = \frac{1}{2}e^{2x}\Big]_0^3 = \frac{1}{2}(e^6 - e^0) \approx 201.21$

(b) $\int_1^2 \frac{1}{x}\,dx = \ln x\Big]_1^2 = \ln 2 - \ln 1 = \ln 2 \approx 0.69$

(c) $\int_1^4 -3\sqrt{x}\,dx = -3\int_1^4 x^{1/2}\,dx$ Rewrite with rational exponent.

$= -3\left[\frac{x^{3/2}}{3/2}\right]_1^4$ Find antiderivative.

$= -2x^{3/2}\Big]_1^4$

$= -2(4^{3/2} - 1^{3/2})$ Apply Fundamental Theorem.

$= -2(8 - 1)$

$= -14$ Simplify.

TRY IT 4

Evaluate each definite integral.

(a) $\int_0^1 e^{4x}\,dx$

(b) $\int_2^5 -\frac{1}{x}\,dx$

STUDY TIP

In Example 4(c), note that the value of a definite integral can be negative.

EXAMPLE 5 **Interpreting Absolute Value**

Evaluate $\int_0^2 |2x - 1|\,dx.$

SOLUTION The region represented by the definite integral is shown in Figure 5.11. From the definition of absolute value, you can write

$$|2x - 1| = \begin{cases} -(2x - 1), & x < \frac{1}{2} \\ 2x - 1, & x \geq \frac{1}{2}. \end{cases}$$

Using Property 3 of definite integrals, you can rewrite the integral as two definite integrals.

$$\int_0^2 |2x - 1|\,dx = \int_0^{1/2} -(2x - 1)\,dx + \int_{1/2}^2 (2x - 1)\,dx$$

$$= \left[-x^2 + x\right]_0^{1/2} + \left[x^2 - x\right]_{1/2}^2$$

$$= \left(-\frac{1}{4} + \frac{1}{2}\right) - (0 + 0) + (4 - 2) - \left(\frac{1}{4} - \frac{1}{2}\right)$$

$$= \frac{5}{2}$$

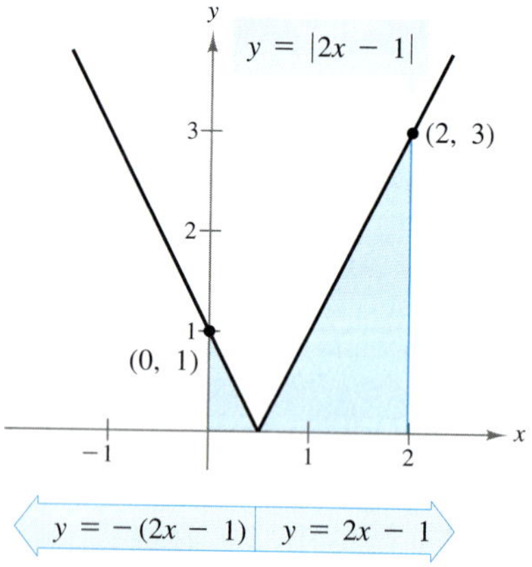

FIGURE 5.11

TRY IT 5

Evaluate $\int_0^5 |x - 2|\,dx.$

Marginal Analysis

You have already studied *marginal analysis* in the context of derivatives and differentials (Sections 2.3 and 3.8). There, you were given a cost, revenue, or profit function, and you used the derivative to approximate the additional cost, revenue, or profit obtained by selling one additional unit. In this section, you will examine the reverse process. That is, you will be given the marginal cost, marginal revenue, or marginal profit and will be asked to use a definite integral to find the exact increase or decrease in cost, revenue, or profit obtained by selling one or several additional units.

For instance, suppose you wanted to find the additional revenue obtained by increasing sales from x_1 to x_2 units. If you knew the revenue function R you could simply subtract $R(x_1)$ from $R(x_2)$. If you didn't know the revenue function, but did know the marginal revenue function, you could still find the additional revenue by using a definite integral, as shown.

$$\int_{x_1}^{x_2} \frac{dR}{dx}\,dx = R(x_2) - R(x_1)$$

EXAMPLE 6 Analyzing a Profit Function

The marginal profit for a product is modeled by $\dfrac{dP}{dx} = -0.0005x + 12.2$.

(a) Find the change in profit when sales increase from 100 to 101 units.

(b) Find the change in profit when sales increase from 100 to 110 units.

SOLUTION

(a) The change in profit obtained by increasing sales from 100 to 101 units is

$$\begin{aligned}\int_{100}^{101} \frac{dP}{dx}\,dx &= \int_{100}^{101} (-0.0005x + 12.2)\,dx \\ &= \Big[-0.00025x^2 + 12.2x\Big]_{100}^{101} \\ &\approx \$12.15.\end{aligned}$$

(b) The change in profit obtained by increasing sales from 100 to 110 units is

$$\begin{aligned}\int_{100}^{110} \frac{dP}{dx}\,dx &= \int_{100}^{110} (-0.0005x + 12.2)\,dx \\ &= \Big[-0.00025x^2 + 12.2x\Big]_{100}^{110} \\ &\approx \$121.48\end{aligned}$$

TRY IT 6

The marginal profit for a product is modeled by

$$\frac{dP}{dx} = -0.0002x + 14.2.$$

(a) Find the change in profit when sales increase from 100 to 101 units.

(b) Find the change in profit when sales increase from 100 to 110 units.

TECHNOLOGY

Symbolic integration utilities can be used to evaluate definite integrals as well as indefinite integrals. If you have access to such a program, try using it to evaluate several of the definite integrals in this section.

Average Value

The *average value* of a function on a closed interval is defined below.

Definition of the Average Value of a Function

If f is continuous on $[a, b]$, then the **average value** of f on $[a, b]$ is

$$\text{Average value of } f \text{ on } [a, b] = \frac{1}{b-a}\int_a^b f(x)\,dx.$$

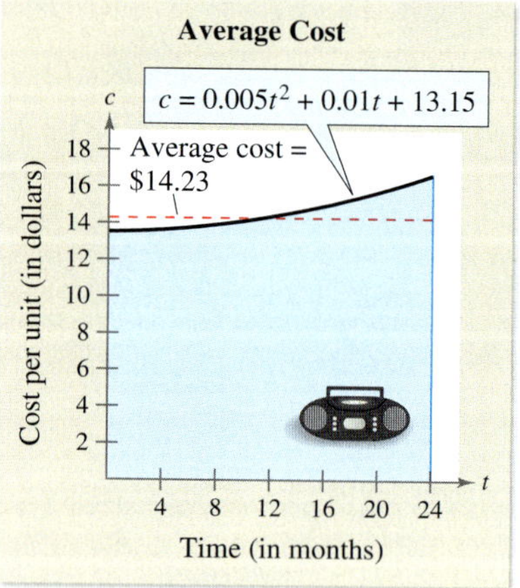

FIGURE 5.12

TRY IT 7

Find the average cost per unit over a two-year period if the cost per unit c of roller blades is given by $c = 0.005t^2 + 0.02t + 12.5$, for $0 \le t \le 24$, where t is the time in months.

In Section 3.5, you studied the effects of production levels on cost using an average cost function. In the next example, you will study the effects of time on cost by using integration to find the average cost.

EXAMPLE 7 Finding the Average Cost

The cost per unit c of producing CD players over a two-year period is modeled by

$$c = 0.005t^2 + 0.01t + 13.15, \qquad 0 \le t \le 24$$

where t is the time in months. Approximate the average cost per unit over the two-year period.

SOLUTION The average cost can be found by integrating c over the interval $[0, 24]$.

$$\begin{aligned}\text{Average cost per unit} &= \frac{1}{24}\int_0^{24}(0.005t^2 + 0.01t + 13.15)\,dt\\ &= \frac{1}{24}\left[\frac{0.005t^3}{3} + \frac{0.01t^2}{2} + 13.15t\right]_0^{24}\\ &= \frac{1}{24}(341.52)\\ &= \$14.23 \qquad \text{(See Figure 5.12.)}\end{aligned}$$

To check the reasonableness of the average value found in Example 7, assume that one unit is produced each month, beginning with $t = 0$ and ending with $t = 24$. When $t = 0$, the cost is

$$\begin{aligned}c &= 0.005(0)^2 + 0.01(0) + 13.15\\ &= \$13.15.\end{aligned}$$

Similarly, when $t = 1$, the cost is

$$\begin{aligned}c &= 0.005(1)^2 + 0.01(1) + 13.15\\ &\approx \$13.17.\end{aligned}$$

Each month, the cost increases, and the average of the 25 costs is

$$\frac{13.15 + 13.17 + 13.19 + 13.23 + \cdots + 16.27}{25} \approx \$14.25.$$

Even and Odd Functions

Several common functions have graphs that are symmetric with respect to the y-axis or the origin, as shown in Figure 5.13. If the graph of f is symmetric with respect to the y-axis, as in Figure 5.13(a), then

$$f(-x) = f(x) \qquad \text{Even function}$$

and f is called an **even** function. If the graph of f is symmetric with respect to the origin, as in Figure 5.13(b), then

$$f(-x) = -f(x) \qquad \text{Odd function}$$

and f is called an **odd** function.

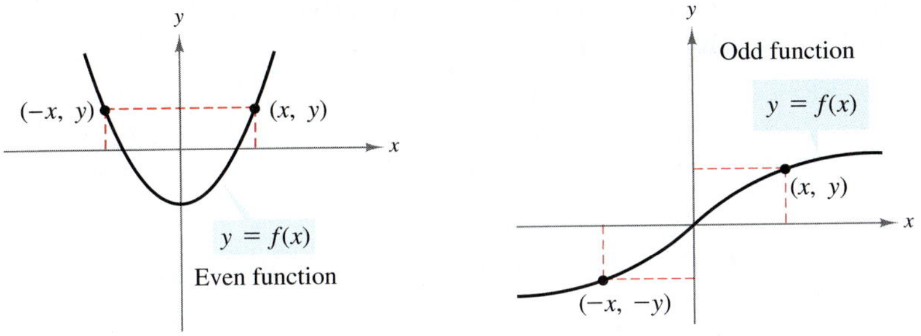

(a) y-axis symmetry (b) Origin symmetry

FIGURE 5.13

Integration of Even and Odd Functions

1. If f is an *even* function, then $\displaystyle\int_{-a}^{a} f(x)\,dx = 2\int_{0}^{a} f(x)\,dx.$

2. If f is an *odd* function, then $\displaystyle\int_{-a}^{a} f(x)\,dx = 0.$

EXAMPLE 8 Integrating Even and Odd Functions

Evaluate each definite integral.

(a) $\displaystyle\int_{-2}^{2} x^2\,dx$ (b) $\displaystyle\int_{-2}^{2} x^3\,dx$

SOLUTION

(a) Because $f(x) = x^2$ is even,

$$\int_{-2}^{2} x^2\,dx = 2\int_{0}^{2} x^2\,dx = 2\left[\frac{x^3}{3}\right]_0^2 = 2\left(\frac{8}{3} - 0\right) = \frac{16}{3}.$$

(b) Because $f(x) = x^3$ is odd,

$$\int_{-2}^{2} x^3\,dx = 0.$$

TRY IT 8

Evaluate each definite integral.

(a) $\displaystyle\int_{-1}^{1} x^4\,dx$

(b) $\displaystyle\int_{-1}^{1} x^5\,dx$

Annuity

A sequence of equal payments made at regular time intervals over a period of time is called an **annuity.** Some examples of annuities are payroll savings plans, monthly home mortgage payments, and individual retirement accounts. The **amount of an annuity** is the sum of the payments plus the interest earned and can be found as shown below.

Amount of an Annuity

If c represents a continuous income function in dollars per year (where t is the time in years), r represents the interest rate compounded continuously, and T represents the term of the annuity in years, then the **amount of an annuity** is

$$\text{Amount of an annuity} = e^{rT}\int_0^T c(t)e^{-rt}\,dt.$$

EXAMPLE 9 Finding the Amount of an Annuity

You deposit \$2000 each year for 15 years in an individual retirement account (IRA) paying 10% interest. How much will you have in your IRA after 15 years?

SOLUTION The income function for your deposit is $c(t) = 2000$. So, the amount of the annuity after 15 years will be

$$\begin{aligned}\text{Amount of an annuity} &= e^{rT}\int_0^T c(t)e^{-rt}\,dt \\ &= e^{(0.10)(15)}\int_0^{15} 2000e^{-0.10t}\,dt \\ &= 2000e^{1.5}\left[-\frac{e^{-0.10t}}{0.10}\right]_0^{15} \\ &\approx \$69{,}633.78.\end{aligned}$$

TRY IT 9

If you deposit \$1000 in a savings account every year, paying 8% interest, how much will be in the account after 10 years?

TAKE ANOTHER LOOK

Using Geometry to Evaluate Definite Integrals

When using the Fundamental Theorem of Calculus to evaluate $\int_a^b f(x)\,dx$, remember that you must first be able to find an antiderivative of $f(x)$. If you are unable to find an antiderivative, you cannot use the Fundamental Theorem. In some cases, you can still evaluate the definite integral. For instance, explain how you can use geometry to evaluate

$$\int_{-2}^{2} \sqrt{4 - x^2}\,dx.$$

Use a symbolic integration utility to verify your answer.

PREREQUISITE REVIEW 5.4

The following warm-up exercises involve skills that were covered in earlier sections. You will use these skills in the exercise set for this section.

In Exercises 1–4, find the indefinite integral.

1. $\int (3x + 7)\, dx$

2. $\int \left(x^{3/2} + 2\sqrt{x}\right) dx$

3. $\int \frac{1}{5x}\, dx$

4. $\int e^{-6x}\, dx$

In Exercises 5 and 6, evaluate the expression when $a = 5$ and $b = 3$.

5. $\left(\frac{a}{5} - a\right) - \left(\frac{b}{5} - b\right)$

6. $\left(6a - \frac{a^3}{3}\right) - \left(6b - \frac{b^3}{3}\right)$

In Exercises 7–10, integrate the marginal function.

7. $\frac{dC}{dx} = 0.02x^{3/2} + 29{,}500$

8. $\frac{dR}{dx} = 9000 + 2x$

9. $\frac{dP}{dx} = 25{,}000 - 0.01x$

10. $\frac{dC}{dx} = 0.03x^2 + 4600$

EXERCISES 5.4

In Exercises 1–8, sketch the region whose area is represented by the definite integral. Then use a geometric formula to evaluate the integral.

1. $\int_0^2 3\, dx$

2. $\int_0^4 2\, dx$

3. $\int_0^5 (x + 1)\, dx$

4. $\int_0^3 (2x + 1)\, dx$

5. $\int_{-2}^3 |x - 1|\, dx$

6. $\int_{-1}^4 |x - 2|\, dx$

7. $\int_{-3}^3 \sqrt{9 - x^2}\, dx$

8. $\int_0^2 \sqrt{4 - x^2}\, dx$

In Exercises 9 and 10, use the values $\int_0^5 f(x)\, dx = 8$ and $\int_0^5 g(x)\, dx = 3$ to evaluate the definite integral.

9. (a) $\int_0^5 [f(x) + g(x)]\, dx$ (b) $\int_0^5 [f(x) - g(x)]\, dx$

(c) $\int_0^5 -4f(x)\, dx$ (d) $\int_0^5 [f(x) - 3g(x)]\, dx$

10. (a) $\int_0^5 2g(x)\, dx$ (b) $\int_5^0 f(x)\, dx$

(c) $\int_5^5 f(x)\, dx$ (d) $\int_0^5 [f(x) - f(x)]\, dx$

In Exercises 11–18, find the area of the region.

11. $y = x - x^2$

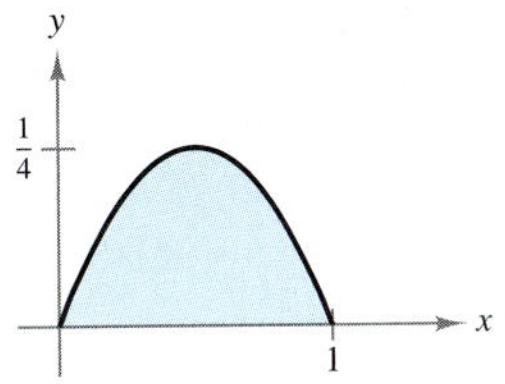

12. $y = 1 - x^4$

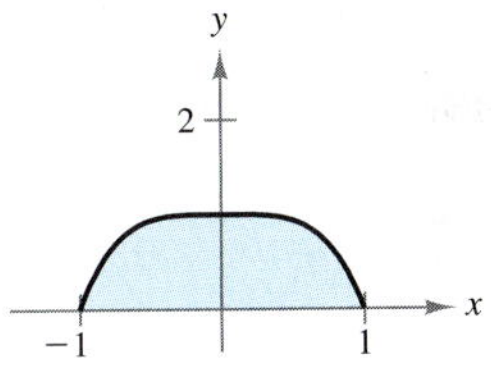

13. $y = \frac{1}{x^2}$

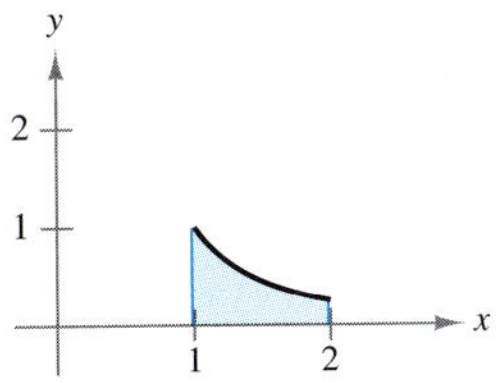

14. $y = \frac{2}{\sqrt{x}}$

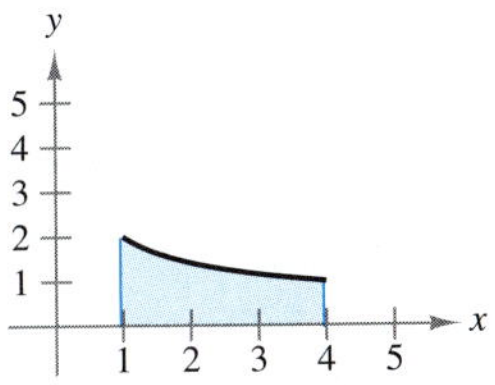

15. $y = 3e^{-x/2}$

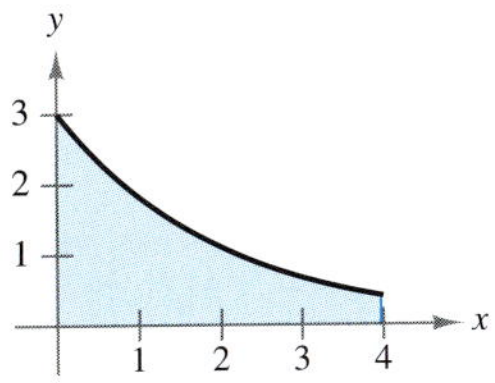

16. $y = 2e^{x/2}$

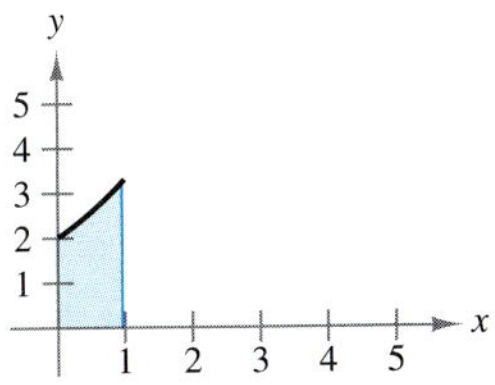

17. $y = \dfrac{x^2 + 4}{x}$

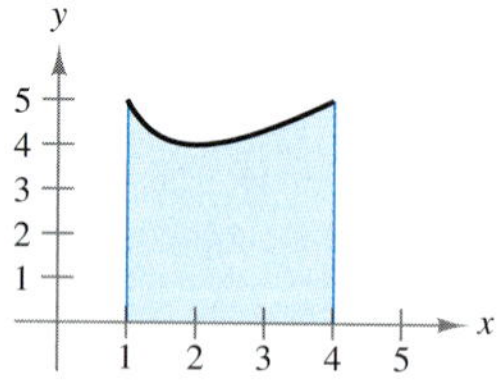

18. $y = \dfrac{x - 2}{x}$

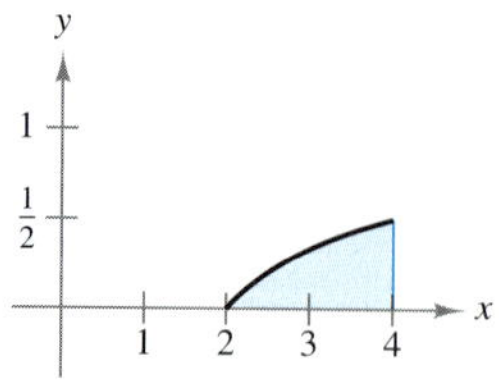

In Exercises 19–42, evaluate the definite integral.

19. $\displaystyle\int_0^1 2x\,dx$

20. $\displaystyle\int_2^7 3v\,dv$

21. $\displaystyle\int_{-1}^0 (2x + 1)\,dx$

22. $\displaystyle\int_2^5 (-3x + 4)\,dx$

23. $\displaystyle\int_{-1}^1 (2t - 1)^2\,dt$

24. $\displaystyle\int_0^1 (1 - 2x)^2\,dx$

25. $\displaystyle\int_0^3 (x - 2)^3\,dx$

26. $\displaystyle\int_2^2 (x - 3)^4\,dx$

27. $\displaystyle\int_{-1}^1 (\sqrt[3]{t} - 2)\,dt$

28. $\displaystyle\int_1^4 \sqrt{\frac{2}{x}}\,dx$

29. $\displaystyle\int_1^4 \frac{2u - 1}{\sqrt{u}}\,du$

30. $\displaystyle\int_0^1 \frac{x - \sqrt{x}}{3}\,dx$

31. $\displaystyle\int_{-1}^0 (t^{1/3} - t^{2/3})\,dt$

32. $\displaystyle\int_0^4 (x^{1/2} + x^{1/4})\,dx$

33. $\displaystyle\int_0^4 \frac{1}{\sqrt{2x + 1}}\,dx$

34. $\displaystyle\int_0^2 \frac{x}{\sqrt{1 + 2x^2}}\,dx$

35. $\displaystyle\int_0^1 e^{-2x}\,dx$

36. $\displaystyle\int_1^2 e^{1-x}\,dx$

37. $\displaystyle\int_1^3 \frac{e^{3/x}}{x^2}\,dx$

38. $\displaystyle\int_{-1}^1 (e^x - e^{-x})\,dx$

39. $\displaystyle\int_0^1 e^{2x}\sqrt{e^{2x} + 1}\,dx$

40. $\displaystyle\int_0^1 \frac{e^{-x}}{\sqrt{e^{-x} + 1}}\,dx$

41. $\displaystyle\int_0^2 \frac{x}{1 + 4x^2}\,dx$

42. $\displaystyle\int_0^1 \frac{e^{2x}}{e^{2x} + 1}\,dx$

In Exercises 43–46, evaluate the definite integral by the most convenient method. Explain your approach.

43. $\displaystyle\int_{-1}^1 |4x|\,dx$

44. $\displaystyle\int_0^3 |2x - 3|\,dx$

45. $\displaystyle\int_0^4 (2 - |x - 2|)\,dx$

46. $\displaystyle\int_{-4}^4 (4 - |x|)\,dx$

In Exercises 47–50, evaluate the definite integral by hand. Then use a symbolic integration utility to evaluate the definite integral. Briefly explain any differences in your results.

47. $\displaystyle\int_{-1}^2 \frac{x}{x^2 - 9}\,dx$

48. $\displaystyle\int_2^3 \frac{x + 1}{x^2 + 2x - 3}\,dx$

49. $\displaystyle\int_0^3 \frac{2e^x}{2 + e^x}\,dx$

50. $\displaystyle\int_1^2 \frac{(2 + \ln x)^3}{x}\,dx$

In Exercises 51–56, evaluate the definite integral by hand. Then use a graphing utility to graph the region whose area is represented by the integral.

51. $\displaystyle\int_1^3 (4x - 3)\,dx$

52. $\displaystyle\int_0^2 (x + 4)\,dx$

53. $\displaystyle\int_0^1 (x - x^3)\,dx$

54. $\displaystyle\int_0^1 \sqrt{x}(1 - x)\,dx$

55. $\displaystyle\int_2^4 \frac{3x^2}{x^3 - 1}\,dx$

56. $\displaystyle\int_0^{\ln 6} \frac{e^x}{2}\,dx$

In Exercises 57–60, find the area of the region bounded by the graphs. Use a graphing utility to graph the region and verify your results.

57. $y = 3x^2 + 1$, $y = 0$, $x = 0$, and $x = 2$

58. $y = 1 + \sqrt{x}$, $y = 0$, $x = 0$, and $x = 4$

59. $y = (x + 5)/x$, $y = 0$, $x = 1$, and $x = 5$

60. $y = 3e^x$, $y = 0$, $x = -2$, and $x = 1$

In Exercises 61–68, use a graphing utility to graph the function over the interval. Find the average value of the function over the interval. Then find all *x*-values in the interval for which the function is equal to its average value.

Function	*Interval*
61. $f(x) = 6 - x^2$	$[-2, 2]$
62. $f(x) = x - 2\sqrt{x}$	$[0, 4]$
63. $f(x) = 5e^{0.2(x-10)}$	$[0, 10]$
64. $f(x) = 2e^{x/4}$	$[0, 4]$
65. $f(x) = x\sqrt{4 - x^2}$	$[0, 2]$
66. $f(x) = \dfrac{1}{(x - 3)^2}$	$[0, 2]$
67. $f(x) = \dfrac{5x}{x^2 + 1}$	$[0, 7]$
68. $f(x) = \dfrac{4x}{x^2 + 1}$	$[0, 1]$

In Exercises 69–72, state whether the function is even, odd, or neither.

69. $f(x) = 3x^4$

70. $g(x) = x^3 - 2x$

71. $g(t) = 2t^5 - 3t^2$

72. $f(t) = 5t^4 + 1$

73. Use the value $\int_0^2 x^2\,dx = \frac{8}{3}$ to evaluate each definite integral. Explain your reasoning.

(a) $\displaystyle\int_{-2}^0 x^2\,dx$ (b) $\displaystyle\int_{-2}^2 x^2\,dx$ (c) $\displaystyle\int_0^2 -x^2\,dx$

74. Use the value $\int_0^2 x^3\,dx = 4$ to evaluate each definite integral. Explain your reasoning.

(a) $\displaystyle\int_{-2}^0 x^3\,dx$ (b) $\displaystyle\int_{-2}^2 x^3\,dx$ (c) $\displaystyle\int_0^2 3x^3\,dx$

Marginal Analysis In Exercises 75–80, find the change in cost C, revenue R, or profit P, for the given marginal. In each case, assume that the number of units x increases by 3 from the specified value of x.

	Marginal	*Number of Units, x*
75.	$\dfrac{dC}{dx} = 2.25$	$x = 100$
76.	$\dfrac{dC}{dx} = \dfrac{20{,}000}{x^2}$	$x = 10$
77.	$\dfrac{dR}{dx} = 48 - 3x$	$x = 12$
78.	$\dfrac{dR}{dx} = 75\left(20 + \dfrac{900}{x}\right)$	$x = 500$
79.	$\dfrac{dP}{dx} = \dfrac{400 - x}{150}$	$x = 200$
80.	$\dfrac{dP}{dx} = 12.5\left(40 - 3\sqrt{x}\right)$	$x = 125$

Annuity In Exercises 81–84, find the amount of an annuity with income function $c(t)$, interest rate r, and term T.

81. $c(t) = \$250, \quad r = 8\%, \quad T = 6$ years

82. $c(t) = \$500, \quad r = 9\%, \quad T = 4$ years

83. $c(t) = \$1500, \quad r = 2\%, \quad T = 10$ years

84. $c(t) = \$2000, \quad r = 3\%, \quad T = 15$ years

Capital Accumulation In Exercises 85–88, you are given the rate of investment dI/dt. Find the capital accumulation over a five-year period by evaluating the definite integral

$$\text{Capital accumulation} = \int_0^5 \frac{dI}{dt}\,dt$$

where t is the time in years.

85. $\dfrac{dI}{dt} = 500$

86. $\dfrac{dI}{dt} = 100t$

87. $\dfrac{dI}{dt} = 500\sqrt{t + 1}$

88. $\dfrac{dI}{dt} = \dfrac{12{,}000t}{(t^2 + 2)^2}$

89. ***Cost*** The total cost of purchasing and maintaining a piece of equipment for x years can be modeled by

$$C = 5000\left(25 + 3\int_0^x t^{1/4}\,dt\right).$$

Find the total cost after (a) 1 year, (b) 5 years, and (c) 10 years.

90. ***Depreciation*** A company purchases a new machine for which the rate of depreciation can be modeled by

$$\frac{dV}{dt} = 10{,}000(t - 6), \quad 0 \le t \le 5$$

where V is the value of the machine after t years. Set up and evaluate the definite integral that yields the total loss of value of the machine over the first 3 years.

91. ***Compound Interest*** A deposit of \$2250 is made in a savings account at an annual interest rate of 12%, compounded continuously. Find the average balance in the account during the first 5 years.

92. ***Mortgage Debt*** The rate of change of mortgage debt outstanding for one- to four-family homes in the United States from 1993 through 2002 can be modeled by

$$\frac{dM}{dt} = 5.4399t^2 + 6603.7e^{-t}$$

where M is the mortgage debt outstanding (in billions of dollars) and $t = 3$ corresponds to 1993. In 1993, the mortgage debt outstanding in the United States was \$3119 billion. *(Source: Board of Governors of the Federal Reserve System)*

(a) Write a model for the debt as a function of t.

(b) What was the average mortgage debt outstanding for 1993 through 2002?

93. ***Medicine*** The velocity v of blood at a distance r from the center of an artery of radius R can be modeled by

$$v = k(R^2 - r^2)$$

where k is a constant. Find the average velocity along a radius of the artery. (Use 0 and R as the limits of integration.)

94. ***Biology*** The rate of change in the number of coyotes $N(t)$ in a population is directly proportional to $650 - N(t)$, where t is time in years.

$$\frac{dN}{dt} = k[650 - N(t)]$$

When $t = 0$, the population is 300, and when $t = 2$, the population has increased to 500.

(a) Find the population function.

(b) Find the average number of coyotes over the first 5 years.

In Exercises 95–98, use a symbolic integration utility to evaluate the definite integral.

95. $\displaystyle\int_3^6 \frac{x}{3\sqrt{x^2 - 8}}\,dx$

96. $\displaystyle\int_{1/2}^1 (x + 1)\sqrt{1 - x}\,dx$

97. $\displaystyle\int_2^5 \left(\frac{1}{x^2} - \frac{1}{x^3}\right)dx$

98. $\displaystyle\int_0^1 x^3(x^3 + 1)^3\,dx$

5.5 THE AREA OF A REGION BOUNDED BY TWO GRAPHS

- Find the areas of regions bounded by two graphs.
- Find consumer and producer surpluses.
- Use the areas of regions bounded by two graphs to solve real-life problems.

Area of a Region Bounded by Two Graphs

With a few modifications, you can extend the use of definite integrals from finding the area of a region *under a graph* to finding the area of a region *bounded by two graphs.* To see how this is done, consider the region bounded by the graphs of f, g, $x = a$, and $x = b$, as shown in Figure 5.14. If the graphs of both f and g lie above the x-axis, then you can interpret the area of the region between the graphs as the area of the region under the graph of g subtracted from the area of the region under the graph of f, as shown in Figure 5.14.

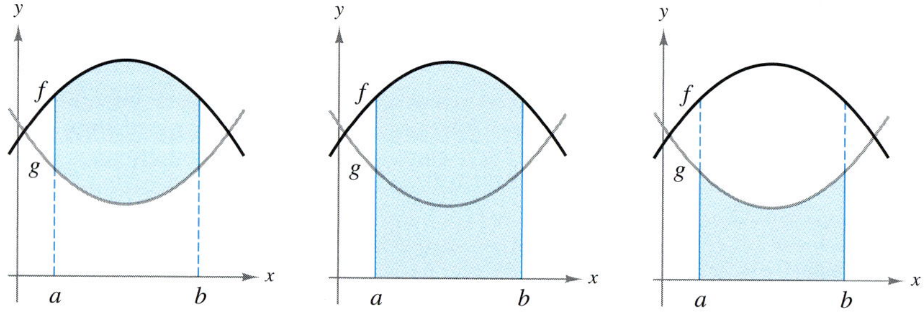

(Area between f and g) = (Area of region under f) − (Area of region under g)

$$\int_a^b [f(x) - g(x)]\, dx = \int_a^b f(x)\, dx - \int_a^b g(x)\, dx$$

FIGURE 5.14

Although Figure 5.14 depicts the graphs of f and g lying above the x-axis, this is not necessary, and the same integrand $[f(x) - g(x)]$ can be used as long as both functions are continuous and $g(x) \le f(x)$ on the interval $[a, b]$.

Area of a Region Bounded by Two Graphs

If f and g are continuous on $[a, b]$ and $g(x) \le f(x)$ for all x in the interval, then the area of the region bounded by the graphs of f, g, $x = a$, and $x = b$ is given by

$$A = \int_a^b [f(x) - g(x)]\, dx.$$

DISCOVERY

Sketch the graph of $f(x) = x^3 - 4x$ and shade in the regions bounded by the graph of f and the x-axis. Write the appropriate integral(s) for this area.

EXAMPLE 1 Finding the Area Bounded by Two Graphs

Find the area of the region bounded by the graphs of

$$y = x^2 + 2 \text{ and } y = x$$

for $0 \le x \le 1$.

SOLUTION Begin by sketching the graphs of both functions, as shown in Figure 5.15. From the figure, you can see that $x \le x^2 + 2$ for all x in $[0, 1]$. So, you can let $f(x) = x^2 + 2$ and $g(x) = x$. Then compute the area as shown.

$$\text{Area} = \int_a^b [f(x) - g(x)]\,dx \quad \text{Area between } f \text{ and } g$$

$$= \int_0^1 [(x^2 + 2) - (x)]\,dx \quad \text{Substitute for } f \text{ and } g.$$

$$= \int_0^1 (x^2 - x + 2)\,dx$$

$$= \left[\frac{x^3}{3} - \frac{x^2}{2} + 2x\right]_0^1 \quad \text{Find antiderivative.}$$

$$= \frac{11}{6} \text{ square units} \quad \text{Apply Fundamental Theorem.}$$

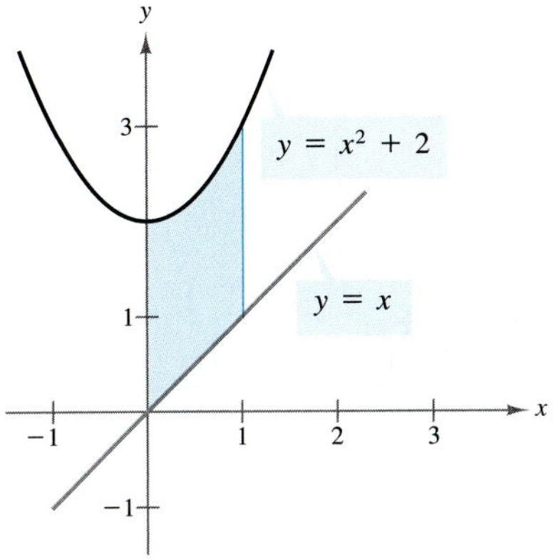

FIGURE 5.15

TRY IT 1

Find the area of the region bounded by the graphs of $y = x^2 + 1$ and $y = x$ for $0 \le x \le 2$. Sketch the region bounded by the graphs.

EXAMPLE 2 Finding the Area Between Intersecting Graphs

Find the area of the region bounded by the graphs of

$$y = 2 - x^2 \text{ and } y = x.$$

SOLUTION In this problem, the values of a and b are not given and you must compute them by finding the points of intersection of the two graphs. To do this, equate the two functions and solve for x. When you do this, you will obtain $x = -2$ and $x = 1$. In Figure 5.16, you can see that the graph of $f(x) = 2 - x^2$ lies above the graph of $g(x) = x$ for all x in the interval $[-2, 1]$.

$$\text{Area} = \int_a^b [f(x) - g(x)]\,dx \quad \text{Area between } f \text{ and } g$$

$$= \int_{-2}^1 [(2 - x^2) - (x)]\,dx \quad \text{Substitute for } f \text{ and } g.$$

$$= \int_{-2}^1 (-x^2 - x + 2)\,dx$$

$$= \left[-\frac{x^3}{3} - \frac{x^2}{2} + 2x\right]_{-2}^1 \quad \text{Find antiderivative.}$$

$$= \frac{9}{2} \text{ square units} \quad \text{Apply Fundamental Theorem.}$$

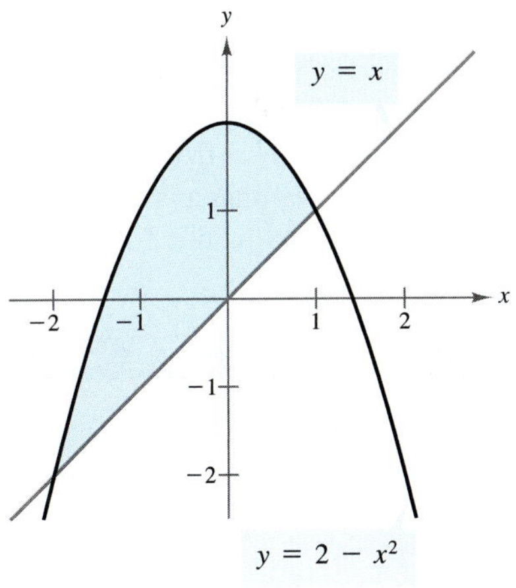

FIGURE 5.16

TRY IT 2

Find the area of the region bounded by the graphs of $y = 3 - x^2$ and $y = 2x$.

EXAMPLE 3 Finding an Area Below the x-Axis

Find the area of the region bounded by the graph of

$$y = x^2 - 3x - 4$$

and the x-axis.

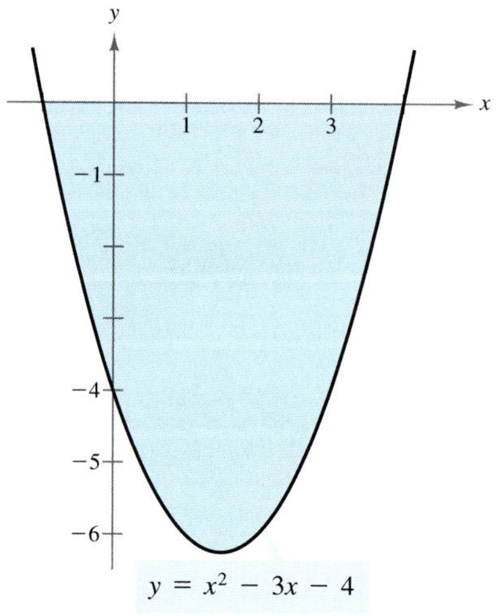

FIGURE 5.17

SOLUTION Begin by finding the x-intercepts of the graph. To do this, set the function equal to zero and solve for x.

$$x^2 - 3x - 4 = 0 \qquad \text{Set function equal to 0.}$$
$$(x - 4)(x + 1) = 0 \qquad \text{Factor.}$$
$$x = 4,\ x = -1 \qquad \text{Solve for } x.$$

From Figure 5.17, you can see that $x^2 - 3x - 4 \le 0$ for all x in the interval $[-1, 4]$. So, you can let $f(x) = 0$ and $g(x) = x^2 - 3x - 4$, and compute the area as shown.

$$\text{Area} = \int_a^b [f(x) - g(x)]\,dx \qquad \text{Area between } f \text{ and } g$$
$$= \int_{-1}^{4} [(0) - (x^2 - 3x - 4)]\,dx \qquad \text{Substitute for } f \text{ and } g.$$
$$= \int_{-1}^{4} (-x^2 + 3x + 4)\,dx$$
$$= \left[-\frac{x^3}{3} + \frac{3x^2}{2} + 4x\right]_{-1}^{4} \qquad \text{Find antiderivative.}$$
$$= \frac{125}{6} \text{ square units} \qquad \text{Apply Fundamental Theorem.}$$

STUDY TIP

When finding the area of a region bounded by two graphs, be sure to use the integrand $[f(x) - g(x)]$. Be sure you realize that you cannot interchange $f(x)$ and $g(x)$. For instance, when solving Example 3, if you subtract $f(x)$ from $g(x)$, you will obtain an answer of $-\frac{125}{6}$, which is not correct.

TRY IT 3

Find the area of the region bounded by the graph of $y = x^2 - x - 2$ and the x-axis.

TECHNOLOGY

Most graphing utilities can display regions that are bounded by two graphs. For instance, to graph the region in Example 3, set the viewing window to $-1 \le x \le 4$ and $-7 \le y \le 1$. Consult your user's manual for specific keystrokes on how to shade the graph. You should obtain the graph shown at the right.

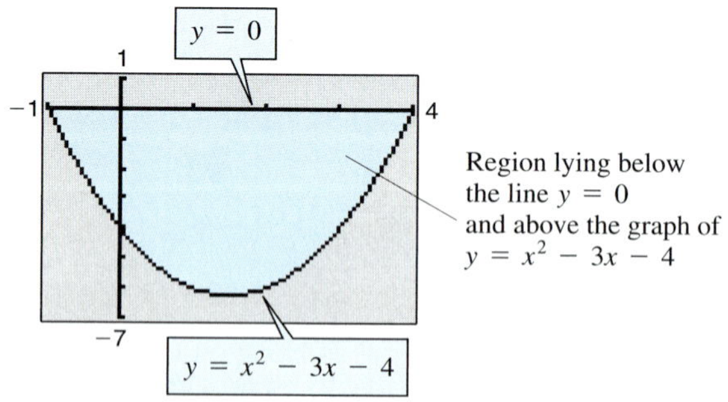

Sometimes two graphs intersect at more than two points. To determine the area of the region bounded by two such graphs, you must find *all* points of intersection and check to see which graph is above the other in each interval determined by the points.

EXAMPLE 4 Using Multiple Points of Intersection

Find the area of the region bounded by the graphs of

$$f(x) = 3x^3 - x^2 - 10x \quad \text{and} \quad g(x) = -x^2 + 2x.$$

SOLUTION To find the points of intersection of the two graphs, set the functions equal to each other and solve for x.

$f(x) = g(x)$	Set $f(x)$ equal to $g(x)$.
$3x^3 - x^2 - 10x = -x^2 + 2x$	Substitute for $f(x)$ and $g(x)$.
$3x^3 - 12x = 0$	Write in standard form.
$3x(x^2 - 4) = 0$	
$3x(x - 2)(x + 2) = 0$	Factor.
$x = 0, x = 2, x = -2$	Solve for x.

These three points of intersection determine two intervals of integration: $[-2, 0]$ and $[0, 2]$. In Figure 5.18, you can see that $g(x) \le f(x)$ in the interval $[-2, 0]$, and that $f(x) \le g(x)$ in the interval $[0, 2]$. So, you must use two integrals to determine the area of the region bounded by the graphs of f and g: one for the interval $[-2, 0]$ and one for the interval $[0, 2]$.

$$\begin{aligned}
\text{Area} &= \int_{-2}^{0} [f(x) - g(x)]\,dx + \int_{0}^{2} [g(x) - f(x)]\,dx \\
&= \int_{-2}^{0} (3x^3 - 12x)\,dx + \int_{0}^{2} (-3x^3 + 12x)\,dx \\
&= \left[\frac{3x^4}{4} - 6x^2\right]_{-2}^{0} + \left[-\frac{3x^4}{4} + 6x^2\right]_{0}^{2} \\
&= (0 - 0) - (12 - 24) + (-12 + 24) - (0 + 0) \\
&= 24
\end{aligned}$$

So, the region has an area of 24 square units.

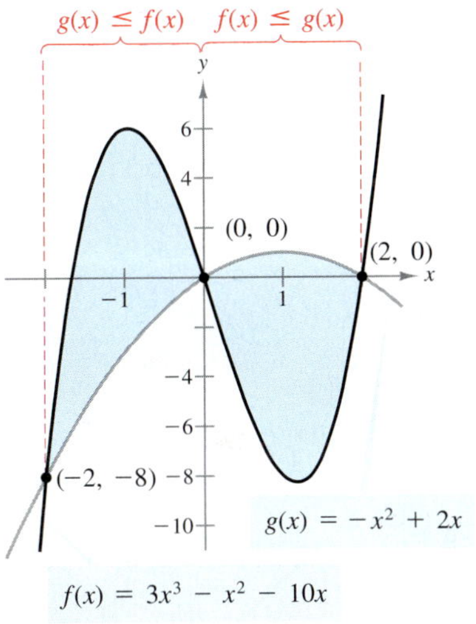

FIGURE 5.18

TRY IT 4

Find the area of the region bounded by the graphs of $f(x) = x^3 + 2x^2 - 3x$ and $g(x) = x^2 + 3x$. Sketch a graph of the region.

STUDY TIP

It is easy to make an error when calculating areas such as that in Example 4. To give yourself some idea about the reasonableness of your solution, you could make a careful sketch of the region on graph paper and then use the grid on the graph paper to approximate the area. Try doing this with the graph shown in Figure 5.18. Is your approximation close to 24 square units?

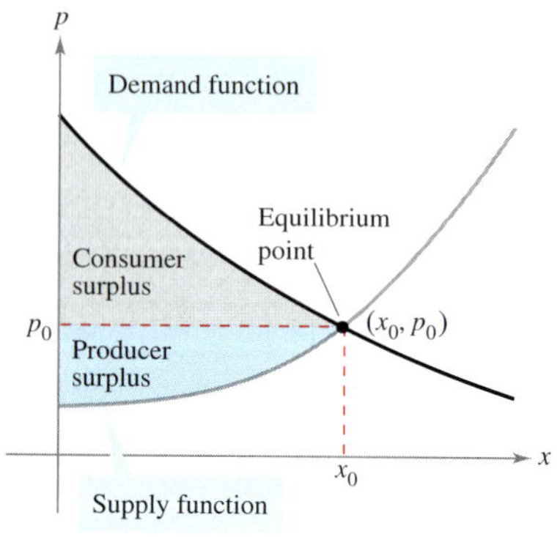

FIGURE 5.19

Consumer Surplus and Producer Surplus

In Section 1.2, you learned that a demand function relates the price of a product to the consumer demand. You also learned that a supply function relates the price of a product to producers' willingness to supply the product. The point (x_0, p_0) at which a demand function $p = D(x)$ and a supply function $p = S(x)$ intersect is the equilibrium point.

Economists call the area of the region bounded by the graph of the demand function, the horizontal line $p = p_0$, and the vertical line $x = 0$ the **consumer surplus.** Similarly, the area of the region bounded by the graph of the supply function, the horizontal line $p = p_0$, and the vertical line $x = 0$ is called the **producer surplus,** as shown in Figure 5.19.

EXAMPLE 5 Finding Surpluses

The demand and supply functions for a product are modeled by

$$\textit{Demand: } p = -0.36x + 9 \quad \text{and} \quad \textit{Supply: } p = 0.14x + 2$$

where x is the number of units (in millions). Find the consumer and producer surpluses for this product.

SOLUTION By equating the demand and supply functions, you can determine that the point of equilibrium occurs when $x = 14$ (million) and the price is $3.96 per unit.

$$\begin{aligned}\text{Consumer surplus} &= \int_0^{14} (\text{demand function} - \text{price})\,dx \\ &= \int_0^{14} [(-0.36x + 9) - 3.96]\,dx \\ &= \Big[-0.18x^2 + 5.04x\Big]_0^{14} \\ &= 35.28\end{aligned}$$

$$\begin{aligned}\text{Producer surplus} &= \int_0^{14} (\text{price} - \text{supply function})\,dx \\ &= \int_0^{14} [3.96 - (0.14x + 2)]\,dx \\ &= \Big[-0.07x^2 + 1.96x\Big]_0^{14} \\ &= 13.72\end{aligned}$$

The consumer surplus and producer surplus are shown in Figure 5.20.

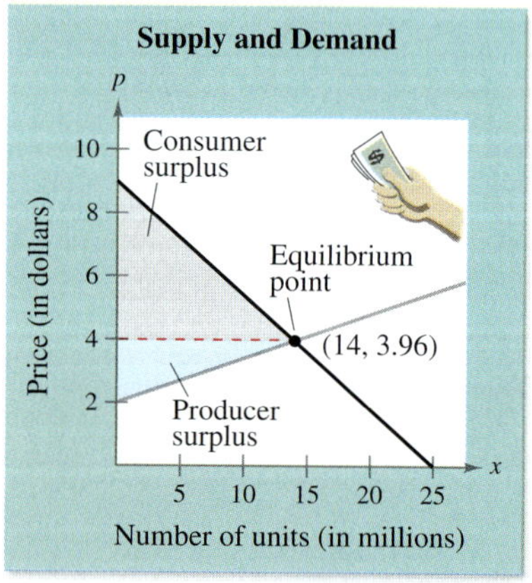

FIGURE 5.20

TRY IT 5

The demand and supply functions for a product are modeled by

$$\textit{Demand: } p = -0.2x + 8 \quad \text{and} \quad \textit{Supply: } p = 0.1x + 2$$

where x is the number of units (in millions). Find the consumer and producer surpluses for this product.

Application

In addition to consumer and producer surpluses, there are many other types of applications involving the area of a region bounded by two graphs. Example 6 shows one of these applications.

EXAMPLE 6 Modeling Petroleum Consumption

In the *Annual Energy Outlook*, the U.S. Energy Information Administration projected the consumption C (in quadrillions of Btu per year) of petroleum to follow the model

$$C_1 = 0.012t^2 + 0.51t + 38.0, \qquad 0 \le t \le 20$$

where $t = 0$ corresponds to January 1, 2000. If the actual consumption more closely followed the model

$$C_2 = 0.019t^2 + 0.37t + 38.2, \qquad 0 \le t \le 20$$

how much petroleum would be saved?

©Michael Schwarz/The Image Works

The three largest sources of petroleum in the United States are Alaska, Texas, and offshore rigs. In 2003, the United States consumed about 39.1 quadrillion Btu of petroleum.

SOLUTION The petroleum saved can be represented as the area of the region between the graphs of C_1 and C_2, as shown in Figure 5.21.

$$\begin{aligned}\text{Petroleum saved} &= \int_0^{20} [C_1 - C_2]\,dt \\ &= \int_0^{20} (-0.007t^2 + 0.14t - 0.2)\,dt \\ &= \left(-\frac{0.007}{3}t^3 + 0.07t^2 - 0.2t\right)\Bigg]_0^{20} \\ &\approx 5.3\end{aligned}$$

So, there would be about 5.3 quadrillion Btu of petroleum saved.

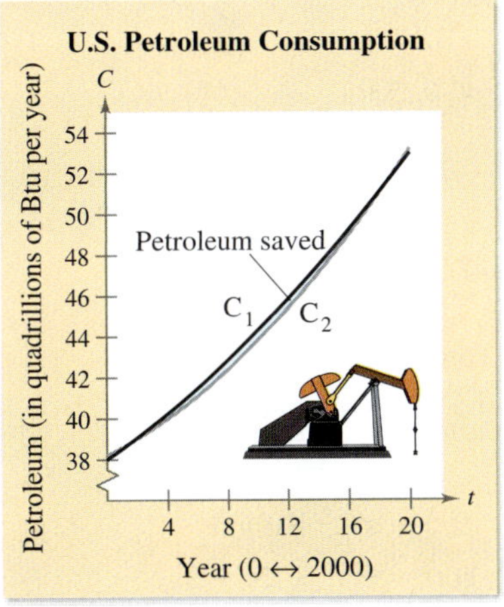

FIGURE 5.21

TRY IT 6

The projected fuel cost C (in millions of dollars per year) for a trucking company from 2005 through 2017 is $C_1 = 5.6 + 2.21t$, $5 \le t \le 17$, where $t = 5$ corresponds to 2005. If the company purchases more efficient truck engines, fuel cost is expected to decrease and to follow the model $C_2 = 4.1 + 1.74t$, $5 \le t \le 17$. How much can the company save with the more efficient engines?

TAKE ANOTHER LOOK

Finding Units for Area

In Example 6, the vertical axis is measured in quadrillions of Btu per year and the horizontal axis is measured in years. Explain why the area of the region shown in Figure 5.21 is measured in quadrillions of Btu.

PREREQUISITE REVIEW 5.5

The following warm-up exercises involve skills that were covered in earlier sections. You will use these skills in the exercise set for this section.

In Exercises 1–4, simplify the expression.

1. $(-x^2 + 4x + 3) - (x + 1)$

2. $(-2x^2 + 3x + 9) - (-x + 5)$

3. $(-x^3 + 3x^2 - 1) - (x^2 - 4x + 4)$

4. $(3x + 1) - (-x^3 + 9x + 2)$

In Exercises 5–10, find the points of intersection of the graphs.

5. $f(x) = x^2 - 4x + 4,\ g(x) = 4$

6. $f(x) = -3x^2,\ g(x) = 6 - 9x$

7. $f(x) = x^2,\ g(x) = -x + 6$

8. $f(x) = \frac{1}{2}x^3,\ g(x) = 2x$

9. $f(x) = x^2 - 3x,\ g(x) = 3x - 5$

10. $f(x) = e^x,\ g(x) = e$

EXERCISES 5.5

In Exercises 1–8, find the area of the region.

1. $f(x) = x^2 - 6x$
$g(x) = 0$

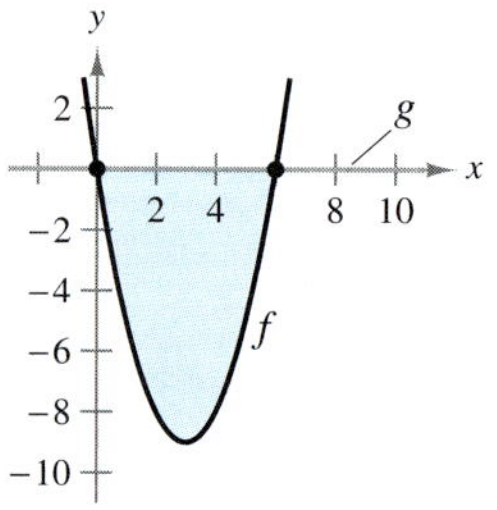

2. $f(x) = x^2 + 2x + 1$
$g(x) = 2x + 5$

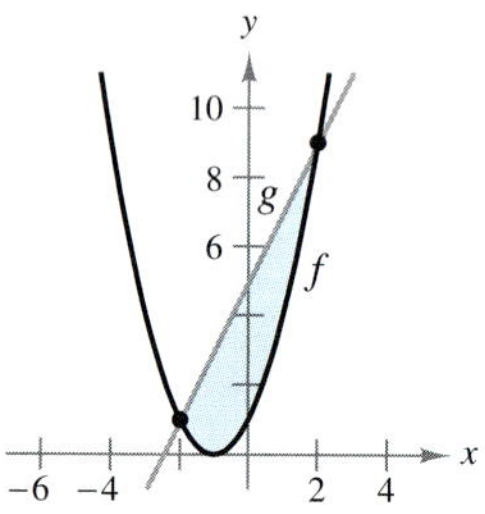

3. $f(x) = x^2 - 4x + 3$
$g(x) = -x^2 + 2x + 3$

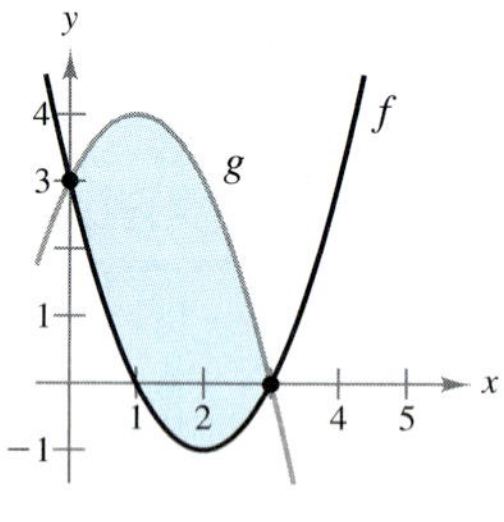

4. $f(x) = x^2$
$g(x) = x^3$

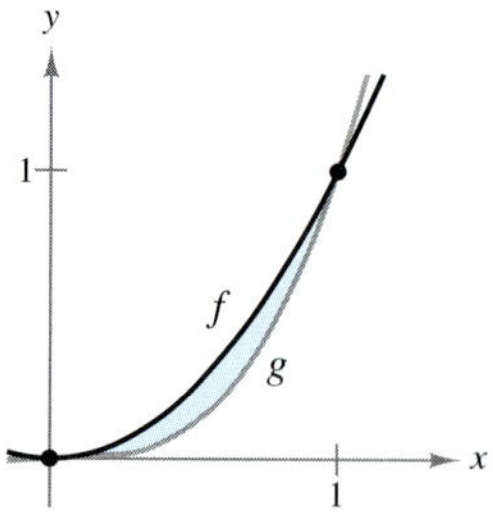

5. $f(x) = 3(x^3 - x)$
$g(x) = 0$

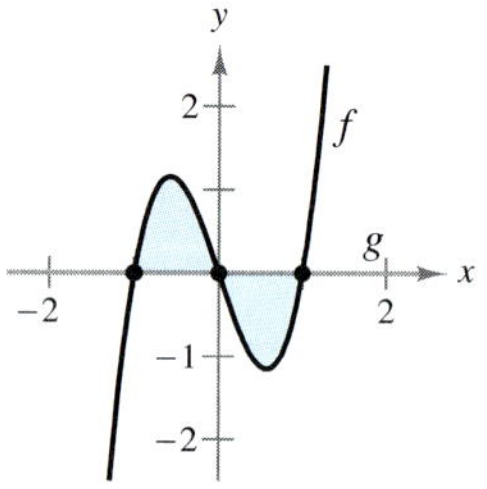

6. $f(x) = (x - 1)^3$
$g(x) = x - 1$

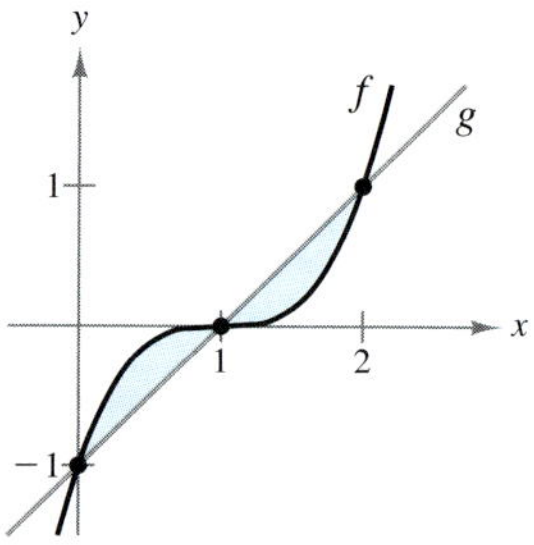

7. $f(x) = e^x - 1$
$g(x) = 0$

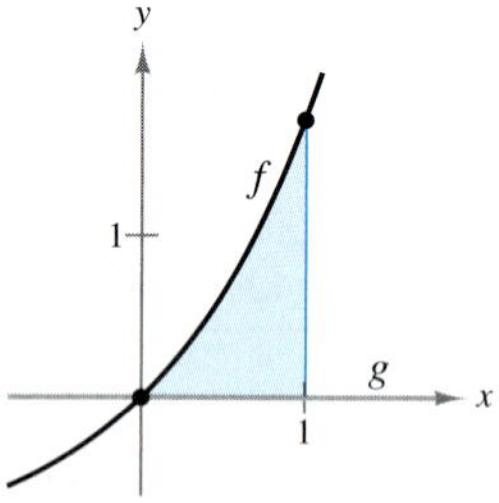

8. $f(x) = -x + 3$
$g(x) = 2x^{-1}$

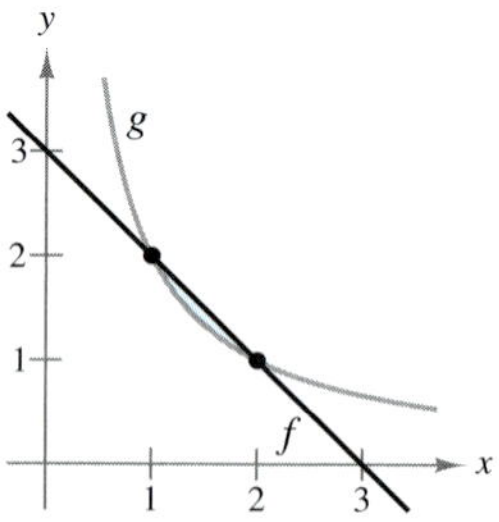

In Exercises 9–14, sketch the region whose area is represented by the definite integral.

9. $\int_0^4 \left[(x + 1) - \frac{1}{2}x\right] dx$

10. $\int_{-1}^1 [(1 - x^2) - (x^2 - 1)] dx$

11. $\int_{-2}^2 [2x^2 - (x^4 - 2x^2)] dx$

12. $\int_{-4}^0 [(x - 6) - (x^2 + 5x - 6)] dx$

13. $\int_{-1}^2 [(y^2 + 2) - 1] dy$

14. $\int_{-2}^3 [(y + 6) - y^2] dy$

In Exercises 15–26, sketch the region bounded by the graphs of the functions and find the area of the region.

15. $y = \dfrac{1}{x^2}, y = 0, x = 1, x = 5$

16. $y = x^3 - 2x + 1, y = -2x, x = 1$

17. $f(x) = \sqrt[3]{x}, g(x) = x$

18. $f(x) = \sqrt{3x} + 1, g(x) = x + 1$

19. $y = x^2 - 4x + 3, y = 3 + 4x - x^2$

20. $y = 4 - x^2, y = x^2$

21. $y = xe^{-x^2}, y = 0, x = 0, x = 1$

22. $y = \dfrac{e^{1/x}}{x^2}, y = 0, x = 1, x = 3$

23. $y = \dfrac{8}{x}, y = x^2, y = 0, x = 1, x = 4$

24. $y = \dfrac{1}{x}, y = x^3, x = \dfrac{1}{2}, x = 1$

25. $f(x) = e^{0.5x}, g(x) = -\dfrac{1}{x}, x = 1, x = 2$

26. $f(x) = \dfrac{1}{x}, g(x) = -e^x, x = \dfrac{1}{2}, x = 1$

In Exercises 27–30, sketch the region bounded by the graphs of the functions and find the area of the region.

27. $f(y) = y^2, g(y) = y + 2$

28. $f(y) = y(2 - y), g(y) = -y$

29. $f(y) = \sqrt{y}, y = 9, x = 0$

30. $f(y) = y^2 + 1, g(y) = 4 - 2y$

In Exercises 31–34, use a graphing utility to graph the region bounded by the graphs of the functions. Write the definite integrals that represent the area of the region. (*Hint:* Multiple integrals may be necessary.)

31. $f(x) = 2x,\ g(x) = 4 - 2x,\ h(x) = 0$

32. $f(x) = x(x^2 - 3x + 3), g(x) = x^2$

33. $y = \dfrac{4}{x}, y = x, x = 1, x = 4$

34. $y = x^3 - 4x^2 + 1, y = x - 3$

In Exercises 35–38, use a graphing utility to graph the region bounded by the graphs of the functions, and find the area of the region.

35. $f(x) = x^2 - 4x, g(x) = 0$

36. $f(x) = 3 - 2x - x^2, g(x) = 0$

37. $f(x) = x^2 + 2x + 1, g(x) = x + 1$

38. $f(x) = -x^2 + 4x + 2, g(x) = x + 2$

In Exercises 39 and 40, use integration to find the area of the triangular region having the given vertices.

39. $(0, 0), (4, 0), (4, 4)$

40. $(0, 0), (4, 0), (6, 4)$

Consumer and Producer Surpluses In Exercises 41–46, find the consumer and producer surpluses.

	Demand Function	*Supply Function*
41.	$p_1(x) = 50 - 0.5x$	$p_2(x) = 0.125x$
42.	$p_1(x) = 300 - x$	$p_2(x) = 100 + x$
43.	$p_1(x) = 200 - 0.02x^2$	$p_2(x) = 100 + x$
44.	$p_1(x) = 1000 - 0.4x^2$	$p_2(x) = 42x$
45.	$p_1(x) = \dfrac{10{,}000}{\sqrt{x + 100}}$	$p_2(x) = 100\sqrt{0.05x + 10}$
46.	$p_1(x) = \sqrt{25 - 0.1x}$	$p_2(x) = \sqrt{9 + 0.1x} - 2$

47. ***Writing*** Describe the characteristics of typical demand and supply functions.

48. ***Writing*** Suppose that the demand and supply functions for a product do not intersect. What can you conclude?

Revenue In Exercises 49 and 50, two models, R_1 and R_2, are given for revenue (in billions of dollars per year) for a large corporation. Both models are estimates of revenues for 2004–2008, with $t = 4$ corresponding to 2004. Which model is projecting the greater revenue? How much more total revenue does that model project over the four-year period?

49. $R_1 = 7.21 + 0.58t, R_2 = 7.21 + 0.45t$

50. $R_1 = 7.21 + 0.26t + 0.02t^2, R_2 = 7.21 + 0.1t + 0.01t^2$

51. *Fuel Cost* The projected fuel cost C (in millions of dollars per year) for an airline company from 2004 through 2010 is $C_1 = 568.5 + 7.15t$, where $t = 4$ corresponds to 2004. If the company purchases more efficient airplane engines, fuel cost is expected to decrease and to follow the model $C_2 = 525.6 + 6.43t$. How much can the company save with the more efficient engines? Explain your reasoning.

52. *Health* An epidemic was spreading such that t weeks after its outbreak it had infected

$$N_1(t) = 0.1t^2 + 0.5t + 150, \quad 0 \le t \le 50$$

people. Twenty-five weeks after the outbreak, a vaccine was developed and administered to the public. At that point, the number of people infected was governed by the model

$$N_2(t) = -0.2t^2 + 6t + 200.$$

Approximate the number of people that the vaccine prevented from becoming ill during the epidemic.

53. *Consumer Trends* For the years 1990 through 2001, the per capita consumption of tomatoes (in pounds per year) in the United States can be modeled by

$$C(t) = \begin{cases} 0.085t^3 - 0.309t^2 + 0.13t + 15.5, \\ 0 \le t \le 4 \\ 0.01515t^4 - 0.5348t^3 + 6.864t^2 - 37.68t + 91.4 \\ 4 < t \le 11 \end{cases}$$

where $t = 0$ corresponds to 1990. *(Source: U.S. Department of Agriculture)*

(a) Use a graphing utility to graph this model.

(b) Suppose the tomato consumption from 1995 through 2001 had continued to follow the model for 1990 through 1994. How many more or fewer pounds of tomatoes would have been consumed from 1995 through 2001?

54. *Consumer and Producer Surpluses* Factory orders for an air conditioner are about 6000 units per week when the price is \$331 and about 8000 units per week when the price is \$303. The supply function is given by $p = 0.0275x$. Find the consumer and producer surpluses. (Assume the demand function is linear.)

55. *Consumer and Producer Surpluses* Repeat Exercise 54 with a demand of about 6000 units per week when the price is \$325 and about 8000 units per week when the price is \$300. Find the consumer and producer surpluses. (Assume the demand function is linear.)

56. *Cost, Revenue, and Profit* The revenue from a manufacturing process (in millions of dollars per year) is projected to follow the model $R = 100$ for 10 years. Over the same period of time, the cost (in millions of dollars per year) is projected to follow the model $C = 60 + 0.2t^2$, where t is the time (in years). Approximate the profit over the 10-year period.

57. *Cost, Revenue, and Profit* Repeat Exercise 56 for revenue and cost models given by $R = 100 + 0.08t$ and $C = 60 + 0.2t^2$.

58. *Lorenz Curve* Economists use *Lorenz curves* to illustrate the distribution of income in a country. Letting x represent the percent of families in a country and y the percent of total income, the model $y = x$ would represent a country in which each family had the same income. The Lorenz curve, $y = f(x)$, represents the actual income distribution. The area between these two models, for $0 \le x \le 100$, indicates the "income inequality" of a country. In 2001, the Lorenz curve for the United States could be modeled by

$$y = (0.00059x^2 + 0.0233x + 1.731)^2, \quad 0 \le x \le 100$$

where x is measured from the poorest to the wealthiest families. Find the income inequality for the United States in 2001. *(Source: U.S. Census Bureau)*

59. *Income Distribution* Using the Lorenz curve in Exercise 58, complete the table, which lists the percent of total income earned by each quintile in the United States in 2001.

Quintile	Lowest	2nd	3rd	4th	Highest
Percent					

BUSINESS CAPSULE

EricCrossan.com

In 1994, wardrobe consultant and personal shopper Marilyn N. Wright started the company Marilyn's Fashions in Newark, Delaware. Wright consults with her 450 American and 250 international clients on their wardrobes, styles, and budgets, then delivers the clothes and accessories to their doorsteps in less than 2 weeks. She used \$5000 in start-up capital while working as a claims adjuster and now brings in \$250,000 in annual revenue.

60. *Research Project* Use your school's library, the Internet, or some other reference source to research a small company similar to that described above. Describe the impact of different factors, such as start-up capital and market conditions, on a company's revenue.

5.6 THE DEFINITE INTEGRAL AS THE LIMIT OF A SUM

- Use the Midpoint Rule to approximate definite integrals.
- Use a symbolic integration utility to approximate definite integrals.

The Midpoint Rule

In Section 5.4, you learned that you cannot use the Fundamental Theorem of Calculus to evaluate a definite integral unless you can find an antiderivative of the integrand. In cases where this cannot be done, you can approximate the value of the integral using an approximation technique. One such technique is called the **Midpoint Rule.** (Two other techniques are discussed in Section 6.5.)

EXAMPLE 1 Approximating the Area of a Plane Region

Use the five rectangles in Figure 5.22 to approximate the area of the region bounded by the graph of $f(x) = -x^2 + 5$, the x-axis, and the lines $x = 0$ and $x = 2$.

SOLUTION You can find the heights of the five rectangles by evaluating f at the midpoint of each of the following intervals.

$$\left[0, \frac{2}{5}\right], \quad \left[\frac{2}{5}, \frac{4}{5}\right], \quad \left[\frac{4}{5}, \frac{6}{5}\right], \quad \left[\frac{6}{5}, \frac{8}{5}\right], \quad \left[\frac{8}{5}, \frac{10}{5}\right]$$

Evaluate f at the midpoints of these intervals.

The width of each rectangle is $\frac{2}{5}$. So, the sum of the five areas is

$$\begin{aligned}
\text{Area} &\approx \frac{2}{5}f\left(\frac{1}{5}\right) + \frac{2}{5}f\left(\frac{3}{5}\right) + \frac{2}{5}f\left(\frac{5}{5}\right) + \frac{2}{5}f\left(\frac{7}{5}\right) + \frac{2}{5}f\left(\frac{9}{5}\right) \\
&= \frac{2}{5}\left[f\left(\frac{1}{5}\right) + f\left(\frac{3}{5}\right) + f\left(\frac{5}{5}\right) + f\left(\frac{7}{5}\right) + f\left(\frac{9}{5}\right)\right] \\
&= \frac{2}{5}\left(\frac{124}{25} + \frac{116}{25} + \frac{100}{25} + \frac{76}{25} + \frac{44}{25}\right) \\
&= \frac{920}{125} \\
&= 7.36.
\end{aligned}$$

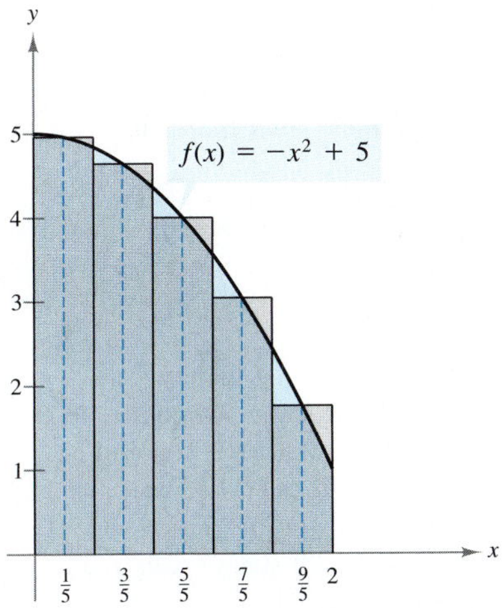

FIGURE 5.22

TRY IT 1

Use four rectangles to approximate the area of the region bounded by the graph of $f(x) = x^2 + 1$, the x-axis, $x = 0$ and $x = 2$.

For the region in Example 1, you can find the exact area with a definite integral. That is,

$$\text{Area} = \int_0^2 (-x^2 + 5)\,dx = \frac{22}{3} \approx 7.33.$$

TECHNOLOGY

The easiest way to use the Midpoint Rule to approximate the definite integral $\int_a^b f(x)\,dx$ is to program it into a computer or programmable calculator. For instance, the pseudocode below will help you write a program to evaluate the Midpoint Rule. (Appendix E lists this program for several models of graphing utilities.)

Program

- *Prompt for value of a.*
- *Input value of a.*
- *Prompt for value of b.*
- *Input value of b.*
- *Prompt for value of n.*
- *Input value of n.*
- *Initialize sum of areas.*
- *Calculate width of subinterval.*
- *Initialize counter.*
- *Begin loop.*
- *Calculate left endpoint.*
- *Calculate right endpoint.*
- *Calculate midpoint of subinterval.*
- *Add area to sum.*
- *Test counter.*
- *End loop.*
- *Display approximation.*

Before executing the program, enter the function. When the program is executed, you will be prompted to enter the lower and upper limits of integration and the number of subintervals you want to use.

With most integrals, you can determine the accuracy of the approximation by using increasingly larger values of n. For example, if you use the program to approximate the value of $\int_0^2 \sqrt{x^3+1}\,dx$, you will obtain an approximation of 3.241, which is accurate to three decimal places.

The approximation procedure used in Example 1 is the **Midpoint Rule.** You can use the Midpoint Rule to approximate *any* definite integral—not just those representing area. The basic steps are summarized below.

Guidelines for Using the Midpoint Rule

To approximate the definite integral $\int_a^b f(x)\,dx$ with the Midpoint Rule, use the steps below.

1. Divide the interval $[a, b]$ into n subintervals, each of width

$$\Delta x = \frac{b-a}{n}.$$

2. Find the midpoint of each subinterval.

$$\text{Midpoints} = \{x_1, x_2, x_3, \ldots, x_n\}$$

3. Evaluate f at each midpoint and form the sum as shown.

$$\int_a^b f(x)\,dx \approx \frac{b-a}{n}[f(x_1) + f(x_2) + f(x_3) + \cdots + f(x_n)]$$

An important characteristic of the Midpoint Rule is that the approximation tends to improve as n increases. The table below shows the approximations for the area of the region described in Example 1 for various values of n. For example, for $n = 10$, the Midpoint Rule yields

$$\int_0^2 (-x^2+5)\,dx \approx \frac{2}{10}\left[f\left(\frac{1}{10}\right) + f\left(\frac{3}{10}\right) + \cdots + f\left(\frac{19}{10}\right)\right]$$
$$= 7.34.$$

n	5	10	15	20	25	30
Approximation	7.3600	7.3400	7.3363	7.3350	7.3344	7.3341

Note that as n increases, the approximation gets closer and closer to the exact value of the integral, which was found to be

$$\frac{22}{3} \approx 7.3333.$$

STUDY TIP

In Example 1, the Midpoint Rule is used to approximate an integral whose exact value can be found with the Fundamental Theorem of Calculus. This was done to illustrate the accuracy of the rule. In practice, of course, you would use the Midpoint Rule to approximate the values of definite integrals for which you cannot find an antiderivative. Examples 2 and 3 illustrate such integrals.

EXAMPLE 2 Using the Midpoint Rule

Use the Midpoint Rule with $n = 5$ to approximate $\int_0^1 \frac{1}{x^2+1}\,dx$.

SOLUTION With $n = 5$, the interval $[0, 1]$ is divided into five subintervals.

$$\left[0, \frac{1}{5}\right], \quad \left[\frac{1}{5}, \frac{2}{5}\right], \quad \left[\frac{2}{5}, \frac{3}{5}\right], \quad \left[\frac{3}{5}, \frac{4}{5}\right], \quad \left[\frac{4}{5}, 1\right]$$

The midpoints of these intervals are $\frac{1}{10}, \frac{3}{10}, \frac{5}{10}, \frac{7}{10}$, and $\frac{9}{10}$. Because each subinterval has a width of $\Delta x = (1 - 0)/5 = \frac{1}{5}$, you can approximate the value of the definite integral as shown.

$$\int_0^1 \frac{1}{x^2+1}\,dx \approx \frac{1}{5}\left(\frac{1}{1.01} + \frac{1}{1.09} + \frac{1}{1.25} + \frac{1}{1.49} + \frac{1}{1.81}\right)$$
$$\approx 0.786$$

The region whose area is represented by the definite integral is shown in Figure 5.23. The actual area of this region is $\pi/4 \approx 0.785$. So, the approximation is off by only 0.001.

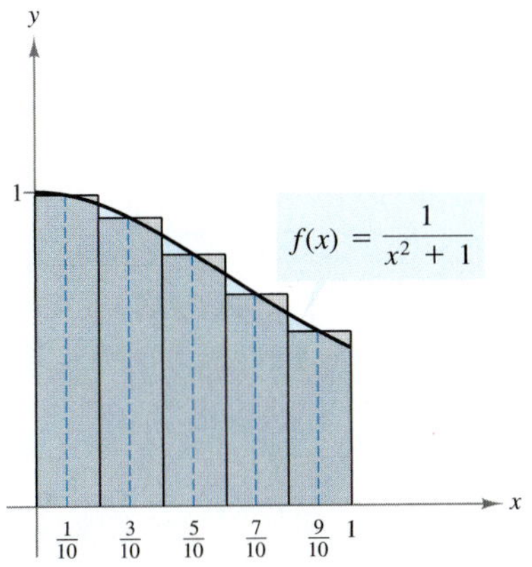

FIGURE 5.23

TRY IT 2

Use the Midpoint Rule with $n = 4$ to approximate the area of the region bounded by the graph of $f(x) = 1/(x^2 + 2)$, the x-axis, and the lines $x = 0$ and $x = 1$.

EXAMPLE 3 Using the Midpoint Rule

Use the Midpoint Rule with $n = 10$ to approximate $\int_1^3 \sqrt{x^2+1}\,dx$.

SOLUTION Begin by dividing the interval $[1, 3]$ into 10 subintervals. The midpoints of these intervals are

$$\frac{11}{10}, \quad \frac{13}{10}, \quad \frac{3}{2}, \quad \frac{17}{10}, \quad \frac{19}{10}, \quad \frac{21}{10}, \quad \frac{23}{10}, \quad \frac{5}{2}, \quad \frac{27}{10}, \quad \text{and} \quad \frac{29}{10}.$$

Because each subinterval has a width of $\Delta x = (3 - 1)/10 = \frac{1}{5}$, you can approximate the value of the definite integral as shown.

$$\int_1^3 \sqrt{x^2+1}\,dx \approx \frac{1}{5}\left[\sqrt{(1.1)^2+1} + \sqrt{(1.3)^2+1} + \cdots + \sqrt{(2.9)^2+1}\right]$$
$$\approx 4.504$$

The region whose area is represented by the definite integral is shown in Figure 5.24. Using techniques that are not within the scope of this course, it can be shown that the actual area is

$$\frac{1}{2}\left[3\sqrt{10} + \ln\left(3 + \sqrt{10}\right) - \sqrt{2} - \ln\left(1 + \sqrt{2}\right)\right] \approx 4.505.$$

So, the approximation is off by only 0.001.

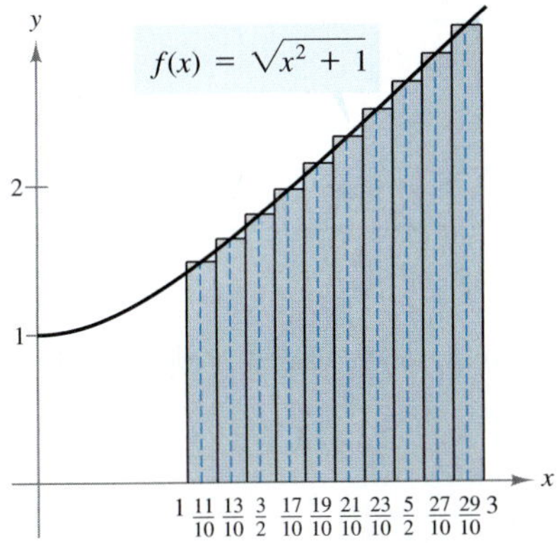

FIGURE 5.24

TRY IT 3

Use the Midpoint Rule with $n = 4$ to approximate the area of the region bounded by the graph of $f(x) = \sqrt{x^2 - 1}$, the x-axis, and the lines $x = 2$ and $x = 4$.

STUDY TIP

The Midpoint Rule is necessary for solving certain real-life problems, such as measuring irregular areas like bodies of water (as seen in Exercise 32 on page 370).

The Definite Integral as the Limit of a Sum

Consider the closed interval $[a, b]$, divided into n subintervals whose midpoints are x_i and whose widths are $\Delta x = (b - a)/n$. In this section, you have seen that the midpoint approximation

$$\int_a^b f(x)\,dx \approx f(x_1)\,\Delta x + f(x_2)\,\Delta x + f(x_3)\,\Delta x + \cdots + f(x_n)\,\Delta x$$
$$= [f(x_1) + f(x_2) + f(x_3) + \cdots + f(x_n)]\,\Delta x$$

becomes better and better as n increases. In fact, the limit of this sum as n approaches infinity is exactly equal to the definite integral. That is,

$$\int_a^b f(x)\,dx = \lim_{n\to\infty} [f(x_1) + f(x_2) + f(x_3) + \cdots + f(x_n)]\,\Delta x.$$

It can be shown that this limit is valid as long as x_i is *any* point in the ith interval.

EXAMPLE 4 Approximating a Definite Integral

Use a computer, programmable calculator, or symbolic integration utility to approximate the definite integral

$$\int_0^1 e^{-x^2}\,dx.$$

SOLUTION Using the program on page 366, with $n = 10, 20, 30, 40,$ and 50, it appears that the value of the integral is approximately 0.7468. If you have access to a computer or calculator with a built-in program for approximating definite integrals, try using it to approximate this integral. When a computer with such a built-in program approximated the integral, it returned a value of 0.746824.

TRY IT 4

Use a computer, programmable calculator, or symbolic integration utility to approximate the definite integral

$$\int_0^1 e^{x^2}\,dx.$$

TAKE ANOTHER LOOK

A Failure of the Midpoint Rule

Suppose you use the Midpoint Rule to approximate the definite integral

$$\int_0^1 \frac{1}{x^2}\,dx.$$

Use a program similar to the one on page 366 to complete the table for $n = 10, 20, 30, 40, 50,$ and 60.

n	10	20	30	40	50	60
Approximation						

Why are the approximations getting larger? Why isn't the Midpoint Rule working?

PREREQUISITE REVIEW 5.6

The following warm-up exercises involve skills that were covered in earlier sections. You will use these skills in the exercise set for this section.

In Exercises 1–6, find the midpoint of the interval.

1. $\left[0, \frac{1}{3}\right]$

2. $\left[\frac{1}{10}, \frac{2}{10}\right]$

3. $\left[\frac{3}{20}, \frac{4}{20}\right]$

4. $\left[1, \frac{7}{6}\right]$

5. $\left[2, \frac{31}{15}\right]$

6. $\left[\frac{26}{9}, 3\right]$

In Exercises 7–10, find the limit.

7. $\displaystyle\lim_{x\to\infty} \frac{2x^2 + 4x - 1}{3x^2 - 2x}$

8. $\displaystyle\lim_{x\to\infty} \frac{4x + 5}{7x - 5}$

9. $\displaystyle\lim_{x\to\infty} \frac{x - 7}{x^2 + 1}$

10. $\displaystyle\lim_{x\to\infty} \frac{5x^3 + 1}{x^3 + x^2 + 4}$

EXERCISES 5.6

In Exercises 1–4, use the Midpoint Rule with $n = 4$ to approximate the area of the region. Compare your result with the exact area obtained with a definite integral.

1. $f(x) = -2x + 3, \quad [0, 1]$

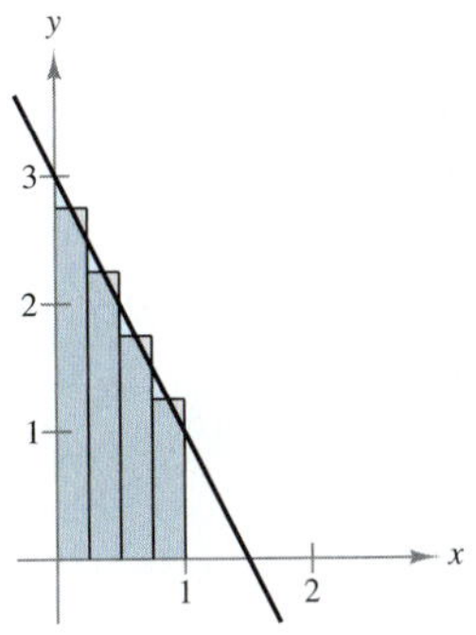

2. $f(x) = \dfrac{1}{x}, \quad [1, 5]$

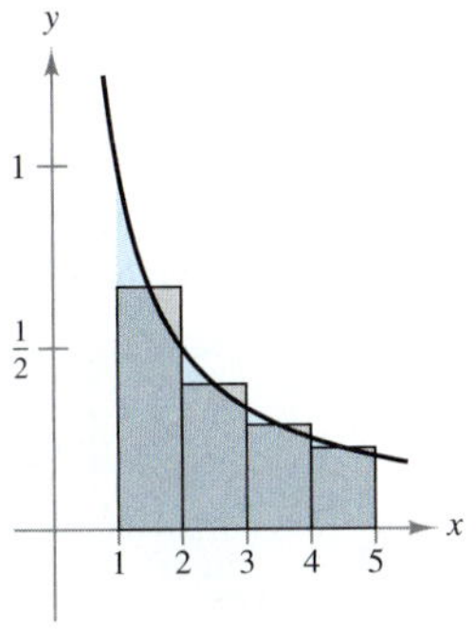

3. $f(x) = \sqrt{x}, \quad [0, 1]$

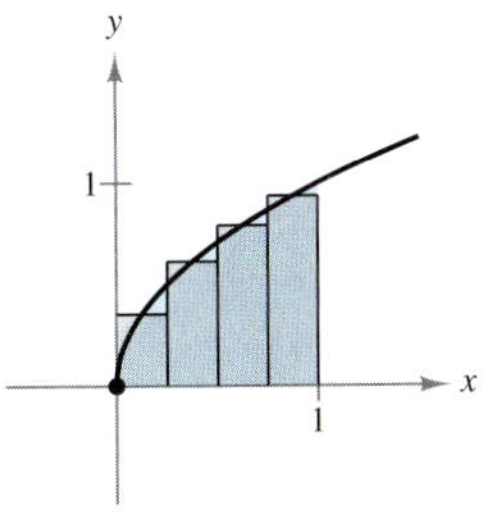

4. $f(x) = 1 - x^2, \quad [-1, 1]$

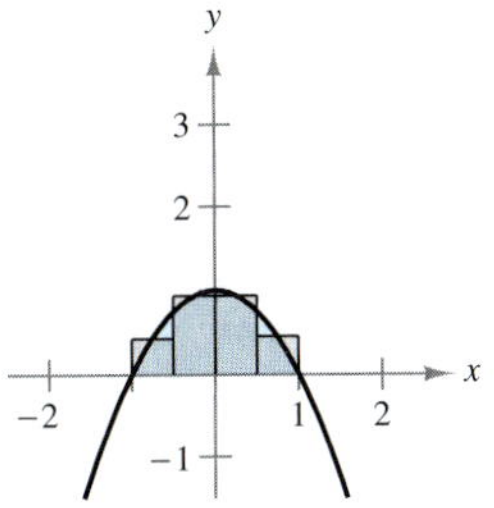

In Exercises 5–12, use the Midpoint Rule with $n = 4$ to approximate the area of the region bounded by the graph of f and the x-axis over the interval. Compare your result with the exact area. Sketch the region.

	Function	*Interval*
5.	$f(x) = x^2 + 2$	$[-1, 1]$
6.	$f(x) = 4 - x^2$	$[0, 2]$
7.	$f(x) = 2x^2$	$[1, 3]$
8.	$f(x) = 2x - x^3$	$[0, 1]$
9.	$f(x) = x^2 - x^3$	$[0, 1]$
10.	$f(x) = x^2 - x^3$	$[-1, 0]$
11.	$f(x) = x(1 - x)^2$	$[0, 1]$
12.	$f(x) = x^2(3 - x)$	$[0, 3]$

In Exercises 13–16, use a program similar to that on page 366 to approximate the area of the region. How large must n be to obtain an approximation that is correct to within 0.01?

13. $\displaystyle\int_0^4 (2x^2 + 3)\,dx$

14. $\displaystyle\int_0^4 (2x^3 + 3)\,dx$

15. $\displaystyle\int_1^2 (2x^2 - x + 1)\,dx$

16. $\displaystyle\int_1^2 (x^3 - 1)\,dx$

In Exercises 17–20, use the Midpoint Rule with $n = 4$ to approximate the area of the region. Compare your result with the exact area obtained with a definite integral.

17. $f(y) = \frac{1}{4}y, \quad [2, 4]$

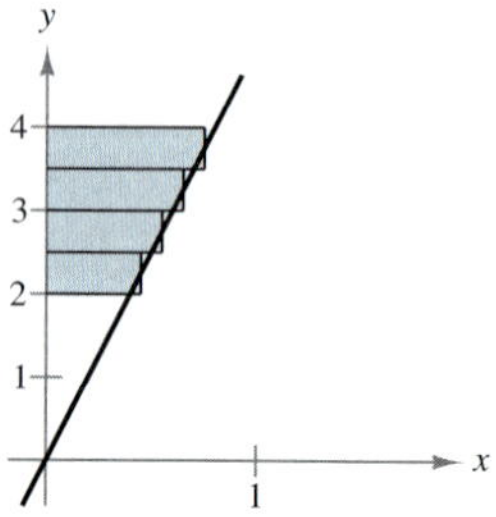

18. $f(y) = 2y, \quad [0, 2]$

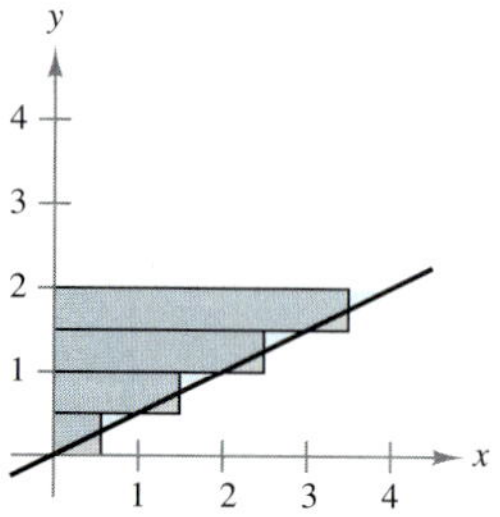

19. $f(y) = y^2 + 1, \quad [0, 4]$

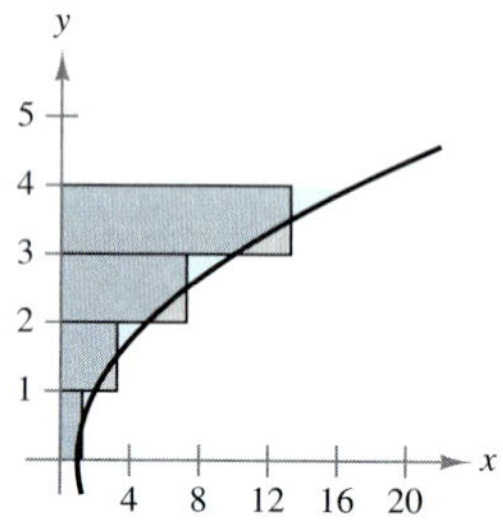

20. $f(y) = 4y - y^2, \quad [0, 4]$

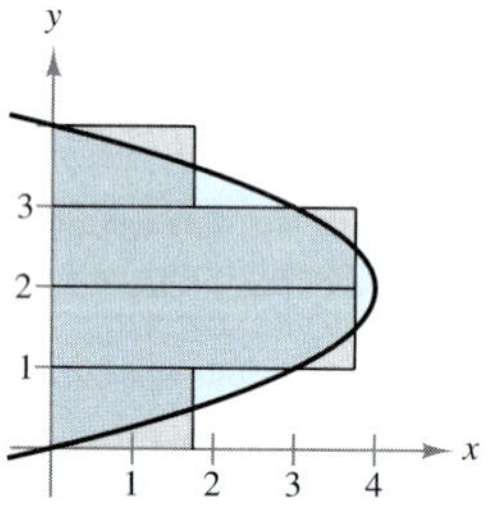

Trapezoidal Rule In Exercises 21 and 22, use the Trapezoidal Rule with $n = 8$ to approximate the definite integral. Compare the result with the exact value and the approximation obtained with $n = 8$ and the Midpoint Rule. Which approximation technique appears to be better? Let f be continuous on $[a, b]$ and let n be the number of equal subintervals (see figure). Then the Trapezoidal Rule for approximating $\int_a^b f(x)\,dx$ is

$$\frac{b-a}{2n}[f(x_0) + 2f(x_1) + \cdots + 2f(x_{n-1}) + f(x_n)].$$

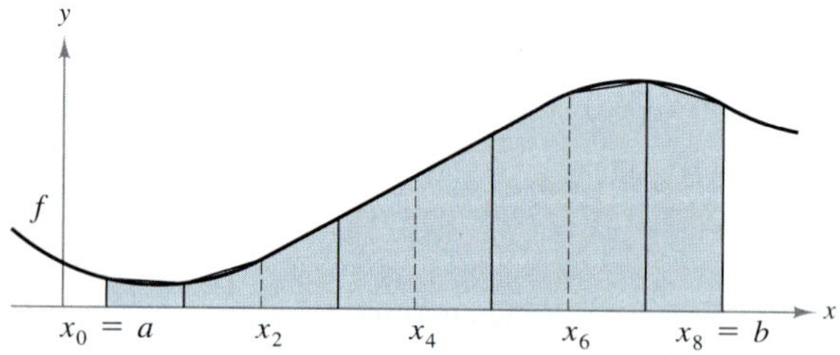

21. $\displaystyle\int_0^2 x^3\,dx$

22. $\displaystyle\int_1^3 \frac{1}{x^2}\,dx$

In Exercises 23–26, use the Trapezoidal Rule with $n = 4$ to approximate the definite integral.

23. $\displaystyle\int_0^2 \frac{1}{x+1}\,dx$

24. $\displaystyle\int_0^4 \sqrt{1+x^2}\,dx$

25. $\displaystyle\int_{-1}^1 \frac{1}{x^2+1}\,dx$

26. $\displaystyle\int_1^5 \frac{\sqrt{x-1}}{x}\,dx$

In Exercises 27 and 28, use a computer or programmable calculator to approximate the definite integral using the Midpoint Rule and the Trapezoidal Rule for $n = 4, 8, 12, 16,$ and 20.

27. $\displaystyle\int_0^4 \sqrt{2+3x^2}\,dx$

28. $\displaystyle\int_0^2 \frac{5}{x^3+1}\,dx$

In Exercises 29 and 30, use the Trapezoidal Rule with $n = 10$ to approximate the area of the region bounded by the graphs of the equations.

29. $y = \sqrt{\dfrac{x^3}{4-x}}, \quad y = 0, \quad x = 3$

30. $y = x\sqrt{\dfrac{4-x}{4+x}}, \quad y = 0, \quad x = 4$

31. *Velocity and Acceleration* The table lists the velocity v (in feet per second) of an accelerating car over a 20-second interval. Use the Trapezoidal Rule to approximate the distance in feet that the car travels during the 20 seconds. (The distance is given by $s = \int_0^{20} v\,dt$.)

Time, t	0	5	10	15	20
Velocity, v	0.0	29.3	51.3	66.0	73.3

32. *Surface Area* To estimate the surface area of a pond, a surveyor takes several measurements, as shown in the figure. Estimate the surface area of the pond using (a) the Midpoint Rule and (b) the Trapezoidal Rule.

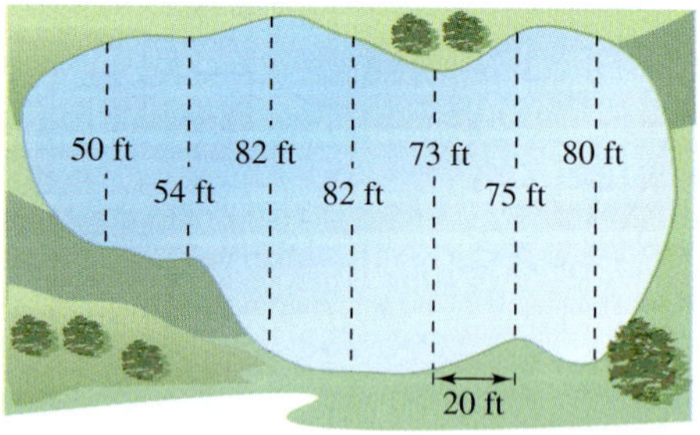

33. *Numerical Approximation* Use the Midpoint Rule and the Trapezoidal Rule with $n = 4$ to approximate π where

$$\pi = \int_0^1 \frac{4}{1+x^2}\,dx.$$

Then use a graphing utility to evaluate the definite integral. Compare all of your results.

5.7 VOLUMES OF SOLIDS OF REVOLUTION

- Use the Disk Method to find volumes of solids of revolution.
- Use the Washer Method to find volumes of solids of revolution with holes.
- Use solids of revolution to solve real-life problems.

The Disk Method

Another important application of the definite integral is its use in finding the volume of a three-dimensional solid. In this section, you will study a particular type of three-dimensional solid—one whose cross sections are similar. You will begin with solids of revolution. These solids, such as axles, funnels, pills, bottles, and pistons, are used commonly in engineering and manufacturing.

As shown in Figure 5.25, a **solid of revolution** is formed by revolving a plane region about a line. The line is called the **axis of revolution.**

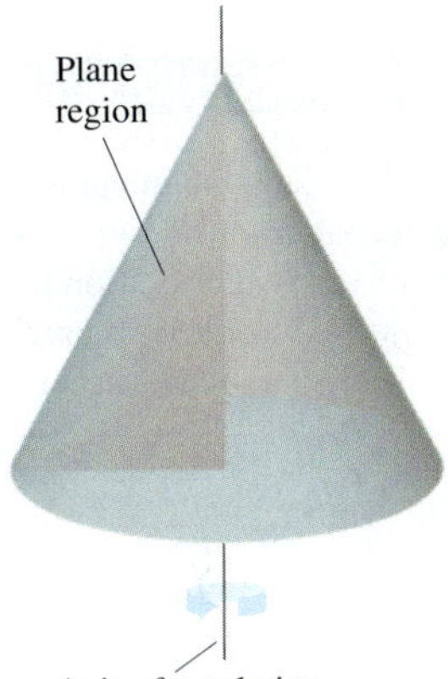

FIGURE 5.25

To develop a formula for finding the volume of a solid of revolution, consider a continuous function f that is nonnegative on the interval $[a, b]$. Suppose that the area of the region is approximated by n rectangles, each of width Δx, as shown in Figure 5.26. By revolving the rectangles about the x-axis, you obtain n circular disks, each with a volume of $\pi[f(x_i)]^2\Delta x$. The volume of the solid formed by revolving the region about the x-axis is approximately equal to the sum of the volumes of the n disks. Moreover, by taking the limit as n approaches infinity, you can see that the exact volume is given by a definite integral. This result is called the **Disk Method.**

The Disk Method

The volume of the solid formed by revolving the region bounded by the graph of f and the x-axis $(a \le x \le b)$ about the x-axis is

$$\text{Volume} = \pi \int_a^b [f(x)]^2\, dx.$$

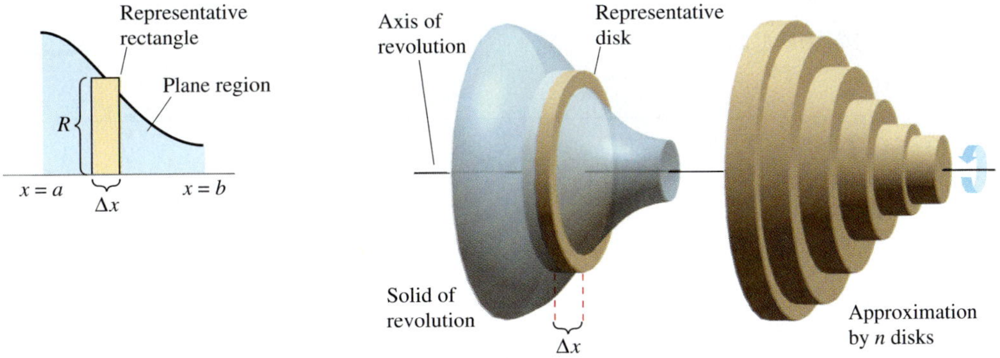

Approximation by n rectangles

FIGURE 5.26

EXAMPLE 1 **Finding the Volume of a Solid of Revolution**

Find the volume of the solid formed by revolving the region bounded by the graph of $f(x) = -x^2 + x$ and the x-axis about the x-axis.

SOLUTION Begin by sketching the region bounded by the graph of f and the x-axis. As shown in Figure 5.27(a), sketch a representative rectangle whose height is $f(x)$ and whose width is Δx. From this rectangle, you can see that the radius of the solid is

$$\text{Radius} = f(x) = -x^2 + x.$$

Using the Disk Method, you can find the volume of the solid of revolution.

$$\begin{aligned}
\text{Volume} &= \pi\int_0^1 [f(x)]^2\,dx && \text{Disk Method} \\
&= \pi\int_0^1 (-x^2 + x)^2\,dx && \text{Substitute for } f(x). \\
&= \pi\int_0^1 (x^4 - 2x^3 + x^2)\,dx && \text{Expand integrand.} \\
&= \pi\left[\frac{x^5}{5} - \frac{x^4}{2} + \frac{x^3}{3}\right]_0^1 && \text{Find antiderivative.} \\
&= \frac{\pi}{30} && \text{Apply Fundamental Theorem.} \\
&\approx 0.105 && \text{Round to three decimal places.}
\end{aligned}$$

So, the volume of the solid is about 0.105 cubic unit.

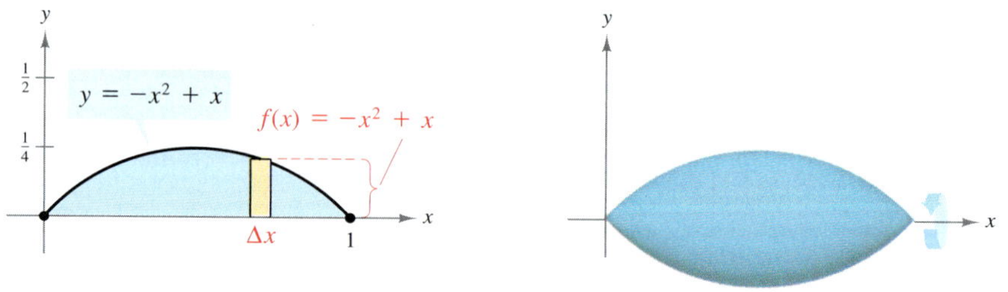

(a) Plane region (b) Solid of revolution

FIGURE 5.27

TECHNOLOGY

Try using the integration capabilities of a graphing utility to verify the solution in Example 1. Consult your user's manual for specific keystrokes.

TRY IT 1

Find the volume of the solid formed by revolving the region bounded by the graph of $f(x) = -x^2 + 4$ and the x-axis about the x-axis.

STUDY TIP

In Example 1, the entire problem was solved *without* referring to the three-dimensional sketch given in Figure 5.27(b). In general, to set up the integral for calculating the volume of a solid of revolution, a sketch of the plane region is more useful than a sketch of the solid, because the radius is more readily visualized in the plane region.

The Washer Method

You can extend the Disk Method to find the volume of a solid of revolution with a *hole*. Consider a region that is bounded by the graphs of f and g, as shown in Figure 5.28(a). If the region is revolved about the x-axis, then the volume of the resulting solid can be found by applying the Disk Method to f and g and subtracting the results.

$$\text{Volume} = \pi\int_a^b [f(x)]^2\,dx - \pi\int_a^b [g(x)]^2\,dx$$

Writing this as a single integral produces the **Washer Method.**

The Washer Method

Let f and g be continuous and nonnegative on the closed interval $[a, b]$, as shown in Figure 5.28(a). If $g(x) \leq f(x)$ for all x in the interval, then the volume of the solid formed by revolving the region bounded by the graphs of f and g $(a \leq x \leq b)$ about the x-axis is

$$\text{Volume} = \pi\int_a^b \{[f(x)]^2 - [g(x)]^2\}\,dx.$$

$f(x)$ is the **outer radius** and $g(x)$ is the **inner radius.**

In Figure 5.28(b), note that the solid of revolution has a hole. Moreover, the radius of the hole is $g(x)$, the inner radius.

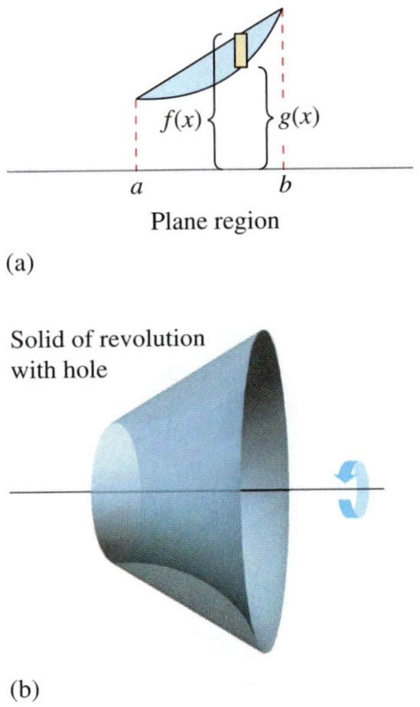

FIGURE 5.28

EXAMPLE 2 **Using the Washer Method**

Find the volume of the solid formed by revolving the region bounded by the graphs of

$$f(x) = \sqrt{25 - x^2} \text{ and } g(x) = 3$$

about the x-axis (see Figure 5.29).

SOLUTION First find the points of intersection of f and g by setting $f(x)$ equal to $g(x)$ and solving for x.

$$f(x) = g(x) \quad \text{Set } f(x) \text{ equal to } g(x).$$

$$\sqrt{25 - x^2} = 3 \quad \text{Substitute for } f(x) \text{ and } g(x).$$

$$25 - x^2 = 9 \quad \text{Square each side.}$$

$$16 = x^2$$

$$\pm 4 = x \quad \text{Solve for } x.$$

Using $f(x)$ as the outer radius and $g(x)$ as the inner radius, you can find the volume of the solid as shown.

$$\text{Volume} = \pi \int_{-4}^{4} \{[f(x)]^2 - [g(x)]^2\}\, dx \quad \text{Washer Method}$$

$$= \pi \int_{-4}^{4} \left[\left(\sqrt{25 - x^2}\right)^2 - (3)^2\right] dx \quad \text{Substitute for } f(x) \text{ and } g(x).$$

$$= \pi \int_{-4}^{4} (16 - x^2)\, dx \quad \text{Simplify.}$$

$$= \pi \left[16x - \frac{x^3}{3}\right]_{-4}^{4} \quad \text{Find antiderivative.}$$

$$= \frac{256\pi}{3} \quad \text{Apply Fundamental Theorem.}$$

$$\approx 268.08 \quad \text{Round to two decimal places.}$$

So, the volume of the solid is about 268.08 cubic inches.

TRY IT 2

Find the volume of the solid formed by revolving the region bounded by the graphs of $f(x) = 5 - x^2$ and $g(x) = 1$ about the x-axis.

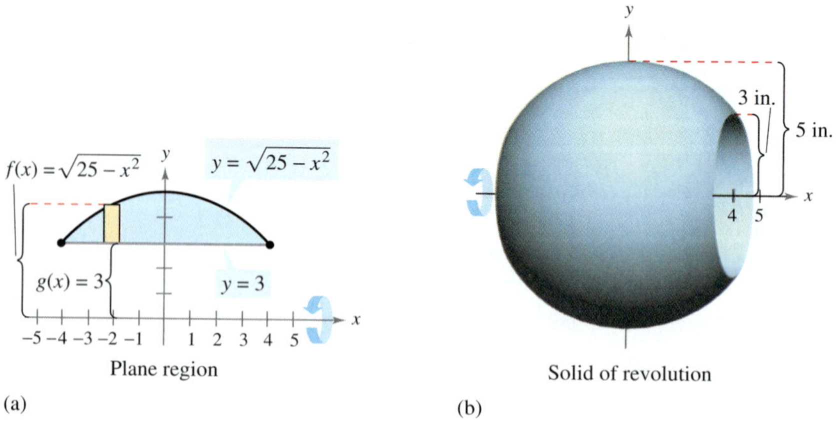

FIGURE 5.29

Application

EXAMPLE 3 Finding a Football's Volume

A regulation-size football can be modeled as a solid of revolution formed by revolving the graph of

$$f(x) = -0.0944x^2 + 3.4, \quad -5.5 \le x \le 5.5$$

about the x-axis, as shown in Figure 5.30. Use this model to find the volume of a football. (In the model, x and y are measured in inches.)

American football, in its modern form, is a twentieth-century invention. In the 1800s a rough, soccer-like game was played with a "round football." In 1905, at the request of President Theodore Roosevelt, the Intercollegiate Athletic Association (which became the NCAA in 1910) was formed. With the introduction of the forward pass in 1906, the shape of the ball was altered to make it easier to grip.

SOLUTION To find the volume of the solid of revolution, use the Disk Method.

$$\text{Volume} = \pi\int_{-5.5}^{5.5} [f(x)]^2\,dx \qquad \text{Disk Method}$$

$$= \pi\int_{-5.5}^{5.5} (-0.0944x^2 + 3.4)^2\,dx \qquad \text{Substitute for } f(x).$$

$$\approx 232 \text{ cubic inches} \qquad \text{Volume}$$

So, the volume of the football is about 232 cubic inches.

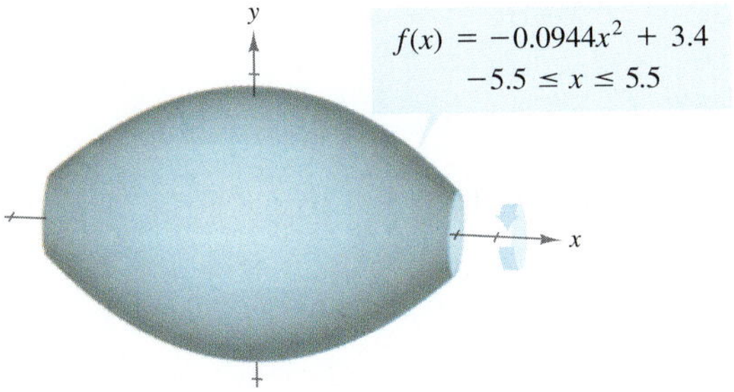

FIGURE 5.30 A football-shaped solid is formed by revolving a parabolic segment about the x-axis.

TRY IT 3

A soup bowl can be modeled as a solid of revolution formed by revolving the graph of

$$f(x) = \sqrt{x} + 1, \quad 0 \le x \le 3$$

about the x-axis. Use this model, where x and y are measured in inches, to find the volume of the soup bowl.

TAKE ANOTHER LOOK

Testing the Reasonableness of an Answer

A football is about 11 inches long and has a diameter of about 7 inches. In Example 3, the volume of a football was approximated to be 232 cubic inches. Explain how you can determine whether this answer is reasonable.

PREREQUISITE REVIEW 5.7

The following warm-up exercises involve skills that were covered in earlier sections. You will use these skills in the exercise set for this section.

In Exercises 1–6, solve for x.

1. $x^2 = 2x$

2. $-x^2 + 4x = x^2$

3. $x = -x^3 + 5x$

4. $x^2 + 1 = x + 3$

5. $-x + 4 = \sqrt{4x - x^2}$

6. $\sqrt{x - 1} = \frac{1}{2}(x - 1)$

In Exercises 7–10, evaluate the integral.

7. $\displaystyle\int_0^2 2e^{2x}\,dx$

8. $\displaystyle\int_{-1}^3 \frac{2x + 1}{x^2 + x + 2}\,dx$

9. $\displaystyle\int_0^2 x\sqrt{x^2 + 1}\,dx$

10. $\displaystyle\int_1^5 \frac{(\ln x)^2}{x}\,dx$

EXERCISES 5.7

In Exercises 1–16, find the volume of the solid formed by revolving the region bounded by the graph(s) of the equation(s) about the x-axis.

1. $y = \sqrt{4 - x^2}$

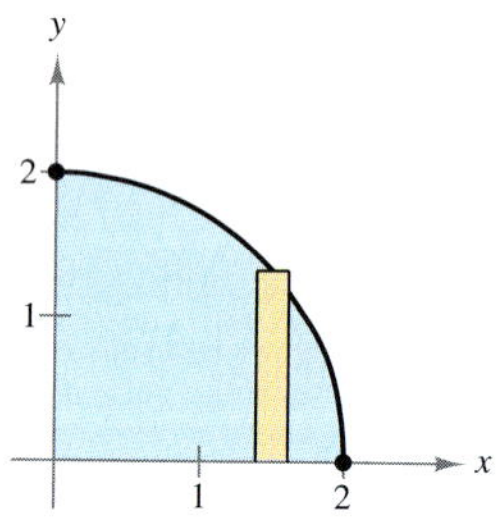

2. $y = x^2$

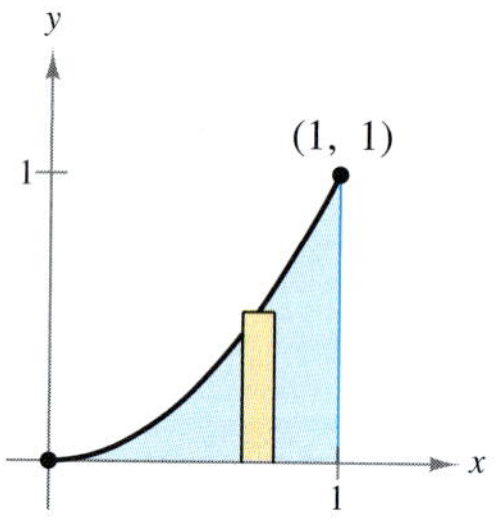

3. $y = \sqrt{x}$

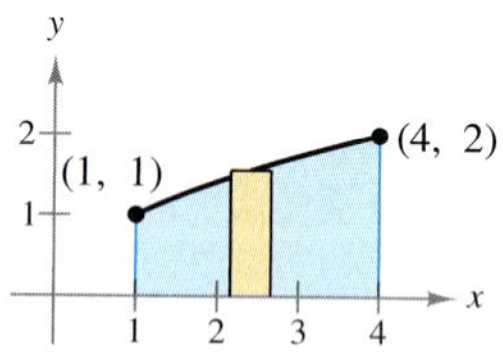

4. $y = \sqrt{4 - x^2}$

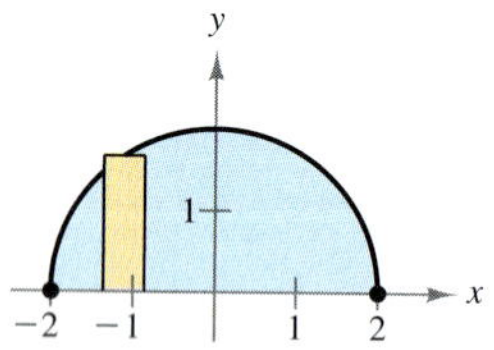

5. $y = 4 - x^2, \quad y = 0$

6. $y = x, \quad y = 0, \quad x = 4$

7. $y = 1 - \frac{1}{4}x^2, \quad y = 0$

8. $y = x^2 + 1, \quad y = 5$

9. $y = -x + 1, \quad y = 0, \quad x = 0$

10. $y = x, \quad y = e^{x-1}, \quad x = 0$

11. $y = \sqrt{x} + 1, \quad y = 0, \quad x = 0, \quad x = 9$

12. $y = \sqrt{x}, \quad y = 0, \quad x = 4$

13. $y = 2x^2, \quad y = 0, \quad x = 2$

14. $y = \dfrac{1}{x}, \quad y = 0, \quad x = 1, \quad x = 3$

15. $y = e^x, \quad y = 0, \quad x = 0, \quad x = 1$

16. $y = x^2, \quad y = 4x - x^2$

In Exercises 17–24, find the volume of the solid formed by revolving the region bounded by the graph(s) of the equation(s) about the y-axis.

17. $y = x^2, \quad y = 4, \quad 0 \le x \le 2$

18. $y = \sqrt{16 - x^2}, \quad y = 0, \quad 0 \le x \le 4$

19. $x = 1 - \frac{1}{2}y, \quad x = 0, \quad y = 0$

20. $x = y(y - 1), \quad x = 0$

21. $y = x^{2/3}$

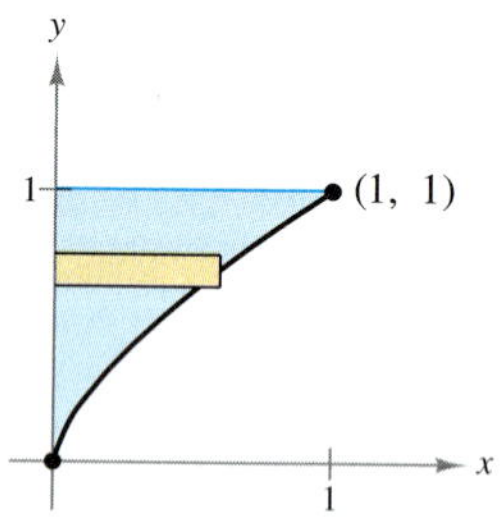

22. $x = -y^2 + 4y$

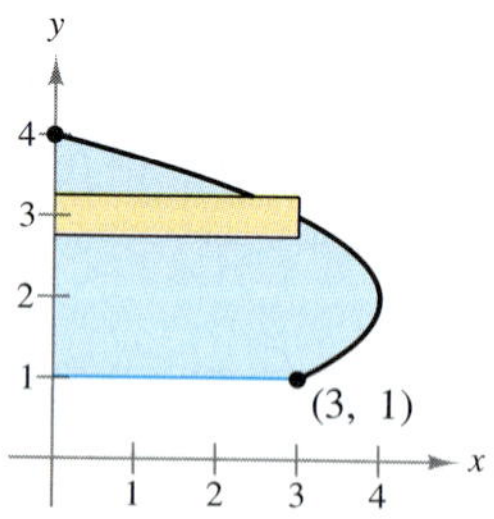

23. $y = \sqrt{4 - x}, \quad y = 0, \quad x = 0$

24. $y = 4, \quad y = 0, \quad x = 2, \quad x = 0$

25. ***Volume*** The line segment from $(0, 0)$ to $(6, 3)$ is revolved about the x-axis to form a cone. What is the volume of the cone?

26. ***Volume*** The line segment from $(0, 0)$ to $(4, 2)$ is revolved about the y-axis to form a cone. What is the volume of the cone?

27. ***Volume*** Use the Disk Method to verify that the volume of a right circular cone is $\frac{1}{3}\pi r^2 h$, where r is the radius of the base and h is the height.

28. ***Volume*** Use the Disk Method to verify that the volume of a sphere of radius r is $\frac{4}{3}\pi r^3$.

29. ***Volume*** The right half of the ellipse

$$9x^2 + 25y^2 = 225$$

is revolved about the y-axis to form an oblate spheroid (shaped like an M&M candy). Find the volume of the spheroid.

30. ***Volume*** The upper half of the ellipse

$$9x^2 + 16y^2 = 144$$

is revolved about the x-axis to form a prolate spheroid (shaped like a football). Find the volume of the spheroid.

31. ***Volume*** A tank on the wing of a jet airplane is modeled by revolving the region bounded by the graph of $y = \frac{1}{8}x^2\sqrt{2 - x}$ and the x-axis about the x-axis, where x and y are measured in meters (see figure). Find the volume of the tank.

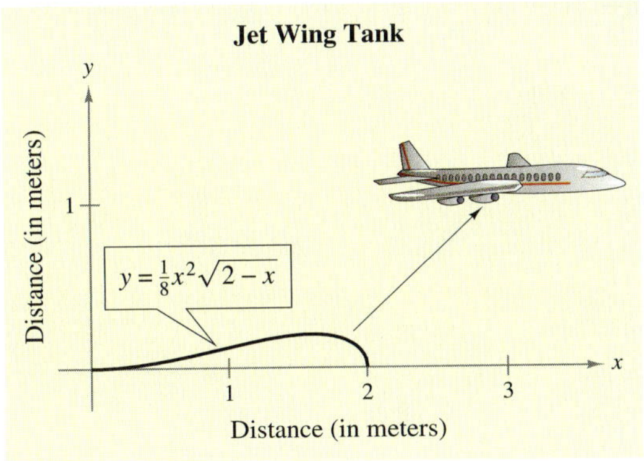

32. ***Volume*** A soup bowl can be modeled as a solid of revolution formed by revolving the graph of

$$y = \sqrt{\frac{x}{2}} + 1, \quad 0 \le x \le 4$$

about the x-axis. Use this model, where x and y are measured in inches, to find the volume of the soup bowl.

33. ***Biology*** A pond is to be stocked with a species of fish. The food supply in 500 cubic feet of pond water can adequately support one fish. The pond is nearly circular, is 20 feet deep at its center, and has a radius of 200 feet. The bottom of the pond can be modeled by

$$y = 20[(0.005x)^2 - 1].$$

(a) How much water is in the pond?

(b) How many fish can the pond support?

34. ***Modeling a Body of Water*** A pond is approximately circular, with a diameter of 400 feet (see figure). Starting at the center, the depth of the water is measured every 25 feet and recorded in the table.

x	0	25	50	75	100
Depth	20	19	19	17	15

x	125	150	175	200
Depth	14	10	6	0

(a) Use a graphing utility to plot the depths and graph the model of the pond's depth, $y = 20 - 0.00045x^2$.

(b) Use the model in part (a) to find the pond's volume.

(c) Use the result of part (b) to approximate the number of gallons of water in the pond ($1 \text{ ft}^3 \approx 7.48$ gal).

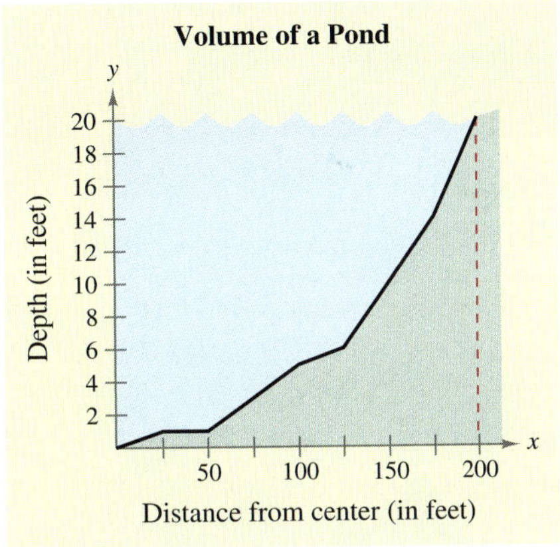

In Exercises 35 and 36, use a program similar to the one on page 366 to approximate the volume of a solid generated by revolving the region bounded by the graphs of the equations about the x-axis.

35. $y = \sqrt[3]{x + 1}, \quad y = 0, \quad x = 0, \quad x = 7$

36. $y = \dfrac{10}{x^2 + 1}, \quad y = 0, \quad x = 0, \quad x = 3$

ALGEBRA REVIEW

"Unsimplifying" an Algebraic Expression

In algebra it is often helpful to write an expression in simplest form. In this chapter, you have seen that the reverse is often true in integration. That is, to fit an integrand to an integration formula, it often helps to "unsimplify" the expression. To do this, you use the same algebraic rules, but your goal is different. Here are some examples.

EXAMPLE 1 Rewriting an Algebraic Expression

Rewrite each algebraic expression as indicated in the example.

(a) $\dfrac{x+1}{\sqrt{x}}$ Example 6, page 322

(b) $x(3-4x^2)^2$ Example 2, page 331

(c) $7x^2\sqrt{x^3+1}$ Example 4, page 332

(d) $5xe^{-x^2}$ Example 3, page 338

SOLUTION

(a) Example 6, page 322

$$\frac{x+1}{\sqrt{x}} = \frac{x}{\sqrt{x}} + \frac{1}{\sqrt{x}} \qquad \text{Rewrite as two fractions.}$$

$$= \frac{x^1}{x^{1/2}} + \frac{1}{x^{1/2}} \qquad \text{Rewrite with rational exponents.}$$

$$= x^{1-1/2} + x^{-1/2} \qquad \text{Properties of exponents}$$

$$= x^{1/2} + x^{-1/2} \qquad \text{Simplify exponent.}$$

(b) Example 2, page 331

$$x(3-4x^2)^2 = \frac{-8}{-8}x(3-4x^2)^2 \qquad \text{Multiply and divide by } -8.$$

$$= \left(-\frac{1}{8}\right)(-8)x(3-4x^2)^2 \qquad \text{Regroup.}$$

$$= \left(-\frac{1}{8}\right)(3-4x^2)^2(-8x) \qquad \text{Regroup.}$$

(c) Example 4, page 332

$$7x^2\sqrt{x^3+1} = 7x^2(x^3+1)^{1/2} \qquad \text{Rewrite with rational exponent.}$$

$$= \frac{3}{3}(7x^2)(x^3+1)^{1/2} \qquad \text{Multiply and divide by 3.}$$

$$= \frac{7}{3}(3x^2)(x^3+1)^{1/2} \qquad \text{Regroup.}$$

$$= \frac{7}{3}(x^3+1)^{1/2}(3x^2) \qquad \text{Regroup.}$$

(d) Example 3, page 338

$$5xe^{-x^2} = \frac{-2}{-2}(5x)e^{-x^2} \qquad \text{Multiply and divide by } -2.$$

$$= \left(-\frac{5}{2}\right)(-2x)e^{-x^2} \qquad \text{Regroup.}$$

$$= \left(-\frac{5}{2}\right)e^{-x^2}(-2x) \qquad \text{Regroup.}$$

EXAMPLE 2 Rewriting an Algebraic Expression

Rewrite each algebraic expression.

(a) $\dfrac{3x^2 + 2x - 1}{x^2}$ (b) $\dfrac{1}{1 + e^{-x}}$

(c) $\dfrac{x^2 + x + 1}{x - 1}$ (d) $\dfrac{x^2 + 6x + 1}{x^2 + 1}$

SOLUTION

(a) $\dfrac{3x^2 + 2x - 1}{x^2} = \dfrac{3x^2}{x^2} + \dfrac{2x}{x^2} - \dfrac{1}{x^2}$ — Example 7(a), page 341 Rewrite as separate fractions.

$= 3 + \dfrac{2}{x} - x^{-2}$ — Properties of exponents.

$= 3 + 2\left(\dfrac{1}{x}\right) - x^{-2}$ — Regroup.

(b) $\dfrac{1}{1 + e^{-x}} = \left(\dfrac{e^x}{e^x}\right)\dfrac{1}{1 + e^{-x}}$ — Example 7(b), page 341 Multiply and divide by e^x.

$= \dfrac{e^x}{e^x + e^x(e^{-x})}$ — Multiply.

$= \dfrac{e^x}{e^x + e^{x-x}}$ — Property of exponents

$= \dfrac{e^x}{e^x + e^0}$ — Simplify exponent.

$= \dfrac{e^x}{e^x + 1}$ — $e^0 = 1$

(c) $\dfrac{x^2 + x + 1}{x - 1} = x + 2 + \dfrac{3}{x - 1}$ — Example 7(c), page 341 Use long division as shown below.

$$
\begin{array}{r}
x + 2 \\
x - 1 \overline{) x^2 + x + 1} \\
\underline{x^2 - x } \\
2x + 1 \\
\underline{2x - 2} \\
3
\end{array}
$$

(d) $\dfrac{x^2 + 6x + 1}{x^2 + 1} = 1 + \dfrac{6x}{x^2 + 1}$ — Bottom of page 340. Use long division as shown below.

$$
\begin{array}{r}
1 \\
x^2 + 1 \overline{) x^2 + 6x + 1} \\
\underline{x^2 + 1} \\
6x
\end{array}
$$

5 CHAPTER SUMMARY AND STUDY STRATEGIES

*After studying this chapter, you should have acquired the following skills. The exercise numbers are keyed to the Review Exercises that begin on page 382. Answers to odd-numbered Review Exercises are given in the back of the text.**

- Use basic integration rules to find indefinite integrals. *(Section 5.1)* — *Review Exercises 1–10*

$$\int k\,dx = kx + C \qquad \int [f(x) - g(x)]\,dx = \int f(x)\,dx - \int g(x)\,dx$$

$$\int kf(x)\,dx = k\int f(x)\,dx \qquad \int x^n\,dx = \frac{x^{n+1}}{n+1} + C, \quad n \neq -1$$

$$\int [f(x) + g(x)]\,dx = \int f(x)\,dx + \int g(x)\,dx$$

- Use initial conditions to find particular solutions of indefinite integrals. *(Section 5.1)* — *Review Exercises 11–14*
- Use antiderivatives to solve real-life problems. *(Section 5.1)* — *Review Exercises 15, 16*
- Use the General Power Rule to find indefinite integrals. *(Section 5.2)* — *Review Exercises 17–24*

$$\int u^n \frac{du}{dx}\,dx = \int u^n\,du = \frac{u^{n+1}}{n+1} + C, \quad n \neq -1$$

- Use the General Power Rule to solve real-life problems. *(Section 5.2)* — *Review Exercises 25, 26*
- Use the Exponential and Log Rules to find indefinite integrals. *(Section 5.3)* — *Review Exercises 27–32*

$$\int e^x\,dx = e^x + C \qquad \int \frac{1}{x}\,dx = \ln|x| + C$$

$$\int e^u \frac{du}{dx}\,dx = \int e^u\,du = e^u + C \qquad \int \frac{du/dx}{u}\,dx = \int \frac{1}{u}\,du = \ln|u| + C$$

- Use a symbolic integration utility to find indefinite integrals. *(Section 5.3)* — *Review Exercises 33, 34*
- Find the areas of regions bounded by the graph of a function and the x-axis. *(Section 5.4)* — *Review Exercises 35–40*
- Use the Fundamental Theorem of Calculus to evaluate definite integrals. *(Section 5.4)* — *Review Exercises 41–52*

$$\int_a^b f(x)\,dx = F(x)\Big]_a^b = F(b) - F(a), \quad \text{where} \quad F'(x) = f(x)$$

- Use definite integrals to solve marginal analysis problems. *(Section 5.4)* — *Review Exercises 53, 54*
- Find average values of functions over closed intervals. *(Section 5.4)* — *Review Exercises 55–58*

$$\text{Average value} = \frac{1}{b-a}\int_a^b f(x)\,dx$$

- Use average values to solve real-life problems. *(Section 5.4)* — *Review Exercises 59–62*
- Find amounts of annuities. *(Section 5.4)* — *Review Exercises 63, 64*
- Use properties of even and odd functions to help evaluate definite integrals. *(Section 5.4)* — *Review Exercises 65–68*

Even function: $f(-x) = f(x)$ *Odd function:* $f(-x) = -f(x)$

* Use a wide range of valuable study aids to help you master the material in this chapter. The *Student Solutions Guide* includes step-by-step solutions to all odd-numbered exercises to help you review and prepare. The *HM mathSpace® Student CD-ROM* helps you brush up on your algebra skills. The *Graphing Technology Guide*, available on the Web at *math.college.hmco.com/students*, offers step-by-step commands and instructions for a wide variety of graphing calculators, including the most recent models.

- Find areas of regions bounded by two (or more) graphs. *(Section 5.5)* — *Review Exercises 69–76*

  $$A = \int_a^b [f(x) - g(x)]\, dx$$

- Find consumer and producer surpluses. *(Section 5.5)* — *Review Exercises 77, 78*
- Use the areas of regions bounded by two graphs to solve real-life problems. *(Section 5.5)* — *Review Exercises 79–82*
- Use the Midpoint Rule to approximate values of definite integrals. *(Section 5.6)* — *Review Exercises 83–86*

  $$\int_a^b f(x)\, dx \approx \frac{b-a}{n}[f(x_1) + f(x_2) + f(x_3) + \cdot \cdot \cdot + f(x_n)]$$

- Use the Disk Method to find volumes of solids of revolution. *(Section 5.7)* — *Review Exercises 87–90*

  $$\text{Volume} = \pi \int_a^b [f(x)]^2\, dx$$

- Use the Washer Method to find volumes of solids of revolution with holes. *(Section 5.7)* — *Review Exercises 91–94*

  $$\text{Volume} = \pi \int_a^b \{[f(x)]^2 - [g(x)]^2\}\, dx$$

- Use solids of revolution to solve real-life problems. *(Section 5.7)* — *Review Exercises 95, 96*
- ***Indefinite and Definite Integrals*** When evaluating integrals, remember that an indefinite integral is a *family of antiderivatives*, each differing by a constant C, whereas a definite integral is a number.
- ***Checking Antiderivatives by Differentiating*** When finding an antiderivative, remember that you can check your result by differentiating. For example, you can check that the antiderivative

  $$\int (3x^3 - 4x)\, dx = \frac{3}{4}x^4 - 2x^2 + C$$

  is correct by differentiating to obtain

  $$\frac{d}{dx}\left[\frac{3}{4}x^4 - 2x^2 + C\right] = 3x^3 - 4x.$$

  Because the derivative is equal to the original integrand, you know that the antiderivative is correct.
- ***Grouping Symbols and the Fundamental Theorem*** When using the Fundamental Theorem of Calculus to evaluate a definite integral, you can avoid sign errors by using grouping symbols. Here is an example.

  $$\int_1^3 (x^3 - 9x)\, dx = \left[\frac{x^4}{4} - \frac{9x^2}{2}\right]_1^3 = \left[\frac{3^4}{4} - \frac{9(3^2)}{2}\right] - \left[\frac{1^4}{4} - \frac{9(1^2)}{2}\right] = \frac{81}{4} - \frac{81}{2} - \frac{1}{4} + \frac{9}{2} = -16$$

Study Tools *Additional resources that accompany this chapter*

- **Algebra Review** (pages 378 and 379)
- **Chapter Summary and Study Strategies** (pages 380 and 381)
- **Review Exercises** (pages 382–385)
- **Sample Post-Graduation Exam Questions** (page 386)
- **Web Exercises** (page 328, Exercise 83; page 364, Exercise 60)
- **Student Solutions Guide**
- **HM mathSpace® Student CD-ROM**
- **Graphing Technology Guide** *(math.college.hmco.com/students)*

5 CHAPTER REVIEW EXERCISES

In Exercises 1–10, find the indefinite integral.

1. $\int 16\, dx$

2. $\int \frac{3}{5}x\, dx$

3. $\int (2x^2 + 5x)\, dx$

4. $\int (5 - 6x^2)\, dx$

5. $\int \frac{2}{3\sqrt[3]{x}}\, dx$

6. $\int 6x^2\sqrt{x}\, dx$

7. $\int \left(\sqrt[3]{x^4} + 3x\right) dx$

8. $\int \left(\frac{4}{\sqrt{x}} + \sqrt{x}\right) dx$

9. $\int \frac{2x^4 - 1}{\sqrt{x}}\, dx$

10. $\int \frac{1 - 3x}{x^2}\, dx$

In Exercises 11–14, find the particular solution, $y = f(x)$, that satisfies the conditions.

11. $f'(x) = 3x + 1, \quad f(2) = 6$

12. $f'(x) = x^{-1/3} - 1, \quad f(8) = 4$

13. $f''(x) = 2x^2, \quad f'(3) = 10, \quad f(3) = 6$

14. $f''(x) = \frac{6}{\sqrt{x}} + 3, \quad f'(1) = 12, \quad f(4) = 56$

15. ***Vertical Motion*** An object is projected upward from the ground with an initial velocity of 80 feet per second.

(a) How long does it take the object to rise to its maximum height?

(b) What is the maximum height?

(c) When is the velocity of the object half of its initial velocity?

(d) What is the height of the object when its velocity is one-half the initial velocity?

16. ***Revenue*** The weekly revenue for a new product has been increasing. The rate of change of the revenue can be modeled by

$$\frac{dR}{dt} = 0.675t^{3/2}, \quad 0 \le t \le 225$$

where t is the time (in weeks). When $t = 0$, $R = 0$.

(a) Find a model for the revenue function.

(b) When will the weekly revenue be \$27,000?

In Exercises 17–24, find the indefinite integral.

17. $\int (1 + 5x)^2\, dx$

18. $\int (x - 6)^{4/3}\, dx$

19. $\int \frac{1}{\sqrt{5x - 1}}\, dx$

20. $\int \frac{4x}{\sqrt{1 - 3x^2}}\, dx$

21. $\int x(1 - 4x^2)\, dx$

22. $\int \frac{x^2}{(x^3 - 4)^2}\, dx$

23. $\int (x^4 - 2x)(2x^3 - 1)\, dx$

24. $\int \frac{\sqrt{x}}{(1 - x^{3/2})^3}\, dx$

25. ***Production*** The output P (in board-feet) of a small sawmill changes according to the model

$$\frac{dP}{dt} = 2t(0.001t^2 + 0.5)^{1/4}, \quad 0 \le t \le 40$$

where t is measured in hours. Find the numbers of board-feet produced in (a) 6 hours and (b) 12 hours.

26. ***Cost*** The marginal cost for a catering service to cater to x people can be modeled by

$$\frac{dC}{dx} = \frac{5x}{\sqrt{x^2 + 1000}}.$$

When $x = 225$, the cost is \$1136.06. Find the costs of catering to (a) 500 people and (b) 1000 people.

In Exercises 27–32, find the indefinite integral.

27. $\int 3e^{-3x}\, dx$

28. $\int (2t - 1)e^{t^2 - t}\, dt$

29. $\int (x - 1)e^{x^2 - 2x}\, dx$

30. $\int \frac{4}{6x - 1}\, dx$

31. $\int \frac{x^2}{1 - x^3}\, dx$

32. $\int \frac{x - 4}{x^2 - 8x}\, dx$

In Exercises 33 and 34, use a symbolic integration utility to find the indefinite integral.

33. $\int \frac{(\sqrt{x} + 1)^2}{\sqrt{x}}\, dx$

34. $\int \frac{e^{5x}}{5 + e^{5x}}\, dx$

In Exercises 35–40, find the area of the region.

35. $f(x) = 4 - 2x$

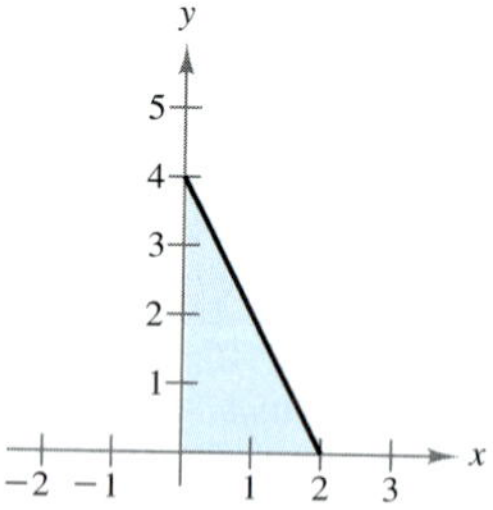

36. $f(x) = 4 - x^2$

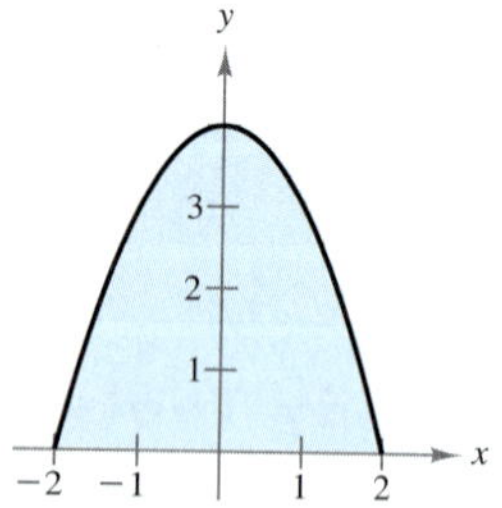

37. $f(y) = (y - 2)^2$

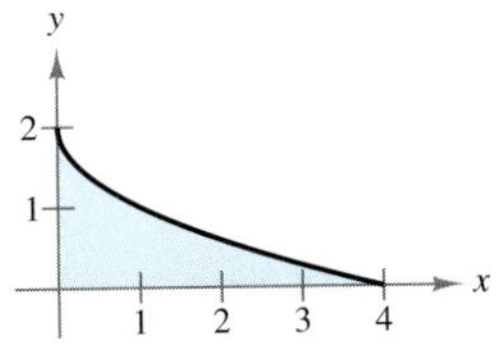

38. $f(x) = \sqrt{9 - x^2}$

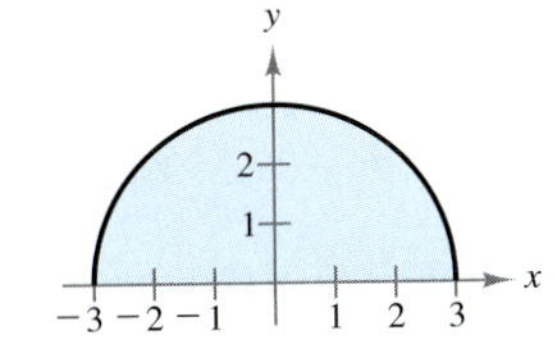

39. $f(x) = \dfrac{2}{x + 1}$

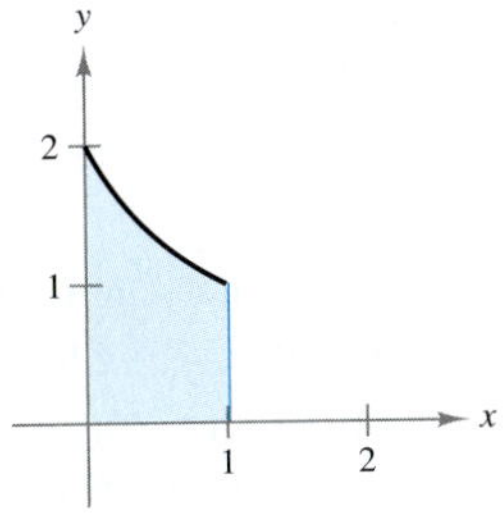

40. $f(x) = 2xe^{x^2 - 4}$

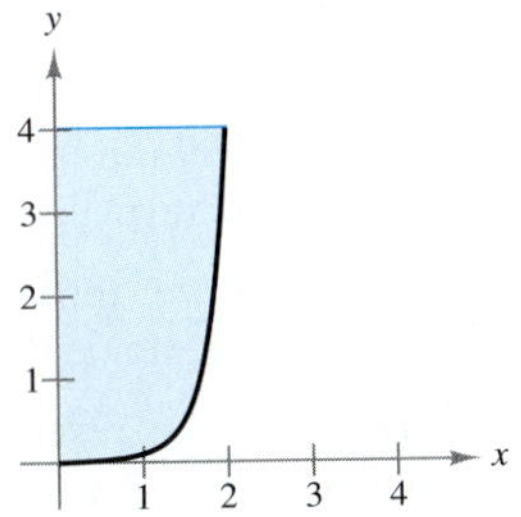

In Exercises 41–52, use the Fundamental Theorem of Calculus to evaluate the definite integral.

41. $\displaystyle\int_0^4 (2 + x)\, dx$

42. $\displaystyle\int_{-1}^1 (t^2 + 2)\, dt$

43. $\displaystyle\int_{-1}^1 (4t^3 - 2t)\, dt$

44. $\displaystyle\int_1^4 2x\sqrt{x}\, dx$

45. $\displaystyle\int_0^3 \frac{1}{\sqrt{1 + x}}\, dx$

46. $\displaystyle\int_3^6 \frac{x}{3\sqrt{x^2 - 8}}\, dx$

47. $\displaystyle\int_1^2 \left(\frac{1}{x^2} - \frac{1}{x^3}\right) dx$

48. $\displaystyle\int_0^1 x^2(x^3 + 1)^3\, dx$

49. $\displaystyle\int_1^3 \frac{(3 + \ln x)}{x}\, dx$

50. $\displaystyle\int_0^{\ln 5} e^{x/5}\, dx$

51. $\displaystyle\int_{-1}^1 3xe^{x^2 - 1}\, dx$

52. $\displaystyle\int_1^3 \frac{1}{x(\ln x + 2)^2}\, dx$

53. ***Cost*** The marginal cost of serving a typical additional client at a law firm can be modeled by

$$\frac{dC}{dx} = 675 + 0.5x$$

where x is the number of clients. How does the cost C change when x increases from 50 to 51 clients?

54. ***Profit*** The marginal profit obtained by selling x dollars of automobile insurance can be modeled by

$$\frac{dP}{dx} = 0.4\left(1 - \frac{5000}{x}\right), \quad x \geq 5000.$$

Find the change in the profit when x increases from $75,000 to $100,000.

In Exercises 55–58, find the average value of the function on the closed interval. Then find all x-values in the interval for which the function is equal to its average value.

55. $f(x) = \dfrac{4}{\sqrt{x - 1}}$, $[5, 10]$

56. $f(x) = \dfrac{20 \ln x}{x}$, $[2, 10]$

57. $f(x) = e^{5-x}$, $[2, 5]$

58. $f(x) = x^3$, $[0, 2]$

59. ***Compound Interest*** An interest-bearing checking account yields 4% interest compounded continuously. If you deposit $500 in such an account, and never write checks, what will the average value of the account be over a period of 2 years? Explain your reasoning.

60. ***Consumer Awareness*** Suppose that the price p of gasoline can be modeled by

$$p = 0.12 + 0.203t - 0.0057t^2$$

where $t = 5$ corresponds to January 1, 1995. Find the cost of gasoline for an automobile that is driven 15,000 miles per year and gets 33 miles per gallon from 1995 through 2000.

61. ***Consumer Trends*** The rates of change of beef prices (in dollars per pound) in the United States from 1998 through 2002 can be modeled by

$$\frac{dB}{dt} = -0.0486t + 0.564$$

where t is the year, with $t = 8$ corresponding to 1998. The price of 1 pound of beef in 2000 was $1.63. *(Source: U.S. Bureau of Labor Statistics)*

(a) Find the price function in terms of the year.

(b) If the price of beef per pound continues to change at this rate, in what year will the price of beef per pound surpass $2.00? Explain your reasoning.

62. ***Medical Science*** The volume V (in liters) of air in the lungs during a five-second respiratory cycle is approximated by the model

$$V = 0.1729t + 0.1522t^2 - 0.0374t^3$$

where t is time in seconds.

(a) Use a graphing utility to graph the equation on the interval $[0, 5]$.

(b) Determine the intervals on which the function is increasing and decreasing.

(c) Determine the maximum volume during the respiratory cycle.

(d) Determine the average volume of air in the lungs during one cycle.

(e) Briefly explain your results for parts (a) through (d).

Annuity In Exercises 63 and 64, find the amount of an annuity with income function $c(t)$, interest rate r, and term T.

63. $c(t) = \$3000$, $r = 6\%$, $T = 5$ years

64. $c(t) = \$1200$, $r = 7\%$, $T = 8$ years

In Exercises 65–68, explain how the given value can be used to evaluate the second integral.

65. $\int_0^2 6x^5\,dx = 64,\quad \int_{-2}^2 6x^5\,dx$

66. $\int_0^3 (x^4 + x^2)\,dx = 57.6,\quad \int_{-3}^3 (x^4 + x^2)\,dx$

67. $\int_1^2 \frac{4}{x^2}\,dx = 2,\quad \int_{-2}^{-1} \frac{4}{x^2}\,dx$

68. $\int_0^1 (x^3 - x)\,dx = -\frac{1}{4},\quad \int_{-1}^0 (x^3 - x)\,dx$

In Exercises 69–74, sketch the region bounded by the graphs of the equations. Then find the area of the region.

69. $y = \frac{1}{x^2},\ y = 4,\ x = 3$

70. $y = 1 - \frac{1}{2}x,\ y = x - 2,\ y = 1$

71. $y = \frac{4}{\sqrt{x+1}},\ y = 0,\ x = 0,\ x = 8$

72. $y = \sqrt{x}(x - 1),\ y = 0$

73. $y = (x - 3)^2,\ y = 8 - (x - 3)^2$

74. $y = 4 - x,\ y = x^2 - 5x + 8,\ x = 0$

In Exercises 75 and 76, use a graphing utility to graph the region bounded by the graphs of the equations. Then find the area of the region.

75. $y = x,\ y = 2 - x^2$

76. $y = x,\ y = x^5$

Consumer and Producer Surpluses In Exercises 77 and 78, find the consumer surplus and producer surplus for the demand and supply functions.

77. Demand function: $p_2(x) = 500 - x$

Supply function: $p_1(x) = 1.25x + 162.5$

78. Demand function: $p_2(x) = \sqrt{100{,}000 - 0.15x^2}$

Supply function: $p_1(x) = \sqrt{0.01x^2 + 36{,}000}$

79. ***Sales*** The sales S (in millions) for Avon from 1994 through 1999 can be modeled by

$$S = -9.1852t^3 + 152.304t^2 - 553.92t + 4625.3,\quad 4 \le t \le 9$$

where $t = 4$ corresponds to 1994. The sales for Avon from 2000 through 2003 can be modeled by

$$S = 79.1833t^3 - 2642.9t^2 + 29{,}569.52t - 104{,}914.8,\quad 9 < t \le 13.$$

If sales for Avon had followed the first model from 1994 through 2003, how much more or less sales would there have been for Avon? *(Source: Avon Products, Inc.)*

80. ***Revenue*** The revenues R (in millions of dollars per year) for AT&T Wireless, Nextel, and Western Wireless from 1999 through 2003 can be modeled by

$$R = -332.500t^2 + 9646.90t - 52{,}416.2 \quad \text{AT\&T Wireless}$$

$$R = -108.643t^2 + 4189.64t - 25{,}469.0 \quad \text{Nextel}$$

$$R = -3.921t^2 + 310.60t - 1896.8 \quad \text{Western Wireless}$$

where $9 \le t \le 13$ corresponds to the five-year period from 1999 through 2003. *(Source: AT&T Wireless Services, Nextel Communications, Inc., and Western Wireless Corp.)*

(a) From 1999 through 2003, how much more was AT&T Wireless' revenue than Nextel's revenue?

(b) From 1999 through 2003, how much more was Nextel's revenue than Western Wireless' revenue?

81. ***Revenue*** The revenues R (in millions of dollars per year) for Time Warner from 1994 through 1999 can be modeled by

$$R = 51.4620t^3 - 819.647t^2 + 4746.19t - 9113.3,\quad 4 \le t \le 9$$

where $t = 4$ corresponds to 1994. From 2000 through 2003, the revenues can be modeled by

$$R = 2107.667t^3 - 74{,}192.86t^2 + 867{,}426.2t - 3{,}328{,}427\quad 9 \le t \le 13.$$

(Source: Time Warner, Inc.)

(a) Use a graphing utility to graph both models over the given ranges for $4 \le t \le 13$.

(b) Using the graphs from part (a), predict whether the revenue would have increased or decreased if the first model had continued to be valid from 2000 through 2003.

(c) Analytically determine how much more or less Time Warner's revenue would have been. (*Hint:* Integrate on the interval [9, 13].)

82. ***Psychology: Sleep Patterns*** The graph on the next page shows three areas, representing awake time, REM (rapid eye movement) sleep time, and non-REM sleep time, over a typical individual's lifetime. Make generalizations about the amount of total sleep, non-REM sleep, and REM sleep an individual gets as he or she gets older. If you wanted to estimate mathematically the amount of non-REM sleep an individual gets between birth and 50 years, how would you do so? How would you mathematically estimate the amount of REM sleep an individual gets during this interval? *(Source: Adapted from Bernstein/Clarke-Stewart/Roy/Wickens,* Psychology, *Fourth Edition)*

Figure for 82

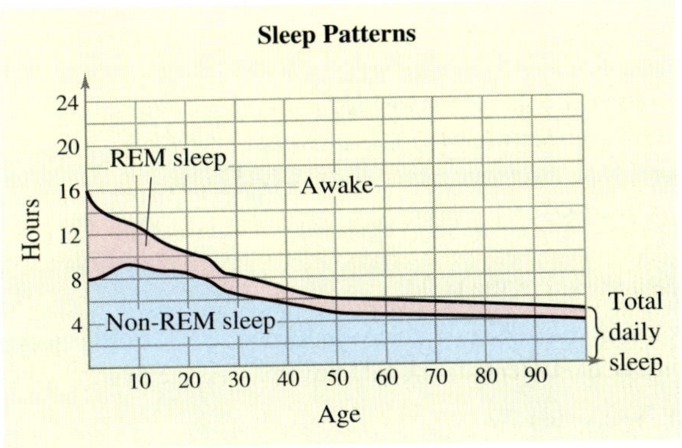

In Exercises 83–86, use the Midpoint Rule with $n = 4$ to approximate the definite integral. Then use a programmable calculator or computer to approximate the definite integral with $n = 20$. Compare the two approximations.

83. $\displaystyle\int_0^2 (x^2 + 1)^2\,dx$ **84.** $\displaystyle\int_{-1}^1 \sqrt{1 - x^2}\,dx$

85. $\displaystyle\int_0^1 \frac{1}{x^2 + 1}\,dx$ **86.** $\displaystyle\int_{-1}^1 e^{3-x^2}\,dx$

In Exercises 87–90, use the Disk Method to find the volume of the solid of revolution formed by revolving the region about the x-axis.

87. $y = \dfrac{1}{\sqrt{x}}$

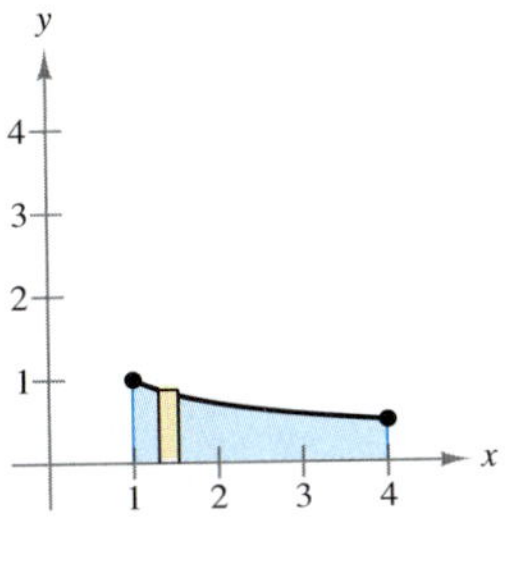

88. $y = \sqrt{16 - x}$

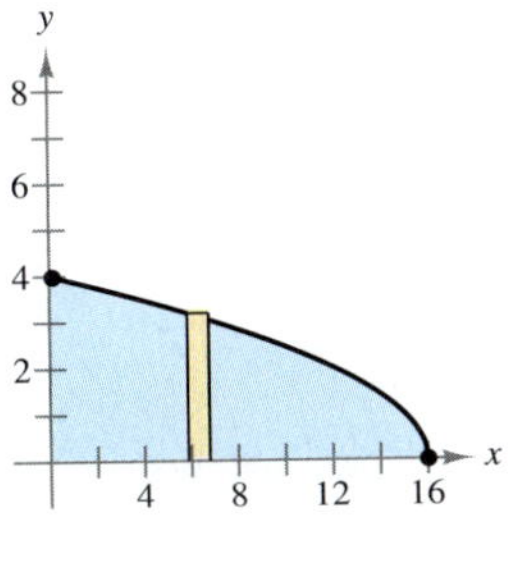

89. $y = e^{1-x}$

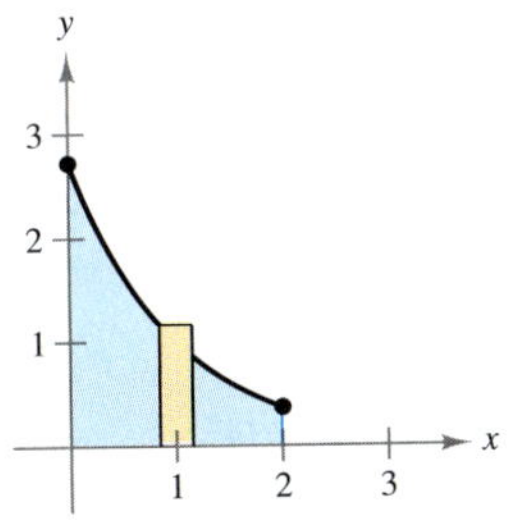

90. $y = \dfrac{1}{x}$

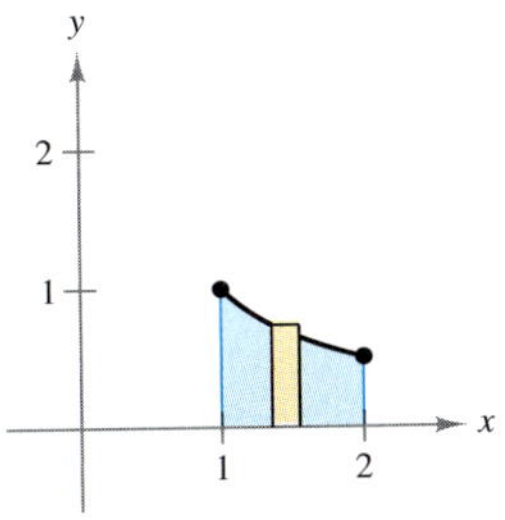

In Exercises 91–94, find the volume of the solid of revolution formed by revolving the region about the x-axis.

91. $y = 2x + 1,\ y = 1,\ x = 2$

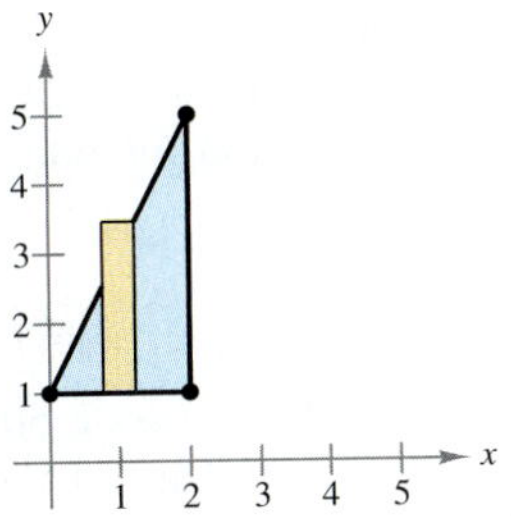

92. $y = \sqrt{x},\ y = 2,\ x = 0$

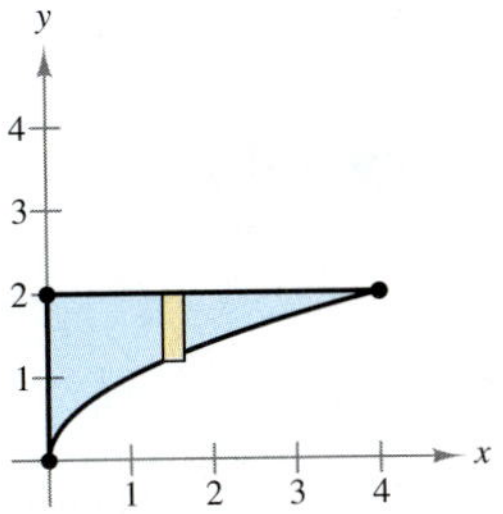

93. $y = x^2,\ y = x^3$

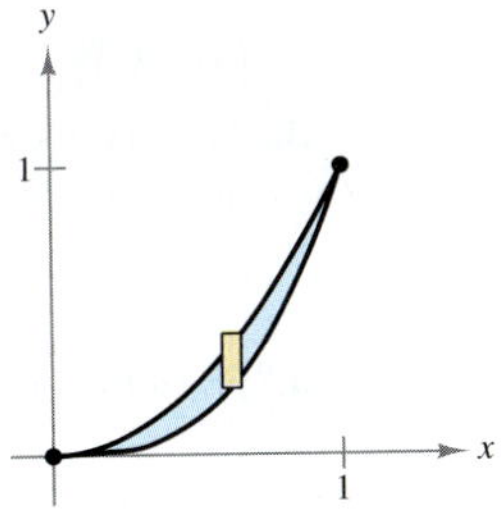

94. $y = 4 - \frac{1}{2}x^2,\ y = 2$

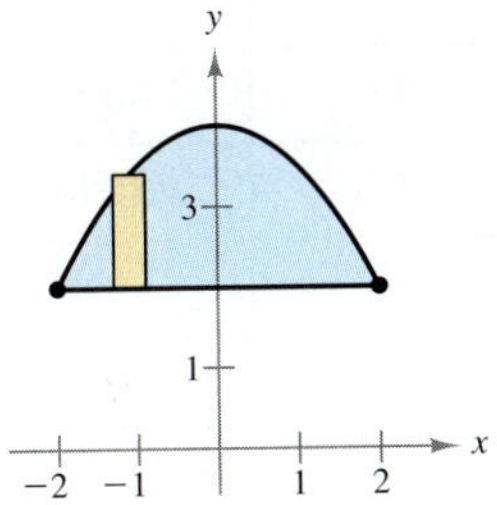

95. *Manufacturing* A manufacturer drills a hole with radius 0.25 inch through the center of a metal sphere whose radius is 1 inch. What is the volume of the resulting ring?

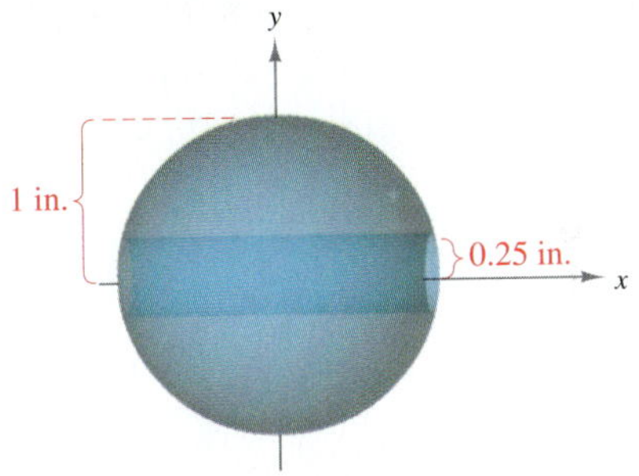

96. *Design* To create a computer design for a funnel, an engineer revolves the region bounded by the lines

$$y = 3 - x, \quad y = 0, \quad \text{and} \quad x = 0$$

about the y-axis, where x and y are measured in feet. Find the volume of the funnel.

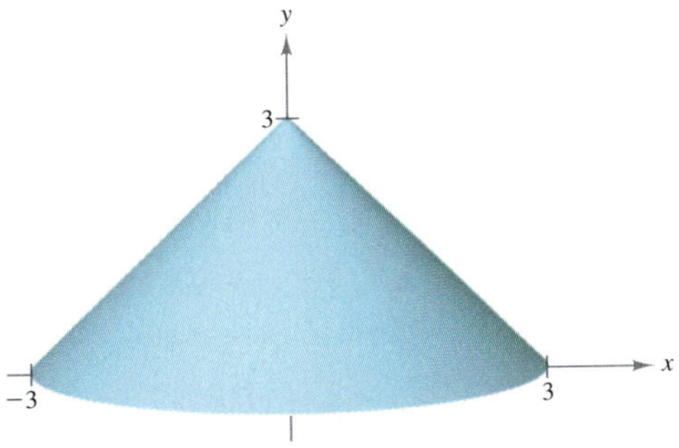

5 SAMPLE POST-GRADUATION EXAM QUESTIONS

CPA
GMAT
GRE
Actuarial
CLAST

The following questions represent the types of questions that appear on certified public accountant (CPA) exams, Graduate Management Admission Tests (GMAT), Graduate Records Exams (GRE), actuarial exams, and College-Level Academic Skills Tests (CLAST). The answers to the questions are given in the back of the book.

Table for 1–3

Number of students	Number of correct answers
12	45 to 50
11	40 to 44
14	35 to 39
10	30 to 34
7	0 to 29

For Questions 1–3, use the data shown in the table.

1. To pass a 50-question exam, a student must correctly answer 75% of the questions. What is the maximum number of students that could have passed the exam?

(a) 23 (b) 47 (c) 30 (d) 37

2. What percent of the class answered 40 or more questions correctly?

(a) 69% (b) 43% (c) 22% (d) 31%

3. The number of students who answered 35 to 39 questions correctly is y times the number who answered 29 or fewer correctly, where y is

(a) $\frac{1}{2}$ (b) 1 (c) 2 d) $\frac{7}{5}$

4. For which of the following statements is $x = -2$ a solution?

I. $x^2 + 4x + 4 \leq 0$

II. $|3x + 6| = 0$

III. $x^2 + 7x + 10 > 0$

(a) I, II, and III (b) I and II (c) II and III (d) I and III

5. The area of the shaded region in the graph is A square units, with A equal to

(a) $\frac{50}{3}$ (b) $\frac{25}{3}$ (c) 50 (d) $\frac{4}{3}$

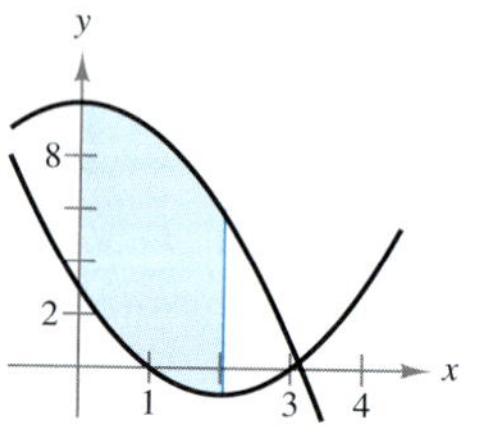

For Questions 6–8, use the data shown below.

56, 58, 54, 54, 59, 56, 55, 57, 56, 62

6. The mean of the data is

(a) 55.5 (b) 56.7 (c) 56.5 (d) 56

7. The median of the data is

(a) 55.5 (b) 56.7 (c) 56.5 (d) 56

8. The mode of the data is

(a) 55.5 (b) 56.7 (c) 56.5 (d) 56

9. If the number 52 were added to these data, which of these would change?

(a) Mean (b) Median (c) Mode (d) All of these

chapter

6 Techniques of Integration

Photodisc/Getty Images

Integration is used to find the present value of an income over several years.

STRATEGIES FOR SUCCESS

WHAT YOU SHOULD LEARN:

- How to find indefinite and definite integrals using integration by substitution
- How to evaluate integrals by parts, by using partial fractions, and by using a table of integrals
- How to use the Trapezoidal Rule and Simpson's Rule to approximate definite integrals
- How to evaluate improper integrals with infinite limits of integration and with infinite integrands

WHY YOU SHOULD LEARN IT:

The various integration techniques can be used in many applications in real life, as demonstrated by the examples below, which represent a small sample of the applications in this chapter.

- Present and Future Value, Exercises 63–74 on page 405
- Health: Epidemic, Exercise 54 on page 414
- Consumer and Producer Surpluses, Exercise 64 on page 425
- Surveying, Exercises 29 and 30 on page 434
- Capitalized Cost, Exercises 41 and 42 on page 445

6.1 INTEGRATION BY SUBSTITUTION

- Use the basic integration formulas to find indefinite integrals.
- Use substitution to find indefinite integrals.
- Use substitution to evaluate definite integrals.
- Use integration to solve real-life problems.

Review of Basic Integration Formulas

Each of the basic integration rules you studied in Chapter 5 was derived from a corresponding differentiation rule. It may surprise you to learn that, although you now have all the necessary tools for *differentiating* algebraic, exponential, and logarithmic functions, your set of tools for *integrating* these functions is by no means complete. The primary objective of this chapter is to develop several techniques that greatly expand the set of integrals to which the basic integration formulas can be applied.

Basic Integration Formulas

1. Constant Rule: $\int k\,dx = kx + C$

2. Simple Power Rule $(n \neq -1)$: $\int x^n\,dx = \frac{x^{n+1}}{n+1} + C$

3. General Power Rule $(n \neq -1)$: $\int u^n \frac{du}{dx}\,dx = \int u^n\,du$

   $= \frac{u^{n+1}}{n+1} + C$

4. Simple Exponential Rule: $\int e^x\,dx = e^x + C$

5. General Exponential Rule: $\int e^u \frac{du}{dx}\,dx = \int e^u\,du$

   $= e^u + C$

6. Simple Log Rule: $\int \frac{1}{x}\,dx = \ln|x| + C$

7. General Log Rule: $\int \frac{du/dx}{u}\,dx = \int \frac{1}{u}\,du$

   $= \ln|u| + C$

As you will see once you work a few integration problems, integration is not nearly as straightforward as differentiation. A major part of any integration problem is determining which basic integration formula (or formulas) to use to solve the problem. This requires remembering the basic formulas, familiarity with various procedures for rewriting integrands in the basic forms, and lots of practice.

Integration by Substitution

There are several techniques for rewriting an integral so that it fits one or more of the basic formulas. One of the most powerful techniques is **integration by substitution.** With this technique, you choose part of the integrand to be u and then rewrite the entire integral in terms of u.

EXAMPLE 1 Integration by Substitution

Use the substitution $u = x + 1$ to find the indefinite integral.

$$\int \frac{x}{(x+1)^2}\,dx$$

SOLUTION From the substitution $u = x + 1$,

$$x = u - 1, \quad \frac{du}{dx} = 1, \quad \text{and} \quad dx = du.$$

By replacing *all* instances of x and dx with the appropriate u-variable forms, you obtain

$$\int \frac{x}{(x+1)^2}\,dx = \int \frac{u-1}{u^2}\,du \qquad \text{Substitute for } x \text{ and } dx.$$

$$= \int \left(\frac{u}{u^2} - \frac{1}{u^2}\right) du \qquad \text{Write as separate fractions.}$$

$$= \int \left(\frac{1}{u} - \frac{1}{u^2}\right) du \qquad \text{Simplify.}$$

$$= \ln|u| + \frac{1}{u} + C \qquad \text{Find antiderivative.}$$

$$= \ln|x+1| + \frac{1}{x+1} + C. \qquad \text{Substitute for } u.$$

STUDY TIP

When you use integration by substitution, you need to realize that your integral should contain just one variable. For instance, the integrals

$$\int \frac{x}{(x+1)^2}\,dx$$

and

$$\int \frac{u-1}{u^2}\,du$$

are in the correct form, but the integral

$$\int \frac{x}{u^2}\,dx$$

is not.

TRY IT 1

Use the substitution $u = x - 2$ to find the indefinite integral.

$$\int \frac{x}{(x-2)^2}\,dx$$

The basic steps for integration by substitution are outlined in the guidelines below.

Guidelines for Integration by Substitution

1. Let u be a function of x (usually part of the integrand).
2. Solve for x and dx in terms of u and du.
3. Convert the entire integral to u-variable form and try to fit it to one or more of the basic integration formulas. If none fits, try a different substitution.
4. After integrating, rewrite the antiderivative as a function of x.

DISCOVERY

Suppose you were asked to evaluate the integrals below. Which one would you choose? Explain your reasoning.

$$\int \sqrt{x^2 + 1}\, dx \quad \text{or}$$

$$\int x\sqrt{x^2 + 1}\, dx$$

EXAMPLE 2 Integration by Substitution

Find $\int x\sqrt{x^2 - 1}\, dx$.

SOLUTION Consider the substitution $u = x^2 - 1$, which produces $du = 2x\, dx$. To create $2x\, dx$ as part of the integral, multiply and divide by 2.

$$\begin{aligned}
\int x\sqrt{x^2 - 1}\, dx &= \frac{1}{2}\int \overbrace{(x^2 - 1)^{1/2}}^{u^{1/2}}\overbrace{2x\, dx}^{du} && \text{Multiply and divide by 2.} \\
&= \frac{1}{2}\int u^{1/2}\, du && \text{Substitute for } x \text{ and } dx. \\
&= \frac{1}{2}\frac{u^{3/2}}{3/2} + C && \text{Power Rule} \\
&= \frac{1}{3}u^{3/2} + C && \text{Simplify.} \\
&= \frac{1}{3}(x^2 - 1)^{3/2} + C && \text{Substitute for } u.
\end{aligned}$$

You can check this result by differentiating.

TRY IT 2

Find $\int x\sqrt{x^2 + 4}\, dx$.

EXAMPLE 3 Integration by Substitution

Find $\int \frac{e^{3x}}{1 + e^{3x}}\, dx$.

SOLUTION Consider the substitution $u = 1 + e^{3x}$, which produces $du = 3e^{3x}\, dx$. To create $3e^{3x}\, dx$ as part of the integral, multiply and divide by 3.

$$\begin{aligned}
\int \frac{e^{3x}}{1 + e^{3x}}\, dx &= \frac{1}{3}\int \overbrace{\frac{1}{1 + e^{3x}}}^{1/u}\overbrace{3e^{3x}\, dx}^{du} && \text{Multiply and divide by 3.} \\
&= \frac{1}{3}\int \frac{1}{u}\, du && \text{Substitute for } x \text{ and } dx. \\
&= \frac{1}{3}\ln|u| + C && \text{Log Rule} \\
&= \frac{1}{3}\ln(1 + e^{3x}) + C && \text{Substitute for } u.
\end{aligned}$$

Note that the absolute value is not necessary in the final answer because the quantity $(1 + e^{3x})$ is positive for all values of x.

TRY IT 3

Find $\int \frac{e^{2x}}{1 + e^{2x}}\, dx$.

EXAMPLE 4 Integration by Substitution

Find the indefinite integral.

$$\int x\sqrt{x-1}\,dx$$

SOLUTION Consider the substitution $u = x - 1$, which produces $du = dx$ and $x = u + 1$.

$$\int x\sqrt{x-1}\,dx = \int (u+1)(u^{1/2})\,du \qquad \text{Substitute for } x \text{ and } dx.$$

$$= \int (u^{3/2} + u^{1/2})\,du \qquad \text{Multiply.}$$

$$= \frac{u^{5/2}}{5/2} + \frac{u^{3/2}}{3/2} + C \qquad \text{Power Rule}$$

$$= \frac{2}{5}(x-1)^{5/2} + \frac{2}{3}(x-1)^{3/2} + C \qquad \text{Substitute for } u.$$

This form of the antiderivative can be further simplified.

$$\frac{2}{5}(x-1)^{5/2} + \frac{2}{3}(x-1)^{3/2} + C = \frac{6}{15}(x-1)^{5/2} + \frac{10}{15}(x-1)^{3/2} + C$$

$$= \frac{2}{15}(x-1)^{3/2}[3(x-1) + 5] + C$$

$$= \frac{2}{15}(x-1)^{3/2}(3x+2) + C$$

You can check this answer by differentiating.

TRY IT 4

Find the indefinite integral.

$$\int x\sqrt{x+2}\,dx$$

Example 4 demonstrates one of the characteristics of integration by substitution. That is, you can often simplify the form of the antiderivative as it exists immediately after resubstitution into x-variable form. So, when working the exercises in this section, don't assume that your answer is incorrect just because it doesn't look exactly like the answer given in the back of the text. You may be able to reconcile the two answers by algebraic simplification.

TECHNOLOGY

If you have access to a symbolic integration utility, try using it to find an antiderivative of $f(x) = x^2\sqrt{x+1}$ and check your answer analytically using the substitution $u = x + 1$. You can also use the utility to solve several of the exercises in this section.

Substitution and Definite Integrals

The fourth step outlined in the guidelines for integration by substitution on page 389 suggests that you convert back to the variable x. To evaluate *definite* integrals, however, it is often more convenient to determine the limits of integration for the variable u. This is often easier than converting back to the variable x and evaluating the antiderivative at the original limits.

EXAMPLE 5 Using Substitution with a Definite Integral

Evaluate $\displaystyle\int_1^5 \frac{x}{\sqrt{2x-1}}\,dx$.

SOLUTION Use the substitution $u = \sqrt{2x-1}$, which implies that $u^2 = 2x - 1$, $x = \frac{1}{2}(u^2+1)$, and $dx = u\,du$. Before substituting, determine the new upper and lower limits of integration.

Lower limit: When $x = 1$, $u = \sqrt{2(1) - 1} = 1$.

Upper limit: When $x = 5$, $u = \sqrt{2(5) - 1} = 3$.

Now, substitute and integrate, as shown.

Integration limits for x — Integration limits for u

$$\int_1^5 \frac{x}{\sqrt{2x-1}}\,dx = \int_1^3 \frac{1}{u}\left(\frac{u^2+1}{2}\right)u\,du$$ Substitute for x, dx, and limits of integration.

$$= \frac{1}{2}\int_1^3 (u^2+1)\,du$$ Simplify.

$$= \frac{1}{2}\left[\frac{u^3}{3} + u\right]_1^3$$ Find antiderivative.

$$= \frac{1}{2}\left[(9+3) - \left(\frac{1}{3}+1\right)\right]$$ Apply Fundamental Theorem.

$$= \frac{16}{3}$$ Simplify.

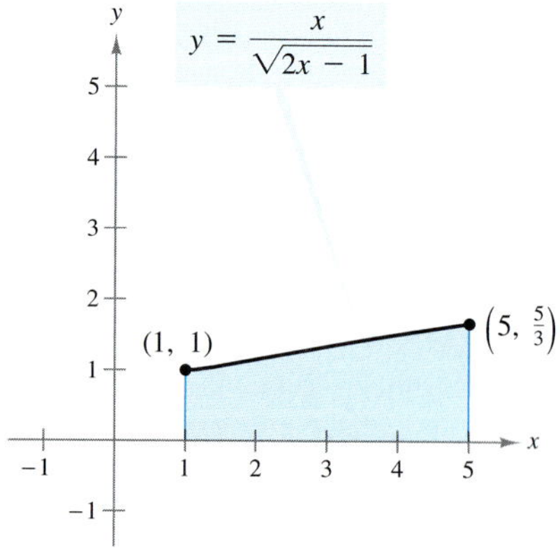

FIGURE 6.1 Region Before Substitution

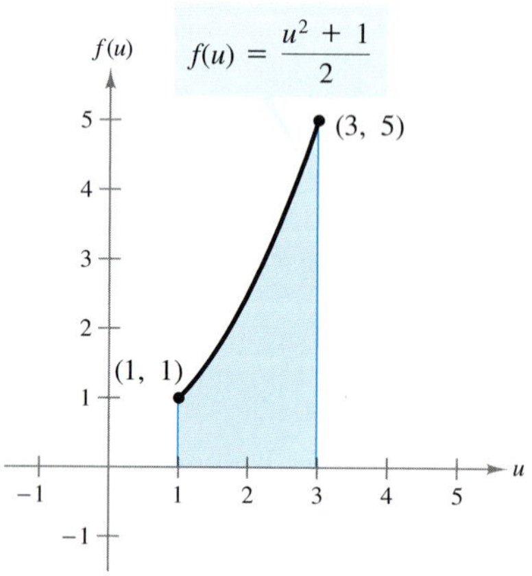

FIGURE 6.2 Region After Substitution

TRY IT 5

Evaluate $\displaystyle\int_0^2 x\sqrt{4x+1}\,dx$.

STUDY TIP

In Example 5, you can interpret the equation

$$\int_1^5 \frac{x}{\sqrt{2x-1}}\,dx = \int_1^3 \frac{1}{u}\left(\frac{u^2+1}{2}\right)u\,du$$

graphically to mean that the two different regions shown in Figures 6.1 and 6.2 have the same area.

Application

Integration can be used to find the probability that an event will occur. In such an application, the real-life situation is modeled by a *probability density function f*, and the probability that x will lie between a and b is represented by

$$P(a \le x \le b) = \int_a^b f(x)\,dx.$$

The probability $P(a \le x \le b)$ must be a number between 0 and 1.

Researchers such as psychologists use definite integrals to represent the probability that an event will occur. For instance, a probability of 0.5 means that an event will occur about 50% of the time.

EXAMPLE 6 Finding a Probability

A psychologist finds that the probability that a participant in a memory experiment will recall between a and b percent (in decimal form) of the material is

$$P(a \le x \le b) = \int_a^b \frac{28}{9}x\sqrt[3]{1-x}\,dx, \quad 0 \le a \le b \le 1.$$

Find the probability that a randomly chosen participant will recall between 0% and 87.5% of the material.

SOLUTION Let $u = \sqrt[3]{1-x}$. Then $u^3 = 1 - x$, $x = 1 - u^3$, and $dx = -3u^2\,du$.

Lower limit: When $x = 0$, $u = \sqrt[3]{1-0} = 1$.

Upper limit: When $x = 0.875$, $u = \sqrt[3]{1-0.875} = 0.5$.

To find the probability, substitute and integrate, as shown.

$$\begin{aligned}\int_0^{0.875} \frac{28}{9}x\sqrt[3]{1-x}\,dx &= \int_1^{1/2} \left[\frac{28}{9}(1-u^3)(u)(-3u^2)\right] du \\ &= \left(\frac{28}{3}\right)\int_1^{1/2} (u^6 - u^3)\,du \\ &= \frac{28}{3}\left[\frac{u^7}{7} - \frac{u^4}{4}\right]_1^{1/2} \\ &\approx 0.865\end{aligned}$$

So, the probability is about 86.5%, as indicated in Figure 6.3.

ALGEBRA REVIEW

For help on the algebra in Example 6, see Example 1(a) in the *Chapter 6 Algebra Review* on page 446.

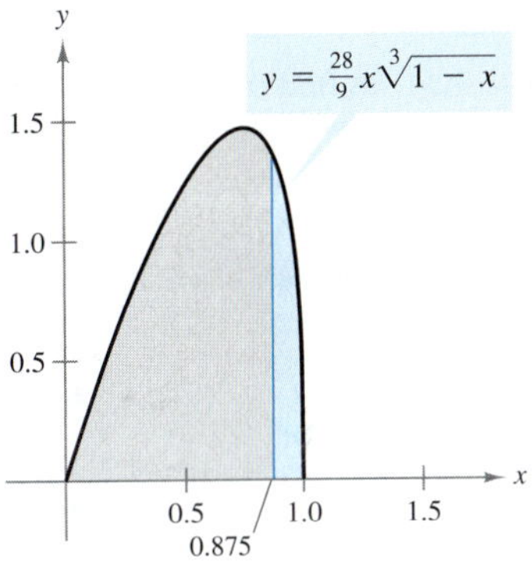

FIGURE 6.3

TRY IT 6

Use Example 6 to find the probability that a participant will recall between 0% and 62.5% of the material.

TAKE ANOTHER LOOK

Probability

In Example 6, explain how you could find a value of b such that $P(0 \le x \le b) = 0.5$.

PREREQUISITE REVIEW 6.1

The following warm-up exercises involve skills that were covered in earlier sections. You will use these skills in the exercise set for this section.

In Exercises 1–8, evaluate the indefinite integral.

1. $\int 5\, dx$

2. $\int \frac{1}{3}\, dx$

3. $\int x^{3/2}\, dx$

4. $\int x^{2/3}\, dx$

5. $\int 2x(x^2 + 1)^3\, dx$

6. $\int 3x^2(x^3 - 1)^2\, dx$

7. $\int 6e^{6x}\, dx$

8. $\int \frac{2}{2x + 1}\, dx$

In Exercises 9–12, simplify the expression.

9. $2x(x - 1)^2 + x(x - 1)$

10. $6x(x + 4)^3 - 3x^2(x + 4)^2$

11. $3(x + 7)^{1/2} - 2x(x + 7)^{-1/2}$

12. $(x + 5)^{1/3} - 5(x + 5)^{-2/3}$

EXERCISES 6.1

In Exercises 1–38, find the indefinite integral.

1. $\int (x - 2)^4\, dx$

2. $\int (x + 5)^{3/2}\, dx$

3. $\int \frac{2}{(t - 9)^2}\, dt$

4. $\int \frac{4}{(1 - t)^3}\, dt$

5. $\int \frac{2t - 1}{t^2 - t + 2}\, dt$

6. $\int \frac{2y^3}{y^4 + 1}\, dy$

7. $\int \sqrt{1 + x}\, dx$

8. $\int (3 + x)^{5/2}\, dx$

9. $\int \frac{12x + 2}{3x^2 + x}\, dx$

10. $\int \frac{6x^2 + 2}{x^3 + x}\, dx$

11. $\int \frac{1}{(5x + 1)^3}\, dx$

12. $\int \frac{1}{(3x + 1)^2}\, dx$

13. $\int \frac{1}{\sqrt{x + 1}}\, dx$

14. $\int \frac{1}{\sqrt{5x + 1}}\, dx$

15. $\int \frac{e^{3x}}{1 - e^{3x}}\, dx$

16. $\int \frac{4e^{2x}}{1 + e^{2x}}\, dx$

17. $\int \frac{2x}{e^{3x^2}}\, dx$

18. $\int \frac{e^{\sqrt{x+1}}}{\sqrt{x + 1}}\, dx$

19. $\int \frac{x^2}{x - 1}\, dx$

20. $\int \frac{2x}{x - 4}\, dx$

21. $\int x\sqrt{x^2 + 4}\, dx$

22. $\int \frac{t}{\sqrt{1 - t^2}}\, dt$

23. $\int e^{5x}\, dx$

24. $\int te^{t^2+1}\, dt$

25. $\int \frac{e^{-x}}{e^{-x} + 2}\, dx$

26. $\int \frac{e^x}{1 + e^x}\, dx$

27. $\int \frac{x}{(x + 1)^4}\, dx$

28. $\int \frac{x^2}{(x + 1)^3}\, dx$

29. $\int \frac{x}{(3x - 1)^2}\, dx$

30. $\int \frac{5x}{(x - 4)^3}\, dx$

31. $\int \frac{1}{\sqrt{t} - 1}\, dt$

32. $\int \frac{1}{\sqrt{x} + 1}\, dx$

33. $\int \frac{2\sqrt{t} + 1}{t}\, dt$

34. $\int \frac{6x + \sqrt{x}}{x}\, dx$

35. $\int \frac{x}{\sqrt{2x + 1}}\, dx$

36. $\int \frac{x^2}{\sqrt{x - 1}}\, dx$

37. $\int t^2\sqrt{1 - t}\, dt$

38. $\int y^2\sqrt[3]{y + 1}\, dy$

In Exercises 39–46, evaluate the definite integral.

39. $\int_0^4 \sqrt{2x + 1}\, dx$

40. $\int_2^4 \sqrt{4x + 1}\, dx$

41. $\int_0^1 3xe^{x^2}\, dx$

42. $\int_0^2 e^{-2x}\, dx$

43. $\int_0^4 \frac{x}{(x + 4)^2}\, dx$

44. $\int_0^1 x(x + 5)^4\, dx$

45. $\int_0^{0.5} x(1 - x)^3\, dx$

46. $\int_0^{0.5} x^2(1 - x)^3\, dx$

In Exercises 47–54, find the area of the region bounded by the graphs of the equations. Then use a graphing utility to graph the region and verify your answer.

47. $y = x\sqrt{x-3},\ y = 0,\ x = 7$

48. $y = x\sqrt{2x+1},\ y = 0,\ x = 4$

49. $y = x^2\sqrt{1-x},\ y = 0,\ x = -3$

50. $y = x^2\sqrt{x+2},\ y = 0,\ x = 7$

51. $y = \dfrac{x^2-1}{\sqrt{2x-1}},\ y = 0,\ x = 1,\ x = 5$

52. $y = \dfrac{2x-1}{\sqrt{x+3}},\ y = 0,\ x = \dfrac{1}{2},\ x = 6$

53. $y = x\sqrt[3]{x+1},\ y = 0,\ x = 0,\ x = 7$

54. $y = x\sqrt[3]{x-2},\ y = 0,\ x = 2,\ x = 10$

In Exercises 55–58, find the area of the region bounded by the graphs of the equations.

55. $y = -x\sqrt{x+2},\ y = 0$

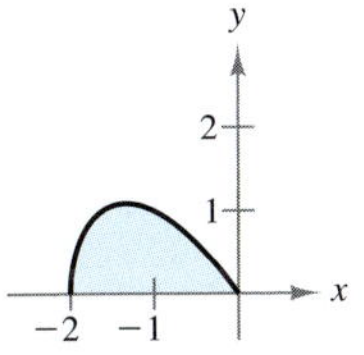

56. $y = x\sqrt[3]{1-x},\ y = 0$

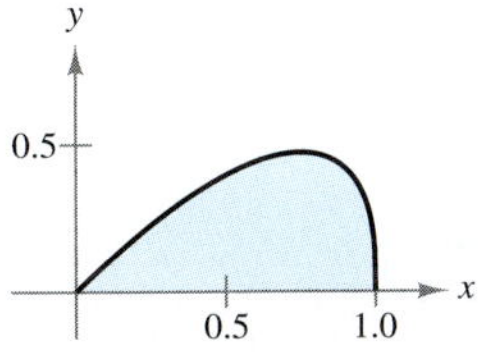

57. $y^2 = x^2(1-x^2)$

(*Hint:* Find the area of the region bounded by $y = x\sqrt{1-x^2}$ and $y = 0$. Then multiply by 4.)

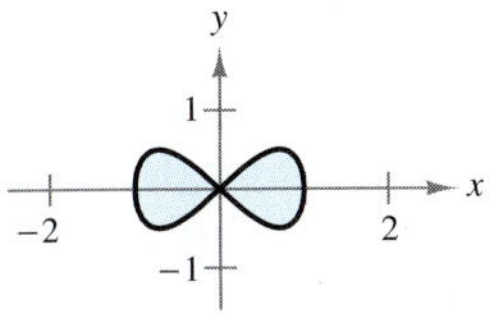

58. $y = 1/(1+\sqrt{x}),$ $y = 0,\ x = 0,\ x = 4$

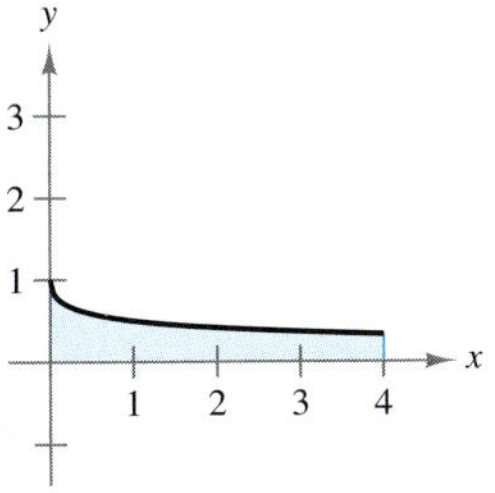

In Exercises 59 and 60, find the volume of the solid generated by revolving the region bounded by the graph(s) of the equation(s) about the x-axis.

59. $y = x\sqrt{1-x^2}$

60. $y = \sqrt{x}(1-x)^2,\ y = 0$

In Exercises 61 and 62, find the average amount by which the function f exceeds the function g on the interval.

61. $f(x) = \dfrac{1}{x+1},\ g(x) = \dfrac{x}{(x+1)^2},\ [0, 1]$

62. $f(x) = x\sqrt{4x+1},\ g(x) = 2\sqrt{x^3},\ [0, 2]$

63. ***Probability*** The probability of recall in an experiment is modeled by

$$P(a \le x \le b) = \int_a^b \frac{15}{4}x\sqrt{1-x}\,dx$$

where x is the percent of recall (see figure).

(a) What is the probability of recalling between 40% and 80%?

(b) What is the median percent recall? That is, for what value of b is $P(0 \le x \le b) = 0.5$?

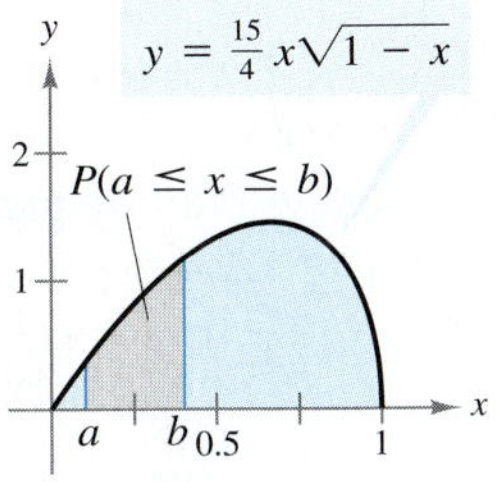

Figure for 63

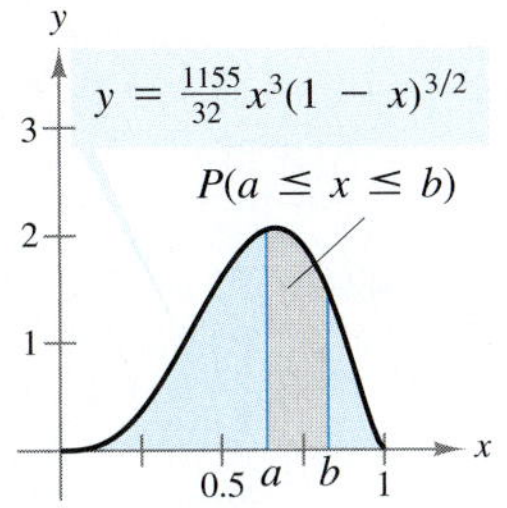

Figure for 64

64. ***Probability*** The probability of finding between a and b percent iron in ore samples is modeled by

$$P(a \le x \le b) = \int_a^b \frac{1155}{32}x^3(1-x)^{3/2}\,dx$$

(see figure). Find the probabilities that a sample will contain between (a) 0% and 25% and (b) 50% and 100% iron.

65. ***Meteorology*** During a two-week period in March in a small town near Lake Erie, the measurable snowfall S (in inches) on the ground can be modeled by

$$S(t) = t\sqrt{14-t}, \quad 0 \le t \le 14$$

where t represents the day.

(a) Use a graphing utility to graph the function.

(b) Find the average amount of snow on the ground during the two-week period.

(c) Find the total snowfall over the two-week period.

66. ***Revenue*** A company sells a seasonal product that generates a daily revenue R (in dollars per day) modeled by

$$R = 0.06t^2(365-t)^{1/2} + 1250, \quad 0 \le t \le 365$$

where t represents the day.

(a) Find the average daily revenue over a period of 1 year.

(b) Describe a product whose seasonal sales pattern resembles the model. Explain your reasoning.

In Exercises 67 and 68, use a program similar to the Midpoint Rule program on page 366 with $n = 10$ to approximate the area of the region bounded by the graph(s) of the equation(s).

67. $y = \sqrt[3]{x}\sqrt{4-x}, \quad y = 0$

68. $y^2 = x^2(1-x^2)$

6.2 INTEGRATION BY PARTS AND PRESENT VALUE

- Use integration by parts to find indefinite and definite integrals.
- Find the present value of future income.

Integration by Parts

In this section, you will study an integration technique called **integration by parts.** This technique is particularly useful for integrands involving the products of algebraic and exponential or logarithmic functions, such as

$$\int x^2e^x\,dx \quad \text{and} \quad \int x\ln x\,dx.$$

Integration by parts is based on the Product Rule for differentiation.

$$\frac{d}{dx}[uv] = u\frac{dv}{dx} + v\frac{du}{dx} \qquad \text{Product Rule}$$

$$uv = \int u\frac{dv}{dx}\,dx + \int v\frac{du}{dx}\,dx \qquad \text{Integrate each side.}$$

$$uv = \int u\,dv + \int v\,du \qquad \text{Write in differential form.}$$

$$\int u\,dv = uv - \int v\,du \qquad \text{Rewrite.}$$

Integration by Parts

Let u and v be differentiable functions of x.

$$\int u\,dv = uv - \int v\,du$$

Note that the formula for integration by parts expresses the original integral in terms of another integral. Depending on the choices for u and dv, it may be easier to evaluate the second integral than the original one.

Guidelines for Integration by Parts

1. Let dv be the most complicated portion of the integrand that fits a basic integration formula. Let u be the remaining factor.
2. Let u be the portion of the integrand whose derivative is a function simpler than u. Let dv be the remaining factor.

STUDY TIP

When using integration by parts, note that you can first choose dv or first choose u. After you choose, however, the choice of the other factor is determined—it must be the remaining portion of the integrand. Also note that dv *must* contain the differential dx of the original integral.

EXAMPLE 1 Integration by Parts

Find $\int xe^x\,dx$.

SOLUTION To apply integration by parts, you must rewrite the original integral in the form $\int u\,dv$. That is, you must break $xe^x\,dx$ into two factors—one "part" representing u and the other "part" representing dv. There are several ways to do this.

$$\int \underbrace{(x)}_{u}\underbrace{(e^x\,dx)}_{dv} \qquad \int \underbrace{(e^x)}_{u}\underbrace{(x\,dx)}_{dv} \qquad \int \underbrace{(1)}_{u}\underbrace{(xe^x\,dx)}_{dv} \qquad \int \underbrace{(xe^x)}_{u}\underbrace{(dx)}_{dv}$$

Following the guidelines, you should choose the first option because $dv = e^x\,dx$ is the most complicated portion of the integrand that fits a basic integration formula *and* because the derivative of $u = x$ is simpler than x.

$$dv = e^x\,dx \quad \Rightarrow \quad v = \int dv = \int e^x\,dx = e^x$$

$$u = x \quad \Rightarrow \quad du = dx$$

With these substitutions, you can apply the integration by parts formula as shown.

$$\int xe^x\,dx = xe^x - \int e^x\,dx \qquad \int u\,dv = uv - \int v\,du$$

$$= xe^x - e^x + C \qquad \text{Integrate } \int e^x\,dx.$$

TRY IT 1

Find $\int xe^{2x}\,dx$.

STUDY TIP

In Example 1, notice that you do not need to include a constant of integration when solving $v = \int e^x\,dx = e^x$. To see why this is true, try replacing e^x by $e^x + C_1$ in the solution.

$$\int xe^x\,dx = x(e^x + C_1) - \int (e^x + C_1)\,dx$$

After integrating, you can see that the terms involving C_1 subtract out.

TECHNOLOGY

If you have access to a symbolic integration utility, try using it to solve several of the exercises in this section. Note that the form of the integral may be slightly different from what you obtain when solving the exercise by hand.

STUDY TIP

To remember the integration by parts formula, you might like to use the "Z" pattern below. The top row represents the original integral, the diagonal row represents uv, and the bottom row represents the new integral.

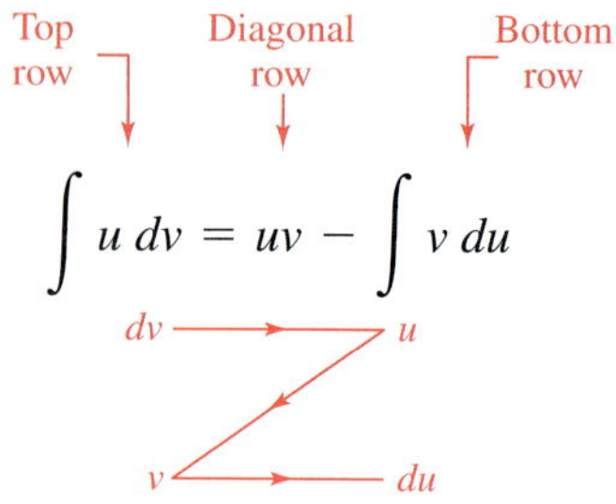

EXAMPLE 2 Integration by Parts

Find $\int x^2 \ln x\,dx$.

SOLUTION For this integral, x^2 is more easily integrated than $\ln x$. Furthermore, the derivative of $\ln x$ is simpler than $\ln x$. So, you should choose $dv = x^2\,dx$.

$$dv = x^2\,dx \quad \Rightarrow \quad v = \int dv = \int x^2\,dx = \frac{x^3}{3}$$

$$u = \ln x \quad \Rightarrow \quad du = \frac{1}{x}\,dx$$

Using these substitutions, apply the integration by parts formula as shown.

$$\int x^2 \ln x\,dx = \frac{x^3}{3}\ln x - \int \left(\frac{x^3}{3}\right)\left(\frac{1}{x}\right)dx \qquad \int u\,dv = uv - \int v\,du$$

$$= \frac{x^3}{3}\ln x - \frac{1}{3}\int x^2\,dx \qquad \text{Simplify.}$$

$$= \frac{x^3}{3}\ln x - \frac{x^3}{9} + C \qquad \text{Integrate.}$$

TRY IT 2

Find $\int x \ln x\,dx$.

EXAMPLE 3 Integrating by Parts with a Single Factor

Find $\int \ln x\,dx$.

SOLUTION This integral is unusual because it has only one factor. In such cases, you should choose $dv = dx$ and choose u to be the single factor.

$$dv = dx \quad \Rightarrow \quad v = \int dv = \int dx = x$$

$$u = \ln x \quad \Rightarrow \quad du = \frac{1}{x}\,dx$$

Using these substitutions, apply the integration by parts formula as shown.

$$\int \ln x\,dx = x \ln x - \int (x)\left(\frac{1}{x}\right)dx \qquad \int u\,dv = uv - \int v\,du$$

$$= x \ln x - \int dx \qquad \text{Simplify.}$$

$$= x \ln x - x + C \qquad \text{Integrate.}$$

TRY IT 3

Differentiate $y = x \ln x - x + C$ to show that it is the antiderivative of $\ln x$.

EXAMPLE 4 Using Integration by Parts Repeatedly

Find $\int x^2e^x\,dx$.

SOLUTION Using the guidelines, notice that the derivative of x^2 becomes simpler, whereas the derivative of e^x does not. So, you should let $u = x^2$ and let $dv = e^x\,dx$.

$$dv = e^x\,dx \quad \Rightarrow \quad v = \int dv = \int e^x\,dx = e^x$$

$$u = x^2 \quad \Rightarrow \quad du = 2x\,dx$$

Using these substitutions, apply the integration by parts formula as shown.

$$\int x^2e^x\,dx = x^2e^x - \int 2xe^x\,dx \qquad \text{First application of integration by parts}$$

To evaluate the new integral on the right, apply integration by parts a second time, using the substitutions below.

$$dv = e^x\,dx \quad \Rightarrow \quad v = \int dv = \int e^x\,dx = e^x$$

$$u = 2x \quad \Rightarrow \quad du = 2\,dx$$

Using these substitutions, apply the integration by parts formula as shown.

$$\int x^2e^x\,dx = x^2e^x - \int 2xe^x\,dx \qquad \text{First application of integration by parts}$$

$$= x^2e^x - \left(2xe^x - \int 2e^x\,dx\right) \qquad \text{Second application of integration by parts}$$

$$= x^2e^x - 2xe^x + 2e^x + C \qquad \text{Integrate.}$$

$$= e^x(x^2 - 2x + 2) + C \qquad \text{Simplify.}$$

You can confirm this result by differentiating.

TRY IT 4

Find $\int x^3e^x\,dx$.

When making repeated applications of integration by parts, be careful not to interchange the substitutions in successive applications. For instance, in Example 4, the first substitutions were $dv = e^x\,dx$ and $u = x^2$. If in the second application you had switched to $dv = 2x\,dx$ and $u = e^x$, you would have reversed the previous integration and returned to the *original* integral.

$$\int x^2e^x\,dx = x^2e^x - \left(x^2e^x - \int x^2e^x\,dx\right)$$

$$= \int x^2e^x\,dx$$

STUDY TIP

Remember that you can check an indefinite integral by differentiating. For instance, in Example 4, try differentiating the antiderivative

$$e^x(x^2 - 2x + 2) + C$$

to check that you obtain the original integrand, x^2e^x.

EXAMPLE 5 **Evaluating a Definite Integral**

Evaluate $\int_1^e \ln x\, dx$.

SOLUTION Integration by parts was used to find the antiderivative of $\ln x$ in Example 3. Using this result, you can evaluate the definite integral as shown.

$$\int_1^e \ln x\, dx = \Big[x \ln x - x\Big]_1^e \quad \text{Use result of Example 3.}$$
$$= (e \ln e - e) - (1 \ln 1 - 1) \quad \text{Apply Fundamental Theorem.}$$
$$= (e - e) - (0 - 1)$$
$$= 1 \quad \text{Simplify.}$$

The area represented by this definite integral is shown in Figure 6.4.

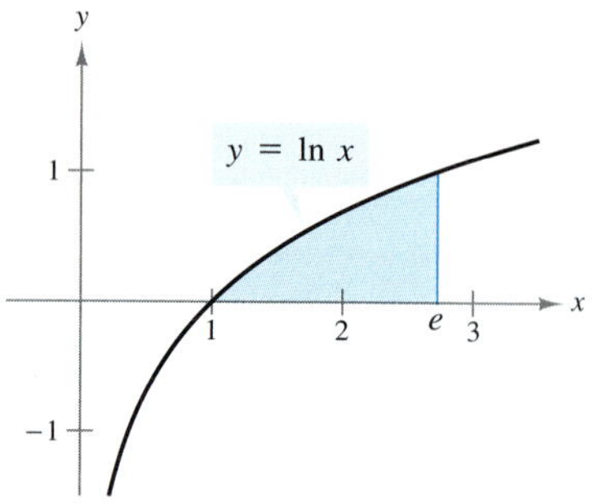

FIGURE 6.4

TRY IT 5

Evaluate $\int_0^1 x^2 e^x\, dx$.

Before starting the exercises in this section, remember that it is not enough to know *how* to use the various integration techniques. You also must know *when* to use them. Integration is first and foremost a problem of recognition—recognizing which formula or technique to apply to obtain an antiderivative. Often, a slight alteration of an integrand will necessitate the use of a different integration technique. Here are some examples.

Integral	*Technique*	*Antiderivative*
$\int x \ln x\, dx$	Integration by parts	$\frac{x^2}{2}\ln x - \frac{x^2}{4} + C$
$\int \frac{\ln x}{x}\, dx$	Power Rule: $\int u^n \frac{du}{dx}\, dx$	$\frac{(\ln x)^2}{2} + C$
$\int \frac{1}{x \ln x}\, dx$	Log Rule: $\int \frac{1}{u}\frac{du}{dx}\, dx$	$\ln\lvert \ln x\rvert + C$

As you gain experience with integration by parts, your skill in determining u and dv will improve. The summary below gives suggestions for choosing u and dv.

Summary of Common Uses of Integration by Parts

1. $\int x^n e^{ax}\, dx$ Let $u = x^n$ and $dv = e^{ax}\, dx$. (Examples 1 and 4)

2. $\int x^n \ln x\, dx$ Let $u = \ln x$ and $dv = x^n\, dx$. (Examples 2 and 3)

Present Value

Recall from Section 4.2 that the present value of a future payment is the amount that would have to be deposited today to produce the future payment. What is the present value of a future payment of $1000 one year from now? Because of inflation, $1000 today buys more than $1000 will buy a year from now. The definition below considers only the effect of inflation.

Present Value

If c represents a continuous income function in dollars per year and the annual rate of inflation is r, then the actual total income over t_1 years is

$$\text{Actual income over } t_1 \text{ years} = \int_0^{t_1} c(t)\,dt$$

and its **present value** is

$$\text{Present value} = \int_0^{t_1} c(t)e^{-rt}\,dt.$$

ALGEBRA REVIEW

According to this definition, if the rate of inflation were 4%, then the present value of $1000 one year from now is just $980.26.

Ignoring inflation, the equation for present value also applies to an interest-bearing account where the annual interest rate r is compounded continuously and c is an income function in dollars per year.

EXAMPLE 6 Finding Present Value

You have just won a state lottery for $1,000,000. You will be paid an annuity of $50,000 a year for 20 years. Assuming an annual inflation rate of 6%, what is the present value of this income?

SOLUTION The income function for your winnings is given by $c(t) = 50{,}000$. So,

$$\text{Actual income} = \int_0^{20} 50{,}000\,dt = \Big[50{,}000t\Big]_0^{20} = \$1{,}000{,}000.$$

Because you do not receive this entire amount now, its present value is

$$\text{Present value} = \int_0^{20} 50{,}000e^{-0.06t}\,dt = \left[\frac{50{,}000}{-0.06}e^{-0.06t}\right]_0^{20} \approx \$582{,}338.$$

This present value represents the amount that the state must deposit now to cover your payments over the next 20 years. This shows why state lotteries are so profitable—for the states!

TRY IT 6

Find the present value of the income from the lottery ticket in Example 6 if the inflation rate is 7%.

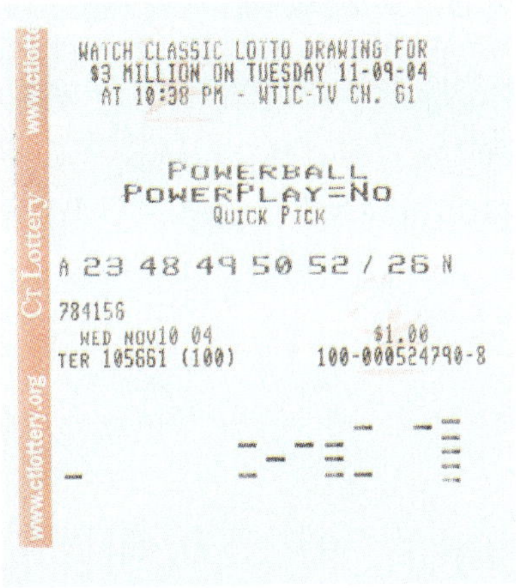

Courtesty of Multi-State Lottery Association

On December 25, 2002, Jack Whittaker won the largest lottery jackpot in the world. He chose to receive a lump sum payment of $170.5 million instead of an annuity that would have paid $314.9 million over a 30-year period. The odds of winning the PowerBall jackpot are 120.5 million to 1.

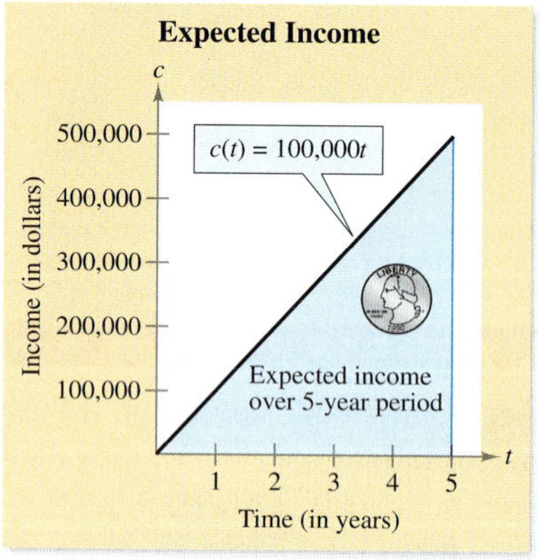

(a)

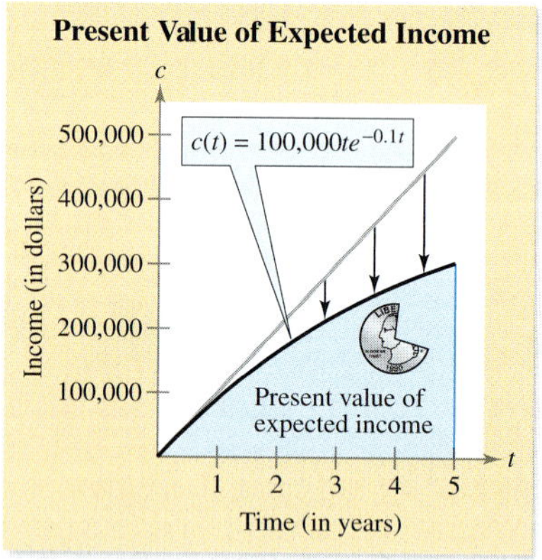

(b)

FIGURE 6.5

EXAMPLE 7 Finding Present Value

A company expects its income during the next 5 years to be given by

$$c(t) = 100{,}000t, \qquad 0 \le t \le 5.$$

See Figure 6.5(a).

Assuming an annual inflation rate of 10%, what is the present value of this income?

SOLUTION The present value is

$$\text{Present value} = \int_0^5 100{,}000te^{-0.1t}\,dt = 100{,}000\int_0^5 te^{-0.1t}\,dt.$$

Using integration by parts, let $dv = e^{-0.1t}\,dt$.

$$dv = e^{-0.1t}\,dt \quad \Rightarrow \quad v = \int dv = \int e^{-0.1t}\,dt = -10e^{-0.1t}$$

$$u = t \quad \Rightarrow \quad du = dt$$

This implies that

$$\begin{aligned}\int te^{-0.1t}\,dt &= -10te^{-0.1t} + 10\int e^{-0.1t}\,dt \\ &= -10te^{-0.1t} - 100e^{-0.1t} \\ &= -10e^{-0.1t}(t + 10).\end{aligned}$$

So, the present value is

$$\begin{aligned}\text{Present value} &= 100{,}000\int_0^5 te^{-0.1t}\,dt \\ &= 100{,}000\Big[-10e^{-0.1t}(t + 10)\Big]_0^5 \\ &\approx \$902{,}040.\end{aligned}$$

See Figure 6.5(b).

TRY IT 7

A company expects its income during the next 10 years to be given by $c(t) = 20{,}000t$, for $0 \le t \le 10$. Assuming an annual inflation rate of 10%, what is the present value of this income?

TAKE ANOTHER LOOK

Present Value

As shown in Figure 6.5, the actual income for the business in Example 7, over the next 5 years, is

$$\text{Actual income} = \int_0^5 100{,}000t\,dt = \$1{,}250{,}000.$$

If you were selling this company for its present value over a five-year period, would it be to your advantage to argue that the inflation rate was lower than 10%? Explain your reasoning.

PREREQUISITE REVIEW 6.2

The following warm-up exercises involve skills that were covered in earlier sections. You will use these skills in the exercise set for this section.

In Exercises 1–6, find $f'(x)$.

1. $f(x) = \ln(x + 1)$

2. $f(x) = \ln(x^2 - 1)$

3. $f(x) = e^{x^3}$

4. $f(x) = e^{-x^2}$

5. $f(x) = x^2e^x$

6. $f(x) = xe^{-2x}$

In Exercises 7–10, find the area between the graphs of f and g.

7. $f(x) = -x^2 + 4,\ g(x) = x^2 - 4$

8. $f(x) = -x^2 + 2,\ g(x) = 1$

9. $f(x) = 4x,\ g(x) = x^2 - 5$

10. $f(x) = x^3 - 3x^2 + 2,\ g(x) = x - 1$

EXERCISES 6.2

In Exercises 1–6, use integration by parts to find the indefinite integral.

1. $\int xe^{3x}\,dx$

2. $\int xe^{-x}\,dx$

3. $\int x^2e^{-x}\,dx$

4. $\int x^2e^{2x}\,dx$

5. $\int \ln 2x\,dx$

6. $\int \ln x^2\,dx$

In Exercises 7–28, find the indefinite integral. (*Hint:* Integration by parts is not required for all the integrals.)

7. $\int e^{4x}\,dx$

8. $\int e^{-2x}\,dx$

9. $\int xe^{4x}\,dx$

10. $\int xe^{-2x}\,dx$

11. $\int xe^{x^2}\,dx$

12. $\int x^2e^{x^3}\,dx$

13. $\int x^2e^x\,dx$

14. $\int \frac{x}{e^x}\,dx$

15. $\int t\ln(t + 1)\,dt$

16. $\int x^3\ln x\,dx$

17. $\int \frac{e^{1/t}}{t^2}\,dt$

18. $\int \frac{1}{x(\ln x)^3}\,dx$

19. $\int x(\ln x)^2\,dx$

20. $\int \ln 3x\,dx$

21. $\int \frac{(\ln x)^2}{x}\,dx$

22. $\int \frac{1}{x\ln x}\,dx$

23. $\int x\sqrt{x - 1}\,dx$

24. $\int \frac{x}{\sqrt{x - 1}}\,dx$

25. $\int x(x + 1)^2\,dx$

26. $\int \frac{x}{\sqrt{2 + 3x}}\,dx$

27. $\int \frac{xe^{2x}}{(2x + 1)^2}\,dx$

28. $\int \frac{x^3e^{x^2}}{(x^2 + 1)^2}\,dx$

In Exercises 29–34, evaluate the definite integral.

29. $\int_0^1 x^2e^x\,dx$

30. $\int_0^2 \frac{x^2}{e^x}\,dx$

31. $\int_1^e x^5\ln x\,dx$

32. $\int_1^e 2x\ln x\,dx$

33. $\int_{-1}^0 \ln(x + 2)\,dx$

34. $\int_0^1 \ln(1 + 2x)\,dx$

In Exercises 35–38, find the area of the region bounded by the graphs of the equations. Then use a graphing utility to graph the region and verify your answer.

35. $y = x^3e^x,\ y = 0,\ x = 0,\ x = 2$

36. $y = (x^2 - 1)e^x,\ y = 0,\ x = -1,\ x = 1$

37. $y = x^2\ln x,\ y = 0,\ x = 1,\ x = e$

38. $y = \frac{\ln x}{x^2},\ y = 0,\ x = 1,\ x = e$

In Exercises 39–42, find the indefinite integral using each specified method. Then write a brief statement explaining which method you prefer.

39. $\int 2x\sqrt{2x-3}\,dx$

(a) By parts, letting $dv = \sqrt{2x-3}\,dx$

(b) By substitution, letting $u = \sqrt{2x-3}$

40. $\int x\sqrt{4+x}\,dx$

(a) By parts, letting $dv = \sqrt{4+x}\,dx$

(b) By substitution, letting $u = \sqrt{4+x}$

41. $\int \frac{x}{\sqrt{4+5x}}\,dx$

(a) By parts, letting $dv = \frac{1}{\sqrt{4+5x}}\,dx$

(b) By substitution, letting $u = \sqrt{4+5x}$

42. $\int x\sqrt{4-x}\,dx$

(a) By parts, letting $dv = \sqrt{4-x}\,dx$

(b) By substitution, letting $u = \sqrt{4-x}$

In Exercises 43 and 44, use integration by parts to verify the formula.

43. $\int x^n \ln x\,dx = \frac{x^{n+1}}{(n+1)^2}[-1+(n+1)\ln x] + C,$

$n \neq -1$

44. $\int x^n e^{ax}\,dx = \frac{x^n e^{ax}}{a} - \frac{n}{a}\int x^{n-1}e^{ax}\,dx$

In Exercises 45–48, use the results of Exercises 43 and 44 to find the indefinite integral.

45. $\int x^2 e^{5x}\,dx$

46. $\int xe^{-3x}\,dx$

47. $\int x^{-2}\ln x\,dx$

48. $\int x^{1/2}\ln x\,dx$

In Exercises 49–52, find the area of the region bounded by the graphs of the given equations.

49. $y = xe^{-x},\ y = 0,\ x = 4$

50. $y = \frac{1}{9}xe^{-x/3},\ y = 0,\ x = 0,\ x = 3$

51. $y = x\ln x,\ y = 0,\ x = e$

52. $y = x^{-3}\ln x,\ y = 0,\ x = e$

53. Given the region bounded by the graphs of $y = 2\ln x$, $y = 0$, and $x = e$ (see figure), find

(a) the area of the region.

(b) the volume of the solid generated by revolving the region about the x-axis.

Figure for 53

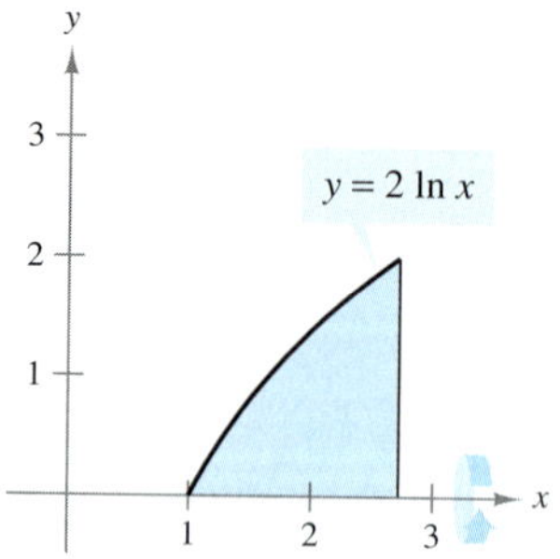

54. Given the region bounded by the graphs of $y = xe^x$, $y = 0$, $x = 0$, and $x = 1$ (see figure), find

(a) the area of the region.

(b) the volume of the solid generated by revolving the region about the x-axis.

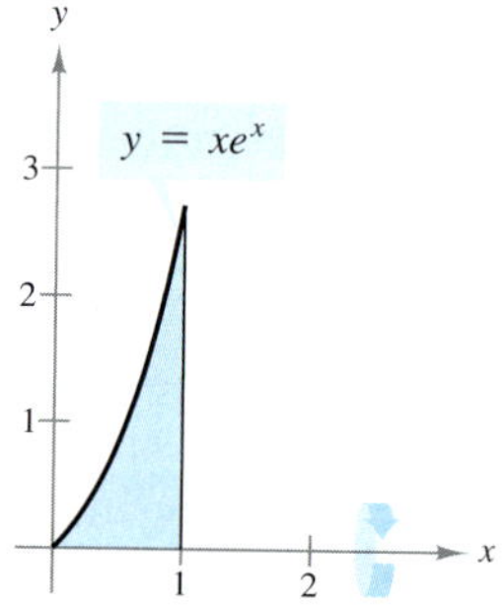

In Exercises 55–58, use a symbolic integration utility to evaluate the integral.

55. $\int_0^2 t^3e^{-4t}\,dt$

56. $\int_1^4 \ln x(x^2+4)\,dx$

57. $\int_0^5 x^4(25-x^2)^{3/2}\,dx$

58. $\int_1^e x^9\ln x\,dx$

59. ***Demand*** A manufacturing company forecasts that the demand x (in units per year) for its product over the next 10 years can be modeled by $x = 500(20 + te^{-0.1t})$ for $0 \le t \le 10$, where t is the time in years.

(a) Use a graphing utility to decide whether the company is forecasting an increase or a decrease in demand over the decade.

(b) According to the model, what is the total demand over the next 10 years?

(c) Find the average annual demand during the 10-year period.

60. ***Capital Campaign*** The board of trustees of a college is planning a five-year capital gifts campaign to raise money for the college. The goal is to have an annual gift income I that is modeled by $I = 2000(375 + 68te^{-0.2t})$ for $0 \le t \le 5$, where t is the time in years.

(a) Use a graphing utility to decide whether the board of trustees expects the gift income to increase or decrease over the five-year period.

(b) Find the expected total gift income over the five-year period.

(c) Determine the average annual gift income over the five-year period. Compare the result with the income given when $t = 3$.

61. *Learning Theory* A model for the ability M of a child to memorize, measured on a scale from 0 to 10, is

$$M = 1 + 1.6t \ln t, \qquad 0 < t \leq 4$$

where t is the child's age in years. Find the average value of this model between

(a) the child's first and second birthdays.

(b) the child's third and fourth birthdays.

62. *Revenue* A company sells a seasonal product. The revenue R (in dollars per year) generated by sales of the product can be modeled by

$$R = 410.5t^2e^{-t/30} + 25{,}000, \qquad 0 \leq t \leq 365$$

where t is the time in days.

(a) Find the average daily receipts during the first quarter, which is given by $0 \leq t \leq 90$.

(b) Find the average daily receipts during the fourth quarter, which is given by $274 \leq t \leq 365$.

(c) Find the total daily receipts during the year.

Present Value In Exercises 63–68, find the present value of the income c (measured in dollars) over t_1 years at the given annual inflation rate r.

63. $c = 5000, \ r = 5\%, \ t_1 = 4$ years

64. $c = 450, \ r = 4\%, \ t_1 = 10$ years

65. $c = 150{,}000 + 2500t, \ r = 4\%, \ t_1 = 10$ years

66. $c = 30{,}000 + 500t, \ r = 7\%, \ t_1 = 6$ years

67. $c = 1000 + 50e^{t/2}, \ r = 6\%, \ t_1 = 4$ years

68. $c = 5000 + 25te^{t/10}, \ r = 6\%, \ t_1 = 10$ years

69. *Present Value* A company expects its income c during the next 4 years to be modeled by

$$c = 150{,}000 + 75{,}000t.$$

(a) Find the actual income for the business over the 4 years.

(b) Assuming an annual inflation rate of 4%, what is the present value of this income?

70. *Present Value* A professional athlete signs a three-year contract in which the earnings can be modeled by

$$c = 300{,}000 + 125{,}000t.$$

(a) Find the actual value of the athlete's contract.

(b) Assuming an annual inflation rate of 5%, what is the present value of the contract?

Future Value In Exercises 71 and 72, find the future value of the income (in dollars) given by $f(t)$ over t_1 years at the annual interest rate of r. If the function f represents a continuous investment over a period of t_1 years at an annual interest rate of r (compounded continuously), then the future value of the investment is given by

$$\text{Future value} = e^{rt_1}\int_0^{t_1} f(t)e^{-rt}\,dt.$$

71. $f(t) = 3000, \ r = 8\%, \ t_1 = 10$ years

72. $f(t) = 3000e^{0.05t}, \ r = 10\%, \ t_1 = 5$ years

73. *Finance: Future Value* Use the equation from Exercises 71 and 72 to calculate the following. *(Source: Adapted from Garman/Forgue,* Personal Finance, *Fifth Edition)*

(a) The future value of \$1200 saved each year for 10 years earning 7% interest.

(b) A person who wishes to invest \$1200 each year finds one investment choice that is expected to pay 9% interest per year and another, riskier choice that may pay 10% interest per year. What is the difference in return (future value) if the investment is made for 15 years?

74. *Consumer Awareness* In 2004, the total cost to attend Pennsylvania State University for 1 year was estimated to be \$19,843. If your grandparents had continuously invested in a college fund according to the model

$$f(t) = 400t$$

for 18 years, at an annual interest rate of 10%, would the fund have grown enough to allow you to cover 4 years of expenses at Pennsylvania State University? *(Source: Pennsylvania State University)*

75. Use a program similar to the Midpoint Rule program on page 366 with $n = 10$ to approximate

$$\int_1^4 \frac{4}{\sqrt{x} + \sqrt[3]{x}}\,dx.$$

76. Use a program similar to the Midpoint Rule program on page 366 with $n = 12$ to approximate the volume of the solid generated by revolving the region bounded by the graphs of

$$y = \frac{10}{\sqrt{x}e^x}, \quad y = 0, \quad x = 1, \text{ and } x = 4$$

about the x-axis.

6.3 PARTIAL FRACTIONS AND LOGISTIC GROWTH

- Use partial fractions to find indefinite integrals.
- Use logistic growth functions to model real-life situations.

Partial Fractions

In Sections 6.1 and 6.2, you studied integration by substitution and by parts. In this section you will study a third technique called **partial fractions.** This technique involves the decomposition of a rational function into the sum of two or more simple rational functions. For instance, suppose you know that

$$\frac{x+7}{x^2-x-6} = \frac{2}{x-3} - \frac{1}{x+2}.$$

Knowing the "partial fractions" on the right side would allow you to integrate the left side as shown.

$$\begin{aligned}\int \frac{x+7}{x^2-x-6}\,dx &= \int \left(\frac{2}{x-3} - \frac{1}{x+2}\right) dx \\ &= 2\int \frac{1}{x-3}\,dx - \int \frac{1}{x+2}\,dx \\ &= 2\ln|x-3| - \ln|x+2| + C\end{aligned}$$

To use this method, you must be able to factor the denominator of the original rational function *and* find the partial fraction decomposition of the function.

Partial Fractions

To find the partial fraction decomposition of the *proper* rational function $p(x)/q(x)$, factor $q(x)$ and write an equation that has the form

$$\frac{p(x)}{q(x)} = (\text{sum of partial fractions}).$$

For each *distinct* linear factor $ax + b$, the right side should include a term of the form

$$\frac{A}{ax+b}.$$

For each *repeated* linear factor $(ax + b)^n$, the right side should include n terms of the form

$$\frac{A_1}{ax+b} + \frac{A_2}{(ax+b)^2} + \cdot\cdot\cdot + \frac{A_n}{(ax+b)^n}.$$

STUDY TIP

Finding the partial fraction decomposition of a rational function is really a *precalculus* topic. Explain how you could verify that

$$\frac{1}{x-1} + \frac{2}{x+2}$$

is the partial fraction decomposition of

$$\frac{3x}{x^2+x-2}.$$

STUDY TIP

A rational function $p(x)/q(x)$ is *proper* if the degree of the numerator is less than the degree of the denominator.

EXAMPLE 1 Finding a Partial Fraction Decomposition

Write the partial fraction decomposition for

$$\frac{x+7}{x^2-x-6}.$$

SOLUTION Begin by factoring the denominator as $x^2 - x - 6 = (x-3)(x+2)$. Then, write the partial fraction decomposition as

$$\frac{x+7}{x^2-x-6} = \frac{A}{x-3} + \frac{B}{x+2}.$$

To solve this equation for A and B, multiply each side of the equation by the least common denominator $(x-3)(x+2)$. This produces the **basic equation** as shown.

$$x + 7 = A(x+2) + B(x-3) \qquad \text{Basic equation}$$

Because this equation is true for all x, you can substitute any convenient values of x into the equation. The x-values that are especially convenient are the ones that make a factor of the least common denominator zero: $x = -2$ and $x = 3$.

Substitute $x = -2$:

$$\begin{aligned} x + 7 &= A(x+2) + B(x-3) && \text{Write basic equation.}\\ -2 + 7 &= A(-2+2) + B(-2-3) && \text{Substitute } -2 \text{ for } x.\\ 5 &= A(0) + B(-5) && \text{Simplify.}\\ -1 &= B && \text{Solve for } B. \end{aligned}$$

Substitute $x = 3$:

$$\begin{aligned} x + 7 &= A(x+2) + B(x-3) && \text{Write basic equation.}\\ 3 + 7 &= A(3+2) + B(3-3) && \text{Substitute 3 for } x.\\ 10 &= A(5) + B(0) && \text{Simplify.}\\ 2 &= A && \text{Solve for } A. \end{aligned}$$

Now that you have solved the basic equation for A and B, you can write the partial fraction decomposition as

$$\frac{x+7}{x^2-x-6} = \frac{2}{x-3} - \frac{1}{x+2}$$

as indicated at the beginning of this section.

ALGEBRA REVIEW

You can check the result in Example 1 by subtracting the partial fractions to obtain the original fraction, as shown in Example 1(b) in the *Chapter 6 Algebra Review*, on page 446.

TRY IT 1

Write the partial fraction decomposition for $\dfrac{x+8}{x^2+7x+12}$.

STUDY TIP

Be sure you see that the substitutions for x in Example 1 are chosen for their convenience in solving for A and B. The value $x = -2$ is selected because it eliminates the term $A(x+2)$, and the value $x = 3$ is chosen because it eliminates the term $B(x-3)$.

TECHNOLOGY

The use of partial fractions depends on the ability to factor the denominator. If this cannot be easily done, then partial fractions should not be used. For instance, consider the integral

$$\int \frac{5x^2 + 20x + 6}{x^3 + 2x^2 + x + 1}\,dx.$$

This integral is only slightly different from that in Example 2, yet it is immensely more difficult to solve. A symbolic integration utility was unable to solve this integral. Of course, if the integral is a definite integral (as is true in many applied problems), then you can use an approximation technique such as the Midpoint Rule.

ALGEBRA REVIEW

You can check the partial fraction decomposition in Example 2 by combining the partial fractions to obtain the original fraction, as shown in Example 1(c) in the *Chapter 6 Algebra Review*, on page 446. Also, for help with the algebra used to simplify the answer, see Example 2(c) on page 447.

EXAMPLE 2 Integrating with Repeated Factors

Find $\displaystyle\int \frac{5x^2 + 20x + 6}{x^3 + 2x^2 + x}\,dx.$

SOLUTION Begin by factoring the denominator as $x(x + 1)^2$. Then, write the partial fraction decomposition as

$$\frac{5x^2 + 20x + 6}{x(x + 1)^2} = \frac{A}{x} + \frac{B}{x + 1} + \frac{C}{(x + 1)^2}.$$

To solve this equation for A, B, and C, multiply each side of the equation by the least common denominator $x(x + 1)^2$.

$$5x^2 + 20x + 6 = A(x + 1)^2 + Bx(x + 1) + Cx \qquad \text{Basic equation}$$

Now, solve for A and C by substituting $x = -1$ and $x = 0$ into the basic equation.

Substitute $x = -1$:

$$\begin{aligned} 5(-1)^2 + 20(-1) + 6 &= A(-1 + 1)^2 + B(-1)(-1 + 1) + C(-1) \\ -9 &= A(0) + B(0) - C \\ 9 &= C \qquad \text{Solve for } C. \end{aligned}$$

Substitute $x = 0$:

$$\begin{aligned} 5(0)^2 + 20(0) + 6 &= A(0 + 1)^2 + B(0)(0 + 1) + C(0) \\ 6 &= A(1) + B(0) + C(0) \\ 6 &= A \qquad \text{Solve for } A. \end{aligned}$$

At this point, you have exhausted the convenient choices for x and have yet to solve for B. When this happens, you can use *any* other x-value along with the known values of A and C.

Substitute $x = 1$, $A = 6$, *and* $C = 9$:

$$\begin{aligned} 5(1)^2 + 20(1) + 6 &= (6)(1 + 1)^2 + B(1)(1 + 1) + (9)(1) \\ 31 &= 6(4) + B(2) + 9(1) \\ -1 &= B \qquad \text{Solve for } B. \end{aligned}$$

Now that you have solved for A, B, and C, you can use the partial fraction decomposition to integrate.

$$\begin{aligned} \int \frac{5x^2 + 20x + 6}{x^3 + 2x^2 + x}\,dx &= \int \left(\frac{6}{x} - \frac{1}{x + 1} + \frac{9}{(x + 1)^2}\right) dx \\ &= 6\ln|x| - \ln|x + 1| + 9\frac{(x + 1)^{-1}}{-1} + C \\ &= \ln\left|\frac{x^6}{x + 1}\right| - \frac{9}{x + 1} + C \end{aligned}$$

TRY IT 2

Find $\displaystyle\int \frac{3x^2 + 7x + 4}{x^3 + 4x^2 + 4x}\,dx.$

You can use the partial fraction decomposition technique outlined in Examples 1 and 2 only with a *proper* rational function—that is, a rational function whose numerator is of lower degree than its denominator. If the numerator is of equal or greater degree, you must divide first. For instance, the rational function

$$\frac{x^3}{x^2 + 1}$$

is improper because the degree of the numerator is greater than the degree of the denominator. Before applying partial fractions to this function, you should divide the denominator into the numerator to obtain

$$\frac{x^3}{x^2 + 1} = x - \frac{x}{x^2 + 1}.$$

EXAMPLE 3 Integrating an Improper Rational Function

Find $\displaystyle\int \frac{x^5 + x - 1}{x^4 - x^3}\,dx.$

SOLUTION This rational function is improper—its numerator has a degree greater than that of its denominator. So, you should begin by dividing the denominator into the numerator to obtain

$$\frac{x^5 + x - 1}{x^4 - x^3} = x + 1 + \frac{x^3 + x - 1}{x^4 - x^3}.$$

Now, applying partial fraction decomposition produces

$$\frac{x^3 + x - 1}{x^3(x - 1)} = \frac{A}{x} + \frac{B}{x^2} + \frac{C}{x^3} + \frac{D}{x - 1}.$$

Multiplying both sides by the least common denominator $x^3(x - 1)$ produces the basic equation.

$$x^3 + x - 1 = Ax^2(x - 1) + Bx(x - 1) + C(x - 1) + Dx^3 \qquad \text{Basic equation}$$

Using techniques similar to those in the first two examples, you can solve for A, B, C, and D to obtain

$$A = 0, \quad B = 0, \quad C = 1, \quad \text{and} \quad D = 1.$$

So, you can integrate as shown.

$$\begin{aligned}\int \frac{x^5 + x - 1}{x^4 - x^3}\,dx &= \int \left(x + 1 + \frac{x^3 + x - 1}{x^4 - x^3}\right) dx \\ &= \int \left(x + 1 + \frac{1}{x^3} + \frac{1}{x - 1}\right) dx \\ &= \frac{x^2}{2} + x - \frac{1}{2x^2} + \ln|x - 1| + C\end{aligned}$$

ALGEBRA REVIEW

You can check the partial fraction decomposition in Example 3 by combining the partial fractions to obtain the original fraction, as shown in Example 2(a) in the *Chapter 6 Algebra Review*, on page 447.

TRY IT 3

Find $\displaystyle\int \frac{x^4 - x^3 + 2x^2 + x + 1}{x^3 + x^2}.$

Logistic Growth Function

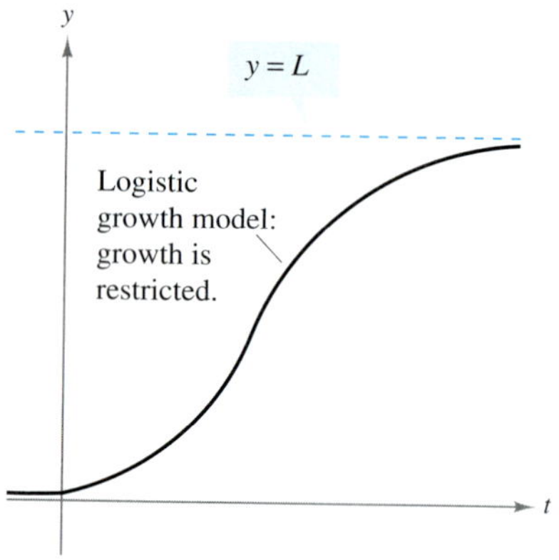

FIGURE 6.6

In Section 4.6, you saw that exponential growth occurs in situations for which the rate of growth is proportional to the quantity present at any given time. That is, if y is the quantity at time t, then

$$\frac{dy}{dt} = ky \qquad \frac{dy}{dt} \text{ is proportional to } y.$$

$$y = Ce^{kt}. \qquad \text{Exponential growth function}$$

Exponential growth is unlimited. As long as C and k are positive, the value of Ce^{kt} can be made arbitrarily large by choosing sufficiently large values of t.

In many real-life situations, however, the growth of a quantity is limited and cannot increase beyond a certain size L, as shown in Figure 6.6. The **logistic growth** model assumes that the rate of growth is proportional to both the quantity y and the difference between the quantity and the limit L. That is

$$\frac{dy}{dt} = ky(L - y). \qquad \frac{dy}{dt} \text{ is proportional to } y \text{ and } (L - y).$$

The solution of this *differential equation* is given in Example 4.

STUDY TIP

The logistic growth model in Example 4 is simplified by assuming that the limit of the quantity y is 1. If the limit were L, then the solution would be

$$y = \frac{L}{1 + be^{-kt}}.$$

In the fourth step of the solution, notice that partial fractions are used to integrate the left side of the equation.

EXAMPLE 4 Deriving the Logistic Growth Function

Solve the equation

$$\frac{dy}{dt} = ky(1 - y).$$

Assume $y > 0$ and $1 - y > 0$.

SOLUTION

$$\frac{dy}{dt} = ky(1 - y) \qquad \text{Write differential equation.}$$

$$\frac{1}{y(1 - y)}\,dy = k\,dt \qquad \text{Write in differential form.}$$

$$\int \frac{1}{y(1 - y)}\,dy = \int k\,dt \qquad \text{Integrate each side.}$$

$$\int \left(\frac{1}{y} + \frac{1}{1 - y}\right) dy = \int k\,dt \qquad \text{Rewrite using partial fractions.}$$

$$\ln y - \ln(1 - y) = kt + C_1 \qquad \text{Find antiderivative.}$$

$$\ln \frac{y}{1 - y} = kt + C_1 \qquad \text{Simplify.}$$

$$\frac{y}{1 - y} = Ce^{kt} \qquad \text{Exponentiate and let } e^{C_1} = C.$$

Solving this equation for y produces

$$y = \frac{1}{1 + be^{-kt}} \qquad \text{Logistic growth function}$$

where $b = 1/C$.

TRY IT 4

Show that if

$$y = \frac{1}{1 + be^{-kt}}, \text{ then}$$

$$\frac{dy}{dt} = ky(1 - y).$$

[*Hint:* First find $ky(1 - y)$ in terms of t, then find dy/dt and show that they are equivalent.]

EXAMPLE 5 Comparing Logistic Growth Functions

Use a graphing utility to investigate the effects of the values of L, b, and k on the graph of

$$y = \frac{L}{1 + be^{-kt}}.$$ Logistic growth function ($L > 0, b > 0, k > 0$)

SOLUTION The value of L determines the horizontal asymptote of the graph to the right. In other words, as t increases without bound, the graph approaches a limit of L (see Figure 6.7).

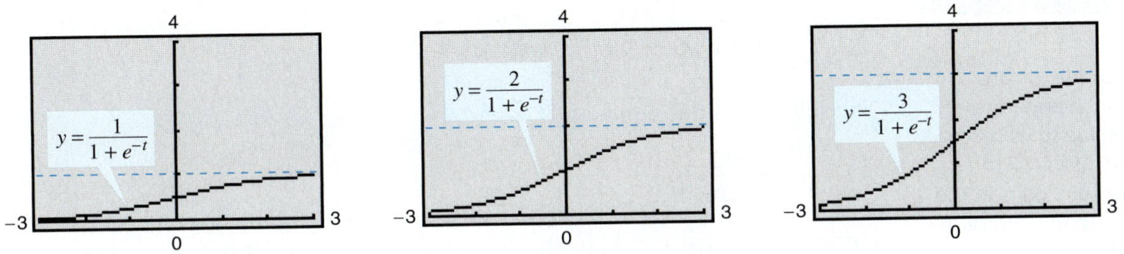

FIGURE 6.7

The value of b determines the point of inflection of the graph. When $b = 1$, the point of inflection occurs when $t = 0$. If $b > 1$, the point of inflection is to the right of the y-axis. If $0 < b < 1$, the point of inflection is to the left of the y-axis (see Figure 6.8).

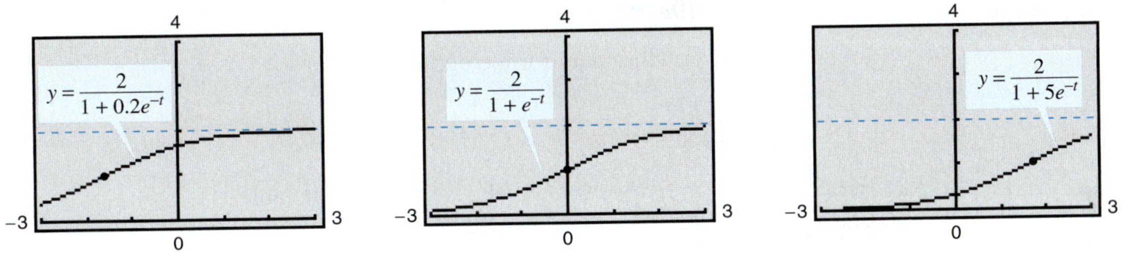

FIGURE 6.8

The value of k determines the rate of growth of the graph. For fixed values of b and L, larger values of k correspond to higher rates of growth (see Figure 6.9).

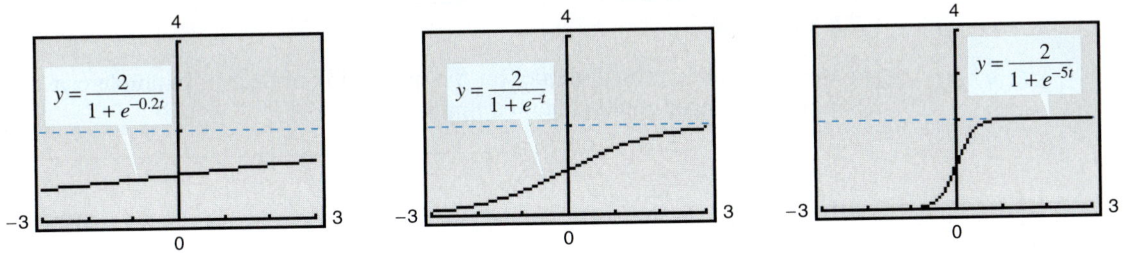

FIGURE 6.9

TRY IT 5

Find the horizontal asymptote of the graph of $y = \frac{4}{1 + 5e^{-6t}}$.

© Galen Rowell/CORBIS

The American peregrine falcon was removed from the endangered species list in 1999 due to its recovery from 324 nesting pairs in North America in 1975 to 1650 pairs in the United States and Canada. The peregrine was put on the endangered species list in 1970 because of the use of the chemical pesticide DDT. The Fish and Wildlife Service, state wildlife agencies, and many other organizations contributed to the recovery by setting up protective breeding programs among other efforts.

EXAMPLE 6 Modeling a Population

The state game commission releases 100 deer into a game preserve. During the first 5 years, the population increases to 432 deer. The commission believes that the population can be modeled by logistic growth with a limit of 2000 deer. Write the logistic growth model for this population. Then use the model to create a table showing the size of the deer population over the next 30 years.

SOLUTION Let y represent the number of deer in year t. Assuming a logistic growth model means that the rate of change in the population is proportional to both y and $(2000 - y)$. That is

$$\frac{dy}{dt} = ky(2000 - y), \qquad 100 \le y \le 2000.$$

The solution of this equation is

$$y = \frac{2000}{1 + be^{-kt}}.$$

Using the fact that $y = 100$ when $t = 0$, you can solve for b.

$$100 = \frac{2000}{1 + be^{-k(0)}} \quad \Rightarrow \quad b = 19$$

Then, using the fact that $y = 432$ when $t = 5$, you can solve for k.

$$432 = \frac{2000}{1 + 19e^{-k(5)}} \quad \Rightarrow \quad k \approx 0.33106$$

So, the logistic growth model for the population is

$$y = \frac{2000}{1 + 19e^{-0.33106t}}. \qquad \text{Logistic growth model}$$

The population, in five-year intervals, is shown in the table.

Time, t	0	5	10	15	20	25	30
Population, y	100	432	1181	1766	1951	1990	1998

TRY IT 6

Write the logistic growth model for the population of deer in Example 6 if the game preserve could contain a limit of 4000 deer.

TAKE ANOTHER LOOK

Logistic Growth

Analyze the graph of the logistic growth function in Example 6. During which years is the *rate of growth* of the herd increasing? During which years is the *rate of growth* of the herd decreasing? How would these answers change if, instead of a limit of 2000 deer, the game preserve could contain a limit of 3000 deer?

PREREQUISITE REVIEW 6.3

The following warm-up exercises involve skills that were covered in earlier sections. You will use these skills in the exercise set for this section.

In Exercises 1–8, factor the expression.

1. $x^2 - 16$

2. $x^2 - 25$

3. $x^2 - x - 12$

4. $x^2 + x - 6$

5. $x^3 - x^2 - 2x$

6. $x^3 - 4x^2 + 4x$

7. $x^3 - 4x^2 + 5x - 2$

8. $x^3 - 5x^2 + 7x - 3$

In Exercises 9–14, rewrite the improper rational expression as the sum of a proper rational expression and a polynomial.

9. $\dfrac{x^2 - 2x + 1}{x - 2}$

10. $\dfrac{2x^2 - 4x + 1}{x - 1}$

11. $\dfrac{x^3 - 3x^2 + 2}{x - 2}$

12. $\dfrac{x^3 + 2x - 1}{x + 1}$

13. $\dfrac{x^3 + 4x^2 + 5x + 2}{x^2 - 1}$

14. $\dfrac{x^3 + 3x^2 - 4}{x^2 - 1}$

EXERCISES 6.3

In Exercises 1–12, write the partial fraction decomposition for the expression.

1. $\dfrac{2(x + 20)}{x^2 - 25}$

2. $\dfrac{3x + 11}{x^2 - 2x - 3}$

3. $\dfrac{8x + 3}{x^2 - 3x}$

4. $\dfrac{10x + 3}{x^2 + x}$

5. $\dfrac{4x - 13}{x^2 - 3x - 10}$

6. $\dfrac{7x + 5}{6(2x^2 + 3x + 1)}$

7. $\dfrac{3x^2 - 2x - 5}{x^3 + x^2}$

8. $\dfrac{3x^2 - x + 1}{x(x + 1)^2}$

9. $\dfrac{x + 1}{3(x - 2)^2}$

10. $\dfrac{3x - 4}{(x - 5)^2}$

11. $\dfrac{8x^2 + 15x + 9}{(x + 1)^3}$

12. $\dfrac{6x^2 - 5x}{(x + 2)^3}$

In Exercises 13–32, find the indefinite integral.

13. $\displaystyle\int \frac{1}{x^2 - 1}\,dx$

14. $\displaystyle\int \frac{9}{x^2 - 9}\,dx$

15. $\displaystyle\int \frac{-2}{x^2 - 16}\,dx$

16. $\displaystyle\int \frac{-4}{x^2 - 4}\,dx$

17. $\displaystyle\int \frac{1}{3x^2 - x}\,dx$

18. $\displaystyle\int \frac{3}{x^2 - 3x}\,dx$

19. $\displaystyle\int \frac{1}{2x^2 + x}\,dx$

20. $\displaystyle\int \frac{5}{x^2 + x - 6}\,dx$

21. $\displaystyle\int \frac{3}{x^2 + x - 2}\,dx$

22. $\displaystyle\int \frac{1}{4x^2 - 9}\,dx$

23. $\displaystyle\int \frac{5 - x}{2x^2 + x - 1}\,dx$

24. $\displaystyle\int \frac{x + 1}{x^2 + 4x + 3}\,dx$

25. $\displaystyle\int \frac{x^2 + 12x + 12}{x^3 - 4x}\,dx$

26. $\displaystyle\int \frac{3x^2 - 7x - 2}{x^3 - x}\,dx$

27. $\displaystyle\int \frac{x + 2}{x^2 - 4x}\,dx$

28. $\displaystyle\int \frac{4x^2 + 2x - 1}{x^3 + x^2}\,dx$

29. $\displaystyle\int \frac{4 - 3x}{(x - 1)^2}\,dx$

30. $\displaystyle\int \frac{x^4}{(x - 1)^3}\,dx$

31. $\displaystyle\int \frac{3x^2 + 3x + 1}{x(x^2 + 2x + 1)}\,dx$

32. $\displaystyle\int \frac{3x}{x^2 - 6x + 9}\,dx$

In Exercises 33–40, evaluate the definite integral.

33. $\displaystyle\int_4^5 \frac{1}{9 - x^2}\,dx$

34. $\displaystyle\int_0^1 \frac{3}{2x^2 + 5x + 2}\,dx$

35. $\displaystyle\int_1^5 \frac{x - 1}{x^2(x + 1)}\,dx$

36. $\displaystyle\int_0^1 \frac{x^2 - x}{x^2 + x + 1}\,dx$

37. $\displaystyle\int_0^1 \frac{x^3}{x^2-2}\,dx$

38. $\displaystyle\int_0^1 \frac{x^3-1}{x^2-4}\,dx$

39. $\displaystyle\int_1^2 \frac{x^3-4x^2-3x+3}{x^2-3x}\,dx$

40. $\displaystyle\int_2^4 \frac{x^4-4}{x^2-1}\,dx$

In Exercises 41–44, find the area of the shaded region.

41. $y = \dfrac{14}{16-x^2}$

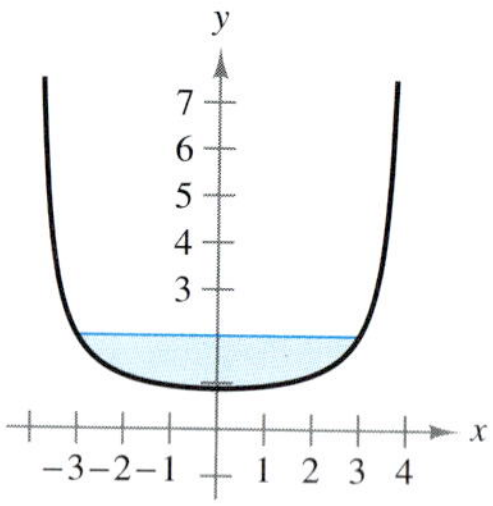

42. $y = \dfrac{-4}{x^2-x-6}$

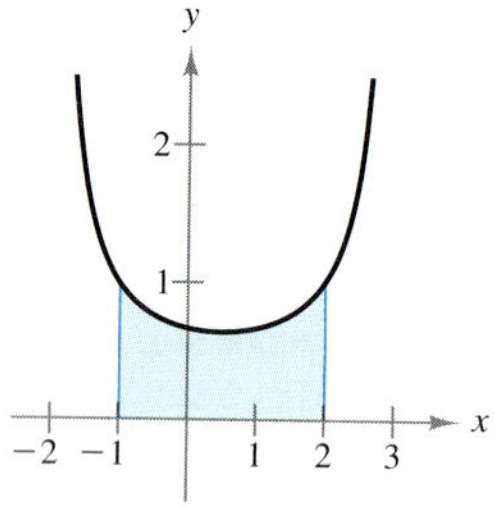

43. $y = \dfrac{x+1}{x^2-x}$

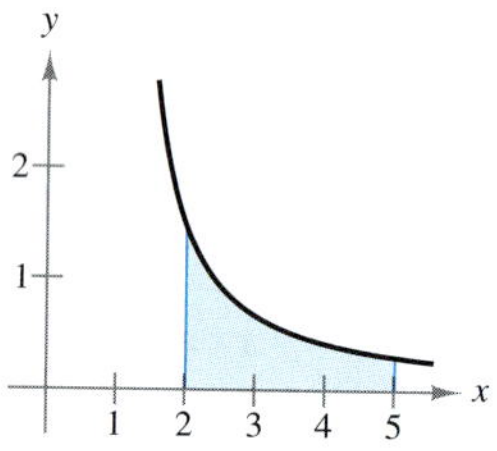

44. $y = \dfrac{x^2+2x-1}{x^2-4}$

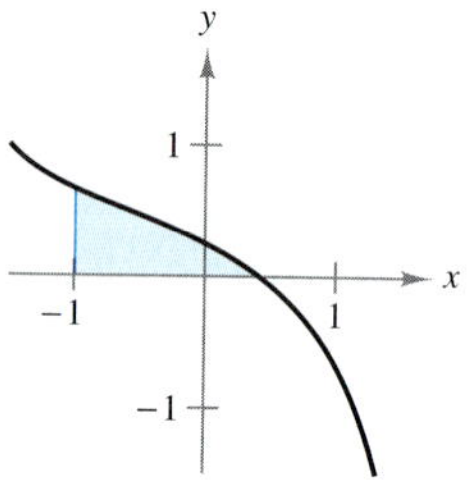

In Exercises 45–48, write the partial fraction decomposition for the rational expression. Check your result algebraically. Then assign a value to the constant a and use a graphing utility to check the result graphically.

45. $\dfrac{1}{a^2-x^2}$

46. $\dfrac{1}{x(x+a)}$

47. $\dfrac{1}{x(a-x)}$

48. $\dfrac{1}{(x+1)(a-x)}$

In Exercises 49–52, use a graphing utility to graph the function. Then find the volume of the solid generated by revolving the region bounded by the graphs of the given equations about the x-axis by using the integration capabilities of a graphing utility and by integrating by hand using partial fraction decomposition.

49. $y = \dfrac{10}{x(x+10)}, \; y = 0, \; x = 1, \; x = 5$

50. $y = \dfrac{-4}{(x+1)(x-4)}, \; y = 0, \; x = 0, \; x = 3$

51. $y = \dfrac{2}{x^2-4}, \; x = 1, \; x = -1, \; y = 0$

52. $y = \dfrac{25x}{x^2+x-6}, \; x = -2, \; x = 0, \; y = 0$

53. ***Biology*** A conservation organization releases 100 animals of an endangered species into a game preserve. The organization believes that the preserve has a capacity of 1000 animals and that the herd will grow according to a logistic growth model. That is, the size y of the herd will follow the equation

$$\int \frac{1}{y(1000-y)}\,dy = \int k\,dt$$

where t is measured in years. Find this logistic curve. (To solve for the constant of integration C and the proportionality constant k, assume $y = 100$ when $t = 0$ and $y = 134$ when $t = 2$.) Use a graphing utility to graph your solution.

54. ***Health: Epidemic*** A single infected individual enters a community of 500 individuals susceptible to the disease. The disease spreads at a rate proportional to the product of the total number infected and the number of susceptible individuals not yet infected. A model for the time it takes for the disease to spread to x individuals is

$$t = 5010\int \frac{1}{(x+1)(500-x)}\,dx$$

where t is the time in hours.

(a) Find the time it takes for 75% of the population to become infected (when $t = 0$, $x = 1$).

(b) Find the number of people infected after 100 hours.

55. ***Marketing*** After test-marketing a new menu item, a fast-food restaurant predicts that sales of the new item will grow according to the model

$$\frac{dS}{dt} = \frac{2t}{(t+4)^2}$$

where t is the time in weeks and S is the sales (in thousands of dollars). Find the sales of the menu item at 10 weeks.

56. ***Biology*** One gram of a bacterial culture is present at time $t = 0$, and 10 grams is the upper limit of the culture's weight. The time required for the culture to grow to y grams is modeled by

$$kt = \int \frac{1}{y(10-y)}\,dy$$

where y is the weight of the culture (in grams) and t is the time in hours.

(a) Verify that the weight of the culture at time t is modeled by

$$y = \frac{10}{1 + 9e^{-10kt}}.$$

Use the fact that $y = 1$ when $t = 0$.

(b) Use the graph to determine the constant k.

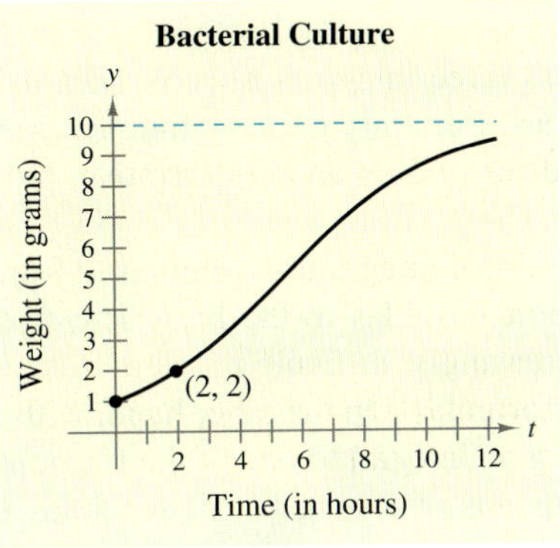

57. *Revenue* The revenue R (in millions of dollars per year) for Symantec Corporation from 1995 through 2003 can be modeled by

$$R = \frac{410t^2 + 28{,}490t + 28{,}080}{-6t^2 + 94t + 100}$$

where $t = 5$ corrresponds to 1995. Find the total revenue from 1995 through 2003. Then find the average revenue during this time period. *(Source: Symantec Corporation)*

58. *Medicine* On a college campus, 50 students return from semester break with a contagious flu virus. The virus has a history of spreading at a rate of

$$\frac{dN}{dt} = \frac{100e^{-0.1t}}{(1 + 4e^{-0.1t})^2}$$

where N is the number of students infected after t days.

(a) Find the model giving the number of students infected with the virus in terms of the number of days since returning from semester break.

(b) If nothing is done to stop the virus from spreading, will the virus spread to infect half the student population of 1000 students? Explain your answer.

59. *Biology* A conservation organization releases 100 animals of an endangered species into a game preserve. The organization believes the population of the species will increase at a rate of

$$\frac{dN}{dt} = \frac{125e^{-0.125t}}{(1 + 9e^{-0.125t})^2}$$

where N is the population and t is the time in months.

(a) Use the fact that $N = 100$ when $t = 0$ to find the population after 2 years.

(b) Find the limiting size of the population as time increases without bound.

60. *Biology: Population Growth* The graph shows the logistic growth curves for two species of the single-celled *Paramecium* in a laboratory culture. During which time intervals is the rate of growth of each species increasing? During which time intervals is the rate of growth of each species decreasing? Which species has a higher limiting population under these conditions? *(Source: Adapted from Levine/Miller,* Biology: Discovering Life, *Second Edition)*

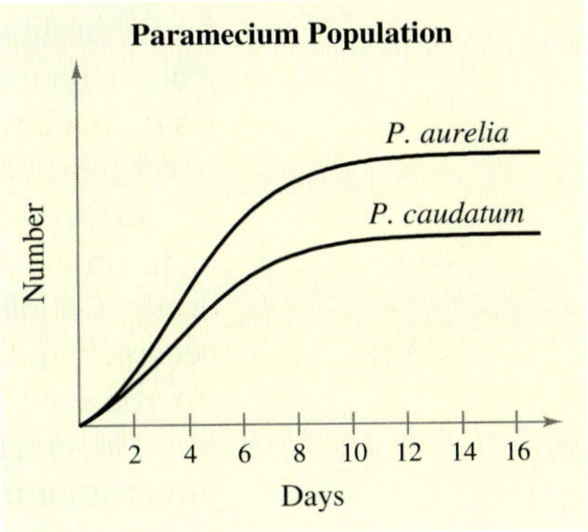

BUSINESS CAPSULE

Courtesy of Susie Wang/Aqua Dessa

While a math communications major at the University of California at Berkeley, Susie Wang began researching the idea of selling natural skin-care products. She used $10,000 to start her company, Aqua Dessa, and uses word-of-mouth as an advertising tactic. Aqua Dessa products are used and sold at spas and exclusive cosmetics counters throughout the United States.

61. *Research Project* Use your school's library, the Internet, or some other reference source to research the opportunity cost of attending graduate school for 2 years to receive a Masters of Business Administration (MBA) degree rather than working for 2 years with a bachelor's degree. Write a short paper describing these costs.

6.4 INTEGRATION TABLES AND COMPLETING THE SQUARE

- Use integration tables to find indefinite integrals.
- Use reduction formulas to find indefinite integrals.
- Use completing the square to find indefinite integrals.

Integration Tables

So far in this chapter, you have studied three integration techniques to be used along with the basic integration formulas. Certainly these techniques and formulas do not cover every possible method for finding an antiderivative, but they do cover most of the important ones.

In this section, you will expand the list of integration formulas to form a table of integrals. As you add new integration formulas to the basic list, two effects occur. On one hand, it becomes increasingly difficult to memorize, or even become familiar with, the entire list of formulas. On the other hand, with a longer list you need fewer techniques for fitting an integral to one of the formulas on the list. The procedure of integrating by means of a long list of formulas is called **integration by tables.** (The table in this section constitutes only a partial listing of integration formulas. Much longer lists exist, some of which contain several hundred formulas.)

Integration by tables should not be considered a trivial task. It requires considerable thought and insight, and it often requires substitution. Many people find a table of integrals to be a valuable supplement to the integration techniques discussed in the first three sections of this chapter. We encourage you to gain competence in the use of integration tables, as well as to continue to improve in the use of the various integration techniques. In doing so, you should find that a combination of techniques and tables is the most versatile approach to integration.

Each integration formula in the table on the next three pages can be developed using one or more of the techniques you have studied. You should try to verify several of the formulas. For instance, Formula 4

$$\int \frac{u}{(a+bu)^2}\,du = \frac{1}{b^2}\left(\frac{a}{a+bu} + \ln|a+bu|\right) + C \qquad \text{Formula 4}$$

can be verified using partial fractions, Formula 17

$$\int \frac{\sqrt{a+bu}}{u}\,du = 2\sqrt{a+bu} + a\int \frac{1}{u\sqrt{a+bu}}\,du \qquad \text{Formula 17}$$

can be verified using integration by parts, and Formula 37

$$\int \frac{1}{1+e^u}\,du = u - \ln(1+e^u) + C \qquad \text{Formula 37}$$

can be verified using substitution.

STUDY TIP

A symbolic integration utility consists, in part, of a database of integration tables. The primary difference between using a symbolic integration utility and using a table of integrals is that with a symbolic integration utility the computer searches through the database to find a fit. With a table of integrals, *you* must do the searching.

In the table of integrals below and on the next two pages, the formulas have been grouped into eight different types according to the form of the integrand.

Forms involving u^n

Forms involving $a + bu$

Forms involving $\sqrt{a + bu}$

Forms involving $\sqrt{u^2 \pm a^2}$

Forms involving $u^2 - a^2$

Forms involving $\sqrt{a^2 - u^2}$

Forms involving e^u

Forms involving $\ln u$

Table of Integrals

Forms involving u^n

1. $\displaystyle\int u^n\,du = \frac{u^{n+1}}{n+1} + C,\quad n \neq -1$

2. $\displaystyle\int \frac{1}{u}\,du = \ln|u| + C$

Forms involving $a + bu$

3. $\displaystyle\int \frac{u}{a+bu}\,du = \frac{1}{b^2}(bu - a\ln|a+bu|) + C$

4. $\displaystyle\int \frac{u}{(a+bu)^2}\,du = \frac{1}{b^2}\left(\frac{a}{a+bu} + \ln|a+bu|\right) + C$

5. $\displaystyle\int \frac{u}{(a+bu)^n}\,du = \frac{1}{b^2}\left[\frac{-1}{(n-2)(a+bu)^{n-2}} + \frac{a}{(n-1)(a+bu)^{n-1}}\right] + C,\quad n \neq 1, 2$

6. $\displaystyle\int \frac{u^2}{a+bu}\,du = \frac{1}{b^3}\left[-\frac{bu}{2}(2a - bu) + a^2\ln|a+bu|\right] + C$

7. $\displaystyle\int \frac{u^2}{(a+bu)^2}\,du = \frac{1}{b^3}\left(bu - \frac{a^2}{a+bu} - 2a\ln|a+bu|\right) + C$

8. $\displaystyle\int \frac{u^2}{(a+bu)^3}\,du = \frac{1}{b^3}\left[\frac{2a}{a+bu} - \frac{a^2}{2(a+bu)^2} + \ln|a+bu|\right] + C$

9. $\displaystyle\int \frac{u^2}{(a+bu)^n}\,du = \frac{1}{b^3}\left[\frac{-1}{(n-3)(a+bu)^{n-3}} + \frac{2a}{(n-2)(a+bu)^{n-2}} - \frac{a^2}{(n-1)(a+bu)^{n-1}}\right] + C,\ n \neq 1, 2, 3$

10. $\displaystyle\int \frac{1}{u(a+bu)}\,du = \frac{1}{a}\ln\left|\frac{u}{a+bu}\right| + C$

11. $\displaystyle\int \frac{1}{u(a+bu)^2}\,du = \frac{1}{a}\left(\frac{1}{a+bu} + \frac{1}{a}\ln\left|\frac{u}{a+bu}\right|\right) + C$

12. $\displaystyle\int \frac{1}{u^2(a+bu)}\,du = -\frac{1}{a}\left(\frac{1}{u} + \frac{b}{a}\ln\left|\frac{u}{a+bu}\right|\right) + C$

13. $\displaystyle\int \frac{1}{u^2(a+bu)^2}\,du = -\frac{1}{a^2}\left[\frac{a+2bu}{u(a+bu)} + \frac{2b}{a}\ln\left|\frac{u}{a+bu}\right|\right] + C$

Table of Integrals (*continued*)

Forms involving $\sqrt{a + bu}$

14. $\displaystyle\int u^n\sqrt{a + bu}\,du = \frac{2}{b(2n + 3)}\left[u^n(a + bu)^{3/2} - na\int u^{n-1}\sqrt{a + bu}\,du\right]$

15. $\displaystyle\int \frac{1}{u\sqrt{a + bu}}\,du = \frac{1}{\sqrt{a}}\ln\left|\frac{\sqrt{a + bu} - \sqrt{a}}{\sqrt{a + bu} + \sqrt{a}}\right| + C, \quad a > 0$

16. $\displaystyle\int \frac{1}{u^n\sqrt{a + bu}}\,du = \frac{-1}{a(n - 1)}\left[\frac{\sqrt{a + bu}}{u^{n-1}} + \frac{(2n - 3)b}{2}\int \frac{1}{u^{n-1}\sqrt{a + bu}}\,du\right], \quad n \neq 1$

17. $\displaystyle\int \frac{\sqrt{a + bu}}{u}\,du = 2\sqrt{a + bu} + a\int \frac{1}{u\sqrt{a + bu}}\,du$

18. $\displaystyle\int \frac{\sqrt{a + bu}}{u^n}\,du = \frac{-1}{a(n - 1)}\left[\frac{(a + bu)^{3/2}}{u^{n-1}} + \frac{(2n - 5)b}{2}\int \frac{\sqrt{a + bu}}{u^{n-1}}\,du\right], \quad n \neq 1$

19. $\displaystyle\int \frac{u}{\sqrt{a + bu}}\,du = -\frac{2(2a - bu)}{3b^2}\sqrt{a + bu} + C$

20. $\displaystyle\int \frac{u^n}{\sqrt{a + bu}}\,du = \frac{2}{(2n + 1)b}\left(u^n\sqrt{a + bu} - na\int \frac{u^{n-1}}{\sqrt{a + bu}}\,du\right)$

Forms involving $\sqrt{u^2 \pm a^2}, \quad a > 0$

21. $\displaystyle\int \sqrt{u^2 \pm a^2}\,du = \frac{1}{2}\left(u\sqrt{u^2 \pm a^2} \pm a^2\ln\left|u + \sqrt{u^2 \pm a^2}\right|\right) + C$

22. $\displaystyle\int u^2\sqrt{u^2 \pm a^2}\,du = \frac{1}{8}\left[u(2u^2 \pm a^2)\sqrt{u^2 \pm a^2} - a^4\ln\left|u + \sqrt{u^2 \pm a^2}\right|\right] + C$

23. $\displaystyle\int \frac{\sqrt{u^2 + a^2}}{u}\,du = \sqrt{u^2 + a^2} - a\ln\left|\frac{a + \sqrt{u^2 + a^2}}{u}\right| + C$

24. $\displaystyle\int \frac{\sqrt{u^2 \pm a^2}}{u^2}\,du = \frac{-\sqrt{u^2 \pm a^2}}{u} + \ln\left|u + \sqrt{u^2 \pm a^2}\right| + C$

25. $\displaystyle\int \frac{1}{\sqrt{u^2 \pm a^2}}\,du = \ln\left|u + \sqrt{u^2 \pm a^2}\right| + C$

26. $\displaystyle\int \frac{1}{u\sqrt{u^2 + a^2}}\,du = -\frac{1}{a}\ln\left|\frac{a + \sqrt{u^2 + a^2}}{u}\right| + C$

27. $\displaystyle\int \frac{u^2}{\sqrt{u^2 \pm a^2}}\,du = \frac{1}{2}\left(u\sqrt{u^2 \pm a^2} \mp a^2\ln\left|u + \sqrt{u^2 \pm a^2}\right|\right) + C$

28. $\displaystyle\int \frac{1}{u^2\sqrt{u^2 \pm a^2}}\,du = \mp\frac{\sqrt{u^2 \pm a^2}}{a^2u} + C$

Table of Integrals (*continued*)

Forms involving $u^2 - a^2, \quad a > 0$

29. $\displaystyle\int \frac{1}{u^2 - a^2}\,du = -\int \frac{1}{a^2 - u^2}\,du = \frac{1}{2a}\ln\left|\frac{u - a}{u + a}\right| + C$

30. $\displaystyle\int \frac{1}{(u^2 - a^2)^n}\,du = \frac{-1}{2a^2(n - 1)}\left[\frac{u}{(u^2 - a^2)^{n-1}} + (2n - 3)\int \frac{1}{(u^2 - a^2)^{n-1}}\,du\right], \quad n \neq 1$

Forms involving $\sqrt{a^2 - u^2}, \quad a > 0$

31. $\displaystyle\int \frac{\sqrt{a^2 - u^2}}{u}\,du = \sqrt{a^2 - u^2} - a\ln\left|\frac{a + \sqrt{a^2 - u^2}}{u}\right| + C$

32. $\displaystyle\int \frac{1}{u\sqrt{a^2 - u^2}}\,du = -\frac{1}{a}\ln\left|\frac{a + \sqrt{a^2 - u^2}}{u}\right| + C$

33. $\displaystyle\int \frac{1}{u^2\sqrt{a^2 - u^2}}\,du = \frac{-\sqrt{a^2 - u^2}}{a^2u} + C$

Forms involving e^u

34. $\displaystyle\int e^u\,du = e^u + C$

35. $\displaystyle\int ue^u\,du = (u - 1)e^u + C$

36. $\displaystyle\int u^ne^u\,du = u^ne^u - n\int u^{n-1}e^u\,du$

37. $\displaystyle\int \frac{1}{1 + e^u}\,du = u - \ln(1 + e^u) + C$

38. $\displaystyle\int \frac{1}{1 + e^{nu}}\,du = u - \frac{1}{n}\ln(1 + e^{nu}) + C$

Forms involving $\ln u$

39. $\displaystyle\int \ln u\,du = u(-1 + \ln u) + C$

40. $\displaystyle\int u\ln u\,du = \frac{u^2}{4}(-1 + 2\ln u) + C$

41. $\displaystyle\int u^n\ln u\,du = \frac{u^{n+1}}{(n + 1)^2}[-1 + (n + 1)\ln u] + C, \quad n \neq -1$

42. $\displaystyle\int (\ln u)^2\,du = u[2 - 2\ln u + (\ln u)^2] + C$

43. $\displaystyle\int (\ln u)^n\,du = u(\ln u)^n - n\int (\ln u)^{n-1}\,du$

TECHNOLOGY

Throughout this section, remember that a symbolic integration utility can be used instead of integration tables. If you have access to such a utility, try using it to find the indefinite integrals in Examples 1 and 2.

EXAMPLE 1 Using Integration Tables

Find $\displaystyle\int \frac{x}{\sqrt{x-1}}\,dx$.

SOLUTION Because the expression inside the radical is linear, you should consider forms involving $\sqrt{a+bu}$, as in Formula 19.

$$\int \frac{u}{\sqrt{a+bu}}\,du = -\frac{2(2a-bu)}{3b^2}\sqrt{a+bu} + C \qquad \text{Formula 19}$$

Using this formula, let $a = -1$, $b = 1$, and $u = x$. Then $du = dx$, and you obtain

$$\int \frac{x}{\sqrt{x-1}}\,dx = -\frac{2(-2-x)}{3}\sqrt{x-1} + C \qquad \text{Substitute values of } a, b, \text{ and } u.$$

$$= \frac{2}{3}(2+x)\sqrt{x-1} + C. \qquad \text{Simplify.}$$

TRY IT 1

Use the integration table to find $\displaystyle\int \frac{x}{\sqrt{2+x}}\,dx$.

EXAMPLE 2 Using Integration Tables

Find $\displaystyle\int x\sqrt{x^4-9}\,dx$.

SOLUTION Because it is not clear which formula to use, you can begin by letting $u = x^2$ and $du = 2x\,dx$. With these substitutions, you can write the integral as shown.

$$\int x\sqrt{x^4-9}\,dx = \frac{1}{2}\int \sqrt{(x^2)^2-9}\,(2x)\,dx \qquad \text{Multiply and divide by 2.}$$

$$= \frac{1}{2}\int \sqrt{u^2-9}\,du \qquad \text{Substitute } u \text{ and } du.$$

Now, it appears that you can use Formula 21.

$$\int \sqrt{u^2-a^2}\,du = \frac{1}{2}\left(u\sqrt{u^2-a^2} - a^2\ln\left|u+\sqrt{u^2-a^2}\right|\right) + C$$

Letting $a = 3$, you obtain

$$\int x\sqrt{x^4-9}\,dx = \frac{1}{2}\int \sqrt{u^2-a^2}\,du$$

$$= \frac{1}{2}\left[\frac{1}{2}\left(u\sqrt{u^2-a^2} - a^2\ln\left|u+\sqrt{u^2-a^2}\right|\right)\right] + C$$

$$= \frac{1}{4}\left(x^2\sqrt{x^4-9} - 9\ln\left|x^2+\sqrt{x^4-9}\right|\right) + C.$$

TRY IT 2

Use the integration table to find

$$\int \frac{\sqrt{x^2+16}}{x}\,dx.$$

EXAMPLE 3 Using Integration Tables

Find $\int \frac{1}{x\sqrt{x+1}}\,dx$.

SOLUTION Considering forms involving $\sqrt{a+bu}$, where $a = 1$, $b = 1$, and $u = x$, you can use Formula 15.

$$\int \frac{1}{u\sqrt{a+bu}}\,du = \frac{1}{\sqrt{a}}\ln\left|\frac{\sqrt{a+bu}-\sqrt{a}}{\sqrt{a+bu}+\sqrt{a}}\right| + C, \qquad a > 0$$

So,

$$\int \frac{1}{x\sqrt{x+1}}\,dx = \int \frac{1}{u\sqrt{a+bu}}\,du = \frac{1}{\sqrt{a}}\ln\left|\frac{\sqrt{a+bu}-\sqrt{a}}{\sqrt{a+bu}+\sqrt{a}}\right| + C$$
$$= \ln\left|\frac{\sqrt{x+1}-1}{\sqrt{x+1}+1}\right| + C.$$

TRY IT 3

Use the integration table to find

$$\int \frac{1}{x^2-4}\,dx.$$

EXAMPLE 4 Using Integration Tables

Evaluate $\int_0^2 \frac{x}{1+e^{-x^2}}\,dx$.

SOLUTION Of the forms involving e^u, Formula 37

$$\int \frac{1}{1+e^u}\,du = u - \ln(1+e^u) + C$$

seems most appropriate. To use this formula, let $u = -x^2$ and $du = -2x\,dx$.

$$\int \frac{x}{1+e^{-x^2}}\,dx = -\frac{1}{2}\int \frac{1}{1+e^{-x^2}}(-2x)\,dx = -\frac{1}{2}\int \frac{1}{1+e^u}\,du$$
$$= -\frac{1}{2}[u - \ln(1+e^u)] + C$$
$$= -\frac{1}{2}[-x^2 - \ln(1+e^{-x^2})] + C$$
$$= \frac{1}{2}[x^2 + \ln(1+e^{-x^2})] + C$$

So, the value of the definite integral is

$$\int_0^2 \frac{x}{1+e^{-x^2}}\,dx = \frac{1}{2}\Big[x^2 + \ln(1+e^{-x^2})\Big]_0^2 \approx 1.66$$

as shown in Figure 6.10.

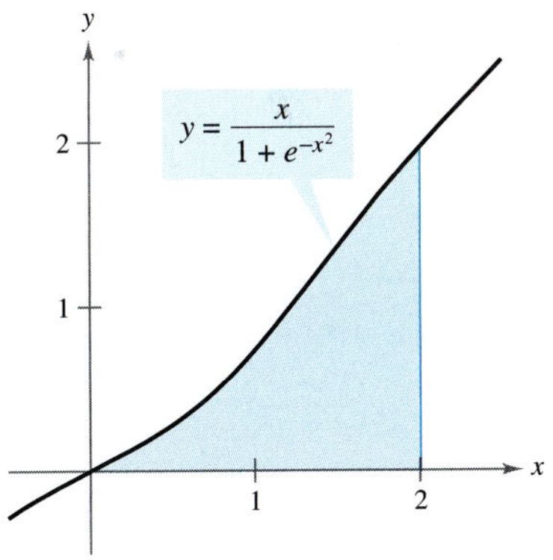

FIGURE 6.10

TRY IT 4

Use the integration table to evaluate $\int_0^1 \frac{x^2}{1+e^{x^3}}\,dx$.

Reduction Formulas

Several of the formulas in the integration table have the form

$$\int f(x)\,dx = g(x) + \int h(x)\,dx$$

where the right side contains an integral. Such integration formulas are called **reduction formulas** because they reduce the original integral to the sum of a function and a simpler integral.

ALGEBRA REVIEW

For help with the algebra in Example 5, see Example 2(b) in the *Chapter 6 Algebra Review* on page 447.

EXAMPLE 5 **Using a Reduction Formula**

Find $\int x^2e^x\,dx$.

SOLUTION Using Formula 36

$$\int u^ne^u\,du = u^ne^u - n\int u^{n-1}e^u\,du$$

you can let $u = x$ and $n = 2$. Then $du = dx$, and you can write

$$\int x^2e^x\,dx = x^2e^x - 2\int xe^x\,dx.$$

Then, using Formula 35

$$\int ue^u\,du = (u - 1)e^u + C$$

you can write

$$\begin{aligned}\int x^2e^x\,dx &= x^2e^x - 2\int xe^x\,dx\\ &= x^2e^x - 2(x - 1)e^x + C\\ &= x^2e^x - 2xe^x + 2e^x + C\\ &= e^x(x^2 - 2x + 2) + C.\end{aligned}$$

TRY IT 5

Use the integration table to find the indefinite integral $\int (\ln x)^2\,dx$.

TECHNOLOGY

You have now studied two ways to find the indefinite integral in Example 5. Example 5 uses an integration table, and Example 4 in Section 6.2 uses integration by parts. A third way would be to use a symbolic integration utility.

Completing the Square

Many integration formulas involve the sum or difference of two squares. You can extend the use of these formulas by an algebraic procedure called **completing the square.** This procedure is demonstrated in Example 6.

EXAMPLE 6 Completing the Square

Find the indefinite integral.

$$\int \frac{1}{x^2 - 4x + 1}\, dx$$

SOLUTION Begin by writing the denominator as the difference of two squares.

$$\begin{aligned} x^2 - 4x + 1 &= (x^2 - 4x + 4) - 4 + 1 \\ &= (x - 2)^2 - 3 \end{aligned}$$

So, you can rewrite the original integral as

$$\int \frac{1}{x^2 - 4x + 1}\, dx = \int \frac{1}{(x-2)^2 - 3}\, dx.$$

Considering $u = x - 2$ and $a = \sqrt{3}$, you can apply Formula 29

$$\int \frac{1}{u^2 - a^2}\, du = \frac{1}{2a} \ln\left|\frac{u - a}{u + a}\right| + C$$

to conclude that

$$\begin{aligned} \int \frac{1}{x^2 - 4x + 1}\, dx &= \int \frac{1}{(x-2)^2 - 3}\, dx \\ &= \int \frac{1}{u^2 - a^2}\, du \\ &= \frac{1}{2a} \ln\left|\frac{u - a}{u + a}\right| + C \\ &= \frac{1}{2\sqrt{3}} \ln\left|\frac{x - 2 - \sqrt{3}}{x - 2 + \sqrt{3}}\right| + C. \end{aligned}$$

TRY IT 6

Find $\int \frac{1}{x^2 - 6x - 1}\, dx.$

TAKE ANOTHER LOOK

Using Integration Tables

Which integration formulas on pages 417–419 would you use to find each indefinite integral? Explain your choice of u for each integral.

a. $\int \frac{e^x}{e^x + 1}\, dx$ b. $\int \frac{1}{e^x + 1}\, dx$ c. $\int \frac{1}{e^{2x} + 1}\, dx$

Use a symbolic integration utility to check that your choices of u were correct.

PREREQUISITE REVIEW 6.4

The following warm-up exercises involve skills that were covered in earlier sections. You will use these skills in the exercise set for this section.

In Exercises 1–4, expand the expression.

1. $(x + 4)^2$

2. $(x - 1)^2$

3. $\left(x + \frac{1}{2}\right)^2$

4. $\left(x - \frac{1}{3}\right)^2$

In Exercises 5–8, write the partial fraction decomposition for the expression.

5. $\dfrac{4}{x(x + 2)}$

6. $\dfrac{3}{x(x - 4)}$

7. $\dfrac{x + 4}{x^2(x - 2)}$

8. $\dfrac{3x^2 + 4x - 8}{x(x - 2)(x + 1)}$

In Exercises 9 and 10, use integration by parts to find the indefinite integral.

9. $\displaystyle\int 2xe^x\,dx$

10. $\displaystyle\int 3x^2 \ln x\,dx$

EXERCISES 6.4

In Exercises 1–8, use the indicated formula from the table of integrals in this section to find the indefinite integral.

1. $\displaystyle\int \frac{x}{(2 + 3x)^2}\,dx$, Formula 4

2. $\displaystyle\int \frac{1}{x(2 + 3x)^2}\,dx$, Formula 11

3. $\displaystyle\int \frac{x}{\sqrt{2 + 3x}}\,dx$, Formula 19

4. $\displaystyle\int \frac{4}{x^2 - 9}\,dx$, Formula 29

5. $\displaystyle\int \frac{2x}{\sqrt{x^4 - 9}}\,dx$, Formula 25

6. $\displaystyle\int x^2\sqrt{x^2 + 9}\,dx$, Formula 22

7. $\displaystyle\int x^3 e^{x^2}\,dx$, Formula 35

8. $\displaystyle\int \frac{x}{1 + e^{x^2}}\,dx$, Formula 37

In Exercises 9–34, use the table of integrals in this section to find the indefinite integral.

9. $\displaystyle\int \frac{1}{x(1 + x)}\,dx$

10. $\displaystyle\int \frac{1}{x(1 + x)^2}\,dx$

11. $\displaystyle\int \frac{1}{x\sqrt{x^2 + 9}}\,dx$

12. $\displaystyle\int \frac{1}{\sqrt{x^2 - 1}}\,dx$

13. $\displaystyle\int \frac{1}{x\sqrt{4 - x^2}}\,dx$

14. $\displaystyle\int \frac{\sqrt{x^2 - 9}}{x^2}\,dx$

15. $\displaystyle\int x \ln x\,dx$

16. $\displaystyle\int x^2(\ln x^3)^2\,dx$

17. $\displaystyle\int \frac{6x}{1 + e^{3x^2}}\,dx$

18. $\displaystyle\int \frac{1}{1 + e^x}\,dx$

19. $\displaystyle\int x\sqrt{x^4 - 4}\,dx$

20. $\displaystyle\int \frac{x}{x^4 - 9}\,dx$

21. $\displaystyle\int \frac{t^2}{(2 + 3t)^3}\,dt$

22. $\displaystyle\int \frac{\sqrt{3 + 4t}}{t}\,dt$

23. $\displaystyle\int \frac{s}{s^2\sqrt{3 + s}}\,ds$

24. $\displaystyle\int \sqrt{3 + x^2}\,dx$

25. $\int \frac{x^2}{(3+2x)^5}\,dx$

26. $\int \frac{1}{x^2\sqrt{x^2-4}}\,dx$

27. $\int \frac{1}{x^2\sqrt{1-x^2}}\,dx$

28. $\int \frac{2x}{(1-3x)^2}\,dx$

29. $\int x^2 \ln x\,dx$

30. $\int xe^{x^2}\,dx$

31. $\int \frac{x^2}{(3x-5)^2}\,dx$

32. $\int \frac{1}{2x^2(2x-1)^2}\,dx$

33. $\int \frac{\ln x}{x(4+3\ln x)}\,dx$

34. $\int (\ln x)^3\,dx$

In Exercises 35–40, use the integration table to find the exact area of the region bounded by the graphs of the equations. Then use a graphing utility to graph the region and approximate the area.

35. $y = \frac{x}{\sqrt{x+1}},\ y = 0,\ x = 8$

36. $y = \frac{2}{1+e^{4x}},\ y = 0,\ x = 0,\ x = 1$

37. $y = \frac{x}{1+e^{x^2}},\ y = 0,\ x = 2$

38. $y = \frac{-e^x}{1-e^{2x}},\ y = 0,\ x = 1,\ x = 2$

39. $y = x^2\sqrt{x^2+4},\ y = 0,\ x = \sqrt{5}$

40. $y = \frac{1}{\sqrt{x}(1+2\sqrt{x})},\ y = 0,\ x = 1,\ x = 4$

In Exercises 41–44, evaluate the definite integral.

41. $\int_0^5 \frac{x}{\sqrt{5+2x}}\,dx$

42. $\int_0^5 \frac{x}{(4+x)^2}\,dx$

43. $\int_0^4 \frac{6}{1+e^{0.5x}}\,dx$

44. $\int_1^4 x\ln x\,dx$

In Exercises 45–48, find the indefinite integral (a) using the integration table and (b) using the specified method.

Integral	*Method*
45. $\int x^2e^x\,dx$	Integration by parts
46. $\int x^4\ln x\,dx$	Integration by parts
47. $\int \frac{1}{x^2(x+1)}\,dx$	Partial fractions
48. $\int \frac{1}{x^2-75}\,dx$	Partial fractions

In Exercises 49–52, complete the square to express each polynomial as the sum or difference of squares.

49. (a) $x^2 + 6x$
(b) $x^2 - 8x + 9$
(c) $x^4 + 2x^2 - 5$
(d) $3 - 2x - x^2$

50. (a) $x^2 + 4x$
(b) $x^2 + 16x - 1$
(c) $x^4 + 8x^2 + 1$
(d) $9x^2 + 36x - 1$

51. (a) $4x^2 + 12x + 15$
(b) $3x^2 - 12x - 9$
(c) $x^2 - 2x$
(d) $9 + 8x - x^2$

52. (a) $16x^2 - 96x + 3$
(b) $x^2 + 4x - 1$
(c) $1 - 8x - x^2$
(d) $6x - x^2$

In Exercises 53–60, complete the square and then use the integration table to find the indefinite integral.

53. $\int \frac{1}{x^2+6x-8}\,dx$

54. $\int \frac{1}{x^2+4x-5}\,dx$

55. $\int \frac{1}{(x-1)\sqrt{x^2-2x+2}}\,dx$

56. $\int \sqrt{x^2-6x}\,dx$

57. $\int \frac{1}{2x^2-4x-6}\,dx$

58. $\int \frac{\sqrt{7-6x-x^2}}{x+3}\,dx$

59. $\int \frac{x}{\sqrt{x^4+2x^2+2}}\,dx$

60. $\int \frac{x\sqrt{x^4+4x^2+5}}{x^2+2}\,dx$

Population Growth In Exercises 61 and 62, use a graphing utility to graph the growth function. Use the table of integrals to find the average value of the growth function over the interval, where N is the size of a population and t is the time in days.

61. $N = \frac{50}{1+e^{4.8-1.9t}},\quad [3, 4]$

62. $N = \frac{375}{1+e^{4.20-0.25t}},\quad [21, 28]$

63. ***Revenue*** The revenue (in dollars per year) for a new product is modeled by

$$R = 10{,}000\left[1 - \frac{1}{(1+0.1t^2)^{1/2}}\right]$$

where t is the time in years. Estimate the total revenue from sales of the product over its first 2 years on the market.

64. ***Consumer and Producer Surpluses*** Find the consumer surplus and the producer surplus for a product with the given demand and supply functions.

Demand: $p = \frac{60}{\sqrt{x^2+81}}$, *Supply:* $p = \frac{x}{3}$

65. ***Profit*** The net profits P (in billions of dollars per year) for Hershey Foods from 2000 through 2003 can be modeled by

$$P = \sqrt{0.04t - 0.3}, \qquad 10 \le t \le 13$$

where t is the time in years, with $t = 10$ corresponding to 2000. Find the average net profit over that time period. *(Source: Hershey Foods Corp.)*

6.5 NUMERICAL INTEGRATION

- Use the Trapezoidal Rule to approximate definite integrals.
- Use Simpson's Rule to approximate definite integrals.
- Analyze the sizes of the errors when approximating definite integrals with the Trapezoidal Rule and Simpson's Rule.

Trapezoidal Rule

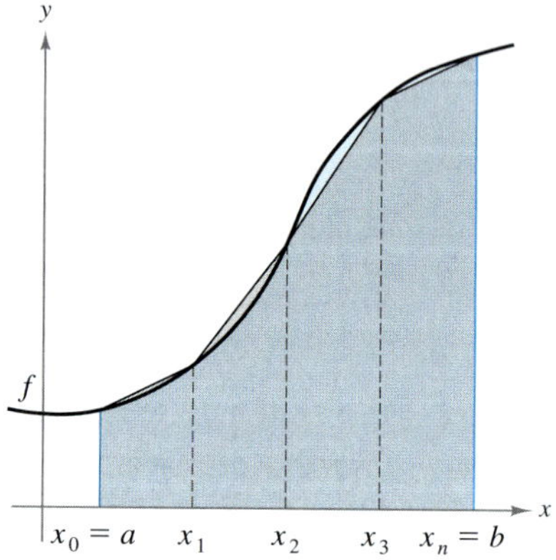

FIGURE 6.11 The area of the region can be approximated using four trapezoids.

In Section 5.6, you studied one technique for approximating the value of a *definite* integral—the Midpoint Rule. In this section, you will study two other approximation techniques: the **Trapezoidal Rule** and **Simpson's Rule.**

To develop the Trapezoidal Rule, consider a function f that is nonnegative and continuous on the closed interval $[a, b]$. To approximate the area represented by $\int_a^b f(x)dx$, partition the interval into n subintervals, each of width

$$\Delta x = \frac{b-a}{n}. \qquad \text{Width of each subinterval}$$

Next, form n trapezoids, as shown in Figure 6.11. As you can see in Figure 6.12, the area of the first trapezoid is

$$\text{Area of first trapezoid} = \left(\frac{b-a}{n}\right)\left[\frac{f(x_0)+f(x_1)}{2}\right].$$

The areas of the other trapezoids follow a similar pattern, and the sum of the n areas is

$$\left(\frac{b-a}{n}\right)\left[\frac{f(x_0)+f(x_1)}{2} + \frac{f(x_1)+f(x_2)}{2} + \cdot\cdot\cdot + \frac{f(x_{n-1})+f(x_n)}{2}\right]$$

$$= \left(\frac{b-a}{2n}\right)[f(x_0)+f(x_1)+f(x_1)+f(x_2)+\cdot\cdot\cdot+f(x_{n-1})+f(x_n)]$$

$$= \left(\frac{b-a}{2n}\right)[f(x_0)+2f(x_1)+2f(x_2)+\cdot\cdot\cdot+2f(x_{n-1})+f(x_n)].$$

Although this development assumes f to be continuous *and* nonnegative on $[a, b]$, the resulting formula is valid as long as f is continuous on $[a, b]$.

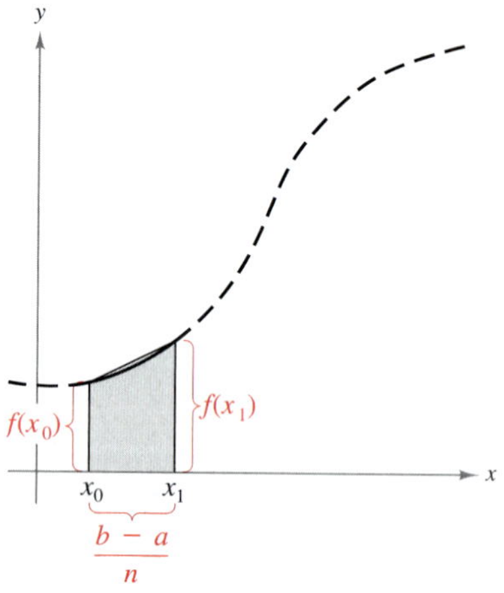

FIGURE 6.12

The Trapezoidal Rule

If f is continuous on $[a, b]$, then

$$\int_a^b f(x)dx \approx \left(\frac{b-a}{2n}\right)[f(x_0)+2f(x_1)+\cdot\cdot\cdot+2f(x_{n-1})+f(x_n)].$$

STUDY TIP

The coefficients in the Trapezoidal Rule have the pattern

1 2 2 2 . . . 2 2 1.

EXAMPLE 1 Using the Trapezoidal Rule

Use the Trapezoidal Rule to approximate

$$\int_0^1 e^x dx.$$

Compare the results for $n = 4$ and $n = 8$.

SOLUTION When $n = 4$, the width of each subinterval is

$$\frac{1-0}{4} = \frac{1}{4}$$

and the endpoints of the subintervals are

$$x_0 = 0, \quad x_1 = \frac{1}{4}, \quad x_2 = \frac{1}{2}, \quad x_3 = \frac{3}{4}, \quad \text{and} \quad x_4 = 1$$

as indicated in Figure 6.13. So, by the Trapezoidal Rule

$$\int_0^1 e^x dx = \frac{1}{8}(e^0 + 2e^{0.25} + 2e^{0.5} + 2e^{0.75} + e^1)$$

$$\approx 1.7272. \qquad \text{Approximation using } n = 4$$

When $n = 8$, the width of each subinterval is

$$\frac{1-0}{8} = \frac{1}{8}$$

and the endpoints of the subintervals are

$$x_0 = 0, \quad x_1 = \frac{1}{8}, \quad x_2 = \frac{1}{4}, \quad x_3 = \frac{3}{8}, \quad x_4 = \frac{1}{2}$$

$$x_5 = \frac{5}{8}, \quad x_6 = \frac{3}{4}, \quad x_7 = \frac{7}{8}, \quad \text{and} \quad x_8 = 1$$

as indicated in Figure 6.14. So, by the Trapezoidal Rule

$$\int_0^1 e^x dx = \frac{1}{16}(e^0 + 2e^{0.125} + 2e^{0.25} + \cdots + 2e^{0.875} + e^1)$$

$$\approx 1.7205. \qquad \text{Approximation using } n = 8$$

Of course, for *this particular* integral, you could have found an antiderivative and used the Fundamental Theorem of Calculus to find the exact value of the definite integral. The exact value is

$$\int_0^1 e^x dx = e - 1 \approx 1.718282. \qquad \text{Exact value}$$

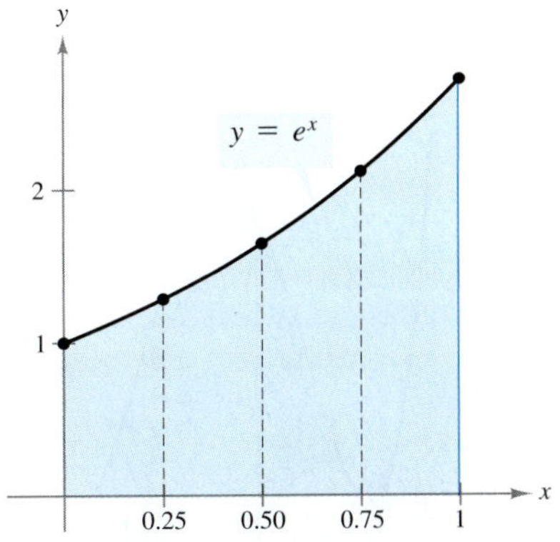

FIGURE 6.13 Four Subintervals

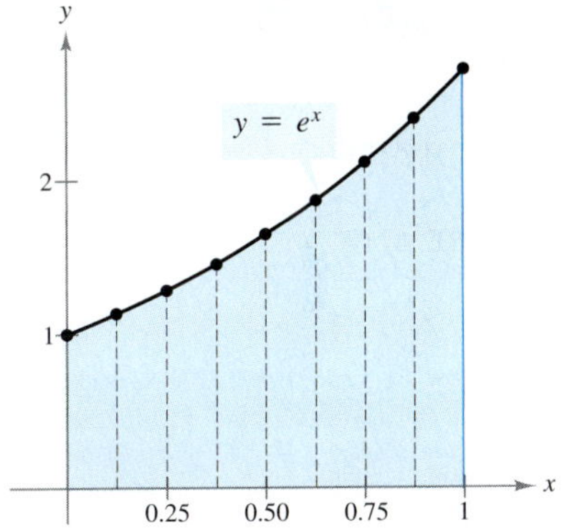

FIGURE 6.14 Eight Subintervals

> **TRY IT 1**
>
> Use the Trapezoidal Rule with $n = 4$ to approximate
>
> $$\int_0^1 e^{2x}\, dx.$$

There are two important points that should be made concerning the Trapezoidal Rule. First, the approximation tends to become more accurate as n increases. For instance, in Example 1, if $n = 16$, the Trapezoidal Rule yields an approximation of 1.7188. Second, although you could have used the Fundamental Theorem of Calculus to evaluate the integral in Example 1, this theorem cannot be used to evaluate an integral as simple as $\int_0^1 e^{x^2} dx$, because e^{x^2} has no elementary function as an antiderivative. Yet the Trapezoidal Rule can be easily applied to this integral.

> **TECHNOLOGY**
>
> A graphing utility can also evaluate a definite integral that does not have an elementary function as an antiderivative. Use the integration capabilities of a graphing utility to approximate the integral $\int_0^1 e^{x^2} dx$.

Simpson's Rule

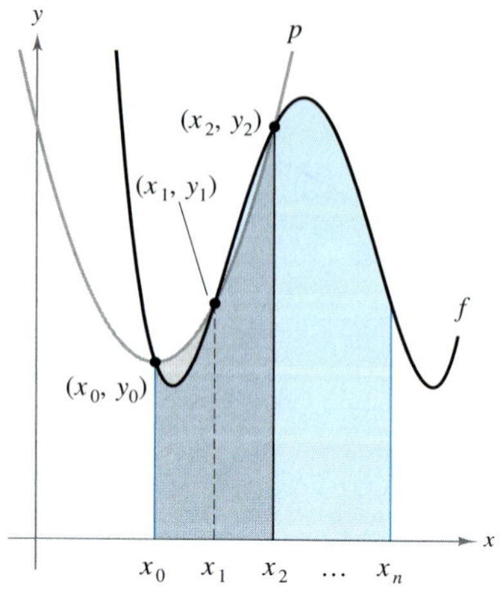

FIGURE 6.15

One way to view the Trapezoidal Rule is to say that on each subinterval, f is approximated by a first-degree polynomial. In Simpson's Rule, f is approximated by a second-degree polynomial on each subinterval.

To develop Simpson's Rule, partition the interval $[a, b]$ into an *even number* n of subintervals, each of width

$$\Delta x = \frac{b-a}{n}.$$

On the subinterval $[x_0, x_2]$, approximate the function f by the second-degree polynomial $p(x)$ that passes through the points

$$(x_0, f(x_0)), \quad (x_1, f(x_1)), \quad \text{and} \quad (x_2, f(x_2))$$

as shown in Figure 6.15. The Fundamental Theorem of Calculus can be used to show that

$$\begin{aligned}\int_{x_0}^{x_2} f(x)\,dx &\approx \int_{x_0}^{x_2} p(x)\,dx \\ &= \left(\frac{x_2 - x_0}{6}\right)\left[p(x_0) + 4p\left(\frac{x_0 + x_2}{2}\right) + p(x_2)\right] \\ &= \frac{2[(b-a)/n]}{6}[p(x_0) + 4p(x_1) + p(x_2)] \\ &= \left(\frac{b-a}{3n}\right)[f(x_0) + 4f(x_1) + f(x_2)].\end{aligned}$$

Repeating this process on the subintervals $[x_{i-2}, x_i]$ produces

$$\int_a^b f(x)\,dx \approx \left(\frac{b-a}{3n}\right)[f(x_0) + 4f(x_1) + f(x_2) + f(x_2) + 4f(x_3) + f(x_4) + \cdots + f(x_{n-2}) + 4f(x_{n-1}) + f(x_n)].$$

By grouping like terms, you can obtain the approximation shown below, which is known as Simpson's Rule. This rule is named after the English mathematician Thomas Simpson (1710–1761).

Simpson's Rule (*n* Is Even)

If f is continuous on $[a, b]$, then

$$\int_a^b f(x)\,dx \approx \left(\frac{b-a}{3n}\right)[f(x_0) + 4f(x_1) + 2f(x_2) + 4f(x_3) + \cdots + 4f(x_{n-1}) + f(x_n)].$$

STUDY TIP

The coefficients in Simpson's Rule have the pattern

1 4 2 4 2 4 . . . 4 2 4 1.

In Example 1, the Trapezoidal Rule was used to estimate the value of

$$\int_0^1 e^x\,dx.$$

The next example uses Simpson's Rule to approximate the same integral.

EXAMPLE 2 Using Simpson's Rule

Use Simpson's Rule to approximate

$$\int_0^1 e^x\,dx.$$

Compare the results for $n = 4$ and $n = 8$.

SOLUTION When $n = 4$, the width of each subinterval is $(1 - 0)/4 = \frac{1}{4}$ and the endpoints of the subintervals are

$$x_0 = 0,\quad x_1 = \frac{1}{4},\quad x_2 = \frac{1}{2},\quad x_3 = \frac{3}{4},\quad \text{and}\quad x_4 = 1$$

as indicated in Figure 6.16. So, by Simpson's Rule

$$\int_0^1 e^x dx = \frac{1}{12}(e^0 + 4e^{0.25} + 2e^{0.5} + 4e^{0.75} + e^1)$$

$$\approx 1.718319. \qquad \text{Approximation using } n = 4$$

When $n = 8$, the width of each subinterval is $(1 - 0)/8 = \frac{1}{8}$ and the endpoints of the subintervals are

$$x_0 = 0,\quad x_1 = \frac{1}{8},\quad x_2 = \frac{1}{4},\quad x_3 = \frac{3}{8},\quad x_4 = \frac{1}{2}$$

$$x_5 = \frac{5}{8},\quad x_6 = \frac{3}{4},\quad x_7 = \frac{7}{8},\quad \text{and}\quad x_8 = 1$$

as indicated in Figure 6.17. So, by Simpson's Rule

$$\int_0^1 e^x dx = \frac{1}{24}(e^0 + 4e^{0.125} + 2e^{0.25} + \cdots + 4e^{0.875} + e^1)$$

$$\approx 1.718284. \qquad \text{Approximation using } n = 8$$

Recall that the exact value of this integral is

$$\int_0^1 e^x dx = e - 1 \approx 1.718282. \qquad \text{Exact value}$$

So, with only eight subintervals, you obtained an approximation that is correct to the nearest 0.000002—an impressive result.

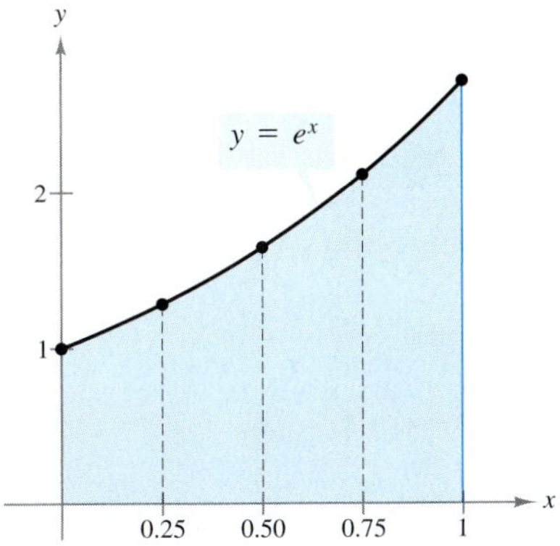

FIGURE 6.16 Four Subintervals

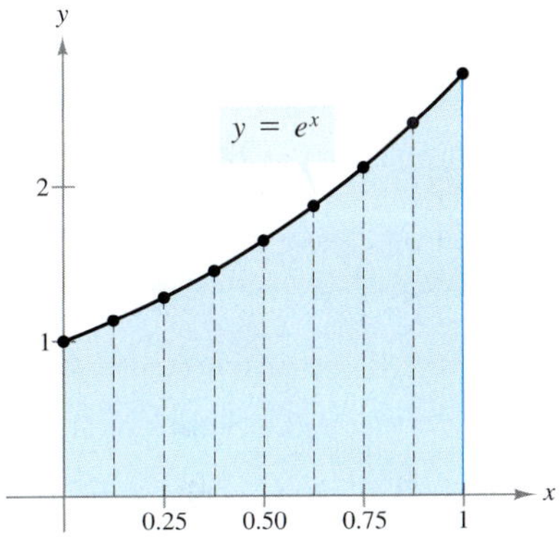

FIGURE 6.17 Eight Subintervals

STUDY TIP

Comparing the results of Examples 1 and 2, you can see that for a given value of n, Simpson's Rule tends to be more accurate than the Trapezoidal Rule.

TRY IT 2

Use Simpson's Rule with $n = 4$ to approximate $\int_0^1 e^{2x}\,dx$.

TECHNOLOGY

Programming Simpson's Rule

In Section 5.6, you saw how to program the Midpoint Rule into a computer or programmable calculator. The pseudocode below can be used to write a program that will evaluate Simpson's Rule. (Appendix E lists this program for several models of graphing utilities.)

Program

- *Prompt for value of a.*
- *Input value of a.*
- *Prompt for value of b.*
- *Input value of b.*
- *Prompt for value of* $n/2$.
- *Input value of* $n/2$.
- *Initialize sum of areas.*
- *Calculate width of subinterval.*
- *Initialize counter.*
- *Begin loop.*
- *Calculate left endpoint.*
- *Calculate right endpoint.*
- *Calculate midpoint of subinterval.*
- *Store left endpoint.*
- *Evaluate* $f(x)$ *at left endpoint.*
- *Store midpoint of subinterval.*
- *Evaluate* $f(x)$ *at midpoint.*
- *Store right endpoint.*
- *Evaluate* $f(x)$ *at right endpoint.*
- *Store Simpson's Rule.*
- *Check value of index.*
- *End loop.*
- *Display approximation.*

Before executing the program, enter the function. When the program is executed, you should be prompted to enter the lower and upper limits of integration, and *half* the number of subintervals you want to use.

To check your steps, try running the program with the integral in Example 2. When $n = 8$ (input 4), you should obtain an approximation of

$$\int_0^1 e^x\,dx \approx 1.718284155.$$

Use your program with $n = 50$ to approximate

$$\int_0^1 e^{x^2}\,dx.$$

STUDY TIP

The Trapezoidal Rule and Simpson's Rule are necessary for solving certain real-life problems, such as approximating the present value of an income. You will see such problems in the exercise set for this section.

Error Analysis

In Examples 1 and 2, you were able to calculate the exact value of the integral and compare that value with the approximations to see how good they were. In practice, you need to have a different way of telling how good an approximation is: such a way is provided in the next result.

Errors in the Trapezoidal Rule and Simpson's Rule

The errors E in approximating $\int_a^b f(x)\,dx$ are as shown.

Trapezoidal Rule: $|E| \le \dfrac{(b-a)^3}{12n^2}\left[\max|f''(x)|\right], \quad a \le x \le b$

Simpson's Rule: $|E| \le \dfrac{(b-a)^5}{180n^4}\left[\max|f^{(4)}(x)|\right], \quad a \le x \le b$

This result indicates that the errors generated by the Trapezoidal Rule and Simpson's Rule have upper bounds dependent on the extreme values of $f''(x)$ and $f^{(4)}(x)$ in the interval $[a, b]$. Furthermore, the bounds for the errors can be made arbitrarily small by *increasing* n. To determine what value of n to choose, consider the steps below.

Trapezoidal Rule

1. Find $f''(x)$.
2. Find the maximum of $|f''(x)|$ on the interval $[a, b]$.
3. Set up the inequality

$$|E| \le \frac{(b-a)^3}{12n^2}\left[\max|f''(x)|\right].$$

4. For an error less than ϵ, solve for n in the inequality

$$\frac{(b-a)^3}{12n^2}\left[\max|f''(x)|\right] < \epsilon.$$

5. Partition $[a, b]$ into n subintervals and apply the Trapezoidal Rule.

Simpson's Rule

1. Find $f^{(4)}(x)$.
2. Find the maximum of $|f^{(4)}(x)|$ on the interval $[a, b]$.
3. Set up the inequality

$$|E| \le \frac{(b-a)^5}{180n^4}\left[\max|f^{(4)}(x)|\right].$$

4. For an error less than ϵ, solve for n in the inequality

$$\frac{(b-a)^5}{180n^4}\left[\max|f^{(4)}(x)|\right] < \epsilon.$$

5. Partition $[a, b]$ into n subintervals and apply Simpson's Rule.

DISCOVERY

How does the error in the Trapezoidal Rule decrease as you increase n, the number of subintervals? In Example 1 on page 427, you approximated the integral $\int_0^1 e^x\,dx$ using the Trapezoidal Rule with four and eight subintervals. Because the exact value of this integral is known to be $e - 1$, the errors are

$n = 4$ error: 0.008918

and

$n = 8$ error: 0.002218.

Notice that the error diminishes by a factor of approximately 4 when the number of subintervals is doubled. Explain how this is consistent with the formula for the Trapezoidal Rule error. What do you think the error would be if you used 16 subintervals in Example 1? (The actual error is 0.0005593, which is approximately 0.002218/4.) How does the error in Simpson's Rule diminish as you increase the number of subintervals?

EXAMPLE 3 **Using the Trapezoidal Rule**

Use the Trapezoidal Rule to estimate the value of

$$\int_0^1 e^{-x^2}\,dx$$

such that the approximation error is less than 0.01.

SOLUTION

1. Begin by finding the second derivative of $f(x) = e^{-x^2}$.

$$f(x) = e^{-x^2}$$
$$f'(x) = -2xe^{-x^2}$$
$$f''(x) = 4x^2e^{-x^2} - 2e^{-x^2}$$
$$= 2e^{-x^2}(2x^2 - 1)$$

2. f'' has only one critical number in the interval $[0, 1]$, and the maximum value of $|f''(x)|$ on this interval is $|f''(0)| = 2$.

3. The error E using the Trapezoidal Rule is bounded by

$$|E| \le \frac{(b-a)^3}{12n^2}(2) = \frac{1}{12n^2}(2) = \frac{1}{6n^2}.$$

4. To ensure that the approximation has an error of less than 0.01, you should choose n such that

$$\frac{1}{6n^2} < 0.01.$$

Solving for n, you can determine that n must be 5 or more.

5. Partition $[0, 1]$ into five subintervals, as shown in Figure 6.18. Then apply the Trapezoidal Rule to obtain

$$\int_0^1 e^{-x^2}\,dx = \frac{1}{10}\left(\frac{1}{e^0} + \frac{2}{e^{0.04}} + \frac{2}{e^{0.16}} + \frac{2}{e^{0.36}} + \frac{2}{e^{0.64}} + \frac{1}{e^1}\right)$$
$$\approx 0.744.$$

So, with an error no larger than 0.01, you know that

$$0.734 \le \int_0^1 e^{-x^2}\,dx \le 0.754.$$

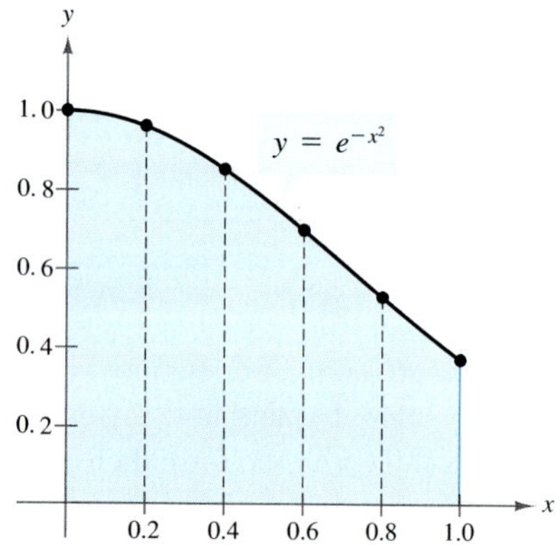

FIGURE 6.18

TRY IT 3

Use the Trapezoidal Rule to estimate the value of

$$\int_0^1 \sqrt{1 + x^2}\,dx$$

such that the approximation error is less than 0.01.

TAKE ANOTHER LOOK

Using the Trapezoidal Rule

In the second step of the solution to Example 3, it is stated that the maximum value of $|f''(x)|$ is 2. Explain why this is true. Verify this result by using a graphing utility to graph f''.

PREREQUISITE REVIEW 6.5

The following warm-up exercises involve skills that were covered in earlier sections. You will use these skills in the exercise set for this section.

In Exercises 1–6, find the indicated derivative.

1. $f(x) = \frac{1}{x}$, $f''(x)$

2. $f(x) = \ln(2x + 1)$, $f^{(4)}(x)$

3. $f(x) = 2 \ln x$, $f^{(4)}(x)$

4. $f(x) = x^3 - 2x^2 + 7x - 12$, $f''(x)$

5. $f(x) = e^{2x}$, $f^{(4)}(x)$

6. $f(x) = e^{x^2}$, $f''(x)$

In Exercises 7 and 8, find the absolute maximum of f on the interval.

7. $f(x) = -x^2 + 6x + 9$, $[0, 4]$

8. $f(x) = \frac{8}{x^3}$, $[1, 2]$

In Exercises 9 and 10, solve for n.

9. $\frac{1}{4n^2} < 0.001$

10. $\frac{1}{16n^4} < 0.0001$

EXERCISES 6.5

In Exercises 1–12, use the Trapezoidal Rule and Simpson's Rule to approximate the value of the definite integral for the indicated value of n. Compare these results with the exact value of the definite integral. Round your answers to four decimal places.

1. $\int_0^2 x^2\,dx,\ n = 4$

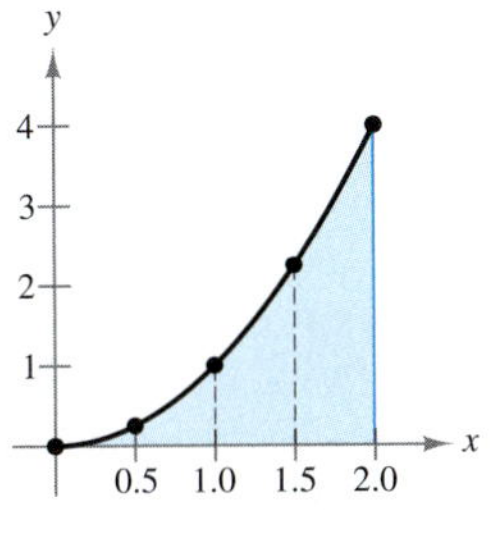

2. $\int_0^1 \left(\frac{x^2}{2} + 1\right) dx,\ n = 4$

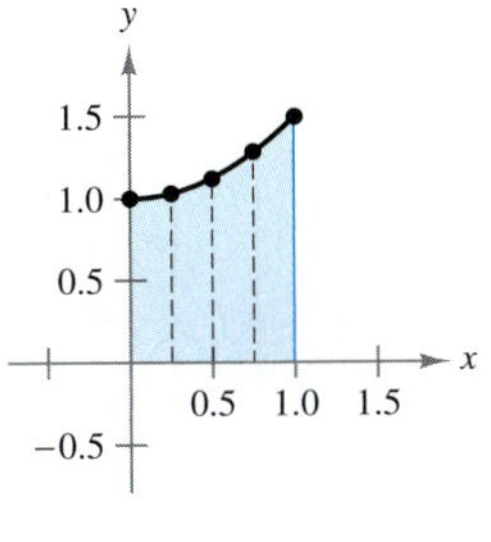

3. $\int_0^2 (x^4 + 1)\,dx,\ n = 4$

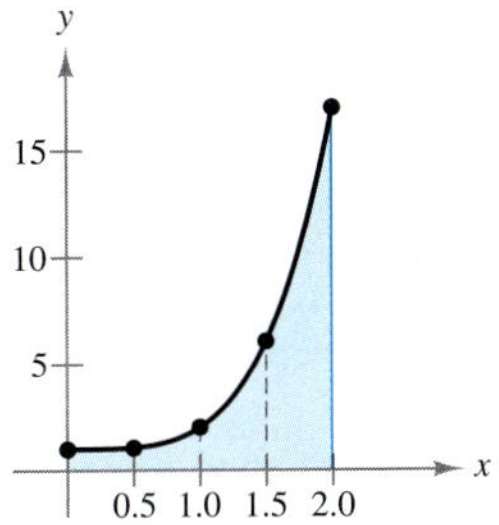

4. $\int_1^2 \frac{1}{x}\,dx,\ n = 4$

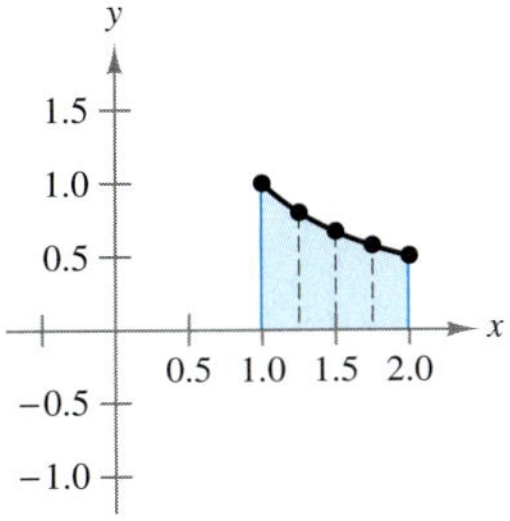

5. $\int_0^2 x^3\,dx,\ n = 8$

6. $\int_1^3 (4 - x^2)\,dx,\ n = 4$

7. $\int_1^2 \frac{1}{x}\,dx,\ n = 8$

8. $\int_1^2 \frac{1}{x^2}\,dx,\ n = 4$

9. $\int_0^4 \sqrt{x}\,dx,\ n = 8$

10. $\int_0^2 \sqrt{1 + x}\,dx,\ n = 4$

11. $\int_0^1 \frac{1}{1 + x}\,dx,\ n = 4$

12. $\int_0^2 x\sqrt{x^2 + 1}\,dx,\ n = 4$

In Exercises 13–20, approximate the integral using (a) the Trapezoidal Rule and (b) Simpson's Rule. (Round your answers to three significant digits.)

Definite Integral	*Subdivisions*
13. $\int_0^1 \frac{1}{1+x^2}\,dx$	$n = 4$
14. $\int_0^2 \frac{1}{\sqrt{1+x^3}}\,dx$	$n = 4$
15. $\int_0^1 \sqrt{1-x^2}\,dx$	$n = 4$
16. $\int_0^1 \sqrt{1-x^2}\,dx$	$n = 8$
17. $\int_0^2 e^{-x^2}\,dx$	$n = 2$
18. $\int_0^2 e^{-x^2}\,dx$	$n = 4$
19. $\int_0^3 \frac{1}{2-2x+x^2}\,dx$	$n = 6$
20. $\int_0^3 \frac{x}{2+x+x^2}\,dx$	$n = 6$

Present Value In Exercises 21 and 22, use a program similar to the Simpson's Rule program on page 430 with $n = 8$ to approximate the present value of the income $c(t)$ over t_1 years at the given annual interest rate r. Then use the integration capabilities of a graphing utility to approximate the present value. Compare the results. (Present value is defined in Section 6.2.)

21. $c(t) = 6000 + 200\sqrt{t},\ r = 7\%,\ t_1 = 4$

22. $c(t) = 200{,}000 + 15{,}000\sqrt[3]{t},\ r = 10\%,\ t_1 = 8$

Marginal Analysis In Exercises 23 and 24, use a program similar to the Simpson's Rule program on page 430 with $n = 4$ to approximate the change in revenue from the marginal revenue function dR/dx. In each case, assume that the number of units sold x increases from 14 to 16.

23. $\frac{dR}{dx} = 5\sqrt{8000 - x^3}$

24. $\frac{dR}{dx} = 50\sqrt{x}\sqrt{20 - x}$

Probability In Exercises 25–28, use a program similar to the Simpson's Rule program on page 430 with $n = 6$ to approximate the indicated normal probability. The standard normal probability density function is

$$f(x) = \frac{1}{\sqrt{2\pi}}e^{-x^2/2}.$$

If x is chosen at random from a population with this density, then the probability that x lies in the interval $[a, b]$ is

$$P(a \le x \le b) = \int_a^b f(x)\,dx.$$

25. $P(0 \le x \le 1)$

26. $P(0 \le x \le 2)$

27. $P(0 \le x \le 4)$

28. $P(0 \le x \le 1.5)$

Surveying In Exercises 29 and 30, use a program similar to the Simpson's Rule program on page 430 to estimate the number of square feet of land in the lot, where x and y are measured in feet, as shown in the figures. In each case, the land is bounded by a stream and two straight roads.

29.

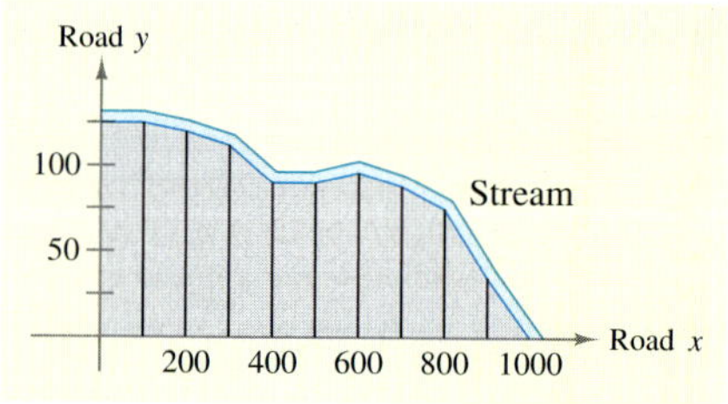

x	0	100	200	300	400	500
y	125	125	120	112	90	90

x	600	700	800	900	1000
y	95	88	75	35	0

30.

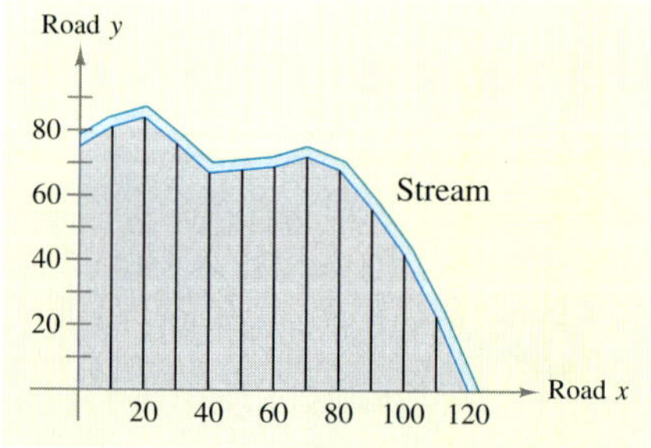

x	0	10	20	30	40	50	60
y	75	81	84	76	67	68	69

x	70	80	90	100	110	120
y	72	68	56	42	23	0

In Exercises 31–34, use the error formulas to find bounds for the error in approximating the integral using (a) the Trapezoidal Rule and (b) Simpson's Rule. (Let $n = 4$.)

31. $\int_0^2 x^4\,dx$ **32.** $\int_0^1 \frac{1}{x+1}\,dx$

33. $\int_0^1 e^{x^3}\,dx$ **34.** $\int_0^1 e^{-x^2}\,dx$

In Exercises 35–38, use the error formulas to find n such that the error in the approximation of the definite integral is less than 0.0001 using (a) the Trapezoidal Rule and (b) Simpson's Rule.

35. $\int_0^1 x^4\,dx$ **36.** $\int_1^3 \frac{1}{x}\,dx$

37. $\int_1^3 e^{2x}\,dx$ **38.** $\int_3^5 \ln x\,dx$

In Exercises 39–42, use the program for Simpson's Rule given on page 430 to approximate the integral. Use $n = 100$.

39. $\int_1^4 x\sqrt{x+4}\,dx$ **40.** $\int_1^4 x^2\sqrt{x+4}\,dx$

41. $\int_2^5 10xe^{-x}\,dx$ **42.** $\int_2^5 10x^2e^{-x}\,dx$

43. Prove that Simpson's Rule is exact when used to approximate the integral of a cubic polynomial function, and demonstrate the result for

$$\int_0^1 x^3\,dx, \qquad n = 2.$$

44. Use a program similar to the Simpson's Rule program on page 430 with $n = 4$ to find the volume of the solid generated by revolving the region bounded by the graphs of

$$y = x\sqrt[3]{x+4}, \quad y = 0, \quad \text{and} \quad x = 4$$

about the x-axis.

In Exercises 45 and 46, use the definite integral below to find the required arc length. If f has a continuous derivative, then the arc length of f between the points $(a, f(a))$ and $(b, f(b))$ is

$$\int_b^a \sqrt{1 + [f'(x)]^2}\,dx.$$

45. *Arc Length* The suspension cable on a bridge that is 400 feet long is in the shape of a parabola whose equation is

$$y = \frac{x^2}{800} \text{ (see figure).}$$

Use a program similar to the Simpson's Rule program on page 430 with $n = 12$ to approximate the length of the cable. Compare this result with the length obtained by using the table of integrals in Section 6.4 to perform the integration.

Figure for 45

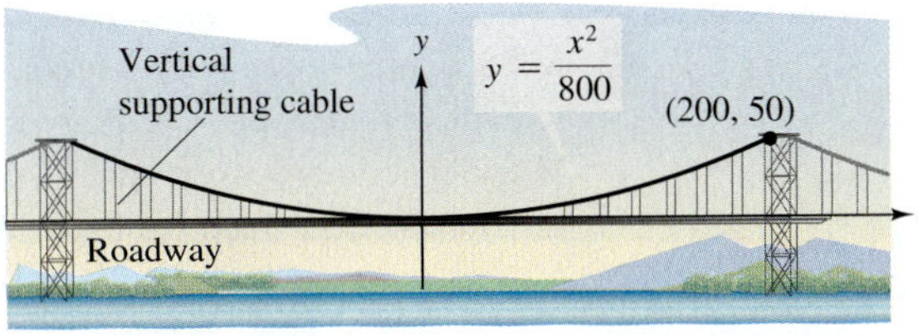

46. *Arc Length* A fleeing hare leaves its burrow (0, 0) and moves due north (up the y-axis). At the same time, a pursuing lynx leaves from 1 yard east of the burrow (1, 0) and always moves toward the fleeing hare (see figure). If the lynx's speed is twice that of the hare's, the equation of the lynx's path is

$$y = \frac{1}{3}(x^{3/2} - 3x^{1/2} + 2).$$

Find the distance traveled by the lynx by integrating over the interval $[0, 1]$.

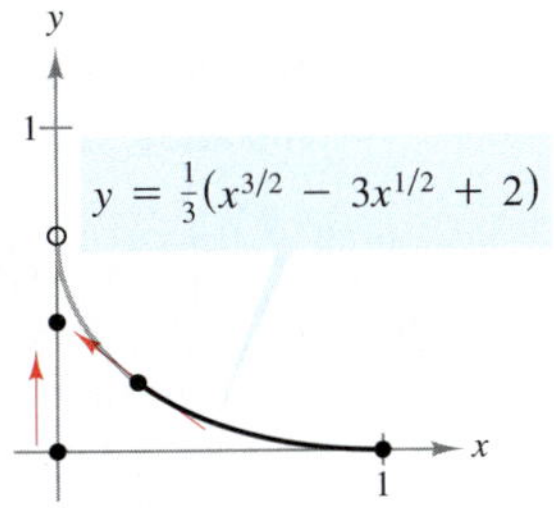

47. *Medicine* A body assimilates a 12-hour cold tablet at a rate modeled by

$$\frac{dC}{dt} = 8 - \ln(t^2 - 2t + 4), \qquad 0 \le t \le 12$$

where dC/dt is measured in milligrams per hour and t is the time in hours. Find the total amount of the drug absorbed into the body during the 12 hours.

48. *Medicine* The concentration M (in grams per liter) of a 6-hour allergy medicine in a body is modeled by

$$M = 12 - 4\ln(t^2 - 4t + 6), \qquad 0 \le t \le 6$$

where t is the time in hours since the allergy medication was taken. Find the average level of concentration in the body over the six-hour period.

49. *Consumer Trends* The rate of change S in the number of subscribers to a newly introduced magazine is modeled by

$$\frac{dS}{dt} = 1000t^2e^{-t}, \qquad 0 \le t \le 6$$

where t is the time in years. Find the total increase in the number of subscribers during the first 6 years.

6.6 IMPROPER INTEGRALS

- Recognize improper integrals.
- Evaluate improper integrals with infinite limits of integration.
- Evaluate improper integrals with infinite integrands.
- Use improper integrals to solve real-life problems.

Improper Integrals

The definition of the definite integral

$$\int_a^b f(x)\,dx$$

includes the requirements that the interval $[a, b]$ be finite and that f be bounded on $[a, b]$. In this section, you will study integrals that do not satisfy these requirements because of one of the conditions below.

1. One or both of the limits of integration are infinite.
2. f has an infinite discontinuity in the interval $[a, b]$.

Integrals having either of these characteristics are called **improper integrals.** For instance, the integrals

$$\int_0^{\infty} e^{-x}\,dx \quad \text{and} \quad \int_{-\infty}^{\infty} \frac{1}{x^2+1}\,dx$$

are improper because one or both limits of integration are infinite, as indicated in Figure 6.19. Similarly, the integrals

$$\int_1^5 \frac{1}{\sqrt{x-1}}\,dx \quad \text{and} \quad \int_{-2}^{2} \frac{1}{(x+1)^2}\,dx$$

are improper because their integrands have an **infinite discontinuity**—that is, they approach infinity somewhere in the interval of integration, as indicated in Figure 6.20.

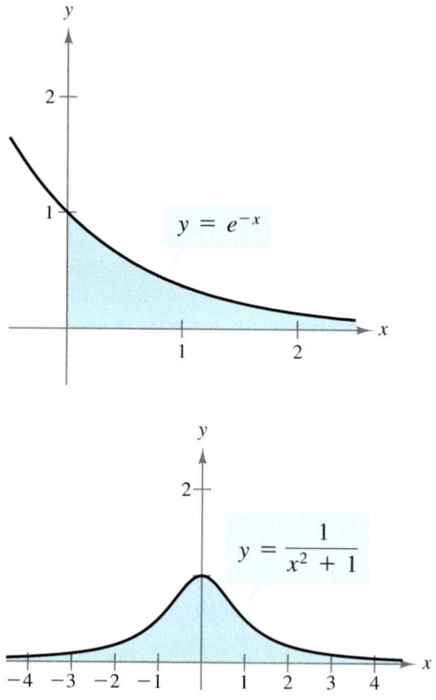

FIGURE 6.19

DISCOVERY

Use a graphing utility to calculate the definite integral $\int_0^b e^{-x}dx$ for $b = 10$ and for $b = 20$. What is the area of the region bounded by the graph of $y = e^{-x}$ and the two coordinate axes?

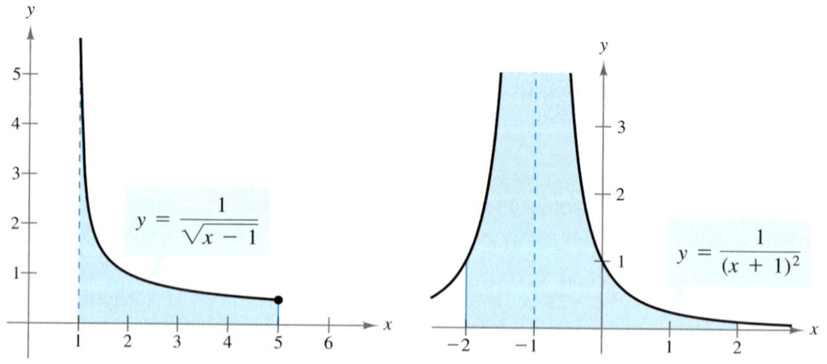

FIGURE 6.20

Integrals with Infinite Limits of Integration

To see how to evaluate an improper integral, consider the integral shown in Figure 6.21. As long as b is a real number that is greater than 1 (no matter how large), this is a definite integral whose value is

$$\int_1^b \frac{1}{x^2}\,dx = \left[-\frac{1}{x}\right]_1^b = -\frac{1}{b} + 1 = 1 - \frac{1}{b}.$$

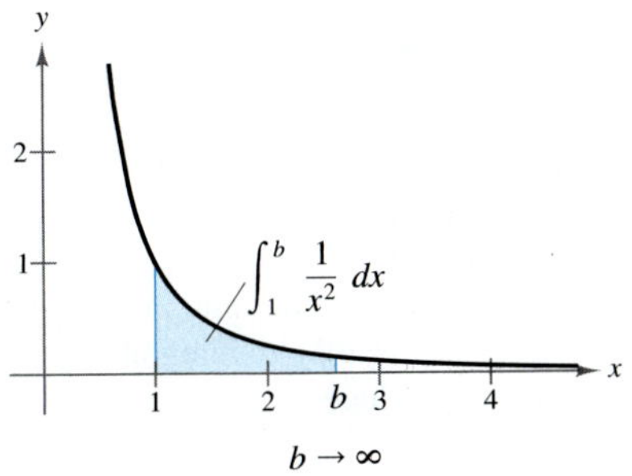

FIGURE 6.21

The table shows the values of this integral for several values of b.

b	2	5	10	100	1000	10,000
$\int_1^b \frac{1}{x^2}\,dx = 1 - \frac{1}{b}$	0.5000	0.8000	0.9000	0.9900	0.9990	0.9999

From this table, it appears that the value of the integral is approaching a limit as b increases without bound. This limit is denoted by the *improper integral* shown below.

$$\int_1^\infty \frac{1}{x^2}\,dx = \lim_{b\to\infty} \int_1^b \frac{1}{x^2}\,dx = \lim_{b\to\infty}\left(1 - \frac{1}{b}\right) = 1$$

Improper Integrals (Infinite Limits of Integration)

1. If f is continuous on the interval $[a, \infty)$, then

$$\int_a^\infty f(x)\,dx = \lim_{b\to\infty} \int_a^b f(x)\,dx.$$

2. If f is continuous on the interval $(-\infty, b]$, then

$$\int_{-\infty}^b f(x)\,dx = \lim_{a\to-\infty} \int_a^b f(x)\,dx.$$

3. If f is continuous on the interval $(-\infty, \infty)$, then

$$\int_{-\infty}^\infty f(x)\,dx = \int_{-\infty}^c f(x)\,dx + \int_c^\infty f(x)\,dx$$

where c is any real number.

In the first two cases, if the limit exists, then the improper integral **converges;** otherwise, the improper integral **diverges.** In the third case, the integral on the left will diverge if either one of the integrals on the right diverges.

TECHNOLOGY

Symbolic integration utilities evaluate improper integrals in much the same way that they evaluate definite integrals. Use a symbolic integration utility to evaluate

$$\int_{-\infty}^{-1} \frac{1}{x^2}\, dx.$$

EXAMPLE 1 Evaluating an Improper Integral

Determine the convergence or divergence of $\displaystyle\int_1^{\infty} \frac{1}{x}\, dx$.

SOLUTION Begin by applying the definition of an improper integral.

$$\int_1^{\infty} \frac{1}{x}\, dx = \lim_{b\to\infty} \int_1^{b} \frac{1}{x}\, dx \qquad \text{Definition of improper integral}$$

$$= \lim_{b\to\infty} \Big[\ln x \Big]_1^b \qquad \text{Find antiderivative.}$$

$$= \lim_{b\to\infty} (\ln b - 0) \qquad \text{Apply Fundamental Theorem.}$$

$$= \infty \qquad \text{Evaluate limit.}$$

Because the limit is infinite, the improper integral diverges.

TRY IT 1

Determine the convergence or divergence of each improper integral.

(a) $\displaystyle\int_1^{\infty} \frac{1}{x^3}\, dx$ (b) $\displaystyle\int_1^{\infty} \frac{1}{\sqrt{x}}\, dx$

As you begin to work with improper integrals, you will find that integrals that appear to be similar can have very different values. For instance, consider the two improper integrals

$$\int_1^{\infty} \frac{1}{x}\, dx = \infty \qquad \text{Divergent integral}$$

and

$$\int_1^{\infty} \frac{1}{x^2}\, dx = 1. \qquad \text{Convergent integral}$$

The first integral diverges and the second converges to 1. Graphically, this means that the areas shown in Figure 6.22 are very different. The region lying between the graph of $y = 1/x$ and the x-axis (for $x \geq 1$) has an *infinite* area, and the region lying between the graph of $y = 1/x^2$ and the x-axis (for $x \geq 1$) has a *finite* area.

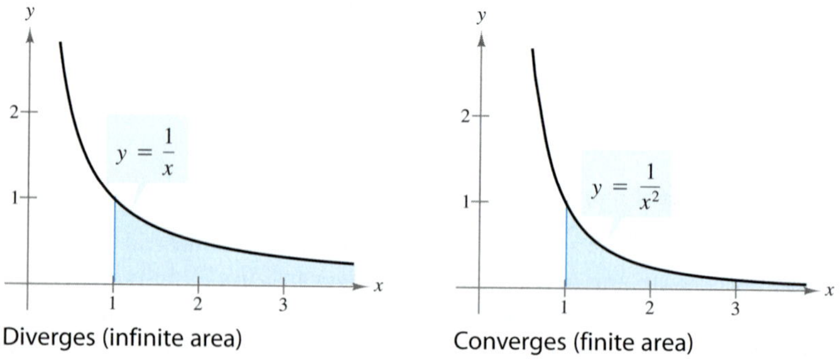

FIGURE 6.22

EXAMPLE 2 Evaluating an Improper Integral

Evaluate the improper integral.

$$\int_{-\infty}^{0} \frac{1}{(1-2x)^{3/2}}\,dx$$

SOLUTION Begin by applying the definition of an improper integral.

$$\begin{aligned}
\int_{-\infty}^{0} \frac{1}{(1-2x)^{3/2}}\,dx &= \lim_{a\to-\infty} \int_{a}^{0} \frac{1}{(1-2x)^{3/2}}\,dx && \text{Definition of improper integral}\\
&= \lim_{a\to-\infty} \left[\frac{1}{\sqrt{1-2x}}\right]_a^0 && \text{Find antiderivative.}\\
&= \lim_{a\to-\infty} \left(1-\frac{1}{\sqrt{1-2a}}\right) && \text{Apply Fundamental Theorem.}\\
&= 1-0 && \text{Evaluate limit.}\\
&= 1 && \text{Simplify.}
\end{aligned}$$

So, the improper integral converges to 1. As shown in Figure 6.23, this implies that the region lying between the graph of $y = 1/(1-2x)^{3/2}$ and the x-axis (for $x \le 0$) has an area of 1 square unit.

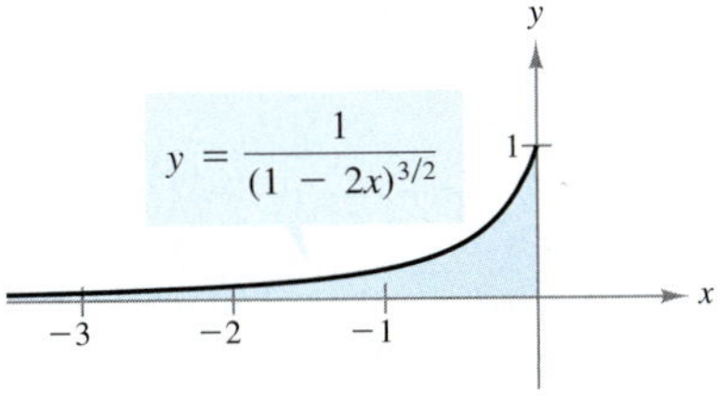

FIGURE 6.23

TRY IT 2

Evaluate the improper integral, if possible.

$$\int_{-\infty}^{0} \frac{1}{(x-1)^2}\,dx$$

EXAMPLE 3 Evaluating an Improper Integral

Evaluate the improper integral.

$$\int_{0}^{\infty} 2xe^{-x^2}\,dx$$

SOLUTION Begin by applying the definition of an improper integral.

$$\begin{aligned}
\int_{0}^{\infty} 2xe^{-x^2}\,dx &= \lim_{b\to\infty} \int_{0}^{b} 2xe^{-x^2}\,dx && \text{Definition of improper integral}\\
&= \lim_{b\to\infty} \left[-e^{-x^2}\right]_0^b && \text{Find antiderivative.}\\
&= \lim_{b\to\infty} \left(-e^{-b^2}+1\right) && \text{Apply Fundamental Theorem.}\\
&= 0+1 && \text{Evaluate limit.}\\
&= 1 && \text{Simplify.}
\end{aligned}$$

So, the improper integral converges to 1. As shown in Figure 6.24, this implies that the region lying between the graph of $y = 2xe^{-x^2}$ and the x-axis (for $x \ge 0$) has an area of 1 square unit.

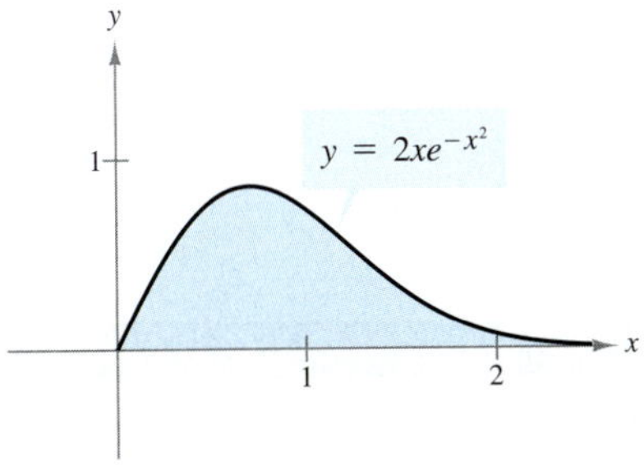

FIGURE 6.24

TRY IT 3

Evaluate the improper integral, if possible.

$$\int_{-\infty}^{0} e^{2x}\,dx$$

Integrals with Infinite Integrands

Improper Integrals (Infinite Integrands)

1. If f is continuous on the interval $[a, b)$ and approaches infinity at b, then

$$\int_a^b f(x)\,dx = \lim_{c\to b^-} \int_a^c f(x)\,dx.$$

2. If f is continuous on the interval $(a, b]$ and approaches infinity at a, then

$$\int_a^b f(x)\,dx = \lim_{c\to a^+} \int_c^b f(x)\,dx.$$

3. If f is continuous on the interval $[a, b]$, except for some c in (a, b) at which f approaches infinity, then

$$\int_a^b f(x)\,dx = \int_a^c f(x)\,dx + \int_c^b f(x)\,dx.$$

In the first two cases, if the limit exists, then the improper integral **converges;** otherwise, the improper integral **diverges.** In the third case, the improper integral on the left diverges if either of the improper integrals on the right diverges.

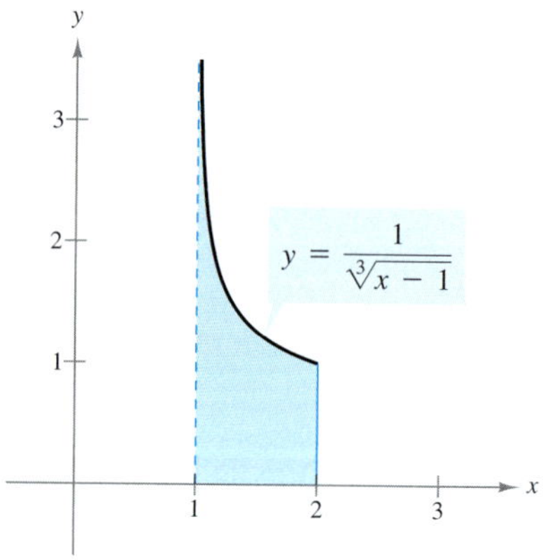

FIGURE 6.25

EXAMPLE 4 **Evaluating an Improper Integral**

Evaluate $\displaystyle\int_1^2 \frac{1}{\sqrt[3]{x-1}}\,dx.$

SOLUTION

$$\int_1^2 \frac{1}{\sqrt[3]{x-1}}\,dx = \lim_{c\to 1^+} \int_c^2 \frac{1}{\sqrt[3]{x-1}}\,dx \qquad \text{Definition of improper integral}$$

$$= \lim_{c\to 1^+} \left[\frac{3}{2}(x-1)^{2/3}\right]_c^2 \qquad \text{Find antiderivative.}$$

$$= \lim_{c\to 1^+} \left[\frac{3}{2} - \frac{3}{2}(c-1)^{2/3}\right] \qquad \text{Apply Fundamental Theorem.}$$

$$= \frac{3}{2} - 0 = \frac{3}{2} \qquad \text{Evaluate limit.}$$

So, the integral converges to $\frac{3}{2}$. This implies that the region shown in Figure 6.25 has an area of $\frac{3}{2}$ square units.

TECHNOLOGY

Use a graphing utility to verify the result of Example 4 by calculating each definite integral.

$$\int_{1.01}^2 \frac{1}{\sqrt[3]{x-1}}\,dx$$

$$\int_{1.001}^2 \frac{1}{\sqrt[3]{x-1}}\,dx$$

$$\int_{1.0001}^2 \frac{1}{\sqrt[3]{x-1}}\,dx$$

TRY IT 4

Evaluate $\displaystyle\int_1^2 \frac{1}{\sqrt{x-1}}\,dx.$

EXAMPLE 5 Evaluating an Improper Integral

Evaluate $\int_1^2 \frac{2}{x^2 - 2x}\, dx.$

SOLUTION

$$\int_1^2 \frac{2}{x^2 - 2x}\, dx = \int_1^2 \left(\frac{1}{x-2} - \frac{1}{x}\right) dx \qquad \text{Use partial fractions.}$$

$$= \lim_{c \to 2^-} \int_1^c \left(\frac{1}{x-2} - \frac{1}{x}\right) dx \qquad \text{Definition of improper integral}$$

$$= \lim_{c \to 2^-} \Big[\ln|x-2| - \ln|x|\Big]_1^c \qquad \text{Find antiderivative.}$$

$$= -\infty \qquad \text{Evaluate limit.}$$

So, you can conclude that the integral diverges. This implies that the region shown in Figure 6.26 has an infinite area.

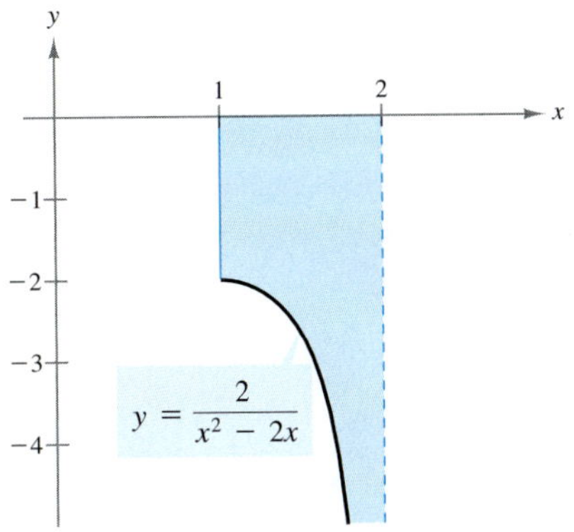

FIGURE 6.26

TRY IT 5

Evaluate $\int_1^3 \frac{3}{x^2 - 3x}\, dx.$

EXAMPLE 6 Evaluating an Improper Integral

Evaluate $\int_{-1}^2 \frac{1}{x^3}\, dx.$

SOLUTION This integral is improper because the integrand has an infinite discontinuity at the interior value $x = 0$, as shown in Figure 6.27. So, you can write

$$\int_{-1}^2 \frac{1}{x^3}\, dx = \int_{-1}^0 \frac{1}{x^3}\, dx + \int_0^2 \frac{1}{x^3}\, dx.$$

By applying the definition of an improper integral, you can show that each of these integrals diverges. So, the original improper integral also diverges.

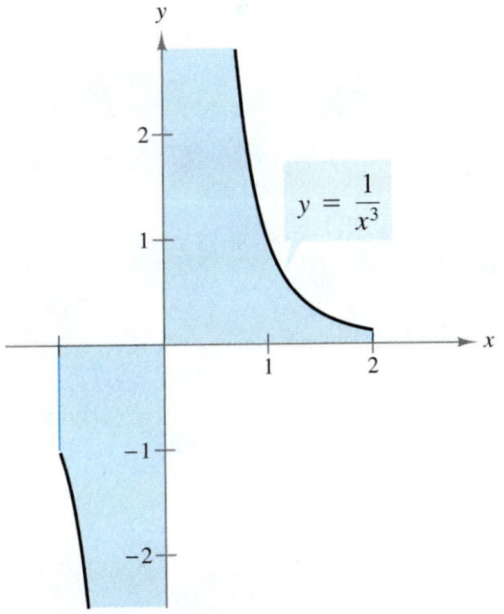

FIGURE 6.27

TRY IT 6

Evaluate $\int_{-1}^1 \frac{1}{x^2}\, dx.$

STUDY TIP

Had you not recognized that the integral in Example 6 was improper, you would have obtained the incorrect result

$$\int_{-1}^2 \frac{1}{x^3}\, dx = \left[-\frac{1}{2x^2}\right]_{-1}^2 = -\frac{1}{8} + \frac{1}{2} = \frac{3}{8}. \qquad \text{Incorrect}$$

Improper integrals in which the integrand has an infinite discontinuity *between* the limits of integration are often overlooked, so keep alert for such possibilities. Even symbolic integrators can have trouble with this type of integral, and can give the same incorrect result.

Applications

In Section 4.3, you studied the graph of the *normal probability density function*

$$f(x) = \frac{1}{\sigma\sqrt{2\pi}} e^{-(x-\mu)^2/2\sigma^2}.$$

This function is used in statistics to represent a population that is normally distributed with a mean of μ and a standard deviation of σ. Specifically, if an outcome x is chosen at random from the population, the probability that x will have a value between a and b is

$$P(a \le x \le b) = \int_a^b \frac{1}{\sigma\sqrt{2\pi}} e^{-(x-\mu)^2/2\sigma^2}\,dx.$$

As shown in Figure 6.28, the probability $P(-\infty < x < \infty)$ is

$$P(-\infty < x < \infty) = \int_{-\infty}^{\infty} \frac{1}{\sigma\sqrt{2\pi}} e^{-(x-\mu)^2/2\sigma^2}\,dx = 1.$$

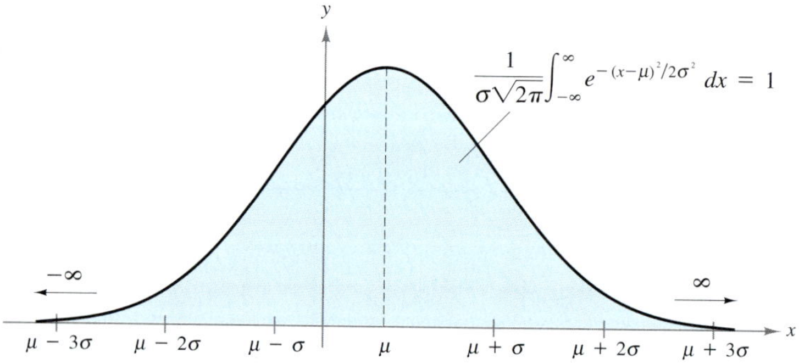

FIGURE 6.28

Many professional basketball players are over $6\frac{1}{2}$ feet tall. If a man is chosen at random from the population, the probability that he is $6\frac{1}{2}$ feet tall or taller is less than half of one percent.

EXAMPLE 7 Finding a Probability

The mean height of American men (between 18 and 24 years old) is 70 inches, and the standard deviation is 3 inches. An 18- to 24-year-old man is chosen at random from the population. What is the probability that he is 6 feet tall or taller? *(Source: U.S. National Center for Health Statistics)*

SOLUTION Using a mean of $\mu = 70$ and a standard deviation of $\sigma = 3$, the probability $P(72 \le x < \infty)$ is given by the improper integral

$$P(72 \le x < \infty) = \int_{72}^{\infty} \frac{1}{3\sqrt{2\pi}} e^{-(x-70)^2/18}\,dx.$$

Using a symbolic integration utility, you can approximate the value of this integral to be 0.252. So, the probability that the man is 6 feet tall or taller is about 25.2%.

TRY IT 7

Use Example 7 to find the probability that an 18- to 24-year-old man chosen at random from the population is 6 feet 6 inches tall or taller.

EXAMPLE 8 Finding Volume

The solid formed by revolving the graph of $f(x) = \frac{1}{x}$ for $1 \le x < \infty$ about the x-axis is called **Gabriel's horn.** (See Figure 6.29.) Find the volume of Gabriel's horn.

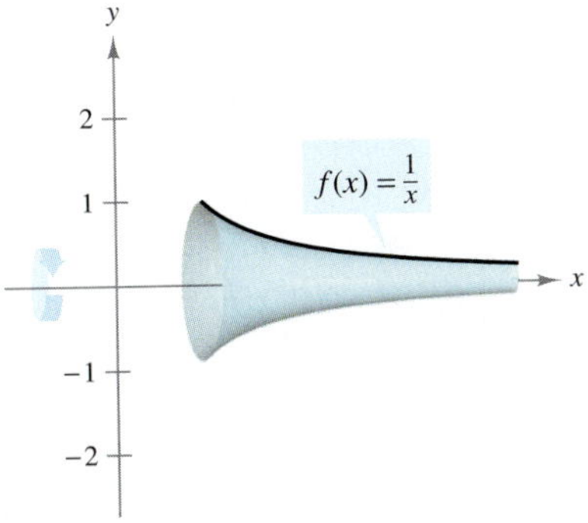

FIGURE 6.29

SOLUTION You can find the volume of the horn using the Disk Method (see Section 5.7).

$$\begin{aligned} \text{Volume} &= \int_1^\infty \pi\left(\frac{1}{x}\right)^2 dx && \text{Disk Method} \\ &= \lim_{b\to\infty} \int_1^b \frac{\pi}{x^2}\, dx && \text{Definition of improper integral} \\ &= \lim_{b\to\infty} \left[-\frac{\pi}{x}\right]_1^b && \text{Find antiderivative.} \\ &= \lim_{b\to\infty} \left(\pi - \frac{\pi}{b}\right) && \text{Apply Fundamental Theorem.} \\ &= \pi && \text{Evaluate limit.} \end{aligned}$$

So, Gabriel's horn has a volume of π cubic units.

TRY IT 8

Use a symbolic integration utility to help you decide which of the improper integrals converge. What can you conclude?

(a) $\displaystyle\int_1^\infty \frac{1}{x^{0.9}}\, dx$

(b) $\displaystyle\int_1^\infty \frac{1}{x^{1.0}}\, dx$

(c) $\displaystyle\int_1^\infty \frac{1}{x^{1.1}}\, dx$

TAKE ANOTHER LOOK

Gabriel's Horn

Example 8 shows that the volume of Gabriel's horn is finite. What is the *area* under the curve $f(x) = 1/x$ and above the x-axis for $x \ge 1$? It can be shown that the surface area of Gabriel's horn is given by the integral

$$2\pi \int_1^\infty \frac{1}{x}\sqrt{1 + \frac{1}{x^4}}\, dx.$$

Use a symbolic integration utility to show that this improper integral diverges. So, Gabriel's horn has finite volume but infinite surface area! How could you paint Gabriel's horn?

PREREQUISITE REVIEW 6.6

The following warm-up exercises involve skills that were covered in earlier sections. You will use these skills in the exercise set for this section.

In Exercises 1–6, find the limit.

1. $\lim_{x\to 2} (2x + 5)$

2. $\lim_{x\to 1} \left(\frac{1}{x} + 2x^2\right)$

3. $\lim_{x\to -4} \frac{x + 4}{x^2 - 16}$

4. $\lim_{x\to 0} \frac{x^2 - 2x}{x^3 + 3x^2}$

5. $\lim_{x\to 1} \frac{1}{\sqrt{x - 1}}$

6. $\lim_{x\to -3} \frac{x^2 + 2x - 3}{x + 3}$

In Exercises 7–10, evaluate the expression (a) when $x = b$ and (b) when $x = 0$.

7. $\frac{4}{3}(2x - 1)^3$

8. $\frac{1}{x - 5} + \frac{3}{(x - 2)^2}$

9. $\ln(5 - 3x^2) - \ln(x + 1)$

10. $e^{3x^2} + e^{-3x^2}$

EXERCISES 6.6

In Exercises 1–14, determine whether or not the improper integral converges. If it does, evaluate the integral.

1. $\int_0^{\infty} e^{-x}\,dx$

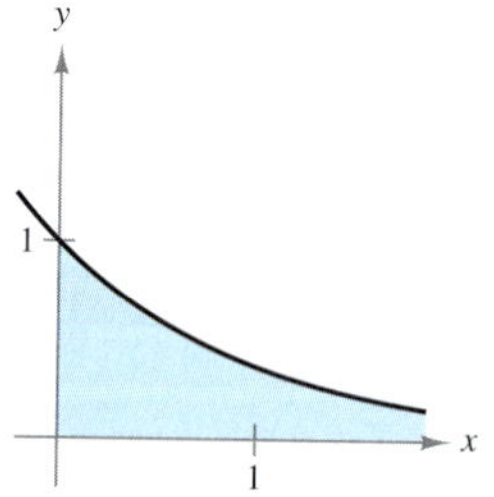

2. $\int_{-\infty}^{0} e^{2x}\,dx$

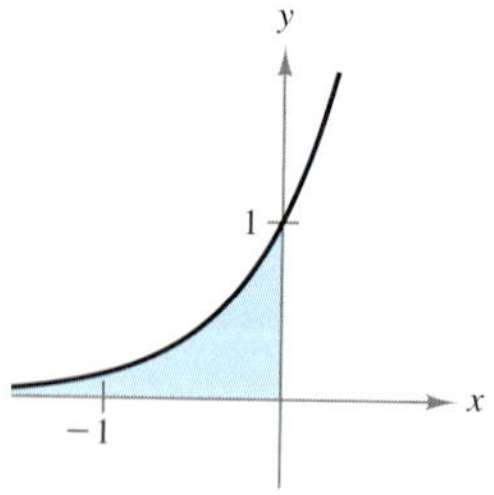

3. $\int_1^{\infty} \frac{1}{x^2}\,dx$

4. $\int_1^{\infty} \frac{1}{\sqrt{x}}\,dx$

5. $\int_0^{\infty} e^{x/3}\,dx$

6. $\int_0^{\infty} \frac{5}{e^{2x}}\,dx$

7. $\int_5^{\infty} \frac{x}{\sqrt{x^2 - 16}}\,dx$

8. $\int_{1/2}^{\infty} \frac{1}{\sqrt{2x - 1}}\,dx$

9. $\int_{-\infty}^{0} e^{-x}\,dx$

10. $\int_{-\infty}^{-1} \frac{1}{x^2}\,dx$

11. $\int_1^{\infty} \frac{e^{\sqrt{x}}}{\sqrt{x}}\,dx$

12. $\int_{-\infty}^{0} \frac{x}{x^2 + 1}\,dx$

13. $\int_{-\infty}^{\infty} 2xe^{-3x^2}\,dx$

14. $\int_{-\infty}^{\infty} x^2e^{-x^3}\,dx$

In Exercises 15–18, determine the divergence or convergence of the improper integral. Evaluate the integral if it converges.

15. $\int_0^4 \frac{1}{\sqrt{x}}\,dx$

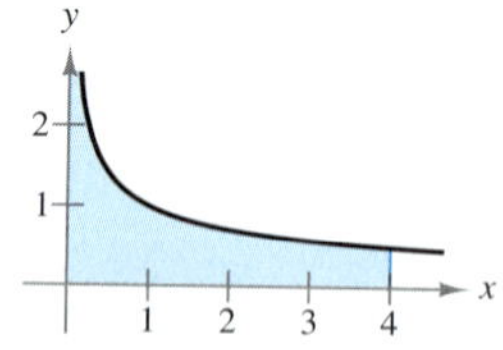

16. $\int_3^4 \frac{1}{\sqrt{x - 3}}\,dx$

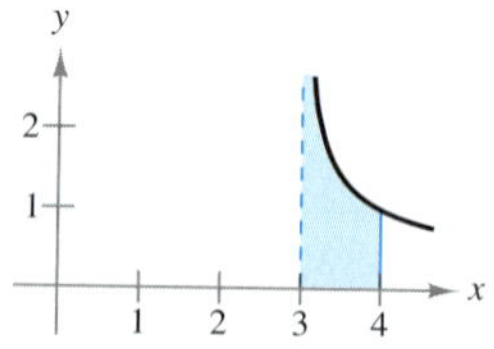

17. $\int_0^2 \frac{1}{(x-1)^{2/3}}\,dx$

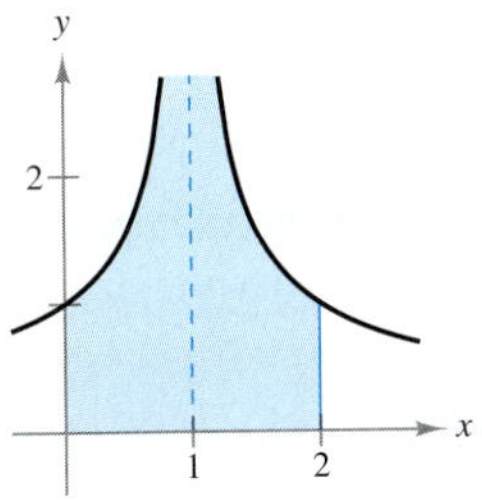

18. $\int_0^2 \frac{1}{(x-1)^2}\,dx$

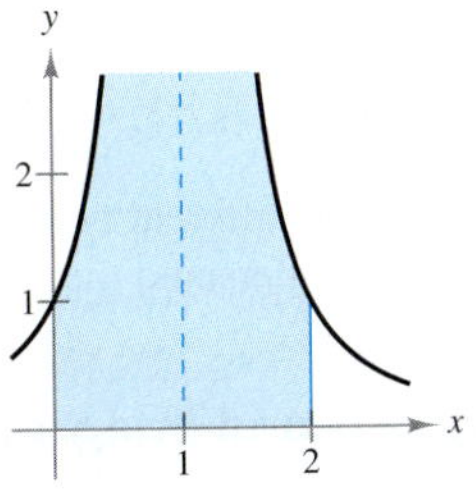

In Exercises 19–28, evaluate the improper integral.

19. $\int_0^1 \frac{1}{1-x}\,dx$

20. $\int_0^{27} \frac{5}{\sqrt[3]{x}}\,dx$

21. $\int_0^9 \frac{1}{\sqrt{9-x}}\,dx$

22. $\int_0^2 \frac{x}{\sqrt{4-x^2}}\,dx$

23. $\int_0^1 \frac{1}{x^2}\,dx$

24. $\int_0^1 \frac{1}{x}\,dx$

25. $\int_0^2 \frac{1}{\sqrt[3]{x-1}}\,dx$

26. $\int_0^2 \frac{1}{(x-1)^{4/3}}\,dx$

27. $\int_3^4 \frac{1}{\sqrt{x^2-9}}\,dx$

28. $\int_3^5 \frac{1}{x^2\sqrt{x^2-9}}\,dx$

In Exercises 29 and 30, (a) find the area of the region bounded by the graphs of the given equations and (b) find the volume of the solid generated by revolving the region about the x-axis.

29. $y = \frac{1}{x^2}, y = 0, x \geq 1$

30. $y = e^{-x}, y = 0, x \geq 0$

In Exercises 31–34, complete the table for the specified values of a and n to demonstrate that

$$\lim_{x\to\infty} x^n e^{-ax} = 0, \qquad a > 0, n > 0.$$

x	1	10	25	50
$x^n e^{-ax}$				

31. $a = 1, n = 1$

32. $a = 2, n = 4$

33. $a = \frac{1}{2}, n = 2$

34. $a = \frac{1}{2}, n = 5$

In Exercises 35–38, use the results of Exercises 31–34 to evaluate the improper integral.

35. $\int_0^\infty x^2 e^{-x}\,dx$

36. $\int_0^\infty (x-1)e^{-x}\,dx$

37. $\int_0^\infty xe^{-2x}\,dx$

38. $\int_0^\infty xe^{-x}\,dx$

39. ***Present Value*** A business is expected to yield a continuous flow of profit at the rate of \$500,000 per year. If money will earn interest at the nominal rate of 9% per year compounded continuously, what is the present value of the business (a) for 20 years and (b) forever? (Present value is defined in Section 6.2.)

40. ***Present Value*** Repeat Exercise 39 for a farm that is expected to produce a profit of \$75,000 per year. Assume that money will earn interest at the nominal rate of 8% compounded continuously. (Present value is defined in Section 6.2.)

Capitalized Cost In Exercises 41 and 42, find the capitalized cost C of an asset (a) for $n = 5$ years, (b) for $n = 10$ years, and (c) forever. The capitalized cost is given by

$$C = C_0 + \int_0^n c(t)e^{-rt}dt$$

where C_0 is the original investment, t is the time in years, r is the annual interest rate compounded continuously, and $c(t)$ is the annual cost of maintenance. [*Hint:* For part (c), see Exercises 31–34.]

41. $C_0 = \$650{,}000,\ c(t) = 25{,}000,\ r = 10\%$

42. $C_0 = \$650{,}000,\ c(t) = \$25{,}000(1 + 0.08t),\ r = 12\%$

43. ***Women's Height*** The mean height of American women between the ages of 25 and 34 is 64.5 inches, and the standard deviation is 2.4 inches. Find the probability that a 25- to 34-year-old woman chosen at random is

(a) between 5 and 6 feet tall.

(b) 5 feet 8 inches or taller.

(c) 6 feet or taller.

(Source: U.S. National Center for Health Statistics)

44. ***Quality Control*** A company manufactures wooden yardsticks. The lengths of the yardsticks are normally distributed with a mean of 36 inches and a standard deviation of 0.2 inch. Find the probability that a yardstick is

(a) longer than 35.5 inches.

(b) longer than 35.9 inches.

ALGEBRA REVIEW

Algebra and Integration Techniques

Integration techniques involve many different algebraic skills. Study the examples in this Algebra Review. Be sure that you understand the algebra used in each step.

EXAMPLE 1 **Algebra and Integration Techniques**

Perform each operation and simplify.

(a) $\frac{28}{9}(1 - u^3)(u)(-3u^2)$ (b) $\frac{2}{x-3} - \frac{1}{x+2}$ (c) $\frac{6}{x} - \frac{1}{x+1} + \frac{9}{(x+1)^2}$

SOLUTION

(a) $\frac{28}{9}(1 - u^3)(u)(-3u^2)$ — Example 6, page 393

$= \frac{(3)(28)}{9}(1 - u^3)(-u^3)$ — Regroup factors.

$= \frac{28}{3}(1 - u^3)(-u^3)$ — Simplify fraction.

$= \frac{28}{3}(u^6 - u^3)$ — Multiply.

(b) $\frac{2}{x-3} - \frac{1}{x+2}$ — Example 1, page 407

$= \frac{2(x+2)}{(x-3)(x+2)} - \frac{(x-3)}{(x-3)(x+2)}$ — Rewrite with common denominator.

$= \frac{2(x+2) - (x-3)}{(x-3)(x+2)}$ — Rewrite as single fraction.

$= \frac{2x + 4 - x + 3}{x^2 - x - 6}$ — Multiply factors.

$= \frac{x + 7}{x^2 - x - 6}$ — Combine like terms.

(c) $\frac{6}{x} - \frac{1}{x+1} + \frac{9}{(x+1)^2}$ — Example 2, page 408

$= \frac{6(x+1)^2}{x(x+1)^2} - \frac{x(x+1)}{x(x+1)^2} + \frac{9x}{x(x+1)^2}$ — Rewrite with common denominator.

$= \frac{6(x+1)^2 - x(x+1) + 9x}{x(x+1)^2}$ — Rewrite as single fraction.

$= \frac{6x^2 + 12x + 6 - x^2 - x + 9x}{x^3 + 2x^2 + x}$ — Multiply factors.

$= \frac{5x^2 + 20x + 6}{x^3 + 2x^2 + x}$ — Combine like terms.

EXAMPLE 2 Algebra and Integration Techniques

Perform each operation and simplify.

(a) $x + 1 + \dfrac{1}{x^3} + \dfrac{1}{x - 1}$

(b) $x^2e^x - 2(x - 1)e^x$

(c) $6 \ln|x| - \ln|x + 1| + 9\dfrac{(x + 1)^{-1}}{-1}$

SOLUTION

(a) $x + 1 + \dfrac{1}{x^3} + \dfrac{1}{x - 1}$ — Example 3, page 409

$= \dfrac{(x + 1)(x^3)(x - 1)}{x^3(x - 1)} + \dfrac{x - 1}{x^3(x - 1)} + \dfrac{x^3}{x^3(x - 1)}$ — Rewrite with common denominator.

$= \dfrac{(x + 1)(x^3)(x - 1) + (x - 1) + x^3}{x^3(x - 1)}$ — Rewrite as single fraction.

$= \dfrac{(x^2 - 1)(x^3) + x - 1 + x^3}{x^3(x - 1)}$ — $(x + 1)(x - 1) = x^2 - 1$

$= \dfrac{x^5 - x^3 + x - 1 + x^3}{x^4 - x^3}$ — Multiply factors.

$= \dfrac{x^5 + x - 1}{x^4 - x^3}$ — Combine like terms.

(b) $x^2e^x - 2(x - 1)e^x$ — Example 5, page 422

$= x^2e^x - 2(xe^x - e^x)$ — Multiply factors.

$= x^2e^x - 2xe^x + 2e^x$ — Multiply factors.

$= e^x(x^2 - 2x + 2)$ — Factor.

(c) $6 \ln|x| - \ln|x + 1| + 9\dfrac{(x + 1)^{-1}}{-1}$ — Example 2, page 408

$= \ln|x|^6 - \ln|x + 1| + 9\dfrac{(x + 1)^{-1}}{-1}$ — $m \ln n = \ln n^m$

$= \ln|x^6| - \ln|x + 1| + 9\dfrac{(x + 1)^{-1}}{-1}$ — Property of absolute value

$= \ln\dfrac{|x^6|}{|x + 1|} + 9\dfrac{(x + 1)^{-1}}{-1}$ — $\ln m - \ln n = \ln\dfrac{m}{n}$

$= \ln\left|\dfrac{x^6}{x + 1}\right| + 9\dfrac{(x + 1)^{-1}}{-1}$ — $\dfrac{|a|}{|b|} = \left|\dfrac{a}{b}\right|$

$= \ln\left|\dfrac{x^6}{x + 1}\right| - 9(x + 1)^{-1}$ — Rewrite sum as difference.

$= \ln\left|\dfrac{x^6}{x + 1}\right| - \dfrac{9}{x + 1}$ — Rewrite with positive exponent.

6 CHAPTER SUMMARY AND STUDY STRATEGIES

*After studying this chapter, you should have acquired the following skills. The exercise numbers are keyed to the Review Exercises that begin on page 450. Answers to odd-numbered Review Exercises are given in the back of the text.**

- Use the basic integration formulas to find indefinite integrals. *(Section 6.1)* — Review Exercises 1–12

  Constant Rule: $\int k\,dx = kx + C$

  Power Rules: $\int x^n\,dx = \frac{x^{n+1}}{n+1} + C, \quad \int u^n \frac{du}{dx}\,dx = \frac{u^{n+1}}{n+1} + C, \quad n \neq -1$

  Exponential Rules: $\int e^x\,dx = e^x + C, \quad \int e^u \frac{du}{dx}\,dx = \int e^u\,du = e^u + C$

  Log Rules: $\int \frac{1}{x}\,dx = \ln|x| + C, \quad \int \frac{du/dx}{u}\,dx = \int \frac{1}{u}\,du = \ln|u| + C$

- Use substitution to find indefinite integrals. *(Section 6.1)* — Review Exercises 13–20
- Use substitution to evaluate definite integrals. *(Section 6.1)* — Review Exercises 21–24
- Use integration to solve real-life problems. *(Section 6.1)* — Review Exercises 25–28
- Use integration by parts to find indefinite integrals. *(Section 6.2)* — Review Exercises 29–32

  $\int u\,dv = uv - \int v\,du$

- Use integration by parts repeatedly to find indefinite integrals. *(Section 6.2)* — Review Exercises 33, 34
- Find the present value of future income. *(Section 6.2)* — Review Exercises 35–42
- Use partial fractions to find indefinite integrals. *(Section 6.3)* — Review Exercises 43–48
- Use logistic growth functions to model real-life situations. *(Section 6.3)* — Review Exercises 49, 50
- Use integration tables to find indefinite and definite integrals. *(Section 6.4)* — Review Exercises 51–56
- Use reduction formulas to find indefinite integrals. *(Section 6.4)* — Review Exercises 57–60
- Use completing the square to find indefinite integrals. *(Section 6.4)* — Review Exercises 61–64
- Use the Trapezoidal Rule to approximate definite integrals. *(Section 6.5)* — Review Exercises 65–68

  $$\int_a^b f(x)\,dx \approx \left(\frac{b-a}{2n}\right)[f(x_0) + 2f(x_1) + \cdots + 2f(x_{n-1}) + f(x_n)]$$

- Use Simpson's Rule to approximate definite integrals. *(Section 6.5)* — Review Exercises 69–72

  $$\int_a^b f(x)\,dx \approx \left(\frac{b-a}{3n}\right)[f(x_0) + 4f(x_1) + 2f(x_2) + 4f(x_3) + \cdots + 4f(x_{n-1}) + f(x_n)]$$

- Analyze the sizes of the errors when approximating definite integrals with the Trapezoidal Rule. *(Section 6.5)* — Review Exercises 73, 74

  $$|E| \le \frac{(b-a)^3}{12n^2}[\max|f''(x)|], \quad a \le x \le b$$

- Analyze the sizes of the errors when approximating definite integrals with Simpson's Rule. *(Section 6.5)* — Review Exercises 75, 76

  $$|E| \le \frac{(b-a)^5}{180n^4}[\max|f^{(4)}(x)|], \quad a \le x \le b$$

* Use a wide range of valuable study aids to help you master the material in this chapter. The *Student Solutions Guide* includes step-by-step solutions to all odd-numbered exercises to help you review and prepare. The *HM mathSpace® Student CD-ROM* helps you brush up on your algebra skills. The *Graphing Technology Guide*, available on the Web at *math.college.hmco.com/students*, offers step-by-step commands and instructions for a wide variety of graphing calculators, including the most recent models.

- Evaluate improper integrals with infinite limits of integration. *(Section 6.6)* — *Review Exercises 77–80*

$$\int_a^{\infty} f(x)\,dx = \lim_{b\to\infty}\int_a^b f(x)\,dx, \qquad \int_{-\infty}^b f(x)\,dx = \lim_{a\to-\infty}\int_a^b f(x)\,dx,$$

$$\int_{-\infty}^{\infty} f(x)\,dx = \int_{-\infty}^{c} f(x)\,dx + \int_c^{\infty} f(x)\,dx$$

- Evaluate improper integrals with infinite integrands. *(Section 6.6)* — *Review Exercises 81–84*

$$\int_a^b f(x)\,dx = \lim_{c\to b^-}\int_a^c f(x)\,dx, \qquad \int_a^b f(x)\,dx = \lim_{c\to a^+}\int_c^b f(x)\,dx,$$

$$\int_a^b f(x)\,dx = \int_a^c f(x)\,dx + \int_c^b f(x)\,dx$$

- Use improper integrals to solve real-life problems. *(Section 6.6)* — *Review Exercises 85–87*
- ***Use a Variety of Approaches*** To be efficient at finding antiderivatives, you need to use a variety of approaches.
  1. Check to see whether the integral fits one of the basic integration formulas—you should have these formulas memorized.
  2. Try an integration technique such as substitution, integration by parts, partial fractions, or completing the square to rewrite the integral in a form that fits one of the basic integration formulas.
  3. Use a table of integrals.
  4. Use a symbolic integration utility.
- ***Use Numerical Integration*** When solving a definite integral, remember that you cannot apply the Fundamental Theorem of Calculus unless you can find an antiderivative of the integrand. This is not always possible—even with a symbolic integration utility. In such cases, you can use a numerical technique such as the Midpoint Rule, the Trapezoidal Rule, or Simpson's Rule to approximate the value of the integral.
- ***Improper Integrals*** When solving integration problems, remember that the symbols used to denote definite integrals are the same as those used to denote improper integrals. Evaluating an improper integral as a definite integral can lead to an incorrect value. For instance, if you evaluated the integral

$$\int_{-2}^{1} \frac{1}{x^2}\,dx$$

as though it were a definite integral, you would obtain a value of $-\frac{3}{2}$. This is not, however, correct. This integral is actually a divergent improper integral. If you have access to a symbolic integration utility, try using it to evaluate this integral—it will probably make the same mistake.

Study Tools *Additional resources that accompany this chapter*

- **Algebra Review** (pages 446 and 447)
- **Chapter Summary and Study Strategies** (pages 448 and 449)
- **Review Exercises** (pages 450–453)
- **Sample Post-Graduation Exam Questions** (page 454)
- **Web Exercise** (page 415, Exercise 61)
- **Student Solutions Guide**
- **HM mathSpace® Student CD-ROM**
- **Graphing Technology Guide** *(math.college.hmco.com/students)*

6 CHAPTER REVIEW EXERCISES

In Exercises 1–12, use a basic integration formula to find the indefinite integral.

1. $\int dt$

2. $\int (x^2 + 2x - 1)\, dx$

3. $\int (x + 5)^3\, dx$

4. $\int \frac{2}{(x-1)^2}\, dx$

5. $\int e^{10x}\, dx$

6. $\int 3xe^{-x^2}\, dx$

7. $\int \frac{1}{5x}\, dx$

8. $\int \frac{2x^3 - x}{x^4 - x^2 + 1}\, dx$

9. $\int x\sqrt{x^2 + 4}\, dx$

10. $\int \frac{1}{\sqrt{2x - 9}}\, dx$

11. $\int \frac{2e^x}{3 + e^x}\, dx$

12. $\int (x^2 - 1)e^{x^3 - 3x}\, dx$

In Exercises 13–20, use substitution to find the indefinite integral.

13. $\int x(x - 2)^3\, dx$

14. $\int x(1 - x)^2\, dx$

15. $\int x\sqrt{x + 1}\, dx$

16. $\int x^2\sqrt{x + 1}\, dx$

17. $\int 2x\sqrt{x - 3}\, dx$

18. $\int \frac{\sqrt{x}}{1 + \sqrt{x}}\, dx$

19. $\int (x + 1)\sqrt{1 - x}\, dx$

20. $\int \frac{x}{x - 1}\, dx$

In Exercises 21–24, use substitution to evaluate the definite integral. Use a symbolic integration utility to verify your answer.

21. $\int_2^3 x\sqrt{x - 2}\, dx$

22. $\int_2^3 x^2\sqrt{x - 2}\, dx$

23. $\int_1^3 x^2(x - 1)^3\, dx$

24. $\int_{-3}^0 x(x + 3)^4\, dx$

25. *Probability* The probability of recall in an experiment is found to be

$$P(a \le x \le b) = \int_a^b \frac{105}{16} x^2\sqrt{1 - x}\, dx$$

where x represents the percent of recall (see figure).

(a) Find the probability that a randomly chosen individual will recall 80% of the material.

(b) What is the median percent recall? That is, for what value of b is it true that $P(0 \le x \le b) = 0.5$?

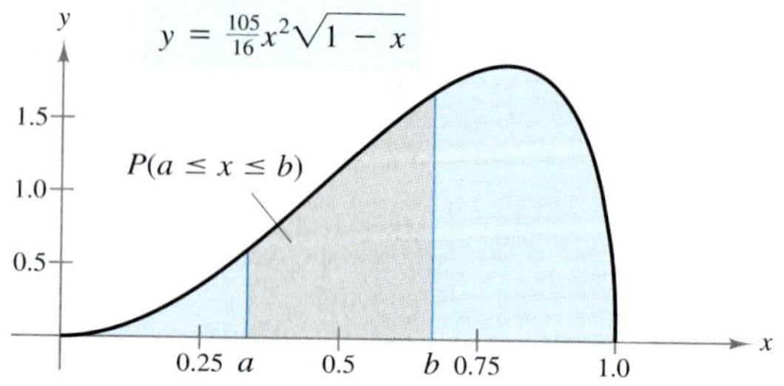

26. *Probability* The probability of locating between a and b percent of oil and gas deposits in a region is

$$P(a \le x \le b) = \int_a^b \frac{140}{27} x^2(1 - x)^{1/3}\, dx$$

(see figure).

(a) Find the probability that between 40% and 60% of the deposits will be found.

(b) Find the probability that between 0% and 50% of the deposits will be found.

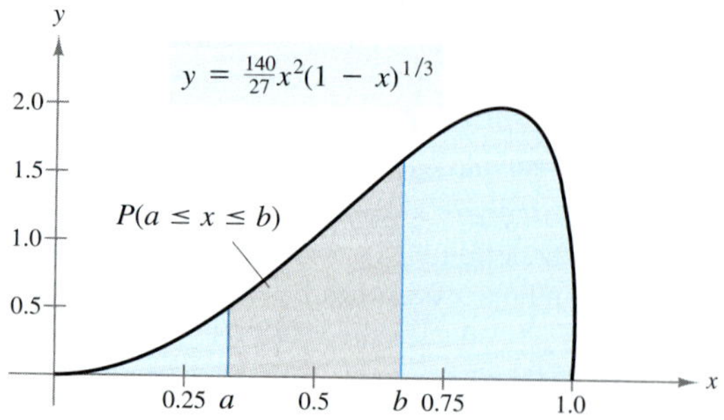

27. *Profit* The net profits P (in millions of dollars per year) for Home Depot from 1997 through 2003 can be modeled by

$$P = -925.08 + 154.753t \ln t, \qquad 7 \le t \le 13$$

where t is the time in years, with $t = 7$ corresponding to 1997. *(Source: The Home Depot, Inc.)*

(a) Find the average net profit for the years 1997 through 2003.

(b) Find the total net profit for the years 1997 through 2003.

28. *Medicine* The effectiveness of a pain-killing drug t hours after ingestion can be modeled by

$$E(t) = te^{-0.4t}$$

where the effectiveness E is measured as a percent.

(a) Use a graphing utility to graph the model over the interval $0 \le t \le 24$.

(b) Over what intervals is the effectiveness increasing and decreasing?

(c) At what time does the medication reach maximum effectiveness?

(d) Determine the average effectiveness of the medication over the 24-hour period.

(e) At what time does the medication have the same effectiveness as the average effectiveness for the entire interval?

In Exercises 29–32, use integration by parts to find the indefinite integral.

29. $\displaystyle\int \frac{\ln x}{\sqrt{x}}\,dx$

30. $\displaystyle\int \sqrt{x}\ln x\,dx$

31. $\displaystyle\int (x-1)e^x\,dx$

32. $\displaystyle\int \ln\left(\frac{x}{x+1}\right)dx$

In Exercises 33 and 34, use integration by parts repeatedly to find the indefinite integral. Use a symbolic integration utility to verify your answer.

33. $\displaystyle\int 2x^2e^{2x}\,dx$

34. $\displaystyle\int (\ln x)^3\,dx$

Present Value In Exercises 35–38, find the present value of the income given by $c(t)$ (measured in dollars) over t_1 years at the given annual inflation rate r.

35. $c(t) = 10{,}000,\ r = 4\%,\ t_1 = 5$ years

36. $c(t) = 20{,}000 + 1500t,\ r = 6\%,\ t_1 = 10$ years

37. $c(t) = 12{,}000t,\ r = 5\%,\ t_1 = 10$ years

38. $c(t) = 10{,}000 + 100e^{t/2},\ r = 5\%,\ t_1 = 5$ years

39. *Economics: Present Value* Calculate the present values of each scenario.

(a) \$1000 per year for 5 years at interest rates of 5%, 10%, and 15%

(b) A lottery ticket that pays \$100,000 per year after taxes over 20 years, assuming an inflation rate of 8%

(Source: Adapted from Boyes/Melvin, Economics, *Third Edition)*

40. *Finance: Present Value* You receive \$1000 at the end of each year for the next 3 years to help with college expenses. Assuming an annual interest rate of 6%, what is the present value of that stream of payments? *(Source: Adapted from Garman/Forgue,* Personal Finance, *Fifth Edition)*

41. *Finance: Present Value* Determine the amount a person planning for retirement would need to deposit today to be able to withdraw \$6000 each year for the next 10 years from an account earning 6% interest. *(Source: Adapted from Garman/Forgue,* Personal Finance, *Fifth Edition)*

42. *Finance: Present Value* A person invests \$50,000 earning 6% interest. If \$6000 is withdrawn each year, use present value to determine how many years it will take for the fund to run out. *(Source: Adapted from Garman/Forgue,* Personal Finance, *Fifth Edition)*

In Exercises 43–48, use partial fractions to find the indefinite integral.

43. $\displaystyle\int \frac{1}{x(x+5)}\,dx$

44. $\displaystyle\int \frac{4x-2}{3(x-1)^2}\,dx$

45. $\displaystyle\int \frac{4x-13}{x^2-3x-10}\,dx$

46. $\displaystyle\int \frac{4x^2-x-5}{x^2(x+5)}\,dx$

47. $\displaystyle\int \frac{x^2}{x^2+2x-15}\,dx$

48. $\displaystyle\int \frac{x^2+2x-12}{x(x+3)}\,dx$

49. *Sales* A new product initially sells 1250 units per week. After 24 weeks, the number of sales increases to 6500. The sales can be modeled by logistic growth with a limit of 10,000 units per week.

(a) Find a logistic growth model for the number of units.

(b) Use the model to complete the table.

Time, t	0	3	6	12	24
Sales, y					

(c) Use the graph shown below to approximate the time t when sales will be 7500.

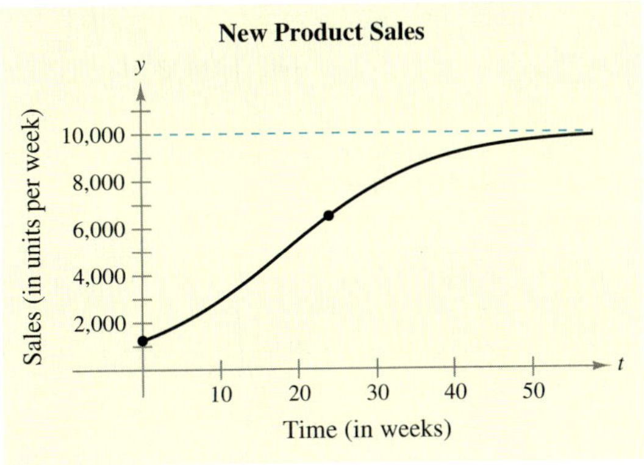

50. *Biology* A conservation society has introduced a population of 300 ring-necked pheasants into a new area. After 5 years, the population has increased to 966. The population can be modeled by logistic growth with a limit of 2700 pheasants.

(a) Find a logistic growth model for the population of ring-necked pheasants.

(b) How many pheasants were present after 4 years?

(c) How long will it take to establish a population of 1750 pheasants?

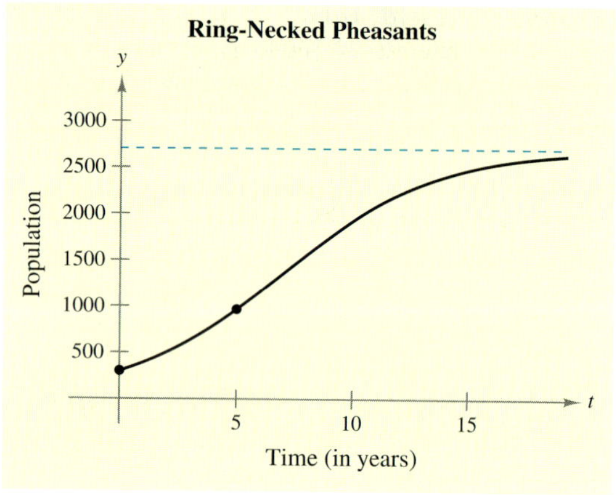

In Exercises 51–56, use the table of integrals in Section 6.4 to evaluate the integral.

51. $\displaystyle\int \frac{\sqrt{x^2+25}}{x}\,dx$

52. $\displaystyle\int \frac{1}{x(4+3x)}\,dx$

53. $\displaystyle\int \frac{1}{x^2-4}\,dx$

54. $\displaystyle\int x(\ln x^2)^2\,dx$

55. $\displaystyle\int_0^3 \frac{x}{\sqrt{1+x}}\,dx$

56. $\displaystyle\int_1^3 \frac{1}{x^2\sqrt{16-x^2}}\,dx$

In Exercises 57–60, use a reduction formula from the table of integrals in Section 6.4 to find the indefinite integral.

57. $\displaystyle\int \frac{\sqrt{1+x}}{x}\,dx$

58. $\displaystyle\int \frac{1}{(x^2-9)^2}\,dx$

59. $\displaystyle\int (x-5)^3e^{x-5}\,dx$

60. $\displaystyle\int (\ln x)^4\,dx$

In Exercises 61–64, complete the square and then use the table of integrals in Section 6.4 to find the indefinite integral.

61. $\displaystyle\int \frac{1}{x^2+4x-21}\,dx$

62. $\displaystyle\int \frac{1}{x^2-8x-52}\,dx$

63. $\displaystyle\int \sqrt{x^2-10x}\,dx$

64. $\displaystyle\int \frac{x}{\sqrt{x^4+6x^2+10}}\,dx$

In Exercises 65–68, use the Trapezoidal Rule to approximate the definite integral.

65. $\displaystyle\int_1^3 \frac{1}{x^2}\,dx,\ n=4$

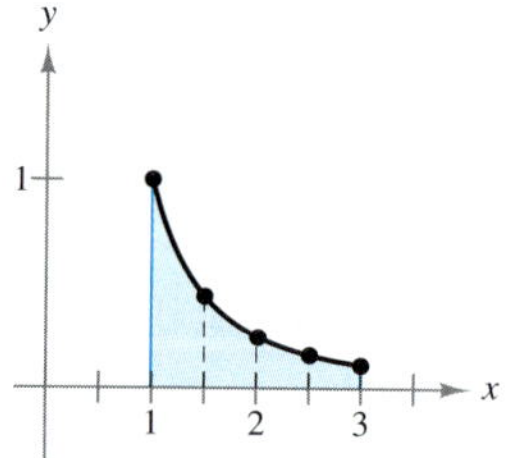

66. $\displaystyle\int_0^2 (x^2+1)\,dx,\ n=4$

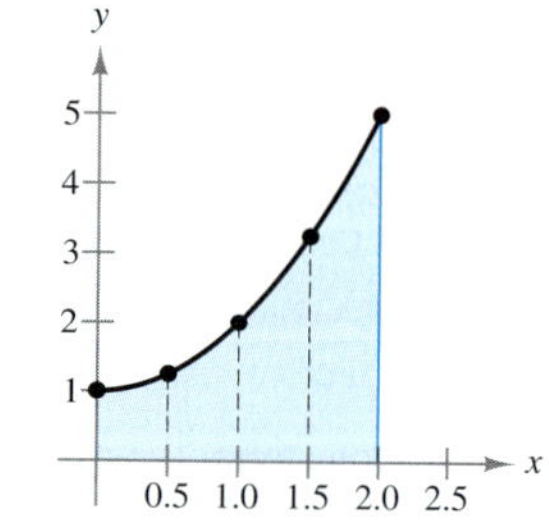

67. $\displaystyle\int_1^2 \frac{1}{1+\ln x}\,dx,\ n=4$

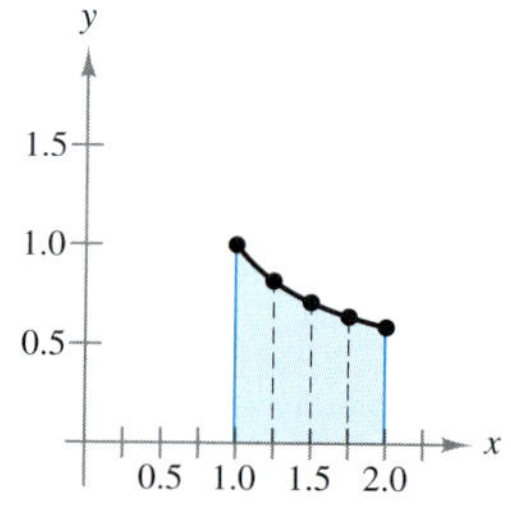

68. $\displaystyle\int_0^2 \frac{1}{\sqrt{1+x^3}}\,dx,\ n=8$

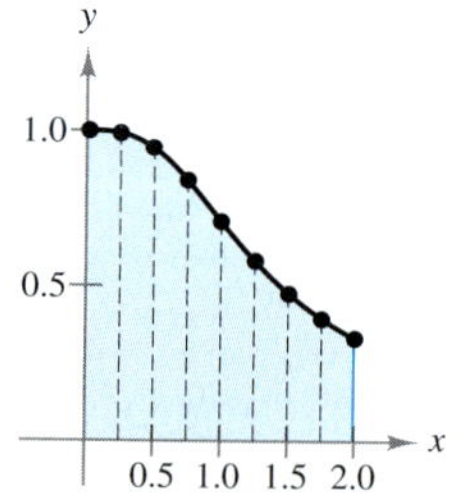

In Exercises 69–72, use Simpson's Rule to approximate the definite integral.

69. $\displaystyle\int_1^2 \frac{1}{x^3}\,dx,\ n=4$

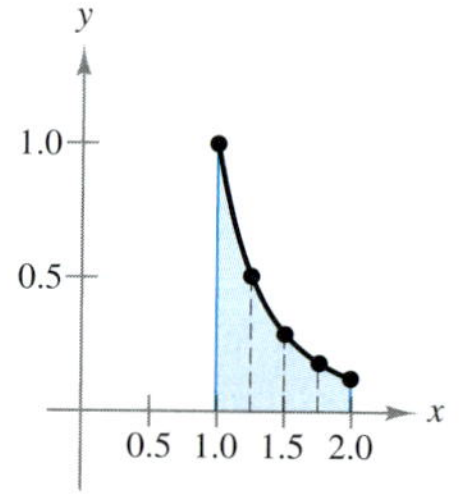

70. $\displaystyle\int_1^2 x^3\,dx,\ n=4$

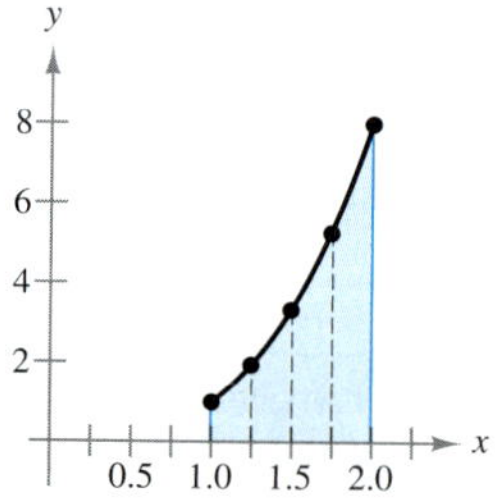

71. $\displaystyle\int_0^1 \frac{x^{3/2}}{2-x^2}\,dx,\ n=4$

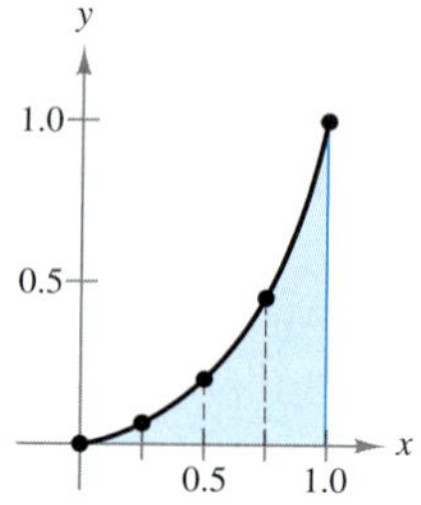

72. $\displaystyle\int_0^1 e^{x^2}\,dx,\ n=6$

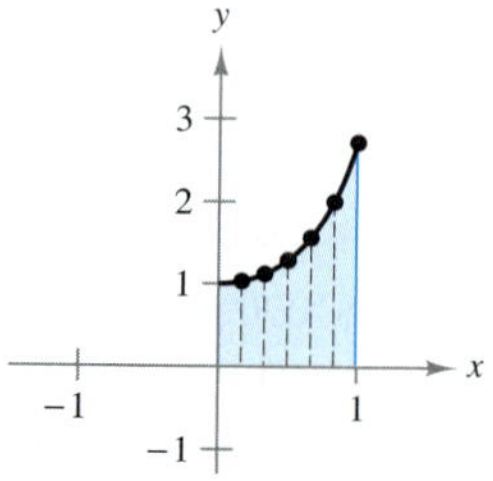

In Exercises 73 and 74, use the error formula to find bounds for the error in approximating the integral using the Trapezoidal Rule.

73. $\int_0^2 e^{2x}\,dx,\ n = 4$

74. $\int_0^2 e^{2x}\,dx,\ n = 8$

In Exercises 75 and 76, use the error formula to find bounds for the error in approximating the integral using Simpson's Rule.

75. $\int_2^4 \frac{1}{x-1}\,dx,\ n = 4$

76. $\int_2^4 \frac{1}{x-1}\,dx,\ n = 8$

In Exercises 77–84, evaluate the improper integral.

77. $\int_0^\infty 4xe^{-2x^2}\,dx$

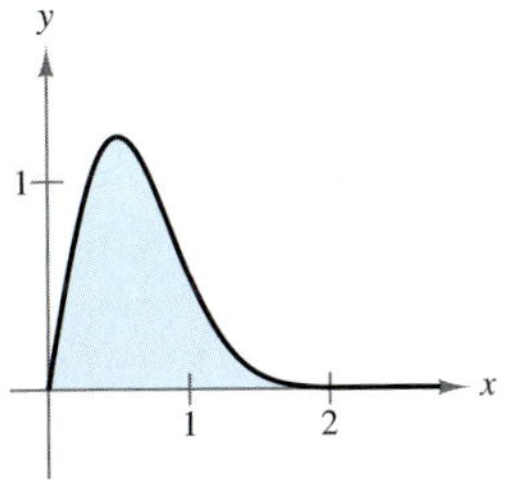

78. $\int_{-\infty}^0 \frac{3}{(1-3x)^{2/3}}\,dx$

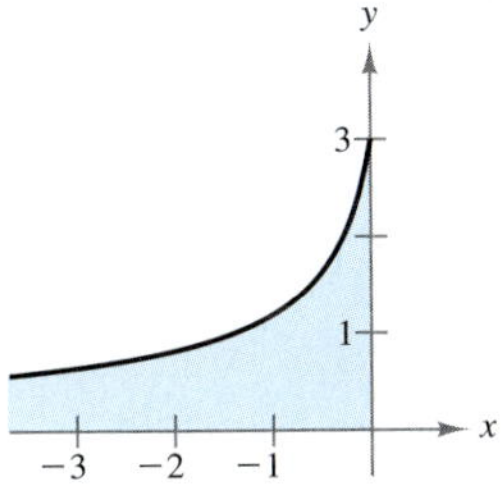

79. $\int_{-\infty}^0 \frac{1}{3x^2}\,dx$

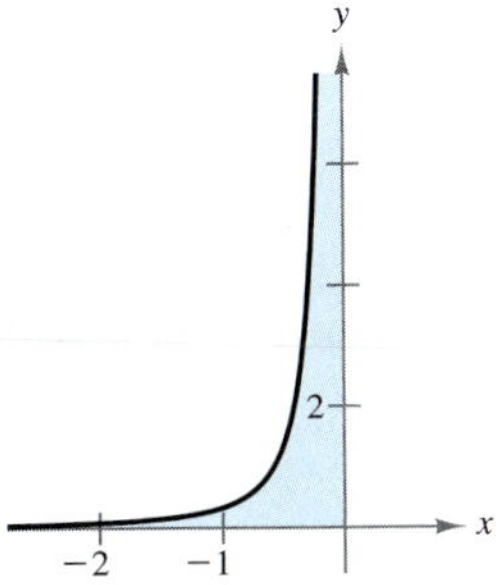

80. $\int_0^\infty 2x^2e^{-x^3}\,dx$

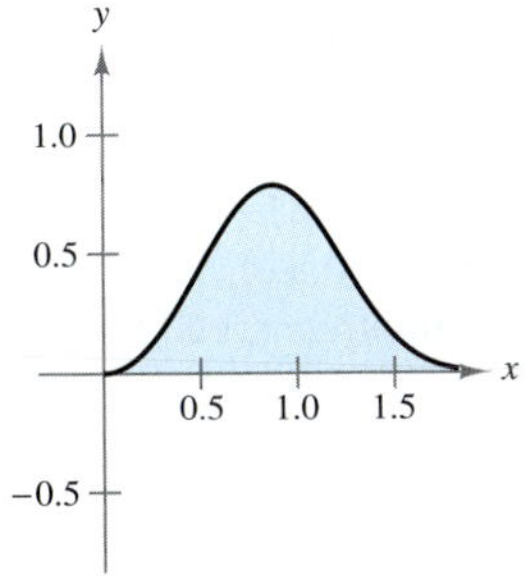

81. $\int_0^4 \frac{1}{\sqrt{4x}}\,dx$

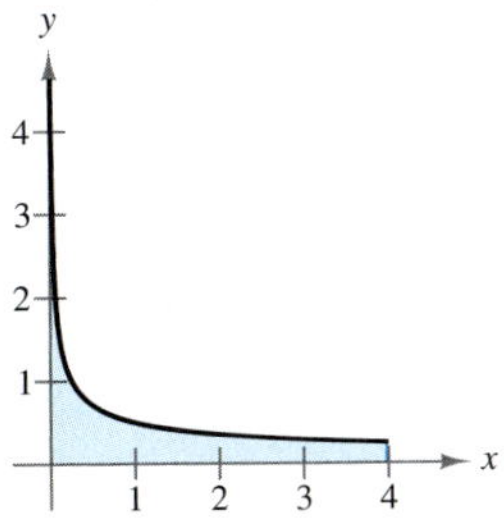

82. $\int_1^2 \frac{x}{16(x-1)^2}\,dx$

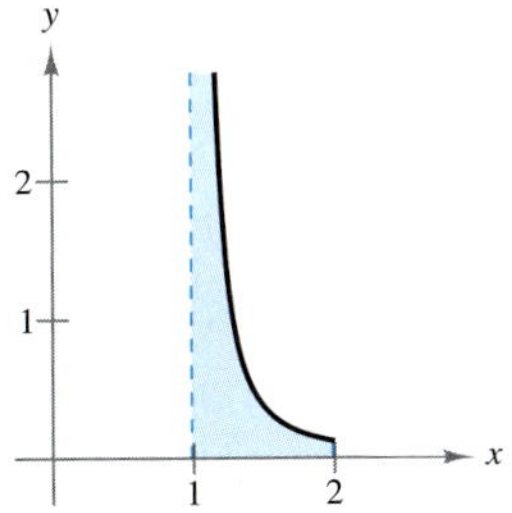

83. $\int_2^3 \frac{1}{\sqrt{x-2}}\,dx$

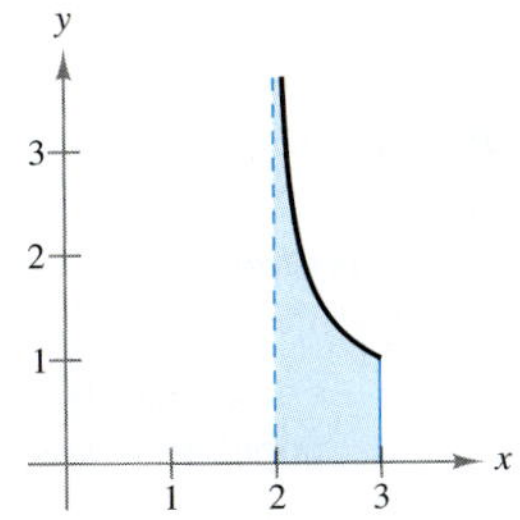

84. $\int_0^2 \frac{x+2}{(x-1)^2}\,dx$

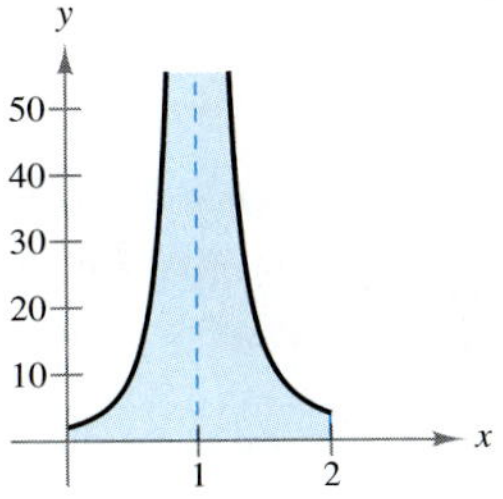

85. ***Present Value*** You are considering buying a franchise that yields a continuous income stream of \$50,000 per year. Find the present value of the franchise (a) for 15 years and (b) forever. Assume that money earns 6% interest per year, compounded continuously.

86. ***Capitalized Cost*** A company invests \$1.5 million in a new manufacturing plant that will cost \$50,000 per year in maintenance. Find the capitalized cost for (a) 20 years and (b) forever. Assume that money earns 6% interest, compounded continuously.

87. ***SAT Scores*** In 2001, the Scholastic Aptitude Test (SAT) math scores for college-bound seniors roughly followed a normal distribution

$$y = 0.0035e^{-(x-514)^2/25{,}538}, \qquad 200 \le x \le 800$$

where x is the SAT score for mathematics. Find the probability that a senior chosen at random had an SAT score (a) between 500 and 650, (b) 650 or better, and (c) 750 or better. *(Source: College Board)*

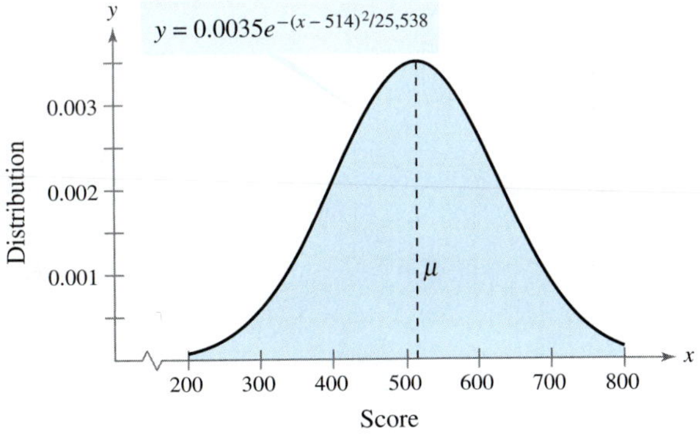

6 SAMPLE POST-GRADUATION EXAM QUESTIONS

CPA
GMAT
GRE
Actuarial
CLAST

The following questions represent the types of questions that appear on certified public accountant (CPA) exams, Graduate Management Admission Tests (GMAT), Graduate Records Exams (GRE), actuarial exams, and College-Level Academic Skills Tests (CLAST). The answers to the questions are given in the back of the book.

For Questions 1–4, the total 2000 population of Florida was 15,982,000 and the total 2002 population was 16,713,000. Also use the data shown in the graphs. *(Source: U.S. Census Bureau)*

Figure for 1–4

Resident Population by Age
State of Florida, 2000

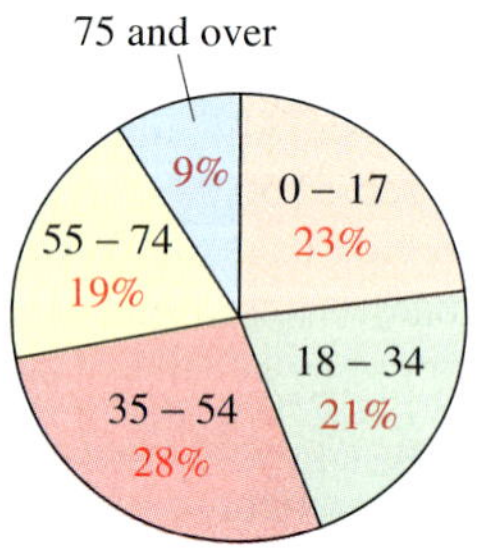

Resident Population by Age
State of Florida, 2002

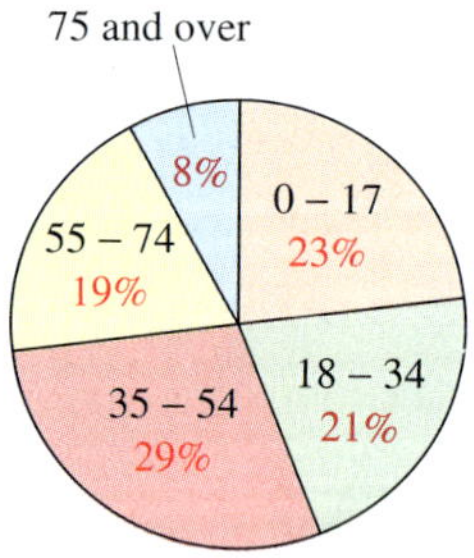

1. Find the number of people aged 75 and over for the year 2000.

(a) 1,438,380 (b) 1,278,560 (c) 1,504,170 (d) 1,337,040

2. Find the increase in population of 35- to 54-year-olds from 2000 to 2002.

(a) 204,680 (b) 448,600 (c) 211,900 (d) 371,810

3. In 2002, how many people were 54 years old or younger?

(a) 12,200,490 (b) 12,033,360 (c) 11,666,860 (d) 11,507,040

4. In what age group did the population and the population percent decrease between 2000 and 2002?

(a) 0–17 (b) 75 and over (c) 18–34 (d) 55–74

5. $\int_0^3 \sqrt{x+1}\, dx =$

(a) $-\frac{14}{3}$ (b) $2\sqrt{3}$ (c) $\frac{14}{3}$ (d) $\frac{15}{2}$

6. $\int_1^6 \frac{x}{\sqrt{x+3}}\, dx =$

(a) $\frac{10}{3}$ (b) $-\frac{20}{3}$ (c) -4 (d) $\frac{20}{3}$

7. $\lim_{x\to\infty} \frac{2x^2 + 5x - 16}{5x^2 - 15x + 48} =$

(a) 0 (b) $\frac{2}{5}$ (c) ∞ (d) $\frac{1}{3}$

8. The city council is planning to construct a new street, as shown in the figure. Construction costs for the new street are $110 per linear foot. What is the projected cost for constructing the new street?

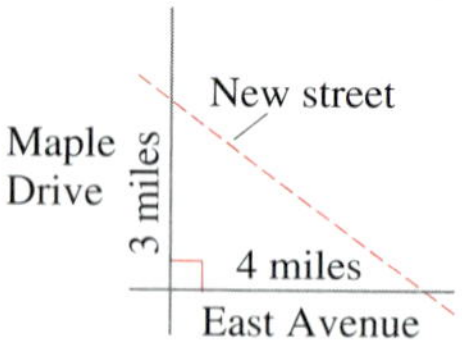

(a) $2,904,000 (b) $4,065,600 (c) $1,904,000 (d) $3,864,000

9. The following information pertains to Varn Co.:

Sales $1,000,000, Variable costs $200,000, Fixed costs $50,000

What is Varn's break-even point in sales dollars?

(a) $40,000 (b) $250,000 (c) $62,500 (d) $200,000

chapter

7 Functions of Several Variables

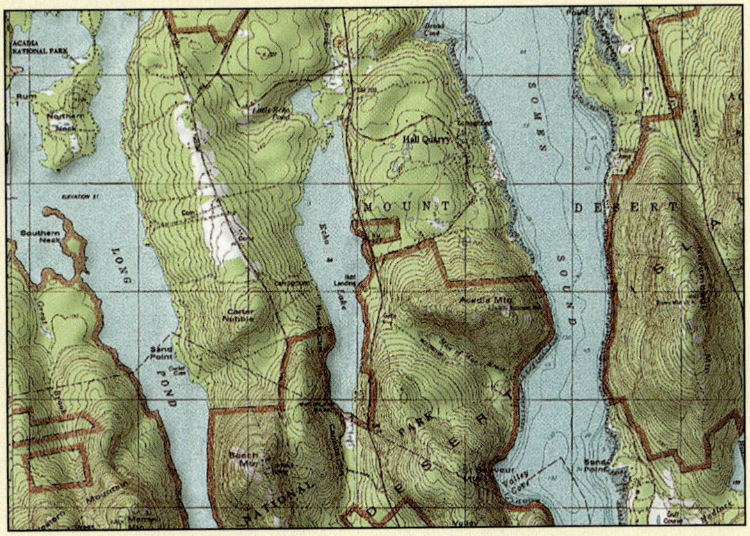

In addition to distance and location, contour maps indicate other variables such as elevation, population, and climate.

STRATEGIES FOR SUCCESS

WHAT YOU SHOULD LEARN:

- How to analyze surfaces and graph functions of two variables on the three-dimensional coordinate system
- How to calculate partial derivatives and find extrema of functions of several variables
- How to use Lagrange multipliers to solve constrained optimization problems
- How to use least squares regression for mathematical modeling
- How to evaluate double integrals and use them to find area and volume

WHY YOU SHOULD LEARN IT:

Functions of several variables have many applications in real life, such as the examples below, which represent a small sample of the applications in this chapter.

- Cobb-Douglas Production Function, Exercises 41 and 42 on page 481
- Geology: A Contour Map, Exercise 49 on page 482
- Complementary and Substitute Products, Exercise 71 on page 493
- Hardy-Weinberg Law, Exercise 45 on page 503
- Investment Strategy, Exercises 49 and 50 on page 513

7.1 THE THREE-DIMENSIONAL COORDINATE SYSTEM

- Plot points in space.
- Find distances between points in space and find midpoints of line segments in space.
- Write the standard forms of the equations of spheres and find the centers and radii of spheres.
- Sketch the coordinate plane traces of surfaces.

The Three-Dimensional Coordinate System

Recall that the Cartesian plane is determined by two perpendicular number lines called the x-axis and the y-axis. These axes together with their point of intersection (the origin) allow you to develop a two-dimensional coordinate system for identifying points in a plane. To identify a point in space, you must introduce a third dimension to the model. The geometry of this three-dimensional model is called **solid analytic geometry.**

DISCOVERY

Describe the location of a point (x, y, z) if $x = 0$. Describe the location of a point (x, y, z) if $x = 0$ and $y = 0$. What can you conclude about the ordered triple (x, y, z) if the point is located on the y-axis? What can you conclude about the ordered triple (x, y, z) if the point is located in the xz-plane?

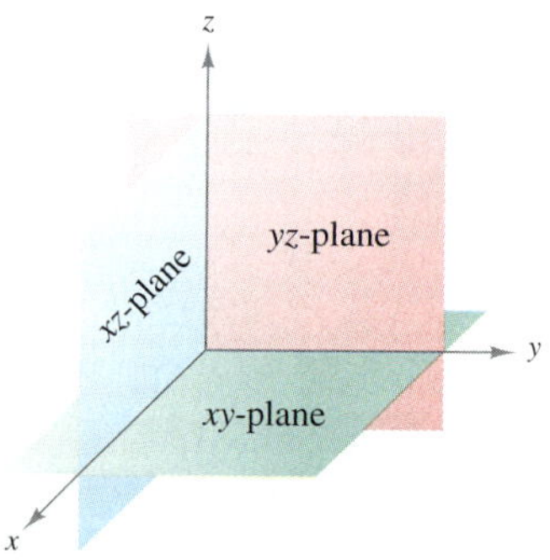

FIGURE 7.1

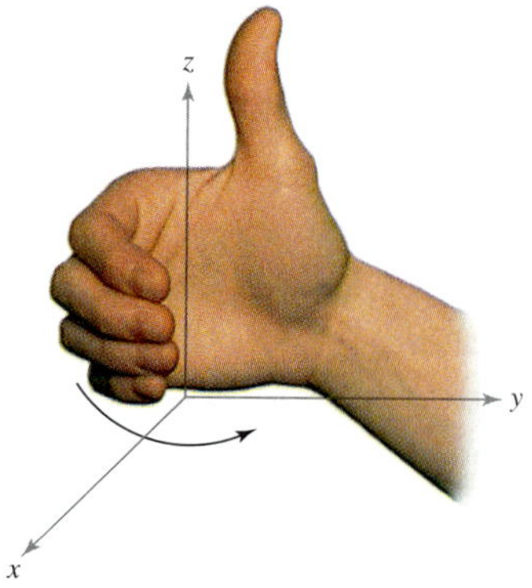

FIGURE 7.2

You can construct a **three-dimensional coordinate system** by passing a z-axis perpendicular to both the x- and y-axes at the origin. Figure 7.1 shows the positive portion of each coordinate axis. Taken as pairs, the axes determine three **coordinate planes:** the ***xy*-plane,** the ***xz*-plane,** and the ***yz*-plane.** These three coordinate planes separate the three-dimensional coordinate system into eight **octants.** The first octant is the one for which all three coordinates are positive. In this three-dimensional system, a point P in space is determined by an ordered triple (x, y, z), where x, y, and z are as shown.

$x =$ directed distance from yz-plane to P

$y =$ directed distance from xz-plane to P

$z =$ directed distance from xy-plane to P

A three-dimensional coordinate system can have either a **left-handed** or a **right-handed** orientation. In this text, you will work exclusively with right-handed systems, as shown in Figure 7.2.

EXAMPLE 1 **Plotting Points in Space**

Plot each point in space.

(a) $(2, -3, 3)$

(b) $(-2, 6, 2)$

(c) $(1, 4, 0)$

(d) $(2, 2, -3)$

SOLUTION To plot the point $(2, -3, 3)$, notice that $x = 2$, $y = -3$, and $z = 3$. To help visualize the point (see Figure 7.3), locate the point $(2, -3)$ in the xy-plane (denoted by a cross). The point $(2, -3, 3)$ lies three units above the cross. The other three points are also shown in the figure.

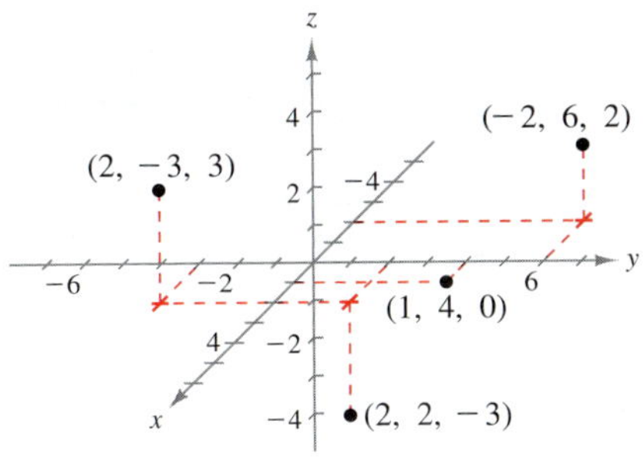

FIGURE 7.3

TRY IT 1

Plot each point on the three-dimensional coordinate system.

(a) $(2, 5, 1)$

(b) $(-2, -4, 3)$

(c) $(4, 0, -5)$

The Distance and Midpoint Formulas

Many of the formulas established for the two-dimensional coordinate system can be extended to three dimensions. For example, to find the distance between two points in space, you can use the Pythagorean Theorem twice, as shown in Figure 7.4. By doing this, you will obtain the formula for the distance between two points in space.

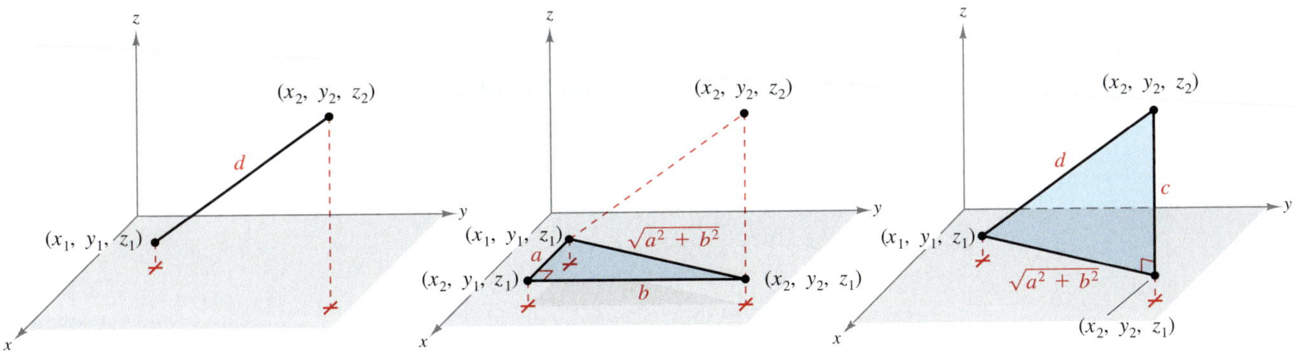

d = distance between two points

$a = |x_2 - x_1|, \quad b = |y_2 - y_1|$

$c = |z_2 - z_1|$

$d = \sqrt{a^2 + b^2 + c^2}$

$= \sqrt{(x_2 - x_1)^2 + (y_2 - y_1)^2 + (z_2 - z_1)^2}$

FIGURE 7.4

Distance Formula in Space

The distance between the points (x_1, y_1, z_1) and (x_2, y_2, z_2) is

$$d = \sqrt{(x_2 - x_1)^2 + (y_2 - y_1)^2 + (z_2 - z_1)^2}.$$

EXAMPLE 2 Finding the Distance Between Two Points

Find the distance between $(1, 0, 2)$ and $(2, 4, -3)$.

SOLUTION

$$\begin{aligned} d &= \sqrt{(x_2 - x_1)^2 + (y_2 - y_1)^2 + (z_2 - z_1)^2} && \text{Write Distance Formula.} \\ &= \sqrt{(2 - 1)^2 + (4 - 0)^2 + (-3 - 2)^2} && \text{Substitute.} \\ &= \sqrt{1 + 16 + 25} && \text{Simplify.} \\ &= \sqrt{42}. && \text{Simplify.} \end{aligned}$$

TRY IT 2

Find the distance between $(2, 3, -1)$ and $(0, 5, 3)$.

Notice the similarity between the Distance Formula in the plane and the Distance Formula in space. The Midpoint Formulas in the plane and in space are also similar.

Midpoint Formula in Space

The midpoint of the line segment joining the points (x_1, y_1, z_1) and (x_2, y_2, z_2) is

$$\text{Midpoint} = \left(\frac{x_1 + x_2}{2}, \frac{y_1 + y_2}{2}, \frac{z_1 + z_2}{2}\right).$$

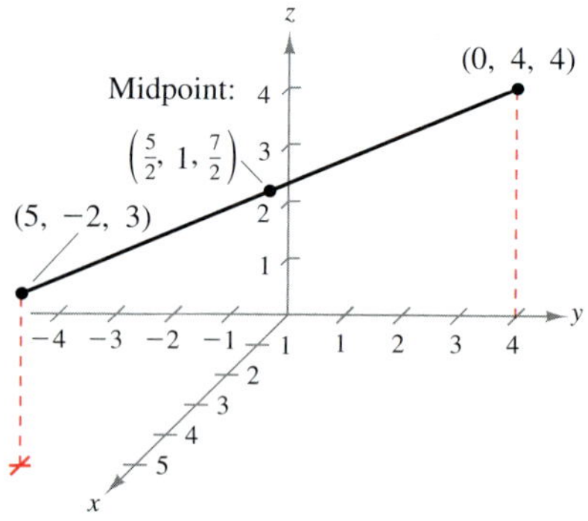

FIGURE 7.5

EXAMPLE 3 Using the Midpoint Formula

Find the midpoint of the line segment joining $(5, -2, 3)$ and $(0, 4, 4)$.

SOLUTION Using the Midpoint Formula, the midpoint is

$$\left(\frac{5 + 0}{2}, \frac{-2 + 4}{2}, \frac{3 + 4}{2}\right) = \left(\frac{5}{2}, 1, \frac{7}{2}\right)$$

as shown in Figure 7.5.

TRY IT 3

Find the midpoint of the line segment joining $(3, -2, 0)$ and $(-8, 6, -4)$.

The Equation of a Sphere

A **sphere** with center at (h, k, l) and radius r is defined to be the set of all points (x, y, z) such that the distance between (x, y, z) and (h, k, l) is r, as shown in Figure 7.6. Using the Distance Formula, this condition can be written as

$$\sqrt{(x - h)^2 + (y - k)^2 + (z - l)^2} = r.$$

By squaring both sides of this equation, you obtain the standard equation of a sphere.

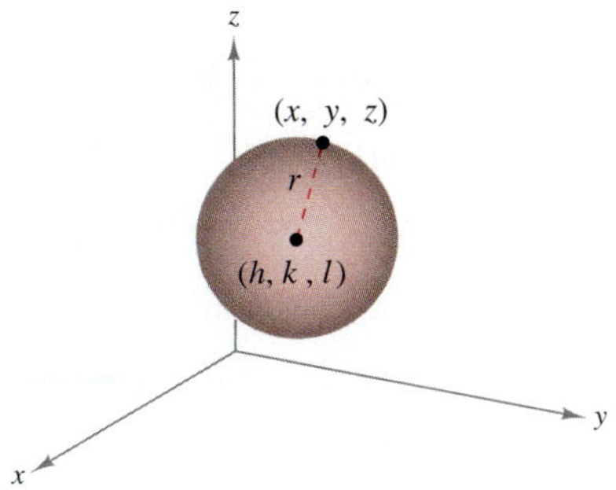

FIGURE 7.6 Sphere: Radius r, Center (h, k, l)

Standard Equation of a Sphere

The **standard equation of a sphere** whose center is (h, k, l) and whose radius is r is

$$(x - h)^2 + (y - k)^2 + (z - l)^2 = r^2.$$

EXAMPLE 4 Finding the Equation of a Sphere

Find the standard equation for the sphere whose center is $(2, 4, 3)$ and whose radius is 3. Does this sphere intersect the xy-plane?

SOLUTION

$$(x - h)^2 + (y - k)^2 + (z - l)^2 = r^2 \quad \text{Write standard equation.}$$

$$(x - 2)^2 + (y - 4)^2 + (z - 3)^2 = 3^2 \quad \text{Substitute.}$$

$$(x - 2)^2 + (y - 4)^2 + (z - 3)^2 = 9 \quad \text{Simplify.}$$

From the graph shown in Figure 7.7, you can see that the center of the sphere lies three units above the xy-plane. Because the sphere has a radius of 3, you can conclude that it does intersect the xy-plane—at the point $(2, 4, 0)$.

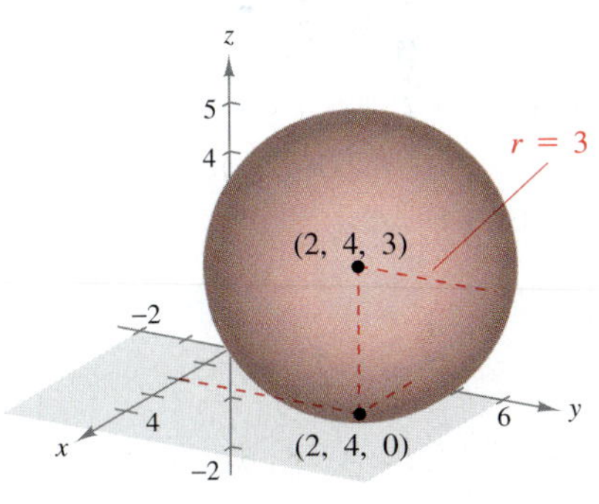

FIGURE 7.7

TRY IT 4

Find the standard equation of the sphere whose center is $(4, 3, 2)$ and whose radius is 5.

EXAMPLE 5 **Finding the Equation of a Sphere**

Find the equation of the sphere that has the points $(3, -2, 6)$ and $(-1, 4, 2)$ as endpoints of a diameter.

SOLUTION By the Midpoint Formula, the center of the sphere is

$$(h, k, l) = \left(\frac{3 + (-1)}{2}, \frac{-2 + 4}{2}, \frac{6 + 2}{2}\right) \quad \text{Apply Midpoint Formula.}$$

$$= (1, 1, 4). \quad \text{Simplify.}$$

By the Distance Formula, the radius is

$$r = \sqrt{(3 - 1)^2 + (-2 - 1)^2 + (6 - 4)^2} \quad \text{Distance Formula using } (3, -2, 6) \text{ and } (1, 1, 4)$$

$$= \sqrt{17}. \quad \text{Simplify.}$$

So, the standard equation of the sphere is

$$(x - h)^2 + (y - k)^2 + (z - l)^2 = r^2 \quad \text{Write formula for a sphere.}$$

$$(x - 1)^2 + (y - 1)^2 + (z - 4)^2 = 17. \quad \text{Substitute.}$$

TRY IT 5

Find the equation of the sphere that has the points $(-2, 5, 7)$ and $(4, 1, -3)$ as endpoints of a diameter.

EXAMPLE 6 **Finding the Center and Radius of a Sphere**

Find the center and radius of the sphere whose equation is

$$x^2 + y^2 + z^2 - 2x + 4y - 6z + 8 = 0.$$

SOLUTION You can obtain the standard equation of the sphere by completing the square. To do this, begin by grouping terms with the same variable. Then add "the square of half the coefficient of each linear term" to each side of the equation. For instance, to complete the square of $(x^2 - 2x)$, add $\left[\frac{1}{2}(-2)\right]^2 = 1$ to each side.

$$x^2 + y^2 + z^2 - 2x + 4y - 6z + 8 = 0$$

$$(x^2 - 2x + \quad) + (y^2 + 4y + \quad) + (z^2 - 6z + \quad) = -8$$

$$(x^2 - 2x + 1) + (y^2 + 4y + 4) + (z^2 - 6z + 9) = -8 + 1 + 4 + 9$$

$$(x - 1)^2 + (y + 2)^2 + (z - 3)^2 = 6$$

So, the center of the sphere is $(1, -2, 3)$, and its radius is $\sqrt{6}$, as shown in Figure 7.8.

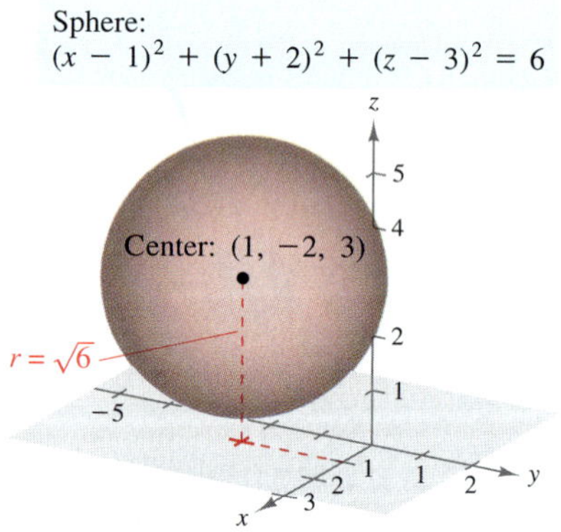

FIGURE 7.8

TRY IT 6

Find the center and radius of the sphere whose equation is

$$x^2 + y^2 + z^2 + 6x - 8y + 2z - 10 = 0.$$

Note in Example 6 that the points satisfying the equation of the sphere are "surface points," not "interior points." In general, the collection of points satisfying an equation involving x, y, and z is called a **surface in space.**

Traces of Surfaces

Finding the intersection of a surface with one of the three coordinate planes (or with a plane parallel to one of the three coordinate planes) helps visualize the surface. Such an intersection is called a **trace** of the surface. For example, the xy-trace of a surface consists of all points that are common to both the surface *and* the xy-plane. Similarly, the xz-trace of a surface consists of all points that are common to both the surface and the xz-plane.

EXAMPLE 7 Finding a Trace of a Surface

Sketch the xy-trace of the sphere whose equation is

$$(x - 3)^2 + (y - 2)^2 + (z + 4)^2 = 5^2.$$

SOLUTION To find the xy-trace of this surface, use the fact that every point in the xy-plane has a z-coordinate of zero. This means that if you substitute $z = 0$ into the original equation, the resulting equation will represent the intersection of the surface with the xy-plane.

$$(x - 3)^2 + (y - 2)^2 + (z + 4)^2 = 5^2 \quad \text{Write original equation.}$$
$$(x - 3)^2 + (y - 2)^2 + (0 + 4)^2 = 25 \quad \text{Let } z = 0 \text{ to find } xy\text{-trace.}$$
$$(x - 3)^2 + (y - 2)^2 + 16 = 25$$
$$(x - 3)^2 + (y - 2)^2 = 9$$
$$(x - 3)^2 + (y - 2)^2 = 3^2 \quad \text{Equation of circle}$$

From this equation, you can see that the xy-trace is a circle of radius 3, as shown in Figure 7.9.

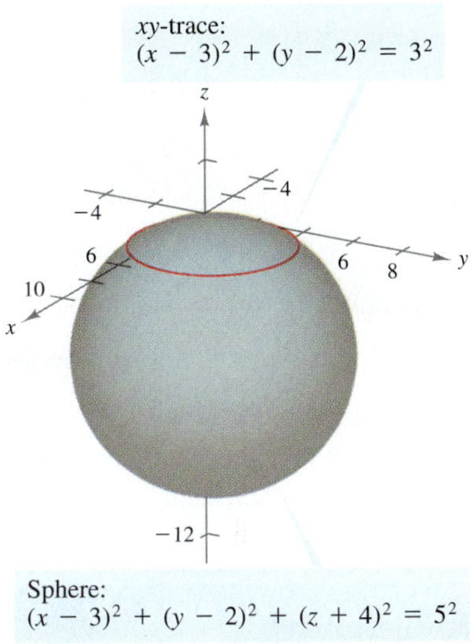

FIGURE 7.9

TRY IT 7

Find the equation of the xy-trace of the sphere whose equation is

$$(x + 1)^2 + (y - 2)^2 + (z + 3)^2 = 5^2.$$

TAKE ANOTHER LOOK

Comparing Two and Three Dimensions

In this section, you saw similarities between formulas in two-dimensional coordinate geometry and formulas in three-dimensional coordinate geometry. In two-dimensional coordinate geometry, the graph of the equation

$$ax + by + c = 0$$

is a line. In three-dimensional coordinate geometry, what is the graph of the equation

$$ax + by + c = 0?$$

Is it a line? Explain your reasoning. Use a three-dimensional graphing utility to verify your result.

PREREQUISITE REVIEW 7.1

The following warm-up exercises involve skills that were covered in earlier sections. You will use these skills in the exercise set for this section.

In Exercises 1–4, find the distance between the points.

1. $(5, 1), (3, 5)$

2. $(2, 3), (-1, -1)$

3. $(-5, 4), (-5, -4)$

4. $(-3, 6), (-3, -2)$

In Exercises 5–8, find the midpoint of the line segment connecting the points.

5. $(2, 5), (6, 9)$

6. $(-1, -2), (3, 2)$

7. $(-6, 0), (6, 6)$

8. $(-4, 3), (2, -1)$

In Exercises 9 and 10, write the standard equation of the circle.

9. Center: $(2, 3)$; radius: 2

10. Endpoints of a diameter: $(4, 0), (-2, 8)$

EXERCISES 7.1

In Exercises 1–4, plot the points on the same three-dimensional coordinate system.

1. (a) $(2, 1, 3)$
(b) $(-1, 2, 1)$

2. (a) $(3, -2, 5)$
(b) $\left(\frac{3}{2}, 4, -2\right)$

3. (a) $(5, -2, 2)$
(b) $(5, -2, -2)$

4. (a) $(0, 4, -5)$
(b) $(4, 0, 5)$

In Exercises 5–8, find the distance between the two points.

5. $(4, 1, 5), (8, 2, 6)$

6. $(-4, -1, 1), (2, -1, 5)$

7. $(-1, -5, 7), (-3, 4, -4)$

8. $(8, -2, 2), (8, -2, 4)$

In Exercises 9–12, find the coordinates of the midpoint of the line segment joining the two points.

9. $(6, -9, 1), (-2, -1, 5)$

10. $(4, 0, -6), (8, 8, 20)$

11. $(-5, -2, 5), (6, 3, -7)$

12. $(0, -2, 5), (4, 2, 7)$

In Exercises 13–16, find (x, y, z).

13.

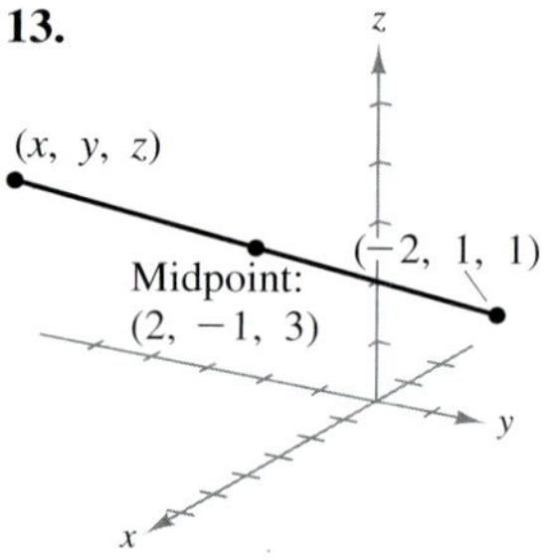

14.

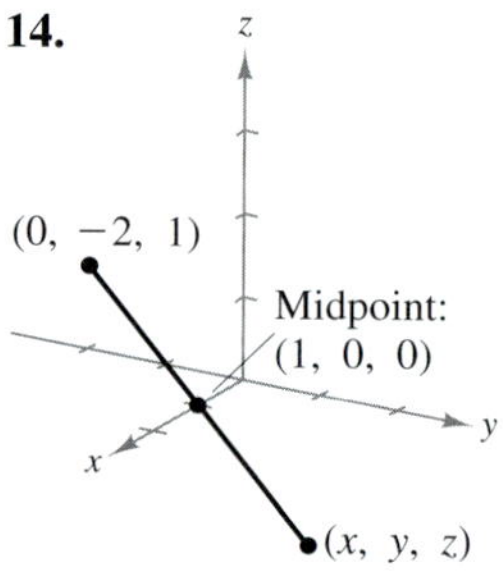

15.

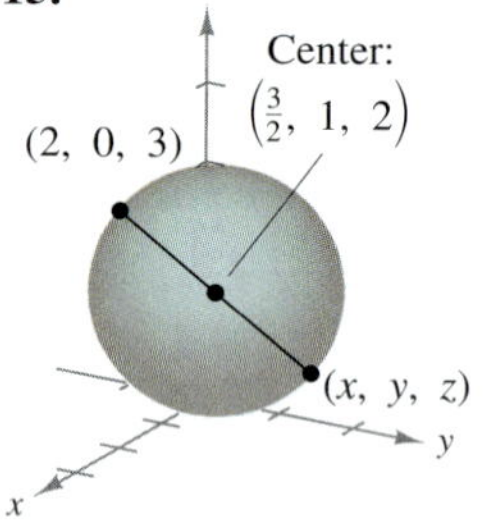

16.

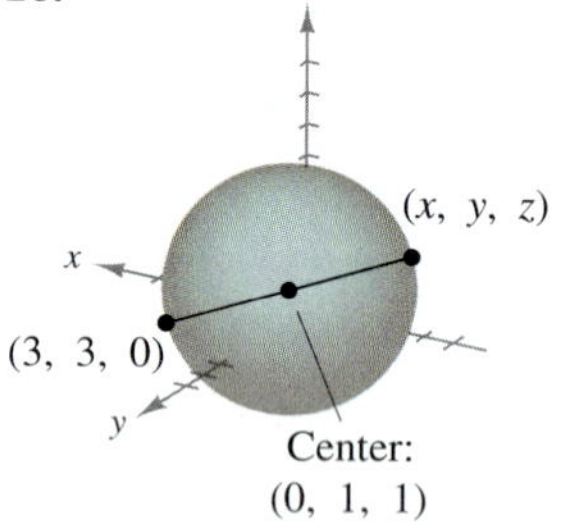

In Exercises 17–20, find the lengths of the sides of the triangle with the given vertices, and determine whether the triangle is a right triangle, an isosceles triangle, or neither of these.

17. $(0, 0, 0), (2, 2, 1), (2, -4, 4)$

18. $(5, 3, 4), (7, 1, 3), (3, 5, 3)$

19. $(-2, 2, 4), (-2, 2, 6), (-2, 4, 8)$

20. $(5, 0, 0), (0, 2, 0), (0, 0, -3)$

In Exercises 21–30, find the standard form of the equation of the sphere.

21.

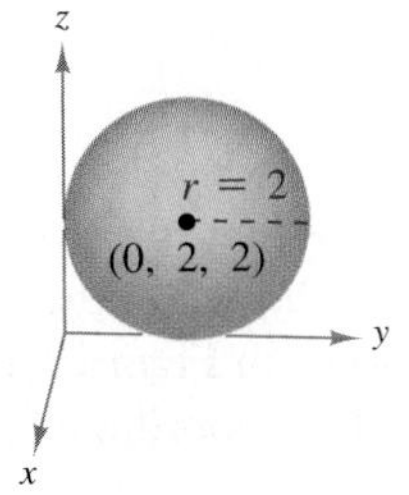

22.

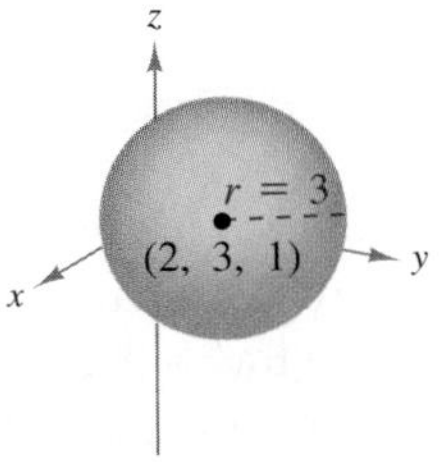

23.

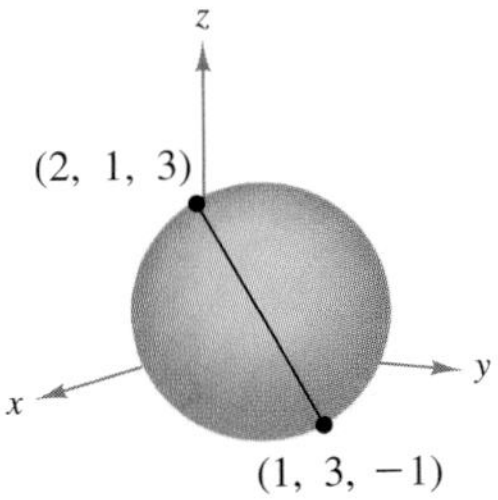

24.

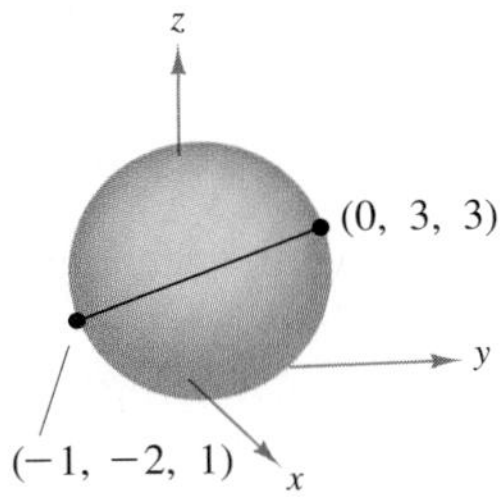

25. Center: $(1, 1, 5)$; radius: 3

26. Center: $(4, -1, 1)$; radius: 5

27. Endpoints of diameter: $(2, 0, 0), (0, 6, 0)$

28. Endpoints of diameter: $(1, 0, 0), (0, 5, 0)$

29. Center: $(-2, 1, 1)$; tangent to the xy-coordinate plane

30. Center: $(1, 2, 0)$; tangent to the yz-coordinate plane

In Exercises 31–36, find the sphere's center and radius.

31. $x^2 + y^2 + z^2 - 5x = 0$

32. $x^2 + y^2 + z^2 - 8y = 0$

33. $x^2 + y^2 + z^2 - 2x + 6y + 8z + 1 = 0$

34. $x^2 + y^2 + z^2 - 4y + 6z + 4 = 0$

35. $2x^2 + 2y^2 + 2z^2 - 4x - 12y - 8z + 3 = 0$

36. $4x^2 + 4y^2 + 4z^2 - 8x + 16y + 11 = 0$

In Exercises 37–40, sketch the xy-trace of the sphere.

37. $(x - 1)^2 + (y - 3)^2 + (z - 2)^2 = 25$

38. $(x + 1)^2 + (y + 2)^2 + (z - 2)^2 = 16$

39. $x^2 + y^2 + z^2 - 6x - 10y + 6z + 30 = 0$

40. $x^2 + y^2 + z^2 - 4y + 2z - 60 = 0$

In Exercises 41 and 42, sketch the yz-trace of the sphere.

41. $x^2 + y^2 + z^2 - 4x - 4y - 6z - 12 = 0$

42. $x^2 + y^2 + z^2 - 6x - 10y + 6z + 30 = 0$

In Exercises 43–46, sketch the trace of the intersection of each plane with the given sphere.

43. $x^2 + y^2 + z^2 = 25$

(a) $z = 3$ (b) $x = 4$

44. $x^2 + y^2 + z^2 = 169$

(a) $x = 5$ (b) $y = 12$

45. $x^2 + y^2 + z^2 - 4x - 6y + 9 = 0$

(a) $x = 2$ (b) $y = 3$

46. $x^2 + y^2 + z^2 - 8x - 6z + 16 = 0$

(a) $x = 4$ (b) $z = 3$

47. ***Geology*** Crystals are classified according to their symmetry. Crystals shaped like cubes are classified as isometric. Suppose you have mapped the vertices of a crystal onto a three-dimensional coordinate system. Determine (x, y, z) if the crystal is isometric.

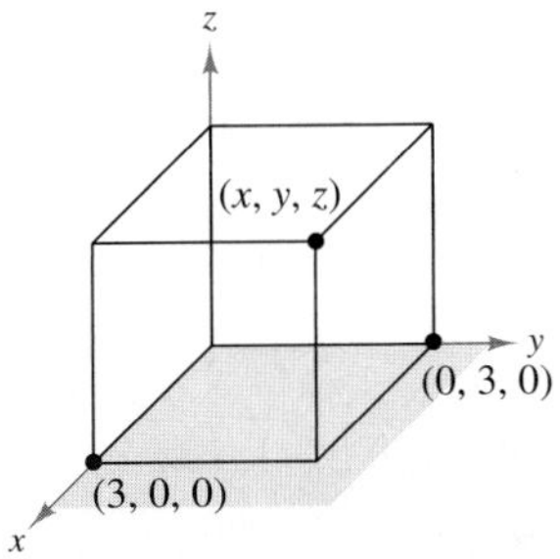

Dr. E.R. Degginger

Halite crystals (rock salt) are classified as isometric.

48. ***Physical Science*** Assume that Earth is a sphere with a radius of 3963 miles. If the center of Earth is placed at the origin of a three-dimensional coordinate system, what is the equation of the sphere? Lines of longitude that run north-south could be represented by what trace(s)? What shape would each of these traces form? Why? Lines of latitude that run east-west could be represented by what trace(s)? Why? What shape would each of these traces form? Why?

7.2 SURFACES IN SPACE

- Sketch planes in space.
- Draw planes in space with different numbers of intercepts.
- Classify quadric surfaces in space.

Equations of Planes in Space

In Section 7.1, you studied one type of surface in space—a sphere. In this section, you will study a second type—a plane in space. The **general equation of a plane** in space is

$$ax + by + cz = d. \qquad \text{General equation of a plane}$$

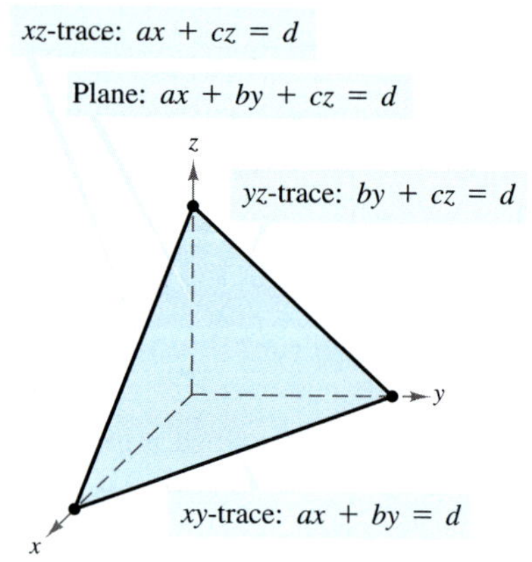

FIGURE 7.10

Note the similarity of this equation to the general equation of a line in the plane. In fact, if you intersect the plane represented by this equation with each of the three coordinate planes, you will obtain traces that are lines, as shown in Figure 7.10.

In Figure 7.10, the points where the plane intersects the three coordinate axes are the x-, y-, and z-intercepts of the plane. By connecting these three points, you can form a triangular region, which helps you visualize the plane in space.

EXAMPLE 1 Sketching a Plane in Space

Find the x-, y-, and z-intercepts of the plane given by

$$3x + 2y + 4z = 12.$$

Then sketch the plane.

SOLUTION To find the x-intercept, let both y and z be zero.

$$3x + 2(0) + 4(0) = 12 \qquad \text{Substitute 0 for } y \text{ and } z.$$

$$3x = 12 \qquad \text{Simplify.}$$

$$x = 4 \qquad \text{Solve for } x.$$

So, the x-intercept is $(4, 0, 0)$. To find the y-intercept, let x and z be zero and conclude that $y = 6$. So, the y-intercept is $(0, 6, 0)$. Similarly, by letting x and y be zero, you can determine that $z = 3$ and that the z-intercept is $(0, 0, 3)$. Figure 7.11 shows the triangular portion of the plane formed by connecting the three intercepts.

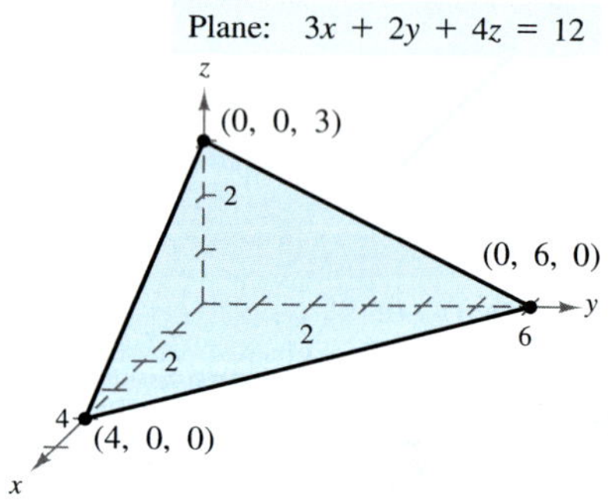

FIGURE 7.11 Sketch Made by Connecting Intercepts: $(4, 0, 0)$, $(0, 6, 0)$, $(0, 0, 3)$

TRY IT 1

Find the x-, y-, and z-intercepts of the plane given by

$$2x + 4y + z = 8.$$

Then sketch the plane.

Drawing Planes in Space

The planes shown in Figures 7.10 and 7.11 have three intercepts. When this occurs, we suggest that you draw the plane by sketching the triangular region formed by connecting the three intercepts.

It is possible for a plane in space to have fewer than three intercepts. This occurs when one or more of the coefficients in the equation $ax + by + cz = d$ is zero. Figure 7.12 shows some planes in space that have only one intercept, and Figure 7.13 shows some that have only two intercepts. In each figure, note the use of dashed lines and shading to give the illusion of three dimensions.

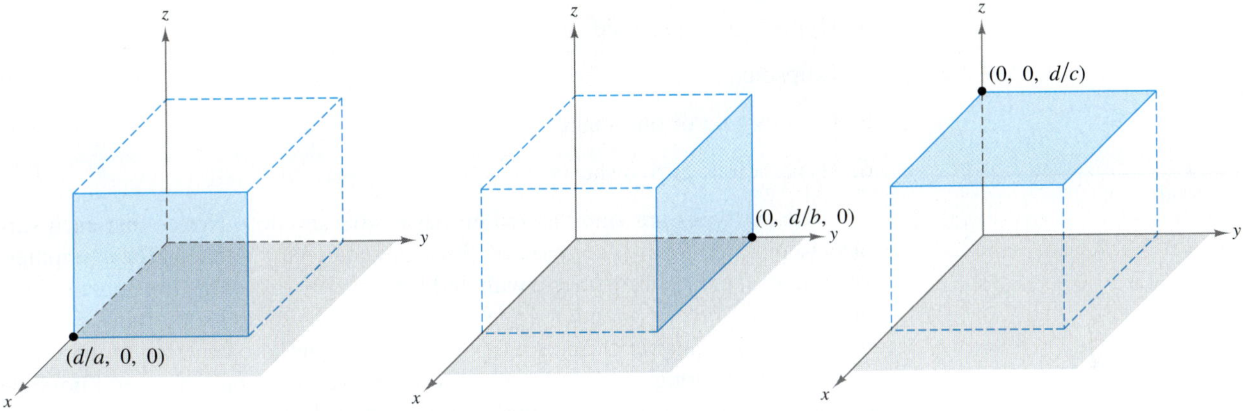

Plane $ax = d$ is parallel to yz-plane. Plane $by = d$ is parallel to xz-plane. Plane $cz = d$ is parallel to xy-plane.

FIGURE 7.12 Planes Parallel to Coordinate Planes

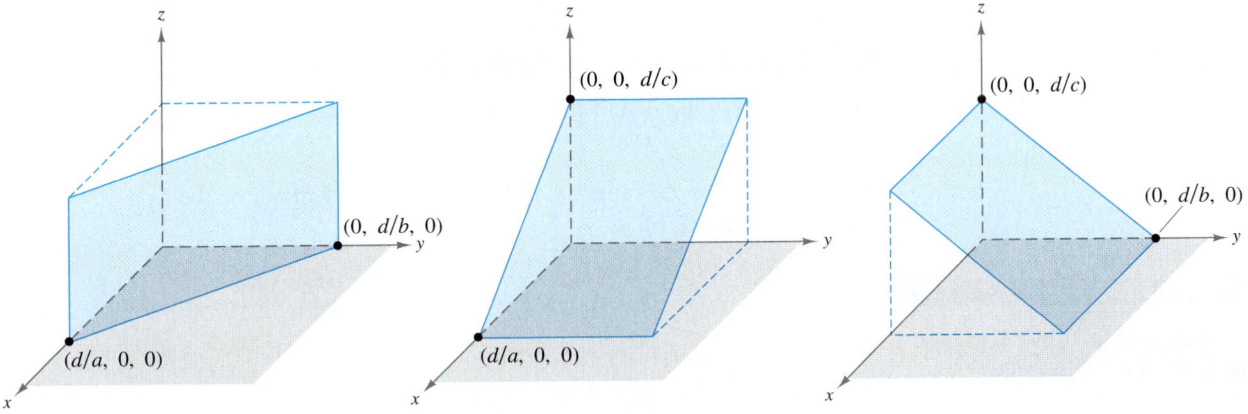

Plane $ax + by = d$ is parallel to z-axis. Plane $ax + cz = d$ is parallel to y-axis. Plane $by + cz = d$ is parallel to x-axis.

FIGURE 7.13 Planes Parallel to Coordinate Axes

DISCOVERY

What is the equation of each plane?

(a) xy-plane (b) xz-plane (c) yz-plane

Quadric Surfaces

A third common type of surface in space is a **quadric surface.** Every quadric surface has an equation of the form

$$Ax^2 + By^2 + Cz^2 + Dx + Ey + Fz + G = 0.$$ Second-degree equation

There are six basic types of quadric surfaces.

1. Elliptic cone
2. Elliptic paraboloid
3. Hyperbolic paraboloid
4. Ellipsoid
5. Hyperboloid of one sheet
6. Hyperboloid of two sheets

The six types are summarized on pages 468 and 469. Notice that each surface is pictured with two types of three-dimensional sketches. The computer-generated sketches use traces with hidden lines to give the illusion of three dimensions. The artist-rendered sketches use shading to create the same illusion.

All of the quadric surfaces on pages 468 and 469 are centered at the origin and have axes along the coordinate axes. Moreover, only one of several possible orientations of each surface is shown. If the surface has a different center or is oriented along a different axis, then its standard equation will change accordingly. For instance, the ellipsoid

$$\frac{x^2}{1^2} + \frac{y^2}{3^2} + \frac{z^2}{2^2} = 1$$

has $(0, 0, 0)$ as its center, but the ellipsoid

$$\frac{(x-2)^2}{1^2} + \frac{(y+1)^2}{3^2} + \frac{(z-4)^2}{2^2} = 1$$

has $(2, -1, 4)$ as its center. A computer-generated graph of the first ellipsoid is shown in Figure 7.14.

DISCOVERY

One way to help visualize a quadric surface is to determine the intercepts of the surface with the coordinate axes. What are the intercepts of the ellipsoid in Figure 7.14?

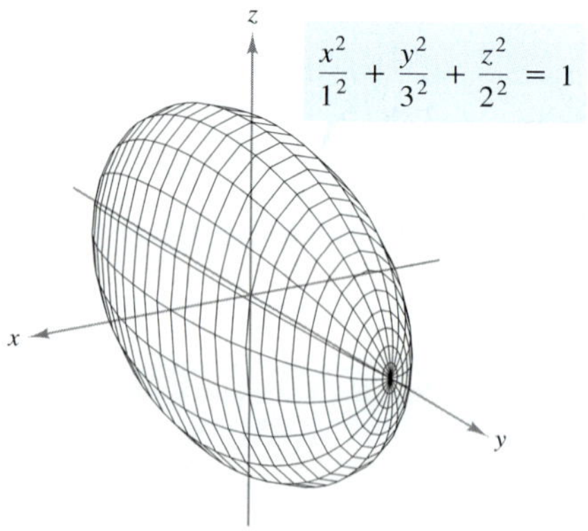

FIGURE 7.14

TECHNOLOGY

Using a Three-Dimensional Graphing Utility

Most three-dimensional graphing utilities represent surfaces by sketching several traces of the surface. The traces are usually taken in equally spaced parallel planes. Depending on the graphing utility, the sketch can be made with one set, two sets, or three sets of traces. For instance, the two sketches shown below use two sets of traces: in one set the traces are parallel to the xz-plane, and in the other set the traces are parallel to the yz-plane.

To sketch the graph of an equation involving x, y, and z with a three-dimensional "function grapher," you must first solve the equation for z. After entering the equation, you need to specify a rectangular viewing box (the three-dimensional analog of a viewing window).

The two sketches shown below were generated by *Derive* for Windows. If you have access to a three-dimensional graphing utility, try using it to duplicate these graphs.

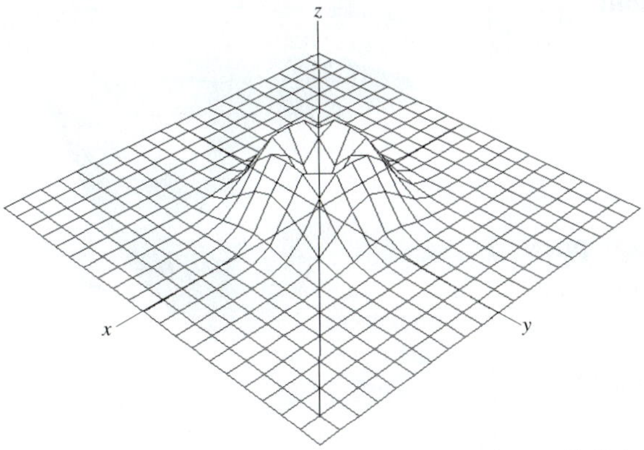

Generated by Derive

Equation: $z = (x^2 + y^2)e^{1-x^2-y^2}$
Grid: 20 traces by 20 traces
Viewing Box: $-5 \le x \le 5$
$-5 \le y \le 5$
$-5 \le z \le 5$

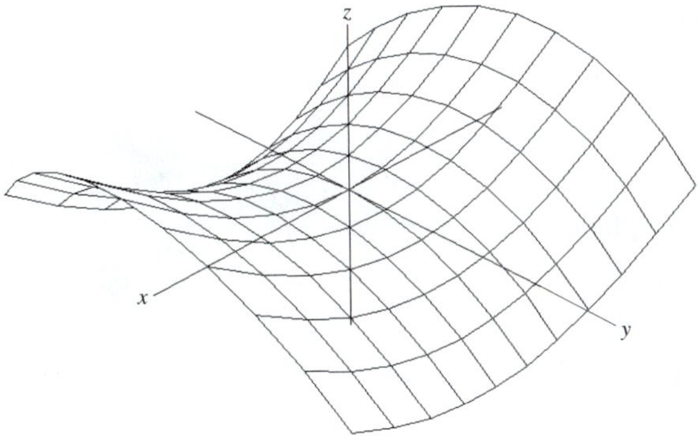

Generated by Derive

Equation: $z = \frac{1}{4}x^2 - \frac{1}{4}y^2$
Grid: 10 traces by 10 traces
Viewing Box: $-4 \le x \le 4$
$-4 \le y \le 4$
$-4 \le z \le 4$

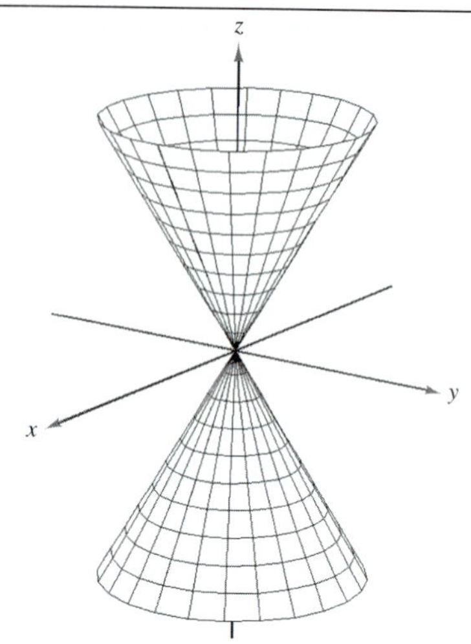

Elliptic Cone

$$\frac{x^2}{a^2} + \frac{y^2}{b^2} - \frac{z^2}{c^2} = 0$$

Trace	*Plane*
Ellipse	Parallel to xy-plane
Hyperbola	Parallel to xz-plane
Hyperbola	Parallel to yz-plane

The axis of the cone corresponds to the variable whose coefficient is negative. The traces in the coordinate planes parallel to this axis are intersecting lines.

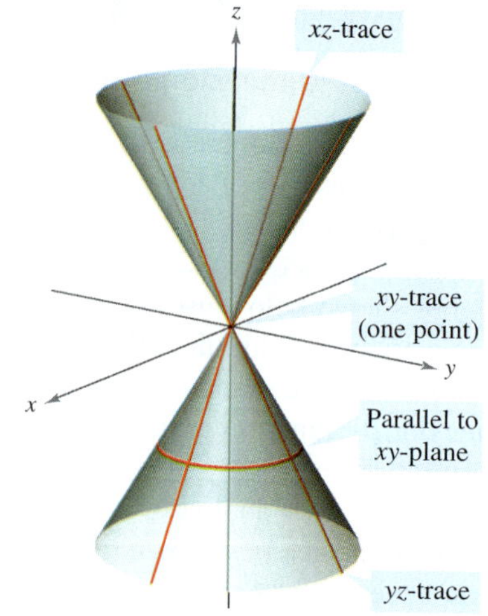

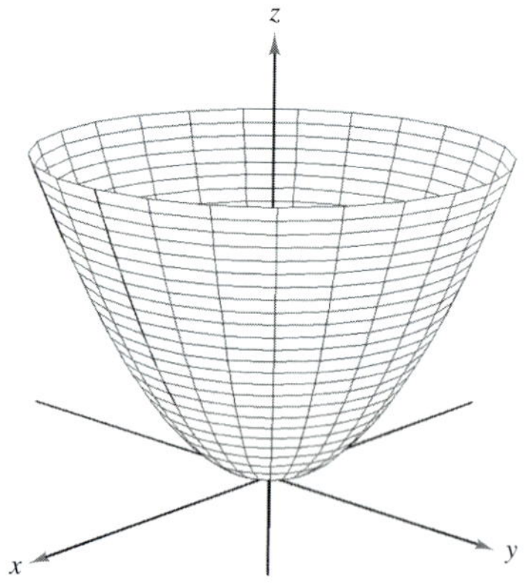

Elliptic Paraboloid

$$z = \frac{x^2}{a^2} + \frac{y^2}{b^2}$$

Trace	*Plane*
Ellipse	Parallel to xy-plane
Parabola	Parallel to xz-plane
Parabola	Parallel to yz-plane

The axis of the paraboloid corresponds to the variable raised to the first power.

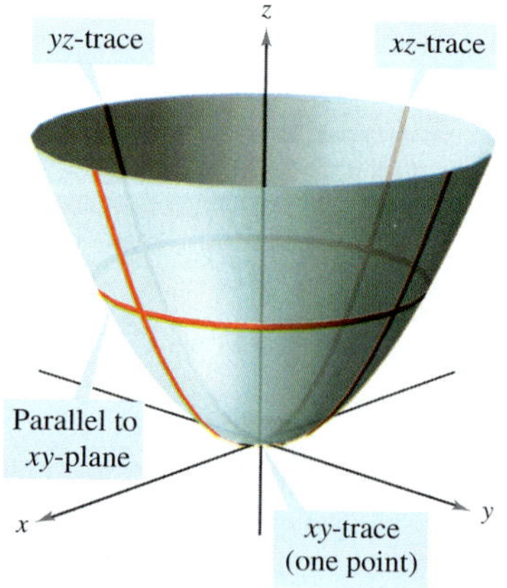

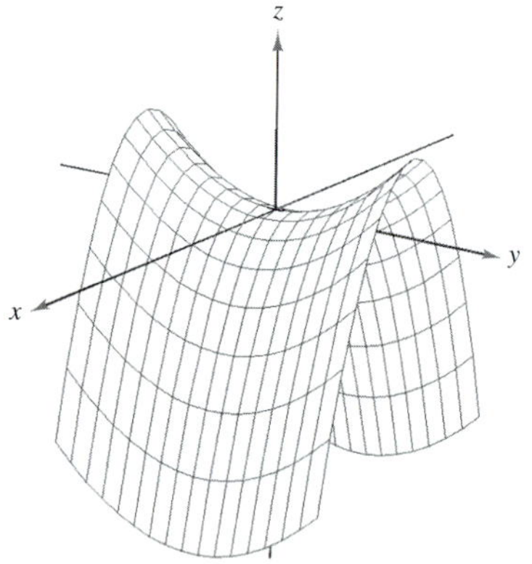

Hyperbolic Paraboloid

$$z = \frac{y^2}{b^2} - \frac{x^2}{a^2}$$

Trace	*Plane*
Hyperbola	Parallel to xy-plane
Parabola	Parallel to xz-plane
Parabola	Parallel to yz-plane

The axis of the paraboloid corresponds to the variable raised to the first power.

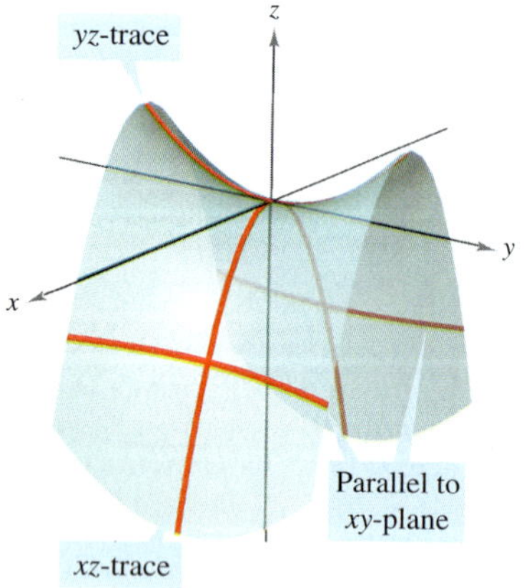

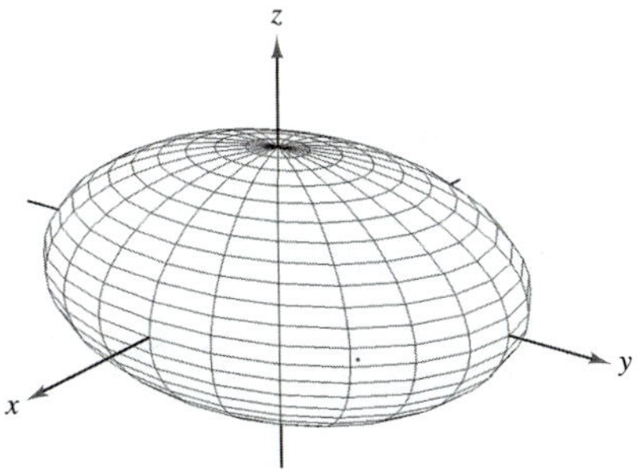

Ellipsoid

$$\frac{x^2}{a^2} + \frac{y^2}{b^2} + \frac{z^2}{c^2} = 1$$

Trace	*Plane*
Ellipse	Parallel to xy-plane
Ellipse	Parallel to xz-plane
Ellipse	Parallel to yz-plane

The surface is a sphere if the coefficients a, b, and c are equal and nonzero.

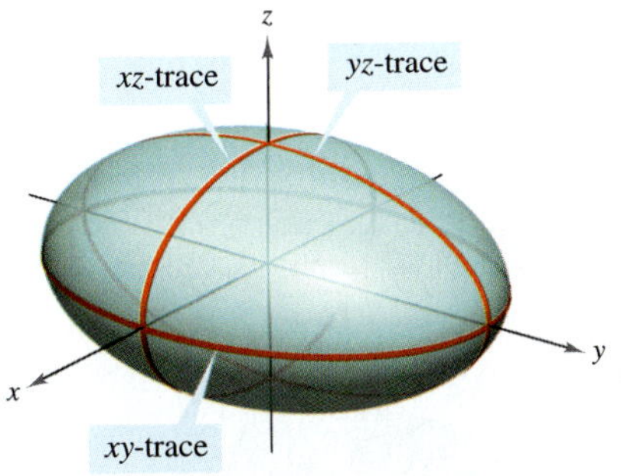

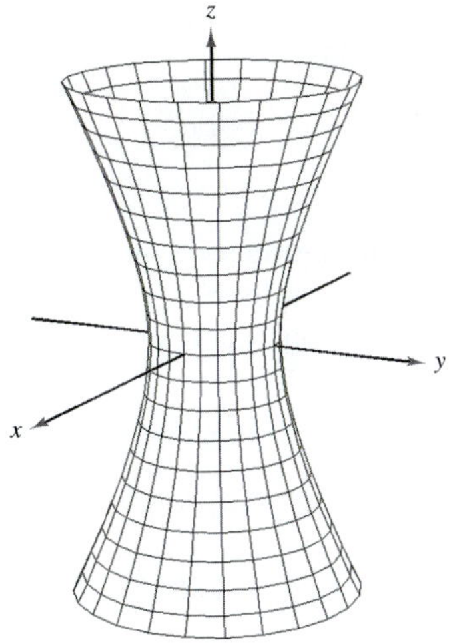

Hyperboloid of One Sheet

$$\frac{x^2}{a^2} + \frac{y^2}{b^2} - \frac{z^2}{c^2} = 1$$

Trace	*Plane*
Ellipse	Parallel to xy-plane
Hyperbola	Parallel to xz-plane
Hyperbola	Parallel to yz-plane

The axis of the hyperboloid corresponds to the variable whose coefficient is negative.

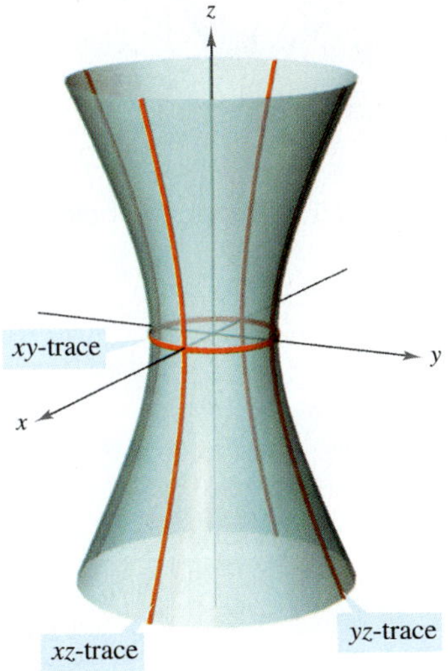

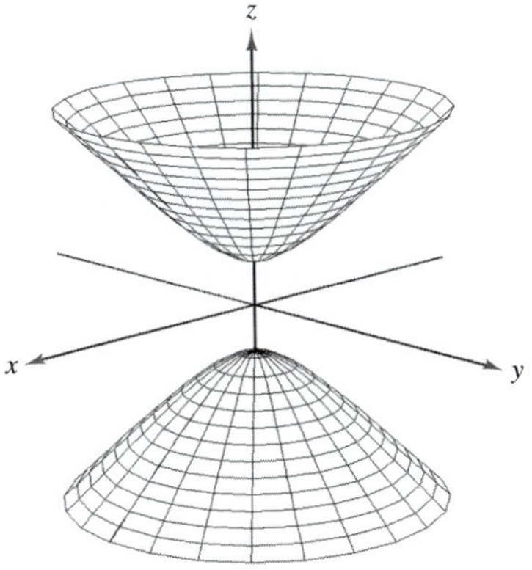

Hyperboloid of Two Sheets

$$\frac{z^2}{c^2} - \frac{x^2}{a^2} - \frac{y^2}{b^2} = 1$$

Trace	*Plane*
Ellipse	Parallel to xy-plane
Hyperbola	Parallel to xz-plane
Hyperbola	Parallel to yz-plane

The axis of the hyperboloid corresponds to the variable whose coefficient is positive. There is no trace in the coordinate plane perpendicular to this axis.

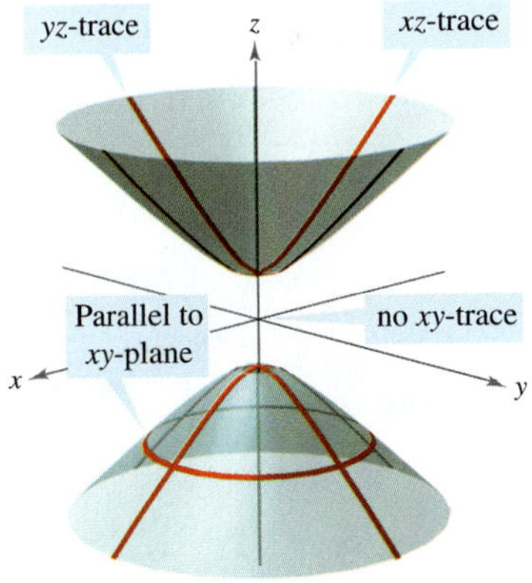

When classifying quadric surfaces, note that the two types of paraboloids have one variable raised to the first power. The other four types of quadric surfaces have equations that are of second degree in *all* three variables.

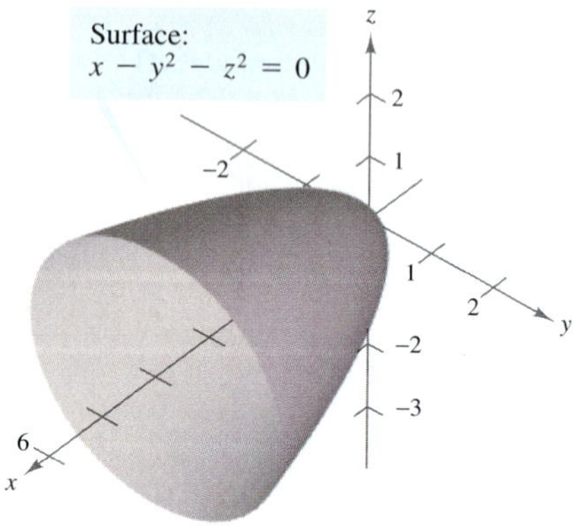

FIGURE 7.15 Elliptic Paraboloid

EXAMPLE 2 Classifying a Quadric Surface

Classify the surface given by $x - y^2 - z^2 = 0$. Describe the traces of the surface in the xy-plane, the xz-plane, and the plane given by $x = 1$.

SOLUTION Because x is raised only to the first power, the surface is a paraboloid whose axis is the x-axis, as shown in Figure 7.15. In standard form, the equation is

$$x = y^2 + z^2.$$

The traces in the xy-plane, the xz-plane, and the plane given by $x = 1$ are as shown.

Trace in xy-plane ($z = 0$):	$x = y^2$	Parabola
Trace in xz-plane ($y = 0$):	$x = z^2$	Parabola
Trace in plane $x = 1$:	$y^2 + z^2 = 1$	Circle

These three traces are shown in Figure 7.16. From the traces, you can see that the surface is an elliptic (or circular) paraboloid. If you have access to a three-dimensional graphing utility, try using it to graph this surface. If you do this, you will discover that sketching surfaces in space is not a simple task—even with a graphing utility.

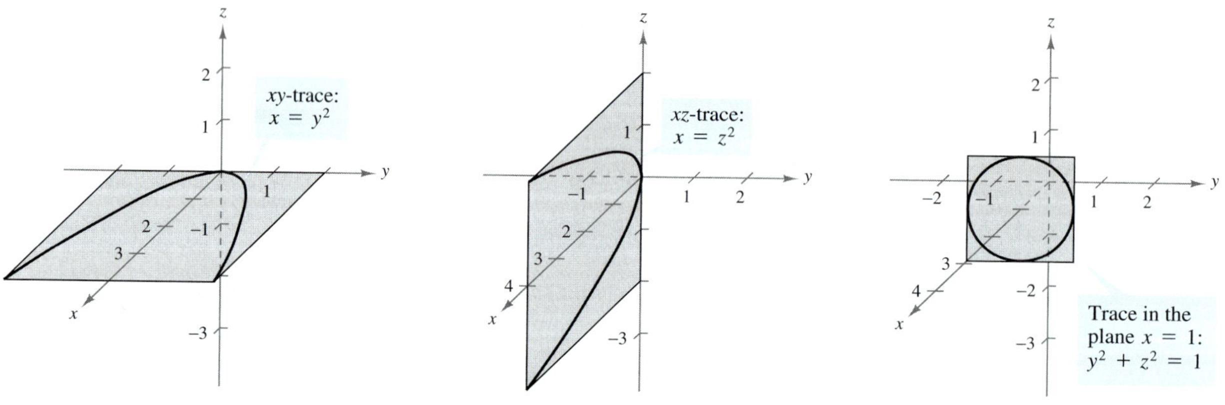

FIGURE 7.16

TRY IT 2

Classify the surface given by $x^2 + y^2 - z^2 = 1$. Describe the traces of the surface in the xy-plane, the yz-plane, the xz-plane, and the plane given by $z = 3$.

EXAMPLE 3 Classifying Quadric Surfaces

Write each quadric surface in standard form and classify each equation.

(a) $x^2 - 4y^2 - 4z^2 - 4 = 0$

(b) $x^2 + 4y^2 + z^2 - 4 = 0$

SOLUTION

(a) The equation $x^2 - 4y^2 - 4z^2 - 4 = 0$ can be written in standard form as

$$\frac{x^2}{4} - y^2 - z^2 = 1. \qquad \text{Standard form}$$

From the standard form, you can see that the graph is a hyperboloid of two sheets, with the x-axis as its axis, as shown in Figure 7.17(a).

(b) The equation $x^2 + 4y^2 + z^2 - 4 = 0$ can be written in standard form as

$$\frac{x^2}{4} + y^2 + \frac{z^2}{4} = 1. \qquad \text{Standard form}$$

From the standard form, you can see that the graph is an ellipsoid, as shown in Figure 7.17(b).

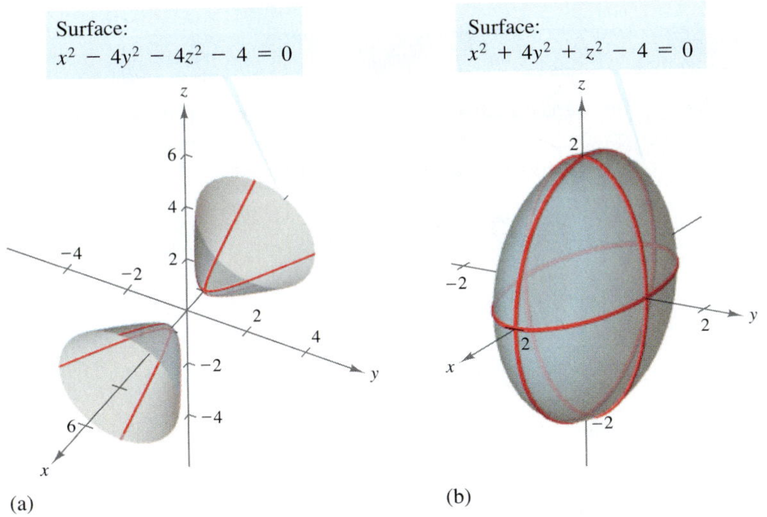

FIGURE 7.17

TRY IT 3

Write each quadric surface in standard form and classify each equation.

(a) $4x^2 + 9y^2 - 36z = 0$

(b) $36x^2 + 16y^2 - 144z^2 = 0$

TAKE ANOTHER LOOK

Classifying Quadric Surfaces

Classify each quadric surface. Use a three-dimensional graphing utility to verify your results.

a. $\dfrac{x^2}{2^2} - \dfrac{y^2}{4^2} + \dfrac{z^2}{3^2} = 0$ b. $\dfrac{x^2}{2^2} + \dfrac{y^2}{4^2} + \dfrac{z^2}{3^2} = 1$ c. $\dfrac{x^2}{2^2} - \dfrac{y^2}{4^2} + \dfrac{z^2}{3^2} = 1$

PREREQUISITE REVIEW 7.2

The following warm-up exercises involve skills that were covered in earlier sections. You will use these skills in the exercise set for this section.

In Exercises 1–4, find the x- and y-intercepts of the function.

1. $3x + 4y = 12$

2. $6x + y = -8$

3. $-2x + y = -2$

4. $-x - y = 5$

In Exercises 5–8, rewrite the expression by completing the square.

5. $x^2 + y^2 + z^2 - 2x - 4y - 6z + 15 = 0$

6. $x^2 + y^2 - z^2 - 8x + 4y - 6z + 11 = 0$

7. $z - 2 = x^2 + y^2 + 2x - 2y$

8. $x^2 + y^2 + z^2 - 6x + 10y + 26z = -202$

In Exercises 9 and 10, write the expression in standard form.

9. $16x^2 - 16y^2 + 16z^2 = 4$

10. $9x^2 - 9y^2 + 9z^2 = 36$

EXERCISES 7.2

In Exercises 1–12, find the intercepts and sketch the graph of the plane.

1. $4x + 2y + 6z = 12$

2. $3x + 6y + 2z = 6$

3. $3x + 3y + 5z = 15$

4. $x + y + z = 3$

5. $2x - y + 3z = 8$

6. $2x - y + z = 4$

7. $z = 3$

8. $y = -4$

9. $y + z = 5$

10. $x + 2y = 8$

11. $x + y - z = 0$

12. $x - 3z = 3$

In Exercises 13–22, determine whether the planes $a_1x + b_1y + c_1z = d_1$ and $a_2x + b_2y + c_2z = d_2$ are parallel, perpendicular, or neither. The planes are parallel if there exists a nonzero constant k such that $a_1 = ka_2$, $b_1 = kb_2$, $c_1 = kc_2$, and perpendicular if $a_1a_2 + b_1b_2 + c_1c_2 = 0$.

13. $5x - 3y + z = 4,\ x + 4y + 7z = 1$

14. $3x + y - 4z = 3,\ -9x - 3y + 12z = 4$

15. $x - 5y - z = 1,\ 5x - 25y - 5z = -3$

16. $x + 3y + 2z = 6,\ 4x - 12y + 8z = 24$

17. $x + 2y = 3,\ 4x + 8y = 5$

18. $x + 3y + z = 7,\ x - 5z = 0$

19. $2x + y = 3,\ 3x - 5z = 0$

20. $2x - z = 1,\ 4x + y + 8z = 10$

21. $x = 6,\ y = -1$

22. $x = -2,\ y = 4$

In Exercises 23–30, find the distance between the point and the plane (see figure). The distance D between a point (x_0, y_0, z_0) and the plane $ax + by + cz + d = 0$ is

$$D = \frac{|ax_0 + by_0 + cz_0 + d|}{\sqrt{a^2 + b^2 + c^2}}.$$

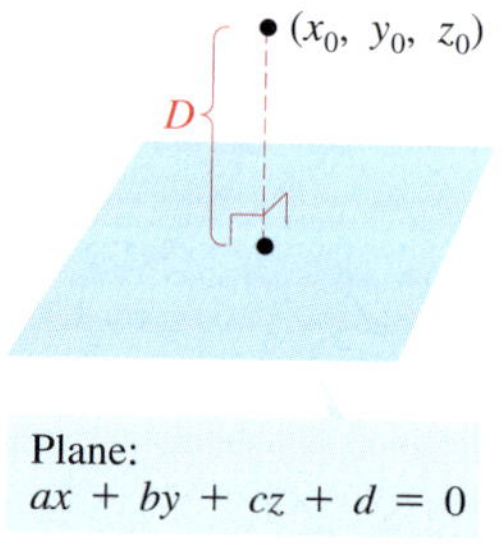

23. $(0, 0, 0),\ 2x + 3y + z = 12$

24. $(0, 0, 0),\ x - 3y + 4z = 6$

25. $(1, 5, -4),\ 3x - y + 2z = 6$

26. $(1, 2, 3),\ 2x - y + z = 4$

27. $(1, 0, -1),\ 2x - 4y + 3z = 12$

28. $(2, -1, 0),\ 3x + 3y + 2z = 6$

29. $(3, 2, -1),\ 2x - 3y + 4z = 24$

30. $(-2, 1, 0),\ 2x + 5y - z = 20$

In Exercises 31–38, match the given equation with the correct graph. [The graphs are labeled (a)–(h).]

31. $\dfrac{x^2}{9} + \dfrac{y^2}{16} + \dfrac{z^2}{9} = 1$

32. $x^2 - \dfrac{4y^2}{15} + z^2 = -\dfrac{4}{15}$

33. $4x^2 + 4y^2 - z^2 = 4$

34. $y^2 = 4x^2 + 9z^2$

35. $4x^2 - 4y + z^2 = 0$

36. $12z = -3y^2 + 4x^2$

37. $4x^2 - y^2 + 4z = 0$

38. $x^2 + y^2 + z^2 = 9$

(a)

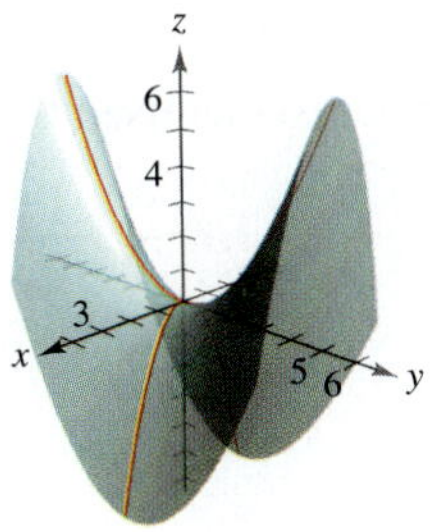

(b)

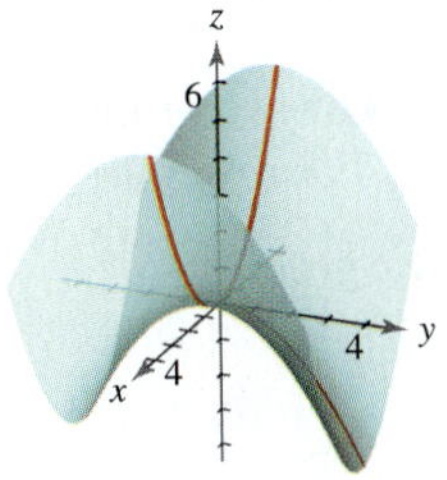

(c)

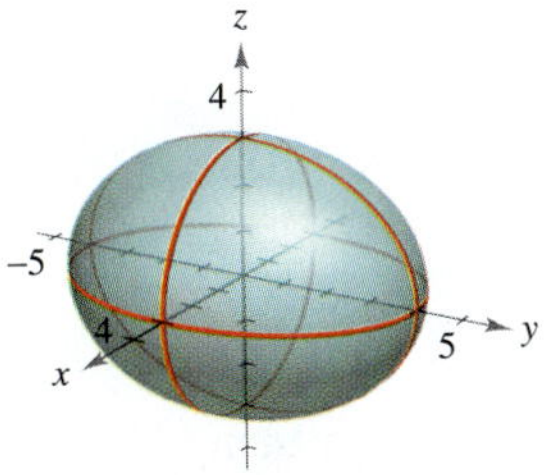

(d)

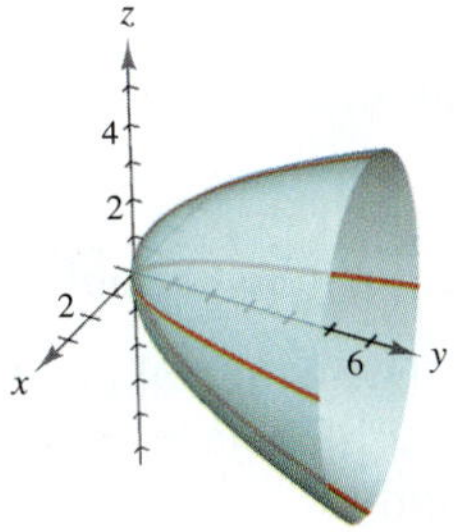

(e)

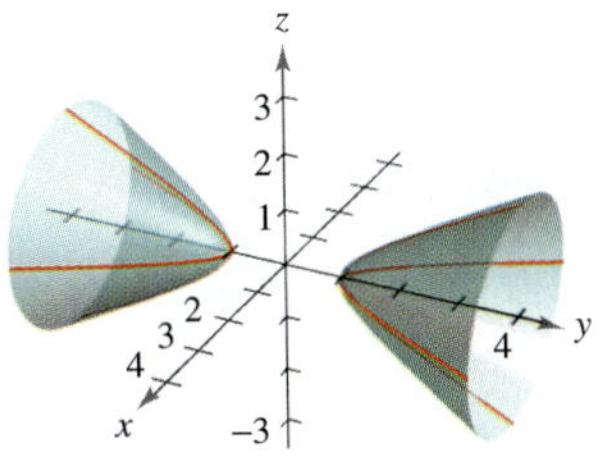

(f)

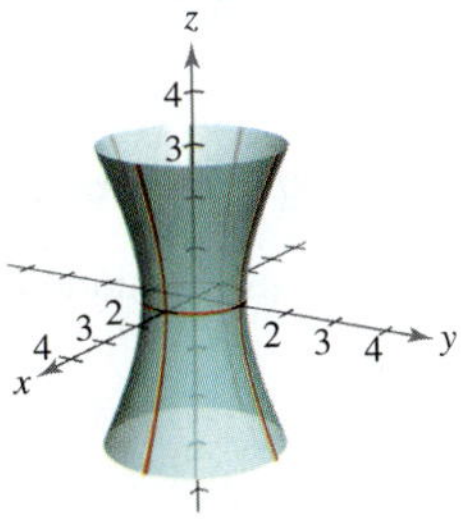

(g)

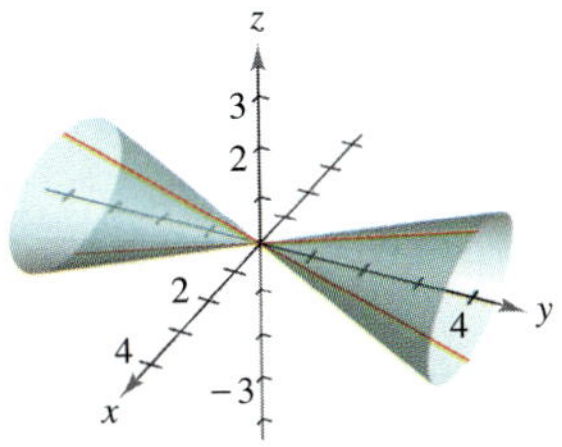

(h)

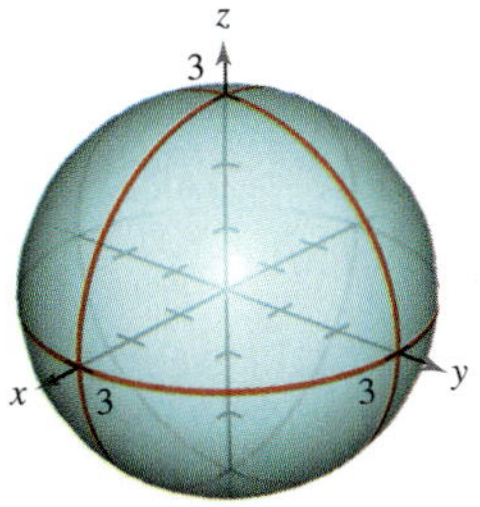

In Exercises 39–42, describe the traces of the surface in the given planes.

Surface	*Planes*
39. $x^2 - y - z^2 = 0$	xy-plane, $y = 1$, yz-plane
40. $y = x^2 + z^2$	xy-plane, $y = 1$, yz-plane
41. $\dfrac{x^2}{4} + y^2 + z^2 = 1$	xy-plane, xz-plane, yz-plane
42. $y^2 + z^2 - x^2 = 1$	xy-plane, xz-plane, yz-plane

In Exercises 43–56, identify the quadric surface.

43. $x^2 + \dfrac{y^2}{4} + z^2 = 1$

44. $\dfrac{x^2}{9} + \dfrac{y^2}{16} + \dfrac{z^2}{16} = 1$

45. $25x^2 + 25y^2 - z^2 = 5$

46. $9x^2 + 4y^2 - 8z^2 = 72$

47. $x^2 - y + z^2 = 0$

48. $z = 4x^2 + y^2$

49. $x^2 - y^2 + z = 0$

50. $z^2 - x^2 - \dfrac{y^2}{4} = 1$

51. $4x^2 - y^2 + 4z^2 = -16$

52. $z^2 = x^2 + \dfrac{y^2}{4}$

53. $z^2 = 9x^2 + y^2$

54. $4y = x^2 + z^2$

55. $3z = -y^2 + x^2$

56. $z^2 = 2x^2 + 2y^2$

In Exercises 57–60, use a three-dimensional graphing utility to graph the function.

57. $z = y^2 - x^2 + 1$

58. $z = x^2 + y^2 + 1$

59. $z = \dfrac{x^2}{2} + \dfrac{y^2}{4}$

60. $z = \frac{1}{12}\sqrt{144 - 16x^2 - 9y^2}$

61. ***Physical Science*** Because of the forces caused by its rotation, Earth is actually an oblate ellipsoid rather than a sphere. The equatorial radius is 3963 miles and the polar radius is 3950 miles. Find an equation of the ellipsoid. Assume that the center of Earth is at the origin and the xy-trace ($z = 0$) corresponds to the equator.

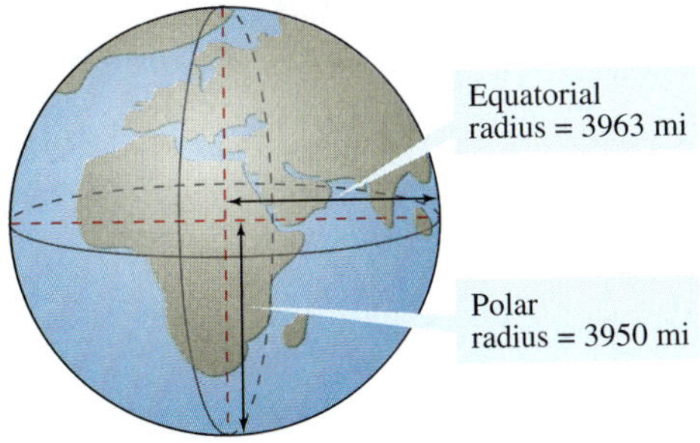

7.3 FUNCTIONS OF SEVERAL VARIABLES

- Evaluate functions of several variables.
- Find the domains and ranges of functions of several variables.
- Read contour maps and sketch level curves of functions of two variables.
- Use functions of several variables to answer questions about real-life situations.

Functions of Several Variables

In the first six chapters of this text, you studied functions of a single independent variable. Many quantities in science, business, and technology, however, are functions not of one, but of two or more variables. For instance, the demand function for a product is often dependent on the price *and* the advertising, rather than on the price alone.

The notation for functions of two or more variables is similar to that used for functions of a single variable. For example,

$$\underbrace{f(x, y)}_{\text{2 variables}} = x^2 + xy \quad \text{and} \quad \underbrace{g(x, y)}_{\text{2 variables}} = e^{x+y}$$

are functions of two variables, and

$$\underbrace{f(x, y, z)}_{\text{3 variables}} = x + 2y - 3z$$

is a function of three variables.

Definition of a Function of Two Variables

Let D be a set of ordered pairs of real numbers. If to each ordered pair (x, y) in D there corresponds a unique real number $z = f(x, y)$, then f is called a **function of x and y.** The set D is the **domain** of f, and the corresponding set of z-values is the **range** of f. Functions of three, four, or more variables are defined similarly.

TRY IT 1

Find the function values of $f(x, y)$.

(a) For $f(x, y) = x^2 + 2xy$, find $f(2, -1)$.

(b) For $f(x, y, z) = \dfrac{2x^2z}{y^3}$, find $f(-3, 2, 1)$.

EXAMPLE 1 **Evaluating Functions of Several Variables**

(a) For $f(x, y) = 2x^2 - y^2$, you can evaluate $f(2, 3)$ as shown.

$$\begin{aligned} f(2, 3) &= 2(2)^2 - (3)^2 \\ &= 8 - 9 \\ &= -1 \end{aligned}$$

(b) For $f(x, y, z) = e^x(y + z)$, you can evaluate $f(0, -1, 4)$ as shown.

$$\begin{aligned} f(0, -1, 4) &= e^0(-1 + 4) \\ &= (1)(3) \\ &= 3 \end{aligned}$$

The Graph of a Function of Two Variables

A function of two variables can be represented graphically as a surface in space by letting $z = f(x, y)$. When sketching the graph of a function of x and y, remember that even though the graph is three-dimensional, the domain of the function is two-dimensional—it consists of the points in the xy-plane for which the function is defined. As with functions of a single variable, unless specifically restricted, the domain of a function of two variables is assumed to be the set of all points (x, y) for which the defining equation has meaning. In other words, to each point (x, y) in the domain of f there corresponds a point (x, y, z) on the surface, and conversely, to each point (x, y, z) on the surface there corresponds a point (x, y) in the domain of f.

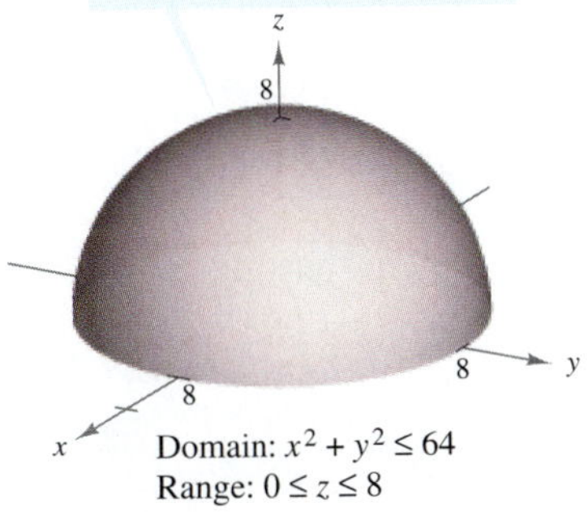

FIGURE 7.18

EXAMPLE 2 Finding the Domain and Range of a Function

Find the domain and range of the function

$$f(x, y) = \sqrt{64 - x^2 - y^2}.$$

SOLUTION Because no restrictions are given, the domain is assumed to be the set of all points for which the defining equation makes sense.

$$64 - x^2 - y^2 \ge 0 \qquad \text{Quantity inside radical must be nonnegative.}$$

$$x^2 + y^2 \le 64 \qquad \text{Domain of the function}$$

So, the domain is the set of all points that lie on or inside the circle given by $x^2 + y^2 = 8^2$. The range of f is the set

$$0 \le z \le 8. \qquad \text{Range of the function}$$

As shown in Figure 7.18, the graph of the function is a hemisphere.

TRY IT 2

Find the domain and range of the function

$$f(x, y) = \sqrt{9 - x^2 - y^2}.$$

TECHNOLOGY

Some three-dimensional graphing utilities can graph equations in x, y, and z. Others are programmed to graph only functions of x and y. A surface in space represents the graph of a function of x and y only if each vertical line intersects the surface at most once. For instance, the surface shown in Figure 7.18 passes this vertical line test, but the surface at the right (drawn by *Mathematica*) does not represent the graph of a function of x and y.

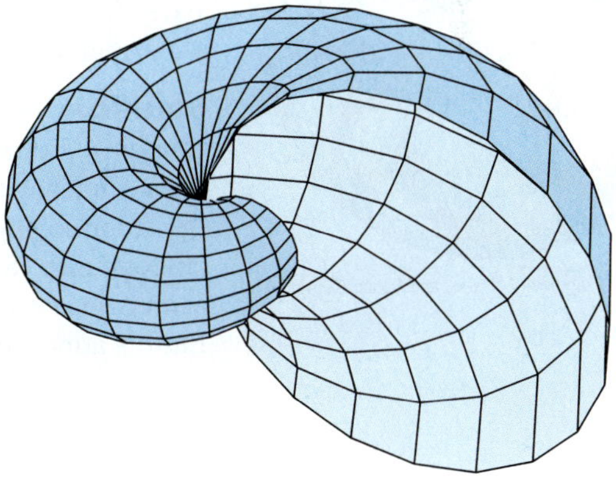

Some vertical lines intersect this surface more than once. So, the surface does not pass the vertical line test and is not a function of x and y.

Contour Maps and Level Curves

A **contour map** of a surface is created by *projecting* traces, taken in evenly spaced planes that are parallel to the xy-plane, onto the xy-plane. Each projection is a **level curve** of the surface.

Contour maps are used to create weather maps, topographical maps, and population density maps. For instance, Figure 7.19(a) shows a graph of a "mountain and valley" surface given by $z = f(x, y)$. Each of the level curves in Figure 7.19(b) represents the intersection of the surface $z = f(x, y)$ with a plane $z = c$, where $c = 828, 830, . . . , 854$.

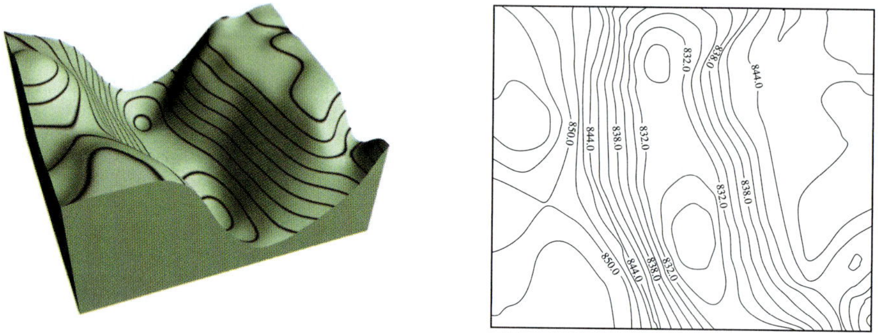

(a) Surface (b) Contour Map

FIGURE 7.19

EXAMPLE 3 Reading a Contour Map

The "contour map" in Figure 7.20 was computer generated using data collected by satellite instrumentation. The map uses color to represent levels of chlorine nitrate in the atmosphere. Chlorine nitrate contributes to the ozone depletion in the Earth's atmosphere. The red areas represent the highest level of chlorine nitrate and the dark blue areas represent the lowest level. Describe the areas that have the highest levels of chlorine nitrate. *(Source: Lockheed Missiles and Space Company)*

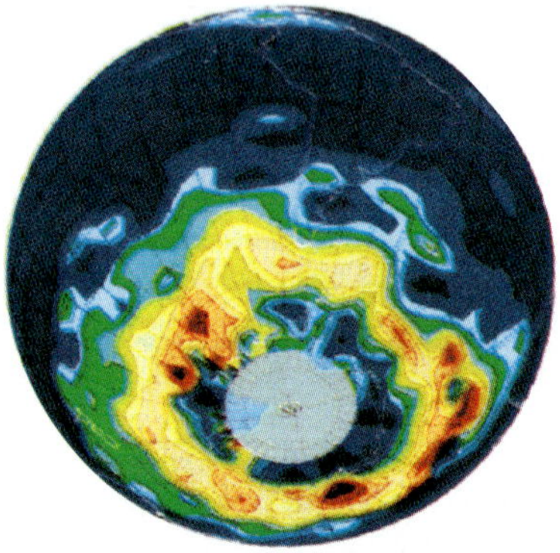

Lockheed Missiles and Space Company

FIGURE 7.20

SOLUTION The highest levels of chlorine nitrate are in the Antarctic Ocean, surrounding Antarctica. Although chlorine nitrate is not itself harmful to ozone, it has a tendency to convert to chlorine monoxide, which *is* harmful to ozone. Once the chlorine nitrate is converted to chlorine monoxide, it no longer shows on the contour map. So, Antarctica itself shows little chlorine nitrate—the nitrate has been converted to monoxide. If you have seen maps showing the "ozone hole" in Earth's atmosphere, you know that the hole occurs over Antarctica.

TRY IT 3

When the level curves of a contour map are close together, is the surface of the contour map steep or nearly level? When the level curves of a contour map are far apart, is the surface of the contour map steep or nearly level?

EXAMPLE 4 Reading a Contour Map

The contour map shown in Figure 7.21 represents the economy in the United States. Discuss the use of color to represent the level curves. *(Source: U.S. Census Bureau)*

SOLUTION You can see from the key that the light yellow regions are mainly used in crop production. The gray areas represent regions that are unproductive. Manufacturing centers are denoted by large red dots and mineral deposits are denoted by small black dots.

One advantage of such a map is that it allows you to "see" the components of the country's economy at a glance. From the map it is clear that the Midwest is responsible for most of the crop production in the United States.

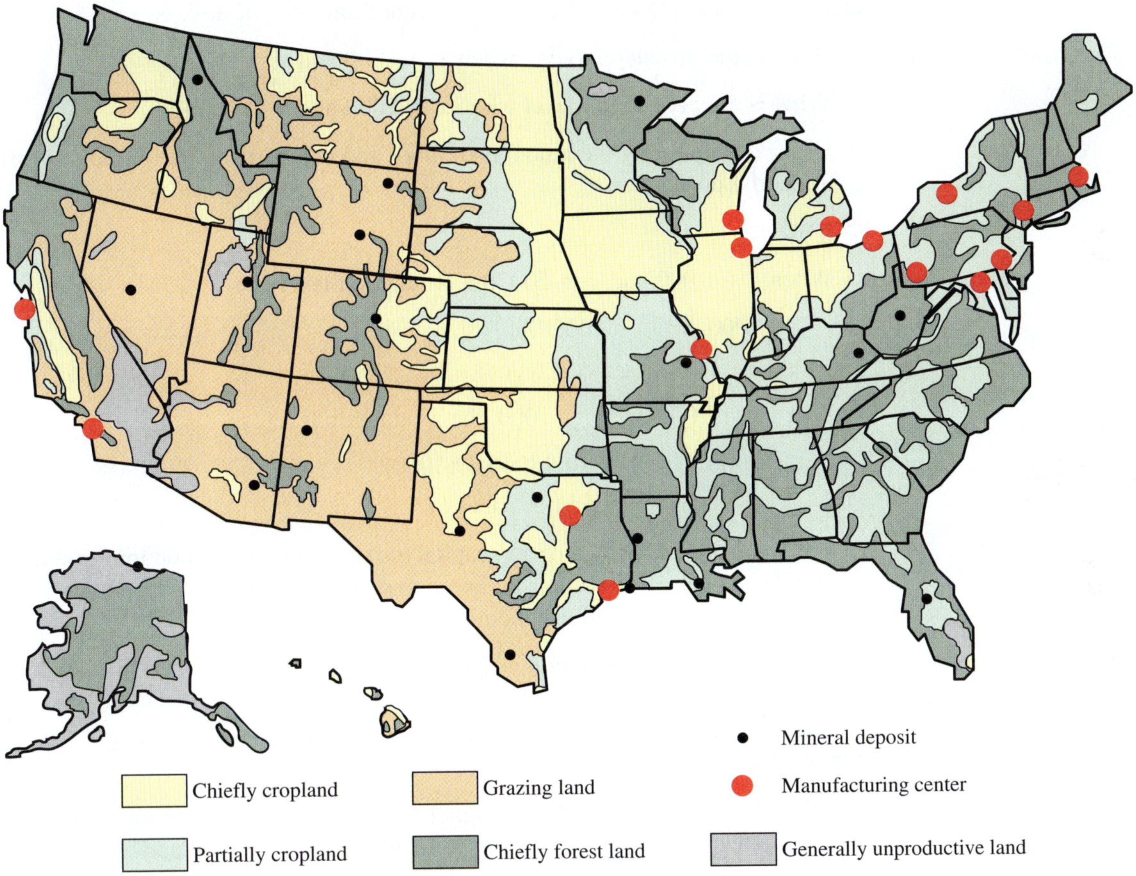

FIGURE 7.21

TRY IT 4

Use Figure 7.21 to describe how Alaska contributes to the U.S. economy. Does Alaska contain any manufacturing centers? Does Alaska contain any mineral deposits?

Applications

The **Cobb-Douglas production function** is used in economics to represent the numbers of units produced by varying amounts of labor and capital. Let x represent the number of units of labor and let y represent the number of units of capital. Then, the number of units produced is modeled by

$$f(x, y) = Cx^a y^{1-a}$$

where C is a constant and $0 < a < 1$.

EXAMPLE 5 Using a Production Function

A manufacturer estimates that its production (measured in units of a product) can be modeled by $f(x, y) = 100x^{0.6}y^{0.4}$ where the labor x is measured in person-hours and the capital y is measured in thousands of dollars.

(a) What is the production level when $x = 1000$ and $y = 500$?

(b) What is the production level when $x = 2000$ and $y = 1000$?

(c) How does doubling the amounts of labor and capital from part (a) to part (b) affect the production?

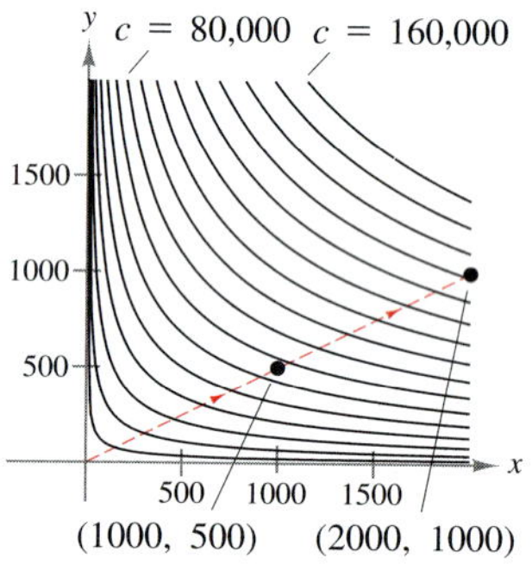

FIGURE 7.22 Level Curves (at Increments of 10,000)

SOLUTION

(a) When $x = 1000$ and $y = 500$, the production level is

$$f(1000, 500) = 100(1000)^{0.6}(500)^{0.4}$$
$$\approx 75{,}786 \text{ units.}$$

(b) When $x = 2000$ and $y = 1000$, the production level is

$$f(2000, 1000) = 100(2000)^{0.6}(1000)^{0.4}$$
$$\approx 151{,}572 \text{ units.}$$

(c) When the amounts of labor and capital are doubled, the production level also doubles. In Exercise 42, you are asked to show that this is characteristic of the Cobb-Douglas production function.

A contour graph of this function is shown in Figure 7.22.

TRY IT 5

Use the Cobb-Douglas production function in Example 5 to find the production levels when $x = 1500$ and $y = 1000$ and when $x = 1000$ and $y = 1500$. Use your results to determine which variable has a greater influence on production.

STUDY TIP

In Figure 7.22, note that the level curves of the function

$$f(x, y) = 100x^{0.6}y^{0.4}$$

occur at increments of 10,000.

EXAMPLE 6 **Finding Monthly Payments**

The monthly payment M for an installment loan of P dollars taken out over t years at an annual interest rate of r is given by

$$M = f(P, r, t) = \frac{\dfrac{Pr}{12}}{1 - \left[\dfrac{1}{1 + (r/12)}\right]^{12t}}.$$

(a) Find the monthly payment for a home mortgage of \$100,000 taken out for 30 years at an annual interest rate of 7%.

(b) Find the monthly payment for a car loan of \$22,000 taken out for 5 years at an annual interest rate of 8%.

SOLUTION

(a) If $P = \$100{,}000$, $r = 0.07$, and $t = 30$, then the monthly payment is

$$M = f(100{,}000, 0.07, 30)$$
$$= \frac{\dfrac{(100{,}000)(0.07)}{12}}{1 - \left[\dfrac{1}{1 + (0.07/12)}\right]^{12(30)}}$$
$$\approx \$665.30.$$

(b) If $P = \$22{,}000$, $r = 0.08$, and $t = 5$, then the monthly payment is

$$M = f(22{,}000, 0.08, 5)$$
$$= \frac{\dfrac{(22{,}000)(0.08)}{12}}{1 - \left[\dfrac{1}{1 + (0.08/12)}\right]^{12(5)}}$$
$$\approx \$446.08.$$

For many Americans, buying a house is the largest single purchase they will ever make. During the 1970s, 1980s, and 1990s, the annual interest rate on home mortgages varied drastically. It was as high as 18% and as low as 5%. Such variations can change monthly payments by hundreds of dollars.

TRY IT 6

(a) Find the monthly payment M for a home mortgage of \$100,000 taken out for 30 years at an annual interest rate of 8%.

(b) Find the total amount of money you will pay for the mortgage.

TAKE ANOTHER LOOK

Monthly Payments

You are taking out a home mortgage for \$120,000, and you are given the options below. Which option would you choose? Explain your reasoning.

a. A fixed annual rate of 8%, over a term of 20 years.

b. A fixed annual rate of 7%, over a term of 30 years.

c. An adjustable annual rate of 7%, over a term of 20 years. The annual rate can fluctuate—each year it is set at 1% above the prime rate.

d. A fixed rate of 7%, over a term of 15 years.

PREREQUISITE REVIEW 7.3

The following warm-up exercises involve skills that were covered in earlier sections. You will use these skills in the exercise set for this section.

In Exercises 1–4, evaluate the function when $x = -3$.

1. $f(x) = 5 - 2x$

2. $f(x) = -x^2 + 4x + 5$

3. $y = \sqrt{4x^2 - 3x + 4}$

4. $y = \sqrt[3]{34 - 4x + 2x^2}$

In Exercises 5–8, find the domain of the function.

5. $f(x) = 5x^2 + 3x - 2$

6. $g(x) = \dfrac{1}{2x} - \dfrac{2}{x+3}$

7. $h(y) = \sqrt{y - 5}$

8. $f(y) = \sqrt{y^2 - 5}$

In Exercises 9 and 10, evaluate the expression.

9. $(476)^{0.65}$

10. $(251)^{0.35}$

EXERCISES 7.3

In Exercises 1–14, find the function values.

1. $f(x, y) = \dfrac{x}{y}$

(a) $f(3, 2)$ (b) $f(-1, 4)$
(c) $f(30, 5)$ (d) $f(5, y)$
(e) $f(x, 2)$ (f) $f(5, t)$

2. $f(x, y) = 4 - x^2 - 4y^2$

(a) $f(0, 0)$ (b) $f(0, 1)$
(c) $f(2, 3)$ (d) $f(1, y)$
(e) $f(x, 0)$ (f) $f(t, 1)$

3. $f(x, y) = xe^y$

(a) $f(5, 0)$ (b) $f(3, 2)$
(c) $f(2, -1)$ (d) $f(5, y)$
(e) $f(x, 2)$ (f) $f(t, t)$

4. $g(x, y) = \ln|x + y|$

(a) $g(2, 3)$ (b) $g(5, 6)$
(c) $g(e, 0)$ (d) $g(0, 1)$
(e) $g(2, -3)$ (f) $g(e, e)$

5. $h(x, y, z) = \dfrac{xy}{z}$

(a) $h(2, 3, 9)$ (b) $h(1, 0, 1)$

6. $f(x, y, z) = \sqrt{x + y + z}$

(a) $f(0, 5, 4)$ (b) $f(6, 8, -3)$

7. $V(r, h) = \pi r^2 h$

(a) $V(3, 10)$ (b) $V(5, 2)$

8. $F(r, N) = 500\left(1 + \dfrac{r}{12}\right)^N$

(a) $F(0.09, 60)$ (b) $F(0.14, 240)$

9. $A(P, r, t) = P\left[\left(1 + \dfrac{r}{12}\right)^{12t} - 1\right]\left(1 + \dfrac{12}{r}\right)$

(a) $A(100, 0.10, 10)$ (b) $A(275, 0.0925, 40)$

10. $A(P, r, t) = Pe^{rt}$

(a) $A(500, 0.10, 5)$ (b) $A(1500, 0.12, 20)$

11. $f(x, y) = \displaystyle\int_x^y (2t - 3)\, dt$

(a) $f(1, 2)$ (b) $f(1, 4)$

12. $g(x, y) = \displaystyle\int_x^y \frac{1}{t}\, dt$

(a) $g(4, 1)$ (b) $g(6, 3)$

13. $f(x, y) = x^2 - 2y$

(a) $f(x + \Delta x, y)$

(b) $\dfrac{f(x, y + \Delta y) - f(x, y)}{\Delta y}$

14. $f(x, y) = 3xy + y^2$

(a) $f(x + \Delta x, y)$

(b) $\dfrac{f(x, y + \Delta y) - f(x, y)}{\Delta y}$

In Exercises 15–18, describe the region R in the xy-coordinate plane that corresponds to the domain of the function, and find the range of the function.

15. $f(x, y) = \sqrt{16 - x^2 - y^2}$

16. $f(x, y) = x^2 + y^2 - 1$

17. $f(x, y) = e^{x/y}$

18. $f(x, y) = \ln(x + y)$

In Exercises 19–28, describe the region R in the xy-coordinate plane that corresponds to the domain of the function.

19. $f(x, y) = \sqrt{9 - 9x^2 - y^2}$

20. $f(x, y) = \sqrt{x^2 + y^2 - 1}$

21. $f(x, y) = \dfrac{x}{y}$

22. $f(x, y) = \dfrac{4y}{x - 1}$

23. $f(x, y) = \dfrac{1}{xy}$

24. $g(x, y) = \dfrac{1}{x - y}$

25. $h(x, y) = x\sqrt{y}$

26. $f(x, y) = \sqrt{xy}$

27. $g(x, y) = \ln(4 - x - y)$

28. $f(x, y) = ye^{1/x}$

In Exercises 29–32, match the graph of the surface with one of the contour maps. [The contour maps are labeled (a)–(d).]

29. $f(x, y) = x^2 + \dfrac{y^2}{4}$

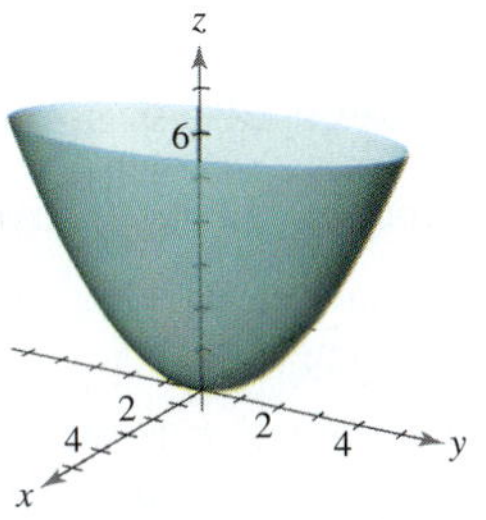

30. $f(x, y) = e^{1-x^2+y^2}$

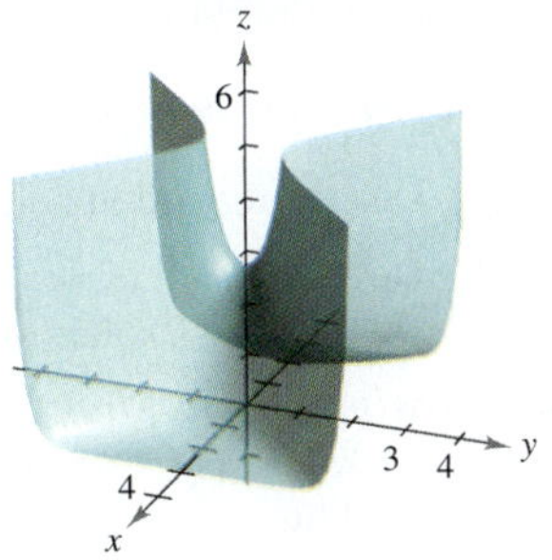

31. $f(x, y) = e^{1-x^2-y^2}$

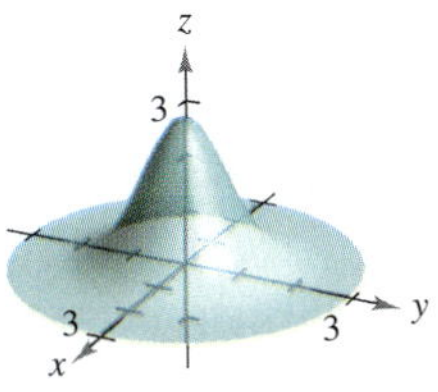

32. $f(x, y) = \ln|y - x^2|$

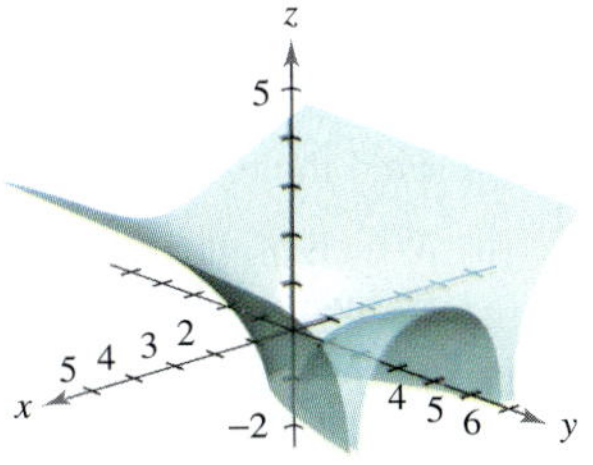

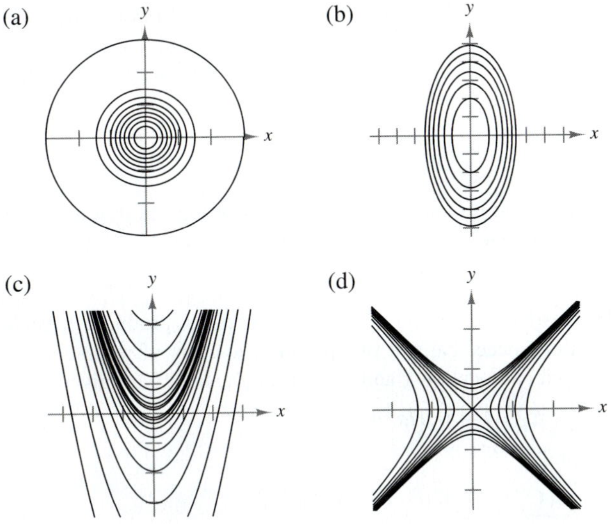

In Exercises 33–40, describe the level curves of the function. Sketch the level curves for the given c-values.

Function	*c-Values*
33. $z = x + y$	$c = -1, 0, 2, 4$
34. $z = 6 - 2x - 3y$	$c = 0, 2, 4, 6, 8, 10$
35. $z = \sqrt{16 - x^2 - y^2}$	$c = 0, 1, 2, 3, 4$
36. $f(x, y) = x^2 + y^2$	$c = 0, 2, 4, 6, 8$
37. $f(x, y) = xy$	$c = \pm 1, \pm 2, \ldots, \pm 6$
38. $z = e^{xy}$	$c = 1, 2, 3, 4, \frac{1}{2}, \frac{1}{3}, \frac{1}{4}$
39. $f(x, y) = \dfrac{x}{x^2 + y^2}$	$c = \pm\dfrac{1}{2}, \pm 1, \pm\dfrac{3}{2}, \pm 2$
40. $f(x, y) = \ln(x - y)$	$c = 0, \pm\frac{1}{2}, \pm 1, \pm\frac{3}{2}, \pm 2$

41. ***Cobb-Douglas Production Function*** A manufacturer estimates the Cobb-Douglas production function to be given by

$$f(x, y) = 100x^{0.75}y^{0.25}.$$

Estimate the production levels when $x = 1500$ and $y = 1000$.

42. ***Cobb-Douglas Production Function*** Use the Cobb-Douglas production function (Example 5) to show that if both the number of units of labor and the number of units of capital are doubled, the production level is also doubled.

43. ***Cost*** A company manufactures two types of woodburning stoves: a freestanding model and a fireplace-insert model. The cost function for producing x freestanding stoves and y fireplace-insert stoves is given by

$$C(x, y) = 27\sqrt{xy} + 195x + 215y + 980.$$

Find the cost when $x = 80$ and $y = 20$.

44. *Forestry* The *Doyle Log Rule* is one of several methods used to determine the lumber yield of a log in board-feet in terms of its diameter d in inches and its length L in feet. The number of board-feet is given by

$$N(d, L) = \left(\frac{d-4}{4}\right)^2 L.$$

(a) Find the number of board-feet of lumber in a log with a diameter of 22 inches and a length of 12 feet.

(b) Find $N(30, 12)$.

45. *Profit* A sporting goods manufacturer produces regulation soccer balls at two plants. The costs of producing x_1 units at location 1 and x_2 units at location 2 are given by

$$C_1(x_1) = 0.02x_1^2 + 4x_1 + 500$$

and

$$C_2(x_2) = 0.05x_2^2 + 4x_2 + 275$$

respectively. If the product sells for \$45 per unit, then the profit function for the product is given by

$$P(x_1, x_2) = 45(x_1 + x_2) - C_1(x_1) - C_2(x_2).$$

Find (a) $P(250, 150)$ and (b) $P(300, 200)$.

46. *Consumer Awareness* The average amount of time that a customer waits in line for service is given by

$$W(x, y) = \frac{1}{x-y}, \qquad y < x$$

where y is the average arrival rate and x is the average service rate (x and y are measured in the number of customers per hour). Evaluate W at each point.

(a) $(15, 10)$ (b) $(12, 9)$ (c) $(12, 6)$ (d) $(4, 2)$

47. *Investment* In 2004, an investment of \$1000 was made in a bond earning 10% compounded annually. The investor pays tax at rate R, and the annual rate of inflation is I. In the year 2014, the value V of the bond in constant 2004 dollars is given by

$$V(I, R) = 1000\left[\frac{1 + 0.10(1-R)}{1+I}\right]^{10}.$$

Use this function of two variables to complete the table.

		Inflation rate, I		
Tax rate, R		0.00	0.03	0.05
	0.00			
	0.28			
	0.35			

48. *Meteorology* Meteorologists measure the atmospheric pressure in millibars. From these observations they create weather maps on which the curves of equal atmospheric pressure (isobars) are drawn (see figure). On the map, the closer the isobars the higher the wind speed. Match points A, B, and C with (a) highest pressure, (b) lowest pressure, and (c) highest wind velocity.

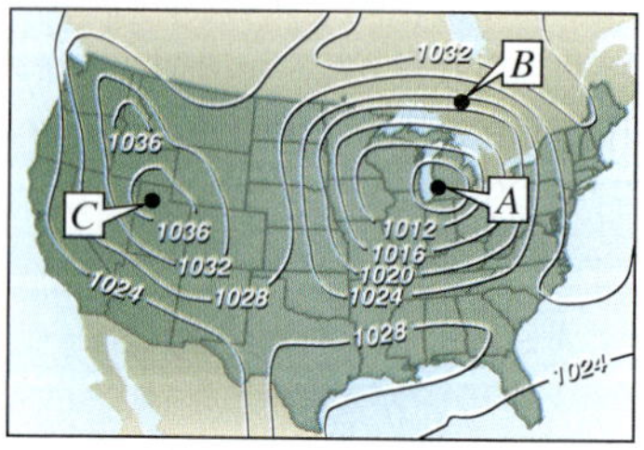

49. *Geology: A Contour Map* The contour map below represents color-coded seismic amplitudes of a fault horizon and a projected contour map, which is used in earthquake studies. *(Source: Adapted from Shipman/Wilson/Todd,* An Introduction to Physical Science, *Tenth Edition)*

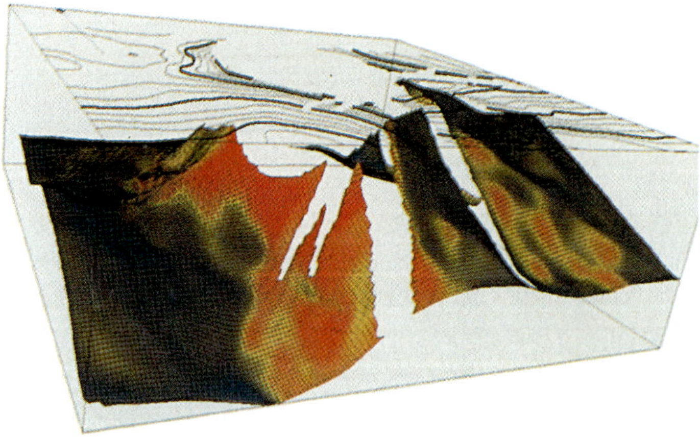

Shipman, *An Introduction to Physical Science* 10/e, 2003, Houghton Mifflin Company

(a) Discuss the use of color to represent the level curves.

(b) Do the level curves correspond to equally spaced amplitudes? Explain your reasoning.

50. *Earnings per Share* The earnings per share for Starbucks Corporation from 1995 through 2003 can be modeled by

$$z = 0.265x - 0.209y + 0.033$$

where x is sales (in billions of dollars) and y is the shareholder's equity (in billions of dollars). *(Source: Starbucks Corporation)*

(a) Find the earnings per share when $x = 8$ and $y = 5$.

(b) Which of the two variables in this model has the greater influence on the earnings per share? Explain your reasoning.

7.4 PARTIAL DERIVATIVES

- Find the first partial derivatives of functions of two variables.
- Find the slopes of surfaces in the x- and y-directions and use partial derivatives to answer questions about real-life situations.
- Find the partial derivatives of functions of several variables.
- Find higher-order partial derivatives.

Functions of Two Variables

Real-life applications of functions of several variables are often concerned with how changes in one of the variables will affect the values of the functions. For instance, an economist who wants to determine the effect of a tax increase on the economy might make calculations using different tax rates while holding all other variables, such as unemployment, constant.

You can follow a similar procedure to find the rate of change of a function f with respect to one of its independent variables. That is, you find the derivative of f with respect to one independent variable, while holding the other variables constant. This process is called **partial differentiation,** and each derivative is called a **partial derivative.** A function of several variables has as many partial derivatives as it has independent variables.

Partial Derivatives of a Function of Two Variables

If $z = f(x, y)$, then the **first partial derivatives of f with respect to x and y** are the functions $\partial z/\partial x$ and $\partial z/\partial y$, defined as shown.

$$\frac{\partial z}{\partial x} = \lim_{\Delta x \to 0} \frac{f(x + \Delta x, y) - f(x, y)}{\Delta x} \qquad y \text{ is held constant.}$$

$$\frac{\partial z}{\partial y} = \lim_{\Delta y \to 0} \frac{f(x, y + \Delta y) - f(x, y)}{\Delta y} \qquad x \text{ is held constant.}$$

STUDY TIP

Note that this definition indicates that partial derivatives of a function of two variables are determined by temporarily considering one variable to be fixed. For instance, if $z = f(x, y)$, then to find $\partial z/\partial x$, you consider y to be constant and differentiate with respect to x. Similarly, to find $\partial z/\partial y$, you consider x to be constant and differentiate with respect to y.

EXAMPLE 1 Finding Partial Derivatives

Find $\partial z/\partial x$ and $\partial z/\partial y$ for the function $z = 3x - x^2y^2 + 2x^3y$.

SOLUTION

$$\frac{\partial z}{\partial x} = 3 - 2xy^2 + 6x^2y \qquad \text{Hold } y \text{ constant and differentiate with respect to } x.$$

$$\frac{\partial z}{\partial y} = -2x^2y + 2x^3 \qquad \text{Hold } x \text{ constant and differentiate with respect to } y.$$

TRY IT 1

Find $\dfrac{\partial z}{\partial x}$ and $\dfrac{\partial z}{\partial y}$ for $z = 2x^2 - 4x^2y^3 + y^4$.

Notation for First Partial Derivatives

The first partial derivatives of $z = f(x, y)$ are denoted by

$$\frac{\partial z}{\partial x} = f_x(x, y) = z_x = \frac{\partial}{\partial x}[f(x, y)]$$

and

$$\frac{\partial z}{\partial y} = f_y(x, y) = z_y = \frac{\partial}{\partial y}[f(x, y)].$$

The values of the first partial derivatives at the point (a, b) are denoted by

$$\left.\frac{\partial z}{\partial x}\right|_{(a,\, b)} = f_x(a, b) \quad \text{and} \quad \left.\frac{\partial z}{\partial y}\right|_{(a,\, b)} = f_y(a, b).$$

TECHNOLOGY

Symbolic differentiation utilities can be used to find partial derivatives of a function of two variables. Try using a symbolic differentiation utility to find the first partial derivatives of the function in Example 2.

EXAMPLE 2 **Finding and Evaluating Partial Derivatives**

Find the first partial derivatives of $f(x, y) = xe^{x^2y}$ and evaluate each at the point $(1, \ln 2)$.

SOLUTION To find the first partial derivative with respect to x, hold y constant and differentiate using the Product Rule.

$$\begin{aligned} f_x(x, y) &= x\frac{\partial}{\partial x}[e^{x^2y}] + e^{x^2y}\frac{\partial}{\partial x}[x] && \text{Apply Product Rule.} \\ &= x(2xy)e^{x^2y} + e^{x^2y} && y \text{ is held constant.} \\ &= e^{x^2y}(2x^2y + 1) && \text{Simplify.} \end{aligned}$$

At the point $(1, \ln 2)$, the value of this derivative is

$$\begin{aligned} f_x(1, \ln 2) &= e^{(1)^2(\ln 2)}[2(1)^2(\ln 2) + 1] && \text{Substitute for } x \text{ and } y. \\ &= 2(2 \ln 2 + 1) && \text{Simplify.} \\ &\approx 4.773. && \text{Use a calculator.} \end{aligned}$$

To find the first partial derivative with respect to y, hold x constant and differentiate to obtain

$$\begin{aligned} f_y(x, y) &= x(x^2)e^{x^2y} && \text{Apply Constant Multiple Rule.} \\ &= x^3e^{x^2y}. && \text{Simplify.} \end{aligned}$$

At the point $(1, \ln 2)$, the value of this derivative is

$$\begin{aligned} f_y(1, \ln 2) &= (1)^3e^{(1)^2(\ln 2)} && \text{Substitute for } x \text{ and } y. \\ &= 2 && \text{Simplify.} \end{aligned}$$

TRY IT 2

Find the first partial derivatives of $f(x, y) = x^2y^3$ and evaluate each at the point $(1, 2)$.

Graphical Interpretation of Partial Derivatives

At the beginning of this course, you studied graphical interpretations of the derivative of a function of a single variable. There, you found that $f'(x_0)$ represents the slope of the tangent line to the graph of $y = f(x)$ at the point (x_0, y_0). The partial derivatives of a function of two variables also have useful graphical interpretations. Consider the function

$z = f(x, y)$. Function of two variables

As shown in Figure 7.23(a), the graph of this function is a surface in space. If the variable y is fixed, say at $y = y_0$, then

$z = f(x, y_0)$ Function of one variable

is a function of one variable. The graph of this function is the curve that is the intersection of the plane $y = y_0$ and the surface $z = f(x, y)$. On this curve, the partial derivative

$f_x(x, y_0)$ Slope in x-direction

represents the slope in the plane $y = y_0$, as shown in Figure 7.23(a). In a similar way, if the variable x is fixed, say at $x = x_0$, then

$z = f(x_0, y)$ Function of one variable

is a function of one variable. Its graph is the intersection of the plane $x = x_0$ and the surface $z = f(x, y)$. On this curve, the partial derivative

$f_y(x_0, y)$ Slope in y-direction

represents the slope in the plane $x = x_0$, as shown in Figure 7.23(b).

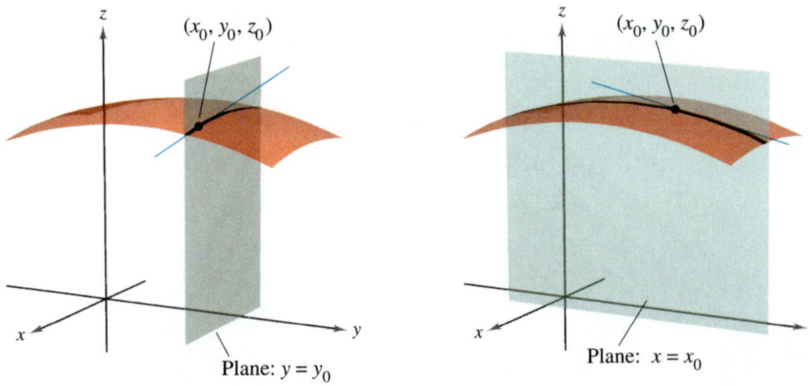

(a) $f_x(x, y_0)$ = slope in x-direction

(b) $f_y(x_0, y)$ = slope in y-direction

FIGURE 7.23

DISCOVERY

How can partial derivatives be used to find *relative extrema* of graphs of functions of two variables?

EXAMPLE 3 Finding Slopes in the x- and y-Directions

Find the slopes of the surface given by

$$f(x, y) = -\frac{x^2}{2} - y^2 + \frac{25}{8}$$

at the point $\left(\frac{1}{2}, 1, 2\right)$ in (a) the x-direction and (b) the y-direction.

SOLUTION

(a) To find the slope in the x-direction, hold y constant and differentiate with respect to x to obtain

$$f_x(x, y) = -x. \qquad \text{Partial derivative with respect to } x$$

At the point $\left(\frac{1}{2}, 1, 2\right)$, the slope in the x-direction is

$$f_x\left(\tfrac{1}{2}, 1\right) = -\tfrac{1}{2} \qquad \text{Slope in } x\text{-direction}$$

as shown in Figure 7.24(a).

(b) To find the slope in the y-direction, hold x constant and differentiate with respect to y to obtain

$$f_y(x, y) = -2y. \qquad \text{Partial derivative with respect to } y$$

At the point $\left(\frac{1}{2}, 1, 2\right)$, the slope in the y-direction is

$$f_y\left(\tfrac{1}{2}, 1\right) = -2 \qquad \text{Slope in } y\text{-direction}$$

as shown in Figure 7.24(b).

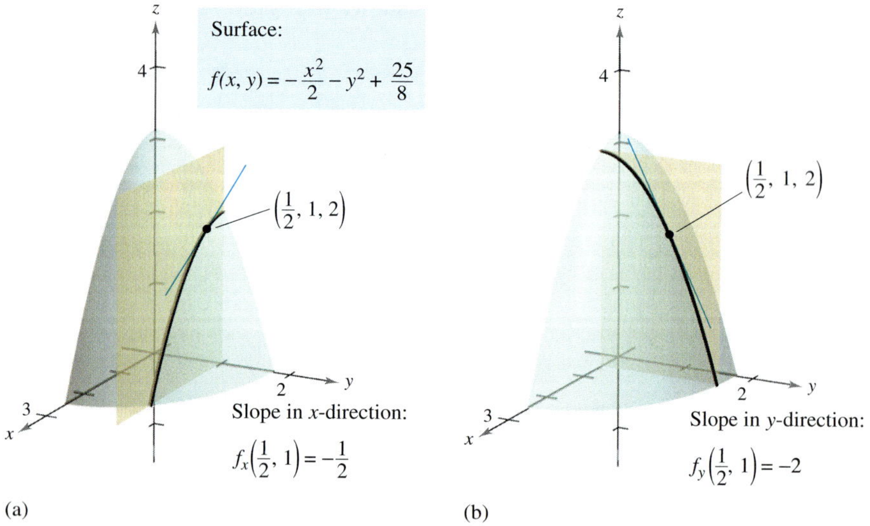

FIGURE 7.24

TRY IT 3

Find the slopes of the surface given by

$$f(x, y) = 4x^2 + 9y^2 + 36$$

at the point $(1, -1, 49)$ in the x-direction and y-direction.

DISCOVERY

Find the partial derivatives f_x and f_y at $(0, 0)$ for the function in Example 3. What are the slopes of f in the x- and y-directions at $(0, 0)$? Describe the shape of the graph of f at this point.

Consumer products in the same market or in related markets can be classified as **complementary** or **substitute products.** If two products have a complementary relationship, an increase in the sale of one product will be accompanied by an increase in the sale of the other product. For instance, DVD players and DVDs have a complementary relationship.

If two products have a substitute relationship, an increase in the sale of one product will be accompanied by a decrease in the sale of the other product. For instance, videocassette recorders and DVD players both compete in the same home entertainment market and you would expect a drop in the price of one to be a deterrent to the sale of the other.

EXAMPLE 4 Examining a Demand Function

The demand functions for two products are represented by

$$x_1 = f(p_1, p_2) \quad \text{and} \quad x_2 = g(p_1, p_2)$$

where p_1 and p_2 are the prices per unit for the two products, and x_1 and x_2 are the numbers of units sold. The graphs of two different demand functions for x_1 are shown below. Use them to classify the products as complementary or substitute products.

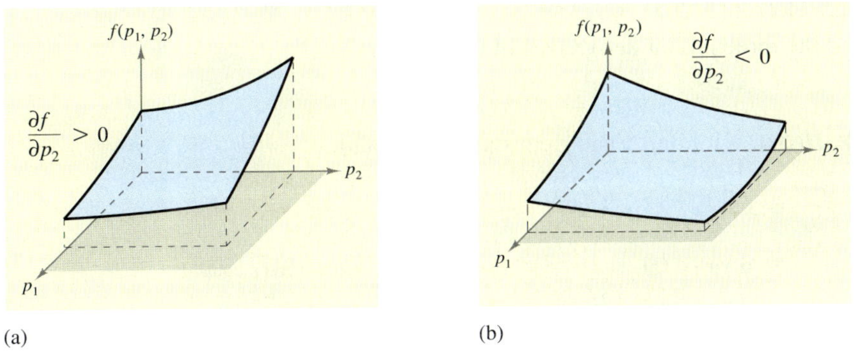

FIGURE 7.25

SOLUTION

(a) Notice that Figure 7.25(a) represents the demand for the *first product*. From the graph of this function, you can see that for a fixed price p_1, an increase in p_2 results in an increase in the demand for the first product. Remember that an increase in p_2 will also result in a decrease in the demand for the second product. So, if $\partial f/\partial p_2 > 0$, the two products have a *substitute* relationship.

(b) Notice that Figure 7.25(b) represents a different demand for the *first product*. From the graph of this function, you can see that for a fixed price p_1, an increase in p_2 results in a decrease in the demand for the first product. Remember that an increase in p_2 will also result in a decrease in the demand for the second product. So, if $\partial f/\partial p_2 < 0$, the two products have a *complementary* relationship.

TRY IT 4

Determine if the demand functions below describe a complementary or a substitute product relationship.

$$x_1 = 100 - 2p_1 + 1.5p_2$$
$$x_2 = 145 + \tfrac{1}{2}p_1 - \tfrac{3}{4}p_2$$

Courtesy of Subway

In 2004, Subway was chosen as the number one franchise by Entrepreneur Magazine. By the end of the year 2003, Subway had a total of 20,941 franchises worldwide. What type of product would be complementary to a Subway sandwich? What type of product would be a substitute?

Functions of Three Variables

The concept of a partial derivative can be extended in a natural way to functions of three or more variables. For instance, the function $w = f(x, y, z)$ has three partial derivatives, each of which is formed by considering two of the variables to be constant. That is, to define the partial derivative of w with respect to x, consider y *and* z to be constant and write

$$\frac{\partial w}{\partial x} = f_x(x, y, z) = \lim_{\Delta x \to 0} \frac{f(x + \Delta x, y, z) - f(x, y, z)}{\Delta x}.$$

To define the partial derivative of w with respect to y, consider x *and* z to be constant and write

$$\frac{\partial w}{\partial y} = f_y(x, y, z) = \lim_{\Delta y \to 0} \frac{f(x, y + \Delta y, z) - f(x, y, z)}{\Delta y}.$$

To define the partial derivative of w with respect to z, consider x *and* y to be constant and write

$$\frac{\partial w}{\partial z} = f_z(x, y, z) = \lim_{\Delta z \to 0} \frac{f(x, y, z + \Delta z) - f(x, y, z)}{\Delta z}.$$

EXAMPLE 5 Finding Partial Derivatives of a Function

Find the three partial derivatives of the function

$$w = xe^{xy+2z}.$$

SOLUTION Holding y and z constant, you obtain

$$\frac{\partial w}{\partial x} = x\frac{\partial}{\partial x}[e^{xy+2z}] + e^{xy+2z}\frac{\partial}{\partial x}[x] \quad \text{Apply Product Rule.}$$

$$= x(ye^{xy+2z}) + e^{xy+2z}(1) \quad \text{Hold } y \text{ and } z \text{ constant.}$$

$$= (xy + 1)e^{xy+2z}. \quad \text{Simplify.}$$

Holding x and z constant, you obtain

$$\frac{\partial w}{\partial y} = x(x)e^{xy+2z} \quad \text{Hold } x \text{ and } z \text{ constant.}$$

$$= x^2e^{xy+2z}. \quad \text{Simplify.}$$

Holding x and y constant, you obtain

$$\frac{\partial w}{\partial z} = x(2)e^{xy+2z} \quad \text{Hold } x \text{ and } y \text{ constant.}$$

$$= 2xe^{xy+2z}. \quad \text{Simplify.}$$

TECHNOLOGY

A symbolic differentiation utility can be used to find the partial derivatives of a function of three or more variables. Try using a symbolic differentiation utility to find the partial derivative $f_y(x, y, z)$ for the function in Example 5.

STUDY TIP

Note that in Example 5 the Product Rule was used only when finding the partial derivative with respect to x. Can you see why?

TRY IT 5

Find the three partial derivatives of the function

$$w = x^2y\ln(xz).$$

Higher-Order Partial Derivatives

As with ordinary derivatives, it is possible to find partial derivatives of second, third, or higher order. For instance, there are four different ways to find a second partial derivative of $z = f(x, y)$.

$$\frac{\partial}{\partial x}\left(\frac{\partial f}{\partial x}\right) = \frac{\partial^2 f}{\partial x^2} = f_{xx}$$ Differentiate twice with respect to x.

$$\frac{\partial}{\partial y}\left(\frac{\partial f}{\partial y}\right) = \frac{\partial^2 f}{\partial y^2} = f_{yy}$$ Differentiate twice with respect to y.

$$\frac{\partial}{\partial y}\left(\frac{\partial f}{\partial x}\right) = \frac{\partial^2 f}{\partial y \partial x} = f_{xy}$$ Differentiate first with respect to x and then with respect to y.

$$\frac{\partial}{\partial x}\left(\frac{\partial f}{\partial y}\right) = \frac{\partial^2 f}{\partial x \partial y} = f_{yx}$$ Differentiate first with respect to y and then with respect to x.

The third and fourth cases are *mixed* partial derivatives. Notice that with the two types of notation for mixed partials, different conventions are used for indicating the order of differentiation. For instance, the partial derivative

$$\frac{\partial}{\partial y}\left(\frac{\partial f}{\partial x}\right) = \frac{\partial^2 f}{\partial y \partial x}$$ Right-to-left order

indicates differentiation with respect to x first, but the partial derivative

$$(f_y)_x = f_{yx}$$ Left-to-right order

indicates differentiation with respect to y first. To remember this, note that in each case you differentiate first with respect to the variable "nearest" f.

EXAMPLE 6 Finding Second Partial Derivatives

Find the second partial derivatives of

$$f(x, y) = 3xy^2 - 2y + 5x^2y^2$$

and determine the value of $f_{xy}(-1, 2)$.

SOLUTION Begin by finding the first partial derivatives.

$$f_x(x, y) = 3y^2 + 10xy^2 \qquad f_y(x, y) = 6xy - 2 + 10x^2y$$

Then, differentiating with respect to x and y produces

$$f_{xx}(x, y) = 10y^2, \qquad f_{yy}(x, y) = 6x + 10x^2$$
$$f_{xy}(x, y) = 6y + 20xy, \qquad f_{yx}(x, y) = 6y + 20xy.$$

Finally, the value of $f_{xy}(x, y)$ at the point $(-1, 2)$ is

$$f_{xy}(-1, 2) = 6(2) + 20(-1)(2) = 12 - 40 = -28.$$

STUDY TIP

Notice in Example 6 that the two mixed partials are equal. This is often the case. In fact, it can be shown that if a function has continuous second partial derivatives, then the order in which the partial derivatives are taken is irrelevant.

TRY IT 6

Find the second partial derivatives of

$$f(x, y) = 4x^2y^2 + 2x + 4y^2.$$

A function of two variables has two first partial derivatives and four second partial derivatives. For a function of three variables, there are three first partials

$$f_x,\ f_y,\quad \text{and}\quad f_z$$

and nine second partials

$$f_{xx},\ f_{xy},\ f_{xz},\ f_{yx},\ f_{yy},\ f_{yz},\ f_{zx},\ f_{zy},\quad \text{and}\quad f_{zz}$$

of which six are mixed partials. To find partial derivatives of order three and higher, follow the same pattern used to find second partial derivatives. For instance, if $z = f(x, y)$, then

$$z_{xxx} = \frac{\partial}{\partial x}\left(\frac{\partial^2 f}{\partial x^2}\right) = \frac{\partial^3 f}{\partial x^3} \quad \text{and} \quad z_{xxy} = \frac{\partial}{\partial y}\left(\frac{\partial^2 f}{\partial x^2}\right) = \frac{\partial^3 f}{\partial y \partial x^2}.$$

EXAMPLE 7 Finding Second Partial Derivatives

Find the second partial derivatives of

$$f(x, y, z) = ye^x + x \ln z.$$

SOLUTION Begin by finding the first partial derivatives.

$$f_x(x, y, z) = ye^x + \ln z, \qquad f_y(x, y, z) = e^x, \qquad f_z(x, y, z) = \frac{x}{z}$$

Then, differentiate with respect to x, y, and z to find the nine second partial derivatives.

$$f_{xx}(x, y, z) = ye^x, \qquad f_{xy}(x, y, z) = e^x, \qquad f_{xz}(x, y, z) = \frac{1}{z}$$

$$f_{yx}(x, y, z) = e^x, \qquad f_{yy}(x, y, z) = 0, \qquad f_{yz}(x, y, z) = 0$$

$$f_{zx}(x, y, z) = \frac{1}{z}, \qquad f_{zy}(x, y, z) = 0, \qquad f_{zz}(x, y, z) = -\frac{x}{z^2}$$

TRY IT 7

Find the second partial derivatives of

$$f(x, y, z) = xe^y + 2xz + y^2.$$

TAKE ANOTHER LOOK

Interpreting a Partial Derivative

The upper half of the sphere shown at the right is given by

$$f(x, y) = \sqrt{1 - x^2 - y^2}.$$

a. Describe the values of $f_x(x, 0)$. Can you find a value of x for which this partial derivative is 1? Use a three-dimensional graphing utility to illustrate your answer graphically.

b. Describe the values of $f_y(0, y)$. Can you find a value of y for which this partial derivative is 1? Use a three-dimensional graphing utility to illustrate your answer graphically.

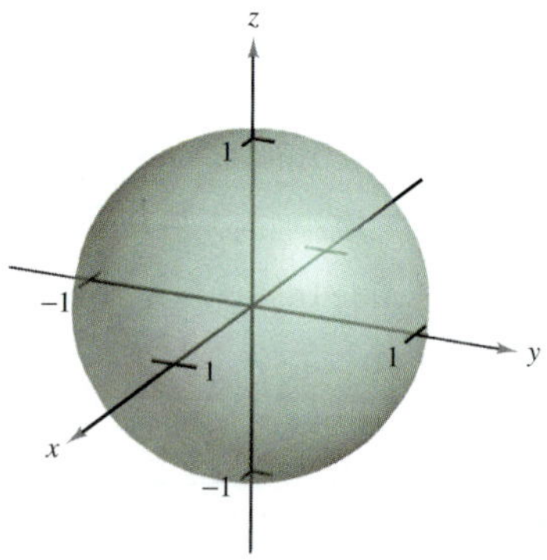

PREREQUISITE REVIEW 7.4

The following warm-up exercises involve skills that were covered in earlier sections. You will use these skills in the exercise set for this section.

In Exercises 1–8, find the derivative of the function.

1. $f(x) = \sqrt{x^2 + 3}$

2. $g(x) = (3 - x^2)^3$

3. $g(t) = te^{2t+1}$

4. $f(x) = e^{2x}\sqrt{1 - e^{2x}}$

5. $f(x) = \ln(3 - 2x)$

6. $u(t) = \ln\sqrt{t^3 - 6t}$

7. $g(x) = \dfrac{5x^2}{(4x - 1)^2}$

8. $f(x) = \dfrac{(x + 2)^3}{(x^2 - 9)^2}$

In Exercises 9 and 10, evaluate the derivative at the point (2, 4).

9. $f(x) = x^2e^{x-2}$

10. $g(x) = x\sqrt{x^2 - x + 2}$

EXERCISES 7.4

In Exercises 1–14, find the first partial derivatives with respect to *x* and with respect to *y*.

1. $f(x, y) = 2x - 3y + 5$

2. $f(x, y) = x^2 - 3y^2 + 7$

3. $f(x, y) = 5\sqrt{x} - 6y^2$

4. $f(x, y) = x^{-1/2} + 4y^{3/2}$

5. $f(x, y) = \dfrac{x}{y}$

6. $z = x\sqrt{y}$

7. $f(x, y) = \sqrt{x^2 + y^2}$

8. $f(x, y) = \dfrac{xy}{x^2 + y^2}$

9. $z = x^2e^{2y}$

10. $z = xe^{x+y}$

11. $h(x, y) = e^{-(x^2+y^2)}$

12. $g(x, y) = e^{x/y}$

13. $z = \ln\dfrac{x - y}{(x + y)^2}$

14. $g(x, y) = \ln\sqrt{x^2 + y^2}$

In Exercises 15–20, let $f(x, y) = 3x^2ye^{x-y}$ and $g(x, y) = 3xy^2e^{y-x}$. Find each of the following.

15. $f_x(x, y)$

16. $f_y(x, y)$

17. $g_x(x, y)$

18. $g_y(x, y)$

19. $f_x(1, 1)$

20. $g_x(-2, -2)$

In Exercises 21–28, evaluate f_x and f_y at the point.

	Function	*Point*
21.	$f(x, y) = 3x^2 + xy - y^2$	$(2, 1)$
22.	$f(x, y) = x^2 - 3xy + y^2$	$(1, -1)$
23.	$f(x, y) = e^{3xy}$	$(0, 4)$
24.	$f(x, y) = e^xy^2$	$(0, 2)$
25.	$f(x, y) = \dfrac{xy}{x - y}$	$(2, -2)$
26.	$f(x, y) = \dfrac{4xy}{\sqrt{x^2 + y^2}}$	$(1, 0)$
27.	$f(x, y) = \ln(x^2 + y^2)$	$(1, 0)$
28.	$f(x, y) = \ln\sqrt{xy}$	$(-1, -1)$

In Exercises 29–32, find the first partial derivatives with respect to *x*, *y*, and *z*.

29. $w = 3x^2y - 5xyz + 10yz^2$

30. $w = \sqrt{x^2 + y^2 + z^2}$

31. $w = \dfrac{xy}{x + y + z}$

32. $w = \dfrac{1}{\sqrt{1 - x^2 - y^2 - z^2}}$

In Exercises 33–38, evaluate w_x, w_y, and w_z at the point.

	Function	*Point*
33.	$w = \sqrt{x^2 + y^2 + z^2}$	$(2, -1, 2)$
34.	$w = \dfrac{xy}{x + y + z}$	$(1, 2, 0)$
35.	$w = \ln\sqrt{x^2 + y^2 + z^2}$	$(3, 0, 4)$
36.	$w = \dfrac{1}{\sqrt{1 - x^2 - y^2 - z^2}}$	$(0, 0, 0)$
37.	$w = 2xz^2 + 3xyz - 6y^2z$	$(1, -1, 2)$
38.	$w = xye^{z^2}$	$(2, 1, 0)$

In Exercises 39–42, find values of x and y such that $f_x(x, y) = 0$ and $f_y(x, y) = 0$ simultaneously.

39. $f(x, y) = x^2 + 4xy + y^2 - 4x + 16y + 3$

40. $f(x, y) = 3x^3 - 12xy + y^3$

41. $f(x, y) = \frac{1}{x} + \frac{1}{y} + xy$

42. $f(x, y) = \ln(x^2 + y^2 + 1)$

In Exercises 43–50, find the slope of the surface at the given point in (a) the x-direction and (b) the y-direction.

	Function	*Point*
43.	$z = 2x - 3y + 5$	$(2, 1, 6)$
44.	$z = xy$	$(1, 2, 2)$
45.	$z = x^2 - 9y^2$	$(3, 1, 0)$
46.	$z = x^2 + 4y^2$	$(2, 1, 8)$
47.	$z = \sqrt{25 - x^2 - y^2}$	$(3, 0, 4)$
48.	$z = \frac{x}{y}$	$(3, 1, 3)$
49.	$z = 4 - x^2 - y^2$	$(1, 1, 2)$

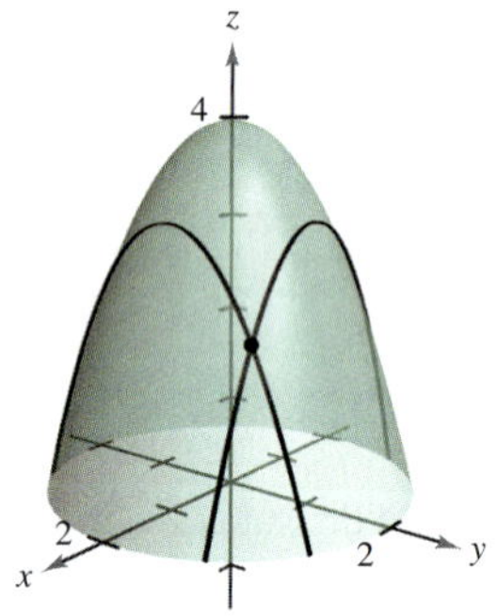

50. $z = x^2 - y^2$ $(-2, 1, 3)$

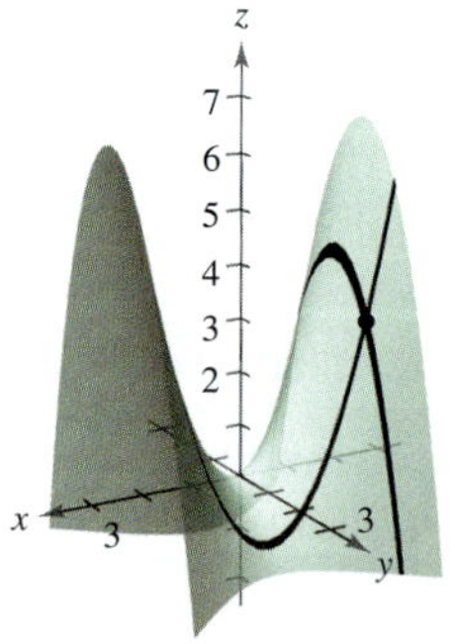

In Exercises 51–54, show that $\partial^2 z/(\partial x \partial y) = \partial^2 z/(\partial y \partial x)$.

51. $z = x^2 - 2xy + 3y^2$

52. $z = x^4 - 3x^2y^2 + y^4$

53. $z = \frac{e^{2xy}}{4x}$

54. $z = \frac{x^2 - y^2}{2xy}$

In Exercises 55–62, find the second partial derivatives

$$\frac{\partial^2 z}{\partial x^2}, \quad \frac{\partial^2 z}{\partial y^2}, \quad \frac{\partial^2 z}{\partial y \partial x}, \quad \text{and} \quad \frac{\partial^2 z}{\partial x \partial y}.$$

55. $z = x^3 - 4y^2$

56. $z = 3x^2 - xy + 2y^3$

57. $z = 4x^3 + 3xy^2 - 4y^3$

58. $z = \sqrt{9 - x^2 - y^2}$

59. $z = \frac{xy}{x - y}$

60. $z = \frac{x}{x + y}$

61. $z = xe^{-y^2}$

62. $z = xe^y + ye^x$

In Exercises 63–66, evaluate the second partial derivatives f_{xx}, f_{xy}, f_{yy}, and f_{yx} at the point.

	Function	*Point*
63.	$f(x, y) = x^4 - 3x^2y^2 + y^2$	$(1, 0)$
64.	$f(x, y) = \sqrt{x^2 + y^2}$	$(0, 2)$
65.	$f(x, y) = \ln(x - y)$	$(2, 1)$
66.	$f(x, y) = x^2e^y$	$(-1, 0)$

67. ***Marginal Cost*** A company manufactures two models of bicycles: a mountain bike and a racing bike. The cost function for producing x mountain bikes and y racing bikes is given by

$$C = 10\sqrt{xy} + 149x + 189y + 675.$$

Find the marginal costs ($\partial C/\partial x$ and $\partial C/\partial y$) when $x = 120$ and $y = 160$.

68. ***Marginal Revenue*** A pharmaceutical corporation has two plants that produce the same over-the-counter medicine. If x_1 and x_2 are the numbers of units produced at plant 1 and plant 2, respectively, then the total revenue for the product is given by

$$R = 200x_1 + 200x_2 - 4x_1^2 - 8x_1x_2 - 4x_2^2.$$

If $x_1 = 4$ and $x_2 = 12$, find the following.

(a) The marginal revenue for plant 1, $\partial R/\partial x_1$

(b) The marginal revenue for plant 2, $\partial R/\partial x_2$

69. *Marginal Productivity* Let $x = 1000$ and $y = 500$ in the Cobb-Douglas production function given by

$$f(x, y) = 100x^{0.6}y^{0.4}.$$

(a) Find the marginal productivity of labor, $\partial f/\partial x$.

(b) Find the marginal productivity of capital, $\partial f/\partial y$.

70. *Marginal Productivity* Repeat Exercise 69 for the production function given by $f(x, y) = 100x^{0.75}y^{0.25}$.

71. *Complementary and Substitute Products* Using the notation of Example 4 in this section, let x_1 and x_2 be the demands for products 1 and 2, respectively, and p_1 and p_2 the prices of products 1 and 2, respectively. Determine whether the demand functions below describe complementary or substitute product relationships.

(a) $x_1 = 150 - 2p_1 - \frac{5}{2}p_2, \quad x_2 = 350 - \frac{3}{2}p_1 - 3p_2$

(b) $x_1 = 150 - 2p_1 + 1.8p_2, \quad x_2 = 350 + \frac{3}{4}p_1 - 1.9p_2$

(c) $x_1 = \dfrac{1000}{\sqrt{p_1p_2}}, \quad x_2 = \dfrac{750}{p_2\sqrt{p_1}}$

72. *Psychology* Early in the twentieth century, an intelligence test called the *Stanford-Binet Test* (more commonly known as the *IQ test*) was developed. In this test, an individual's mental age M is divided by the individual's chronological age C and the quotient is multiplied by 100. The result is the individual's IQ.

$$IQ(M, C) = \frac{M}{C} \times 100$$

Find the partial derivatives of IQ with respect to M and with respect to C. Evaluate the partial derivatives at the point $(12, 10)$ and interpret the result. *(Source: Adapted from Bernstein/Clark-Stewart/Roy/Wickens,* Psychology, *Fourth Edition)*

73. *Education* Let N be the number of applicants to a university, p the charge for food and housing at the university, and t the tuition. Suppose that N is a function of p and t such that $\partial N/\partial p < 0$ and $\partial N/\partial t < 0$. How would you interpret the fact that both partials are negative?

74. *Chemistry* The temperature at any point (x, y) in a steel plate is given by

$$T = 500 - 0.6x^2 - 1.5y^2$$

where x and y are measured in meters. At the point $(2, 3)$, find the rate of change of the temperature with respect to the distance moved along the plate in the directions of the x- and y-axes.

75. *Chemistry* A measure of what hot weather feels like to two average persons is the Apparent Temperature Index. A model for this index is

$$A = 0.885t - 78.7h + 1.20th + 2.70$$

where A is the apparent temperature, t is the air temperature, and h is the relative humidity in decimal form. *(Source:* The UMAP Journal*)*

(a) Find $\partial A/\partial t$ and $\partial A/\partial h$ when $t = 90°\text{F}$ and $h = 0.80$.

(b) Which has a greater effect on A, air temperature or humidity? Explain your reasoning.

76. *Marginal Utility* The utility function $U = f(x, y)$ is a measure of the utility (or satisfaction) derived by a person from the consumption of two goods x and y. Suppose the utility function is given by

$$U = -5x^2 + xy - 3y^2.$$

(a) Determine the marginal utility of good x.

(b) Determine the marginal utility of good y.

(c) When $x = 2$ and $y = 3$, should a person consume one more unit of good x or one more unit of good y? Explain your reasoning.

(d) Use a three-dimensional graphing utility to graph the function. Interpret the marginal utilities of goods x and y graphically.

BUSINESS CAPSULE

© Charles Nix/www.nixphoto.com

Fred and Richard Calloway of Augusta, Georgia, cofounded Male Care, which provides barber, dry cleaning, and car wash services. Among the many advertising techniques used by the Calloways to attract new clients are coupons, customer referrals, and radio advertising. They also feel that their prime location provides them with a strong customer base. Eighty percent of their 1800 monthly clients are repeat customers.

77. *Research Project* Use your school's library, the Internet, or some other reference source to research a company that increased the demand for its product by creative advertising. Write a paper about the company. Use graphs to show how a change in demand is related to a change in the marginal utility of a product or service.

7.5 EXTREMA OF FUNCTIONS OF TWO VARIABLES

- Understand the relative extrema of functions of two variables.
- Use the First-Partials Test to find the relative extrema of functions of two variables.
- Use the Second-Partials Test to find the relative extrema of functions of two variables.
- Use relative extrema to answer questions about real-life situations.

Relative Extrema

Earlier in the text, you learned how to use derivatives to find the relative minimum and relative maximum values of a function of a single variable. In this section, you will learn how to use partial derivatives to find the relative minimum and relative maximum values of a function of two variables.

Relative Extrema of a Function of Two Variables

Let f be a function defined on a region containing (x_0, y_0). The number $f(x_0, y_0)$ is a **relative maximum** of f if there is a circular region R centered at (x_0, y_0) such that

$$f(x, y) \leq f(x_0, y_0) \qquad f(x_0, y_0) \text{ is a relative maximum.}$$

for all (x, y) in R. The number $f(x_0, y_0)$ is a **relative minimum** of f if there is a circular region R centered at (x_0, y_0) such that

$$f(x, y) \geq f(x_0, y_0) \qquad f(x_0, y_0) \text{ is a relative minimum.}$$

for all (x, y) in R.

Graphically, you can think of a point on a surface as a relative maximum if it is at least as high as all "nearby" points on the surface, and you can think of a point on a surface as a relative minimum if it is at least as low as all "nearby" points on the surface. Several relative maxima and relative minima are shown in Figure 7.26.

As in single-variable calculus, you need to distinguish between relative extrema and absolute extrema of a function of two variables. The number $f(x_0, y_0)$ is an absolute maximum of f in the region R if it is greater than or equal to all other function values in the region. For instance, the function $f(x, y) = -(x^2 + y^2)$ graphs as a paraboloid, opening downward, with vertex at $(0, 0, 0)$. The number $f(0, 0) = 0$ is an absolute maximum of the function over the entire xy-plane. An absolute minimum of f in a region is defined similarly.

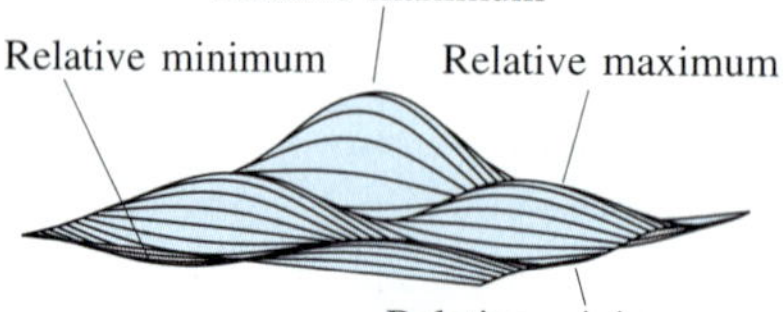

FIGURE 7.26 Relative Extrema

The First-Partials Test for Relative Extrema

To locate the relative extrema of a function of two variables, you can use a procedure that is similar to the First-Derivative Test used for functions of a single variable.

First-Partials Test for Relative Extrema

If $f(x_0, y_0)$ is a relative extremum of f on an open region R in the xy-plane, and the first partial derivatives of f exist in R, then

$$f_x(x_0, y_0) = 0$$

and

$$f_y(x_0, y_0) = 0$$

as shown in Figure 7.27.

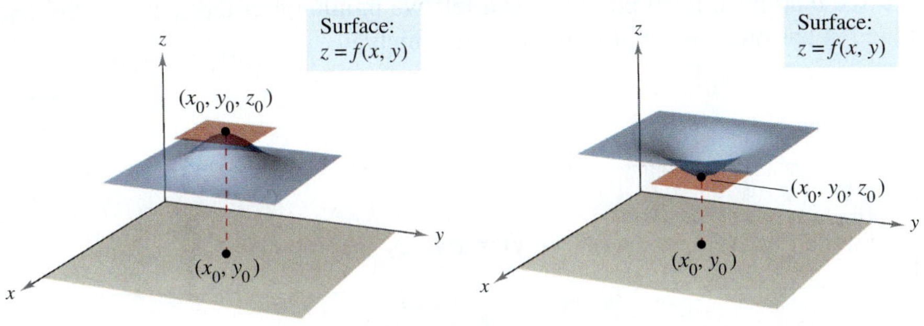

Relative Maximum   Relative Minimum

FIGURE 7.27

STUDY TIP

An *open* region in the xy-plane is similar to an open interval on the real number line. For instance, the region R consisting of the interior of the circle $x^2 + y^2 = 1$ is an open region. If the region R consists of the interior of the circle *and* the points on the circle, then it is a *closed* region.

A point (x_0, y_0) is a **critical point** of f if $f_x(x_0, y_0)$ or $f_y(x_0, y_0)$ is undefined or if

$$f_x(x_0, y_0) = 0 \quad \text{and} \quad f_y(x_0, y_0) = 0. \qquad \text{Critical point}$$

The First-Partials Test states that if the first partial derivatives exist, then you need only examine values of $f(x, y)$ at critical points to find the relative extrema. As is true for a function of a single variable, however, the critical points of a function of two variables do not always yield relative extrema. For instance, the point $(0, 0)$ is a critical point of the surface shown in Figure 7.28, but $f(0, 0)$ is not a relative extremum of the function. Such points are called **saddle points** of the function.

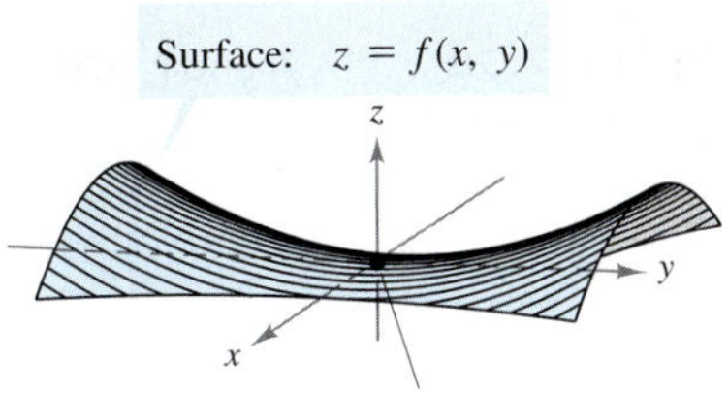

Saddle point at $(0, 0, 0)$:
$f_x(0, 0) = f_y(0, 0) = 0$

FIGURE 7.28

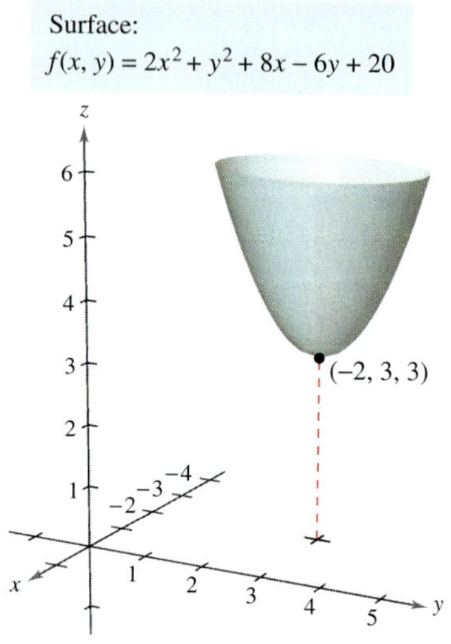

FIGURE 7.29

EXAMPLE 1 Finding Relative Extrema

Find the relative extrema of

$$f(x, y) = 2x^2 + y^2 + 8x - 6y + 20.$$

SOLUTION Begin by finding the first partial derivatives of f.

$$f_x(x, y) = 4x + 8 \quad \text{and} \quad f_y(x, y) = 2y - 6$$

Because these partial derivatives are defined for all points in the xy-plane, the only critical points are those for which both first partial derivatives are zero. To locate these points, set $f_x(x, y)$ and $f_y(x, y)$ equal to 0, and solve the resulting system of equations.

$$4x + 8 = 0 \qquad \text{Set } f_x(x, y) \text{ equal to 0.}$$

$$2y - 6 = 0 \qquad \text{Set } f_y(x, y) \text{ equal to 0.}$$

The solution of this system is $x = -2$ and $y = 3$. So, the point $(-2, 3)$ is the only critical number of f. From the graph of the function, shown in Figure 7.29, you can see that this critical point yields a relative minimum of the function. So, the function has only one relative extremum, which is

$$f(-2, 3) = 3. \qquad \text{Relative minimum}$$

TRY IT 1

Find the relative extrema of $f(x, y) = x^2 + 2y^2 + 16x - 8y + 8$.

Example 1 shows a relative minimum occurring at one type of critical point—the type for which both $f_x(x, y)$ and $f_y(x, y)$ are zero. The next example shows a relative maximum that occurs at the other type of critical point—the type for which either $f_x(x, y)$ or $f_y(x, y)$ is undefined.

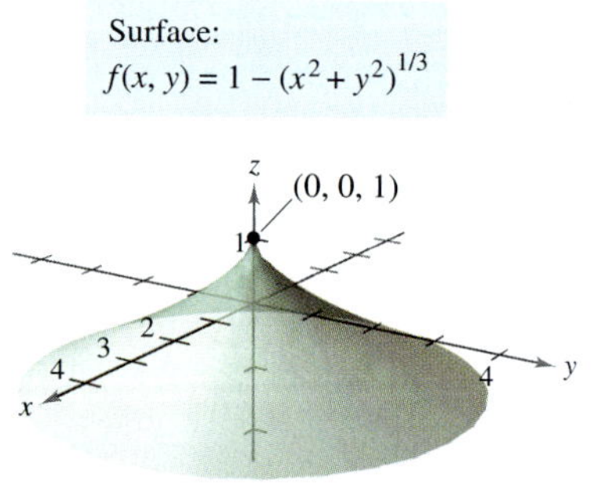

FIGURE 7.30 $f_x(x, y)$ and $f_y(x, y)$ are undefined at (0, 0).

EXAMPLE 2 Finding Relative Extrema

Find the relative extrema of

$$f(x, y) = 1 - (x^2 + y^2)^{1/3}.$$

SOLUTION Begin by finding the first partial derivatives of f.

$$f_x(x, y) = -\frac{2x}{3(x^2 + y^2)^{2/3}} \quad \text{and} \quad f_y(x, y) = -\frac{2y}{3(x^2 + y^2)^{2/3}}$$

These partial derivatives are defined for all points in the xy-plane *except* the point (0, 0). So, (0, 0) is a critical point of f. Moreover, this is the only critical point, because there are no other values of x and y for which either partial is undefined or for which both partials are zero. From the graph of the function, shown in Figure 7.30, you can see that this critical point yields a relative maximum of the function. So, the function has only one relative extremum, which is

$$f(0, 0) = 1. \qquad \text{Relative maximum}$$

TRY IT 2

Find the relative extrema of

$$f(x, y) = \sqrt{1 - \frac{x^2}{16} - \frac{y^2}{4}}.$$

TECHNOLOGY

Classifying Extrema with a Graphing Utility

The First-Partials Test tells you that the relative extrema of a function $z = f(x, y)$ can occur only at the critical points of the function. The test does not, however, give you a method for determining whether a critical point actually yields a relative minimum, a relative maximum, or neither. Once you have located the critical points of a function, you can use technology to find the relative extrema graphically. Four examples of surfaces are shown below. Try using the graphs and the given critical numbers to classify the relative extrema.

Surface: $f(x, y) = x^2 - 2x + y^2 + 4y + 5$

Critical point: $(1, -2)$

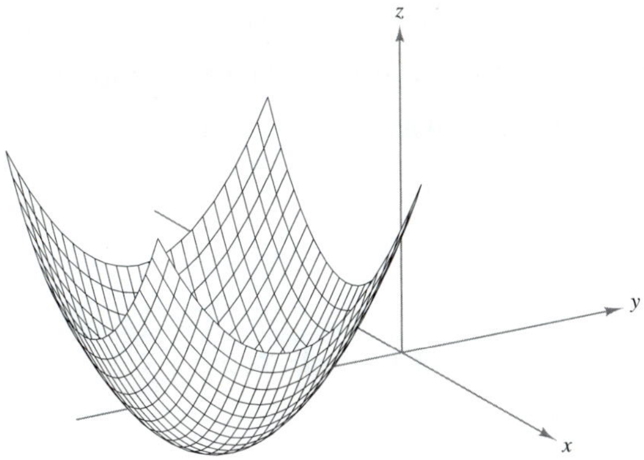

Generated by Mathematica

Surface: $f(x, y) = \sqrt{x^2 + y^2}$

Critical point: $(0, 0)$

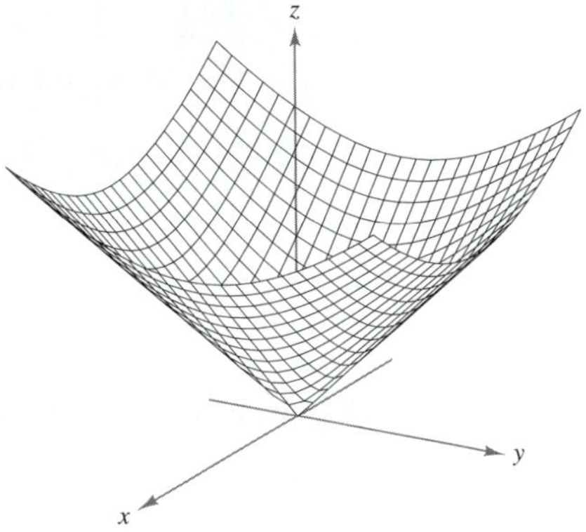

Generated by Mathematica

Surface: $f(x, y) = \dfrac{1}{1 + x^2 + y^2}$

Critical point: $(0, 0)$

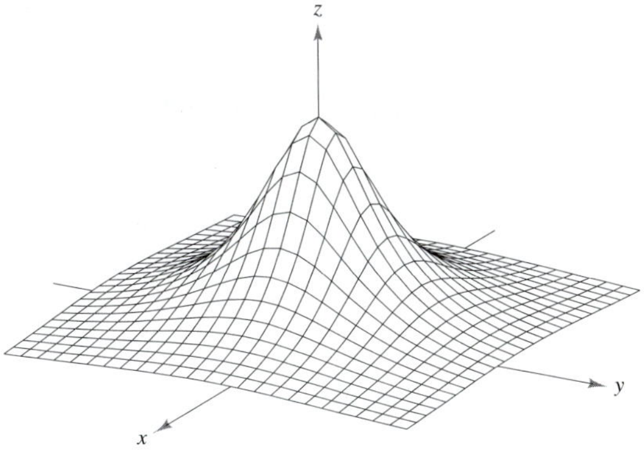

Generated by Mathematica

Surface: $f(x, y) = 8xye^{-x^2-y^2}$

Critical points: $(0, 0), \left(\pm\frac{1}{\sqrt{2}}, \pm\frac{1}{\sqrt{2}}\right), \left(\pm\frac{1}{\sqrt{2}}, \pm\frac{1}{\sqrt{2}}\right)$

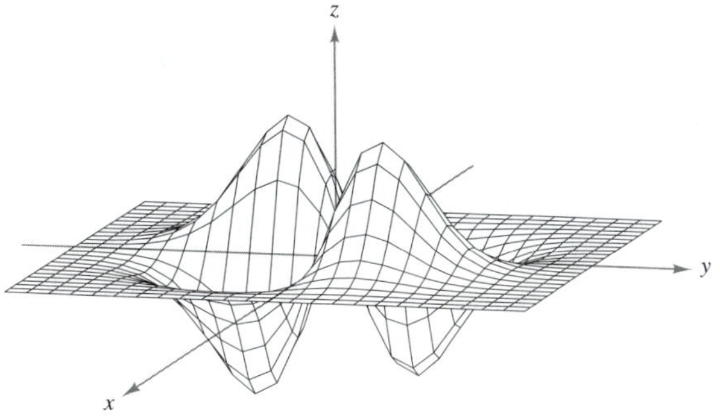

Generated by Mathematica

STUDY TIP

If $d > 0$, then $f_{xx}(a, b)$ and $f_{yy}(a, b)$ must have the same sign. So, you can replace $f_{xx}(a, b)$ with $f_{yy}(a, b)$ in the first two parts of the test.

ALGEBRA REVIEW

For help in solving the system of equations

$$y - x^3 = 0$$
$$x - y^3 = 0$$

in Example 3, see Example 1(a) in the *Chapter 7 Algebra Review*, on page 540.

The Second-Partials Test for Relative Extrema

For functions such as those in Examples 1 and 2, you can determine the *type* of extrema at the critical points by sketching the graph of the function. For more complicated functions, a graphical approach is not so easy to use. The **Second-Partials Test** is an analytical test that can be used to determine whether a critical number yields a relative minimum, a relative maximum, or neither.

Second-Partials Test for Relative Extrema

If f has continuous first and second partial derivatives in an open region and there exists a point (a, b) in the region such that $f_x(a, b) = 0$ and $f_y(a, b) = 0$, then the quantity

$$d = f_{xx}(a, b)f_{yy}(a, b) - [f_{xy}(a, b)]^2$$

can be used as shown.

1. $f(a, b)$ is a relative minimum if $d > 0$ and $f_{xx}(a, b) > 0$.
2. $f(a, b)$ is a relative maximum if $d > 0$ and $f_{xx}(a, b) < 0$.
3. $(a, b, f(a, b))$ is a saddle point if $d < 0$.
4. The test gives no information if $d = 0$.

EXAMPLE 3 **Applying the Second-Partials Test**

Find the relative extrema and saddle points of $f(x, y) = xy - \frac{1}{4}x^4 - \frac{1}{4}y^4$.

SOLUTION Begin by finding the critical points of f. Because $f_x(x, y) = y - x^3$ and $f_y(x, y) = x - y^3$ are defined for all points in the xy-plane, the only critical points are those for which both first partial derivatives are zero. By solving the equations $y - x^3 = 0$ and $x - y^3 = 0$ simultaneously, you can determine that the critical points are $(1, 1)$, $(-1, -1)$, and $(0, 0)$. Furthermore, because

$$f_{xx}(x, y) = -3x^2, \quad f_{yy}(x, y) = -3y^2, \quad \text{and} \quad f_{xy}(x, y) = 1$$

you can use the quantity $d = f_{xx}(a, b)f_{yy}(a, b) - [f_{xy}(a, b)]^2$ to classify the critical points as shown.

Critical Point	d	$f_{xx}(x, y)$	*Conclusion*
$(1, 1)$	$(-3)(-3) - 1 = 8$	-3	Relative maximum
$(-1, -1)$	$(-3)(-3) - 1 = 8$	-3	Relative maximum
$(0, 0)$	$(0)(0) - 1 = -1$	0	Saddle point

The graph of f is shown in Figure 7.31.

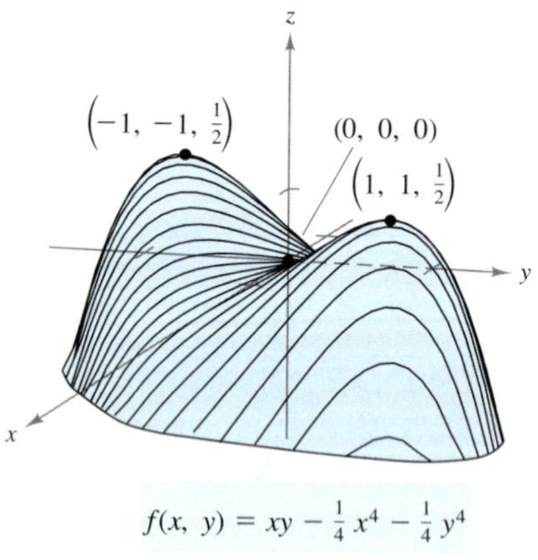

FIGURE 7.31

TRY IT 3

Find the relative extrema and saddle points of $f(x, y) = \dfrac{y^2}{16} - \dfrac{x^2}{4}$.

Applications of Extrema

EXAMPLE 4 Finding a Maximum Profit

A company makes two substitute products whose demand functions are given by

$$x_1 = 200(p_2 - p_1) \quad \text{Demand for product 1}$$
$$x_2 = 500 + 100p_1 - 180p_2 \quad \text{Demand for product 2}$$

where p_1 and p_2 are the prices per unit (in dollars) and x_1 and x_2 are the numbers of units sold. The costs of producing the two products are \$0.50 and \$0.75 per unit, respectively. Find the prices that will yield a maximum profit.

SOLUTION The cost and revenue functions are as shown.

$$C = 0.5x_1 + 0.75x_2 \quad \text{Write cost function.}$$
$$= 0.5(200)(p_2 - p_1) + 0.75(500 + 100p_1 - 180p_2) \quad \text{Substitute.}$$
$$= 375 - 25p_1 - 35p_2 \quad \text{Simplify.}$$
$$R = p_1x_1 + p_2x_2 \quad \text{Write revenue function.}$$
$$= p_1(200)(p_2 - p_1) + p_2(500 + 100p_1 - 180p_2) \quad \text{Substitute.}$$
$$= -200p_1^2 - 180p_2^2 + 300p_1p_2 + 500p_2 \quad \text{Simplify.}$$

This implies that the profit function is

$$P = R - C \quad \text{Write profit function.}$$
$$= -200p_1^2 - 180p_2^2 + 300p_1p_2 + 500p_2 - (375 - 25p_1 - 35p_2)$$
$$= -200p_1^2 - 180p_2^2 + 300p_1p_2 + 25p_1 + 535p_2 - 375.$$

The maximum profit occurs when the two first partial derivatives are zero.

$$\frac{\partial P}{\partial p_1} = -400p_1 + 300p_2 + 25 = 0$$
$$\frac{\partial P}{\partial p_2} = 300p_1 - 360p_2 + 535 = 0$$

By solving this system simultaneously, you can conclude that the solution is $p_1 = \$3.14$ and $p_2 = \$4.10$. From the graph of the function shown in Figure 7.32, you can see that this critical number yields a maximum. So, the maximum profit is

$$P = P(3.14, 4.10)$$
$$= \$761.48.$$

TRY IT 4

Find the prices that will yield a maximum profit for the products in Example 4 if the costs of producing the two products are \$0.75 and \$0.50 per unit, respectively.

STUDY TIP

In Example 4, you can check that the two products are substitutes by observing that x_1 increases as p_2 increases and x_2 increases as p_1 increases.

ALGEBRA REVIEW

For help in solving the system of equations in Example 4, see Example 1(b) in the *Chapter 7 Algebra Review*, on page 540.

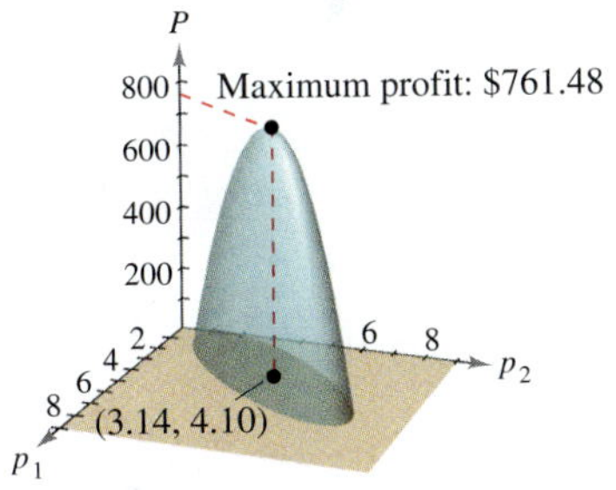

FIGURE 7.32

STUDY TIP

In Example 4, to convince yourself that the maximum profit is \$761.48, try substituting other prices into the profit function. For each pair of prices, you will obtain a profit that is less than \$761.48. For instance, if $p_1 = \$2$ and $p_2 = \$3$, then the profit is $P = P(2, 3) = \$660.00$.

ALGEBRA REVIEW

For help in solving the system of equations

$$y(24 - 12x - 4y) = 0$$
$$x(24 - 6x - 8y) = 0$$

in Example 5, see Example 2(a) in the *Chapter 7 Algebra Review*, on page 541.

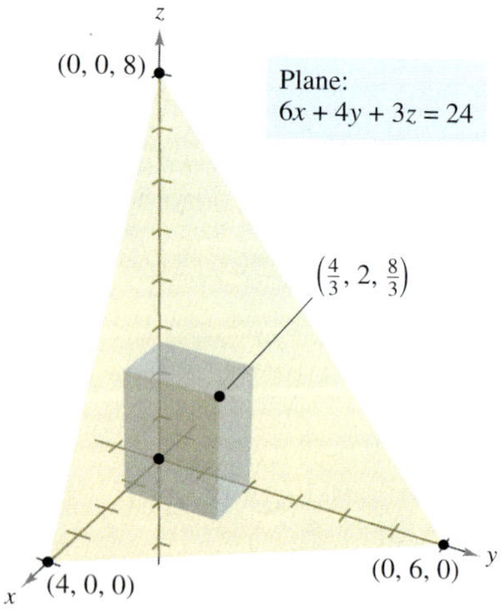

FIGURE 7.33

EXAMPLE 5 **Finding a Maximum Volume**

Consider all possible rectangular boxes that are resting on the xy-plane with one vertex at the origin and the opposite vertex in the plane $6x + 4y + 3z = 24$, as shown in Figure 7.33. Of all such boxes, which has the greatest volume?

SOLUTION Because one vertex of the box lies in the plane given by $6x + 4y + 3z = 24$ or $z = \frac{1}{3}(24 - 6x - 4y)$, you can write the volume of the box as

$$\begin{aligned} V &= xyz && \text{Volume = (width)(length)(height)} \\ &= xy\left(\tfrac{1}{3}\right)(24 - 6x - 4y) && \text{Substitute for } z. \\ &= \tfrac{1}{3}(24xy - 6x^2y - 4xy^2). && \text{Simplify.} \end{aligned}$$

To find the critical numbers, set the first partial derivatives equal to zero.

$$\begin{aligned} V_x &= \tfrac{1}{3}(24y - 12xy - 4y^2) && \text{Partial with respect to } x \\ &= \tfrac{1}{3}y(24 - 12x - 4y) = 0 && \text{Factor and set equal to 0.} \\ V_y &= \tfrac{1}{3}(24x - 6x^2 - 8xy) && \text{Partial with respect to } y \\ &= \tfrac{1}{3}x(24 - 6x - 8y) = 0 && \text{Factor and set equal to 0.} \end{aligned}$$

The four solutions of this system are $(0, 0)$, $(0, 6)$, $(4, 0)$, and $\left(\frac{4}{3}, 2\right)$. Using the Second-Partials Test, you can determine that the maximum volume occurs when the width is $x = \frac{4}{3}$ and the length is $y = 2$. For these values, the height of the box is

$$z = \tfrac{1}{3}\left[24 - 6\left(\tfrac{4}{3}\right) - 4(2)\right] = \tfrac{8}{3}.$$

So, the maximum volume is

$$V = xyz = \left(\tfrac{4}{3}\right)(2)\left(\tfrac{8}{3}\right) = \tfrac{64}{9} \text{ cubic units.}$$

TRY IT 5

Find the maximum volume of a box that is resting on the xy-plane with one vertex at the origin and the opposite vertex in the plane $2x + 4y + z = 8$.

TAKE ANOTHER LOOK

Using the Second-Partials Test

In Example 5, explain why you can disregard the three critical numbers $(0, 0)$, $(0, 6)$, and $(4, 0)$. Then use the second partial derivatives

$$V_{xx}(x, y) = -4y$$
$$V_{yy}(x, y) = -\tfrac{8}{3}x$$
$$V_{xy}(x, y) = \tfrac{1}{3}(24 - 12x - 8y)$$

to explain why the critical number $\left(\frac{4}{3}, 2\right)$ yields a maximum.

PREREQUISITE REVIEW 7.5

The following warm-up exercises involve skills that were covered in earlier sections. You will use these skills in the exercise set for this section.

In Exercises 1–8, solve the system of equations.

1. $\begin{cases} 5x = 15 \\ 3x - 2y = 5 \end{cases}$

2. $\begin{cases} \frac{1}{2}y = 3 \\ -x + 5y = 19 \end{cases}$

3. $\begin{cases} x + y = 5 \\ x - y = -3 \end{cases}$

4. $\begin{cases} x + y = 8 \\ 2x - y = 4 \end{cases}$

5. $\begin{cases} 2x - y = 8 \\ 3x - 4y = 7 \end{cases}$

6. $\begin{cases} 2x - 4y = 14 \\ 3x + y = 7 \end{cases}$

7. $\begin{cases} x^2 + x = 0 \\ 2yx + y = 0 \end{cases}$

8. $\begin{cases} 3y^2 + 6y = 0 \\ xy + x + 2 = 0 \end{cases}$

In Exercises 9–14, find all first and second partial derivatives of the function.

9. $z = 4x^3 - 3y^2$

10. $z = 2x^5 - y^3$

11. $z = x^4 - \sqrt{xy} + 2y$

12. $z = 2x^2 - 3xy + y^2$

13. $z = ye^{xy^2}$

14. $z = xe^{xy}$

EXERCISES 7.5

In Exercises 1–4, find any critical points and relative extrema of the function.

1. $f(x, y) = x^2 - y^2 + 4x - 8y - 11$

2. $f(x, y) = x^2 + y^2 + 2x - 6y + 6$

3. $f(x, y) = \sqrt{x^2 + y^2 + 1}$

4. $f(x, y) = \sqrt{25 - (x - 2)^2 - y^2}$

In Exercises 5–20, examine each function for relative extrema and saddle points.

5. $f(x, y) = (x - 1)^2 + (y - 3)^2$

6. $f(x, y) = 9 - (x - 3)^2 - (y + 2)^2$

7. $f(x, y) = 2x^2 + 2xy + y^2 + 2x - 3$

8. $f(x, y) = -x^2 - 5y^2 + 8x - 10y - 13$

9. $f(x, y) = -5x^2 + 4xy - y^2 + 16x + 10$

10. $f(x, y) = x^2 + 6xy + 10y^2 - 4y + 4$

11. $f(x, y) = 3x^2 + 2y^2 - 12x - 4y + 7$

12. $f(x, y) = -3x^2 - 2y^2 + 3x - 4y + 5$

13. $f(x, y) = x^2 - y^2 + 4x - 4y - 8$

14. $f(x, y) = x^2 - 3xy - y^2$

15. $f(x, y) = xy$

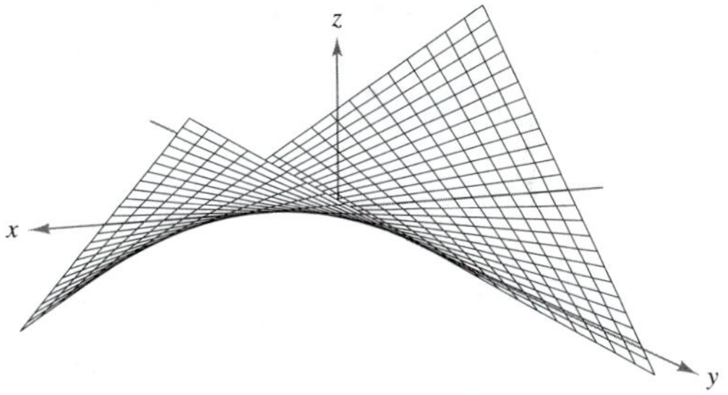

16. $f(x, y) = 12x + 12y - xy - x^2 - y^2$

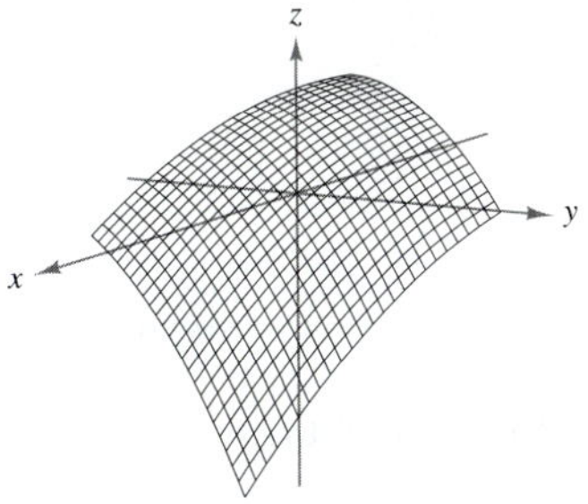

17. $f(x, y) = (x^2 + 4y^2)e^{1-x^2-y^2}$

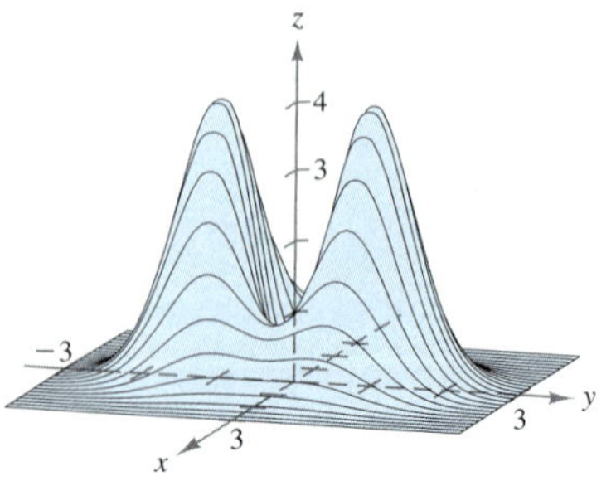

18. $f(x, y) = e^{-(x^2+y^2)}$

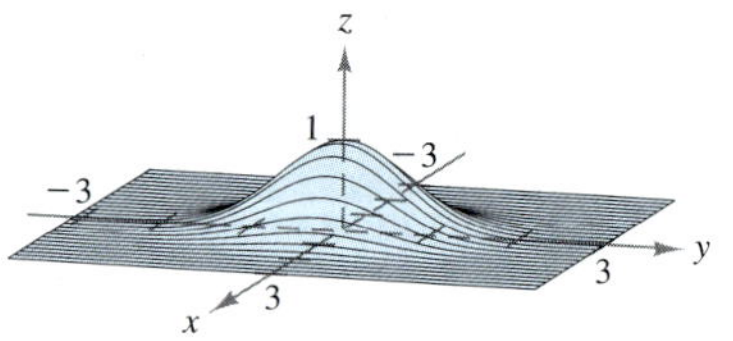

19. $f(x, y) = e^{xy}$

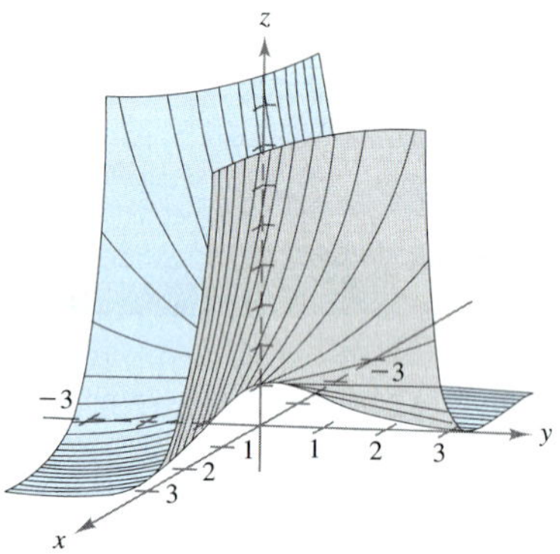

20. $f(x, y) = -\dfrac{4x}{x^2 + y^2 + 1}$

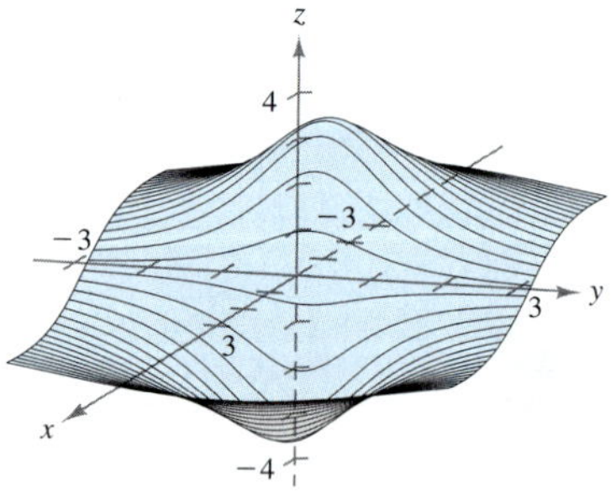

In Exercises 21–24, determine whether there is a relative maximum, a relative minimum, a saddle point, or insufficient information to determine the nature of the function $f(x, y)$ at the critical point (x_0, y_0).

21. $f_{xx}(x_0, y_0) = 16$
$f_{yy}(x_0, y_0) = 4$
$f_{xy}(x_0, y_0) = 8$

22. $f_{xx}(x_0, y_0) = -4$
$f_{yy}(x_0, y_0) = -6$
$f_{xy}(x_0, y_0) = 3$

23. $f_{xx}(x_0, y_0) = -7$
$f_{yy}(x_0, y_0) = 4$
$f_{xy}(x_0, y_0) = 9$

24. $f_{xx}(x_0, y_0) = 20$
$f_{yy}(x_0, y_0) = 8$
$f_{xy}(x_0, y_0) = 9$

In Exercises 25–30, find the critical points and test for relative extrema. List the critical points for which the Second-Partials Test fails.

25. $f(x, y) = (xy)^2$

26. $f(x, y) = \sqrt{x^2 + y^2}$

27. $f(x, y) = x^3 + y^3$

28. $f(x, y) = x^3 + y^3 - 3x^2 + 6y^2 + 3x + 12y + 7$

29. $f(x, y) = x^{2/3} + y^{2/3}$

30. $f(x, y) = (x^2 + y^2)^{2/3}$

In Exercises 31 and 32, find the critical points of the function and, from the form of the function, determine whether each critical point is a relative maximum or a relative minimum.

31. $f(x, y, z) = (x - 1)^2 + (y + 3)^2 + z^2$

32. $f(x, y, z) = 6 - [x(y + 2)(z - 1)]^2$

In Exercises 33–36, find three positive numbers x, y, and z that satisfy the given conditions.

33. The sum is 30 and the product is maximum.

34. The sum is 32 and $P = xy^2z$ is maximum.

35. The sum is 30 and the sum of the squares is minimum.

36. The sum is 1 and the sum of the squares is minimum.

37. ***Revenue*** A company manufactures two products. The total revenue from x_1 units of product 1 and x_2 units of product 2 is given by

$$R = -5x_1^2 - 8x_2^2 - 2x_1x_2 + 42x_1 + 102x_2.$$

Find x_1 and x_2 so as to maximize the revenue.

38. ***Revenue*** A retail outlet sells two competitive products, the prices of which are p_1 and p_2. Find p_1 and p_2 so as to maximize the total revenue

$$R = 500p_1 + 800p_2 + 1.5p_1p_2 - 1.5p_1^2 - p_2^2.$$

Revenue In Exercises 39 and 40, find p_1 and p_2 so as to maximize the total revenue $R = x_1p_1 + x_2p_2$ for a retail outlet that sells two competitive products with the given demand functions.

39. $x_1 = 1000 - 2p_1 + p_2$, $x_2 = 1500 + 2p_1 - 1.5p_2$

40. $x_1 = 1000 - 4p_1 + 2p_2$, $x_2 = 900 + 4p_1 - 3p_2$

41. ***Profit*** A corporation manufactures a high-performance automobile engine product at two locations. The cost functions for producing x_1 units at location 1 and x_2 units at location 2 are given by

$$C_1 = 0.05x_1^2 + 15x_1 + 5400$$

and

$$C_2 = 0.03x_2^2 + 15x_2 + 6100$$

respectively. The demand function for the product is given by

$$p = 225 - 0.4(x_1 + x_2)$$

and so the total revenue function is given by

$$R = [225 - 0.4(x_1 + x_2)](x_1 + x_2).$$

Find the production levels at the two locations that will maximize the profit

$$P = R - C_1 - C_2.$$

42. ***Cost*** The material for constructing the base of an open box costs 1.5 times as much as the material for constructing the sides. Find the dimensions of the box of largest volume that can be made for a fixed amount of money C.

43. ***Volume*** Find the dimensions of a rectangular package of largest volume that may be sent by a shipping company assuming that the sum of the length and the girth (perimeter of a cross section) cannot exceed 144 inches.

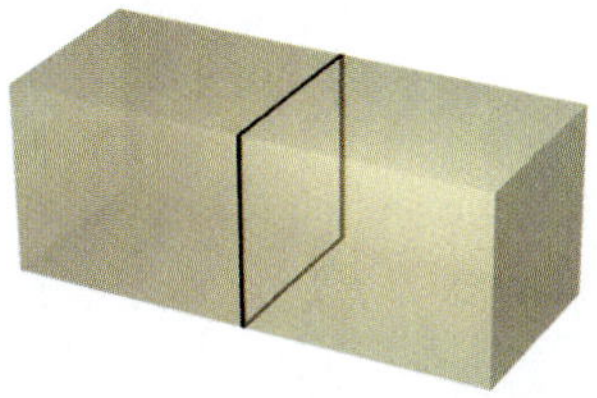

44. ***Volume*** Show that a rectangular box of given volume and minimum surface area is a cube.

45. ***Hardy-Weinberg Law*** Common blood types are determined genetically by the three alleles A, B, and O. (An allele is any of a group of possible mutational forms of a gene.) A person whose blood type is AA, BB, or OO is homozygous. A person whose blood type is AB, AO, or BO is heterozygous. The Hardy-Weinberg Law states that the proportion P of heterozygous individuals in any given population is modeled by

$$P(p, q, r) = 2pq + 2pr + 2qr$$

where p represents the percent of allele A in the population, q represents the percent of allele B in the population, and r represents the percent of allele O in the population. Use the fact that $p + q + r = 1$ (the sum of the three must equal 100%) to show that the maximum proportion of heterozygous individuals in any population is $\frac{2}{3}$.

46. ***Biology*** A lake is to be stocked with smallmouth and largemouth bass. Let x represent the number of smallmouth bass and let y represent the number of largemouth bass in the lake. The weight of each fish is dependent on the population densities. After a six-month period, the weight of a single smallmouth bass is given by

$$W_1 = 3 - 0.002x - 0.005y$$

and the weight of a single largemouth bass is given by

$$W_2 = 4.5 - 0.003x - 0.004y.$$

Assuming that no fish die during the six-month period, how many smallmouth and largemouth bass should be stocked in the lake so that the *total* weight T of bass in the lake is a maximum?

47. ***Medicine*** In order to treat a certain bacterial infection, a combination of two drugs is being tested. Studies have shown that the duration of the infection in laboratory tests can be modeled by

$$D(x, y) = x^2 + 2y^2 - 18x - 24y + 2xy + 120$$

where x is the dose in hundreds of milligrams of the first drug and y is the dose in hundreds of milligrams of the second drug. Determine the partial derivatives of D with respect to x and with respect to y. Find the amount of each drug necessary to minimize the duration of the infection.

True or False? In Exercises 48–51, determine whether the statement is true or false. If it is false, explain why or give an example that shows it is false.

48. If $d > 0$ and $f_x(a, b) < 0$, then $f(a, b)$ is a relative minimum.

49. A saddle point always occurs at a critical point.

50. If $f(x, y)$ has a relative maximum (x_0, y_0, z_0), then $f_x(x_0, y_0) = f_y(x_0, y_0) = 0$.

51. The function

$$f(x, y) = \sqrt[3]{x^2 + y^2}$$

has a relative maximum at the origin.

7.6 LAGRANGE MULTIPLIERS

- Use Lagrange multipliers with one constraint to find extrema of functions of several variables and to answer questions about real-life situations.
- Use Lagrange multipliers with two constraints to find extrema of functions of several variables.

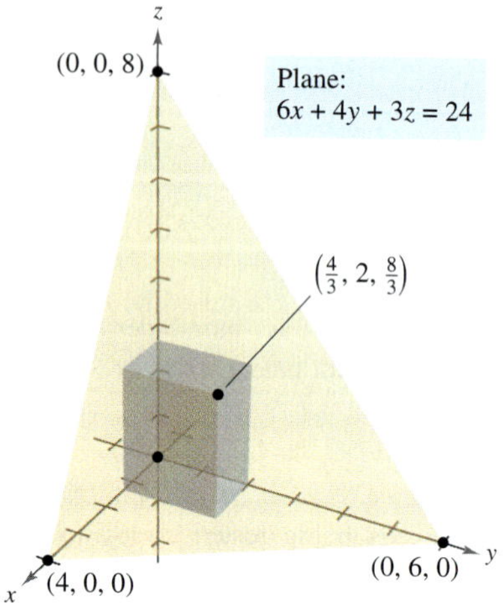

FIGURE 7.34

Lagrange Multipliers with One Constraint

In Example 5 in Section 7.5, you were asked to find the dimensions of the rectangular box of maximum volume that would fit in the first octant beneath the plane

$$6x + 4y + 3z = 24$$

as shown again in Figure 7.34. Another way of stating this problem is to say that you are asked to find the maximum of

$$V = xyz \qquad \text{Objective function}$$

subject to the constraint

$$6x + 4y + 3z - 24 = 0. \qquad \text{Constraint}$$

This type of problem is called a **constrained optimization** problem. In Section 7.5, you answered this question by solving for z in the constraint equation and then rewriting V as a function of two variables.

In this section, you will study a different (and often better) way to solve constrained optimization problems. This method involves the use of variables called **Lagrange multipliers,** named after the French mathematician Joseph Louis Lagrange (1736–1813).

Method of Lagrange Multipliers

If $f(x, y)$ has a maximum or minimum subject to the constraint $g(x, y) = 0$, then it will occur at one of the critical numbers of the function F defined by

$$F(x, y, \lambda) = f(x, y) - \lambda g(x, y).$$

The variable λ (the lowercase Greek letter lambda) is called a **Lagrange multiplier.** For functions of three variables, F has the form

$$F(x, y, z, \lambda) = f(x, y, z) - \lambda g(x, y, z).$$

The method of Lagrange multipliers gives you a way of finding critical points but does not tell you whether these points yield minima, maxima, or neither. To make this distinction, you must rely on the context of the problem.

EXAMPLE 1 **Using Lagrange Multipliers: One Constraint**

Find the maximum of

$V = xyz$ Objective function

subject to the constraint

$6x + 4y + 3z - 24 = 0.$ Constraint

SOLUTION First, let $f(x, y, z) = xyz$ and $g(x, y, z) = 6x + 4y + 3z - 24$. Then, define a new function F as

$$\begin{aligned} F(x, y, z, \lambda) &= f(x, y, z) - \lambda g(x, y, z) \\ &= xyz - \lambda(6x + 4y + 3z - 24). \end{aligned}$$

To find the critical numbers of F, set the partial derivatives of F with respect to x, y, z, and λ equal to zero and obtain

$$\begin{aligned} F_x(x, y, z, \lambda) &= yz - 6\lambda = 0 \\ F_y(x, y, z, \lambda) &= xz - 4\lambda = 0 \\ F_z(x, y, z, \lambda) &= xy - 3\lambda = 0 \\ F_\lambda(x, y, z, \lambda) &= -6x - 4y - 3z + 24 = 0. \end{aligned}$$

Solving for λ in the first equation and substituting into the second and third equations produces the values below.

$$xz - 4\left(\frac{yz}{6}\right) = 0 \quad \Rightarrow \quad y = \frac{3}{2}x$$

$$xy - 3\left(\frac{yz}{6}\right) = 0 \quad \Rightarrow \quad z = 2x$$

Next, substitute these values for y and z into the equation $F_\lambda(x, y, z, \lambda) = 0$ and solve for x.

$$\begin{aligned} F_\lambda(x, y, z, \lambda) &= 0 \\ -6x - 4\left(\tfrac{3}{2}x\right) - 3(2x) + 24 &= 0 \\ -18x &= -24 \\ x &= \tfrac{4}{3} \end{aligned}$$

Using this x-value, you can conclude that the critical values are $x = \frac{4}{3}$, $y = 2$, and $z = \frac{8}{3}$, which implies that the maximum is

$V = xyz$ Write objective function.

$= \left(\frac{4}{3}\right)(2)\left(\frac{8}{3}\right)$ Substitute values of x, y, and z.

$= \frac{64}{9}$ cubic units. Maximum volume

STUDY TIP

Example 1 shows how Lagrange multipliers can be used to solve the same problem that was solved in Example 5 in Section 7.5.

ALGEBRA REVIEW

The most difficult aspect of many Lagrange multiplier problems is the complicated algebra needed to solve the system of equations arising from $F(x, y, \lambda) = f(x, y) - \lambda g(x, y)$. There is no general way to proceed in every case, so you should study the examples carefully, and refer to the *Chapter 7 Algebra Review* on pages 540 and 541.

TRY IT 1

Find the maximum volume of $V = xyz$ subject to the constraint $2x + 4y + z - 8 = 0$.

Maximillian Stock LTD/Phototake

For many industrial applications, a simple robot can cost more than a year's wages and benefits for one employee. So, manufacturers must carefully balance the amount of money spent on labor and capital.

EXAMPLE 2 Finding a Maximum Production Level

A manufacturer's production is modeled by the Cobb-Douglas function

$$f(x, y) = 100x^{3/4}y^{1/4} \qquad \text{Objective function}$$

where x represents the units of labor and y represents the units of capital. Each labor unit costs \$150 and each capital unit costs \$250. The total expenses for labor and capital cannot exceed \$50,000. Find the maximum production level.

SOLUTION Because total labor and capital expenses cannot exceed \$50,000, the constraint is

$$150x + 250y = 50{,}000 \qquad \text{Constraint}$$
$$150x + 250y - 50{,}000 = 0. \qquad \text{Write in standard form.}$$

To find the maximum production level, begin by writing the function

$$F(x, y, \lambda) = 100x^{3/4}y^{1/4} - \lambda(150x + 250y - 50{,}000).$$

Next, set the partial derivatives of this function equal to zero.

$$F_x(x, y, \lambda) = 75x^{-1/4}y^{1/4} - 150\lambda = 0 \qquad \text{Equation 1}$$
$$F_y(x, y, \lambda) = 25x^{3/4}y^{-3/4} - 250\lambda = 0 \qquad \text{Equation 2}$$
$$F_\lambda(x, y, \lambda) = -150x - 250y + 50{,}000 = 0 \qquad \text{Equation 3}$$

The strategy for solving such a system must be customized to the particular system. In this case, you can solve for λ in the first equation, substitute into the second equation, solve for x, substitute into the third equation, and solve for y.

$$75x^{-1/4}y^{1/4} - 150\lambda = 0 \qquad \text{Equation 1}$$
$$\lambda = \tfrac{1}{2}x^{-1/4}y^{1/4} \qquad \text{Solve for } \lambda.$$
$$25x^{3/4}y^{-3/4} - 250\left(\tfrac{1}{2}\right)x^{-1/4}y^{1/4} = 0 \qquad \text{Substitute in Equation 2.}$$
$$25x - 125y = 0 \qquad \text{Multiply by } x^{1/4}y^{3/4}.$$
$$x = 5y \qquad \text{Solve for } x.$$
$$-150(5y) - 250y + 50{,}000 = 0 \qquad \text{Substitute in Equation 3.}$$
$$-1000y = -50{,}000 \qquad \text{Simplify.}$$
$$y = 50 \qquad \text{Solve for } y.$$

Using this value for y, it follows that $x = 5(50) = 250$. So, the maximum production level of

$$f(250, 50) = 100(250)^{3/4}(50)^{1/4} \qquad \text{Substitute for } x \text{ and } y.$$
$$\approx 16{,}719 \text{ units} \qquad \text{Maximum production}$$

occurs when $x = 250$ units of labor and $y = 50$ units of capital.

TRY IT 2

In Example 2, suppose that each labor unit costs \$200 and each capital unit costs \$250. Find the maximum production level if labor and capital cannot exceed \$50,000.

Economists call the Lagrange multiplier obtained in a production function the **marginal productivity of money.** For instance, in Example 2, the marginal productivity of money when $x = 250$ and $y = 50$ is

$$\lambda = \frac{1}{2}x^{-1/4}y^{1/4} = \frac{1}{2}(250)^{-1/4}(50)^{1/4} \approx 0.334.$$

This means that if one additional dollar is spent on production, approximately 0.334 additional unit of the product can be produced.

EXAMPLE 3 Finding a Maximum Production Level

In Example 2, suppose that \$70,000 is available for labor and capital. What is the maximum number of units that can be produced?

SOLUTION You could rework the entire problem, as demonstrated in Example 2. However, because the only change in the problem is the availability of additional money to spend on labor and capital, you can use the fact that the marginal productivity of money is

$$\lambda \approx 0.334.$$

Because an additional \$20,000 is available and the maximum production in Example 2 was 16,719 units, you can conclude that the maximum production is now

$$16{,}719 + (0.334)(20{,}000) \approx 23{,}400 \text{ units.}$$

Try using the procedure demonstrated in Example 2 to confirm this result.

TRY IT 3

In Example 3, suppose that \$80,000 is available for labor and capital. What is the maximum number of units that can be produced?

TECHNOLOGY

You can use a three-dimensional graphing utility to confirm graphically the results of Examples 2 and 3. Begin by graphing the surface $f(x, y) = 100x^{3/4}y^{1/4}$. Then graph the vertical plane given by $150x + 250y = 50{,}000$. As shown at the right, the maximum production level corresponds to the highest point on the intersection of the surface and the plane.

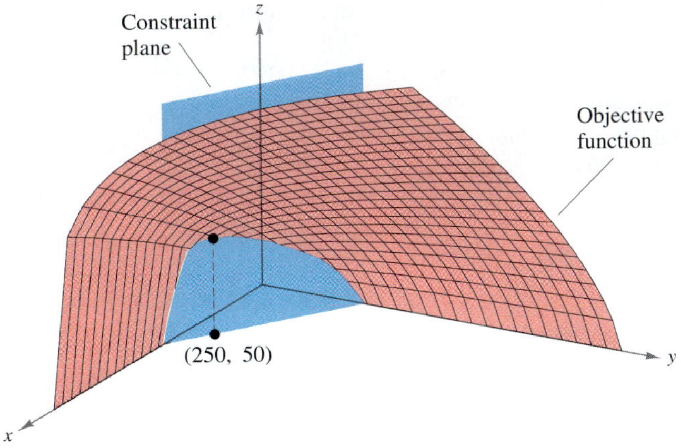

Intersection of the surface $z = 100x^{3/4}y^{1/4}$ and the plane $150x + 250y = 50{,}000$

TECHNOLOGY

You can use a spreadsheet to solve constrained optimization problems. Spreadsheet software programs have a built-in algorithm that finds absolute extrema of functions. Be sure you enter each constraint and the objective function into the spreadsheet. You should also enter initial values of the variables you are working with. Try using a spreadsheet to solve the problem in Example 4. What is your result? (Consult the user's manual of a spreadsheet software program for specific instructions on how to solve a constrained optimization problem.)

EXAMPLE 4 Optimizing with Four Variables

Find the minimum of the function

$$f(x, y, z, w) = x^2 + y^2 + z^2 + w^2 \qquad \text{Objective function}$$

subject to the constraint

$$3x + 2y - 4z + w - 3 = 0. \qquad \text{Constraint}$$

SOLUTION Begin by forming the function

$$F(x, y, z, w, \lambda) = x^2 + y^2 + z^2 + w^2 - \lambda(3x + 2y - 4z + w - 3).$$

Next, set the five partial derivatives of F equal to zero.

$$F_x(x, y, z, w, \lambda) = 2x - 3\lambda = 0 \qquad \text{Equation 1}$$
$$F_y(x, y, z, w, \lambda) = 2y - 2\lambda = 0 \qquad \text{Equation 2}$$
$$F_z(x, y, z, w, \lambda) = 2z + 4\lambda = 0 \qquad \text{Equation 3}$$
$$F_w(x, y, z, w, \lambda) = 2w - \lambda = 0 \qquad \text{Equation 4}$$
$$F_\lambda(x, y, z, w, \lambda) = -3x - 2y + 4z - w + 3 = 0 \qquad \text{Equation 5}$$

For this system, you can solve the first four equations in terms of λ.

$$2x - 3\lambda = 0 \quad \Longrightarrow \quad x = \tfrac{3}{2}\lambda$$
$$2y - 2\lambda = 0 \quad \Longrightarrow \quad y = \lambda$$
$$2z + 4\lambda = 0 \quad \Longrightarrow \quad z = -2\lambda$$
$$2w - \lambda = 0 \quad \Longrightarrow \quad w = \tfrac{1}{2}\lambda$$

Substituting these values into the fifth equation allows you to solve for λ.

$$-3x - 2y + 4z - w + 3 = 0$$
$$-3\left(\tfrac{3}{2}\lambda\right) - 2(\lambda) + 4(-2\lambda) - \left(\tfrac{1}{2}\lambda\right) = -3$$
$$-15\lambda = -3$$
$$\lambda = \tfrac{1}{5}$$

So, the critical numbers are

$$x = \tfrac{3}{10}, \quad y = \tfrac{1}{5}, \quad z = -\tfrac{2}{5}, \quad \text{and} \quad w = \tfrac{1}{10}$$

and the minimum value of f subject to the given constraint is

$$f\left(\tfrac{3}{10}, \tfrac{1}{5}, -\tfrac{2}{5}, \tfrac{1}{10}\right) = \left(\tfrac{3}{10}\right)^2 + \left(\tfrac{1}{5}\right)^2 + \left(-\tfrac{2}{5}\right)^2 + \left(\tfrac{1}{10}\right)^2 \qquad \text{Substitute.}$$
$$= \tfrac{9}{100} + \tfrac{1}{25} + \tfrac{4}{25} + \tfrac{1}{100} \qquad \text{Simplify.}$$
$$= \tfrac{3}{10}. \qquad \text{Minimum value}$$

So, the minimum value of $f(x, y, z, w)$ is $\tfrac{3}{10}$.

DISCOVERY

In Example 4, what would be the minimum value of the objective function $f(x, y, z, w) = x^2 + y^2 + z^2 + w^2$ if there were no constraint? What would be the maximum value of f in this case? How do you know that the answer to Example 4 is the minimum value of f subject to the given constraint, and not the maximum value?

TRY IT 4

Find the minimum of the function

$$f(x, y, z, w) = x^2 + y^2 + z^2 + w^2$$

subject to the constraint $2x + 3y + z - 4w - 15 = 0$.

In Example 4 in Section 7.5, you found the maximum profit for two substitute products whose demand functions are given by

$x_1 = 200(p_2 - p_1)$ Demand for product 1

$x_2 = 500 + 100p_1 - 180p_2.$ Demand for product 2

With this model, the total demand, $x_1 + x_2$, is completely determined by the prices p_1 and p_2. In many real-life situations, this assumption is too simplistic; regardless of the prices of the substitute brands, the annual total demands for some products, such as toothpaste, are relatively constant. In such situations, the total demand is **limited,** and variations in price do not affect the total demand as much as they affect the market share of the substitute brands.

EXAMPLE 5 Finding a Maximum Profit

A company makes two substitute products whose demand functions are given by

$x_1 = 200(p_2 - p_1)$ Demand for product 1

$x_2 = 500 + 100p_1 - 180p_2$ Demand for product 2

where p_1 and p_2 are the prices per unit (in dollars) and x_1 and x_2 are the numbers of units sold. The costs of producing the two products are \$0.50 and \$0.75 per unit, respectively. The total demand is limited to 200 units per year. Find the prices that will yield a maximum profit.

SOLUTION From Example 4 in Section 7.5, the profit function is modeled by

$$P = -200p_1^2 - 180p_2^2 + 300p_1p_2 + 25p_1 + 535p_2 - 375.$$

The total demand for the two products is

$$\begin{aligned} x_1 + x_2 &= 200(p_2 - p_1) + 500 + 100p_1 - 180p_2 \\ &= -100p_1 + 20p_2 + 500. \end{aligned}$$

Because the total demand is limited to 200 units,

$-100p_1 + 20p_2 + 500 = 200.$ Constraint

Using Lagrange multipliers, you can determine that the maximum profit occurs when $p_1 = \$3.94$ and $p_2 = \$4.69$. This corresponds to an annual profit of \$712.21.

TRY IT 5

In Example 5, suppose the total demand is limited to 250 units per year. Find the prices that will yield a maximum profit.

STUDY TIP

The constrained optimization problem in Example 5 is represented graphically in Figure 7.35. The graph of the objective function is a paraboloid and the graph of the constraint is a vertical plane. In the "unconstrained" optimization problem on page 499, the maximum profit occurred at the vertex of the paraboloid. In this "constrained" problem, however, the maximum profit corresponds to the highest point on the curve that is the intersection of the paraboloid and the vertical "constraint" plane.

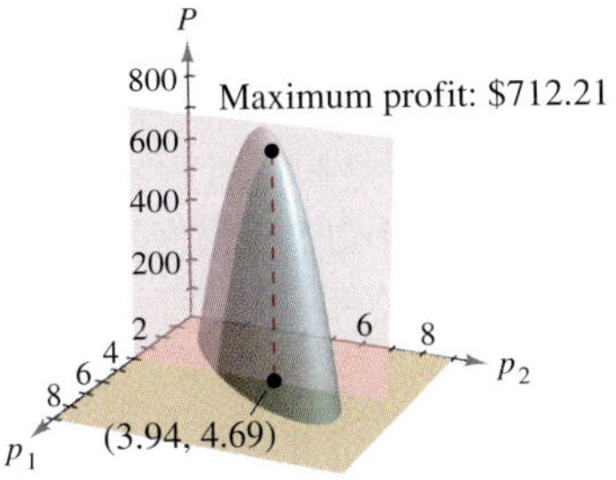

FIGURE 7.35

Lagrange Multipliers with Two Constraints

In Examples 1 through 5, each of the optimization problems contained only one constraint. When an optimization problem has two constraints, you need to introduce a second Lagrange multiplier. The customary symbol for this second multiplier is μ, the Greek letter mu.

EXAMPLE 6 Using Lagrange Multipliers: Two Constraints

Find the minimum value of

$$f(x, y, z) = x^2 + y^2 + z^2 \qquad \text{Objective function}$$

subject to the constraints

$$x + y - 3 = 0 \qquad \text{Constraint 1}$$
$$x + z - 5 = 0. \qquad \text{Constraint 2}$$

SOLUTION Begin by forming the function

$$F(x, y, z, \lambda, \mu) = x^2 + y^2 + z^2 - \lambda(x + y - 3) - \mu(x + z - 5).$$

Next, set the five partial derivatives equal to zero, and solve the resulting system of equations for x, y, and z.

$$F_x(x, y, z, \lambda, \mu) = 2x - \lambda - \mu = 0 \qquad \text{Equation 1}$$
$$F_y(x, y, z, \lambda, \mu) = 2y - \lambda = 0 \qquad \text{Equation 2}$$
$$F_z(x, y, z, \lambda, \mu) = 2z - \mu = 0 \qquad \text{Equation 3}$$
$$F_\lambda(x, y, z, \lambda, \mu) = -x - y + 3 = 0 \qquad \text{Equation 4}$$
$$F_\mu(x, y, z, \lambda, \mu) = -x - z + 5 = 0 \qquad \text{Equation 5}$$

Solving this system of equations produces $x = \frac{8}{3}$, $y = \frac{1}{3}$, and $z = \frac{7}{3}$. So, the minimum value of $f(x, y, z)$ is

$$f\left(\frac{8}{3}, \frac{1}{3}, \frac{7}{3}\right) = \left(\frac{8}{3}\right)^2 + \left(\frac{1}{3}\right)^2 + \left(\frac{7}{3}\right)^2$$
$$= \frac{38}{3}.$$

TRY IT 6

Find the minimum value of $f(x, y, z) = x^2 + y^2 + z^2$ subject to the constraints

$$x + y - 2 = 0$$
$$x + z - 4 = 0.$$

TAKE ANOTHER LOOK

Solving a System of Equations

Explain how you would solve the system of equations shown in Example 6. Illustrate your description by actually solving the system.

Suppose that in Example 6 you had been asked to find the maximum value of $f(x, y, z)$ subject to the two constraints. How would your answer have been different? Use a three-dimensional graphing utility to verify your conclusion.

PREREQUISITE REVIEW 7.6

The following warm-up exercises involve skills that were covered in earlier sections. You will use these skills in the exercise set for this section.

In Exercises 1–6, solve the system of linear equations.

1. $\begin{cases} 4x - 6y = 3 \\ 2x + 3y = 2 \end{cases}$

2. $\begin{cases} 6x - 6y = 5 \\ -3x - y = 1 \end{cases}$

3. $\begin{cases} 5x - y = 25 \\ x - 5y = 15 \end{cases}$

4. $\begin{cases} 4x - 9y = 5 \\ -x + 8y = -2 \end{cases}$

5. $\begin{cases} 2x - y + z = 3 \\ 2x + 2y + z = 4 \\ -x + 2y + 3z = -1 \end{cases}$

6. $\begin{cases} -x - 4y + 6z = -2 \\ x - 3y - 3z = 4 \\ 3x + y + 3z = 0 \end{cases}$

In Exercises 7–10, find all first partial derivatives.

7. $f(x, y) = x^2y + xy^2$

8. $f(x, y) = 25(xy + y^2)^2$

9. $f(x, y, z) = x(x^2 - 2xy + yz)$

10. $f(x, y, z) = z(xy + xz + yz)$

EXERCISES 7.6

In Exercises 1–12, use Lagrange multipliers to find the given extremum. In each case, assume that x and y are positive.

	Objective Function	*Constraint*
1.	Maximize $f(x, y) = xy$	$x + y = 10$
2.	Maximize $f(x, y) = xy$	$2x + y = 4$
3.	Minimize $f(x, y) = x^2 + y^2$	$x + y - 4 = 0$
4.	Minimize $f(x, y) = x^2 + y^2$	$-2x - 4y + 5 = 0$
5.	Maximize $f(x, y) = x^2 - y^2$	$y - x^2 = 0$
6.	Minimize $f(x, y) = x^2 - y^2$	$x - 2y + 6 = 0$
7.	Maximize $f(x, y) = 3x + xy + 3y$	$x + y = 25$
8.	Maximize $f(x, y) = 3x + y + 10$	$x^2y = 6$
9.	Maximize $f(x, y) = \sqrt{6 - x^2 - y^2}$	$x + y - 2 = 0$
10.	Minimize $f(x, y) = \sqrt{x^2 + y^2}$	$2x + 4y - 15 = 0$
11.	Maximize $f(x, y) = e^{xy}$	$x^2 + y^2 - 8 = 0$
12.	Minimize $f(x, y) = 2x + y$	$xy = 32$

In Exercises 13–18, use Lagrange multipliers to find the given extremum. In each case, assume that x, y, and z are positive.

13. Minimize $f(x, y, z) = 2x^2 + 3y^2 + 2z^2$

Constraint: $x + y + z - 24 = 0$

14. Maximize $f(x, y, z) = xyz$

Constraint: $x + y + z - 6 = 0$

15. Minimize $f(x, y, z) = x^2 + y^2 + z^2$

Constraint: $x + y + z = 1$

16. Minimize $f(x, y) = x^2 - 8x + y^2 - 12y + 48$

Constraint: $x + y = 8$

17. Maximize $f(x, y, z) = x + y + z$

Constraint: $x^2 + y^2 + z^2 = 1$

18. Maximize $f(x, y, z) = x^2y^2z^2$

Constraint: $x^2 + y^2 + z^2 = 1$

In Exercises 19 and 20, use Lagrange multipliers with the objective function

$$f(x, y, z, w) = 2x^2 + y^2 + z^2 + 2w^2$$

and with the given constraints to find the given extremum. In each case, assume that x, y, z, and w are nonnegative.

19. Maximize $f(x, y, z, w)$

Constraint: $2x + 2y + z + w = 2$

20. Maximize $f(x, y, z, w)$

Constraint: $x + y + 2z + 2w = 4$

In Exercises 21–24, use Lagrange multipliers to find the given extremum of f subject to two constraints. In each case, assume that x, y, and z are nonnegative.

21. Maximize $f(x, y, z) = xyz$

Constraints: $x + y + z = 24,\ x - y + z = 12$

22. Minimize $f(x, y, z) = x^2 + y^2 + z^2$

Constraints: $x + 2z = 4,\ x + y = 8$

23. Maximize $f(x, y, z) = xyz$

Constraints: $x^2 + z^2 = 5,\ x - 2y = 0$

24. Maximize $f(x, y, z) = xy + yz$

Constraints: $x + 2y = 6,\ x - 3z = 0$

In Exercises 25 and 26, use a spreadsheet to find the given extremum. In each case, assume that x, y, and z are nonnegative.

25. Maximize $f(x, y, z) = xyz$

Constraints: $x + 3y = 6,\ x - 2z = 0$

26. Minimize $f(x, y, z) = x^2 + y^2 + z^2$

Constraints: $x + 2y = 8,\ x + z = 4$

In Exercises 27–30, find three positive numbers x, y, and z that satisfy the given conditions.

27. The sum is 120 and the product is maximum.

28. The sum is 120 and the sum of the squares is minimum.

29. The sum is S and the product is maximum.

30. The sum is S and the sum of the squares is minimum.

In Exercises 31–34, find the minimum distance from the curve or surface to the given point. (*Hint:* Start by minimizing the square of the distance.)

31. Line: $x + 2y = 5,\ (0, 0)$

Minimize $d^2 = x^2 + y^2$

32. Circle: $(x - 4)^2 + y^2 = 4,\ (0, 10)$

Minimize $d^2 = x^2 + (y - 10)^2$

33. Plane: $x + y + z = 1,\ (2, 1, 1)$

Minimize $d^2 = (x - 2)^2 + (y - 1)^2 + (z - 1)^2$

34. Cone: $z = \sqrt{x^2 + y^2},\ (4, 0, 0)$

Minimize $d^2 = (x - 4)^2 + y^2 + z^2$

35. ***Volume*** Find the dimensions of the rectangular package of largest volume subject to the constraint that the sum of the length and the girth cannot exceed 108 inches (see figure). (*Hint:* Maximize $V = xyz$ subject to the constraint $x + 2y + 2z = 108$.)

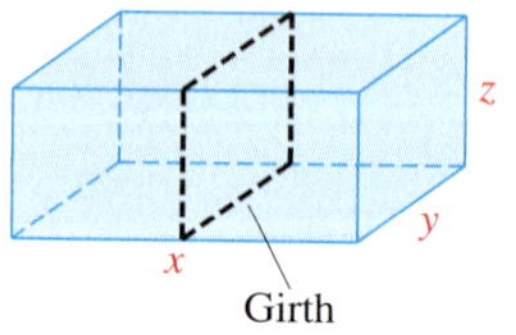

36. ***Cost*** In redecorating an office, the cost for new carpeting is five times the cost of wallpapering a wall. Find the dimensions of the largest office that can be redecorated for a fixed cost C (see figure). (*Hint:* Maximize $V = xyz$ subject to $5xy + 2xz + 2yz = C$.)

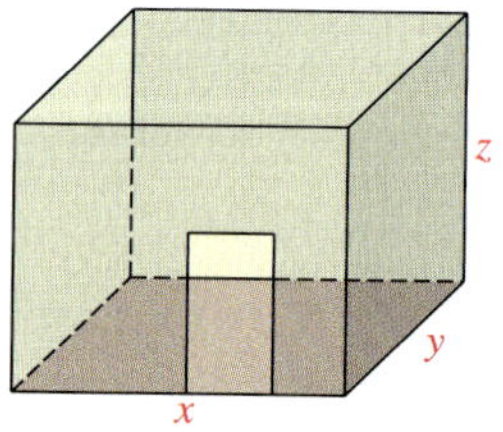

37. ***Cost*** A cargo container (in the shape of a rectangular solid) must have a volume of 480 cubic feet. Use Lagrange multipliers to find the dimensions of the container of this size that has a minimum cost, if the bottom will cost \$5 per square foot to construct and the sides and top will cost \$3 per square foot to construct.

38. ***Cost*** A manufacturer has an order for 1000 units of fine paper that can be produced at two locations. Let x_1 and x_2 be the numbers of units produced at the two plants. Find the number of units that should be produced at each plant to minimize the cost if the cost function is given by

$$C = 0.25x_1^2 + 25x_1 + 0.05x_2^2 + 12x_2.$$

39. ***Cost*** A manufacturer has an order for 2000 units of all-terrain vehicle tires that can be produced at two locations. Let x_1 and x_2 be the numbers of units produced at the two plants. The cost function is modeled by

$$C = 0.25x_1^2 + 10x_1 + 0.15x_2^2 + 12x_2.$$

Find the number of units that should be produced at each plant to minimize the cost.

40. ***Hardy-Weinberg Law*** Repeat Exercise 45 in Section 7.5 using Lagrange multipliers—that is, maximize

$$P(p, q, r) = 2pq + 2pr + 2qr$$

subject to the constraint

$$p + q + r = 1.$$

41. ***Least-Cost Rule*** The production function for a company is given by

$$f(x, y) = 100x^{0.25}y^{0.75}$$

where x is the number of units of labor and y is the number of units of capital. Suppose that labor costs \$48 per unit, capital costs \$36 per unit, and management sets a production goal of 20,000 units.

(a) Find the numbers of units of labor and capital needed to meet the production goal while minimizing the cost.

(b) Show that the conditions of part (a) are met when

$$\frac{\text{Marginal productivity of labor}}{\text{Marginal productivity of capital}} = \frac{\text{unit price of labor}}{\text{unit price of capital}}.$$

This proportion is called the *Least-Cost Rule* (or Equimarginal Rule).

42. ***Least-Cost Rule*** Repeat Exercise 41 for the production function given by

$$f(x, y) = 100x^{0.6}y^{0.4}.$$

43. ***Production*** The production function for a company is given by

$$f(x, y) = 100x^{0.25}y^{0.75}$$

where x is the number of units of labor and y is the number of units of capital. Suppose that labor costs \$48 per unit and capital costs \$36 per unit. The total cost of labor and capital is limited to \$100,000.

(a) Find the maximum production level for this manufacturer.

(b) Find the marginal productivity of money.

(c) Use the marginal productivity of money to find the maximum number of units that can be produced if \$125,000 is available for labor and capital.

44. ***Production*** Repeat Exercise 43 for the production function given by

$$f(x, y) = 100x^{0.06}y^{0.04}.$$

45. ***Biology*** A microbiologist must prepare a culture medium in which to grow a certain type of bacteria. The percent of salt contained in this medium is given by

$$S = 12xyz$$

where x, y, and z are the nutrient solutions to be mixed in the medium. For the bacteria to grow, the medium must be 13% salt. Nutrient solutions x, y, and z cost \$1, \$2, and \$3 per liter, respectively. How much of each nutrient solution should be used to minimize the cost of the culture medium?

46. ***Biology*** Repeat Exercise 45 for a salt-content model of

$$S = 0.01x^2y^2z^2.$$

47. ***Construction*** A rancher plans to use an existing stone wall and the side of a barn as a boundary for two adjacent rectangular corrals. Fencing for the perimeter costs \$10 per foot. To separate the corrals, a fence that costs \$4 per foot will divide the region. The total area of the two corrals is to be 6000 square feet.

(a) Use Lagrange multipliers to find the dimensions that will minimize the cost of the fencing.

(b) What is the minimum cost?

48. ***Area*** Use Lagrange multipliers to show that the maximum area of a rectangle with dimensions x and y and a given perimeter P is $\frac{1}{16}P^2$.

49. ***Investment Strategy*** An investor is considering three different stocks in which to invest \$300,000. The average annual dividends for the stocks are

General Mills (G)	2.5%
Pepsico, Inc. (P)	1.4%
Sara Lee (S)	3.1%.

The amount invested in Pepsico, Inc. must follow the equation

$$3000(G) - 3000(S) + P^2 = 0.$$

How much should be invested in each stock to yield a maximum of dividends?

50. ***Investment Strategy*** An investor is considering three different stocks in which to invest \$20,000. The average annual dividends for the stocks are

General Motors (G)	5.2%
Campbell Soup (C)	2.7%
Kellogg Co. (K)	3.2%.

The amount invested in Campbell Soup must follow the equation

$$1000(K) - 1000(G) + C^2 = 0.$$

How much should be invested in each stock to yield a maximum of dividends?

51. ***Research Project*** Use your school's library, the Internet, or some other reference source to write a paper about two different types of available investment options. Find examples of each type and find the data about their dividends for the past 10 years. What are the similarities and differences between the two types?

7.7 LEAST SQUARES REGRESSION ANALYSIS

- Find the sum of the squared errors for mathematical models.
- Find the least squares regression lines for data.
- Find the least squares regression quadratics for data.

Measuring the Accuracy of a Mathematical Model

When seeking a mathematical model to fit real-life data, you should try to find a model that is both as *simple* and as *accurate* as possible. For instance, a simple linear model for the points shown in Figure 7.36(a) is

$$f(x) = 1.8566x - 5.0246. \quad \text{Linear model}$$

However, Figure 7.36(b) shows that by choosing a slightly more complicated quadratic model

$$g(x) = 0.1996x^2 - 0.7281x + 1.3749 \quad \text{Quadratic model}$$

you can obtain significantly greater accuracy.

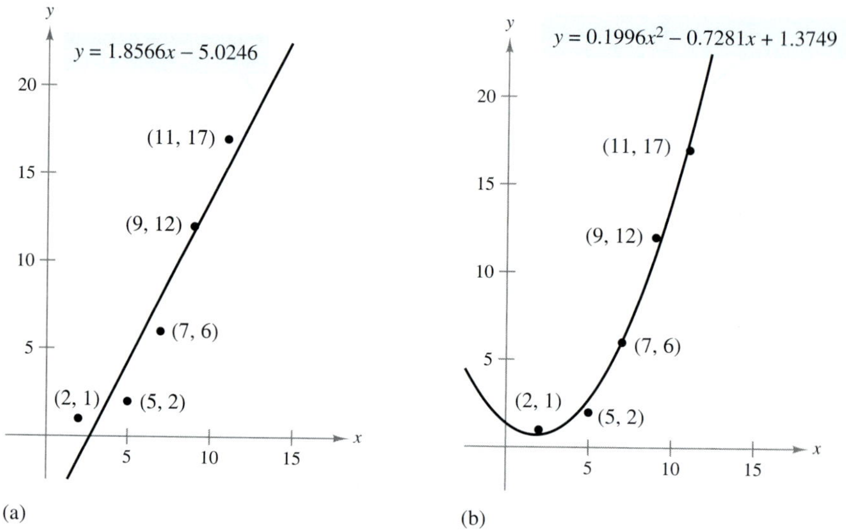

FIGURE 7.36

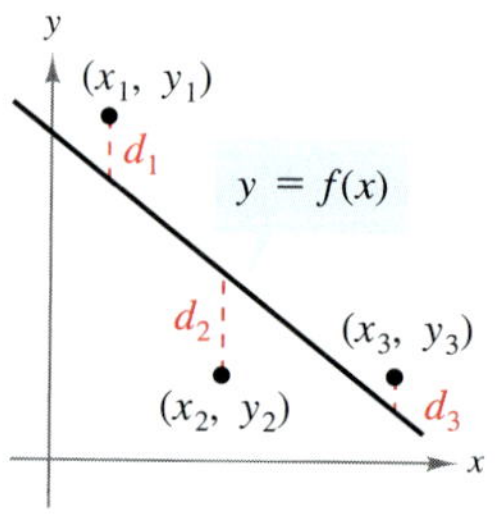

$S = d_1^2 + d_2^2 + d_3^2$

FIGURE 7.37

To measure how well the model $y = f(x)$ fits a collection of points, sum the squares of the differences between the actual y-values and the model's y-values. This sum is called the **sum of the squared errors** and is denoted by S. Graphically, S is the sum of the squares of the vertical distances between the graph of f and the given points in the plane, as shown in Figure 7.37. If the model is a perfect fit, then $S = 0$. However, when a perfect fit is not feasible, you should use a model that minimizes S.

Definition of the Sum of the Squared Errors

The **sum of the squared errors** for the model $y = f(x)$ with respect to the points $(x_1, y_1), (x_2, y_2), \ldots, (x_n, y_n)$ is given by

$$S = [f(x_1) - y_1]^2 + [f(x_2) - y_2]^2 + \cdots + [f(x_n) - y_n]^2.$$

EXAMPLE 1 Finding the Sum of the Squared Errors

Find the sum of the squared errors for the linear and quadratic models

$f(x) = 1.8566x - 5.0246$ Linear model

$g(x) = 0.1996x^2 - 0.7281x + 1.3749$ Quadratic model

(see Figures 7.35 and 7.36) with respect to the points

$(2, 1), (5, 2), (7, 6), (9, 12), (11, 17).$

SOLUTION Begin by evaluating each model at the given x-values, as shown in the table.

x	2	5	7	9	11
Actual y-values	1	2	6	12	17
Linear model, $f(x)$	-1.3114	4.2584	7.9716	11.6848	15.3980
Quadratic model, $g(x)$	0.7171	2.7244	6.0586	10.9896	17.5174

For the linear model f, the sum of the squared errors is

$$\begin{aligned} S &= (-1.3114 - 1)^2 + (4.2584 - 2)^2 + (7.9716 - 6)^2 \\ &\quad + (11.6848 - 12)^2 + (15.3980 - 17)^2 \\ &\approx 16.9959. \end{aligned}$$

Similarly, the sum of the squared errors for the quadratic model g is

$$\begin{aligned} S &= (0.7171 - 1)^2 + (2.7244 - 2)^2 + (6.0586 - 6)^2 \\ &\quad + (10.9896 - 12)^2 + (17.5174 - 17)^2 \\ &\approx 1.8968. \end{aligned}$$

STUDY TIP

In Example 1, note that the sum of the squared errors for the quadratic model is less than the sum of the squared errors for the linear model, which confirms that the quadratic model is a better fit.

TRY IT 1

Find the sum of the squared errors for the linear and quadratic models

$f(x) = 2.85x - 6.1$

$g(x) = 0.1964x^2 + 0.4929x - 0.6$

with respect to the points (2, 1), (4, 5), (6, 9), (8, 16), (10, 24). Then decide which model is a better fit.

Least Squares Regression Line

The sum of the squared errors can be used to determine which of several models is the best fit for a collection of data. In general, if the sum of the squared errors of f is less than the sum of the squared errors of g, then f is said to be a better fit for the data than g. In regression analysis, you consider all possible models of a certain type. The one that is defined to be the best-fitting model is the one with the least sum of the squared errors. Example 2 shows how to use the optimization techniques described in Section 7.5 to find the best-fitting linear model for a collection of data.

ALGEBRA REVIEW

For help in solving the system of equations in Example 2, see Example 2(b) in the *Chapter 7 Algebra Review* on page 541.

EXAMPLE 2 Finding the Best Linear Model

Find the values of a and b such that the linear model

$$f(x) = ax + b$$

has a minimum sum of the squared errors for the points

$$(-3, 0), (-1, 1), (0, 2), (2, 3).$$

SOLUTION The sum of the squared errors is

$$\begin{aligned} S &= [f(x_1) - y_1]^2 + [f(x_2) - y_2]^2 + [f(x_3) - y_3]^2 + [f(x_4) - y_4]^2 \\ &= (-3a + b - 0)^2 + (-a + b - 1)^2 + (b - 2)^2 + (2a + b - 3)^2 \\ &= 14a^2 - 4ab + 4b^2 - 10a - 12b + 14. \end{aligned}$$

To find the values of a and b for which S is a minimum, you can use the techniques described in Section 7.5. That is, find the partial derivatives of S.

$$\frac{\partial S}{\partial a} = 28a - 4b - 10 \qquad \text{Differentiate with respect to } a.$$

$$\frac{\partial S}{\partial b} = -4a + 8b - 12 \qquad \text{Differentiate with respect to } b.$$

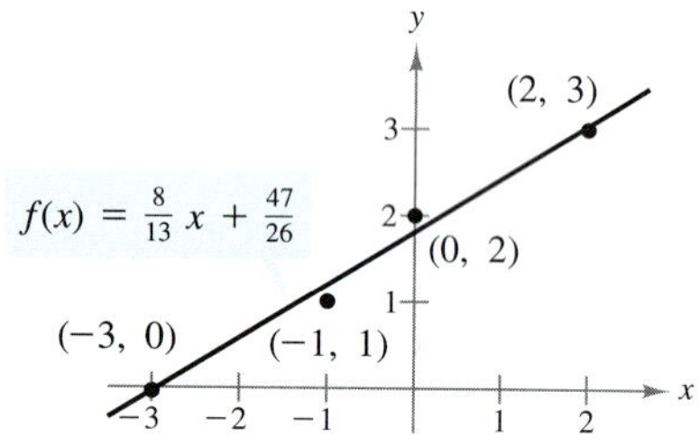

FIGURE 7.38

Next, set each partial derivative equal to zero.

$$28a - 4b - 10 = 0 \qquad \text{Set } \partial S/\partial a \text{ equal to 0.}$$

$$-4a + 8b - 12 = 0 \qquad \text{Set } \partial S/\partial b \text{ equal to 0.}$$

The solution of this system of linear equations is

$$a = \frac{8}{13} \quad \text{and} \quad b = \frac{47}{26}.$$

So, the best-fitting linear model for the given points is

$$f(x) = \frac{8}{13}x + \frac{47}{26}.$$

The graph of this model is shown in Figure 7.38.

TRY IT 2

Find the values of a and b such that the linear model $f(x) = ax + b$ has a minimum sum of the squared errors for the points $(-2, 0)$, $(0, 2)$, $(2, 5)$, $(4, 7)$.

The line in Example 2 is called the **least squares regression line** for the given data. The solution shown in Example 2 can be generalized to find a formula for the least squares regression line, as shown below. Consider the linear model

$$f(x) = ax + b$$

and the points $(x_1, y_1), (x_2, y_2), \ldots, (x_n, y_n)$. The sum of the squared errors is

$$S = \sum_{i=1}^{n}[f(x_i) - y_i]^2 = \sum_{i=1}^{n}(ax_i + b - y_i)^2.$$

To minimize S, set the partial derivatives $\partial S/\partial a$ and $\partial S/\partial b$ equal to zero and solve for a and b. The results are summarized below.

The Least Squares Regression Line

The **least squares regression line** for the points

$$(x_1, y_1), (x_2, y_2), \ldots, (x_n, y_n)$$

is $y = ax + b$, where

$$a = \frac{n\sum_{i=1}^{n}x_iy_i - \sum_{i=1}^{n}x_i\sum_{i=1}^{n}y_i}{n\sum_{i=1}^{n}x_i^2 - \left(\sum_{i=1}^{n}x_i\right)^2} \quad \text{and} \quad b = \frac{1}{n}\left(\sum_{i=1}^{n}y_i - a\sum_{i=1}^{n}x_i\right).$$

In the formula for the least squares regression line, note that if the x-values are symmetrically spaced about zero, then

$$\sum_{i=1}^{n}x_i = 0$$

and the formulas for a and b simplify to

$$a = \frac{n\sum_{i=1}^{n}x_iy_i}{n\sum_{i=1}^{n}x_i^2} \quad \text{and} \quad b = \frac{1}{n}\sum_{i=1}^{n}y_i.$$

Note also that only the *development* of the least squares regression line involves partial derivatives. The *application* of this formula is simply a matter of computing the values of a and b—a task that is performed much more simply on a calculator or a computer than by hand.

DISCOVERY

Graph the three points $(2, 2)$, $(2, 1)$, and $(2.1, 1.5)$ and visually estimate the least squares regression line for these data. Now use the formulas on this page or a graphing utility to show that the equation of the line is actually $y = 1.5$. In general, the least squares regression line for "nearly vertical" data can be quite unusual. Show that by interchanging the roles of x and y, you can obtain a better linear approximation.

EXAMPLE 3 Modeling Hourly Wage

The average hourly wages y (in dollars per hour) for production workers in manufacturing industries from 1994 through 2002 are shown in the table. Find the least squares regression line for the data and use the result to estimate the average hourly wage in 2006. *(Source: U.S. Bureau of Labor Statistics)*

Year	1994	1995	1996	1997	1998	1999	2000	2001	2002
y	12.07	12.37	12.78	13.17	13.49	13.91	14.38	14.76	15.29

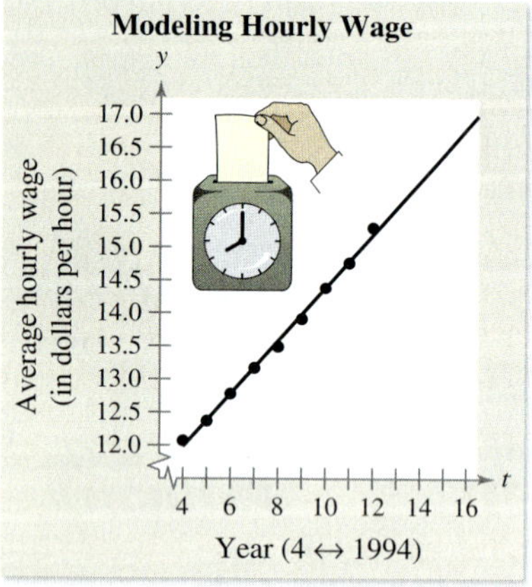

FIGURE 7.39

SOLUTION Let t represent the year, with $t = 4$ corresponding to 1994. Then, you need to find the linear model that best fits the points

$$(4, 12.07), (5, 12.37), (6, 12.78), (7, 13.17), (8, 13.49),$$
$$(9, 13.91), (10, 14.38), (11, 14.76), (12, 15.29).$$

Using a calculator with a built-in least squares regression program, you can determine that the best-fitting line is

$$y = 10.38 + 0.400t.$$

With this model, you can estimate the 2006 average hourly wage, using $t = 16$, to be

$$y = 10.38 + 0.400(16) = 16.78 \text{ per hour.}$$

This result is shown graphically in Figure 7.39.

TRY IT 3

The numbers of cellular phone subscribers y (in thousands) for the years 1998 through 2002 are shown in the table. Find the least squares regression line for the data and use the result to estimate the number of subscribers in 2006. Let t represent the year, with $t = 8$ corresponding to 1998.
(Source: Cellular Telecommunications & Internet Association)

Year	1998	1999	2000	2001	2002
y	69,209	86,047	109,478	128,375	140,766

TECHNOLOGY

Most graphing utilities and spreadsheet software programs have a built-in linear regression program. When you run such a program, the "r-value" gives a measure of how well the model fits the data. The closer the value of $|r|$ is to 1, the better the fit. For the data in Example 3, $r \approx 0.998$, which implies that the model is a very good fit. Use a spreadsheet software program to find the least squares regression line and compare your results to Example 3. (See Appendix E for specific keystrokes on how to find a least squares regression line using a graphing utility. Consult the user's manual of a spreadsheet software program for specific instructions.)

Least Squares Regression Quadratic

When using regression analysis to model data, remember that the least squares regression line provides only the best *linear* model for a set of data. It does not necessarily provide the best possible model. For instance, in Example 1, you saw that the quadratic model was a better fit than the linear model.

Regression analysis can be performed with many different types of models, such as exponential or logarithmic models. The following development shows how to find the best-fitting quadratic model for a collection of data points. Consider a quadratic model of the form

$$f(x) = ax^2 + bx + c.$$

The sum of the squared errors for this model is

$$S = \sum_{i=1}^{n}[f(x_i) - y_i]^2 = \sum_{i=1}^{n}(ax_i^2 + bx_i + c - y_i)^2.$$

To find the values of a, b, and c that minimize S, set the three partial derivatives, $\partial S/\partial a$, $\partial S/\partial b$, and $\partial S/\partial c$, equal to zero.

$$\frac{\partial S}{\partial a} = \sum_{i=1}^{n} 2x_i^2(ax_i^2 + bx_i + c - y_i) = 0$$

$$\frac{\partial S}{\partial b} = \sum_{i=1}^{n} 2x_i(ax_i^2 + bx_i + c - y_i) = 0$$

$$\frac{\partial S}{\partial c} = \sum_{i=1}^{n} 2(ax_i^2 + bx_i + c - y_i) = 0$$

By expanding this system, you obtain the result given in the summary below.

Least Squares Regression Quadratic

The **least squares regression quadratic** for the points

$$(x_1, y_1), (x_2, y_2), \ldots, (x_n, y_n)$$

is $y = ax^2 + bx + c$, where a, b, and c are the solutions of the system of equations below.

$$a\sum_{i=1}^{n}x_i^4 + b\sum_{i=1}^{n}x_i^3 + c\sum_{i=1}^{n}x_i^2 = \sum_{i=1}^{n}x_i^2y_i$$

$$a\sum_{i=1}^{n}x_i^3 + b\sum_{i=1}^{n}x_i^2 + c\sum_{i=1}^{n}x_i = \sum_{i=1}^{n}x_iy_i$$

$$a\sum_{i=1}^{n}x_i^2 + b\sum_{i=1}^{n}x_i + cn = \sum_{i=1}^{n}y_i$$

TECHNOLOGY

Most graphing utilities have a built-in program for finding the least squares regression quadratic. This program works just like the program for the least squares line. You should use this program to verify your solutions to the exercises.

EXAMPLE 4 Modeling the Number of Newspapers

The numbers y of daily morning newspapers in the United States from 1992 through 2002 are shown in the table. Find the least squares regression quadratic for the data and use the result to estimate the number of daily morning newspapers in 2006. *(Source: Editor & Publisher Co.)*

Year	1992	1993	1994	1995	1996	1997	1998	1999	2000	2001	2002
y	596	623	635	656	686	705	721	736	766	776	776

SOLUTION Let t represent the year, with $t = 2$ corresponding to 1992. Then, you need to find the quadratic model that best fits the points

$$(2, 596), (3, 623), (4, 635), (5, 656), (6, 686), (7, 705),$$
$$(8, 721), (9, 736), (10, 766), (11, 776), (12, 776).$$

Using a calculator with a built-in least squares regression program, you can determine that the best-fitting quadratic is $y = -0.57t^2 + 27.1t + 542$. With this model, you can estimate the number of daily morning newspapers in 2006, using $t = 16$, to be

$$y = -0.57(16)^2 + 27.1(16) + 542 \approx 830.$$

This result is shown graphically in Figure 7.40.

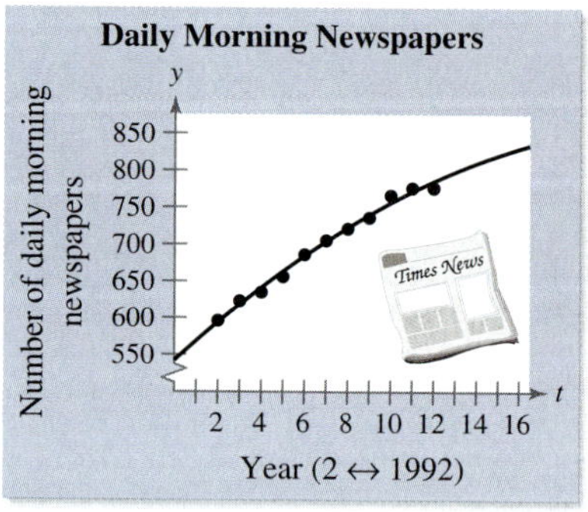

FIGURE 7.40

TRY IT 4

The per capita expenditures for health services and supplies in dollars in the United States for selected years are listed in the table. Find the least squares regression quadratic for the data and use the result to estimate the per capita expenditure for health care in 2006. Let t represent the year, with $t = 5$ corresponding to 1995. *(Source: U.S. Centers for Medicare and Medicaid Services)*

Year	1995	1996	1997	1998	1999	2000	2001
Expenditure	3576	3721	3870	4039	4244	4502	4851

TAKE ANOTHER LOOK

Least Squares Regression Analysis

Most graphing utilities have several built-in least squares regression programs. Try entering the points from Example 4 in a graphing utility. Then run one of the regression programs for the points. The quadratic model in Example 4 has a correlation coefficient of $r \approx 0.996$. Can you find a model, such as an exponential model or power model

$y = ab^x$ Exponential model $\qquad$ $y = ax^b$ Power model

that has a better correlation coefficient?

PREREQUISITE REVIEW 7.7

The following warm-up exercises involve skills that were covered in earlier sections. You will use these skills in the exercise set for this section.

In Exercises 1 and 2, evaluate the expression.

1. $(2.5 - 1)^2 + (3.25 - 2)^2 + (4.1 - 3)^2$

2. $(1.1 - 1)^2 + (2.08 - 2)^2 + (2.95 - 3)^2$

In Exercises 3 and 4, find the partial derivatives of S.

3. $S = a^2 + 6b^2 - 4a - 8b - 4ab + 6$

4. $S = 4a^2 + 9b^2 - 6a - 4b - 2ab + 8$

In Exercises 5–10, evaluate the sum.

5. $\sum_{i=1}^{5} i$

6. $\sum_{i=1}^{6} 2i$

7. $\sum_{i=1}^{4} \frac{1}{i}$

8. $\sum_{i=1}^{3} i^2$

9. $\sum_{i=1}^{6} (2 - i)^2$

10. $\sum_{i=1}^{5} (30 - i^2)$

EXERCISES 7.7

In Exercises 1–4, (a) use the method of least squares to find the least squares regression line and (b) calculate the sum of the squared errors.

1.

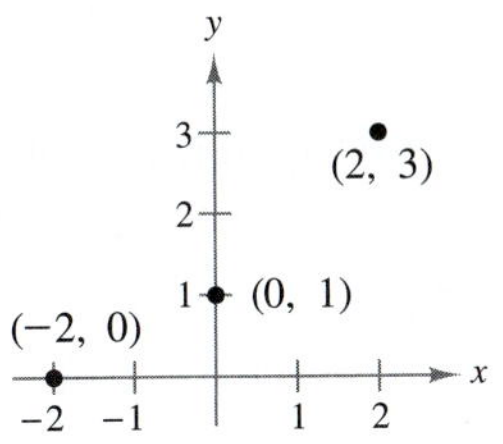

2.

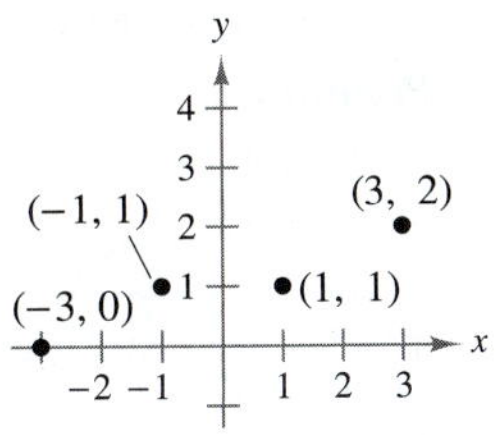

3.

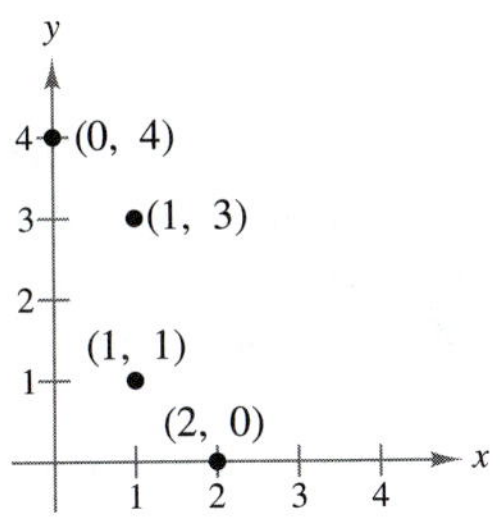

4.

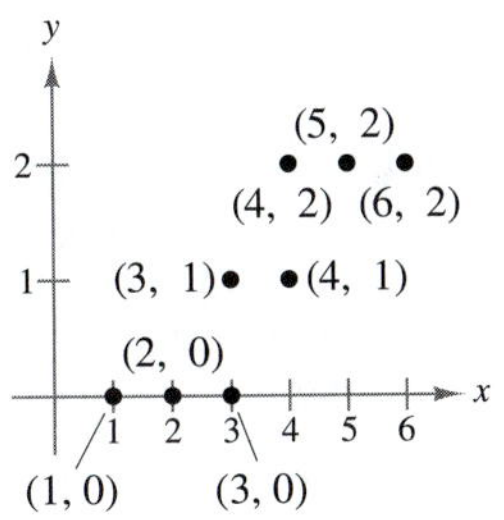

In Exercises 5–8, use partial derivatives to find the values of a and b such that the linear model $f(x) = ax + b$ has a minimum sum of the squared errors for the given points.

5. $(-2, -1), (0, 0), (2, 3)$

6. $(-3, 0), (-1, 1), (1, 1), (3, 2)$

7. $(-2, 4), (-1, 1), (0, -1), (1, -3)$

8. $(-5, -3), (-4, -2), (-2, -1), (-1, 1)$

In Exercises 9–18, use a graphing utility with a least squares regression program to find the least squares regression line for the given points.

9. $(-2, 0), (-1, 1), (0, 1), (1, 2), (2, 3)$

10. $(-4, -1), (-2, 0), (2, 4), (4, 5)$

11. $(-2, 2), (2, 6), (3, 7)$

12. $(-5, 1), (1, 3), (2, 3), (2, 5)$

13. $(-3, 4), (-1, 2), (1, 1), (3, 0)$

14. $(-10, 10), (-5, 8), (3, 6), (7, 4), (5, 0)$

15. $(0, 0), (1, 1), (3, 4), (4, 2), (5, 5)$

16. $(1, 0), (3, 3), (5, 6)$

17. $(0, 6), (4, 3), (5, 0), (8, -4), (10, -5)$

18. $(5, 2), (0, 0), (2, 1), (7, 4), (10, 6), (12, 6)$

In Exercises 19–22, use a graphing utility with a least squares regression program to find the least squares regression quadratic for the given points. Then plot the points and graph the least squares quadratic.

19. $(-2, 0), (-1, 0), (0, 1), (1, 2), (2, 5)$

20. $(-4, 5), (-2, 6), (2, 6), (4, 2)$

21. $(1, 0), (2, 1), (3, 7), (4, 13)$

22. $(0, 10), (1, 9), (2, 6), (3, 0)$

In Exercises 23–26, use a graphing utility with a least squares regression program to find linear and quadratic models for the data. State which model best fits the data.

23. $(-4, 1), (-3, 2), (-2, 2), (-1, 4), (0, 6), (1, 8), (2, 9)$

24. $(-1, -4), (0, -3), (1, -3), (2, 0), (4, 5), (6, 9), (9, 3)$

25. $(0, 769), (1, 677), (2, 601), (3, 543), (4, 489), (5, 411)$

26. $(1, 10.3), (2, 14.2), (3, 18.9), (4, 23.7), (5, 29.1), (6, 35)$

27. ***Demand*** A store manager wants to know the demand for a product as a function of price. The daily sales for three different prices of the product are listed in the table.

Price, x	\$1.00	\$1.25	\$1.50
Demand, y	450	375	330

(a) Use a graphing utility with a least squares regression program to find the least squares regression line for the data.

(b) Estimate the demand when the price is \$1.40.

(c) What price will create a demand of 500 products?

28. ***Demand*** A hardware retailer wants to know the demand for a tool as a function of price. The monthly sales for four different prices of the tool are listed in the table.

Price, x	\$25	\$30	\$35	\$40
Demand, y	82	75	67	55

(a) Use a graphing utility with a least squares regression program to find the least squares regression line for the data.

(b) Estimate the demand when the price is \$32.95.

(c) What price will create a demand of 83 tools?

29. ***Agriculture*** A farmer used four test plots to determine the relationship between wheat yield in bushels per acre and the amount of fertilizer in hundreds of pounds per acre. The results are listed in the table.

Fertilizer, x	1.0	1.5	2.0	2.5
Yield, y	35	44	50	56

(a) Use a spreadsheet software program with a least squares regression program to find the least squares regression line for the data.

(b) Estimate the yield for a fertilizer application of 160 pounds per acre.

30. ***Medical Science*** After contamination by a carcinogen, people in different geographic regions were assigned an exposure index that represented the degree of contamination. Use the data listed in the table and a spreadsheet software program with a least squares regression program to find a least squares regression line to estimate the mortality per 100,000 people for a given exposure.

Exposure, x	1.35	2.67	3.93	5.14	7.43
Mortality, y	118.5	135.2	167.3	197.6	204.7

31. ***Medical Science*** To study the number of infant deaths per 1000 live births in the United States, a medical researcher obtains the data listed in the table. *(Source: U.S. National Center for Health Statistics)*

Year	1950	1960	1970	1980	1990	2000
Deaths, y	29.2	26.0	20.0	12.6	9.2	6.9

(a) Use a graphing utility with a least squares regression program to find the least squares regression line for the data and use this line to estimate the number of infant deaths in 2010. Let $t = 0$ represent 1970.

(b) Find the least squares regression quadratic for the data and use the model to estimate the number of infant deaths in 2010.

32. ***Population Growth*** The table gives the approximate world population (in billions) for six different years. *(Source: U.S. Census Bureau)*

Year	1800	1850	1900	1950	1990	2000
Time, t	-2	-1	0	1	1.8	2
Population, y	0.8	1.1	1.6	2.4	5.3	6.1

(a) During the 1800s, population growth was almost linear. Use a graphing utility with a least squares regression program to find a least squares regression line for those years and use the line to estimate the population in 1875.

(b) Use a graphing utility with a least squares regression program to find a least squares regression quadratic for the data from 1850 through 2000 and use the model to estimate the population in the year 2010.

(c) Even though the rate of growth of the population has begun to decline, most demographers believe the population size will pass the 8 billion mark sometime in the next 30 years. What do you think?

33. ***Engineering*** After the development of a new turbocharger for an automobile engine, the experimental data listed in the table were obtained for speed in miles per hour at two-second intervals.

Time, x	0	2	4	6	8	10
Speed, y	0	15	30	50	65	70

(a) Use a graphing utility with a least squares regression program to find a least squares regression quadratic for the data.

(b) Use the model to estimate the speed after 5 seconds.

34. ***Revenue*** The revenue (in millions of dollars) for Earthlink from 1997 through 2003 is shown in the table. *(Source: Earthlink, Inc.)*

Year	1997	1998	1999	2000	2001	2002	2003
Revenue	133.4	290.6	670.4	986.6	1244.9	1357.4	1401.9

(a) Use a graphing utility to create a scatter plot of the data. Let $t = 7$ correspond to 1997.

(b) Use a graphing utility with a least squares regression program to find an appropriate model for the data.

(c) Explain why you chose the type of model that you created in part (b).

In Exercises 35–38, use a graphing utility with a least squares regression program to find any model that best fits the data points.

35. (1, 13), (2, 16.5), (4, 24), (5, 28), (8, 39), (11, 50.25), (17, 72), (20, 85)

36. (1, 5.5), (3, 7.75), (6, 15.2), (8, 23.5), (11, 46), (15, 110)

37. (1, 1.5), (2.5, 8.5), (5, 13.5), (8, 16.7), (9, 18), (20, 22)

38. (0, 0.5), (1, 7.6), (3, 60), (4.2, 117), (5, 170), (7.9, 380)

In Exercises 39–42, plot the points and determine whether the data have positive, negative, or no correlation (see graphs at top of next column). Then use a graphing utility to find the value of r and confirm your result. The number r is called the *correlation coefficient*. It is a measure of how well the model fits the data. Correlation coefficients vary between -1 and 1, and the closer $|r|$ is to 1, the better the model.

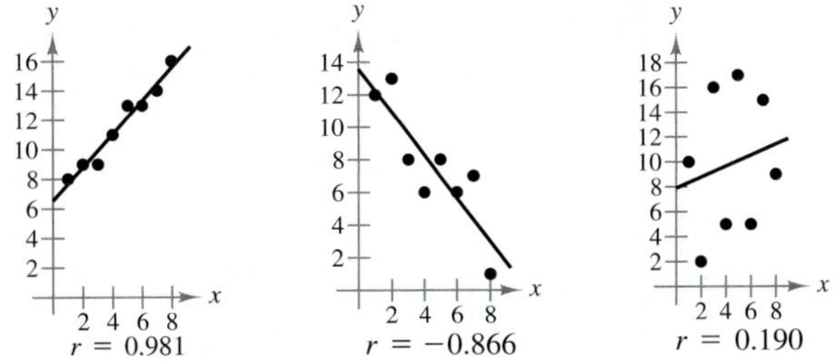

Positive Correlation   Negative Correlation   No Correlation

39. (1, 4), (2, 6), (3, 8), (4, 11), (5, 13), (6, 15)

40. (1, 7.5), (2, 7), (3, 7), (4, 6), (5, 5), (6, 4.9)

41. (1, 3), (2, 6), (3, 2), (4, 3), (5, 9), (6, 1)

42. (0.5, 2), (0.75, 1.75), (1, 3), (1.5, 3.2), (2, 3.7), (2.6, 4)

43. ***Finance: Median Income*** In the table below are the median income levels for various age levels in the United States. Use the data and a graphing utility with a least squares regression program to find the least squares regression quadratic for the data and use the resulting model to estimate the median income for someone who is 28 years old. *(Source: U.S. Census Bureau)*

Age level	Median income
20	\$28,200
30	\$45,100
40	\$53,300
50	\$58,000
60	\$45,900
70	\$23,100

True or False? In Exercises 44–47, determine whether the statement is true or false. If it is false, explain why or give an example that shows it is false.

44. Data that are modeled by $y = 3.29x - 4.17$ have a negative correlation.

45. Data that are modeled by $y = -0.238x + 25$ have a negative correlation.

46. If the correlation coefficient is $r \approx -0.98781$, the model is a good fit.

47. A correlation coefficient of $r \approx 0.201$ implies that the data have no correlation.

7.8 DOUBLE INTEGRALS AND AREA IN THE PLANE

- Evaluate double integrals.
- Use double integrals to find the areas of regions.

Double Integrals

In Section 7.4, you learned that it is meaningful to differentiate functions of several variables by differentiating with respect to one variable at a time while holding the other variables constant. It should not be surprising to learn that you can *integrate* functions of two or more variables using a similar procedure. For instance, if you are given the partial derivative

$$f_x(x, y) = 2xy \qquad \text{Partial with respect to } x$$

then, by holding y constant, you can integrate with respect to x to obtain

$$\int f_x(x, y)\,dx = f(x, y)$$
$$= x^2y + C(y).$$

This procedure is called **partial integration with respect to *x*.** Note that the "constant of integration" $C(y)$ is assumed to be a function of y, because y is fixed during integration with respect to x. Similarly, if you are given the partial derivative

$$f_y(x, y) = x^2 + 2 \qquad \text{Partial with respect to } y$$

then, by holding x constant, you can integrate with respect to y to obtain

$$\int f_y(x, y)\,dy = f(x, y)$$
$$= x^2y + 2y + C(x).$$

In this case, the "constant of integration" $C(x)$ is assumed to be a function of x, because x is fixed during integration with respect to y.

To evaluate a definite integral of a function of two or more variables, you can apply the Fundamental Theorem of Calculus to one variable while holding the others constant, as shown.

$$\int_1^{2y} 2xy\,dx = x^2y\Big]_1^{2y} = (2y)^2y - (1)^2y$$

x is the variable of integration and y is fixed.

Replace x by the limits of integration.

$$= 4y^3 - y.$$

The result is a function of y.

Note that you omit the constant of integration, just as you do for a definite integral of a function of one variable.

EXAMPLE 1 Finding Partial Integrals

Find each partial integral.

(a) $\displaystyle\int_1^x (2x^2y^{-2} + 2y)\,dy$ (b) $\displaystyle\int_y^{5y} \sqrt{x - y}\,dx$

SOLUTION

(a) $$\int_1^x (2x^2y^{-2} + 2y)\,dy = \left[\frac{-2x^2}{y} + y^2\right]_1^x$$ Hold x constant.

$$= \left(\frac{-2x^2}{x} + x^2\right) - \left(\frac{-2x^2}{1} + 1\right)$$

$$= 3x^2 - 2x - 1$$

(b) $$\int_y^{5y} \sqrt{x - y}\,dx = \left[\frac{2}{3}(x - y)^{3/2}\right]_y^{5y}$$ Hold y constant.

$$= \frac{2}{3}[(5y - y)^{3/2} - (y - y)^{3/2}]$$

$$= \frac{16}{3}y^{3/2}$$

TRY IT 1

Find each partial integral.

(a) $\displaystyle\int_1^x (4xy + y^3)\,dy$

(b) $\displaystyle\int_y^{y^2} \frac{1}{x + y}\,dx$

In Example 1(a), note that the definite integral defines a function of x and can *itself* be integrated. An "integral of an integral" is called a **double integral.** With a function of two variables, there are two types of double integrals.

$$\int_a^b \int_{g_1(x)}^{g_2(x)} f(x, y)\,dy\,dx = \int_a^b \left[\int_{g_1(x)}^{g_2(x)} f(x, y)\,dy\right] dx$$

$$\int_a^b \int_{g_1(y)}^{g_2(y)} f(x, y)\,dx\,dy = \int_a^b \left[\int_{g_1(y)}^{g_2(y)} f(x, y)\,dx\right] dy$$

STUDY TIP

Notice that the difference between the two types of double integrals is the order in which the integration is performed, $dy\,dx$ or $dx\,dy$.

EXAMPLE 2 Evaluating a Double Integral

Evaluate $\displaystyle\int_1^2 \int_0^x (2xy + 3)\,dy\,dx.$

SOLUTION

$$\int_1^2 \int_0^x (2xy + 3)\,dy\,dx = \int_1^2 \left[\int_0^x (2xy + 3)\,dy\right] dx$$

$$= \int_1^2 \left[xy^2 + 3y\right]_0^x dx$$

$$= \int_1^2 (x^3 + 3x)\,dx$$

$$= \left[\frac{x^4}{4} + \frac{3x^2}{2}\right]_1^2$$

$$= \left(\frac{2^4}{4} + \frac{3(2^2)}{2}\right) - \left(\frac{1^4}{4} + \frac{3(1^2)}{2}\right) = \frac{33}{4}$$

TECHNOLOGY

A symbolic integration utility can be used to evaluate double integrals. To do this, you need to enter the integrand, then integrate twice—once with respect to one of the variables and then with respect to the other variable. Use a symbolic integration utility to evaluate the double integral in Example 2.

TRY IT 2

Evaluate the double integral.

$$\int_1^2 \int_0^x (5x^2y - 2)\,dy\,dx$$

Finding Area with a Double Integral

One of the simplest applications of a double integral is finding the area of a plane region. For instance, consider the region R that is bounded by

$$a \le x \le b \quad \text{and} \quad g_1(x) \le y \le g_2(x).$$

Using the techniques described in Section 5.5, you know that the area of R is

$$\int_a^b [g_2(x) - g_1(x)]\, dx.$$

This same area is also given by the double integral

$$\int_a^b \int_{g_1(x)}^{g_2(x)} dy\, dx$$

because

$$\int_a^b \int_{g_1(x)}^{g_2(x)} dy\, dx = \int_a^b \Big[y \Big]_{g_1(x)}^{g_2(x)} dx$$
$$= \int_a^b [g_2(x) - g_1(x)]\, dx.$$

Figure 7.41 shows the two basic types of plane regions whose areas can be determined by a double integral.

Determining Area in the Plane by Double Integrals

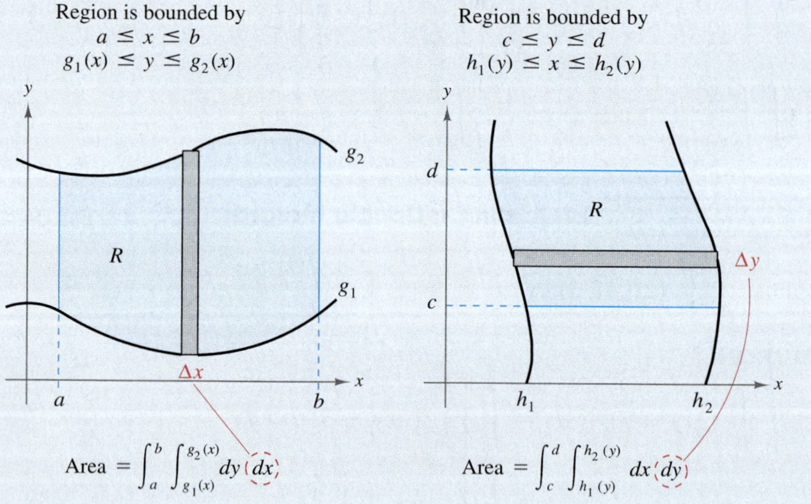

FIGURE 7.41

STUDY TIP

In Figure 7.41, note that the horizontal or vertical orientation of the narrow rectangle indicates the order of integration. The "outer" variable of integration always corresponds to the width of the rectangle. Notice also that the outer limits of integration for a double integral are constant, whereas the inner limits may be functions of the outer variable.

EXAMPLE 3 **Finding Area by a Double Integral**

Use a double integral to find the area of the rectangular region shown in Figure 7.42.

SOLUTION The bounds for x are $1 \le x \le 5$ and the bounds for y are $2 \le y \le 4$. So, the area of the region is

$$\int_1^5 \int_2^4 dy\,dx = \int_1^5 \Big[y \Big]_2^4 dx \qquad \text{Integrate with respect to } y.$$

$$= \int_1^5 (4-2)\,dx \qquad \text{Apply Fundamental Theorem of Calculus.}$$

$$= \int_1^5 2\,dx \qquad \text{Simplify.}$$

$$= \Big[2x \Big]_1^5 \qquad \text{Integrate with respect to } x.$$

$$= 10 - 2 \qquad \text{Apply Fundamental Theorem of Calculus.}$$

$$= 8 \text{ square units.} \qquad \text{Simplify.}$$

You can confirm this by noting that the rectangle is two units by four units.

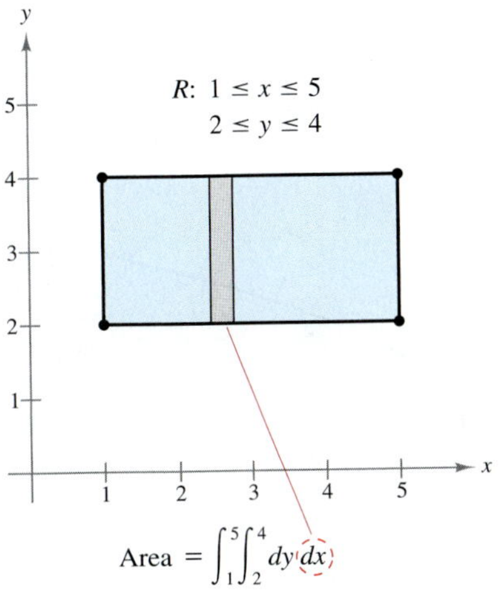

FIGURE 7.42

TRY IT 3

Use a double integral to find the area of the rectangular region shown in Example 3 by integrating with respect to x and then y.

EXAMPLE 4 **Finding Area by a Double Integral**

Use a double integral to find the area of the region bounded by the graphs of $y = x^2$ and $y = x^3$.

SOLUTION As shown in Figure 7.43, the two graphs intersect when $x = 0$ and $x = 1$. Choosing x to be the outer variable, the bounds for x are $0 \le x \le 1$ and the bounds for y are $x^3 \le y \le x^2$. This implies that the area of the region is

$$\int_0^1 \int_{x^3}^{x^2} dy\,dx = \int_0^1 \Big[y \Big]_{x^3}^{x^2} dx \qquad \text{Integrate with respect to } y.$$

$$= \int_0^1 (x^2 - x^3)\,dx \qquad \text{Apply Fundamental Theorem of Calculus.}$$

$$= \left[\frac{x^3}{3} - \frac{x^4}{4} \right]_0^1 \qquad \text{Integrate with respect to } x.$$

$$= \frac{1}{3} - \frac{1}{4} \qquad \text{Apply Fundamental Theorem of Calculus.}$$

$$= \frac{1}{12} \text{ square unit.} \qquad \text{Simplify.}$$

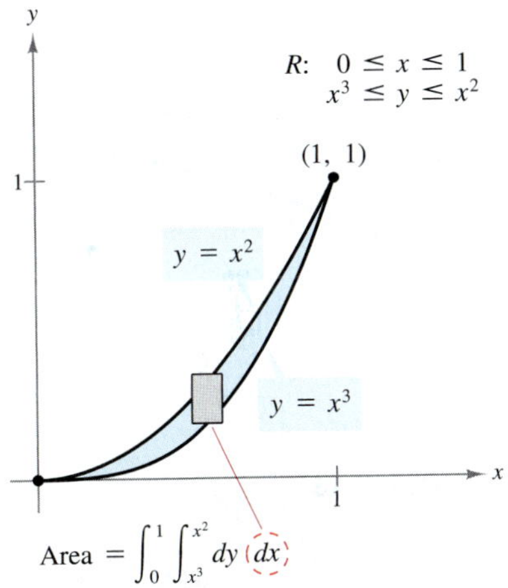

FIGURE 7.43

TRY IT 4

Use a double integral to find the area of the region bounded by the graphs of $y = 2x$ and $y = x^2$.

In setting up double integrals, the most difficult task is likely to be determining the correct limits of integration. This can be simplified by making a sketch of the region R and identifying the appropriate bounds for x and y.

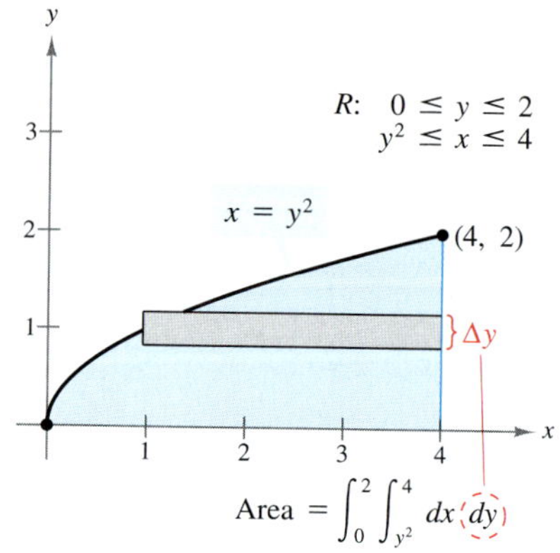

FIGURE 7.44

EXAMPLE 5 Changing the Order of Integration

For the double integral

$$\int_0^2 \int_{y^2}^4 dx\,dy$$

(a) sketch the region R whose area is represented by the integral,

(b) rewrite the integral so that x is the outer variable, and

(c) show that both orders of integration yield the same value.

SOLUTION

(a) From the limits of integration, you know that

$$y^2 \le x \le 4 \qquad \text{Variable bounds for } x$$

which means that the region R is bounded on the left by the parabola $x = y^2$ and on the right by the line $x = 4$. Furthermore, because

$$0 \le y \le 2 \qquad \text{Constant bounds for } y$$

you know that the region lies above the x-axis, as shown in Figure 7.44.

(b) If you interchange the order of integration so that x is the outer variable, then x will have constant bounds of integration given by $0 \le x \le 4$. Solving for y in the equation $x = y^2$ implies that the bounds for y are $0 \le y \le \sqrt{x}$, as shown in Figure 7.45. So, with x as the outer variable, the integral can be written as

$$\int_0^4 \int_0^{\sqrt{x}} dy\,dx.$$

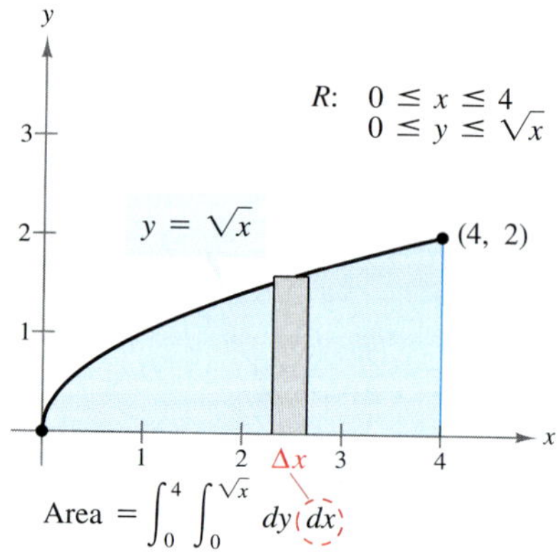

FIGURE 7.45

(c) Both integrals yield the same value.

$$\int_0^2 \int_{y^2}^4 dx\,dy = \int_0^2 \Big[x\Big]_{y^2}^4 dy = \int_0^2 (4 - y^2)\,dy = \left[4y - \frac{y^3}{3}\right]_0^2 = \frac{16}{3}$$

$$\int_0^4 \int_0^{\sqrt{x}} dy\,dx = \int_0^4 \Big[y\Big]_0^{\sqrt{x}} dx = \int_0^4 \sqrt{x}\,dx = \left[\frac{2}{3}x^{3/2}\right]_0^4 = \frac{16}{3}$$

STUDY TIP

To designate a double integral or an area of a region without specifying a particular order of integration, you can use the symbol

$$\int\!\!\int_R dA$$

where $dA = dx\,dy$ or $dA = dy\,dx$.

TRY IT 5

For the double integral $\int_0^2 \int_{2y}^4 dx\,dy$,

(a) sketch the region R whose area is represented by the integral,

(b) rewrite the integral so that x is the outer variable, and

(c) show that both orders of integration yield the same result.

EXAMPLE 6 Finding Area by a Double Integral

Use a double integral to calculate the area denoted by

$$\int_R\int dA$$

where R is the region bounded by $y = x$ and $y = x^2 - x$.

SOLUTION Begin by sketching the region R, as shown in Figure 7.46. From the sketch, you can see that vertical rectangles of width dx are more convenient than horizontal ones. So, x is the outer variable of integration and its constant bounds are $0 \le x \le 2$. This implies that the bounds for y are $x^2 - x \le y \le x$, and the area is given by

$$\int_R\int dA = \int_0^2 \int_{x^2-x}^{x} dy\,dx \qquad \text{Substitute bounds for region.}$$

$$= \int_0^2 \Big[y\Big]_{x^2-x}^{x} dx \qquad \text{Integrate with respect to } y.$$

$$= \int_0^2 [x - (x^2 - x)]\,dx \qquad \text{Apply Fundamental Theorem of Calculus.}$$

$$= \int_0^2 (2x - x^2)\,dx \qquad \text{Simplify.}$$

$$= \left[x^2 - \frac{x^3}{3}\right]_0^2 \qquad \text{Integrate with respect to } x.$$

$$= 4 - \frac{8}{3} \qquad \text{Apply Fundamental Theorem of Calculus.}$$

$$= \frac{4}{3} \text{ square units.} \qquad \text{Simplify.}$$

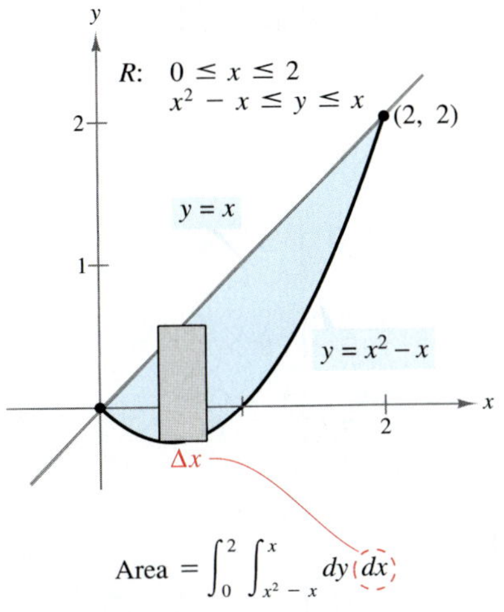

FIGURE 7.46

TRY IT 6

Use a double integral to calculate the area denoted by $\int_R\int dA$ where R is the region bounded by $y = 2x + 3$ and $y = x^2$.

As you are working the exercises for this section, you should be aware that the primary uses of double integrals will be discussed in Section 7.9. Double integrals by way of areas in the plane have been introduced so that you can gain practice in finding the limits of integration. When setting up a double integral, remember that your first step should be to sketch the region R. After doing this, you have two choices of integration orders: $dx\,dy$ or $dy\,dx$.

TAKE ANOTHER LOOK

Sketching Regions in the Plane

Sketch the region represented by the double integral.

a. $\int_0^4 \int_0^4 dx\,dy$ b. $\int_1^4 \int_0^5 dy\,dx$ c. $\int_{-1}^4 \int_2^6 dy\,dx$

What can you conclude about a region of integration if the inside and outside limits of integration are constants?

PREREQUISITE REVIEW 7.8

The following warm-up exercises involve skills that were covered in earlier sections. You will use these skills in the exercise set for this section.

In Exercises 1–12, evaluate the definite integral.

1. $\int_0^1 dx$

2. $\int_0^2 3\,dy$

3. $\int_1^4 2x^2\,dx$

4. $\int_0^1 2x^3\,dx$

5. $\int_1^2 (x^3 - 2x + 4)\,dx$

6. $\int_0^2 (4 - y^2)\,dy$

7. $\int_1^2 \frac{2}{7x^2}\,dx$

8. $\int_1^4 \frac{2}{\sqrt{x}}\,dx$

9. $\int_0^2 \frac{2x}{x^2 + 1}\,dx$

10. $\int_2^e \frac{1}{y - 1}\,dy$

11. $\int_0^2 xe^{x^2+1}\,dx$

12. $\int_0^1 e^{-2y}\,dy$

In Exercises 13–16, sketch the region bounded by the graphs of the equations.

13. $y = x,\ y = 0,\ x = 3$

14. $y = x,\ y = 3,\ x = 0$

15. $y = 4 - x^2,\ y = 0,\ x = 0$

16. $y = x^2,\ y = 4x$

EXERCISES 7.8

In Exercises 1–10, evaluate the partial integral.

1. $\int_0^x (2x - y)\,dy$

2. $\int_x^{x^2} \frac{y}{x}\,dy$

3. $\int_1^{2y} \frac{y}{x}\,dx$

4. $\int_0^{e^y} y\,dx$

5. $\int_0^{\sqrt{9-x^2}} x^2y\,dy$

6. $\int_{x^2}^{\sqrt{x}} (x^2 + y^2)\,dy$

7. $\int_{e^y}^{y} \frac{y \ln x}{x}\,dx$

8. $\int_{-\sqrt{1-y^2}}^{\sqrt{1-y^2}} (x^2 + y^2)\,dx$

9. $\int_0^{x^3} ye^{-y/x}\,dy$

10. $\int_y^3 \frac{xy}{\sqrt{x^2 + 1}}\,dx$

In Exercises 11–24, evaluate the double integral.

11. $\int_0^2 \int_0^1 (x - y)\,dy\,dx$

12. $\int_0^2 \int_0^2 (6 - x^2)\,dy\,dx$

13. $\int_0^4 \int_0^3 xy\,dy\,dx$

14. $\int_0^1 \int_0^x \sqrt{1 - x^2}\,dy\,dx$

15. $\int_0^1 \int_0^{\sqrt{1-y^2}} (x + y)\,dx\,dy$

16. $\int_0^2 \int_{3y^2-6y}^{2y-y^2} 3y\,dx\,dy$

17. $\int_1^2 \int_0^4 (x^2 - 2y^2 + 1)\,dx\,dy$

18. $\int_0^1 \int_y^{2y} (1 + 2x^2 + 2y^2)\,dx\,dy$

19. $\int_0^2 \int_0^{\sqrt{1-y^2}} -5xy\,dx\,dy$

20. $\int_0^4 \int_0^x \frac{2}{(x + 1)(y + 1)}\,dy\,dx$

21. $\int_0^2 \int_0^{4-x^2} x^3\,dy\,dx$

22. $\int_0^a \int_0^{a-x} (x^2 + y^2)\,dy\,dx$

23. $\int_0^\infty \int_0^\infty e^{-(x+y)/2}\,dy\,dx$

24. $\int_0^\infty \int_0^\infty xye^{-(x^2+y^2)}\,dx\,dy$

In Exercises 25–32, sketch the region R whose area is given by the double integral. Then change the order of integration and show that both orders yield the same area.

25. $\int_0^1 \int_0^2 dy\, dx$

26. $\int_1^2 \int_2^4 dx\, dy$

27. $\int_0^1 \int_{2y}^2 dx\, dy$

28. $\int_0^4 \int_0^{\sqrt{x}} dy\, dx$

29. $\int_0^2 \int_{x/2}^1 dy\, dx$

30. $\int_0^4 \int_{\sqrt{x}}^2 dx\, dy$

31. $\int_0^1 \int_{y^2}^{\sqrt[3]{y}} dx\, dy$

32. $\int_{-2}^2 \int_0^{4-y^2} dx\, dy$

In Exercises 33 and 34, evaluate the double integral. Note that it is necessary to change the order of integration.

33. $\int_0^3 \int_y^3 e^{x^2}\, dx\, dy$

34. $\int_0^2 \int_x^2 e^{-y^2}\, dy\, dx$

In Exercises 35–40, use a double integral to find the area of the specified region.

35.

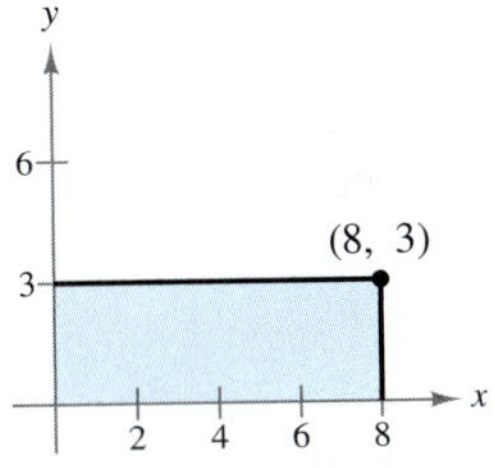

36.

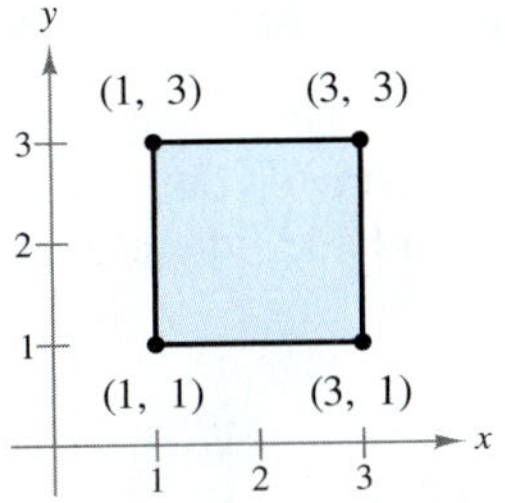

37.

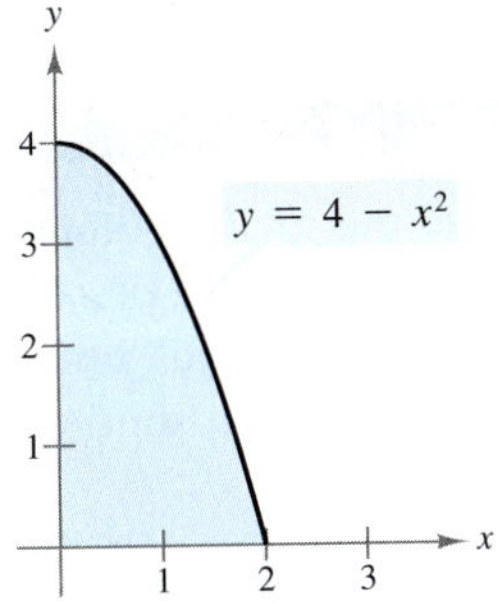

38.

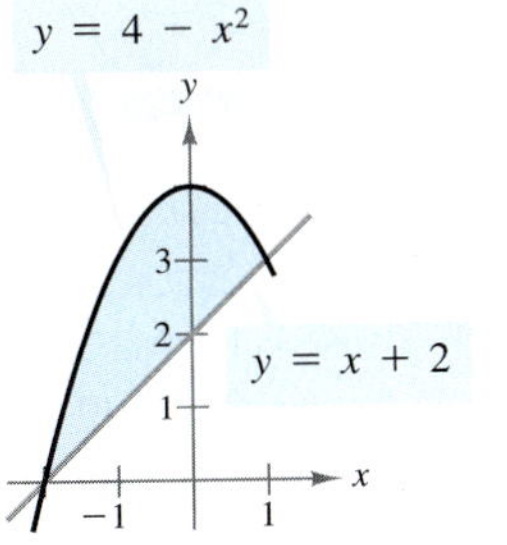

39.

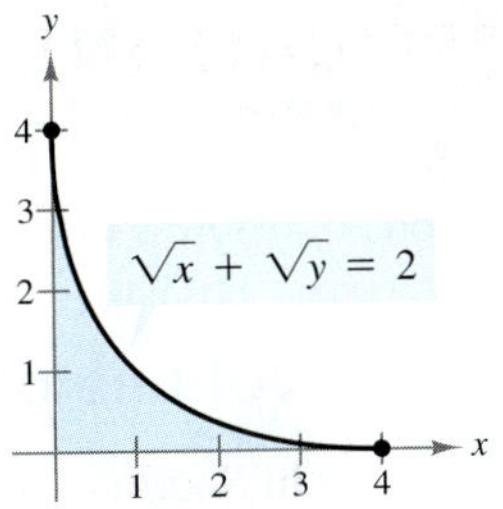

40.

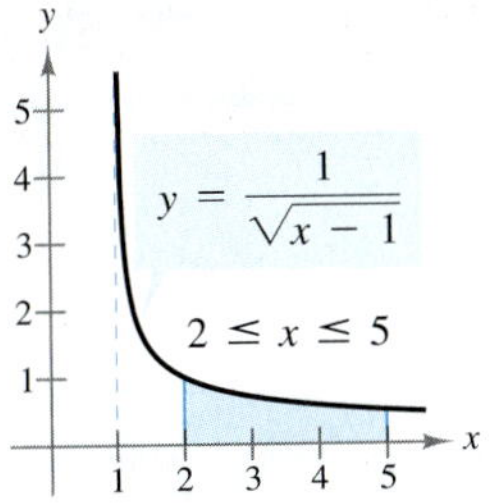

In Exercises 41–46, use a double integral to find the area of the region bounded by the graphs of the equations.

41. $y = 25 - x^2,\ y = 0$

42. $y = x^{3/2},\ y = x$

43. $5x - 2y = 0,\ x + y = 3,\ y = 0$

44. $xy = 9,\ y = x,\ y = 0,\ x = 9$

45. $y = x,\ y = 2x,\ x = 2$

46. $y = x^2 + 2x + 1,\ y = 3(x + 1)$

In Exercises 47–54, use a symbolic integration utility to evaluate the double integral.

47. $\int_0^1 \int_0^2 e^{-x^2 - y^2}\, dx\, dy$

48. $\int_0^2 \int_{x^2}^{2x} (x^3 + 3y^2)\, dy\, dx$

49. $\int_1^2 \int_0^x e^{xy}\, dy\, dx$

50. $\int_1^2 \int_y^{2y} \ln(x + y)\, dx\, dy$

51. $\int_0^1 \int_x^1 \sqrt{1 - x^2}\, dy\, dx$

52. $\int_0^3 \int_0^{x^2} \sqrt{x}\sqrt{1 + x}\, dy\, dx$

53. $\int_0^2 \int_{\sqrt{4-x^2}}^{4-x^2/4} \frac{xy}{x^2 + y^2 + 1}\, dy\, dx$

54. $\int_0^4 \int_0^y \frac{2}{(x + 1)(y + 1)}\, dx\, dy$

True or False? In Exercises 55 and 56, determine whether the statement is true or false. If it is false, explain why or give an example that shows it is false.

55. Changing the order of integration will sometimes change the value of a double integral.

56. $\int_2^5 \int_1^6 x\, dy\, dx = \int_1^6 \int_2^5 x\, dx\, dy$

7.9 APPLICATIONS OF DOUBLE INTEGRALS

- Use double integrals to find the volumes of solids.
- Use double integrals to find the average values of real-life models.

Volume of a Solid Region

In Section 7.8, you used double integrals as an alternative way to find the area of a plane region. In this section, you will study the primary uses of double integrals: to find the volume of a solid region and to find the average value of a function.

Consider a function $z = f(x, y)$ that is continuous and nonnegative over a region R. Let S be the solid region that lies between the xy-plane and the surface

$$z = f(x, y) \qquad \text{Surface lying above the } xy\text{-plane}$$

directly above the region R, as shown in Figure 7.47. You can find the volume of S by integrating $f(x, y)$ over the region R.

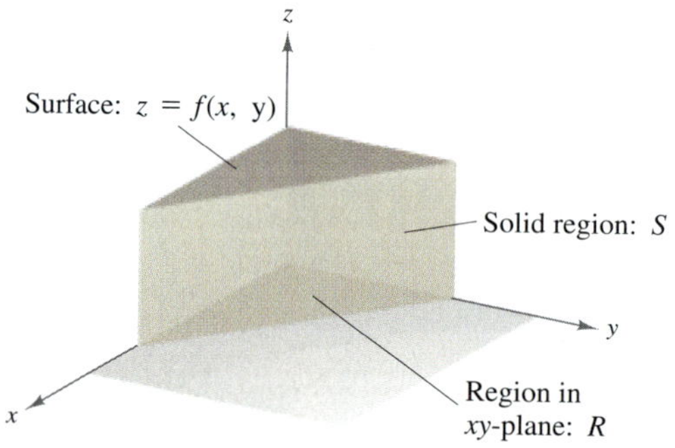

FIGURE 7.47

Determining Volume by Double Integrals

If R is a bounded region in the xy-plane and f is continuous and nonnegative over R, then the **volume of the solid** region between the surface $z = f(x, y)$ and R is given by the double integral

$$\int_R\int f(x, y)\, dA$$

where $dA = dx\, dy$ or $dA = dy\, dx$.

DISCOVERY

A solid has an elliptical (not a circular) base. Consider the region bounded by the circular paraboloid $z = a^2 - x^2 - y^2$ for $a > 0$ and the xy-plane. How many ways do you know for finding the volume of this solid? For instance, you could use the Disk Method to find the volume as a solid of revolution. Does each method involve integration?

EXAMPLE 1 Finding the Volume of a Solid

Find the volume of the solid region bounded in the first octant by the plane

$$z = 2 - x - 2y.$$

SOLUTION

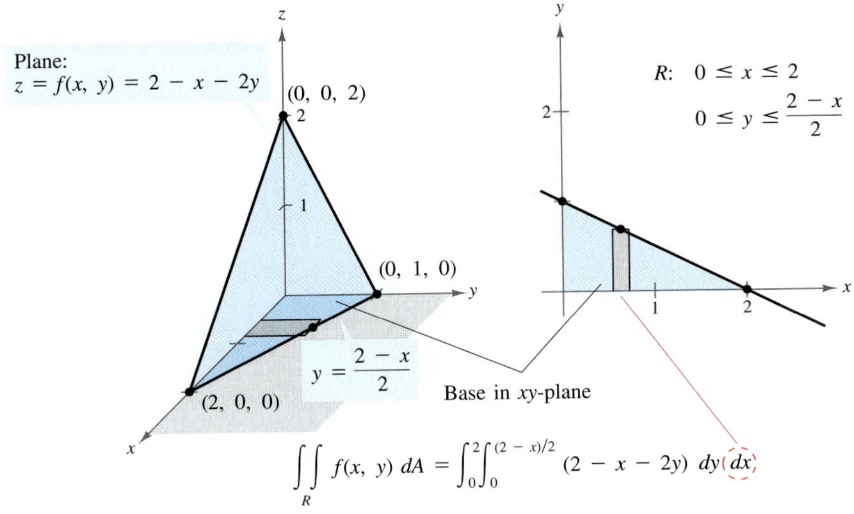

FIGURE 7.48

To set up the double integral for the volume, it is helpful to sketch both the solid region and the plane region R in the xy-plane. In Figure 7.48, you can see that the region R is bounded by the lines $x = 0$, $y = 0$, and $y = \frac{1}{2}(2 - x)$. One way to set up the double integral is to choose x as the outer variable. With that choice, the constant bounds for x are $0 \le x \le 2$ and the variable bounds for y are $0 \le y \le \frac{1}{2}(2 - x)$. So, the volume of the solid region is

$$\begin{aligned}
V &= \int_0^2 \int_0^{(2-x)/2} (2 - x - 2y)\, dy\, dx \\
&= \int_0^2 \Big[(2 - x)y - y^2\Big]_0^{(2-x)/2} dx \\
&= \int_0^2 \left\{(2 - x)\left(\frac{1}{2}\right)(2 - x) - \left[\frac{1}{2}(2 - x)\right]^2\right\} dx \\
&= \frac{1}{4}\int_0^2 (2 - x)^2\, dx \\
&= \left[-\frac{1}{12}(2 - x)^3\right]_0^2 \\
&= \frac{2}{3} \text{ cubic unit.}
\end{aligned}$$

STUDY TIP

Example 1 uses $dy\,dx$ as the order of integration. Try using the other order, $dx\,dy$, as indicated in Figure 7.49, to find the volume of the region. Do you get the same result as in Example 1?

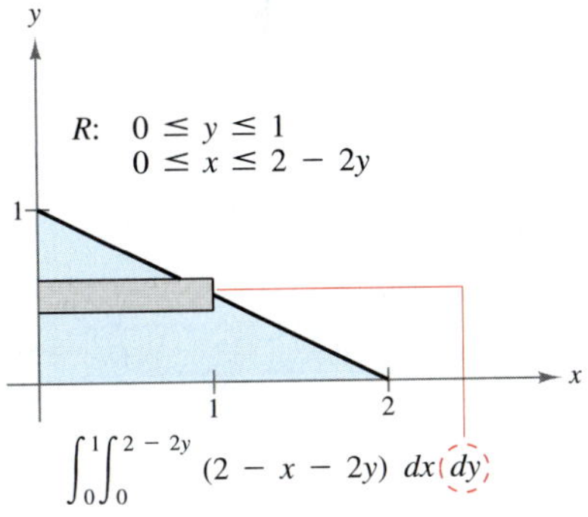

FIGURE 7.49

TRY IT 1

Find the volume of the solid region bounded in the first octant by the plane $z = 4 - 2x - y$.

In Example 1, the order of integration was arbitrary. Although the example used x as the outer variable, you could just as easily have used y as the outer variable. The next example describes a situation in which one order of integration is more convenient than the other.

EXAMPLE 2 Comparing Different Orders of Integration

Find the volume under the surface $f(x, y) = e^{-x^2}$ bounded by the xz-plane and the planes $y = x$ and $x = 1$, as shown in Figure 7.50.

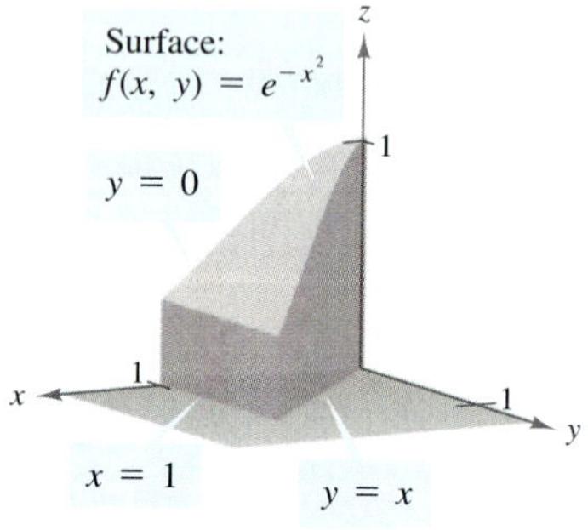

FIGURE 7.50

SOLUTION

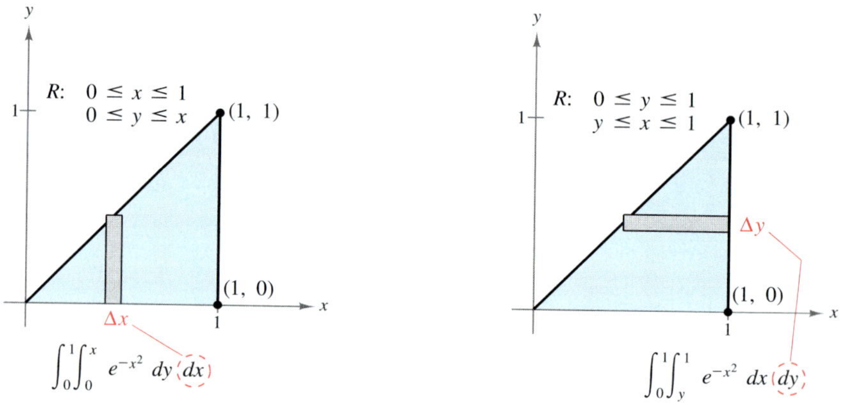

FIGURE 7.51

In the xy-plane, the bounds of region R are the lines $y = 0$, $x = 1$, and $y = x$. The two possible orders of integration are indicated in Figure 7.51. If you attempt to evaluate the two double integrals shown in the figure, you will discover that the one on the right involves finding the antiderivative of e^{-x^2}, which you know is not an elementary function. The integral on the left, however, can be evaluated more easily, as shown.

$$\begin{aligned} V &= \int_0^1 \int_0^x e^{-x^2}\, dy\, dx \\ &= \int_0^1 \left[e^{-x^2} y \right]_0^x dx \\ &= \int_0^1 x e^{-x^2}\, dx \\ &= \left[-\frac{1}{2} e^{-x^2} \right]_0^1 \\ &= -\frac{1}{2}\left(\frac{1}{e} - 1\right) \approx 0.316 \text{ cubic unit} \end{aligned}$$

TECHNOLOGY

Use a symbolic integration utility to evaluate the double integral in Example 2.

TRY IT 2

Find the volume under the surface $f(x, y) = e^{x^2}$, bounded by the xz-plane and the planes $y = 2x$ and $x = 1$.

Guidelines for Finding the Volume of a Solid

1. Write the equation of the surface in the form $z = f(x, y)$ and sketch the solid region.
2. Sketch the region R in the xy-plane and determine the order and limits of integration.
3. Evaluate the double integral

$$\int_R\int f(x, y)\, dA$$

using the order and limits determined in the second step.

The first step above suggests that you sketch the three-dimensional solid region. This is a good suggestion, but it is not always feasible and is not as important as making a sketch of the two-dimensional region R.

EXAMPLE 3 **Finding the Volume of a Solid**

Find the volume of the solid bounded above by the surface

$$f(x, y) = 6x^2 - 2xy$$

and below by the plane region R shown in Figure 7.52.

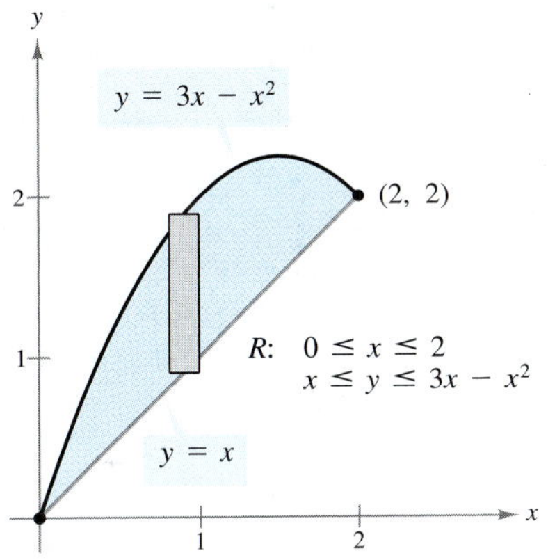

FIGURE 7.52

SOLUTION Because the region R is bounded by the parabola $y = 3x - x^2$ and the line $y = x$, the limits for y are $x \le y \le 3x - x^2$. The limits for x are $0 \le x \le 2$, and the volume of the solid is

$$\begin{aligned}
V &= \int_0^2 \int_x^{3x-x^2} (6x^2 - 2xy)\, dy\, dx \\
&= \int_0^2 \Big[6x^2y - xy^2 \Big]_x^{3x-x^2} dx \\
&= \int_0^2 [(18x^3 - 6x^4 - 9x^3 + 6x^4 - x^5) - (6x^3 - x^3)]\, dx \\
&= \int_0^2 (4x^3 - x^5)\, dx \\
&= \left[x^4 - \frac{x^6}{6} \right]_0^2 \\
&= \frac{16}{3} \text{ cubic units.}
\end{aligned}$$

TRY IT 3

Find the volume of the solid bounded above by the surface $f(x, y) = 4x^2 + 2xy$ and below by the plane region bounded by $y = x^2$ and $y = 2x$.

A *population density function* $p = f(x, y)$ is a model that describes density (in people per square unit) of a region. To find the population of a region R, evaluate the double integral

$$\int_R \int f(x, y)\, dA.$$

EXAMPLE 4 Finding the Population of a Region

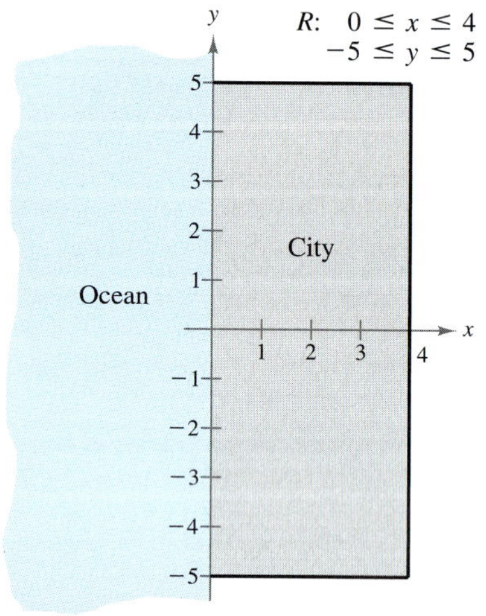

FIGURE 7.53

The population density (in people per square mile) of the city shown in Figure 7.53 can be modeled by

$$f(x, y) = \frac{50{,}000}{x + |y| + 1}$$

where x and y are measured in miles. Approximate the city's population. What is the city's average population density?

SOLUTION Because the model involves the absolute value of y, it follows that the population density is symmetrical about the x-axis. So, the population in the first quadrant is equal to the population in the fourth quadrant. This means that you can find the total population by doubling the population in the first quadrant.

$$\begin{aligned}
\text{Population} &= 2\int_0^4 \int_0^5 \frac{50{,}000}{x + y + 1}\, dy\, dx \\
&= 100{,}000 \int_0^4 \Big[\ln(x + y + 1)\Big]_0^5 dx \\
&= 100{,}000 \int_0^4 [\ln(x + 6) - \ln(x + 1)]\, dx \\
&= 100{,}000\Big[(x + 6)\ln(x + 6) - (x + 6) - \\
&\qquad (x + 1)\ln(x + 1) + (x + 1)\Big]_0^4 \\
&= 100{,}000\Big[(x + 6)\ln(x + 6) - (x + 1)\ln(x + 1) - 5\Big]_0^4 \\
&= 100{,}000[10\ln(10) - 5\ln(5) - 5 - 6\ln(6) + 5] \\
&\approx 422{,}810 \text{ people}
\end{aligned}$$

So, the city's population is about 422,810. Because the city covers a region 4 miles wide and 10 miles long, its area is 40 square miles. So, the average population density is

$$\begin{aligned}
\text{Average population density} &= \frac{422{,}810}{40} \\
&\approx 10{,}570 \text{ people per square mile.}
\end{aligned}$$

TRY IT 4

In Example 4, what integration technique was used to integrate

$$\int [\ln(x + 6) - \ln(x + 1)]\, dx?$$

Average Value of a Function over a Region

Average Value of a Function over a Region

If f is integrable over the plane region R with area A, then its **average value** over R is

$$\text{Average value} = \frac{1}{A}\int_R\int f(x, y)\,dA.$$

EXAMPLE 5 Finding Average Profit

A manufacturer determines that the profit for selling x units of one product and y units of a second product is modeled by

$$P = -(x - 200)^2 - (y - 100)^2 + 5000.$$

The weekly sales for product 1 vary between 150 and 200 units, and the weekly sales for product 2 vary between 80 and 100 units. Estimate the average weekly profit for the two products.

SOLUTION Because $150 \le x \le 200$ and $80 \le y \le 100$, you can estimate the weekly profit to be the average of the profit function over the rectangular region shown in Figure 7.54. Because the area of this rectangular region is $(50)(20) = 1000$, it follows that the average profit V is

$$\begin{aligned} V &= \frac{1}{1000}\int_{150}^{200}\int_{80}^{100}[-(x - 200)^2 - (y - 100)^2 + 5000]\,dy\,dx \\ &= \frac{1}{1000}\int_{150}^{200}\left[-(x - 200)^2y - \frac{(y - 100)^3}{3} + 5000y\right]_{80}^{100}dx \\ &= \frac{1}{1000}\int_{150}^{200}\left[-20(x - 200)^2 - \frac{292{,}000}{3}\right]dx \\ &= \frac{1}{3000}\Big[-20(x - 200)^3 + 292{,}000x\Big]_{150}^{200} \\ &\approx \$4033. \end{aligned}$$

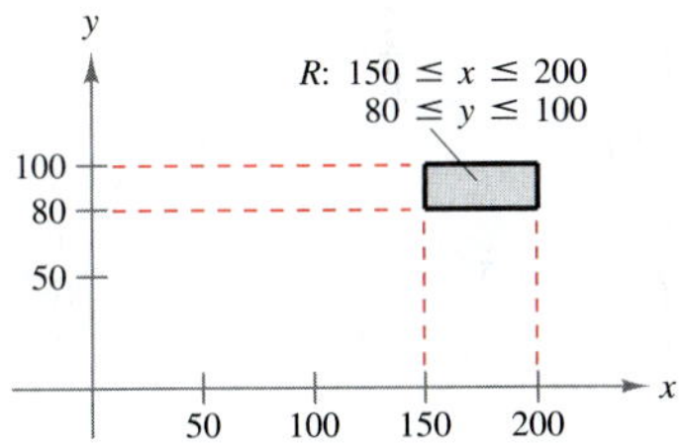

FIGURE 7.54

TRY IT 5

Find the average value of $f(x, y) = 4 - \frac{1}{2}x - \frac{1}{2}y$ over the region $0 \le x \le 2$ and $0 \le y \le 2$.

TAKE ANOTHER LOOK

Finding an Average Value

What is the average value of the constant function $f(x, y) = 100$ over a region R? Explain your reasoning.

PREREQUISITE REVIEW 7.9

The following warm-up exercises involve skills that were covered in earlier sections. You will use these skills in the exercise set for this section.

In Exercises 1–4, sketch the region that is described.

1. $0 \le x \le 2,\ 0 \le y \le 1$

2. $1 \le x \le 3,\ 2 \le y \le 3$

3. $0 \le x \le 4,\ 0 \le y \le 2x - 1$

4. $0 \le x \le 2,\ 0 \le y \le x^2$

In Exercises 5–10, evaluate the double integral.

5. $\displaystyle\int_0^1 \int_1^2 dy\, dx$

6. $\displaystyle\int_0^3 \int_1^3 dx\, dy$

7. $\displaystyle\int_0^1 \int_0^x x\, dy\, dx$

8. $\displaystyle\int_0^4 \int_1^y y\, dx\, dy$

9. $\displaystyle\int_1^3 \int_x^{x^2} 2\, dy\, dx$

10. $\displaystyle\int_0^1 \int_x^{-x^2+2} dy\, dx$

EXERCISES 7.9

In Exercises 1–8, sketch the region of integration and evaluate the double integral.

1. $\displaystyle\int_0^2 \int_0^1 (3x + 4y)\, dy\, dx$

2. $\displaystyle\int_0^3 \int_0^1 (2x + 6y)\, dy\, dx$

3. $\displaystyle\int_0^1 \int_y^{\sqrt{y}} x^2y^2\, dx\, dy$

4. $\displaystyle\int_0^6 \int_{y/2}^3 (x + y)\, dx\, dy$

5. $\displaystyle\int_0^1 \int_0^{\sqrt{1-x^2}} y\, dy\, dx$

6. $\displaystyle\int_0^2 \int_0^{4-x^2} xy^2\, dy\, dx$

7. $\displaystyle\int_{-a}^a \int_{-\sqrt{a^2-x^2}}^{\sqrt{a^2-x^2}} dy\, dx$

8. $\displaystyle\int_0^a \int_0^{\sqrt{a^2-x^2}} dy\, dx$

In Exercises 9–12, set up the integral for both orders of integration and use the more convenient order to evaluate the integral over the region R.

9. $\displaystyle\int_R\int xy\, dA$

R: rectangle with vertices at $(0, 0)$, $(0, 5)$, $(3, 5)$, $(3, 0)$

10. $\displaystyle\int_R\int x\, dA$

R: semicircle bounded by $y = \sqrt{25 - x^2}$ and $y = 0$

11. $\displaystyle\int_R\int \frac{y}{x^2 + y^2}\, dA$

R: triangle bounded by $y = x$, $y = 2x$, $x = 2$

12. $\displaystyle\int_R\int \frac{y}{1 + x^2}\, dA$

R: region bounded by $y = 0$, $y = \sqrt{x}$, $x = 4$

In Exercises 13 and 14, evaluate the double integral. Note that it is necessary to change the order of integration.

13. $\displaystyle\int_0^1 \int_{y/2}^{1/2} e^{-x^2}\, dx\, dy$

14. $\displaystyle\int_0^{\ln 10} \int_{e^x}^{10} \frac{1}{\ln y}\, dy\, dx$

In Exercises 15–26, use a double integral to find the volume of the specified solid.

15.

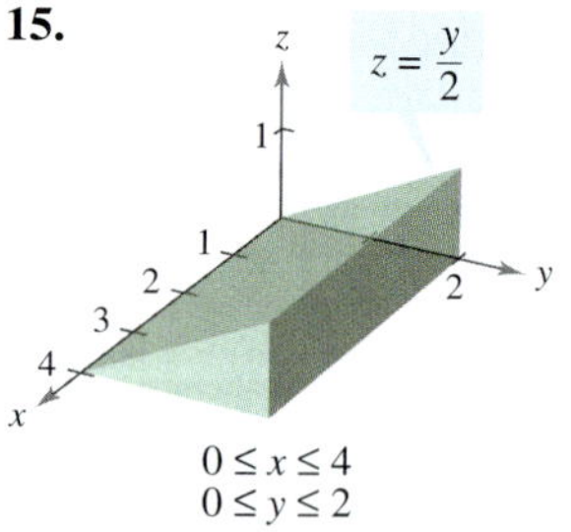

16.

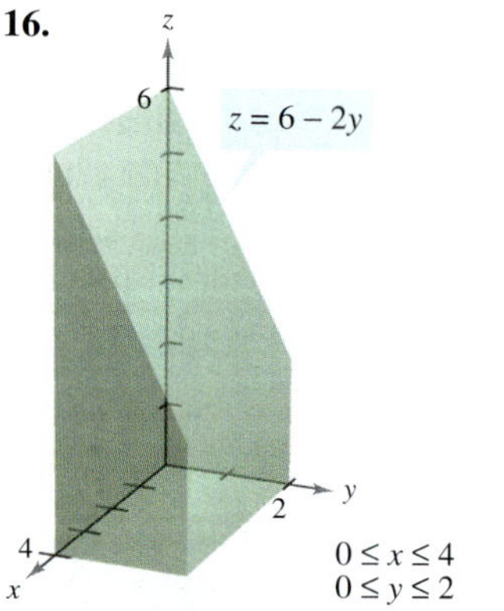

17.

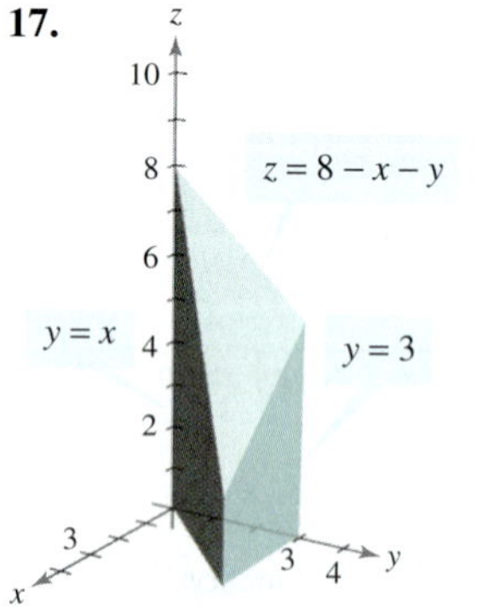

18.

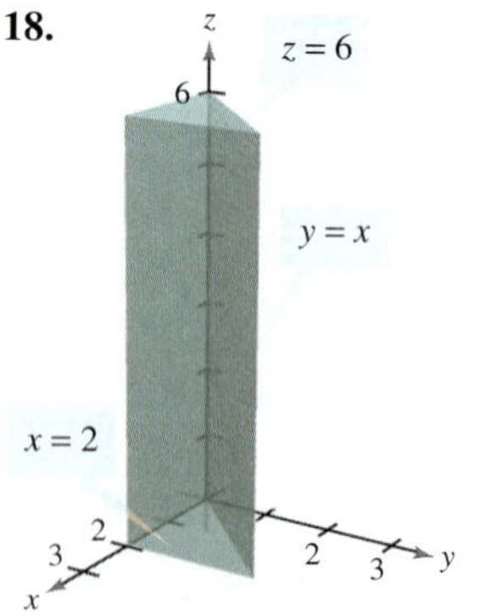

19.

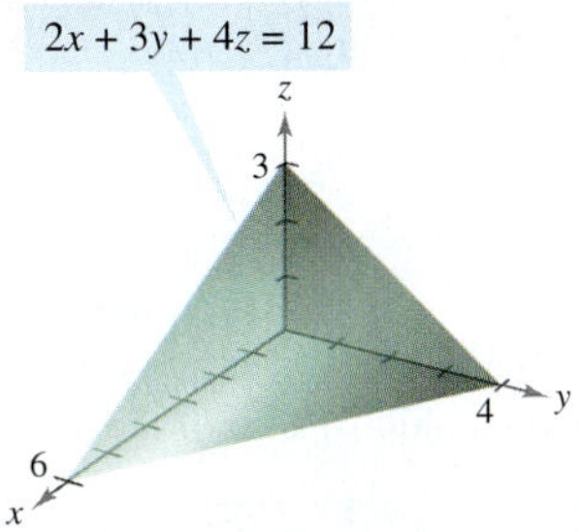

20.

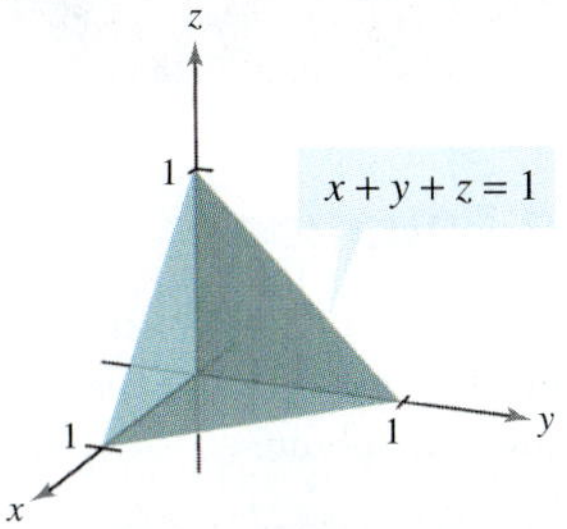

21.

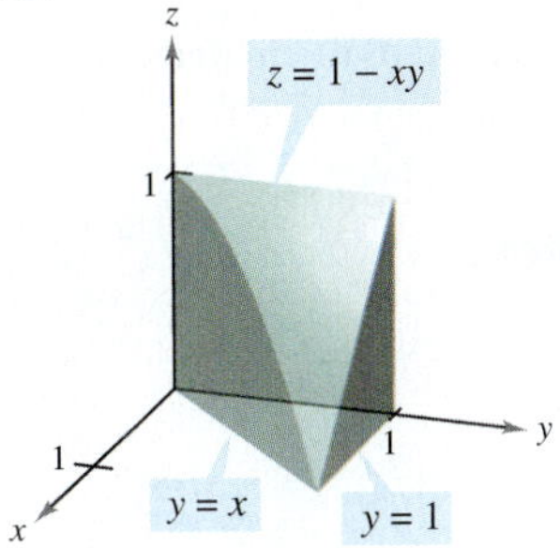

22.

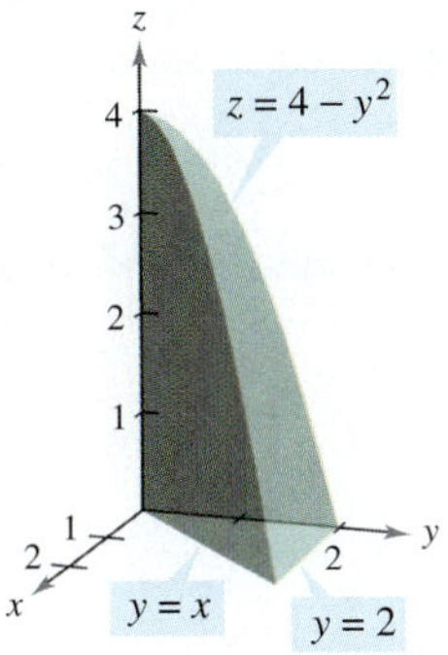

23.

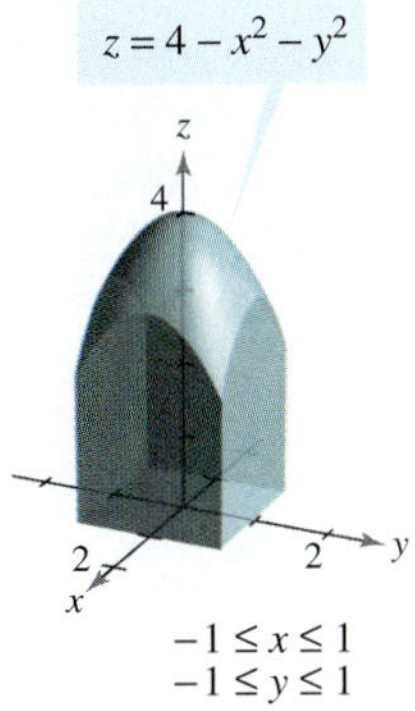

24.

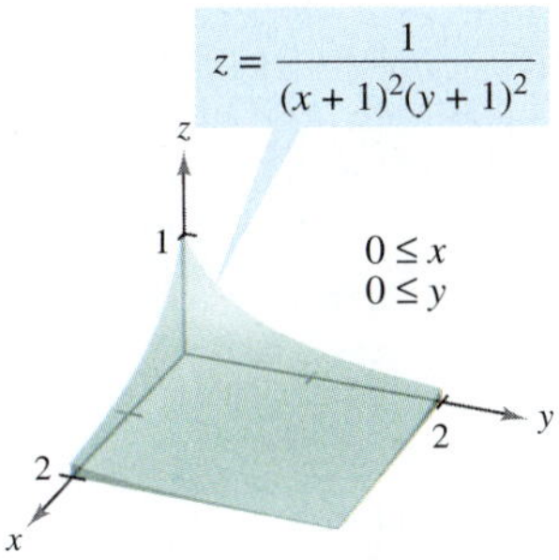

25.

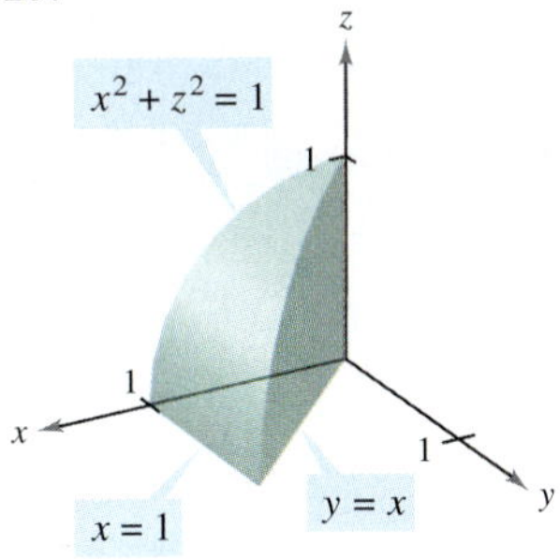

26.

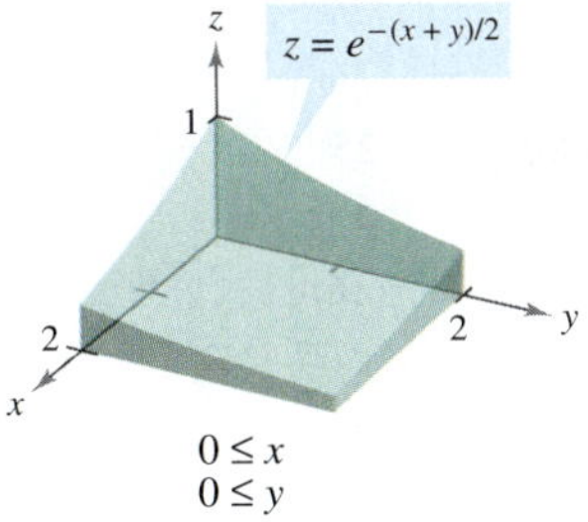

In Exercises 27–30, use a double integral to find the volume of the solid bounded by the graphs of the equations.

27. $z = xy,\ z = 0,\ y = 0,\ y = 4,\ x = 0,\ x = 1$

28. $z = x,\ z = 0,\ y = x,\ y = 0,\ x = 0,\ x = 4$

29. $z = x^2,\ z = 0,\ x = 0,\ x = 2,\ y = 0,\ y = 4$

30. $z = x + y,\ x^2 + y^2 = 4$ (first octant)

31. ***Population Density*** The population density (in people per square mile) for a coastal town can be modeled by

$$f(x, y) = \frac{120{,}000}{(2 + x + y)^3}$$

where x and y are measured in miles. What is the population inside the rectangular area defined by the vertices (0, 0), (2, 0), (0, 2), and (2, 2)?

32. ***Population Density*** The population density (in people per square mile) for a coastal town on an island can be modeled by

$$f(x, y) = \frac{5000xe^y}{1 + 2x^2}$$

where x and y are measured in miles. What is the population inside the rectangular area defined by the vertices (0, 0), (4, 0), (0, −2), and (4, −2)?

In Exercises 33–36, find the average value of $f(x, y)$ over the region R.

	Integral	*Region R*
33.	$f(x, y) = x$	Rectangle with vertices (0, 0), (4, 0), (4, 2), (0, 2)
34.	$f(x, y) = xy$	Rectangle with vertices (0, 0), (4, 0), (4, 2), (0, 2)
35.	$f(x, y) = x^2 + y^2$	Square with vertices (0, 0), (2, 0), (2, 2), (0, 2)
36.	$f(x, y) = e^{x+y}$	Triangle with vertices (0, 0), (0, 1), (1, 1)

37. ***Average Revenue*** A company sells two products whose demand functions are given by

$$x_1 = 500 - 3p_1 \quad \text{and} \quad x_2 = 750 - 2.4p_2.$$

So, the total revenue is given by

$$R = x_1p_1 + x_2p_2.$$

Estimate the average revenue if the price p_1 varies between \$50 and \$75 and the price p_2 varies between \$100 and \$150.

38. ***Average Weekly Profit*** A firm's weekly profit in marketing two products is given by

$$P = 192x_1 + 576x_2 - x_1^2 - 5x_2^2 - 2x_1x_2 - 5000$$

where x_1 and x_2 represent the numbers of units of each product sold weekly. Estimate the average weekly profit if x_1 varies between 40 and 50 units and x_2 varies between 45 and 50 units.

ALGEBRA REVIEW

Solving Systems of Equations

Nonlinear System in Two Variables

$$\begin{cases} 4x + 3y = 6 \\ x^2 - y = 4 \end{cases}$$

Linear System in Three Variables

$$\begin{cases} -x + 2y + 4z = 2 \\ 2x - y + z = 0 \\ 6x + 2z = 3 \end{cases}$$

Three of the sections in this chapter (7.5, 7.6, and 7.7) involve solutions of systems of equations. These systems can be linear or nonlinear, as shown at the left.

There are many techniques for solving a system of linear equations. Two of the more common ones are listed here.

1. *Substitution*: Solve for one of the variables in one of the equations and substitute the value into another equation.
2. *Elimination*: Add multiples of one equation to a second equation to eliminate a variable in the second equation.

EXAMPLE 1 Solving a System of Equations

Solve each system of equations.

(a) $\begin{cases} y - x^3 = 0 \\ x - y^3 = 0 \end{cases}$

(b) $\begin{cases} -400p_1 + 300p_2 = -25 \\ 300p_1 - 360p_2 = -535 \end{cases}$

SOLUTION

(a) Example 3, page 498

$\begin{cases} y - x^3 = 0 \\ x - y^3 = 0 \end{cases}$	Equation 1 Equation 2
$y = x^3$	Solve for y in Equation 1.
$x - (x^3)^3 = 0$	Substitute x^3 for y in Equation 2.
$x - x^9 = 0$	$(x^m)^n = x^{mn}$
$x(x - 1)(x + 1)(x^2 + 1)(x^4 + 1) = 0$	Factor.
$x = 0$	Set factors equal to zero.
$x = 1$	Set factors equal to zero.
$x = -1$	Set factors equal to zero.

(b) Example 4, page 499

$\begin{cases} -400p_1 + 300p_2 = -25 \\ 300p_1 - 360p_2 = -535 \end{cases}$	Equation 1 Equation 2
$p_2 = \frac{1}{12}(16p_1 - 1)$	Solve for p_2 in Equation 1.
$300p_1 - 360\left(\frac{1}{12}\right)(16p_1 - 1) = -535$	Substitute for p_2 in Equation 2.
$300p_1 - 30(16p_1 - 1) = -535$	Multiply factors.
$-180p_1 = -565$	Combine like terms.
$p_1 = \frac{113}{36} \approx 3.14$	Divide each side by -180.
$p_2 = \frac{1}{12}\left[16\left(\frac{113}{36}\right) - 1\right]$	Find p_2 by substituting p_1.
$p_2 \approx 4.10$	Solve for p_2.

EXAMPLE 2 Solving a System of Equations

Solve each system of equations.

(a) $\begin{cases} y(24 - 12x - 4y) = 0 \\ x(24 - 6x - 8y) = 0 \end{cases}$ (b) $\begin{cases} 28a - 4b = 10 \\ -4a + 8b = 12 \end{cases}$

SOLUTION

(a) Example 5, page 500

Before solving this system of equations, factor 4 out of the first equation and factor 2 out of the second equation.

$$\begin{cases} y(24 - 12x - 4y) = 0 \\ x(24 - 6x - 8y) = 0 \end{cases}$$ Original Equation 1 / Original Equation 2

$$\begin{cases} y(4)(6 - 3x - y) = 0 \\ x(2)(12 - 3x - 4y) = 0 \end{cases}$$ Factor 4 out of Equation 1. / Factor 2 out of Equation 2.

$$\begin{cases} y(6 - 3x - y) = 0 \\ x(12 - 3x - 4y) = 0 \end{cases}$$ Equation 1 / Equation 2

In each equation, either factor can be 0, so you obtain four different linear systems. For the first system, substitute $y = 0$ into the second equation to obtain $x = 4$.

$$\begin{cases} y = 0 \\ 12 - 3x - 4y = 0 \end{cases}$$ $(4, 0)$ is a solution.

You can solve the second system by the method of elimination.

$$\begin{cases} 6 - 3x - y = 0 \\ 12 - 3x - 4y = 0 \end{cases}$$ $\left(\frac{4}{3}, 2\right)$ is a solution.

The third system is already solved.

$$\begin{cases} y = 0 \\ x = 0 \end{cases}$$ $(0, 0)$ is a solution.

You can solve the last system by substituting $x = 0$ into the first equation to obtain $y = 6$.

$$\begin{cases} 6 - 3x - y = 0 \\ x = 0 \end{cases}$$ $(0, 6)$ is a solution.

(b) Example 2, page 516

$$\begin{cases} 28a - 4b = 10 \\ -4a + 8b = 12 \end{cases}$$ Equation 1 / Equation 2

$-2a + 4b = 6$ Divide Equation 2 by 2.

$26a = 16$ Add new equation to Equation 1.

$a = \frac{8}{13}$ Divide each side by 26.

$28\left(\frac{8}{13}\right) - 4b = 10$ Substitute for a in Equation 1.

$b = \frac{47}{26}$ Solve for b.

7 CHAPTER SUMMARY AND STUDY STRATEGIES

*After studying this chapter, you should have acquired the following skills. The exercise numbers are keyed to the Review Exercises that begin on page 544. Answers to odd-numbered Review Exercises are given in the back of the text.**

- Plot points in space. *(Section 7.1)* — *Review Exercises 1, 2*
- Find the distance between two points in space. *(Section 7.1)* — *Review Exercises 3, 4*

  $d = \sqrt{(x_2 - x_1)^2 + (y_2 - y_1)^2 + (z_2 - z_1)^2}$
- Find the midpoints of line segments in space. *(Section 7.1)* — *Review Exercises 5, 6*

  $\text{Midpoint} = \left(\frac{x_1 + x_2}{2}, \frac{y_1 + y_2}{2}, \frac{z_1 + z_2}{2}\right)$
- Write the standard forms of the equations of spheres. *(Section 7.1)* — *Review Exercises 7–10*

  $(x - h)^2 + (y - k)^2 + (z - l)^2 = r^2$
- Find the centers and radii of spheres. *(Section 7.1)* — *Review Exercises 11, 12*
- Sketch the coordinate plane traces of spheres. *(Section 7.1)* — *Review Exercises 13, 14*
- Sketch planes in space. *(Section 7.2)* — *Review Exercises 15–18*
- Classify quadric surfaces in space. *(Section 7.2)* — *Review Exercises 19–26*
- Evaluate functions of several variables. *(Section 7.3)* — *Review Exercises 27, 28*
- Find the domains and ranges of functions of several variables. *(Section 7.3)* — *Review Exercises 29, 30*
- Sketch the level curves of functions of two variables. *(Section 7.3)* — *Review Exercises 31–34*
- Use functions of several variables to answer questions about real-life situations. *(Section 7.3)* — *Review Exercises 35–40*
- Find the first partial derivatives of functions of several variables. *(Section 7.4)* — *Review Exercises 41–50*

  $\frac{\partial z}{\partial x} = \lim_{\Delta x \to 0} \frac{f(x + \Delta x, y) - f(x, y)}{\Delta x}$

  $\frac{\partial z}{\partial y} = \lim_{\Delta y \to 0} \frac{f(x, y + \Delta y) - f(x, y)}{\Delta y}$
- Find the slopes of surfaces in the x- and y-directions. *(Section 7.4)* — *Review Exercises 51–54*
- Find the second partial derivatives of functions of several variables. *(Section 7.4)* — *Review Exercises 55–58*
- Use partial derivatives to answer questions about real-life situations. *(Section 7.4)* — *Review Exercises 59–62*
- Find the relative extrema of functions of two variables. *(Section 7.5)* — *Review Exercises 63–70*
- Use relative extrema to answer questions about real-life situations. *(Section 7.5)* — *Review Exercises 71, 72*
- Use Lagrange multipliers to find extrema of functions of several variables. *(Section 7.6)* — *Review Exercises 73–78*
- Use a spreadsheet to find the indicated extremum. *(Section 7.6)* — *Review Exercises 79, 80*
- Use Lagrange multipliers to answer questions about real-life situations. *(Section 7.6)* — *Review Exercises 81, 82*

* Use a wide range of valuable study aids to help you master the material in this chapter. The *Student Solutions Guide* includes step-by-step solutions to all odd-numbered exercises to help you review and prepare. The *HM mathSpace® Student CD-ROM* helps you brush up on your algebra skills. The *Graphing Technology Guide*, available on the Web at *math.college.hmco.com/students*, offers step-by-step commands and instructions for a wide variety of graphing calculators, including the most recent models.

- Find the least squares regression lines, $y = ax + b$, for data and calculate the sum of the squared errors for data. *(Section 7.7)* — *Review Exercises 83, 84*

$$a = \left[n\sum_{i=1}^{n} x_i y_i - \sum_{i=1}^{n} x_i \sum_{i=1}^{n} y_i\right] \bigg/ \left[n\sum_{i=1}^{n} x_i^2 - \left(\sum_{i=1}^{n} x_i\right)^2\right], \qquad b = \frac{1}{n}\left(\sum_{i=1}^{n} y_i - a\sum_{i=1}^{n} x_i\right)$$

- Use least squares regression lines to model real-life data. *(Section 7.7)* — *Review Exercises 85, 86*
- Find the least squares regression quadratics for data. *(Section 7.7)* — *Review Exercises 87, 88*
- Evaluate double integrals. *(Section 7.8)* — *Review Exercises 89–92*
- Use double integrals to find the areas of regions. *(Section 7.8)* — *Review Exercises 93–96*
- Use double integrals to find the volumes of solids. *(Section 7.9)* — *Review Exercises 97, 98*

$$\text{Volume} = \int_R\int f(x, y)\, dA$$

- Use double integrals to find the average values of real-life models. *(Section 7.9)* — *Review Exercises 99, 100*

$$\text{Average value} = \frac{1}{A}\int_R\int f(x, y)\, dA$$

- ***Comparing Two Dimensions with Three Dimensions*** Many of the formulas and techniques in this chapter are generalizations of formulas and techniques used in earlier chapters in the text. Here are several examples.

Two-Dimensional Coordinate System	Three-Dimensional Coordinate System
Distance Formula $d = \sqrt{(x_2 - x_1)^2 + (y_2 - y_1)^2}$	*Distance Formula* $d = \sqrt{(x_2 - x_1)^2 + (y_2 - y_1)^2 + (z_2 - z_1)^2}$
Midpoint Formula $\text{Midpoint} = \left(\frac{x_1 + x_2}{2}, \frac{y_1 + y_2}{2}\right)$	*Midpoint Formula* $\text{Midpoint} = \left(\frac{x_1 + x_2}{2}, \frac{y_1 + y_2}{2}, \frac{z_1 + z_2}{2}\right)$
Equation of Circle $(x - h)^2 + (y - k)^2 = r^2$	*Equation of Sphere* $(x - h)^2 + (y - k)^2 + (z - l)^2 = r^2$
Equation of Line $ax + by = c$	*Equation of Plane* $ax + by + cz = d$
Derivative of $y = f(x)$ $\frac{dy}{dx} = \lim_{\Delta x \to 0} \frac{f(x + \Delta x) - f(x)}{\Delta x}$	*Partial Derivative of $z = f(x, y)$* $\frac{\partial z}{\partial x} = \lim_{\Delta x \to 0} \frac{f(x + \Delta x, y) - f(x, y)}{\Delta x}$
Area of Region $A = \int_a^b f(x)\, dx$	*Volume of Region* $V = \int_R\int f(x, y)\, dA$

Study Tools *Additional resources that accompany this chapter*

7 CHAPTER REVIEW EXERCISES

In Exercises 1 and 2, plot the points.

1. $(2, -1, 4), (-1, 3, -3)$

2. $(1, -2, -3), (-4, -3, 5)$

In Exercises 3 and 4, find the distance between the two points.

3. $(0, 0, 0), (2, 5, 9)$

4. $(-4, 1, 5), (1, 3, 7)$

In Exercises 5 and 6, find the midpoint of the line segment joining the two points.

5. $(2, 6, 4), (-4, 2, 8)$

6. $(5, 0, 7), (-1, -2, 9)$

In Exercises 7–10, find the standard form of the equation of the sphere.

7. Center: $(0, 1, 0)$; radius: 5

8. Center: $(4, -5, 3)$; radius: 10

9. Diameter endpoints: $(3, 4, 0), (5, 8, 2)$

10. Diameter endpoints: $(-2, 5, 1), (4, -3, 3)$

In Exercises 11 and 12, find the center and radius of the sphere.

11. $x^2 + y^2 + z^2 + 4x - 2y - 8z + 5 = 0$

12. $x^2 + y^2 + z^2 + 4y - 10z - 7 = 0$

In Exercises 13 and 14, sketch the *xy*-trace of the sphere.

13. $(x + 2)^2 + (y - 1)^2 + (z - 3)^2 = 25$

14. $(x - 1)^2 + (y + 3)^2 + (z - 6)^2 = 72$

In Exercises 15–18, find the intercepts and sketch the graph of the plane.

15. $x + 2y + 3z = 6$

16. $2y + z = 4$

17. $6x + 3y - 6z = 12$

18. $4x - y + 2z = 8$

In Exercises 19–26, identify the surface.

19. $x^2 + y^2 + z^2 - 2x + 4y - 6z + 5 = 0$

20. $16x^2 + 16y^2 - 9z^2 = 0$

21. $x^2 + \dfrac{y^2}{16} + \dfrac{z^2}{9} = 1$

22. $-x^2 + \dfrac{y^2}{16} + \dfrac{z^2}{9} = 1$

23. $z = \dfrac{x^2}{9} + y^2$

24. $-4x^2 + y^2 + z^2 = 4$

25. $z = \sqrt{x^2 + y^2}$

26. $z = 9x + 3y - 5$

In Exercises 27 and 28, find the function values.

27. $f(x, y) = xy^2$

(a) $f(2, 3)$ (b) $f(0, 1)$

(c) $f(-5, 7)$ (d) $f(-2, -4)$

28. $f(x, y) = \dfrac{x^2}{y}$

(a) $f(6, 9)$ (b) $f(8, 4)$

(c) $f(t, 2)$ (d) $f(r, r)$

In Exercises 29 and 30, describe the region *R* in the *xy*-plane that corresponds to the domain of the function. Then find the range of the function.

29. $f(x, y) = \sqrt{1 - x^2 - y^2}$

30. $f(x, y) = \dfrac{1}{x + y}$

In Exercises 31–34, describe the level curves of the function. Sketch the level curves for the given *c*-values.

31. $z = 10 - 2x - 5y,\ c = 0, 2, 4, 5, 10$

32. $z = \sqrt{9 - x^2 - y^2},\ c = 0, 1, 2, 3$

33. $z = (xy)^2,\ c = 1, 4, 9, 12, 16$

34. $z = 2e^{xy},\ c = 1, 2, 3, 4, 5$

35. ***Meteorology*** The contour map shown below represents the average yearly precipitation for Iowa. *(Source: U.S. National Oceanic and Atmospheric Administration)*

(a) Discuss the use of color to represent the level curves.

(b) Which part of Iowa receives the most precipitation?

(c) Which part of Iowa receives the least precipitation?

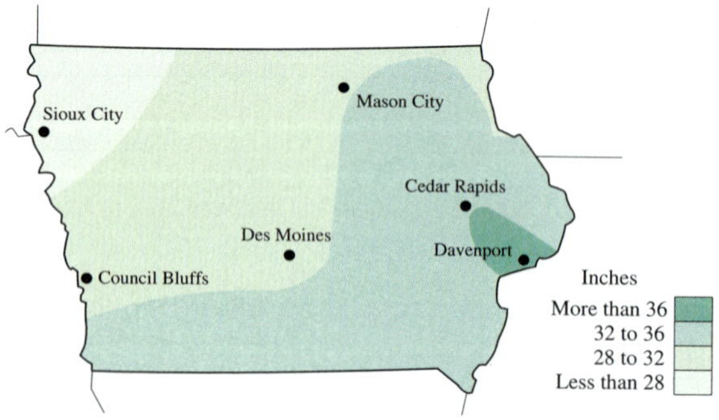

36. ***Population Density*** The contour map below represents the population density of New York. *(Source: U.S. Bureau of Census)*

(a) Discuss the use of color to represent the level curves.

(b) Do the level curves correspond to equally spaced population densities?

(c) Describe how to obtain a more detailed contour map.

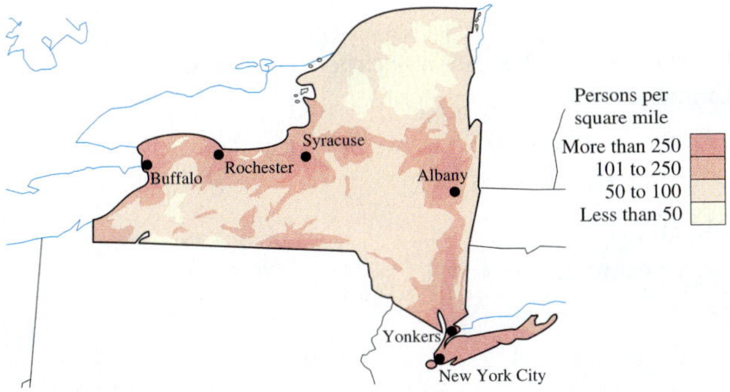

37. ***Chemistry*** The acidity of rainwater is measured in units called pH, and smaller pH values are increasingly acidic. The map shows the curves of equal pH and gives evidence that downwind of heavily industrialized areas the acidity has been increasing. Using the level curves on the map, determine the direction of the prevailing winds in the northeastern United States.

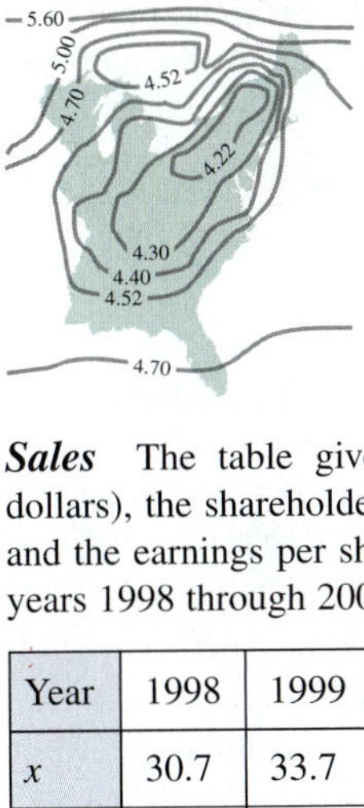

38. ***Sales*** The table gives the revenues x (in billions of dollars), the shareholder's equity y (in billions of dollars), and the earnings per share z (in dollars) for Target for the years 1998 through 2003. *(Source: Target Corp.)*

Year	1998	1999	2000	2001	2002	2003
x	30.7	33.7	36.9	39.9	43.9	48.2
y	5.3	5.9	6.5	7.9	9.4	11.1
z	1.02	1.27	1.38	1.56	1.81	2.01

A model for these data is

$$z = f(x, y) = 0.061x - 0.017y - 0.748.$$

(a) Use a graphing utility and the model to approximate z for the given values of x and y.

(b) Which of the two variables in this model has the greater influence on shareholder's equity?

(c) Simplify the expression for $f(x, 45)$ and interpret its meaning in the context of the problem.

39. ***Equation of Exchange*** Economists use an equation of exchange to express the relation among money, prices, and business transactions. This equation can be written as

$$P = \frac{MV}{T}$$

where M is the money supply, V is the velocity of circulation, T is the total number of transactions, and P is the price level. Find P when $M = \$2500$, $V = 6$, and $T = 6000$.

40. ***Biomechanics*** The Froude number F, defined as

$$F = \frac{v^2}{gl}$$

where v represents velocity, g represents gravitational acceleration, and l represents stride length, is an example of a "similarity criterion." Find the Froude number of a rabbit for which velocity is 2 meters per second, gravitational acceleration is 3 meters per second squared, and stride length is 0.75 meter.

In Exercises 41–50, find the first partial derivatives.

41. $f(x, y) = x^2y + 3xy + 2x - 5y$

42. $f(x, y) = 5x^2 + 4xy - 3y^2$

43. $z = 6x^2\sqrt{y} + 3\sqrt{xy} - 7xy$

44. $z = (xy + 2x + 4y)^2$

45. $f(x, y) = \ln(2x + 3y)$

46. $f(x, y) = \ln\sqrt{2x + 3y}$

47. $f(x, y) = x^2e^y - y^2e^x$

48. $f(x, y) = x^2e^{-2y}$

49. $w = xyz^2$

50. $w = xyz + 2xy - 9xz + 4yz - y^2z + 4z^2$

In Exercises 51–54, find the slope of the surface at the indicated point in (a) the x-direction and (b) the y-direction.

51. $z = 3x - 4y + 9$, $(3, 2, 10)$

52. $z = 4x^2 - y^2$, $(2, 4, 0)$

53. $z = 8 - x^2 - y^2$, $(1, 2, 3)$

54. $z = x^2 - y^2$, $(5, -4, 9)$

In Exercises 55–58, find all second partial derivatives.

55. $f(x, y) = x^3 - 4xy^2 + y^3$

56. $f(x, y) = \dfrac{y}{x + y}$

57. $f(x, y) = \sqrt{64 - x^2 - y^2}$

58. $f(x, y) = x^2e^{-y^2}$

59. ***Marginal Cost*** A company manufactures two models of skis: cross-country skis and downhill skis. The cost function for producing x cross-country skis and y downhill skis is given by

$$C = 15(xy)^{1/3} + 99x + 139y + 2293.$$

Find the marginal costs when $x = 250$ and $y = 175$.

60. ***Marginal Revenue*** At a baseball stadium, souvenir caps are sold at two locations. If x_1 and x_2 are the numbers of baseball caps sold at location 1 and location 2, respectively, then the total revenue for the caps is modeled by

$$R = 15x_1 + 16x_2 - \frac{1}{10}x_1^2 - \frac{1}{10}x_2^2 - \frac{1}{100}x_1x_2.$$

Given that $x_1 = 50$ and $x_2 = 40$, find the marginal revenues at location 1 and at location 2.

61. ***Medical Science*** The surface area A of an average human body in square centimeters can be approximated by the model

$$A(w, h) = 101.4w^{0.425}h^{0.725}$$

where w is the weight in pounds and h is the height in inches.

(a) Determine the partial derivatives of A with respect to w and with respect to h.

(b) Evaluate dA/dw at $(180, 70)$. Explain your result.

62. ***Medicine*** In order to treat a certain bacterial infection, a combination of two drugs is being tested. Studies have shown that the duration of the infection in laboratory tests can be modeled by

$$D(x, y) = x^2 + 2y^2 - 18x - 24y + 2xy + 120$$

where x is the dose in hundreds of milligrams of the first drug and y is the dose in hundreds of milligrams of the second drug. Evaluate $D(5, 2.5)$ and $D(7.5, 8)$ and interpret your results.

In Exercises 63–70, find any critical points and relative extrema of the function.

63. $f(x, y) = x^2 + 2xy + y^2$

64. $f(x, y) = x^3 - 3xy + y^2$

65. $f(x, y) = x^2 + 6xy + 3y^2 + 6x + 8$

66. $f(x, y) = x + y^2 - e^x$

67. $f(x, y) = x^3 + y^2 - xy$

68. $f(x, y) = y^2 + xy + 3y - 2x + 5$

69. $f(x, y) = x^3 + y^3 - 3x - 3y + 2$

70. $f(x, y) = y^2 - x^2$

71. ***Revenue*** A company manufactures and sells two products. The demand functions for the products are given by

$$p_1 = 100 - x_1$$

and

$$p_2 = 200 - 0.5x_2.$$

(a) Find the total revenue functions for x_1 and x_2.

(b) Find x_1 and x_2 such that the revenue is maximized.

(c) What is the maximum revenue?

72. ***Profit*** A company manufactures a product at two locations. The costs of manufacturing x_1 units at plant 1 and x_2 units at plant 2 are modeled by

$$C_1 = 0.03x_1^2 + 4x_1 + 300 \quad \text{and}$$

$$C_2 = 0.05x_2^2 + 7x_2 + 175$$

respectively. If the product sells for \$10 per unit, find x_1 and x_2 such that the profit, $P = 10(x_1 + x_2) - C_1 - C_2$, is maximized.

In Exercises 73–78, locate any extrema of the function by using Lagrange multipliers.

73. $f(x, y) = x^2y$

Constraint: $x + 2y = 2$

74. $f(x, y) = x^2 + y^2$

Constraint: $x + y = 4$

75. $f(x, y, z) = xyz$

Constraint: $x + 2y + z - 4 = 0$

76. $f(x, y, z) = xz + yz$

Constraint: $x + y + z = 6$

77. $f(x, y, z) = x^2 + y^2 + z^2$

Constraints: $x + z = 6,\ y + z = 8$

78. $f(x, y, z) = xyz$

Constraints: $x + y + z = 32,\ x - y + z = 0$

In Exercises 79 and 80, use a spreadsheet to find the indicated extremum. In each case, assume that x, y, and z are nonnegative.

79. Maximize $f(x, y, z) = xy$

Constraints: $x^2 + y^2 = 16,\ x - 2z = 0$

80. Minimize $f(x, y, z) = x^2 + y^2 + z^2$

Constraints: $x - 2z = 4,\ x + y = 8$

81. ***Maximum Production Level*** The production function for a manufacturer is given by $f(x, y) = 4x + xy + 2y$. Assume that the total amount available for labor x and capital y is \$2000 and that units of labor and capital cost \$20 and \$4, respectively. Find the maximum production level for this manufacturer.

82. ***Minimum Cost*** A manufacturer has an order for 1500 units that can be produced at two locations. Let x_1 and x_2 be the numbers of units produced at the two locations. Find the number that should be produced at each location to meet the order and minimize cost if the cost function is given by

$$C = 0.20x_1^2 + 10x_1 + 0.15x_2^2 + 12x_2.$$

In Exercises 83 and 84, (a) use the method of least squares to find the least squares regression line and (b) calculate the sum of the squared errors.

83. $(-2, -3), (-1, -1), (1, 2), (3, 2)$

84. $(-3, -1), (-2, -1), (0, 0), (1, 1), (2, 1)$

85. ***Agriculture*** An agronomist used four test plots to determine the relationship between the wheat yield (in bushels per acre) and the amount of fertilizer (in hundreds of pounds per acre). The results are listed in the table.

Fertilizer, x	1.0	1.5	2.0	2.5
Yield, y	32	41	48	53

(a) Use a graphing utility with a least squares regression program to find the least squares regression line for the data.

(b) Estimate the yield for a fertilizer application of 20 pounds per acre.

86. ***Work Force*** The table gives the percents and numbers (in millions) of women in the work force for selected years. *(Source: U.S. Bureau of Labor Statistics)*

Year	1970	1975	1980	1985
Percent, x	43.3	46.3	51.5	54.5
Number, y	31.5	37.5	45.5	51.1

Year	1990	1995	2000
Percent, x	57.5	58.9	59.9
Number, y	56.8	60.9	66.3

(a) Use a spreadsheet software program with a least squares regression program to find the least squares regression line for the data.

(b) According to this model, approximately how many women enter the labor force for each one-point increase in the percent of women in the labor force?

In Exercises 87 and 88, find the least squares regression quadratic for the given points. Plot the points and sketch the least squares regression quadratic.

87. $(-1, 9), (0, 7), (1, 5), (2, 6), (4, 23)$

88. $(0, 10), (2, 9), (3, 7), (4, 4), (5, 0)$

In Exercises 89–92, evaluate the double integral.

89. $\displaystyle\int_0^1 \int_0^{1+x} (3x + 2y)\, dy\, dx$

90. $\displaystyle\int_{-2}^2 \int_0^4 (x - y^2)\, dx\, dy$

91. $\displaystyle\int_1^2 \int_1^{2y} \frac{x}{y^2}\, dx\, dy$

92. $\displaystyle\int_0^4 \int_0^{\sqrt{16-x^2}} 2x\, dy\, dx$

In Exercises 93–96, use a double integral to find the area of the region.

93.

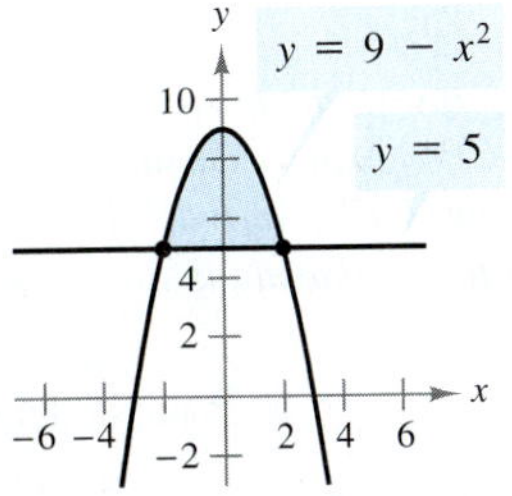

94.

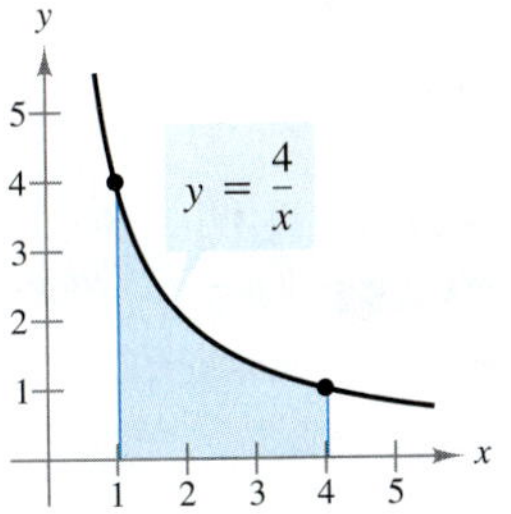

95.

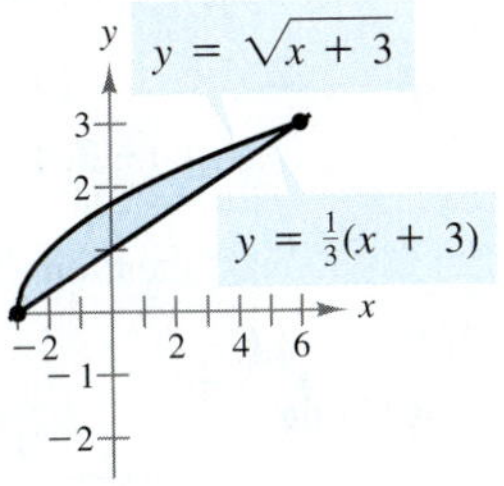

96.

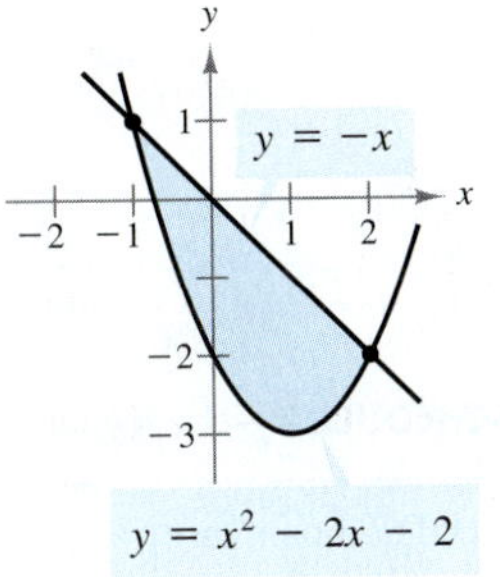

97. Find the volume of the solid bounded by the graphs of $z = (xy)^2$, $z = 0$, $y = 0$, $y = 4$, $x = 0$, and $x = 4$.

98. Find the volume of the solid bounded by the graphs of $z = x + y$, $z = 0$, $x = 0$, $x = 3$, $y = x$, and $y = 0$.

99. ***Average Elevation*** In a triangular coastal area, the elevation in miles above sea level at the point (x, y) is modeled by

$$f(x, y) = 0.25 - 0.025x - 0.01y$$

where x and y are measured in miles (see figure). Find the average elevation of the triangular area.

Figure for 99

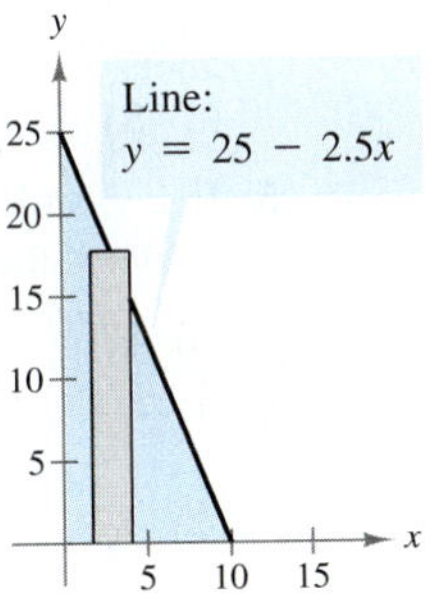

Figure for 100

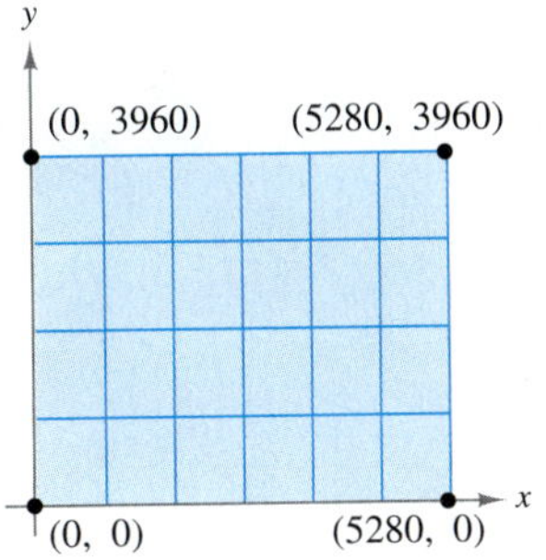

100. ***Real Estate*** The average value of real estate (in dollars per square foot) for a rectangular section of a city is given by

$$f(x, y) = 2.5x^{3/2}y^{3/4}$$

(see figure). Find the average value of real estate for this section.

7 SAMPLE POST-GRADUATION EXAM QUESTIONS

CPA
GMAT
GRE
Actuarial
CLAST

The following questions represent the types of questions that appear on certified public accountant (CPA) exams, Graduate Management Admission Tests (GMAT), Graduate Records Exams (GRE), actuarial exams, and College-Level Academic Skills Tests (CLAST). The answers to the questions are given in the back of the book.

1. What is the derivative of $f(x, y) = y^2(x + y)^3$ with respect to y?

(a) $6y(x + y)^2$ (b) $y(x + y)^2(2x + 5y)$

(c) $3y^2(x + y)^2$ (d) $y(x + y)^2(5x + 2y)$

2. Let $f(x, y) = x^2 + y^2 + 6x - 2y + 4$. At which point does f have a relative minimum?

(a) $(-3, 1, -13)$ (b) $(-3, 1, -6)$ (c) $(3, 1, -2)$ (d) $(-3, -1, -2)$

For Questions 3 and 4, use the following excerpts from the 2003 Tax Rate Schedules.

SCHEDULE X—Use if your filing status is **Single**

If the amount on Form 1040, line 40, is over—	But not over—	Enter on Form 1040, line 41	of the amount over—
$0	$7000	..............**10%**	**$0**
7000	28,400	**$700.00 + 15%**	**7000**
28,400	68,800	**3910.00 + 25%**	**28,400**
68,800	143,500	**14,010.00 + 28%**	**68,800**
143,500	311,950	**34,926.00 + 33%**	**143,500**
311,950	....	**90,514.50 + 35%**	**311,950**

SCHEDULE Y-1—Use if your filing status is **Married filing jointly** or **Qualifying widow(er)**

If the amount on Form 1040, line 40, is over—	But not over—	Enter on Form 1040, line 41	of the amount over—
$0	$14,000	..............**10%**	**$0**
14,000	56,800	**$1400.00 + 15%**	**14,000**
56,800	114,650	**7820.00 + 25%**	**56,800**
114,650	174,700	**22,282.50 + 28%**	**114,650**
174,700	311,950	**39,096.50 + 33%**	**174,700**
311,950	....	**84,389.00 + 35%**	**311,950**

3. The tax for a married couple filing jointly whose amount on Form 1040, line 40, is $125,480 is

(a) $29,880.40 (b) $55,339.10 (c) $25,314.90 (d) $54,384.50

4. The tax for a single person whose amount on Form 1040, line 40, is $1,000,000 is

(a) $199,697.00 (b) $325,206.50 (c) $343,630.08 (d) $331,332.00

5. If x, y, and z are chosen from the three numbers $\frac{1}{3}$, -2, and 4, what is the largest possible value of the expression $x^2 + z/y^2$?

(a) 126 (b) 40 (c) 24 (d) 72

6. If $xz = 4y$, then

$$\frac{x^3}{2}$$

equals

(a) $\frac{2y^3}{z^3}$ (b) $\frac{16y^2}{z^3}$ (c) $\frac{32y^3}{z^3}$ (d) $\frac{64y^3}{z^3}$

7. Mave Co. calculated the following ratios for one of its profit centers: gross margin 31%, return on sales 26%, capital turnover 0.5 time.

What is Mave's return on investment for this profit center?

(a) 8.5% (b) 13% (c) 15.5% (d) 26%

APPENDICES

Appendices C, D, and E are located on the website that accompanies this text at *college.hmco.com*.

A ALTERNATIVE INTRODUCTION TO THE FUNDAMENTAL THEOREM OF CALCULUS

- Approximate the areas of regions using Riemann sums.
- Evaluate definite integrals.

In this appendix, a summation process is used to provide an alternative development of the definite integral. It is intended that this supplement follow Section 5.3 in the text. If used, this appendix should replace the material preceding Example 2 in Section 5.4. Example 1 below shows how the area of a region in the plane can be approximated by the use of rectangles.

EXAMPLE 1 **Using Rectangles to Approximate the Area of a Region**

Use the four rectangles indicated in Figure A.1 to approximate the area of the region lying between the graph of

$$f(x) = \frac{x^2}{2}$$

and the x-axis, between $x = 0$ and $x = 4$.

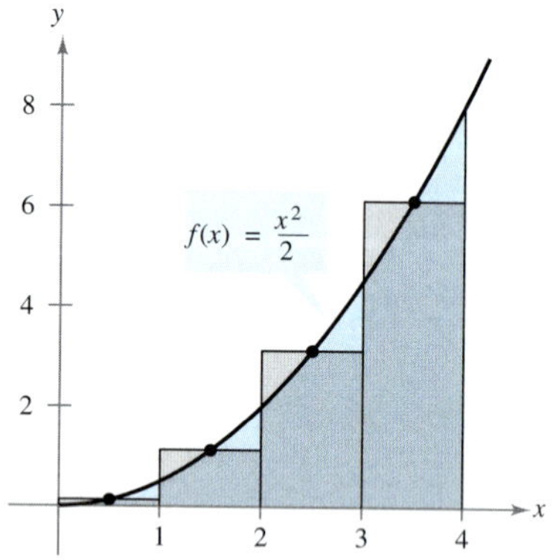

FIGURE A.1

SOLUTION You can find the heights of the rectangles by evaluating the function f at each of the midpoints of the subintervals

$$[0, 1],\quad [1, 2],\quad [2, 3],\quad [3, 4].$$

Because the width of each rectangle is 1, the sum of the areas of the four rectangles is

$$\begin{aligned} S &= \overbrace{(1)}^{\text{width}}\, \overbrace{f\left(\frac{1}{2}\right)}^{\text{height}} + \overbrace{(1)}^{\text{width}}\, \overbrace{f\left(\frac{3}{2}\right)}^{\text{height}} + \overbrace{(1)}^{\text{width}}\, \overbrace{f\left(\frac{5}{2}\right)}^{\text{height}} + \overbrace{(1)}^{\text{width}}\, \overbrace{f\left(\frac{7}{2}\right)}^{\text{height}} \\ &= \frac{1}{8} + \frac{9}{8} + \frac{25}{8} + \frac{49}{8} \\ &= \frac{84}{8} \\ &= 10.5. \end{aligned}$$

So, you can approximate the area of the region to be 10.5 square units.

STUDY TIP

The approximation technique used in Example 1 is called the **Midpoint Rule.** The Midpoint Rule is discussed further in Section 5.6.

The procedure shown in Example 1 can be generalized. Let f be a continuous function defined on the closed interval $[a, b]$. To begin, partition the interval into n subintervals, each of width $\Delta x = (b - a)/n$, as shown.

$$a = x_0 < x_1 < x_2 < \cdot \cdot \cdot < x_{n-1} < x_n = b$$

In each subinterval $[x_{i-1}, x_i]$ choose an arbitrary point c_i and form the sum

$$S = f(c_1)\,\Delta x + f(c_2)\,\Delta x + \cdot \cdot \cdot + f(c_{n-1})\,\Delta x + f(c_n)\,\Delta x.$$

This type of summation is called a **Riemann sum,** and is often written using summation notation as shown below.

$$S = \sum_{i=1}^{n} f(c_i)\,\Delta x, \quad x_{i-1} \le c_i \le x_i$$

For the Riemann sum in Example 1, the interval is $[a, b] = [0, 4]$, the number of subintervals is $n = 4$, the width of each interval is $\Delta x = 1$, and the point c_i in each subinterval is its midpoint. So, you can write the approximation in Example 1 as

$$\begin{aligned} S &= \sum_{i=1}^{n} f(c_i)\,\Delta x \\ &= \sum_{i=1}^{4} f(c_i)(1) \\ &= \frac{1}{8} + \frac{9}{8} + \frac{25}{8} + \frac{49}{8} \\ &= \frac{84}{8}. \end{aligned}$$

EXAMPLE 2 Using a Riemann Sum to Approximate Area

Use a Riemann sum to approximate the area of the region bounded by the graph of $f(x) = -x^2 + 2x$ and the x-axis, for $0 \le x \le 2$. In the Riemann sum, let $n = 6$ and choose c_i to be the left endpoint of each subinterval.

SOLUTION Subdivide the interval $[0, 2]$ into six subintervals, each of width

$$\begin{aligned} \Delta x &= \frac{2 - 0}{6} \\ &= \frac{1}{3} \end{aligned}$$

as shown in Figure A.2. Because c_i is the left endpoint of each subinterval, the Riemann sum is given by

$$\begin{aligned} S &= \sum_{i=1}^{n} f(c_i)\,\Delta x \\ &= \left[f(0) + f\left(\frac{1}{3}\right) + f\left(\frac{2}{3}\right) + f(1) + f\left(\frac{4}{3}\right) + f\left(\frac{5}{3}\right)\right]\left(\frac{1}{3}\right) \\ &= \left[0 + \frac{5}{9} + \frac{8}{9} + 1 + \frac{8}{9} + \frac{5}{9}\right]\left(\frac{1}{3}\right) \\ &= \frac{35}{27} \text{ square units.} \end{aligned}$$

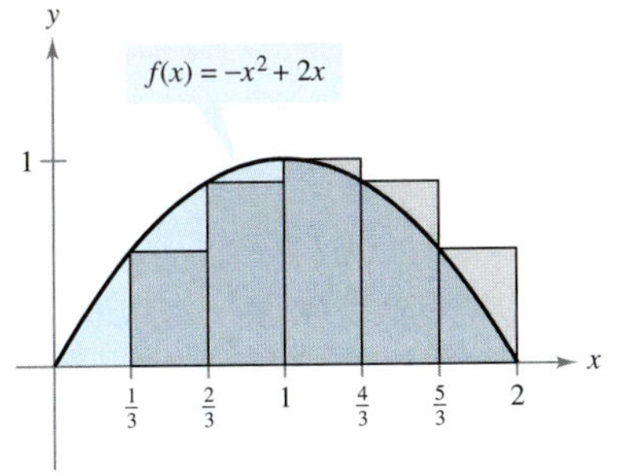

FIGURE A.2

Example 2 illustrates an important point. If a function f is continuous and nonnegative over the interval $[a, b]$, then the Riemann sum

$$S = \sum_{i=1}^{n} f(c_i)\,\Delta x$$

can be used to approximate the area of the region bounded by the graph of f and the x-axis, between $x = a$ and $x = b$. Moreover, for a given interval, as the number of subintervals increases, the approximation to the actual area will improve. This is illustrated in the next two examples by using Riemann sums to approximate the area of a triangle.

EXAMPLE 3 **Approximating the Area of a Triangle**

Use a Riemann sum to approximate the area of the triangular region bounded by the graph of $f(x) = 2x$ and the x-axis, $0 \le x \le 3$. Use a partition of six subintervals and choose c_i to be the left endpoint of each subinterval.

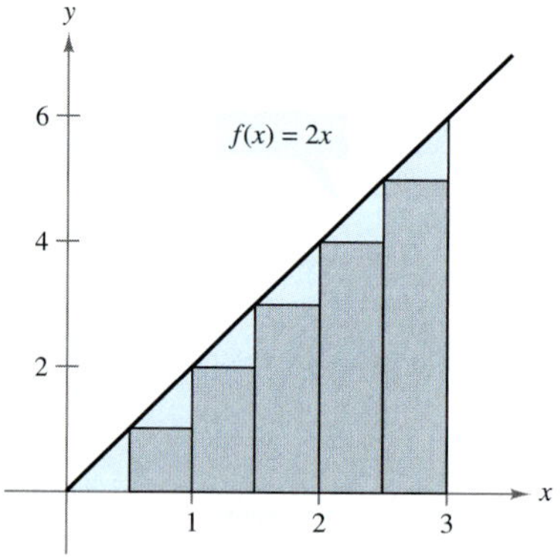

FIGURE A.3

SOLUTION Subdivide the interval $[0, 3]$ into six subintervals, each of width

$$\Delta x = \frac{3-0}{6} = \frac{1}{2}$$

as shown in Figure A.3. Because c_i is the left endpoint of each subinterval, the Riemann sum is given by

$$\begin{aligned} S &= \sum_{i=1}^{n} f(c_i)\,\Delta x \\ &= \left[f(0) + f\left(\frac{1}{2}\right) + f(1) + f\left(\frac{3}{2}\right) + f(2) + f\left(\frac{5}{2}\right)\right]\left(\frac{1}{2}\right) \\ &= [0 + 1 + 2 + 3 + 4 + 5]\left(\frac{1}{2}\right) \\ &= \frac{15}{2} \text{ square units.} \end{aligned}$$

The approximations in Examples 2 and 3 are called **left Riemann sums,** because c_i was chosen to be the left endpoint of each subinterval. If the right endpoints had been used in Example 3, the **right Riemann sum** would have been $\frac{21}{2}$. Note that the exact area of the triangular region in Example 3 is

$$\text{Area} = \frac{1}{2}(\text{base})(\text{height}) = \frac{1}{2}(3)(6) = 9 \text{ square units.}$$

So, the left Riemann sum gives an approximation that is less than the actual area, and the right Riemann sum gives an approximation that is greater than the actual area.

In Example 4, you will see that the approximation improves as the number of subintervals increases.

TECHNOLOGY

Most graphing utilities are able to sum the first n terms of a sequence. Try using a graphing utility to verify the right Riemann sum in Example 3.

EXAMPLE 4 Increasing the Number of Subintervals

Let

$$f(x) = 2x, \quad 0 \le x \le 3.$$

Use a computer to determine the left and right Riemann sums for $n = 10$, $n = 100$, and $n = 1000$ subintervals.

SOLUTION A basic computer program for this problem is as shown.

```
10   INPUT; N
20   DELTA=3/N
30   LSUM=0: RSUM=0
40   FOR I=1 TO N
50   LC=(I-1)*DELTA: RC=I*DELTA
60   LSUM=LSUM+2*LC*DELTA: RSUM=RSUM+2*RC*DELTA
70   NEXT
80   PRINT "LEFT RIEMANN SUM:"; LSUM
90   PRINT "RIGHT RIEMANN SUM:"; RSUM
100  END
```

Running this program for $n = 10$, $n = 100$, and $n = 1000$ gave the results shown in the table.

n	Left Riemann sum	Right Riemann sum
10	8.100	9.900
100	8.910	9.090
1000	8.991	9.009

From the results of Example 4, it appears that the Riemann sums are approaching the limit 9 as n approaches infinity. It is this observation that motivates the definition of a **definite integral.** In this definition, consider the partition of $[a, b]$ into n subintervals of equal width $\Delta x = (b - a)/n$ as shown.

$$a = x_0 < x_1 < x_2 < \cdot \cdot \cdot < x_{n-1} < x_n = b$$

Moreover, consider c_i to be an arbitrary point in the ith subinterval $[x_{i-1}, x_i]$. To say that the number of subintervals n approaches infinity is equivalent to saying that the width, Δx, of the subintervals approaches zero.

Definition of Definite Integral

If f is a continuous function defined on the closed interval $[a, b]$, then the **definite integral of f on $[a, b]$** is

$$\int_a^b f(x)\,dx = \lim_{\Delta x \to 0} \sum_{i=1}^{n} f(c_i)\,\Delta x$$

$$= \lim_{n \to \infty} \sum_{i=1}^{n} f(c_i)\,\Delta x.$$

If f is continuous and nonnegative on the interval $[a, b]$, then the definite integral of f on $[a, b]$ gives the area of the region bounded by the graph of f, the x-axis, and the vertical lines $x = a$ and $x = b$.

Evaluation of a definite integral by its limit definition can be difficult. However, there are times when a definite integral can be solved by recognizing that it represents the area of a common type of geometric figure.

EXAMPLE 5 The Areas of Common Geometric Figures

Sketch the region corresponding to each of the definite integrals. Then evaluate each definite integral using a geometric formula.

(a) $\int_1^3 4\,dx$

(b) $\int_0^3 (x + 2)\,dx$

(c) $\int_{-2}^2 \sqrt{4 - x^2}\,dx$

SOLUTION A sketch of each region is shown in Figure A.4 (page A7).

(a) The region associated with this definite integral is a rectangle of height 4 and width 2. Moreover, because the function $f(x) = 4$ is continuous and nonnegative on the interval $[1, 3]$, you can conclude that the area of the rectangle is given by the definite integral. So, the value of the definite integral is

$$\int_1^3 4\,dx = 4(2) = 8 \text{ square units.}$$

(b) The region associated with this definite integral is a trapezoid with an altitude of 3 and parallel bases of lengths 2 and 5. The formula for the area of a trapezoid is $\frac{1}{2}h(b_1 + b_2)$, and so you have

$$\int_0^3 (x + 2)\,dx = \frac{1}{2}(3)(2 + 5)$$

$$= \frac{21}{2} \text{ square units.}$$

(c) The region associated with this definite integral is a semicircle of radius 2. So, the area is $\frac{1}{2}\pi r^2$, and you have

$$\int_{-2}^{2} \sqrt{4 - x^2}\, dx = \frac{1}{2}\pi(2^2) = 2\pi \text{ square units.}$$

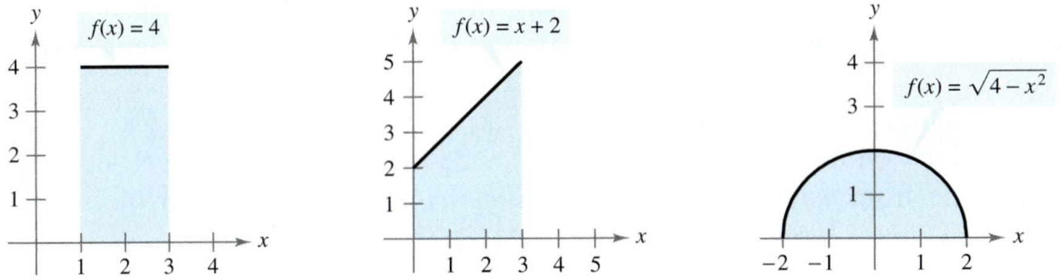

(a) $\int_1^3 4\, dx$
Rectangle

(b) $\int_0^3 (x + 2)\, dx$
Trapezoid

(c) $\int_{-2}^{2} \sqrt{4 - x^2}\, dx$
Semicircle

FIGURE A.4

For some simple functions, it is possible to evaluate definite integrals by the Riemann sum definition. In the next example, you will use the fact that the sum of the first n integers is given by the formula

$$1 + 2 + \cdot \cdot \cdot + n = \sum_{i=1}^{n} i = \frac{n(n+1)}{2}$$

See Exercise 29.

to compute the area of the triangular region in Examples 3 and 4.

EXAMPLE 6 Evaluating a Definite Integral by Its Definition

Evaluate $\int_0^3 2x\, dx$.

SOLUTION Let $\Delta x = (b - a)/n = 3/n$, and choose c_i to be the right endpoint of each subinterval, $c_i = 3i/n$. Then you have

$$\begin{aligned}
\int_0^3 2x\, dx &= \lim_{\Delta x \to 0} \sum_{i=1}^{n} f(c_i)\Delta x \\
&= \lim_{n\to\infty} \sum_{i=1}^{n} 2\left(i\frac{3}{n}\right)\left(\frac{3}{n}\right) \\
&= \lim_{n\to\infty} \frac{18}{n^2} \sum_{i=1}^{n} i \\
&= \lim_{n\to\infty} \left(\frac{18}{n^2}\right)\left(\frac{n(n+1)}{2}\right) \\
&= \lim_{n\to\infty} \left(9 + \frac{9}{n}\right).
\end{aligned}$$

This limit can be evaluated in the same way that you calculated horizontal asymptotes in Section 3.6. In particular, as n approaches infinity, you see that $9/n$ approaches 0, and the limit above is 9. So, you can conclude that

$$\int_0^3 2x\, dx = 9.$$

From Example 6, you can see that it can be difficult to evaluate the definite integral of even a simple function by using Riemann sums. A computer can help in calculating these sums for large values of n, but this procedure would only give an approximation of the definite integral. Fortunately, the **Fundamental Theorem of Calculus** provides a technique for evaluating definite integrals using antiderivatives, and for this reason it is often thought to be the most important theorem in calculus. In the remainder of this appendix, you will see how derivatives and integrals are related via the Fundamental Theorem of Calculus.

To simplify the discussion, assume that f is a continuous nonnegative function defined on the interval $[a, b]$. Let $A(x)$ be the area of the region under the graph of f from a to x, as indicated in Figure A.5. The area under the shaded region in Figure A.6 is $A(x + \Delta x) - A(x)$.

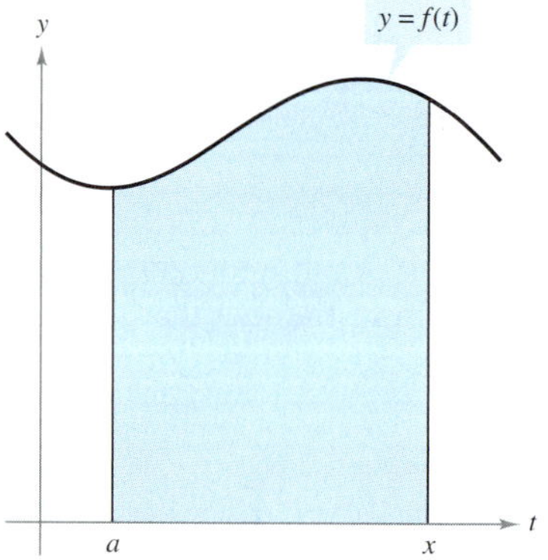

FIGURE A.5

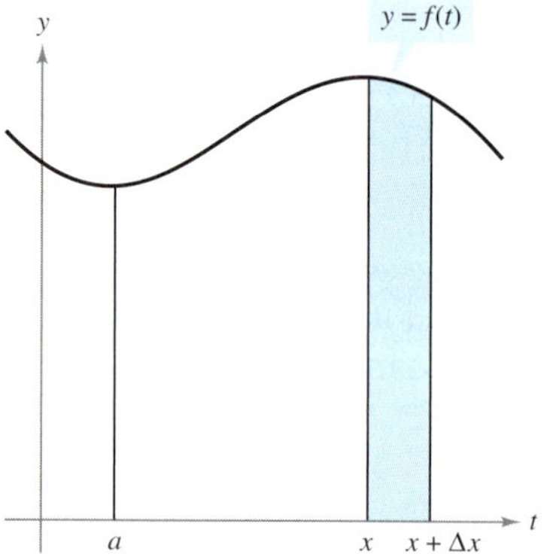

FIGURE A.6

If Δx is small, then this area is approximated by the area of the rectangle of height $f(x)$ and width Δx. So, you have

$$A(x + \Delta x) - A(x) \approx f(x)\,\Delta x.$$

Dividing by Δx produces

$$f(x) \approx \frac{A(x + \Delta x) - A(x)}{\Delta x}.$$

By taking the limit as Δx approaches 0, you can see that

$$\begin{aligned} f(x) &= \lim_{\Delta x \to 0} \frac{A(x + \Delta x) - A(x)}{\Delta x} \\ &= A'(x) \end{aligned}$$

and you can establish the fact that the area function $A(x)$ is an antiderivative of f. Although it was assumed that f is continuous and nonnegative, this development is valid if the function f is simply continuous on the closed interval $[a, b]$. This result is used in the proof of the Fundamental Theorem of Calculus.

Fundamental Theorem of Calculus

If f is a continuous function on the closed interval $[a, b]$, then

$$\int_a^b f(x)dx = F(b) - F(a)$$

where F is any function such that $F'(x) = f(x)$.

PROOF From the discussion above, you know that

$$\int_a^x f(x)\,dx = A(x)$$

and in particular

$$A(a) = \int_a^a f(x)\,dx = 0$$

and

$$A(b) = \int_a^b f(x)\,dx.$$

If F is *any* antiderivative of f, then you know that F differs from A by a constant. That is, $A(x) = F(x) + C$. So

$$\begin{aligned}\int_a^b f(x)\,dx &= A(b) - A(a)\\ &= [F(b) + C] - [F(a) + C]\\ &= F(b) - F(a).\end{aligned}$$

You are now ready to continue Section 5.4, on page 346, just after the statement of the Fundamental Theorem of Calculus.

APPENDIX A EXERCISES

In Exercises 1–6, use the left Riemann sum and the right Riemann sum to approximate the area of the region using the indicated number of subintervals.

1. $y = \sqrt{x}$

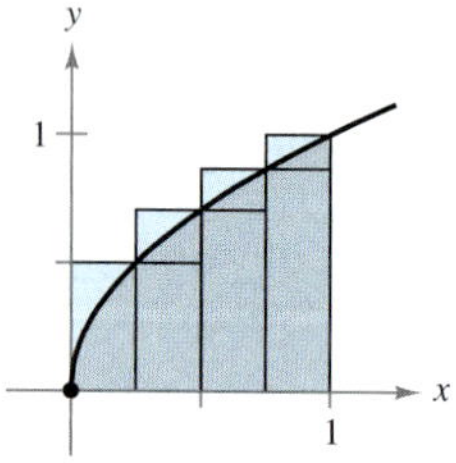

2. $y = \sqrt{x} + 1$

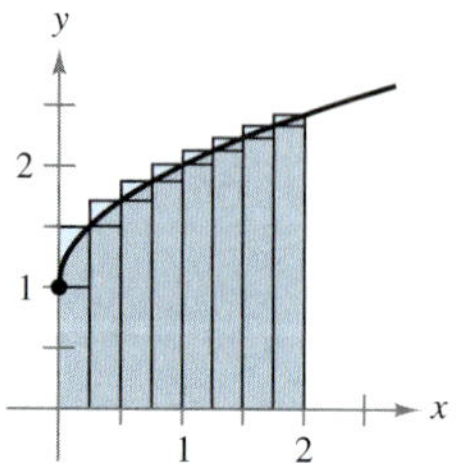

3. $y = \dfrac{1}{x}$

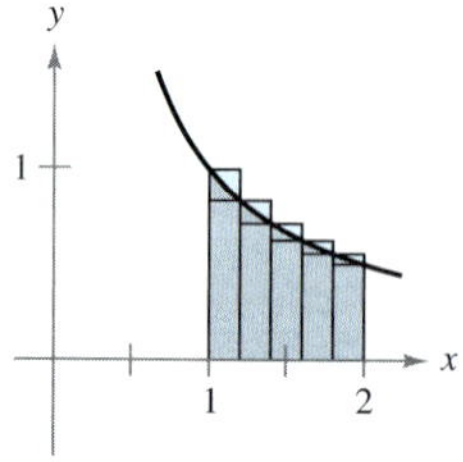

4. $y = \dfrac{1}{x-2}$

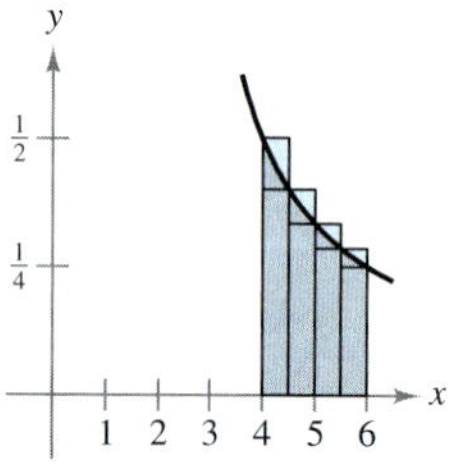

5. $y = \sqrt{1 - x^2}$

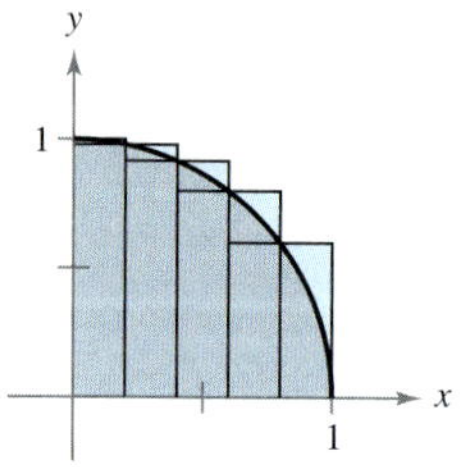

6. $y = \sqrt{x + 1}$

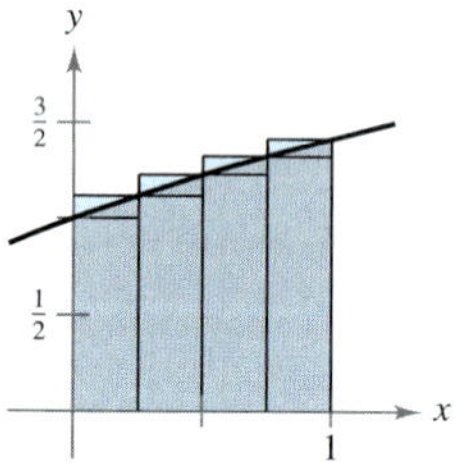

7. Repeat Exercise 1 using the midpoint Riemann sum.

8. Repeat Exercise 2 using the midpoint Riemann sum.

9. Consider a triangle of area 2 bounded by the graphs of $y = x$, $y = 0$, and $x = 2$.

(a) Sketch the graph of the region.

(b) Divide the interval $[0, 2]$ into n equal subintervals and show that the endpoints are

$$0 < 1\left(\frac{2}{n}\right) < \cdots < (n-1)\left(\frac{2}{n}\right) < n\left(\frac{2}{n}\right).$$

(c) Show that the left Riemann sum is

$$S_L = \sum_{i=1}^{n} \left[(i-1)\left(\frac{2}{n}\right)\right]\left(\frac{2}{n}\right).$$

(d) Show that the right Riemann sum is

$$S_R = \sum_{i=1}^{n} \left[i\left(\frac{2}{n}\right)\right]\left(\frac{2}{n}\right).$$

(e) Complete the table below.

n	5	10	50	100
Left sum, S_L				
Right sum, S_R				

(f) Show that $\lim\limits_{n\to\infty} S_L = \lim\limits_{n\to\infty} S_R = 2$.

10. Consider a trapezoid of area 4 bounded by the graphs of $y = x$, $y = 0$, $x = 1$, and $x = 3$.

(a) Sketch the graph of the region.

(b) Divide the interval $[1, 3]$ into n equal subintervals and show that the endpoints are

$$1 < 1 + 1\left(\frac{2}{n}\right) < \cdots < 1 + (n-1)\left(\frac{2}{n}\right) < 1 + n\left(\frac{2}{n}\right).$$

(c) Show that the left Riemann sum is

$$S_L = \sum_{i=1}^{n} \left[1 + (i-1)\left(\frac{2}{n}\right)\right]\left(\frac{2}{n}\right).$$

(d) Show that the right Riemann sum is

$$S_R = \sum_{i=1}^{n} \left[1 + i\left(\frac{2}{n}\right)\right]\left(\frac{2}{n}\right).$$

(e) Complete the table below.

n	5	10	50	100
Left sum, S_L				
Right sum, S_R				

(f) Show that $\lim\limits_{n\to\infty} S_L = \lim\limits_{n\to\infty} S_R = 4$.

In Exercises 11–18, set up a definite integral that yields the area of the region. (Do not evaluate the integral.)

11. $f(x) = 3$

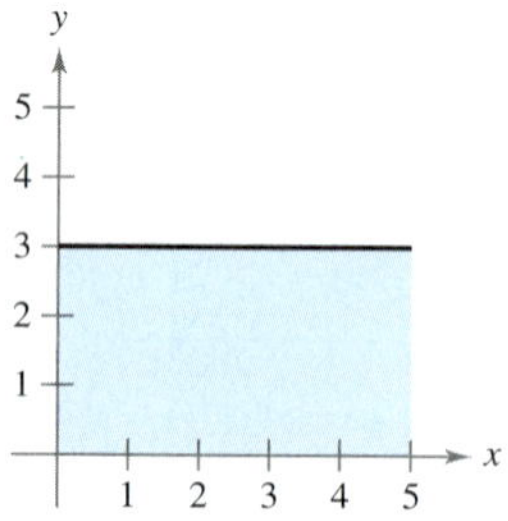

12. $f(x) = 4 - 2x$

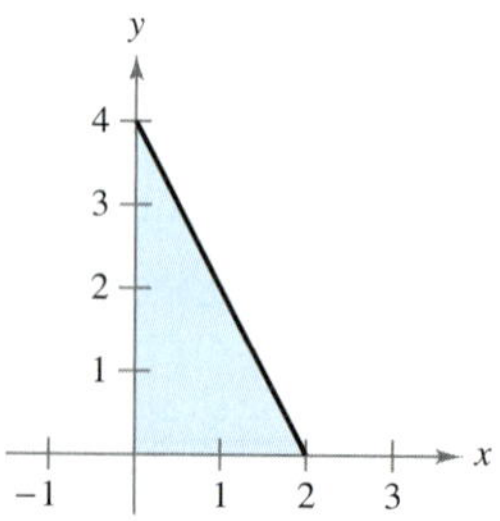

13. $f(x) = 4 - |x|$

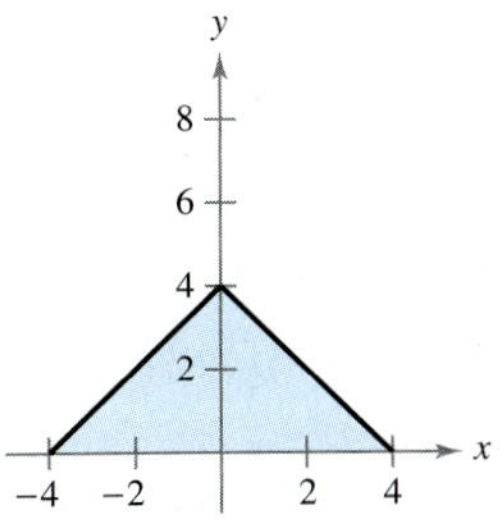

14. $f(x) = x^2$

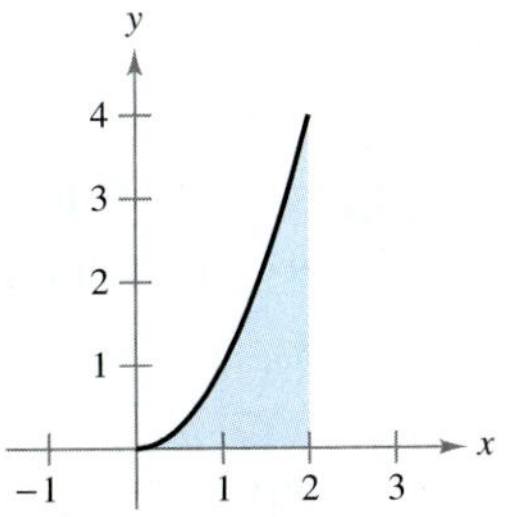

15. $f(x) = 4 - x^2$

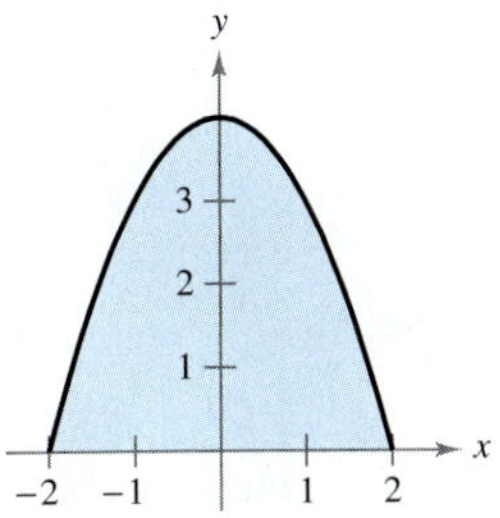

16. $f(x) = \dfrac{1}{x^2 + 1}$

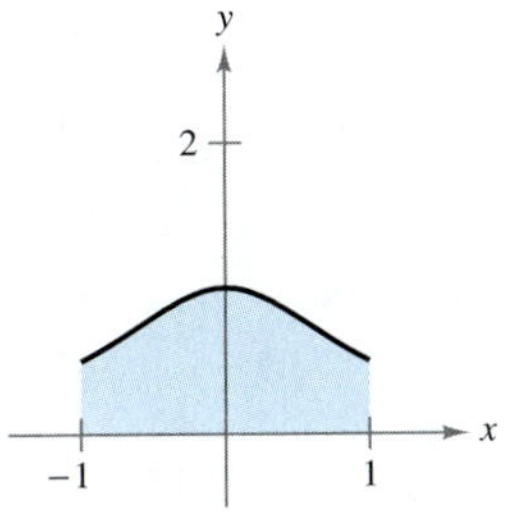

17. $f(x) = \sqrt{x + 1}$

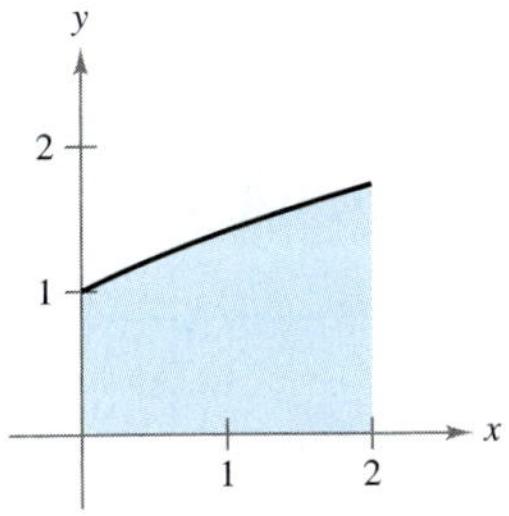

18. $f(x) = (x^2 + 1)^2$

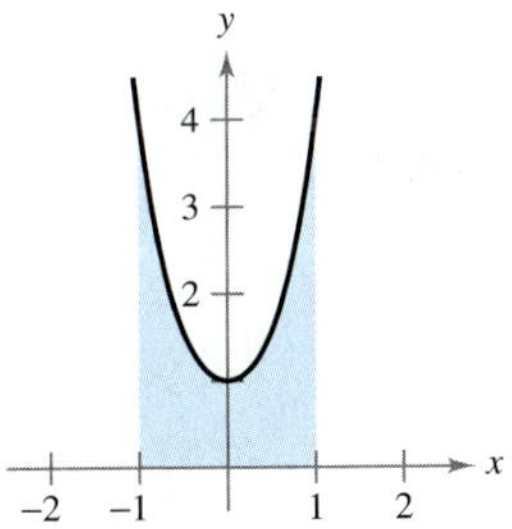

In Exercises 19–28, sketch the region whose area is given by the definite integral. Then use a geometric formula to evaluate the integral ($a > 0, r > 0$).

19. $\displaystyle\int_0^3 4\,dx$

20. $\displaystyle\int_{-a}^{a} 4\,dx$

21. $\displaystyle\int_0^4 x\,dx$

22. $\displaystyle\int_0^4 \frac{x}{2}\,dx$

23. $\displaystyle\int_0^2 (2x + 5)\,dx$

24. $\displaystyle\int_0^5 (5 - x)\,dx$

25. $\displaystyle\int_{-1}^{1} (1 - |x|)\,dx$

26. $\displaystyle\int_{-a}^{a} (a - |x|)\,dx$

27. $\displaystyle\int_{-3}^{3} \sqrt{9 - x^2}\,dx$

28. $\displaystyle\int_{-r}^{r} \sqrt{r^2 - x^2}\,dx$

29. Show that $\displaystyle\sum_{i=1}^{n} i = \frac{n(n+1)}{2}$. (*Hint:* Add the two sums below.)

$$S = 1 + 2 + 3 + \cdot\cdot\cdot + (n - 2) + (n - 1) + n$$
$$S = n + (n - 1) + (n - 2) + \cdot\cdot\cdot + 3 + 2 + 1$$

30. Use the Riemann sum definition of the definite integral and the result of Exercise 29 to evaluate $\int_1^2 x\,dx$.

In Exercises 31 and 32, use the figure to fill in the blank with the symbol <, >, or =.

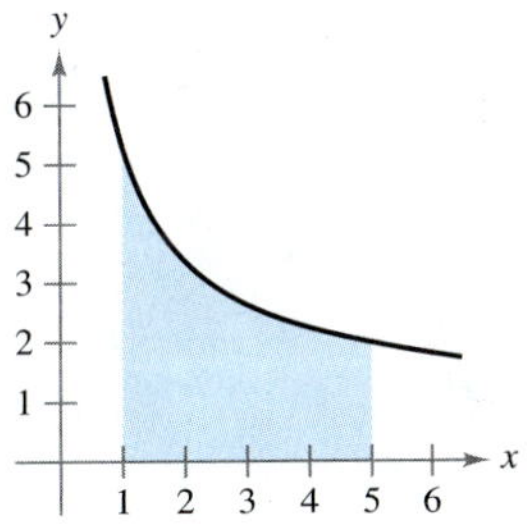

31. The interval $[1, 5]$ is partitioned into n subintervals of equal width Δx, and x_i is the left endpoint of the ith subinterval.

$$\sum_{i=1}^{n} f(x_i)\,\Delta x \quad \square \quad \int_1^5 f(x)\,dx$$

32. The interval $[1, 5]$ is partitioned into n subintervals of equal width Δx, and x_i is the right endpoint of the ith subinterval.

$$\sum_{i=1}^{n} f(x_i)\,\Delta x \quad \square \quad \int_1^5 f(x)\,dx$$

B FORMULAS

B.1 DIFFERENTIATION AND INTEGRATION FORMULAS

Use differentiation and integration tables to supplement differentiation and integration techniques.

Differentiation Formulas

1. $\frac{d}{dx}[cu] = cu'$
2. $\frac{d}{dx}[u \pm v] = u' \pm v'$
3. $\frac{d}{dx}[uv] = uv' + vu'$
4. $\frac{d}{dx}\left[\frac{u}{v}\right] = \frac{vu' - uv'}{v^2}$
5. $\frac{d}{dx}[c] = 0$
6. $\frac{d}{dx}[u^n] = nu^{n-1}u'$
7. $\frac{d}{dx}[x] = 1$
8. $\frac{d}{dx}[\ln u] = \frac{u'}{u}$
9. $\frac{d}{dx}[e^u] = e^u u'$
10. $\frac{d}{dx}[\sin u] = (\cos u)u'$
11. $\frac{d}{dx}[\cos u] = -(\sin u)u'$
12. $\frac{d}{dx}[\tan u] = (\sec^2 u)u'$
13. $\frac{d}{dx}[\cot u] = -(\csc^2 u)u'$
14. $\frac{d}{dx}[\sec u] = (\sec u \tan u)u'$
15. $\frac{d}{dx}[\csc u] = -(\csc u \cot u)u'$

Integration Formulas

Forms Involving u^n

1. $\int u^n \, du = \frac{u^{n+1}}{n+1} + C, \quad n \neq -1$
2. $\int \frac{1}{u} \, du = \ln|u| + C$

Forms Involving $a + bu$

3. $\int \frac{u}{a + bu} \, du = \frac{1}{b^2}(bu - a \ln|a + bu|) + C$
4. $\int \frac{u}{(a + bu)^2} \, du = \frac{1}{b^2}\left(\frac{a}{a + bu} + \ln|a + bu|\right) + C$
5. $\int \frac{u}{(a + bu)^n} \, du = \frac{1}{b^2}\left[\frac{-1}{(n-2)(a + bu)^{n-2}} + \frac{a}{(n-1)(a + bu)^{n-1}}\right] + C, \quad n \neq 1, 2$
6. $\int \frac{u^2}{a + bu} \, du = \frac{1}{b^3}\left[-\frac{bu}{2}(2a - bu) + a^2 \ln|a + bu|\right] + C$
7. $\int \frac{u^2}{(a + bu)^2} \, du = \frac{1}{b^3}\left(bu - \frac{a^2}{a + bu} - 2a \ln|a + bu|\right) + C$

8. $\displaystyle\int \frac{u^2}{(a+bu)^3}\,du = \frac{1}{b^3}\left[\frac{2a}{a+bu} - \frac{a^2}{2(a+bu)^2} + \ln|a+bu|\right] + C$

9. $\displaystyle\int \frac{u^2}{(a+bu)^n}\,du = \frac{1}{b^3}\left[\frac{-1}{(n-3)(a+bu)^{n-3}} + \frac{2a}{(n-2)(a+bu)^{n-2}} - \frac{a^2}{(n-1)(a+bu)^{n-1}}\right] + C, \quad n \neq 1, 2, 3$

10. $\displaystyle\int \frac{1}{u(a+bu)}\,du = \frac{1}{a}\ln\left|\frac{u}{a+bu}\right| + C$

11. $\displaystyle\int \frac{1}{u(a+bu)^2}\,du = \frac{1}{a}\left(\frac{1}{a+bu} + \frac{1}{a}\ln\left|\frac{u}{a+bu}\right|\right) + C$

12. $\displaystyle\int \frac{1}{u^2(a+bu)}\,du = -\frac{1}{a}\left(\frac{1}{u} + \frac{b}{a}\ln\left|\frac{u}{a+bu}\right|\right) + C$

13. $\displaystyle\int \frac{1}{u^2(a+bu)^2}\,du = -\frac{1}{a^2}\left[\frac{a+2bu}{u(a+bu)} + \frac{2b}{a}\ln\left|\frac{u}{a+bu}\right|\right] + C$

Forms Involving $\sqrt{a+bu}$

14. $\displaystyle\int u^n\sqrt{a+bu}\,du = \frac{2}{b(2n+3)}\left[u^n(a+bu)^{3/2} - na\int u^{n-1}\sqrt{a+bu}\,du\right]$

15. $\displaystyle\int \frac{1}{u\sqrt{a+bu}}\,du = \frac{1}{\sqrt{a}}\ln\left|\frac{\sqrt{a+bu}-\sqrt{a}}{\sqrt{a+bu}+\sqrt{a}}\right| + C, \quad a > 0$

16. $\displaystyle\int \frac{1}{u^n\sqrt{a+bu}}\,du = \frac{-1}{a(n-1)}\left[\frac{\sqrt{a+bu}}{u^{n-1}} + \frac{(2n-3)b}{2}\int \frac{1}{u^{n-1}\sqrt{a+bu}}\,du\right], \quad n \neq 1$

17. $\displaystyle\int \frac{\sqrt{a+bu}}{u}\,du = 2\sqrt{a+bu} + a\int \frac{1}{u\sqrt{a+bu}}\,du$

18. $\displaystyle\int \frac{\sqrt{a+bu}}{u^n}\,du = \frac{-1}{a(n-1)}\left[\frac{(a+bu)^{3/2}}{u^{n-1}} + \frac{(2n-5)b}{2}\int \frac{\sqrt{a+bu}}{u^{n-1}}\,du\right], \quad n \neq 1$

19. $\displaystyle\int \frac{u}{\sqrt{a+bu}}\,du = -\frac{2(2a-bu)}{3b^2}\sqrt{a+bu} + C$

20. $\displaystyle\int \frac{u^n}{\sqrt{a+bu}}\,du = \frac{2}{(2n+1)b}\left(u^n\sqrt{a+bu} - na\int \frac{u^{n-1}}{\sqrt{a+bu}}\,du\right)$

Forms Involving $u^2 - a^2, a > 0$

21. $\displaystyle\int \frac{1}{u^2-a^2}\,du = -\int \frac{1}{a^2-u^2}\,du = \frac{1}{2a}\ln\left|\frac{u-a}{u+a}\right| + C$

22. $\displaystyle\int \frac{1}{(u^2-a^2)^n}\,du = \frac{-1}{2a^2(n-1)}\left[\frac{u}{(u^2-a^2)^{n-1}} + (2n-3)\int \frac{1}{(u^2-a^2)^{n-1}}\,du\right], \quad n \neq 1$

Integration Formulas (Continued)

Forms Involving $\sqrt{u^2 \pm a^2}, a > 0$

23. $\displaystyle\int \sqrt{u^2 \pm a^2}\, du = \frac{1}{2}\left(u\sqrt{u^2 \pm a^2} \pm a^2 \ln\left|u + \sqrt{u^2 \pm a^2}\right|\right) + C$

24. $\displaystyle\int u^2\sqrt{u^2 \pm a^2}\, du = \frac{1}{8}\left[u(2u^2 \pm a^2)\sqrt{u^2 \pm a^2} - a^4 \ln\left|u + \sqrt{u^2 \pm a^2}\right|\right] + C$

25. $\displaystyle\int \frac{\sqrt{u^2 + a^2}}{u}\, du = \sqrt{u^2 + a^2} - a \ln\left|\frac{a + \sqrt{u^2 + a^2}}{u}\right| + C$

26. $\displaystyle\int \frac{\sqrt{u^2 \pm a^2}}{u^2}\, du = \frac{-\sqrt{u^2 \pm a^2}}{u} + \ln\left|u + \sqrt{u^2 \pm a^2}\right| + C$

27. $\displaystyle\int \frac{1}{\sqrt{u^2 \pm a^2}}\, du = \ln\left|u + \sqrt{u^2 \pm a^2}\right| + C$

28. $\displaystyle\int \frac{1}{u\sqrt{u^2 + a^2}}\, du = \frac{-1}{a} \ln\left|\frac{a + \sqrt{u^2 + a^2}}{u}\right| + C$

29. $\displaystyle\int \frac{u^2}{\sqrt{u^2 \pm a^2}}\, du = \frac{1}{2}\left(u\sqrt{u^2 \pm a^2} \mp a^2 \ln\left|u + \sqrt{u^2 \pm a^2}\right|\right) + C$

30. $\displaystyle\int \frac{1}{u^2\sqrt{u^2 \pm a^2}}\, du = \mp \frac{\sqrt{u^2 \pm a^2}}{a^2 u} + C$

31. $\displaystyle\int \frac{1}{(u^2 \pm a^2)^{3/2}}\, du = \frac{\pm u}{a^2\sqrt{u^2 \pm a^2}} + C$

Forms Involving $\sqrt{a^2 - u^2}, a > 0$

32. $\displaystyle\int \frac{\sqrt{a^2 - u^2}}{u}\, du = \sqrt{a^2 - u^2} - a \ln\left|\frac{a + \sqrt{a^2 - u^2}}{u}\right| + C$

33. $\displaystyle\int \frac{1}{u\sqrt{a^2 - u^2}}\, du = \frac{-1}{a} \ln\left|\frac{a + \sqrt{a^2 - u^2}}{u}\right| + C$

34. $\displaystyle\int \frac{1}{u^2\sqrt{a^2 - u^2}}\, du = \frac{-\sqrt{a^2 - u^2}}{a^2 u} + C$

35. $\displaystyle\int \frac{1}{(a^2 - u^2)^{3/2}}\, du = \frac{u}{a^2\sqrt{a^2 - u^2}} + C$

Forms Involving e^u

36. $\displaystyle\int e^u\, du = e^u + C$

37. $\int ue^u\,du = (u - 1)e^u + C$

38. $\int u^n e^u\,du = u^n e^u - n\int u^{n-1}e^u\,du$

39. $\int \frac{1}{1 + e^u}\,du = u - \ln(1 + e^u) + C$

40. $\int \frac{1}{1 + e^{nu}}\,du = u - \frac{1}{n}\ln(1 + e^{nu}) + C$

Forms Involving $\ln u$

41. $\int \ln u\,du = u(-1 + \ln u) + C$

42. $\int u \ln u\,du = \frac{u^2}{4}(-1 + 2\ln u) + C$

43. $\int u^n \ln u\,du = \frac{u^{n+1}}{(n + 1)^2}[-1 + (n + 1)\ln u] + C, \quad n \neq -1$

44. $\int (\ln u)^2\,du = u[2 - 2\ln u + (\ln u)^2] + C$

45. $\int (\ln u)^n\,du = u(\ln u)^n - n\int (\ln u)^{n-1}\,du$

Forms Involving $\sin u$ or $\cos u$

46. $\int \sin u\,du = -\cos u + C$

47. $\int \cos u\,du = \sin u + C$

48. $\int \sin^2 u\,du = \frac{1}{2}(u - \sin u \cos u) + C$

49. $\int \cos^2 u\,du = \frac{1}{2}(u + \sin u \cos u) + C$

50. $\int \sin^n u\,du = -\frac{\sin^{n-1} u \cos u}{n} + \frac{n - 1}{n}\int \sin^{n-2} u\,du$

51. $\int \cos^n u\,du = \frac{\cos^{n-1} u \sin u}{n} + \frac{n - 1}{n}\int \cos^{n-2} u\,du$

52. $\int u \sin u\,du = \sin u - u\cos u + C$

Integration Formulas (Continued)

53. $\displaystyle\int u \cos u \, du = \cos u + u \sin u + C$

54. $\displaystyle\int u^n \sin u \, du = -u^n \cos u + n \int u^{n-1} \cos u \, du$

55. $\displaystyle\int u^n \cos u \, du = u^n \sin u - n \int u^{n-1} \sin u \, du$

56. $\displaystyle\int \frac{1}{1 \pm \sin u} \, du = \tan u \mp \sec u + C$

57. $\displaystyle\int \frac{1}{1 \pm \cos u} \, du = -\cot u \pm \csc u + C$

58. $\displaystyle\int \frac{1}{\sin u \cos u} \, du = \ln|\tan u| + C$

Forms Involving $\tan u$, $\cot u$, $\sec u$, or $\csc u$

59. $\displaystyle\int \tan u \, du = -\ln|\cos u| + C$

60. $\displaystyle\int \cot u \, du = \ln|\sin u| + C$

61. $\displaystyle\int \sec u \, du = \ln|\sec u + \tan u| + C$

62. $\displaystyle\int \csc u \, du = \ln|\csc u - \cot u| + C$

63. $\displaystyle\int \tan^2 u \, du = -u + \tan u + C$

64. $\displaystyle\int \cot^2 u \, du = -u - \cot u + C$

65. $\displaystyle\int \sec^2 u \, du = \tan u + C$

66. $\displaystyle\int \csc^2 u \, du = -\cot u + C$

67. $\displaystyle\int \tan^n u \, du = \frac{\tan^{n-1} u}{n-1} - \int \tan^{n-2} u \, du, \quad n \neq 1$

68. $\displaystyle\int \cot^n u \, du = -\frac{\cot^{n-1} u}{n-1} - \int \cot^{n-2} u \, du, \quad n \neq 1$

69. $\displaystyle\int \sec^n u \, du = \frac{\sec^{n-2} u \tan u}{n-1} + \frac{n-2}{n-1} \int \sec^{n-2} u \, du, \quad n \neq 1$

70. $\displaystyle\int \csc^n u \, du = -\frac{\csc^{n-2} u \cot u}{n-1} + \frac{n-2}{n-1} \int \csc^{n-2} u \, du, \quad n \neq 1$

71. $\displaystyle\int \frac{1}{1 \pm \tan u}\,du = \frac{1}{2}(u \pm \ln|\cos u \pm \sin u|) + C$

72. $\displaystyle\int \frac{1}{1 \pm \cot u}\,du = \frac{1}{2}(u \mp \ln|\sin u \pm \cos u|) + C$

73. $\displaystyle\int \frac{1}{1 \pm \sec u}\,du = u + \cot u \mp \csc u + C$

74. $\displaystyle\int \frac{1}{1 \pm \csc u}\,du = u - \tan u \pm \sec u + C$

B.2 FORMULAS FROM BUSINESS AND FINANCE

Summary of business and finance formulas

Formulas from Business

Basic Terms

x = number of units produced (or sold)

p = price per unit

R = total revenue from selling x units

C = total cost of producing x units

$\overline{C}$ = average cost per unit

P = total profit from selling x units

Basic Equations

$$R = xp \qquad \overline{C} = \frac{C}{x} \qquad P = R - C$$

Typical Graphs of Supply and Demand Curves

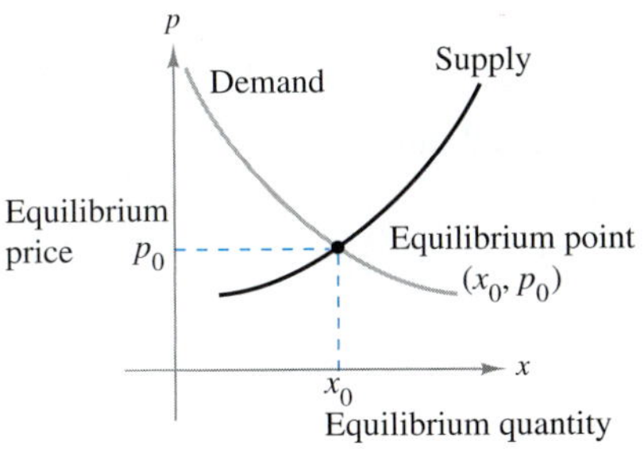

Supply curves increase as price increases and demand curves decrease as price increases. The equilibrium point occurs when the supply and demand curves intersect.

Formulas from Business (Continued)

Demand Function: $p = f(x) =$ price required to sell x units

$$\eta = \frac{p/x}{dp/dx} = \text{price elasticity of demand}$$

(If $|\eta| < 1$, the demand is inelastic. If $|\eta| > 1$, the demand is elastic.)

Typical Graphs of Revenue, Cost, and Profit Functions

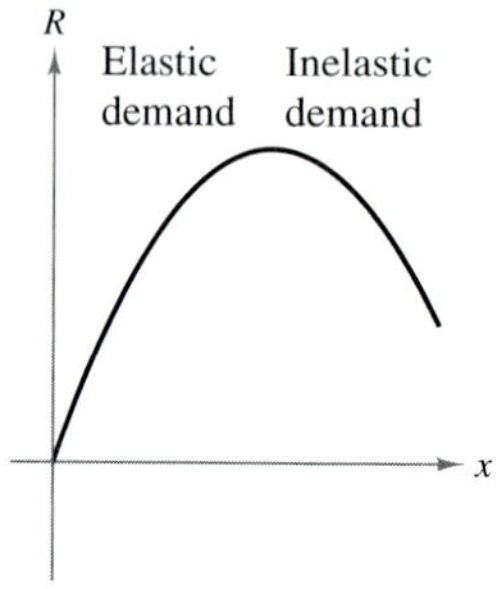

Revenue Function

The low prices required to sell more units eventually result in a decreasing revenue.

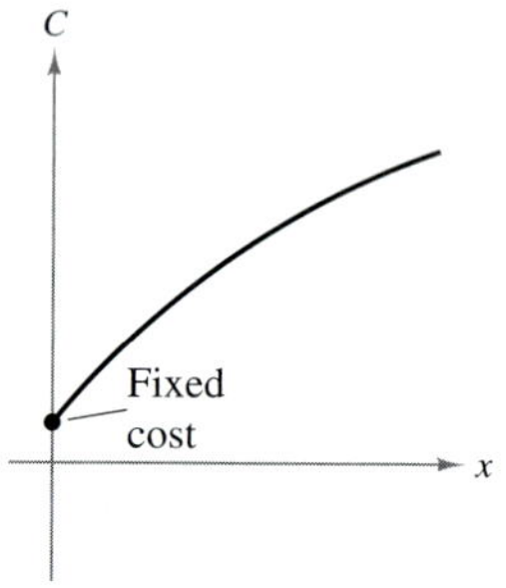

Cost Function

The total cost to produce x units includes the fixed cost.

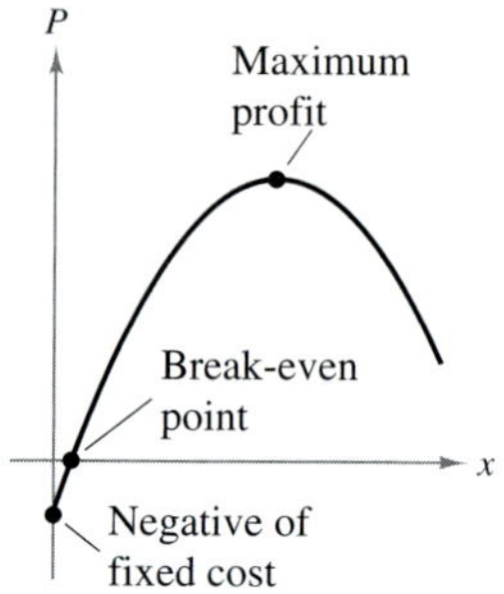

Profit Function

The break-even point occurs when $R = C$.

Marginals

$$\frac{dR}{dx} = \text{marginal revenue}$$

$\approx$ the *extra* revenue from selling one additional unit

$$\frac{dC}{dx} = \text{marginal cost}$$

$\approx$ the *extra* cost of producing one additional unit

$$\frac{dP}{dx} = \text{marginal profit}$$

$\approx$ the *extra* profit from selling one additional unit

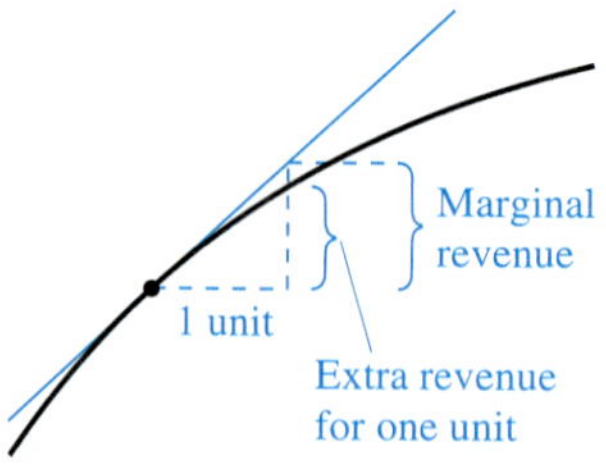

Revenue Function

Formulas from Finance

Basic Terms

$P =$ amount of deposit
$r =$ interest rate
$n =$ number of times interest is compounded per year
$t =$ number of years
$A =$ balance after t years

Compound Interest Formulas

1. Balance when interest is compounded n times per year

$$A = P\left(1 + \frac{r}{n}\right)^{nt}$$

2. Balance when interest is compounded continuously

$$A = Pe^{rt}$$

Effective Rate of Interest

$$r_{eff} = \left(1 + \frac{r}{n}\right)^{n} - 1$$

Present Value of a Future Investment

$$P = \frac{A}{\left(1 + \frac{r}{n}\right)^{nt}}$$

Balance of an Increasing Annuity After n Deposits of P per Year for t Years

$$A = P\left[\left(1 + \frac{r}{n}\right)^{nt} - 1\right]\left(1 + \frac{n}{r}\right)$$

Initial Deposit for a Decreasing Annuity with n Withdrawals of W per Year for t Years

$$P = W\left(\frac{n}{r}\right)\left\{1 - \left[\frac{1}{1 + (r/n)}\right]^{nt}\right\}$$

Monthly Installment M for a Loan of P Dollars over t Years at r% Interest

$$M = P\left\{\frac{r/12}{1 - \left[\frac{1}{1 + (r/12)}\right]^{12t}}\right\}$$

Amount of an Annuity

$$e^{rT}\int_0^T c(t)e^{-rt}\,dt$$

$c(t)$ is the continuous income function in dollars per year and T is the term of the annuity in years.

Answers to Selected Exercises

CHAPTER 0

SECTION 0.1 (page 0-7)

1. Rational **3.** Irrational **5.** Rational

7. Rational **9.** Irrational

11. (a) Yes (b) No (c) Yes (d) No

13. (a) Yes (b) No (c) No (d) Yes

15. $x \ge 12$

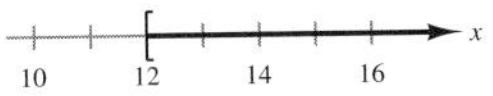

17. $x < -\frac{1}{2}$

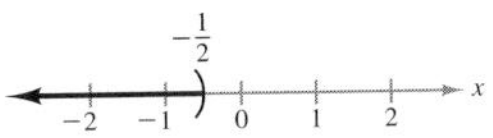

19. $x > 1$

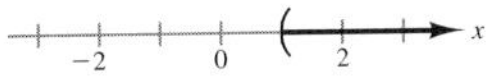

21. $-\frac{1}{2} < x < \frac{7}{2}$

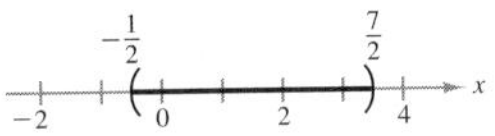

23. $-\frac{3}{4} < x < -\frac{1}{4}$

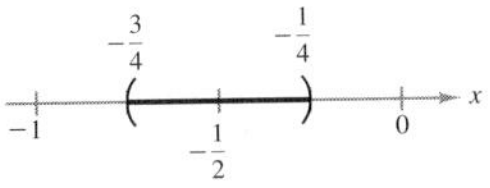

25. $x > 6$

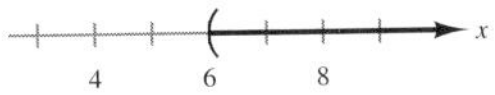

27. $-\frac{3}{2} < x < 2$

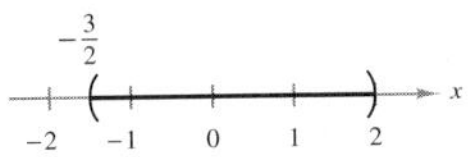

29.

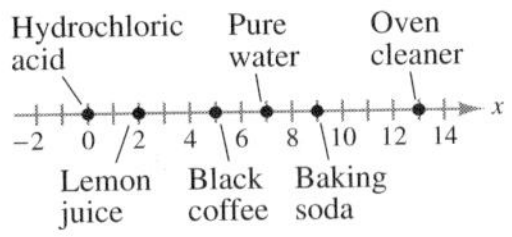

31. $x \ge 36$ units **33.** $285.71 < x < 428.57$ miles

35. (a) False (b) True (c) True (d) False

SECTION 0.2 (page 0-12)

1. (a) -51 (b) 51 (c) 51

3. (a) -14.99 (b) 14.99 (c) 14.99

5. (a) $-\frac{128}{75}$ (b) $\frac{128}{75}$ (c) $\frac{128}{75}$ **7.** $|x| \le 2$

9. $|x| > 2$ **11.** $|x - 4| \le 2$ **13.** $|x - 2| > 2$

15. $|x - 4| < 2$ **17.** $|y - a| \le 2$

19. $-5 < x < 5$

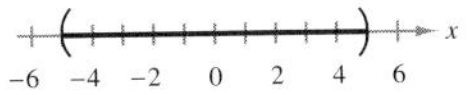

21. $x < -6$ or $x > 6$

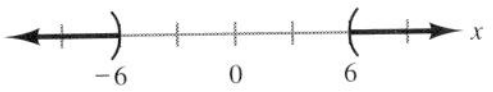

23. $-7 < x < 3$

25. $x \le -7$ or $x \ge 13$

27. $x < 6$ or $x > 14$

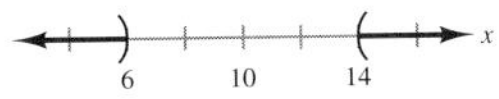

29. $4 < x < 5$

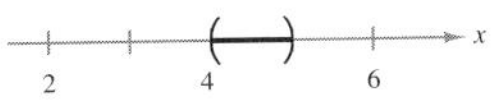

31. $a - b \le x \le a + b$

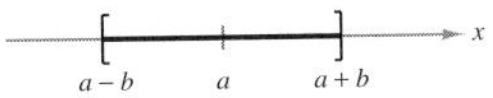

33. $\dfrac{a - 8b}{3} < x < \dfrac{a + 8b}{3}$

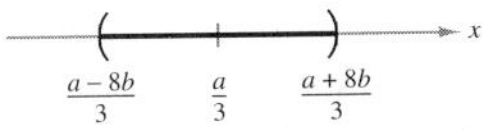

35. 14 **37.** 1.25 **39.** $\frac{1}{8}$ **41.** $|M - 1083.4| < 0.2$

43. $65.8 \le h \le 71.2$ **45.** $175{,}000 \le x \le 225{,}000$

47. (a) $|4750 - E| \le 500$, $|4750 - E| \le 237.50$
(b) At variance

49. (a) $|20{,}000 - E| \le 500$, $|20{,}000 - E| \le 1000$
(b) At variance

SECTION 0.3 (page 0-18)

1. -24 **3.** $\frac{1}{2}$ **5.** 3 **7.** 44 **9.** 5 **11.** 9

13. $\frac{1}{2}$ **15.** $\frac{1}{4}$ **17.** 908.3483 **19.** -5.3601

21. $\dfrac{3}{4y^{14}}$ **23.** $10x^4$ **25.** $7x^5$

27. $\frac{4}{3}(x + y)^5,\ x \ne -y$ **29.** $3x,\ x > 0$

31. (a) $2\sqrt{2}$ (b) $3\sqrt{2}$

33. (a) $2x\sqrt[3]{2x^2}$ (b) $2|x|z\sqrt[4]{2z}$

35. (a) $\dfrac{2x^3z}{y}\sqrt[3]{\dfrac{18z^2}{y}}$ (b) $2(3x + 5)^3\sqrt{3(3x + 5)}$

37. $2x(2x^2 - 3)$ **39.** $(2x^3 + 1)/x^{1/2}$

41. $3(x + 1)^{1/2}(x + 2)(x - 1)$

43. $\dfrac{2(x - 1)^2}{(x + 1)^2}$ **45.** $\dfrac{(x^2 + 1)(3x^2 - 2x + 1)}{\sqrt{x - 1}}$

47. $x \ge 1$ **49.** $(-\infty, \infty)$ **51.** $(-\infty, 1) \cup (1, \infty)$

53. $x \ne 4,\ x \ge -2$ **55.** $1 \le x \le 5$ **57.** \$19,121.84

59. \$11,345.46 **61.** $\dfrac{\sqrt{2}}{2}\pi$ seconds or about 2.22 seconds

SECTION 0.4 (page 0-24)

1. $\dfrac{1}{2}, -\dfrac{1}{3}$ **3.** $\dfrac{3}{2}$ **5.** $-2 \pm \sqrt{3}$ **7.** $\dfrac{-3 \pm \sqrt{41}}{4}$

9. $(x - 2)^2$ **11.** $(2x + 1)^2$ **13.** $(x + 2)(x - 1)$

15. $(3x - 2)(x - 1)$ **17.** $(x - 2y)^2$

19. $(3 + y)(3 - y)(9 + y^2)$ **21.** $(x - 2)(x^2 + 2x + 4)$

23. $(y + 4)(y^2 - 4y + 16)$ **25.** $(x - 3)(x^2 + 3x + 9)$

27. $(x - 4)(x - 1)(x + 1)$ **29.** $(2x - 3)(x^2 + 2)$

31. $(x - 2)(2x^2 - 1)$ **33.** $(x + 4)(x - 4)(x^2 + 1)$

35. 0, 5 **37.** ± 3 **39.** $\pm\sqrt{3}$ **41.** 0, 6

43. $-2, 1$ **45.** $2, 3$ **47.** -4 **49.** ± 2 **51.** $1, \pm 2$

53. $(-\infty, -2] \cup [2, \infty)$ **55.** $(-\infty, 3] \cup [4, \infty)$

57. $(x + 1)(x^2 - 4x - 2)$ **59.** $(x - 1)(2x^2 + x - 1)$

61. $-2, -1, 4$ **63.** $1, 2, 3$ **65.** $-\frac{2}{3}, -\frac{1}{2}, 3$

67. 4 **69.** 2000 units **71.** 3.4×10^{-5}

SECTION 0.5 *(page 0-32)*

1. $\dfrac{x+5}{x-1}$ **3.** $\dfrac{5x-1}{x^2+2}$ **5.** $-\dfrac{x}{x^2-4}$ **7.** $\dfrac{2}{x-3}$

9. $\dfrac{(A+C)x^2 - (A - B - 2C)x - (2A + 2B - C)}{(x+1)^2(x-2)}$

11. $\dfrac{(A+B)x^2 - (6B - C)x + 3(A - 2C)}{(x-6)(x^2+3)}$

13. $-\dfrac{(x-1)^2}{x(x^2+1)}$ **15.** $-\dfrac{x^2+3}{(x+1)(x-2)(x-3)}$

17. $\dfrac{x+2}{(x+1)^{3/2}}$ **19.** $-\dfrac{3t}{2\sqrt{1+t}}$ **21.** $\dfrac{x(x^2+2)}{(x^2+1)^{3/2}}$

23. $\dfrac{2}{x^2\sqrt{x^2+2}}$ **25.** $\dfrac{1}{2\sqrt{x}(x+1)^{3/2}}$ **27.** $\dfrac{3x+4}{2(x+2)^{1/2}}$

29. $\dfrac{3x(x+2)}{(2x+3)^{3/2}}$ **31.** $\dfrac{\sqrt{6}}{2}$ **33.** $\dfrac{x\sqrt{x-4}}{x-4}$

35. $\dfrac{49\sqrt{x^2-9}}{x+3}$ **37.** $\dfrac{\sqrt{14}+2}{2}$ **39.** $\dfrac{x(5+\sqrt{3})}{11}$

41. $\sqrt{6} - \sqrt{5}$ **43.** $\sqrt{x} - \sqrt{x-2}$

45. $\dfrac{4 - 3x^2}{x^4(4 - x^2)^{3/2}}$ **47.** $232.68

CHAPTER 1

SECTION 1.1 *(page 8)*

Prerequisite Review

1. $3\sqrt{5}$ **2.** $2\sqrt{5}$ **3.** $\frac{1}{2}$ **4.** -2 **5.** $5\sqrt{3}$

6. $-\sqrt{2}$ **7.** $x = -3, x = 9$

8. $y = -8, y = 4$ **9.** $x = 19$ **10.** $y = 1$

1. (a) $a = 4, b = 3, c = 5$

(b) $4^2 + 3^2 = 5^2$

3. (a) $a = 10, b = 3, c = \sqrt{109}$

(b) $10^2 + 3^2 = (\sqrt{109})^2$

5. (a) $a = 4, b = 5, c = \sqrt{41}$

(b) $4^2 + 5^2 = (\sqrt{41})^2$

7. (a)

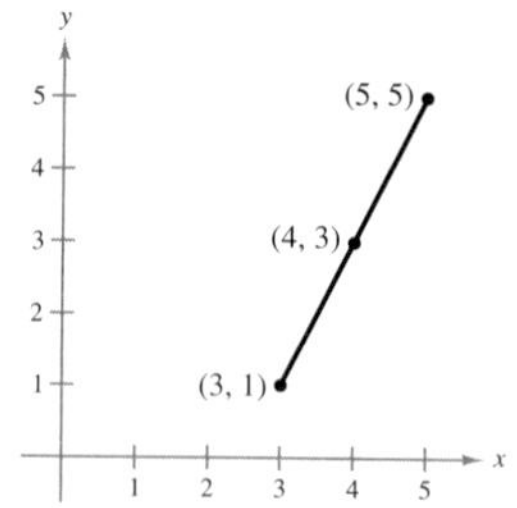

(b) $d = 2\sqrt{5}$

(c) Midpoint: $(4, 3)$

9. (a)

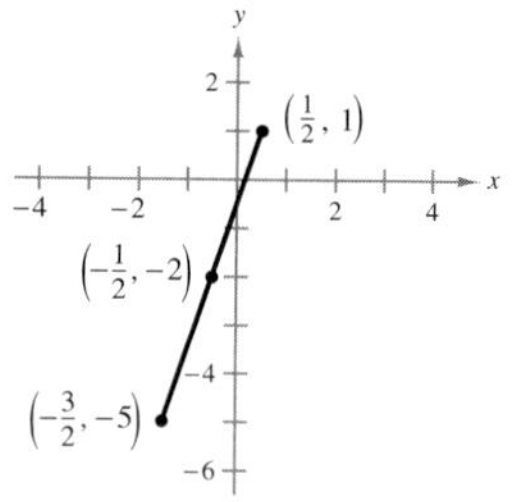

(b) $d = 2\sqrt{10}$

(c) Midpoint: $\left(-\frac{1}{2}, -2\right)$

11. (a)

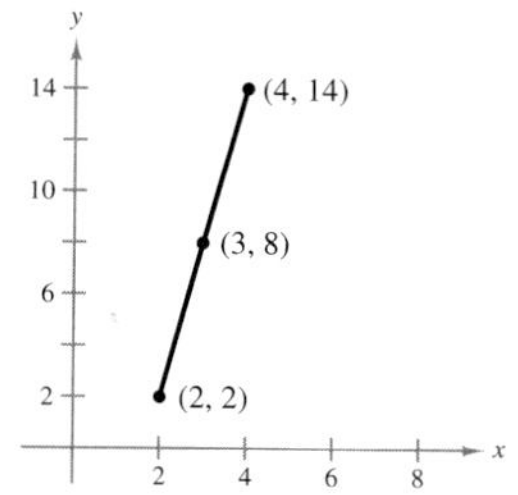

(b) $d = 2\sqrt{37}$ (c) Midpoint: $(3, 8)$

13. (a)

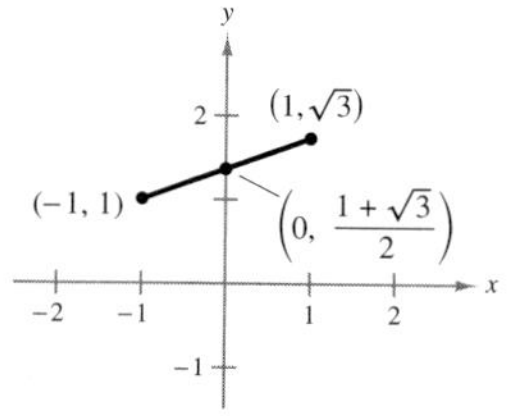

(b) $d = \sqrt{8 - 2\sqrt{3}}$ (c) Midpoint: $\left(0, \dfrac{1 + \sqrt{3}}{2}\right)$

15. $d_1 = \sqrt{45}, d_2 = \sqrt{20},$
$d_3 = \sqrt{65}$
$d_1^2 + d_2^2 = d_3^2$

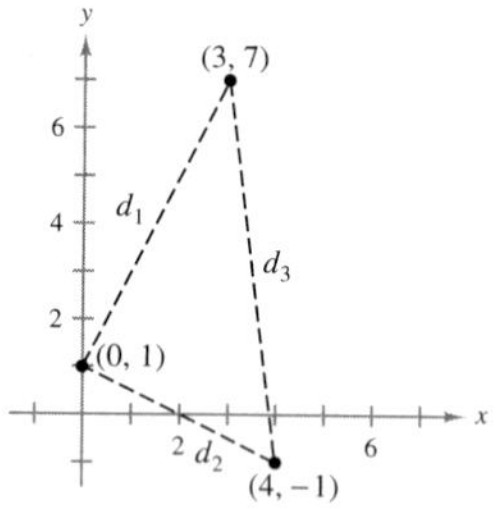

17. $d_1 = d_2 = d_3 = d_4$
$= \sqrt{5}$

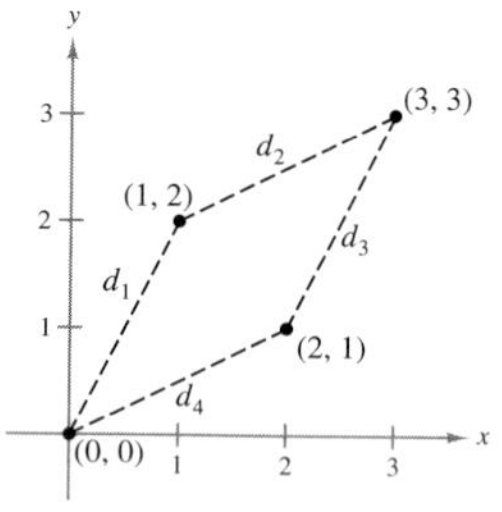

19. Collinear, because

$d_1 + d_2 = d_3$

$2\sqrt{5} + \sqrt{5} = 3\sqrt{5}$

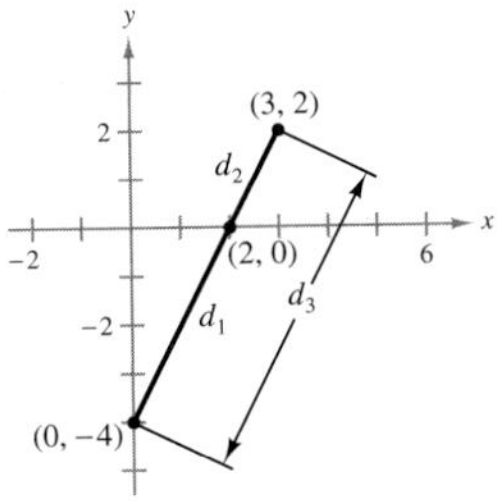

21. Not collinear, because

$d_1 + d_2 \neq d_3$

$d_1 = \sqrt{18}, d_2 = \sqrt{41}$

$d_3 = \sqrt{113}$

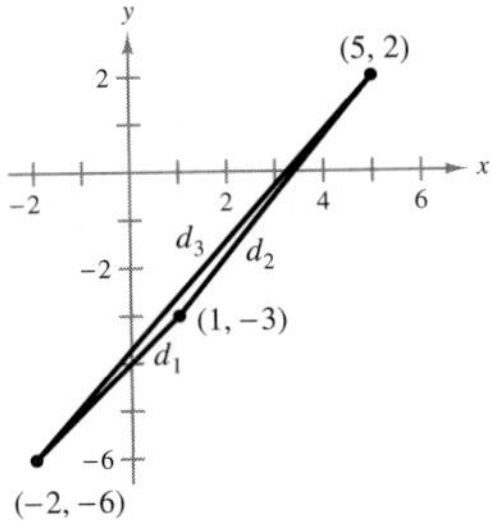

23. $x = 4, -2$ **25.** $y = \pm\sqrt{55}$

27. $\left(\frac{3x_1 + x_2}{4}, \frac{3y_1 + y_2}{4}\right), \left(\frac{x_1 + x_2}{2}, \frac{y_1 + y_2}{2}\right),$

$\left(\frac{x_1 + 3x_2}{4}, \frac{y_1 + 3y_2}{4}\right)$

29. (a) $\left(\frac{7}{4}, -\frac{7}{4}\right), \left(\frac{5}{2}, -\frac{3}{2}\right), \left(\frac{13}{4}, -\frac{5}{4}\right)$

(b) $\left(-\frac{3}{2}, -\frac{9}{4}\right), \left(-1, -\frac{3}{2}\right), \left(-\frac{1}{2}, -\frac{3}{4}\right)$

31. (a) 16.76 feet (b) 1341.04 square feet

33. Answers will vary. Sample answer:

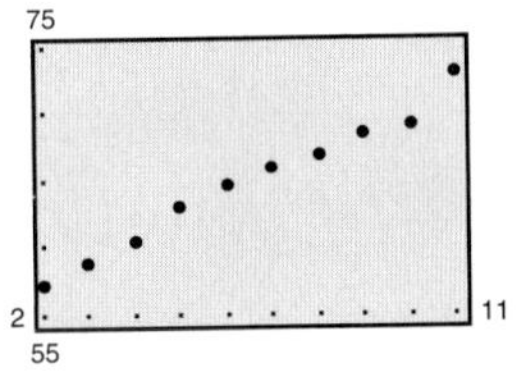

The number of subscribers appears to be increasing linearly.

35. (a) 10,400 (b) 8900 (c) 8500 (d) 10,500

37. (a) \$85 thousand (b) \$100 thousand

(c) \$122 thousand (d) \$159 thousand

39. (a) Revenue: \$25,172 million

Profit: \$890.7 million

(b) Actual 2001 revenue: \$24,623 million

Actual 2001 profit: \$885.6 million

(c) Yes, the increase in revenue per year is $\approx$ \$3667 million.

Yes, the increase in profit per year is $\approx$ \$133 million.

(d) Expenses for 1999: \$17,214.9 million

Expenses for 2001: \$23,737.4 million

Expenses for 2003: \$31,347.7 million

(e) Answers will vary.

41. (a) $(-1, 2), (1, 1), (2, 3)$

(b)

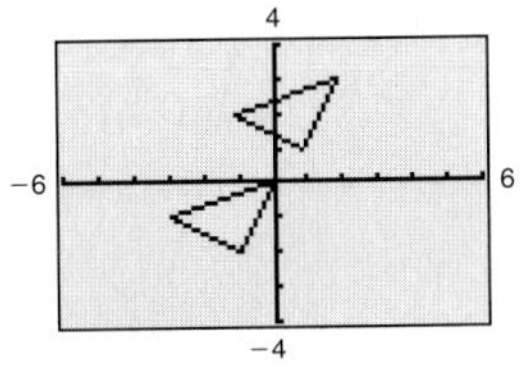

43. (a)

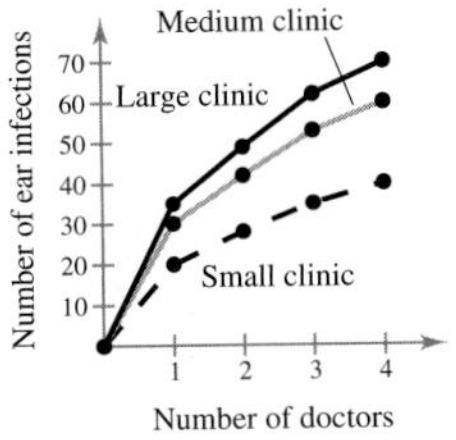

(b) The larger the clinic, the more patients a doctor can treat.

SECTION 1.2 *(page 21)*

Prerequisite Review

1. $y = \frac{1}{5}(x + 12)$ **2.** $y = x - 15$

3. $y = \dfrac{1}{x^3 + 2}$

4. $y = \pm\sqrt{x^2 + x - 6} = \pm\sqrt{(x + 3)(x - 2)}$

5. $y = -1 \pm \sqrt{9 - (x - 2)^2}$

6. $y = 5 \pm \sqrt{81 - (x + 6)^2}$ **7.** $x^2 - 4x + 4$

8. $x^2 + 6x + 9$ **9.** $x^2 - 5x + \frac{25}{4}$

10. $x^2 + 3x + \frac{9}{4}$ **11.** $(x - 2)(x - 1)$

12. $(x + 3)(x + 2)$ **13.** $\left(y - \frac{3}{2}\right)^2$ **14.** $\left(y - \frac{7}{2}\right)^2$

1. (a) Not a solution point (b) Solution point

(c) Solution point

3. (a) Solution point (b) Not a solution point

(c) Not a solution point

5. (a) Not a solution point (b) Solution point

(c) Solution point

7. e **8.** b **9.** c **10.** f **11.** a **12.** d

13. $(0, -3), \left(\frac{3}{2}, 0\right)$ **15.** $(0, -2), (-2, 0), (1, 0)$

17. $(0, 0), (-3, 0), (3, 0)$ **19.** $(-2, 0), (0, 2)$

21. $(0, 0)$

23.

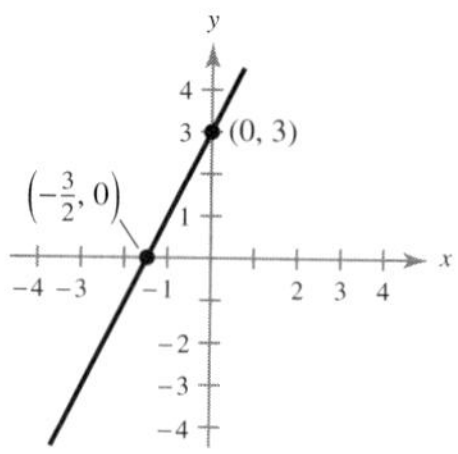

25.

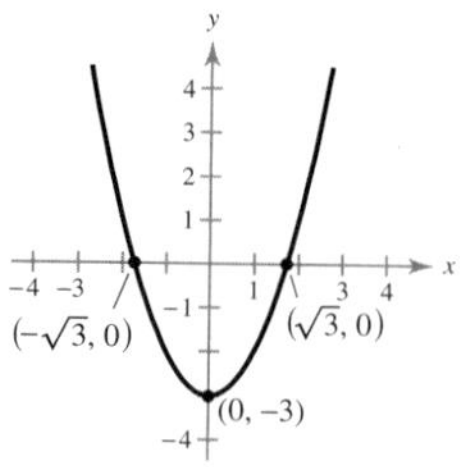

27.

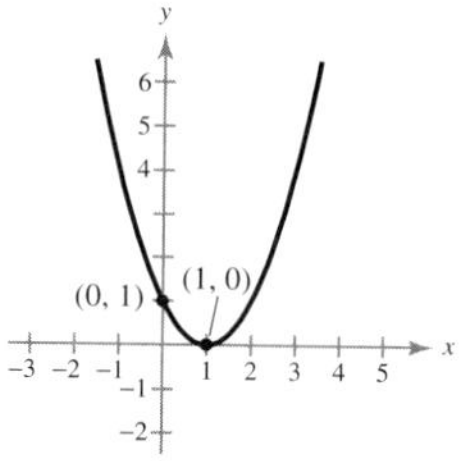

29.

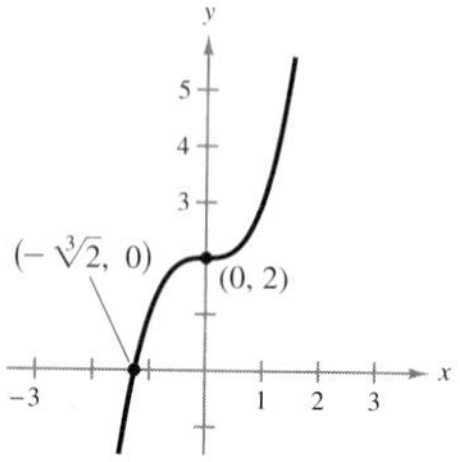

31.

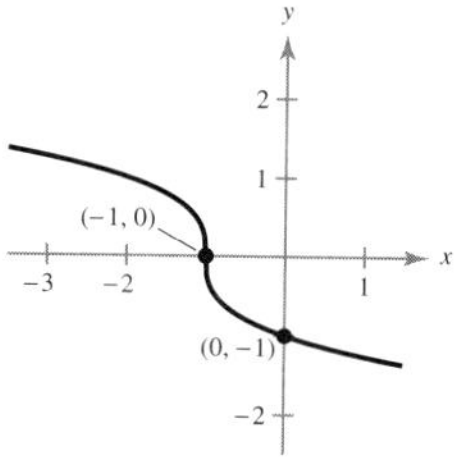

33.

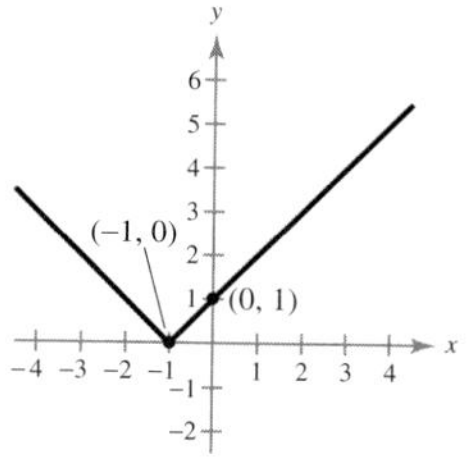

35.

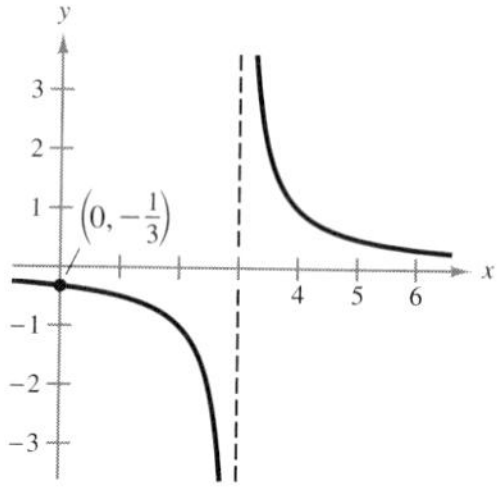

37.

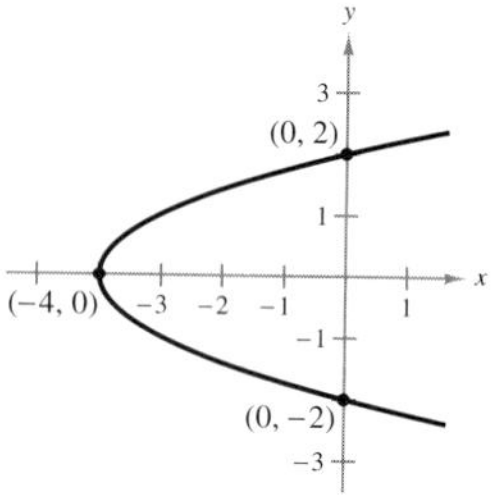

39. $x^2 + y^2 - 9 = 0$ **41.** $x^2 + y^2 - 4x + 2y - 11 = 0$

43. $x^2 + y^2 + 2x - 4y = 0$ **45.** $x^2 + y^2 - 6y = 0$

47. $(x - 1)^2 + (y + 3)^2 = 4$

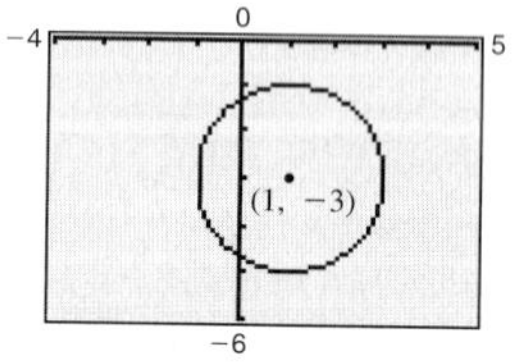

49. $(x + 2)^2 + (y + 3)^2 = 16$

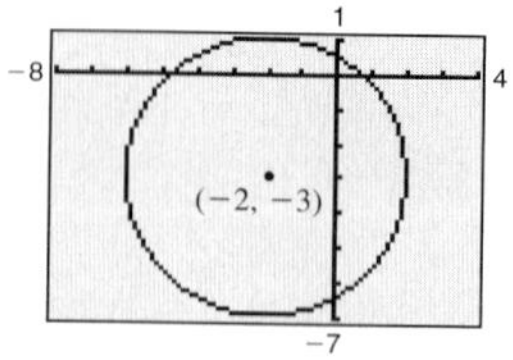

51. $\left(x - \frac{1}{2}\right)^2 + \left(y - \frac{1}{2}\right)^2 = 2$

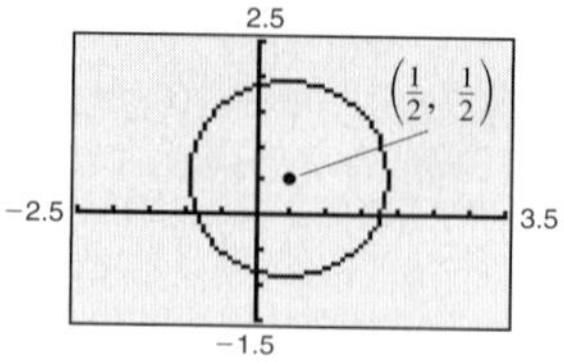

53. $\left(x + \frac{1}{2}\right)^2 + \left(y + \frac{5}{4}\right)^2 = \frac{9}{4}$

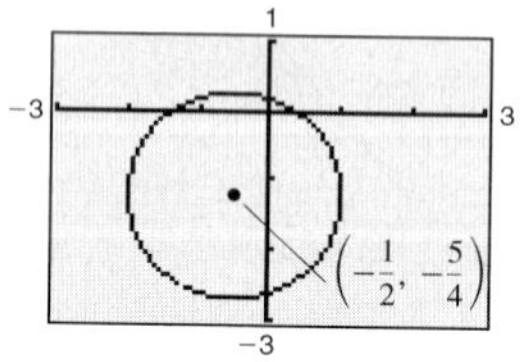

55. (1, 1) **57.** (3, 4), (5, 0)

59. $(0, 0), \left(\sqrt{2}, 2\sqrt{2}\right), \left(-\sqrt{2}, -2\sqrt{2}\right)$

61. $(-1, 0), (0, 1), (1, 0)$

63. (a) $C = 11.8x + 15{,}000$; $R = 19.3x$ (b) 2000 units

(c) 2134 units

65.

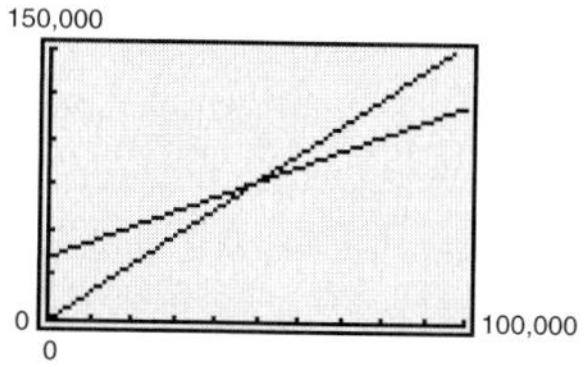

50,000 units

67.

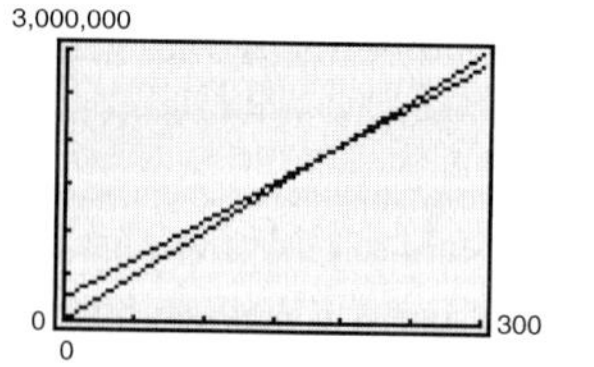

193 units

69. (15, 120)

71. (a) The model is good. Explanations will vary.

(b) $11,547 million

73. (a)

Year	1997	1998	1999
Salary	488.18	512.28	535.56

Year	2000	2001	2004
Salary	558.07	579.84	641.06

(b) Answers will vary.

(c) $678.79; answers will vary.

75.

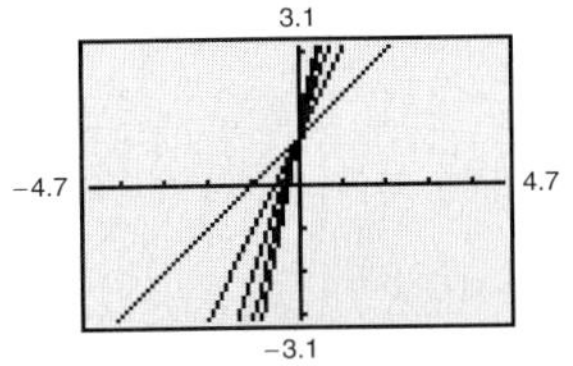

The greater the value of c, the steeper the line.

77.

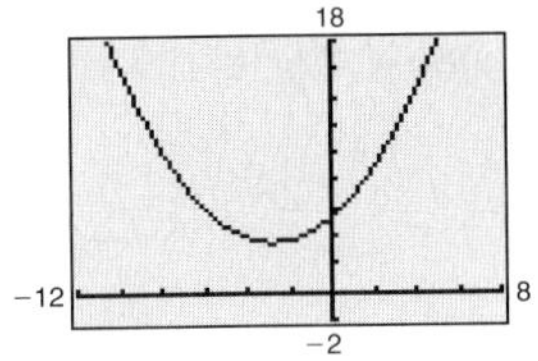

(0, 5.36)

79.

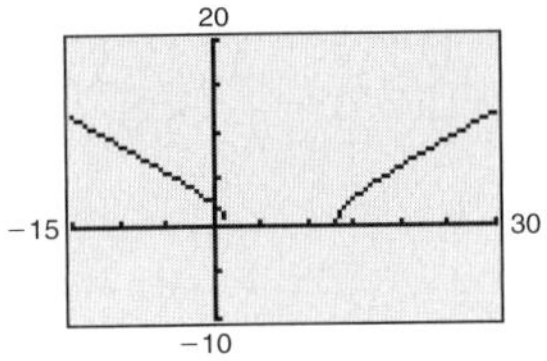

(1.4780, 0), (12.8553, 0), (0, 2.3875)

81.

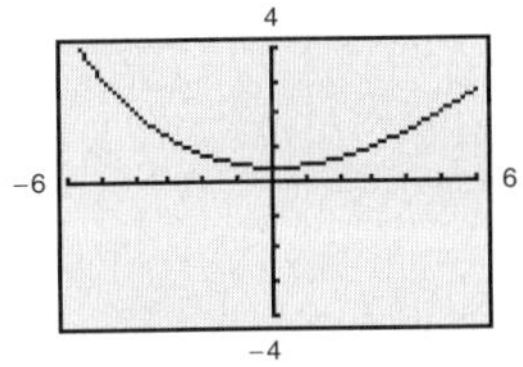

(0, 0.4167)

SECTION 1.3 *(page 33)*

Prerequisite Review

1. -1 **2.** $-\frac{7}{3}$ **3.** $\frac{1}{3}$ **4.** $-\frac{7}{6}$

5. $y = 4x + 7$ **6.** $y = 3x - 7$

7. $y = 3x - 10$ **8.** $y = -x - 7$

9. $y = 7x - 17$ **10.** $y = \frac{2}{3}x + \frac{5}{3}$

1. 1 **3.** 0

5.

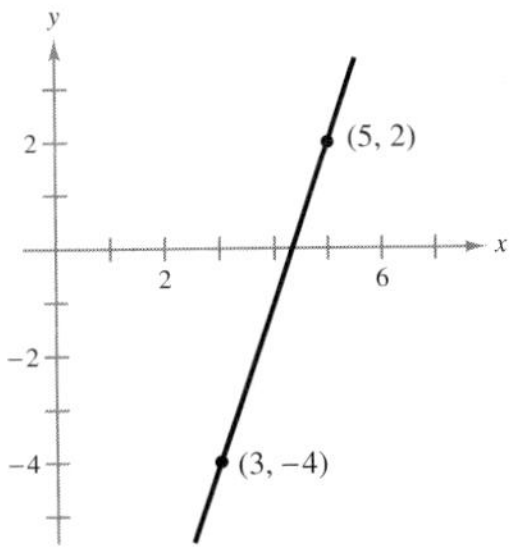

$m = 3$

7.

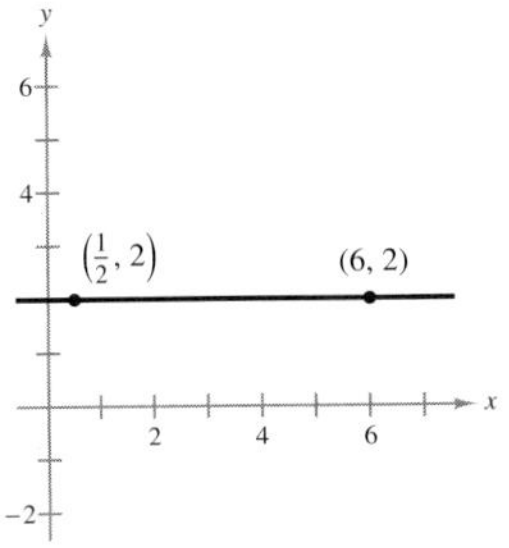

$m = 0$

9.

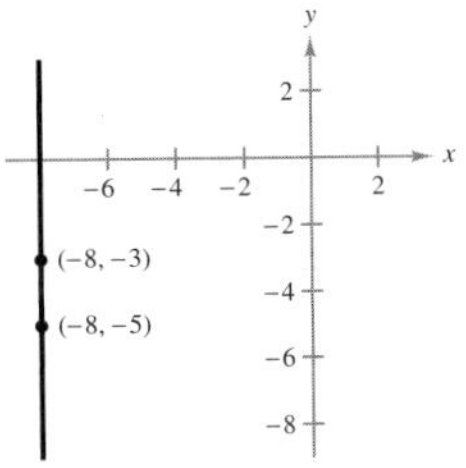

m is undefined.

11.

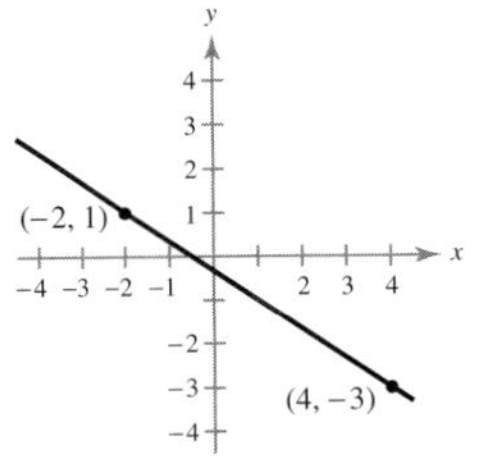

$m = -\frac{2}{3}$

13.

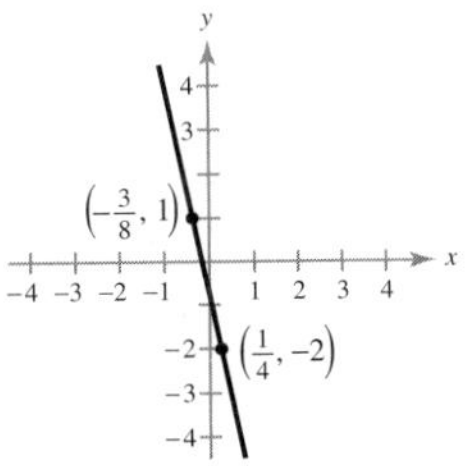

$m = -\frac{24}{5}$

15.

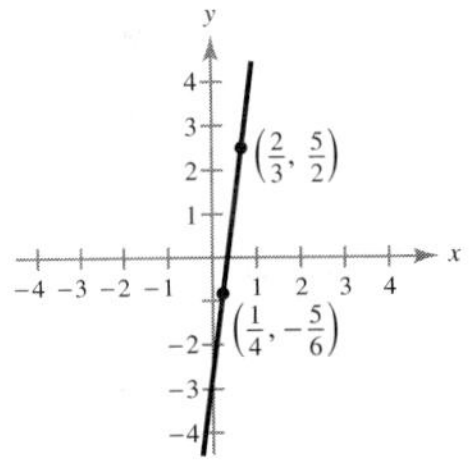

$m = 8$

17. (0, 1), (1, 1), (3, 1) **19.** (3, −6), (9, −2), (12, 0)

21. (0, 10), (2, 4), (3, 1) **23.** (−8, 0), (−8, 2), (−8, 3)

25. $m = -\frac{1}{5}$, (0, 4) **27.** $m = \frac{7}{5}$, (0, −3)

29. $m = 3$, (0, −15) **31.** m is undefined; no y-intercept.

33. $m = 0$, (0, 4)

35. $y = 2x - 5$

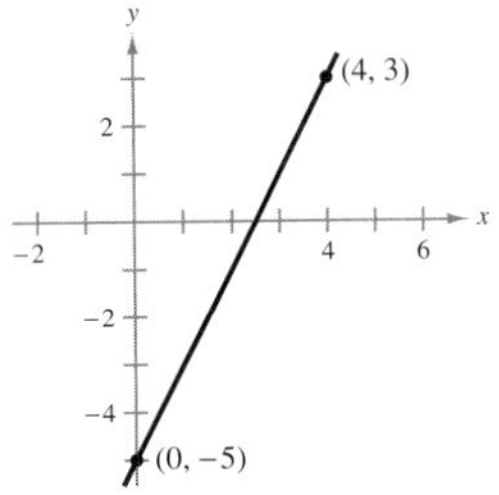

37. $3x + y = 0$

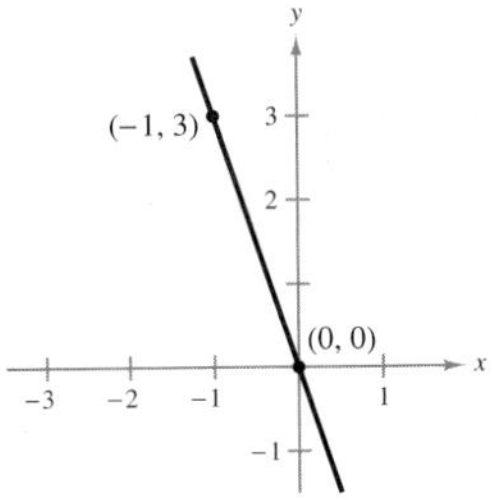

39. $x - 2 = 0$

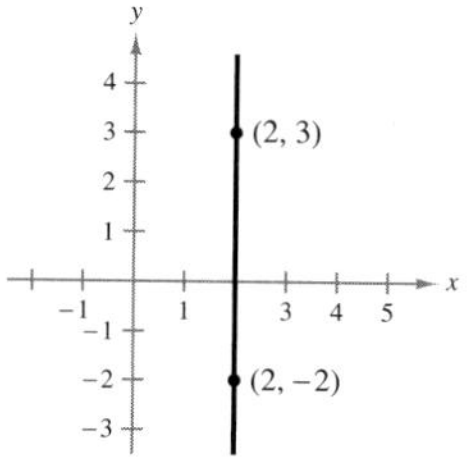

41. $y + 1 = 0$

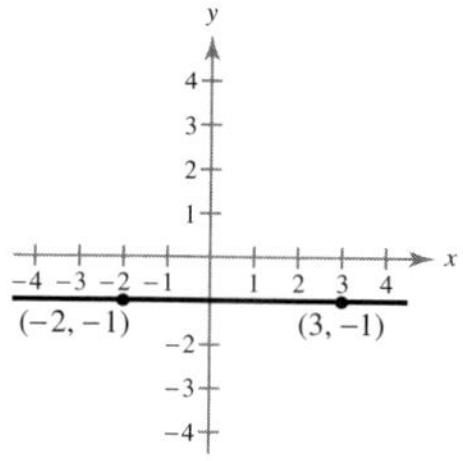

43. $3x - 6y + 7 = 0$

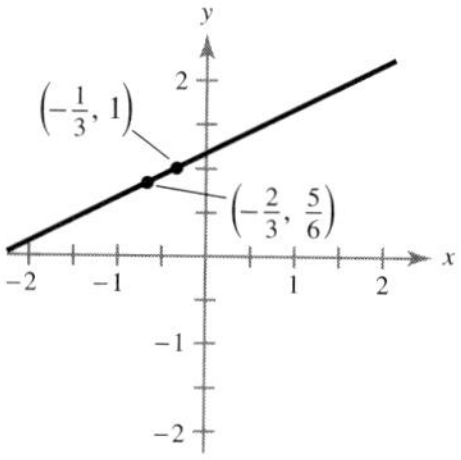

45. $4x - y + 6 = 0$

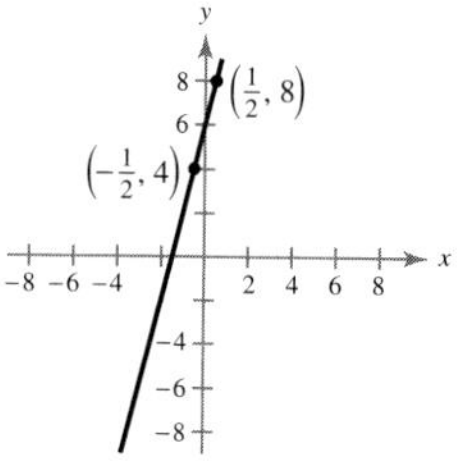

47. $3x - 4y + 12 = 0$

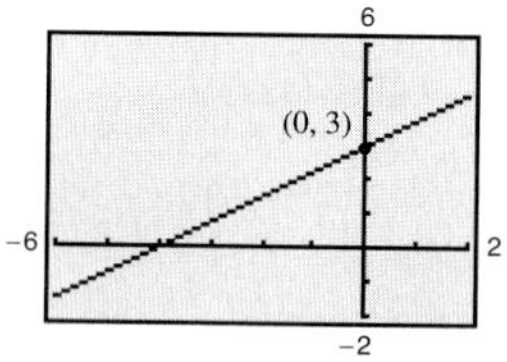

49. $x + 1 = 0$

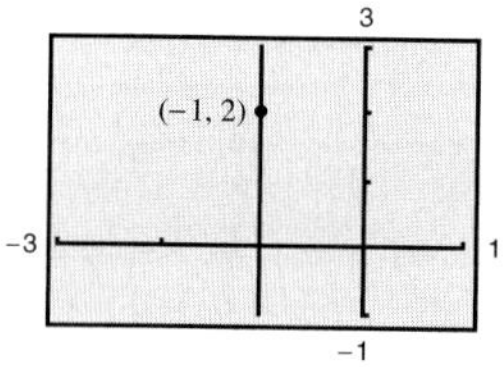

51. $y - 7 = 0$

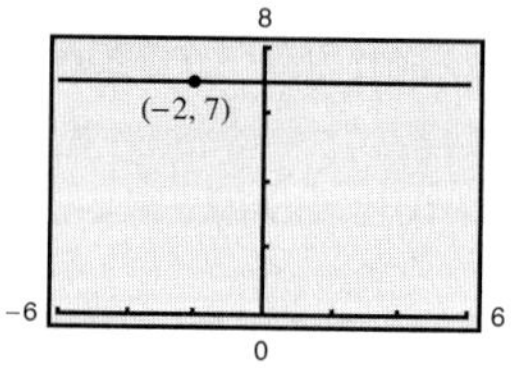

53. $y = -4x - 2$

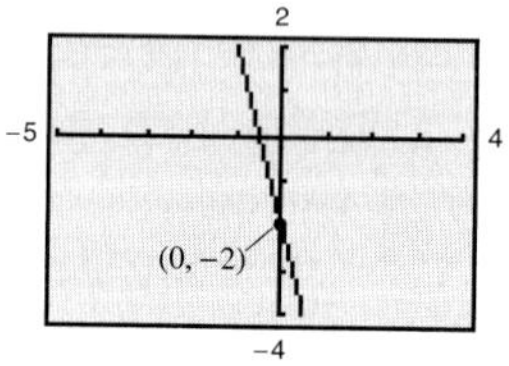

55. $9x - 12y + 8 = 0$

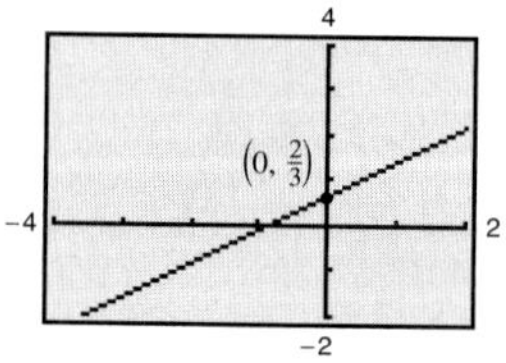

57. The points are not collinear. Explanations of methods will vary.

59. $x - 3 = 0$ **61.** $y + 10 = 0$

63. (a) $x + y + 1 = 0$ (b) $x - y + 5 = 0$

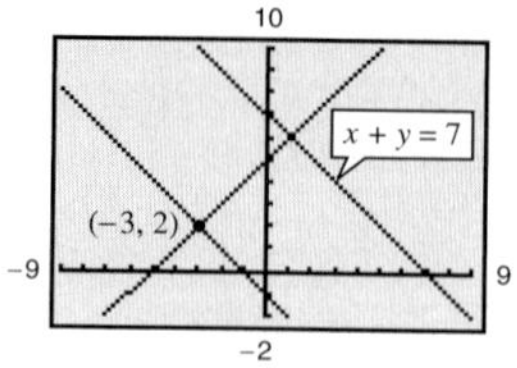

65. (a) $6x + 8y - 3 = 0$ (b) $96x - 72y + 127 = 0$

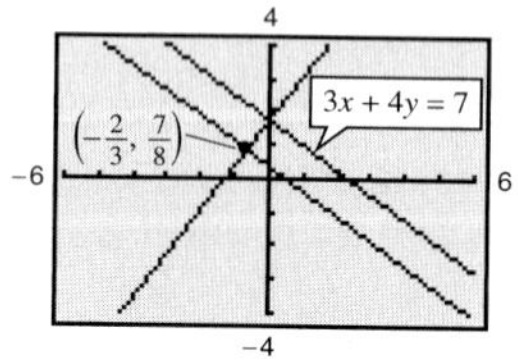

67. (a) $y = 0$ (b) $x + 1 = 0$

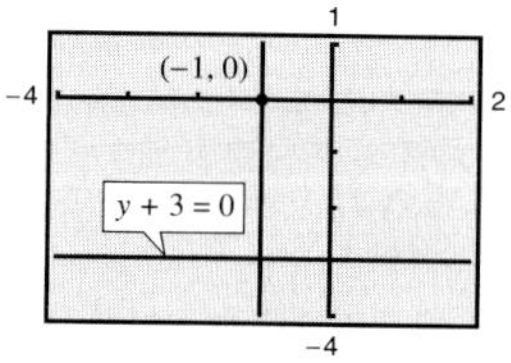

69. (a) $x - 1 = 0$ (b) $y - 1 = 0$

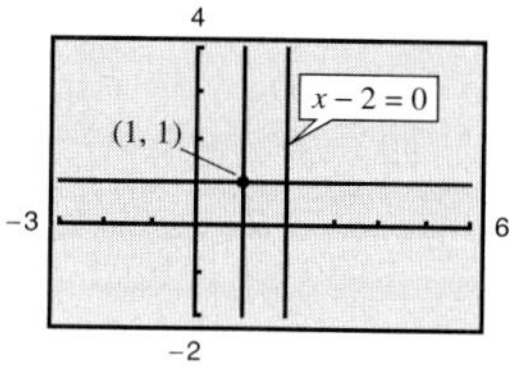

71.

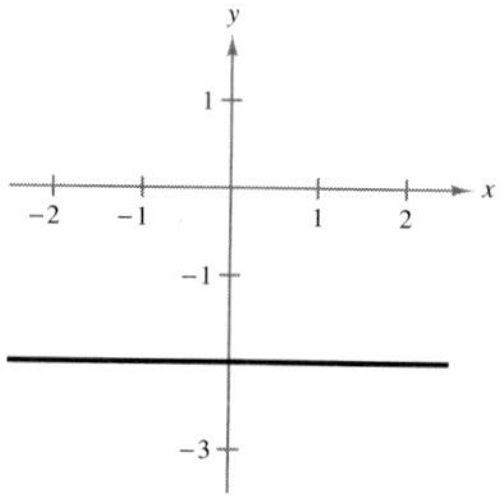

73.

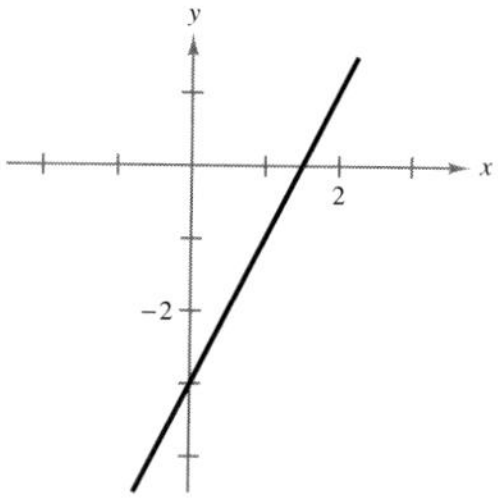

75.

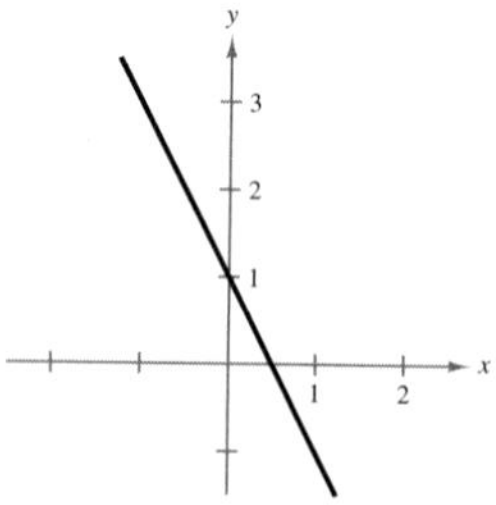

77.

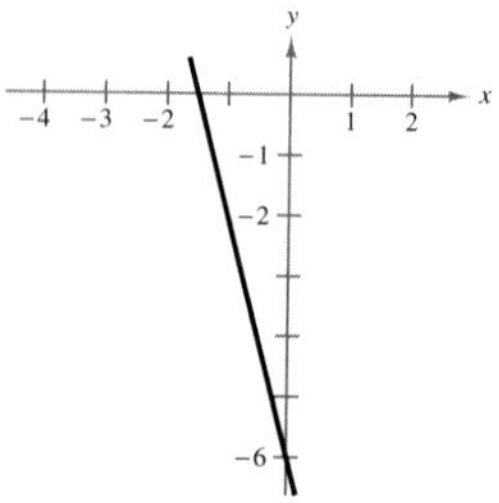

79. (a) $y = 49.4t + 3514.2$; The slope $m = 49.4$ tells you that the population increases 49.4 thousand each year.

(b) 3958.8 thousand (3,958,800)

(c) 4057.6 thousand (4,057,600)

(d) 1999: 3975 thousand (3,975,000);

2001: 4062 thousand (4,062,000)

The estimates were close to the actual populations.

(e) The model could possibly be used to predict the population in 2006 if the population continues to grow at the same linear rate.

81. $F = \frac{9}{5}C + 32$ or $C = \frac{5}{9}F - \frac{160}{9}$

83. $C = 0.34x + 150$

85. (a) $C = \frac{8}{5}A - 45$ or $A = \frac{5}{8}C + \frac{225}{8}$

(b) $F = \frac{72}{25}A - 49$ or $A = \frac{25}{72}F + \frac{1225}{72}$

(c) 75° (d) $C = 92.6°$; $F = 198.68°$ (e) $A = 56.25°$

87. (a) $y = 1025 - 205t, 0 \le t \le 5$

(b)

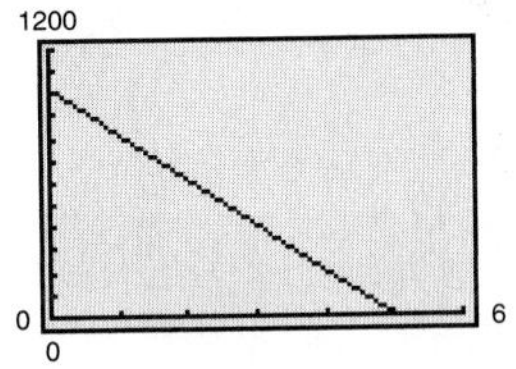

(c) \$410 (d) $t = 2.07$ years

89. (a) $x = -\frac{1}{15}p + \frac{226}{3}$ (b) 45 units (c) 49 units

91. (a) $Y = 437t + 3878$ (b) \$7811 billion

(c) \$9122 billion

(d) 1999: \$7786.5 billion

2002: \$8929.1 billion

93.

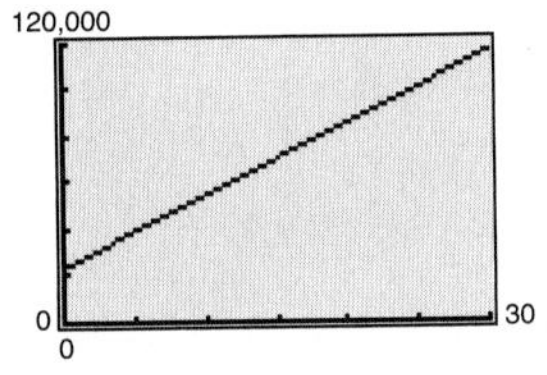

$x \le 24$ units

95.

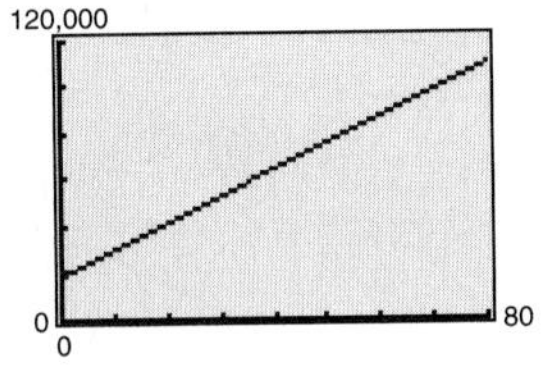

$x \le 70$ units

97.

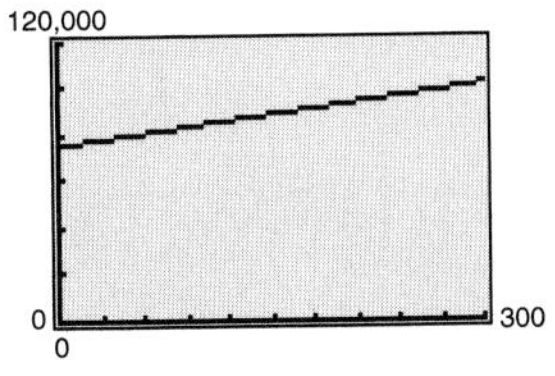

$x \le 275$ units

99.

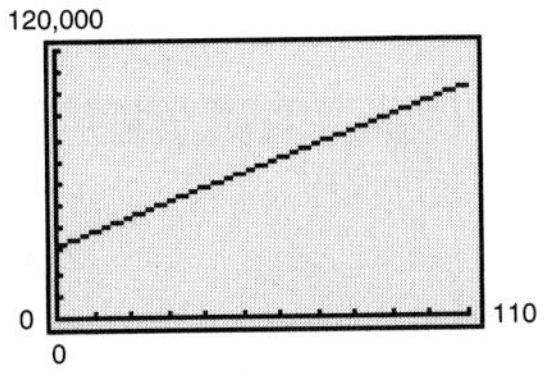

$x \le 104$ units

101.

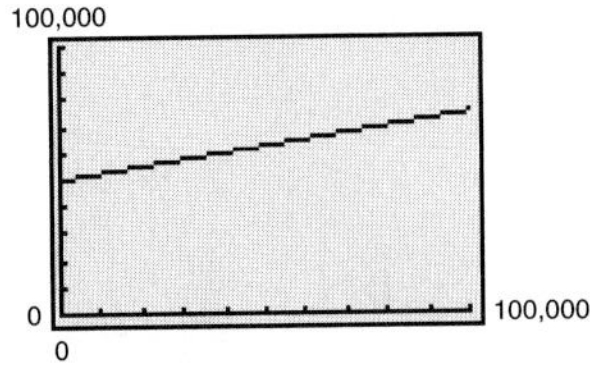

$x \le 200{,}000$ units

SECTION 1.4 *(page 45)*

Prerequisite Review

1. 20 **2.** 10 **3.** $x^2 + x - 6$

4. $x^3 + 9x^2 + 26x + 30$ **5.** $\frac{1}{x}$ **6.** $\frac{2x - 1}{x}$

7. $y = -2x + 17$ **8.** $y = \frac{6}{5}x^2 + \frac{1}{5}$

9. $y = 3 \pm \sqrt{5 + (x + 1)^2}$ **10.** $y = \pm\sqrt{4x^2 + 2}$

11. $y = 2x + \frac{1}{2}$ **12.** $y = \frac{x^3}{2} + \frac{1}{2}$

1. y is not a function of x. **3.** y is a function of x.

5. y is a function of x. **7.** y is not a function of x.

9.

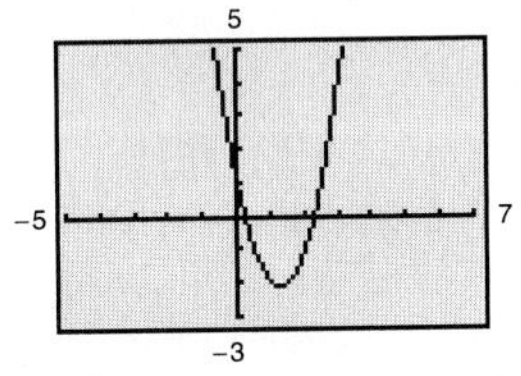

Domain: $(-\infty, \infty)$

Range: $[-2.125, \infty)$

11.

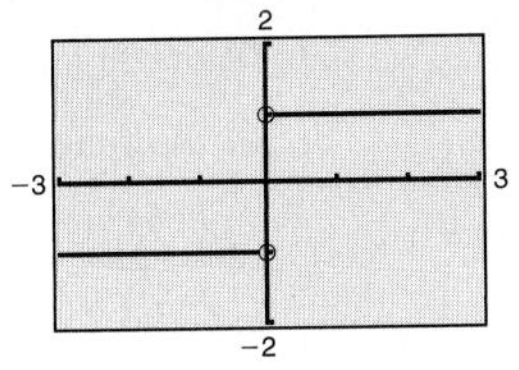

Domain: $(-\infty, 0) \cup (0, \infty)$

Range: $y = -1$ or $y = 1$

13.

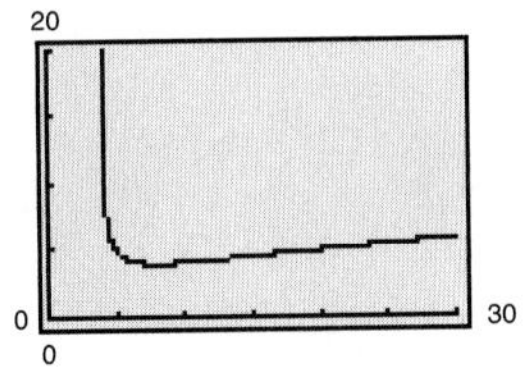

Domain: $(4, \infty)$

Range: $[4, \infty)$

15.

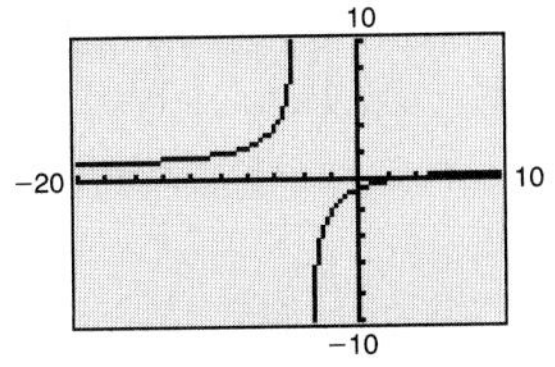

Domain: $(-\infty, -4) \cup (-4, \infty)$

Range: $(-\infty, 1) \cup (1, \infty)$

17. Domain: $(-\infty, \infty)$
Range: $(-\infty, \infty)$

19. Domain: $(-\infty, \infty)$
Range: $(-\infty, 4]$

21. (a) -3 (b) -9 (c) $2x - 5$ (d) $2x + 2\Delta x - 3$

23. (a) $\frac{1}{2}$ (b) 4 (c) $\frac{1}{x+4}$ (d) $-\frac{\Delta x}{x(x + \Delta x)}$

25. $2x + \Delta x - 4, \quad \Delta x \neq 0$

27. $\frac{1}{\sqrt{x + \Delta x + 3} + \sqrt{x + 3}}, \quad \Delta x \neq 0$

29. $-\frac{1}{(x + \Delta x - 2)(x - 2)}, \quad \Delta x \neq 0$

31. y is not a function of x.

33. y is a function of x.

35. (a) $2x$ (b) $10x - 25$ (c) $\frac{2x - 5}{5}$ (d) 5 (e) 5

37. (a) $x^2 + x$ (b) $(x^2 + 1)(x - 1) = x^3 - x^2 + x - 1$
(c) $\frac{x^2 + 1}{x - 1}$ (d) $x^2 - 2x + 2$ (e) x^2

39. (a) $\frac{x + 1}{x^2}$ (b) $\frac{1}{x^3}$ (c) $x, x \neq 0$
(d) $x^2, x \neq 0$ (e) $x^2, x \neq 0$

41. (a) 0 (b) 0 (c) -1 (d) $\sqrt{15}$
(e) $\sqrt{x^2 - 1}$ (f) $x - 1, x \geq 0$

43. The data fit the function (b), $g(x) = cx^2$, with $c = -2$.

45. The data fit the function (d), $r(x) = c/x$, with $c = 32$.

47. $f(g(x)) = 5\left(\frac{x - 1}{5}\right) + 1 = x$

$g(f(x)) = \frac{5x + 1 - 1}{5} = x$

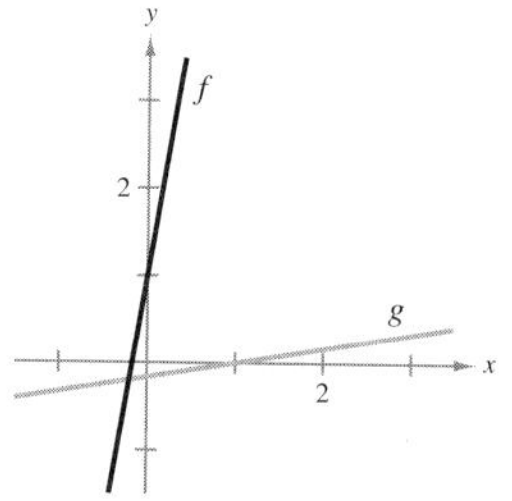

49. $f(g(x)) = 9 - \left(\sqrt{9 - x}\right)^2 = 9 - (9 - x) = x$
$g(f(x)) = \sqrt{9 - (9 - x^2)} = \sqrt{x^2} = x$

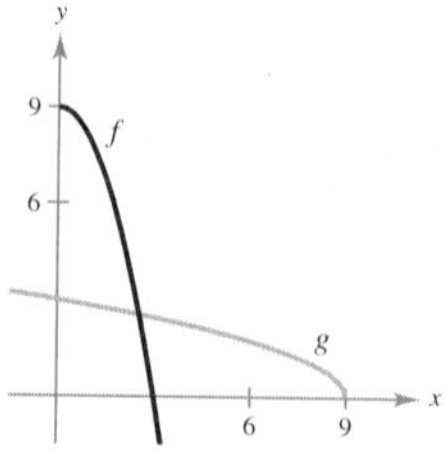

51. $f(x) = 2x - 3,\ f^{-1}(x) = \frac{x + 3}{2}$

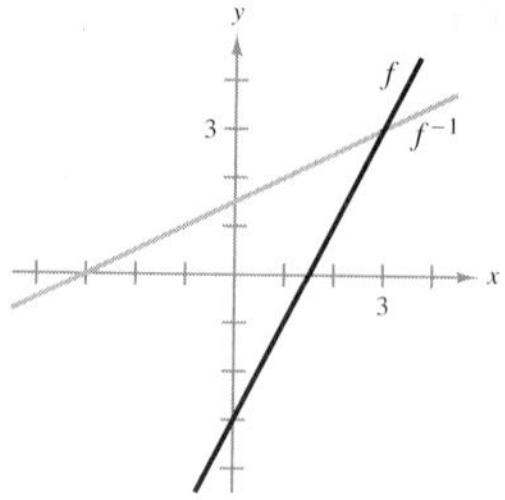

53. $f(x) = x^5,\ f^{-1}(x) = \sqrt[5]{x}$

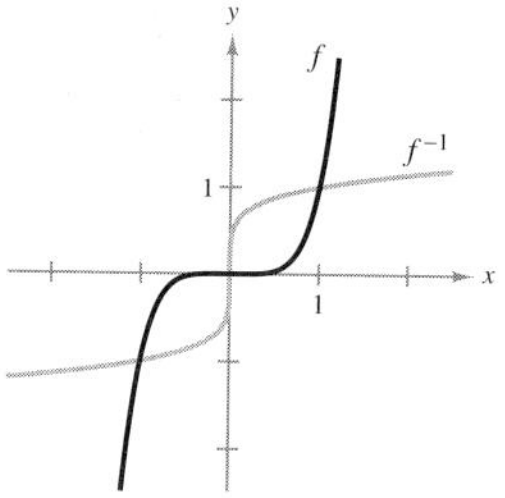

55. $f(x) = \sqrt{9 - x^2}, 0 \leq x \leq 3$
$f^{-1}(x) = \sqrt{9 - x^2}, 0 \leq x \leq 3$

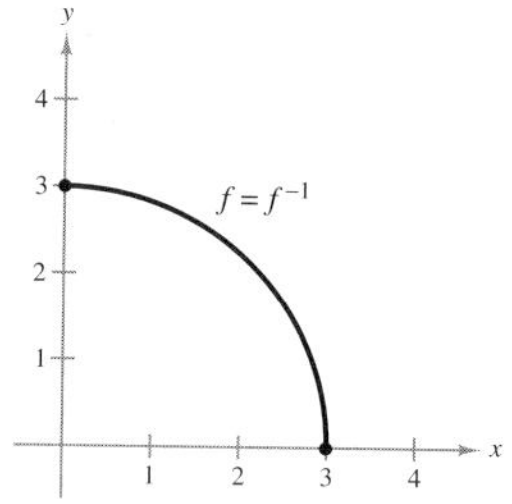

57. $f(x) = x^{2/3}, x \geq 0$
$f^{-1}(x) = x^{3/2}, x \geq 0$

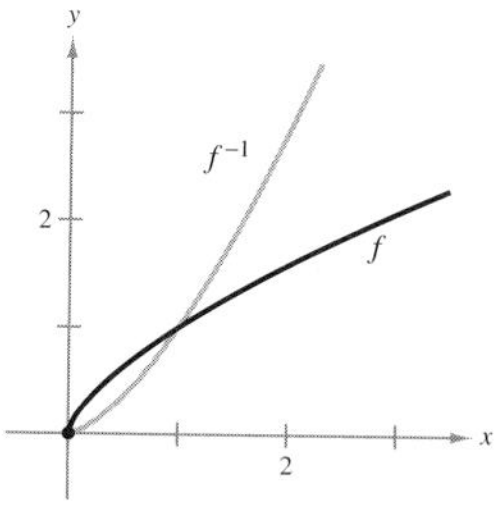

59.

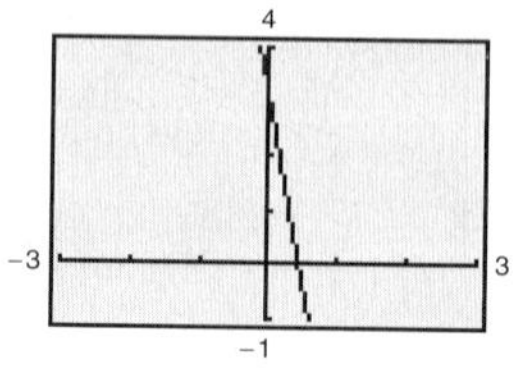

$f(x)$ is one-to-one. $f^{-1}(x) = \frac{3 - x}{7}$

61.

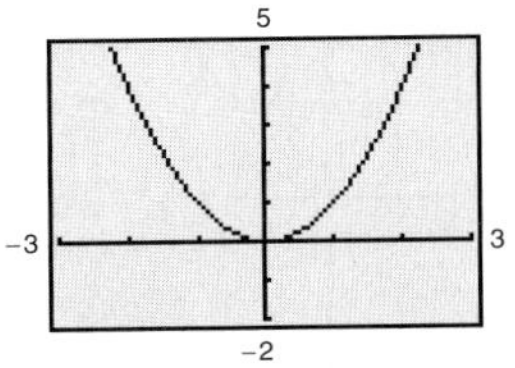

$f(x)$ is not one-to-one.

63.

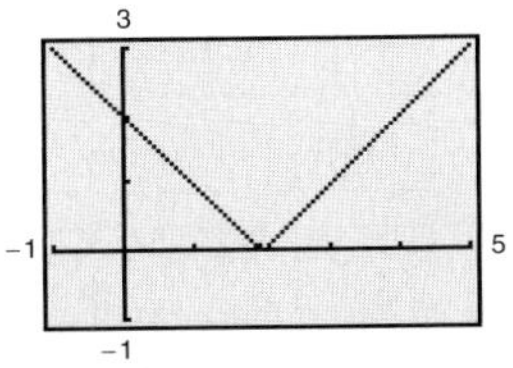

$f(x)$ is not one-to-one.

65. (a)

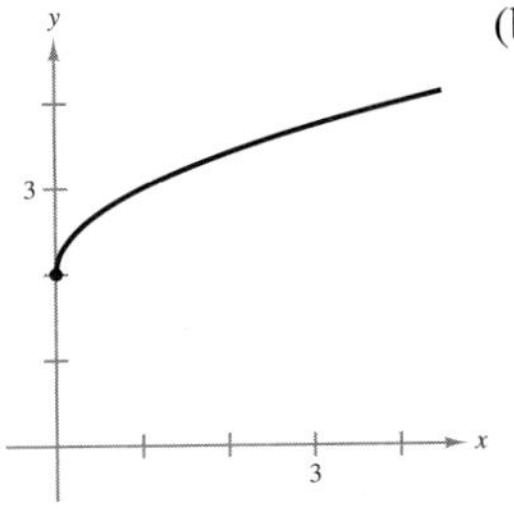

(b)

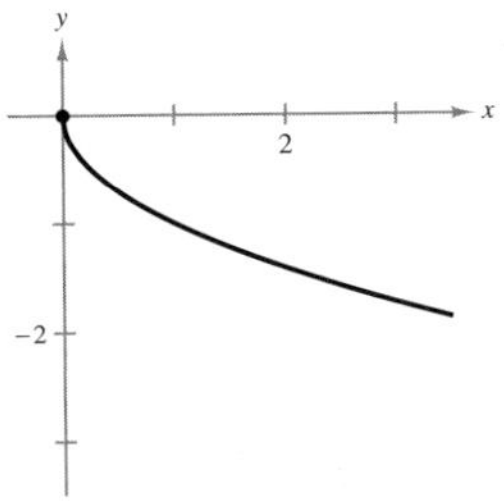

(c)

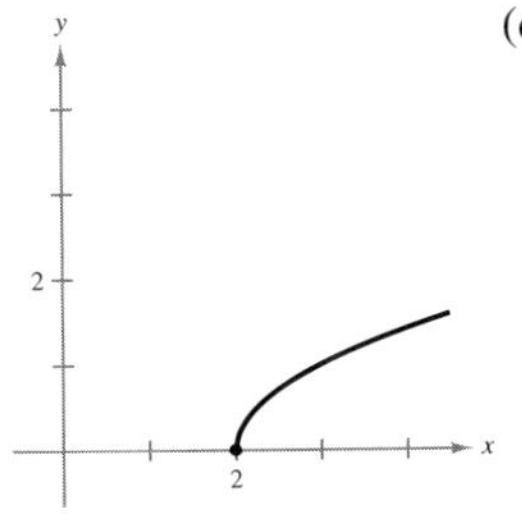

(d)

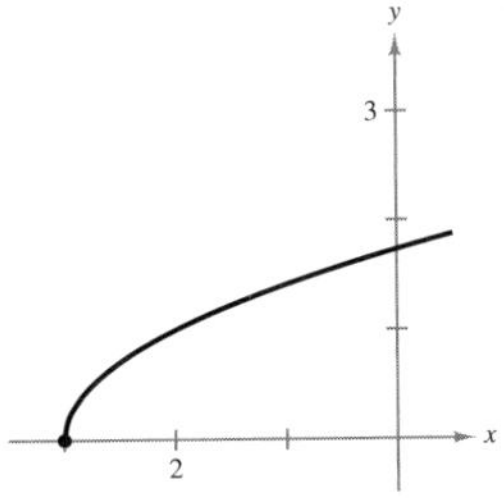

(e)

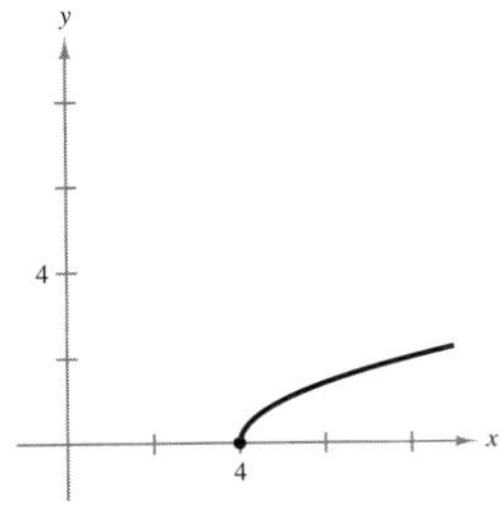

(f)

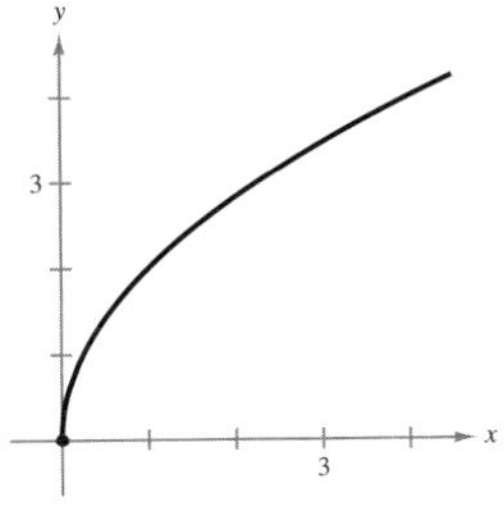

67. (a) $y = (x + 3)^2$ (b) $y = x^2 + 3$
(c) $y = -(x - 3)^2 + 6$ (d) $y = -(x + 6)^2 - 3$

69. $R_T = R_1 + R_2 = -0.8t^2 - 7.22t + 734,$
$t = 0, 1, \ldots, 6$

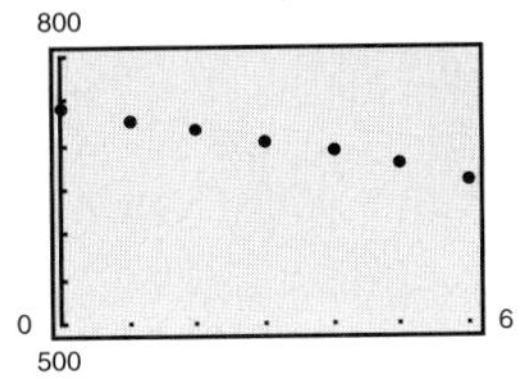

71. (a) $x = \dfrac{1475}{p} - 100$ (b) 47.5 units

73. $C(x(t)) = 2800t + 375$

C is the weekly cost in terms of t hours of manufacturing.

75. (a) $p = \begin{cases} 90, & 0 \le x \le 100 \\ 91 - 0.01x, & 100 < x \le 1600 \\ 75, & x > 1600 \end{cases}$

(b) $P = \begin{cases} 30x, & 0 \le x \le 100 \\ 31x - 0.01x^2, & 100 < x \le 1600 \\ 15x, & x > 1600 \end{cases}$

77. (a) $R = rn = [8 - 0.05(n - 80)]n$

(b)

n	90	100	110	120
R	675	700	715	720

n	130	140	150
R	715	700	675

(c) Answers will vary.

79.

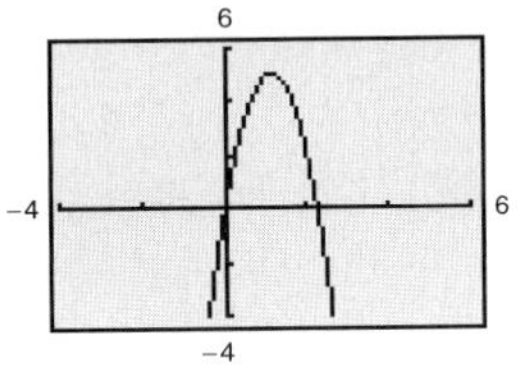

Zeros: $x = 0, \frac{9}{4}$
$f(x)$ is not one-to-one.

81.

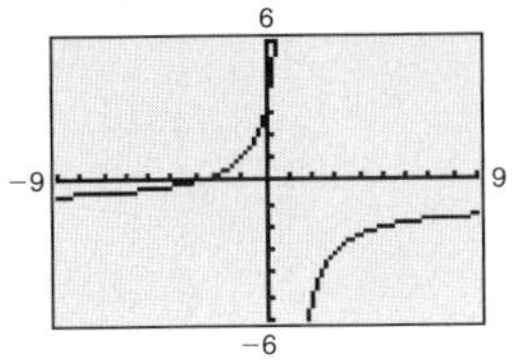

Zero: $t = -3$
$g(t)$ is one-to-one.

83.

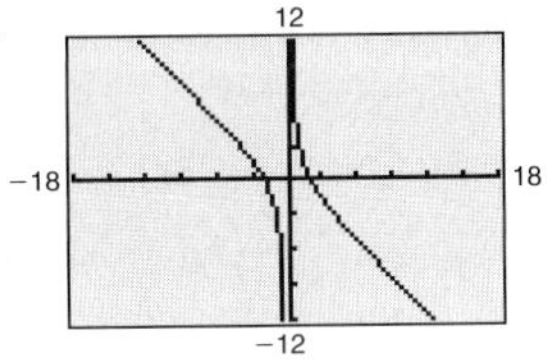

Zeros: $x = \pm 2$
$f(x)$ is not one-to-one.

85.

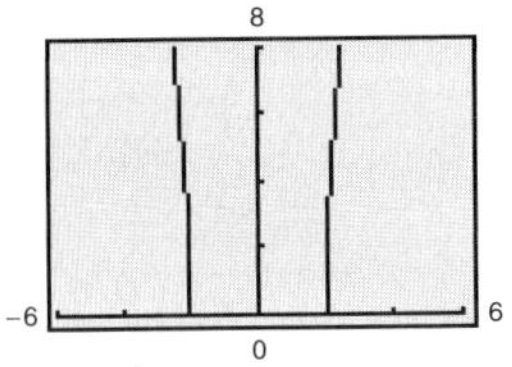

Zero: ± 2
$g(t)$ is not one-to-one.

87. Answers will vary.

SECTION 1.5 *(page 58)*

Prerequisite Review

1. (a) 7 (b) $c^2 - 3c + 3$
(c) $x^2 + 2xh + h^2 - 3x - 3h + 3$

2. (a) -4 (b) 10 (c) $3t^2 + 4$

3. h **4.** 4

5. Domain: $(-\infty, 0) \cup (0, \infty)$
Range: $(-\infty, 0) \cup (0, \infty)$

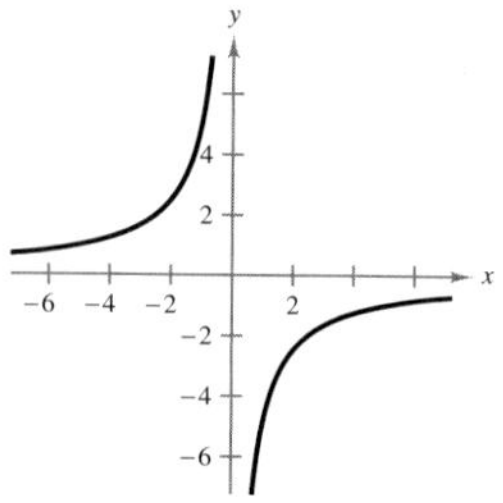

6. Domain: $[-5, 5]$
Range: $[0, 5]$

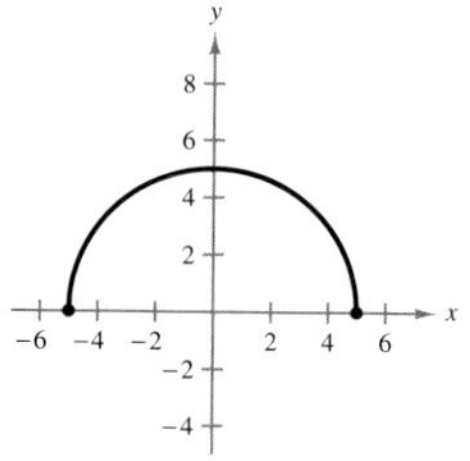

7. Domain: $(-\infty, \infty)$
Range: $[0, \infty)$

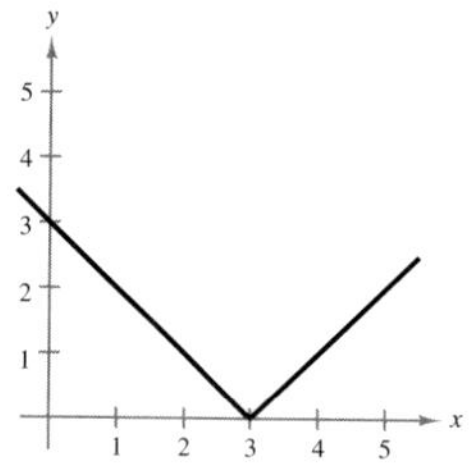

8. Domain: $(-\infty, 0) \cup (0, \infty)$
Range: $-1, 1$

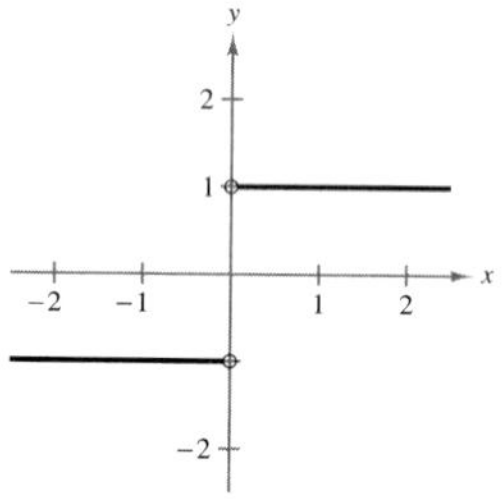

9. y is not a function of x.

10. y is a function of x.

1.

x	1.9	1.99	1.999	2
$f(x)$	13.5	13.95	13.995	?

x	2.001	2.01	2.1
$f(x)$	14.005	14.05	14.5

$\lim_{x\to 2} (5x + 4) = 14$

3.

x	1.9	1.99	1.999	2
$f(x)$	0.2564	0.2506	0.2501	?

x	2.001	2.01	2.1
$f(x)$	0.2499	0.2494	0.2439

$\lim_{x\to 2} \frac{x-2}{x^2-4} = \frac{1}{4}$

5.

x	−0.1	−0.01	−0.001	0
$f(x)$	0.2911	0.2889	0.2887	?

x	0.001	0.01	0.1
$f(x)$	0.2887	0.2884	0.2863

$\lim_{x\to 0} \frac{\sqrt{x+3}-\sqrt{3}}{x} \approx 0.2887$

$\left(\text{The actual limit is } \frac{1}{2\sqrt{3}}.\right)$

7.

x	−0.5	−0.1	−0.01	−0.001	0
$f(x)$	−0.0714	−0.0641	−0.0627	−0.0625	?

$$\lim_{x\to 0^-} \frac{\frac{1}{x+4} - \frac{1}{4}}{x} = -\frac{1}{16}$$

9. (a) 1 (b) 3

11. (a) 1 (b) 3

13. (a) 12 (b) 27 (c) $\frac{1}{3}$

15. (a) 4 (b) 48 (c) 256

17. (a) 1 (b) 1 (c) 1

19. (a) 0 (b) 0 (c) 0

21. (a) 3 (b) −3 (c) Limit does not exist.

23. 16 **25.** −7 **27.** 0 **29.** 2 **31.** −2

33. $-\frac{3}{4}$ **35.** $\frac{35}{9}$ **37.** $\frac{1}{3}$ **39.** $-\frac{1}{20}$ **41.** −2

43. Limit does not exist. **45.** $\frac{1}{10}$ **47.** 12

49. Limit does not exist.

51. −1 **53.** 2 **55.** $\frac{1}{2\sqrt{x+2}}$ **57.** $2t - 5$

59.

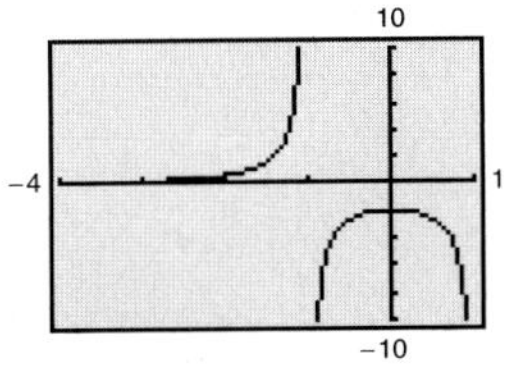

x	0	0.5	0.9	0.99
$f(x)$	−2	−2.67	−10.53	−100.5

x	0.999	0.9999	1
$f(x)$	−1000.5	−10,000.5	Undefined

$-\infty$

61.

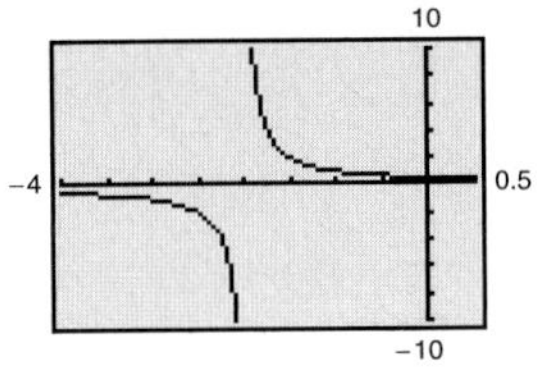

x	−3	−2.5	−2.1	−2.01
$f(x)$	−1	−2	−10	−100

x	−2.001	−2.0001	−2
$f(x)$	−1000	−10,000	Undefined

$-\infty$

63.

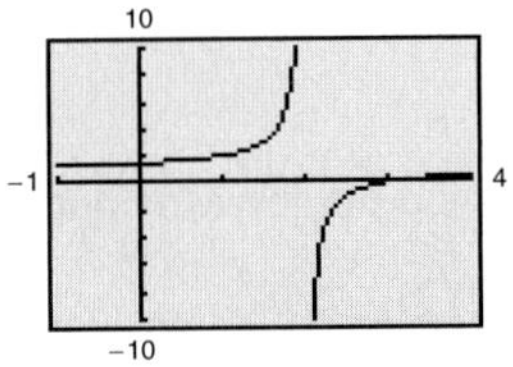

Limit does not exist.

65.

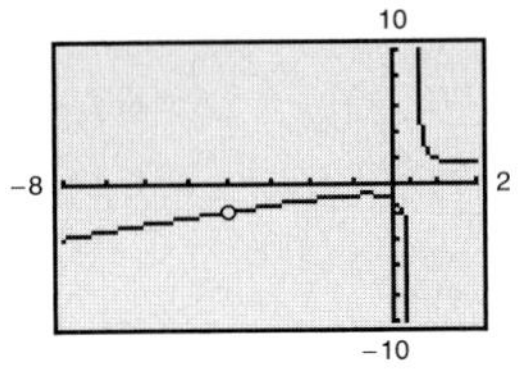

$-\frac{17}{9} \approx -1.8889$

67. (a)

x	−0.01	−0.001	−0.0001	0
$f(x)$	2.732	2.720	2.718	Undefined

x	0.0001	0.001	0.01
$f(x)$	2.718	2.717	2.705

$\lim_{x \to 0} (1 + x)^{1/x} \approx 2.718$

(b)

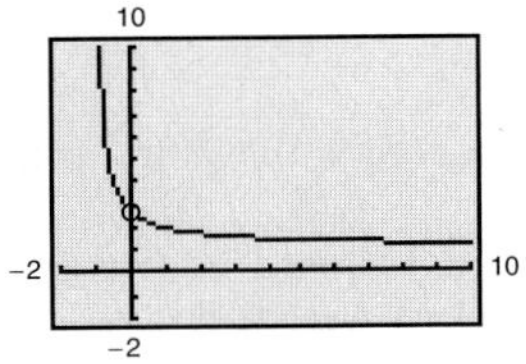

(c) Domain: $(-1, 0) \cup (0, \infty)$

Range: $(1, e) \cup (e, \infty)$

69. (a) \$25,000 (b) 80%

(c) ∞; The cost function increases without bound as x approaches 100 from the left. Therefore, according to the model, it is not possible to remove 100% of the pollutants.

71. (a)

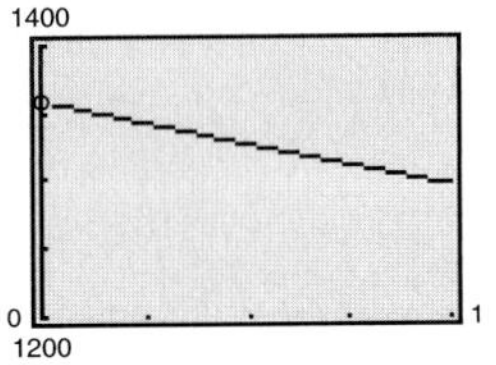

(b) For $x = 0.25$, $A \approx \$1342.53$.

For $x = \frac{1}{365}$, $A \approx \$1358.95$.

(c) $\lim_{x \to 0^+} 500(1 + 0.1x)^{10/x} = 500e \approx \$1359.14$; continuous compounding

SECTION 1.6 *(page 69)*

Prerequisite Review

1. $\dfrac{x + 4}{x - 8}$ **2.** $\dfrac{x + 1}{x - 3}$ **3.** $\dfrac{x + 2}{2(x - 3)}$ **4.** $\dfrac{x - 4}{x - 2}$

5. $x = 0, -7$ **6.** $x = -5, 1$ **7.** $x = -\frac{2}{3}, -2$

8. $x = 0, 3, -8$ **9.** 13 **10.** −1

1. Continuous; The function is a polynomial.

3. Not continuous $(x \neq \pm 2)$

5. Continuous; The rational function's domain is the set of real numbers.

7. Not continuous ($x \neq 3$ and $x \neq 5$)

9. Not continuous ($x \neq \pm 2$)

11. $(-\infty, 0)$ and $(0, \infty)$ **13.** $(-\infty, -1)$ and $(-1, \infty)$

15. $(-\infty, \infty)$ **17.** $(-\infty, -1)$ and $(-1, 1)$ and $(1, \infty)$

19. $(-\infty, \infty)$ **21.** $(-\infty, 4)$ and $(4, 5)$ and $(5, \infty)$

23. Continuous on all intervals $\left(\frac{c}{2}, \frac{c}{2} + \frac{1}{2}\right)$ where c is an integer.

25. $(-\infty, \infty)$ **27.** $(-\infty, 2]$ and $(2, \infty)$

29. $(-\infty, -1)$ and $(-1, \infty)$

31. Continuous on all intervals $(c, c + 1)$, where c is an integer.

33. $(1, \infty)$ **35.** Continuous

37. Nonremovable discontinuity at $x = 2$

39.

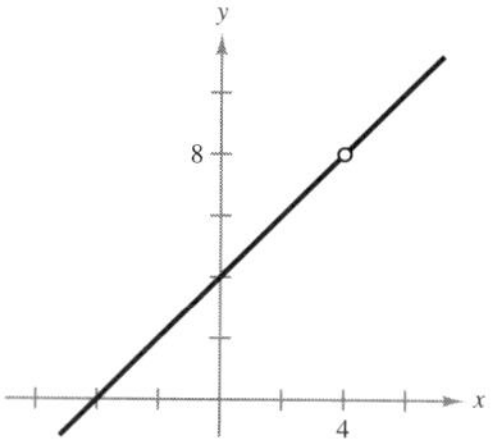

Continuous on $(-\infty, 4)$ and $(4, \infty)$

41.

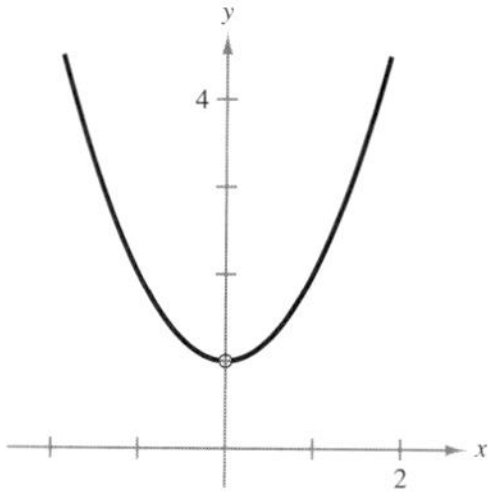

Continuous on $(-\infty, 0)$ and $(0, \infty)$

43.

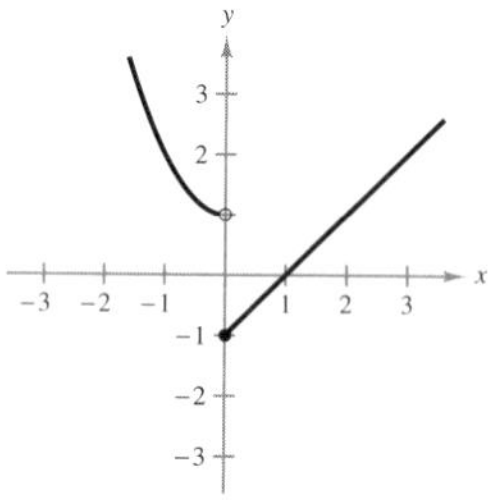

Continuous on $(-\infty, 0)$ and $(0, \infty)$

45. $a = 2$

47.

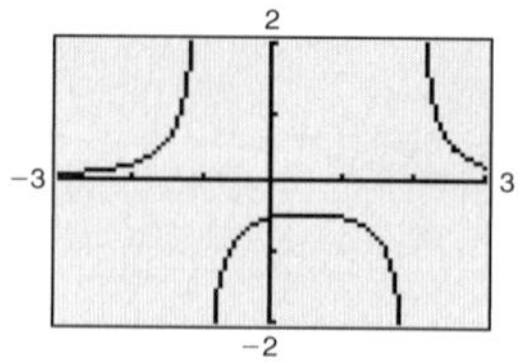

Not continuous at $x = 2$ and $x = -1$

49.

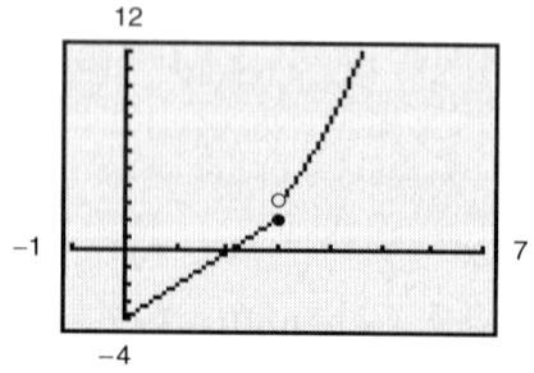

Not continuous at $x = 3$

51.

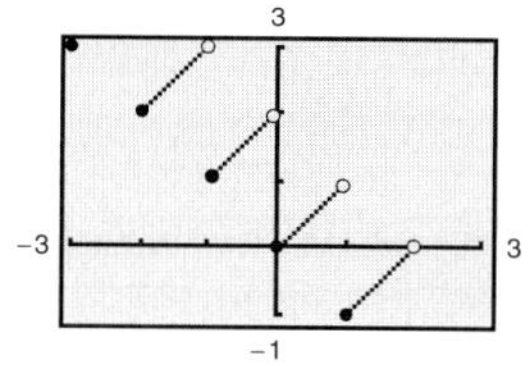

Not continuous at all integers c

53. $(-\infty, \infty)$

55. Continuous on all intervals $\left(\frac{c}{2}, \frac{c+1}{2}\right)$, where c is an integer.

57.

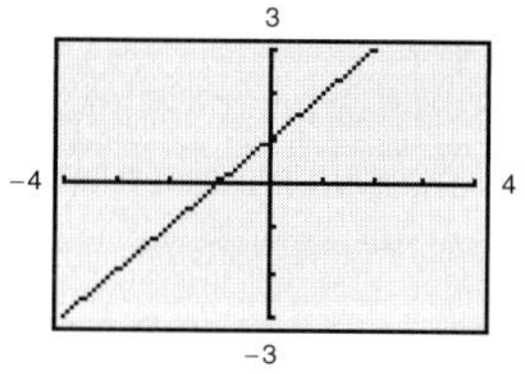

The graph of $f(x) = \dfrac{x^2 + x}{x}$ appears to be continuous on $[-4, 4]$, but f is not continuous at $x = 0$.

59. (a)

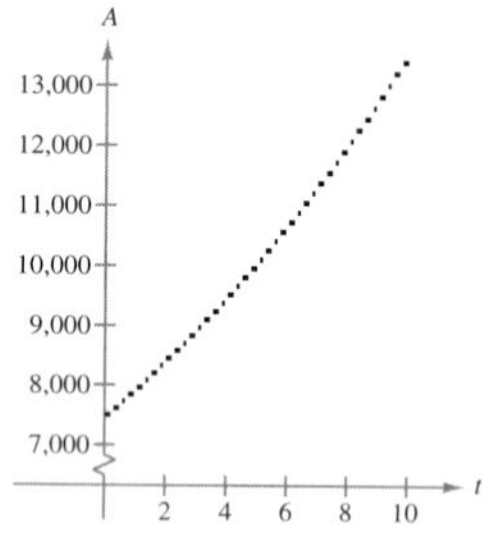

The graph has nonremovable discontinuities at $t = \frac{1}{4}, \frac{1}{2}, \frac{3}{4}, 1, \frac{5}{4}, . . .$

(b) \$11,379.17

61. $C = 9.80 - 2.50[\![1 - x]\!]$

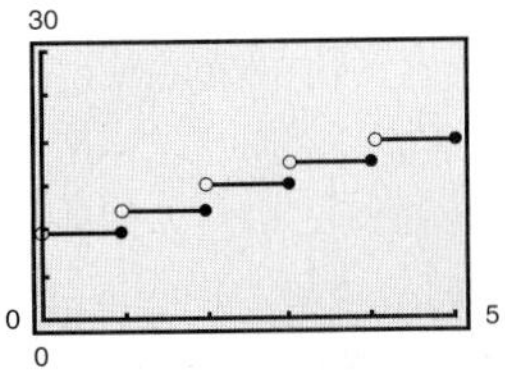

C is not continuous at $x = 1, 2, 3, \ldots$

63. (a)

$$C(t) = \begin{cases} 1.04, & 0 < t \le 2 \\ 1.04 + 0.36[\![t - 1]\!], & t > 2, t \text{ is not an integer.} \\ 1.04 + 0.36(t - 2), & t > 2, t \text{ is an integer.} \end{cases}$$

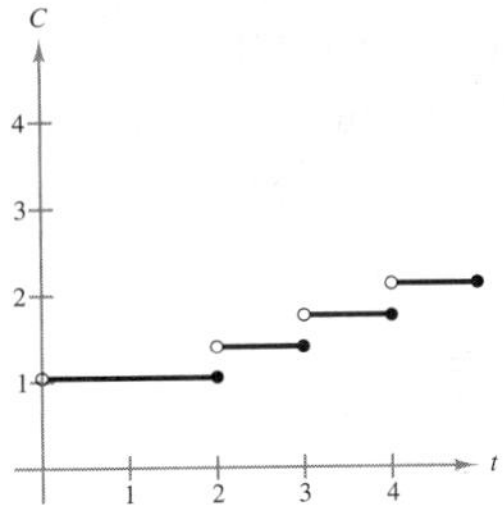

C is not continuous at $t = 2, 3, 4, \ldots$

(b) \$3.56

65. (a)

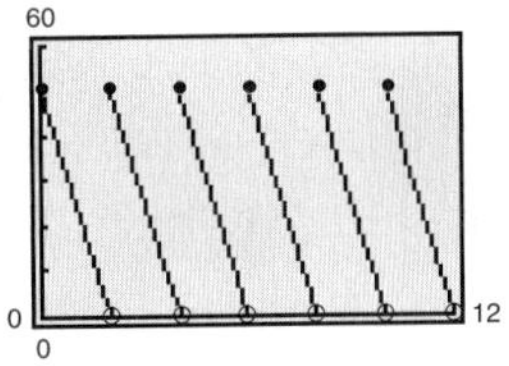

Nonremovable discontinuities at $t = 2, 4, 6, 8, \ldots$

(b) Every 2 months

67.

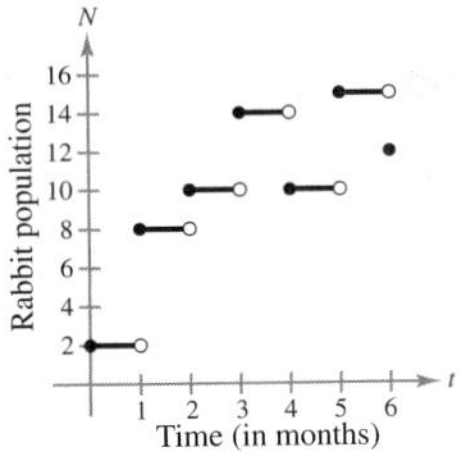

Nonremovable discontinuities at $t = 1, 2, 3, 4, 5$, and 6

REVIEW EXERCISES FOR CHAPTER 1

(page 76)

1. a **2.** c **3.** b **4.** d **5.** $\sqrt{29}$ **7.** $3\sqrt{2}$

9. (7, 4) **11.** (−8, 6)

13. The tallest bars in the graph represent revenues. The middle bars represent costs. The bars on the left in each group represent profits, because $P = R - C$.

15. (4, 7), (5, 8), (8, 10)

17.

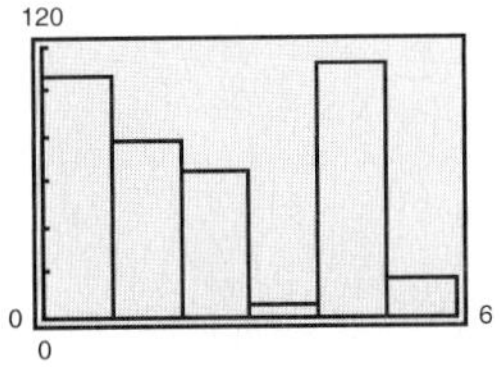

19.

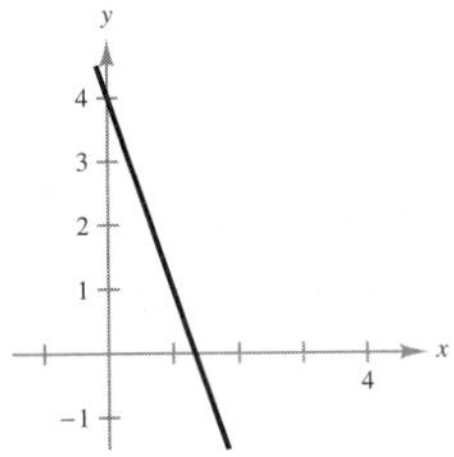

21.

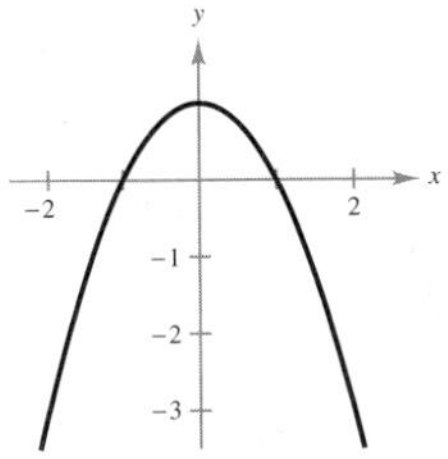

23.

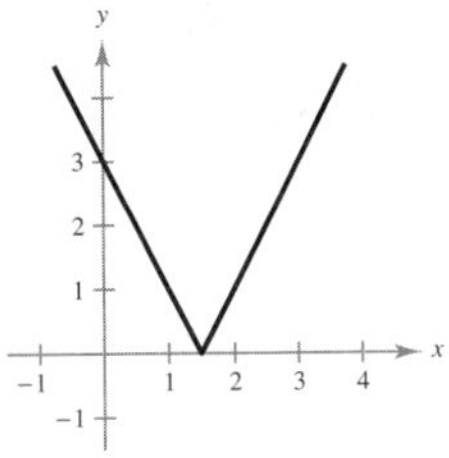

25.

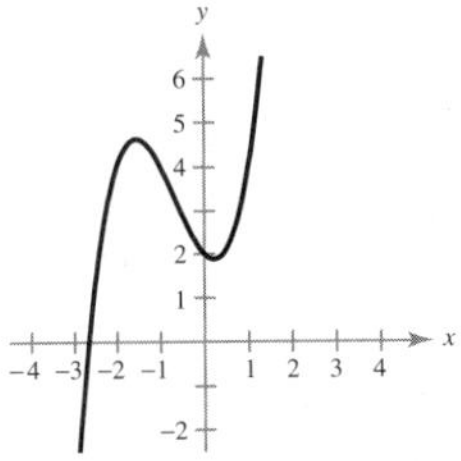

27.

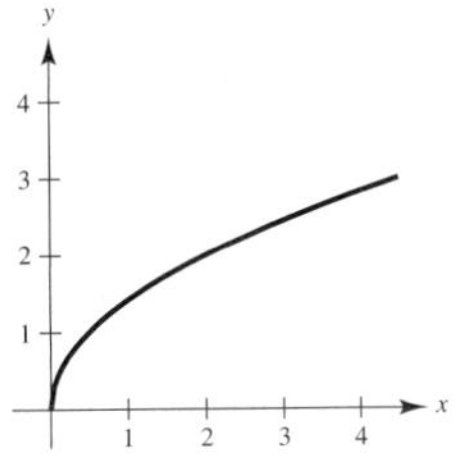

29. (0, 1), (1, 0), (−1, 0) **31.** $(x - 2)^2 + (y + 1)^2 = 73$

33. $(x + 5)^2 + (y + 2)^2 = 36$

Center: $(-5, -2)$

Radius: 6

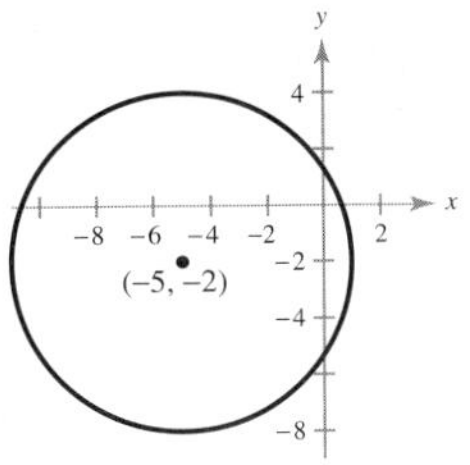

35. (2, 1), (−1, −2) **37.** (0, 0), (1, 1)

39. (a) $C = 6000 + 6.50x$

$R = 13.90x$

(b) ≈ 811 units

41. Slope: -3

y-intercept: $(0, -2)$

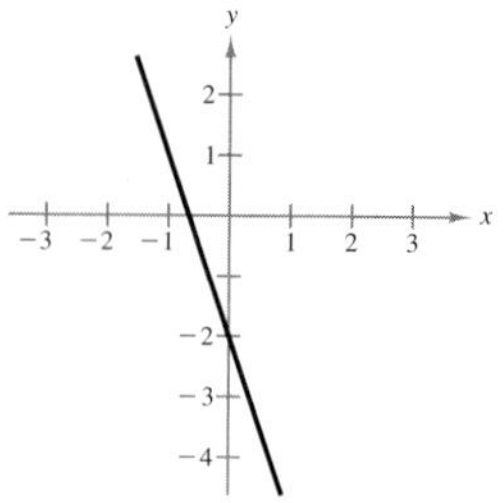

43. Slope: 0 (horizontal line)

y-intercept: $\left(0, -\frac{5}{3}\right)$

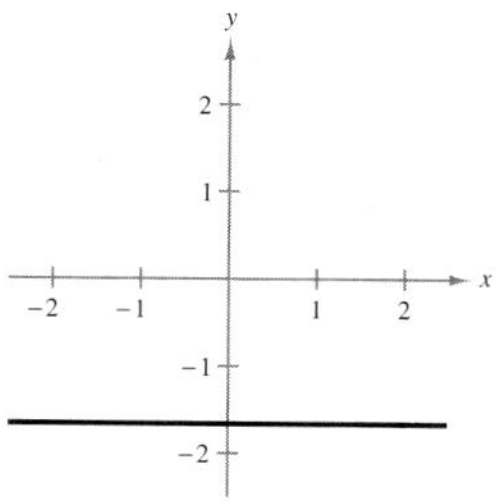

45. Slope: $-\frac{2}{5}$

y-intercept: $(0, -1)$

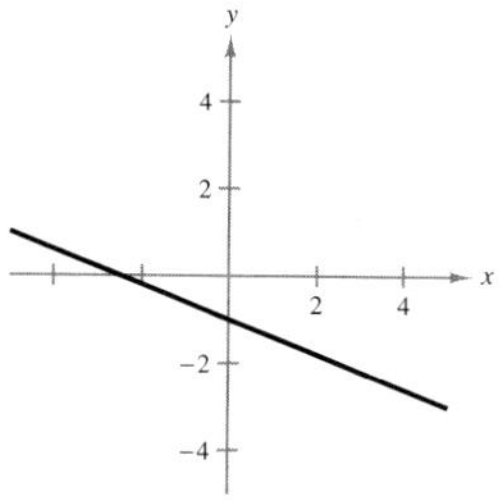

47. $\frac{6}{7}$ **49.** $\frac{20}{21}$

51. $y = -2x + 5$

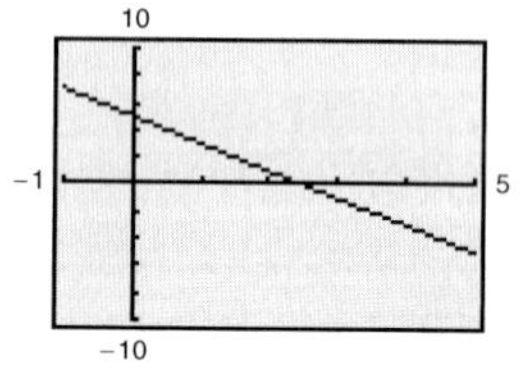

53. (a) $7x - 8y + 69 = 0$ (b) $2x + y = 0$

(c) $2x + y = 0$ (d) $2x + 3y - 12 = 0$

55. (a) $x = -10p + 1070$ (b) 725 units (c) 650 units

57. y is a function of x. **59.** y is not a function of x.

61. (a) 7 (b) $3x + 7$ (c) $10 + 3\Delta x$

63.

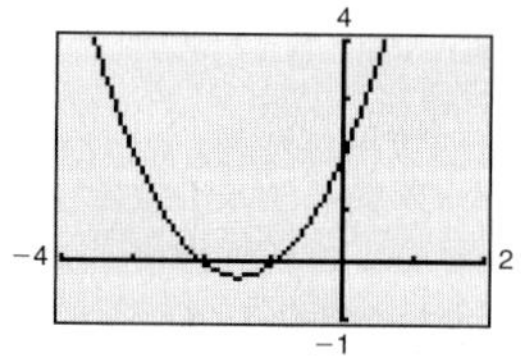

Domain: $(-\infty, \infty)$

Range: $\left[-\frac{1}{4}, \infty\right)$

65.

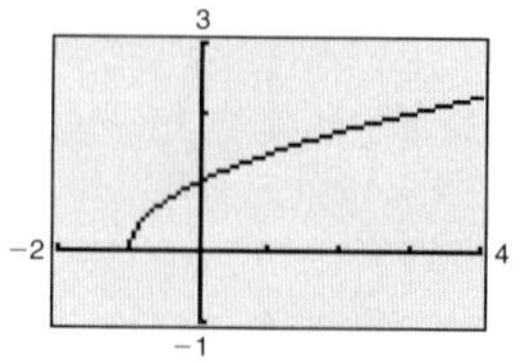

Domain: $[-1, \infty)$

Range: $[0, \infty)$

67.

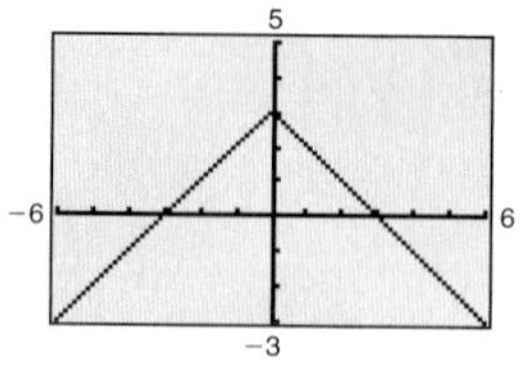

Domain: $(-\infty, \infty)$

Range: $(-\infty, 3]$

69. (a) $x^2 + 2x$ (b) $x^2 - 2x + 2$

(c) $2x^3 - x^2 + 2x - 1$ (d) $\dfrac{1 + x^2}{2x - 1}$

(e) $4x^2 - 4x + 2$ (f) $2x^2 + 1$

71. $f^{-1}(x) = \frac{2}{3}x$

73. $f(x)$ does not have an inverse function.

75. 7 **77.** 49 **79.** $\frac{10}{3}$ **81.** -2

83. $-\frac{1}{4}$ **85.** $-\infty$ **87.** Limit does not exist.

89. $-\frac{1}{16}$ **91.** $3x^2 - 1$ **93.** 0.5774

95. False, limit does not exist.

97. False, limit does not exist.

99. False, limit does not exist.

101. $(-\infty, -4)$ and $(-4, \infty)$ **103.** $(-\infty, -1)$ and $(-1, \infty)$

105. Continuous on all intervals $(c, c + 1)$, where c is an integer

107. $(-\infty, 0)$ and $(0, \infty)$ **109.** $a = 2$

111. (a)

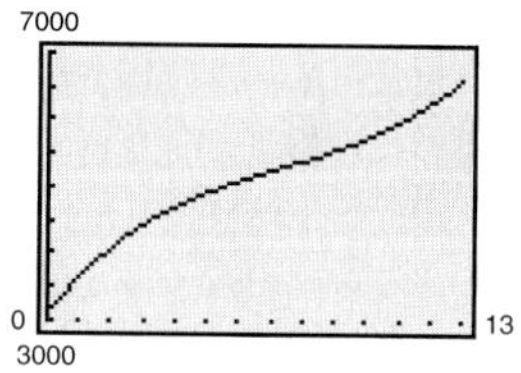

(b)

t	0	1	2	3	4
Debt	3206.3	3598.2	4001.8	4351.0	4643.3
Model	3103.6	3644.1	4078.9	4424.7	4697.8

t	5	6	7	8	9
Debt	4920.6	5181.5	5369.2	5478.2	5605.5
Model	4914.8	5092.2	5246.5	5394.1	5551.7

t	10	11	12	13
Debt	5628.7	5769.9	6198.4	6752.0
Model	5735.6	5962.4	6248.6	6610.8

(c) $10,137.2 billion

113. 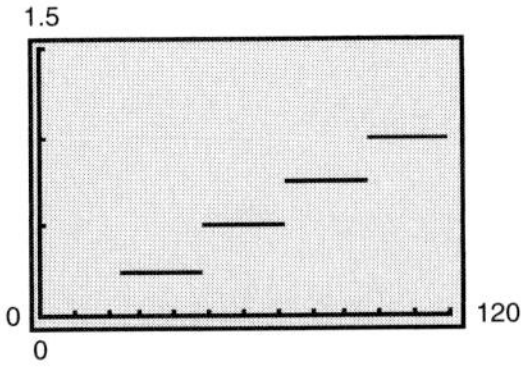

(a) Discontinuous at $x = 24n$, where n is a positive integer

(b) \$15.50

SAMPLE POST-GRAD EXAM QUESTIONS

(page 80)

1. d **2.** d **3.** b **4.** a **5.** e

6. c **7.** e **8.** b **9.** b

CHAPTER 2

SECTION 2.1 *(page 90)*

Prerequisite Review

1. $x = 2$ **2.** $y = 2$ **3.** $2x$ **4.** $3x^2$

5. $\frac{1}{x^2}$ **6.** $2x$ **7.** $(-\infty, 1) \cup (1, \infty)$

8. $(-\infty, \infty)$ **9.** $(-\infty, 0) \cup (0, \infty)$

10. $(-\infty, -4) \cup (-4, 3) \cup (3, \infty)$

1. 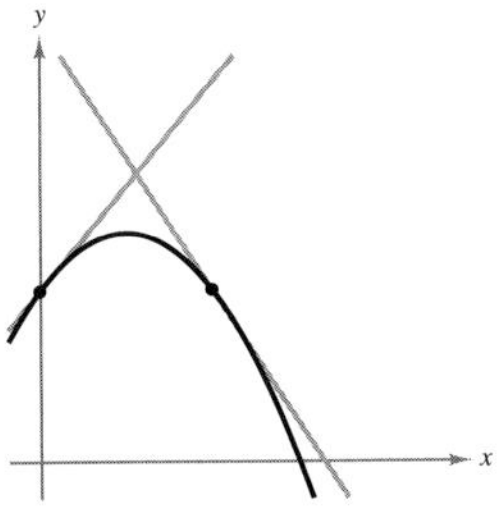

3. 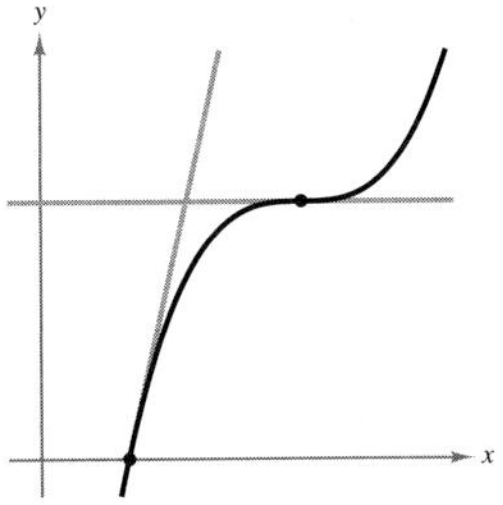

5. $m = 1$ **7.** $m = 0$ **9.** $m = -\frac{1}{3}$

11. 1997: $m \approx 250$
2000: $m \approx 200$
2002: $m \approx 0$

13. $t = 1$: $m \approx 65$
$t = 8$: $m \approx 0$
$t = 12$: $m \approx -1000$

15. $f(x) = 3$
$f(x + \Delta x) = 3$
$f(x + \Delta x) - f(x) = 0$
$\dfrac{f(x + \Delta x) - f(x)}{\Delta x} = 0$
$\lim\limits_{\Delta x \to 0} \dfrac{f(x + \Delta x) - f(x)}{\Delta x} = 0$

17. $f(x) = -5x + 3$
$f(x + \Delta x) = -5x - 5\Delta x + 3$
$f(x + \Delta x) - f(x) = -5\Delta x$
$\dfrac{f(x + \Delta x) - f(x)}{\Delta x} = -5$
$\lim\limits_{\Delta x \to 0} \dfrac{f(x + \Delta x) - f(x)}{\Delta x} = -5$

19. $f(x) = x^2 - 4$
$f(x + \Delta x) = x^2 + 2x\Delta x + (\Delta x)^2 - 4$
$f(x + \Delta x) - f(x) = 2x\Delta x + (\Delta x)^2$
$\dfrac{f(x + \Delta x) - f(x)}{\Delta x} = 2x + \Delta x$
$\lim\limits_{\Delta x \to 0} \dfrac{f(x + \Delta x) - f(x)}{\Delta x} = 2x$

21. $h(t) = \sqrt{t - 1}$
$h(t + \Delta t) = \sqrt{t + \Delta t - 1}$
$h(t + \Delta t) - h(t) = \sqrt{t + \Delta t - 1} - \sqrt{t - 1}$
$\dfrac{h(t + \Delta t) - h(t)}{\Delta t} = \dfrac{1}{\sqrt{t + \Delta t - 1} + \sqrt{t - 1}}$
$\lim\limits_{\Delta t \to 0} \dfrac{h(t + \Delta t) - h(t)}{\Delta t} = \dfrac{1}{2\sqrt{t - 1}}$

23. $f(t) = t^3 - 12t$
$f(t + \Delta t) = t^3 + 3t^2\Delta t + 3t(\Delta t)^2$
$\quad + (\Delta t)^3 - 12t - 12\Delta t$
$f(t + \Delta t) - f(t) = 3t^2\Delta t + 3t(\Delta t)^2 + (\Delta t)^3 - 12\Delta t$
$\dfrac{f(t + \Delta t) - f(t)}{\Delta t} = 3t^2 + 3t\Delta t + (\Delta t)^2 - 12$
$\lim\limits_{\Delta t \to 0} \dfrac{f(t + \Delta t) - f(t)}{\Delta t} = 3t^2 - 12$

25. $f(x) = \dfrac{1}{x + 2}$
$f(x + \Delta x) = \dfrac{1}{x + \Delta x + 2}$
$f(x + \Delta x) - f(x) = \dfrac{-\Delta x}{(x + \Delta x + 2)(x + 2)}$
$\dfrac{f(x + \Delta x) - f(x)}{\Delta x} = \dfrac{-1}{(x + \Delta x + 2)(x + 2)}$
$\lim\limits_{\Delta x \to 0} \dfrac{f(x + \Delta x) - f(x)}{\Delta x} = -\dfrac{1}{(x + 2)^2}$

27. $f'(x) = -2$
$f'(2) = -2$

29. $f'(x) = 0$
$f'(0) = 0$

31. $f'(x) = 2x$
$f'(2) = 4$

33. $f'(x) = 3x^2 - 1$
$f'(2) = 11$

35. $f'(x) = -\dfrac{1}{\sqrt{1 - 2x}}$

$f'(-4) = -\dfrac{1}{3}$

37. $y = 2x - 2$

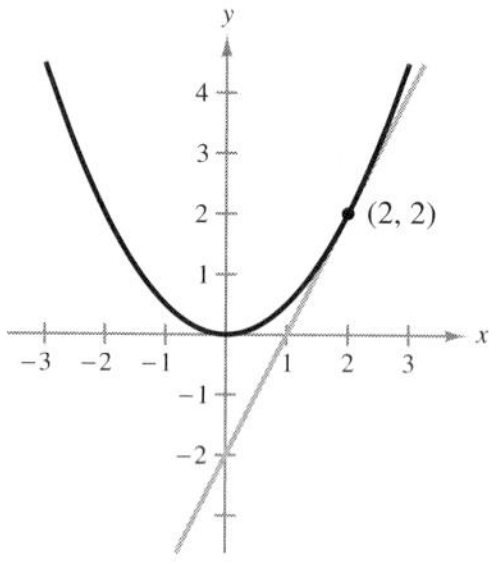

39. $y = -6x - 3$

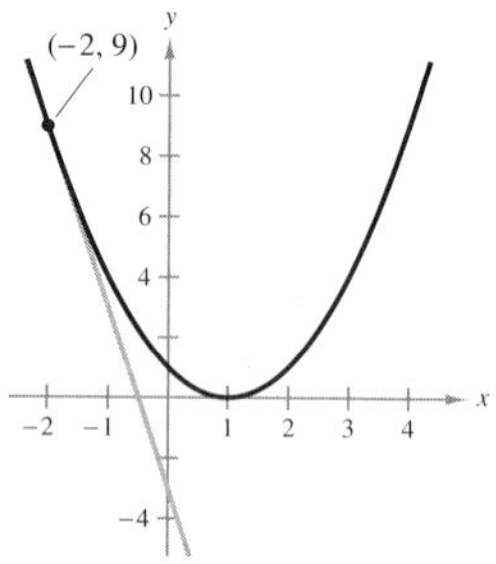

41. $y = \dfrac{x}{4} + 2$

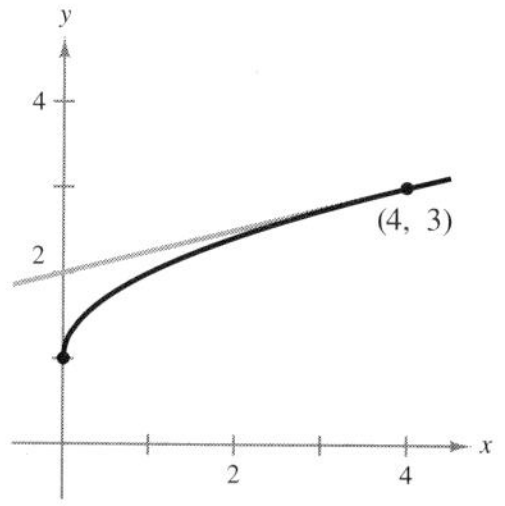

43. $y = -x + 2$

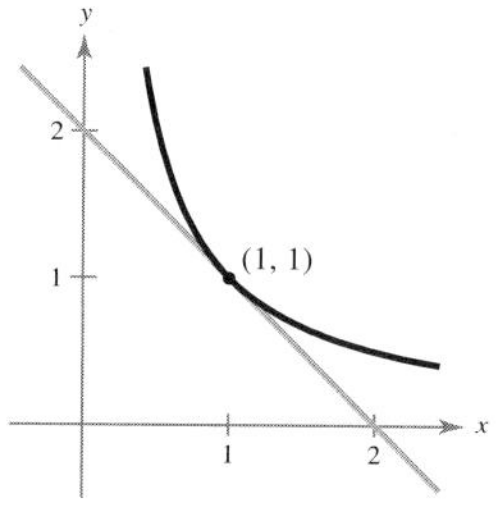

45. $y = -x + 1$

47. $y = -6x + 8$

$y = -6x - 8$

49. $x \neq -3$ (node) **51.** $x \neq 3$ (cusp) **53.** $x > 1$

55. $x \neq 0$ (nonremovable discontinuity)

57.

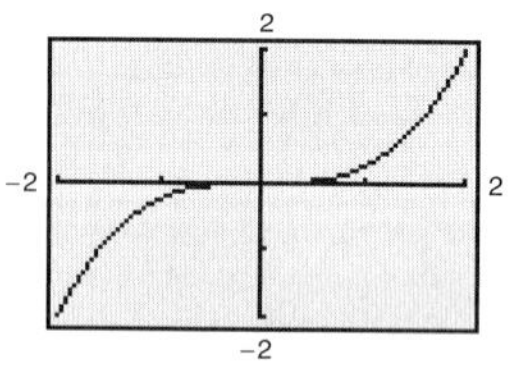

$f'(x) = \frac{3}{4}x^2$

x	-2	$-\frac{3}{2}$	-1	$-\frac{1}{2}$
$f(x)$	-2	-0.8438	-0.25	-0.0313
$f'(x)$	3	1.6875	0.75	0.1875

x	0	$\frac{1}{2}$	1	$\frac{3}{2}$	2
$f(x)$	0	0.0313	0.25	0.8438	2
$f'(x)$	0	0.1875	0.75	1.6875	3

59.

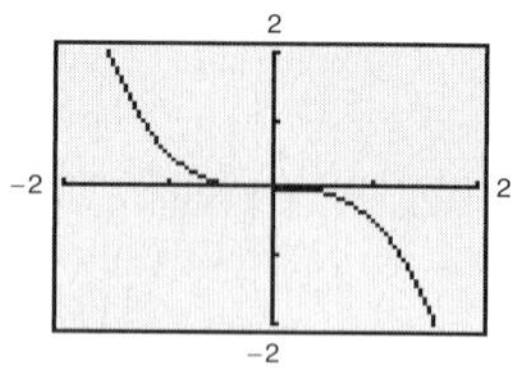

$f'(x) = -\frac{3}{2}x^2$

x	-2	$-\frac{3}{2}$	-1	$-\frac{1}{2}$
$f(x)$	4	1.6875	0.5	0.0625
$f'(x)$	-6	-3.375	-1.5	-0.375

x	0	$\frac{1}{2}$	1	$\frac{3}{2}$	2
$f(x)$	0	-0.0625	-0.5	-1.6875	-4
$f'(x)$	0	-0.375	-1.5	-3.375	-6

61. $f'(x) = 2x - 4$

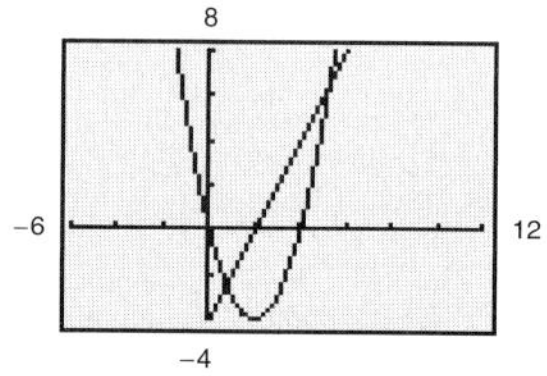

The x-intercept of the derivative indicates a point of horizontal tangency for f.

63. $f'(x) = 3x^2 - 3$

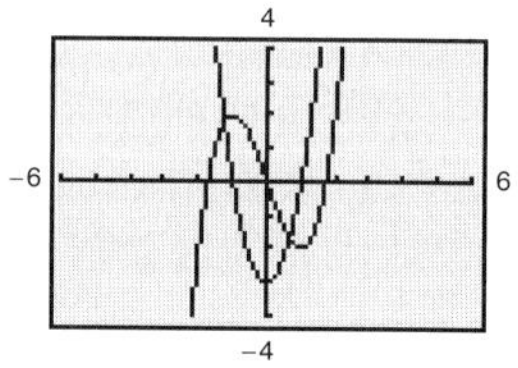

The x-intercepts of the derivative indicate points of horizontal tangency for f.

65. Answers will vary. Sample answer: $f(x) = -x$

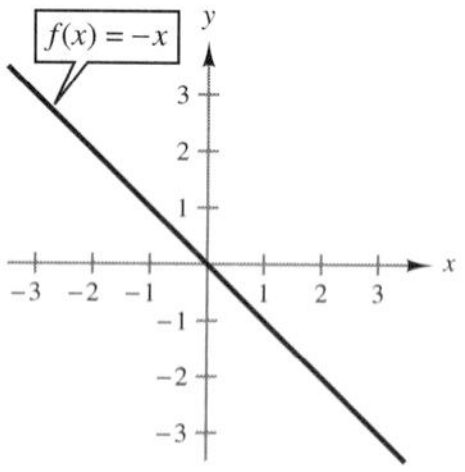

$f'(x) = -1$

67.

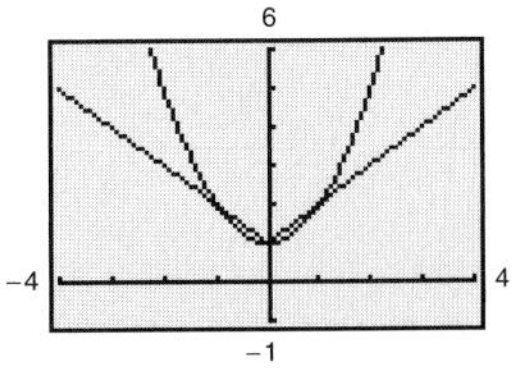

The graph of f is smooth at $(0, 1)$, but the graph of g has a sharp point at $(0, 1)$. The function g is not differentiable at $x = 0$.

69. False. $f(x) = |x|$ is continuous, but not differentiable, at $x = 0$.

71. True

SECTION 2.2 *(page 102)*

Prerequisite Review

1. (a) 8 (b) 16 (c) $\frac{1}{2}$

2. (a) $\frac{1}{36}$ (b) $\frac{1}{32}$ (c) $\frac{1}{64}$

3. $4x(3x^2 + 1)$ **4.** $\frac{3}{2}x^{1/2}(x^{3/2} - 1)$ **5.** $\frac{1}{4x^{3/4}}$

6. $x^2 - \frac{1}{x^{1/2}} + \frac{1}{3x^{2/3}}$ **7.** $0, -\frac{2}{3}$

8. $0, \pm 1$ **9.** $-10, 2$ **10.** $-2, 12$

1. (a) 2 (b) $\frac{1}{2}$ **3.** (a) -1 (b) $-\frac{1}{3}$ **5.** 0

7. 4 **9.** $2x + 4$ **11.** $-6t + 2$ **13.** $3t^2 - 2$

15. $\frac{16}{3}t^{1/3}$ **17.** $\frac{2}{\sqrt{x}}$ **19.** $-\frac{8}{x^3} + 4x$

21. Function: $y = \frac{1}{4x^3}$

Rewrite: $y = \frac{1}{4}x^{-3}$

Differentiate: $y' = \frac{-3}{4}x^{-4}$

Simplify: $y' = -\frac{3}{4x^4}$

23. Function: $y = \frac{1}{(4x)^3}$

Rewrite: $y = \frac{1}{64}x^{-3}$

Differentiate: $y' = -\frac{3}{64}x^{-4}$

Simplify: $y' = -\frac{3}{64x^4}$

25. Function: $y = \frac{\sqrt{x}}{x}$

Rewrite: $y = x^{-1/2}$

Differentiate: $y' = -\frac{1}{2}x^{-3/2}$

Simplify: $y' = -\frac{1}{2x^{3/2}}$

27. -1 **29.** -2 **31.** 4 **33.** $2x + \frac{4}{x^2} + \frac{6}{x^3}$

35. $2x - 2 + \frac{8}{x^5}$ **37.** $3x^2 + 1$ **39.** $6x^2 + 16x - 1$

41. $\frac{2x^3 - 6}{x^3}$ **43.** $\frac{4x^3 - 2x - 10}{x^3}$ **45.** $\frac{4}{5x^{1/5}} + 1$

47. $y = 2x - 2$ **49.** $y = \frac{8}{15}x + \frac{22}{15}$

51. $(0, -1), \left(-\frac{\sqrt{6}}{2}, \frac{5}{4}\right), \left(\frac{\sqrt{6}}{2}, \frac{5}{4}\right)$ **53.** $(-5, -12.5)$

55. (a)

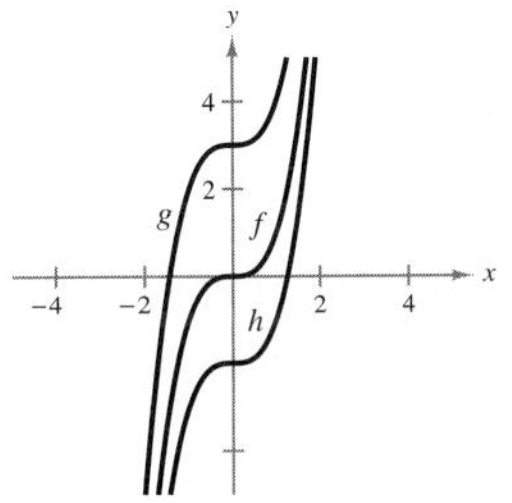

(b) $f'(1) = g'(1)$
$= h'(1)$
$= 3$

(c)

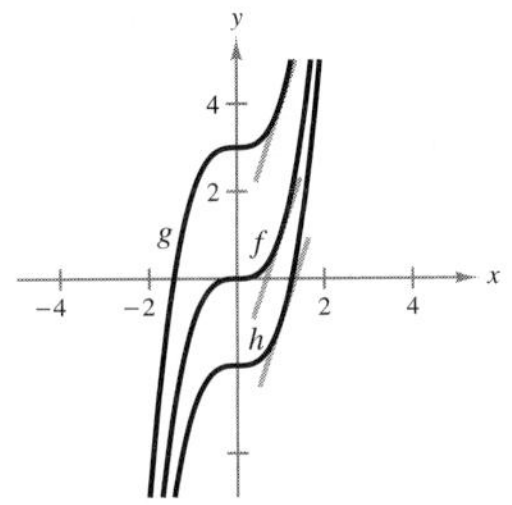

57. (a) 3 (b) 6 (c) -3 (d) 6

59. (a) 1998: 401.1
2001: -79.1
2003: 516.2
(b) The results are similar.
(c) Millions of dollars per year per year

61. $p = 0.40x - 250$
$p' = 0.40$

63. 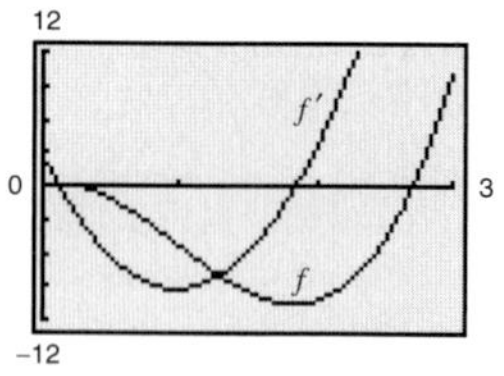

(0.11, 0.14), (1.84, −10.49)

65. False. Let $f(x) = x$ and $g(x) = x + 1$.

SECTION 2.3 *(page 116)*

Prerequisite Review

1. 3 **2.** −7 **3.** $y' = 8x - 2$

4. $y' = -9t^2 + 4t$ **5.** $s' = -32t + 24$

6. $y' = -32x + 54$ **7.** $A' = -\frac{3}{5}r^2 + \frac{3}{5}r + \frac{1}{2}$

8. $y' = 2x^2 - 4x + 7$ **9.** $y' = 12 - \dfrac{x}{2500}$

10. $y' = 74 - \dfrac{3x^2}{10{,}000}$

1. (a) \$11 billion per year (b) \$7 billion per year
(c) \$6 billion per year (d) \$16 billion per year
(e) \$9.5 billion per year (f) \$10.4 billion per year

3. 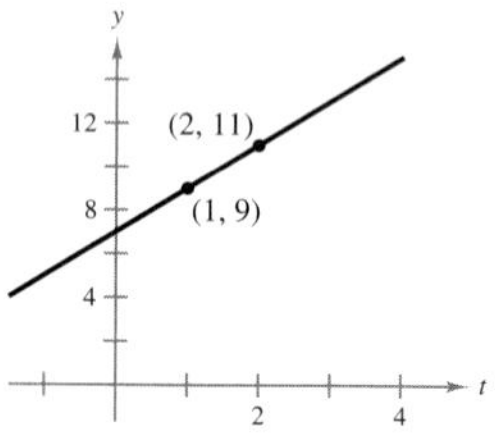

Average rate: 2

Instantaneous rates: $f'(1) = 2, f'(2) = 2$

5. 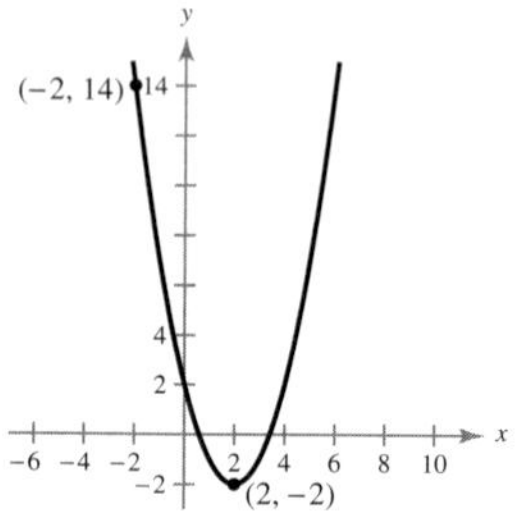

Average rate: −4

Instantaneous rates: $h'(-2) = -8, h'(2) = 0$

7. 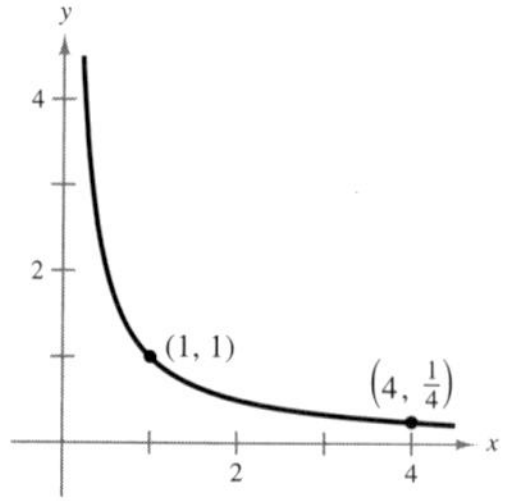

Average rate: $-\frac{1}{4}$

Instantaneous rates: $f'(1) = -1, f'(4) = -\frac{1}{16}$

9. 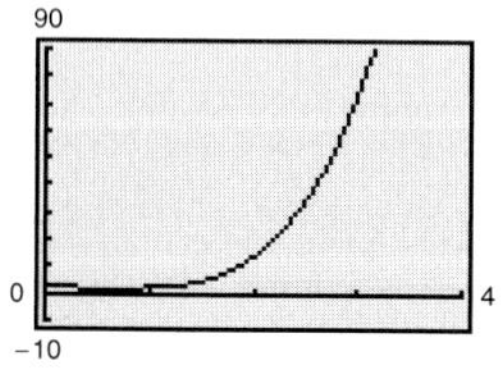

Average rate: 36

Instantaneous rates: $g'(1) = 2, g'(3) = 102$

11. (a) −500

The number of visitors to the park is decreasing at an average rate of 500 hundred thousand people per month from September to December.

(b) Answers will vary. The instantaneous rate of change at $t = 8$ is approximately 0.

13. (a) Average rate: $\frac{11}{27}$

Instantaneous rates: $E'(0) = \frac{1}{3}, E'(1) = \frac{4}{9}$

(b) Average rate: $\frac{11}{27}$

Instantaneous rates: $E'(1) = \frac{4}{9}, E'(2) = \frac{1}{3}$

(c) Average rate: $\frac{5}{27}$

Instantaneous rates: $E'(2) = \frac{1}{3}, E'(3) = 0$

(d) Average rate: $-\frac{7}{27}$

Instantaneous rates: $E'(3) = 0, E'(4) = -\frac{5}{9}$

15. (a) −80 feet per second

(b) $s'(2) = -64$ feet per second,
$s'(3) = -96$ feet per second

(c) $\dfrac{\sqrt{555}}{4} \approx 5.89$ seconds

(d) $-8\sqrt{555} \approx -188.5$ feet per second

17. 1.47 dollars **19.** $470 - 0.5x$ dollars, $0 \le x \le 940$

21. $50 - x$ dollars **23.** $-18x^2 + 16x + 200$ dollars

25. $-4x + 72$ dollars **27.** $-0.0005x + 12.2$ dollars

29. (a) \$0.58 (b) \$0.60

(c) The results are nearly the same.

31. (a) \$4.95 (b) \$5.00

(c) The results are nearly the same.

33. (a)

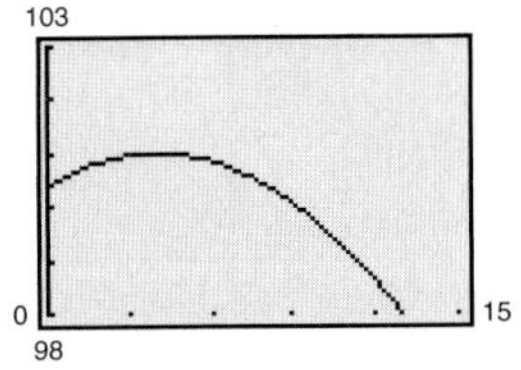

(b) For $t < 4$, increasing; for $t > 4$, decreasing; shows when fever is going up and down.

(c) $T(0) = 100.4$

$T(4) = 101$

$T(8) = 100.4$

$T(12) = 98.6$

(d) $T'(t) = -0.075t + 0.3$

The rate of change of temperature

(e) $T'(0) = 0.3$

$T'(4) = 0$

$T'(8) = -0.3$

$T'(12) = -0.6$

35. (a) $R = 5x - 0.001x^2$

(b) $P = -0.001x^2 + 3.5x - 35$

(c)

x	600	1200	1800	2400	3000
dR/dx	3.8	2.6	1.4	0.2	-1
dP/dx	2.3	1.1	-0.1	-1.3	-2.5
P	1705	2725	3025	2605	1465

37. (a) $P = -0.0025x^2 + 1.7x - 20$

(b)

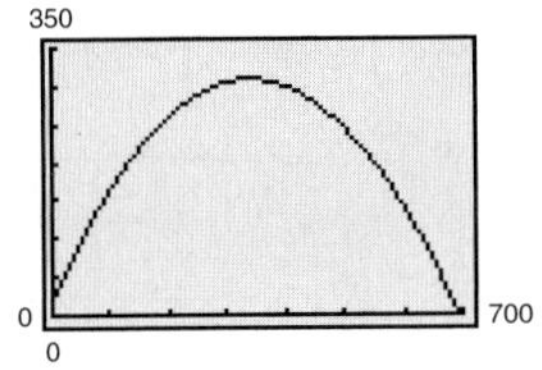

When $x = 200$, slope is positive.

When $x = 400$, slope is negative.

(c) $P'(200) = 0.7$

$P'(400) = -0.3$

39. (a) $P = -\dfrac{1}{6000}x^2 + 11.8x - 85{,}000$

(b)

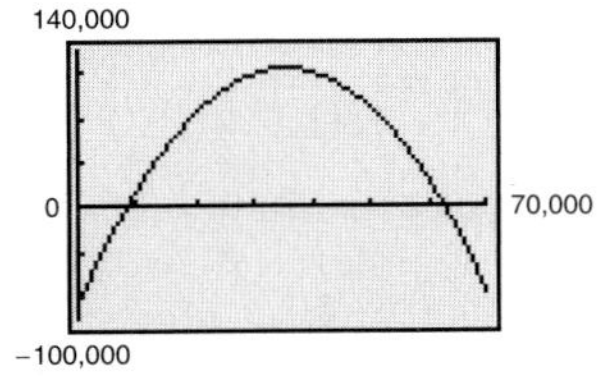

When $x = 18{,}000$, slope is positive.

When $x = 36{,}000$, slope is negative.

(c) $P'(18{,}000) = 5.8$ dollars

$P'(36{,}000) = -0.2$ dollars

41. (a) \$0.33 per unit (b) \$0.13 per unit

(c) \$0 per unit (d) \$−0.08 per unit

$p'(2500) = 0$ indicates that $x = 2500$ is the optimal value of x. So, $p = \dfrac{50}{\sqrt{x}} = \dfrac{50}{\sqrt{2500}} = \$1.00$.

43. (a) ≈ 4.7 miles per year

(b) ≈ 3.5 miles per year

(c) ≈ 3.85 miles per year

45. (a) The rate of change of the number of gallons of gasoline sold when the price is \$1.479 per gallon.

(b) In general, the rate of change when $p = 1.479$ should be negative.

47.

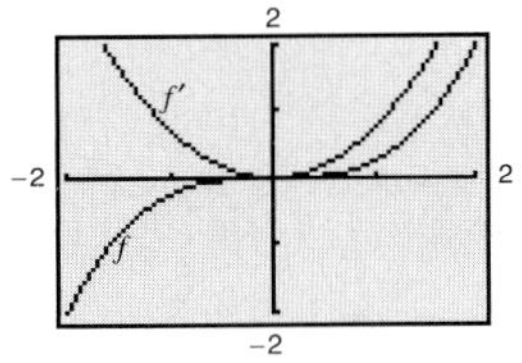

f has a horizontal tangent at $x = 0$.

49. The population in each phase is increasing. During the acceleration phase the population's growth is the greatest, so the slopes of the tangent lines are greater than the slopes of the tangent lines during the lag phase and the deceleration phase. Possible reasons for the changing rates could be seasonal growth and food supplies.

SECTION 2.4 *(page 128)*

Prerequisite Review

1. $2(3x^2 + 7x + 1)$ **2.** $4x^2(6 - 5x^2)$

3. $8x^2(x^2 + 2)^3 + (x^2 + 4)$

4. $(2x)(2x + 1)[2x + (2x + 1)^3]$

5. $\dfrac{23}{(2x + 7)^2}$ **6.** $-\dfrac{x^2 + 8x + 4}{(x^2 - 4)^2}$

7. $-\dfrac{2(x^2 + x - 1)}{(x^2 + 1)^2}$ 8. $\dfrac{4(3x^4 - x^3 + 1)}{(1 - x^4)^2}$

9. $\dfrac{4x^3 - 3x^2 + 3}{x^2}$ 10. $\dfrac{x^2 - 2x + 4}{(x - 1)^2}$

11. 11 12. 0 13. $-\frac{1}{4}$ 14. $\frac{17}{4}$

1. $f'(1) = 13$ 3. $f'(0) = 0$ 5. $g'(4) = 11$

7. $h'(6) = -5$ 9. $f'(3) = \frac{3}{4}$ 11. $g'(6) = -11$

13. $f'(1) = \frac{2}{5}$

15. Function: $y = \dfrac{x^2 + 2x}{x}$

Rewrite: $y = x + 2,\ x \neq 0$

Differentiate: $y' = 1,\ x \neq 0$

Simplify: $y' = 1,\ x \neq 0$

17. Function: $y = \dfrac{7}{3x^3}$

Rewrite: $y = \dfrac{7}{3}x^{-3}$

Differentiate: $y' = -7x^{-4}$

Simplify: $y' = -\dfrac{7}{x^4}$

19. Function: $y = \dfrac{4x^2 - 3x}{8\sqrt{x}}$

Rewrite: $y = \dfrac{1}{2}x^{3/2} - \dfrac{3}{8}x^{1/2}$

Differentiate: $y' = \dfrac{3}{4}x^{1/2} - \dfrac{3}{16}x^{-1/2}$

Simplify: $y' = \dfrac{3}{4}\sqrt{x} - \dfrac{3}{16\sqrt{x}}$

21. Function: $y = \dfrac{x^2 - 4x + 3}{x - 1}$

Rewrite: $y = x - 3,\ x \neq 1$

Differentiate: $y' = 1,\ x \neq 1$

Simplify: $y' = 1,\ x \neq 1$

23. $10x^4 + 12x^3 - 3x^2 - 18x - 15$

25. $12t^2(2t^3 - 1)$

27. $\dfrac{5}{6x^{1/6}} + \dfrac{1}{x^{2/3}}$ 29. $-\dfrac{5}{(2x - 3)^2}$

31. $\dfrac{2}{(x + 1)^2},\ x \neq 1$ 33. $\dfrac{x^2 + 2x - 1}{(x + 1)^2}$

35. $\dfrac{3s^2 - 2s - 5}{2s^{3/2}}$ 37. $\dfrac{2x^3 + 11x^2 - 8x - 17}{(x + 4)^2}$

39. $y = 5x - 2$

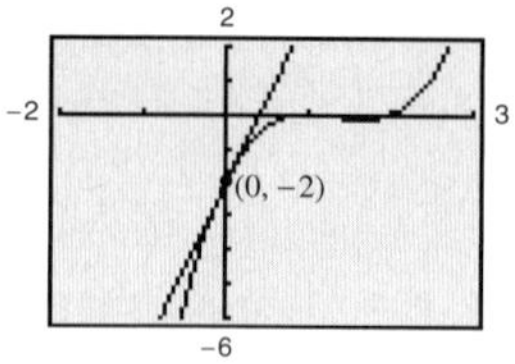

41. $y = \frac{3}{4}x - \frac{5}{4}$

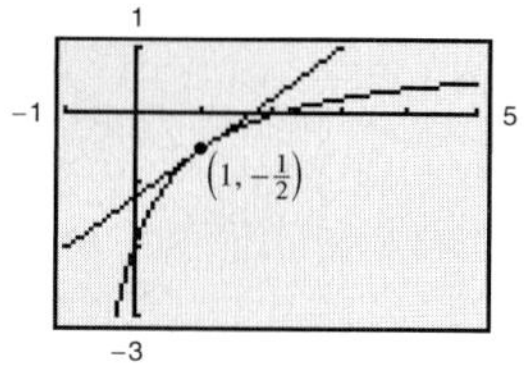

43. $y = -16x - 5$

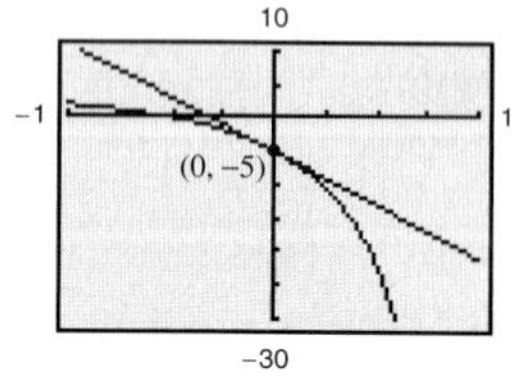

45. (0, 0), (2, 4) 47. (0, 0), $(\sqrt[3]{-4}, -2.117)$

49.

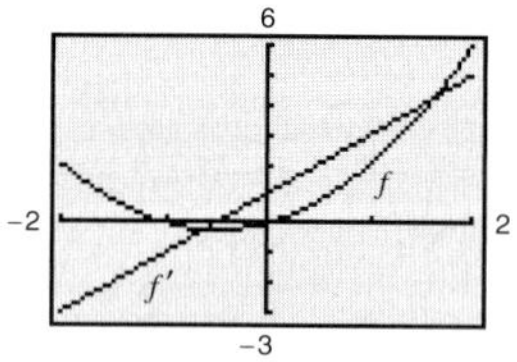

51.

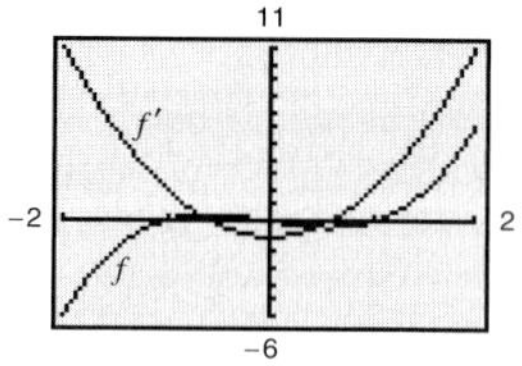

53. -1.87 55. (a) -0.480 (b) 0.120 (c) 0.015

57. 31.55 bacteria per hour

59. (a) $p = \dfrac{4000}{\sqrt{x}}$ (b) $C = 250x + 10{,}000$

(c) $P = 4000\sqrt{x} - 250x - 10{,}000$

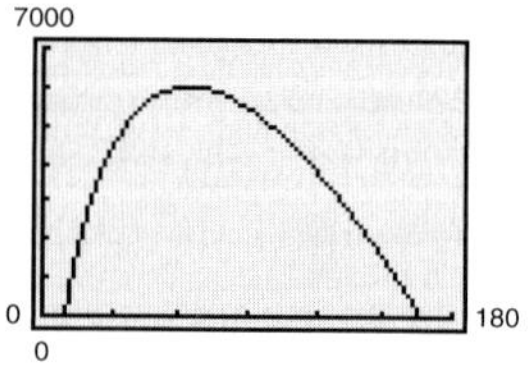

$500 per unit

61. (a)

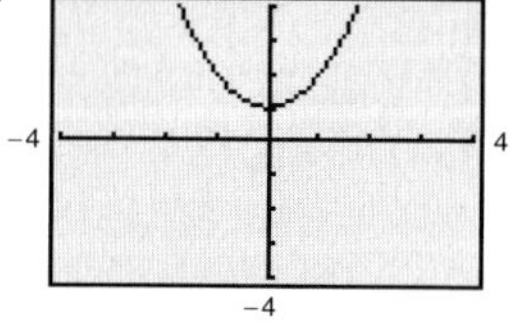

(b)

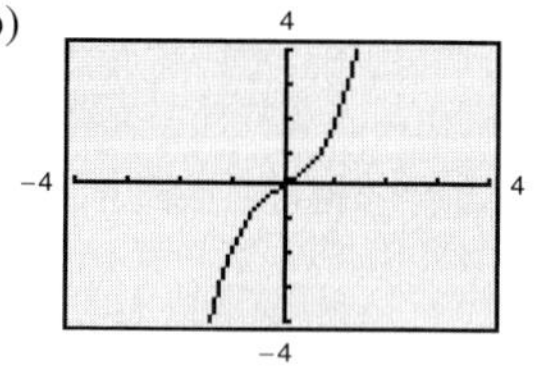

(c)

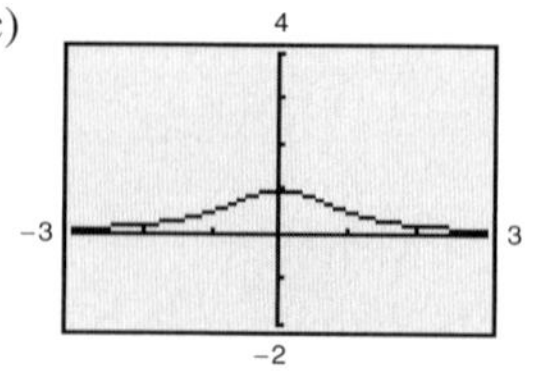

The graph of (c) would most likely represent a demand function. As the number of units increases, demand is likely to decrease, not increase as in (a) and (b).

63. (a) -38.125

(b) -10.37

(c) -3.80

Increasing the order size reduces the cost per item.

65. $\dfrac{dP}{dt} = -\dfrac{1.25(40{,}900 - 8360t + 403t^2)}{(2500 - 515t + 28t^2)^2}$

$P'(5) = -0.0294$

$P'(7) = -0.0373$

$P'(9) = 0.1199$

$P'(11) = 0.0577$

The rate of change in price at year t

SECTION 2.5 *(page 138)*

Prerequisite Review

1. $(1 - 5x)^{2/5}$ **2.** $(2x - 1)^{3/4}$

3. $(4x^2 + 1)^{-1/2}$ **4.** $(x - 6)^{-1/3}$

5. $x^{1/2}(1 - 2x)^{-1/3}$ **6.** $(2x)^{-1}(3 - 7x)^{3/2}$

7. $(x - 2)(3x^2 + 5)$ **8.** $(x - 1)(5\sqrt{x} - 1)$

9. $(x^2 + 1)^2(4 - x - x^3)$

10. $(3 - x^2)(x - 1)(x^2 + x + 1)$

$y = f(g(x))$	$u = g(x)$	$y = f(u)$
1. $y = (6x - 5)^4$	$u = 6x - 5$	$y = u^4$
3. $y = (4 - x^2)^{-1}$	$u = 4 - x^2$	$y = u^{-1}$
5. $y = \sqrt{5x - 2}$	$u = 5x - 2$	$y = \sqrt{u}$
7. $y = \dfrac{1}{3x + 1}$	$u = 3x + 1$	$y = u^{-1}$

9. c **11.** b **13.** a **15.** c **17.** $6(2x - 7)^2$

19. $-6(4 - 2x)^2$ **21.** $6x(6 - x^2)(2 - x^2)$

23. $\dfrac{4x}{3(x^2 - 9)^{1/3}}$ **25.** $\dfrac{1}{2\sqrt{t + 1}}$ **27.** $\dfrac{4t + 5}{2\sqrt{2t^2 + 5t + 2}}$

29. $\dfrac{6x}{(9x^2 + 4)^{2/3}}$ **31.** $\dfrac{27}{4(2 - 9x)^{3/4}}$ **33.** $\dfrac{4x^2}{(4 - x^3)^{7/3}}$

35. $y = 216x - 378$ **37.** $y = \frac{8}{3}x - \frac{7}{3}$

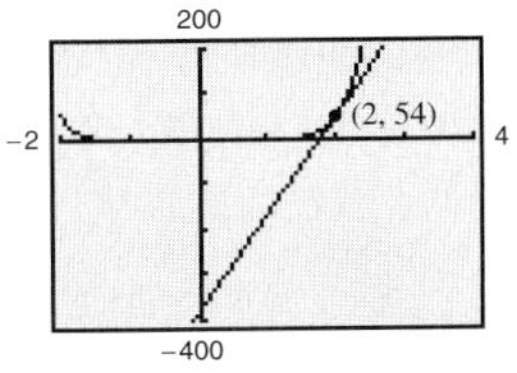

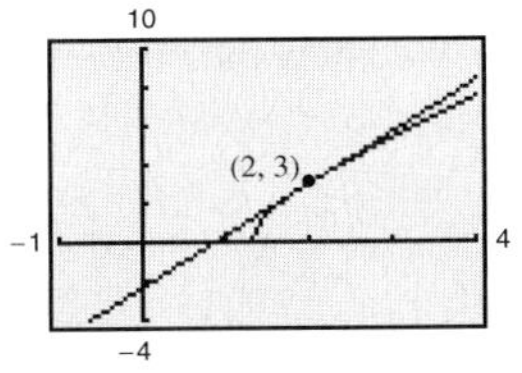

39. $y = x - 1$

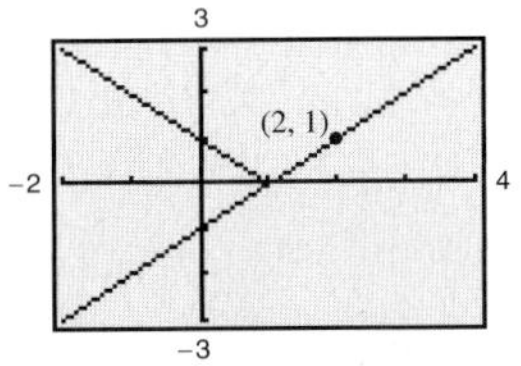

41. $f'(x) = \dfrac{1 - 3x^2 - 4x^{3/2}}{2\sqrt{x}(x^2 + 1)^2}$

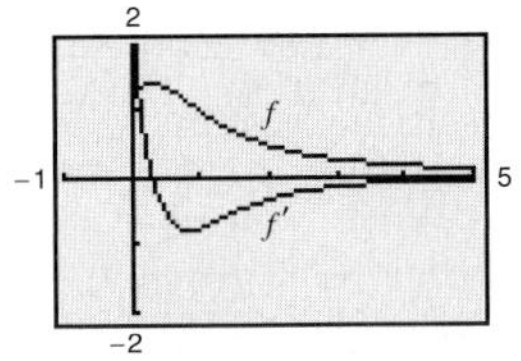

The zero of $f'(x)$ corresponds to the point on the graph of $f(x)$ where the tangent line is horizontal.

43. $f'(x) = -\dfrac{\sqrt{(x + 1)/x}}{2x(x + 1)}$

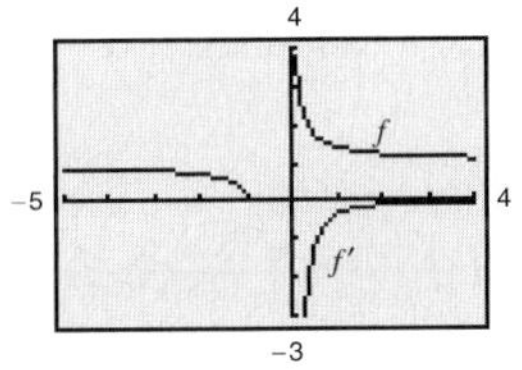

$f'(x)$ has no zeros.

45. $-\dfrac{1}{(x - 2)^2}$ **47.** $\dfrac{8}{(t + 2)^3}$ **49.** $-\dfrac{2(2x - 3)}{(x^2 - 3x)^3}$

51. $-\dfrac{2t}{(t^2 - 2)^2}$ **53.** $27(x - 3)^2(4x - 3)$

55. $\dfrac{3(x + 1)}{\sqrt{2x + 3}}$ **57.** $\dfrac{t(5t - 8)}{2\sqrt{t - 2}}$

59. $-\dfrac{3}{4x^{3/2}\sqrt{3 - 2x}}$ **61.** $\dfrac{x}{\sqrt{x^2 + 1}} - \dfrac{x}{\sqrt{x^2 - 1}}$

63. $\dfrac{2(6 - 5x)(5x^2 - 12x + 5)}{(x^2 - 1)^3}$

65. $y = \frac{8}{3}t + 4$ **67.** $y = -6t - 14$

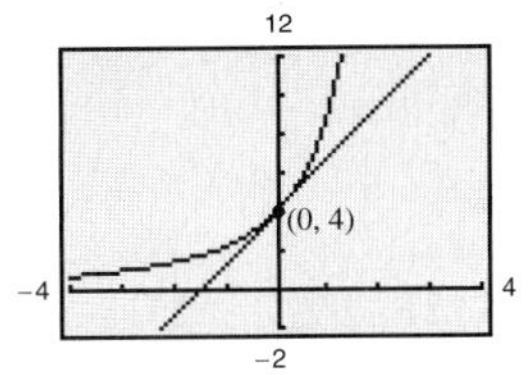

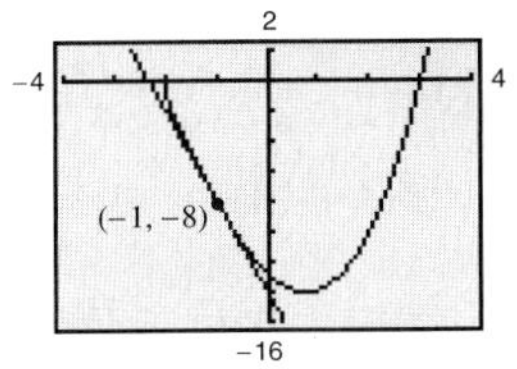

69. $y = -2x + 7$

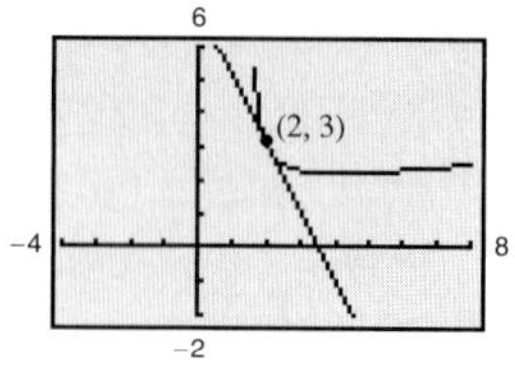

71. (a) \$74.00 per 1% (b) \$81.59 per 1%
(c) \$89.94 per 1%

73.

t	0	1	2	3	4
$\frac{dN}{dt}$	0	177.78	44.44	10.82	3.29

The rate of growth of N is decreasing.

75. (a) $V = \dfrac{10{,}000}{\sqrt[3]{t + 1}}$

(b) $-\$1322.83$ per year

(c) $-\$524.97$ per year

77. False. $y' = \frac{1}{2}(1 - x)^{-1/2}(-1) = -\frac{1}{2}(1 - x)^{-1/2}$

SECTION 2.6 *(page 145)*

Prerequisite Review

1. $t = 0, \frac{3}{2}$ **2.** $t = -2, 7$ **3.** $t = -2, 10$

4. $t = \dfrac{9 \pm 3\sqrt{10{,}249}}{32}$ **5.** $\dfrac{dy}{dx} = 6x^2 + 14x$

6. $\dfrac{dy}{dx} = 8x^3 + 18x^2 - 10x - 15$

7. $\dfrac{dy}{dx} = \dfrac{2x(x + 7)}{(2x + 7)^2}$ **8.** $\dfrac{dy}{dx} = -\dfrac{6x^2 + 10x + 15}{(2x^2 - 5)^2}$

9. Domain: $(-\infty, \infty)$
Range: $[-4, \infty)$

10. Domain: $[7, \infty)$
Range: $[0, \infty)$

1. 0 **3.** 2 **5.** $2t - 8$ **7.** $\dfrac{9}{2t^4}$

9. $18(2 - x^2)(5x^2 - 2)$ **11.** $\dfrac{4}{(x - 1)^3}$

13. $12x^2 + 24x + 16$ **15.** $60x^2 - 72x$

17. $120x + 360$ **19.** $-\dfrac{9}{2x^5}$ **21.** 260 **23.** $-\dfrac{1}{648}$

25. -126 **27.** $4x$ **29.** $\dfrac{2}{x^2}$ **31.** 2

33. $f''(x) = 6(x - 3) = 0$ when $x = 3$.

35. $f''(x) = 2(3x + 4) = 0$ when $x = -\frac{4}{3}$.

37. $f''(x) = \dfrac{x(2x^2 - 3)}{(x^2 - 1)^{3/2}} = 0$ when $x = \pm\dfrac{\sqrt{6}}{2}$

39. $f''(x) = \dfrac{2x(x + 3)(x - 3)}{(x^2 + 3)^3}$
$= 0$ when $x = 0$ or $x = \pm 3$.

41. (a) $s(t) = -16t^2 + 144t$

(b) $v(t) = -32t + 144$
$a(t) = -32$

(c) 4.5 seconds; 324 feet

(d) $v(9) = -144$ feet per second, which is the same speed as the initial velocity

43.

t	0	10	20	30	40	50	60
$\frac{ds}{dt}$	0	45	60	67.5	72	75	77.1
$\frac{d^2s}{dt^2}$	9	2.25	1	0.56	0.36	0.25	0.18

As time increases, velocity increases and acceleration decreases.

45. $f(x) = x^2 - 6x + 6$
$f'(x) = 2x - 6$
$f''(x) = 2$

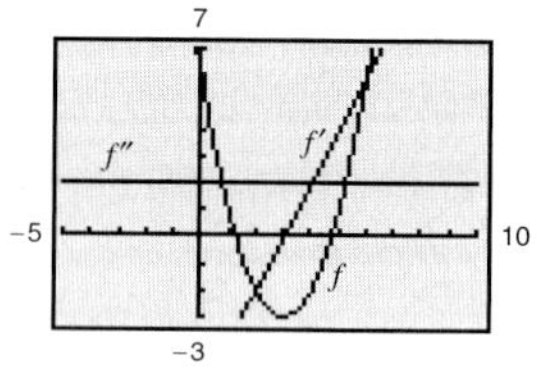

The degrees of the successive derivatives decrease by 1.

47.

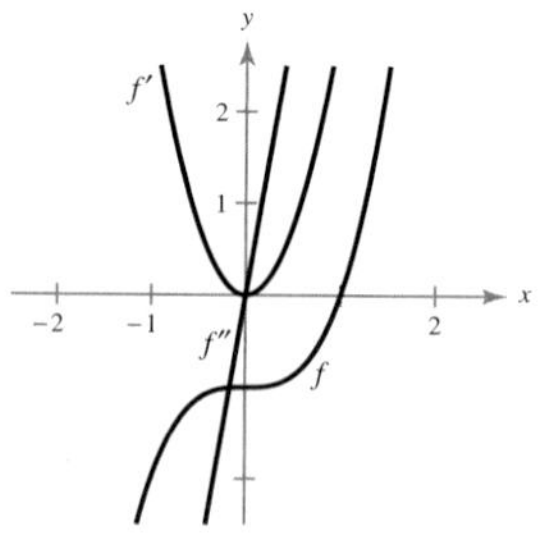

We know the degrees of the successive derivatives decrease by 1.

49. (a)

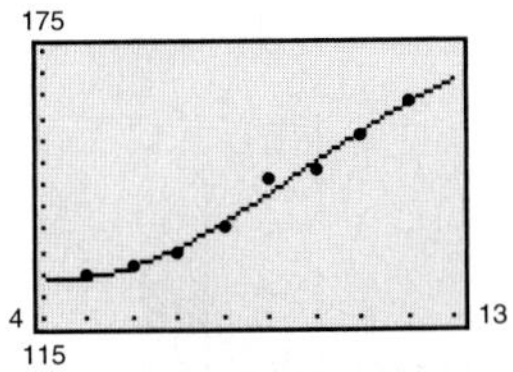

(b) $y' = -0.2484t^2 + 4.886t - 17.06$

$y'' = -0.4968t + 4.886$

(c) $y' > 0$ on $[5, 12]$ (d) 1999 ($t = 9.835$)

(e) The first derivative is used to show that the price of homes is increasing in (c), and the second derivative is used to show the maximum increase in (d).

51. False. The product rule is

$[f(x)g(x)]' = f'(x)g(x) + f(x)g'(x)$.

53. True **55.** True **57.** $xf^{(n)}(x) + nf^{(n-1)}(x)$

SECTION 2.7 *(page 152)*

Prerequisite Review

1. $y = x^2 - 2x$ **2.** $y = \dfrac{x-3}{4}$

3. $y = 1, x \neq -6$ **4.** $y = -4, x \neq \pm\sqrt{3}$

5. $y = \pm\sqrt{5 - x^2}$ **6.** $y = \pm\sqrt{6 - x^2}$ **7.** $\frac{8}{3}$

8. $-\frac{1}{2}$ **9.** $\frac{5}{7}$ **10.** 1

1. $-\dfrac{y}{x}$ **3.** $-\dfrac{x}{y}$ **5.** $\dfrac{xy^2}{2 - x^2y}$ **7.** $\dfrac{y}{8y - x}$

9. $-\dfrac{1}{10y - 2}$ **11.** $\dfrac{1}{2}$ **13.** $-\dfrac{x}{y}, 0$

15. $-\dfrac{y}{x+1}, -\dfrac{1}{4}$ **17.** $\dfrac{y - 3x^2}{2y - x}, \dfrac{1}{2}$ **19.** $\dfrac{1 - 3x^2y^3}{3x^3y^2 - 1}, -1$

21. $-\sqrt{\dfrac{y}{x}}, -\dfrac{5}{4}$ **23.** $-\sqrt[3]{\dfrac{y}{x}}, -\dfrac{1}{2}$ **25.** 3

27. 0 **29.** $-\dfrac{\sqrt{5}}{3}$ **31.** $-\dfrac{x}{y}, \dfrac{4}{3}$ **33.** $\dfrac{1}{2y}, -\dfrac{1}{2}$

35. At (5, 12): $5x + 12y - 169 = 0$

At (−12, 5): $12x - 5y + 169 = 0$

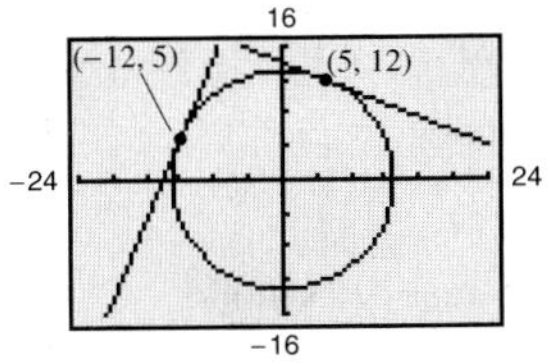

37. At $(1, \sqrt{5})$: $15x - 2\sqrt{5}y - 5 = 0$

At $(1, -\sqrt{5})$: $15x + 2\sqrt{5}y - 5 = 0$

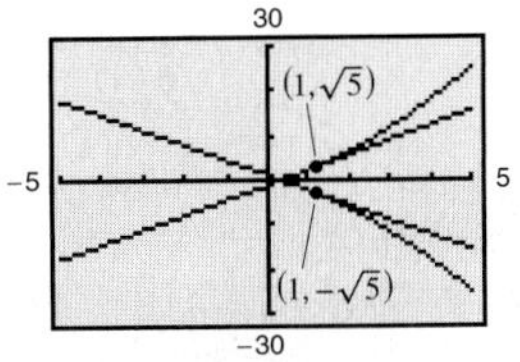

39. At (0, 2): $y = 2$

At (2, 0): $x = 2$

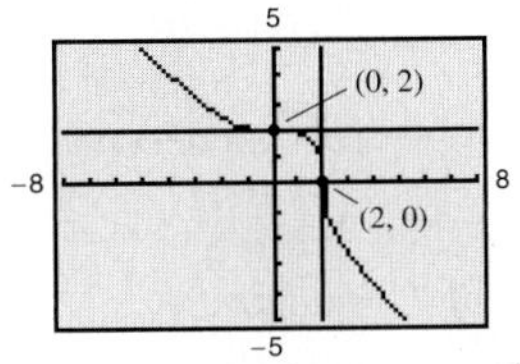

41. $\dfrac{1}{0.024x^3 + 0.04x}$ **43.** $-\dfrac{4xp}{2p^2 + 1}$

45. (a) −2

(b)

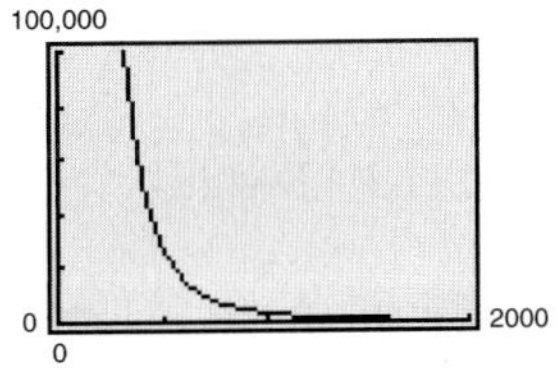

As more labor is used, less capital is available.

As more capital is used, less labor is available.

SECTION 2.8 *(page 160)*

Prerequisite Review

1. $A = \pi r^2$ **2.** $V = \frac{4}{3}\pi r^3$ **3.** $S = 6s^2$

4. $V = s^3$ **5.** $V = \frac{1}{3}\pi r^2 h$ **6.** $A = \frac{1}{2}bh$

7. $-\dfrac{x}{y}$ **8.** $\dfrac{2x - 3y}{3x}$ **9.** $-\dfrac{2x + y}{x + 2}$

10. $-\dfrac{y^2 - y + 1}{2xy - 2y - x}$

1. (a) 62 (b) $\frac{32}{85}$ **3.** (a) $-\frac{5}{8}$ (b) $\frac{3}{2}$

5. (a) 24π square inches per minute

(b) 96π square inches per minute

7. If $\dfrac{dr}{dt}$ is constant, $\dfrac{dA}{dt} = 2\pi r \dfrac{dr}{dt}$ and so is proportional to r.

9. (a) $\frac{5}{\pi}$ feet per minute (b) $\frac{5}{4\pi}$ feet per minute

11. (a) 112.5 dollars per week

(b) 7500 dollars per week

(c) 7387.5 dollars per week

13. (a) 9 cubic centimeters per second

(b) 900 cubic centimeters per second

15. (a) -12 centimeters per minute

(b) 0 centimeters per minute

(c) 4 centimeters per minute

(d) 12 centimeters per minute

17. (a) $-\frac{7}{12}$ foot per second (b) $-\frac{3}{2}$ feet per second

(c) $-\frac{48}{7}$ feet per second

19. (a) -750 miles per hour (b) 20 minutes

21. -8.33 feet per second **23.** $\approx 188.5\ \text{ft}^3$ per minute

25. 5 units per week

REVIEW EXERCISES FOR CHAPTER 2

(page 166)

1. -2 **3.** 0

5. $t = 7$: slope $\approx$ \$5500 million per year per year (sales are increasing)

$t = 10$: slope $\approx$ \$7500 million per year per year (sales are increasing)

$t = 12$: slope $\approx$ \$3500 million per year per year (sales are increasing)

7. $t = 0$: slope ≈ 180
$t = 4$: slope ≈ -70
$t = 6$: slope ≈ -900

9. $-3; -3$ **11.** $2x - 4; -2$

13. $\frac{1}{2\sqrt{x+9}}; \frac{1}{4}$ **15.** $-\frac{1}{(x-5)^2}; -1$

17. -5 **19.** 0 **21.** $\frac{1}{6}$ **23.** -5 **25.** 1 **27.** 0

29. $y = -\frac{4}{3}t + 2$

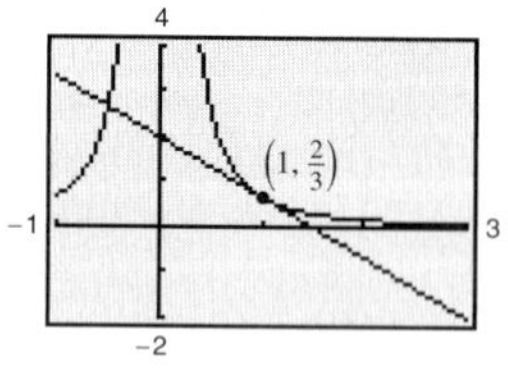

31. $y = 2x + 2$

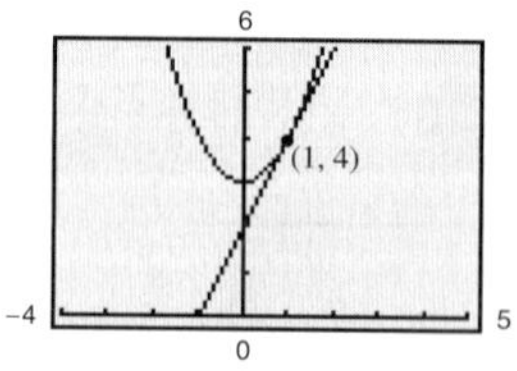

33. $y = -34x - 27$

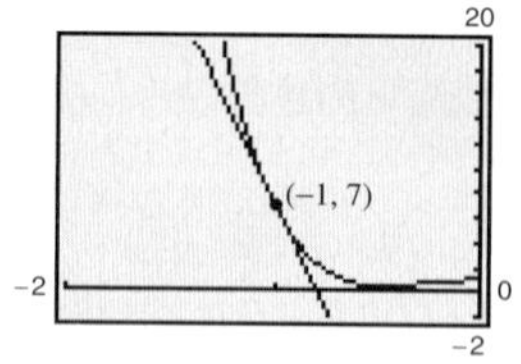

35. $y = x - 1$

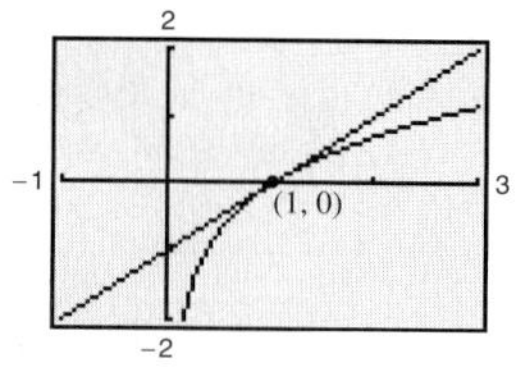

37. $y = -2x + 6$

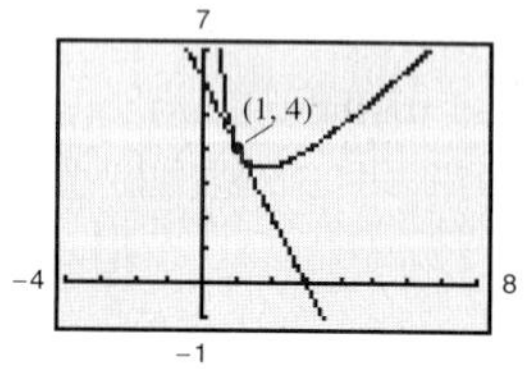

39. Average rate of change: 4

Instantaneous rate of change when $x = 0$: 3

Instantaneous rate of change when $x = 1$: 5

41. (a) \$6976 million per year

(b) 1998: \$7317 million per year

2002: \$3876 million per year

(c) Sales were increasing in 1998 and 2002, and grew at a rate of \$6976 million over the period 1998–2002.

43. (a) $P'(t) = -0.004236t^3 + 0.09045t^2 - 0.57t + 1.007$

(b) $-\$0.004$ per pound in 1997

\$0.116 per pound in 2000

$-\$0.128$ per pound in 2002

(c)

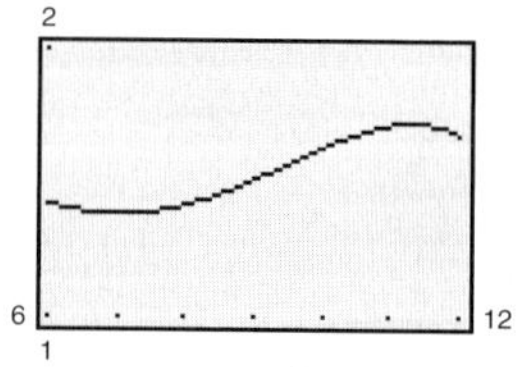

The price is increasing from 1997 to 2001, and decreasing from 1996 to 1997 and from 2001 to 2002.

(d) Negative slope: $6 < t < 7$ and $11 < t < 12$

Positive slope: $7 < t < 11$

(e) When the price increases, the slope is positive.

When the price decreases, the slope is negative.

45. (a) $s(t) = -16t^2 + 276$ (b) -32 feet per second

(c) $t = 2$: -64 feet per second

$t = 3$: -96 feet per second

(d) 4.15 second (e) 132.8 feet per second

47. $R = 27.50x$

$C = 15x + 2500$

$P = 12.50x - 2500$

49. $\dfrac{dC}{dx} = 320$ **51.** $\dfrac{dC}{dx} = \dfrac{1.275}{\sqrt{x}}$

53. $\dfrac{dR}{dx} = 200 - \dfrac{2}{5}x$ **55.** $\dfrac{dR}{dx} = \dfrac{35(x-4)}{2(x-2)^{3/2}}$

57. $\dfrac{dP}{dx} = -0.0006x^2 + 12x - 1$ **59.** $15x^2(1 - x^2)$

61. $16x^3 - 33x^2 + 12x$ **63.** $\dfrac{2(3 + 5x - 3x^2)}{(x^2+1)^2}$

65. $30x(5x^2 + 2)^2$ **67.** $-\dfrac{1}{(x+1)^{3/2}}$ **69.** $\dfrac{2x^2+1}{\sqrt{x^2+1}}$

71. $80x^4 - 24x^2 + 1$ **73.** $18x^5(x+1)(2x+3)^2$

75. $x(x-1)^4(7x-2)$ **77.** $\dfrac{3(9t+5)}{2\sqrt{3t+1}(1-3t)^3}$

79. (a) $t = 1$: -6.63 $t = 3$: -6.5

$t = 5$: -4.33 $t = 10$: -1.36

(b)

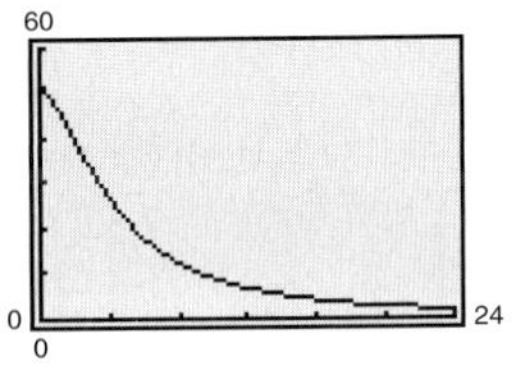

The rate of decrease is approaching zero.

81. 6 **83.** $-\dfrac{120}{x^6}$ **85.** $\dfrac{35x^{3/2}}{2}$ **87.** $\dfrac{2}{x^{2/3}}$

89. (a) $s(t) = -16t^2 + 5t + 30$ (b) 1.534 seconds

(c) -44.09 feet per second

(d) -32 feet per second squared

91. $-\dfrac{2x+3y}{3(x+y^2)}$ **93.** $\dfrac{2x-8}{2y-9}$ **95.** $y = \dfrac{1}{3}x + \dfrac{1}{3}$

97. $y = \frac{4}{3}x + \frac{2}{3}$ **99.** $\frac{1}{64}$ feet per minute

SAMPLE POST-GRAD EXAM QUESTIONS

(page 170)

1. c **2.** e **3.** e **4.** c **5.** c **6.** a

CHAPTER 3

SECTION 3.1 *(page 179)*

Prerequisite Review

1. $x = 0, x = 8$ **2.** $x = 0, x = 24$ **3.** $x = \pm 5$

4. $x = 0$ **5.** $(-\infty, 3) \cup (3, \infty)$ **6.** $(-\infty, 1)$

7. $(-\infty, -2) \cup (-2, 5) \cup (5, \infty)$ **8.** $(-\sqrt{3}, \sqrt{3})$

9. $x = -2$: -6 **10.** $x = -2$: 60

$x = 0$: 2 $x = 0$: -4

$x = 2$: -6 $x = 2$: 60

11. $x = -2$: $-\frac{1}{3}$ **12.** $x = -2$: $\frac{1}{18}$

$x = 0$: 1 $x = 0$: $-\frac{1}{8}$

$x = 2$: 5 $x = 2$: $-\frac{3}{2}$

1. $f'(-1) = -\frac{8}{25}$ **3.** $f'(-3) = -\frac{2}{3}$

$f'(0) = 0$ $f'(-2)$ is undefined.

$f'(1) = \frac{8}{25}$ $f'(-1) = \frac{2}{3}$

5. Increasing on $(-\infty, -1)$

Decreasing on $(-1, \infty)$

7. Increasing on $(-1, 0)$ and $(1, \infty)$

Decreasing on $(-\infty, -1)$ and $(0, 1)$

9. No critical numbers

Increasing on $(-\infty, \infty)$

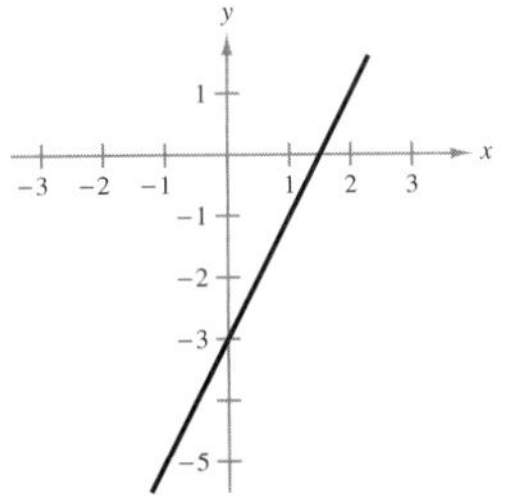

11. Critical number: $x = 1$

Increasing on $(-\infty, 1)$

Decreasing on $(1, \infty)$

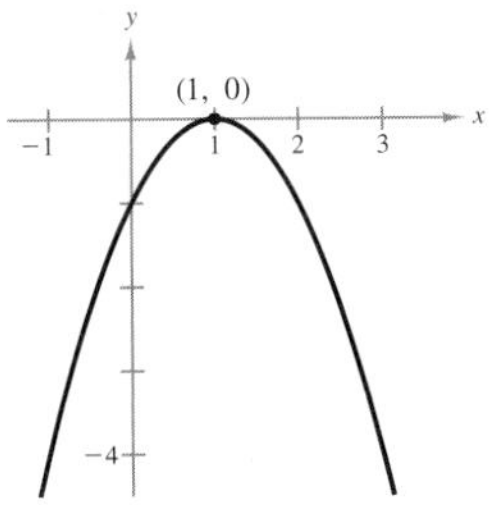

13. Critical number: $x = \frac{5}{2}$

Decreasing on $\left(-\infty, \frac{5}{2}\right)$

Increasing on $\left(\frac{5}{2}, \infty\right)$

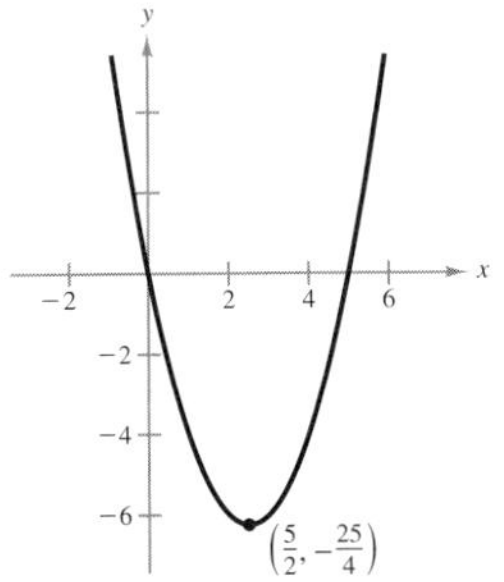

15. Critical numbers: $x = 0, x = 4$

Increasing on $(-\infty, 0)$ and $(4, \infty)$

Decreasing on $(0, 4)$

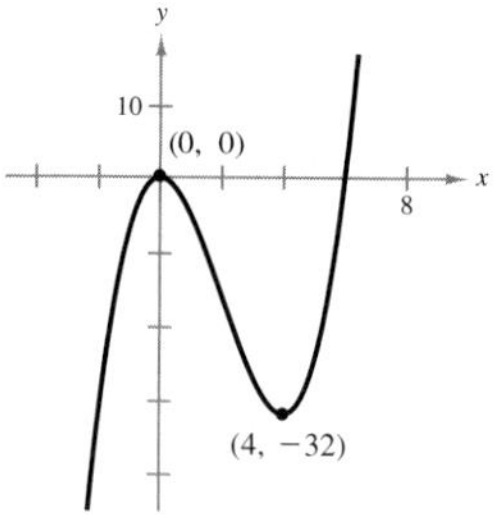

17. Critical numbers: $x = -1, x = 1$

Decreasing on $(-\infty, -1)$

Increasing on $(1, \infty)$

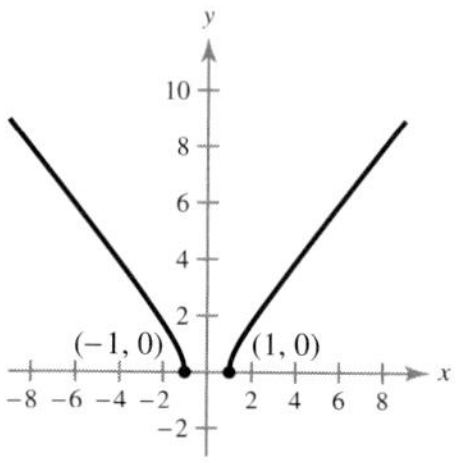

19. Critical number: $x = 1$

Increasing on $(-\infty, 1)$

Decreasing on $(1, \infty)$

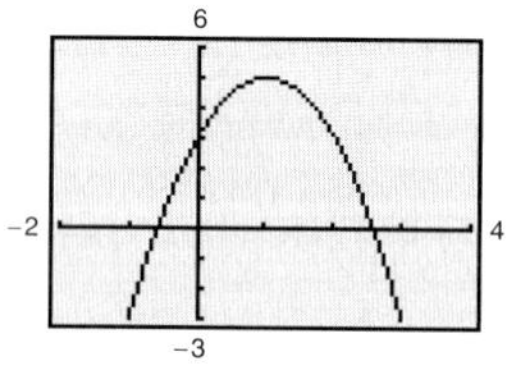

21. Critical numbers: $x = -1, x = -\frac{5}{3}$

Increasing on $\left(-\infty, -\frac{5}{3}\right)$ and $(-1, \infty)$

Decreasing on $\left(-\frac{5}{3}, -1\right)$

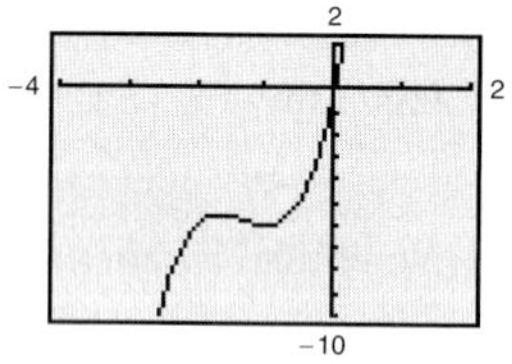

23. Critical numbers: $x = -1, x = -\frac{2}{3}$

Decreasing on $\left(-1, -\frac{2}{3}\right)$

Increasing on $\left(-\frac{2}{3}, \infty\right)$

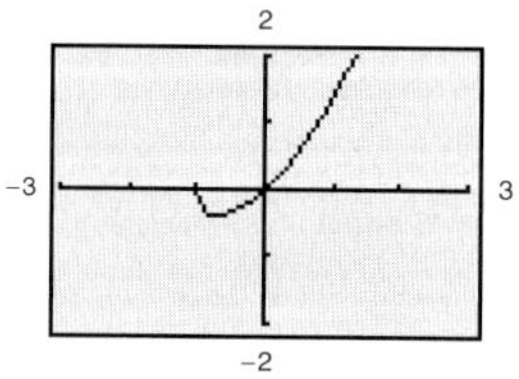

25. Critical numbers: $x = 0, x = \frac{3}{2}$

Decreasing on $\left(-\infty, \frac{3}{2}\right)$

Increasing on $\left(\frac{3}{2}, \infty\right)$

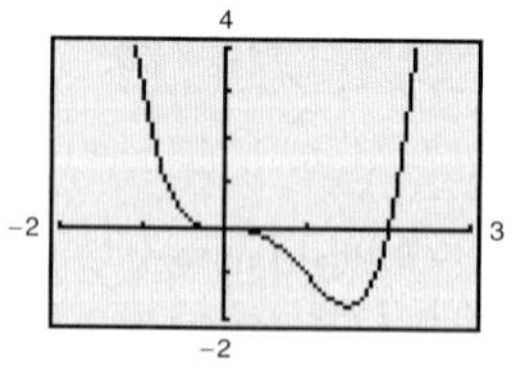

27. Critical numbers: $x = 2, x = -2$

Decreasing on $(-\infty, -2)$ and $(2, \infty)$

Increasing on $(-2, 2)$

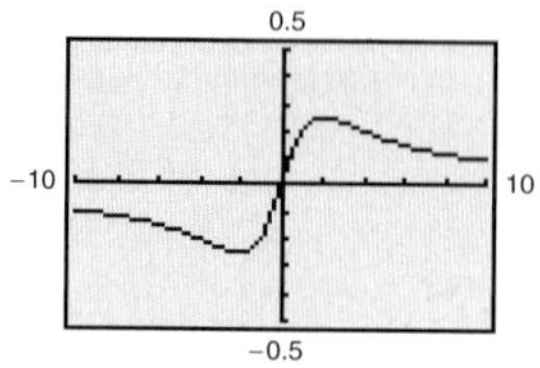

29. No critical numbers

Discontinuities: $x = \pm 4$

Increasing on $(-\infty, -4)$, $(-4, 4)$, and $(4, \infty)$

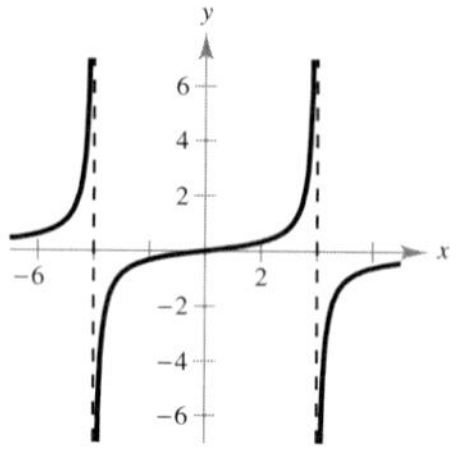

31. Critical number: $x = 0$

Discontinuity: $x = 0$

Increasing on $(-\infty, 0)$

Decreasing on $(0, \infty)$

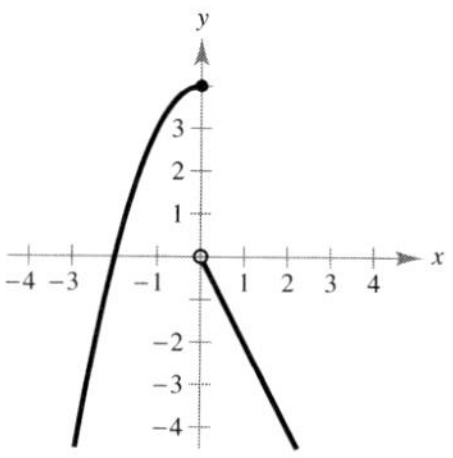

33. Critical number: $x = 1$

No discontinuity, but the function is not differentiable at $x = 1$.

Increasing on $(-\infty, 1)$

Decreasing on $(1, \infty)$

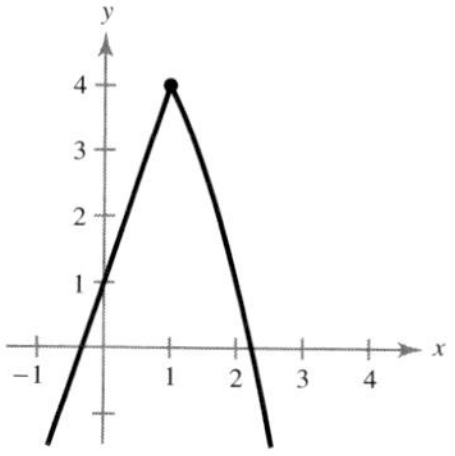

35. (a) Decreasing on $[1, 4.10)$

Increasing on $(4.10, \infty)$

(b)

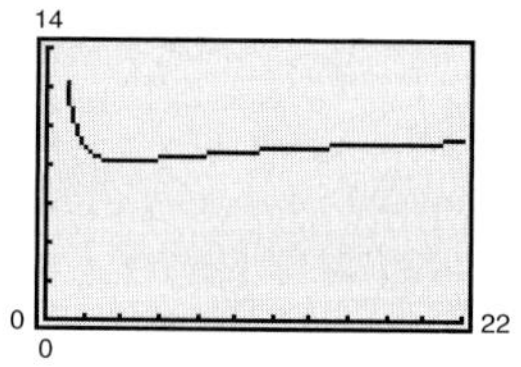

(c) $C = 9$ (or \$900) when $x = 2$ and $x = 15$. Use an order size of $x = 4$, which will minimize the cost C.

37. Rising on $(0, 3)$

Falling on $(3, 6)$

39. (a)

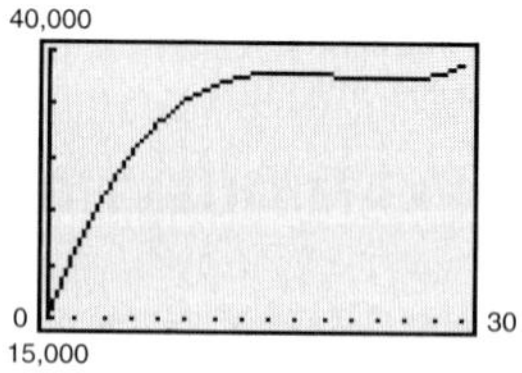

Increasing from 1970 to early 1987 and from mid-1994 to 2000

Decreasing from early 1987 to mid-1994

(b) $y' = 8.229t^2 - 343.1t + 3462.3 = 0$;

Critical numbers: $x = 24.57, x = 17.12$;

Therefore the model is increasing from 1970 to early 1987 and from mid-1994 to 2000 and decreasing from early 1987 to mid-1994.

SECTION 3.2 *(page 189)*

Prerequisite Review

1. $0, \pm\frac{1}{2}$ **2.** $-2, 5$ **3.** 1 **4.** $0, 125$

5. $-4 \pm \sqrt{17}$ **6.** $1 \pm \sqrt{5}$

7. Negative **8.** Positive **9.** Positive

10. Negative **11.** Increasing **12.** Decreasing

1. Relative maximum: $(1, 5)$

3. Relative minimum: $(3, -9)$

5. Relative maximum: $\left(\frac{2}{3}, \frac{28}{9}\right)$

Relative minimum: $(1, 3)$

7. No relative extrema

9. Relative maximum: $(0, 15)$

Relative minimum: $(4, -17)$

11. Relative minimum: $\left(\frac{3}{2}, -\frac{27}{16}\right)$

13.

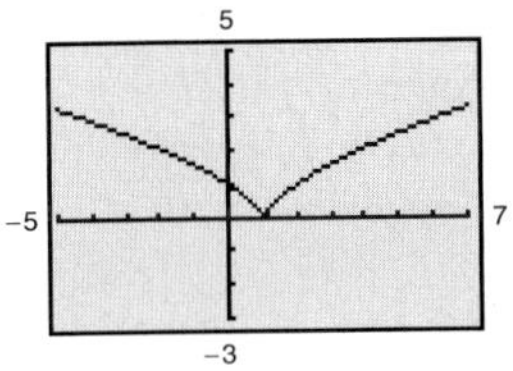

Relative minimum: $(1, 0)$

15.

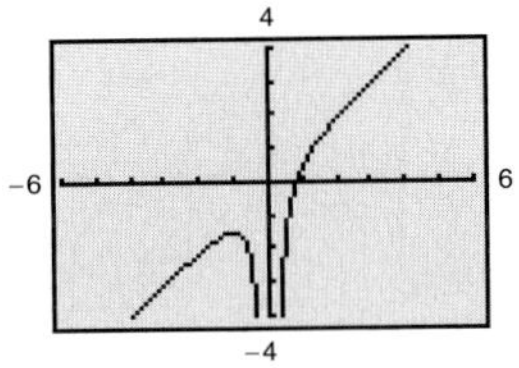

Relative maximum: $\left(-1, -\frac{3}{2}\right)$

17.

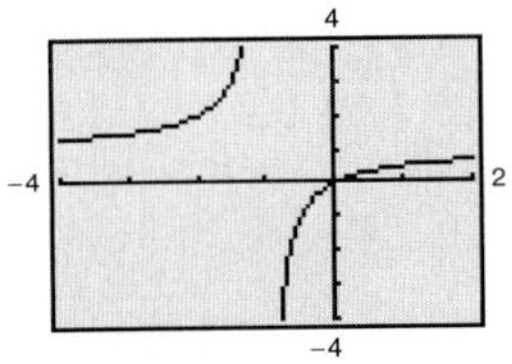

No relative extrema

19. Minimum: $(2, 2)$

Maximum: $(-1, 8)$

21. Maximum: $(0, 5)$

Minimum: $(3, -13)$

23. Minima: $(-1, -4), (2, -4)$

Maxima: $(0, 0), (3, 0)$

25. Maximum: $(2, 1)$

Minimum: $\left(0, \frac{1}{3}\right)$

27. Maximum: $(-1, 5)$

Minimum: $(0, 0)$

29. Maximum: $(-7, 4)$

Minimum: $(1, 0)$

31. Maximum: $(5, 7)$

Minimum: $(2.69, -5.55)$

33. Maximum: $(2, 2.\overline{6})$

Minima: $(0, 0), (3, 0)$

35. Minimum: $(0, 0)$

Maximum: $(1, 2)$

37. Maximum: $\left(2, \frac{1}{2}\right)$

Minimum: $(0, 0)$

39. $\frac{40\sqrt{3}}{3}$ **41.** 360 **43.** 82 units **45.** \$2.10

47. (a) Population tends to increase each year, so the minimum population occurred in 1790 and the maximum population occurred in 2000.

(b) Maximum population: 278.968 million

Minimum population: 3.775 million

(c) The minimum population was about 3.775 million in 1790 and the maximum population was about 278.968 million in 2000.

SECTION 3.3 *(page 198)*

Prerequisite Review

1. $f''(x) = 48x^2 - 54x$

2. $g''(s) = 12s^2 - 18s + 2$

3. $g''(x) = 56x^6 + 120x^4 + 72x^2 + 8$

4. $f''(x) = \dfrac{4}{9(x-3)^{2/3}}$ **5.** $h''(x) = \dfrac{190}{(5x-1)^3}$

6. $f''(x) = -\dfrac{42}{(3x+2)^3}$ **7.** $x = \pm\dfrac{\sqrt{3}}{3}$

8. $x = 0, 3$ **9.** $t = \pm 4$ **10.** $x = 0, \pm 5$

1. Concave upward on $(-\infty, \infty)$

3. Concave upward on $\left(-\infty, -\frac{1}{2}\right)$

Concave downward on $\left(-\frac{1}{2}, \infty\right)$

5. Concave upward on $(-\infty, -2)$ and $(2, \infty)$

Concave downward on $(-2, 2)$

7. Concave upward on $(-\infty, 2)$

Concave downward on $(2, \infty)$

9. Relative maximum: $(3, 9)$

11. Relative maximum: $(1, 3)$

Relative minimum: $\left(\frac{7}{3}, \frac{49}{27}\right)$

13. Relative minimum: $(0, -3)$

15. Relative minimum: $(0, 1)$ **17.** No relative extrema

19. Relative maximum: $(0, 0)$

Relative minima: $(-0.5, -0.052), (1, -0.\overline{3})$

21. Relative maximum: $(2, 9)$

Relative minimum: $(0, 5)$

23. Sign of $f'(x)$ on $(0, 2)$ is positive.

Sign of $f''(x)$ on $(0, 2)$ is positive.

25. Sign of $f'(x)$ on $(0, 2)$ is negative.

Sign of $f''(x)$ on $(0, 2)$ is negative.

27. (3, 0) **29.** (1, 0), (3, −16)

31. No inflection points **33.** $\left(\frac{3}{2}, -\frac{1}{16}\right)$, (2, 0)

35.

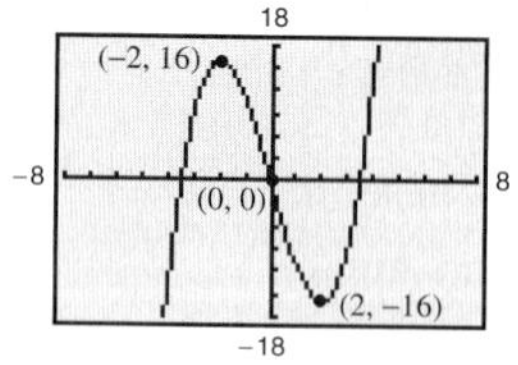

Relative maximum: (−2, 16)

Relative minimum: (2, −16)

Point of inflection: (0, 0)

37.

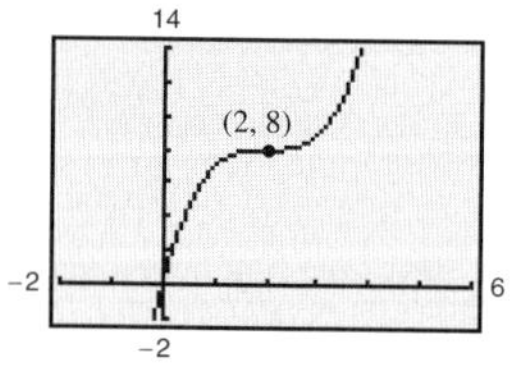

No relative extrema

Point of inflection: (2, 8)

39.

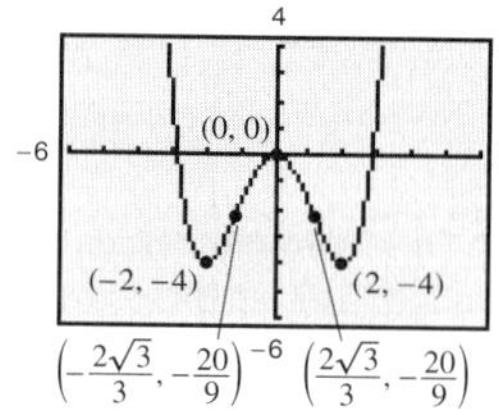

Relative maximum: (0, 0)

Relative minima: (±2, −4)

Points of inflection: $\left(\pm\frac{2\sqrt{3}}{3}, -\frac{20}{9}\right)$

41.

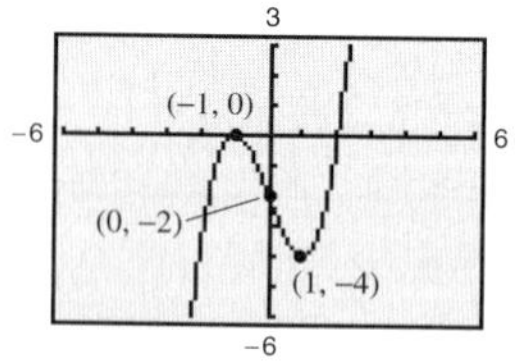

Relative maximum: (−1, 0)

Relative minimum: (1, −4)

Point of inflection: (0, −2)

43.

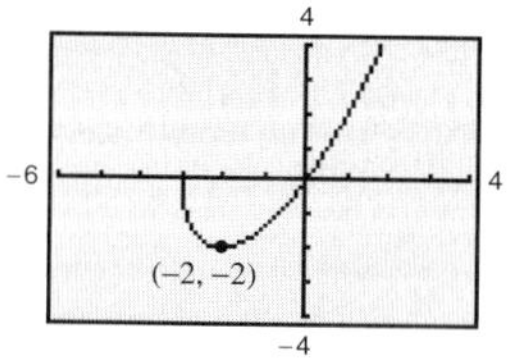

Relative minimum: (−2, −2)

No inflection points

45.

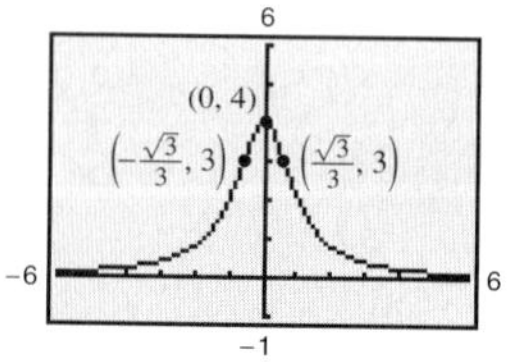

Relative maximum: (0, 4)

Points of inflection: $\left(\pm\frac{\sqrt{3}}{3}, 3\right)$

47.

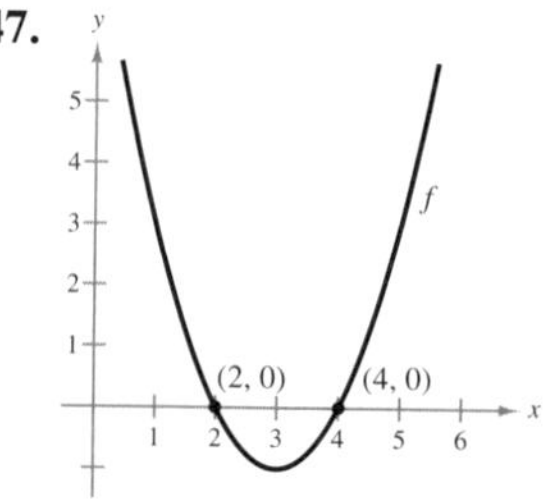

49.

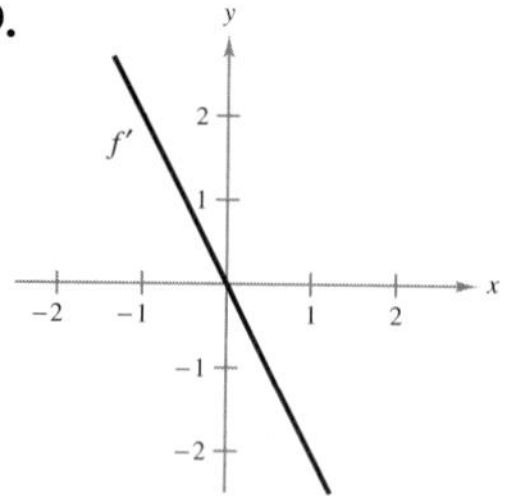

(a) f': Positive on $(-\infty, 0)$

f: Increasing on $(-\infty, 0)$

(b) f': Negative on $(0, \infty)$

f: Decreasing on $(0, \infty)$

(c) f': Not increasing

f: Not concave upward

(d) f': Decreasing on $(-\infty, \infty)$

f: Concave downward on $(-\infty, \infty)$

51. (200, 320) **53.** 100 units **55.** 8:30 P.M.

57. $\sqrt{3} \approx 1.732$ years

59.

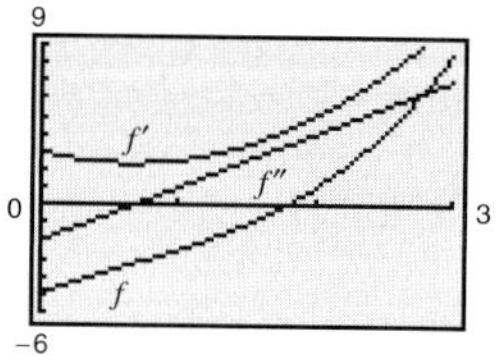

Relative minimum: (0, −5)

Relative maximum: (3, 8.5)

Point of inflection: $\left(\frac{2}{3}, -3.2963\right)$

61.

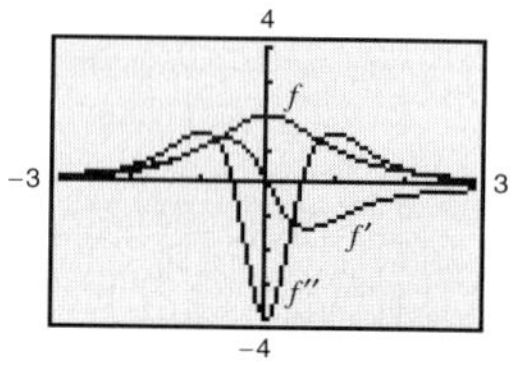

Relative maximum: (0, 2)

Points of inflection: (0.58, 1.5), (−0.58, 1.5)

63. (a) Relative maximum: (2, 2150)

Relative minimum: (1, 2050)

Absolute maximum: (0, 2250)

Absolute minimum: (7.5, 1740)

The market opened at the maximum for the day and closed at the minimum. At approximately 9:30 A.M., the market started to recover after falling, and at approximately 10:30 A.M., the market started to fall again.

(b) (4, 2010);

At approximately 12:30 P.M., the market began to fall at a greater rate.

65. (a) At $t = 8$, 256 people will be infected.

(b) At $t = 4$, the virus will be spreading most rapidly.

(c)

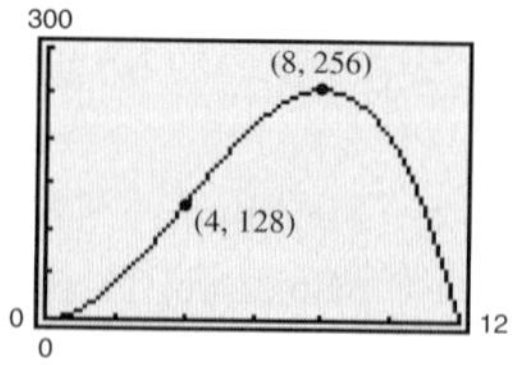

SECTION 3.4 (page 207)

Prerequisite Review

1. $x + \frac{1}{2}y = 12$ **2.** $2xy = 24$ **3.** $xy = 24$
4. $\sqrt{(x_2 - x_1)^2 + (y_2 - y_1)^2} = 10$
5. $x = -3$ **6.** $x = -\frac{2}{3}, 1$ **7.** $x = \pm 5$
8. $x = 4$ **9.** $x = \pm 1$ **10.** $x = \pm 3$

1. 55, 55 **3.** 18, 9 **5.** $\sqrt{192}, \sqrt{192}$ **7.** 1
9. $l = w = 25$ meters **11.** $l = w = 8$ feet
13. $x = 25$ feet, $y = \frac{100}{3}$ feet
15. (a) Proof
(b) $V_1 = 99$ cubic inches
$V_2 = 125$ cubic inches
$V_3 = 117$ cubic inches
(c) 5 inches × 5 inches × 5 inches
17. Rectangular portion: $\frac{16}{\pi + 4} \times \frac{32}{\pi + 4}$ feet
19. 1.056 cubic feet
21. Plant 18 trees to yield 1296 apples.
23. $2\sqrt{15} + 4$ inches × $\sqrt{15} + 2$ inches
25. (a) $L = \sqrt{x^2 + 4 + \frac{8}{x - 1} + \frac{4}{(x - 1)^2}}, \quad x > 1$
(b)

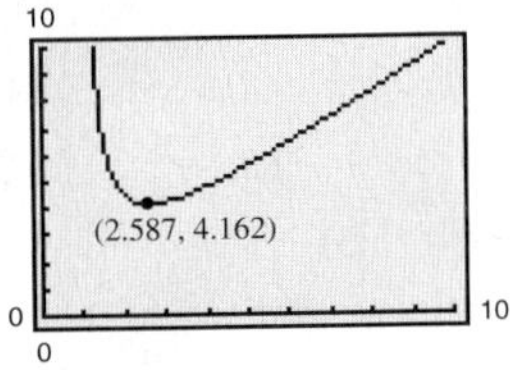

Minimum when $x \approx 2.587$
(c) (0, 0), (2, 0), (0, 4)
27. $\sqrt{2}r \times \frac{\sqrt{2}}{2}r$ **29.** $\frac{4\pi r^3}{3\sqrt{3}}$ **31.** $\left(\pm\sqrt{\frac{5}{2}}, \frac{7}{2}\right)$
33. 18 inches × 18 inches × 36 inches
35. Radius of circle: $\frac{8}{\pi + 4}$
Side of square: $\frac{16}{\pi + 4}$
37. $x = 1$ mile
39. Harvesting in the 6th week will yield 140 bushels for a maximum value of $490.

41.

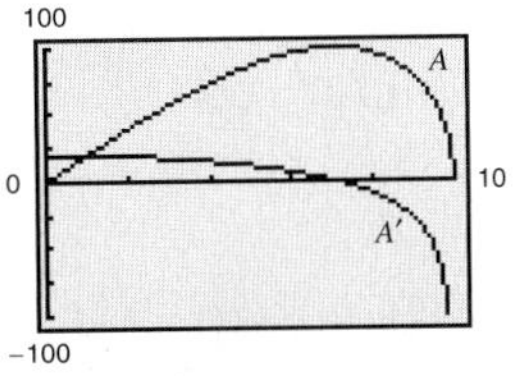

Length $= 10\sqrt{2} \approx 14.14$
Width $= 5\sqrt{2} \approx 7.07$

SECTION 3.5 (page 217)

Prerequisite Review

1. 1 **2.** $\frac{6}{5}$ **3.** 2 **4.** $\frac{1}{2}$
5. $\frac{dC}{dx} = 1.2 + 0.006x$ **6.** $\frac{dP}{dx} = 0.02x + 11$
7. $\frac{dR}{dx} = 14 - \frac{x}{1000}$ **8.** $\frac{dR}{dx} = 3.4 - \frac{x}{750}$
9. $\frac{dP}{dx} = -1.4x + 7$ **10.** $\frac{dC}{dx} = 4.2 + 0.003x^2$

1. 2000 units **3.** 200 units **5.** 80 units
7. 50 units **9.** $60 **11.** $69.68
13. 3 units
$\overline{C}(3) = 17; \frac{dC}{dx} = 4x + 5$; when $x = 3, \frac{dC}{dx} = 17$

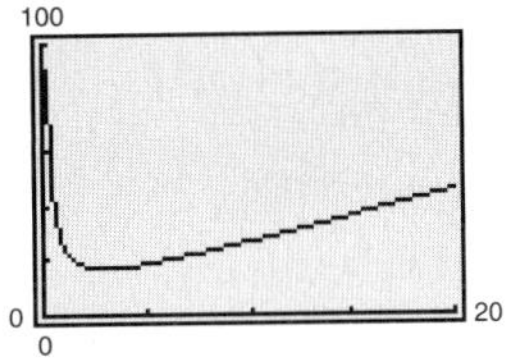

15. (a) $80 (b) $45.93
17. The maximum profit occurs when $s = 10$ (or $10,000).
The point of diminishing returns occurs at $s = \frac{35}{6}$ (or $5833.33).
19. 200 radios **21.** $50
23. C = cost under water + cost on land
$= 8\sqrt{0.25 + x^2} + 6(6 - x)$

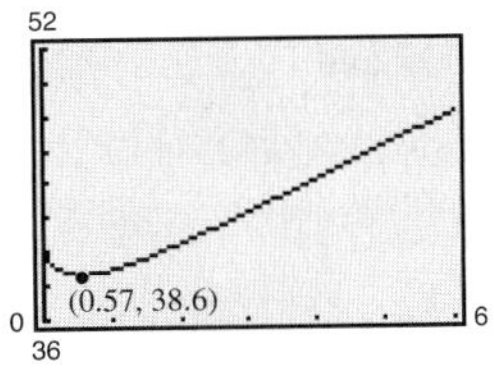

The line should run from the power station to a point across the river approximately 0.57 mile downstream.
(Exact: $3/(2\sqrt{7})$ mile)

25. 77.46 miles per hour

27. $-\frac{17}{3}$, elastic

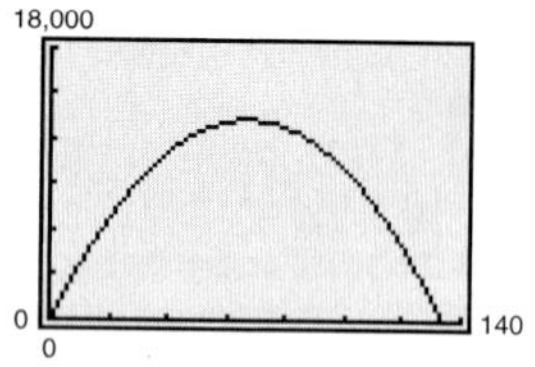

Elastic: $\left(0, \frac{200}{3}\right) \approx (0, 66.7)$

Inelastic: $\left(\frac{200}{3}, \frac{400}{3}\right) \approx (66.7, 133.3)$

29. $-\frac{9997}{3}$, elastic

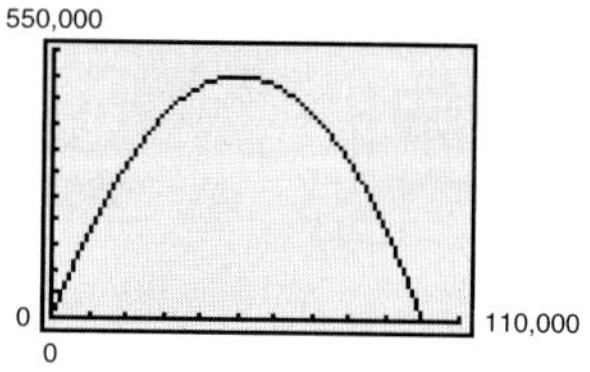

Elastic: $(0, 50{,}000)$

Inelastic: $(50{,}000, 100{,}000)$

31. $-\frac{3}{2}$, elastic

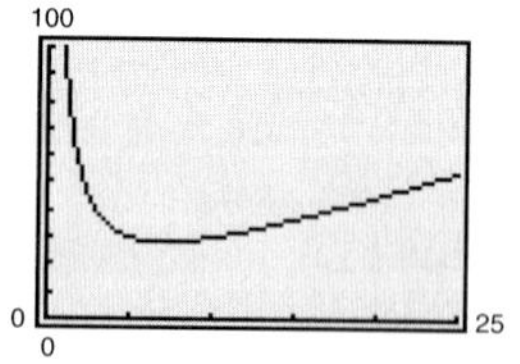

Elastic: $\left(5\sqrt{2}, \infty\right) \approx (7.1, \infty)$

Inelastic: $\left(0, 5\sqrt{2}\right) \approx (0, 7.1)$

33. (a) -2.48% (b) -0.496

(c) -0.5; The results are approximately the same.

(d) $R = p^3 - 20p^2 + 100p$, $x = \frac{400}{9}$ units, $p = \$3.33$

35. (a) $-\frac{11}{14}$ (b) $x = 500$ units, $p = \$10$

(c) Answers will vary.

37. No; when $p = 5$, $x = 350$ and $\eta = -\frac{5}{7}$.

Because $|\eta| = \frac{5}{7} < 1$, demand is inelastic.

39. Proof

41. (a) Revenue was greatest in 2001 and least in 1994.

(b) Revenue was increasing at the greatest rate in 1998 and decreasing at the greatest rate in 2003.

(c)

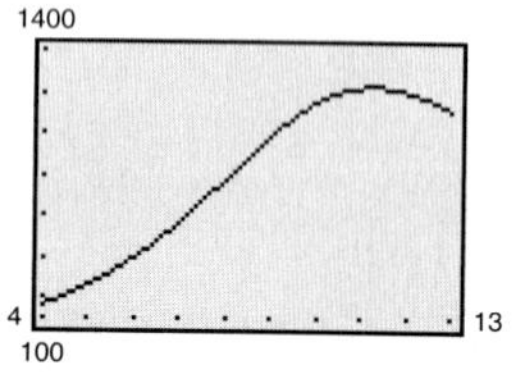

43. Answers will vary.

SECTION 3.6 *(page 228)*

Prerequisite Review

1. 3 **2.** 1 **3.** -11 **4.** 4 **5.** $-\frac{1}{4}$

6. -2 **7.** 0 **8.** 1

9. $\overline{C} = \dfrac{150}{x} + 3$

$\dfrac{dC}{dx} = 3$

10. $\overline{C} = \dfrac{1900}{x} + 1.7 + 0.002x$

$\dfrac{dC}{dx} = 1.7 + 0.004x$

11. $\overline{C} = 0.005x + 0.5 + \dfrac{1375}{x}$

$\dfrac{dC}{dx} = 0.01x + 0.5$

12. $\overline{C} = \dfrac{760}{x} + 0.05$

$\dfrac{dC}{dx} = 0.05$

1. Vertical asymptote: $x = 0$

Horizontal asymptote: $y = 1$

3. Vertical asymptotes: $x = -1, x = 2$

Horizontal asymptote: $y = 1$

5. Vertical asymptote: none

Horizontal asymptote: $y = \frac{3}{2}$

7. Vertical asymptotes: $x = \pm 2$

Horizontal asymptote: $y = \frac{1}{2}$

9. f **10.** b **11.** c **12.** a **13.** e **14.** d

15. ∞ **17.** $-\infty$ **19.** $-\infty$ **21.** $-\infty$ **23.** $\frac{2}{3}$

25. 0 **27.** $-\infty$ **29.** ∞ **31.** 5

33.

x	10^0	10^1	10^2	10^3
$f(x)$	2.000	0.348	0.101	0.032

x	10^4	10^5	10^6
$f(x)$	0.010	0.003	0.001

$\lim_{x \to \infty} \dfrac{x + 1}{x\sqrt{x}} = 0$

35.

x	10^0	10^1	10^2	10^3
$f(x)$	0	49.5	49.995	49.99995

x	10^4	10^5	10^6
$f(x)$	50.0	50.0	50.0

$$\lim_{x\to\infty} \frac{x^2 - 1}{0.02x^2} = 50$$

37.

x	-10^6	-10^4	-10^2	10^0
$f(x)$	-2	-2	-1.9996	0.8944

x	10^2	10^4	10^6
$f(x)$	1.9996	2	2

$$\lim_{x\to-\infty} \frac{2x}{\sqrt{x^2+4}} = -2,\ \lim_{x\to\infty} \frac{2x}{\sqrt{x^2+4}} = 2$$

39.

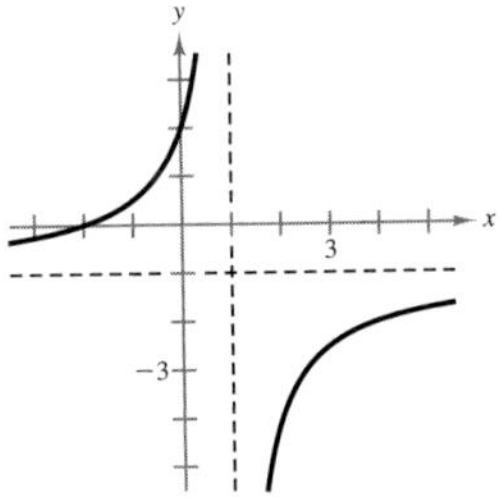

41.

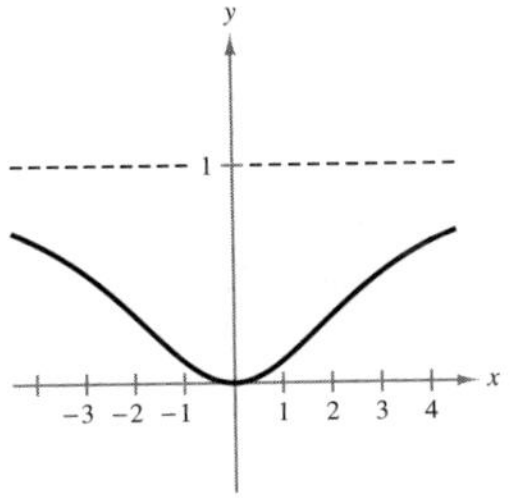

43.

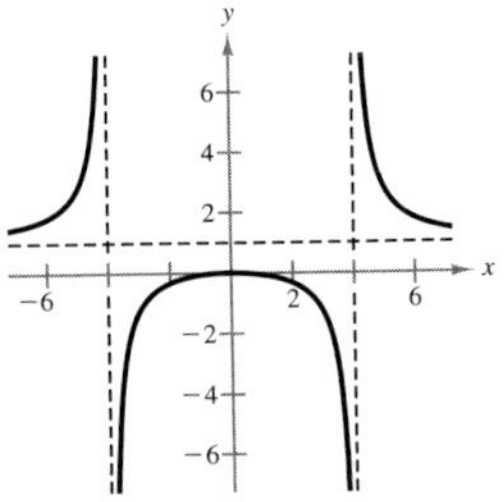

45.

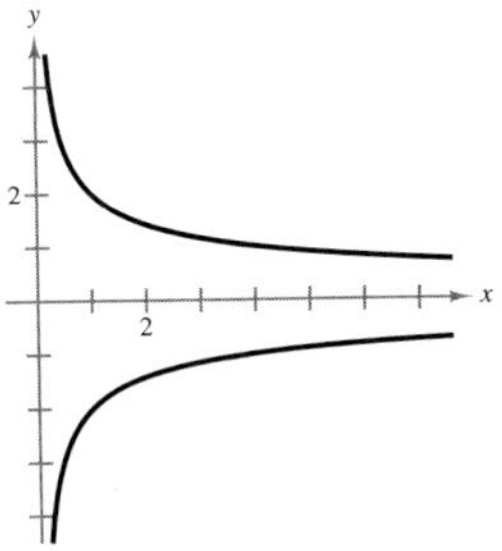

47.

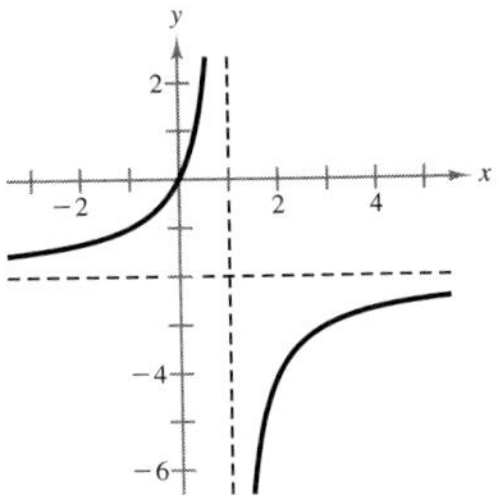

49.

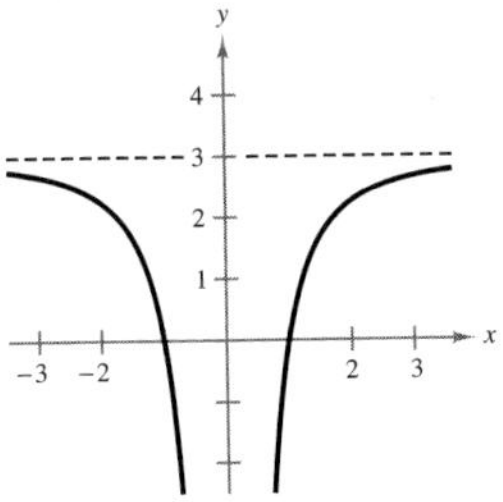

51.

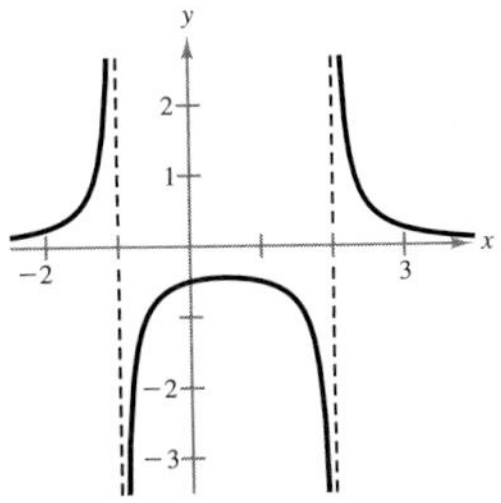

53.

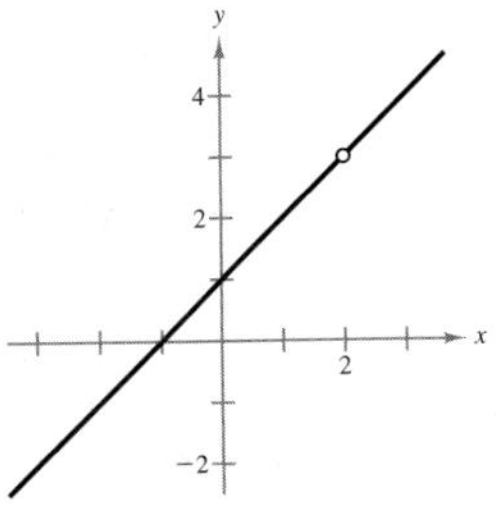

55.

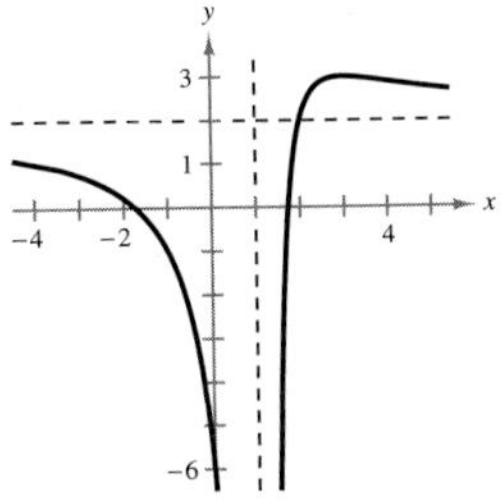

57. (a) $\overline{C} = 1.35 + \dfrac{4570}{x}$ (b) \$47.05, \$5.92 (c) \$1.35

59. (a) 25%; \$176 million; 50%; \$528 million; 75%; \$1584 million

(b) ∞

(c)

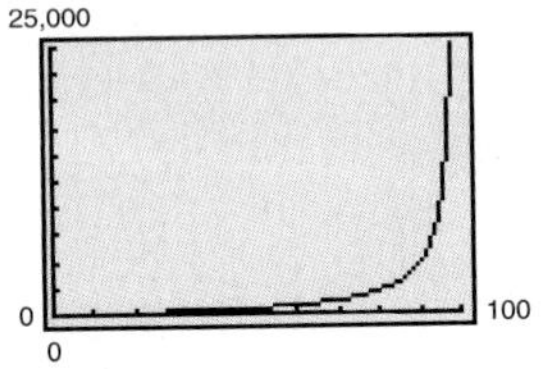

61. a

63. (a) 5 years: 153 elk (b) 400 elk
10 years: 215 elk
25 years: 294 elk

65. (a) $\overline{P} = 50 - \dfrac{1000}{x}$ (b) \$40, \$48, \$49 (c) \$50

SECTION 3.7 *(page 238)*

Prerequisite Review

1. Vertical asymptote: $x = 0$
   Horizontal asymptote: $y = 0$
2. Vertical asymptote: $x = 2$
   Horizontal asymptote: $y = 0$
3. Vertical asymptote: $x = -3$
   Horizontal asymptote: $y = 40$
4. Vertical asymptotes: $x = 1, x = 3$
   Horizontal asymptote: $y = 1$

5. Decreasing on $(-\infty, -2)$
Increasing on $(-2, \infty)$

6. Increasing on $(-\infty, -4)$
Decreasing on $(-4, \infty)$

7. Increasing on $(-\infty, -1)$ and $(1, \infty)$
Decreasing on $(-1, 1)$

8. Decreasing on $(-\infty, 0)$ and $\left(\sqrt[3]{2}, \infty\right)$
Increasing on $\left(0, \sqrt[3]{2}\right)$

9. Increasing on $(-\infty, 1)$ and $(1, \infty)$

10. Decreasing on $(-\infty, -3)$ and $\left(\frac{1}{3}, \infty\right)$
Increasing on $\left(-3, \frac{1}{3}\right)$

1.
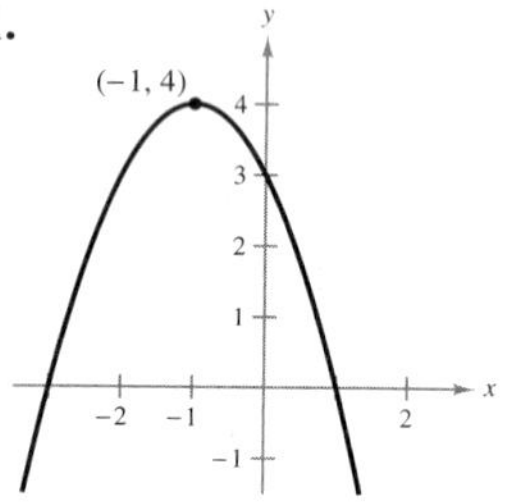

3.
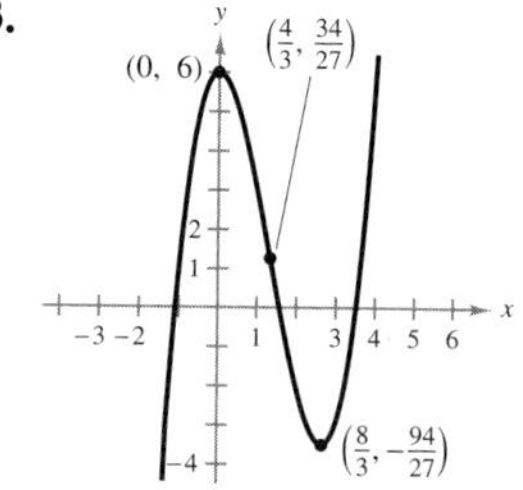

5.
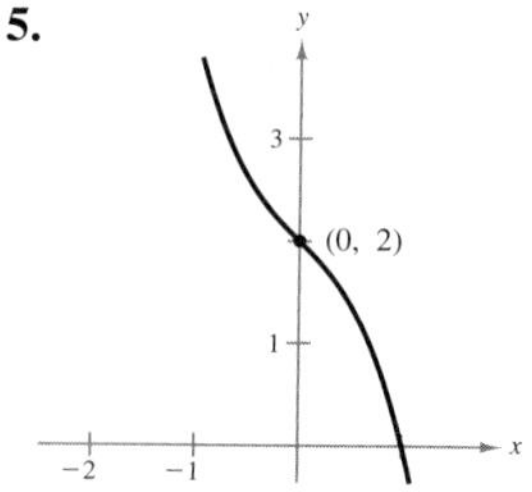

7.
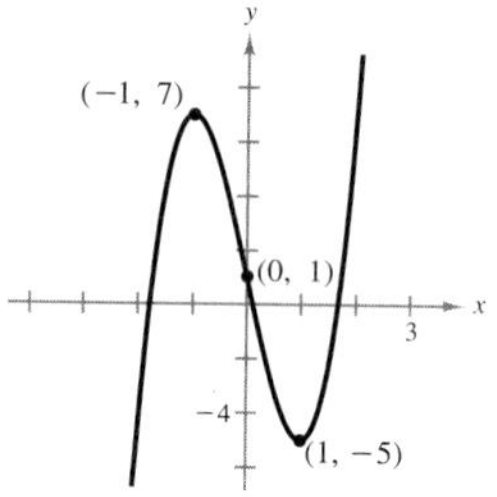

9.
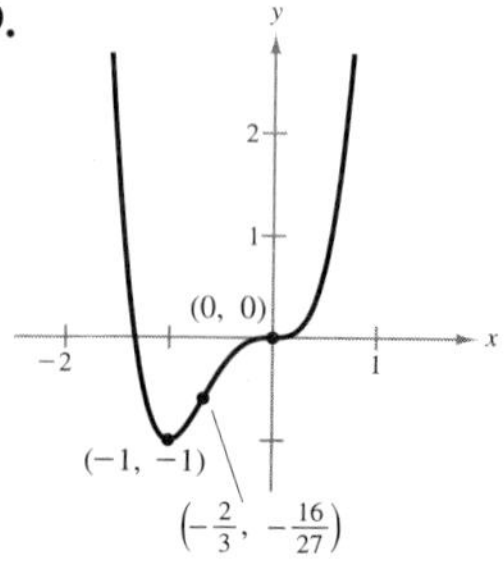

11.
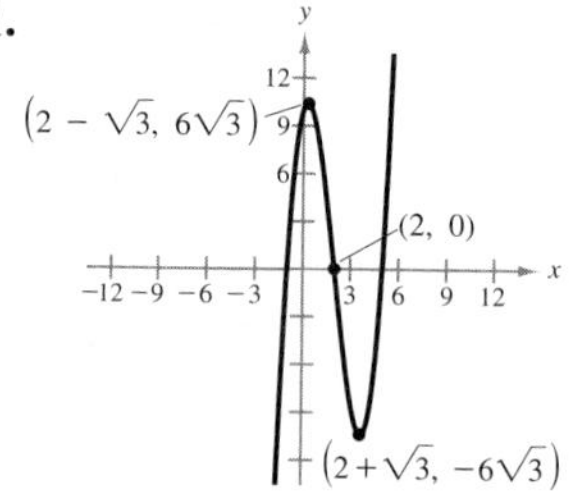

13.
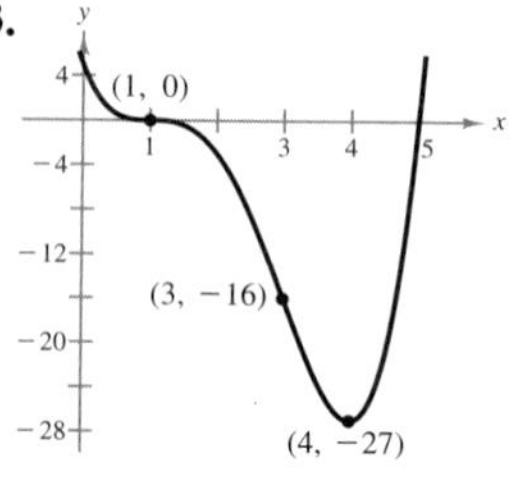

15.
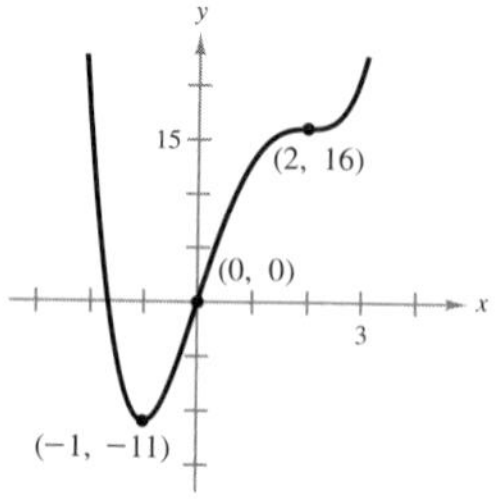

17.
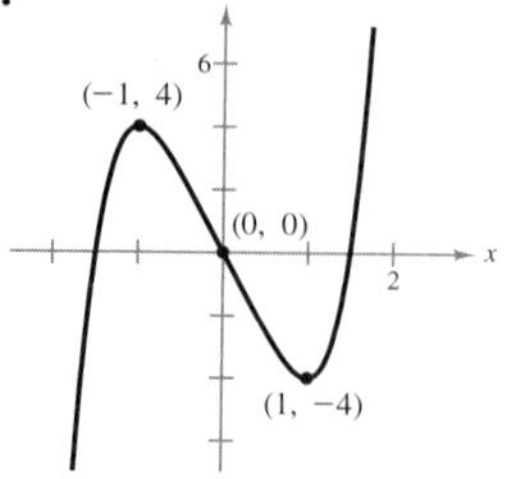

19.
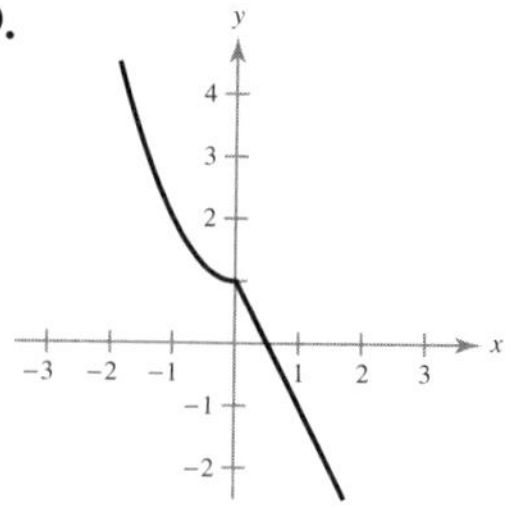

21.
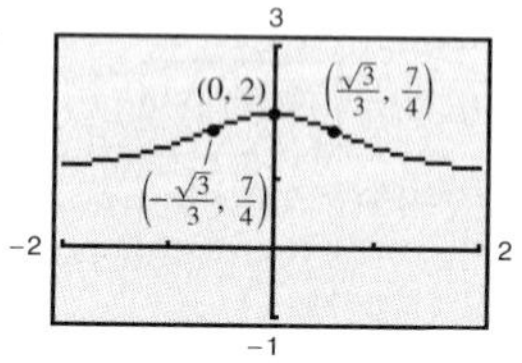

23.
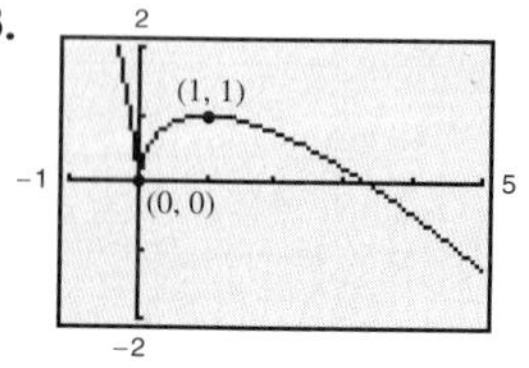

25.
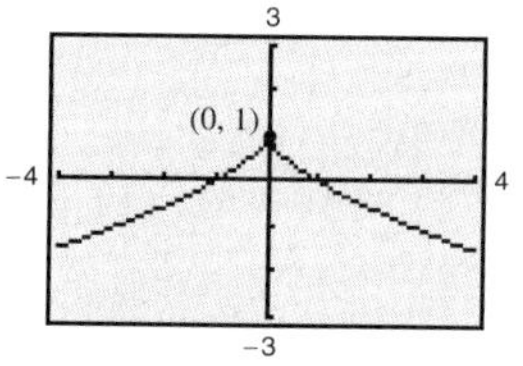

27.
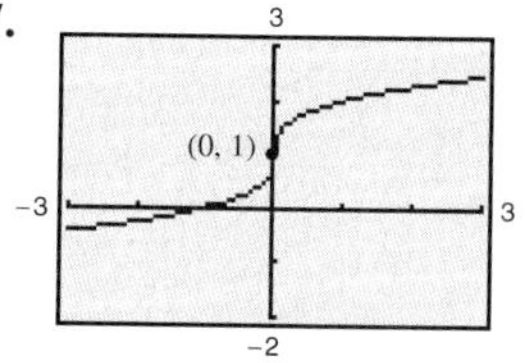

29.
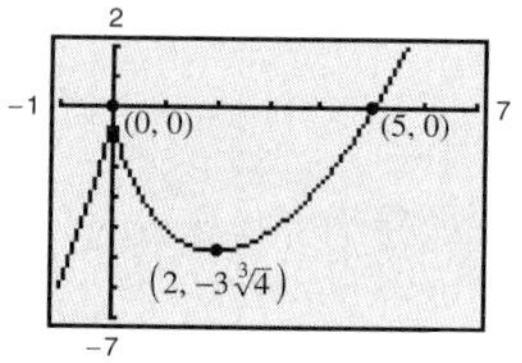

31.
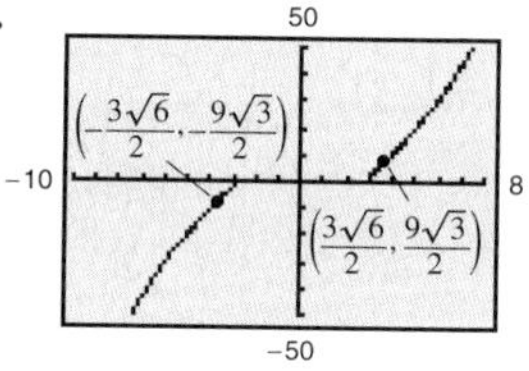

33.
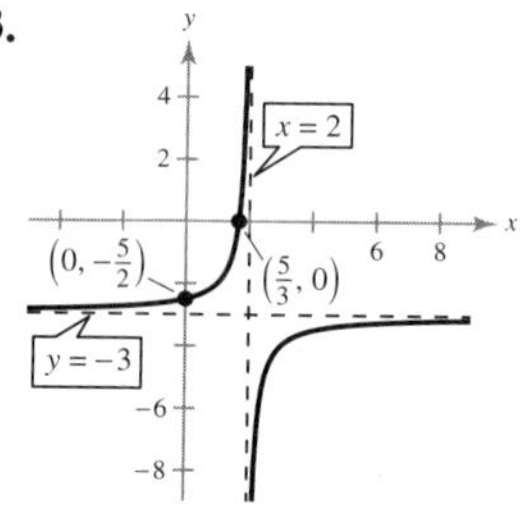

Domain: $(-\infty, 2) \cup (2, \infty)$

35.
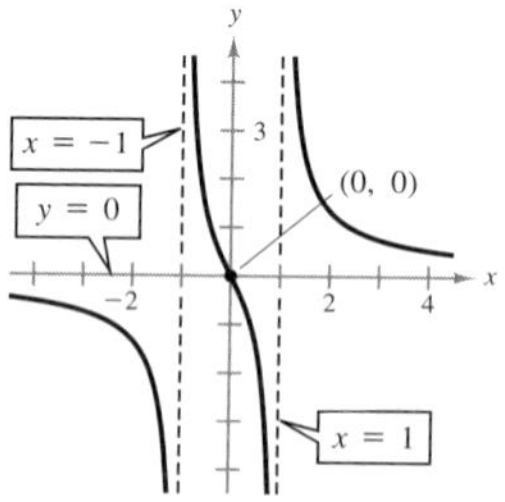

Domain: $(-\infty, -1) \cup (-1, 1) \cup (1, \infty)$

37.

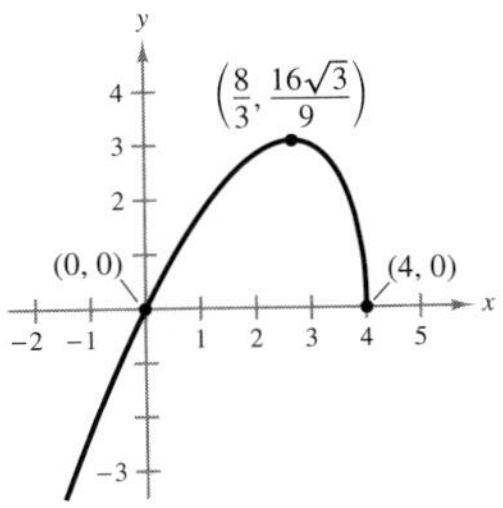

Domain: $(-\infty, 4]$

39.

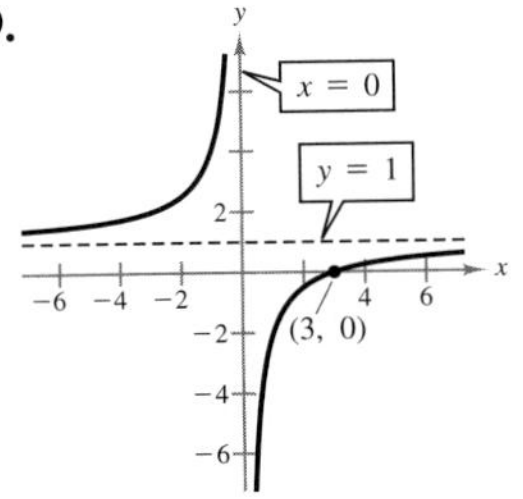

Domain: $(-\infty, 0) \cup (0, \infty)$

41.

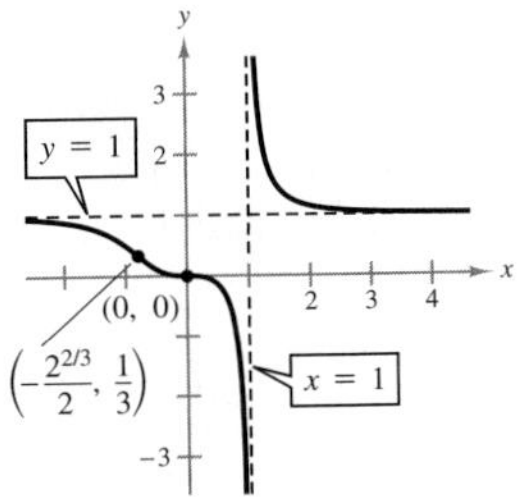

Domain: $(-\infty, 1) \cup (1, \infty)$

43. Answers will vary.

Sample answer: $f(x) = -x^3 + x^2 + x + 1$

45. Answers will vary. Sample answer: $f(x) = x^3 + 1$

47. Answers will vary.

Sample answer:

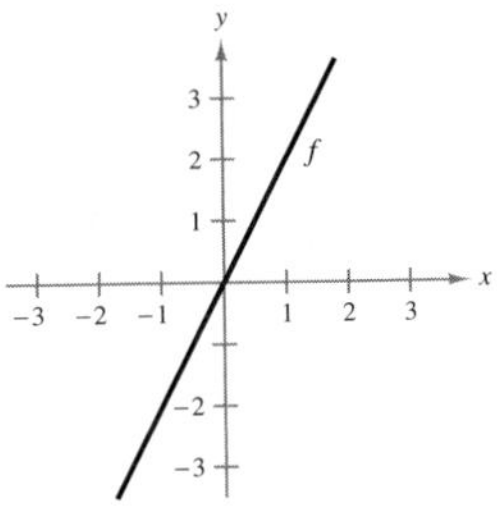

49. Answers will vary.

Sample answer:

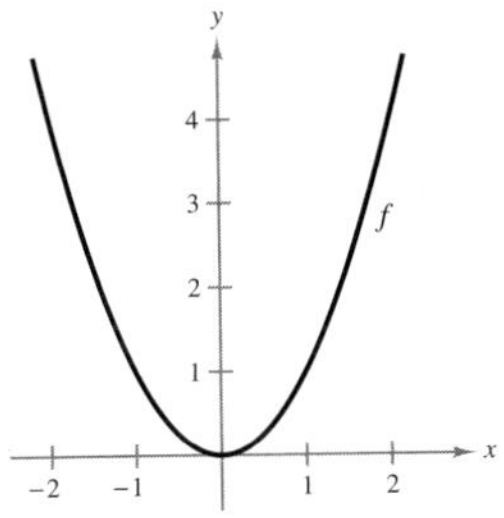

51. Answers will vary. Sample answer:

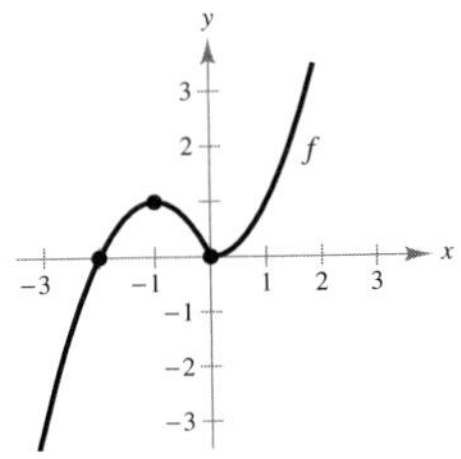

53. (a) $C = 0.36s + \dfrac{900}{s}, \quad 40 \le s \le 65$

(b)

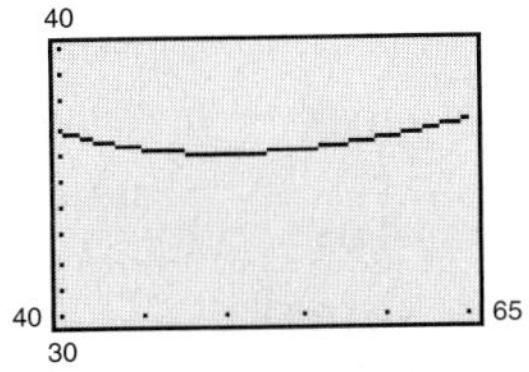

Most economical speed $\approx$ 50 miles per hour

55.

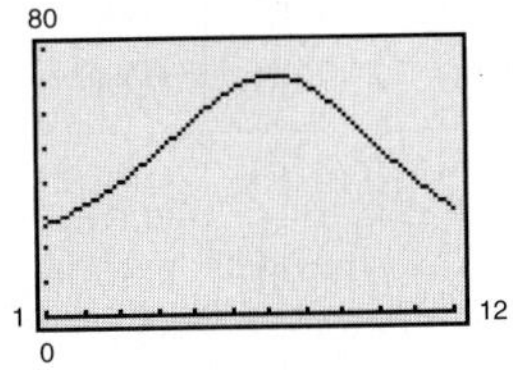

Absolute minimum: $(1, 27.23)$

Absolute maximum: $(7.13, 71.23)$

The monthly normal temperature reaches a minimum of 27.23°F in January and a maximum of 71.23°F at the beginning of July.

57.

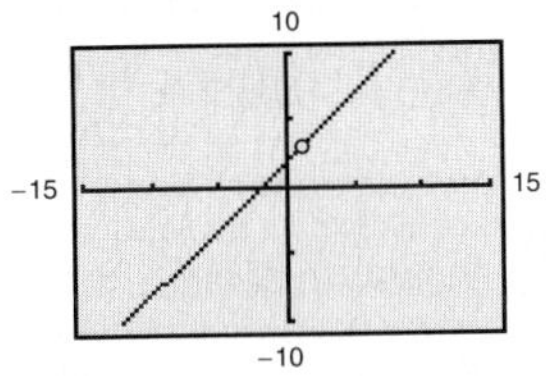

The rational function has the common factor of $x - 1$ in the numerator and denominator. At $x = 1$, there is a hole in the graph, not a vertical asymptote.

SECTION 3.8 *(page 246)*

Prerequisite Review

1. $\dfrac{dC}{dx} = 0.18x$ **2.** $\dfrac{dC}{dx} = 0.15$

3. $\dfrac{dR}{dx} = 1.25 + 0.03\sqrt{x}$ **4.** $\dfrac{dR}{dx} = 15.5 - 3.1x$

5. $\dfrac{dP}{dx} = -\dfrac{0.01}{\sqrt[3]{x^2}} + 1.4$ **6.** $\dfrac{dP}{dx} = -0.04x + 25$

7. $\dfrac{dA}{dx} = \dfrac{\sqrt{3}}{2}x$ **8.** $\dfrac{dA}{dx} = 12x$ **9.** $\dfrac{dC}{dr} = 2\pi$

10. $\dfrac{dP}{dw} = 4$ **11.** $\dfrac{dS}{dr} = 8\pi r$ **12.** $\dfrac{dP}{dx} = 2 + \sqrt{2}$

13. $A = \pi r^2$ **14.** $A = x^2$

15. $V = x^3$ **16.** $V = \frac{4}{3}\pi r^3$

1. $dy = 6x\,dx$ **3.** $dy = 12(4x - 1)^2\,dx$

5. $dy = \dfrac{x}{\sqrt{x^2 + 1}}\,dx$ **7.** 0.1005 **9.** -0.013245

11. $dy = 0.3$
$\Delta y = 0.331$

13. $dy = -0.04$
$\Delta y \approx -0.0394$

15.

$dx = \Delta x$	dy	Δy	$\Delta y - dy$	$\dfrac{dy}{\Delta y}$
1.000	4.000	5.0000	1.0000	0.8000
0.500	2.000	2.2500	0.2500	0.8889
0.100	0.400	0.4100	0.0100	0.9756
0.010	0.040	0.0401	0.0001	0.9975
0.001	0.004	0.0040	0.0000	1.0000

17.

$dx = \Delta x$	dy	Δy	$\Delta y - dy$	$\dfrac{dy}{\Delta y}$
1.000	−0.25000	−0.13889	0.11111	1.79999
0.500	−0.12500	−0.09000	0.03500	1.38889
0.100	−0.02500	−0.02324	0.00176	1.07573
0.010	−0.00250	−0.00248	0.00002	1.00806
0.001	−0.00025	−0.00025	0.00000	1.00000

19.

$dx = \Delta x$	dy	Δy	$\Delta y - dy$	$\dfrac{dy}{\Delta y}$
1.000	0.14865	0.12687	−0.02178	1.17167
0.500	0.07433	0.06823	−0.00610	1.08940
0.100	0.01487	0.01459	−0.00028	1.01919
0.010	0.00149	0.00148	−0.00001	1.00676
0.001	0.00015	0.00015	0.00000	1.00000

21. $y = 28x + 37$
For $\Delta x = -0.01$, $f(x + \Delta x) = -19.281302$ and
$y(x + \Delta x) = -19.28$
For $\Delta x = 0.01$, $f(x + \Delta x) = -18.721298$ and
$y(x + \Delta x) = -18.72$

23. $y = x$
For $\Delta x = -0.01$, $f(x + \Delta x) = -0.009999$ and
$y(x + \Delta x) = -0.01$
For $\Delta x = 0.01$, $f(x + \Delta x) = 0.009999$ and
$y(x + \Delta x) = 0.01$

25. A53(a) $\Delta p = -0.25 = dp$ (b) $\Delta p = -0.25 = dp$

27. \$5.20

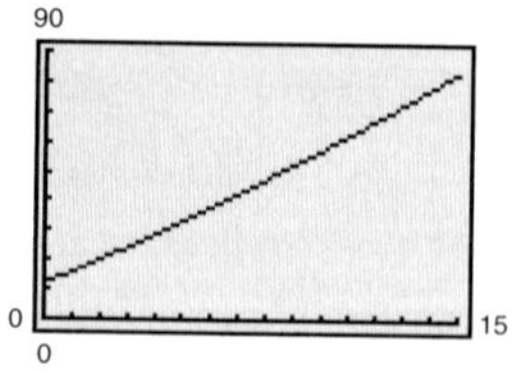

29. \$7.50

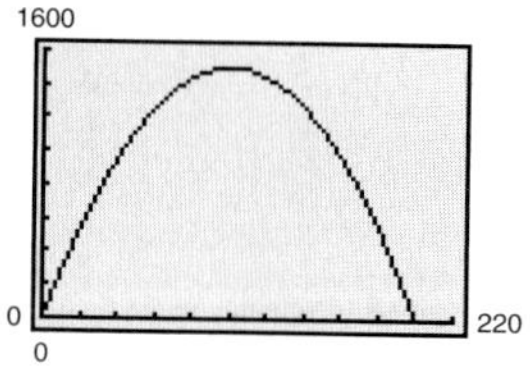

31. $-\$1250$

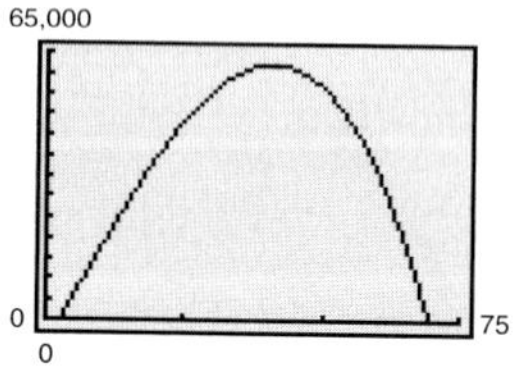

33. $R = -\frac{1}{3}x^2 + 100x$; \$6

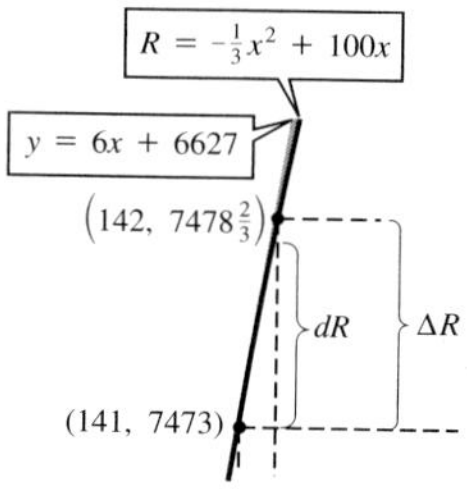

35. $P = -\dfrac{1}{2000}x^2 + 23x - 275{,}000$; $-\$5$

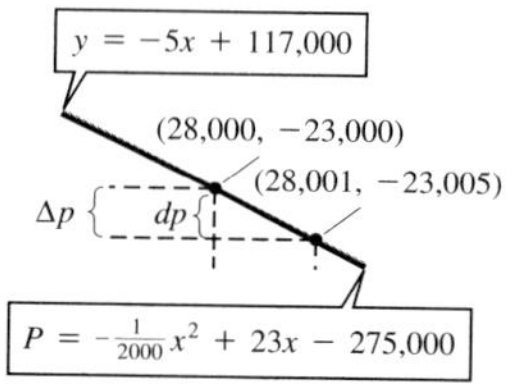

37. (a) $dA = 2x\,\Delta x$, $\Delta A = 2x\,\Delta x + (\Delta x)^2$
(b) and (c)

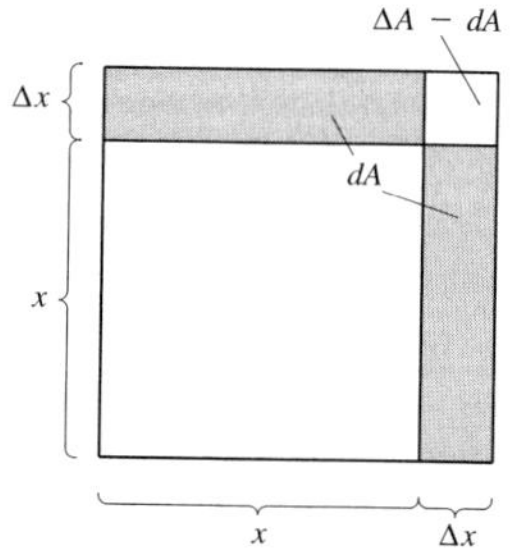

39. $\pm\frac{5}{2}\pi$ square inches, $\pm\frac{1}{40}$

41. $\pm 2.88\pi$ cubic inches, ± 0.01 **43.** True

REVIEW EXERCISES FOR CHAPTER 3

(page 252)

1. $x = 1$ **3.** $x = 0, x = 1$

5. Increasing on $\left(-\frac{1}{2}, \infty\right)$
Decreasing on $\left(-\infty, -\frac{1}{2}\right)$

7. Increasing on $(-\infty, 3)$ and $(3, \infty)$

9. (a) (1.48, 7.28) (b) (1, 1.48), (7.28, 12)

(c) Normal monthly temperature is rising from early January to July.

Normal monthly temperature is decreasing in early January and from early July to December.

(d)

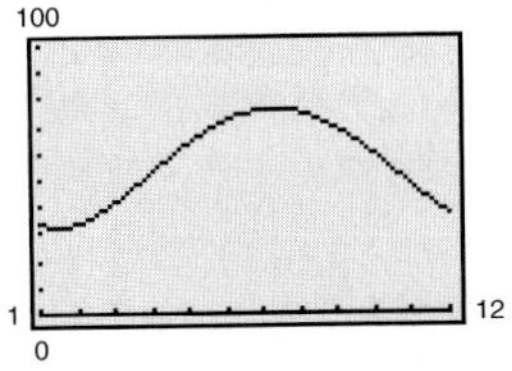

11. Relative maximum: $(0, -2)$

Relative minimum: $(1, -4)$

13. Relative minimum: $(8, -52)$

15. Relative maxima: $(-1, 1), (1, 1)$

Relative minimum: $(0, 0)$

17. Relative maximum: $(0, 6)$

19. Relative maximum: $(0, 0)$

Relative minimum: $(4, 8)$

21. Maximum: $(0, 6)$

Minimum: $\left(-\frac{5}{2}, -\frac{1}{4}\right)$

23. Maxima: $(-2, 17), (4, 17)$

Minima: $(-4, -15), (2, -15)$

25. Maximum: $(1, 3)$

Minimum: $\left(3, 4\sqrt{3} - 9\right)$

27. Maximum: $(2, 26)$

Minimum: $(1, -1)$

29. Maximum: $(1, 1)$

Minimum: $(-1, -1)$

31.

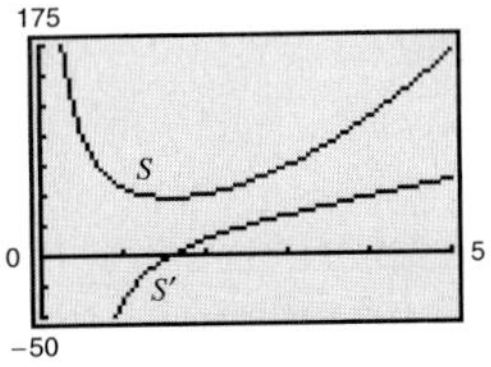

$r \approx 1.58$ inches

33. Concave upward on $(2, \infty)$

Concave downward on $(-\infty, 2)$

35. Concave upward on $\left(-\frac{2}{\sqrt{3}}, \frac{2}{\sqrt{3}}\right)$

Concave downward on $\left(-\infty, -\frac{2}{\sqrt{3}}\right)$ and $\left(\frac{2}{\sqrt{3}}, \infty\right)$

37. $(0, 0), (4, -128)$

39. $(0, 0), (1.0652, 4.5244), (2.5348, 3.5246)$

41. Relative maximum: $\left(-\sqrt{3}, 6\sqrt{3}\right)$

Relative minimum: $\left(\sqrt{3}, -6\sqrt{3}\right)$

43. Relative maximum: $(-4, 0)$

Relative minimum: $(-2, -108)$

45. $\left(50, 166\frac{2}{3}\right)$

47. 13, 13

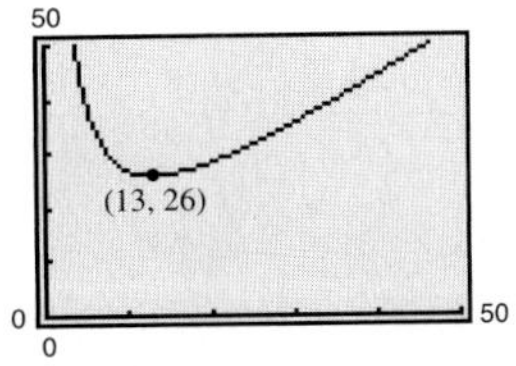

49. (a) Absolute maximum: (4.23, 1764.29)

Absolute minimum: (30, 1485)

(b) 1989

(c) The maximum number of daily newspapers in circulation was 1764.29 million in 1974 and the minimum number was 1485 million in 2000.

The year 1989 was when circulation was changing at the greatest rate.

51. $x = \frac{137}{9} \approx 15.2$ years

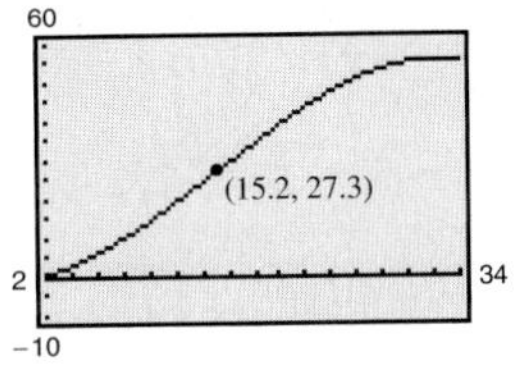

53. $s'(r) = -2cr$

$-2cr = 0 \Rightarrow r = 0$

$s''(r) = -2c < 0$ for all r

Therefore, $r = 0$ yields a maximum value of s.

55. $N = 85$ (maximizes revenue)

57. 125 units

59. Elastic: (0, 75)

Inelastic: (75, 150)

Demand is of unit elasticity when $x = 75$.

61. Elastic: (0, 200)

Inelastic: (200, 300)

Demand is of unit elasticity when $x = 200$.

63. Vertical asymptote: $x = 4$

Horizontal asymptote: $y = 2$

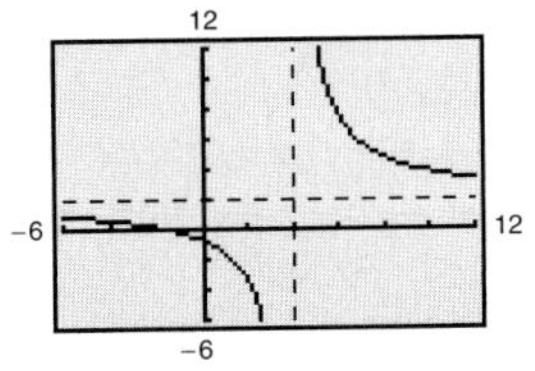

65. Vertical asymptote: $x = 0$

Horizontal asymptotes: $y = \pm 3$

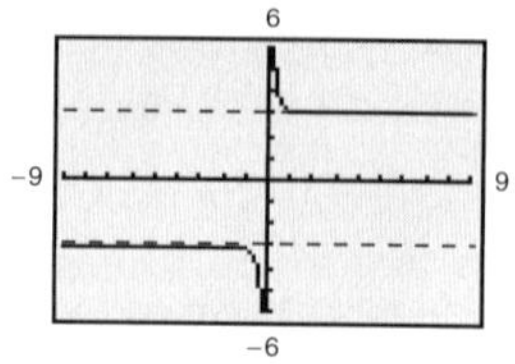

67. Vertical asymptotes: $x = 1, x = 4$

Horizontal asymptote: $y = 0$

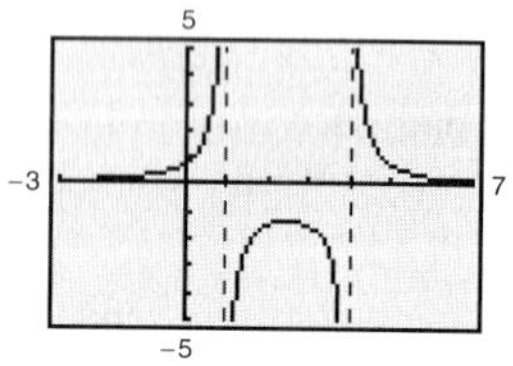

69. $-\infty$ **71.** ∞ **73.** $\frac{5}{2}$ **75.** $-\infty$

77. (a)

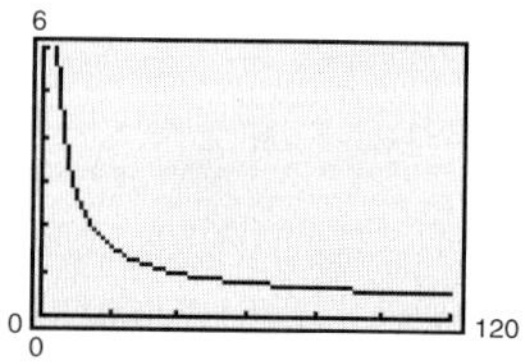

(b) $\lim_{s \to \infty} T = 0.37$

79.

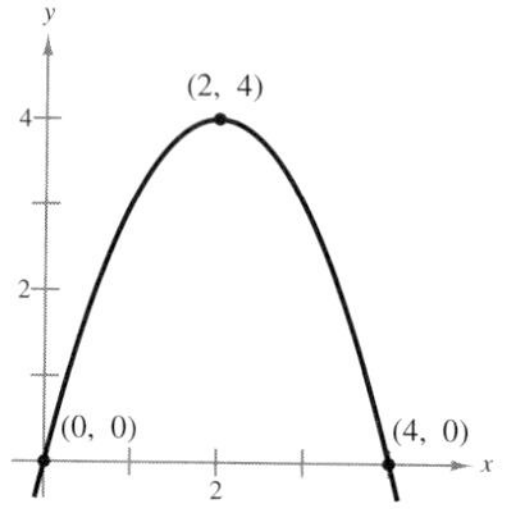

Domain: $(-\infty, \infty)$

81.

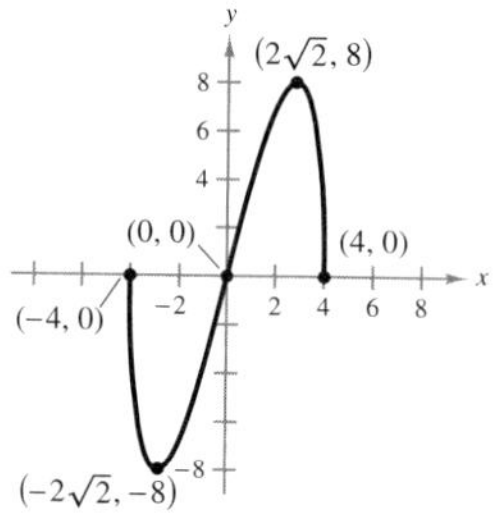

Domain: $[-4, 4]$

83.

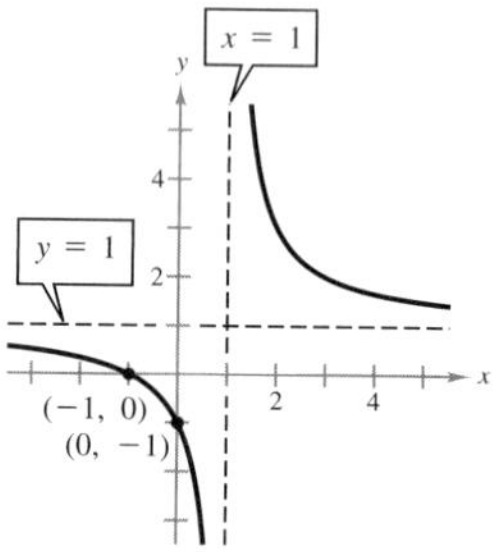

Domain: $(-\infty, 1) \cup (1, \infty)$

85.

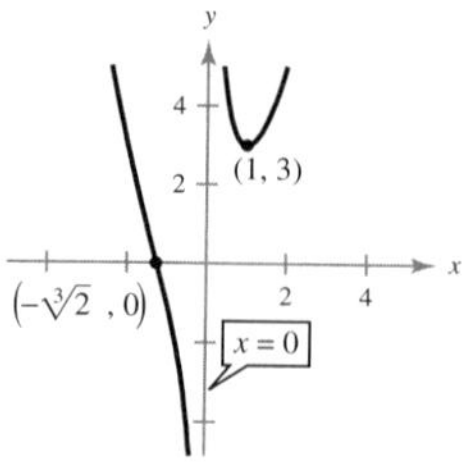

Domain: $(-\infty, 0) \cup (0, \infty)$

87. $dy = 12x\,dx$ **89.** $dy = \dfrac{5}{3x^{4/3}}\,dx$

91. \$800 **93.** \$15.25

95. (a)

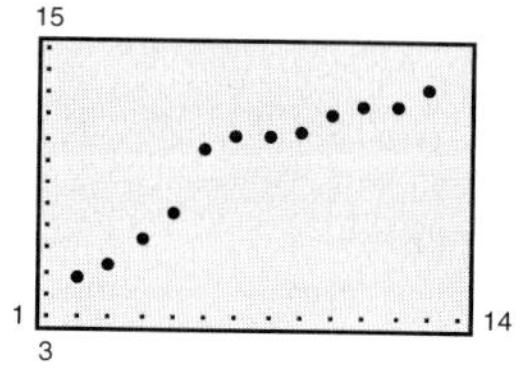

(b) Appears to be a positive correlation

(c)

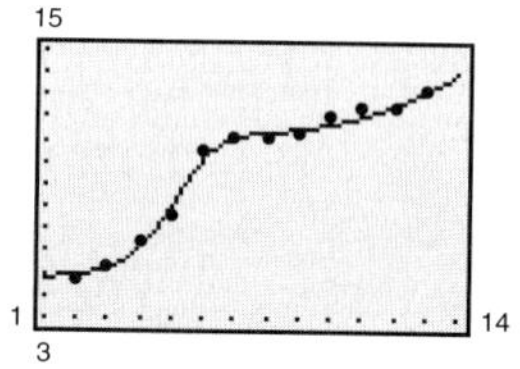

(d) Increasing: 1992 to 2003

(e) Increasing: 1992 to 1995 and 1998 to 2003

Decreasing: 1995 to 1998

(f) Answers will vary.

97. $dS = \pm 1.8\pi$ square inches

$dV = \pm 8.1\pi$ cubic inches

99.

Quantity of output	Price	Total revenue	Marginal revenue
1	14.00	14.00	10.00
2	12.00	24.00	6.00
3	10.00	30.00	4.00
4	8.50	34.00	1.00
5	7.00	35.00	−2.00
6	5.50	33.00	

(a) $R = -1.43x^2 + 13.8x + 2$

(b) $\frac{dR}{dx} = -2.86x + 13.8$;

10.94, 8.08, 5.22, 2.36, -0.50, -3.36;

The model is a fairly good estimate.

(c) About 5 units of output: (4.83, 35.29)

SAMPLE POST-GRAD EXAM QUESTIONS

(page 256)

1. b **2.** a **3.** b **4.** d **5.** b

CHAPTER 4

SECTION 4.1 *(page 262)*

Prerequisite Review

1. Horizontal shift to the left two units
2. Reflection about the x-axis
3. Vertical shift down one unit
4. Reflection about the y-axis
5. Horizontal shift to the right one unit
6. Vertical shift up two units
7. Nonremovable discontinuity at $x = -4$
8. Continuous on $(-\infty, \infty)$
9. Discontinuous at $x = \pm 1$
10. Continuous on $(-\infty, \infty)$

11. 5 **12.** $\frac{4}{3}$ **13.** $-9, 1$ **14.** $2 \pm 2\sqrt{2}$

15. $1, -5$ **16.** $\frac{1}{2}, 1$

1. (a) 625 (b) 9 (c) $16\sqrt{2}$
(d) 9 (e) 125 (f) 4

3. (a) 3125 (b) $\frac{1}{5}$ (c) 625 (d) $\frac{1}{125}$

5. (a) $\frac{1}{5}$ (b) 27 (c) 5 (d) 4096

7. (a) 4 (b) $\frac{\sqrt{2}}{2} \approx 0.707$ (c) $\frac{1}{8}$ (d) $\frac{\sqrt{2}}{8} \approx 0.177$

9. (a) 0.907 (b) 348.912 (c) 1.796 (d) 1.308

11. 4 **13.** -2 **15.** 2 **17.** 16 **19.** e

20. c **21.** a **22.** f **23.** d **24.** b

25.

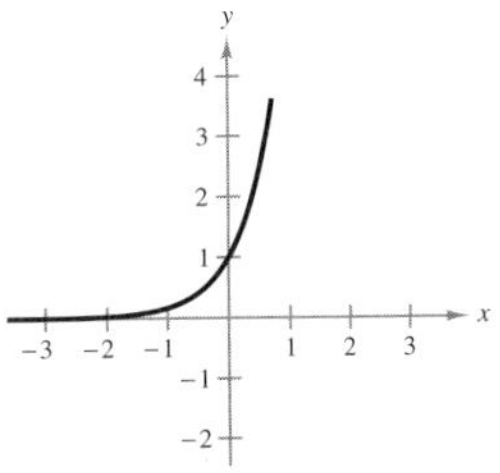

27.

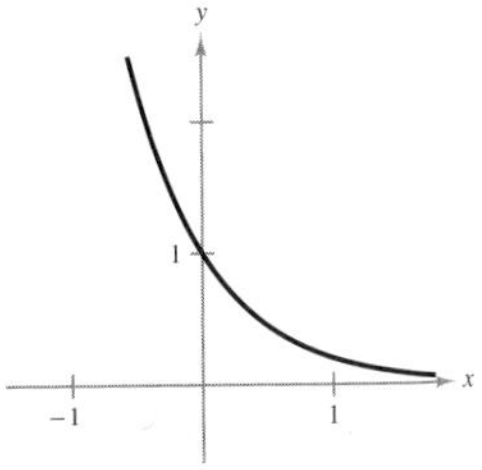

29.

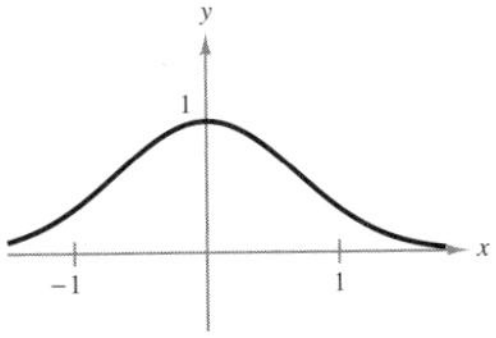

31.

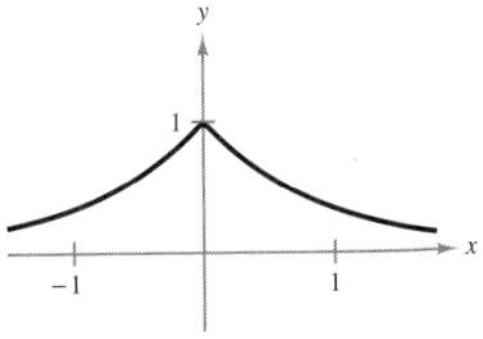

33.

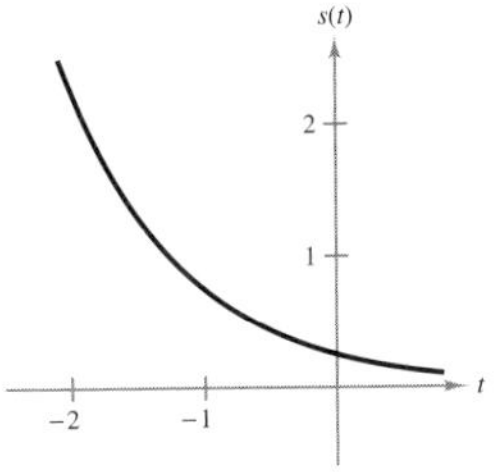

35. (a) $P(16) \approx 303.15$ million
(b) $P(22) \approx 325.25$ million

37. (a) $V(5) \approx \$80{,}634.95$ (b) $V(20) \approx \$161{,}269.89$

39. $V(t) = 16{,}000\left(\frac{3}{4}\right)^t$

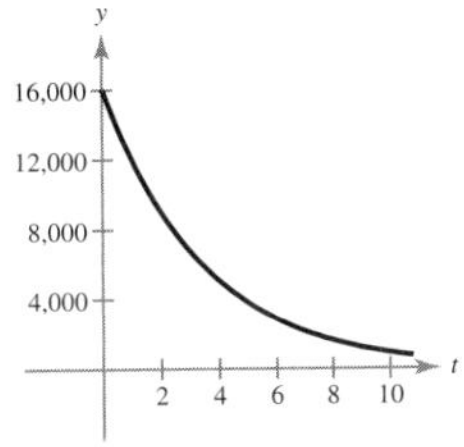

$V(4) = 16{,}000\left(\frac{3}{4}\right)^4 \approx \$5062.50$

41. (a)

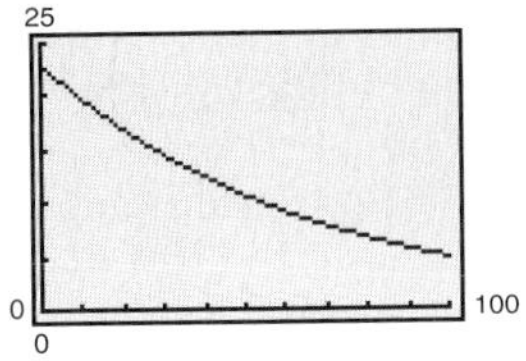

(b) $y(75) \approx 7.24$ grams (c) $y = 1$ for $t \approx 203.56$ years

SECTION 4.2 *(page 270)*

Prerequisite Review

1. Continuous on $(-\infty, \infty)$
2. Discontinuous for $x = \pm 2$
3. Discontinuous for $x = \pm\sqrt{3}$
4. Removable discontinuity at $x = 4$

5. 0 **6.** 0 **7.** 4 **8.** $\frac{1}{2}$ **9.** $\frac{3}{2}$
10. 6 **11.** 0 **12.** 0

1. (a) e^7 (b) e^{12} (c) $\dfrac{1}{e^6}$ (d) 1

3. (a) e^5 (b) $e^{5/2}$ (c) e^6 (d) e^7

5. $-\frac{1}{3}$ **7.** 9 **9.** $\pm e$ **11.** $\sqrt[3]{3}e$ **13.** f

14. e **15.** d **16.** b **17.** c **18.** a

19.

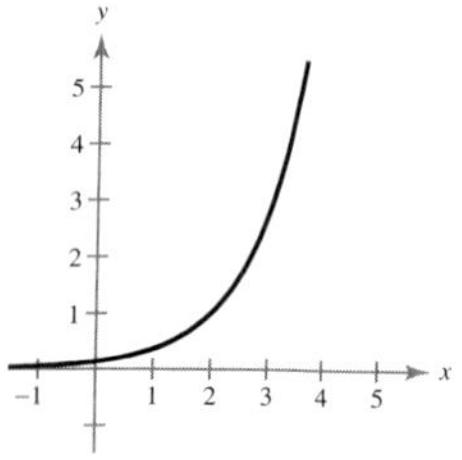

21.

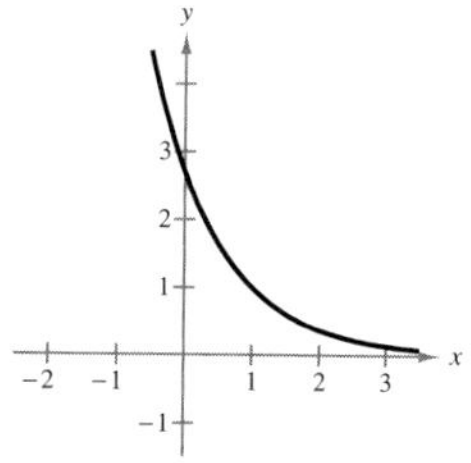

23.

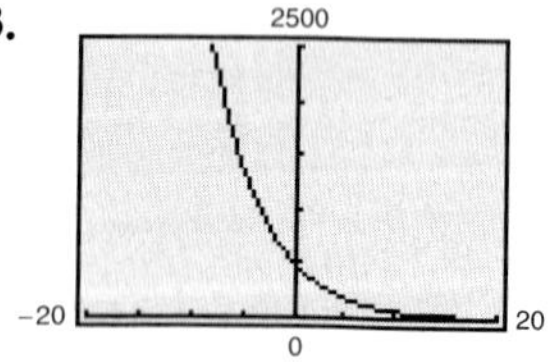

25.

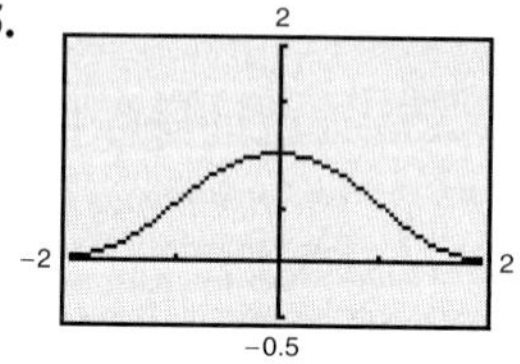

27.

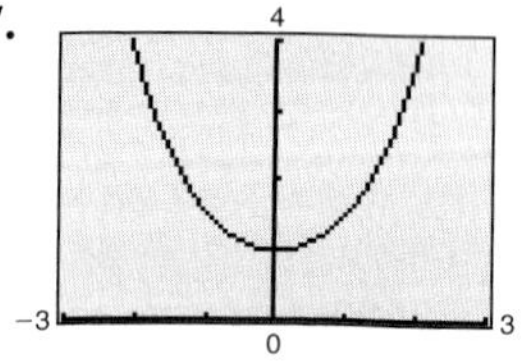

No horizontal asymptotes

Continuous on the entire real number line

29.

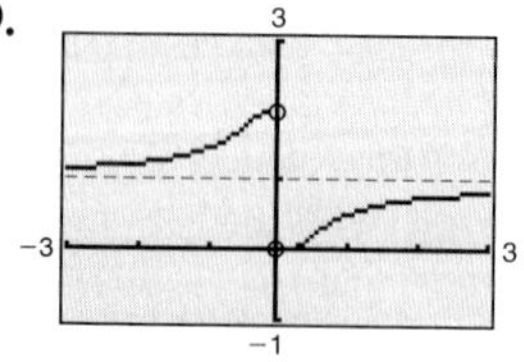

Horizontal asymptote: $y = 1$

Discontinuous at $x = 0$

31.

n	1	2	4	12
A	1343.92	1346.86	1348.35	1349.35

n	365	Continuous compounding
A	1349.84	1349.86

33.

n	1	2	4	12
A	3262.04	3290.66	3305.28	3315.15

n	365	Continuous compounding
A	3319.95	3320.12

35.

t	1	10	20
P	96,078.94	67,032.00	44,932.90

t	30	40	50
P	30,119.42	20,189.65	13,533.53

37.

t	1	10	20
P	95,132.82	60,716.10	36,864.45

t	30	40	50
P	22,382.66	13,589.88	8251.24

39. (a) 9% (b) 9.2% (c) 9.31% (d) 9.38%

41. \$12,500 **43.** \$8751.92

45. (a) \$849.53 (b) \$421.12

$\lim_{x\to\infty} p = 0$

47. (a) 0.1535 (b) 0.4866 (c) 0.8111

49. (a) 1995: \$4510.69 million; 2000: \$5719.90 million
2003: \$6595.92 million

(b) Yes. There is a positive correlation between sales and time in years.

(c) 2011

51. (a)

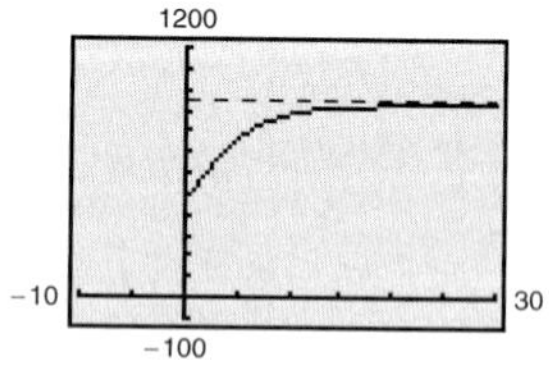

(b) Yes, $\lim_{t\to\infty} \dfrac{925}{1 + e^{-0.3t}} = 925$

(c) $\lim_{t\to\infty} \dfrac{1000}{1 + e^{-0.3t}} = 1000$

Models similar to this logistic growth model where $y = \dfrac{a}{1 + be^{-ct}}$ have a limit of a as $t \to \infty$.

53. (a) 0.731 (b) 11 (c) Yes, $\lim_{n\to\infty} \dfrac{0.83}{1 + e^{-0.2n}} = 0.83$

SECTION 4.3 *(page 279)*

Prerequisite Review

1. $\frac{1}{2}e^x(2x^2 - 1)$ **2.** $\dfrac{e^x(x + 1)}{x}$ **3.** $e^x(x - e^x)$

4. $e^{-x}(e^{2x} - x)$ **5.** $-\dfrac{6}{7x^3}$ **6.** $6x - \dfrac{1}{6}$

7. $6(2x^2 - x + 6)$ **8.** $\dfrac{t + 2}{2t^{3/2}}$

9. Relative maximum: $\left(-\dfrac{4\sqrt{3}}{3}, \dfrac{16\sqrt{3}}{9}\right)$

Relative minimum: $\left(\dfrac{4\sqrt{3}}{3}, -\dfrac{16\sqrt{3}}{9}\right)$

10. Relative maximum: $(0, 5)$

Relative minima: $(-1, 4), (1, 4)$

1. 3 **3.** -1 **5.** $4e^{4x}$ **7.** $-2xe^{-x^2}$

9. $\dfrac{2}{x^3}e^{-1/x^2}$ **11.** $e^{4x}(4x^2 + 2x + 4)$

13. $-\dfrac{6(e^x - e^{-x})}{(e^x + e^{-x})^4}$ **15.** $xe^x + e^x + 4e^{-x}$

17. $y = -2x + 1$ **19.** $y = \dfrac{4}{e^2}$ **21.** $y = 24x + 8$

23. $\dfrac{dy}{dx} = \dfrac{1}{2}(-x - 1 - 2y)$ $\left(\text{Equivalently, } \dfrac{dy}{dx} = -\dfrac{1}{2}\right)$

25. $\dfrac{dy}{dx} = \dfrac{e^{-x}(x^2 - 2x) + y}{4y - x}$

27. $6(3e^{3x} + 2e^{-2x})$ **29.** $5e^{-x} - 50e^{-5x}$

31.

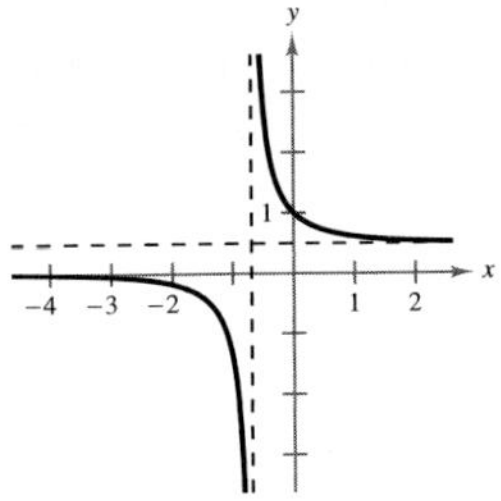

No relative extrema

No points of inflection

Horizontal asymptote to the right: $y = \frac{1}{2}$

Horizontal asymptote to the left: $y = 0$

Vertical asymptote: $x \approx -0.693$

33.

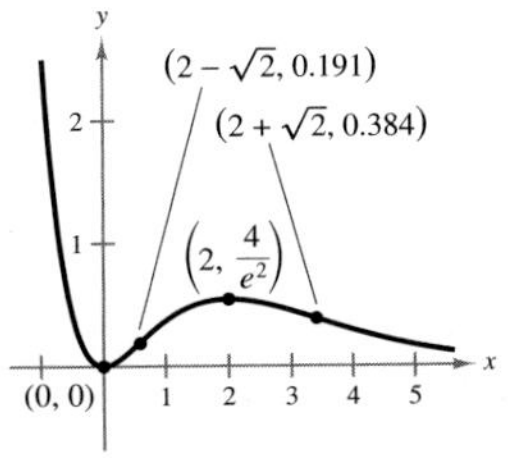

Relative minimum: $(0, 0)$

Relative maximum: $(2, 4e^{-2})$

Points of inflection: $(2 - \sqrt{2}, 0.191), (2 + \sqrt{2}, 0.384)$

35.

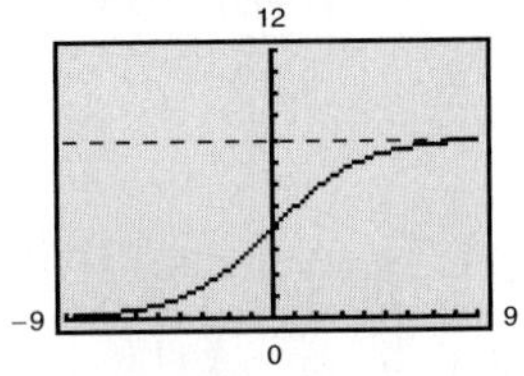

Asymptotes: $y = 0, y = 8$

37. (a)

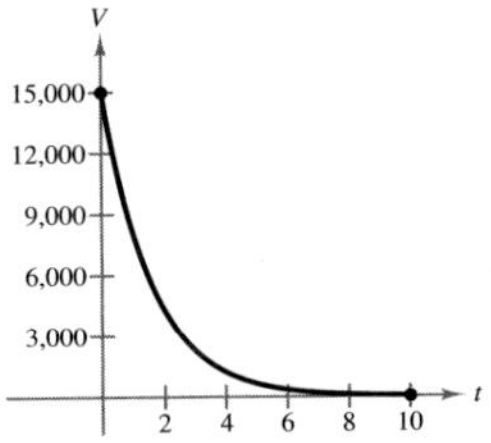

(b) $-\$5028.84$/year (c) $-\$406.89$/year

(d) $v = -1497.2t + 15{,}000$

(e) In the exponential function, the initial rate of depreciation is greater than in a linear model. The linear model has a constant rate of depreciation.

39. (a)

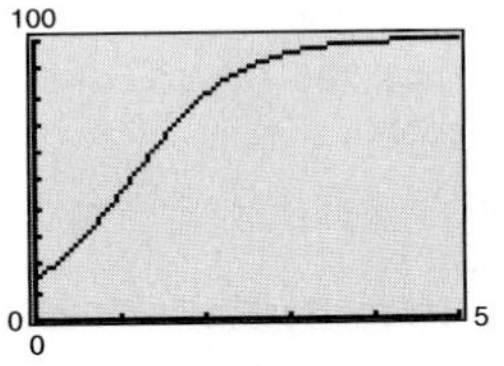

(b) 80.3%

(c) $x \approx 1.1$ or approximately 1100 egg masses

41. (a) \$433.31/year (b) \$890.22/year

(c) \$21,839.26/year

43. $t = 5$: 14.44

$t = 10$: 3.63

$t = 25$: 0.58

45. (a) $f(x) = \dfrac{1}{12.5\sqrt{2\pi}}e^{-(x-650)^2/2(12.5)^2}$

$= \dfrac{1}{12.5\sqrt{2\pi}}e^{-(x-650)^2/312.5}$

(b)

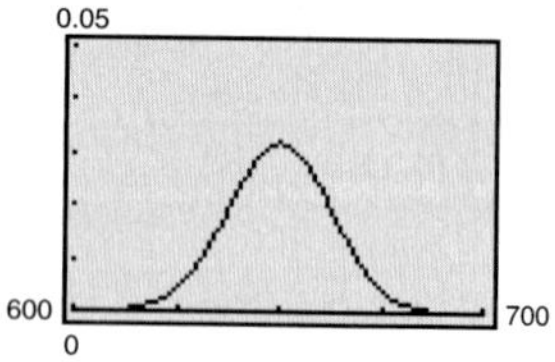

(c) $f'(x) = \dfrac{-4\sqrt{2}(x-650)e^{-2(x-650)^2/624}}{15625\sqrt{\pi}}$

(d) Answers will vary.

47.

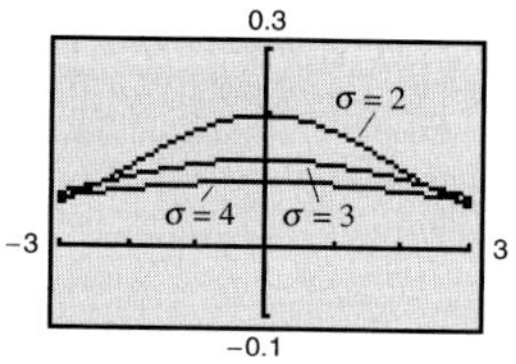

As σ increases, the graph becomes flatter.

49. (a) $\dfrac{dh}{dt} = -80e^{-1.6t} - 20$

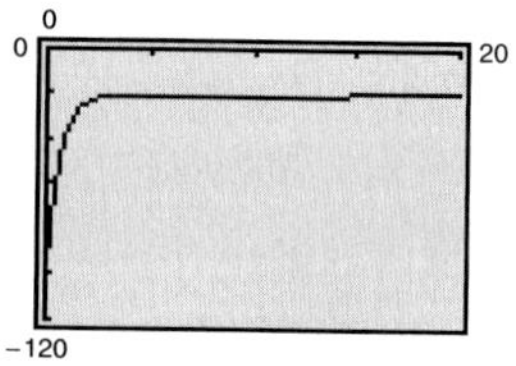

(b) $-100, -36.15, -20.03, -20.00, -20.00$

(c) The values in (b) are rates of descent in feet per second. As time increases, the rate is approximately constant at -20 feet/second.

SECTION 4.4 *(page 287)*

Prerequisite Review

1. $\frac{1}{4}$ **2.** 64 **3.** 3^6 **4.** $\left(\frac{2}{3}\right)^3$ **5.** 1

6. $81e^4$ **7.** $\dfrac{e^3}{2}$ **8.** $\dfrac{125}{8e^3}$ **9.** $x > -4$

10. Any real number x **11.** $x < -1$ or $x > 1$

12. $x > 5$ **13.** \$3462.03 **14.** \$3374.65

1. $e^{0.6931\ldots} = 2$ **3.** $e^{-1.6094\ldots} = 0.2$ **5.** $\ln 1 = 0$

7. $\ln(0.0498\ldots) = -3$ **9.** c **10.** d

11. b **12.** a

13.

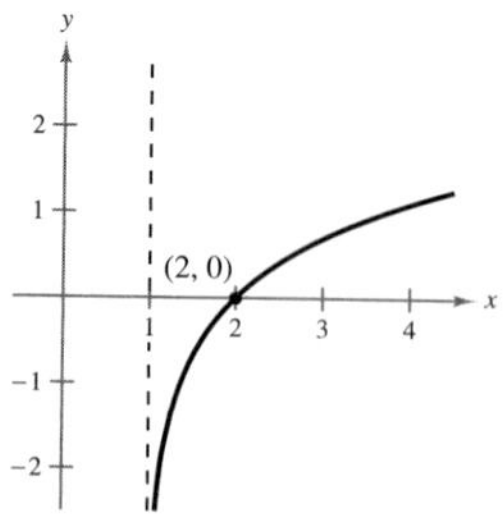

15.

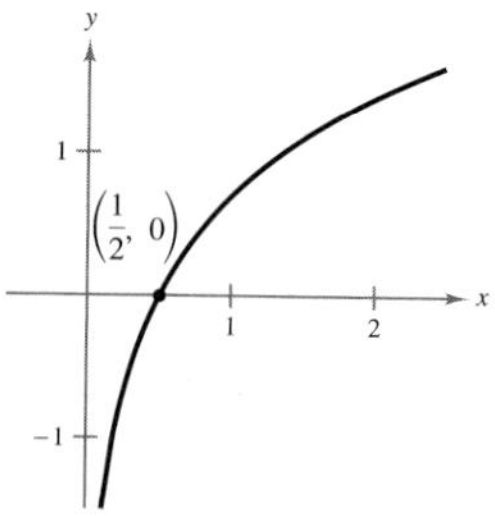

17.

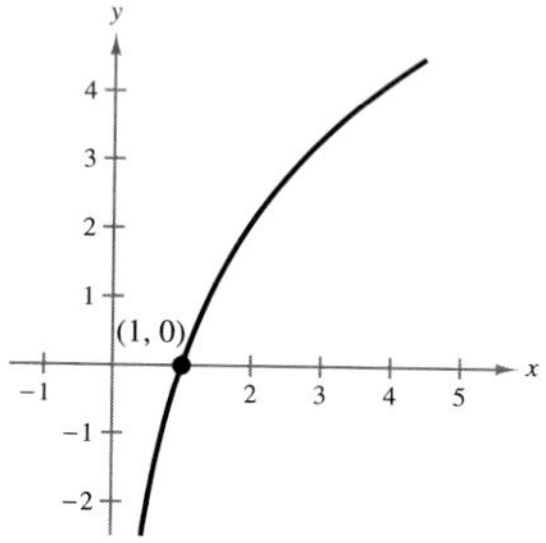

19. Answers will vary.

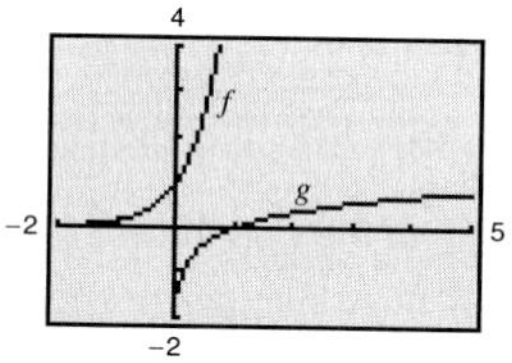

21. Answers will vary.

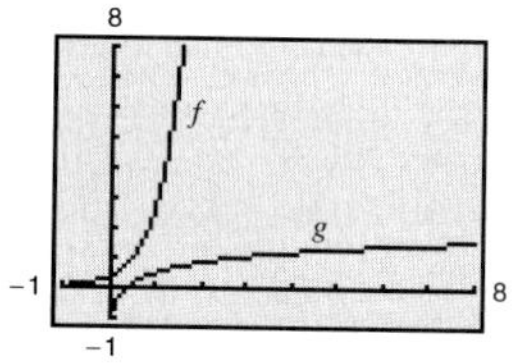

23. x^2 **25.** $5x + 2$ **27.** $2x - 1$

29. (a) 1.7917 (b) 0.4055 (c) 4.3944 (d) 0.5493

31. $\ln 2 - \ln 3$ **33.** $\ln x + \ln y + \ln z$

35. $\frac{1}{2}\ln(x^2 + 1)$ **37.** $\ln z + 2\ln(z - 1)$

39. $\ln 3 + \ln x + \ln(x + 1) - 2\ln(2x + 1)$

41. $\ln \dfrac{x-2}{x+2}$ **43.** $\ln \dfrac{x^3y^2}{z^4}$ **45.** $\ln\left[\dfrac{x(x+3)}{x+4}\right]^3$

47. $\ln\left[\dfrac{x(x^2+1)}{x+1}\right]^{3/2}$ **49.** $\ln \dfrac{(x+1)^{1/3}}{(x-1)^{2/3}}$

51. $x = 4$ **53.** $x = 1$ **55.** $x = \ln 4 - 1 \approx 0.3863$

57. $t = \dfrac{\ln 7 - \ln 3}{-0.2} \approx -4.2365$

59. $x = \frac{1}{2}\left(1 + \ln\frac{3}{2}\right) \approx 0.7027$

61. $x = -100\ln\frac{3}{4} \approx 28.7682$

63. $x = \dfrac{\ln 15}{2\ln 5} \approx 0.8413$ **65.** $t = \dfrac{\ln 2}{\ln 1.07} \approx 10.2448$

67. $t = \dfrac{\ln 3}{12\ln[1 + (0.07/12)]} \approx 15.740$

69. (a) 14.21 years (b) 13.89 years
(c) 13.86 years (d) 13.86 years

71. $t \approx -12{,}194$ years

73. (a) $P(20) \approx 1{,}681{,}900$ (b) 2011

75. 9395 years **77.** 12,484 years

79. (a) 80 (b) 57.5 (c) 10 months

81.

x	y	$\dfrac{\ln x}{\ln y}$	$\ln \dfrac{x}{y}$	$\ln x - \ln y$
1	2	0	-0.6931	-0.6931
3	4	0.7925	-0.2877	-0.2877
10	5	1.4307	0.6931	0.6931
4	0.5	-2.0000	2.0794	2.0794

83.

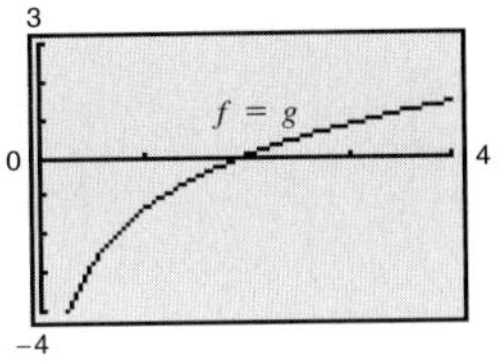

85. False. $f(x) = \ln x$ is undefined for $x \le 0$.

87. False. $f\left(\dfrac{x}{2}\right) = f(x) - f(2)$ **89.** False. $u = v^2$

SECTION 4.5 *(page 296)*

Prerequisite Review

1. $2 \ln(x + 1)$ **2.** $\ln x + \ln(x + 1)$

3. $\ln x - \ln(x + 1)$ **4.** $3[\ln x - \ln(x - 3)]$

5. $\ln 4 + \ln x + \ln(x - 7) - 2 \ln x$

6. $3 \ln x + \ln(x + 1)$

7. $-\dfrac{y}{x + 2y}$ **8.** $\dfrac{3 - 2xy + y^2}{x(x - 2y)}$

9. $-12x + 2$ **10.** $-\dfrac{6}{x^4}$

1. 3 **3.** 2 **5.** $\dfrac{2}{x}$ **7.** $\dfrac{2x}{x^2 + 3}$ **9.** $\dfrac{2(x^3 - 1)}{x(x^3 - 4)}$

11. $\dfrac{3}{x}(\ln x)^5$ **13.** $x(1 + \ln x^2)$ **15.** $\dfrac{2x^2 - 1}{x(x^2 - 1)}$

17. $\dfrac{1}{x(x + 1)}$ **19.** $\dfrac{2}{3(x^2 - 1)}$ **21.** $-\dfrac{4}{x(4 + x^2)}$

23. $e^{-x}\left(\dfrac{1}{x} - \ln x\right)$ **25.** $\dfrac{e^x - e^{-x}}{e^x + e^{-x}}$ **27.** $e^{x(\ln 2)}$

29. $\dfrac{1}{\ln 4} \ln x$ **31.** 5.585 **33.** -0.631

35. -2.134 **37.** $(\ln 3)3^x$ **39.** $\dfrac{1}{x \ln 2}$

41. $(2 \ln 4)4^{2x-3}$ **43.** $\dfrac{2x + 6}{(x^2 + 6x) \ln 10}$

45. $2^x(1 + x \ln 2)$ **47.** $y = x - 1$

49. $y = \dfrac{1}{3 \ln 3}x - \dfrac{2}{9 \ln 3} + 2$ or $y = 0.303x + 1.798$

51. $\dfrac{2xy}{3 - 2y^2}$ **53.** $\dfrac{y(1 - 6x^2)}{1 + y}$ **55.** $\dfrac{1}{2x}$ **57.** $(\ln 5)^2 5^x$

59. $\dfrac{d\beta}{dI} = \dfrac{10}{(\ln 10)I}$; for $I = 10^{-4}$, the rate of change is about 43,429.4 decibels per watt per square centimeters.

61. 2, $y = 2x - 1$ **63.** $-\dfrac{8}{5}$, $y = -\dfrac{8}{5}x - 4$

65. $\dfrac{1}{\ln 2}$, $y = \dfrac{1}{\ln 2}x - \dfrac{1}{\ln 2}$

67.

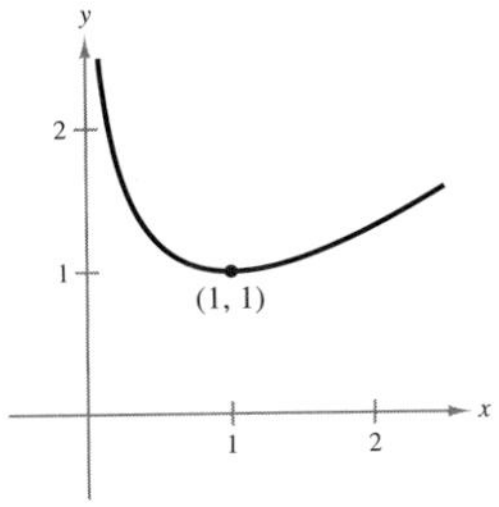

Relative minimum: $(1, 1)$

69.

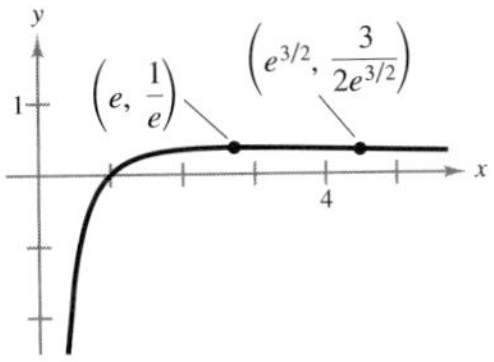

Relative maximum: $\left(e, \dfrac{1}{e}\right)$

Point of inflection: $\left(e^{3/2}, \dfrac{3}{2e^{3/2}}\right)$

71.

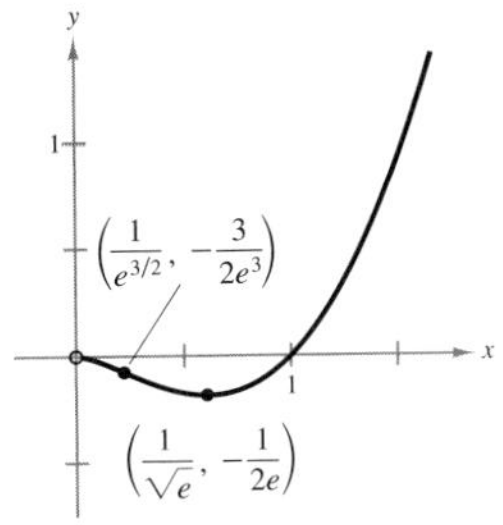

Relative minimum: $\left(\dfrac{1}{\sqrt{e}}, -\dfrac{1}{2e}\right)$

Point of inflection: $\left(\dfrac{1}{e^{3/2}}, -\dfrac{3}{2e^3}\right)$

73. $-\frac{1}{p}, -\frac{1}{10}$ **75.** $-\frac{1000p}{(p^2+1)[\ln(p^2+1)]^2}, -4.65$

77. $p = 1000e^{-x}$

$\frac{dp}{dx} = -1000e^{-x}$

At $p = 10$, rate of change $= -10$.

$\frac{dp}{dx}$ and $\frac{dx}{dp}$ are reciprocals of each other.

79. (a) $\overline{C} = \frac{500 + 300x - 300 \ln x}{x}$

(b) Minimum of 279.15 at $e^{8/3}$

(c)

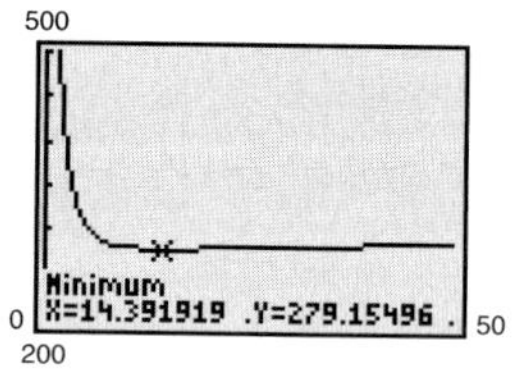

81. (a)

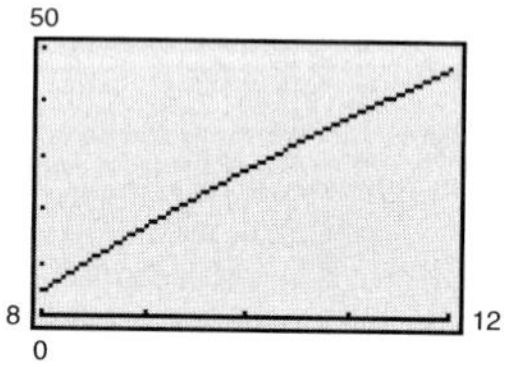

(b) \$27.56 billion (c) 10.33

83. Answers will vary.

SECTION 4.6 *(page 305)*

Prerequisite Review

1. $-\frac{1}{4}\ln 2$ **2.** $\frac{1}{5}\ln\frac{10}{3}$ **3.** $-\frac{\ln(25/16)}{0.01}$

4. $-\frac{\ln(11/16)}{0.02}$ **5.** $7.36e^{0.23t}$ **6.** $1.296e^{0.072t}$

7. $-33.6e^{-1.4t}$ **8.** $-0.025e^{-0.001t}$ **9.** 4

10. 12 **11.** $2x + 1$ **12.** $x^2 + 1$

1. $y = 2e^{0.1014t}$ **3.** $y = 4e^{-0.4159t}$

5. $y = 0.6687e^{0.4024t}$ **7.** $y = 10e^{2t}$, exponential growth

9. $y = 30e^{-4t}$, exponential decay

11. *Amount after 1000 years:* 6.48 grams

Amount after 10,000 years: 0.13 gram

13. *Initial quantity:* 6.73 grams

Amount after 1000 years: 5.96 grams

15. *Initial quantity:* 2.16 grams

Amount after 10,000 years: 1.62 grams

17. 68% **19.** 15,642 years

21. $k_1 = \frac{\ln 4}{12} \approx 0.1155$, so $y_1 = 5e^{0.1155t}$.

$k_2 = \frac{1}{6}$, so $y_2 = 5(2)^{t/6}$

Explanations will vary.

23. (a) 1350 (b) $\frac{5 \ln 2}{\ln 3} \approx 3.15$ hours

(c) No. Answers will vary.

25. *Time to double:* 5.78 years

Amount after 10 years: \$3320.12

Amount after 25 years: \$20,085.54

27. *Annual rate:* 8.94%

Amount after 10 years: \$1833.67

Amount after 25 years: \$7009.86

29. *Annual rate:* 9.50%

Time to double: 7.30 years

Amount after 25 years: \$5375.51

31. (a) Answers will vary. (b) 6.17%

33.

Number of compoundings/yr	4	12
Effective yield	5.095%	5.116%

Number of compoundings/yr	365	Continuous
Effective yield	5.127%	5.127%

35. Answers will vary.

37. (a) \$1034.08 million (b) \$628.25 million

(c)

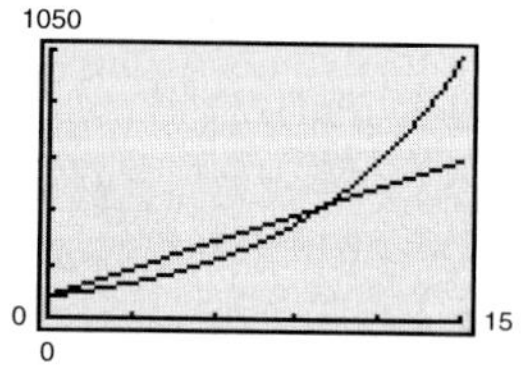

$t = 0$ corresponds to 1993.

Answers will vary.

39. (a) $C = 30$

$k = \ln\left(\frac{1}{6}\right) \approx -1.7918$

(b) $30e^{-0.35836} = 20.9646$ or 20,965 units

(c)

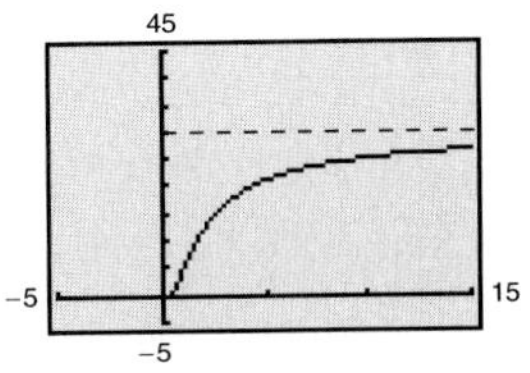

41. About 36 days **43.** \$496,806

45. (a) $C = \frac{625}{64}$

$k = \frac{1}{100} \ln \frac{4}{5}$

(b) $x = 448$ units; $P = \$3.59$

47. 2046

REVIEW EXERCISES FOR CHAPTER 4

(page 312)

1. 8 **3.** 125 **5.** 1 **7.** $\frac{1}{6}$ **9.** 4

11. 16 **13.** $\frac{1}{2}$ **15.** $e\sqrt[3]{3}$

17. (a) 1995: $R(5) \approx \$524.04$ million

1998: $R(8) \approx \$636.58$ million

2001: $R(11) \approx \$773.30$ million

(b) Answers will vary.

19.

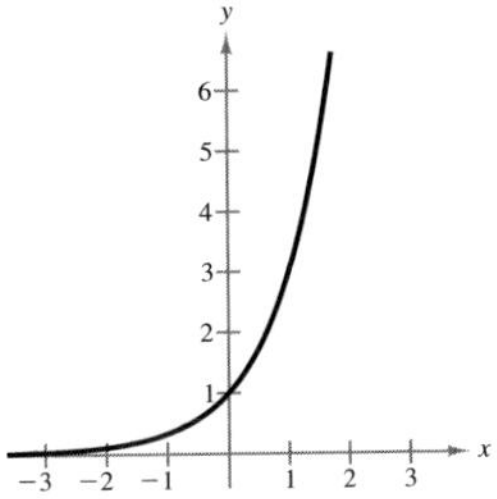

21.

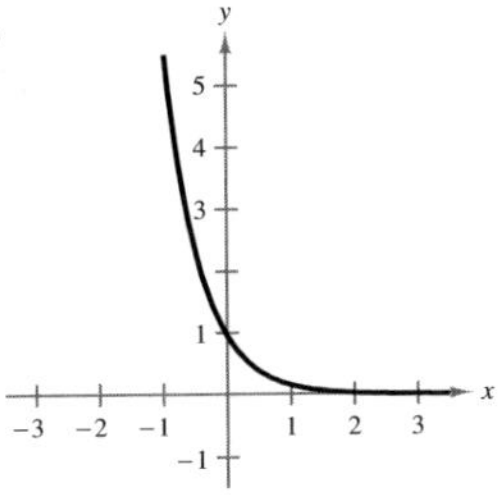

23.

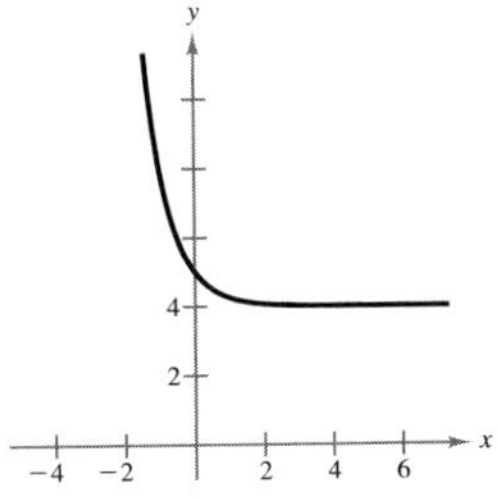

25.

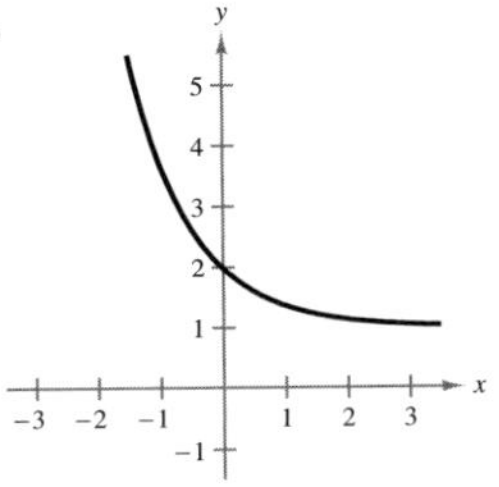

27.

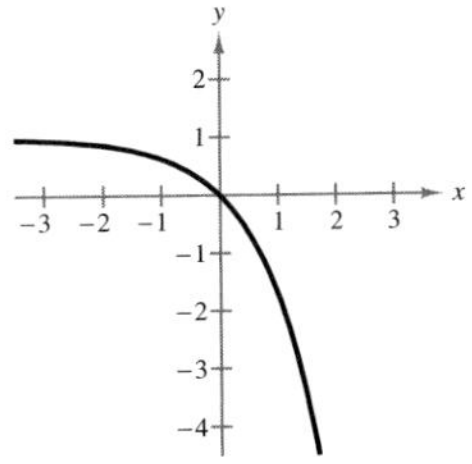

29. \$7500
Explanations will vary.

31. (a) $2e \approx 5.4366$ (b) $2e^{-1/2} \approx 1.2131$

(c) $2e^9 \approx 16{,}206.168$

33. (a) $12e^{-3.4} \approx 0.4005$ (b) $12e^{-10} \approx 0.0005$

(c) $12e^{-20} \approx 2.4734 \times 10^{-8}$

35. (a)

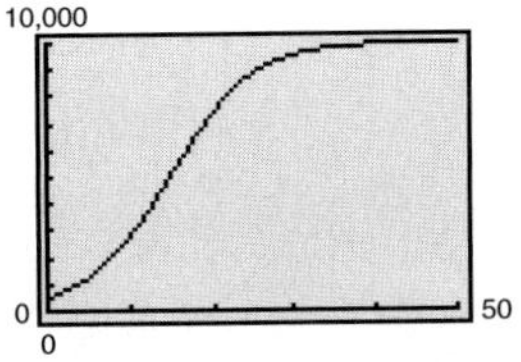

(b) $P \approx 1049$ fish

(c) Yes, P approaches 10,000 fish as t approaches ∞.

(d) The population is increasing most rapidly at the inflection point, which occurs around $t = 15$ months.

37.

n	1	2	4	12
A	\$1216.65	\$1218.99	\$1220.19	\$1221.00

n	365	Continuous compounding
A	\$1221.39	\$1221.40

39. b **41.** (a) 6.14% (b) 6.17% **43.** \$9889.50

45. 1980: $P(0) = 24.196$ million

1995: $P(15) \approx 31.675$ million

2000: $P(20) = 32.700$ million

47. $8xe^{x^2}$ **49.** $\frac{1 - 2x}{e^{2x}}$ **51.** $4e^{2x}$ **53.** $\frac{-10e^{2x}}{(1 + e^{2x})^2}$

55.

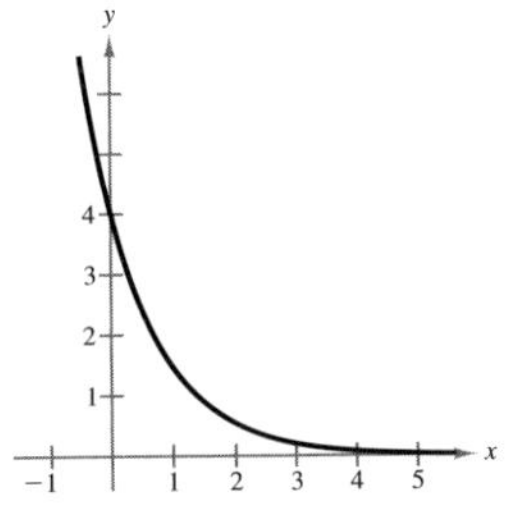

No relative extrema

No points of inflection

Horizontal asymptote: $y = 0$

57.

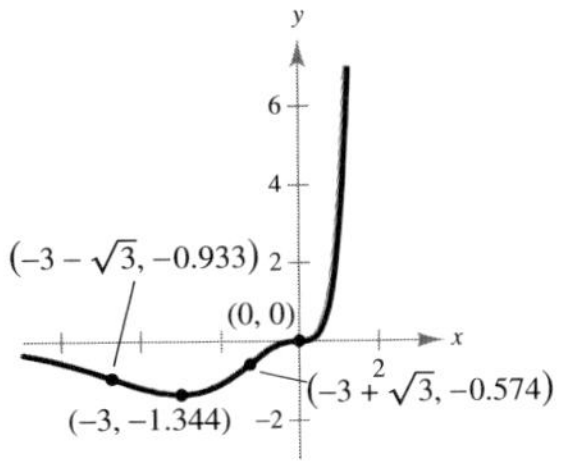

Relative minimum: $(-3, -1.344)$

Inflection points: $(0, 0)$, $(-3 + \sqrt{3}, -0.574)$, and $(-3 - \sqrt{3}, -0.933)$

Horizontal asymptote: $y = 0$

59.

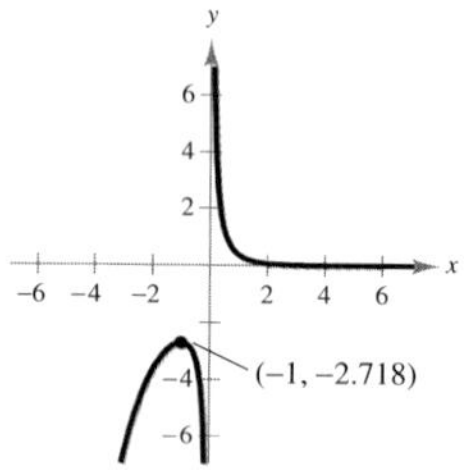

Relative maximum: $(-1, -2.718)$

Horizontal asymptote: $y = 0$

Vertical asymptote: $x = 0$

61.

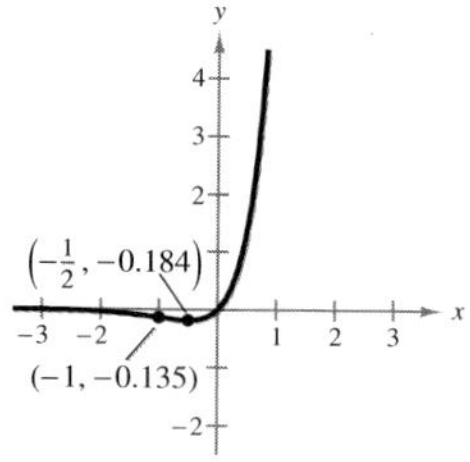

Relative minimum: $\left(-\frac{1}{2}, -\frac{1}{2e}\right)$

Inflection point: $\left(-1, -\frac{1}{e^2}\right)$

Horizontal asymptote: $y = 0$

63. $e^{2.4849} \approx 12$ **65.** $\ln 4.4817 \approx 1.5$

67.

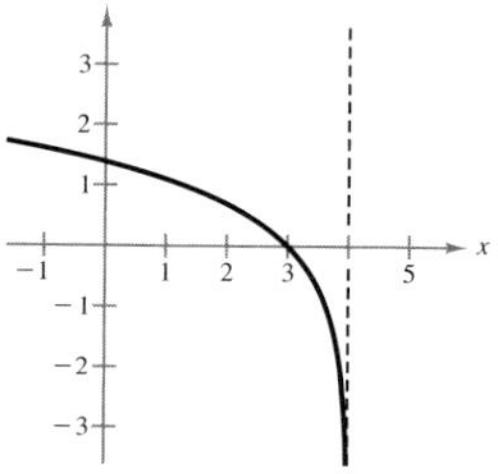

69.

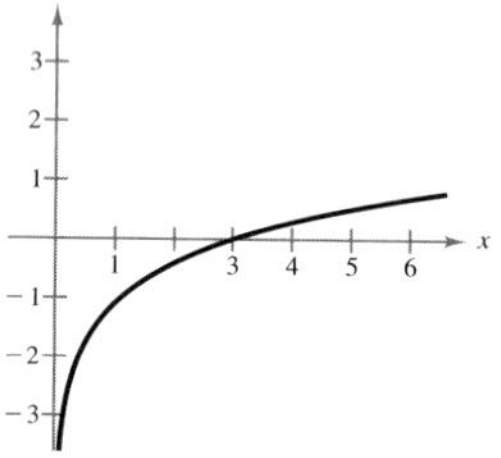

71. $\ln x + \frac{1}{2}\ln(x - 1)$ **73.** $2 \ln x - 3 \ln(x + 1)$

75. $3[\ln(1 - x) - \ln 3 - \ln x]$ **77.** 3

79. $e^{3e^{-1}} \approx 3.0151$ **81.** 1

83. $\frac{1}{2}(\ln 6 + 1) \approx 1.3959$

85. $\dfrac{3 + \sqrt{13}}{2} \approx 3.3028$ **87.** $-\dfrac{\ln(0.25)}{1.386} \approx 1.0002$

89. $\dfrac{\ln 1.1}{\ln 1.21} = 0.5$ **91.** $100 \ln\left(\dfrac{25}{4}\right) \approx 183.2581$

93. (a)

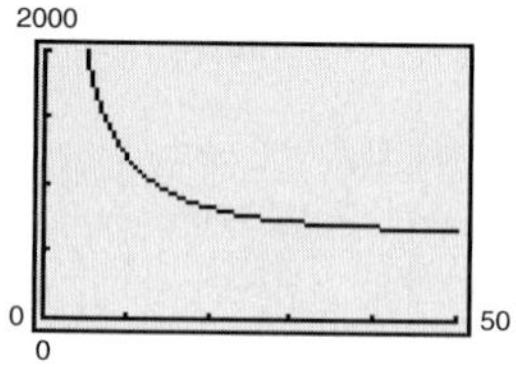

(b) A 30-year term has a smaller monthly payment, but it takes more time to pay off than a 20-year term.

95. $\dfrac{2}{x}$ **97.** $\dfrac{1}{x} + \dfrac{1}{x - 1} - \dfrac{1}{x - 2} = \dfrac{x^2 - 4x + 2}{x(x - 2)(x - 1)}$

99. 2 **101.** $\dfrac{1 - 3 \ln x}{x^4}$ **103.** $\dfrac{4x}{3(x^2 - 2)}$

105. $\dfrac{2}{x} + \dfrac{1}{2(x + 1)}$ **107.** $\dfrac{1}{1 + e^x}$

109.

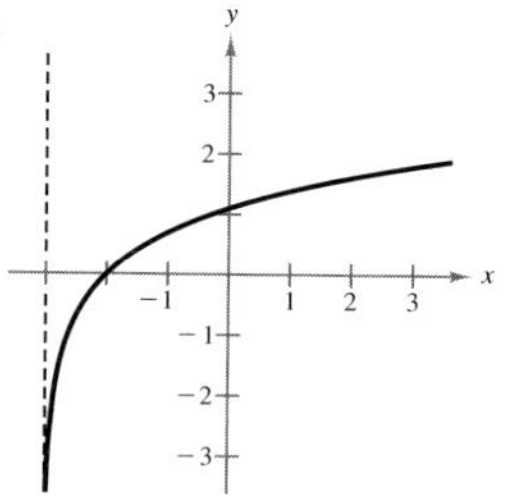

No relative extrema

No points of inflection

111.

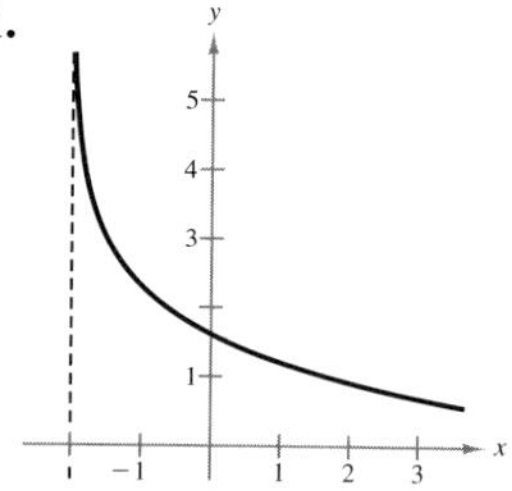

No relative extrema

No points of inflection

113. 2 **115.** 0 **117.** 1.431 **119.** 1.500

121. $\dfrac{2}{(2x - 1) \ln 3}$ **123.** $-\dfrac{2}{x \ln 2}$

125. (a)

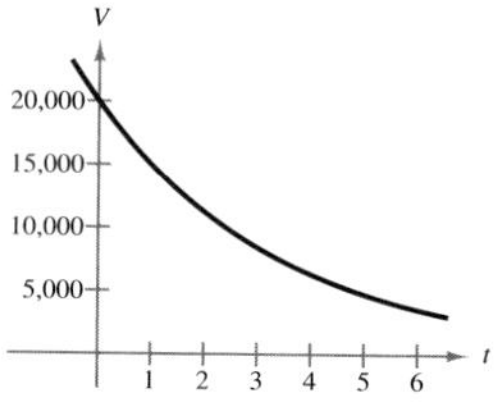

$t = 2$: \$11,250

(b) $t = 1$: -4315.23 dollars/year

$t = 4$: -1820.49 dollars/year

(c) $t \approx 4.8$ years

127. $A = 500e^{-0.01277t}$ **129.** 27.9 years

131. \$896.10 million

SAMPLE POST-GRAD EXAM QUESTIONS

(page 316)

1. d **2.** b **3.** c **4.** a **5.** c

6. d **7.** a **8.** b **9.** d

CHAPTER 5

SECTION 5.1 *(page 326)*

Prerequisite Review

1. $x^{-1/2}$ **2.** $(2x)^{4/3}$ **3.** $5^{1/2}x^{3/2} + x^{5/2}$
4. $x^{-1/2} + x^{-2/3}$ **5.** $(x + 1)^{5/2}$ **6.** $x^{1/6}$
7. -12 **8.** -10 **9.** 14 **10.** 14

1–7. Answers will vary.

9. $6x + C$

$$\frac{d}{dx}[6x + C] = 6$$

11. $\frac{5}{3}t^3 + C$

$$\frac{d}{dt}\left[\frac{5}{3}t^3 + C\right] = 5t^2$$

13. $-\frac{5}{2x^2} + C$

$$\frac{d}{dx}\left[-\frac{5}{2x^2} + C\right] = 5x^{-3}$$

15. $u + C$

$$\frac{d}{du}[u + C] = 1$$

17. $et + C$

$$\frac{d}{dt}[et + C] = e$$

19. $\frac{2}{5}y^{5/2} + C$

$$\frac{d}{dy}\left[\frac{2}{5}y^{5/2} + C\right] = y^{3/2}$$

	Rewrite	Integrate	Simplify
21.	$\int x^{1/3}\,dx$	$\frac{x^{4/3}}{4/3} + C$	$\frac{3}{4}x^{4/3} + C$
23.	$\int x^{-3/2}\,dx$	$\frac{x^{-1/2}}{-1/2} + C$	$-\frac{2}{\sqrt{x}} + C$
25.	$\int (x^3 + 3x)\,dx$	$\frac{x^4}{4} + \frac{3x^2}{2} + C$	$\frac{x^2}{4}(x^2 + 6) + C$
27.	$\frac{1}{2}\int x^{-3}\,dx$	$\frac{1}{2}\left(\frac{x^{-2}}{-2}\right) + C$	$-\frac{1}{4x^2} + C$

29.

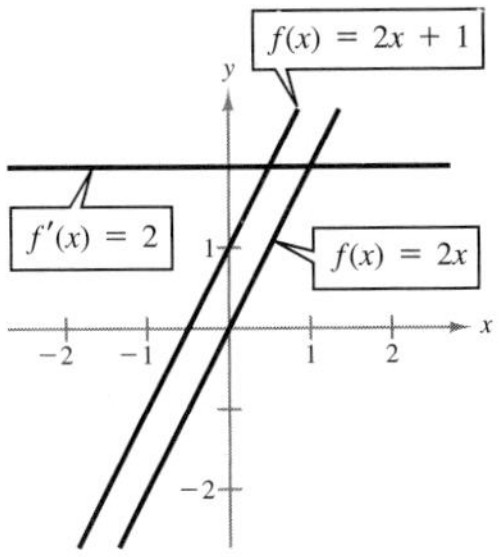

31.

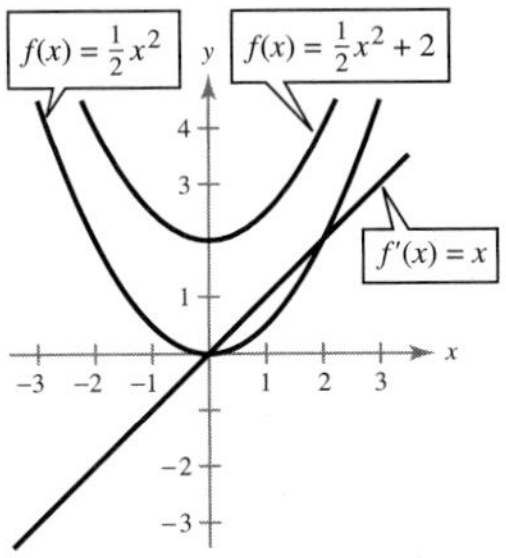

33. $\frac{1}{4}x^4 + 2x + C$ **35.** $\frac{3}{4}x^{4/3} - \frac{3}{4}x^{2/3} + C$

37. $\frac{3}{5}x^{5/3} + x + C$

39. $-\frac{1}{9x^3} + C$ **41.** $2x - \frac{1}{2x^2} + C$

43. $\frac{3}{4}u^4 + \frac{1}{2}u^2 + C$ **45.** $2x^3 - \frac{11}{2}x^2 + 5x + C$

47. $\frac{2}{7}y^{7/2} + C$ **49.** $f(x) = 2x^{3/2} + 3x - 1$

51. $f(x) = 2x^3 - 3x^2$ **53.** $f(x) = -\frac{1}{x^2} + \frac{1}{x} + \frac{1}{2}$

55. $y = -\frac{5}{2}x^2 - 2x + 2$ **57.** $f(x) = 4x^{3/2} - 10x + 10$

59. $f(x) = x^2 + x + 4$ **61.** $f(x) = \frac{9}{4}x^{4/3}$

63. $C = 85x + 5500$ **65.** $C = \frac{1}{10}\sqrt{x} + 4x + 750$

67. $R = 225x - \frac{3}{2}x^2, p = 225 - \frac{3}{2}x$

69. $R = 225x + x^2 - \frac{1}{3}x^3, p = 225 + x - \frac{1}{3}x^2$

71. $P = -9x^2 + 1650x$ **73.** $P = -12x^2 + 805x + 68$

75. 56.25 feet **77.** $v_0 = 40\sqrt{22} \approx 187.617$ feet/second

79. (a) $C = x^2 - 12x + 125$ (b) \$2025

$$\overline{C} = x - 12 + \frac{125}{x}$$

(c) \$125 is fixed.

\$1900 is variable.

Examples will vary.

81. (a) $M = 0.212t^3 - 14.24t^2 + 632.7t + 44{,}608$ (in thousands)

(b) 60,700,000; Yes, this seems reasonable.

83. Answers will vary.

SECTION 5.2 *(page 335)*

Prerequisite Review

1. $\frac{1}{2}x^4 + x + C$ **2.** $\frac{3}{2}x^2 + \frac{2}{3}x^{3/2} - 4x + C$

3. $-\frac{1}{x} + C$ **4.** $-\frac{1}{6t^2} + C$

5. $\frac{4}{7}t^{7/2} + \frac{2}{5}t^{5/2} + C$ **6.** $\frac{4}{5}x^{5/2} - \frac{2}{3}x^{3/2} + C$

7. $\frac{5x^3 - 4}{2x} + C$ **8.** $\frac{-6x^2 + 5}{3x^3} + C$

9. $\frac{1}{5}x^5 + \frac{2}{3}x^3 + x + C$

10. $\frac{1}{7}x^7 - \frac{4}{5}x^5 + \frac{1}{2}x^4 + \frac{4}{3}x^3 - 2x^2 + x + C$

11. $-\frac{5(x - 2)^4}{16}$ **12.** $-\frac{1}{12(x - 1)^2}$

13. $9(x^2 + 3)^{2/3}$ **14.** $-\frac{5}{(1 - x^3)^{1/2}}$

$\int u^n \frac{du}{dx}\,dx$	u	$\frac{du}{dx}$
1. $\int (5x^2+1)^2(10x)\,dx$	$5x^2+1$	$10x$
3. $\int \sqrt{1-x^2}\,(-2x)\,dx$	$1-x^2$	$-2x$
5. $\int \left(4+\frac{1}{x^2}\right)^5\left(\frac{-2}{x^3}\right)dx$	$4+\frac{1}{x^2}$	$-\frac{2}{x^3}$
7. $\int (1+\sqrt{x})^3\left(\frac{1}{2\sqrt{x}}\right)dx$	$1+\sqrt{x}$	$\frac{1}{2\sqrt{x}}$

9. $\frac{1}{5}(1+2x)^5+C$ **11.** $\frac{2}{3}(5x^2-4)^{3/2}+C$

13. $\frac{1}{5}(x-1)^5+C$ **15.** $\frac{1}{16}(x^2-1)^8+C$

17. $-\frac{1}{3(1+x^3)}+C$ **19.** $-\frac{1}{2(x^2+2x-3)}+C$

21. $\sqrt{x^2-4x+3}+C$ **23.** $-\frac{15}{8}(1-u^2)^{4/3}+C$

25. $4\sqrt{1+y^2}+C$ **27.** $-3\sqrt{2t+3}+C$

29. $-\frac{1}{2}\sqrt{1-x^4}+C$ **31.** $-\frac{1}{24}\left(1+\frac{4}{t^2}\right)^3+C$

33. $\frac{1}{6}(x^3+3x)^2+C$ **35.** $\frac{1}{48}(6x^2-1)^4+C$

37. $-\frac{2}{45}(2-3x^3)^{5/2}+C$ **39.** $\sqrt{x^2+25}+C$

41. $\frac{2}{3}\sqrt{x^3+3x+4}+C$

43. (a) $\frac{1}{6}(2x-1)^3+C_1=\frac{4}{3}x^3-2x^2+x+C_2$

(b) Answers differ by a constant: $C_2=C_1-\frac{1}{6}$

(c) Answers will vary.

45. (a) $\frac{(x^2-1)^3}{6}+C_1=\frac{1}{6}x^6-\frac{1}{2}x^4+\frac{1}{2}x^2+C_2$

(b) Answers differ by a constant: $C_2=C_1-\frac{1}{6}$

(c) Answers will vary.

47. $f(x)=\frac{1}{3}[5-(1-x^2)^{3/2}]$

49. (a) $C=8\sqrt{x+1}+18$

(b)

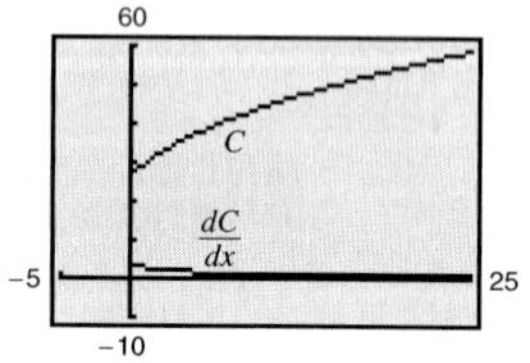

51. $x=\frac{1}{3}(p^2-25)^{3/2}+24$ **53.** $x=\frac{6000}{\sqrt{p^2-16}}+3000$

55. (a) $h=\sqrt{17.6t^2+1}+5$ (b) 26 inches

57. (a) $Q=(x-19{,}999)^{0.95}+19{,}999$

(b)

x	20,000	50,000	100,000	150,000
Q	20,000	37,916.56	65,491.59	92,151.16
$x-Q$	0	12,083.44	34,508.41	57,848.84

(c)

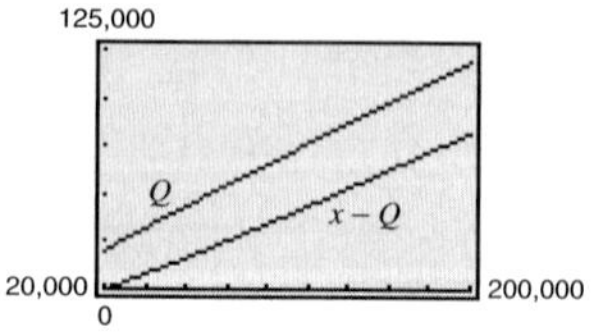

59. $-\frac{2}{3}x^{3/2}+\frac{2}{3}(x+1)^{3/2}+C$

SECTION 5.3 *(page 342)*

Prerequisite Review

1. $\left(\frac{5}{2},\infty\right)$ **2.** $(-\infty,2)\cup(3,\infty)$

3. $x+2-\frac{2}{x+2}$ **4.** $x-2+\frac{1}{x-4}$

5. $x+8+\frac{2x-4}{x^2-4x}$ **6.** $x^2-x-4+\frac{20x+22}{x^2+5}$

7. $\frac{1}{4}x^4-\frac{1}{x}+C$ **8.** $\frac{1}{2}x^2+2x+C$

9. $\frac{1}{2}x^2-\frac{4}{x}+C$ **10.** $-\frac{1}{x}-\frac{3}{2x^2}+C$

1. $e^{2x}+C$ **3.** $\frac{1}{4}e^{4x}+C$ **5.** $-\frac{9}{2}e^{-x^2}+C$

7. $\frac{5}{3}e^{x^3}+C$ **9.** $\frac{1}{3}e^{x^3+3x^2-1}+C$ **11.** $-5e^{2-x}+C$

13. $\ln|x+1|+C$ **15.** $-\frac{1}{2}\ln|3-2x|+C$

17. $\frac{2}{3}\ln|3x+5|+C$

19. $\ln\sqrt{x^2+1}+C$ **21.** $\frac{1}{3}\ln|x^3+1|+C$

23. $\frac{1}{2}\ln|x^2+6x+7|+C$ **25.** $\ln|\ln x|+C$

27. $\ln|1-e^{-x}|+C$ **29.** $-\frac{1}{2}e^{2/x}+C$ **31.** $2e^{\sqrt{x}}+C$

33. $\frac{1}{2}e^{2x}-4e^x+4x+C$ **35.** $-\ln(1+e^{-x})+C$

37. $-2\ln|5-e^{2x}|+C$ **39.** $e^x+2x-e^{-x}+C$

41. $-\frac{2}{3}(1-e^x)^{3/2}+C$ **43.** $-\frac{1}{x-1}+C$

45. $2e^{2x-1}+C$ **47.** $\frac{1}{4}x^2-4\ln|x|+C$

49. $2\ln(e^x+1)+C$ **51.** $\frac{1}{2}x^2+3x+8\ln|x-1|+C$

53. $\ln|e^x+x|+C$

55. $f(x)=\frac{1}{2}x^2+5x+8\ln|x-1|-8$

57. (a) $P(t) = 1000[1 + \ln(1 + 0.25t)^{12}]$

(b) $P(3) \approx 7715$ (c) $t \approx 6$ days

59. (a) $p = -50e^{-x/500} + 45.06$

(b)

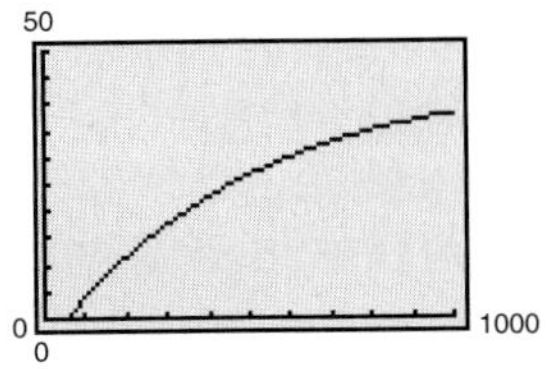

The price increases as the demand increases.

(c) 387

61. (a) $S = 37{,}452.86e^{0.07t} + 37{,}606.58$ (in dollars)

(b) \$107,928.47

63. False. $\ln x^{1/2} = \frac{1}{2}\ln x$

SECTION 5.4 *(page 353)*

Prerequisite Review

1. $\frac{3}{2}x^2 + 7x + C$ **2.** $\frac{2}{5}x^{5/2} + \frac{4}{3}x^{3/2} + C$

3. $\frac{1}{5}\ln|x| + C$ **4.** $-\frac{1}{6e^{6x}} + C$ **5.** $-\frac{8}{5}$

6. $-\frac{62}{3}$ **7.** $C = 0.008x^{5/2} + 29{,}500x + C$

8. $R = x^2 + 9000x + C$

9. $P = 25{,}000x - 0.005x^2 + C$

10. $C = 0.01x^3 + 4600x + C$

1.

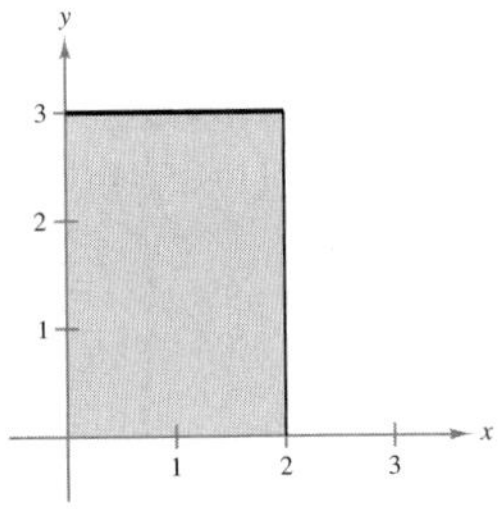

Area = 6

3.

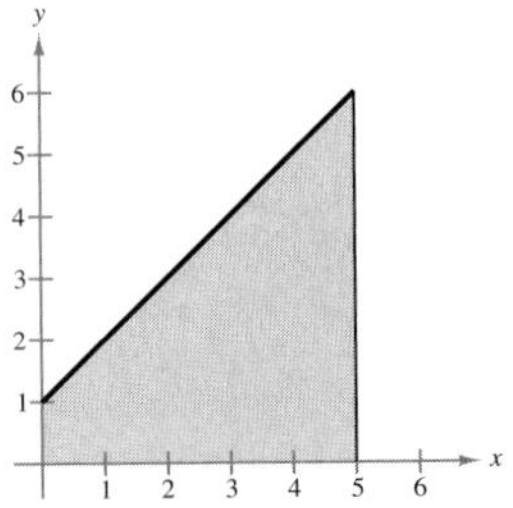

Area = $\frac{35}{2}$

5.

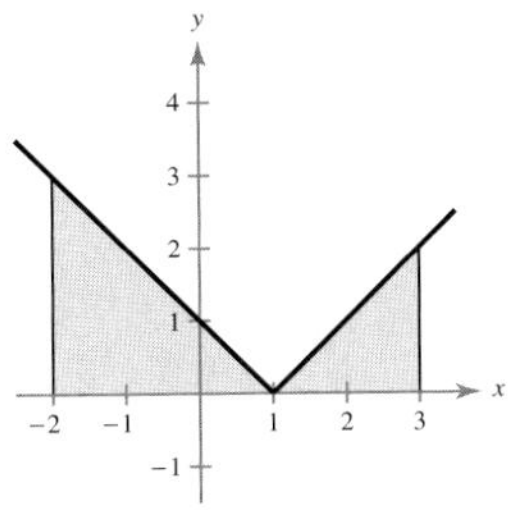

Area = $\frac{13}{2}$

7.

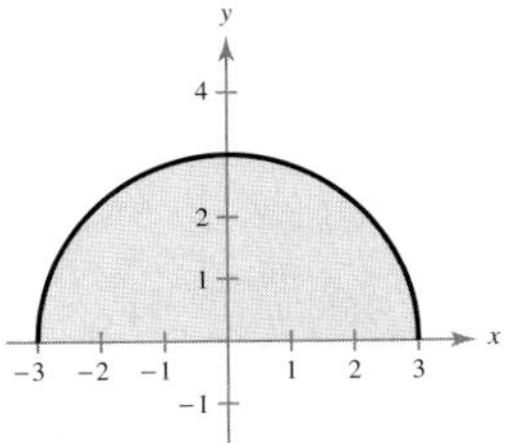

Area = $\frac{9\pi}{2}$

9. (a) 11 (b) 5 (c) -32 (d) -1

11. $\frac{1}{6}$ **13.** $\frac{1}{2}$ **15.** $6\left(1 - \frac{1}{e^2}\right)$ **17.** $8\ln 2 + \frac{15}{2}$

19. 1 **21.** 0 **23.** $\frac{14}{3}$ **25.** $-\frac{15}{4}$ **27.** -4

29. $\frac{22}{3}$ **31.** $-\frac{27}{20}$ **33.** 2 **35.** $\frac{1}{2}(1 - e^{-2}) \approx 0.432$

37. $\frac{e^3 - e}{3} \approx 5.789$ **39.** $\frac{1}{3}\left[(e^2 + 1)^{3/2} - 2\sqrt{2}\right] \approx 7.157$

41. $\frac{1}{8}\ln 17 \approx 0.354$ **43.** 4 **45.** 4

47. $\frac{1}{2}\ln 5 - \frac{1}{2}\ln 8 \approx -0.235$

49. $2\ln(2 + e^3) - 2\ln 3 \approx 3.993$

51. Area = 10

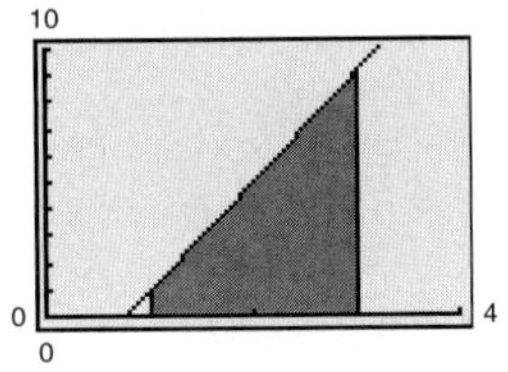

53. Area = $\frac{1}{4}$

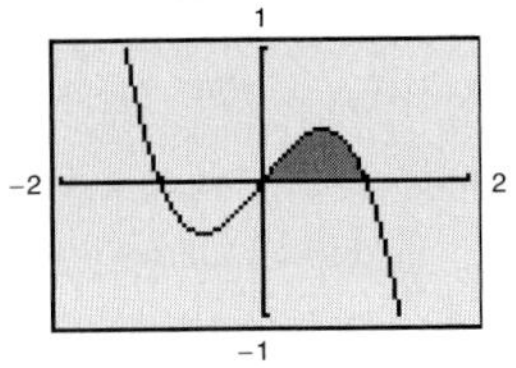

55. Area = $\ln 9$

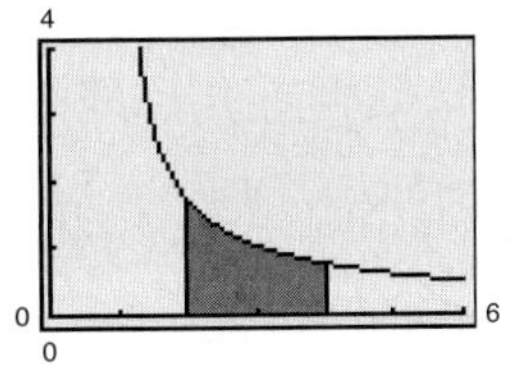

57. 10 **59.** $4 + 5\ln 5 \approx 12.047$

61.

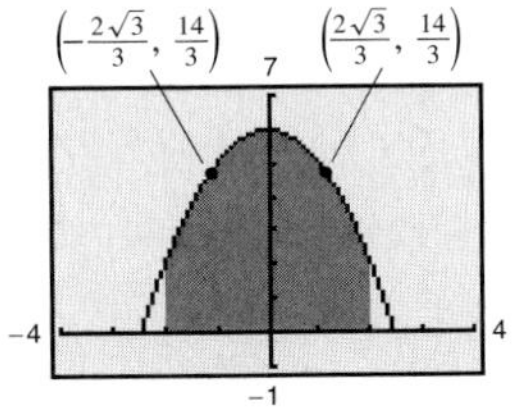

Average = $\frac{14}{3}$

$x = \pm\frac{2\sqrt{3}}{3} \approx \pm 1.155$

63.

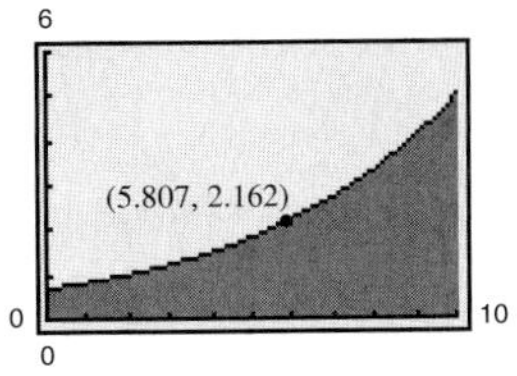

Average = $\frac{5}{2} - \frac{5}{2}e^{-2}$

$x = 5\ln\left(\frac{e^2 - 1}{2}\right) \approx 5.807$

65.

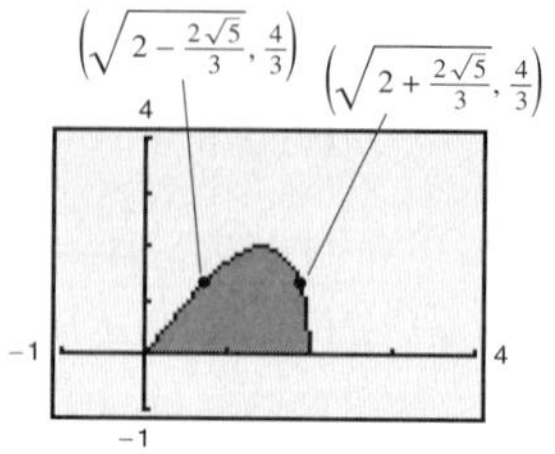

Average $= \dfrac{4}{3}$

$x = \sqrt{2 + \dfrac{2\sqrt{5}}{3}} \approx 1.868$

$x = \sqrt{2 - \dfrac{2\sqrt{5}}{3}} \approx 0.714$

67.

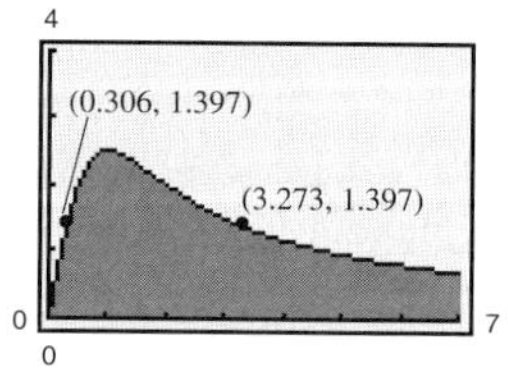

Average $= \frac{5}{14} \ln 50$

$x \approx 0.306$

$x \approx 3.273$

69. Even **71.** Neither odd nor even

73. (a) $\frac{8}{3}$ (b) $\frac{16}{3}$ (c) $-\frac{8}{3}$

Explanations will vary.

75. \$6.75 **77.** \$22.50 **79.** \$3.97 **81.** \$1925.23

83. \$16,605.21 **85.** \$2500 **87.** \$4565.65

89. (a) \$137,000 (b) \$214,720.93 (c) \$338,393.53

91. \$3082.95 **93.** $\dfrac{2kR^2}{3}$ **95.** $\dfrac{2}{3}\sqrt{7} - \dfrac{1}{3}$ **97.** $\dfrac{39}{200}$

SECTION 5.5 *(page 362)*

Prerequisite Review

1. $-x^2 + 3x + 2$ **2.** $-2x^2 + 4x + 4$

3. $-x^3 + 2x^2 + 4x - 5$ **4.** $x^3 - 6x - 1$

5. (0, 4), (4, 4) **6.** (1, −3), (2, −12)

7. (−3, 9), (2, 4) **8.** (−2, −4), (0, 0), (2, 4)

9. (1, −2), (5, 10) **10.** (1, e)

1. 36 **3.** 9 **5.** $\frac{3}{2}$ **7.** $e - 2$

9.

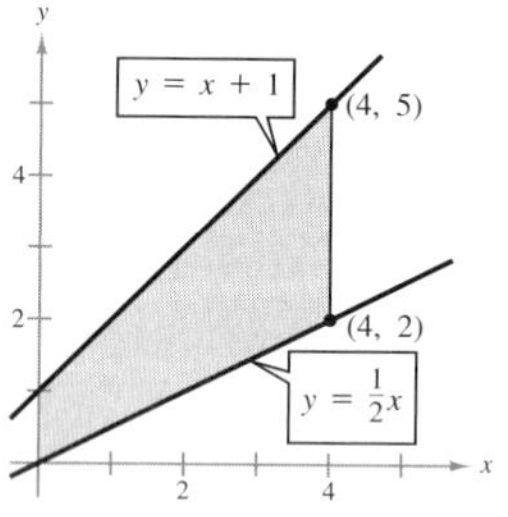

11.

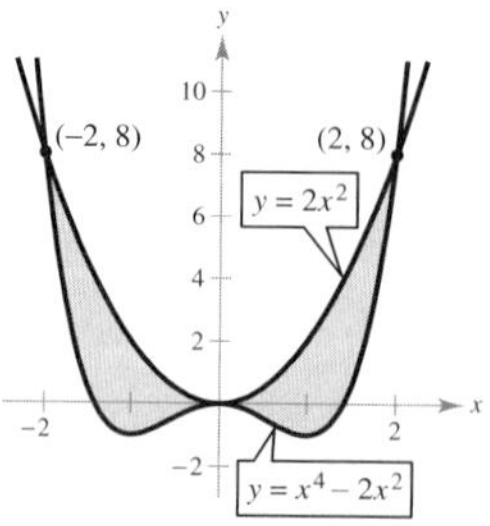

13.

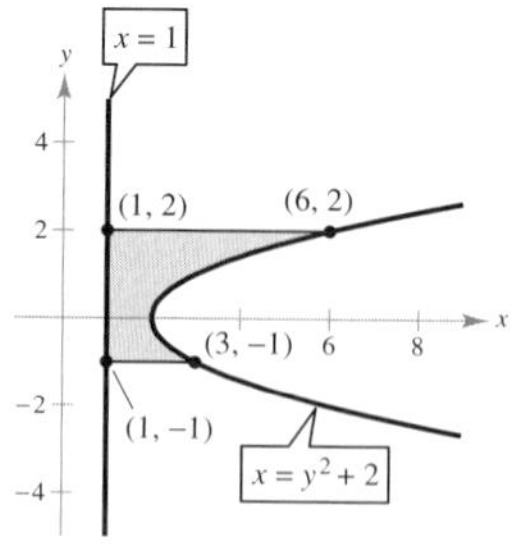

15.

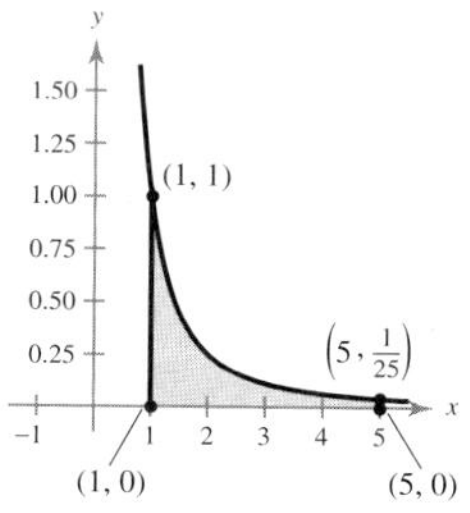

Area $= \frac{4}{5}$

17.

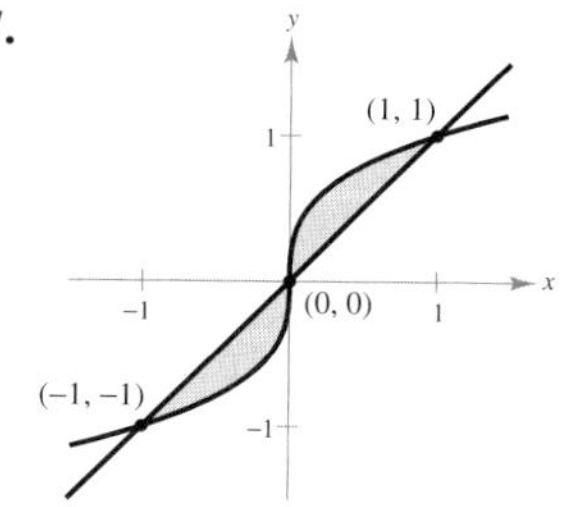

Area $= \frac{1}{2}$

19.

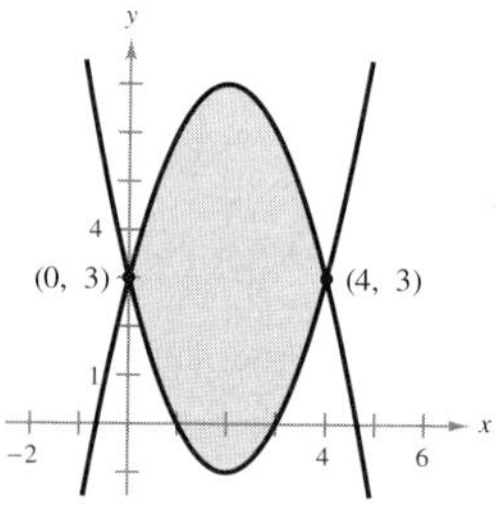

Area $= \frac{64}{3}$

21.

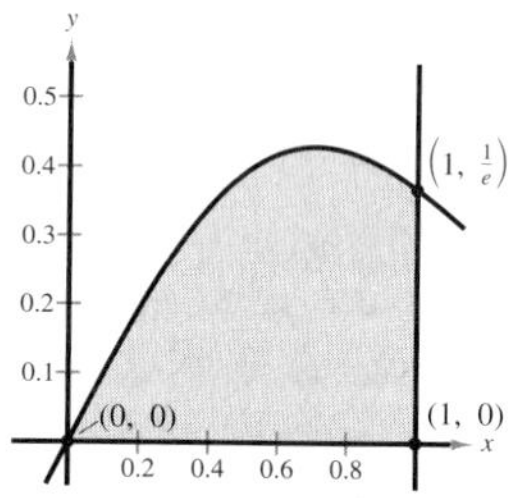

Area $= -\frac{1}{2}e^{-1} + \frac{1}{2}$

23.

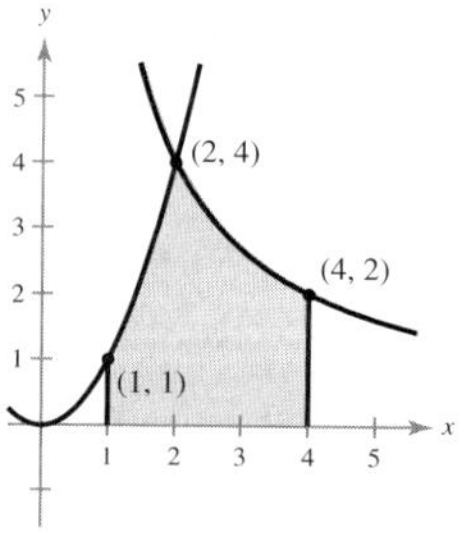

Area $= \frac{7}{3} + 8 \ln 2$

25.

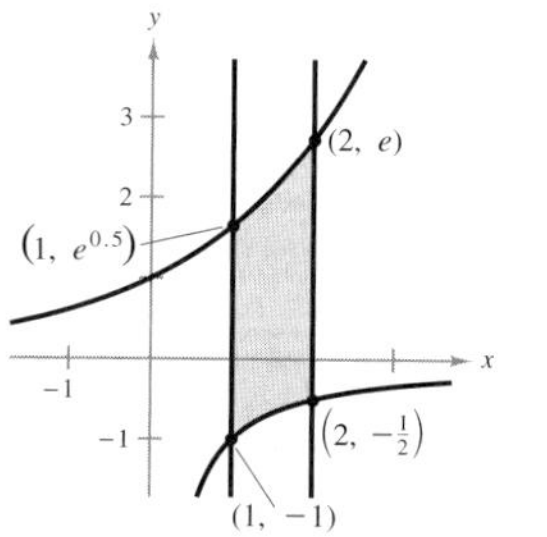

Area $= (2e + \ln 2) - 2e^{1/2}$

27.

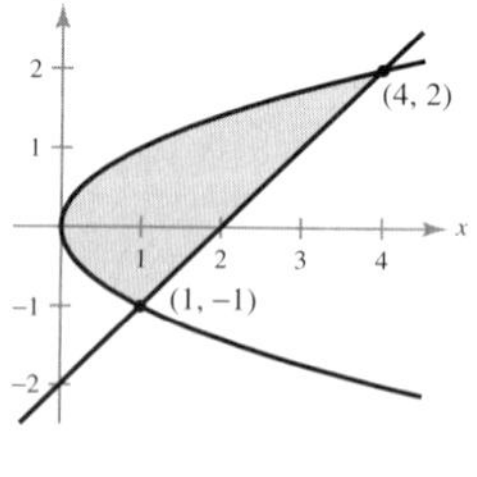

Area $= \frac{9}{2}$

29.

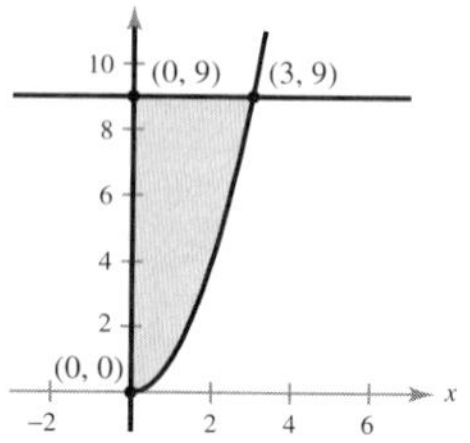

Area $= 18$

31.

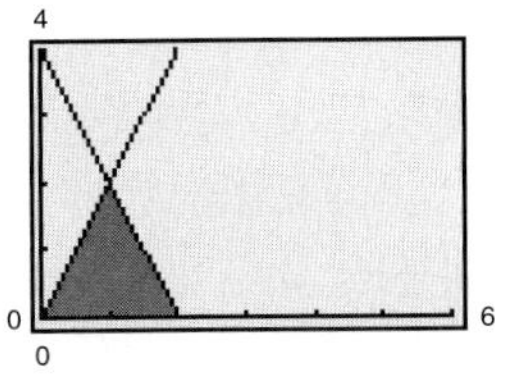

$$\text{Area} = \int_0^1 2x\,dx + \int_1^2 (4 - 2x)\,dx$$

33.

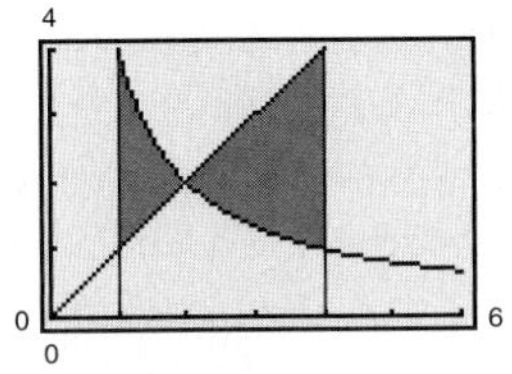

$$\text{Area} = \int_1^2 \left(\frac{4}{x} - x\right) dx + \int_2^4 \left(x - \frac{4}{x}\right) dx$$

35.

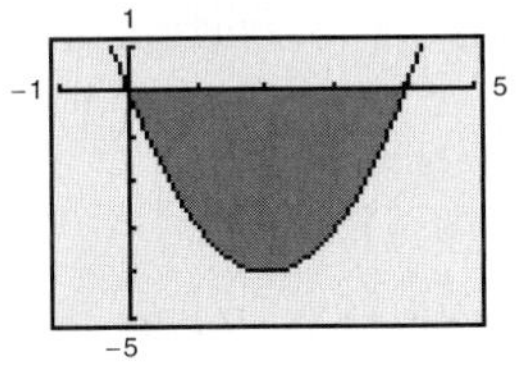

Area $= \frac{32}{3}$

37.

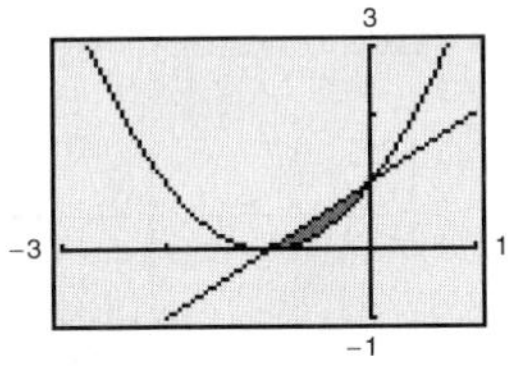

Area $= \frac{1}{6}$

39. 8

41. Consumer surplus = 1600

Producer surplus = 400

43. Consumer surplus ≈ 1666.67

Producer surplus = 1250

45. Consumer surplus = 50,000

Producer surplus ≈ 25,497

47. A typical demand function is decreasing, whereas a typical supply function is increasing.

49. R_1, \$3.12 billion

51. \$287.64 million; Explanations will vary.

53. (a)

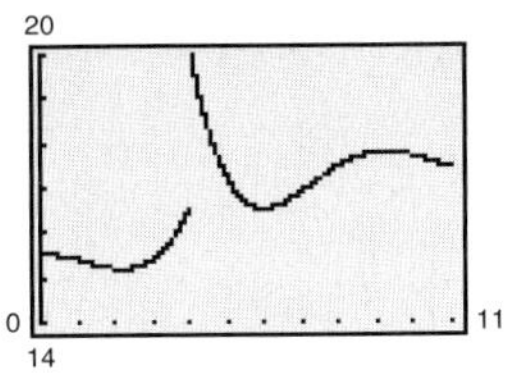

(b) 169.2 pounds more

55. Consumer surplus = \$625,000

Producer surplus = \$1,375,000

57. \$337.33 million

59.

Quintile	Lowest	2nd	3rd	4th	Highest
Percent	2.92	7.09	14.58	26.74	44.89

SECTION 5.6 *(page 369)*

Prerequisite Review

1. $\frac{1}{6}$ **2.** $\frac{3}{20}$ **3.** $\frac{7}{40}$ **4.** $\frac{13}{12}$ **5.** $\frac{61}{30}$ **6.** $\frac{53}{18}$
7. $\frac{2}{3}$ **8.** $\frac{4}{7}$ **9.** 0 **10.** 5

1. Midpoint Rule: 2

Exact area: 2

3. Midpoint Rule: 0.6730

Exact area: $\frac{2}{3} \approx 0.6667$

5. Midpoint Rule: 4.6250

Exact area: $\frac{14}{3} = 4.\overline{6}$

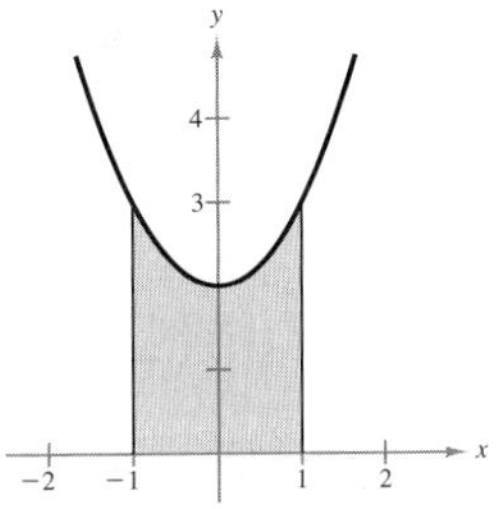

7. Midpoint Rule: 17.2500

Exact area: $\frac{52}{3} = 17.\overline{3}$

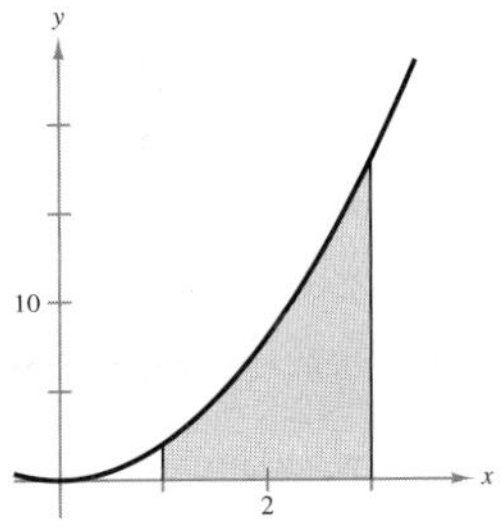

9. Midpoint Rule: 0.0859

Exact area: $\frac{1}{12} = 0.08\overline{3}$

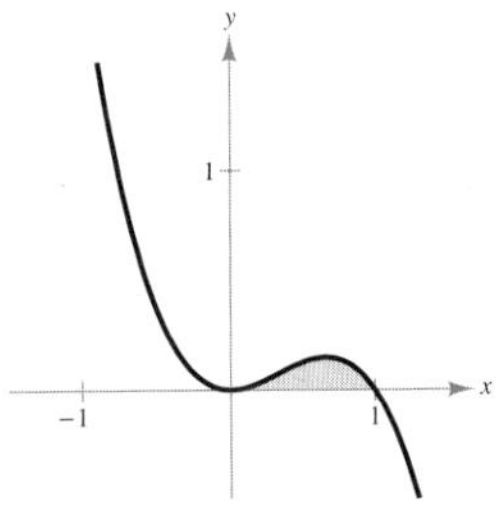

11. Midpoint Rule: 0.0859

Exact area: $\frac{1}{12} = 0.08\overline{3}$

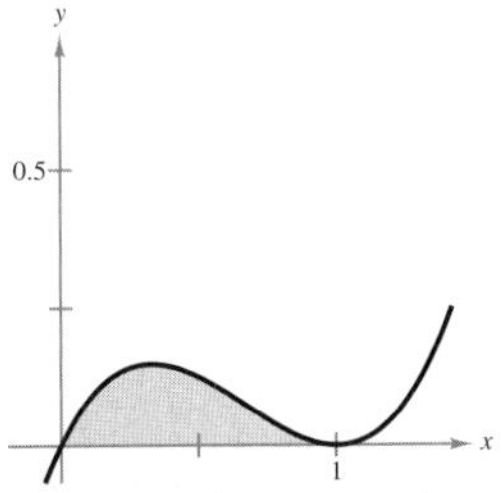

13. Area ≈ 54.6667,

$n = 31$

15. Area ≈ 4.16,

$n = 5$

17. Midpoint Rule: 1.5

Exact area: 1.5

19. Midpoint Rule: 25

Exact area: $\frac{76}{3} = 25.\overline{3}$

21. Exact: 4

Trapezoidal Rule: 4.0625

Midpoint Rule: 3.9688

Midpoint Rule is better in this example.

23. 1.1167 **25.** 1.55

27.

n	Midpoint Rule	Trapezoidal Rule
4	15.3965	15.6055
8	15.4480	15.5010
12	15.4578	15.4814
16	15.4613	15.4745
20	15.4628	15.4713

29. 4.8103 **31.** 916.25 feet

33. Midpoint Rule: $\pi \approx 3.146801$

Trapezoidal Rule: $\pi \approx 3.131176$

Graphing utility: $\pi \approx 3.141593$

SECTION 5.7 *(page 376)*

Prerequisite Review

1. 0, 2 **2.** 0, 2 **3.** 0, 2, -2 **4.** -1, 2

5. 2, 4 **6.** 1, 5 **7.** $e^4 - 1$ **8.** ln 7

9. $\frac{5\sqrt{5}}{3} - \frac{1}{3}$ **10.** $\frac{(\ln 5)^3}{3}$

1. $\frac{16\pi}{3}$ **3.** $\frac{15\pi}{2}$ **5.** $\frac{512\pi}{15}$ **7.** $\frac{32\pi}{15}$ **9.** $\frac{\pi}{3}$

11. $\frac{171\pi}{2}$ **13.** $\frac{128\pi}{5}$ **15.** $\frac{\pi}{2}(e^2 - 1)$ **17.** 8π

19. $\frac{2\pi}{3}$ **21.** $\frac{\pi}{4}$ **23.** $\frac{256\pi}{15}$ **25.** 18π

27. $V = \pi \int_0^h \left(\frac{r}{h}x\right)^2 dx = \frac{1}{3}\pi r^2 h$ **29.** 100π **31.** $\frac{\pi}{30}$

33. (a) 1,256,637 cubic feet (b) 2513 fish

35. 58.434

REVIEW EXERCISES FOR CHAPTER 5
(page 382)

1. $16x + C$ **3.** $\frac{2}{3}x^3 + \frac{5}{2}x^2 + C$ **5.** $x^{2/3} + C$

7. $\frac{3}{7}x^{7/3} + \frac{3}{2}x^2 + C$ **9.** $\frac{4}{9}x^{9/2} - 2\sqrt{x} + C$

11. $f(x) = \frac{3}{2}x^2 + x - 2$ **13.** $f(x) = \frac{1}{6}x^4 - 8x + \frac{33}{2}$

15. (a) 2.5 seconds (b) 100 feet

(c) 1.25 seconds (d) 75 feet

17. $x + 5x^2 + \frac{25}{3}x^3 + C$ or $\frac{1}{15}(1 + 5x)^3 + C_1$

19. $\frac{2}{5}\sqrt{5x - 1} + C$ **21.** $\frac{1}{2}x^2 - x^4 + C$

23. $\frac{1}{4}(x^4 - 2x)^2 + C$

25. (a) 30.5 board-feet (b) 125.2 board-feet

27. $-e^{-3x} + C$ **29.** $\frac{1}{2}e^{x^2 - 2x} + C$

31. $-\frac{1}{3}\ln|1 - x^3| + C$ **33.** $\frac{2}{3}x^{3/2} + 2x + 2x^{1/2} + C$

35. $A = 4$ **37.** $A = \frac{8}{3}$ **39.** $A = 2 \ln 2$

41. 16 **43.** 0 **45.** 2 **47.** $\frac{1}{8}$ **49.** 3.899

51. 0 **53.** Increases $700.25

55. Average value: $\frac{8}{5}$, $x = \frac{29}{4}$

57. Average value: $\frac{1}{3}(-1 + e^3) \approx 6.362$, $x \approx 3.150$

59. $520.54; Explanations will vary.

61. (a) $B = -0.0243t^2 + 0.564t - 1.58$

(b) The price of beef per pound does not surpass $2.00. The highest price is $1.69 during 2001, and after that the prices are decreasing.

63. $17,492.94

65. $\int_{-2}^{2} 6x^5 \, dx = 0$ (Odd function)

67. $\int_{-2}^{-1} \frac{4}{x^2} \, dx = \int_{1}^{2} \frac{4}{x^2} \, dx = 2$ (Symmetric about y-axis)

69.

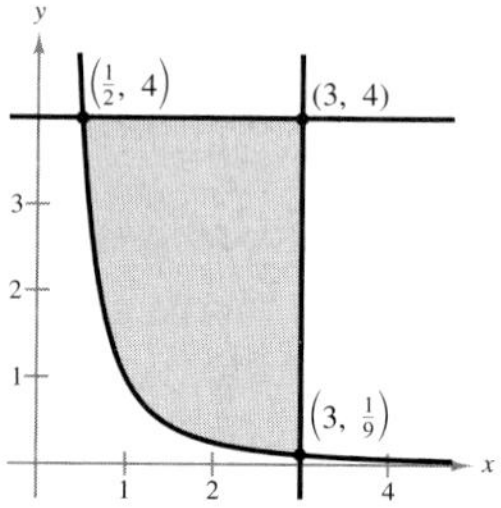

Area $= \frac{25}{3}$

71.

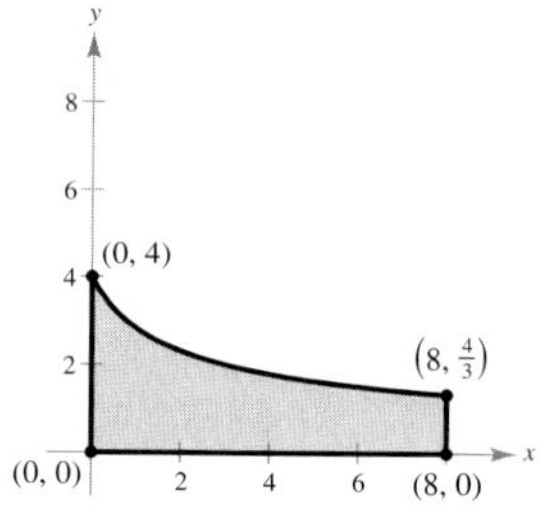

Area = 16

73.

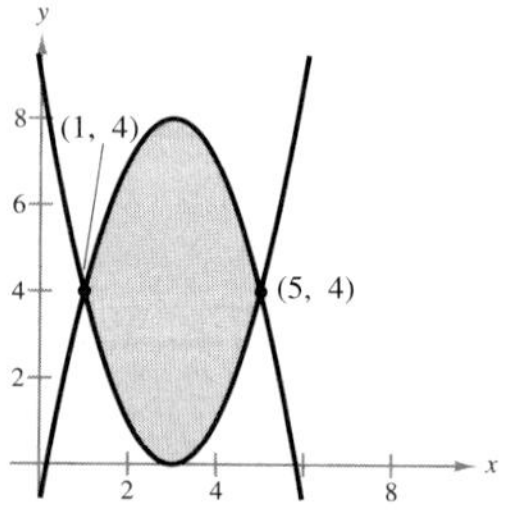

Area $= \frac{64}{3}$

75.

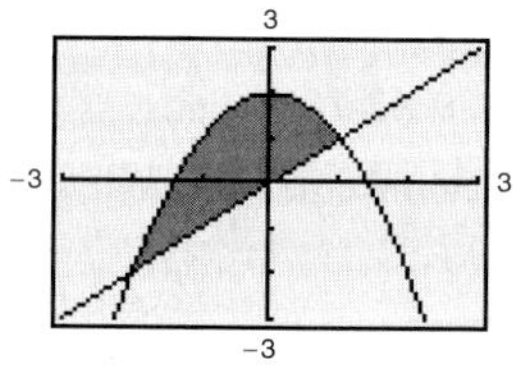

Area $= \frac{9}{2}$

77. Consumer surplus: 11,250

Producer surplus: 14,062.5

79. $5511 million less

81. (a)

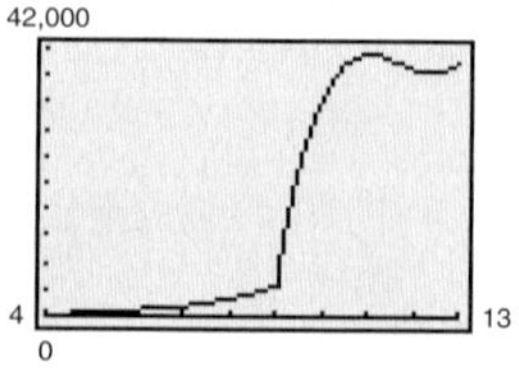

(b) Decreased

(c) $85,834 million less

83. $n = 4$: 13.3203
$n = 20$: 13.7167

85. $n = 4$: 0.7867
$n = 20$: 0.7855

87. $\pi \ln 4 \approx 4.355$ **89.** $\frac{\pi}{2}(e^2 - e^{-2}) \approx 11.394$

91. $\frac{56\pi}{3}$ **93.** $\frac{2\pi}{35}$ **95.** $\frac{5\pi}{16}\sqrt{15}$

SAMPLE POST-GRAD EXAM QUESTIONS

(page 386)

1. d **2.** b **3.** c **4.** b **5.** a
6. b **7.** d **8.** d **9.** a

CHAPTER 6

SECTION 6.1 *(page 394)*

Prerequisite Review

1. $5x + C$ **2.** $\frac{1}{3}x + C$ **3.** $\frac{2}{5}x^{5/2} + C$
4. $\frac{3}{5}x^{5/3} + C$ **5.** $\frac{1}{4}(x^2 + 1)^4 + C$
6. $\frac{(x^3 - 1)^3}{3} + C$ **7.** $e^{6x} + C$
8. $\ln|2x + 1| + C$ **9.** $x(x - 1)(2x - 1)$
10. $3x(x + 4)^2(x + 8)$
11. $(x + 21)(x + 7)^{-1/2}$ **12.** $x(x + 5)^{-2/3}$

1. $\frac{1}{5}(x - 2)^5 + C$ **3.** $\frac{2}{9 - t} + C$
5. $\ln|t^2 - t + 2| + C$ **7.** $\frac{2}{3}(1 + x)^{3/2} + C$
9. $\ln(3x^2 + x)^2 + C$ **11.** $-\frac{1}{10(5x + 1)^2} + C$
13. $2\sqrt{x + 1} + C$ **15.** $-\frac{1}{3}\ln|1 - e^{3x}| + C$
17. $-\frac{1}{3}e^{-3x^2} + C$ **19.** $\frac{1}{2}x^2 + x + \ln|x - 1| + C$
21. $\frac{1}{3}(x^2 + 4)^{3/2} + C$ **23.** $\frac{1}{5}e^{5x} + C$
25. $-\ln|e^{-x} + 2| + C$ **27.** $\frac{-1}{2(x + 1)^2} + \frac{1}{3(x + 1)^3} + C$
29. $\frac{1}{9}\left(\ln|3x - 1| - \frac{1}{3x - 1}\right) + C$
31. $2(\sqrt{t} - 1) + 2\ln|\sqrt{t} - 1| + C$
33. $4\sqrt{t} + \ln|t| + C$ **35.** $\frac{1}{3}(x - 1)\sqrt{2x + 1} + C$
37. $\{-\frac{2}{105}(1 - t)^{3/2}[35 - 42(1 - t) + 15(1 - t)^2]\} + C =$
$-\frac{2}{105}(15t^2 + 12t + 8)(1 - t)^{3/2} + C$
39. $\frac{26}{3}$ **41.** $\frac{3}{2}(e - 1) \approx 2.577$
43. $\ln 2 - \frac{1}{2} \approx 0.193$ **45.** $\frac{13}{320}$

47. Area $= \frac{144}{5}$

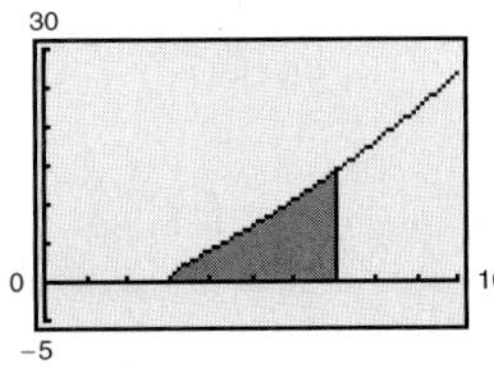

49. Area $= \frac{1696}{105}$

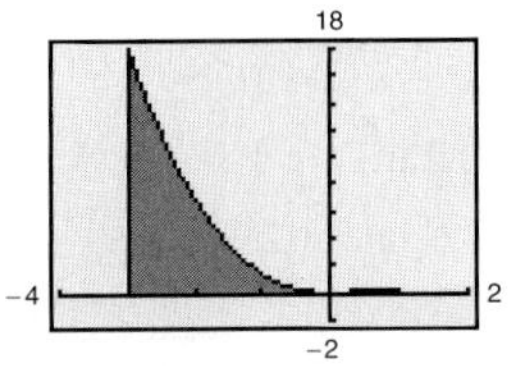

51. Area $= \frac{224}{15}$

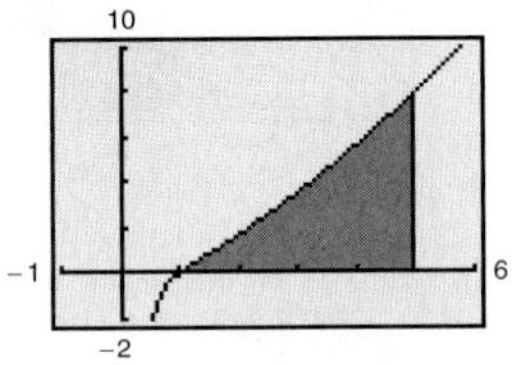

53. Area $= \frac{1209}{28}$

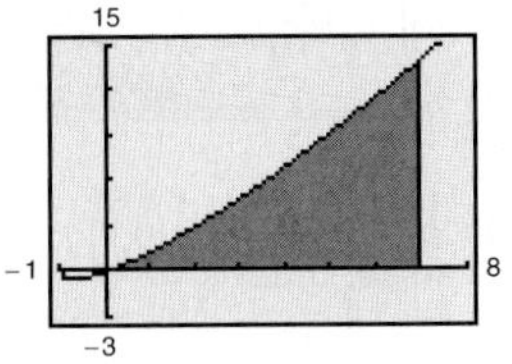

55. $\frac{16}{15}\sqrt{2}$ **57.** $\frac{4}{3}$ **59.** $\frac{4\pi}{15} \approx 0.838$ **61.** $\frac{1}{2}$
63. (a) 0.547 (b) 0.586
65. (a)

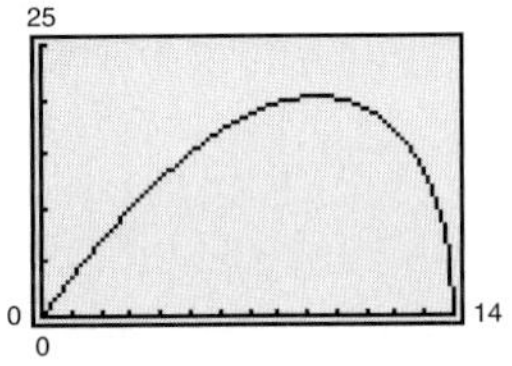

(b) About 13.97 inches (c) About 195.56 inches
67. 5.885

SECTION 6.2 *(page 403)*

Prerequisite Review

1. $\frac{1}{x + 1}$ **2.** $\frac{2x}{x^2 - 1}$ **3.** $3x^2e^{x^3}$
4. $-2xe^{-x^2}$ **5.** $e^x(x^2 + 2x)$ **6.** $e^{-2x}(1 - 2x)$
7. $\frac{64}{3}$ **8.** $\frac{4}{3}$ **9.** 36 **10.** 8

1. $\frac{1}{3}xe^{3x} - \frac{1}{9}e^{3x} + C$ **3.** $-x^2e^{-x} - 2xe^{-x} - 2e^{-x} + C$
5. $x \ln 2x - x + C$ **7.** $\frac{1}{4}e^{4x} + C$
9. $\frac{e^{4x}}{16}(4x - 1) + C$ **11.** $\frac{1}{2}e^{x^2} + C$
13. $x^2e^x - 2e^x x + 2e^x + C$
15. $\frac{1}{2}t^2 \ln|t + 1| - \frac{1}{2}\ln|t + 1| - \frac{1}{4}(t - 1)^2 + C$
17. $-e^{1/t} + C$ **19.** $\frac{x^2}{2}(\ln x)^2 - \frac{x^2}{2}\ln x + \frac{x^2}{4} + C$
21. $\frac{1}{3}(\ln x)^3 + C$ **23.** $\frac{2}{15}(x - 1)^{3/2}(3x + 2) + C$

25. $\frac{1}{4}x^4 + \frac{2}{3}x^3 + \frac{1}{2}x^2 + C$ **27.** $\frac{e^{2x}}{4(2x+1)} + C$

29. $e - 2 \approx 0.718$ **31.** $\frac{5}{36}e^6 + \frac{1}{36} \approx 56.060$

33. $2 \ln 2 - 1 \approx 0.386$

35. Area $= 2e^2 + 6$ **37.** Area $= \frac{1}{9}(2e^3 + 1)$

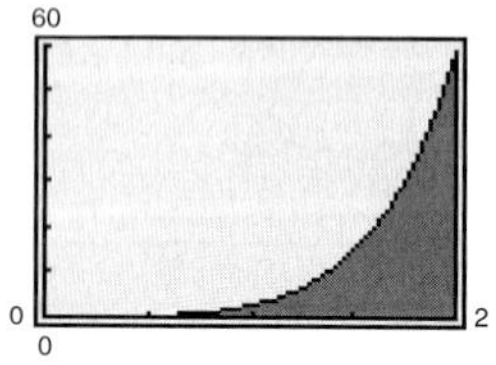

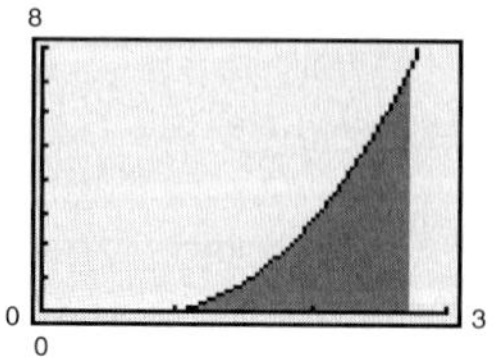

39. $\frac{2}{5}(2x-3)^{3/2}(x+1) + C$

41. $\frac{2}{75}\sqrt{4+5x}(5x-8) + C$ **43.** Proof

45. $\frac{e^{5x}}{125}(25x^2 - 10x + 2) + C$ **47.** $-\frac{1}{x} - \frac{\ln x}{x} + C$

49. $1 - 5e^{-4} \approx 0.908$ **51.** $\frac{1}{4}(e^2 + 1) \approx 2.097$

53. (a) 2 (b) $4\pi(e-2) \approx 9.026$

55. $\frac{3}{128} - \frac{379}{128}e^{-8} \approx 0.022$

57. $\frac{1,171,875}{256}\pi \approx 14,381.070$

59.

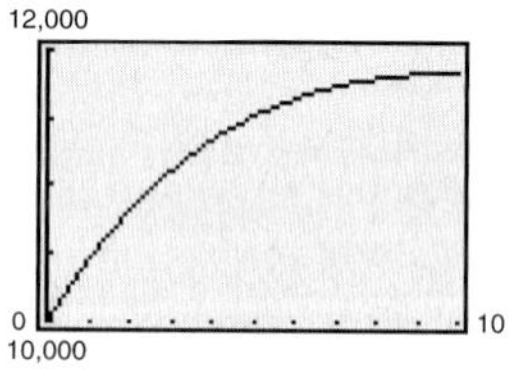

(a) Increase

(b) 113,212 units (c) 11,321 units per year

61. (a) $3.2 \ln 2 - 0.2 \approx 2.018$

(b) $12.8 \ln 4 - 7.2 \ln 3 - 1.8 \approx 8.035$

63. \$18,126.92 **65.** \$1,332,474.72 **67.** \$4103.07

69. (a) \$1,200,000 (b) \$1,094,142.26

71. \$45,957.78 **73.** (a) \$17,378.62 (b) \$3681.26

75. 4.254

SECTION 6.3 *(page 413)*

Prerequisite Review

1. $(x-4)(x+4)$ **2.** $(x-5)(x+5)$

3. $(x-4)(x+3)$ **4.** $(x-2)(x+3)$

5. $x(x-2)(x+1)$ **6.** $x(x-2)^2$

7. $(x-2)(x-1)^2$ **8.** $(x-3)(x-1)^2$

9. $\frac{1}{x-2} + x$ **10.** $-\frac{1}{1-x} + 2x - 2$

11. $-\frac{2}{x-2} + x^2 - x - 2$

12. $-\frac{4}{x+1} + x^2 - x + 3$

13. $\frac{6}{x-1} + x + 4, \quad x \neq -1$

14. $\frac{1}{x+1} + x + 3 \quad x \neq 1$

1. $\frac{5}{x-5} - \frac{3}{x+5}$ **3.** $\frac{9}{x-3} - \frac{1}{x}$ **5.** $\frac{1}{x-5} + \frac{3}{x+2}$

7. $\frac{3}{x} - \frac{5}{x^2}$ **9.** $\frac{1}{3(x-2)} + \frac{1}{(x-2)^2}$

11. $\frac{8}{x+1} - \frac{1}{(x+1)^2} + \frac{2}{(x+1)^3}$ **13.** $\frac{1}{2}\ln\left|\frac{x-1}{x+1}\right| + C$

15. $\frac{1}{4}\ln\left|\frac{x+4}{x-4}\right| + C$ **17.** $\ln\left|\frac{3x-1}{x}\right| + C$

19. $\ln\left|\frac{x}{2x+1}\right| + C$ **21.** $\ln\left|\frac{x-1}{x+2}\right| + C$

23. $\frac{3}{2}\ln|2x-1| - 2\ln|x+1| + C$

25. $5\ln|x-2| - \ln|x+2| - 3\ln|x| + C$

27. $\frac{1}{2}(3\ln|x-4| - \ln|x|) + C$

29. $-3\ln|x-1| - \frac{1}{x-1} + C$

31. $\ln|x| + 2\ln|x+1| + \frac{1}{x+1} + C$

33. $\frac{1}{6}\ln\frac{4}{7} \approx -0.093$ **35.** $-\frac{4}{5} + 2\ln\frac{5}{3} \approx 0.222$

37. $\frac{1}{2} - \ln 2 \approx -0.193$ **39.** $4\ln 2 + \frac{1}{2} \approx 3.273$

41. $12 - \frac{7}{2}\ln 7 \approx 5.189$ **43.** $5\ln 2 - \ln 5 \approx 1.856$

45. $\frac{1}{2a}\left(\frac{1}{a+x} + \frac{1}{a-x}\right)$ **47.** $\frac{1}{a}\left(\frac{1}{x} + \frac{1}{a-x}\right)$

49.

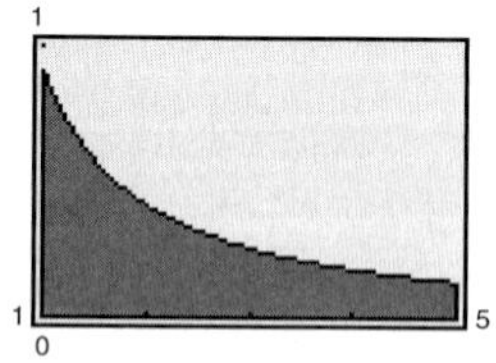

$\frac{\pi}{165}\left[136 - 33\ln\frac{11}{3}\right] \approx 1.7731$

51.

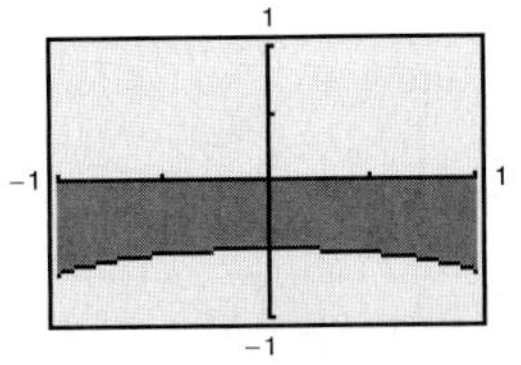

$\pi\left(\frac{1}{3} + \frac{1}{4}\ln 3\right) \approx 1.9100$

53. $y = \dfrac{1000}{1 + 9e^{-0.1656t}}$

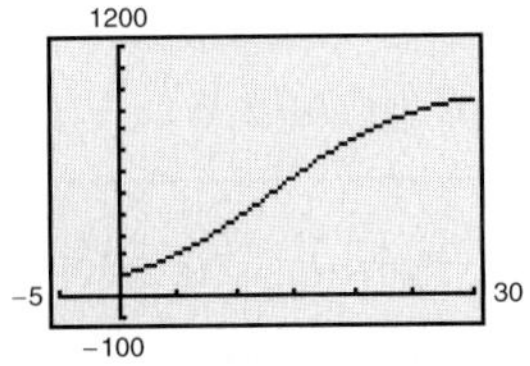

55. \$1.077 thousand **57.** \$6188.4 million; \$773.6 million

59. (a) 103 (b) 200 **61.** Answers will vary.

SECTION 6.4 *(page 424)*

Prerequisite Review

1. $x^2 + 8x + 16$ **2.** $x^2 - 2x + 1$

3. $x^2 + x + \frac{1}{4}$ **4.** $x^2 - \frac{2}{3}x + \frac{1}{9}$

5. $\dfrac{2}{x} - \dfrac{2}{x+2}$ **6.** $-\dfrac{3}{4x} + \dfrac{3}{4(x-4)}$

7. $\dfrac{3}{2(x-2)} - \dfrac{2}{x^2} - \dfrac{3}{2x}$ **8.** $-\dfrac{3}{x+1} + \dfrac{2}{x-2} + \dfrac{4}{x}$

9. $2e^x(x-1) + C$ **10.** $x^3 \ln x - \dfrac{x^3}{3} + C$

1. $\dfrac{1}{9}\left(\dfrac{2}{2+3x} + \ln|2+3x|\right) + C$

3. $\dfrac{2(3x-4)}{27}\sqrt{2+3x} + C$ **5.** $\ln\left(x^2 + \sqrt{x^4 - 9}\right) + C$

7. $\dfrac{1}{2}(x^2 - 1)e^{x^2} + C$ **9.** $\ln\left|\dfrac{x}{1+x}\right| + C$

11. $-\dfrac{1}{3}\ln\left|\dfrac{3 + \sqrt{x^2+9}}{x}\right| + C$

13. $-\dfrac{1}{2}\ln\left|\dfrac{2 + \sqrt{4 - x^2}}{x}\right| + C$

15. $\frac{1}{4}x^2(-1 + 2\ln x) + C$ **17.** $3x^2 - \ln(1 + e^{3x^2}) + C$

19. $\frac{1}{4}\left(x^2\sqrt{x^4 - 4} - 4\ln\left|x^2 + \sqrt{x^4 - 4}\right|\right) + C$

21. $\dfrac{1}{27}\left[\dfrac{4}{2+3t} - \dfrac{4}{2(2+3t)^2} + \ln|2+3t|\right] + C$

23. $\dfrac{1}{\sqrt{3}}\ln\left|\dfrac{\sqrt{3+s} - \sqrt{3}}{\sqrt{3+s} + \sqrt{3}}\right| + C$

25. $\dfrac{1}{8}\left[\dfrac{-1}{2(3+2x)^2} + \dfrac{6}{3(3+2x)^3} - \dfrac{9}{4(3+2x)^4}\right] + C$

27. $-\dfrac{\sqrt{1-x^2}}{x} + C$ **29.** $\dfrac{1}{9}x^3(-1 + 3\ln x) + C$

31. $\dfrac{1}{27}\left(3x - \dfrac{25}{3x-5} + 10\ln|3x-5|\right) + C$

33. $\frac{1}{9}(3\ln x - 4\ln|4 + 3\ln x|) + C$

35. Area $= \frac{40}{3}$

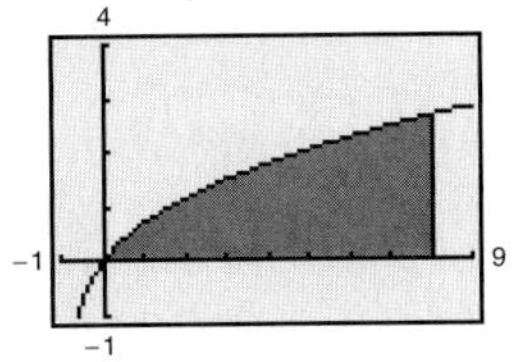

Area $= 13.\overline{3}$

37. Area $= \dfrac{1}{2}\left[4 + \ln\left(\dfrac{2}{1+e^4}\right)\right]$

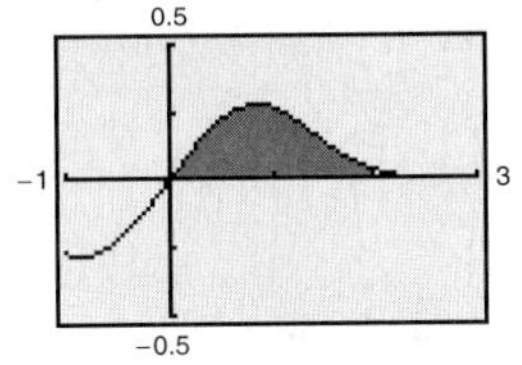

Area ≈ 0.3375

39. Area $= \frac{1}{4}\left[21\sqrt{5} - 8\ln\left(\sqrt{5} + 3\right) + 8\ln 2\right]$

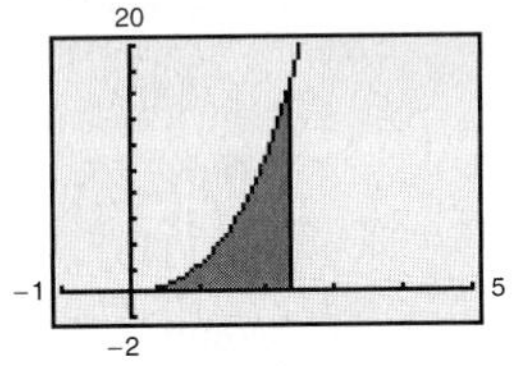

Area ≈ 9.8145

41. $\dfrac{5\sqrt{5}}{3}$ **43.** $12\left(2 + \ln\left|\dfrac{2}{1+e^2}\right|\right) \approx 6.7946$

45. $(x^2 - 2x + 2)e^x + C$ **47.** $-\left(\dfrac{1}{x} + \ln\left|\dfrac{x}{x+1}\right|\right) + C$

49. (a) $(x+3)^2 - 9$ (b) $(x-4)^2 - 7$
(c) $(x^2+1)^2 - 6$ (d) $4 - (x+1)^2$

51. (a) $4\left(x + \frac{3}{2}\right)^2 + 6$ (b) $3(x-2)^2 - 21$
(c) $(x-1)^2 - 1$ (d) $25 - (x-4)^2$

53. $\dfrac{1}{2\sqrt{17}}\ln\left|\dfrac{x + 3 - \sqrt{17}}{x + 3 + \sqrt{17}}\right| + C$

55. $-\ln\left|\dfrac{1 + \sqrt{x^2 - 2x + 2}}{x - 1}\right| + C$

57. $\dfrac{1}{8}\ln\left|\dfrac{x-3}{x+1}\right| + C$

59. $\frac{1}{2}\ln\left|x^2 + 1 + \sqrt{x^4 + 2x^2 + 2}\right| + C$

61.

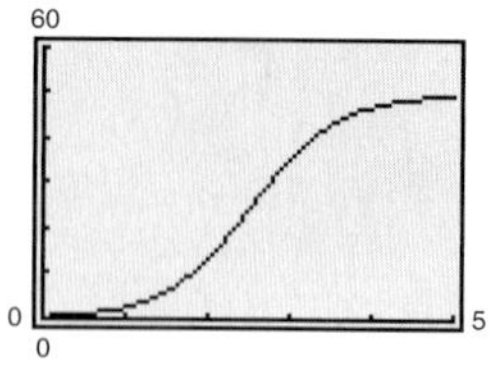

Average value: 42.58

63. \$1138.43 **65.** \$0.40 billion per year

SECTION 6.5 *(page 433)*

Prerequisite Review

1. $\frac{2}{x^3}$ **2.** $-\frac{96}{(2x+1)^4}$ **3.** $-\frac{12}{x^4}$ **4.** $6x - 4$

5. $16e^{2x}$ **6.** $e^{x^2}(4x^2 + 2)$ **7.** $(3, 18)$

8. $(1, 8)$ **9.** $n < -5\sqrt{10},\ n > 5\sqrt{10}$

10. $n < -5, n > 5$

	Exact value	*Trapezoidal Rule*	*Simpson's Rule*
1.	2.6667	2.7500	2.6667
3.	8.4000	9.0625	8.4167
5.	4.0000	4.0625	4.0000
7.	0.6931	0.6941	0.6932
9.	5.3333	5.2650	5.3046
11.	0.6931	0.6970	0.6933

13. (a) 0.783 (b) 0.785

15. (a) 0.749 (b) 0.771

17. (a) 0.877 (b) 0.830

19. (a) 1.880 (b) 1.890

21. \$21,831.20 **23.** \$678.36 **25.** 0.3413 = 34.13%

27. 0.4999 = 49.99% **29.** 89,500 square feet

31. (a) 2 (b) $\frac{2^5}{180(4^4)}(24) \approx 0.017$

33. (a) $\frac{5e}{64} \approx 0.212$ (b) $\frac{13e}{1024} \approx 0.035$

35. (a) $n = 101$ (b) $n = 8$

37. (a) $n = 3280$ (b) $n = 60$

39. 19.5215 **41.** 3.6558

43. Exact value: $\int_0^1 x^3\,dx = \frac{x^4}{4}\Big]_0^1 = \frac{1}{4}$

Simpson's Rule: $\int_0^1 x^3\,dx = \frac{1}{6}\left[0^3 + 4\left(\frac{1}{2}\right)^3 + 1^3\right] = \frac{1}{4}$

45. 416.1 feet

47. 58.876 milligrams (Simpson's Rule with $n = 100$)

49. 1876 subscribers (Simpson's Rule with $n = 100$)

SECTION 6.6 *(page 444)*

Prerequisite Review

1. 9 **2.** 3 **3.** $-\frac{1}{8}$ **4.** Limit does not exist.

5. Limit does not exist. **6.** -4

7. (a) $\frac{32}{3}b^3 - 16b^2 + 8b - \frac{4}{3}$ (b) $-\frac{4}{3}$

8. (a) $\frac{b^2 - b - 11}{(b-2)^2(b-5)}$ (b) $\frac{11}{20}$

9. (a) $\ln\left(\frac{5 - 3b^2}{b + 1}\right)$ (b) $\ln 5 \approx 1.609$

10. (a) $e^{-3b^2}(e^{6b^2} + 1)$ (b) 2

1. 1 **3.** 1 **5.** Diverges **7.** Diverges

9. Diverges **11.** Diverges **13.** 0 **15.** 4

17. 6 **19.** Diverges **21.** 6 **23.** Diverges

25. 0 **27.** $\ln(4 + \sqrt{7}) - \ln 3 \approx 0.7954$

29. (a) 1 (b) $\frac{\pi}{3}$

31.

x	1	10	25	50
xe^{-x}	0.3679	0.0005	0.0000	0.0000

33.

x	1	10	25	50
$x^2e^{-(1/2)x}$	0.6065	0.6738	0.0023	0.0000

35. 2 **37.** $\frac{1}{4}$ **39.** (a) \$4,637,228 (b) \$5,555,556

41. (a) \$748,367.34 (b) \$808,030.14 (c) \$900,000.00

43. (a) 0.9687 (b) 0.0724 (c) 0.0009

REVIEW EXERCISES FOR CHAPTER 6
(page 450)

1. $t + C$ **3.** $\frac{(x+5)^4}{4} + C$ **5.** $\frac{1}{10}e^{10x} + C$

7. $\frac{1}{5}\ln|x| + C$ **9.** $\frac{1}{3}(x^2 + 4)^{3/2} + C$

11. $2\ln(3 + e^x) + C$ **13.** $\frac{(x-2)^5}{5} + \frac{(x-2)^4}{2} + C$

15. $\frac{2}{15}(x + 1)^{3/2}(3x - 2) + C$

17. $\frac{4}{5}(x - 3)^{3/2}(x + 2) + C$

19. $-\frac{2}{15}(1 - x)^{3/2}(3x + 7) + C$

21. $\frac{26}{15}$ **23.** $\frac{412}{15}$ **25.** (a) 0.696 (b) 0.693

27. (a) \$2661.667 million (b) \$15,970.002 million

29. $2\sqrt{x}\ln x - 4\sqrt{x} + C$ **31.** $xe^x - 2e^x + C$

33. $x^2e^{2x} - xe^{2x} + \frac{1}{2}e^{2x} + C$ **35.** \$45,317.31

37. \$432,979.25

39. (a) \$4423.98, \$3934.69, \$3517.56 (b) \$997,629.35

41. \$45,118.84 **43.** $\frac{1}{5}\ln\left|\frac{x}{x+5}\right| + C$

45. $\ln|x-5| + 3\ln|x+2| + C$

47. $x - \frac{25}{8}\ln|x+5| + \frac{9}{8}\ln|x-3| + C$

49. (a) $y = \dfrac{10,000}{1 + 7e^{-0.106873t}}$

(b)

Time, t	0	3	6	12	24
Sales, y	1250	1645	2134	3400	6500

(c) $t \approx 28$ weeks

51. $\sqrt{x^2+25} - 5\ln\left|\dfrac{5+\sqrt{x^2+25}}{x}\right| + C$

53. $\dfrac{1}{4}\ln\left|\dfrac{x-2}{x+2}\right| + C$ **55.** $\dfrac{8}{3}$

57. $2\sqrt{1+x} + \ln\left|\dfrac{\sqrt{1+x}-1}{\sqrt{1+x}+1}\right| + C$

59. $(x-5)^3e^{x-5} - 3(x-5)^2e^{x-5} + 6(x-6)e^{x-5} + C$

61. $\dfrac{1}{10}\ln\left|\dfrac{x-3}{x+7}\right| + C$

63. $\frac{1}{2}\Big[(x-5)\sqrt{(x-5)^2-25}$
$- 25\ln\left|(x-5) + \sqrt{(x-5)^2-25}\right|\Big] + C$

65. 0.705 **67.** 0.741 **69.** 0.376 **71.** 0.289

73. 9.0997 **75.** 0.017 **77.** 1 **79.** Diverges

81. 2 **83.** 2 **85.** (a) \$494,525.28 (b) \$833,333.33

87. (a) 0.431 (b) 0.108 (c) 0.013

SAMPLE POST-GRAD EXAM QUESTIONS

(page 454)

1. a **2.** d **3.** a **4.** b **5.** c

6. d **7.** b **8.** a **9.** b

CHAPTER 7

SECTION 7.1 *(page 462)*

Prerequisite Review

1. $2\sqrt{5}$ **2.** 5 **3.** 8 **4.** 8 **5.** (4, 7)

6. (1, 0) **7.** (0, 3) **8.** (−1, 1)

9. $(x-2)^2 + (y-3)^2 = 4$

10. $(x-1)^2 + (y-4)^2 = 25$

1.

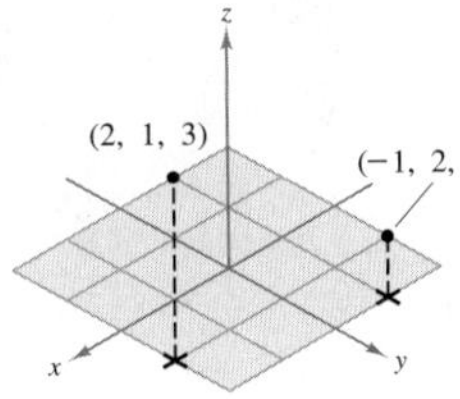

3.

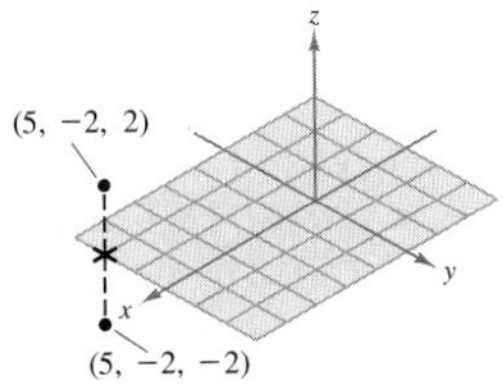

5. $3\sqrt{2}$ **7.** $\sqrt{206}$ **9.** (2, −5, 3)

11. $\left(\frac{1}{2}, \frac{1}{2}, -1\right)$ **13.** (6, −3, 5) **15.** (1, 2, 1)

17. 3, $3\sqrt{5}$, 6; right triangle

19. 2, $2\sqrt{5}$, $2\sqrt{2}$; neither right nor isosceles

21. $x^2 + (y-2)^2 + (z-2)^2 = 4$

23. $\left(x - \frac{3}{2}\right)^2 + (y-2)^2 + (z-1)^2 = \frac{21}{4}$

25. $(x-1)^2 + (y-1)^2 + (z-5)^2 = 9$

27. $(x-1)^2 + (y-3)^2 + z^2 = 10$

29. $(x+2)^2 + (y-1)^2 + (z-1)^2 = 1$

31. Center: $\left(\frac{5}{2}, 0, 0\right)$ **33.** Center: (1, −3, −4)
Radius: $\frac{5}{2}$ Radius: 5

35. Center: (1, 3, 2)
Radius: $\dfrac{5\sqrt{2}}{2}$

37.

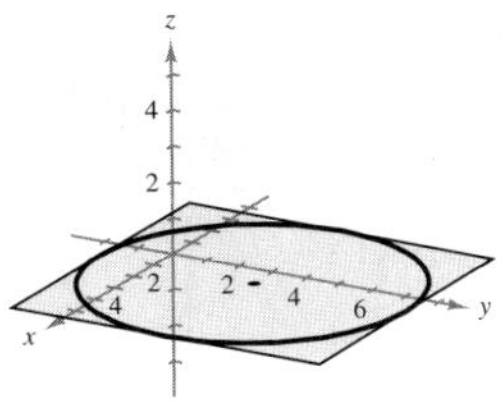

39.

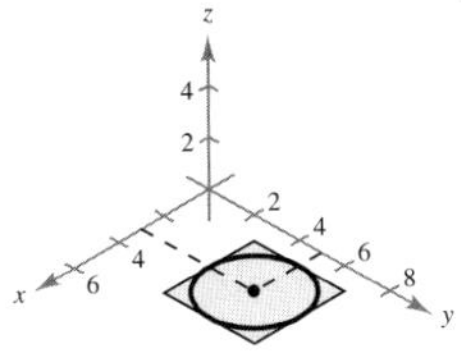

41.

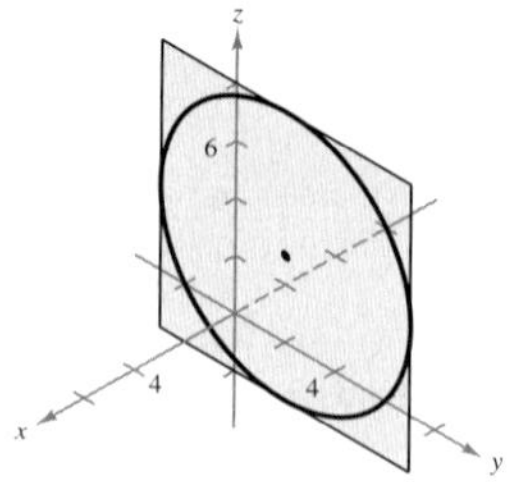

43. (a)

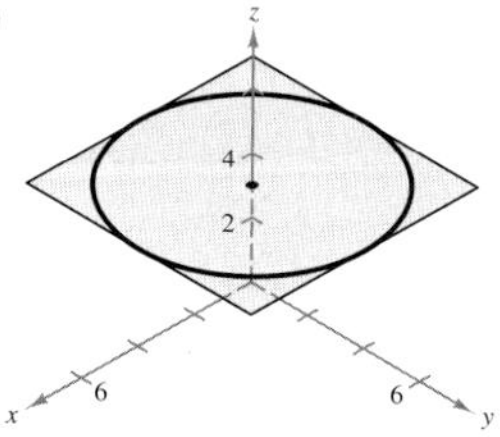

(b)

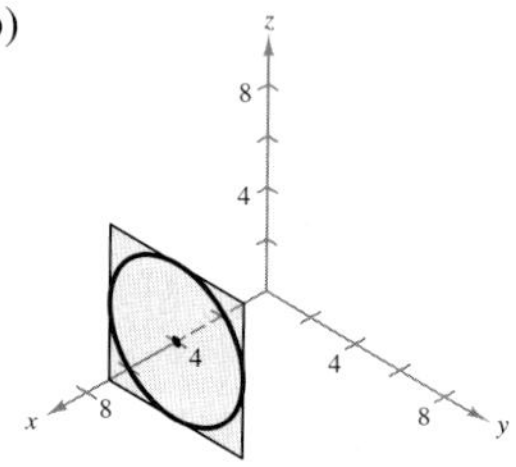

45. (a)

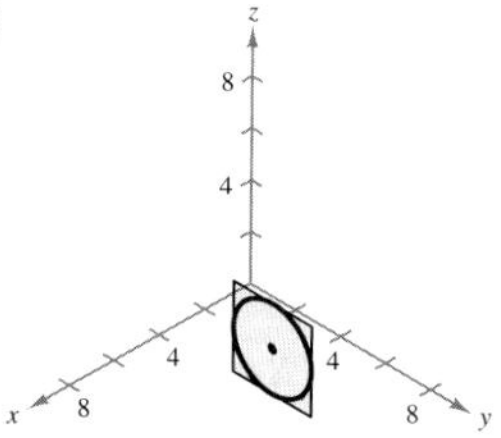

(b)

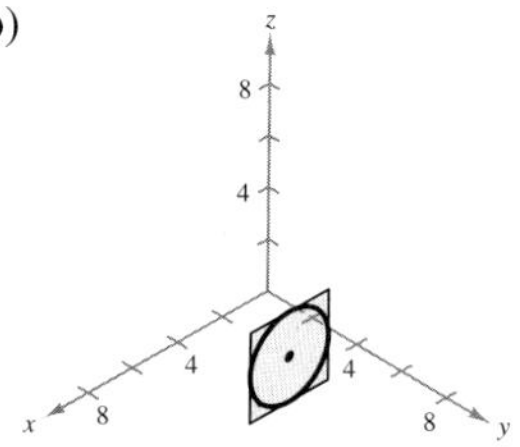

47. (3, 3, 3)

SECTION 7.2 *(page 472)*

Prerequisite Review

1. (4, 0), (0, 3) **2.** $\left(-\frac{4}{3}, 0\right)$, (0, −8)

3. (1, 0), (0, −2) **4.** (−5, 0), (0, −5)

5. $(x-1)^2 + (y-2)^2 + (z-3)^2 + 1 = 0$

6. $(x-4)^2 + (y+2)^2 - (z+3)^2 = 0$

7. $(x+1)^2 + (y-1)^2 - z = 0$

8. $(x-3)^2 + (y+5)^2 + (z+13)^2 = 1$

9. $x^2 - y^2 + z^2 = \frac{1}{4}$ **10.** $x^2 - y^2 + z^2 = 4$

1.

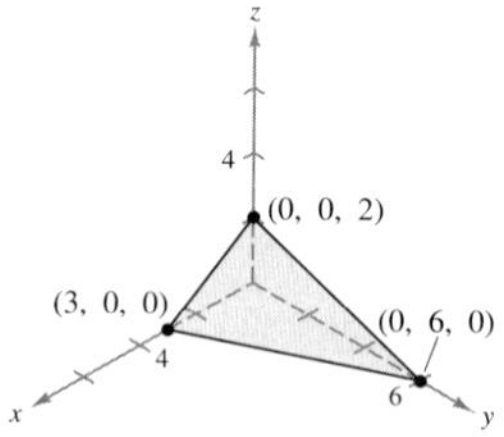

3.

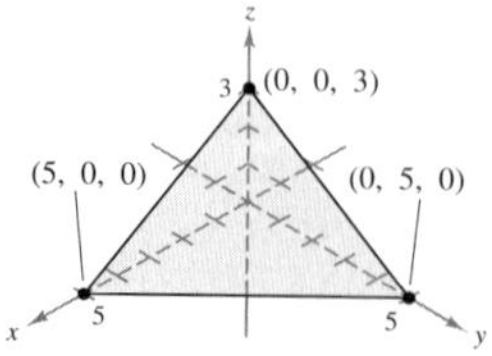

5.

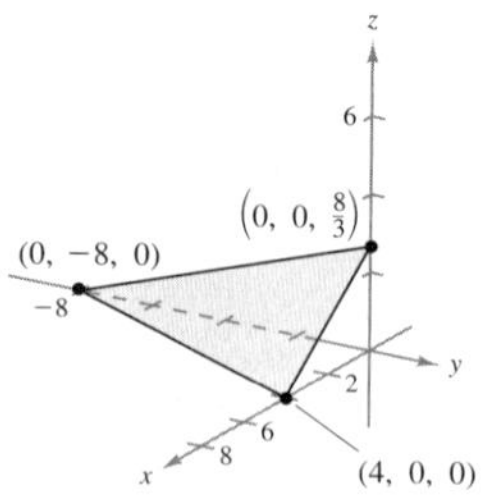

7.

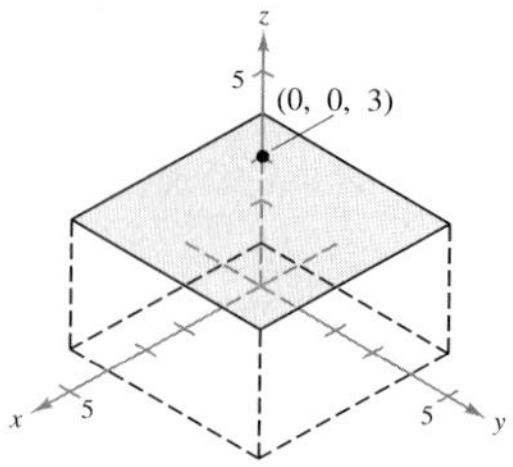

9.

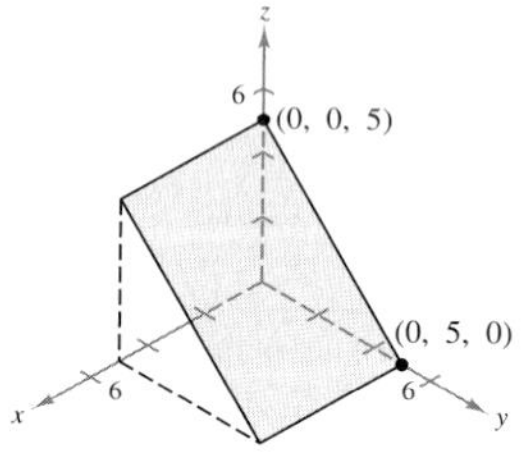

11.

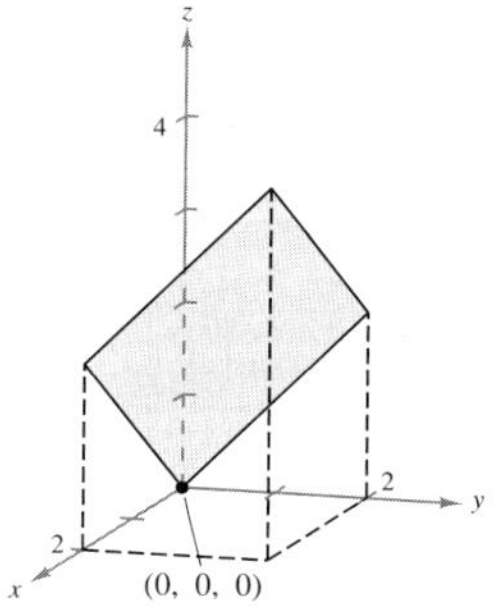

13. Perpendicular **15.** Parallel **17.** Parallel

19. Neither parallel nor perpendicular **21.** Perpendicular

23. $\dfrac{6\sqrt{14}}{7}$ **25.** $\dfrac{8\sqrt{14}}{7}$ **27.** $\dfrac{13\sqrt{29}}{29}$ **29.** $\dfrac{28\sqrt{29}}{29}$

31. c **32.** e **33.** f **34.** g

35. d **36.** b **37.** a **38.** h

39. Trace in xy-plane ($z = 0$): $y = x^2$ (parabola)

Trace in plane $y = 1$: $x^2 - z^2 = 1$ (hyperbola)

Trace in yz-plane ($x = 0$): $y = -z^2$ (parabola)

41. Trace in xy-plane ($z = 0$): $\dfrac{x^2}{4} + y^2 = 1$ (ellipse)

Trace in xz-plane ($y = 0$): $\dfrac{x^2}{4} + z^2 = 1$ (ellipse)

Trace in yz-plane ($x = 0$): $y^2 + z^2 = 1$ (circle)

43. Ellipsoid **45.** Hyperboloid of one sheet

47. Elliptic paraboloid **49.** Hyperbolic paraboloid

51. Hyperboloid of two sheets **53.** Elliptic cone

55. Hyperbolic paraboloid

57.

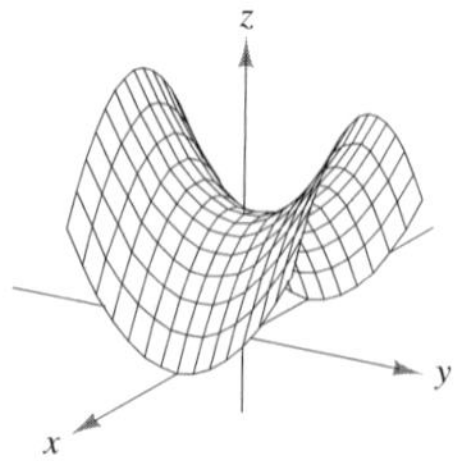

59.

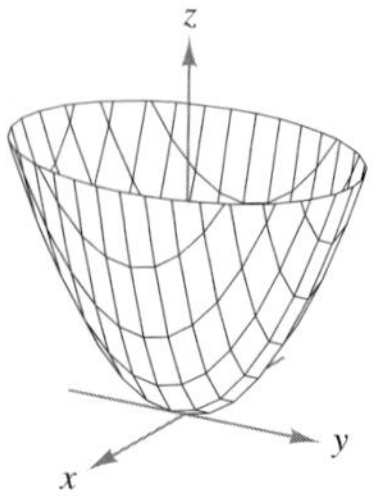

61. $\dfrac{x^2}{3963^2} + \dfrac{y^2}{3963^2} + \dfrac{z^2}{3950^2} = 1$

SECTION 7.3 *(page 480)*

Prerequisite Review

1. 11 **2.** -16 **3.** 7

4. 4 **5.** $(-\infty, \infty)$

6. $(-\infty, -3) \cup (-3, 0) \cup (0, \infty)$

7. $[5, \infty)$ **8.** $\left(-\infty, -\sqrt{5}\right] \cup \left[\sqrt{5}, \infty\right)$

9. 55.0104 **10.** 6.9165

1. (a) $\frac{3}{2}$ (b) $-\frac{1}{4}$ (c) 6 (d) $\frac{5}{y}$ (e) $\frac{x}{2}$ (f) $\frac{5}{t}$

3. (a) 5 (b) $3e^2$ (c) $2e^{-1}$
(d) $5e^y$ (e) xe^2 (f) te^t

5. (a) $\frac{2}{3}$ (b) 0 **7.** (a) 90π (b) 50π

9. (a) \$20,655 (b) \$1,397,673 **11.** (a) 0 (b) 6

13. (a) $x^2 + 2x\,\Delta x + (\Delta x)^2 - 2y$ (b) -2, $\Delta y \neq 0$

15. Domain: all points (x, y) inside and on the circle
$x^2 + y^2 = 16$
Range: $[0, 4]$

17. Domain: all points (x, y) such that $y \neq 0$
Range: $(0, \infty)$

19. All points inside and on the ellipse $9x^2 + y^2 = 9$

21. All points (x, y) such that $y \neq 0$

23. All points (x, y) such that $x \neq 0$ nor $y \neq 0$

25. All points (x, y) such that $y \geq 0$

27. The half-plane below the line $y = -x + 4$

29. b **30.** d **31.** a **32.** c

33. The level curves are parallel lines.

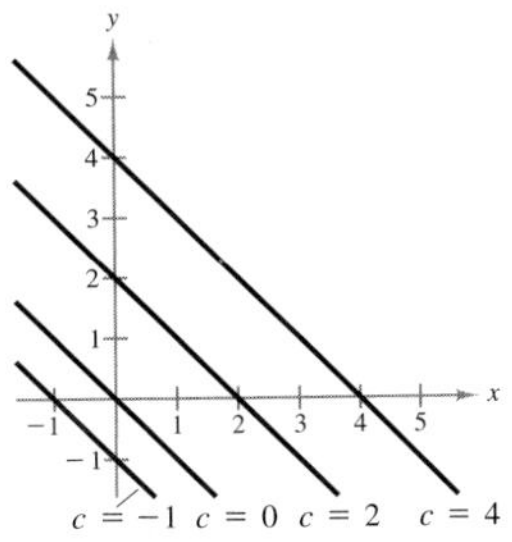

35. The level curves are circles.

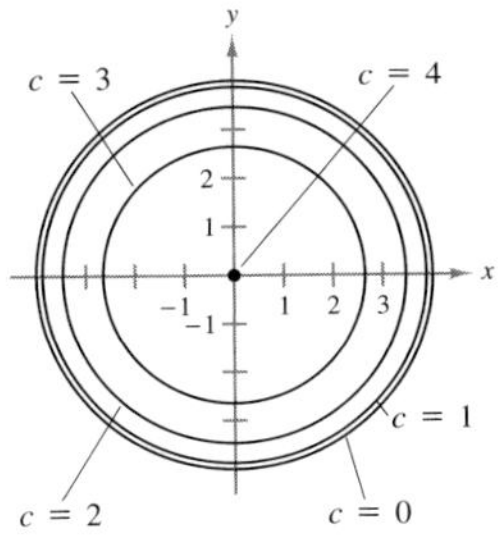

37. The level curves are hyperbolas.

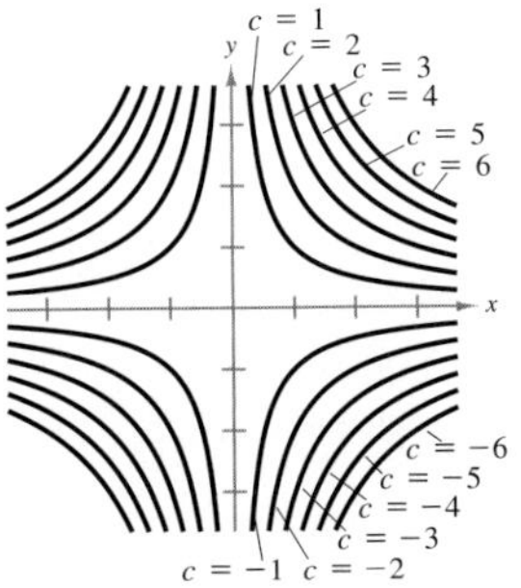

39. The level curves are circles.

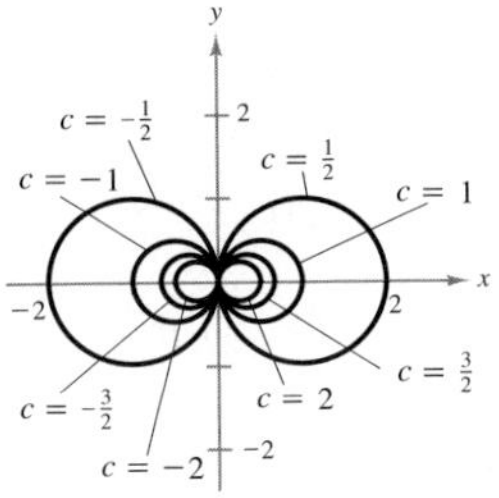

41. 135,540 units **43.** \$21,960

45. (a) \$13,250.00 (b) \$15,925.00

47.

R \ I	0	0.03	0.05
0	\$2593.74	\$1929.99	\$1592.33
0.28	\$2004.23	\$1491.34	\$1230.42
0.35	\$1877.14	\$1396.77	\$1152.40

49. (a) The different colors represent various amplitudes.
(b) No, the level curves are uneven and sporadically spaced.

SECTION 7.4 *(page 491)*

Prerequisite Review

1. $\dfrac{x}{\sqrt{x^2 + 3}}$ **2.** $-6x(3 - x^2)^2$ **3.** $e^{2t+1}(2t + 1)$

4. $\dfrac{e^{2x}(2 - 3e^{2x})}{\sqrt{1 - e^{2x}}}$ **5.** $-\dfrac{2}{3 - 2x}$ **6.** $\dfrac{3(t^2 - 2)}{2t(t^2 - 6)}$

7. $-\dfrac{10x}{(4x - 1)^3}$ **8.** $-\dfrac{(x + 2)^2(x^2 + 8x + 27)}{(x^2 - 9)^3}$

9. $f'(2) = 8$ **10.** $g'(2) = \frac{7}{2}$

1. $f_x(x, y) = 2$
$f_y(x, y) = -3$

3. $f_x(x, y) = \dfrac{5}{2\sqrt{x}}$
$f_y(x, y) = -12y$

5. $f_x(x, y) = \frac{1}{y}$

$f_y(x, y) = -\frac{x}{y^2}$

7. $f_x(x, y) = \frac{x}{\sqrt{x^2 + y^2}}$

$f_y(x, y) = \frac{y}{\sqrt{x^2 + y^2}}$

9. $\frac{\partial z}{\partial x} = 2xe^{2y}$

$\frac{\partial z}{\partial y} = 2x^2e^{2y}$

11. $h_x(x, y) = -2xe^{-(x^2+y^2)}$

$h_y(x, y) = -2ye^{-(x^2+y^2)}$

13. $\frac{\partial z}{\partial x} = \frac{3y - x}{x^2 - y^2}$

$\frac{\partial z}{\partial y} = \frac{y - 3x}{x^2 - y^2}$

15. $f_x(x, y) = 3xye^{x-y}(2 + x)$

17. $g_x(x, y) = 3y^2e^{y-x}(1 - x)$ **19.** 9

21. $f_x(x, y) = 6x + y,\ 13;\ f_y(x, y) = x - 2y,\ 0$

23. $f_x(x, y) = 3ye^{3xy},\ 12;\ f_y(x, y) = 3xe^{3xy},\ 0$

25. $f_x(x, y) = -\frac{y^2}{(x - y)^2}, -\frac{1}{4}$

$f_y(x, y) = \frac{x^2}{(x - y)^2}, \frac{1}{4}$

27. $f_x(x, y) = \frac{2x}{x^2 + y^2}, 2$

$f_y(x, y) = \frac{2y}{x^2 + y^2}, 0$

29. $w_x = 6xy - 5yz$

$w_y = 3x^2 - 5xz + 10z^2$

$w_z = -5xy + 20yz$

31. $w_x = \frac{y(y + z)}{(x + y + z)^2}$

$w_y = \frac{x(x + z)}{(x + y + z)^2}$

$w_z = -\frac{xy}{(x + y + z)^2}$

33. $w_x = \frac{x}{\sqrt{x^2 + y^2 + z^2}}, \frac{2}{3}$

$w_y = \frac{y}{\sqrt{x^2 + y^2 + z^2}}, -\frac{1}{3}$

$w_z = \frac{z}{\sqrt{x^2 + y^2 + z^2}}, \frac{2}{3}$

35. $w_x = \frac{x}{x^2 + y^2 + z^2}, \frac{3}{25}$

$w_y = \frac{y}{x^2 + y^2 + z^2}, 0$

$w_z = \frac{z}{x^2 + y^2 + z^2}, \frac{4}{25}$

37. $w_x = 2z^2 + 3yz,\ 2$

$w_y = 3xz - 12yz,\ 30$

$w_z = 4xz + 3xy - 6y^2,\ -1$

39. $(-6, 4)$ **41.** $(1, 1)$

43. (a) 2 (b) -3 **45.** (a) 6 (b) -18

47. (a) $-\frac{3}{4}$ (b) 0 **49.** (a) -2 (b) -2

51. $\frac{\partial^2 z}{\partial x \partial y} = \frac{\partial^2 z}{\partial y \partial x} = -2$ **53.** $\frac{\partial^2 z}{\partial x \partial y} = \frac{\partial^2 z}{\partial y \partial x} = ye^{2xy}$

55. $\frac{\partial^2 z}{\partial x^2} = 6x$

$\frac{\partial^2 z}{\partial y^2} = -8$

$\frac{\partial^2 z}{\partial y \partial x} = \frac{\partial^2 z}{\partial x \partial y} = 0$

57. $\frac{\partial^2 z}{\partial x^2} = 24x$

$\frac{\partial^2 z}{\partial y^2} = 6x - 24y$

$\frac{\partial^2 z}{\partial y \partial x} = \frac{\partial^2 z}{\partial x \partial y} = 6y$

59. $\frac{\partial^2 z}{\partial x^2} = \frac{2y^2}{(x - y)^3}$

$\frac{\partial^2 z}{\partial y^2} = \frac{2x^2}{(x - y)^3}$

$\frac{\partial^2 z}{\partial y \partial x} = \frac{\partial^2 z}{\partial x \partial y}$

$= -\frac{2xy}{(x - y)^3}$

61. $\frac{\partial^2 z}{\partial x^2} = 0$

$\frac{\partial^2 z}{\partial y^2} = 2xe^{-y^2}(2y^2 - 1)$

$\frac{\partial^2 z}{\partial y \partial x} = \frac{\partial^2 z}{\partial x \partial y} = -2ye^{-y^2}$

63. $f_{xx}(x, y) = 12x^2 - 6y^2,\ 12$

$f_{xy}(x, y) = -12xy,\ 0$

$f_{yy}(x, y) = -6x^2 + 2,\ -4$

$f_{yx}(x, y) = -12xy,\ 0$

65. $f_{xx}(x, y) = -\frac{1}{(x - y)^2}, -1$

$f_{xy}(x, y) = \frac{1}{(x - y)^2}, 1$

$f_{yy}(x, y) = -\frac{1}{(x - y)^2}, -1$

$f_{yx}(x, y) = \frac{1}{(x - y)^2}, 1$

67. At (120, 160), $\frac{\partial C}{\partial x} \approx 154.77$

At (120, 160), $\frac{\partial C}{\partial y} \approx 193.33$

69. (a) $f_x(x, y) = 60\left(\frac{y}{x}\right)^{0.4}$, $f_x(1000, 500) \approx 45.47$

(b) $f_y(x, y) = 40\left(\frac{x}{y}\right)^{0.6}$, $f_x(1000, 500) \approx 60.63$

71. (a) Complementary (b) Substitute
(c) Complementary

73. An increase in either price will cause a decrease in the number of applicants.

75. (a) At $t = 90°$ and $h = 0.80$, $\frac{\partial A}{\partial t} = 1.845$.

At $t = 90°$ and $h = 0.80$, $\frac{\partial A}{\partial h} = 29.3$.

(b) The humidity has a greater effect since the coefficient of h is greater.

77. Answers will vary.

SECTION 7.5 *(page 501)*

Prerequisite Review

1. $(3, 2)$ **2.** $(11, 6)$ **3.** $(1, 4)$ **4.** $(4, 4)$

5. $(5, 2)$ **6.** $(3, -2)$ **7.** $(0, 0), (-1, 0)$

8. $(-2, 0), (2, -2)$

9. $\frac{\partial z}{\partial x} = 12x^2$ $\quad \frac{\partial^2 z}{\partial y^2} = -6$

$\frac{\partial z}{\partial y} = -6y$ $\quad \frac{\partial^2 z}{\partial x \partial y} = 0$

$\frac{\partial^2 z}{\partial x^2} = 24x$ $\quad \frac{\partial^2 z}{\partial y \partial x} = 0$

10. $\frac{\partial z}{\partial x} = 10x^4$ $\quad \frac{\partial^2 z}{\partial y^2} = -6y$

$\frac{\partial z}{\partial y} = -3y^2$ $\quad \frac{\partial^2 z}{\partial x \partial y} = 0$

$\frac{\partial^2 z}{\partial x^2} = 40x^3$ $\quad \frac{\partial^2 z}{\partial y \partial x} = 0$

11. $\frac{\partial z}{\partial x} = 4x^3 - \frac{\sqrt{xy}}{2x}$ $\quad \frac{\partial^2 z}{\partial y^2} = \frac{\sqrt{xy}}{4y^2}$

$\frac{\partial z}{\partial y} = -\frac{\sqrt{xy}}{2y} + 2$ $\quad \frac{\partial^2 z}{\partial x \partial y} = -\frac{\sqrt{xy}}{4xy}$

$\frac{\partial^2 z}{\partial x^2} = 12x^2 + \frac{\sqrt{xy}}{4x^2}$ $\quad \frac{\partial^2 z}{\partial y \partial x} = -\frac{\sqrt{xy}}{4xy}$

12. $\frac{\partial z}{\partial x} = 4x - 3y$ $\quad \frac{\partial^2 z}{\partial y^2} = 2$

$\frac{\partial z}{\partial y} = 2y - 3x$ $\quad \frac{\partial^2 z}{\partial x \partial y} = -3$

$\frac{\partial^2 z}{\partial x^2} = 4$ $\quad \frac{\partial^2 z}{\partial y \partial x} = -3$

13. $\frac{\partial z}{\partial x} = y^3 e^{xy^2}$ $\quad \frac{\partial^2 z}{\partial y^2} = 4x^2y^3e^{xy^2} + 6xye^{xy^2}$

$\frac{\partial z}{\partial y} = 2xy^2e^{xy^2} + e^{xy^2}$ $\quad \frac{\partial^2 z}{\partial x \partial y} = 2xy^4e^{xy^2} + 3y^2e^{xy^2}$

$\frac{\partial^2 z}{\partial x^2} = y^5e^{xy^2}$ $\quad \frac{\partial^2 z}{\partial y \partial x} = 2xy^4e^{xy^2} + 3y^2e^{xy^2}$

14. $\frac{\partial z}{\partial x} = e^{xy}(xy + 1)$ $\quad \frac{\partial^2 z}{\partial y^2} = x^3e^{xy}$

$\frac{\partial z}{\partial y} = x^2 e^{xy}$ $\quad \frac{\partial^2 z}{\partial x \partial y} = xe^{xy}(xy + 2)$

$\frac{\partial^2 z}{\partial x^2} = ye^{xy}(xy + 2)$ $\quad \frac{\partial^2 z}{\partial y \partial x} = xe^{xy}(xy + 2)$

1. Critical point: $(-2, -4)$
No relative extrema
$(-2, -4, 1)$ is a saddle point.

3. Critical point: $(0, 0)$
Relative minimum: $(0, 0, 1)$

5. Relative minimum: $(1, 3, 0)$

7. Relative minimum: $(-1, 1, -4)$

9. Relative maximum: $(8, 16, 74)$

11. Relative minimum: $(2, 1, -7)$

13. Saddle point: $(-2, -2, -8)$

15. Saddle point: $(0, 0, 0)$

17. Relative maxima: $(0, \pm 1, 4)$
Relative minimum: $(0, 0, 0)$
Saddle points: $(\pm 1, 0, 1)$

19. Saddle point: $(0, 0, 1)$

21. Insufficient information

23. $f(x_0, y_0)$ is a saddle point.

25. Relative minima: $(a, 0, 0), (0, b, 0)$
Second-Partials Test fails at $(a, 0)$ and $(0, b)$.

27. Saddle point: $(0, 0, 0)$
Second-Partials Test fails at $(0, 0)$.

29. Relative minimum: $(0, 0, 0)$
Second-Partials Test fails at $(0, 0)$.

31. Relative minimum: $(1, -3, 0)$

33. 10, 10, 10 **35.** 10, 10, 10

37. $x_1 = 3, x_2 = 6$ **39.** $p_1 = 2500, p_2 = 3000$

41. $x_1 \approx 94, x_2 \approx 157$

43. 48 inches × 24 inches × 24 inches

45. Proof

47. $D_x(x, y) = 2x - 18 + 2y$
$D_y(x, y) = 4y - 24 + 2x$
To minimize the duration of the infection, 600 mg of the first drug and 300 mg of the second drug are necessary.

49. True

51. False. The origin is a minimum.

SECTION 7.6 (page 511)

Prerequisite Review

1. $\left(\frac{7}{8}, \frac{1}{12}\right)$ **2.** $\left(-\frac{1}{24}, -\frac{7}{8}\right)$ **3.** $\left(\frac{55}{12}, -\frac{25}{12}\right)$

4. $\left(\frac{22}{23}, -\frac{3}{23}\right)$ **5.** $\left(\frac{5}{3}, \frac{1}{3}, 0\right)$ **6.** $\left(\frac{14}{19}, -\frac{10}{19}, -\frac{32}{57}\right)$

7. $f_x = 2xy + y^2$, $f_y = x^2 + 2xy$ **8.** $f_x = 50y^2(x + y)$, $f_y = 50y(x + y)(x + 2y)$

9. $f_x = 3x^2 - 4xy + yz$, $f_y = -2x^2 + xz$, $f_z = xy$ **10.** $f_x = yz + z^2$, $f_y = xz + z^2$, $f_z = xy + 2xz + 2yz$

1. $f(5, 5) = 25$ **3.** $f(2, 2) = 8$

5. $f\left(\frac{\sqrt{2}}{2}, \frac{1}{2}\right) = \frac{1}{4}$ **7.** $f\left(\frac{25}{2}, \frac{25}{2}\right) = 231.25$

9. $f(1, 1) = 2$ **11.** $f(2, 2) = e^4$ **13.** $f(9, 6, 9) = 432$

15. $f\left(\frac{1}{3}, \frac{1}{3}, \frac{1}{3}\right) = \frac{1}{3}$ **17.** $f\left(\frac{1}{\sqrt{3}}, \frac{1}{\sqrt{3}}, \frac{1}{\sqrt{3}}\right) = \sqrt{3}$

19. $f\left(\frac{4}{15}, \frac{8}{15}, \frac{4}{15}, \frac{2}{15}\right) = \frac{8}{15}$ **21.** $f(9, 6, 9) = 486$

23. $f\left(\sqrt{\frac{10}{3}}, \frac{1}{2}\sqrt{\frac{10}{3}}, \sqrt{\frac{5}{3}}\right) = \frac{5\sqrt{15}}{9}$

25. $x = 4, y = \frac{2}{3}, z = 2$ **27.** 40, 40, 40 **29.** $\frac{S}{3}, \frac{S}{3}, \frac{S}{3}$

31. $\sqrt{5}$ **33.** $\sqrt{3}$

35. 36 inches × 18 inches × 18 inches

37. Length = width = $\sqrt[3]{360} \approx 7.1$ feet

Height = $\frac{480}{360^{2/3}} \approx 9.5$ feet

39. $x_1 = 752.5, x_2 = 1247.5$

To minimize cost, let $x_1 = 753$ units and $x_2 = 1247$ units.

41. (a) $x = 50\sqrt{2} \approx 71$ (b) Answers will vary.

$y = 200\sqrt{2} \approx 283$

43. (a) $f\left(\frac{3125}{6}, \frac{6250}{3}\right) \approx 147{,}314$ (b) 1.473 (c) 184,142

45. $x = \sqrt[3]{0.065} \approx 0.402$ liter

$y = \frac{1}{2}\sqrt[3]{0.065} \approx 0.201$ liter

$z = \frac{1}{3}\sqrt[3]{0.065} \approx 0.134$ liter

47. (a) 50 feet × 120 feet (b) \$2400

49. Stock G: \$138,333.33

Stock P: \$7000.00

Stock S: \$154,666.67

51. Answers will vary.

SECTION 7.7 (page 521)

Prerequisite Review

1. 5.0225 **2.** 0.0189

3. $S_a = 2a - 4 - 4b$, $S_b = 12b - 8 - 4a$ **4.** $S_a = 8a - 6 - 2b$, $S_b = 18b - 4 - 2a$

5. 15 **6.** 42 **7.** $\frac{25}{12}$

8. 14 **9.** 31 **10.** 95

1. (a) $y = \frac{3}{4}x + \frac{4}{3}$ (b) $\frac{1}{6}$

3. (a) $y = -2x + 4$ (b) 2

5. $y = x + \frac{2}{3}$ **7.** $y = -2.3x - 0.9$ **9.** $y = \frac{7}{10}x + \frac{7}{5}$

11. $y = x + 4$ **13.** $y = -\frac{13}{20}x + \frac{7}{4}$

15. $y = \frac{37}{43}x + \frac{7}{43}$ **17.** $y = -\frac{175}{148}x + \frac{945}{148}$

19. $y = \frac{3}{7}x^2 + \frac{6}{5}x + \frac{26}{35}$ **21.** $y = 1.25x^2 - 1.75x + 0.25$

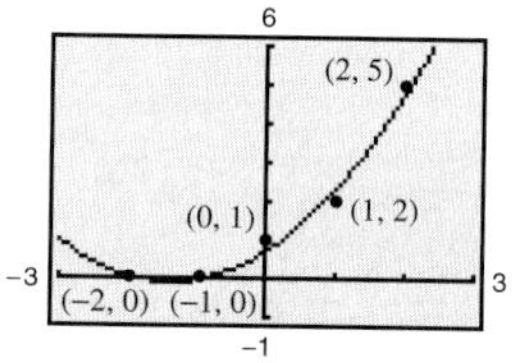

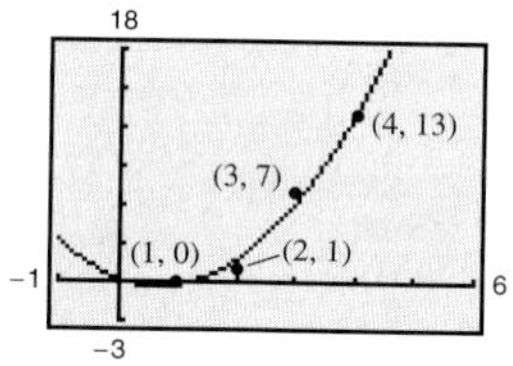

23. Linear: $y = 1.4x + 6$

Quadratic: $y = 0.12x^2 + 1.7x + 6$

The quadratic model is a better fit.

25. Linear: $y = -68.9x + 754$

Quadratic: $y = 2.82x^2 - 83.0x + 763$

The quadratic model is a better fit.

27. (a) $y = -240x + 685$ (b) 349 (c) \$0.77

29. (a) $y = 13.8x + 22.1$ (b) 44.18 bushels/acre

31. (a) $y = -0.48t + 19.74$; In 2010, $y \approx 0.54$ deaths

(b) $y = 0.0027t^2 - 0.51t + 19.03$

In 2010, $y \approx 2.95$ deaths

33. (a) $y = -\frac{25}{112}x^2 + \frac{541}{56}x - \frac{25}{14}$ (b) 40.9 miles per hour

35. Linear: $y = 3.757x + 9.03$

Quadratic: $y = 0.006x^2 + 3.625x + 9.43$

Either model is a good fit for the data.

37. Quadratic: $y = -0.087x^2 + 2.82x + 0.4$

39.

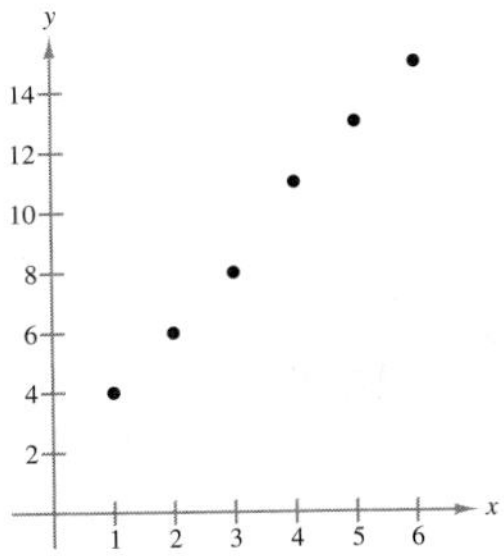

Positive correlation, $r = 0.9981$

41.

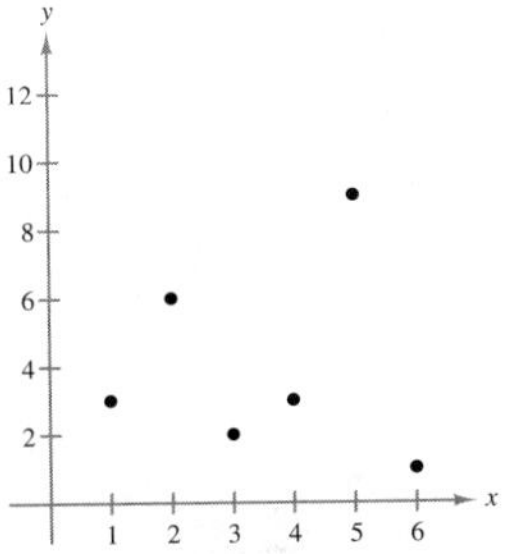

No correlation, $r = 0$

43. $y = -49.95t^2 + 4442.6t - 41{,}941$, where $t = 20$ represents 20-year-olds; $\approx \$43{,}291$

45. True **47.** True

SECTION 7.8 *(page 530)*

Prerequisite Review

1. 1 **2.** 6 **3.** 42 **4.** $\frac{1}{2}$ **5.** $\frac{19}{4}$

6. $\frac{16}{3}$ **7.** $\frac{1}{7}$ **8.** 4 **9.** $\ln 5$ **10.** $\ln|e - 1|$

11. $\frac{e}{2}(e^4 - 1) \approx 72.8474$ **12.** $\frac{1}{2}\left(1 - \frac{1}{e^2}\right)$

13.

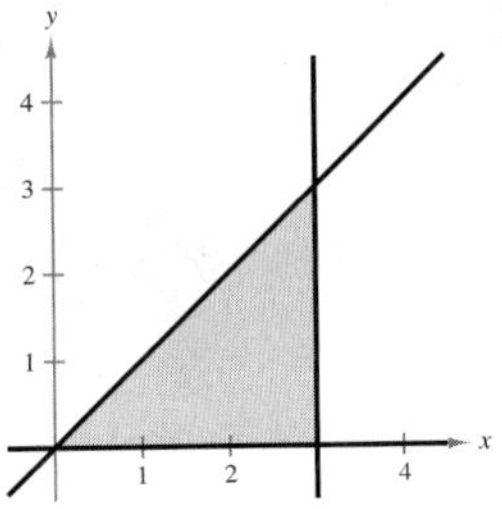

14.

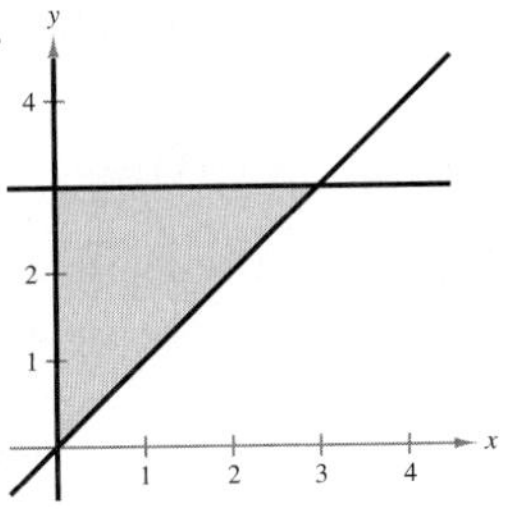

15.

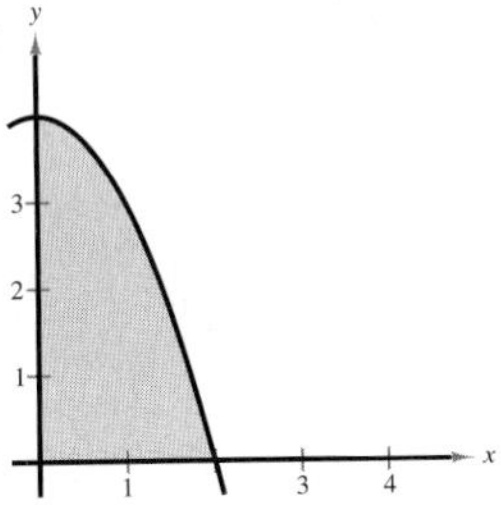

16.

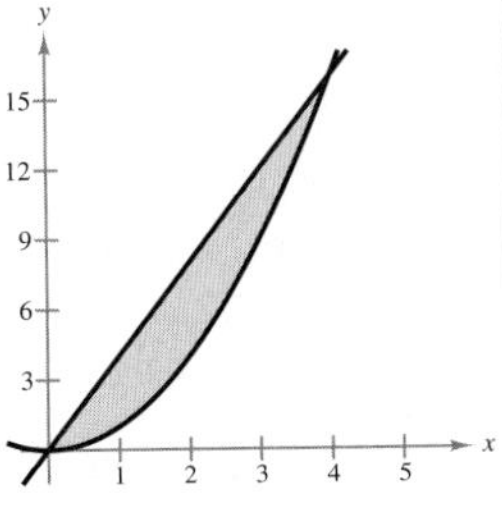

1. $\frac{3x^2}{2}$ **3.** $y \ln|2y|$ **5.** $\frac{x^2}{2}(9 - x^2)$

7. $\frac{1}{2}y[(\ln y)^2 - y^2]$ **9.** $x^2(1 - e^{-x^2} - x^2e^{-x^2})$ **11.** 1

13. 36 **15.** $\frac{2}{3}$ **17.** $\frac{20}{3}$ **19.** 5 **21.** $\frac{16}{3}$ **23.** 4

25.

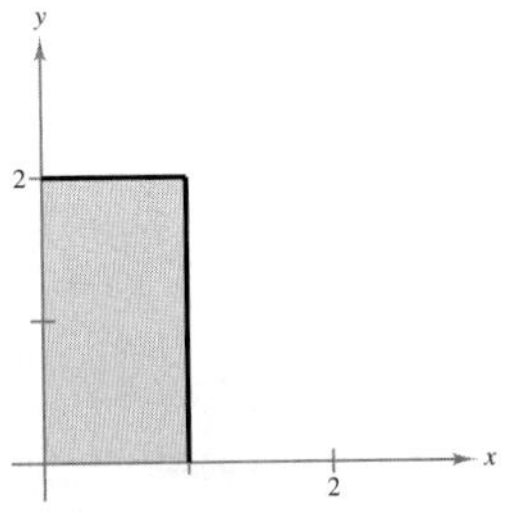

$$\int_0^1 \int_0^2 dy\,dx = \int_0^2 \int_0^1 dx\,dy = 2$$

27.

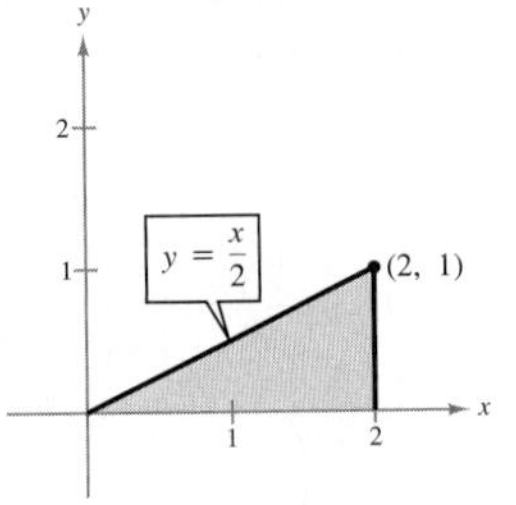

$$\int_0^1 \int_{2y}^2 dx\,dy = \int_0^2 \int_0^{x/2} dy\,dx = 1$$

29.

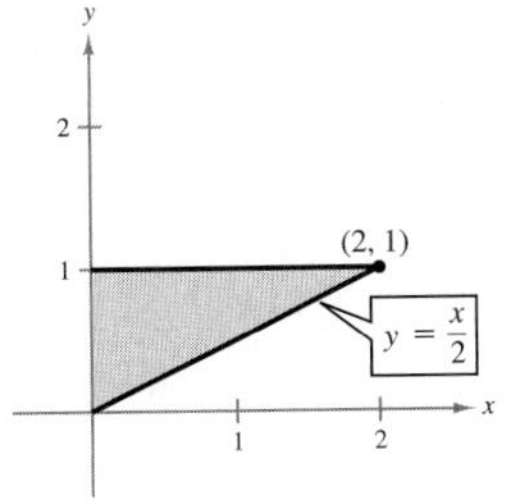

$$\int_0^2 \int_{x/2}^1 dy\,dx = \int_0^1 \int_0^{2y} dx\,dy = 1$$

31.

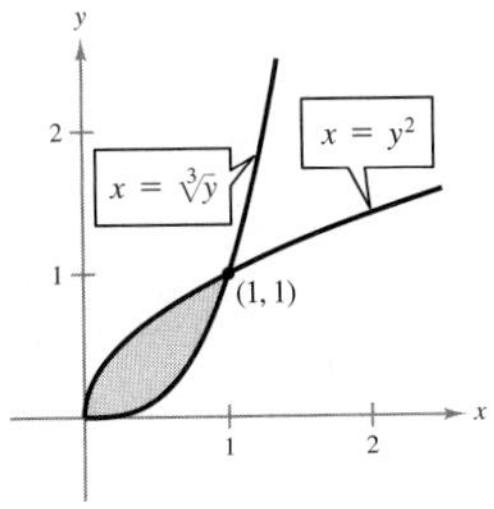

$$\int_0^1 \int_{y^2}^{\sqrt[3]{y}} dx\,dy = \int_0^1 \int_{x^3}^{\sqrt{x}} dy\,dx = \frac{5}{12}$$

33. $\frac{1}{2}(e^9 - 1) \approx 4051.042$ **35.** 24 **37.** $\frac{16}{3}$

39. $\frac{8}{3}$ **41.** $\frac{500}{3}$ **43.** $\frac{45}{14}$ **45.** 2 **47.** 0.6588

49. 8.1747 **51.** 0.4521 **53.** 1.1190

55. False, because $dA = dy\,dx = dx\,dy$, it doesn't matter in what order the integration is performed.

SECTION 7.9 *(page 538)*

Prerequisite Review

1.

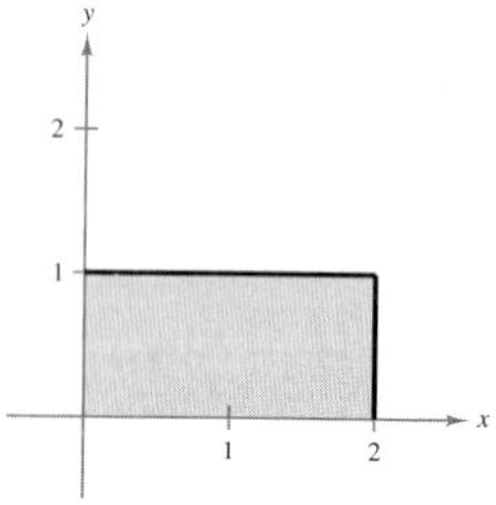

2.

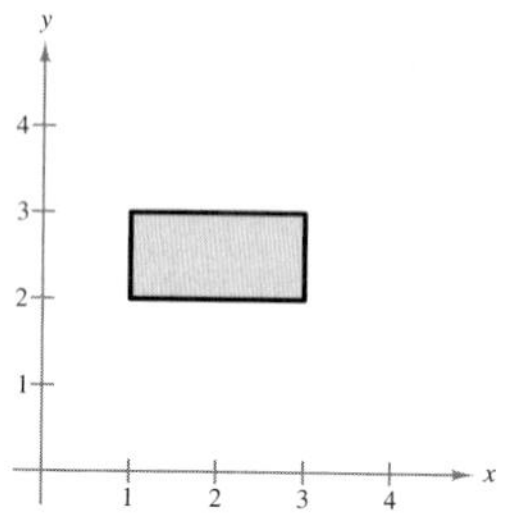

3.

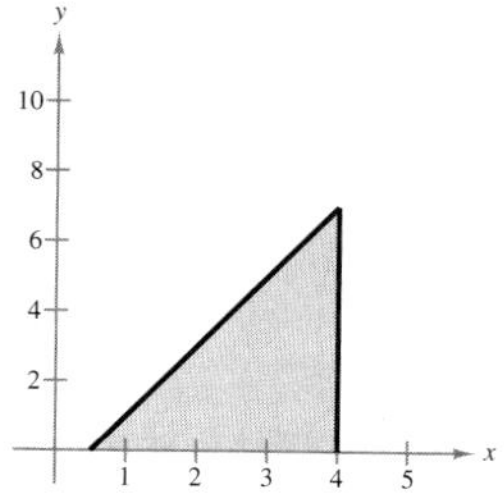

4.

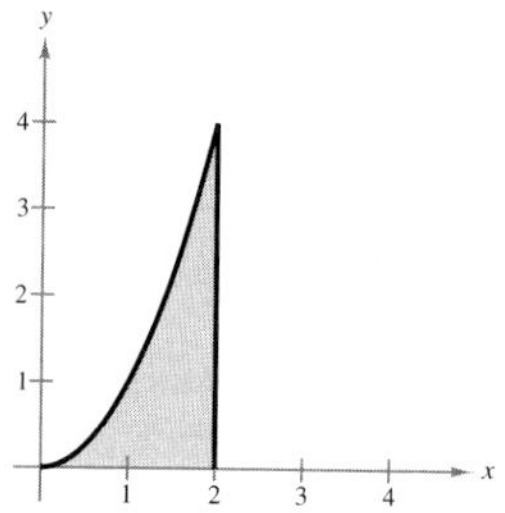

5. 1 **6.** 6 **7.** $\frac{1}{3}$ **8.** $\frac{40}{3}$
9. $\frac{28}{3}$ **10.** $\frac{7}{6}$

1.

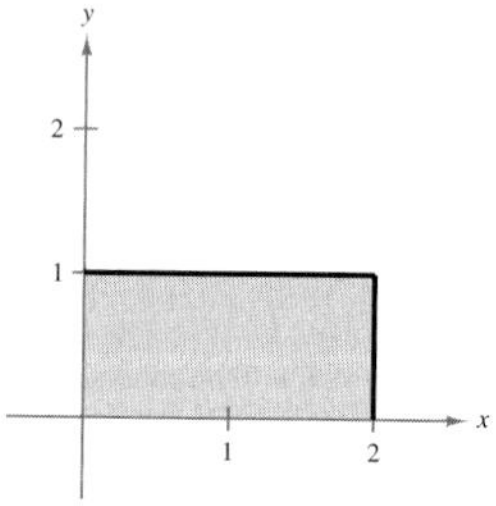

10

3.

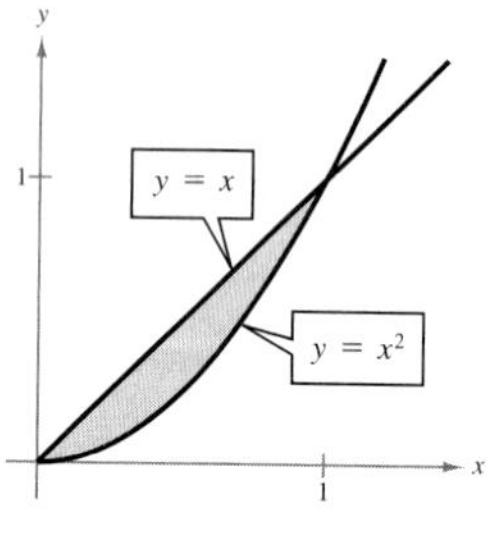

$\frac{1}{54}$

5.

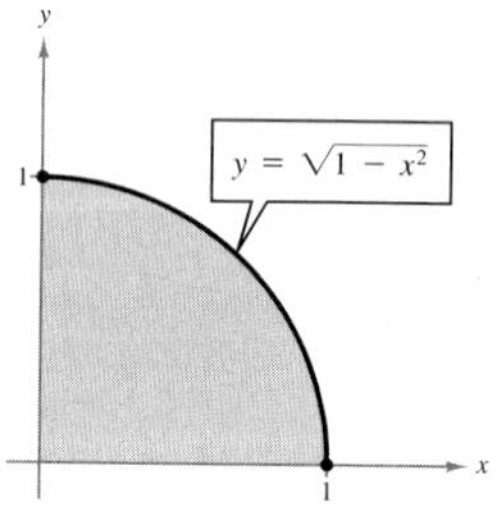

$\frac{1}{3}$

7.

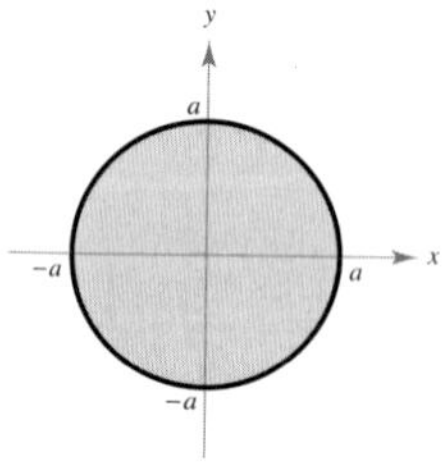

πa^2

9. $\displaystyle\int_0^3\int_0^5 xy\,dy\,dx = \int_0^5\int_0^3 xy\,dx\,dy = \frac{225}{4}$

11. $\displaystyle\int_0^2\int_x^{2x} \frac{y}{x^2 + y^2}\,dy\,dx = \int_0^2\int_{y/2}^{y} \frac{y}{x^2 + y^2}\,dx\,dy + \int_2^4\int_{y/2}^{2} \frac{y}{x^2 + y^2}\,dx\,dy = \ln\frac{5}{2}$

13. $\displaystyle\int_0^{1/2}\int_0^{2x} e^{-x^2}\,dy\,dx = 0.2212$ **15.** 4

17. 22.5 **19.** 12 **21.** $\frac{3}{8}$ **23.** $\frac{40}{3}$
25. $\frac{1}{3}$ **27.** 4 **29.** $\frac{32}{3}$ **31.** 10,000
33. 2 **35.** $\frac{8}{3}$ **37.** \$75,125

REVIEW EXERCISES FOR CHAPTER 7
(page 544)

1.

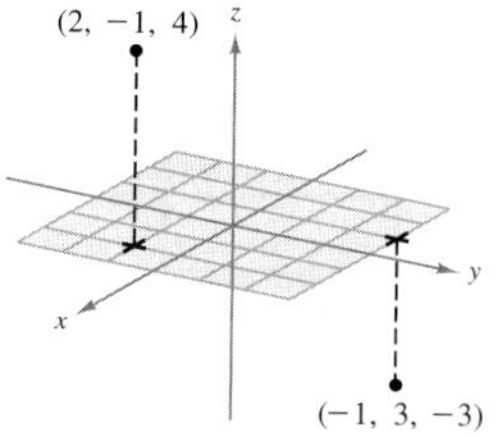

3. $\sqrt{110}$ **5.** $(-1, 4, 6)$

7. $x^2 + (y - 1)^2 + z^2 = 25$

9. $(x - 4)^2 + (y - 6)^2 + (z - 1)^2 = 6$

11. Center: $(-2, 1, 4)$; radius: 4

13.

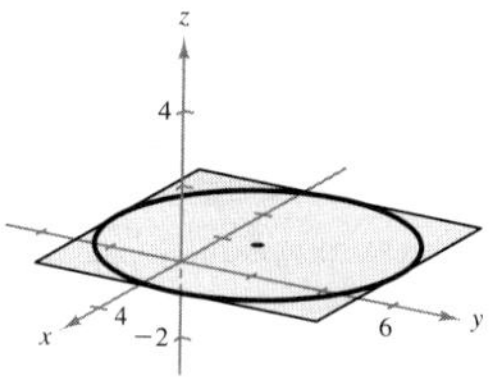

15.

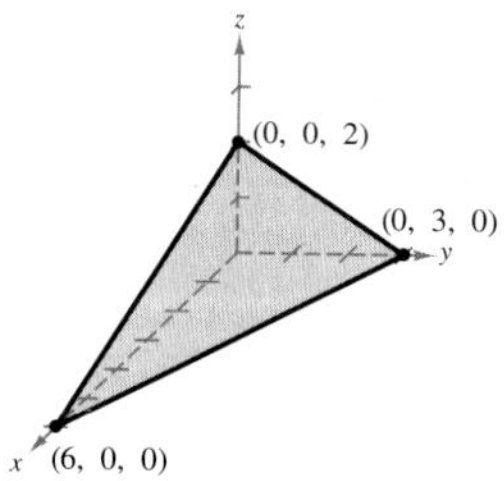

17.

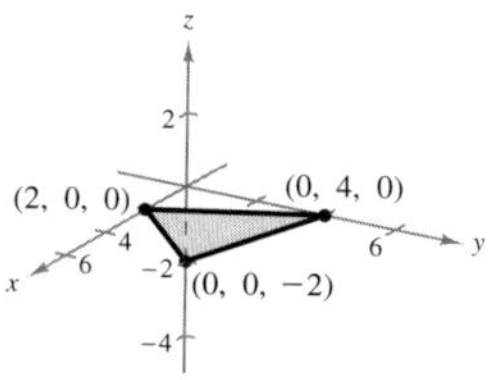

19. Sphere **21.** Ellipsoid **23.** Elliptic Paraboloid

25. Top half of a circular cone

27. (a) 18 (b) 0 (c) −245 (d) −32

29. The domain is the set of all points inside or on the circle $x^2 + y^2 = 1$, and the range is $[0, 1]$.

31. The level curves are lines of slope $-\frac{2}{5}$.

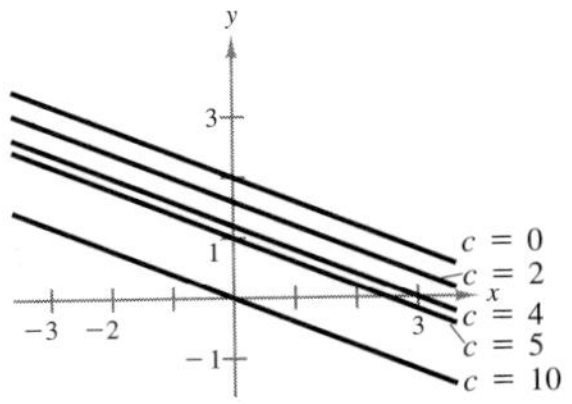

33. The level curves are hyperbolas.

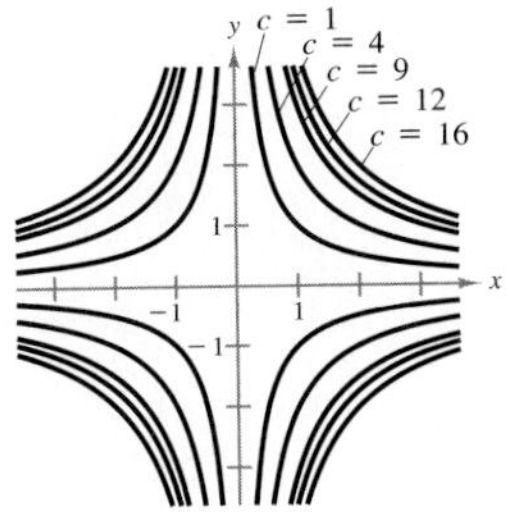

35. (a) As the color darkens from light green to dark green, the average yearly precipitation increases.

(b) The small eastern portion containing Davenport

(c) The northwestern portion containing Sioux City

37. Southwest **39.** \$2.50 **41.** $f_x = 2xy + 3y + 2$

$f_y = x^2 + 3x - 5$

43. $z_x = 12x\sqrt{y} + \dfrac{3}{2}\sqrt{\dfrac{y}{x}} - 7y$

$z_y = \dfrac{3x^2}{\sqrt{y}} + \dfrac{3}{2}\sqrt{\dfrac{x}{y}} - 7x$

45. $f_x = \dfrac{2}{2x + 3y}$

$f_y = \dfrac{3}{2x + 3y}$

47. $f_x = 2xe^y - y^2e^x$

$f_y = x^2e^y - 2ye^x$

49. $w_x = yz^2$

$w_y = xz^2$

$w_z = 2xyz$

51. (a) $z_x = 3$ (b) $z_y = -4$

53. (a) $z_x = -2x$
At $(1, 2, 3)$, $z_x = -2$.
(b) $z_y = -2y$
At $(1, 2, 3)$, $z_y = -4$.

55. $f_{xx} = 6x$

$f_{yy} = -8x + 6y$

$f_{xy} = f_{yx} = -8y$

57. $f_{xx} = \dfrac{y^2 - 64}{(64 - x^2 - y^2)^{3/2}}$

$f_{yy} = \dfrac{x^2 - 64}{(64 - x^2 - y^2)^{3/2}}$

$f_{xy} = f_{yx} = \dfrac{-xy}{(64 - x^2 - y^2)^{3/2}}$

59. $C_x(250, 175) \approx 99.70$

$C_y(250, 175) \approx 140.01$

61. (a) $A_w = 43.095w^{-0.575}h^{0.725}$

$A_h = 73.515w^{0.425}h^{-0.275}$

(b) ≈ 47.35;
The surface area of an average human body increases approximately 47.35 square centimeters per pound for a human who weighs 180 pounds and is 70 inches tall.

63. Relative minimum: $(x, -x, 0)$

65. Saddle point: $\left(\frac{3}{2}, -\frac{3}{2}, \frac{25}{2}\right)$

67. Relative minimum: $\left(\frac{1}{6}, \frac{1}{12}, -\frac{1}{432}\right)$

Saddle point: $(0, 0, 0)$

69. Relative minimum: $(1, 1, -2)$

Relative maximum: $(-1, -1, 6)$

Saddle points: $(1, -1, 2)$, $(-1, 1, 2)$

71. (a) $R = -x_1^2 - \frac{1}{2}x_2^2 + 100x_1 + 200x_2$

(b) $x_1 = 50, x_2 = 200$ (c) \$22,500.00

73. At $\left(\frac{4}{3}, \frac{1}{3}\right)$, the relative maximum is $\frac{16}{27}$.

At $(0, 1)$, the relative minimum is 0.

75. At $\left(\frac{4}{3}, \frac{2}{3}, \frac{4}{3}\right)$, the relative maximum is $\frac{32}{27}$.

77. At $\left(\frac{4}{3}, \frac{10}{3}, \frac{14}{3}\right)$, the relative minimum is $34\frac{2}{3}$.

79. $x = 2\sqrt{2}, y = 2\sqrt{2}, z = \sqrt{2}$

81. $f(49.4, 253) \approx 13{,}202$

83. (a) $y = \frac{60}{59}x - \frac{15}{59}$ (b) 2.746

85. (a) $y = 14x + 19$ (b) 21.8 bushels/acre

87. $y = 1.71x^2 - 2.57x + 5.56$

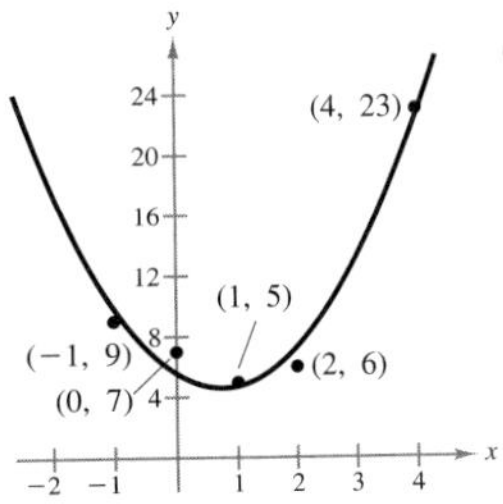

89. $\frac{29}{6}$ **91.** $\frac{7}{4}$

93. $\displaystyle\int_{-2}^{2}\int_{5}^{9-x^2} dy\,dx = \int_{5}^{9}\int_{-\sqrt{9-y}}^{\sqrt{9-y}} dx\,dy = \frac{32}{3}$

95. $\displaystyle\int_{-3}^{6}\int_{1/3(x+3)}^{\sqrt{x+3}} dy\,dx = \int_{0}^{3}\int_{3y-3}^{y^2-3} dx\,dy = \frac{9}{2}$

97. $\frac{4096}{9}$ **99.** 0.0833 mile

SAMPLE POST-GRAD EXAM QUESTIONS

(page 548)

1. b **2.** b **3.** c **4.** d

5. a **6.** c **7.** b

APPENDIX A *(page A10)*

1. Left Riemann sum: 0.518
Right Riemann sum: 0.768

3. Left Riemann sum: 0.746
Right Riemann sum: 0.646

5. Left Riemann sum: 0.859
Right Riemann sum: 0.659

7. Midpoint Rule: 0.673

9. (a)

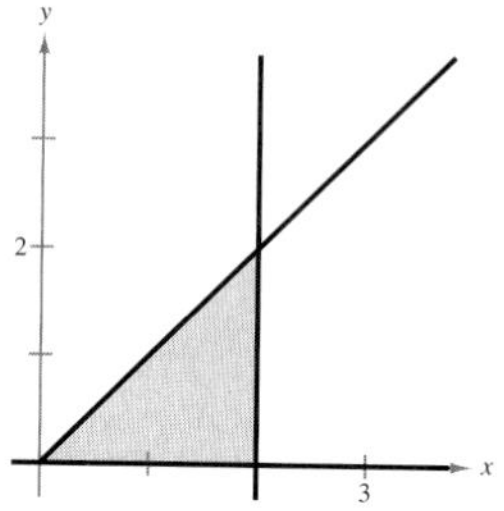

(b) Answers will vary.
(c) Answers will vary.
(d) Answers will vary.

(e)

n	5	10	50	100
Left sum, S_L	1.6	1.8	1.96	1.98
Right sum, S_R	2.4	2.2	2.04	2.02

(f) Answers will vary.

11. $\int_0^5 3\,dx$

13. $\int_{-4}^4 (4 - |x|)\,dx = \int_{-4}^0 (4 + x)\,dx + \int_0^4 (4 - x)\,dx$

15. $\int_{-2}^2 (4 - x^2)\,dx$ **17.** $\int_0^2 \sqrt{x + 1}\,dx$

19.

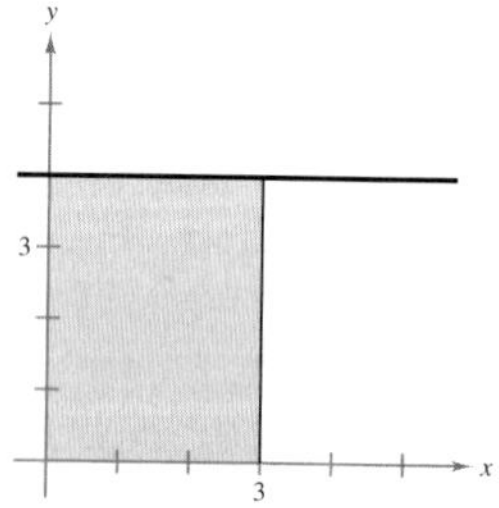

Rectangle

$A = 12$

21.

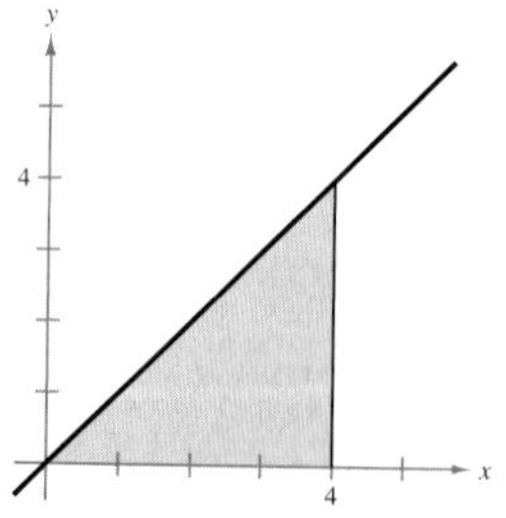

Triangle

$A = 8$

23.

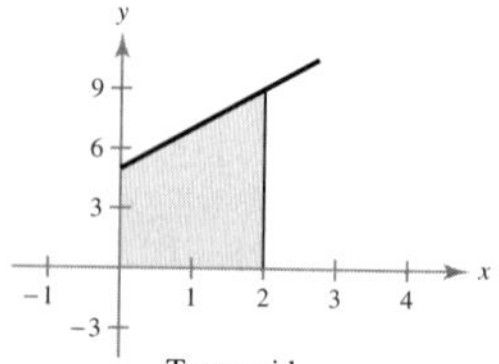

Trapezoid

$A = 14$

25.

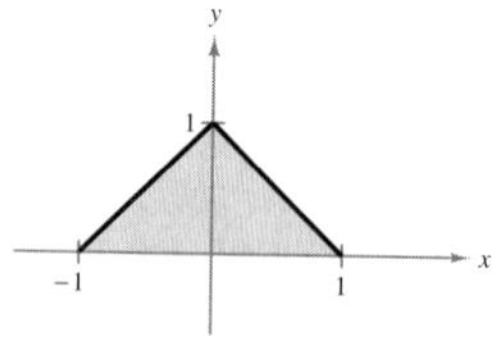

Triangle

$A = 1$

27.

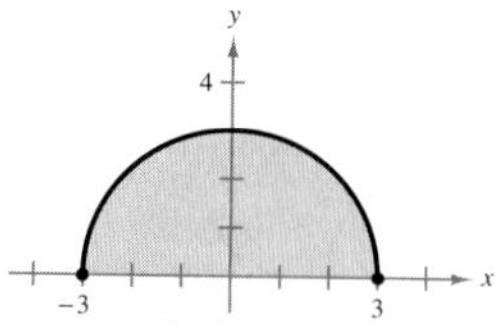

Semicircle

$A = \frac{9\pi}{2}$

29. Answers will vary. **31.** >

Answers to Try Its

CHAPTER 0

SECTION 0.1

Try It 1 $x < 5$ or $(-\infty, 5)$

Try It 2 $x < -2$ or $x > 5$; $(-\infty, -2) \cup (5, \infty)$

Try It 3 $200 \le x \le 400$; so the daily production levels during the month varied between a low of 200 units and a high of 400 units.

SECTION 0.2

Try It 1 8; 8; -8

Try It 2 $2 \le x \le 10$

Try It 3 $\$4027.50 \le C \le \$11{,}635$

SECTION 0.3

Try It 1 $\frac{4}{9}$

Try It 2 8

Try It 3 (a) $3x^6$ (b) $8x^{7/2}$ (c) $4x^{4/3}$

Try It 4 (a) $x(x^2 - 2)$ (b) $2x^{1/2}(1 + 4x)$

Try It 5 $\dfrac{(3x - 1)^{3/2}(13x - 2)}{(x + 2)^{1/2}}$

Try It 6 $\dfrac{10(x - 4)}{(x + 1)^3}$

Try It 7 (a) $[2, \infty)$ (b) $(2, \infty)$ (c) $(-\infty, \infty)$

SECTION 0.4

Try It 1 (a) $\dfrac{-2 \pm \sqrt{2}}{2}$ (b) 4 (c) No real zeros

Try It 2 (a) $x = -3$ and $x = 5$ (b) $x = -1$

(c) $x = \frac{3}{2}$ and $x = 2$

Try It 3 $(-\infty, -2] \cup [1, \infty)$

Try It 4 $-1, \frac{1}{2}, 2$

SECTION 0.5

Try It 1 (a) $\dfrac{x^2 + 2}{x}$ (b) $\dfrac{3x + 1}{(x + 1)(2x + 1)}$

Try It 2 (a) $\dfrac{3x + 4}{(x + 2)(x - 2)}$ (b) $-\dfrac{x + 1}{3x(x + 2)}$

Try It 3

(a) $\dfrac{(A + B + C)x^2 + (A + 3B)x + (-2A + 2B - C)}{(x + 1)(x - 1)(x + 2)}$

(b) $\dfrac{(A + C)x^2 + (-A + B + 2C)x + (-2A - 2B + C)}{(x + 1)^2(x - 2)}$

Try It 4 (a) $\dfrac{3x + 8}{4(x + 2)^{3/2}}$ (b) $\dfrac{1}{\sqrt{x^2 + 4}}$

Try It 5 $\dfrac{\sqrt{x^2 + 4}}{x^2}$

Try It 6 (a) $\dfrac{5\sqrt{2}}{4}$ (b) $\dfrac{x + 2}{4\sqrt{x + 2}}$ (c) $\dfrac{\sqrt{6} + \sqrt{3}}{3}$

(d) $\dfrac{\sqrt{x + 2} - \sqrt{x}}{2}$

CHAPTER 1

SECTION 1.1

Try It 1

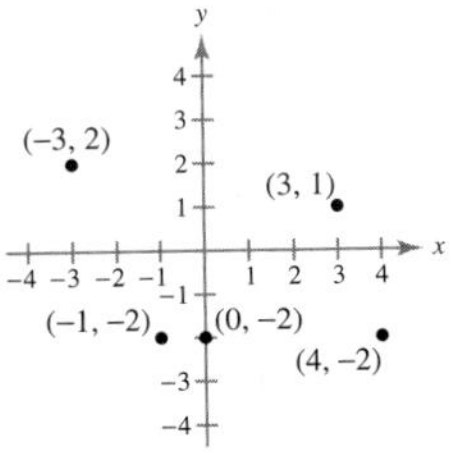

Try It 2

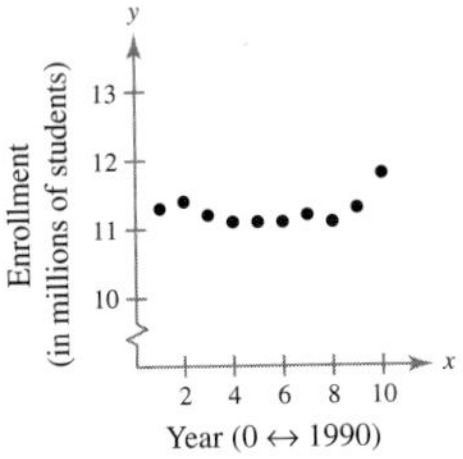

Try It 3 5

Try It 4 $d_1 = \sqrt{20}, d_2 = \sqrt{45}, d_3 = \sqrt{65}$

$d_1^2 + d_2^2 = 20 + 45 = 65 = d_3^2$

Try It 5 25 yards

Try It 6 $(-2, 5)$

Try It 7 \$4.56 billion

Try It 8 $(-1, -4), (1, -2), (1, 2), (-1, 0)$

SECTION 1.2

Try It 1

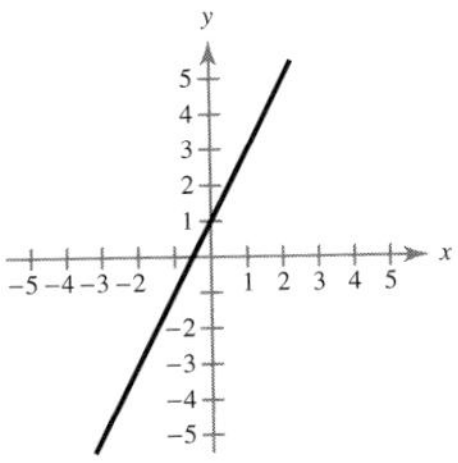

Try It 2

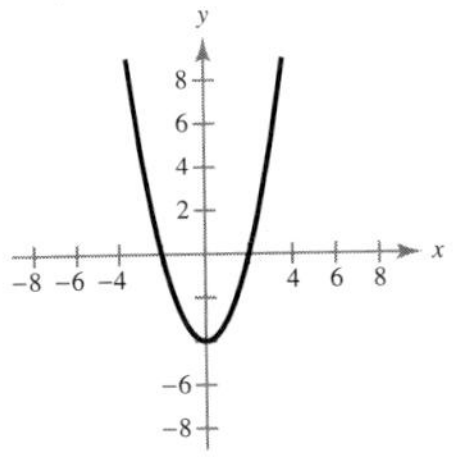

Try It 3 (a) x-intercepts: $(3, 0), (-1, 0)$
y-intercept: $(0, -3)$
(b) x-intercept: $(-4, 0)$
y-intercepts: $(0, 2), (0, -2)$

Try It 4 $(x + 2)^2 + (y - 1)^2 = 25$

Try It 5 $(x - 2)^2 + (y + 1)^2 = 4$

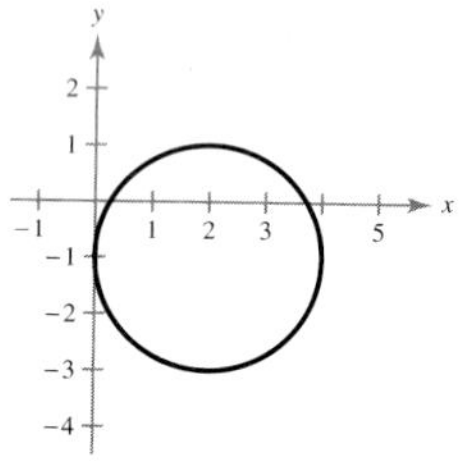

Try It 6 12,500 units

Try It 7 4 million units at \$122/unit

Try It 8 The projection obtained from the model is \$7962.02 million, which is close to the *Value Line* projection.

SECTION 1.3

Try It 1

(a)

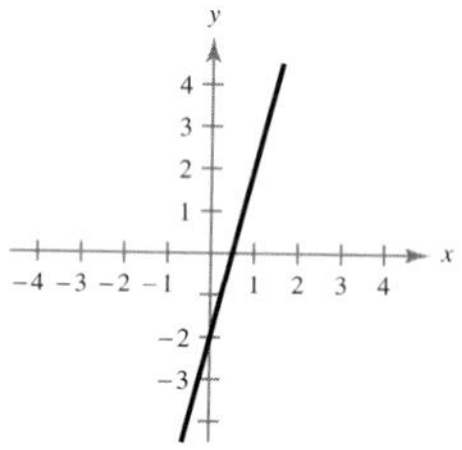

(b)

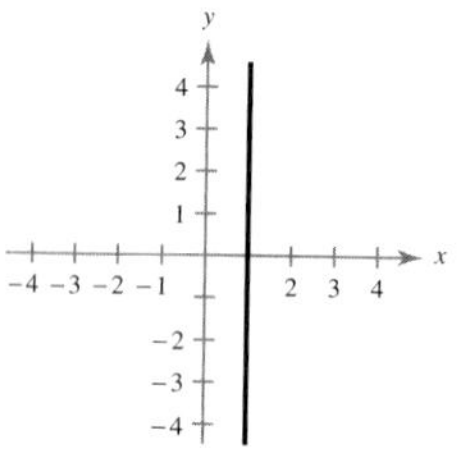

Try It 2 Yes, $\frac{27}{312} \approx 0.08654 > \frac{1}{12} = 0.08\overline{3}$.

Try It 3 The y-intercept $(0, 875)$ tells you that the original value of the copier is \$875. The slope of $m = -175$ tells you that the value decreases \$175/year.

Try It 4 (a) 2 (b) $-\frac{1}{2}$

Try It 5 $y = 2x + 4$

Try It 6 $S = 1.72t + 16.49$; \$19.93

Try It 7 (a) $y = \frac{1}{2}x$ (b) $y = -2x + 5$

Try It 8 $V = -1375t + 12{,}000$

SECTION 1.4

Try It 1 (a) Yes, $y = x - 1$. (b) No, $y = \pm\sqrt{4 - x^2}$.
(c) No, $y = \pm\sqrt{2 - x}$. (d) Yes, $y = x^2$.

Try It 2 (a) Domain: $[-1, \infty)$ (b) Domain: $(-\infty, \infty)$
Range: $[0, \infty)$ Range: $[0, \infty)$

Try It 3 $f(0) = 1, f(1) = -3, f(4) = -3$
No, f is not one-to-one.

Try It 4 (a) $x^2 + 2x\,\Delta x + (\Delta x)^2 - 2x - 2\,\Delta x + 3$
(b) $2x + \Delta x - 2, \Delta x \neq 0$

Try It 5 (a) $2x^2 + 5$ (b) $4x^2 + 4x + 3$

Try It 6 (a) $f^{-1}(x) = 5x$ (b) $f^{-1}(x) = \frac{1}{3}(x - 2)$

Try It 7 $f^{-1}(x) = \sqrt{x - 2}$

Try It 8
$$f(x) = x^2 + 4$$
$$y = x^2 + 4$$
$$x = y^2 + 4$$
$$x - 4 = y^2$$
$$\pm\sqrt{x - 4} = y$$

SECTION 1.5

Try It 1 6

Try It 2 (a) 4 (b) Does not exist (c) 4

Try It 3 5

Try It 4 12

Try It 5 7

Try It 6 $\frac{1}{4}$

Try It 7 (a) -1 (b) 1

Try It 8 1

Try It 9 $\lim_{x \to 1^-} f(x) = 8$ and $\lim_{x \to 1^+} f(x) = 10$
$\lim_{x \to 1^-} f(x) \neq \lim_{x \to 1^+} f(x)$

Try It 10 Does not exist

SECTION 1.6

Try It 1 (a) f is continuous on the entire real line.
(b) f is continuous on the entire real line.

Try It 2 (a) f is continuous on $(-\infty, 1)$ and $(1, \infty)$.
(b) f is continuous on $(-\infty, 2)$ and $(2, \infty)$.
(c) f is continuous on the entire real line.

Try It 3 f is continuous on $[2, \infty)$.

Try It 4 f is continuous on $[-1, 5]$.

Try It 5

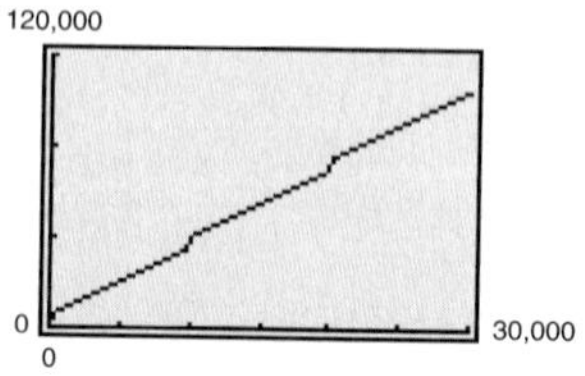

Try It 6 $A = 10{,}000(1 + 0.02)^{[\![4t]\!]}$

CHAPTER 2

SECTION 2.1

Try It 1 3

Try It 2 For the months on the graph to the left of July, the tangent lines have positive slopes. For the months to the right of July, the tangent lines have negative slopes. The average daily temperature is increasing prior to July and decreasing after July.

Try It 3 4

Try It 4 2

Try It 5 $m = 8x$

At $(0, 1)$, $m = 0$.

At $(1, 5)$, $m = 8$.

Try It 6 $2x - 5$

Try It 7 $-\dfrac{4}{t^2}$

SECTION 2.2

Try It 1 (a) 0 (b) 0 (c) 0 (d) 0

Try It 2 (a) $4x^3$ (b) $-\dfrac{3}{x^4}$ (c) $2w$ (d) $-\dfrac{1}{t^2}$

Try It 3 $f'(x) = 3x^2$

$m = f'(-1) = 3;$

$m = f^{-1}(0) = 0;$

$m = f^{-1}(1) = 3$

Try It 4 (a) $8x$ (b) $\dfrac{8}{\sqrt{x}}$

Try It 5 (a) $\frac{1}{4}$ (b) $-\frac{2}{5}$

Try It 6 (a) $-\dfrac{9}{2x^3}$ (b) $-\dfrac{9}{8x^3}$

Try It 7 (a) $\dfrac{\sqrt{5}}{2\sqrt{x}}$ (b) $\dfrac{1}{3x^{2/3}}$

Try It 8 -1

Try It 9 $y = -x + 2$

Try It 10 $R'(8) \approx \$0.76$/year

SECTION 2.3

Try It 1 (a) $0.5\overline{6}$ milligrams per milliliter/minute

(b) 0 milligrams per milliliter/minute

(c) -1.5 milligrams per milliliter/minute

Try It 2 (a) -16 feet/second (b) -48 feet/second

(c) -80 feet/second

Try It 3 When $t = 1.75$, $h'(1.75) = -56$ feet/second.

When $t = 2$, $h'(2) = -64$ feet/second.

Try It 4 $h = -16t^2 + 16t + 12$

$v = h' = -32t + 16$

Try It 5 When $x = 100$, $\dfrac{dP}{dx} = \$16$/unit.

Actual gain $= \$16.06$

Try It 6 $p = 11 - \dfrac{x}{2000}$

Try It 7 Revenue: $R = 2000x - 4x^2$

Marginal revenue: $\dfrac{dR}{dx} = 2000 - 8x$

Try It 8 $\dfrac{dP}{dx} = \$1.44$/unit

Actual increase in profit $\approx \$1.44$

SECTION 2.4

Try It 1 $-27x^2 + 12x + 24$

Try It 2 $\dfrac{2x^2 - 1}{x^2}$

Try It 3 (a) $18x^2 + 30x$ (b) $12x + 15$

Try It 4 $-\dfrac{22}{(5x - 2)^2}$

Try It 5 $y = \frac{8}{25}x - \frac{4}{5}$;

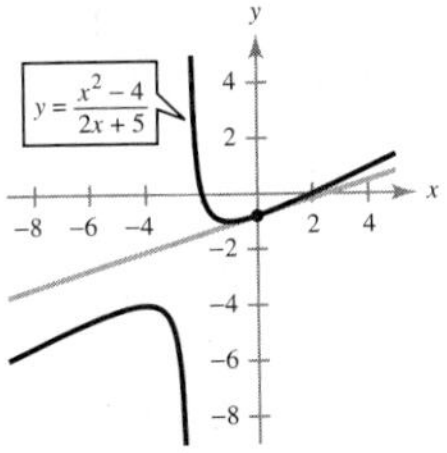

Try It 6 $\dfrac{-3x^2 + 4x + 8}{x^2(x + 4)^2}$

Try It 7 (a) $\frac{2}{5}x + \frac{4}{5}$ (b) $3x^3$

Try It 8 $\dfrac{2x^2 - 4x}{(x - 1)^2}$

Try It 9

t	0	1	2	3	4	5	6	7
$\dfrac{dP}{dt}$	0	-50	-16	-6	-2.77	-1.48	-0.88	-0.56

As t increases, the rate at which the blood pressure drops decreases.

SECTION 2.5

Try It 1 (a) $u = g(x) = x + 1$

$y = f(u) = \dfrac{1}{\sqrt{u}}$

(b) $u = g(x) = x^2 + 2x + 5$

$y = f(u) = u^3$

Try It 2 $6x^2(x^3 + 1)$

Try It 3 $4(2x + 3)(x^2 + 3x)^3$

Try It 4 $y = \frac{1}{3}x + \frac{8}{3}$

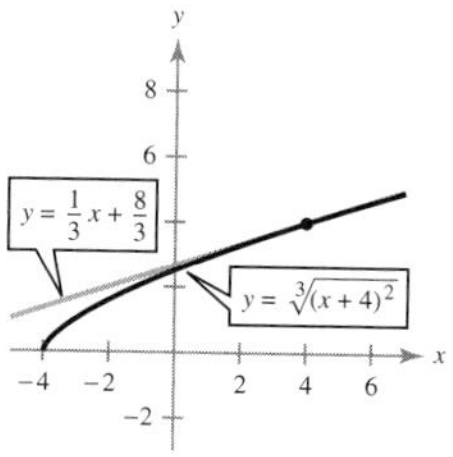

Try It 5 (a) $-\dfrac{8}{(2x + 1)^2}$ (b) $-\dfrac{6}{(x - 1)^4}$

Try It 6 $\dfrac{x(3x^2 + 2)}{\sqrt{x^2 + 1}}$

Try It 7 $-\dfrac{12(x + 1)}{(x - 5)^3}$

Try It 8 About \$2.65/year

SECTION 2.6

Try It 1 $f'(x) = 18x^2 - 4x,\ f''(x) = 36x - 4,$

$f'''(x) = 36,\ f^{(4)}(x) = 0$

Try It 2 18

Try It 3 $\dfrac{120}{x^6}$

Try It 4 $s(t) = -16t^2 + 64t + 80$

$v(t) = s'(t) = -32t + 64$

$a(t) = v'(t) = s''(t) = -32$

Try It 5 -9.8 meters/seconds squared

Try It 6

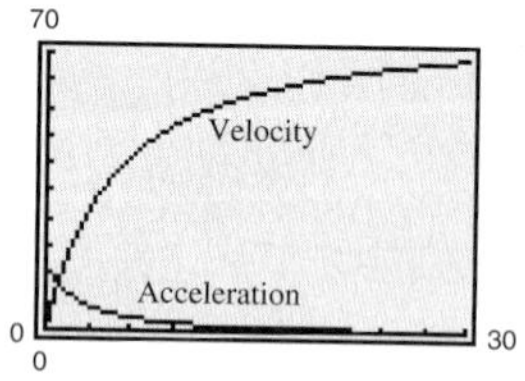

Acceleration approaches zero.

SECTION 2.7

Try It 1 $-\dfrac{2}{x^3}$

Try It 2 (a) $12x^2$ (b) $6y\dfrac{dy}{dx}$ (c) $1 + 5\dfrac{dy}{dx}$

(d) $y^3 + 3xy^2\dfrac{dy}{dx}$

Try It 3 $\frac{3}{4}$

Try It 4 $\dfrac{dy}{dx} = \dfrac{x + 2}{-y + 1}$

Try It 5 $\frac{5}{9}$

Try It 6 $\dfrac{dx}{dp} = -\dfrac{1000}{xp^2}$

SECTION 2.8

Try It 1 9

Try It 2 $12\pi \approx 37.7$ squared feet/second

Try It 3 $72\pi \approx 226.2$ squared inches/minute

Try It 4 \$1500/day

Try It 5 \$28,400/week

CHAPTER 3

SECTION 3.1

Try It 1 $f'(x) = 4x^3$

$f'(x) < 0$ if $x < 0$; therefore, f is decreasing on $(-\infty, 0)$.

$f'(x) > 0$ if $x > 0$; therefore, f is increasing on $(0, \infty)$.

Try It 2 $\dfrac{dW}{dt} = 0.198t + 0.17 > 0$ if $t > 0$, which implies that the consumption of bottled water was increasing from 1990 through 1999.

Try It 3 Increasing on $(-\infty, -2)$ and $(2, \infty)$

Decreasing on $(-2, 2)$

Try It 4 Increasing on $(0, \infty)$

Decreasing on $(-\infty, 0)$

Try It 5 Because $f'(x) = -3x^2 = 0$ when $x = 0$ and because f is decreasing on $(-\infty, 0) \cup (0, \infty)$, f is decreasing on $(-\infty, \infty)$.

Try It 6 $(0, 3000)$

SECTION 3.2

Try It 1 Relative maximum at $(-1, 5)$

Relative minimum at $(1, -3)$

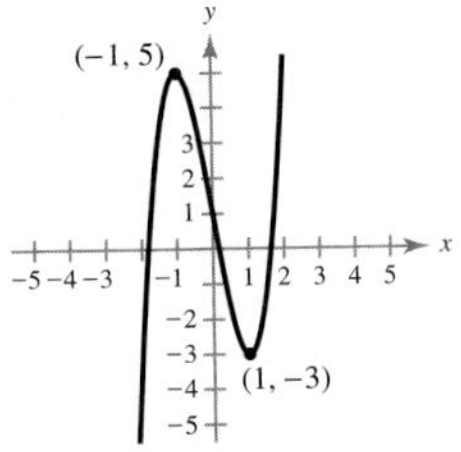

Try It 2 Relative minimum at $(3, -27)$

Try It 3 Relative maximum at $(1, 1)$

Relative minimum at $(0, 0)$

Try It 4 Absolute maximum at $(0, 10)$

Absolute minimum at $(4, -6)$

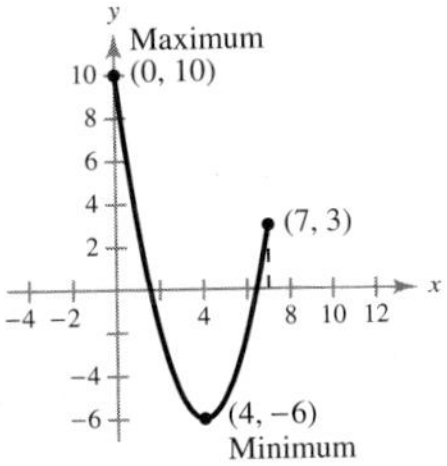

Try It 5

x (units)	24,000	24,200	24,300	24,400
P (profit)	\$24,760	\$24,766	\$24,767.50	\$24,768

x (units)	24,500	24,600	24,800	25,000
P (profit)	\$24,767.50	\$24,766	\$24,760	\$24,750

SECTION 3.3

Try It 1 (a) $f'' = -4$; because $f''(x) < 0$ for all x, f is concave downward for all x.

(b) $f''(x) = \dfrac{1}{2x^{3/2}}$; because $f''(x) > 0$ for all $x > 0$, f is concave upward for all $x > 0$.

Try It 2 Because $f''(x) > 0$ for $x < -2$ and $x > 2$, f is concave upward on $(-\infty, -2)$ and $(2, \infty)$.

Because $f''(x) < 0$ for $-2 < x < 2$, f is concave downward on $(-2, 2)$.

Try It 3 f is concave upward on $(-\infty, 0)$ and $(1, \infty)$.

f is concave downward on $(0, 1)$.

Points of inflection: $(0, 1)$, $(1, 0)$

Try It 4 Relative minimum: $(3, -26)$

Try It 5 Point of diminishing returns: $x = \$150$ thousand

SECTION 3.4

Try It 1

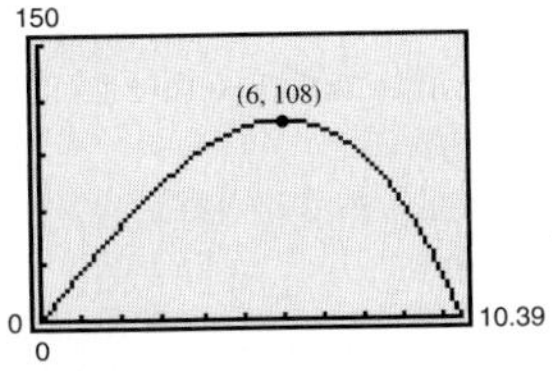

Maximum volume = 108 cubic inches

Try It 2 $x = 6$, $y = 12$

Try It 3 $\left(\sqrt{\frac{1}{2}}, \frac{7}{2}\right)$ and $\left(-\sqrt{\frac{1}{2}}, \frac{7}{2}\right)$

Try It 4 8 inches by 12 inches

SECTION 3.5

Try It 1 125 units yield a maximum revenue of \$1,562,500.

Try It 2 400 units

Try It 3 \$6.25/unit

Try It 4 \$4.00

Try It 5 Demand is elastic when $0 < x < 1156/9$.

Demand is inelastic when $1156/9 < x < 289$.

Demand is of unit elasticity when $x = 1156/9$.

SECTION 3.6

Try It 1 (a) $\displaystyle\lim_{x\to 2^-} \frac{1}{x-2} = -\infty$; $\displaystyle\lim_{x\to 2^+} \frac{1}{x-2} = \infty$

(b) $\displaystyle\lim_{x\to -3^-} \frac{-1}{x+3} = \infty$; $\displaystyle\lim_{x\to -3^+} \frac{-1}{x+3} = -\infty$

Try It 2 $x = 0$, $x = 4$

Try It 3 $x = 3$

Try It 4 $\displaystyle\lim_{x\to 2^-} \frac{x^2-4x}{x-2} = \infty$; $\displaystyle\lim_{x\to 2^+} \frac{x^2-4x}{x-2} = -\infty$

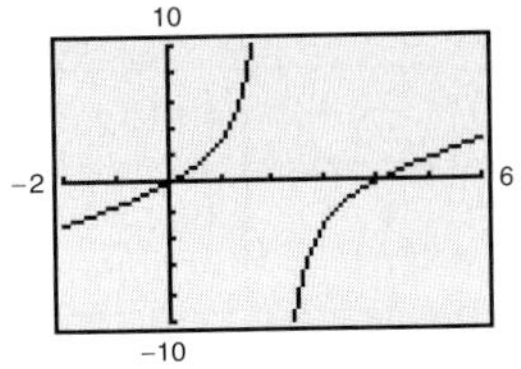

Try It 5 2

Try It 6 (a) $y = 0$ (b) $y = \frac{1}{2}$

(c) No horizontal asymptote

Try It 7 $C = 0.75x + 25{,}000$

$$\overline{C} = 0.75 + \frac{25{,}000}{x}$$

$$\lim_{x\to\infty} \overline{C} = \$0.75/\text{unit}$$

Try It 8 No, the cost function is not defined at $p = 100$, which implies that it is not possible to remove 100% of the pollutants.

SECTION 3.7

Try It 1

	$f(x)$	$f'(x)$	$f''(x)$	Shape of graph
x in $(-\infty, -1)$		$-$	$+$	Decreasing, concave upward
$x = -1$	-32	0	$+$	Relative minimum
x in $(-1, 1)$		$+$	$+$	Increasing, concave upward
$x = 1$	-16	$+$	0	Point of inflection
x in $(1, 3)$		$+$	$-$	Increasing, concave downward
$x = 3$	0	0	$-$	Relative maximum
x in $(3, \infty)$		$-$	$-$	Decreasing, concave downward

Try It 2

	$f(x)$	$f'(x)$	$f''(x)$	Shape of graph
x in $(-\infty, 0)$		$-$	$+$	Decreasing, concave upward
$x = 0$	5	0	0	Point of inflection
x in $(0, 2)$		$-$	$-$	Decreasing, concave downward
$x = 2$	-11	$-$	0	Point of inflection
x in $(2, 3)$		$-$	$+$	Decreasing, concave upward
$x = 3$	-22	0	$+$	Relative minimum
x in $(3, \infty)$		$+$	$+$	Increasing, concave upward

Try It 3

	$f(x)$	$f'(x)$	$f''(x)$	Shape of graph
x in $(-\infty, 0)$		$+$	$-$	Increasing, concave downward
$x = 0$	0	0	$-$	Relative maximum
x in $(0, 1)$		$-$	$-$	Decreasing, concave downward
$x = 1$	Undef.	Undef.	Undef.	Vertical asymptote
x in $(1, 2)$		$-$	$+$	Decreasing, concave upward
$x = 2$	4	0	$+$	Relative minimum
x in $(2, \infty)$		$+$	$+$	Increasing, concave upward

Try It 4

	$f(x)$	$f'(x)$	$f''(x)$	Shape of graph
x in $(-\infty, -1)$		$+$	$+$	Increasing, concave upward
$x = -1$	Undef.	Undef.	Undef.	Vertical asymptote
x in $(-1, 0)$		$+$	$-$	Increasing, concave downward
$x = 0$	-1	0	$-$	Relative maximum
x in $(0, 1)$		$-$	$-$	Decreasing, concave downward
$x = 1$	Undef.	Undef.	Undef.	Vertical asymptote
x in $(1, \infty)$		$-$	$+$	Decreasing, concave upward

Try It 5

	$f(x)$	$f'(x)$	$f''(x)$	Shape of graph
x in $(0, 1)$		$-$	$+$	Decreasing, concave upward
$x = 1$	-4	0	$+$	Relative minimum
x in $(1, \infty)$		$+$	$+$	Increasing, concave upward

SECTION 3.8

Try It 1 $dy = 0.32$; $\Delta y = 0.32240801$

Try It 2 $dR = \$5$; $\Delta R = \$4.94$

Try It 3 $dP = \$10.96$; $\Delta P = \$10.98$

Try It 4 (a) $dy = 12x^2\,dx$ (b) $dy = \frac{2}{3}\,dx$

(c) $dy = (6x - 2)\,dx$ (d) $dy = -\dfrac{2}{x^3}\,dx$

Try It 5 $S = 1.96\pi$ squared inches ≈ 6.1575 squared inches

$dS = \pm 0.056\pi$ squared inches $\approx \pm 0.1759$ squared inches

CHAPTER 4

SECTION 4.1

Try It 1 (a) 243 (b) 3 (c) 64

(d) 8 (e) $\frac{1}{2}$ (f) $\sqrt{10}$

Try It 2 (a) 5.453×10^{-13} (b) 1.621×10^{-13}

(c) 2.629×10^{-14}

Try It 3

x	-3	-2	-1	0	1	2	3
$f(x)$	$\frac{1}{125}$	$\frac{1}{25}$	$\frac{1}{5}$	1	5	25	125

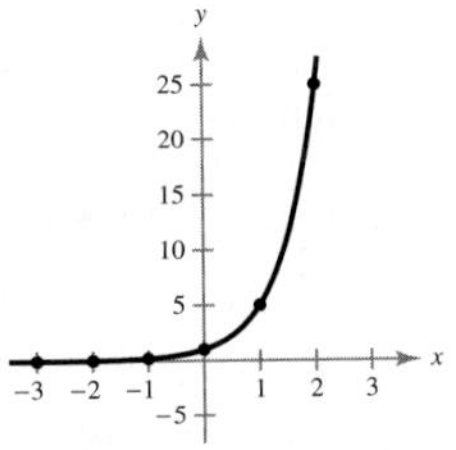

Try It 4

x	-3	-2	-1	0	1	2	3
$f(x)$	9	5	3	2	$\frac{3}{2}$	$\frac{5}{4}$	$\frac{9}{8}$

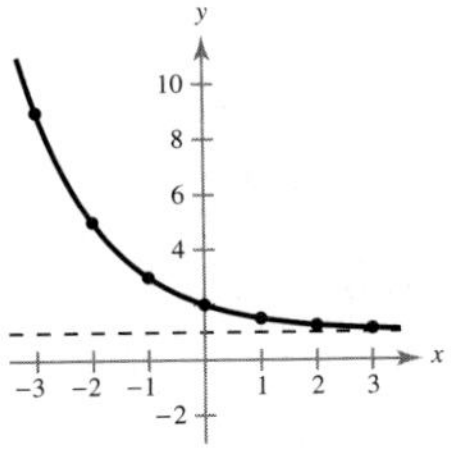

Horizontal asymptote: $y = 1$

SECTION 4.2

Try It 1

x	-2	-1	0	1	2
$f(x)$	$e^2 \approx 7.389$	$e \approx 2.718$	1	$\frac{1}{e} \approx 0.368$	$\frac{1}{e^2} \approx 0.135$

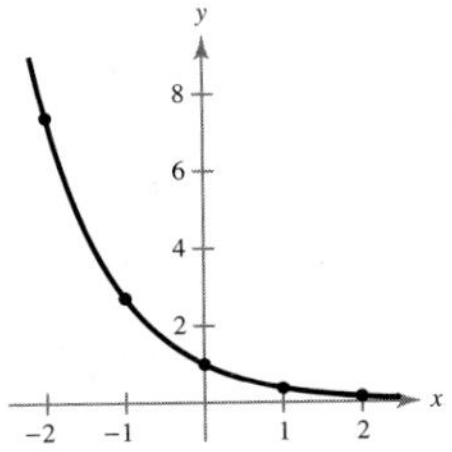

Try It 2 After 0 hours, $y = 1.25$ grams.

After 1 hour, $y \approx 1.338$ grams.

After 10 hours, $y \approx 1.498$ grams.

$$\lim_{t\to\infty} \frac{1.50}{1 + 0.2e^{-0.5t}} = 1.50 \text{ grams.}$$

Try It 3 (a) \$4870.38 (b) \$4902.71

(c) \$4918.66 (d) \$4919.21

All else being equal, the more often interest is compounded, the greater the balance.

Try It 4 (a) 7.12% (b) 7.25%

Try It 5 \$16,712.90

SECTION 4.3

Try It 1 At $(0, 1)$, $y = x + 1$.

At $(1, e)$, $y = ex$.

Try It 2 (a) $3e^{3x}$ (b) $-\dfrac{6x^2}{e^{2x^3}}$ (c) $8xe^{x^2}$ (d) $-\dfrac{2}{e^{2x}}$

Try It 3 (a) $xe^x(x + 2)$ (b) $\frac{1}{2}(e^x - e^{-x})$

(c) $\dfrac{e^x(x - 2)}{x^3}$ (d) $e^x(x^2 + 2x - 1)$

Try It 4

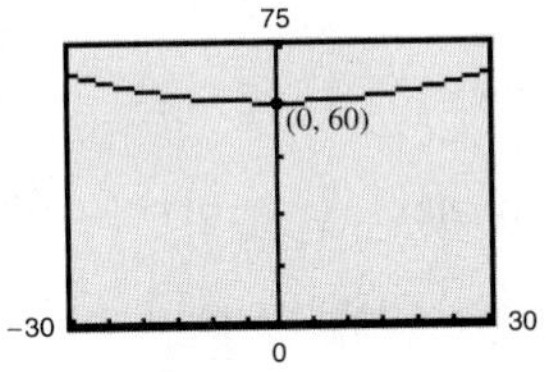

Try It 5 \$18.39/unit (80,000 units)

Try It 6

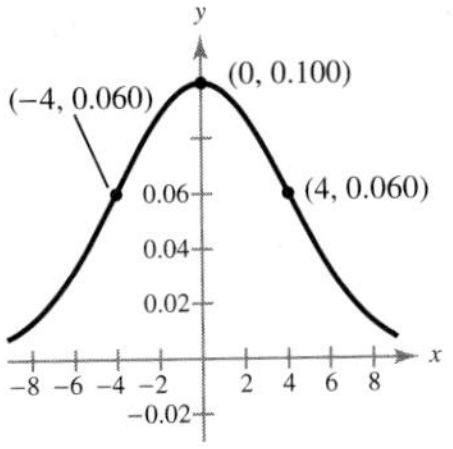

Points of inflection: $(-4, 0.060)$, $(4, 0.060)$

SECTION 4.4

Try It 1

x	-1.5	-1	-0.5	0	0.5	1
$f(x)$	-0.693	0	0.405	0.693	0.916	1.099

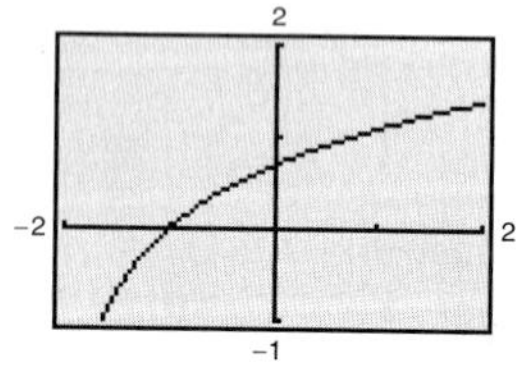

Try It 2 (a) 3 (b) $x + 1$

Try It 3 (a) $\ln 2 - \ln 5$ (b) $\frac{1}{3}\ln(x + 2)$
(c) $\ln x - \ln 5 - \ln y$ (d) $\ln x + 2\ln(x + 1)$

Try It 4 (a) $\ln x^4y^3$ (b) $\ln \dfrac{x + 1}{(x + 3)^2}$

Try It 5 (a) $\ln 6$ (b) $5 \ln 5$

Try It 6 (a) e^4 (b) e^3

Try It 7 7.9 years

SECTION 4.5

Try It 1 $\dfrac{1}{x}$

Try It 2 (a) $\dfrac{2x}{x^2 - 4}$ (b) $x(1 + 2\ln x)$ (c) $\dfrac{2\ln x - 1}{x^3}$

Try It 3 $\dfrac{1}{3(x + 1)}$

Try It 4 $\dfrac{2}{x} + \dfrac{x}{x^2 + 1}$

Try It 5 Relative minimum: $(2, 2 - 2\ln 2) \approx (2, 0.6137)$

Try It 6 $\dfrac{dp}{dt} = -1.3\%/\text{month}$

The average score would decrease at a greater rate than the model in Example 6.

Try It 7 (a) 4 (b) -2 (c) -5 (d) 3

Try It 8 (a) 2.322 (b) 2.631 (c) 3.161 (d) -0.5

Try It 9

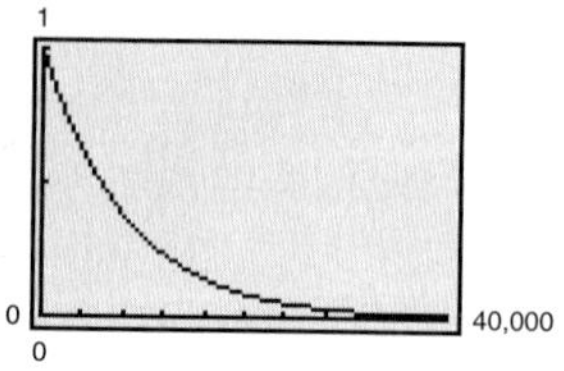

As time increases, the derivative approaches 0. The rate of change of the amount of carbon isotopes is proportional to the amount present.

SECTION 4.6

Try It 1 About 2113.7 years

Try It 2 $y = 25e^{0.6931t}$

Try It 3 $r = \frac{1}{8}\ln 2 \approx 0.0866$ or 8.66%

Try It 4 About 12.42 months

CHAPTER 5

SECTION 5.1

Try It 1 (a) $\displaystyle\int 3\,dx = 3x + C$ (b) $\displaystyle\int 2x\,dx = x^2 + C$
(c) $\displaystyle\int 9t^2\,dt = 3t^3 + C$

Try It 2 (a) $5x + C$ (b) $-r + C$ (c) $2t + C$

Try It 3 $\frac{5}{2}x^2 + C$

Try It 4 (a) $-\dfrac{1}{x} + C$ (b) $\dfrac{3}{4}x^{4/3} + C$

Try It 5 (a) $\frac{1}{2}x^2 + 4x + C$ (b) $x^4 - \frac{5}{2}x^2 + 2x + C$

Try It 6 $\frac{2}{3}x^{3/2} + 4x^{1/2} + C$

Try It 7 General solution: $F(x) = 2x^2 + 2x + C$
Particular solution: $F(x) = 2x^2 + 2x + 4$

Try It 8 $s(t) = -16t^2 + 32t + 48$. The ball hits the ground 3 seconds after it is thrown, with a velocity of -64 feet per second.

Try It 9 $C = -0.01x^2 + 28x + 12.01$
$C(200) = \$5212.01$

SECTION 5.2

Try It 1 (a) $\dfrac{(x^3 + 6x)^3}{3} + C$ (b) $\dfrac{2}{3}(x^2 - 2)^{3/2} + C$

Try It 2 $\frac{1}{36}(3x^4 + 1)^3 + C$

Try It 3 $2x^9 + \frac{12}{5}x^5 + 2x + C$

Try It 4 $\frac{5}{3}(x^2 + 1)^{3/2} + C$

Try It 5 $-\frac{1}{3}(1 - 2x)^{3/2} + C$

Try It 6 Approximately \$30,045

SECTION 5.3

Try It 1 (a) $3e^x + C$ (b) $e^{5x} + C$ (c) $e^x - \dfrac{x^2}{2} + C$

Try It 2 $\frac{1}{2}e^{2x+3} + C$

Try It 3 $2e^{x^2} + C$

Try It 4 (a) $2 \ln|x| + C$ (b) $\ln|x^3| + C$
(c) $\ln|2x + 1| + C$

Try It 5 $\frac{1}{4} \ln|4x + 1| + C$

Try It 6 $\frac{3}{2} \ln(x^2 + 4) + C$

Try It 7 (a) $4x - 3 \ln|x| - \dfrac{2}{x} + C$
(b) $2 \ln(1 + e^x) + C\,dx$
(c) $\dfrac{x^2}{2} + x + 3 \ln|x + 1| + C$

SECTION 5.4

Try It 1 $\frac{1}{2}(3)(12) = 18$

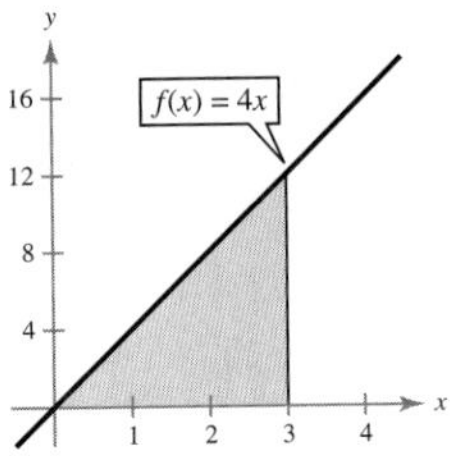

Try It 2 $\frac{22}{3}$ square units

Try It 3 68

Try It 4 (a) $\frac{1}{4}(e^4 - 1) \approx 13.3995$
(b) $-\ln 5 + \ln 2 \approx -0.9163$

Try It 5 $\frac{13}{2}$

Try It 6 (a) About \$14.18 (b) \$141.79

Try It 7 \$13.70

Try It 8 (a) $\frac{2}{5}$ (b) 0

Try It 9 About \$15,319.26

SECTION 5.5

Try It 1 $\frac{8}{3}$ square units

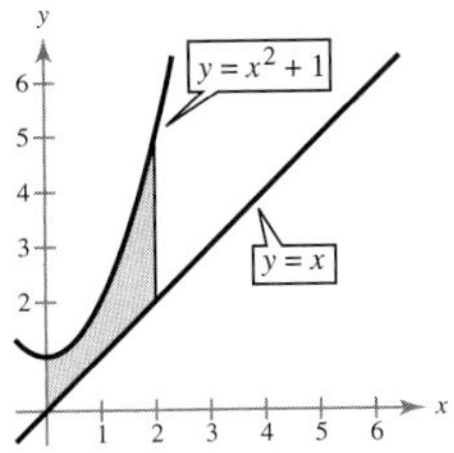

Try It 2 $\frac{32}{3}$ square units

Try It 3 $\frac{9}{2}$ square units

Try It 4 $\frac{253}{12}$ square units

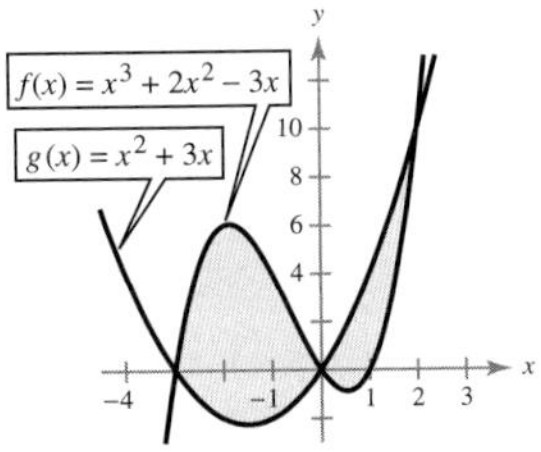

Try It 5 Consumer surplus: 40
Producer surplus: 20

Try It 6 The company can save \$80.04 million.

SECTION 5.6

Try It 1 $\frac{37}{8}$ square units

Try It 2 0.436 square unit

Try It 3 5.642 square units

Try It 4 ≈ 1.463

SECTION 5.7

Try It 1 $\dfrac{512\pi}{15}$ cubic units

Try It 2 $\dfrac{832\pi}{15}$ cubic units

Try It 3 About 45 cubic inches

CHAPTER 6

SECTION 6.1

Try It 1 $\ln|x - 2| - \dfrac{2}{x - 2} + C$

Try It 2 $\frac{1}{3}(x^2 + 4)^{3/2} + C$

Try It 3 $\frac{1}{2} \ln|1 + e^{2x}| + C$

Try It 4 $\frac{2}{15}(x + 2)^{3/2}(3x - 4) + C$

Try It 5 $\frac{149}{30}$

Try It 6 50.4%

SECTION 6.2

Try It 1 $\frac{1}{2}xe^{2x} - \frac{1}{4}e^{2x} + C$

Try It 2 $\dfrac{x^2}{2} \ln x - \dfrac{1}{4}x^2 + C$

Try It 3 $\dfrac{d}{dx}[x \ln x - x + C] = x\left(\dfrac{1}{x}\right) + \ln x - 1$
$= \ln x$

Try It 4 $e^x(x^3 - 3x^2 + 6x - 6) + C$

Try It 5 $e - 2$

Try It 6 \$538,145

Try It 7 \$528,482

SECTION 6.3

Try It 1 $\dfrac{5}{x+3} - \dfrac{4}{x+4}$

Try It 2 $\ln|x(x+2)^2| + \dfrac{1}{x+2} + C$

Try It 3 $\dfrac{1}{2}x^2 - 2x - \dfrac{1}{x} + 4\ln|x+1| + C$

Try It 4 $ky(1-y) = \dfrac{kbe - kt}{(1+be^{-kt})^2}$

$$y = (1 + be^{-kt})^{-1}$$

$$\frac{dy}{dt} = \frac{kbe^{-kt}}{(1+be^{-kt})^2}$$

Therefore, $\dfrac{dy}{dt} = ky(1-y)$

Try It 5 $y = 4$

Try It 6 $y = \dfrac{4000}{1 + 39e^{-0.31045t}}$

SECTION 6.4

Try It 1 $\frac{2}{3}(x-4)\sqrt{2+x} + C$ (Formula 19)

Try It 2 $\sqrt{x^2+16} - 4\ln\left|\dfrac{4+\sqrt{x^2+16}}{x}\right| + C$
(Formula 23)

Try It 3 $\dfrac{1}{4}\ln\left|\dfrac{x-2}{x+2}\right| + C$ (Formula 29)

Try It 4 $\frac{1}{3}[1 - \ln(1+e) + \ln 2] \approx 0.12663$ (Formula 37)

Try It 5 $x(\ln x)^2 + 2x - 2x\ln x + C$ (Formula 42)

Try It 6 $\dfrac{\sqrt{10}}{20}\ln\left|\dfrac{x-3-\sqrt{10}}{x-3+\sqrt{10}}\right| + C$ (Formula 29)

SECTION 6.5

Try It 1 3.2608

Try It 2 3.1956

Try It 3 1.154

SECTION 6.6

Try It 1 (a) Converges; $\frac{1}{2}$ (b) Diverges

Try It 2 1

Try It 3 $\frac{1}{2}$

Try It 4 2

Try It 5 Diverges

Try It 6 Diverges

Try It 7 0.0038 or $\approx 0.4\%$

Try It 8 (c) Converges, $\displaystyle\int_1^{\infty} \frac{1}{x^p}\,dx$ diverges for $0 < p \le 1$ and converges for $p > 1$.

CHAPTER 7

SECTION 7.1

Try It 1

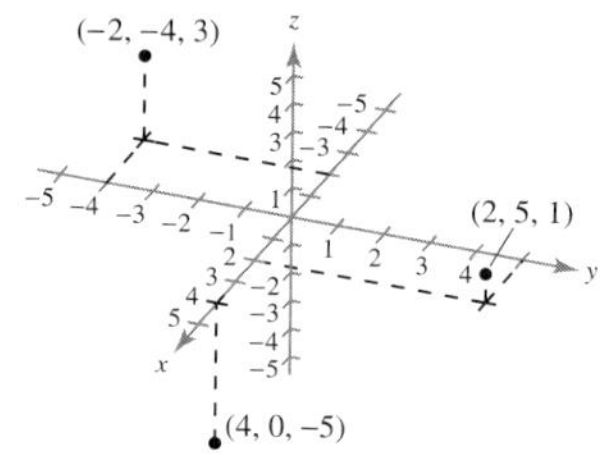

Try It 2 $2\sqrt{6}$

Try It 3 $\left(-\frac{5}{2}, 2, -2\right)$

Try It 4 $(x-4)^2 + (y-3)^2 + (z-2)^2 = 25$

Try It 5 $(x-1)^2 + (y-3)^2 + (z-2)^2 = 38$

Try It 6 Center: $(-3, 4, -1)$; radius: 6

Try It 7 $(x+1)^2 + (y-2)^2 = 16$

SECTION 7.2

Try It 1 x-intercept: $(4, 0, 0)$;
y-intercept: $(0, 2, 0)$;
z-intercept: $(0, 0, 8)$

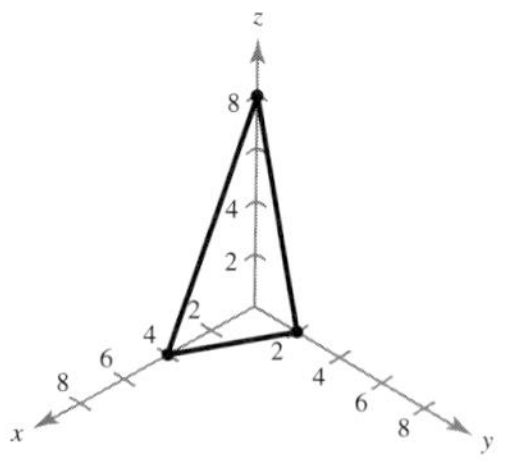

Try It 2 Hyperboloid of one sheet
xy-trace: circle, $x^2 + y^2 = 1$; yz-trace: hyperbola, $y^2 - z^2 = 1$; xz-trace: hyperbola, $x^2 - z^2 = 1$;
$z = 3$ trace: circle, $x^2 + y^2 = 10$

Try It 3 (a) $\dfrac{x^2}{9} + \dfrac{y^2}{4} = z$; elliptic paraboloid

(b) $\dfrac{x^2}{4} + \dfrac{y^2}{9} - z^2 = 0$; elliptic cone

SECTION 7.3

Try It 1 (a) 0 (b) $\frac{9}{4}$

Try It 2 Domain: $x^2 + y^2 \le 9$

Range: $0 \le z \le 3$

Try It 3 Steep; nearly level

Try It 4 Alaska is mainly used for forest land. Alaska does not contain any manufacturing centers, but it does contain a mineral deposit of petroleum.

Try It 5 $f(1500, 1000) \approx 127{,}542$ units

$f(1000, 1500) \approx 117{,}608$ units

x, person-hours, has a greater effect on production.

Try It 6 (a) $M = \$733.76$/month

(b) Total paid $= (30 \times 12) \times 733.76 = \$264{,}153.60$

SECTION 7.4

Try It 1 $\dfrac{\partial z}{\partial x} = 4x - 8xy^3$

$\dfrac{\partial z}{\partial y} = -12x^2y^2 + 4y^3$

Try It 2 $f_x(x, y) = 2xy^3;\ f_x(1, 2) = 16$

$f_y(x, y) = 3x^2y^2;\ f_y(1, 2) = 12$

Try It 3 In the x-direction: $f_x(1, -1, 49) = 8$

In the y-direction: $f_y(1, -1, 49) = -18$

Try It 4 Substitute product relationship

Try It 5 $\dfrac{\partial w}{\partial x} = xy + 2xy \ln(xz)$

$\dfrac{\partial w}{\partial y} = x^2 \ln xz$

$\dfrac{\partial w}{\partial z} = \dfrac{x^2y}{z}$

Try It 6 $f_{xx} = 8y^2$

$f_{yy} = 8x^2 + 8$

$f_{xy} = 16xy$

$f_{yx} = 16xy$

Try It 7

$f_{xx} = 0$	$f_{xy} = e^y$	$f_{xz} = 2$
$f_{yx} = e^y$	$f_{yy} = xe^y + 2$	$f_{yz} = 0$
$f_{zx} = 2$	$f_{zy} = 0$	$f_{zz} = 0$

SECTION 7.5

Try It 1 $f(-8, 2) = -64$: relative minimum

Try It 2 $f(0, 0) = 1$: relative maximum

Try It 3 $f(0, 0) = 0$: saddle point

Try It 4 $P(3.11, 3.81) = \$744.81$ maximum profit

Try It 5 $V\left(\frac{4}{3}, \frac{2}{3}, \frac{8}{3}\right) = \frac{64}{27}$ cubic units

SECTION 7.6

Try It 1 $V\left(\frac{4}{3}, \frac{2}{3}, \frac{8}{3}\right) = \frac{64}{27}$ cubic units

Try It 2 $f(187.5, 50) \approx 13{,}474$ units

Try It 3 $\approx 26{,}740$ units

Try It 4 $f(x, y, z, w) = \frac{15}{2}$

Try It 5 $P(3.35, 4.26) = \$758.08$ maximum profit

Try It 6 $f(2, 0, 2) = 8$

SECTION 7.7

Try It 1 For $f(x)$, $S \approx 9.1$.

For $g(x)$, $S \approx 0.47515$.

The quadratic model is a better fit.

Try It 2 $f(x) = \frac{6}{5}x + \frac{23}{10}$

Try It 3 $y = 18{,}544.2t - 78{,}667$

In 2006, $y \approx 218{,}040{,}000$ subscribers

Try It 4 $y = 19.49t^2 - 106.1t + 3638$

In 2006, $y \approx 6930$ dollars.

SECTION 7.8

Try It 1 (a) $\frac{1}{4}x^4 + 2x^3 - 2x - \frac{1}{4}$ (b) $\ln|y^2 + y| - \ln|2y|$

Try It 2 $\frac{25}{2}$

Try It 3 $\displaystyle\int_2^4\int_1^5 dx\,dy = 8$

Try It 4 $\frac{4}{3}$

Try It 5 (a)

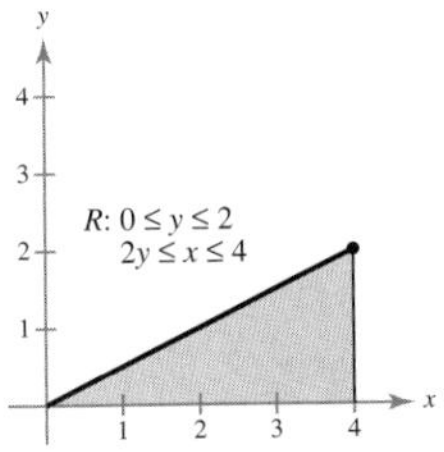

(b) $\displaystyle\int_0^4\int_0^{x/2} dy\,dx$ (c) $\displaystyle\int_0^2\int_{2y}^4 dx\,dy = 4 = \int_0^4\int_0^{x/2} dy\,dx$

Try It 6 $\displaystyle\int_{-1}^3\int_{x^2}^{2x+3} dy\,dx = \frac{32}{3}$

SECTION 7.9

Try It 1 $\frac{16}{3}$

Try It 2 $e - 1$

Try It 3 $\frac{176}{15}$

Try It 4 Integration by parts

Try It 5 3

Index

E

F

G

H

I

C DIFFERENTIAL EQUATIONS

C.1 SOLUTIONS OF DIFFERENTIAL EQUATIONS

Find general solutions of differential equations. • Find particular solutions of differential equations.

General Solution of a Differential Equation

A **differential equation** is an equation involving a differentiable function and one or more of its derivatives. For instance,

$$y' + 2y = 0 \qquad \text{Differential equation}$$

is a differential equation. A function $y = f(x)$ is a **solution** of a differential equation if the equation is satisfied when y and its derivatives are replaced by $f(x)$ and its derivatives. For instance,

$$y = e^{-2x} \qquad \text{Solution of differential equation}$$

is a solution of the differential equation shown above. To see this, substitute for y and $y' = -2e^{-2x}$ in the original equation.

$$\begin{aligned} y' + 2y &= -2e^{-2x} + 2(e^{-2x}) && \text{Substitute for } y \text{ and } y'. \\ &= 0 \end{aligned}$$

In the same way, you can show that $y = 2e^{-2x}$, $y = -3e^{-2x}$, and $y = \frac{1}{2}e^{-2x}$ are also solutions of the differential equation. In fact, each function given by

$$y = Ce^{-2x} \qquad \text{General solution}$$

where C is a real number, is a solution of the equation. This family of solutions is called the **general solution** of the differential equation.

EXAMPLE 1 **Checking Solutions**

Show that

(a) $y = Ce^x$ and (b) $y = Ce^{-x}$

are solutions of the differential equation $y'' - y = 0$.

SOLUTION

(a) Because $y' = Ce^x$ and $y'' = Ce^x$, it follows that

$$\begin{aligned} y'' - y &= Ce^x - Ce^x \\ &= 0. \end{aligned}$$

So, $y = Ce^x$ is a solution.

(b) Because $y' = -Ce^{-x}$ and $y'' = Ce^{-x}$, it follows that

$$\begin{aligned} y'' - y &= Ce^{-x} - Ce^{-x} \\ &= 0. \end{aligned}$$

So, $y = Ce^{-x}$ is also a solution.

Particular Solutions and Initial Conditions

A **particular solution** of a differential equation is any solution that is obtained by assigning specific values to the constants in the general equation.*

Geometrically, the general solution of a differential equation is a family of graphs called **solution curves.** For instance, the general solution of the differential equation $xy' - 2y = 0$ is

$y = Cx^2.$ General solution

Figure A.7 shows several solution curves of this differential equation.

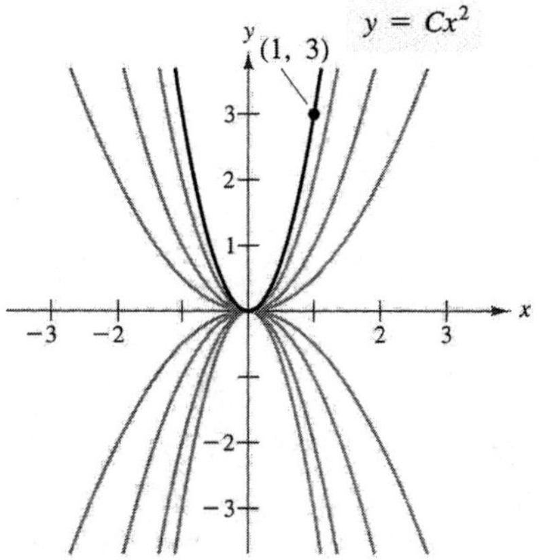

FIGURE A.7

Particular solutions of a differential equation are obtained from **initial conditions** placed on the unknown function and its derivatives. For instance, in Figure A.7, suppose you want to find the particular solution whose graph passes through the point $(1, 3)$. This initial condition can be written as

$y = 3$ when $x = 1.$ Initial condition

Substituting these values into the general solution produces $3 = C(1)^2$, which implies that $C = 3$. So, the particular solution is

$y = 3x^2.$ Particular solution

EXAMPLE 2 Finding a Particular Solution

Verify that

$y = Cx^3$ General solution

is a solution of the differential equation $xy' - 3y = 0$ for any value of C. Then find the particular solution determined by the initial condition

$y = 2$ when $x = -3.$ Initial condition

SOLUTION The derivative of $y = Cx^3$ is $y' = 3Cx^2$. Substituting into the differential equation produces

$$\begin{aligned} xy' - 3y &= x(3Cx^2) - 3(Cx^3) \\ &= 0. \end{aligned}$$

So, $y = Cx^3$ is a solution for any value of C. To find the particular solution, substitute $x = -3$ and $y = 2$ into the general solution to obtain

$$2 = C(-3)^3 \quad \text{or} \quad C = -\frac{2}{27}.$$

This implies that the particular solution is

$y = -\frac{2}{27}x^3.$ Particular solution

*Some differential equations have solutions other than those given by their general solutions. These are called **singular solutions.** In this brief discussion of differential equations, singular solutions will not be discussed.

EXAMPLE 3 **Finding a Particular Solution**

You are working in the marketing department of a company that is producing a new cereal product to be sold nationally. You determine that a maximum of 10 million units of the product could be sold in a year. You hypothesize that the rate of growth of the sales x (in millions of units) is proportional to the difference between the maximum sales and the current sales. As a differential equation, this hypothesis can be written as

$$\frac{dx}{dt} = k(10 - x), \qquad 0 \le x \le 10.$$

Rate of change of x — is propor-tional to — the difference between 10 and x.

The general solution of this differential equation is

$x = 10 - Ce^{-kt}$ General solution

where t is the time in years. After 1 year, 250,000 units have been sold. Sketch the graph of the sales function over a 10-year period.

SOLUTION Because the product is new, you can assume that $x = 0$ when $t = 0$. So, you have *two* initial conditions.

$x = 0$ when $t = 0$ First initial condition

$x = 0.25$ when $t = 1$ Second initial condition

Substituting the first initial condition into the general solution produces

$0 = 10 - Ce^{-k(0)}$

which implies that $C = 10$. Substituting the second initial condition into the general solution produces

$0.25 = 10 - 10e^{-k(1)}$

which implies that $k = \ln \frac{40}{39} \approx 0.0253$. So, the particular solution is

$x = 10 - 10e^{-0.0253t}$. Particular solution

The table shows the annual sales during the first 10 years, and the graph of the solution is shown in Figure A.8.

t	1	2	3	4	5	6	7	8	9	10
x	0.25	0.49	0.73	0.96	1.19	1.41	1.62	1.83	2.04	2.24

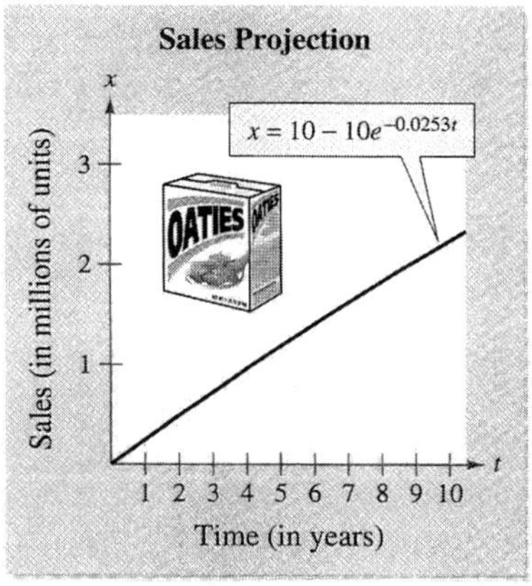

FIGURE A.8

In the first three examples in this section, each solution was given in explicit form, such as $y = f(x)$. Sometimes you will encounter solutions for which it is more convenient to write the solution in implicit form, as shown in Example 4.

EXAMPLE 4 Sketching Graphs of Solutions

Verify that

$$2y^2 - x^2 = C \qquad \text{General solution}$$

is a solution of the differential equation

$$2yy' - x = 0.$$

Then sketch the particular solutions represented by $C = 0$, $C = \pm 1$, and $C = \pm 4$.

SOLUTION To verify the given solution, differentiate each side with respect to x.

$$2y^2 - x^2 = C \qquad \text{Given general solution}$$
$$4yy' - 2x = 0 \qquad \text{Differentiate with respect to } x.$$
$$2yy' - x = 0 \qquad \text{Divide each side by 2.}$$

Because the third equation is the given differential equation, you can conclude that

$$2y^2 - x^2 = C$$

is a solution. The particular solutions represented by $C = 0$, $C = \pm 1$, and $C = \pm 4$ are shown in Figure A.9.

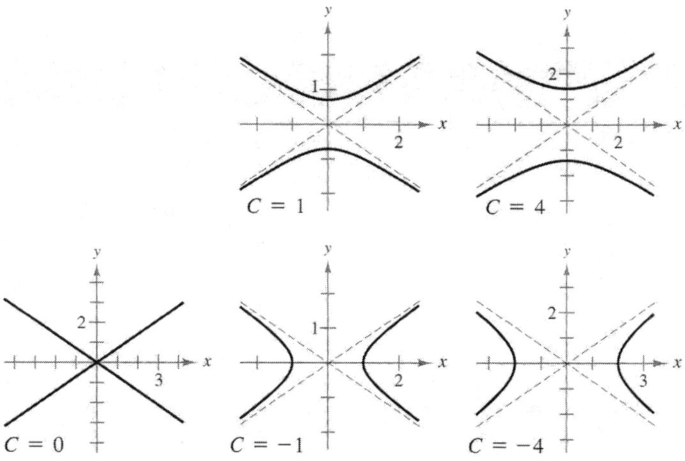

FIGURE A.9 Graphs of Five Particular Solutions

TAKE ANOTHER LOOK

Writing a Differential Equation

Write a differential equation that has the family of circles

$$x^2 + y^2 = C$$

as a general solution.

PREREQUISITE REVIEW C.1

The following warm-up exercises involve skills that were covered in earlier sections. You will use these skills in the exercise set for this section.

In Exercises 1–4, find the first and second derivatives of the function.

1. $y = 3x^2 + 2x + 1$

2. $y = -2x^3 - 8x + 4$

3. $y = -3e^{2x}$

4. $y = -3e^{x^2}$

In Exercises 5–8, use implicit differentiation to find *dy/dx*.

5. $x^2 + y^2 = 2x$

6. $2x - y^3 = 4y$

7. $xy^2 = 3$

8. $3xy + x^2y^2 = 10$

In Exercises 9 and 10, solve for *k*.

9. $0.5 = 9 - 9e^{-k}$

10. $14.75 = 25 - 25e^{-2k}$

EXERCISES C.1

In Exercises 1–10, verify that the function is a solution of the differential equation.

Solution	Differential Equation
1. $y = x^3 + 5$	$y' = 3x^2$
2. $y = 2x^3 - x + 1$	$y' = 6x^2 - 1$
3. $y = e^{-2x}$	$y' + 2y = 0$
4. $y = 3e^{x^2}$	$y' - 2xy = 0$
5. $y = 2x^3$	$y' - \frac{3}{x}y = 0$
6. $y = 4x^2$	$y' - \frac{2}{x}y = 0$
7. $y = x^2$	$x^2y'' - 2y = 0$
8. $y = \frac{1}{x}$	$xy'' + 2y' = 0$
9. $y = 2e^{2x}$	$y'' - y' - 2y = 0$
10. $y = e^{x^3}$	$y'' - 3x^2y' - 6xy = 0$

In Exercises 11–28, verify that the function is a solution of the differential equation for any value of *C*.

Solution	Differential Equation
11. $y = \frac{1}{x} + C$	$\frac{dy}{dx} = -\frac{1}{x^2}$
12. $y = \sqrt{4 - x^2} + C$	$\frac{dy}{dx} = -\frac{x}{\sqrt{4 - x^2}}$
13. $y = Ce^{4x}$	$\frac{dy}{dx} = 4y$
14. $y = Ce^{-4x}$	$\frac{dy}{dx} = -4y$
15. $y = Ce^{-t/3} + 7$	$3\frac{dy}{dt} + y - 7 = 0$
16. $y = Ce^{-t} + 10$	$y' + y - 10 = 0$
17. $y = Cx^2 - 3x$	$xy' - 3x - 2y = 0$
18. $y = x \ln x^2 + 2x^{3/2} + Cx$	$y' - \frac{y}{x} = 2 + \sqrt{x}$
19. $y = x^2 + 2x + \frac{C}{x}$	$xy' + y = x(3x + 4)$
20. $y = C_1 + C_2e^x$	$y'' - y' = 0$

Solution	*Differential Equation*
21. $y = C_1e^{x/2} + C_2e^{-2x}$	$2y'' + 3y' - 2y = 0$
22. $y = C_1e^{4x} + C_2e^{-x}$	$y'' - 3y' - 4y = 0$
23. $y = \dfrac{bx^4}{4-a} + Cx^a$	$y' - \dfrac{ay}{x} = bx^3$
24. $y = \dfrac{x^3}{5} - x + C\sqrt{x}$	$2xy' - y = x^3 - x$
25. $y = \dfrac{2}{1 + Ce^{x^2}}$	$y' + 2xy = xy^2$
26. $y = Ce^{x-x^2}$	$y' + (2x - 1)y = 0$
27. $y = x \ln x + Cx + 4$	$x(y' - 1) - (y - 4) = 0$
28. $y = x(\ln x + C)$	$x + y - xy' = 0$

In Exercises 29–32, use implicit differentiation to verify that the equation is a solution of the differential equation for any value of C.

Solution	*Differential Equation*
29. $x^2 + y^2 = Cy$	$y' = \dfrac{2xy}{x^2 - y^2}$
30. $y^2 + 2xy - x^2 = C$	$(x + y)y' - x + y = 0$
31. $x^2 + xy = C$	$x^2y'' - 2(x + y) = 0$
32. $x^2 - y^2 = C$	$y^3y'' + x^2 - y^2 = 0$

In Exercises 33–36, determine whether the function is a solution of the differential equation $y^{(4)} - 16y = 0$.

33. $y = e^{-2x}$

34. $y = 5 \ln x$

35. $y = \dfrac{4}{x}$

36. $y = 4e^{2x}$

In Exercises 37–40, determine whether the function is a solution of the differential equation $y''' - 3y' + 2y = 0$.

37. $y = \frac{2}{9}xe^{-2x}$

38. $y = 4e^x + \frac{2}{9}xe^{-2x}$

39. $y = xe^x$

40. $y = x \ln x$

In Exercises 41–48, verify that the general solution satisfies the differential equation. Then find the particular solution that satisfies the initial condition.

41. General solution: $y = Ce^{-2x}$
Differential equation: $y' + 2y = 0$
Initial condition: $y = 3$ when $x = 0$

42. General solution: $2x^2 + 3y^2 = C$
Differential equation: $2x + 3yy' = 0$
Initial condition: $y = 2$ when $x = 1$

43. General solution: $y = C_1 + C_2 \ln|x|,\ x > 0$
Differential equation: $xy'' + y' = 0$
Initial condition: $y = 5$ and $y' = 0.5$ when $x = 1$

44. General solution: $y = C_1x + C_2x^3$
Differential equation: $x^2y'' - 3xy' + 3y = 0$
Initial condition: $y = 0$ and $y' = 4$ when $x = 2$

45. General solution: $y = C_1e^{4x} + C_2e^{-3x}$
Differential equation: $y'' - y' - 12y = 0$
Initial condition: $y = 5$ and $y' = 6$ when $x = 0$

46. General solution: $y = Ce^{x-x^2}$
Differential equation: $y' + (2x - 1)y = 0$
Initial condition: $y = 2$ when $x = 1$

47. General solution: $y = e^{2x/3}(C_1 + C_2x)$
Differential equation: $9y'' - 12y' + 4y = 0$
Initial condition: $y = 4$ when $x = 0$
$y = 0$ when $x = 3$

48. General solution: $y = \left(C_1 + C_2x + \frac{1}{12}x^4\right)e^{2x}$
Differential equation: $y'' - 4y' + 4y = x^2e^{2x}$
Initial condition: $y = 2$ and $y' = 1$ when $x = 0$

In Exercises 49–52, the general solution of the differential equation is given. Use a graphing utility to graph the particular solutions that correspond to the indicated values of C.

General Solution	*Differential Equation*	*C-Values*
49. $y = Cx^2$	$xy' - 2y = 0$	$1, 2, 4$
50. $4y^2 - x^2 = C$	$4yy' - x = 0$	$0, \pm1, \pm4$
51. $y = C(x + 2)^2$	$(x + 2)y' - 2y = 0$	$0, \pm1, \pm2$
52. $y = Ce^{-x}$	$y' + y = 0$	$0, \pm1, \pm2$

In Exercises 53–60, use integration to find the general solution of the differential equation.

53. $\dfrac{dy}{dx} = 3x^2$

54. $\dfrac{dy}{dx} = \dfrac{1}{1 + x}$

55. $\dfrac{dy}{dx} = \dfrac{x + 3}{x}$

56. $\dfrac{dy}{dx} = \dfrac{x - 2}{x}$

57. $\dfrac{dy}{dx} = \dfrac{1}{x^2 - 1}$

58. $\dfrac{dy}{dx} = \dfrac{x}{1 + x^2}$

59. $\dfrac{dy}{dx} = x\sqrt{x - 3}$

60. $\dfrac{dy}{dx} = xe^x$

In Exercises 61–64, you are shown the graphs of some of the solutions of the differential equation. Find the particular solution whose graph passes through the indicated point.

61. $y^2 = Cx^3$

$2xy' - 3y = 0$

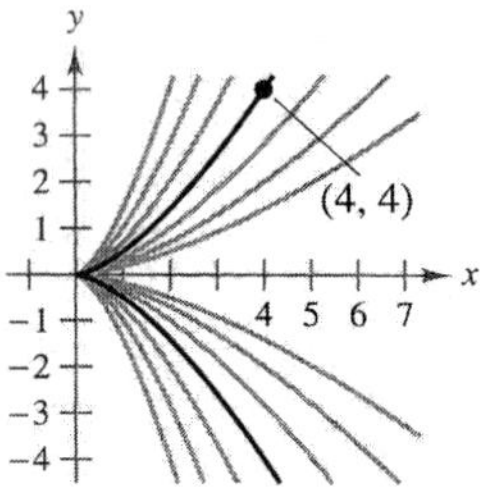

62. $2x^2 - y^2 = C$

$yy' - 2x = 0$

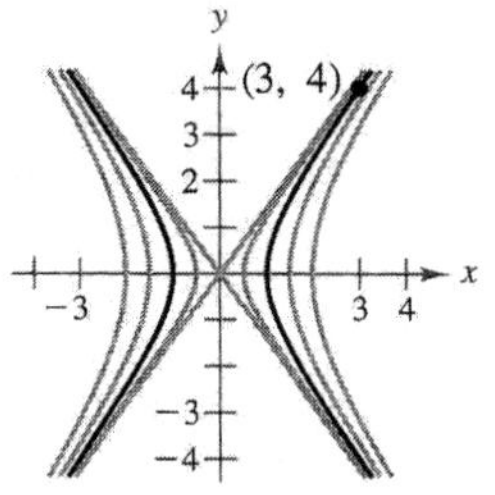

63. $y = Ce^x$

$y' - y = 0$

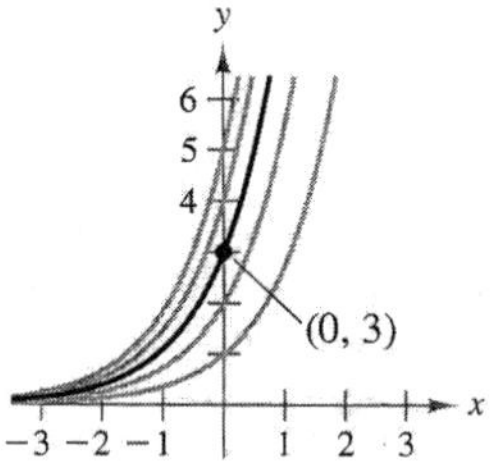

64. $y^2 = 2Cx$

$2xy' - y = 0$

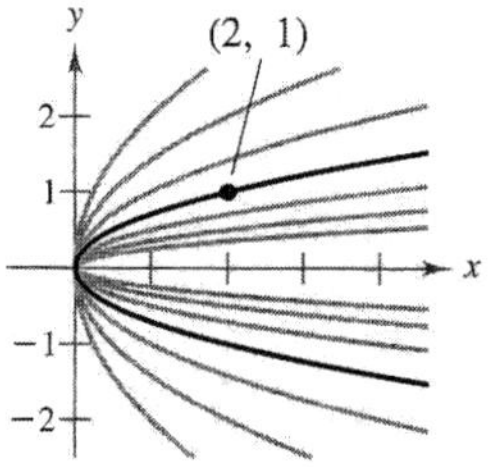

65. ***Biology*** The limiting capacity of the habitat of a wildlife herd is 750. The growth rate dN/dt of the herd is proportional to the unutilized opportunity for growth, as described by the differential equation

$$\frac{dN}{dt} = k(750 - N).$$

The general solution of this differential equation is

$$N = 750 - Ce^{-kt}.$$

When $t = 0$, the population of the herd is 100. After 2 years, the population has grown to 160.

(a) Write the population function N as a function of t.

(b) Use a graphing utility to graph the population function.

(c) What is the population of the herd after 4 years?

66. ***Investment*** The rate of growth of an investment is proportional to the amount in the investment at any time t. That is,

$$\frac{dA}{dt} = kA.$$

The initial investment is \$1000, and after 10 years the balance is \$3320.12. The general solution is

$$A = Ce^{kt}.$$

What is the particular solution?

67. ***Marketing*** You are working in the marketing department of a computer software company. Your marketing team determines that a maximum of 30,000 units of a new product can be sold in a year. You hypothesize that the rate of growth of the sales x is proportional to the difference between the maximum sales and the current sales. That is,

$$\frac{dx}{dt} = k(30{,}000 - x).$$

The general solution of this differential equation is

$$x = 30{,}000 - Ce^{-kt}$$

where t is the time in years. During the first year, 2000 units are sold. Complete the table showing the numbers of units sold in subsequent years.

Year, t	2	4	6	8	10
Units, x					

68. ***Marketing*** In Exercise 67, suppose that the maximum annual sales are 50,000 units. How does this change the sales shown in the table?

69. ***Safety*** Assume that the rate of change in the number of miles s of road cleared per hour by a snowplow is inversely proportional to the depth h of the snow. This rate of change is described by the differential equation

$$\frac{ds}{dh} = \frac{k}{h}.$$

Show that

$$s = 25 - \frac{13}{\ln 3} \ln \frac{h}{2}$$

is a solution of this differential equation.

70. Show that $y = a + Ce^{k(1-b)t}$ is a solution of the differential equation

$$y = a + b(y - a) + \left(\frac{1}{k}\right)\left(\frac{dy}{dt}\right)$$

where k is a constant.

71. The function $y = Ce^{kx}$ is a solution of the differential equation

$$\frac{dy}{dx} = 0.07y.$$

Is it possible to determine C or k from the information given? If so, find its value.

True or False? In Exercises 72 and 73, determine whether the statement is true or false. If it is false, explain why or give an example that shows it is false.

72. A differential equation can have more than one solution.

73. If $y = f(x)$ is a solution of a differential equation, then $y = f(x) + C$ is also a solution.

C.2 SEPARATION OF VARIABLES

Use separation of variables to solve differential equations. • Use differential equations to model and solve real-life problems.

Separation of Variables

The simplest type of differential equation is one of the form $y' = f(x)$. You know that this type of equation can be solved by integration to obtain

$$y = \int f(x)\,dx.$$

In this section, you will learn how to use integration to solve another important family of differential equations—those in which the variables can be separated. This technique is called **separation of variables.**

Separation of Variables

If f and g are continuous functions, then the differential equation

$$\frac{dy}{dx} = f(x)g(y)$$

has a general solution of

$$\int \frac{1}{g(y)}\,dy = \int f(x)\,dx + C.$$

TECHNOLOGY

You can use a symbolic integration utility to solve a separable variables differential equation. Use a symbolic integration utility to solve the differential equation

$$y' = \frac{x}{y^2 + 1}.$$

Essentially, the technique of separation of variables is just what its name implies. For a differential equation involving x and y, you separate the x variables to one side and the y variables to the other. After separating variables, integrate each side to obtain the general solution. Here is an example.

EXAMPLE 1 **Solving a Differential Equation**

Find the general solution of

$$\frac{dy}{dx} = \frac{x}{y^2 + 1}.$$

SOLUTION Begin by separating variables, then integrate each side.

$$\frac{dy}{dx} = \frac{x}{y^2 + 1} \qquad \text{Differential equation}$$

$$(y^2 + 1)\,dy = x\,dx \qquad \text{Separate variables.}$$

$$\int (y^2 + 1)\,dy = \int x\,dx \qquad \text{Integrate each side.}$$

$$\frac{y^3}{3} + y = \frac{x^2}{2} + C \qquad \text{General solution}$$

EXAMPLE 2 Solving a Differential Equation

Find the general solution of

$$\frac{dy}{dx} = \frac{x}{y}.$$

SOLUTION Begin by separating variables, then integrate each side.

$$\frac{dy}{dx} = \frac{x}{y} \qquad \text{Differential equation}$$

$$y\,dy = x\,dx \qquad \text{Separate variables.}$$

$$\int y\,dy = \int x\,dx \qquad \text{Integrate each side.}$$

$$\frac{y^2}{2} = \frac{x^2}{2} + C_1 \qquad \text{Find antiderivatives.}$$

$$y^2 = x^2 + C \qquad \text{Multiply each side by 2.}$$

So, the general solution is $y^2 = x^2 + C$. Note that C_1 is used as a temporary constant of integration in anticipation of multiplying each side of the equation by 2 to produce the constant C.

STUDY TIP

After finding the general solution of a differential equation, you should use the techniques demonstrated in Section C.1 to check the solution. For instance, in Example 2 you can check the solution by differentiating the equation $y^2 = x^2 + C$ to obtain $2yy' = 2x$ or $y' = x/y$.

EXAMPLE 3 Solving a Differential Equation

Find the general solution of

$$e^y \frac{dy}{dx} = 2x.$$

Use a graphing utility to graph several solutions.

SOLUTION Begin by separating variables, then integrate each side.

$$e^y \frac{dy}{dx} = 2x \qquad \text{Differential equation}$$

$$e^y\,dy = 2x\,dx \qquad \text{Separate variables.}$$

$$\int e^y\,dy = \int 2x\,dx \qquad \text{Integrate each side.}$$

$$e^y = x^2 + C \qquad \text{Find antiderivatives.}$$

By taking the natural logarithm of each side, you can write the general solution as

$$y = \ln(x^2 + C). \qquad \text{General solution}$$

The graphs of the particular solutions given by $C = 0$, $C = 5$, $C = 10$, and $C = 15$ are shown in Figure A.10.

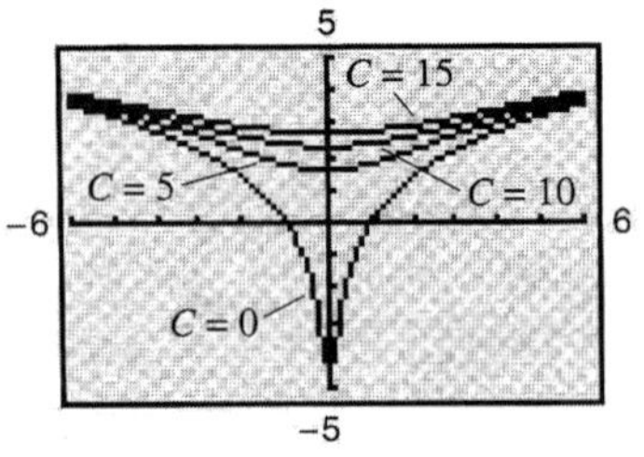

FIGURE A.10

EXAMPLE 4 Finding a Particular Solution

Solve the differential equation

$$xe^{x^2} + yy' = 0$$

subject to the initial condition $y = 1$ when $x = 0$.

SOLUTION

$$xe^{x^2} + yy' = 0 \qquad \text{Differential equation}$$

$$y\frac{dy}{dx} = -xe^{x^2} \qquad \text{Subtract } xe^{x^2} \text{ from each side.}$$

$$y\,dy = -xe^{x^2}\,dx \qquad \text{Separate variables.}$$

$$\int y\,dy = \int -xe^{x^2}\,dx \qquad \text{Integrate each side.}$$

$$\frac{y^2}{2} = -\frac{1}{2}e^{x^2} + C_1 \qquad \text{Find antiderivatives.}$$

$$y^2 = -e^{x^2} + C \qquad \text{Multiply each side by 2.}$$

To find the particular solution, substitute the initial condition values to obtain

$$(1)^2 = -e^{(0)^2} + C.$$

This implies that $1 = -1 + C$, or $C = 2$. So, the particular solution that satisfies the initial condition is

$$y^2 = -e^{x^2} + 2. \qquad \text{Particular solution}$$

EXAMPLE 5 Solving a Differential Equation

Example 3 in Section C.1 uses the differential equation

$$\frac{dx}{dt} = k(10 - x)$$

to model the sales of a new product. Solve this differential equation.

SOLUTION

$$\frac{dx}{dt} = k(10 - x) \qquad \text{Differential equation}$$

$$\frac{1}{10 - x}\,dx = k\,dt \qquad \text{Separate variables.}$$

$$\int \frac{1}{10 - x}\,dx = \int k\,dt \qquad \text{Integrate each side.}$$

$$-\ln(10 - x) = kt + C_1 \qquad \text{Find antiderivatives.}$$

$$\ln(10 - x) = -kt - C_1 \qquad \text{Multiply each side by } -1.$$

$$10 - x = e^{-kt - C_1} \qquad \text{Exponentiate each side.}$$

$$x = 10 - Ce^{-kt} \qquad \text{Solve for } x.$$

STUDY TIP

In Example 5, the context of the original model indicates that $(10 - x)$ is positive. So, when you integrate $1/(10 - x)$, you can write $-\ln(10 - x)$, rather than $-\ln|10 - x|$.

Also note in Example 5 that the solution agrees with the one that was given in Example 3 in Section C.1.

Applications

EXAMPLE 6 Modeling National Income

Let y represent the national income, let a represent the income spent on necessities, and let b represent the percent of the remaining income spent on luxuries. A commonly used economic model that relates these three quantities is

$$\frac{dy}{dt} = k(1 - b)(y - a)$$

where t is the time in years. Assume that b is 75%, and solve the resulting differential equation.

Corporate profits in the United States are closely monitored by New York City's Wall Street executives. Corporate profits, however, represent only about 10.5% of the national income. In 2003, the national income was more than \$9.5 trillion. Of this, about 65% was employee compensation.

SOLUTION Because b is 75%, it follows that $(1 - b)$ is 0.25. So, you can solve the differential equation as shown.

$$\frac{dy}{dt} = k(0.25)(y - a) \qquad \text{Differential equation}$$

$$\frac{1}{y - a}\,dy = 0.25k\,dt \qquad \text{Separate variables.}$$

$$\int \frac{1}{y - a}\,dy = \int 0.25k\,dt \qquad \text{Integrate each side.}$$

$$\ln(y - a) = 0.25kt + C_1 \qquad \text{Find antiderivatives, given } y - a > 0.$$

$$y - a = Ce^{0.25kt} \qquad \text{Exponentiate each side.}$$

$$y = a + Ce^{0.25kt} \qquad \text{Add } a \text{ to each side.}$$

The graph of this solution is shown in Figure A.11. In the figure, note that the national income is spent in three ways.

(National income) = (necessities) + (luxuries) + (capital investment)

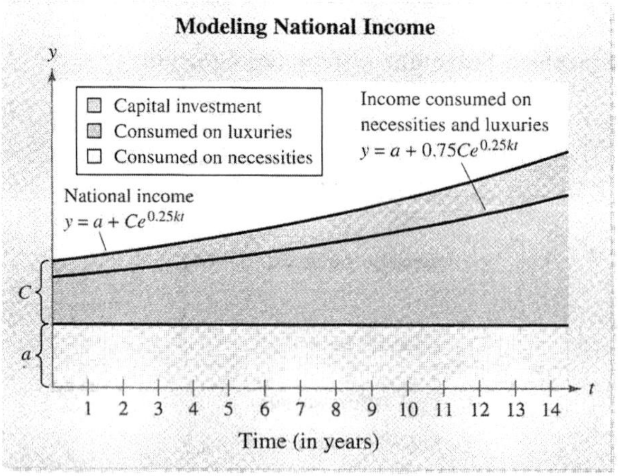

FIGURE A.11

EXAMPLE 7 **Using Graphical Information**

Find the equation of the graph that has the characteristics listed below.

1. At each point (x, y) on the graph, the slope is $-x/2y$.
2. The graph passes through the point $(2, 1)$.

SOLUTION Using the information about the slope of the graph, you can write the differential equation

$$\frac{dy}{dx} = -\frac{x}{2y}.$$

Using the point on the graph, you can determine the initial condition $y = 1$ when $x = 2$.

$$\frac{dy}{dx} = -\frac{x}{2y} \qquad \text{Differential equation}$$

$$2y\,dy = -x\,dx \qquad \text{Separate variables.}$$

$$\int 2y\,dy = \int -x\,dx \qquad \text{Integrate each side.}$$

$$y^2 = -\frac{x^2}{2} + C_1 \qquad \text{Find antiderivatives.}$$

$$2y^2 = -x^2 + C \qquad \text{Multiply each side by 2.}$$

$$x^2 + 2y^2 = C \qquad \text{Simplify.}$$

Applying the initial condition yields

$$(2)^2 + 2(1)^2 = C$$

which implies that $C = 6$. So, the equation that satisfies the two given conditions is

$$x^2 + 2y^2 = 6. \qquad \text{Particular solution}$$

As shown in Figure A.12, the graph of this equation is an ellipse.

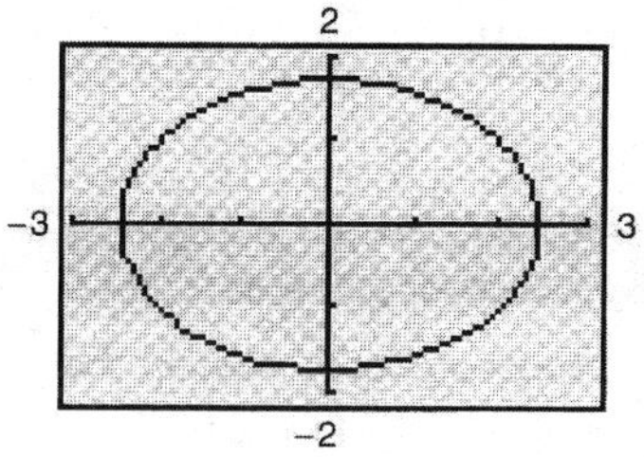

FIGURE A.12

TAKE ANOTHER LOOK

Classifying Differential Equations

In which of the differential equations can the variables be separated?

a. $\dfrac{dy}{dx} = \dfrac{3x}{y}$

b. $\dfrac{dy}{dx} = \dfrac{3x}{y} + 1$

c. $x^2 \dfrac{dy}{dx} = \dfrac{3x}{y}$

d. $\dfrac{dy}{dx} = \dfrac{3x + y}{y}$

PREREQUISITE REVIEW C.2

The following warm-up exercises involve skills that were covered in earlier sections. You will use these skills in the exercise set for this section.

In Exercises 1–6, find the indefinite integral and check your result by differentiating.

1. $\int x^{3/2}\,dx$

2. $\int (t^3 - t^{1/3})\,dt$

3. $\int \frac{2}{x-5}\,dx$

4. $\int \frac{y}{2y^2+1}\,dy$

5. $\int e^{2y}\,dy$

6. $\int xe^{1-x^2}\,dx$

In Exercises 7–10, solve the equation for C or k.

7. $(3)^2 - 6(3) = 1 + C$

8. $(-1)^2 + (-2)^2 = C$

9. $10 = 2e^{2k}$

10. $(6)^2 - 3(6) = e^{-k}$

EXERCISES C.2

In Exercises 1–6, decide whether the variables in the differential equation can be separated.

1. $\frac{dy}{dx} = \frac{x}{y+3}$

2. $\frac{dy}{dx} = \frac{x+1}{x}$

3. $\frac{dy}{dx} = \frac{1}{x} + 1$

4. $\frac{dy}{dx} = \frac{x}{x+y}$

5. $\frac{dy}{dx} = x - y$

6. $x\frac{dy}{dx} = \frac{1}{y}$

In Exercises 7–26, use separation of variables to find the general solution of the differential equation.

7. $\frac{dy}{dx} = 2x$

8. $\frac{dy}{dx} = \frac{1}{x}$

9. $3y^2\frac{dy}{dx} = 1$

10. $\frac{dy}{dx} = x^2y$

11. $(y+1)\frac{dy}{dx} = 2x$

12. $(1+y)\frac{dy}{dx} - 4x = 0$

13. $y' - xy = 0$

14. $y' - y = 5$

15. $\frac{dy}{dt} = \frac{e^t}{4y}$

16. $e^y\frac{dy}{dt} = 3t^2 + 1$

17. $\frac{dy}{dx} = \sqrt{1-y}$

18. $\frac{dy}{dx} = \sqrt{\frac{x}{y}}$

19. $(2+x)y' = 2y$

20. $y' = (2x-1)(y+3)$

21. $xy' = y$

22. $y' - y(x+1) = 0$

23. $y' = \frac{x}{y} - \frac{x}{1+y}$

24. $\frac{dy}{dx} = \frac{x^2+2}{3y^2}$

25. $e^x(y'+1) = 1$

26. $yy' - 2xe^x = 0$

In Exercises 27–32, use the initial condition to find the particular solution of the differential equation.

Differential Equation	*Initial Condition*
27. $yy' - e^x = 0$	$y = 4$ when $x = 0$
28. $\sqrt{x} + \sqrt{y}\,y' = 0$	$y = 4$ when $x = 1$
29. $x(y+4) + y' = 0$	$y = -5$ when $x = 0$
30. $\frac{dy}{dx} = x^2(1+y)$	$y = 3$ when $x = 0$
31. $dP - 6P\,dt = 0$	$P = 5$ when $t = 0$
32. $dT + k(T-70)\,dt = 0$	$T = 140$ when $t = 0$

In Exercises 33 and 34, find an equation for the graph that passes through the point and has the specified slope. Then graph the equation.

33. Point: $(-1, 1)$

Slope: $y' = \dfrac{6x}{5y}$

34. Point: $(8, 2)$

Slope: $y' = \dfrac{2y}{3x}$

Velocity In Exercises 35 and 36, solve the differential equation to find velocity v as a function of time t if $v = 0$ when $t = 0$. The differential equation models the motion of two people on a toboggan after consideration of the force of gravity, friction, and air resistance.

35. $12.5\dfrac{dv}{dt} = 43.2 - 1.25v$

36. $12.5\dfrac{dv}{dt} = 43.2 - 1.75v$

Chemistry: Newton's Law of Cooling In Exercises 37–39, use Newton's Law of Cooling, which states that the rate of change in the temperature T of an object is proportional to the difference between the temperature T of the object and the temperature T_0 of the surrounding environment. This is described by the differential equation $dT/dt = k(T - T_0)$.

37. A steel ingot whose temperature is 1500°F is placed in a room whose temperature is a constant 90°F. One hour later, the temperature of the ingot is 1120°F. What is the ingot's temperature 5 hours after it is placed in the room?

38. A room is kept at a constant temperature of 70°F. An object placed in the room cools from 350°F to 150°F in 45 minutes. How long will it take for the object to cool to a temperature of 80°F?

39. Food at a temperature of 70°F is placed in a freezer that is set at 0°F. After 1 hour, the temperature of the food is 48°F.

(a) Find the temperature of the food after it has been in the freezer 6 hours.

(b) How long will it take the food to cool to a temperature of 10°F?

40. ***Biology: Cell Growth*** The rate of growth of a spherical cell with volume V is proportional to its surface area S. For a sphere, the surface area and volume are related by $S = kV^{2/3}$. So, a model for the cell's growth is

$$\frac{dV}{dt} = kV^{2/3}.$$

Solve this differential equation.

41. ***Learning Theory*** The management of a factory has found that a worker can produce at most 30 units per day. The number of units N per day produced by a new employee will increase at a rate proportional to the difference between 30 and N. This is described by the differential equation

$$\frac{dN}{dt} = k(30 - N)$$

where t is the time in days. Solve this differential equation.

42. ***Sales*** The rate of increase in sales S (in thousands of units) of a product is proportional to the current level of sales and inversely proportional to the square of the time t. This is described by the differential equation

$$\frac{dS}{dt} = \frac{kS}{t^2}$$

where t is the time in years. The saturation point for the market is 50,000 units. That is, the limit of S as $t \to \infty$ is 50. After 1 year, 10,000 units have been sold. Find S as a function of the time t.

43. ***Economics: Pareto's Law*** According to the economist Vilfredo Pareto (1848–1923), the rate of decrease of the number of people y in a stable economy having an income of at least x dollars is directly proportional to the number of such people and inversely proportional to their income x. This is modeled by the differential equation

$$\frac{dy}{dx} = -k\frac{y}{x}.$$

Solve this differential equation.

44. ***Economics: Pareto's Law*** In 2001, 15.2 million people in the United States earned more than \$75,000 and 90.9 million people earned more than \$25,000 (see figure). Assume that Pareto's Law holds and use the result of Exercise 43 to determine the number of people (in millions) who earned (a) more than \$20,000 and (b) more than \$100,000. *(Source: U.S. Census Bureau)*

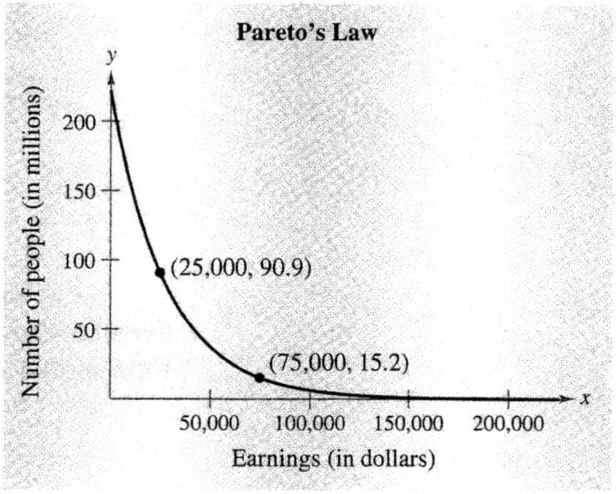

C.3 FIRST-ORDER LINEAR DIFFERENTIAL EQUATIONS

Solve first-order linear differential equations. • Use first-order linear differential equations to model and solve real-life problems.

First-Order Linear Differential Equations

Definition of a First-Order Linear Differential Equation

A **first-order linear differential equation** is an equation of the form

$$y' + P(x)y = Q(x)$$

where P and Q are functions of x. An equation that is written in this form is said to be in **standard form.**

STUDY TIP

The term "first-order" refers to the fact that the highest-order derivative of y in the equation is the first derivative.

To solve a linear differential equation, write it in standard form to identify the functions $P(x)$ and $Q(x)$. Then integrate $P(x)$ and form the expression

$$u(x) = e^{\int P(x)\,dx} \qquad \text{Integrating factor}$$

which is called an **integrating factor.** The general solution of the equation is

$$y = \frac{1}{u(x)}\int Q(x)u(x)\,dx. \qquad \text{General solution}$$

EXAMPLE 1 **Solving a Linear Differential Equation**

Find the general solution of

$$y' + y = e^x.$$

SOLUTION For this equation, $P(x) = 1$ and $Q(x) = e^x$. So, the integrating factor is

$$\begin{aligned} u(x) &= e^{\int dx} \qquad \text{Integrating factor} \\ &= e^x. \end{aligned}$$

This implies that the general solution is

$$\begin{aligned} y &= \frac{1}{e^x}\int e^x(e^x)\,dx \\ &= e^{-x}\left(\frac{1}{2}e^{2x} + C\right) \\ &= \frac{1}{2}e^x + Ce^{-x}. \qquad \text{General solution} \end{aligned}$$

In Example 1, the differential equation was given in standard form. For equations that are not written in standard form, you should first convert to standard form so that you can identify the functions $P(x)$ and $Q(x)$.

EXAMPLE 2 Solving a Linear Differential Equation

Find the general solution of

$$xy' - 2y = x^2.$$

Assume $x > 0$.

SOLUTION Begin by writing the equation in standard form.

$$y' - \left(\frac{2}{x}\right)y = x \qquad \text{Standard form, } y' + P(x)y = Q(x)$$

In this form, you can see that $P(x) = -2/x$ and $Q(x) = x$. So,

$$\begin{aligned}\int P(x)\,dx &= -\int \frac{2}{x}\,dx \\ &= -2\ln x \\ &= -\ln x^2\end{aligned}$$

which implies that the integrating factor is

$$\begin{aligned}u(x) &= e^{\int P(x)\,dx} \\ &= e^{-\ln x^2} \\ &= \frac{1}{x^2}. \qquad \text{Integrating factor}\end{aligned}$$

This implies that the general solution is

$$\begin{aligned}y &= \frac{1}{u(x)}\int Q(x)u(x)\,dx && \text{Form of general solution} \\ &= \frac{1}{1/x^2}\int x\left(\frac{1}{x^2}\right)dx && \text{Substitute.} \\ &= x^2\int \frac{1}{x}\,dx && \text{Simplify.} \\ &= x^2(\ln x + C). && \text{General solution}\end{aligned}$$

DISCOVERY

Solve for y' in the differential equation in Example 2. Use this equation for y' to determine the slopes of y at the points $(1, 0)$ and $(e^{-1/2}, -1/2e)$. Now graph the particular solution $y = x^2 \ln x$ and estimate the slopes at $x = 1$ and $x = e^{-1/2}$. What happens to the slope of y as x approaches zero?

TECHNOLOGY

From Example 2, you can see that it can be difficult to solve a linear differential equation. Fortunately, the task is greatly simplified by symbolic integration utilities. Use a symbolic integration utility to find the particular solution of the differential equation in Example 2, given the initial condition $y = 1$ when $x = 1$.

Guidelines for Solving a Linear Differential Equation

1. Write the equation in standard form
$$y' + P(x)y = Q(x).$$
2. Find the integrating factor
$$u(x) = e^{\int P(x)\,dx}.$$
3. Evaluate the integral below to find the general solution.
$$y = \frac{1}{u(x)}\int Q(x)u(x)\,dx$$

Application

EXAMPLE 3 **Finding a Balance**

You are setting up a "continuous annuity" trust fund. For 20 years, money is continuously transferred from your checking account to the trust fund at the rate of \$1000 per year (about \$2.74 per day). The account earns 8% interest, compounded continuously. What is the balance in the account after 20 years?

SOLUTION Let A represent the balance after t years. The balance increases in two ways: with interest *and* with additional deposits. The rate at which the balance is changing can be modeled by

$$\frac{dA}{dt} = \underbrace{0.08A}_{\text{Interest}} + \underbrace{1000}_{\text{Deposits}}.$$

In standard form, this linear differential equation is

$$\frac{dA}{dt} - 0.08A = 1000 \qquad \text{Standard form}$$

which implies that $P(t) = -0.08$ and $Q(t) = 1000$. The general solution is

$$A = -12{,}500 + Ce^{0.08t}. \qquad \text{General solution}$$

Because $A = 0$ when $t = 0$, you can determine that $C = 12{,}500$. So, the revenue after 20 years is

$$\begin{aligned} A &= -12{,}500 + 12{,}500e^{0.08(20)} \\ &\approx -12{,}500 + 61{,}912.91 \\ &= \$49{,}412.91. \end{aligned}$$

TAKE ANOTHER LOOK

Why an Integrating Factor Works

When both sides of the first-order linear differential equation

$$y' + P(x)y = Q(x)$$

are multiplied by the integrating factor $e^{\int P(x)\,dx}$, you obtain

$$y'e^{\int P(x)\,dx} + P(x)e^{\int P(x)\,dx}y = Q(x)e^{\int P(x)\,dx}.$$

Show that the left side is the derivative of $ye^{\int P(x)\,dx}$, which implies that the general solution is given by

$$ye^{\int P(x)\,dx} = \int Q(x)e^{\int P(x)\,dx}\,dx.$$

PREREQUISITE REVIEW C.3

The following warm-up exercises involve skills that were covered in earlier sections. You will use these skills in the exercise set for this section.

In Exercises 1–4, simplify the expression.

1. $e^{-x}(e^{2x} + e^{x})$

2. $\frac{1}{e^{-x}}(e^{-x} + e^{2x})$

3. $e^{-\ln x^3}$

4. $e^{2\ln x + x}$

In Exercises 5–10, find the indefinite integral.

5. $\int e^{x}(2 + e^{-2x})\,dx$

6. $\int e^{2x}(xe^{x} + 1)\,dx$

7. $\int \frac{1}{2x + 5}\,dx$

8. $\int \frac{x + 1}{x^2 + 2x + 3}\,dx$

9. $\int (4x - 3)^2\,dx$

10. $\int x(1 - x^2)^2\,dx$

EXERCISES C.3

In Exercises 1–6, write the linear differential equation in standard form.

1. $x^3 - 2x^2y' + 3y = 0$

2. $y' - 5(2x - y) = 0$

3. $xy' + y = xe^{x}$

4. $xy' + y = x^3y$

5. $y + 1 = (x - 1)y'$

6. $x = x^2(y' + y)$

In Exercises 7–18, solve the differential equation.

7. $\frac{dy}{dx} + 3y = 6$

8. $\frac{dy}{dx} + 5y = 15$

9. $\frac{dy}{dx} + y = e^{-x}$

10. $\frac{dy}{dx} + 3y = e^{-3x}$

11. $\frac{dy}{dx} + \frac{y}{x} = 3x + 4$

12. $\frac{dy}{dx} + \frac{2y}{x} = 3x + 1$

13. $y' + 5xy = x$

14. $y' + 5y = e^{5x}$

15. $(x - 1)y' + y = x^2 - 1$

16. $xy' + y = x^2 + 1$

17. $x^3y' + 2y = e^{1/x^2}$

18. $xy' + y = x^2 \ln x$

In Exercises 19–22, solve for *y* in two ways.

19. $y' + y = 4$

20. $y' + 10y = 5$

21. $y' - 2xy = 2x$

22. $y' + 4xy = x$

In Exercises 23–26, match the differential equation with its solution.

Differential Equation	*Solution*
23. $y' - 2x = 0$	(a) $y = Ce^{x^2}$
24. $y' - 2y = 0$	(b) $y = -\frac{1}{2} + Ce^{x^2}$
25. $y' - 2xy = 0$	(c) $y = x^2 + C$
26. $y' - 2xy = x$	(d) $y = Ce^{2x}$

In Exercises 27–34, find the particular solution that satisfies the initial condition.

Differential Equation	*Initial Condition*
27. $y' + y = 6e^{x}$	$y = 3$ when $x = 0$
28. $y' + 2y = e^{-2x}$	$y = 4$ when $x = 1$
29. $xy' + y = 0$	$y = 2$ when $x = 2$
30. $y' + y = x$	$y = 4$ when $x = 0$
31. $y' + 3x^2y = 3x^2$	$y = 6$ when $x = 0$
32. $y' + (2x - 1)y = 0$	$y = 2$ when $x = 1$
33. $xy' - 2y = -x^2$	$y = 5$ when $x = 1$
34. $x^2y' - 4xy = 10$	$y = 10$ when $x = 1$

35. ***Sales*** The rate of change (in thousands of units) in sales S is modeled by

$$\frac{dS}{dt} = 0.2(100 - S) + 0.2t$$

where t is the time in years. Solve this differential equation and use the result to complete the table.

t	0	1	2	3	4	5	6	7	8	9	10
S	0										

36. ***Sales*** The rate of change in sales S is modeled by

$$\frac{dS}{dt} = k_1(L - S) + k_2t$$

where t is the time in years and $S = 0$ when $t = 0$. Solve this differential equation for S as a function of t.

Elasticity of Demand In Exercises 37 and 38, find the demand function $p = f(x)$. Recall from Section 3.5 that the price elasticity of demand was defined as $\eta = (p/x)/(dp/dx)$.

37. $\eta = 1 - \dfrac{400}{3x}$, $p = 340$ when $x = 20$

38. $\eta = 1 - \dfrac{500}{3x}$, $p = 2$ when $x = 100$

Supply and Demand In Exercises 39 and 40, use the demand and supply functions to find the price p as a function of time t. Begin by setting $D(t)$ equal to $S(t)$ and solving the resulting differential equation. Find the general solution, and then use the initial condition to find the particular solution.

39. $D(t) = 480 + 5p(t) - 2p'(t)$ Demand function

$S(t) = 300 + 8p(t) + p'(t)$ Supply function

$p(0) = \$75.00$ Initial condition

40. $D(t) = 4000 + 5p(t) - 4p'(t)$ Demand function

$S(t) = 2800 + 7p(t) + 2p'(t)$ Supply function

$p(0) = \$1000.00$ Initial condition

41. ***Investment*** A brokerage firm opens a new real estate investment plan for which the earnings are equivalent to continuous compounding at the rate of r. The firm estimates that deposits from investors will create a net cash flow of Pt dollars, where t is the time in years. The rate of change in the total investment A is modeled by

$$\frac{dA}{dt} = rA + Pt.$$

(a) Solve the differential equation and find the total investment A as a function of t. Assume that $A = 0$ when $t = 0$.

(b) Find the total investment A after 10 years given that $P = \$500{,}000$ and $r = 9\%$.

42. ***Investment*** Let $A(t)$ be the amount in a fund earning interest at the annual rate of r, compounded continuously. If a continuous cash flow of P dollars per year is withdrawn from the fund, then the rate of decrease of A is given by the differential equation

$$\frac{dA}{dt} = rA - P$$

where $A = A_0$ when $t = 0$.

(a) Solve this equation for A as a function of t.

(b) Use the result of part (a) to find A when $A_0 =$ \$2,000,000, $r = 7\%$, $P = \$250{,}000$, and $t = 5$ years.

(c) Find A_0 if a retired person wants a continuous cash flow of \$40,000 per year for 20 years. Assume that the person's investment will earn 8%, compounded continuously.

43. ***Velocity*** A booster rocket carrying an observation satellite is launched into space. The rocket and satellite have mass m and are subject to air resistance proportional to the velocity v at any time t. A differential equation that models the velocity of the rocket and satellite is

$$m\frac{dv}{dt} = -mg - kv$$

where g is the acceleration due to gravity. Solve the differential equation for v as a function of t.

44. ***Health*** An infectious disease spreads through a large population according to the model

$$\frac{dy}{dt} = \frac{1 - y}{4}$$

where y is the percent of the population exposed to the disease, and t is the time in years.

(a) Solve this differential equation, assuming $y(0) = 0$.

(b) Find the number of years it takes for half of the population to have been exposed to the disease.

(c) Find the percentage of the population that has been exposed to the disease after 4 years.

45. ***Research Project*** Use your school's library, the Internet, or some other reference source to find an article in a scientific or business journal that uses a differential equation to model a real-life situation. Write a short paper describing the situation. If possible, describe the solution of the differential equation.

C.4 APPLICATIONS OF DIFFERENTIAL EQUATIONS

Use differential equations to model and solve real-life problems.

EXAMPLE 1 Modeling Advertising Awareness

The new cereal product from Example 3 in Section C.1 is introduced through an advertising campaign to a population of 1 million potential customers. The rate at which the population hears about the product is assumed to be proportional to the number of people who are not yet aware of the product. By the end of 1 year, half of the population has heard of the product. How many will have heard of it by the end of 2 years?

SOLUTION Let y be the number (in millions) of people at time t who have heard of the product. This means that $(1 - y)$ is the number of people who have not heard of it, and dy/dt is the rate at which the population hears about the product. From the given assumption, you can write the differential equation as shown.

$$\frac{dy}{dt} = k(1 - y)$$

Rate of change of y — is proportional to — the difference between 1 and y.

Using separation of variables *or* a symbolic integration utility, you can find the general solution to be

$$y = 1 - Ce^{-kt}. \qquad \text{General solution}$$

To solve for the constants C and k, use the initial conditions. That is, because $y = 0$ when $t = 0$, you can determine that $C = 1$. Similarly, because $y = 0.5$ when $t = 1$, it follows that $0.5 = 1 - e^{-k}$, which implies that

$$k = \ln 2 \approx 0.693.$$

So, the particular solution is

$$y = 1 - e^{-0.693t}. \qquad \text{Particular solution}$$

This model is shown graphically in Figure A.13. Using the model, you can determine that the number of people who have heard of the product after 2 years is

$$y = 1 - e^{-0.693(2)}$$
$$\approx 0.75 \text{ or } 750{,}000 \text{ people.}$$

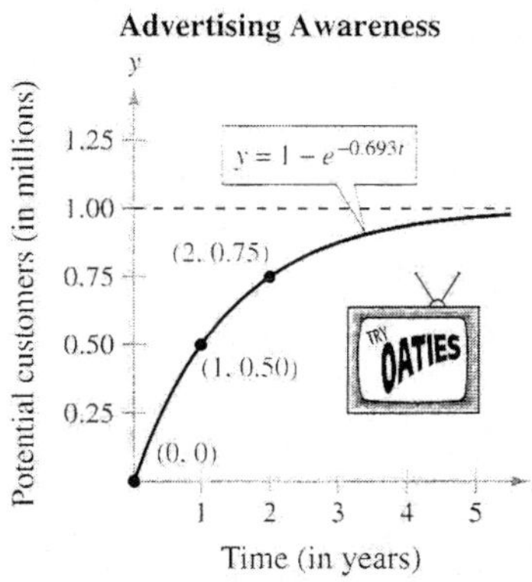

FIGURE A.13

EXAMPLE 2 Modeling a Chemical Reaction

During a chemical reaction, substance A is converted into substance B at a rate that is proportional to the square of the amount of A. When $t = 0$, 60 grams of A is present, and after 1 hour ($t = 1$), only 10 grams of A remains unconverted. How much of A is present after 2 hours?

SOLUTION Let y be the unconverted amount of substance A at any time t. From the given assumption about the conversion rate, you can write the differential equation as shown.

$$\frac{dy}{dt} = ky^2$$

Rate of change of y / is proportional to / the square of y.

Using separation of variables *or* a symbolic integration utility, you can find the general solution to be

$$y = \frac{-1}{kt + C}.$$ General solution

To solve for the constants C and k, use the initial conditions. That is, because $y = 60$ when $t = 0$, you can determine that $C = -\frac{1}{60}$. Similarly, because $y = 10$ when $t = 1$, it follows that

$$10 = \frac{-1}{k - (1/60)}$$

which implies that $k = -\frac{1}{12}$. So, the particular solution is

$$y = \frac{-1}{(-1/12)t - (1/60)}$$ Substitute for k and C.

$$= \frac{60}{5t + 1}.$$ Particular solution

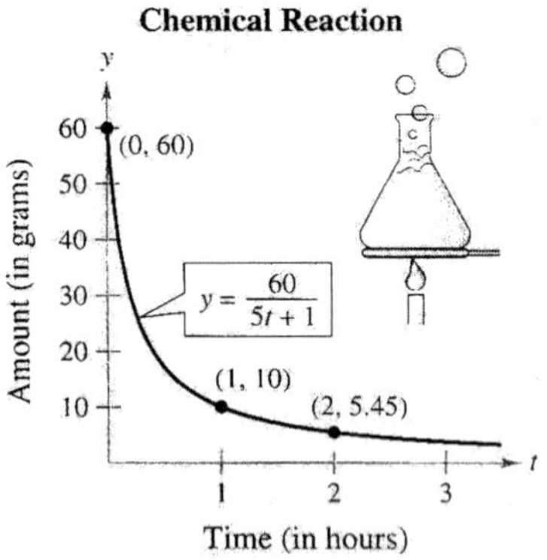

FIGURE A.14

Using the model, you can determine that the unconverted amount of substance A after 2 hours is

$$y = \frac{60}{5(2) + 1}$$

$$\approx 5.45 \text{ grams}.$$

In Figure A.14, note that the chemical conversion is occurring rapidly during the first hour. Then, as more and more of substance A is converted, the conversion rate slows down.

STUDY TIP

In Example 2, the rate of conversion was assumed to be proportional to the *square* of the unconverted amount. How would the result change if the rate of conversion were assumed to be proportional to the unconverted amount?

Earlier in the text, you studied two models for population growth: *exponential growth*, which assumes that the rate of change of y is proportional to y, and *logistic growth*, which assumes that the rate of change of y is proportional to y and $L - y$, where L is the population limit.

The next example describes a third type of growth model called a **Gompertz growth model.** This model assumes that the rate of change of y is proportional to y and the natural log of L/y, where L is the population limit.

EXAMPLE 3 Modeling Population Growth

A population of 20 wolves has been introduced into a national park. The forest service estimates that the maximum population the park can sustain is 200 wolves. After 3 years, the population is estimated to be 40 wolves. If the population follows a Gompertz growth model, how many wolves will there be 10 years after their introduction?

During the second half of the twentieth century, wolves disappeared from most of the middle and northern areas of the United States. Recently, however, wolf populations have been reappearing in several northern national parks.

SOLUTION Let y be the number of wolves at any time t. From the given assumption about the rate of growth of the population, you can write the differential equation as shown.

$$\frac{dy}{dt} = ky \ln \frac{200}{y}$$

Rate of change of y — is proportional to — the product of y and — the log of the ratio of 200 and y.

Using separation of variables *or* a symbolic integration utility, you can find the general solution to be

$$y = 200e^{-Ce^{-kt}}. \qquad \text{General solution}$$

To solve for the constants C and k, use the initial conditions. That is, because $y = 20$ when $t = 0$, you can determine that

$$C = \ln 10$$
$$\approx 2.3026.$$

Similarly, because $y = 40$ when $t = 3$, it follows that

$$40 = 200e^{-2.3026e^{-3k}}$$

which implies that $k \approx 0.1194$. So, the particular solution is

$$y = 200e^{-2.3026e^{-0.1194t}}. \qquad \text{Particular solution}$$

Using the model, you can estimate the wolf population after 10 years to be

$$y = 200e^{-2.3026e^{-0.1194(10)}}$$
$$\approx 100 \text{ wolves}.$$

In Figure A.15, note that after 10 years the population has reached about half of the estimated maximum population. Try checking the growth model to see that it yields $y = 20$ when $t = 0$ and $y = 40$ when $t = 3$.

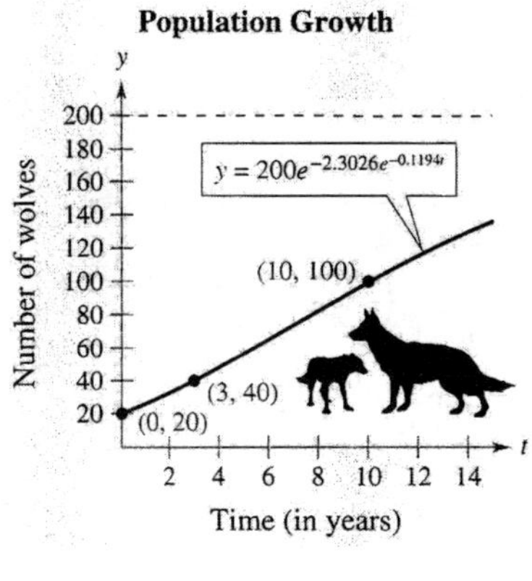

FIGURE A.15

In genetics, a commonly used hybrid selection model is based on the differential equation

$$\frac{dy}{dt} = ky(1 - y)(a - by).$$

In this model, y represents the portion of the population that has a certain characteristic and t represents the time (measured in generations). The numbers a, b, and k are constants that depend on the genetic characteristic that is being studied.

EXAMPLE 4 **Modeling Hybrid Selection**

You are studying a population of beetles to determine how quickly characteristic D will pass from one generation to the next. At the beginning of your study ($t = 0$), you find that half the population has characteristic D. After four generations ($t = 4$), you find that 80% of the population has characteristic D. Use the hybrid selection model above with $a = 2$ and $b = 1$ to find the percent of the population that will have characteristic D after 10 generations.

SOLUTION Using $a = 2$ and $b = 1$, the differential equation for the hybrid selection model is

$$\frac{dy}{dt} = ky(1 - y)(2 - y).$$

Using separation of variables *or* a symbolic integration utility, you can find the general solution to be

$$\frac{y(2 - y)}{(1 - y)^2} = Ce^{2kt}. \qquad \text{General solution}$$

To solve for the constants C and k, use the initial conditions. That is, because $y = 0.5$ when $t = 0$, you can determine that $C = 3$. Similarly, because $y = 0.8$ when $t = 4$, it follows that

$$\frac{0.8(1.2)}{(0.2)^2} = 3e^{8k}$$

which implies that

$$k = \frac{1}{8}\ln 8 \approx 0.2599.$$

So, the particular solution is

$$\frac{y(2 - y)}{(1 - y)^2} = 3e^{0.5199t}. \qquad \text{Particular solution}$$

Using the model, you can estimate the percent of the population that will have characteristic D after 10 generations to be given by

$$\frac{y(2 - y)}{(1 - y)^2} = 3e^{0.5199(10)}.$$

Using a symbolic algebra utility, you can solve this equation for y to obtain $y \approx 0.96$. The graph of the model is shown in Figure A.16.

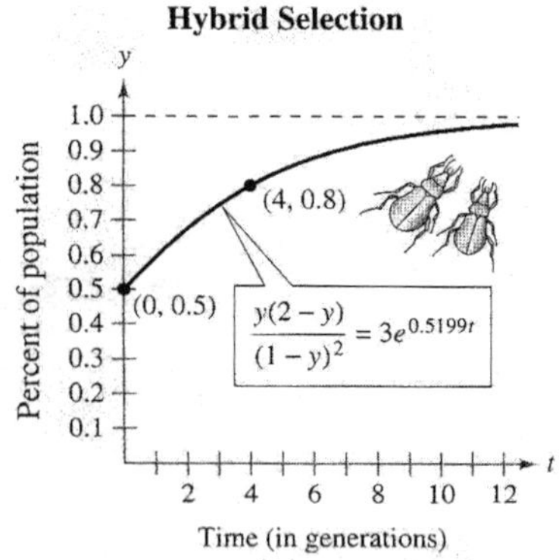

FIGURE A.16

EXAMPLE 5 Modeling a Chemical Mixture

A tank contains 40 gallons of a solution composed of 90% water and 10% alcohol. A second solution containing half water and half alcohol is added to the tank at the rate of 4 gallons per minute. At the same time, the tank is being drained at the rate of 4 gallons per minute, as shown in Figure A.17. Assuming that the solution is stirred constantly, how much alcohol will be in the tank after 10 minutes?

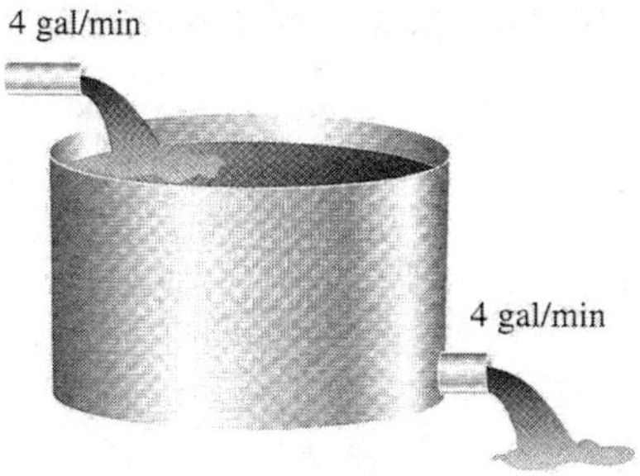

FIGURE A.17

SOLUTION Let y be the number of gallons of alcohol in the tank at any time t. The *percent* of alcohol in the 40-gallon tank at any time is $y/40$. Moreover, because 4 gallons of solution is being drained each minute, the rate of change of y is

$$\frac{dy}{dt} = -4\left(\frac{y}{40}\right) + 2$$

Rate of change of y — is equal to the amount of alcohol draining out — plus the amount of alcohol entering.

where 2 represents the number of gallons of alcohol entering each minute in the 50% solution. In standard form, this linear differential equation is

$$y' + \frac{1}{10}y = 2. \qquad \text{Standard form}$$

Using an integrating factor *or* a symbolic integration utility, you can find the general solution to be

$$y = 20 + Ce^{-t/10}. \qquad \text{General solution}$$

Because $y = 4$ when $t = 0$, you can conclude that $C = -16$. So, the particular solution is

$$y = 20 - 16e^{-t/10}. \qquad \text{Particular solution}$$

Using this model, you can determine that the amount of alcohol in the tank when $t = 10$ is

$$y = 20 - 16e^{-(10)/10}$$
$$\approx 14.1 \text{ gallons.}$$

TAKE ANOTHER LOOK

Chemical Mixture

Sketch the particular solution obtained in Example 5. Describe the rate of change of the amount of alcohol in the tank. Does the amount approach 0 as t increases? Explain your reasoning.

PREREQUISITE REVIEW C.4

The following warm-up exercises involve skills that were covered in earlier sections. You will use these skills in the exercise set for this section.

In Exercises 1–4, use separation of variables to find the general solution of the differential equation.

1. $\dfrac{dy}{dx} = 3x$

2. $2y\dfrac{dy}{dx} = 3$

3. $\dfrac{dy}{dx} = 2xy$

4. $\dfrac{dy}{dx} = \dfrac{x - 4}{4y^3}$

In Exercises 5–8, use an integrating factor to solve the first-order linear differential equation.

5. $y' + 2y = 4$

6. $y' + 2y = e^{-2x}$

7. $y' + xy = x$

8. $xy' + 2y = x^2$

In Exercises 9 and 10, write the equation that models the statement.

9. The rate of change of y with respect to x is proportional to the square of x.

10. The rate of change of x with respect to t is proportional to the difference of x and t.

EXERCISES C.4

In Exercises 1–6, assume that the rate of change of y is proportional to y. Solve the resulting differential equation $dy/dx = ky$ and find the particular solution that passes through the points.

1. $(0, 1), (3, 2)$

2. $(0, 4), (1, 6)$

3. $(0, 4), (4, 1)$

4. $(0, 60), (5, 30)$

5. $(2, 2), (3, 4)$

6. $(1, 4), (2, 1)$

7. ***Investment*** The rate of growth of an investment is proportional to the amount A of the investment at any time t. An investment of \$2000 increases to a value of \$2983.65 in 5 years. Find its value after 10 years.

8. ***Population Growth*** The rate of change of the population of a city is proportional to the population P at any time t. In 1998, the population was 400,000, and the constant of proportionality was 0.015. Estimate the population of the city in the year 2005.

9. ***Sales Growth*** The rate of change in sales S (in thousands of units) of a new product is proportional to the difference between L and S (in thousands of units) at any time t. When $t = 0$, $S = 0$. Write and solve the differential equation for this sales model.

10. ***Sales Growth*** Use the result of Exercise 9 to write S as a function of t if (a) $L = 100$, $S = 25$ when $t = 2$, and (b) $L = 500$, $S = 50$ when $t = 1$.

In Exercises 11–14, the rate of change of y is proportional to the product of y and the difference of L and y. Solve the resulting differential equation $dy/dx = ky(L - y)$ and find the particular solution that passes through the points for the given value of L.

11. $L = 20$; $(0, 1), (5, 10)$

12. $L = 100$; $(0, 10), (5, 30)$

13. $L = 5000$; $(0, 250), (25, 2000)$

14. $L = 1000$; $(0, 100), (4, 750)$

15. ***Biology*** At any time t, the rate of growth of the population N of deer in a state park is proportional to the product of N and $L - N$, where $L = 500$ is the maximum number of deer the park can maintain. When $t = 0$, $N = 100$, and when $t = 4$, $N = 200$. Write N as a function of t.

16. ***Sales Growth*** The rate of change in sales S (in thousands of units) of a new product is proportional to the product of S and $L - S$. L (in thousands of units) is the estimated maximum level of sales, and $S = 10$ when $t = 0$. Write and solve the differential equation for this sales model.

Learning Theory In Exercises 17 and 18, assume that the rate of change in the proportion P of correct responses after n trials is proportional to the product of P and $L - P$, where L is the limiting proportion of correct responses.

17. Write and solve the differential equation for this learning theory model.

18. Use the solution of Exercise 17 to write P as a function of n, and then use a graphing utility to graph the solution.

(a) $L = 1.00$

$P = 0.50$ when $n = 0$

$P = 0.85$ when $n = 4$

(b) $L = 0.80$

$P = 0.25$ when $n = 0$

$P = 0.60$ when $n = 10$

Chemical Reaction In Exercises 19 and 20, use the chemical reaction model in Example 2 to find the amount y as a function of t, and use a graphing utility to graph the function.

19. $y = 45$ grams when $t = 0$; $y = 4$ grams when $t = 2$

20. $y = 75$ grams when $t = 0$; $y = 12$ grams when $t = 1$

In Exercises 21 and 22, use the Gompertz growth model described in Example 3 to find the growth function, and sketch its graph.

21. $L = 500$; $y = 100$ when $t = 0$; $y = 150$ when $t = 2$

22. $L = 5000$; $y = 500$ when $t = 0$; $y = 625$ when $t = 1$

23. ***Biology*** A population of eight beavers has been introduced into a new wetlands area. Biologists estimate that the maximum population the wetlands can sustain is 60 beavers. After 3 years, the population is 15 beavers. If the population follows a Gompertz growth model, how many beavers will be in the wetlands after 10 years?

24. ***Biology*** A population of 30 rabbits has been introduced into a new region. It is estimated that the maximum population the region can sustain is 400 rabbits. After 1 year, the population is estimated to be 90 rabbits. If the population follows a Gompertz growth model, how many rabbits will be present after 3 years?

Biology In Exercises 25 and 26, use the hybrid selection model in Example 4 to find the percent of the population that has the indicated characteristic.

25. You are studying a population of mayflies to determine how quickly characteristic A will pass from one generation to the next. At the start of the study, half the population has characteristic A. After four generations, 75% of the population has characteristic A. Find the percent of the population that will have characteristic A after 10 generations. (Assume $a = 2$ and $b = 1$.)

26. A research team is studying a population of snails to determine how quickly characteristic B will pass from one generation to the next. At the start of the study, 40% of the snails have characteristic B. After five generations, 80% of the population has characteristic B. Find the percent of the population that will have characteristic B after eight generations. (Assume $a = 2$ and $b = 1$.)

27. ***Chemical Reaction*** In a chemical reaction, a compound changes into another compound at a rate proportional to the unchanged amount, according to the model

$$\frac{dy}{dt} = ky.$$

(a) Solve the differential equation.

(b) If the initial amount of the original compound is 20 grams, and the amount remaining after 1 hour is 16 grams, when will 75% of the compound have been changed?

28. ***Chemical Mixture*** A 100-gallon tank is full of a solution containing 25 pounds of a concentrate. Starting at time $t = 0$, distilled water is admitted to the tank at the rate of 5 gallons per minute, and the well-stirred solution is withdrawn at the same rate.

(a) Find the amount Q of the concentrate in the solution as a function of t. (*Hint:* $Q' + Q/20 = 0$)

(b) Find the time when the amount of concentrate in the tank reaches 15 pounds.

29. ***Chemical Mixture*** A 200-gallon tank is half full of distilled water. At time $t = 0$, a solution containing 0.5 pound of concentrate per gallon enters the tank at the rate of 5 gallons per minute, and the well-stirred mixture is withdrawn at the same rate. Find the amount Q of concentrate in the tank after 30 minutes. (*Hint:* $Q' + Q/20 = \frac{5}{2}$)

30. ***Safety*** Assume that the rate of change in the number of miles s of road cleared per hour by a snowplow is inversely proportional to the depth h of snow. That is,

$$\frac{ds}{dh} = \frac{k}{h}.$$

Find s as a function of h if $s = 25$ miles when $h = 2$ inches and $s = 12$ miles when $h = 6$ inches $(2 \le h \le 15)$.

31. ***Chemistry*** A wet towel hung from a clothesline to dry loses moisture through evaporation at a rate proportional to its moisture content. If after 1 hour the towel has lost 40% of its original moisture content, after how long will it have lost 80%?

32. ***Biology*** Let x and y be the sizes of two internal organs of a particular mammal at time t. Empirical data indicate that the relative growth rates of these two organs are equal, and can be modeled by

$$\frac{1}{x}\frac{dx}{dt} = \frac{1}{y}\frac{dy}{dt}.$$

Use this differential equation to write y as a function of x.

33. ***Population Growth*** When predicting population growth, demographers must consider birth and death rates as well as the net change caused by the difference between the rates of immigration and emigration. Let P be the population at time t and let N be the net increase per unit time due to the difference between immigration and emigration. So, the rate of growth of the population is given by

$$\frac{dP}{dt} = kP + N, \quad N \text{ is constant.}$$

Solve this differential equation to find P as a function of time.

34. ***Meteorology*** The barometric pressure y (in inches of mercury) at an altitude of x miles above sea level decreases at a rate proportional to the current pressure according to the model

$$\frac{dy}{dx} = -0.2y$$

where $y = 29.92$ inches when $x = 0$. Find the barometric pressure (a) at the top of Mt. St. Helens (8364 feet) and (b) at the top of Mt. McKinley (20,320 feet).

35. ***Investment*** A large corporation starts at time $t = 0$ to invest part of its receipts at a rate of P dollars per year in a fund for future corporate expansion. Assume that the fund earns r percent interest per year compounded continuously. So, the rate of growth of the amount A in the fund is given by

$$\frac{dA}{dt} = rA + P$$

where $A = 0$ when $t = 0$. Solve this differential equation for A as a function of t.

Investment In Exercises 36–38, use the result of Exercise 35.

36. Find A for each situation.

(a) $P = \$100{,}000$, $r = 12\%$, and $t = 5$ years

(b) $P = \$250{,}000$, $r = 15\%$, and $t = 10$ years

37. Find P if the corporation needs \$120,000,000 in 8 years and the fund earns $16\frac{1}{4}\%$ interest compounded continuously.

38. Find t if the corporation needs \$800,000 and it can invest \$75,000 per year in a fund earning 13% interest compounded continuously.

Medical Science In Exercises 39–41, a medical researcher wants to determine the concentration C (in moles per liter) of a tracer drug injected into a moving fluid. Solve this problem by considering a single-compartment dilution model (see figure). Assume that the fluid is continuously mixed and that the volume of fluid in the compartment is constant.

Figure for 39–41

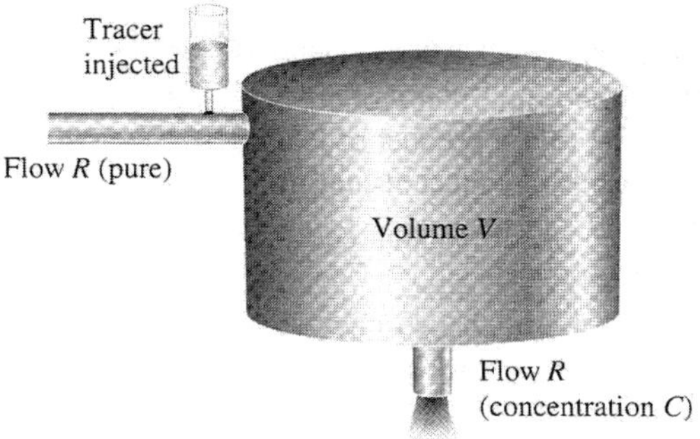

39. If the tracer is injected instantaneously at time $t = 0$, then the concentration of the fluid in the compartment begins diluting according to the differential equation

$$\frac{dC}{dt} = \left(-\frac{R}{V}\right)C, \quad C = C_0 \text{ when } t = 0.$$

(a) Solve this differential equation to find the concentration as a function of time.

(b) Find the limit of C as $t \to \infty$.

40. Use the solution of the differential equation in Exercise 39 to find the concentration as a function of time, and use a graphing utility to graph the function.

(a) $V = 2$ liters, $R = 0.5$ L/min, and $C_0 = 0.6$ mol/L

(b) $V = 2$ liters, $R = 1.5$ L/min, and $C_0 = 0.6$ mol/L

41. In Exercises 39 and 40, it was assumed that there was a single initial injection of the tracer drug into the compartment. Now consider the case in which the tracer is continuously injected (beginning at $t = 0$) at the rate of Q mol/min. Considering Q to be negligible compared with R, use the differential equation

$$\frac{dC}{dt} = \frac{Q}{V} - \left(\frac{R}{V}\right)C, \quad C = 0 \text{ when } t = 0.$$

(a) Solve this differential equation to find the concentration as a function of time.

(b) Find the limit of C as $t \to \infty$.

APPENDIX A (page A10)

1. Left Riemann sum: 0.518
   Right Riemann sum: 0.768
3. Left Riemann sum: 0.746
   Right Riemann sum: 0.646
5. Left Riemann sum: 0.859
   Right Riemann sum: 0.659
7. Midpoint Rule: 0.673
9. (a)

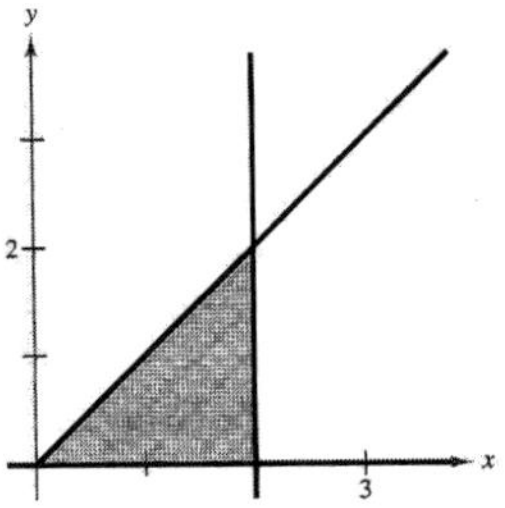

(b) Answers will vary.
(c) Answers will vary.
(d) Answers will vary.

(e)

n	5	10	50	100
Left sum, S_L	1.6	1.8	1.96	1.98
Right sum, S_R	2.4	2.2	2.04	2.02

(f) Answers will vary.

11. $\int_0^5 3\,dx$

13. $\int_{-4}^4 (4 - |x|)\,dx = \int_{-4}^0 (4 + x)\,dx + \int_0^4 (4 - x)\,dx$

15. $\int_{-2}^2 (4 - x^2)\,dx$ 17. $\int_0^2 \sqrt{x + 1}\,dx$

19.

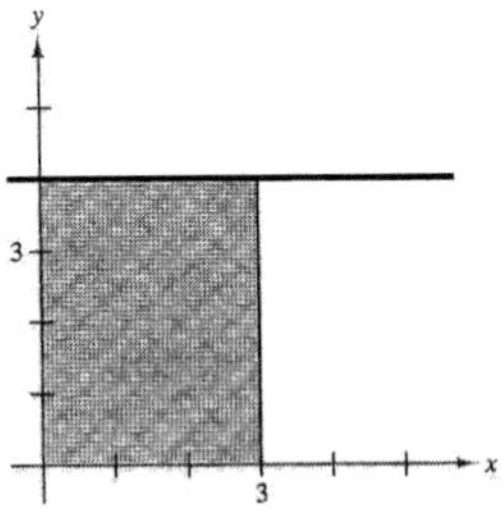

Rectangle

$A = 12$

21.

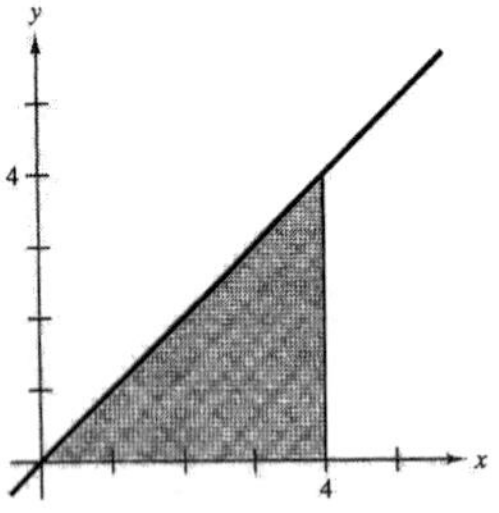

Triangle

$A = 8$.

23.

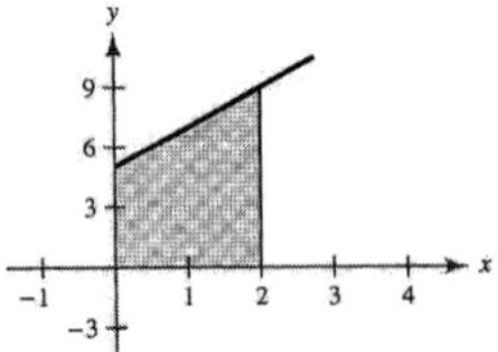

Trapezoid

$A = 14$

25.

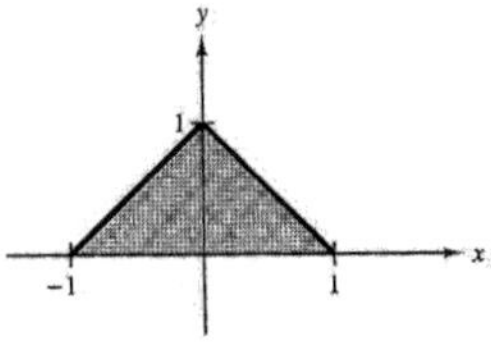

Triangle

$A = 1$

27.

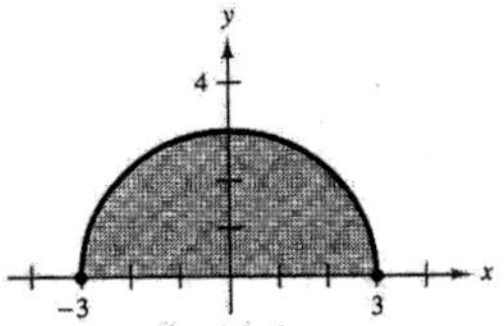

Semicircle

$A = \frac{9\pi}{2}$

29. Answers will vary. 31. >

APPENDIX C

Section C.1 (page A24)

Prerequisite Review

1. $y' = 6x + 2$
   $y'' = 6$
2. $y' = -6x^2 - 8$
   $y'' = -12x$
3. $y' = -6e^{2x}$
   $y'' = -12e^{2x}$
4. $y' = -6xe^{x^2}$
   $y'' = -6e^{x^2}(2x^2 + 1)$
5. $\frac{1 - x}{y}$ 6. $\frac{2}{3y^2 + 4}$ 7. $-\frac{y}{2x}$
8. $-\frac{y}{x}$ 9. $k = 2 \ln 3 - \ln \frac{17}{2} \approx 0.0572$
10. $k = \ln 10 - \frac{\ln 41}{2} \approx 0.4458$

1. $y = 3x^2$
3. $y' = -2e^{-2x}$ and $y' + 2y = -2e^{-2x} + 2(e^{-2x}) = 0$
5. $y' = 6x^2$ and $y' - \frac{3}{x}y = 6x^2 - \frac{3}{x}(2x^3) = 0$
7. $y'' = 2$ and $x^2y'' - 2y = x^2(2) - 2(x^2) = 0$
9. $y' = 4e^{2x}$, $y'' = 8e^{2x}$, and
   $y'' - y' - 2y = 8e^{2x} - 4e^{2x} - 2(2e^{2x}) = 0$
11. $\frac{dy}{dx} = -\frac{1}{x^2}$ 13. $\frac{dy}{dx} = 4Ce^{4x} = 4y$
15. $\frac{dy}{dt} = -\frac{1}{3}Ce^{-t/3}$ and
    $3\frac{dy}{dt} + y - 7 = 3\left(-\frac{1}{3}Ce^{-t/3}\right) + (Ce^{-t/3} + 7) - 7 = 0$
17. $xy' - 3x - 2y = x(2Cx - 3) - 3x - 2(Cx^2 - 3x) = 0$
19. $xy' + y = x\left(2x + 2 - \frac{C}{x^2}\right) + \left(x^2 + 2x + \frac{C}{x}\right)$
    $= x(3x + 4)$

21. $2y'' + 3y' - 2y = 2\left(\frac{1}{4}C_1e^{x/2} + 4C_2e^{-2x}\right)$
$+ 3\left(\frac{1}{2}C_1e^{x/2} - 2C_2e^{-2x}\right)$
$- 2(C_1e^{x/2} + C_2e^{-2x}) = 0$

23. $y' - \dfrac{ay}{x} = \left(\dfrac{4bx^3}{4-a} + aCx^{a-1}\right) - \dfrac{a}{x}\left(\dfrac{bx^4}{4-a} + Cx^a\right)$
$= bx^3$

25. $y' + 2xy = -\dfrac{4Cxe^{x^2}}{(1 - Ce^{x^2})^2} + 2x\left(\dfrac{2}{1 + Ce^{x^2}}\right) = xy^2$

27. $y' = \ln x + 1 + C$ and
$x(y' - 1) - (y - 4) = x(\ln x + 1 + C - 1)$
$- (x \ln x + Cx + 4 - 4) = 0$

29. $2x + 2yy' = Cy'$
$y' = \dfrac{2x}{C - 2y} = \dfrac{2xy}{Cy - 2y^2}$
$= \dfrac{2xy}{(x^2 + y^2) - 2y^2} = \dfrac{2xy}{x^2 - y^2}$

31. $x + y = \dfrac{C}{x}$
$y'' = \dfrac{2C}{x^3}$
$x^2y'' - 2(x + y) = \dfrac{2C}{x} - \dfrac{2C}{x} = 0$

33. Solution **35.** Not a solution **37.** Not a solution

39. Solution **41.** $y = 3e^{-2x}$ **43.** $y = 5 + \ln\sqrt{|x|}$

45. $y = 3e^{4x} + 2e^{-3x}$ **47.** $y = \frac{4}{3}(3 - x)e^{2x/3}$

49.

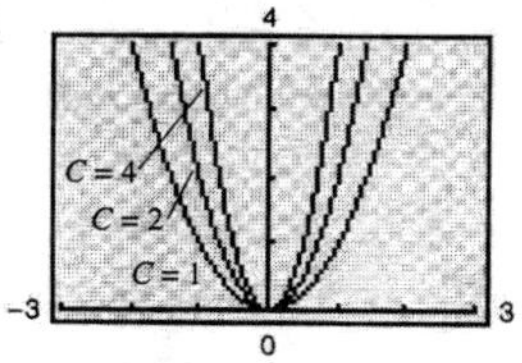

51.

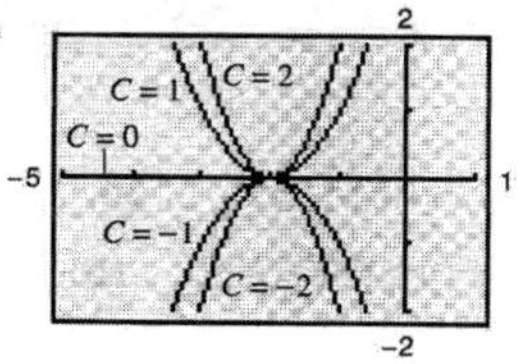

53. $y = x^3 + C$ **55.** $y = x + 3\ln|x| + C$

57. $y = \dfrac{1}{2}\ln\left(\dfrac{x - 1}{x + 1}\right) + C$

59. $y = \frac{2}{5}(x - 3)^{3/2}(x + 2) + C$

61. $y^2 = \frac{1}{4}x^3$ **63.** $y = 3e^x$

65. (a) $N = 750 - 650e^{-0.0484t}$

(b)

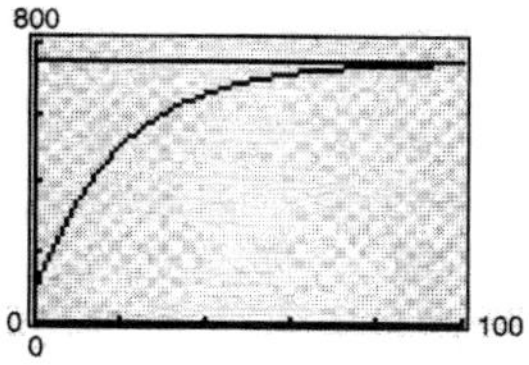

(c) $N \approx 214$

67.

Year, t	2	4	6	8	10
Units, x	3867	7235	10,169	12,725	14,951

69. Because

$$\frac{ds}{dh} = -\frac{13}{\ln 3}\left(\frac{1/2}{h/2}\right) = -\frac{13}{\ln 3}\frac{1}{h}, \text{ and } -\frac{13}{\ln 3}$$

is a constant, we can conclude that the equation is a solution of $ds/dh = k/h$ where $k = -13/(\ln 3)$.

71. $k = 0.07$

73. False. From Example 1, $y = e^x$ is a solution of $y'' - y = 0$, but $y = e^x + 1$ is not.

Section C.2 *(page A32)*

Prerequisite Review

1. $\frac{2}{5}x^{5/2} + C$ **2.** $\frac{1}{4}t^4 - \frac{3}{4}t^{4/3} + C$

3. $2\ln|x - 5| + C$ **4.** $\frac{1}{4}\ln|2y^2 + 1| + C$

5. $\frac{1}{2}e^{2y} + C$ **6.** $-\frac{1}{2}e^{1-x^2} + C$ **7.** $C = -10$

8. $C = 5$ **9.** $k = \dfrac{\ln 5}{2} \approx 0.8047$

10. $k = -2\ln 3 - \ln 2 \approx -2.8904$

1. Yes
$(y + 3)\,dy = x\,dx$

3. Yes
$dy = \left(\dfrac{1}{x} + 1\right)dx$

5. No. The variables cannot be separated.

7. $y = x^2 + C$ **9.** $y = \sqrt[3]{x + C}$

11. $C = 2x^2 - (y + 1)^2$ **13.** $y = Ce^{x^2/2}$

15. $y^2 = \frac{1}{2}e^t + C$

17. $y = 1 - \left(C - \dfrac{x}{2}\right)^2$

19. $y = C(2 + x)^2$ **21.** $y = Cx$

23. $3y^2 + 2y^3 = 3x^2 + C$ **25.** $y = -e^{-x} - x + C$

27. $y^2 = 2e^x + 14$ **29.** $y = -4 - e^{-x^2/2}$

31. $P = 5e^{6t}$

33. $5y^2 = 6x^2 - 1$ or $6x^2 - 5y^2 = 1$

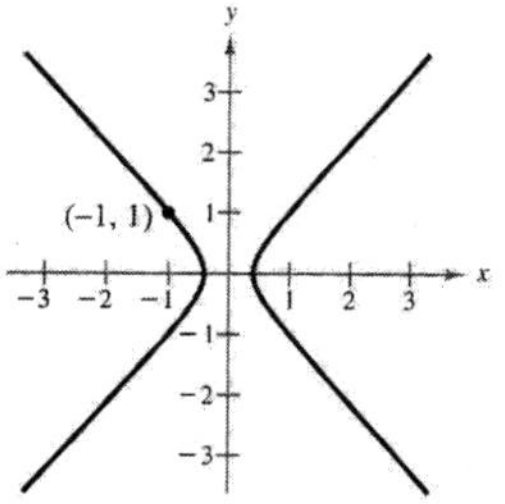

35. $v = 34.56(1 - e^{-0.1t})$ **37.** $T \approx 383.298°\text{F}$

39. (a) $T \approx 7.277°\text{F}$ (b) $t \approx 5.158$ hours

41. $N = 30 + Ce^{-kt}$ **43.** $y = Cx^{-k}$

Section C.3 *(page A37)*

Prerequisite Review

1. $e^x + 1$ **2.** $e^{3x} + 1$ **3.** $\dfrac{1}{x^3}$ **4.** x^2e^x

5. $2e^x - e^{-x} + C$ **6.** $e^{3x}\left(\dfrac{x}{3} - \dfrac{1}{9}\right) + \dfrac{1}{2}e^{2x} + C$

7. $\frac{1}{2}\ln|2x + 5| + C$ **8.** $\frac{1}{2}\ln|x^2 + 2x + 3| + C$

9. $\frac{1}{12}(4x - 3)^3 + C$ **10.** $\frac{1}{6}(x^2 - 1)^3 + C$

1. $y' + \dfrac{-3}{2x^2}y = \dfrac{x}{2}$ **3.** $y' + \dfrac{1}{x}y = e^x$

5. $y' + \dfrac{1}{1 - x}y = \dfrac{1}{x - 1}$ **7.** $y = 2 + Ce^{-3x}$

9. $y = e^{-x}(x + C)$ **11.** $y = x^2 + 2x + \dfrac{C}{x}$

13. $y = \dfrac{1}{5} + Ce^{-(5/2)x^2}$ **15.** $y = \dfrac{x^3 - 3x + C}{3(x - 1)}$

17. $y = e^{1/x^2}\left(-\dfrac{1}{2x^2} + C\right)$ **19.** $y = Ce^{-x} + 4$

21. $y = Ce^{x^2} - 1$

23. c **24.** d **25.** a **26.** b

27. $y = 3e^x$ **29.** $xy = 4$

31. $y = 1 + 5e^{-x^3}$ **33.** $y = x^2(5 - \ln|x|)$

35. $S = t + 95(1 - e^{-t/5})$

t	0	1	2	3	4	5
S	0	18.22	33.32	45.86	56.31	65.05

t	6	7	8	9	10
S	72.39	78.57	83.82	88.30	92.14

37. $p = 400 - 3x$ **39.** $p = 15(4 + e^{-t})$

41. (a) $A = \dfrac{P}{r^2}(rt - 1 + e^{-rt})$ (b) $A \approx \$18{,}924{,}053.07$

43. $v = -\dfrac{gm}{k} + Ce^{-kt/m}$ **45.** Answers will vary.

Section C.4 *(page A44)*

Prerequisite Review

1. $y = \frac{3}{2}x^2 + C$ **2.** $y^2 = 3x + C$

3. $y = Ce^{x^2}$ **4.** $y^4 = \frac{1}{2}(x - 4)^2 + C$

5. $y = 2 + Ce^{-2x}$ **6.** $y = xe^{-2x} + Ce^{-2x}$

7. $y = 1 + Ce^{-x^2/2}$ **8.** $y = \frac{1}{4}x^2 + Cx^{-2}$

9. $\dfrac{dy}{dx} = Cx^2$ **10.** $\dfrac{dx}{dt} = C(x - t)$

1. $y = e^{(x\ln 2)/3} \approx e^{0.2310x}$ **3.** $y = 4e^{-(x\ln 4)/4} \approx 4e^{-0.3466x}$

5. $y = \frac{1}{2}e^{(\ln 2)x} \approx \frac{1}{2}e^{0.6931x}$ **7.** \$4451.08

9. $S = L(1 - e^{-kt})$ **11.** $y = \dfrac{20}{1 + 19e^{-0.5889x}}$

13. $y = \dfrac{5000}{1 + 19e^{-0.10156x}}$ **15.** $N = \dfrac{500}{1 + 4e^{-0.2452t}}$

17. $\dfrac{dP}{dn} = kP(L - P)$, $P = \dfrac{CL}{e^{-Lkn} + C}$

19. $y = \dfrac{360}{8 + 41t}$ **21.** $y = 500e^{-1.6094e^{-0.1451t}}$

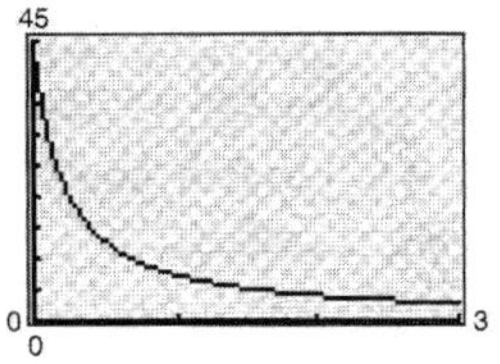

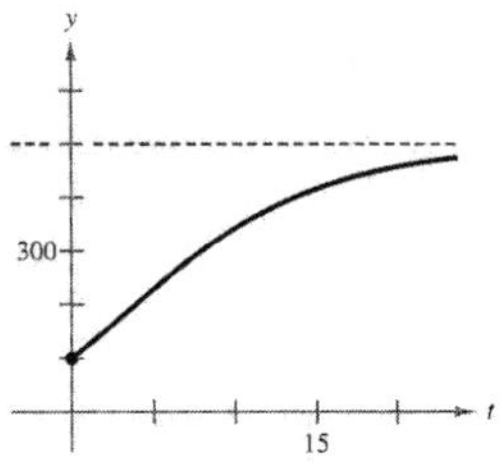

23. 34 beavers **25.** 92%

27. (a) $y = Ce^{kt}$ (b) ≈ 6.2 hours

29. 38.843 pounds per gallon **31.** ≈ 3.15 hours

33. $P = Ce^{kt} - \dfrac{N}{k}$ **35.** $A = \dfrac{P}{r}(e^{rt} - 1)$

37. \$7,305,295.15 **39.** (a) $C = C_0e^{-Rt/V}$ (b) 0

41. (a) $C(t) = \dfrac{Q}{R}(1 - e^{-Rt/V})$ (b) $\dfrac{Q}{R}$

APPLICATIONS *(continued from front endsheets)*

Physical Sciences

General